The World of Iron

The World of Iron

Edited by Jane Humphris and Thilo Rehren

Archetype Publications

First published 2013 by Archetype Publications Ltd

Archetype Publications Ltd
c/o International Academic Projects
1 Birdcage Walk
London SW1H 9JJ
www.archetype.co.uk

ISBN: 978-1-904982-975

British Library Cataloguing in Publication Data
A catalogue record for this book is available from the British Library.

Front cover illustration: BSE image of iron slag excavated from within a mid-first millennium AD iron smelting furnace in Southern Rwanda. Image width 800μm
Photograph: Jane Humphris

Typeset by Sandra Hall
Printed and bound in Estonia by Print Best

Contents

Theoretical approaches to technology

East Asia

Foreword

The idea for the World of Iron Conference (WIC) was conceived in 2004 during many hours of slag processing in the UCL Institute of Archaeology Wolfson Archaeological Science Laboratories. At the time, Jane Humphris was working on iron smelting remains from the Buganda Kingdom (Uganda) for her Masters' dissertation, while Xander Veldhuijzen was investigating some of the earliest iron smelting in the Levant for his PhD, both working under the supervision of Professor Thilo Rehren. Meanwhile, several major conferences were taking place within a year or two of each other, all dedicated to aspects of early iron in Europe – clearly demonstrating the strong appetite among the academic community to explore the impact of this metal from a technological as well as societal perspective, but for our taste a bit narrow in geographical limit. The concept for a global iron conference grew over the next few years until 2007, when Xander was a Leverhulme postdoctoral fellow at the Institute and Jane was carrying out her PhD research on Rwandan material. They decided to try to pull together the ideas they had been discussing and create a conference aimed specifically at the non-European evidence – which to our knowledge had never been presented and discussed in a coherent manner. Thilo supported the idea and momentum quickly started to gather.

The initial aim of the conference was to create a forum for the presentation and discussion of the latest research concerning the inception, adoption, expansion and impact of prehistoric iron production, specifically outside Europe. Through a combination of thematic and regional sessions, the scope of the conference would allow for case studies from around the world to be presented side-by-side with theoretical and scientific papers. It was hoped that this inclusive, international academic forum would stimulate the exchange of ideas and experiences, and provide the opportunity to encourage new research and future collaborations. We felt that these aims were best achieved by bringing together both well-established and young academics from around the world to create a globally comparative perspective. It also became clear very quickly that this rather ambitious concept required considerable financial support; few of the scholars we wanted to attract to present their research and contribute to the conference discussions had access to the sort of conference funding usually available for European academics.

A year and a half later, from Monday 16th February to Friday 20th February 2009, the conference took place at the Natural History Museum in London. The Regional Sessions brought together scholars and research from four key regions around the world (Africa; the Indian Subcontinent; West and Central Asia; and East Asia) to discuss the latest anthropological, archaeological and metallurgical research in the context of region-specific considerations. The Themed Sessions (Invention, Inspiration and Innovation; Theoretical Approaches to Technology; Scientific Approaches to Technology; Environmental Considerations) incorporated the latest research being carried out in all regions, including Europe.

In total, 69 papers were presented orally and 35 posters were displayed during the conference, which was attended by close to 150 participants. Within these conference proceedings, fifty-three papers have been selected from across the oral and poster presentations, following an open invitation to all active participants, an anonymous peer-reviewing process and sympathetic editing of the submitted manuscripts. Almost all of this work was done by Jane; without her efforts, patience and persistence this book would not have been completed (writes Thilo). It is hoped that this volume will not only illustrate the wealth of the global archaeometallurgical record, but also highlight the variety of approaches available to us as archaeologists and archaeometallurgists, and the innovative means by which these approaches are being applied to the diverse archaeometallurgical records around the world. It is also hoped that the papers in this book provide an overview of the state of research in their particular geographic and academic fields, and as such provide a useful platform for considering how research should and could move forward from this point.

The book has been laid out in the order that the sessions were presented during the conference. Where possible the session chairs (listed below) have provided the first paper of each section of the book, which either summarise the session or provide a more detailed insight into the particular session focus. The first session (Africa) was chaired by Bertram Mapunda, Professor at the Archaeology Unit of the Department of History, University of Dar es Salaam in Tanzania. Bertram began researching iron working in Southern

Tanzania in 1990, and in addition to archaeometallurgy he has a particular interest in cultural heritage and the history of East Africa, an area throughout which he has worked extensively. Bertram has published his own research into African iron production in a number of seminal papers. He continues to be actively involved in research and to provide academic support as a supervisor to a number of Masters' and PhD students from Tanzania and beyond. During the Africa session, archaeometallurgical research from across the continent was presented by academics working in both African and non-African institutions. A key theme during this session was the presentation and discussion of early dates for iron production from research carried out in countries including Nigeria, Togo and Central Africa, highlighting the continuing desire to understand the origins of this technology within the continent. Particular processes within the *chaîne opératoire* were also discussed, drawing on case studies from Nigeria, Botswana, Kenya and Tanzania, while a summary of the recent work carried out by students from the UCL Institute of Archaeology in Eastern Africa was also provided.

The Indian Subcontinent session was chaired by Professor Sharada Srinivasan of the National Institute of Advanced Studies in Bangalore, India. Sharada has been involved in the field of archaeometallurgy since the early 1990s, and is particularly interested in the relationship between the scientific fields of archaeology and art. She has been involved in the study of bronze statuary and early iron and steel making in the subcontinent, and ethnoarchaeological studies of foundry practices related to metalworking and craft traditions. The papers presented in this session broadly paralleled those presented in the Africa session in terms of themes. These themes included early iron production, the latest discoveries of ancient Indian iron industries, fascinating ethnographic studies and papers discussing the classic topic of wootz steel. There were also a number of papers and posters detailing the innovative and exciting work being carried out by Gillian Juleff and her team on the linear furnaces of Sri Lanka. Sharada, an expert in classical Indian dance, very kindly agreed to provide the conference audience with an Indian dance performance entitled 'The Singing Pillars of Hampi' during the lunch break on Thursday. This performance was thoroughly enjoyed by everyone.

The thematic session on Invention, Innovation and Inspiration took place on Tuesday afternoon, and was included to allow for an exploration of the concept of invention, and also to provide a stage for discussions concerning the role of the innovator and matters concerning technological evolution. Dr Gillian Juleff, senior lecturer at the University of Exeter, whose work on early iron production has shed much light on social and environmental stimuli for innovation from Sri Lanka to Exmoor in the United Kingdom, offered her wealth of experience to chair this session. Gillian's research in Sri Lanka in the early 1990's led to the discovery of a previously unknown first millennium AD iron industry based on a wind-powered smelting technology. The experimental re-enactment of this technology, published in the journal Nature, brought wide recognition to the study of ancient metallurgy and has influenced a new generation of archaeometallurgists. Papers presented in this session considered the inception and development of iron production in Africa, India, China and Europe. This session also included a live musical performance by Peter Halkon (to the delight of the audience), who sang a traditional British folk song during his paper which explored the role of blacksmiths in England in the past.

On Wednesday morning, Professor Thilo Rehren stepped in at the last minute to chair the West and Central Asia session. Now director of the new UCL Qatar campus in Doha, Thilo drew on his extensive experience and interest in pyrotechnology, metallurgical processes and analytical methods to ensure the session and the following discussion were stimulating. Xander Veldhuijzen's paper on slag inclusion analysis began this session, reflecting the particularly technical archaeometallurgical approach to the archaeology of iron production in this region. Ancient mining in Jordan, metal workshops in Turkey, Israel and Palestine, and iron production in Kazakhstan and Turkmenistan were also discussed in this regional session.

Professor Pierre Lemonnier of the CNRS in France, whose influential work on the role of technology in society has provided an academic springboard for many academics studying technologies of the past, chaired the session that dealt with Theoretical Approaches to Technology. Topics presented in this session included variability, epistemology, symbolism, the role of ethnography, evolutionary theory, and a number of theoretical methods for studying technology in the past. Hence, wider anthropological issues as well as specific case studies were explored during this session.

The focus of the final regional session, which took place on Thursday morning, was recent archaeometallurgical work being carried out in East Asia, an area with which UCL has a particular connection, having launched the International Centre for Chinese Heritage and Archaeology (ICCHA) jointly with Peking University in 2003. Professor Han Rubin, former director of the Institute of Historical Metallurgy and Materials, University of Science and Technology in Beijing, brought her far-reaching knowledge of the history of metals in China to this session, for which she acted as chair. Iron production locations from across the region and from a variety of different periods were presented. As well as a number of case studies from China, papers also focused on recent work in Korea and Thailand.

The penultimate session was dedicated to Scientific Approaches to Iron in Archaeology, and allowed for the latest and most pioneering scientific research methods to be presented. Professor Vincent Serneels from the Department of Geosciences at the University of Fribourg in Switzerland brought his archaeometric specialism to chairing this session. Vincent's diverse research background utilising a variety of analytical techniques to study material including Roman glass mosaic tesserae, Alpine soapstone, and recently leading a major multi-faceted investigation into iron production in Mali, meant he was well suited to the role of chair for this session. Topics presented included geoprospection, the study of iron blooms, and the identification of smithing workshops. The role of iron in Medieval French monuments, slag inclusion analysis, measuring variation and compositional fluid dynamics were also discussed within papers presented during this session.

The final session of the conference, on Friday morning, was that of Environmental Considerations. This session was chaired by Professor Donald B Wagner of the Nordic Institute of Asian Studies in Copenhagen, Denmark, who has published extensively on the history of science and technology, particularly of the Far East. An especially pertinent session in today's world, when human interaction with and impact on the natural environment is such an important theme in everyday life, this session explored the possible effects on the environment and past landscapes that the fuel-thirsty iron production industries may have had. Both the ways in which this impact can be studied, and particular case studies focused on Exmoor, Snowdonia, Tanzania and Mali, were some of the issues explored during this session.

In addition to the sessions, a number of extra-curricular activities were organised for the participants. A wine reception was held at the UCL Institute of Archaeology on the Sunday before the conference began, to allow for early registration and a chance for participants to meet each other ahead of the proceedings. During the lunch breaks, two short presentations of recent ethno-experimental smelting in Rwanda (a film by Jane Humphris) and Sri Lanka (by Gill Juleff) were given. Conference sponsor Tata Steel hosted a drinks reception in the Earth Science Foyer of the Natural History Museum on the Tuesday evening which was attended by a number of senior Tata Steel Directors. Dr Mukherjee, Director of Innovation and Technology at Tata Steel, gave a speech outlining why Tata Steel had decided to be one of the conference sponsors and the company's vision for investment in such worthwhile endeavors as academic research. On the Thursday evening another sponsor, Rio Tinto supported a drinks reception and buffet which was held in the Earth Treasury Gallery of the Natural History Museum. Housing the museum's extensive and beautiful collection of minerals, rocks and gems, this gallery formed a fabulous backdrop to a very enjoyable evening. The conference finished on Friday afternoon back at the UCL Institute of Archaeology, with a final gathering for farewell drinks.

Of course, none of the above would have been possible without the generous support of the companies and institutions who charitably supported the conference in a variety of ways. Thanks to the outstanding kindness of the Natural History Museum, the use of this fantastic venue was provided at no cost, which meant that significant bursaries could be offered to support many scholars from overseas to attend the conference. The efforts of Professor Andy Fleet (head of the Earth Science Department) and an extensive and efficient event management team ensured that the conference ran smoothly and was enjoyed by all. Following numerous letters and applications submitted by the conference organisers to try to raise funds, in the end Tata Steel, Rio Tinto, the Wenner-Gren Foundation, the UCL Institute of Archaeology and the Institute for Archaeo-Metallurgical Studies (IAMS) all supported this venture. UCL Qatar provided financial support towards the publication costs and the dissemination of books to the authors. We are incredibly thankful to these companies and institutions for recognising the potential for such a conference and supporting the venture both financially and logistically. Without their most generously given funding, few of the non-European participants would have been able to attend and contribute.

As the conference drew nearer, a number of staff and students from UCL provided invaluable logistical help, from binding abstract books into the small hours of Sunday night, to helping register participants and taking photographs during the event. While there is not room to mention everyone, special thanks must go to Louise Iles, Tim Hatton, Eleanor Blakelock, Miljana Radivojević, Wenli Zhou, Thomas Thondhlana, Edwinus Lyaya, Janice Li, Kat Manning, Tessa Dickinson, Oli Pryce and Thomas Birch. We would also like to thank our scientific committee and all those who assisted in the choosing of abstracts for both the conference and the proceedings, and who participated in the peer reviewing process. We would like to thank Raul Carstocea for the diligent manner in which he approached the copy editing of the contributions for this book, and Stuart Laidlaw for his help with the formatting of images. Finally, we would like to offer our sincere thanks to the authors of the papers included in this volume. Because of your hard work and patience, we think we have all created a volume that can form a key text in the field of ferrous archaeometallurgy for years to come.

We hope that you enjoy the papers in this volume.

Jane Humphris and Thilo Rehren

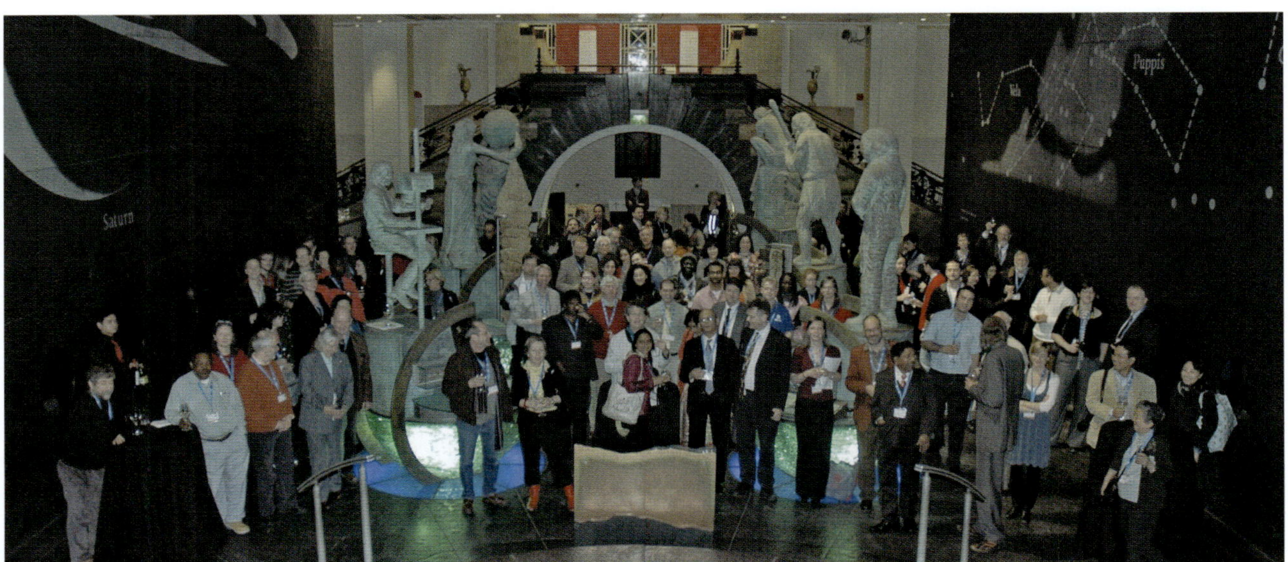

Conference Sponsors

Institute for Archaeo-Metallurgical Studies

For the past forty years, the Institute for Archaeo-Metallurgical Studies (IAMS) at the UCL Institute of Archaeology has supported research into past mining and metallurgy, and has conducted some of the most significant projects of the times, including those at Timna in Israel, and Rio Tinto in Spain. Whilst continually striving to expand our knowledge of metallurgy from ancient times to the recent past through innovative research, teaching archaeo-metallurgy has always been a core purpose of IAMS. The IAMS teaching programme is integrated into the UCL Institute of Archaeology undergraduate and post graduate courses, and covers a wide range of topics including mining and quarrying, and extractive metallurgy. Every year IAMS hosts an intensive two-week summer school, providing lectures covering a wide range of topics presented by leading academics of the field.

Throughout its existence, IAMS has supported numerous visiting researchers and PhD students, thanks to the outstanding generosity of its trustees and associated mining companies. Former IAMS students are now holding leading academic positions in India, Africa, Europe and the US, and continue to shape the future of the discipline.

The World of Iron Conference perfectly fitted into to remit of IAMS, with its goals of inclusive, global participation in order to encourage and promote the investigation and discussion of iron production throughout the ages and across the world. As such, IAMS was delighted to be able to support this venture financially, and to help with the logistics of the event.

Natural History Museum

With a world-class mineralogy collection of 405,000 specimens including rocks, ores and meteorites from around the globe, the Natural History Museum in London provided the perfect backdrop for the World of Iron conference. The Museum was incredibly supportive towards the conference, hosting it at the museum at no charge, thus providing an extremely generous level of support towards this academic venture. The conference was held in the Flett Lecture theatre, situated next to the Earth Hall with its giant metallic Earth sculpture and showcases of moon rock and colourful minerals, which provided an inspiring entrance to the venue every day for the participants. In addition, two receptions were held at the museum after hours, one in the Earth Hall and one in the Earth's Treasury, a gallery filled with minerals, gems and rocks.

Rio Tinto

Rio Tinto was founded in 1873 to reopen and work the ancient copper mines in southern Spain. By the end of the twentieth century Rio Tinto was mining many different materials including iron, copper, aluminium, as well as minerals such as borates and zircon, in diverse locations all around the world. In 2006 the company was integral to the development of a new and innovative method of iron making (HIsmelt).

Today Rio Tinto is one of the world's leading mining companies, with the concept of sustainable development at the centre of its business policy. While continuing to strive for the highest levels of mining and procurement performance, Rio Tinto is committed to comprehensive and extensive social and environmental policies. In addition, the company holds technology and innovation as key components of its future strategy, and encourages academic excellence in this field to promote the development of mining and production techniques.

As a company founded on ancient mines, and one with an extensive global history, Rio Tinto has long supported the Institute for Archaeometallurgical Studies (IAMS) at UCL in recognition of the significant academic value of the research conducted into ancient metallurgy performed by students and staff supported by IAMS. Rio Tinto was very happy to support the World of Iron Conference, as a venture striving to bring researches together from around the world to discuss and present findings on the diverse and pioneering metallurgy of the past.

Tata Steel

Tata Steel was pleased to support the World of Iron Conference 2009 at the Natural History Museum, London. Tata Steel recognises the importance and significance of the development of iron smelting through the ages, and the societal impact on our world today. The World of Iron Conference 2009, organised by the UCL Institute of Archaeology, was an opportunity for leading academics from academic institutions both within and outside of Europe to convene and consolidate recent research undertaken globally in this exciting area of archaeology. The research papers and findings presented by delegates at the Conference have been collated into this book for posterity, to enable shared understanding of the developments and increased knowledge of this area of archaeology.

Tata Steel celebrated its centenary in 2007 after being established in 1907 as Asia's first integrated private sector steel company. Tata Steel Group today is a global steelmaker with production plants in India, South East Asia, the UK and mainland Europe. After the 2007 acquisition of Corus, Europe's second largest steel producer, the combined enterprise has a crude steel capacity of over 28 million tonnes and over 82,000 employees across four continents. The Group provides innovative products and solutions to the construction, automotive, packaging, mechanical engineering and other markets worldwide.

Tata Steel Group's vision is to become the world steel benchmark for Value Creation and Corporate Citizenship. Tata Steel was pleased to support this conference showing our commitment to these values and recognising the importance and significance of the development of iron smelting through the ages, and the societal impact on our world today.

UCL Institute of Archaeology

The Institute of Archaeology has long been committed to global archaeology and is widely recognised as the only archaeology department in the UK that has the capacity to pursue a global agenda. It also has a long and illustrious tradition of archaeometallurgy, going back to the work of Tylecote in the 1970s. Thus the 'World of Iron' conference, organised by current students and staff, demonstrates the Institute's continuing major contribution to this vibrant field and fits perfectly into its research agenda; in doing so it makes an important contribution to the realisation of the Institute's long-term strategic goals. For these reasons the Institute was very happy to provide the venue for receptions at the beginning and end of the conference.

UCL Qatar

UCL Qatar is a partnership between University College London (UCL), Qatar Foundation and Qatar Museums Authority. Based at Education City as a partner of Hamad bin Khalifa University, UCL Qatar offers postgraduate degree programmes in the Archaeology of the Arab and Islamic World, Conservation Studies, and Museum and Gallery Practice. Academic research of the highest standard is also a core focus, with local and international excavations and research projects based and managed in Doha. In addition to world-class teaching and facilities, the campus houses one of the most comprehensively equipped analytical and conservation laboratories in the region, within which a wide spectrum of archaeological materials can be analysed, preserved and stabilised using state-of-the-art apparatus and techniques. Community outreach and education is also a priority within both the teaching curriculum and fieldwork practice.

UCL Qatar are delighted to contribute funding to support the publishing of the World of Iron, a resource that will provide a key text for students, academics and the public alike for many years. UCL Qatar are also happy to endorse the distribution of copies of the book to the authors.

Wenner Gren Foundation

The Wenner-Gren Foundation is a private operating foundation dedicated to the advancement of anthropology throughout the world. It is one of the major funding sources for international anthropological research and is actively engaged with the anthropological community through its varied grant, fellowship, networking, conference and symposia programs. The Foundation works to support all branches of anthropology and closely related disciplines concerned with human biological and cultural origins, development, and variation.

The Wenner-Gren Foundation has three major goals – to support significant and innovative anthropological research into humanity's biological and cultural origins, development and variation, to foster the creation of an international community of research scholars in anthropology, and to provide leadership at the forefronts of the discipline. Therefore, the Wenner-Gren foundation was very happy to support the World of Iron Conference. With the aim of bringing together academics from around the world; the remit of discussing and presenting the latest research into the history, technologies and impacts of iron production; and the creation of this volume as a conference proceedings, we feel the World of Iron Conference was a success that we are proud to have contributed towards as a sponsor.

Africa

Iron and its influence on the prehistoric site of Lejja

Pamela Eze–Uzomaka

ABSTRACT: Archaeological investigations at the prehistoric iron smelting site of Lejja, Nigeria, made possible the study of some large scale iron smelting debris – evidence of a once thriving iron smelting community. The main village square in Lejja, contains over 800 slag blocks weighing between 34 and 57 kg. Furnaces, tuyères and slag are visible on the surface in many places. The site has been radiocarbon dated to approximately 2000 BC. Iron smelting is currently extinct in Lejja, and there are no active iron smiths in the entire community.

For a community to experience such a profuse production of iron, the people must have interacted with situations and circumstances which impacted their life and modified their social existence. This raises some important questions: Why did smelting become extinct in the area? Are there traces of the past in the present life of the inhabitants of Lejja? What influenced them to embark on such an extensive iron production in prehistory?

Keywords: Iron smelting, slag, radiocarbon dates, ethnography, politics, taboos.

Evidence of iron smelting in Lejja

Intensive smelting of iron took place at the site of Lejja, in south east Nigeria during prehistoric periods. This statement is substantiated by the extensive iron smelting debris left behind in Lejja. The debris could point not only to an extensive iron smelting period in the history of the site, but could even represent the remains of a once thriving industry. Iron smelting often involved the community as a whole, and its effects were usually far reaching; from changing the status and living standards of the smelters to the actual development of some African cultures (Holl 2000, p.5). The socio–cultural effects on the indigenous peoples must have been far reaching and smelting must have occupied a central position in their everyday existence.

In the Nsukka area, there are several towns and communities that were intensively involved in iron smelting in prehistory. Smelting sites include Opi, Aku, Obimo, Obukpa, Owerre Elu, Lejja, etc.

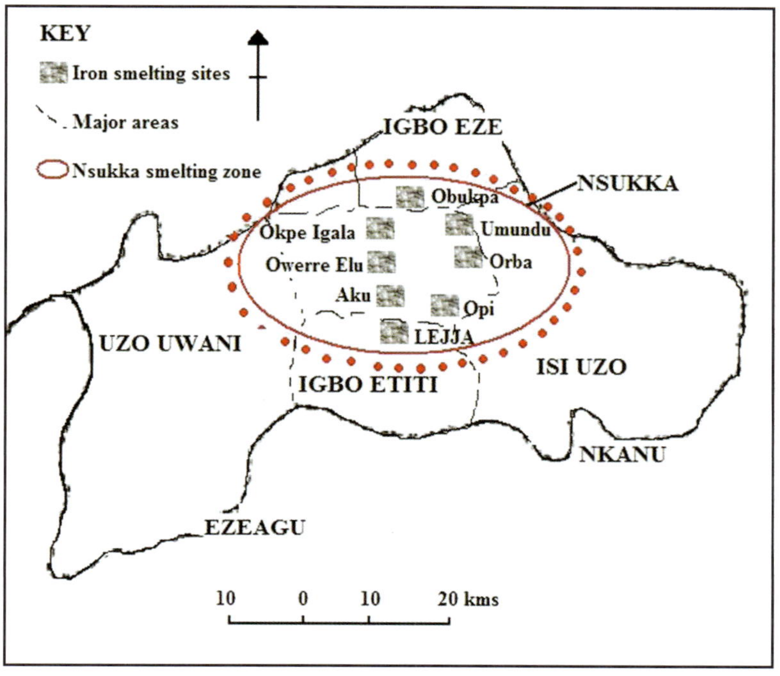

Figure 1: Map of the Nsukka area showing Lejja and some other iron smelting sites in the area.

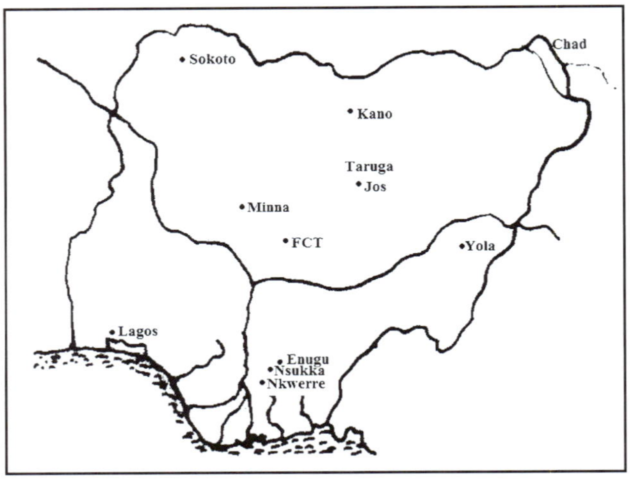

Figure 2: Map of Nigeria showing some major towns.

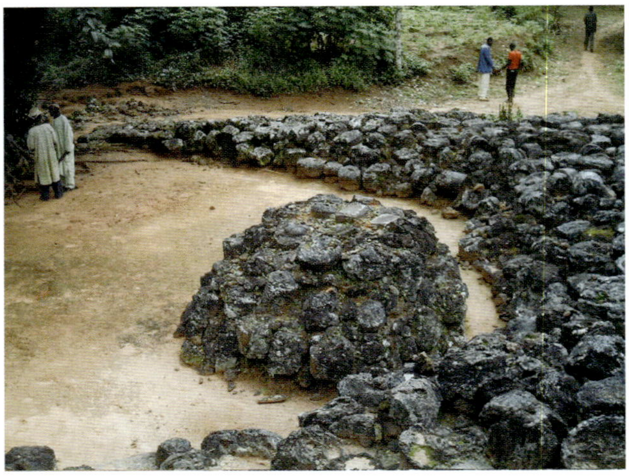

Figure 3: The slag blocks at Otobo Dunoka, Lejja.

Otobo Dunoka, the Lejja village square, is the site that presents the greatest concentration of slag blocks. The slag blocks are placed in rows, and they are often used as seats for village meetings (see Fig. 3). There are over 800 blocks of slag in this square, and the ends of some of them are either conical or spherical in shape. The blocks look alike, but are of different sizes. The diameters of the slag blocks range between 31 and 56 cm, with heights ranging from 22 to 36 cm. Some of the slag blocks are broken, but the unbroken ones have an average weight of between 34 and 57 kg (Okafor 1993, p.439).

The slag blocks at the village square are still intact because they are durable, and also because of the traditional taboos associated with tampering with, or removing them from the village square (Eze-Uzomaka 2000, p.102).

This village square (Otobo Dunoka) in Lejja (Fig. 3) is quite extraordinary. The cylindrical slag blocks are placed in a semi-circular arrangement and serve as seats during meetings, for both men and women. In the same square, there are also other monuments, like the small mound at the entrance of the village square. This mound is considered to be a sacred mound and is dedicated to the local deities. Other village squares in other parts of Lejja have cylindrical slag blocks, but in very limited numbers compared to the Otobo Dunoka. Towards the back of the village square, there is a hall whose foundation is lined with these cylindrical blocks. Some of the stones visible on the sides of the wall are the 'iron stones' found on the roads as one moves around Lejja.

In Taruga, Nigeria, the furnaces discovered included free standing vertical shafts of medium height over slag pits dug into the soil and rock. The only openings in the shafts were for tuyères positioned radially around them or near ground level (Tylecote 1975, p.5, Schmidt 1981, p.32). This suggests that the tuyères were inserted in the furnace for pre–heating and one observes some parallels between this Taruga style and that of the Early Iron Age in north–western Tanzania. The low shaft furnace discovered in Lejja (Trench 1) was also a free standing vertical shaft on top of a slag pit dug into the soil. The seven openings (tuyère ports) were positioned radially around it, where the ground level should have been many years ago.

Radiocarbon dates obtained from Lejja were much older than similar results from iron smelting sites in other parts of the Nsukka area, as well as from other sites in Nigeria as a whole (Table 1). The first set of dates was acquired from Trenches 1 and 2, and the last set of dates was obtained from Trench 4, at a depth of 140cm.

Detailed archaeological reports of all recent excavations carried out in Lejja, including the finds, as well as the collection and interpretation of samples, have been published (Eze-Uzomaka, 2009; 2010)

Table 1: Radiocarbon dates from a few sites in Nigeria.

Iron Smelting Sites – Nigeria	Radiocarbon dates (bp)	Lab. No.	References
Taruga	2230 ± 120	11459	Calvocoressi & David 1979: 26
	2390 ± 140	12960	ibid
	2250 ± 100	13400	ibid
	2042 ± 126	BM532	ibid
	2269 ± 116	BM533	ibid
	2541 ± 104	BM938	ibid
	2488 ± 84	BM940	ibid
	2541 ± 104	BM941	ibid
	2291 ± 123	BM942	ibid
Nok	2160 ± 95	14913	ibid
Rafin Ndoko	2020 ± 75	N2585	ibid
Rop Rock Shelter	1975 ± 125	1406	ibid
Opi–Nsukka	2305 ± 90	OxA–3201	Okafor 1993: 437
	2170 ± 80	OxA2691	ibid
	2080 ± 90	OxA3200	ibid
Owerre Elu	1060 ± 60	OxA2738	ibid
	570 ± 60	OxA2739	ibid
Orba	300 ± 90	OxA2693	ibid
	215 ± 100	OxA2694	ibid
	295 ± 85	OxA2695	ibid
Umundu	200 ± 80	OxA2688	ibid
	205 ± 80	OxA2689	ibid
	130 ± 80	OxA2690	ibid
Lejja	4 005 ± 40	Ua–34415	Eze–Uzomaka 2008
	2 370 ± 40	Ua–34417	ibid
	1 715 ± 35	Ua–34416	ibid
	3 445 ± 40	Ua–37422	Eze–Uzomaka 2009
	2 050 ± 30	Ua–37421	ibid

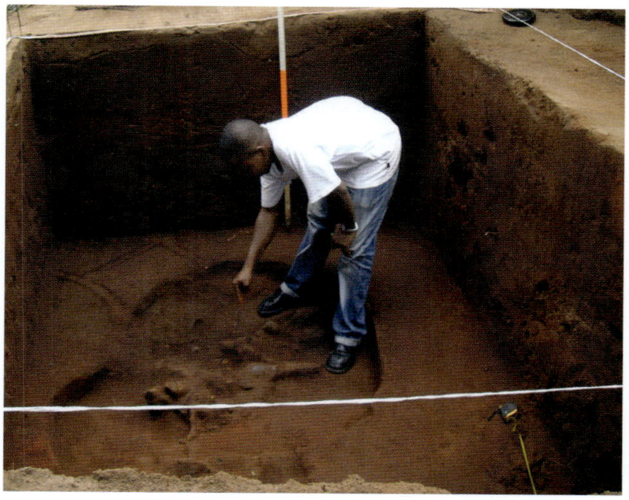

Figure 4: Excavation of Trench 4, Lejja.

Effects of iron smelting on the Lejja community

Although the initial objective was to examine the archaeological remains of Lejja, as the excavations continued, it became imperative to examine the ethnography of the people in more detail. Technological systems of any kind are a fundamental part of the societies in which they operate. "The advent of technologies may thereby create new and unsuspected social demands or satisfy pre-existing ones" (Holl 2000, p.20).

Influence on environment

One of the local people in Lejja explained that the techniques that the Nsukka people used to adapt to their own environment were diverse. He claimed that the environment in which people lived in 'the days of old' (prehistoric times) was usually the forest, and, as a result, their forefathers were constantly facing danger from large carnivorous animals and had to devise tools to protect themselves in their environment. As a result of the very real dangers they faced living in the forest, they must have lived in groups, and, as new settlers arrived, they preferred to stay close together in clusters. This strategy led to an increase in the density of the population. In addition, the use of iron tools not only helped people increase their food production and nourish a large population (that was continually expanding following migrations from arid regions) but they also provided support for life in the forest. The presence of iron ore in many parts of the Nsukka area must have led to an occupational specialisation which may have intensified and spread to other places.

It is also known that in the early agricultural period, iron implements were used to improve the processes of planting and harvesting. De Barros (1985) conducted a study on the Bassar and observed that the production and use of iron implements "increased food production through the use of more efficient bush clearing tools, tools which permitted the clearing of forest and increased productivity in areas where stone was not readily available for stone axes; increased population densities as a result of increased food production; increased specialisation and social differentiation, especially the formation of iron working castes or classes; larger and more stable communities due to both increased specialisation and the existence of food surpluses; increased trade due to increased specialisation; and the 'embryonic rise' of modern politics, 'the politics of class differentiation'" (de Barros 1985, p.467).

The smelters interacted with their environment, since the iron ore, charcoal, clay, and other materials added in the construction of a furnace are all part of the physical environment which is inadvertently decimated by constant use. In societies which were focussed on intensive large scale production, nearly all able–bodied freemen were involved in either smelting or smithing, and most, if not all of the farming was carried out by the iron workers' wives and children, as well as slaves (de Barros 1988).

It has been speculated that the smelting process was often carried out away from the rest of the community. However, in Lejja and in other parts of the Nsukka area (see Fig. 1), furnaces were found buried very close to the locations where people live at present, and also in places overgrown by the forest. Whether the people lived in the thick forest areas in the past is yet to be determined.

Political implications of iron smelting

In Lejja, mythical stories have been created around the iron smelters, emphasising their godlike importance and thereby heightening their political significance. They were said to have influenced many wars and local skirmishes by supplying iron weapons to warring parties. Smelting affected the politics of Lejja in the past and was controlled by clans. It was not a general trade, like smithing. Only the members of some clans had the right to smelt iron, as well as rule the community as political leaders. In the past, the Chief or King of Lejja had to be an iron smelter. The present King of Lejja (*Eze Lejja*) is supposed to be a descendant of a lineage of past smelters. The symbol of the collective authority of the town (*Arua*), which is an equivalent of the modern day mace, is made of iron. This mace is moulded from a collection of pieces of iron contributed from the different villages in Lejja. When the mace is old and needs to be replaced, different villages must contribute a piece of iron each to replace it. This aspect is of significance in Lejja, because the act of moulding all the pieces of iron together showed that all the villages were united to each other in the making of the mace (C.C. Opata, personal communication, 2008).

C.C. Opata also provided an insight into the political implications of the slag arrangement in the main village square of Lejja. On top of the northern edge of the slag embankment there is a large cylindrical slag block named *Eze mkpume* – king of stones. This specific slag block cannot be touched by anyone, since it is a symbol of the authority of the ancestral father of Lejja. The taboo on touching the slag block is also an act of respect shown towards the elders. Behind this particular slag block is a special slag block which is used by the king of Lejja as a seat during social gatherings. On the left hand side of the king's seat there is a slag block usually occupied by the next high chief, known as Eze-Uwani – King of Uwani. The next slag block is meant for another high chief, known as Eze Akaibute – King of Akaibute. Both are high chiefs in their own right, but ancestrally subject to the King of Lejja. The village square is arranged in such a fashion that men are seated on the left hand side of the King of Lejja, while the women sit on the other side, near the masquerade house that is shaped like a very big furnace.

This masquerade house (Fig. 5) is known in the local language as Okiti Akpuriagedege. The people believe that the shape of the masquerade house in the Otobo Dunoka village square is important to their history as an iron smelting clan. The sitting arrangement is also significant and culturally relevant to the people. The seat of the King of Lejja behind this 'king of stones' places him in an ancestral protective position over the other chiefs, as well as the rest of the people gathered there. Finally, it is important to note that anyone (except the three most senior elders – Eze-Lejja, Eze-Uwani and Eze-Akaibute) who wishes to speak during the meeting must move to the centre of the village square, where there are no cylindrical slag blocks, and stand there to speak to the rest of the people.

Among the Tiv of Nigeria, iron portends power and prestige. "The tie between politics and iron working is especially strong in the Congo basin and seems to diminish in the periphery" (de Maret 1985: 75). Outside the Bantu world, among the Tiv, semi-Bantus of Nigeria, iron connotes power and prestige (East 1939, p.139). "Further north east at the border between Nigeria and Cameroon, among those societies where the chief's power is sacred – as among the Sukur, Margi and Ndjeny – blacksmiths play an important part, by forging the symbols of political power and by their role during the funerals of chiefs" (de Maret 1985: 75–6). Many chiefs, even though they are not smiths themselves, still kept a complete set of smith's tools, including tongs, anvil, poker, hammer, bellows and tuyère (East, 1939, p.139). "In Central Africa, the blacksmith is a public figure, close to both the symbolic and political powers." (de Maret 1985, p.75). In the oral tradition of the Kuba, a crucial episode known as "the hammers and the anvil" links iron metallurgy to the foundation of kingship (Vansina 1978, pp.49–51). The craft is linked in several instances with royalty and "it was even said that from the early days of the kingdom, all men of the royal lineage had to know the art of smithing" (Vansina 1978, p.182). One question still remains unanswered: did iron smelting lead to the appearance of kings or did the kings establish iron smelting in Lejja?

Economic control and power

Monetary affluence and an increase in the standard of living could have been some of the results of the large scale smelting carried out in Lejja in the past. From the ethno-historical information presently available, it is evident that iron ore as well as artefacts were important commodities exchanged within local networks in the southern forested area (Essomba 1993). Iron goods had a high value, since most of the early money (*manilas*) was made of iron, and later copper. The smith was undoubtedly a wealthy man. Smelting and forging were well rewarded, and those who worked with iron were inadvertently placed at the core of socio-economic circuits (de Maret 1985, p.76). One aspect which can be inferred from the cylindrical slag blocks left behind is that the Lejja smelters must have had excess production of iron, and this may have

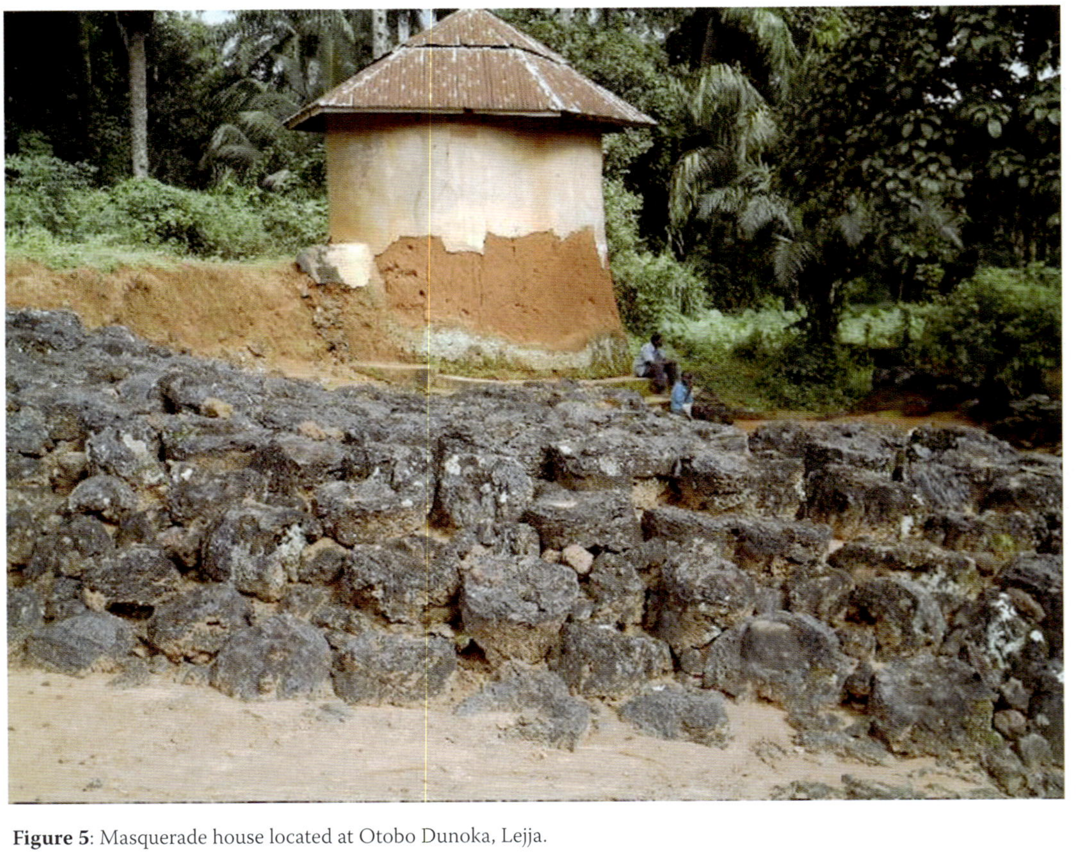

Figure 5: Masquerade house located at Otobo Dunoka, Lejja.

led to extensive trade to far and distant places, sustained over a long period of time. If this was the case, it is therefore not surprising that iron smelting seemed to coincide with the rise of prestigious leadership.

Although smelting did give rise to material comfort for the smelters, it was not always a matter of wealth and affluence; more important perhaps was the hold on power, which bordered on the divine, creating an aura around the craftsmen, who became traditional custodians of power. However, "in traditional African societies, investiture of leaders, be they chiefs or kings or whatever, is also a process of transmutation by which the paramount attains the aura of semi divine. Almost everywhere, smiths play a prominent role in the investiture process, sanctioning by their attainments as transmuters, the elevation of society's principal mediators" (Holl 2000, p.77). Only the affluent in society could take on this task and carry it through.

Socio-religious and mystical implications

Smelting also had religious implications, and thereby affected the belief system of prehistoric people. Ironworkers became experts in rituals meant to encourage good production and to ward off evil spirits, including songs and prayers, the giving of medicines, and even sacrifices. The latter were usually put in the furnace itself or buried under the base of the furnace. Examples of these date back as far as the Early Iron Age in Tanzania and Rwanda (Schmidt 1997, in Childs and Herbert 2005, p.293). In Lejja and in other towns in the Nsukka area, masquerade houses are built very close to the village squares (Otobo) where slag blocks are packed. This is significant because masquerades are regarded as 'masked spirits of the ancestors'. The masquerade houses are located in seemingly direct relation to the cylindrical slag blocks stacked in the village squares. Furthermore, during the excavations in Lejja in 2007, we could not find ritual materials buried at the base of the furnaces. It was therefore postulated that the masquerade houses could have served as places for the performance of rituals, carried out before, during, or after the smelt. It could also be that sacrifices were offered beside the furnace and not underneath the smelt, in which case they would not be preserved in any form.

The *Oji* is a set of iron gongs placed together on an iron rod which was used by diviners and masked spirits, medicine men, to ward off evil spirits and create awareness of the presence of a diviner. Masked spirits of the ancestors often hold it in their hands when walking. When the *Oji* is struck on the ground, it is said to create a magnetic field around itself. This magnetic field is thought to have the ability to absorb all the evil in the area, so that if the masked spirit is moving forward, there is a radius of influence nullifying the influence of evil. As a result, the *Oji* is normally feared by people. People believe that the *Oji* has mystical powers and could be used for enchantment (C.C. Opata, personal communication, 2008).

Some cultures associated sexual images with iron production. Smelting is symbolically intertwined with fertility, and sometimes the production of the bloom is compared to conception (smelting) and birth (the actual reception of the bloom). The furnaces are also often extravagantly adorned to resemble a woman, the 'mother of the bloom' (Childs and Herbert, 2005). The smelting process is carried out entirely by men and often away from the village. It is thought that if women were to touch any of the materials or even be present, this could jeopardise the success of the production.

Taboos were part of the religious implications of smelting. It was said that iron smelters created and maintained an aura around themselves. In Otobo Dunoka, Lejja, the village square where the cylindrical slag blocks are packed, no woman undergoing her menstrual cycle is allowed to pass through the centre of the square. If this happens, the woman is said to experience persistent bleeding and an unending menstrual flow. The bleeding is said to continue unless palm oil is used to propitiate one of the idols (Odegwo). Odegwo is an idol erected in the square (Fig. 6). It looks like a triangular pyramid of moulded clay. In spite of the stories surrounding this pyramidal structure, studies are presently being carried out to examine whether this structure had anything to do with astrology and star gazing in the past.

In some other parts of Nigeria, there were many strict taboos surrounding the processes of smelting and smithing. A study carried out in Nkwerre (see Fig. 2), in south–eastern Nigeria, enumerates some taboos associated with iron smelting and smithing. For example, a menstruating woman could not enter the workshop of a blacksmith, or she would damage whatever the smith was fashioning. It was considered an offence to fan the smithing fires by blowing with the mouth, as this would ruin the load. Again, it was taboo for one to swear falsely on Ogadazu. Ogadazu was the god of iron in Nkwerre, and was represented by the anvil upon which iron tools are fashioned. It was therefore believed that the anvil itself was a supernatural being, although to all appearances it is an inanimate object. Death or ill luck awaited the person who swore on it. It was also taboo to enter the smith's workshop through the rear; this was considered to bring bad luck (Eze-Uzomaka, 2000, p.125). The smiths believed that their success or failure in business lay with the strict observance of these rules and regulations.

A study of other cultures also describes ceremonies that feature the personification of the hammer and the anvil. Among the Luba of Central Africa, one of the important aspects of the enthronement of the King of Luba is the

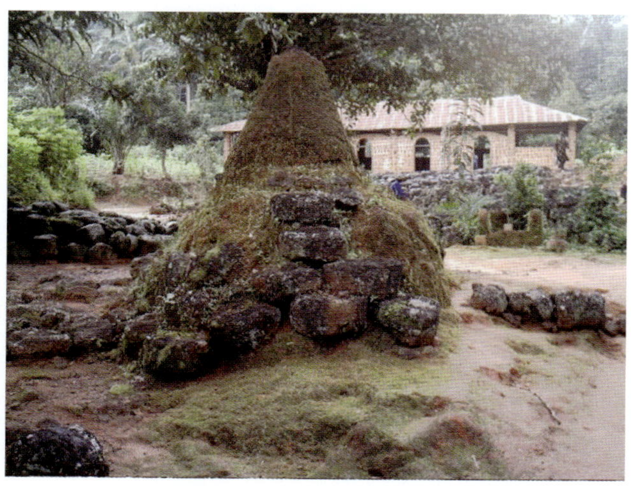

Figure 6: The idol known as *Odegwo* in the Lejja Village Square.

'striking of the anvil'. This ceremony is symbolic, in the sense that when the king is about to be crowned, a revered elder in the community beats the knees of the king at the same time as a blacksmith strikes his anvil. It is at this moment that the king receives the spirit of the kingdom (de Maret 1985, p.80). This transforms the person of the king into a sacred one. The King of Tio was also a sacred king who was associated with an anvil. He was called *Ngaandzunnu*, "owner of the anvil" (Vansina 1973, p.374).

Religious beliefs also forbade women from smelting. Women were said to be weak and would inadvertently destroy the smelt. In research carried out in Nkwerre, Nigeria, questions on the gender of the supreme deity generated a mixture of surprise and amusement from the respondents. Apparently, it had never occurred to any of the respondents to think of the supreme deity as anything but of the male gender. One respondent replied in controlled amusement: "What a question. The Supreme Being is definitely a man. God has never been regarded as female in the past. The important thing is that God is. But then we know that the person who made the whole world cannot be a woman" (Eze-Uzomaka, 2000, p.125). In Lejja and in other African societies, women were typically involved in other duties. They were usually the potters, and carried out other menial tasks. The wives of iron workers were also potters (Herbert, 1993, p.203).

Iron smelters and smiths held different social status depending on their culture. In some cultures, the skills of smelting are often passed down in the family and would be associated with great social status (sometimes even considered as witch doctors) within their community. Their powerful knowledge allowed them to produce materials on which the entire community relied. In some communities, they were believed to have such strong supernatural powers that they were regarded as highly as the king or chief. An excavation at the royal tomb of King Rujugira (Great Lakes, Eastern Africa), for example, revealed two iron anvils placed at his head (Childs and Herbert, 2005, p.288), suggesting their importance and powerful significance to the people.

The end of iron smelting in Lejja

There is no living iron smelter in present day Lejja. There are no smiths either. The trade has completely disappeared. What caused the demise of such a vast smelting industry in Lejja? A number of reasons could be given for this occurrence based on studies carried out in other places. These could easily be applied to Lejja, although such an application should be approached with caution. For example, the influx of European iron goods into Africa must have presented a challenge to indigenous smelters. Cheap steel produced in blast furnaces in Europe was imported and promoted by the colonial governments at the time (Flint, 1974, p.387). The smelters obviously came under a lot of pressure, either to produce greater quantities of iron and to do this more efficiently using the charcoal and ore available to them (Goucher 1981), or to quit.

Many other factors may have played their part in the termination of iron smelting, including the depletion of the forest. This was partly related to the use of charcoal in iron working, which entailed massive cutting of trees for the preparation of charcoal (Shaw, 1985). In turn, this decimation of trees led to deforestation and the extinction of some tree species (Shaw, 1981). In Toumra, today's blacksmiths have to walk for two days to acquire the wood needed for iron forging, while 25 years ago they claim that these trees were growing around the village (Haaland 1985, p.63). In addition to firewood collection having therefore become an inconvenience, people also left the villages for a number of reasons other than deforestation (e.g. very little annual rainfall).

Other factors also played an important part in the decline of the iron working tradition: socio-cultural and economic factors related to labour, price, ritual value of local iron, and the social organisation of iron working groups. The production of iron required a high level of skill and a heavy investment of time. As such, this labour intensive industry was seriously affected by the shortage of labour as a result of politico-economic developments (Oguagha 1982).

Finally, it must be added that the lack of strict regard for taboos and rituals which resulted from the influx of the Christian religion may also have played a part in the gradual decline of iron smelting in Lejja. Christianity encouraged the young people to go to school and seek 'prestigious' white collar jobs instead of continuing in the trade of their fathers. Moreover, in Lejja, the King of Dunoka is a paramount chief and is supposed to be a descendant of a lineage of smelters; however, during our research, we learned that he no longer believes in ancestral creeds and would not perform the ritualistic practices of his forefathers. This was related to a lack of in-depth knowledge, by the king, of the myths and legends that should regulate his actions at present. Although he is King of Lejja, the rituals that should be performed by him in his role as king have been overtaken by the Christian religion.

Conclusion

In the present study, I have not been able to fully address some of the questions which may arise concerning the socio–cultural impact of iron smelting on the people of Lejja. The answers to these questions may have to come from inferences made from the study of other cultures where iron smelting occurred in the past 50 years or less, and where there are some survivors who witnessed or even participated in the smelting of iron.

During the study, the connections between politics and metallurgy became clearer. Although smelting may have ceased in Lejja a long time ago, evidence shows that the link between the dynastic lineage of Lejja and smelting still remains, and is evident in their cultural make-up. Taboos are still recounted by the people, but they no longer command the same fear as they used to in the past.

One of the observations made during my research in Lejja was the inclusion of slag blocks at the bases of buildings and during the construction of the foundations of houses. This occurred frequently in many places in Lejja. The people of

Lejja explained that as far as they could tell, there was no significance in the practice. Since slag was in abundance everywhere and was durable in nature, they used it to make the house foundations strong.

The influence of iron working can be observed all around the Lejja site. From the foregoing, it is probable that the demise of iron smelting in the area must have drastically affected the social and economic existence of the people. Iron smelting was linked to royalty in several other kingdoms in Africa. Further research should therefore be undertaken to examine the tie between politics and iron working in the past as well as the tangible economic benefits that may have been lost after the demise of the industry in Lejja.

References

Anozie, F.N., Ray, K. And Haslan J., 1993. Preliminary archaeological studies of early iron smelting in Igboland, SE Nigeria. In: B. Andah, P. de Maret and R. Soper eds. *The Proceedings of the 9th Congress of the Pan–African Association of Prehistory and Related Studies, Jos, Nigeria*. Ibadan: Rex Charles Publications.

Childs, S.T. and Herbert, E.W., 2005. Metallurgy and its Consequences. In: A. Brower Stahl ed. *African archaeology: a critical introduction*. Oxford: Blackwell, pp.276–300.

de Barros, P., 1985. *The Bassar: large–scale iron producers of the West African savanna*. PhD thesis, University of California at Los Angeles. Michigan: University Microfilms International, Ann Arbor.

de Barros, P., 1988. Societal repercussions of the rise of large–scale traditional iron production: a West African example. *African Archaeological Review*, 6, pp.91–113.

de Maret, P., 1985. The smith's myth and the origin of leadership in Central Africa. In: R. Haaland and P. Shinnie eds. *African iron working, ancient and traditional*. Oslo: Norwegian University Press, pp.73–5.

East, R., 1939. *Akiga's story. The Tiv tribe as seen by one of its members*. London and New York: Published for the International Institute of African Languages and Cultures by the Oxford University Press.

Essomba, J.M., 1993. *Le fer dans le passé des sociétiés du sud Cameroun*. Paris: L'Harmattan.

Eze-Uzomaka, P.I., 2000. *Museums, Archaeologists and Indigenous People*. British Archaeological Reports International Series 904. Oxford: Archaeopress.

Eze-Uzomaka, P.I. 2009 Iron Age Archaeology in Lejja, Nigeria. In Dimensions of African Archaeology: Studies in the African Past. Pwiti, G. Radimilahy, C. and Macamo S. (eds) Dar es Salaam. E & D Vision Pub. Ltd. Vol. 7; 2009. pp 41–51.

Eze-Uzomaka, P.I. 2010 Excavation of Amaovoko: A Further Study of the Lejja Iron Smelting Culture. In Studies in the African Past. The Journal of African Archaeology Network. Chami, F.A. and Radimilahy, C. (eds.) Dar es Salaam. E & D Vision Pub. Ltd. Vol. 8; 2010. pp 178–191.

Flint, J.E., 1974. Economic change in West Africa in the 19th century. In: J.F.A. Ajayi and M. Crowther eds. *History of West Africa, Vol. 2*. London: Longman, pp.387–91.

Goucher, C.L., 1981. Iron is iron 'til it is rust: trade and ecology in the decline of West African iron–smelting. *Journal of African History*, 22, pp.179–89.

Haaland, R., 1985. Iron Production, its Socio–Cultural Context and Ecological Implications. In: R. Haaland and P. Shinnie eds. *African iron working, ancient and traditional*. Oslo: Norwegian University Press, pp.50–72.

Herbert, E.W., 1993. *Iron, Gender and Power: Rituals of Transformation in African Societies*. Bloomington: Indiana University Press.

Holl, A.F.C., 2000. Metals and Precolonial African Society. Toward an Archaeology of Innovation, Process and Transformation. In: J.O. Vogel ed. *Ancient African metallurgy. the socio–cultural context*. Oxford: AltaMira Press, pp. 1–83.

Oguagha, P.A., 1982. The 'conquest hypothesis' in Igbo–Igala relations: a re-examination. *West African Journal of Archaeology*, 12, pp.55–61.

Okafor, E.E., 1993. New evidence on early iron smelting from south-eastern Nigeria. In: T. Shaw, P. Sinclair, B. Andah and A. Okpoko eds. *The Archaeology of Africa: Food, Metals and Towns*. London: Routledge, pp.432–48.

Schmidt, P.R., 1981. *The Origins of Iron Smelting in Africa: A Complex Technology in Tanzania*. Research Papers in Anthropology Vol. 1. Providence, RI: Brown University.

Schmidt, P.R., 1997. *Iron Technology in East Africa. Symbolism, Science and Archaeology*. Oxford: James Currey Publishers.

Shaw, T., 1981. The Nok sculptures of Nigeria. *Scientific American*, 244, pp.154–66.

Shaw, T., 1985. The Prehistory of West Africa. In: J.F.A. Ajayi and M. Crowther eds. *History of West Africa*. Harlow: Longman, pp.48–86.

Tylecote, R.F., 1975. The Origin of Iron Smelting in Africa. *West African Journal of Archaeology*, 5, pp.1–10.

Vansina, J., 1973. *The Tio Kingdom of the Middle Congo, 1880–1892*. London: Oxford University Press for the International African Institute.

Vansina, J., 1978. *The Children of Woot: A History of the Kuba People*. Madison: University of Wisconsin Press.

Author's address

* Head of Department of Archaeology, University of Nigeria, Nsukka, Nigeria
drpamela26@yahoo.com

A comparison of early and later iron age societies in the Bassar region of Togo

Philip de Barros

ABSTRACT: The Bassar region of northern Togo has been the locus of two major ironworking societies over the last 2500 years. An Early Iron Age society is documented at the large ironworking centre of Dekpassanware and its satellite hamlets, dating from cal BC 420 to cal AD 130. After an apparent hiatus of a millennium, extensive ironworking reappeared after cal AD 1265 and thrived well into the 20th century. Enough data is now available on both periods to make a worthwhile comparison focused on: 1) regional settlement patterns; 2) intra-site spatial organisation in terms of habitation and industrial areas; 3) smelting and smithing technologies; 4) iron production levels; 5) diet; 6) material culture; and 7) probable ethnicity of the inhabitants. The data is derived from ethnographic, archaeological, and analytical research carried out by the author between 1981 and 2009, along with past and on-going specialist studies of botanical, faunal, and metallurgical remains.

Keywords: Ironworking, Early Iron Age vs. Later Iron Age, settlement patterns, technology, production, diet, material culture, ethnicity.

Introduction

The Bassar region is located in northern Togo, about 10° north of the Equator (Fig. 1).

An annual rainfall of 55 inches produces a mixed woodland-savannah habitat heavily impacted by past ironworking activities (deforestation), and past and current horticultural practices. The region contains nearly pure hematite ores (69% iron) near Bandjeli and numerous iron ore hills to the east (35-45% iron). The archaeological study of the region is based primarily on an extensive survey of slag mound sites identified by local farmers and ironworkers, and a stratified, systematic unaligned sample survey of a 14 x 3 km^2 area stretching from a major iron ore source in the east (*Bidjilib*) to the Katcha River in the west (Fig. 2). An area of nearly 10km^2 was intensively surveyed to obtain information on site size and settlement patterns in the context of a dated ceramic sequence. The survey recorded both Later Iron Age (LIA) and what would eventually be identified as Early Iron Age (EIA) sites, such as Dekpassanware (de Barros 1985, 1986, 1988, 2003).

Two major ironworking societies have occupied the Bassar region over the last 2500 years. An EIA society was centred on the 28 ha village and industrial site of Dekpassanware and its satellite hamlets, dating from 2500 to 2000 BP. After an apparent hiatus of a millennium, ironworking reappeared during the Later Iron Age, from about 650 BP to the mid-20th century (de Barros 1986, 2003). These EIA and LIA societies are compared below in terms of their temporal ranges, settlement patterns, intra site spatial organisation, ironworking technology, iron production levels, diet, material culture, and probable ethnicity of their members.

Temporal range

The EIA deposits at Dekpassanware have yielded radiocarbon dates ranging from cal BC 420 to cal AD 130. These are directly underlain by a Late Stone Age (LSA) component that dates from cal BC 1360 to 1050 and cal BC 830 to 400 (Table 1).

The LIA in Bassar is dated from cal BC 1265 to the 1950s, during which time Dekpassanware was reoccupied from cal AD 1280 to 1615. The LIA dating is based on ethnography, ethnohistory, and 16 calibrated, charcoal-based radiocarbon dates from the Bassar region (de Barros, 1986). Research carried out thus far has not found evidence for ironworking during the long hiatus between these ironworking occupations. This period is long enough that it cannot be easily explained by a simple ecological collapse of the early iron industry, followed by a period of reforestation, and the subsequent reintroduction of ironworking (Goucher, 1984; de Barros, 2000). This hiatus is likely due to the insufficient intensive sample survey of this vast region, whose core area – centred on the communities of Bassar, Kabu, Bandjeli, and Bitchabe – is minimally 625 km^2. Other ironworking sites or centres may exist during this hiatus in areas not yet surveyed and/or dated.

Regional settlement patterns

As noted above, Dekpassanware is underlain in places by terminal LSA deposits. The LSA regional settlement pattern, based on lithic and lithic-ceramic scatters recorded during the intensive survey, consists of dispersed hamlets and activity areas ranging from <0.5 to <3 ha in size (de

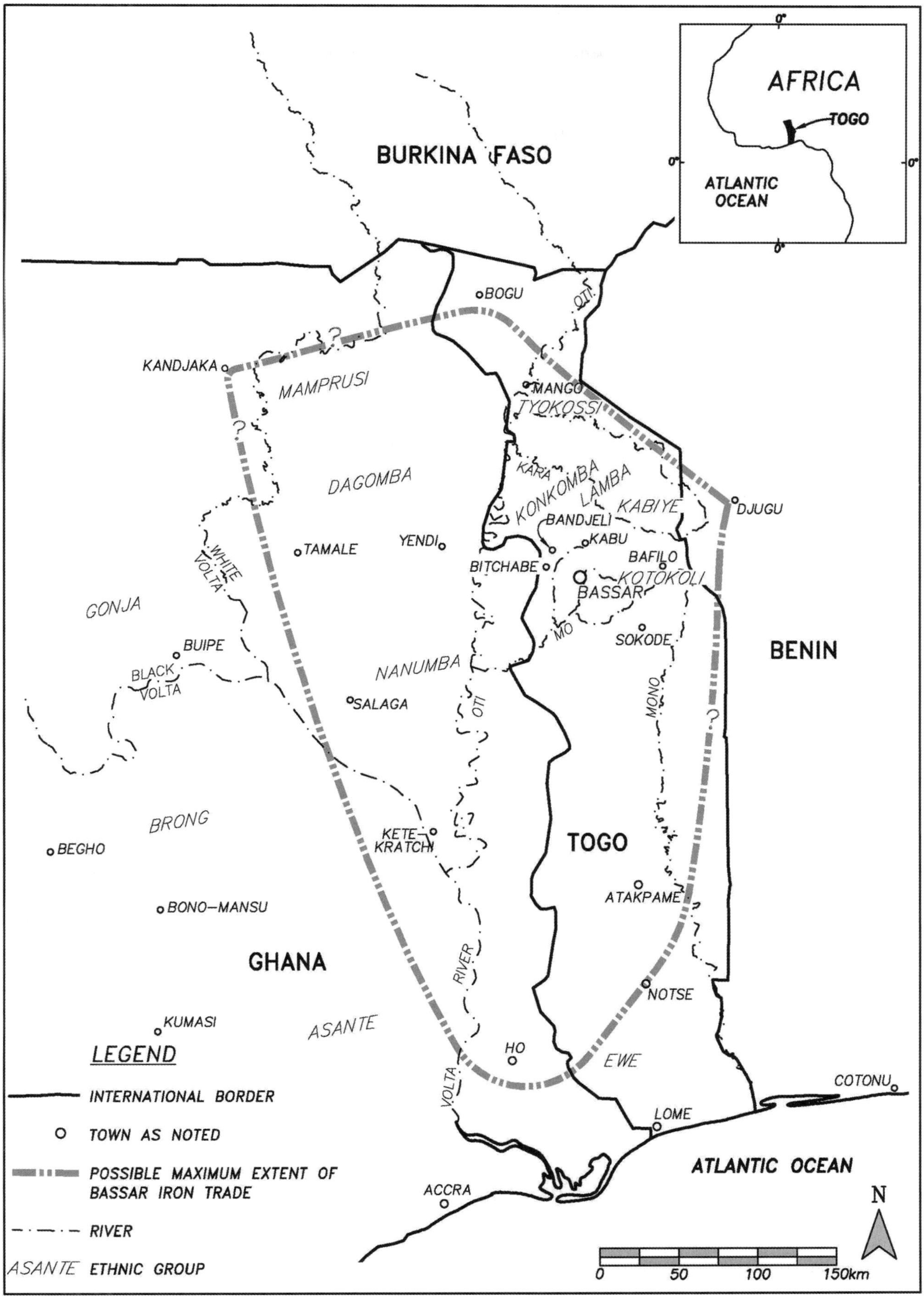

Figure 1: The Bassar region in Togo showing the maximum extent of Bassar iron trade in the LIA.

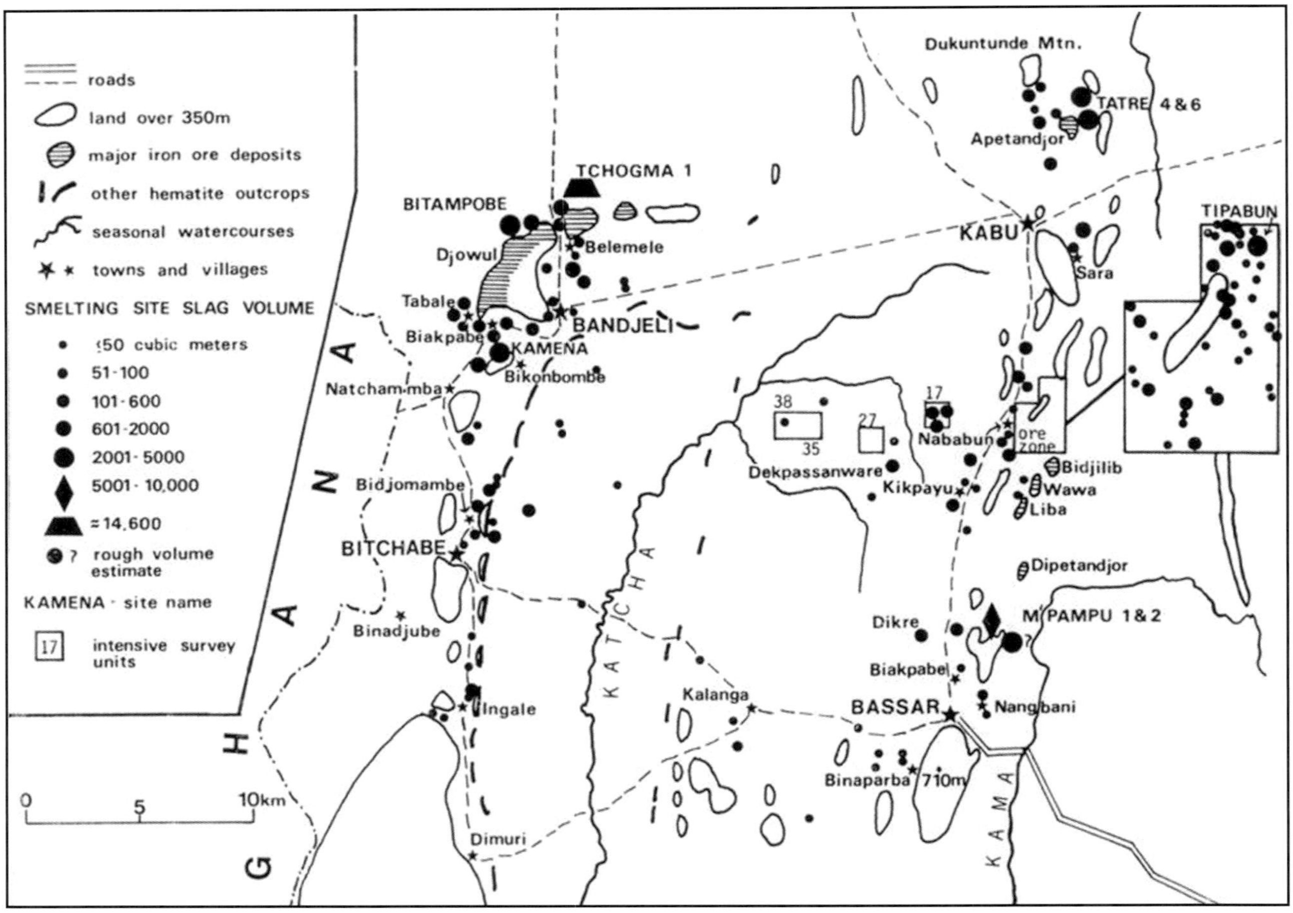

Figure 2: Bassar region smelting sites recorded during extensive and intensive surveys. Major ore deposits, intensive survey units, and the EIA site of Dekpassanware are also shown.

Barros 1985, pp.376-444). The EIA settlement pattern, also documented by survey, includes the 28 ha ironworking centre of Dekpassanware and five 1-2 ha satellite hamlets located 3-7 km away (Fig. 3). Some of these hamlets have ironworking debris (slag and tuyère fragments) and some do not. Perhaps reflecting its importance as a major centre, Dekpassanware is situated on a high gentle rise from which virtually the entire core of the Bassar region is visible (Fig. 2).

At the onset of the LIA (13-14th centuries AD), intensive survey results indicate a relatively dispersed pattern of sites averaging about 3 ha in size, with a range of <1 to 3.5 ha. As iron production increasingly became the dominant regional economic activity (late 16th century onwards), settlements shift eastward, aggregating in the *Bidjilib* iron ore zone. Regional iron production increases in the order of 300-450% during this period (de Barros, 1986). By the 17-18th centuries, the settlement pattern consisted of two large villages of 7 and 9.5 ha, respectively, and at least eight satellite hamlets averaging 3 ha, with a range between 1 and 5 ha (de Barros, 1988; Figs. 4-6). During the late 18th and 19th centuries, Dagomba slave raiding from the west and Tyokossi slave raids from the north led to the near total abandonment of the Bassar peneplain and the establishment of refuge sites near major mountains or hills. When the Germans arrived in the 1890s, they encountered regional iron specialisation along the western Bandjeli-Bitchabe-Dimuri axis – with smelting focused at Bandjeli, smithing north and south of Bitchabe, and charcoal making at Dimuri. This regional specialisation developed due to the economies of scale associated with large-scale production, and to regional resource variation. Mount Djowul near Bandjeli contains nearly pure hematite ores with little phosphorous or other impurities; the Bitchabe region is rich in high quality quartzite for making stone anvils and hammers; and Dimuri is located adjacent to the remaining forest in the region to the south. The eastern Kabu-Bassar axis consisted of the village chiefdom of Kabu (founded ca. 1860) to the north, with its specialist ironworking village at Sara, and the larger Bassar chiefdom to the south (founded ca. 1780), with its specialist smelting, smithing, and pottery villages near Mt. Bassar (de Barros, 1988; Fig. 7). Thus, during both the EIA and LIA, the pattern of central villages with satellite hamlets arises as ironworking becomes a major activity. This low-level settlement hierarchy may or may not have been associated with chiefdoms or some kind of informal 'big man' system (de Barros, 1988, 2001; Dugast, 1988). Finally, while the introduction of ironworking resulted in significant increases in regional population densities and maximum site size, as well as in the aggregation of settlements in the iron ore zone during the LIA large scale production, it did not lead to any significant increase in political centralisation during either period (de Barros, 1988). This conclusion is based on the small size of settlements and on the lack of evidence for any developed site size hierarchy or social stratification.

Table 1: Charcoal-based radiocarbon dates from Dekpassanware, in the Bassar region of northern Togo.

Beta Lab No.	Conventional C14 Date	Calibrated Date (2 sigma)	Unit	Site Area	10-cm Level	Associated Ironworking Debris	AMS or Extended Counting
Eastern Zone of Site: Units 23, 24, and 45 (2002-03, 2008) [Units 23 and 45 are directly adjacent]							
173469	2320 ± 60 BP	BC 520-350 and BC 310-210	23	Rubbish Dump	6	Iron ore	
173470	2030 ± 80 BP	BC 340-320 and BC 210-AD 130	23	Rubbish Dump	9	Slag	
169559	2300 ± 60 BP	BC 420-200	23	Rubbish Dump	10		
173471	2580 ± 60 BP	BC 830-740 and BC 710-530	23	Rubbish Dump	14		
252672	2210 ± 40 BP	BC 390-170	45	Rubbish Dump	15 147 cm	Slag (1) in Level 16	AMS
252674	2970 ± 40 BP	BC 1360-1350 & BC 1310-1050	45	Sparse debitage & ceramics	18 174 cm		AMS
169560	2260 ± 50 BP	BC 400-190	24	Residential Area	17		
Western Zone of Site (95 m west of Unit 23): Units 36 and 42 (2002-03; 2008 [directly adjacent units]							
169562	2150 ± 70 BP	BC 380-10	36	Rubbish Dump	8	Slag and ore	
173472	2110 ± 90 BP	BC 380-AD 70	36	Rubbish Dump	10	Slag	Ext Ct
173473	2210 ± 40 BP	BC 380-170	36	Rubbish Dump	12	Level 11: slag & ore	
169563	2520 ± 80 BP	BC 820-400	36	Rubbish Dump	13		
173474	2310 ± 80 BP	BC 750-700 & BC 540-190	36	Rubbish Dump	14		
252673	2200 ± 40 BP	BC 380-170	42	Rubbish Dump	15 150 cm	Slag	AMS
Industrial Zone 2 (160 m west of Unit 36): Unit 29 (2003)							
173468	2150 ± 80 BP	BC 390-AD 20	29	Adjacent to Industrial Zone	8	Slag & tuyere frags	Ext Ct
Later Iron Age Reoccupation - Large Slag Mounds: Unit 32 (2002) & unnumbered unit (1982)							
169561	640 ± 40 BP	AD 1280-1410	32	Slag Mound No. 5	5	Slag & furnace frags	
5352	330 ± 30 BP	AD 1470-1615 (1 sigma)	NA	Slag Mound No. 1	multiple levels	Slag & furnace frags	

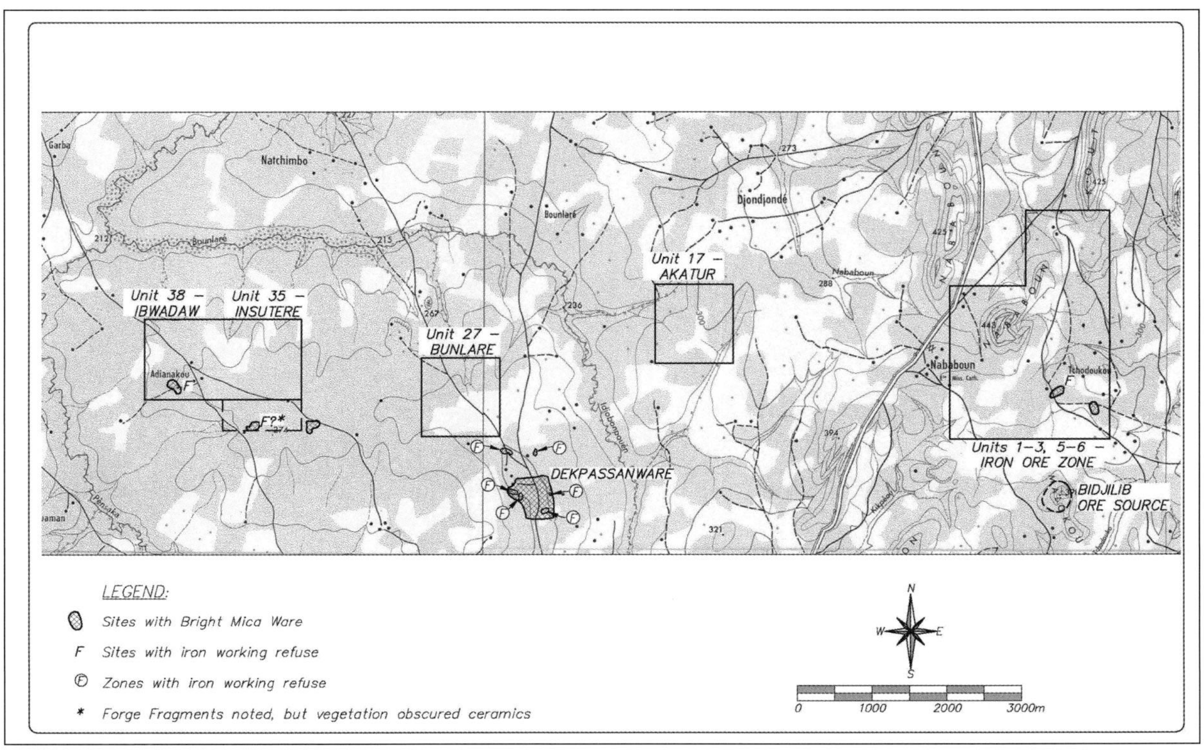

Figure 3: EIA settlement pattern, ca. 400 BC to 100 AD, with Dekpassanware and satellite hamlets.

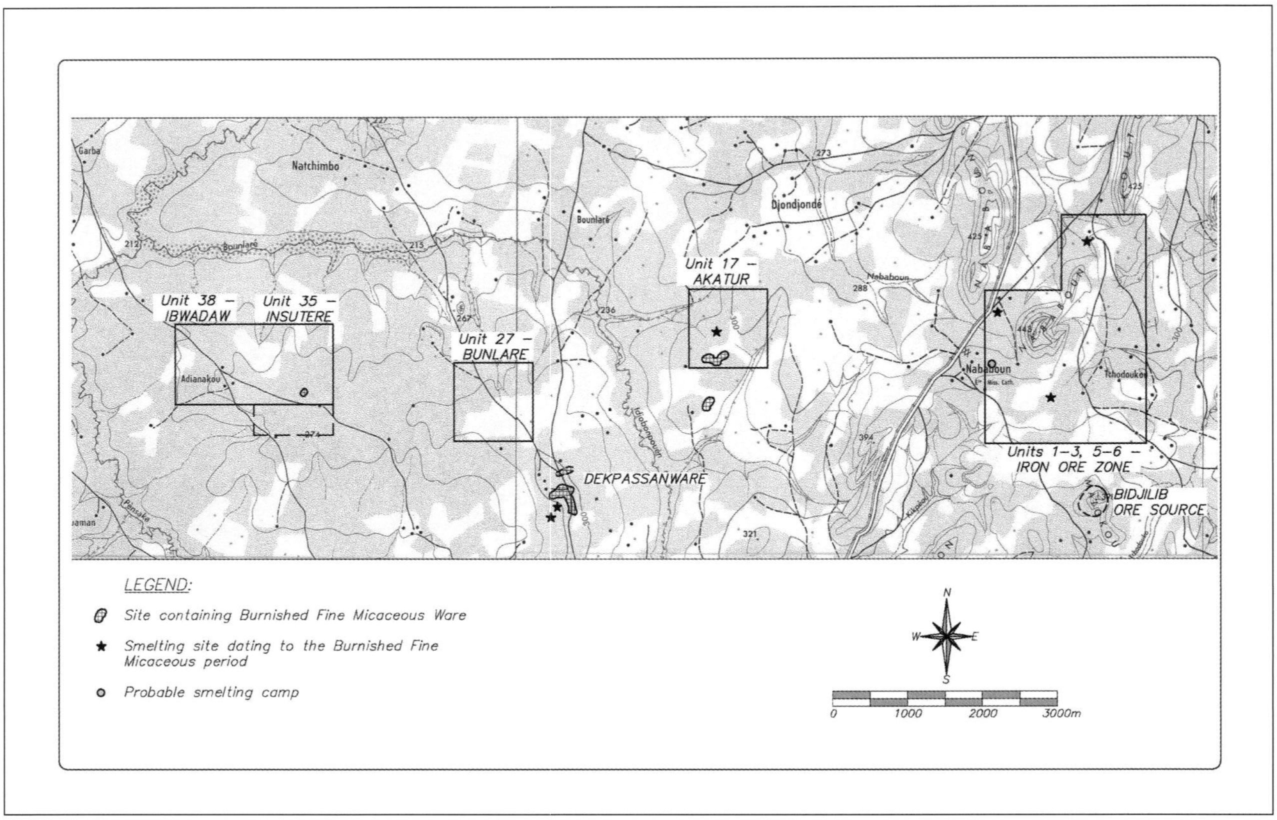

Figure 4: LIA settlement pattern, ca. 1300 to 1500 AD, showing habitation and smelting sites.

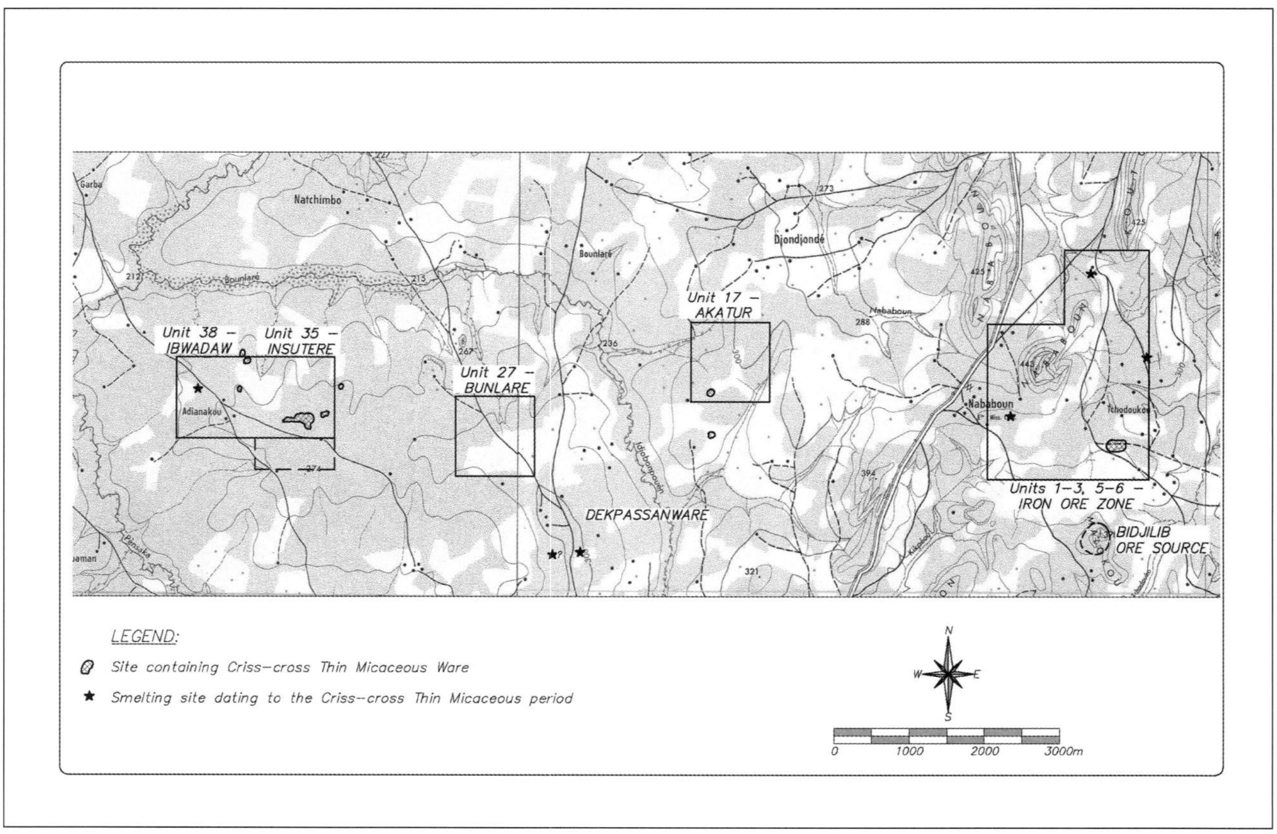

Figure 5: LIA settlement pattern, ca. 1500 to 1600 AD, showing habitation and smelting sites.

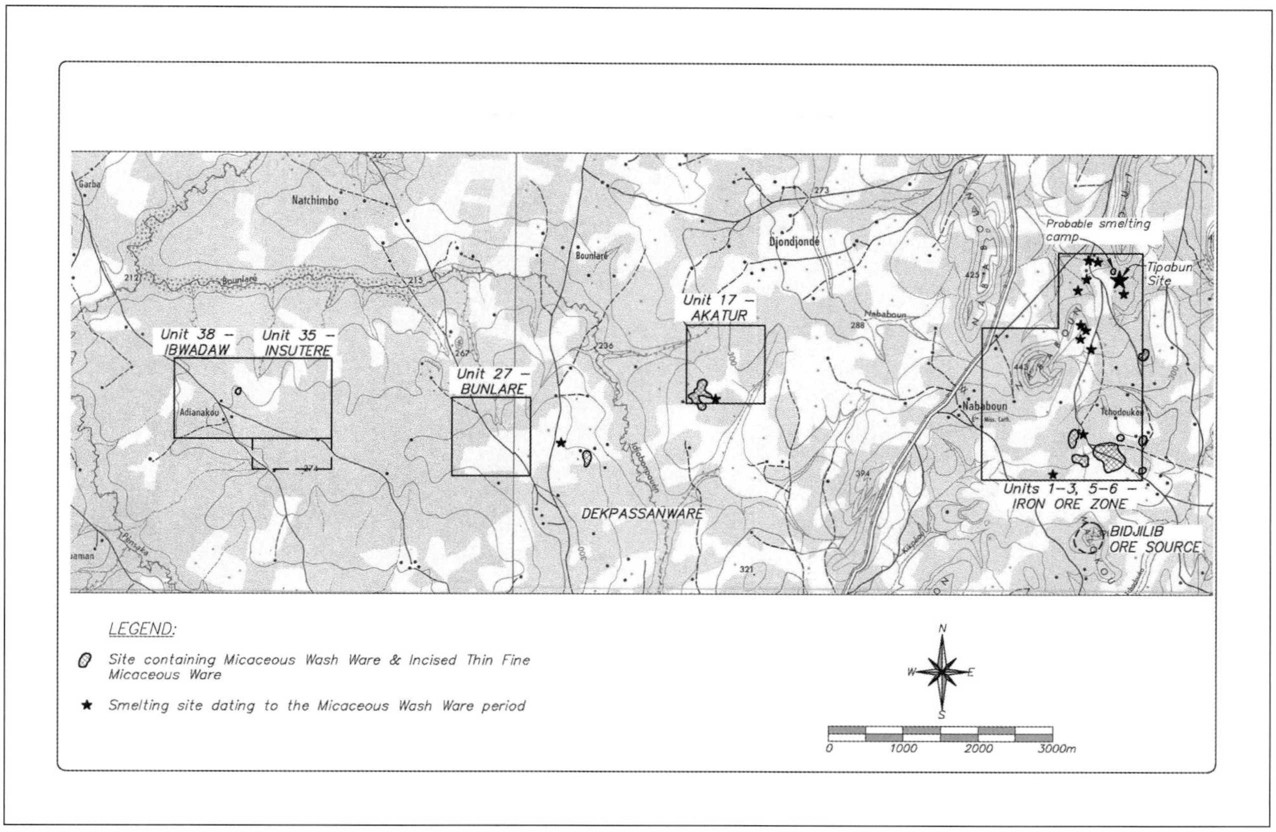

Figure 6: LIA settlement pattern, ca. 1600 to 1800 AD, showing habitation and smelting sites.

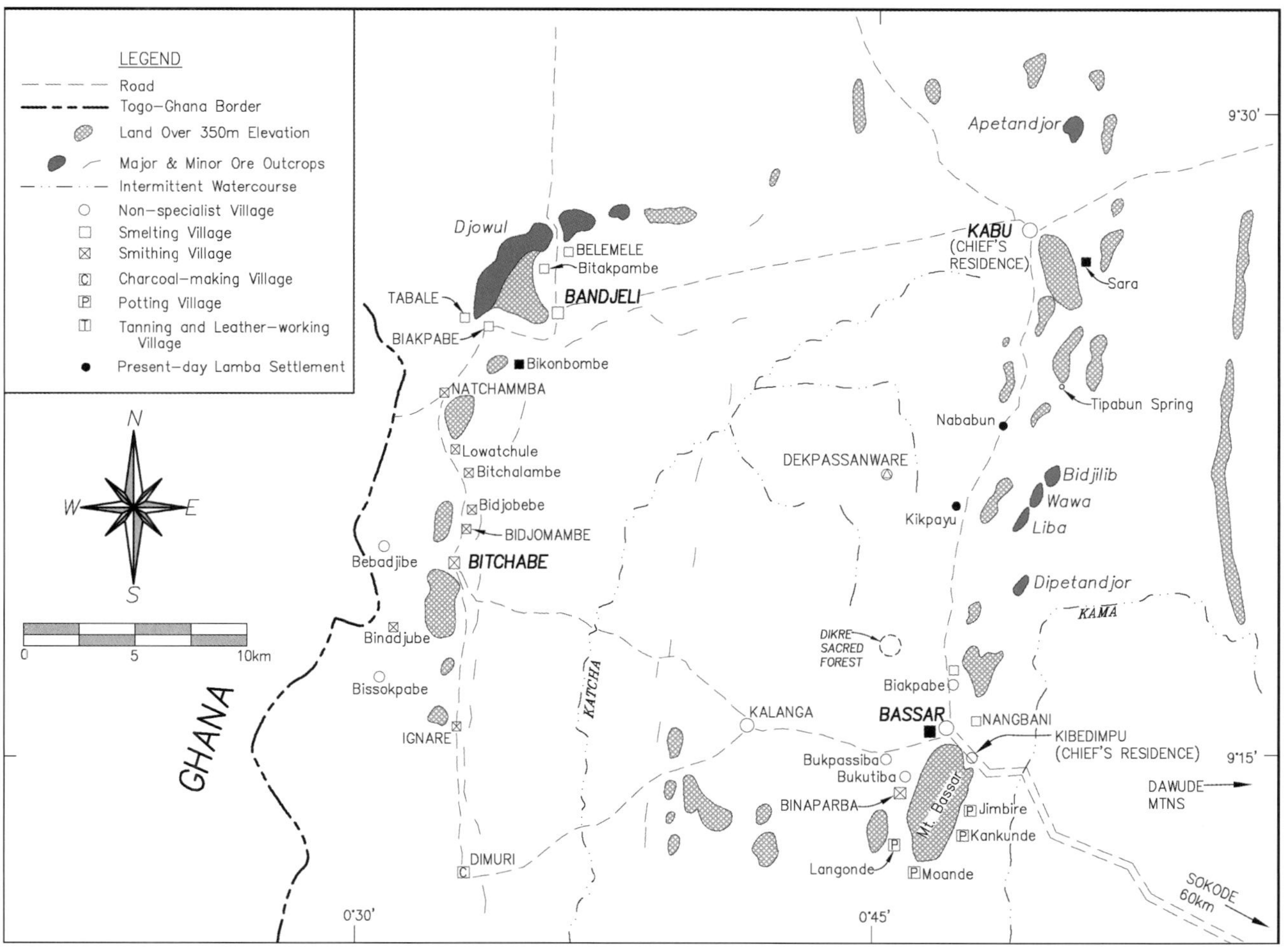

Figure 7: Bassar region specialist and non-specialist villages when the Germans arrived in the 1890s. The EIA site of Dekpassanware is also shown.

Spatial organisation of habitation and ironworking areas

In West Africa, including the Bassar region, several factors affected the position of the site where smelting took place. These include: the location of quality ores, proximity to water, appropriate wood fuel, and refractory clay sources for furnace and tuyère construction, and avoidance of habitation areas. Reasons for the latter included sexual taboos regarding women's participation in or proximity to smelting operations; fear of witchcraft and evil spirits; protection of ironworking secrets from espionage; and/or the avoidance of fire hazards (Goucher, 1984; Herbert, 1993; de Barros, 2000). Aside from iron bloom crushing, Bassar women did not participate in traditional smithing, but cultural taboos apparently did not lead to a significant spatial isolation of such activities. At the EIA site of Dekpassanware, where both smelting and smithing took place, four of the six industrial zones – which contain slag, ceramics, broken tuyères, stone hammer fragments, and in one case, stationary stone anvils and iron bloom crushing mortars (see I-1 in Fig. 8) – are on the periphery of residential areas; however, the other two zones are respectively 280 and 330 m to the north (Fig. 8). Since the industrial areas are generally similar in their contents, there apparently was no strict requirement that ironworking areas be located at a considerable distance from habitation areas. In fact, at the smaller hamlets with ironworking debris (slag, tuyères), there is no physical separation between habitation and ironworking areas; however, it is possible that only smithing took place at these sites.

At present, little is known about EIA smelting activities near the major iron ore deposits to the east (Fig. 2). This is partially because the intensive survey avoided dating small or surface slag deposits, to avoid charcoals introduced by rodent and agricultural disturbance (de Barros, 1985, pp.523-526); as a result, the two EIA hamlets in the *Bidjilib* ore zone have not been excavated or dated (Fig. 3). To verify the locations of forge activities, soil samples were collected from the vicinity of probable stone anvil sites and within industrial zones; these were then sifted for hammer scales and microspatter (*battitures*). These signatures of smithing activity were found in abundance in industrial areas 1, 2, and 4, including near the stationary anvils; traces were also found near a possible portable anvil. Samples from industrial areas 3 and 6 have not yet been examined. Future studies will systematically sample residential areas to check for forge locations, as is known ethnographically for the LIA (de Barros, 1985, pp.166-184).

Archaeological and ethnographic data from LIA sites in the Bassar region exhibit a pattern of distinct, often major, separation between smelting and residential areas (de Barros, 1985, 2001). In a few cases, smelting was carried out a short distance away from residential areas, but these may have been smelting camps (de Barros, 1985, 1988; Figs. 4 and 6). During the LIA reoccupation of Dekpassanware, the two smelting slag zones were located 95 and 240m from residential areas, respectively (Fig. 8). During the slave raiding of the late 18th and 19th centuries, which led to major population shifts (including the abandonment of Dekpassanware), major smelting sites continued to be set well apart from habitation sites, although smaller ones were occasionally close by (de Barros, 2001). However, after the end of smelting in the mid-20th century, people began to build houses near abandoned smelting sites, which sometimes renders interpretations of the archaeological record difficult.

During the LIA, smithing consisted of three operations – crushing iron bloom in stone mortars, purifying the bloom, and manufacturing iron tools, the latter two in the forge (Goucher, 1984; de Barros, 1985; Dugast 1986). Mortar locations were associated with good stone outcrops, which might have been somewhat removed from habitation areas, whereas forge activities were conducted either near the mortars and/or more commonly within or near habitation

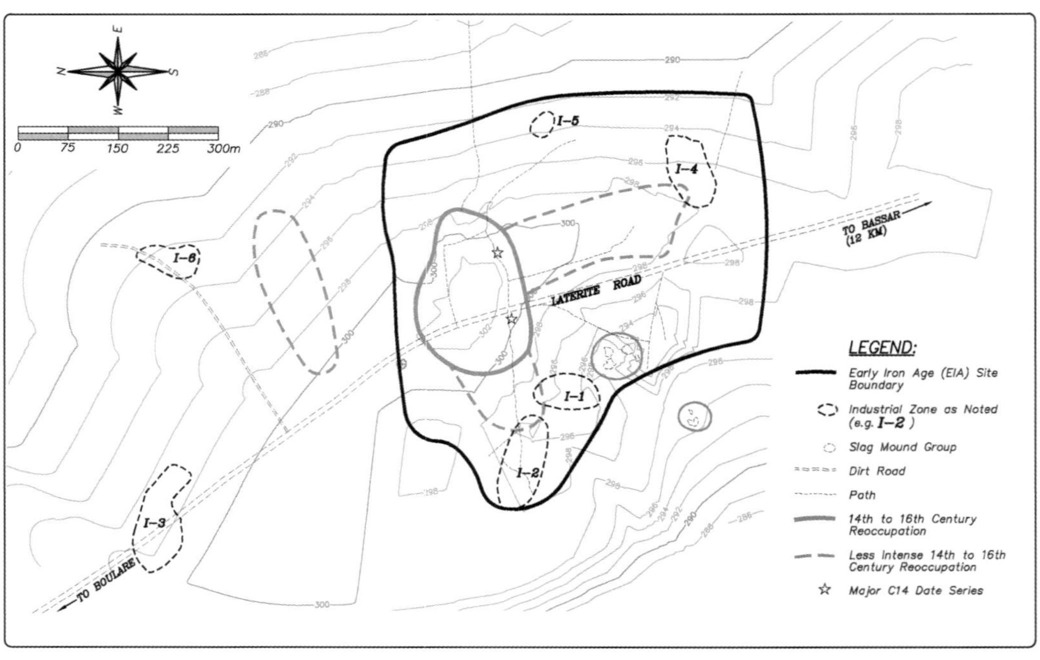

Figure 8: Dekpassanware and vicinity, showing EIA and LIA components.

areas. The intensive survey identified smithing debris (tuyères) associated only with EIA sites. This aspect is puzzling, and suggests that regional specialisation in smelting and smithing may date back to the early part of the LIA, with smithing perhaps carried out in the Bitchabe-Bidjomambe area, which was not intensively surveyed. Ethnographic and archaeological data obtained from the Bitchabe zone in 2002 suggest stone mortars were generally located near habitation areas. Prior to the slave raiding period, archaeological data suggest forges were grouped in large smithing zones near residential areas, but after contact smithing was more likely to be directly adjacent or within residential areas. Additional studies are necessary to clarify these patterns.

Ironworking technology

Based on the physical appearance and preliminary chemical analyses of ore and slag samples recovered from excavations at Dekpassanware, iron ores were probably mined during the EIA from the iron-rich hills and mountains of the region. There is no evidence for the use of lateritic ores. Little is known about the furnaces used, but the paucity of furnace structural debris suggests they were small and bellows-driven. A search for furnace emplacements on the surface or in shallow pits has not yet been successful. The smithing technology resembles that of the LIA, with conical tuyères, stone anvils, stone hammers, smithing furnace bottoms, and stone iron bloom crushing mortars (de Barros, 2006). The metallurgical analysis of an iron bracelet from an EIA burial indicates it is of good quality (low slag content), though somewhat heterogeneous in carbon content (0.1 to 0.6%), with evidence of folding but no quenching (David Killick, personal communication, 13 February 2009; Fig. 9). Initial chemical analysis of ores from both EIA and LIA components at Dekpassanware indicates that relatively high quality ores were being smelted during both periods (Vincent Serneels, personal communication, 12 February 2009).

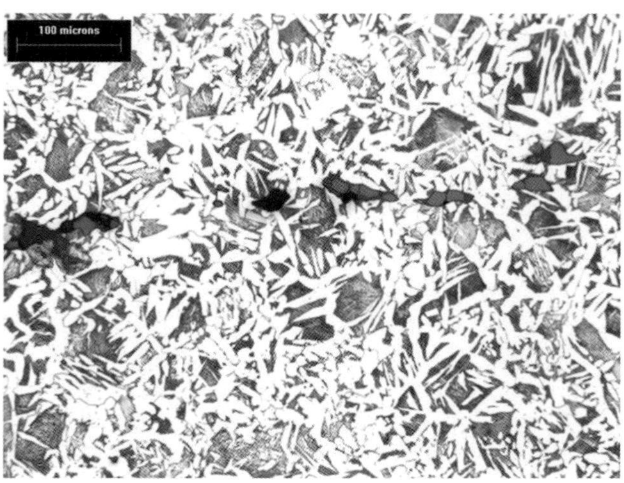

Figure 9: 200x etched slide of EIA bracelet, Dekpassanware: 1) slag inclusions: fat grey blobs of wüstite (FeO) in glass, in a 0.4% carbon steel; 2) white areas: ferrite (pure iron); 3) grey areas in fine structure: pearlite (eutectic of ferrite and iron carbide). Eutectic structure indicates that the bracelet was slowly air cooled, not water quenched.

Abundant ethnographic evidence, including data from replicative smelts (Goucher, 1984; Hahn, 1997), is available for the LIA (de Barros, 1985, pp.134-185; 1986). Ore was mined from iron-rich hills and mountains using iron picks with wooden handles, usually by women. Wood charcoal was made locally or imported by women, and highly refractory clays were obtained locally. Furnaces and tewels (circular liners for draft holes) were constructed by men. Tall, more efficient induced draft furnaces were used, which operated without bellows or tuyères. Smelting took 2 days in the Bandjeli area and 3-6 days in the eastern region, due to lower quality ores requiring more charcoal fuel and larger capacity furnaces. A typical iron bloom weighed 30kg. The blacksmith or his wife crushed the bloom in stone mortars using large spherical stone pounders or smaller hammer stones made of quartzite in order to remove as much slag as possible before forging. The crushed bloom, now free of powdered slag, was reworked in the forge to eliminate additional impurities and to obtain the desired carbon content (Goucher, 1984). Iron was worked into pill-shaped ingots (*lukopile*) that were later fashioned primarily into circular hoe blades, but also into axes, picks, arrow points, knives, bracelets, machetes, nails, and musical gongs. The all-male blacksmith team consisted of a master smith who manipulated the iron with iron tongs and later carried out the finishing work using metal hammers; a stone hammer man who wielded four different types of large quartzite hammers for rough-out work; and a third man who worked the bellows (Dugast, 1986). The forge was made of clay, often reinforced at the lip with a large everted jar rim. Tuyères (made from clay mixed with straw) were used to direct air from the concertina bellows (made of wood and animal skins) into the forge. Slag was not recycled for iron production or other major uses. Ongoing metallurgical analyses of iron bloom, tools, and bracelets by David Killick at the University of Arizona will help determine to what degree there was continuity and/or change between the EIA and LIA ironworking technologies.

Iron production levels

At the EIA site of Dekpassanware, all indicators suggest production for local and regional needs. Slag deposits are generally level with the surface (rarely 10-20 cm high) and range between 30 and 60 cm in depth. More importantly, the entire 28 ha site has only one bloom crushing mortar complex, with only seven mortars. Smelting activity at the site is suggested by the presence of: 1) subsurface ore nodules and nodules broken up in preparation for smelting; 2) two slag deposits that contain mostly heavy slag with few smithing furnace bottoms (Tylecote, 1962, p.190 and 3) two tuyères from these deposits completely plugged at one end by slag. However, the high density of furnace smithing bottoms in most excavated deposits, the virtual lack of furnace lining, the paucity of fired clay furnace fragments, and the inability thus far to identify furnace emplacements suggest the dominant ironworking activity at Dekpassanware was smithing. Smelting probably also took place closer to the *Bidjilib-Wawa-Liba* ore sources to the east, but this remains to be demonstrated (Fig. 2).

Patterns in production and use appear to change during the LIA. From the late 13th to the mid-16th century, iron was produced for both local and regional consumption, and was traded with adjacent ethnic groups; however, from the mid-16th to the early 20th century, production data (slag mound volume, large iron bloom crushing mortar sites) indicate large-scale production for long-distance trade (de Barros, 1985, 1986). During this time, production levels increased in the order of 300-450%. The induced draft furnaces used varied in size from 2-2.5 m in the Bandjeli area to 2.5-3.5+ m along the Kabu-Bassar axis – due to the important differences in iron ore quality noted above. The estimated annual regional production during the LIA was 7-20 tonnes from the 14th to the mid-16th century; 28-81 tonnes from the mid-16th century to ca. 1800 AD; 60-135 tonnes during the 19th century; and 150-200 tonnes from ca. 1900-1915 at Bandjeli alone. The latter figure is based on actual bloom weights measured at markets in the Bandjeli region by German colonial officials. The former figures are based on the conversion of iron slag mound volumes to iron production (de Barros, 1985, pp.317-364; 1986). The Bassar region was one of most important iron production centres in sub-Saharan Africa, with a volume of over 80,000 m^3 of slag, often found in very large mounds (Fig. 10). Bassar compares favourably with other large ironworking centres at Mema (Mali), Babungo (Cameroon), and Meroe (Sudan) (de Barros, 1986). This large-scale production is thought to be due to the rise of the Dagomba, Gonja, and Mamprusi states within the Volta Basin in the 15th and 16th centuries, which would have increased the demand for iron hoes and weapons, and which also helped stimulate long distance trade into the Middle Volta Basin, as did the rise of Bono-Mansu and its Asante successor state (de Barros, 1986).

Diet

The available evidence for the LSA deposits at Dekpassanware is limited, consisting primarily of faunal remains that suggest that the consumption of domesticated cattle was not significant. Faunal analyses by Murrey (2003) and Linseele (2008), and the study of charred seeds and wood from flotation and excavation samples by Eichhorn and Kahlheber (2002) and Eichhorn (2008) indicate the EIA diet at Dekpassanware consisted of domesticated pearl millet (*Pennisetum glaucum*) and cowpeas (*Vigna unguiculata*) or black-eyed peas, as well as domesticated West African shorthorn cattle (*Bos taurus brachyceros*; cf. Rege, et al., 1994), whose remains dominate the faunal sample. Small amounts of goat/sheep and guinea fowl (*Numida meleagris*) were also recovered – as opposed to chicken, which probably arrived in West Africa around 100-500 AD (Macdonald and Blench, 2000). Yams are inferred based on current cultivation patterns and the assumption of its use (Harlan, et al., 1976; Sowunmi, 1985). Attempts to extract yam starch from excavated ground stone tools are ongoing, carried out by Amanda Logan.

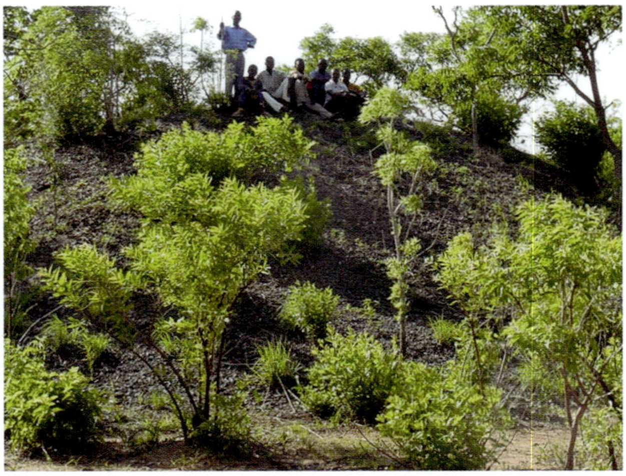

Figure 10: Large 5m+ high slag mound near the LIA habitation site of Titur, north of Bandjeli.

Domesticated sorghum has not yet been documented prior to the 1st century AD at 10° north latitude (Kahlheber and Neumann, 2007). Moreover, except for one unripened grain of what might be sorghum (Stefanie Kahlheber, personal communication, 3 September 2008), this plant was not recovered at Dekpassanware. The project area is also within the zone of wild fonio, which extends from central Nigeria into northern Benin and Togo (Ruskin, 1996), but no wild or domesticated fonio was recovered either. Although the sample size is small, the shorthorn cattle at Dekpassanware may be dwarf shorthorn (Linseele, 2008), although today's dwarf shorthorn cattle are found only in more humid and forested environments than those typical of Bassar's savannah-woodland. Wild animal foods are surprisingly uncommon as compared to sites in Ghana (Gautier and Van Neer, 2005), but there is some evidence for hare (*Lepus sp.*), antelope, duiker (*Cephalophus* or *Sylvicapra sp.*), and rodents – probably the cane rat or grasscutter (*Thryonomys swinderianus*). A fragment of giant land snail (*Achatina sp.*) was also recovered. Some bones could be either of cattle or savannah buffalo (*Syncerus cafer*), but the lack of secure buffalo identifications suggests they are all of cattle (Linseele 2008). Charcoal studies also suggest the presence of a park savannah at this time, with probable karité, néré, and possibly baobab (Eichhorn 2008).

During the LIA, early observations by Frobenius (1913) and Klose (1899) suggest the following for the precolonial LIA diet: yams; millet and sorghum or 'gros mil' (*sorghum bicolor*), although the date of introduction of sorghum is not known; beans (cowpeas); groundnuts (*Arachis hypogaea*); sesame seeds, and taro and tobacco at the house; and domesticated cattle, dwarf goats, sheep, guinea fowl (*Numida meleagris*), and chicken (*Gallus gallus*). Wild animal foods included antelope, hare, leopard (*Panthera pardus*), bush pig (*Potamochoerus porcus*), monkeys, and cane rat. Wild plant foods included karité (*Butyrospermum parkii*), néré (*Parkia biglobosa*), baobab (*Adansonia digitata*) seeds and greens, palm nuts, and wild leaves and roots. Hippos were occasionally hunted, and the cultivation of okra and red peppers is likely.

Aside from slag mounds, excavated data for the LIA are limited to the upper 50cm at Dekpassanware. Surprisingly, faunal remains are not common, and only sparse remains of domesticated cattle are present. This major difference between the EIA and LIA ironworking occupations at Dekpassanware is not easily explained given the ethnohistoric data emphasising the ownership of cattle. However, given the limited sample size of the LIA deposits, it would be presumptuous to assume that EIA populations depended more upon cattle for food than did LIA populations.

Other aspects of culture

The evidence from Dekpassanware suggests that ironworking did not completely replace flaked stone technology. A few quartz flakes, core fragments, and scrapers were recovered from EIA levels in residential areas, although such material was virtually absent in industrial areas. It is not clear whether LSA artefacts were introduced via rodent disturbance or whether flaked stone tools continued to be used. Since flaked stone fire starters are still used today, it is likely that remnants of flaked stone technology persisted. Ground stone tools (manos, metates, hand stones), hammer stones, polished pebbles, fire-altered rock, and abundant ceramics are present. Most ceramics were made from clays derived from rock types found east of the Bassar region (de Barros, 1985, p.543), indicating non-local manufacture. Clear quartz, rose quartz, chalcedony, chert, and shell beads are present in small quantities (de Barros, 2008). Ceremonial items, such as a bird-like object made of volcanic tuff and a rasp-like object (non-Kintampo), are also present. Iron bracelets (and a finger ring) of different types are common, along with fragments of knives, hoes, and iron wire. Several burials are known at the site, some excavated and some found by a local farmer digging a borrow pit. Excavations in 2002 identified what is either a communal or familial burial containing at least four adults and one juvenile, the latter with a large jar base fragment in direct contact with its skull. Two adults and one child wore iron anklets, bracelets or necklets.

Information on LIA material culture is more limited. It is based primarily on the extensive and intensive regional surveys and associated surface collections, the excavation of slag mounds, and material recovered from the upper 50 cm at Dekpassanware. Ethnographic data indicate that while cemeteries existed, burials were not communal or familial because adults and children were buried separately, and grave goods did not generally accompany the deceased (Cornevin 1962; Tigon 2002; Stéphan Dugast, personal communication, 12 March 2009). No LIA burials have been excavated. During the LIA, ceramics were even more abundant than during the EIA. Based on resemblances to ethnographic forms and present-day ceramic trading patterns, ceramics were probably imported from the neighbouring Losso or possibly Nawuda (Niamtugu) to the northeast, the Kabiye to the east, the Kotokoli to the southeast, and the Konkomba from the north, the latter primarily for the Bandjeli area (de Barros, 1985, pp.541-593). Calabash millet beer drinking bowls appear to have been replaced by fine, thin-walled, smudged and decorated imported ceramic drinking bowls. Recovered geometric potsherds are probable remnants of potsherd pavements, based on the discovery of eroded pavement remnants at a late 18th century site. After ca. 1800, slave raiding radically altered Bassar settlement patterns, and imported ceramics declined drastically in quantity, with new, local clay sources subsequently developed near Mount Bassar (de Barros, 2001; Fig. 7).

Probable ethnic affiliation

Attempting to infer ethnic affiliation from the archaeological past is often very problematic. However, ethnographic, ceramic, technological, and mortuary data strongly suggest the EIA inhabitants of Dekpassanware were related to the present-day Kabiye (and/or Lamba). Oral traditions collected by Froelich and Alexandre (1960) suggest the region was formerly occupied by the Lama, ancestors of the present-day Kabiye and Lamba. The paste of the dominant Bright Mica Ware contains minerals derived from gneisses and granites from the Precambrian shield exposed in the Kabiye region to the east (de Barros, 1985, p.573). Moreover, the vessel forms strongly resemble those of Kabiye potters today, including the frequent use of hollow ring bases and groove and incision decorative motifs. In addition, the Kabiye who claim to be indigenous – stating that their founding ancestors descended from the sky – are also the only other ironworking group in Togo who use large stone hammers for making iron tools (Posnansky and de Barros 1980; Hahn, 1997). Finally, Bassar mortuary patterns do not include communal or familial tombs with adults and children buried in the same area, whereas the Kabiye do have burial chambers for clan or family burials (Posnansky and de Barros, 1980, pp.20-26). Local informants also stated the Bassar do not normally bury their dead with grave goods, whereas the Dekpassanware EIA burials frequently contain iron bracelets, anklets, and necklets worn by the deceased.

During the LIA, ethnographic, archaeological, and metallurgical data suggest that induced draft furnaces were imported into the Bassar region, perhaps from the north, and that the Bassar represent a mixture of indigenous peoples, with larger numbers of later ironworking (and non-ironworking) migrants attracted to the rich ironworking centres of the region (de Barros, 1985; Dugast, 1992). Cornevin (1962), de Barros (2000), and Dugast (1987) have documented the multiple origins of Bassar clans: Dagomba, Konkomba, Gurma, Mossi, Gonja, Kotokoli, Lamba, Kabiye, Tyokossi, and others. But who were the indigenous peoples of the Bassar region at the onset of the LIA? While the evidence strongly suggests Lama-Kabiye peoples 2000 years ago, the evidence is ambiguous for the LIA. Various traditions suggest possibly indigenous Lamba at Kankundi, possibly indigenous Gurma at Byakpabe, and possibly indigenous Gonja (Nataka clan) at Nangbani and formerly at the sacred forest of Dikre (Dugast, 1987; de Barros, in press). These data could support the hypotheses of either a long-term development of the Bassar from ancient indigenous inhabitants and/or the replacement and/or absorption of ancient indigenous peoples of another origin (Stéphan Dugast, personal communication, 12 March 2009).

Conclusion

The EIA-LIA comparison suggests the following: 1) the presence of ironworking during the EIA and LIA correlates with the rise of a low-level settlement hierarchy, but surprisingly without any dramatic increase in political centralisation (see Figs. 3-6; de Barros, 1988); 2) smelting areas tend to be spatially separated from habitation areas, but this is not necessarily the case for smithing areas; 3) whereas EIA furnaces were probably bellows-driven, the large-scale production of the LIA was associated with the introduction of large, induced draft furnaces; 4) a relatively good quality carbon steel was already intentionally produced during the EIA; 5) millet, cowpeas, and shorthorn cattle were important in the EIA diet, but the relative absence of cattle in LIA deposits remains to be explained; 6) the EIA population appears linked to Lama-Kabiye (gurunsi) peoples, whereas the LIA one is linked to a mixture of indigenous Lama and/ or other paragurma-speaking peoples and later immigrant groups, together called the Bassar; and, 7) further research is needed to determine whether the millennium-long gap between the EIA and LIA is real or due to sampling issues.

Acknowledgements

Special thanks are due to Merrick Posnansky for his constant encouragement; to Stéphan Dugast for his deep knowledge and understanding of Bassar culture; to Bossman Murrey, Veerle Linseele, David Killick, Barbara Eichhorn, Stephanie Kahlheber and Vincent Serneels for specialist studies; to A. Gnon-Konde, Yao Nambou, Kokou Tagareb, Yao Adjankli, and other informants from Bandjeli, Bassar, Bitchabe, and Kabu; to Dola Aguigah for providing students for training and to Dovi Kuevi for logistical support.

References

Cornevin, R., 1962. *Les Bassari du Nord-Togo*. Paris: Berger-Levrault.

de Barros, P., 1985. *The Bassar: large-scale iron producers of the West African savanna*. PhD thesis, University of California at Los Angeles. Michigan: University Microfilms International, Ann Arbor.

de Barros, P., 1986. Bassar: a quantified, chronologically controlled, regional approach to a traditional iron production centre in West Africa. *Africa*, 56, pp.148-73.

de Barros, P., 1988. Societal repercussions of the rise of large-scale traditional iron production: a West African example. *African Archaeological Review*, 6, pp.91-113.

de Barros, P., 2000. Iron metallurgy: sociocultural context. In: J. Vogel ed. *Ancient African metallurgy*. Walnut Creek, California: AltaMira Press, pp.147-98.

de Barros, P., 2001. The effect of the slave trade on the Bassar ironworking society. In: C.R. DeCorse ed. *West Africa during the Atlantic slave trade: archaeological perspectives*. London: Leicester University Press, pp.59-80.

de Barros, P., 2003. Recent early Iron Age research in Bassar, Togo. *Nyame Akuma*, 59, pp.76-8.

de Barros, P., 2006. The surprising Early Iron Age site of Dekpassanware (Togo). Paper presented at the *Society for Africanist Archaeologists 18th Biennial Conference, Calgary, 23-26 June 2006*.

de Barros, P., 2008. Dekpassanware: Early Iron Age site in the Bassar region of northern Togo. Paper presented at the colloquium *Cultural developments and technological innovations in 1st millennium BC/AD West Africa, Ouagadougou, 12-14 March 2008*.

de Barros, P., in press. The rise of the Bassar chiefdom in the context of Africa's internal frontier. In: J.C. Monroe and A. Ogundiran eds. *The politics of landscape in Atlantic West Africa*. Cambridge University Press.

Dugast, S., 1986. La pince et le soufflet: deux techniques de forge traditionnelle au Nord-Togo. *Journal des Africanistes*, 56, pp.29-53.

Dugast, S., 1987. L'Agglomération de Bassar: classement selon le clan fondateur; classement par village. Unpublished paper.

Dugast, S., 1988. Déterminations économiques *versus* fondements symboliques: la chefferie de Bassar. *Cahiers d'Etudes africaines*, 28, pp.265-80.

Dugast, S., 1992. *Rites et organization sociale: l'agglomération de Bassar au Nord-Togo*. PhD Thesis, École des hautes études en sciences sociales, Paris.

Eichhorn, B., 2008. Charcoal analysis (Dekpassanware excavations in Togo): preliminary report for 2008 samples. Unpublished paper.

Eichhorn, B. and Kahlheber, S., 2002. Comments on the identification of archaeobotanical remains of Dekpassanware, Togo. Unpublished paper.

Frobenius, L., 1913. *Und Afrika sprach III: unter unsträflichen Aethiopien*. Berlin: Bita, Deutsches Berlagshaus.

Froelich, J.C. and Alexandre, P., 1960. Histoire traditionnelle des Kotokoli et Bi-Tchambi du Nord-Togo. *Bulletin de l'Institut Fondamental d'Afrique Noire*, 22, pp.228-74.

Gautier, A. and Van Neer, W., 2005. The continuous exploitation of wild animal resources in the archaeozoological record of Ghana. *Journal of African Archaeology*, 3, pp. 195-212.

Goucher, C.L., 1984. The iron industry of Bassar, Togo: an interdisciplinary investigation of African technological history. PhD thesis, University of California, Los Angeles.

Hahn, H.P., 1997. *Techniques de métallurgie au Nord-Togo*. Lomé: Collection 'Patrimoines', Presses de l'Université du Bénin.

Harlan, J.R., de Wet, M.J. and Stemler, A.B.L. eds., 1976. *Origins of plant domestication*. The Hague: Mouton.

Herbert, E.W., 1993. *Iron, gender, and power: rituals of transformation in African societies*. Bloomington: Indiana University Press.

Kahlheber, S. and Neumann, K., 2007. The development of plant cultivation in semi-arid West Africa. In: T.P. Denham, J. Iriarte and L. Vrydaghs eds. *Rethinking agriculture: archaeological and ethno-archaeological perspectives*. London: University College Press, pp.318-44.

Klose, H., 1899. *Togo unter deutscher flagge: reisebilder und betrachtungen*. Berlin: Reimer.

Linseele, V., 2008. Faunal analysis, Dekpassanware. Unpublished paper.

MacDonald, K.C. and Blench, R.M., 2000. Chickens. In: K.F. Kiple and K.C. Ornelas eds. *The world history of food, Vol. I*. Cambridge: Cambridge University Press, pp.496-9.

Murrey, B., 2003. Faunal analysis, Dekpassanware, Bassar region, Togo. Unpublished paper.

Posnansky, M. and P.L.F. de Barros, 1980. An archaeological reconnaissance of Togo, August 1979. Self-published in book form.

Rege, J.E.O., Aboagye, G.S. and Tawah, C.L., 1994. *Shorthorn cattle of West and Central Africa I: origin, distribution, classification and population statistics*. New York: Food and Agricultural Organization, United Nations.

Ruskin, F.R. ed., 1996. *Lost crops of Africa, Vol. I, grains*. Washington DC: National Academy Press.

Sowunmi, M.A., 1985. The beginnings of agriculture in West Africa: botanical evidence. *Current Anthropology*, 26, pp. 127-9.

Tigon, D., 2002. Système et mode d'enterrement chez les Bassar. Oral traditions from N. Djayo, at Bidjomombe, near Bitchabe, 13 June 2002.

Tylecote, D., 1962. *Metallurgy in archaeology. A prehistory of metallurgy in the British Isles*. London: Edward Arnold.

Author's address

- 13730 Via Cima Bella, San Diego, CA 92129
pdebarros@palomar.edu

Our iron smelting ^{14}C dates from Central Africa: from a plain appointment to a full blown relationship

Bernard Clist

ABSTRACT: This paper reviews the ^{14}C dates associated with early iron smelting in Central Africa before 1900 BP.

All of the archaeological sites from which these early dates have been obtained are critically examined. For example, each dated sample is checked for its stratigraphic integrity, its degree of association with dated artefacts, and its associated ceramic sequence.

The series of 'good' 14C dates obtained from this re-appraisal is then presented at the regional level in order to provide a new paradigm of the expansion of iron metallurgy through Central Africa, from Cameroon to Congo.

The work presented here follows several published papers and book chapters, and is based on research within the region by the author which began in 1982.

Keywords: Central Africa, Sub-Saharan Africa, Iron Age, Neolithic, radiocarbon dating, iron smelting, village expansion, iron smelting expansion.

Introduction

Back in 1977, at the time that the first synthesis of radiocarbon dates from Central Africa was published in the *Journal of African History*, the oldest date for iron metallurgy was around the 4th century AD (Maret, et al., 1977; David, 1980). Following the excavations carried out in the Upemba depression of Zaïre (now the Democratic Republic of Congo, DRC), we glimpsed that from the 6th century AD onward, iron was widely known and used in the south-eastern reaches of the Congo river basin. This was later pushed back in time to the 3rd century AD, in the Lubumbashi area (DRC) (Maret, 1978, 1982, 1985; Anciaux de Faveaux et Maret, 1984).

In 1985 the laboratory results from the Obobogo archaeological site (Cameroon), which had been excavated in 1983, enabled us to jump to 2120 and 2310 BP (Maret, 1985). There iron slag was found in deep refuse pits, which had been dug from the village level which was later disturbed by fields and buildings.

The evidence unearthed in Cameroon was quickly confirmed by fieldwork carried out in Gabon. In 1985, the earliest iron in Gabon was cited at 1850 BP (Clist, et al., 1986; Clist, 1989a). Soon, however, a series of radiocarbon dates closely associated with archaeological evidence identified an Early Iron Age period in which iron smelting bracketed between 2350 and 2050 BP around the country (Digombe, et al., 1985; Schmidt, et al., 1985; Clist, 1989b; regional approach in Lanfranchi and Clist, 1991).

It must be stressed that the Gabonese data demonstrates an unequal distribution of early iron smelting and iron use across the country. While in the north-west of Gabon there is no evidence for iron production around 2300 BP, evidence brought to light during excavations led by Peter Schmidt (Digombe, et al., 1987; Clist, 1995, 1997, 2005) suggests that smelting existed on quite a large scale in the south-east, in the Haut-Ogooué province at Moanda.

Before progressing to consider the modern picture of early iron smelting in Central Africa it is important to mention two methodological issues: the context of the dates, and the identification of the associated archaeological data. H.-P. Wotzka has been able to state very recently that, "Collecting dates is tantamount to running into continual trouble, to do first with missing, incomplete, inaccurate or distorted contextual information [...] considerable time is wasted trying to track down or cross-check information which every researcher would have submitted along with their samples". (Wotzka, 2006, pp.273-274).

In an effort to resolve these issues I propose a revised, regional approach to radiocarbon dates which relate to the earliest evidence of iron smelting in Central Africa. I believe that this new work will be of major interest as a result of the dynamics of archaeological research carried out since 2005 in Cameroon and Gabon, and due to certain publications aiming to increase to c. 3000 BP the beginning of iron smelting in Africa (e.g. Zangato, 2007).

When investigating the history of iron smelting in Central Africa, there are three types of sites which need to be considered. The first type contains the remains of furnaces and direct evidence of smelting, and clearly pinpoints iron smelting centres; the second contains iron smelting residues such as tuyères and iron slag, either in stratigraphic layers or refuse dumps, which suggests iron smelting was carried out in close proximity to the excavation site. The final type of site includes those where iron objects are found, suggesting the possibility that iron implements had been imported from smelting centres elsewhere.

The corpus of radiocarbon dates and the diffusion of iron smelting

Recently, all the sites with archaeological finds relating to iron smelting were reviewed (Clist, 2005). The resultant corpus of radiocarbon dates was then discussed in close relationship with their local context: degree of association with artefacts, stratigraphy, various typological or laboratory analyses of artefacts, etc.

This work has been updated to incorporate data found up to December 2008 and a new catalogue of early iron smelting sites from Central Africa before 1900 BP has been published (Clist, 2012). Within this catalogue explanations will be included to help the reader understand why a specific date has been put aside; or why it must be used with the utmost caution (the reader is referred to this catalogue as no further critical discussion can be advanced within the scope of this paper).

The maps in Figs. 1-6 show the geographical dispersal of those archaeological sites which show evidence of iron smelting, associated to calibrated time frames **(1)**:

- before 500 BC: iron smelting is known in western Centrafrique, perhaps also to the north-east of Yaoundé in Cameroon at Zili and in Rwanda/Burundi on some rare sites (Fig. 1);
- between 500 and 400 BC: iron smelting centers are set up in Centrafrique, in and around Yaoundé (Cameroon **(2)**), in Rwanda/Burundi and in Gabon at Moanda (Fig. 2). The extension of iron smelting to the south was likely made possible by the dry climatic spell which opened up savannah corridors through the equatorial forest and caused heavy forest fragmentation around 650 BC (see Clist, 2006a; Maley, 2004; and Fig. 7); coupled with the still open environment and the presence of communities of villagers, who settled in the western parts of the equatorial forest and throughout south-east Gabon. Settled from 1200 cal BC in southern Cameroon (Obobogo site at Yaoundé) the successive waves of advance of these peoples can be traced to Gabon, where the earliest evidence is today dated to 600-500 BC with the Okala Tradition (Clist, 2005, 2006b). The Gabonese villagers had not been established in the savannahs of Haut Ogooué long before they discovered or were introduced to iron smelting.

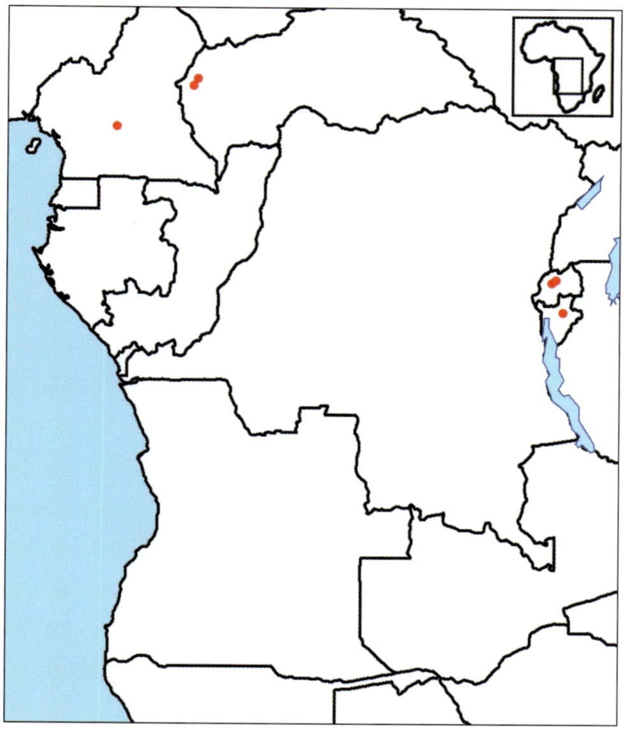

Figure 1: Iron smelting sites before 500 cal BC.

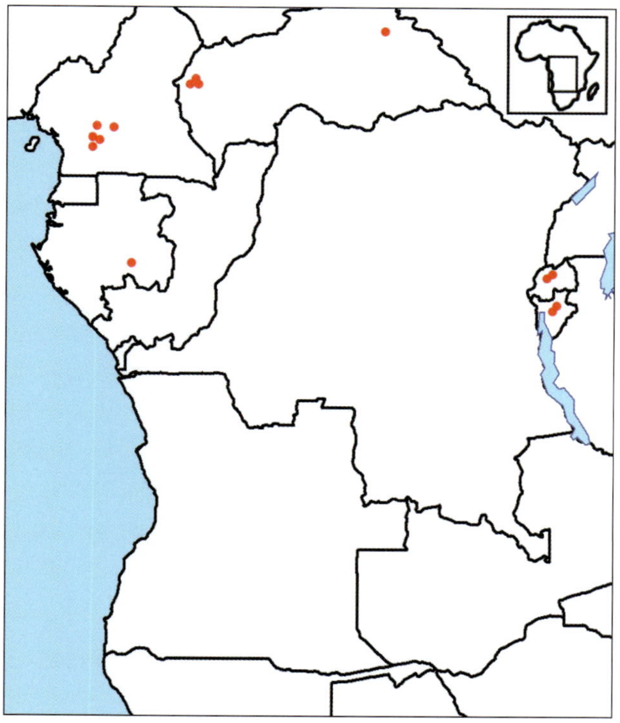

Figure 2: Iron smelting sites from 500 to 400 cal BC.

- 400 to 300 BC: other smelting sites have been found deep in the forests of southern Cameroon and northern Gabon (Fig. 3). The use of iron implements amongst Neolithic villagers was slowly becoming more widespread; perhaps some areas only imported iron implements based on economic ties established generations earlier
- 300 to 150 BC: iron smelting was still expanding to villages at this point and local production tended to be the general rule. Iron smelting became common along the Atlantic coast of Congo (Fig. 4);

- 150 BC to 100 AD: the metallurgists expanded to the Angolan border, the lower Congo and upstream along the major rivers of the Congo basin; deep into the equatorial forest (Figs. 5 and 6).

From Fig. 8, using the Gbabiri dates in Centrafrique (group B of dates), it is possible to infer that iron smelting started around c. 800 BC in the north-west of Central Africa (Zangato, 1999, 2007; Clist, 2005).

The Oboui site and other 3000 BP dates

The Oboui site excavations in Centrafrique have raised the possibility of the origin of iron smelting in Africa being significantly earlier than previous evidence has suggested. The calibrated plot of the six ^{14}C dates obtained for an early occurrence of iron working at that site has been set up against all other acceptable dates (Fig. 8). Excluding Pa-2196 (much younger), the remaining Oboui dates fall roughly between 2150-2000 cal BC (Fig. 8, group A); they are in strict isolation

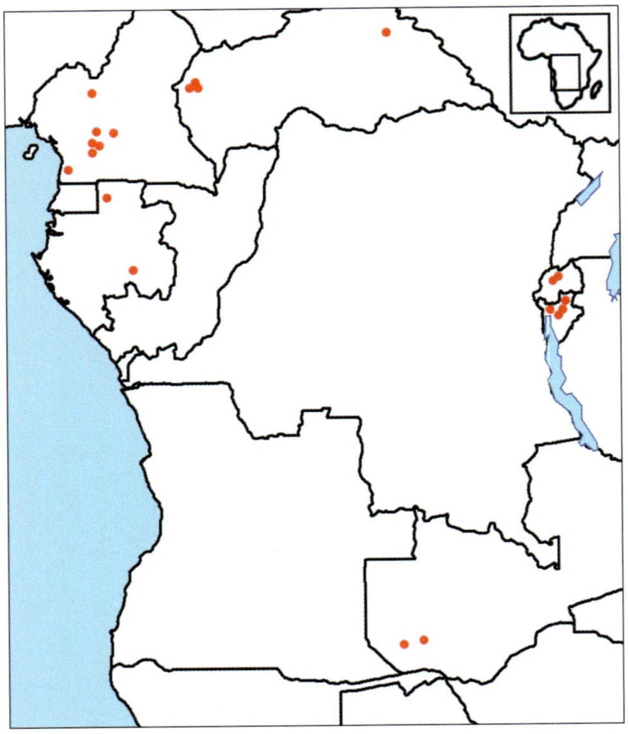

Figure 3: Iron smelting sites from 400 to 300 cal BC.

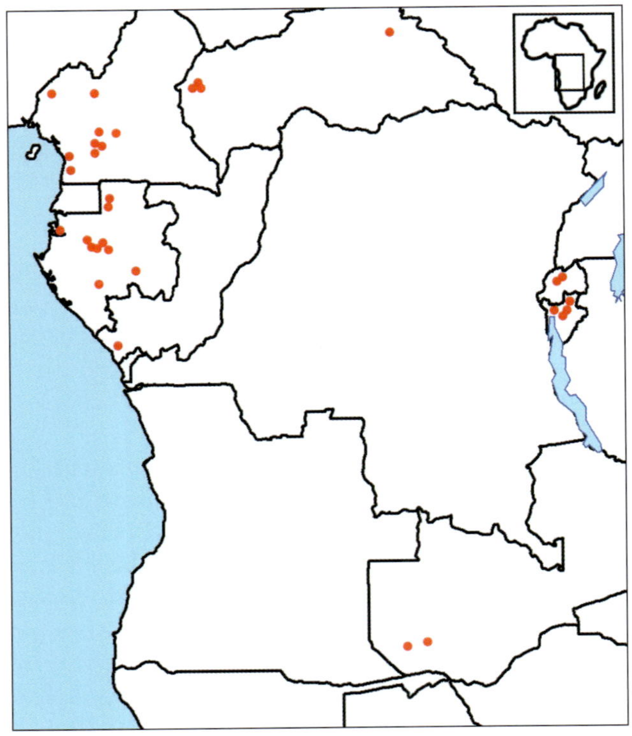

Figure 4: Iron smelting sites from 300 to 150 cal BC.

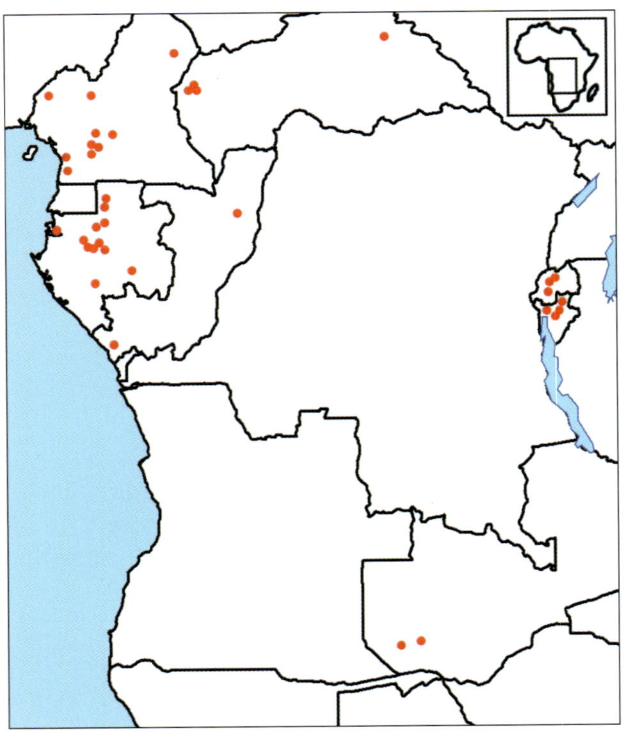

Figure 5: Iron smelting sites from 150 to 10 cal BC.

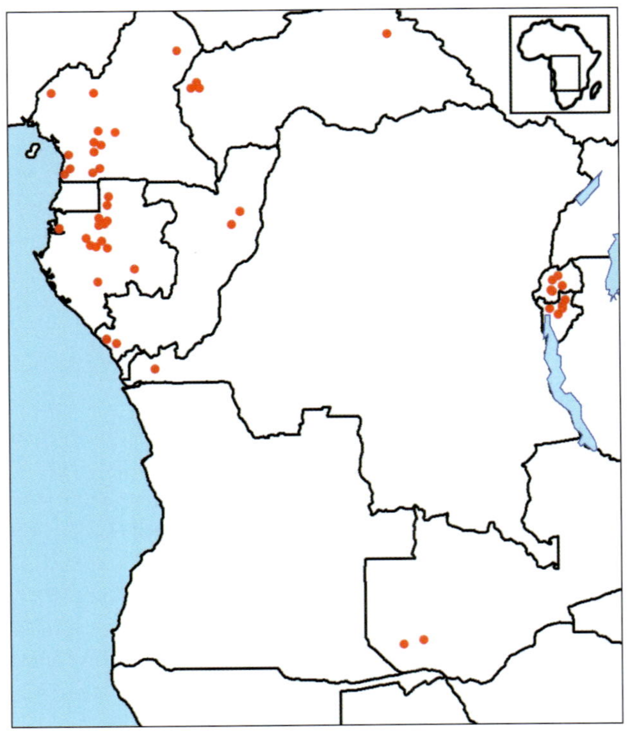

Figure 6: Iron smelting sites from 10 cal BC to 100 AD.

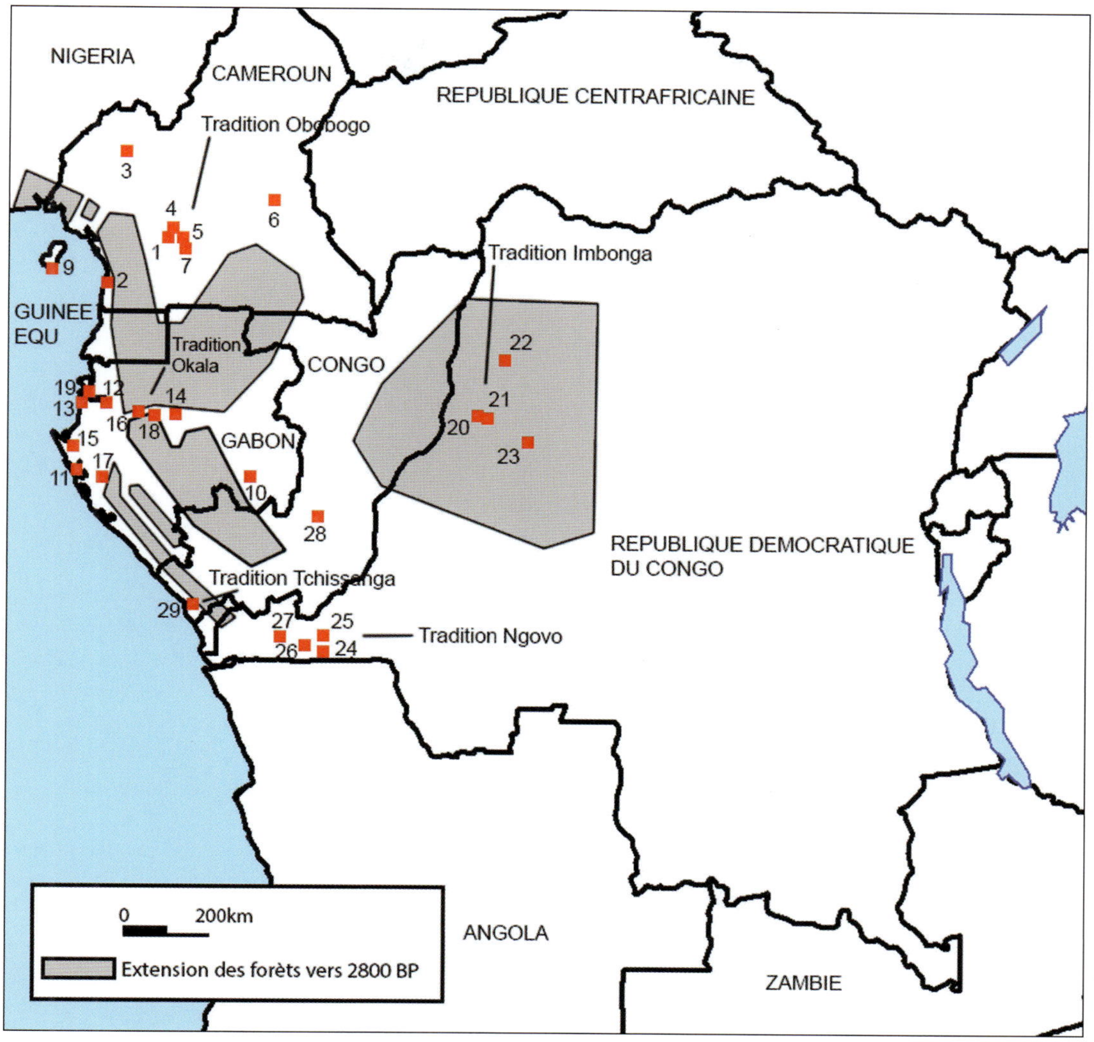

Figure 7: Theoretical extension of the western part of the equatorial forest around 3000 BP (greyed zones) and distribution map of the main villagers' traditions. (**see Note 3**)

from the rest of the corpus. Over 1000 cal years separate them from the earliest dates at Gbabiri, which lies only a few kilometres away (Fig. 8, group B).

Further c. 3000 BP dates within the bibliography of Central Africa have been investigated for comparison to the very early dates from Oboui; however, as several seminal questions on the Oboui excavation results remain unanswered (see Clist, 2012) and, due to the isolated nature of its calibrated dates, it seems pertinent for the time being to disregard the early c. 3600 BP 14C dates from the Oboui site.

The Lv-1880 3390+/-100 BP date from the Ogrotoulou cave in Centrafrique is one of a short series of fairly divergent dates coming from a single layer, 20cm thick, of mixed archaeological material. This single date must thus be used with caution (see Moga, 2008; Clist 2012). Other dates from the neighbourhood of Oboui have also been listed and all come from pre-*tazunu* layers at the Balimbé 2 site 68, Balimbé 2 site 21, and the Dokoko site 22 (Fig. 7). No iron artefacts or traces of iron production are associated to the dated samples (see Zangato, 1999, 2000, 2007, Zangato & Holl, 2010). Later, again from the Ogrotoulou cave, we have two ¹⁴C dates which overlap around 2350 BP (Moga, 2008: Lv-1879, 2360±60 BP, and Lv-5948, 2350±50 BP).

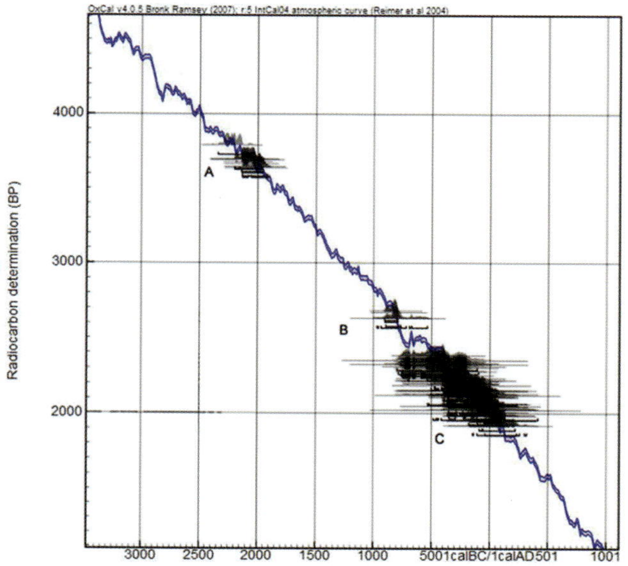

Figure 8: Calibrated dates for early iron smelting in Central Africa (using OxCal v.4.0.5).

In the south-east of the region, several dates from Rwanda and Burundi relate to this time range. One is referred to two papers by Belgian excavator Van Grunderbeek (Van Grunderbeek, 1992; Van Grunderbeek, et al., 2001), although all the 3000 BP dates have been discarded by her based on various considerations. The remaining "good" early ^{14}C dates are the following: Kabacusi Hv-121232815 +/- 165 BP, Gasiza 1 Hv-11143 2635 +/- 95 BP, Mirama 3a Hv-11142 2480 +/- 85 BP, Karuriye Hv-12925 2260 +/-110 BP, Rwiyange 1 Hv-12926 2250 +/- 125 BP, Muganza 1 Hv-12125 2245 +/- 65 BP, Rurembo GrN-5752 2180 +/- 50 BP, Mutwarubona 1 Ly-2267 2020 +/- 330 BP. All these dates fall between c. 2600 and 2000 BP. M.-C. Van Grunderbeek refers to the Kabacusi and Gasiza dates as evidence of the earliest stages of iron smelting in the Great Lakes area, excluding all the other earlier dated samples (Van Grunderbeek, et al., 2001, pp.276-277). In her account, it is the Gasiza 1 date which must be used to pinpoint the first known smelts (i.e. circa 750 cal BC).

At the Katuruka site in Tanzania, excavated by P. Schmidt, three early dates cluster around 2500 BP (N-895 2500 +/- 115 BP, RL-406 2470 +/- 110 BP, N-890 2400 +/- 115 BP) (see Clist, 1987). No other site is earlier than circa 600 cal BC.

No other dates from Central Africa are available in the 4000-3000 BP time range. To find similar alternative data it is necessary to turn to West Africa.

In Nigeria, at the Nsukka site, the earliest reported dates are 2305+/- 90 BP, 2170 +/- 80 BP, and 2080 +/- 90 BP (Okafor and Phillips, 1992; Okafor, 2004). By using the 2305 BP date, iron smelting in central Nigeria can be calibrated to around 400 cal BC. In Igboland, to the South-East, E. Okafor speaks of a circa 760 cal BC date. At the Lejja site, a single 4000 BP date on charcoal was obtained from charcoal imbedded in the crevices of an iron slag (Eje-Uzomaka **(4)**). Other ^{14}C dates from Lejja are earlier, although in all of these cases it is necessary to wait for the final site publication to understand the local context.

In Kano State in 1984, surveys led to thermoluminescence (TL) dating of three bowl furnaces (Darling, 1985). **(5)**. Both the Matafanda and Fitola sites are of very early origin. These results must be used cautiously however as they come from luminescence analyses carried out 25 years ago, after which time nothing else has been published about the sites' context. In addition, the dates of the bowl furnace sites are isolated.

In Burkina Faso, at Douroula, the earliest evidence for iron smelting is not earlier than 460 cal BC (Millogo, 2000, p.66).

In Senegal, amongst a small series of interesting archaeological sites, a deposit of iron tools and a few hints of local iron production on two mounds excavated at the site of Walaldé have produced calibrated dates between 800-200 BC (Deme and MacIntosh, 2006, pp.327-328 and 341-344). The earliest occupation phase appears to document a period of transitional iron use, with some worked stone in evidence (800-550 cal BC) (iron-using agropastoralists?). Smelting, forging slags and tuyères are in evidence in considerable quantities in the later phase (400-200 cal BC). It is possible to state that iron has been known there since at least 550 cal BC, as dated in the earliest levels.

In Mauritania, fieldwork carried out in 2000 dated one iron smelting furnace around 760-400 cal BC, or 582 cal BC (MacDonald et al., 2009, p.36; GX-30818-AMS 2440+/-40 BP).

Sadly, it has become fashionable over the last few years to use the ancient dating of metallurgical processes from the Termit area in Niger (Person et Quechon, 2001, 2004; Quechon, 2001, 2004) to support claims of autochthonous development of iron smelting (see for example Bocoum, 2004). Although contradicted by some (e.g. Killick, 2004; Alpern, 2005), the Niger data is still called upon to force the hypothesis of an independent development of iron technology in sub-Saharan Africa. G. Quéchon presented his position as follows: "... iron and copper objects appeared in Termit in approximately 1500 BC, and the first known smelting furnaces date from approximately 800 BC "(Quéchon, 2004, p.112). The major issues with Niger concern the quality of the ^{14}C dates and their degree of association to metallurgy. All the artefacts originate from unstratified settlements. They lie today on the surface of wind deflated sands. Even today, one can find up to "[...] 41 windy days on a 45 days fieldwork period" (Quéchon, 2001, p.248, translation B. Clist). An archaeologist who has specialised in West Africa, Anne Haour, has stated: "no definitive statement can be made on the occurrence of metal implements on surface settlements . . . other than to wish that these had been recovered in stratigraphically secure contexts. Skepticism . . . is the wisest approach" (Haour, 2003, pp.217-218).

The only more acceptable dataset for early metallurgy in Niger comes from the Do Dimmi 15 archaeological site, where some 22 iron smelting bowl furnaces were studied. Six radiocarbon dates bracket metallurgy between 3000 and 2000 BP (Quéchon, 1995, pp.310-311; Killick, 2004, p.104 and **(6)**). Once again however, the same stratigraphic problem outlined earlier for the Termit site is found here. At the time of the excavation, the excavators doubted the real association between the dated charcoal and the site itself (Quéchon and Roset, 1974, p.97). In addition, the regimented grouping of so many bowl furnaces in a relatively small area demonstrated by photographic documentation allegedly points to a short-lived episode rather than the far-ranging ^{14}C dates (Killick, 2004, p.105). One may add to this the fact that the oldest dates were obtained from the same laboratory, in Dakar.

Conclusion

It is now possible to suggest a circa 800 cal BC period when iron smelting may have started in Centrafrique and in Cameroon (Zili and Gbabiri sites). At 600 cal BC, iron tools were certainly produced locally in Cameroon at a small handful of sites. Iron in Centrafrique was not produced earlier than c. 500 cal BC.

Up to c. 1200 cal BC hunter-gatherers roamed the landscape; their material culture has enabled us to distinguish between several Late Stone Age industries (see Lanfranchi and Clist, 1991, Late Stone Age chapters for the latest synthesis). From 1200 to 400 cal BC, villagers settled the southern savannahs of Cameroon, the equatorial forest and, eventually, the northern regions of the savannahs south of this forest; a slow and irregular expansion of the Forest co-Traditions from North to South along the Atlantic coastline and inland through the forest block can be suggested (Clist, 2005, chap.8, pp.693-763). (Fig. 7). There existed a period when villagers used polished stone adzes and axes, cultivated small fields and relied heavily on oil palm and various other fruit trees. This is a genuine Neolithic culture.

From 600 to 400 cal BC, iron smelting began in places such as Zili in Cameroon and Moanda in Gabon, places which then exported finished iron products. Other centres were set up closer to those communities which already used iron tools (Figs. 1-7). During this transitional phase, when Neolithic villagers first acquired finished products from distant iron smelting specialists through already established trade routes, hunter-gatherer groups still existed. This aspect thus points toward a mosaic of interacting cultures which started to flourish around 3000 years ago in Central Africa and which persisted late into the Iron Age (Clist, 1991, 2006a, 2006b).

What then of the surrounding areas of Central Africa? Interestingly enough, iron was well established in the western Great Lakes area to the east, i.e. in Rwanda, Burundi, and west Tanzania around 600 cal BC. Moving to the north-west however, no clear picture of iron-smelting emerges. Central Nigeria has smelts going on from 400 cal BC onward, with the south-eastern region only showing evidence from around 760 cal BC. In Burkina Faso this is dated around 460 cal BC while in Senegal dates are known from around c. 550 cal BC, closer to those from south-eastern Mauritania, c. 580 cal BC one.

All of these dates from West and Central Africa fall within the 800-400 cal BC range, where a ^{14}C date can mean various things due to the discrepancies of the calibration scale and of the ^{14}C atmospheric contents, giving rise to a 400 years range. Some have called this period the "no date land" or "Hallstatt plateau" (Clist, 2005, pp.129-138).

Other evidence must be called for to differentiate contrasting data and to reinforce our paradigm. AMS dating (7), wood identification of dated charcoals, detailed ceramic sequence of the excavated site, detailed analysis of other artefacts or ecofacts, detailed presentation of stratigraphic and spatial contexts of finds: all of these methods must now be systemically used to help us to resolve the debate about the origins and dating of iron metallurgy. Far more interestingly, they must be used to discern the differing social ways in which the forging and use of iron dispersed throughout the region came to be accepted and transformed by villagers. For the time being however, this remains wishful thinking, limited by the state of our data sets.

Acknowledgements

For reviewing the manuscript and/or strengthening some arguments and/or suggesting new material, in alphabetical order - Raul Carstocea, Peter Darling, Jane Humphris, Eric Huysecom, David Killick, Kevin McDonald, Scott McEachern, Conny Meister, Silja Meyer, AnnMarie van der Heyden.

Notes

1. Using "Online CalPal" at http://www.calpal-online.de/index.html. "Online CalPal" was used throughout this paper and especially for its conclusions.
2. None of the ^{14}C dates from the well-known Oliga site in Cameroon have been used in Figures 1-6 and 8. I discussed the data at length a few years ago (Clist, 2005, pp. 771-772). S. Alpern, reviewing the pros and cons of the origin of iron smelting in Africa, synthesised my standpoint on Oliga: "[B. Clist] believes the data come from what archeologists deem a 'secondary deposit', a a natural pit at the base of a hillside down which rain has washed remains of an iron-smelting furnace, including slag and tuyères, along with charcoal from various periods besides that of the furnace" (Alpern, 2005, pp.76-77).
3. Cameroon - 1: Avoh, 2: Kribi (sites of Bwambé, Eboundja, Malongo), 3: Njinikejem, 4: Nkang, 5: Nkométou, 6: Ouélé Maroua, 7: Yaoundé (sites of Ndindan, Obobogo, Okolo); Centrafrique – 8: Balimbé ; Equatorial-Guinea – 9: Bioko island; Gabon – 10: Franceville, 11: Ikengué, 12: Kango, 13: Libreville (sites of Okala, Cap Estérias), 14: Lopé (sites of Lopé 2, Lopé 12, Okanda 1, Okanda 6, Toubé 1), 15: Mbilapé, 16: Ndjolé, 17: Ofoubou (sites of Ofoubou 5, Ofoubou 16), 18: Otoumbi 13, 19: Rivière Denis; Democratic Republic of Congo – 20: Bokele, 21: Bokuma, 22: Boso Njafo, 23: Imbonga, 24: Kwimba, 25: Ngovo/ Dimba, 26: Ntadi Ntadi, 27: Sakuzi; Congo – 28: Djambala, 29: Tchissanga et Tandou Youmbi.
4. See Ezo-Uzomaka, communication at the SAFA meeting in Frankfurt, Germany, 8-11 September 2008: http://cohesion.rice.edu/centersandinst/safa/emplibrary/AlphabeticalProceedings2008.pdf
5. Roni site: DurTL-45-2AS , 380 +/- 320 AD; Matafanda site: DurTL-57-3AS, 1400 +/- 50 BC; Fitola site: DurTL-57-2AS, 2400 +/- 1100 BC (Darling, 1985).
6. Do Dimni 15: Dak-147: 2924 +/- 120 BP; Dak-145: 2628+/- 120 BP; Pa-288: 2500 +/- 70 BP; UPS-?: 2065 +/- 60 BP.
7. A new dating method entailing the dating of ceramics by way of their progressive building up of water contents (rehydroxylation dating) - still in its experimental stage - may lead to major breakthroughs in Sub-Saharan Africa. See for example Wilson et al. (2009).

References

Alpern, S.B., 2005. Did they or didn't they invent it? Iron in Sub-Saharan Africa. *History in Africa*, 32, pp.41-94.

Anciaux de Faveaux, A. and Maret, P. de, 1984. Premières datations pour la fonte du cuivre au Shaba (Zaïre). *Bulletin de la Société Royale Belge d'Anthropologie et de Préhistoire*, 95, pp.5-20.

Bocoum, H. ed., 2004. *The origins of Iron Metallurgy in Africa: new light on its antiquity*. Paris: UNESCO.

Clist, B., 1987. A critical reappraisal of the chronological framework of Urewe Early Iron Age Industry. *Muntu*, 6, pp.35-62.

Clist, B., 1989a. Bilan des premiers travaux du Département d'Archéologie du CICIBA: missions du premier semestre 1985. In: *Actes du colloque international "les peuples bantu, migrations, expansion et identité culturelle", Libreville, 1-6 Avril 1985, Vol.1.* Paris: L'Harmattan/CICIBA, pp.103-10.

Clist, B., 1989b. Archaeology in Gabon, 1886-1988. *The African Archaeological Review*, 7, pp.59-95.

Clist, B., 1991. Synthèse régionale sur l'Age du Fer Ancien. in Lanfranchi, R. and Clist, B. eds. *Aux origines de l'Afrique Centrale*. Ministère de la Coopération et du Développement / Centre Culturel français de Libreville / Sépia, Paris, pp.225-226.

Clist, B., 1995. *Gabon: 100.000 ans d'Histoire*. Paris: Centre Culturel français Saint-Exupéry / Sépia.

Clist, B., 1997. Le site d'Okala, Province de l'Estuaire, Gabon et son importance pour la compréhension du passage à la sédentarisation

en Afrique Centrale. *Comptes rendus de l'Académie des Sciences de Paris, série Sciences de la terre et des planètes*, 325, pp.151-6.
Clist, B., 2005. *Des premiers villages aux premiers européens autour de l'estuaire du Gabon: quatre millénaires d'interactions entre l'homme et son milieu*. PhD Thesis, Université Libre de Bruxelles.
Clist, B., 2006a. Mais où se sont taillées nos pierres en Afrique Centrale entre 7.000 et 2.000 bp? In: H.-P. Wotzka, ed. *Grundlegungen. Beiträge zur europäischen und afrikanischen Archäologie für Manfred K.H. Eggert*. Tübingen: Francke Attempto Verlag GmbH and Co. KG, pp.291-302.
Clist, B., 2006b. Coexistences matérielles entre 6000 et 20 cal BC en Afrique centrale: une mosaïque culturelle. In: L. Astruc, F. Bon, V. Lea, P.-Y. Milcent and S. Phillibert eds. *Normes techniques et pratiques sociales. De la simplicité des outillages pré- et protohistoriques, XXVIe rencontres internationales d'archéologie et d'histoire d'Antibes*. Antibes: Editions APDCA, pp.377-83.
Clist, B., 2012. Vers une réduction des préjugés et la fonte des antagonismes: un bilan de l'expansion de la métallurgie du fer en Afrique sud-saharienne. *Journal of African Archaeology*, 10, 1, pp.71-84.
Clist, B., Oslisly, R. and Peyrot, B., 1986. La métallurgie du fer au Gabon: premiers éléments de synthèse. *Muntu*, 4-5, pp.47-55.
Darling, P., 1985. Fieldwork surveys in and around Kano State, Nigeria, 1982-1985. *Nyame Akuma*, 27, pp.39-41.
David, N., 1980. Early bantu expansion in the context of Central African prehistory: 4000-1 BC. In: L. Bouquiaux ed. *L'expansion bantoue (= Actes du Colloque international du CNRS, Viviers, France, 1977), Vol. II*. Paris: SELAF, pp.609-47.
Deme, A. and McIntosh, S.K., 2006. Excavations at Walaldé: new light on the settlement of the Middle Senegal Valley by iron-using peoples. *Journal of African Archaeology*, 4(2), pp.317-47.
Digombe, L., Schmidt, P., Locko, M., Mouleingui Boukossou, V. and Mombo, J.B., 1985. Radiocarbon dates for the Iron Age in Gabon. *Current Anthropology*, 26(4), p.516.
Digombe, L., Schmidt, P., Mouleingui Boukossou, V., Mombo, J.B. and Locko, M., 1987. L'Age du Fer Ancien au Gabon. *L'Anthropologie*, 91(2), pp.711-7.
Haour, A., 2003. One Hundred Years of Archaeology in Niger. *Journal of World Prehistory*, 17, pp.217-8.
Killick, D., 2004. Review essay: what do we know about African iron working? *Journal of African Archaeology*, 2(1), pp.97-112.
Lanfranchi, R. and Clist, B. eds., 1991. *Aux origines de l'Afrique Centrale*. Paris: Centers Culturels français d'Afrique Centrale/ CICIBA.
MacDonald, K.C., Vernet, R., Martinon-Torres, M. and Fuller, D., 2009. Dhar Néma: from early agriculture to metallurgy in southeastern Mauritania. *Azania: archaeological research in Africa*, 44(1), pp.3-48.
Maley, J., 2004. Les variations de la végétation et des paléoenvironnements du Domaine Forestier africain au cours du Quaternaire récent. In: J. Renault-Miskovsky and A.M. Semah, eds. *L'évolution de la Végétation depuis deux millions d'années*. Paris: Publications Artcom - Errance, pp.143-78.
Maret, P. de, 1978. *Chronologie de l'âge du fer dans la dépression de l'Upemba en République du Zaïre*. PhD Thesis manuscript, Université Libre de Bruxelles.
Maret, P. de, 1982. New survey of archaeological research and dates from West-Central and North-Central Africa. *Journal of African History*, 23, pp.1-15.
Maret, P. de, 1985. Recent archaeological research and dates from Central Africa. *Journal of African History*, 26(2-3), pp.129-48.
Maret, P. de, Van Noten, F. and Cahen, D., 1977. Radiocarbon dates from West Central Africa: a synthesis. *Journal of African History*, 18(4), pp.481-505.
Millogo, A.K., 2000. Les premières métallurgies. In: R. Vernet, ed. *L'archéologie en Afrique de l'Ouest*. Saint Maur: Editions Sépia, pp.59-70.
Moga, J., 2008. La métallurgie du fer en Centrafrique: premiers éléments de synthèse. *Revue Centre Africaine d'Anthropologie*, 1, pp.1-9.
Okafor, E.E., 2004. Twenty-five centuries of bloomer iron smelting in Nigeria. In: H. Bocoum, ed. *The origins of Iron Metallurgy in Africa: new light on its antiquity*. Paris: UNESCO, pp.43-54.
Okafor, E.E. and Phillips, P., 1992. New 14C ages from Nsukka, Nigeria and the origins of African metallurgy. *Antiquity*, 66, pp.686-8.
Person, A. and Quechon, G., 2001. Données chronométriques et chronologiques de la métallurgie du fer à Termit: matériaux graphiques pour l'étude des Ages anciens du fer. *Mediterranean Archaeology*, 14, pp.255-62.
Person, A. and Quechon, G., 2004. Chronometric and chronological data on metallurgy at Termit: graphs for the study of the ancient Iron Ages. In: H. Bocoum, ed. *The origins of Iron Metallurgy in Africa: new light on its antiquity*. Paris: UNESCO, pp.119-26.
Quechon, G., 1995. La fin du néolithique et les débuts de la métallurgie dans le massif de Termit (Niger): éléments de méthodologie. In: A. Marliac, ed. *Milieux, sociétés et archéologues*. Paris: ORSTOM-Karthala, pp.303-12.
Quechon, G., 2001. Les datations de la métallurgie du fer à Termit (Niger): leur fiabilité, leur signification. *Mediterranean Archaeology*, 14, pp.247-53.
Quechon, G., 2004. Iron metallurgy datings from Termit (Niger): their reliability and significance. In: H. Bocoum, ed. *The origins of Iron Metallurgy in Africa: new light on its antiquity*. Paris: UNESCO, pp.109-18.
Quechon, G. and Roset, J.-P., 1974. Prospection archéologique du massif de Termit (Niger). *Cahiers ORSTOM, série Sciences Humaines*, 11(1), pp.85-104.
Schmidt, P., Digombe, L., Locko, M. and Mouleingui, V., 1985. Newly dated Iron Age sites in Gabon. *Nyame Akuma*, 26, pp.16-8.
Van Grunderbeek, M.-C., 1992. Essai de délimitation chronologique de l'Age du Fer Ancien au Burundi, au Rwanda et dans la region des Grands Lacs. *Azania*, 27, pp.53-80.
Van Grunderbeek, M.-C., Roche, E. and Doutrelepont, H., 2001. Un type de fourneau de fonte du fer associé à la culture Urewe (Age du Fer Ancien), au Rwanda et au Burundi. *Mediterranean Archaeology*, 14 (= *Proceedings of the First International Colloquium on the archaeology of Africa and the Mediterranean basin held at the Museum of Natural History in Geneva, 4-7 June 1999*), pp.271-97.
Wilson, M.A., Carter, M.A., Hall, C., Hoff, W.D., Ince, I., Savage, S.D., McKay, B. and Betts, I.M., 2009. Dating fired-clay ceramics using long-term power law rehydroxylation kinetics. *Proceedings of the Royal Society* (A), 465(2108), pp.2407-15.
Wotzka, H.-P., 2006. Records of activity: radiocarbon and the structure of Iron Age settlement in Central Africa. In: H.-P. Wotzka, ed. *Grundlegungen. Beiträge zur europäischen und afrikanischen Archäologie für Manfred K.H. Eggert*, Tübingen: Francke Attempto Verlag GmbH and Co. KG, pp.271-89.
Zangato, E., 1999. *Sociétés préhistoriques et mégalithes dans le Nord-Ouest de la République Centrafricaine*, Cambridge Monographs in African Archaeology 46, British Archaeological Reports International Series 768. Oxford: Archaeopress.
Zangato, E. 2000. *Les occupations néolithiques dans le Nord-Ouest de la République centrafricaine*. Collection Préhistoires, Volume 3, Éd. M. Mergoil, Montagnac.
Zangato, E., 2007. *Les ateliers d'Oboui. Premières communautés métallurgistes dans le nord-ouest du Centrafrique*. Paris: Editions Recherche sur les Civilisations.
Zangato, E. and Holl, A.F.C. 2010. On the Iron Front: New Evidence from North-Central Africa. *Journal of African Archaeology*, 8 (1), pp.7-23.

Author's address

- KongoKing Research Group
  Department of Languages and Cultures, Ghent University,
  B-9000 Ghent, Belgium
  bernardolivier.clist@ugent.be

Yoruba iron metallurgy: raw materials, routine and rituals

O. Akin Ige

ABSTRACT: Over the centuries, iron has been both the substance and agent of transformation that promoted the acquisition of wealth and war amongst the Yoruba of south-western Nigeria. Amongst the members of this ethno-linguistic group iron technology is considered as deriving from divine inspiration, brought to humans by cultural deities.

Ethnographic and archaeological research at several excavated iron-working sites in south-western Nigeria has shown that the raw materials for iron smelting consisted of weathered igneous rocks combined with ilmenite rich sands which, when smelted, resulted in the accumulation of titanium-rich slags. There is evidence to suggest that the location of these raw materials and the mining techniques and furnaces used to smelt them – and even the dress code and the discipline of those who carried these processes out – were all aspects that were directed by the oracles. Altars dedicated to *Ogun* (the god of iron) and a wide array of temples, still intact today, attest the strength that *Ogun* – and consequently, iron – was perceived to invest in the daily lives of the Yoruba people.

Using the Ile-Ife and Oyo areas of south-western Nigeria as case studies, this paper will link the various cultural rituals referenced above with some of the routine processes involved in iron metallurgy among the Yoruba people.

Keywords: Yoruba, smelting, raw materials, ritual, routine, *Ogun.*

Introduction

The name 'Yoruba' is applied to an ethnic and linguistic group numbering about 30 million people, who occupy a large area extending through the Kwara, Kogi, Edo, Ekiti, Ondo, Osun, Oyo, Ogun, and Lagos states of Nigeria; and thence to the republics of Benin, Togo, Ghana, Liberia, Sudan, and Sierra Leone (Fig. 1). In addition to a common language the Yoruba are united by their shared culture and an oral tradition which traces their origins to Ile-Ife, an ancient city in south-western Nigeria (Adediran, 1992; Olomola, 1992).

Although the Yoruba's ancestors eventually dispersed to establish settlements in their present homelands across West Africa, Ile-Ife continued to be regarded as their spiritual home and home of their divinities, including the fiery god *Ogun*. Due to the belief in the divine nature of iron in Yoruba culture sacrifices were made to Ogun – who was believed to control all aspects of iron metallurgy and usage – before the lighting of the smelting furnace to protect the processes that transformed the raw materials for daily and ritualistic use.

Many Yoruba deities are represented holding iron or metalwork, as illustrated in Figs. 2a-d (Adepegba, 1991). The iron staff of *Erinle*, god of the forest, is topped with a crested bird and surrounded by a circle of farmers' tools, hunters' weapons, and reptiles (Fig. 2b). *Orisa-oko*, god of agriculture, holds a heavy, sword-like flat iron staff, engraved with abstract patterns and stylised animals (Fig. 2c). These staffs were kept in shrines to various deities (Fig. 2d), as symbols of authority (Adepegba, 1991) and also functioned as ritual symbols, presented in ceremony to particular titled men.

Ethnographic and archaeometric data suggests that the Yoruba operated hundreds of iron mining and smelting sites, with some sources (Adesina, 1985; Ige and Rehren, 2003) suggesting that iron smelting was directly linked to the migratory and economic activities of the ancient Yoruba; thus

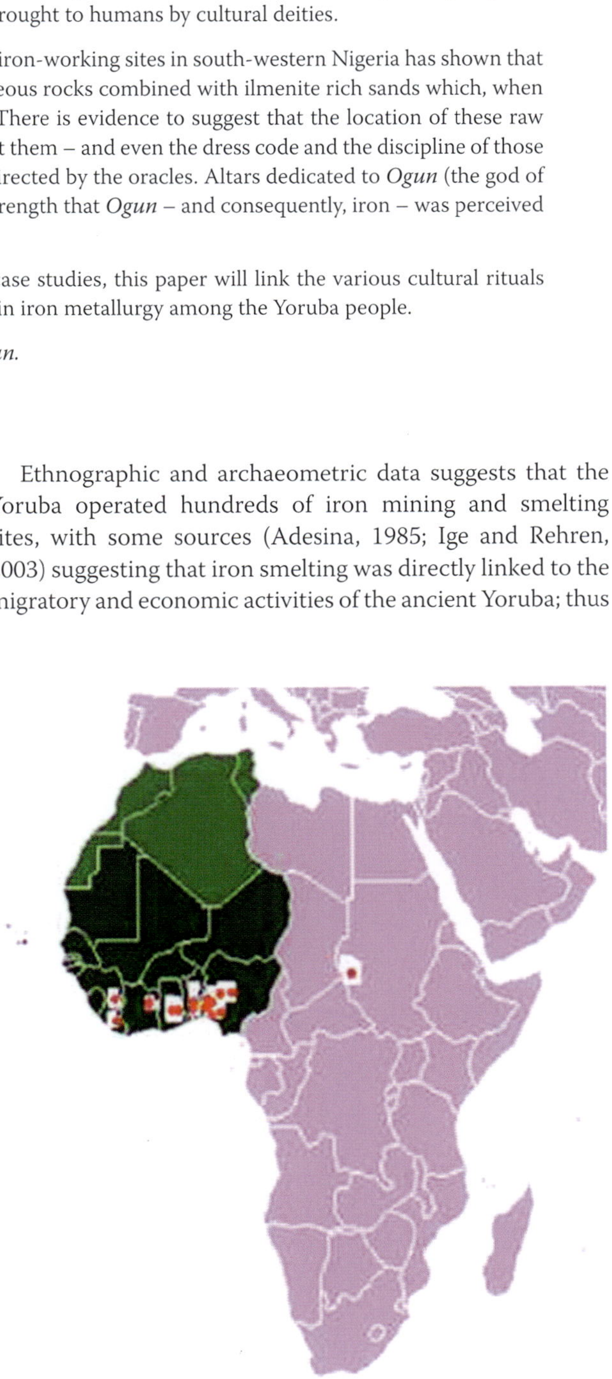

Figure 1: Map showing Yoruba areas of West Africa.

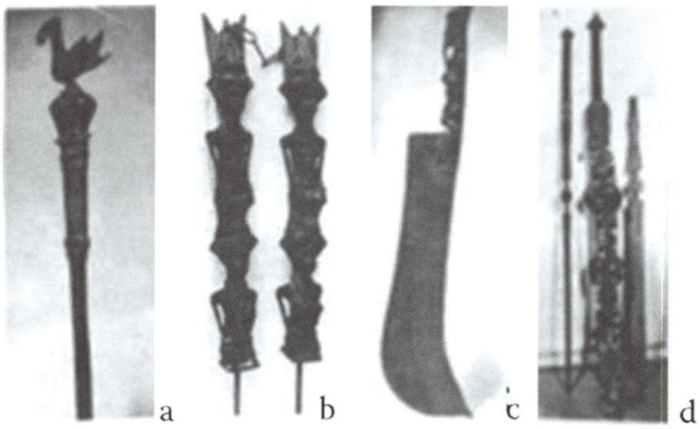

Figure 2: Examples of Yoruba deities represented with iron or metal works.

relating their practice to diffusionist theories advanced for the origin of iron working in West Africa.

One school of thought (Tylecote, et al., 1971) suggests that iron was introduced into West Africa from Carthage, either across the Sahara or by sea. Another (Andah, 1979) suggests that it came from Egypt via Meroe and the Sudan.

Historical and archaeological data collected in some parts of south-western Nigeria (Adesina, 1985) show that the knowledge of mining and metallurgy was mostly confined to royalty. As they spread across the land and towards the north, they founded kingdoms and empires. This produced a flourishing of iron working, due to the need to acquire weapons to conquer more lands, leading to the establishment of powerful kingdoms, such as the Old Oyo Empire. As soon as the Fulani jihadists defeated Old Oyo around 1830, a decline in iron production ensued (Willett, 1960; Soper and Darling, 1980). The fall of the Old Oyo Empire drove the Yoruba back to the original homelands in Ife. As they migrated homeward to Ife, a reduction in iron production seems to have occurred, which must have contributed significantly to the eventual discontinuation of the art of smelting during the early-mid twentieth century. Other possible contributing factors include the instability caused by intertribal wars, the appearance of European scrap iron, and perhaps the agrarian revolution that altered the commercial leaning of the local people, or a combination of all these factors. Lack of raw materials can be ruled out as a possibility, as there is still an abundance of various grades of lateritic ironstone in almost all known Yoruba towns. However, as reported by other authors (e.g. Miller et al., 1994), local production in many African countries declined sharply with European colonisation, as imported iron assumed increasing economic importance.

Raw materials in Yoruba iron metallurgy

Slag dumps occur at several historical sites in south-western Nigeria, especially around Ife, Ibadan, and Oyo: the most powerful Yoruba empires during the 18th and 19th centuries. These important archaeometallurgical sites were identified and characterised by their immense slag deposits, although, as elsewhere on the continent, the hard-wearing slag is being quarried away for use in road foundations and the building of houses.

Across the Ile-Ife and Oyo regions the underlying geology is comprised of igneous and metamorphic rocks (Ige and Asubiojo, 1991) (Fig. 3). The igneous rocks consist of granites and some basic rocks (iron amphibole-rich rocks), while the metamorphic rocks consist of gneisses and schists of basic to ultrabasic compositions (Elueze, 2000). The basic rocks are transformed by intense tropical weathering into large deposits of ilmenite rich sand. In addition, these weathering products contain other iron minerals, such as ($Fe_2O_3.nH_2O$) hematite (Fe_2O_3) and goethite ($Fe_2O_3.H_2O$). Each stage of the weathering process appears to have led to a rapid accumulation of iron oxides.

The Yoruba have a variety of words for these stones. The gravel-stones, known as *idaro* or *oko* (meaning 'bullet-like ores'), are of three types with several variants. The name *Sagodo* is given to a variant of *idaro* ore, which is picked up from the ground after a downpour. The *Sagodo* stone has been identified as the lateritic iron concretions rich in goethite and obtained from the weathering of iron-rich amphibolites. The *agunwinwin* stone is a variant of *oko* and has been identified as accumulated ilmenite-rich, fine-grained black stone. The ilmenite, titanium-rich sand was blended with these lateritic rocks by the ancient smelters. According to oral evidence provided by the *Oba* (monarch) of *Modakeke*, iron smelters prospected for the productive mines by observing bits of the iron-rich gravel on the surface, inferring that more easily mined iron materials abound in the ground. Test pits were then dug to locate the most productive mines and the highest qualities of ores After these ores were discovered, men began to mine while the women and children took the mined ore to the processing location. Remnants of intense surface and underground mining activities are still visible today, with collapsed structures and underground tunnels marking the location of the ancient mines (Fig. 4). Modern excavations showed that the iron ore deposits that were mined in the past could be up to two metres thick in some locations (Ige and Rehren, 2003). Due to the difficulty and labour involved in

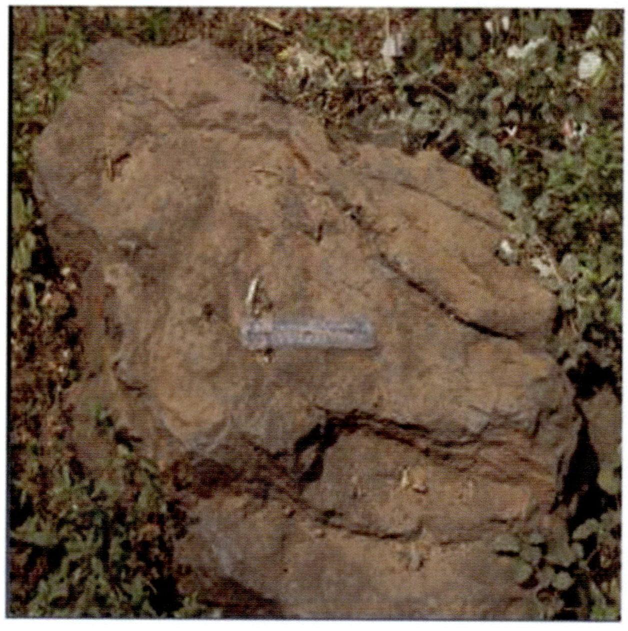

Figure 3: Lateritized amphibole rock.

transporting large quantities of ore for great distances, iron masters built their furnaces close to the ore deposits.

Excavations revealed that most of the furnaces had been completely destroyed and only their bases remained (Fig. 5). Usually the excavations include three layers. Layer 1 (0-15cm) consists of the top soil, while Layer 2, measured at a depth of 15-40cm, is composed of gritty and dark soil. The finds identified in this layer are potsherds, ores, and some slags. Layer 3 was between 40 and 75cm in depth and consists of homogenous loose and light brown soils containing potsherds, laterite-coated pieces of iron slags, tuyères, and furnace lining. Below this level was the bedrock that consists of titanium-rich amphibolite. Ceramic evidence from these three layers indicates that the iron industries belong to one cultural unit.

Laboratory analyses

Several samples of ores, black sand and slags have been subjected to metallographic and compositional analyses. The chemical analysis was performed at the UCL Institute of Archaeology Wolfson Archaeological Science Laboratories using X-ray Fluorescence Spectroscopy on homogenous powder pellets and polished sections were analysed using Scanning Electron Microscopy (SEM-EDX). Analytical details and comprehensive data have been presented elsewhere (Ige and Rehren, 2003). Analysis of the ore and black sand confirms that the materials are high in iron. The analysed titanium-bearing, unweathered basic rocks are characterised by high contents of Fe_{total} and Ti, as well as by high Fe/Mg ratios. Matching the high modal amounts of ilmenite (Fig. 6) in these rocks, the TiO_2 contents of 4.21 to 5.97 wt% were markedly high. Analytical data obtained on slags show that FeO values range between 59% and 65%, SiO_2 between 17 and 22%, Al_2O_3 between 5 and 9.4%, TiO_2 between 6.3 and 11%, MnO 0.1-0.2%, CaO 0.7-1.5%, P_2O_5 0.3-1.3%, and K_2O between 0.2% and 0.78%. The analysed lateritic iron ores were typically massive fragments and reddish to brownish in colour in situ. Under the microscope, they appear to consist of an aggregation of goethite and limonite in a groundmass of decomposed amphiboles. The ilmenite sands contain between 46.52 and 58.29% iron by weight and between 25.76 and 48.23% TiO_2 by weight. In addition, the sands also contain small amounts of chromium (about 1.5%). Modal analyses show that the iron content (as hematite and magnetite) ranges between 37 and 53%, while the ilmenite content ranges between 35 and 52%. Some analytical data obtained from the analysis of rocks, ore and sands is presented in Table 1, while slag data is presented in Table 2.

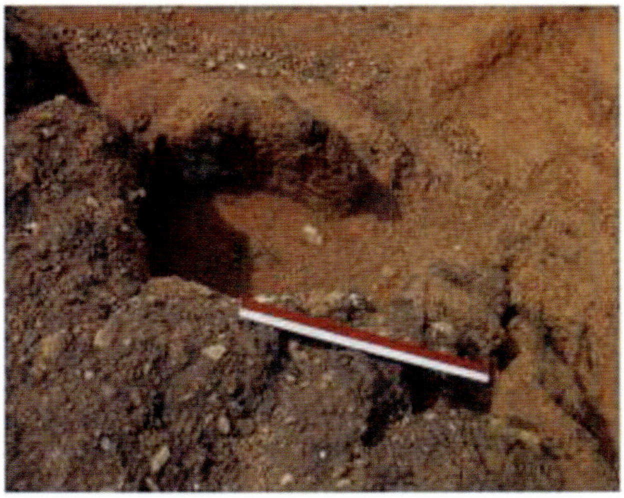

Figure 5: Furnace remains.

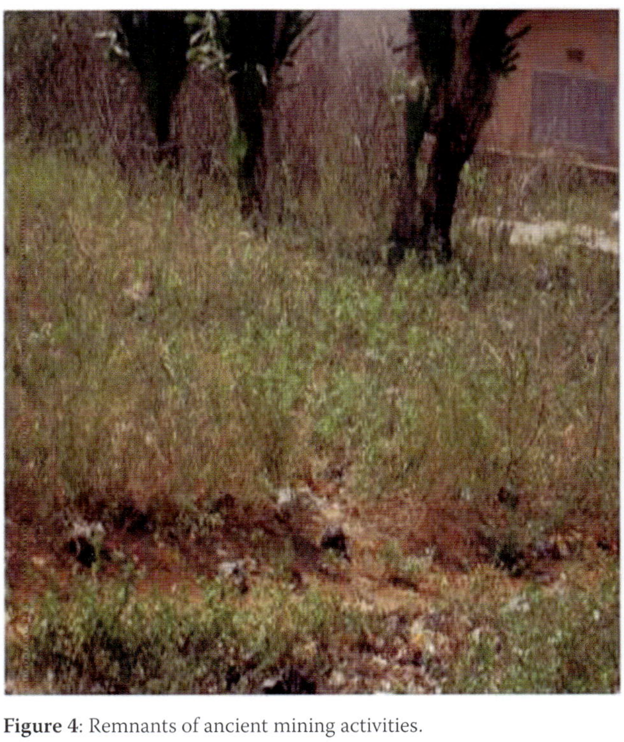

Figure 4: Remnants of ancient mining activities.

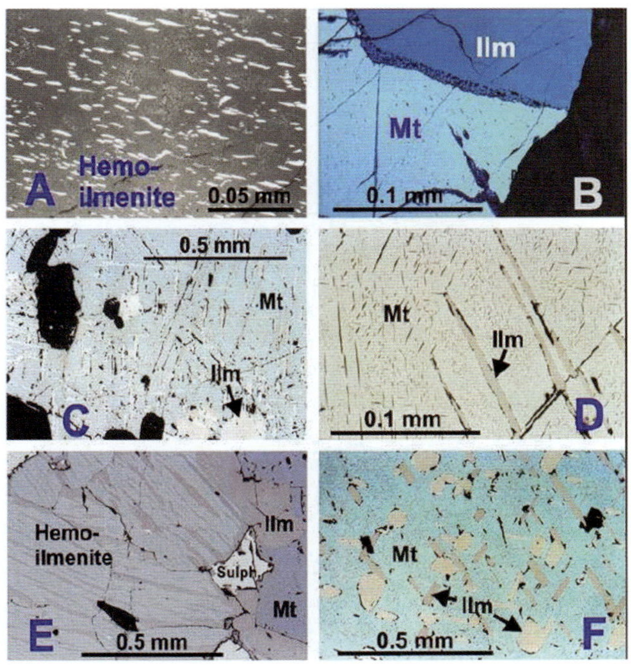

Figure 6: (A) Ilmenite (dark brown) with very fine hematite exsolution lamellae (light grey) (B) ilmenite with magnetite contact (C) titanomagnetite (light grey) with closely spaced ilmenite lamellae and abundant euhedral apatite (dark gey) (D) titanomagnetite (light grey) with several generations of fine and very fine spinel exsolutionilmenite.

Mineralogical studies show that the phases present in the slags consist of light grey fayalite matrix, cubic spinel, and dark glass. The spinel is markedly zoned, with alumina-rich core and titanium-rich border. A characteristic skeletal and elongated structure of fayalite, indicating rapid cooling, was identified. The marked absence of wüstite was noted, which may be an indication of the efficiency of the furnace (Tylecote, et al., 1971) (Fig. 7).

Table 1: Data obtained from the chemical analysis of rocks, ores and sand.

	Fresh Rock	Ilmenite Sand	Ore 1	Ore 2
SiO_2	40.81	36.66	2.33	5.23
TiO_2	4.74	10.77	0.84	1.88
Al_2O_3	4.47	10.35	4.28	1.44
Fe_2O_3	17.44	40.41	82.22	73.02
Cr_2O_3	0.31	1.31	1.54	1.56
NaO	0.1	0.22	0.66	0.61
MnO	0.32	0.57	0.52	0.2
MgO	6.57	2.12	0.11	0.11
CaO	8.91	1.1	0.2	0.1
K_2O	0.51	0.01	0	0

Table 2: Data obtained from the chemical analysis of slag samples.

WT%	Slag 1	Slag 2	Slag 3	Slag 4	Slag 5
FeO	59.42	65.17	61.29	65.35	63.05
Al_2O_3	9.48	7.04	6.7	5.31	5.94
MgO	0.31	0.5	0.3	0.39	0.32
SiO_2	20.93	20.08	17.54	20.01	21.65
TiO_2	6.3	6.6	11.12	7.72	7.73
MnO	0.11	0.18	0.06	0.31	0.24
P_2O_5	1.32	0.95	0.34	0.63	0.65
CaO	1.41	0.11	1.54	0.17	0.7
K_2O	0.78	0.04	0.68	0.18	0.02

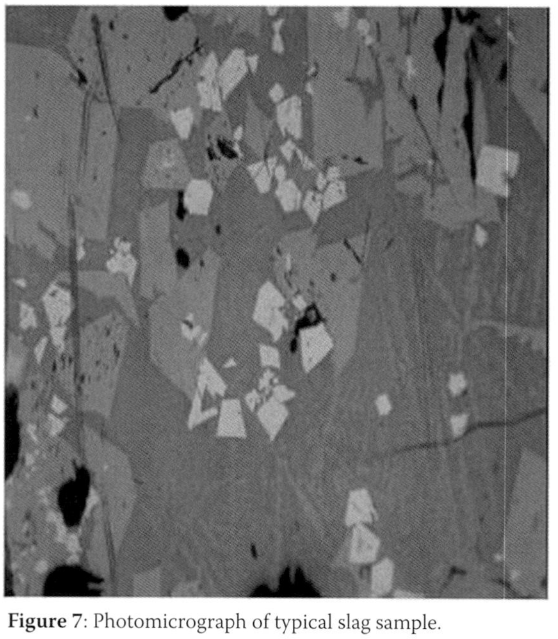

Figure 7: Photomicrograph of typical slag sample.

The routine

Evidence shows that the secrets of iron and iron smelting have been repeatedly learned by human beings in many places and at many times, yet different groups have developed different methods to smelt iron. In all processes that led to the fashioning of weapons and utensils from iron, the Yoruba followed specific, traditional routines. As mentioned above, prospecting for suitable ores was carried out mostly by visual means, which have been repeatedly utilised for many generations. Furnaces were always constructed on hillsides in the path of the wind, allowing self-draughting to take place; this was considered as the 'midnight intervention by gods' in fanning the flames. From ethnographic data, it is clear that there was a chief smelter who usually directed the smelting process.

Just like smelting, blacksmithing was widely practiced among the Yoruba and most villages had their own blacksmiths. Unlike iron smelting however, blacksmithing was generally a hereditary craft, with fathers in the families that practiced it transferring their skills and knowledge to their sons (Folarin, 1987). In some cases, the blacksmith was different from the iron smelter. It was customary for an individual who wanted farm implements or other items to provide the iron to the blacksmith. In some rare circumstances, however, the blacksmith could provide the iron and the person had to pay more for the product. Apart from this, the blacksmith also made various farm implements, which were used to stimulate intra and inter-group trade, although at this stage this trade had not yet been fully developed.

The blacksmith routinely hammered out a piece of iron until it became flat, subsequently working the iron into a lump again, and thereby driving out a lot of the slag and other impurities from the bloom iron. The process was repeated enough times so that he could hammer out the extra carbon and be left with a softer, more malleable wrought iron which could be manipulated into various shapes. The best finished objects were hard and tough and able to maintain a good edge capable of withstanding the rigors of extended use.

Iron smelting and blacksmithing were carried out on a part-time basis. Because most smelters and blacksmiths were also full-time farmers, iron working only took place during the months preceding the rains.

The rituals

Among the Yoruba, iron metallurgy was accompanied by various elaborate rituals and taboos which regulated the smelting and smithing of iron objects. *Ogun*, the furious god of iron, was believed to have introduced iron working to men, and consequently prayers and offerings had to be made to him before the processes of iron smelting and smithing. The use of the resultant weapons was also associated with this deity. The *ogun* tree (*Peregun, Dracaena fragram* or *Dracaena mannii*; Fig. 8) was always planted at the entrance of smelting areas and besides smithing sites, as well as at *Ogun* shrines. The name of this tree is literally translatable as the 'appeaser for the *Ogun*', the god of iron, and it is found at all ancient mining and smelting sites across the Yoruba land. This tree is so common at the entrance to mining and smelting locations that archaeologists may in the future be able to use these trees as guides when searching for ancient iron production locations in south-western Nigeria.

The basic rituals of appeasing *Ogun*, the god of iron, were associated with the processes of prospecting for iron ore, smelting and smithing. Before a family went out in search of ironstones and during the smelting process, it had to make peace with itself and the god. It was also important to make peace with neighbours and seek forgiveness for offences. Master smelters controlled this magic, organised the long months of strenuous labour necessary to bring the smelting to fruition, and imposed on their workers certain prohibitions from ordinary activities such as sexual contact, to 'frame' the smelting and endorse it as a special event. They also knew and performed invocations, dances, songs, praises, exhortations, jokes, aphorisms, and other esoteric activities thought necessary for rendering the entire process of smelting a success.

The smith transformed, using his secret of transmutations, iron blooms into indispensable tools and implements used in daily life and rituals. Thus, iron was used for the production of tools, household implements, weapons, agricultural implements, beautifully forged ornaments, sculptures, musical instruments, ceremonial objects, and masks. Iron was also used to fashion objects used in the worship of gods associated with agriculture and medicine, like the staffs of *Orisa oko*, god of agriculture, and *Osanyin*, god of medicine.

Some swords produced by powerful blacksmiths were thought to have magical or spiritual powers and were handed down from one generation of warriors to the next, a treasured heirloom and a focal point of family sagas in military societies. Mysticism developed around fine weapons that withstood the rigours of battle, including the alleged blessings and protection they conferred on the warriors who used them (Onabajo, 1986).

Some blacksmiths fashioned objects mainly for traditional religious cults. These cults, such as *Ogboni*, *Osun*, *Orisa-Oko* and *Sango*, commissioned various insignia of office from the smiths. One of the most remarkable images used by The *Ogboni* is the *Edan* male and female figure. Other items made by the ironmasters were prestige items used by elites, such as those used as protective talisman known as *Oruka elelo*, or rings worn as protection against evil spirits.

Ritual was also involved in recruiting apprentices to the practice of iron working. In most cases the *Ifa* oracle was consulted for information on the type of occupation a child should follow in order to prosper. Any child who took to learning the craft of iron working did so under the express instruction of the *Ifa* oracle.

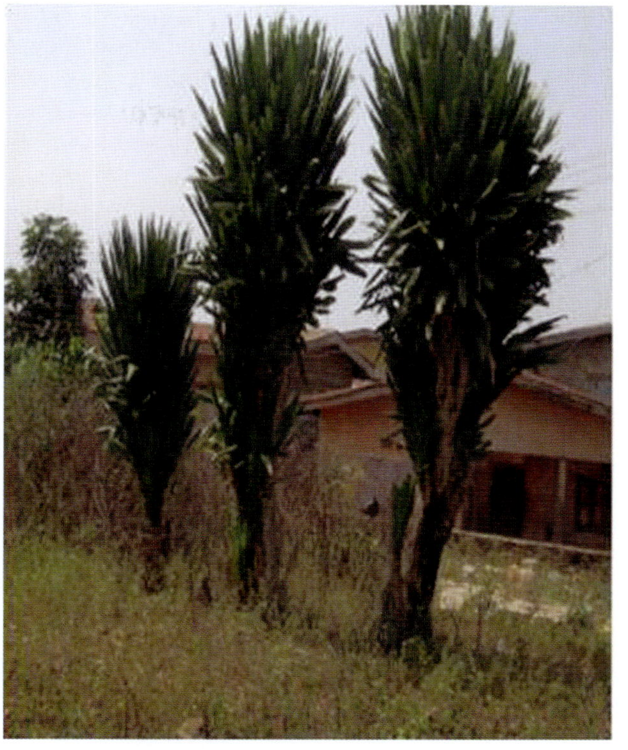

Figure 8: The *ogun* tree.

Discussion

The metallurgical processes in the Yoruba area of south-western Nigeria were carried out using lateritic ore blended with titanium- and aluminum-rich sand. These operations produced slags rich in titanium and aluminum spinel, in the range of the ulvöspinel-hercynite series. The most difficult task for a modern scientist studying ancient Yoruba iron metallurgy lies in the fact that it is notoriously difficult to relate individual iron smelting sites to specific iron ore deposits. Frequently, the low-grade and spatially limited deposits were already exhausted in ancient times, so that modern investigators are either forced to draw conclusions from those deposits still available or from deposits formed only recently. In this work, the samples were from sites that can be linked to mining sites. In the Ile-Ife area, many massive underground mines have been discovered.

According to field and laboratory analyses, iron processing at all of the sites studied in south-western Nigeria is similar in that it took place in self-draughting, slag-tapping furnaces This type of furnace takes advantage of the lower density of preheated air to that of the outside air. The hot air rises, resulting in the low pressure in the furnace, which in turn causes more air to be drawn in through the tuyères below. In the area studied for the present paper, iron smelting took place in bloomery furnaces constructed from either kaolinite/illite clay or sillimanite-quartz brick with clay cement. The mineralogical analyses of the ores and slags show that most of the ores used in the smelting operations were mined from the area and that routine blending of the lateritic ores with ilmenite-rich ores is indicative of sound scientific prospecting and exploration.

Okafor (1992) noted that bloomery iron smelting in West Africa is distinguished by three different characteristics: direct production of medium carbon steel from the furnace; pre-heated draught arrangement; and the development of shaft furnaces with self-induced draught. In spite of these distinctive achievements and this level of technical sophistication, the iron working industries gradually declined at different rates in different areas, when presented with the challenged posed by cheap, imported iron from the blast furnaces of Europe. This occurred mostly around the turn of the century, when European colonisers prohibited local African industries such as iron smelting in order to undermine African self-sufficiency (Schmidt, 1996).

Earlier technology, ideology, magic, and social organisation of ironworking became obsolete in most parts of the Yoruba land. Scrap metal from Europe represented a new source for iron objects during the colonial period, causing the Yoruba blacksmiths to adapt their skills in order to keep the knowledge of iron production alive. The Yoruba blacksmiths thus came to transform the scrap iron into many useful utensils and weapons. Since they are often felt to be imbued with mystical powers of transformation, they may be associated with fertility, charming, rainmaking, divination, and spells

This great power of iron is indicated by its deification among many West African populations, whose shrines are nowadays decorated with iron objects. Thus, iron remains of great value in much of the contemporary Yoruba land.

Conclusion

There is no doubt that the ancient Yoruba were highly successful in mining and metallurgy, as evident from data obtained from the ancient mining and smelting sites studied and from the quality of the objects produced. Beneath the cloud of mythical secrecy in the Yoruba mining and metallurgy lie sound scientific sophisticated patterns of mining and smelting of iron.

They smelted in furnaces that were constructed in the path of the wind, with tuyeres positioned to assist in the creation of a draught through the furnace to facilitate the smelting process.

The various disciplinary measures imposed on miners and smelters (e.g. sexual abstinence) are not mere rituals but are in conformity with the principles of conservation of energy – a requirement for such daring and dangerous operations. Sacrifices were made to appease the gods.

Acknowledgements

The analyses were carried out at UCL Institute of Archaeology, London. I am grateful to the Commonwealth Science Council and the Royal Society of Britain for grants to visit London and the Obafemi Awolowo University Research Committee for research grant No 11-812-AVB. Immense thanks are due to Prof Thilo Rehren and Dr Xander Veldhuijzen for their help in laboratory work. The organisers of WIC 2009 are thanked for support that enabled me to attend the conference.

References

Adediran, B., 1992. The Early Beginnings of the Ife State. In: I.A. Akinjogbin ed. *The Cradle of a Race (Ife from the Beginning to 1980)*. Port Harcourt: Sunray Publications Lagos, pp.77-95.

Adepegba, C.O., 1991. *Yoruba metal sculpture*. Ibadan: Ibadan University Press.

Adesina, M.O., 1985. *Reconnaissance survey of pre-colonial iron working sites in the central part Oyo state: Iponrin via Ajase as a case study*. MSc Dissertation, University of Ibadan.

Andah, B.W., 1979. Iron Age Beginnings in West Africa: Reflections and Suggestions. *West African Journal of Archaeology*, 9, pp.135-50.

Elueze, A.A., 2000. Compositional appraisal and petrotectonic significance of the Imelu banded ferruginous rock in the Ilesha schist belt, southwestern Nigeria. *Journal of Mining and Geology*, 36, pp.9-18.

Folarin, A., 1987. Notes on creative experimentation, Nigerian contemporary arts example. *Nigerian Magazine* 4: pp.66-8.

Ige, O.A. and Asubiojo, O.I., 1991. Trace element geochemistry and petrogenesis of some meta-ultramafites in Apomu and Ife-Ilesha area of southwestern Nigeria. *Chemical Geology*, 91, pp.19-32.

Ige, O.A. and Rehren, Th., 2003. Black sand and iron stone: iron smelting in Modakeke, Ife, south western Nigeria. *Institute of Archaeometallurgical Studies*, 23, pp.15-20.

Miller, D. and van der Merwe, N.J. 1994. Early metalworking in sub-Saharan Africa: a review of recent research. *Journal of African History*, 35. pp.1-36.

Okafor, E., 1992. *New evidence on early iron smelting from southeastern Nigeria*. PhD Thesis, University of Manchester.

Olomola, I., 1992. Ife before Oduduwa. In: I.A. Akinjogbin ed. *The Cradle of a Race (Ife from the Beginning to 1980)*. Port Harcourt: Sunray Publications Lagos, pp.51-61.

Onabajo, O., 1986. Archaeological reconnaissance at the site of Kiriji Ekiti Parapo Awr Camp. In: *National Conference on Centenary Kiriji Peace Treaty*. Ife: University of Ife, Nigeria, pp.12-4.

Schmidt, P. ed., 1996. *The Culture and Technology of African Iron Production*. Gainesville: University Press of Florida.

Soper, R. and Darling, P., 1980. The Walls of Old Oyo. *West African Journal of Archaeology*, 10, pp.61-81.

Tylecote, R.F., Austin, J.N. and Wraith, A.E., 1971. The Mechanism of the Bloomery Process in Shaft Furnaces. *Journal of the Iron and Steel Institute*, 209, pp.342-63.

Willett, F., 1960. Investigation at Old Oyo 1956-7: An Interim Report. *Journal of Historical Society of Nigeria*, 2, pp.59-77.

Author's address

- Natural History Museum Obafemi Awolowo University, Ile-Ife, Nigeria
  Email: oige@oauife.edu.ng, akin_ige@live.com

Mining and moving specular haematite in Botswana, ca. 200-1300 AD

Edwin N. Wilmsen[1,2,3], Alec C. Campbell[4], George A. Brook[5], Lawrence H. Robbins[6], and Michael Murphy[7]

ABSTRACT: In this paper, we examine the Early and Middle Iron Age (ca. 200-1300 AD) appeal of specular haematite, or specularite (Fe_20_3): an iron oxide of steel grey or black colour and with a metallic luster; which was a valued cosmetic in southern Africa from very early times well into the colonial period.

We have investigated more than twenty specularite mines and fourteen archaeological sites in the Tsodilo Hills in northwest Botswana, as well as two additional mines in southeastern Botswana. The scale of specularite production found in these locations in addition to its discovery elsewhere - small slabs of specularite were found at sites 250km to the west and 400km to the east of the Tsodilo Hills – implies that it was distributed beyond local populations during the time under consideration, as the small slabs mentioned above have no known source closer than Tsodilo.

As instrumental neutron activation analysis (INAA) has enabled the provenance of mine samples to be accurately determined, the work in progress which follows will attempt to determine if these specimens did indeed come from the Tsodilo Hills.

Keywords: Specularite, Tsodilo mining, discriminant function analysis, Botswana, Iron Age archaeology.

Introduction

In southern Africa there is little evidence, other than quarrying, to suggest that mining indigenous hard rock iron ore for metallurgical purposes occurred prehistorically; it would seem that enough magnetite and goethite existed on or near the surface for the purposes of smelting. It would appear however, that a considerable amount of hard rock mining for the purposes of adornment took place – particularly for specularite.

Specular haematite, or specularite (Fe_20_3), an iron oxide with steel grey to black colour; specular (i.e., metallic) lustre; platy or tabular habit (like micas) and a red streak; was a valued cosmetic in southern Africa from very early times until well into the colonial period, and formidable expenditures of labour were invested in its acquisition.

At mines such as Doornfontein (Beaumont and Boshier, 1974) and Blinkklipkop in the Northern Cape, mining began pre 800 AD (Thackeray, et al., 1983). At the Tsodilo Hills in northwestern Botswana (Fig. 1), hard rock adit mining for specularite was carried out on a massive scale from an indeterminate early date. Evidence for occasional mining in the Hills dates to before ca. cal. 5300±160 BP at Rhino Cave (Robbins, et al., 1996), although there is no evidence to suggest the scope of such early endeavours. A series of radiocarbon dates indicate that mining at Tsodilo, rather than quarrying, was intensive during the period ca. 750-1025 cal. AD (Robbins, et al., 1998). Using tunnels and inclined shafts, the considerable labour and large scale of specularite production at Tsodilo indicates that specularite was traded or distributed beyond the local population (Kiehn, et al., 2007, p.464). Archaeological finds of ostrich eggshells filled with specularite in the Northern Cape (Humphreys, 1974) and specularite filled pottery found in southern Namibia (Jacobson, 1977) support this theory and provide evidence for its prehistoric transport. The location of the Namibian pottery ware, for example, has no known source of specularite within more than 100 kilometres (Beaumont, 1973).

The distribution of specularite has been documented in more recent times. In the 1820s, Burchell (1967 [1822-1824], p.256) commented that the ore from Blinkklipkop, "constitutes in some degree an article of barter with the more distant tribes... the use of it extends over at least five degrees of latitude." John Campbell (1835, p.112) wrote: "Blink Mountain [Blinkklipkop] is a kind of Mecca to the nations around, who are constantly making pilgrimages to it, to obtain fresh supplies [of specularite]." Burchell (1967 [1822-1824], p.256) observed that the Batlhaping ground it with grease and then smeared it: "generally over the body, but chiefly on the head... staining them of a dark red or ferruginous color... the hair is so much loaded and clotted with an accumulation of it, that the clots exhibit the appearance of lumps of mineral". To Metheun (1846, p.95) it looked as if, "they wore metal skull-caps." The use of specularite was not restricted to Bantu-speaking peoples native to the area but was also employed by peoples speaking Khoisan languages. Baines (1864, pp.111, 125), for example, described the hair of Ngamiland San as, "felted with red clay and grease into ...

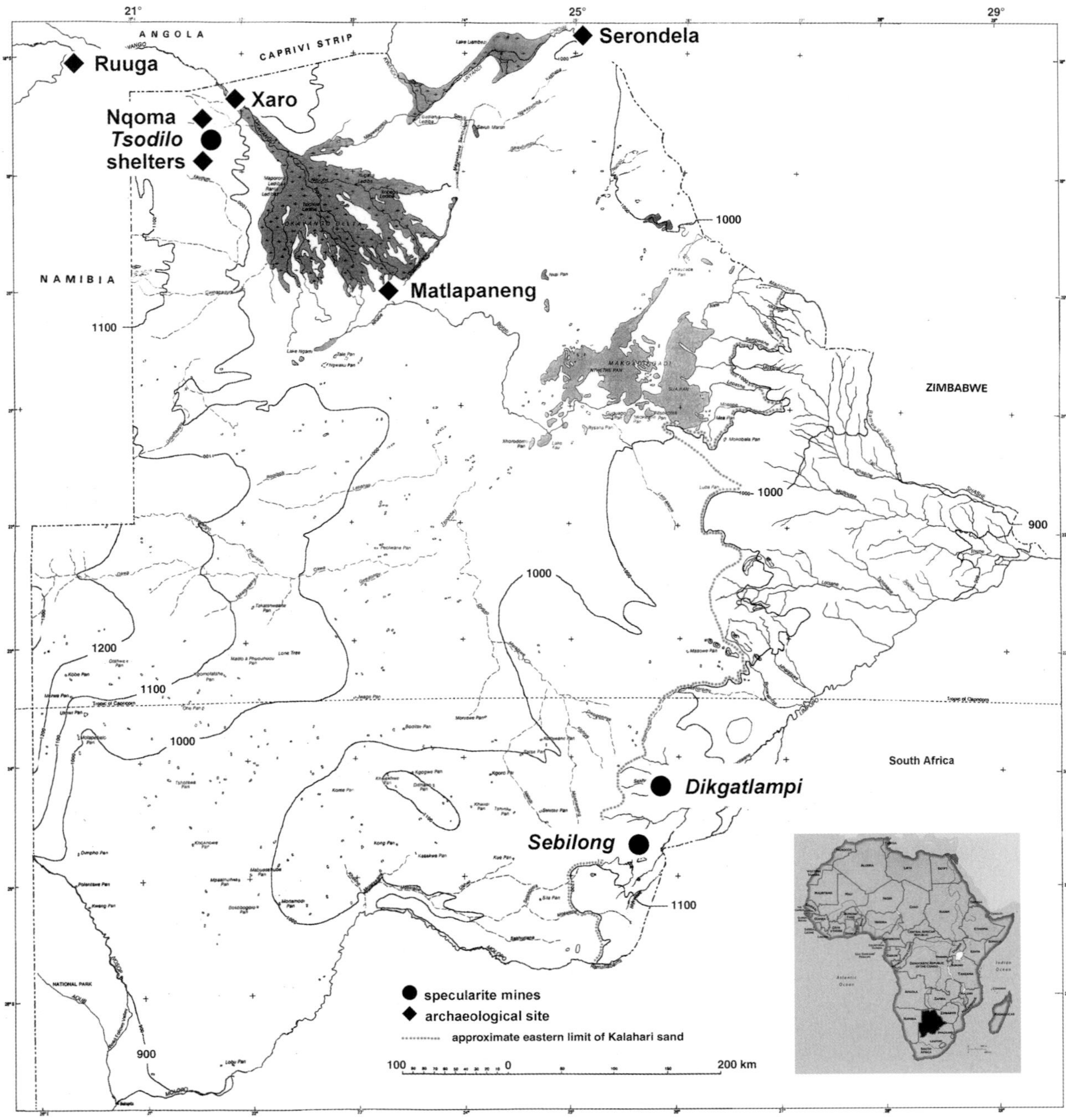

Figure 1: Location of mine and site areas.

a kind of Scotch (Glengarry) bonnet ... to which ornaments of various kinds, beads, bits of ostrich egg-shell ... &c., were attached." An identical hair-dressing practice is depicted on prehistoric rock paintings at many southern African locations.

In this paper we confine our attention to the Tsodilo Hills and to that critical period called the Early Iron Age (EIA, ca. 200-900 AD), when ceramics, metallurgy, and domesticated food production were introduced into southern Africa, and to the Middle Iron Age (MIA, ca. 900-1300 AD) when those economies had become entrenched and complicated further by the emergence of stratified societies.

The first evidence (Zhizo glass beads and Indo-Pacific marine shells) for the region's partcipation in intercontinental trade in the final centuries of the EIA was found at Nqoma, a stratified site on a plateau of the 'Female' Hill at Tsodilo. Excavation of the site in 1985 by Denbow and Wilmsen (1986; Wilmsen, 2011) recovered masses of specularite in both fragmented and pulverised form, although the origin of this material was unknown at the time. In 1991 however, Campbell and Robbins recognised that what had previously been thought to be natural caves were actually ancient mines, and initiated a programme of intensive survey and excavation, investigating more than twenty mines in the area (Robbins, et al., 1998) (Fig. 2). Brook and Kiehn (Kiehn, et al., 2007) subsequently tested the feasibility of geochemically fingerprinting sources using instrumental neutron activation analysis (INAA), by comparing Tsodilo specularite with that from two mines (Sebilong and Dikgatlampi) in southeastern Botswana (Fig. 1). These tests demonstrated that the INAA method is robust enough to be used in determining the

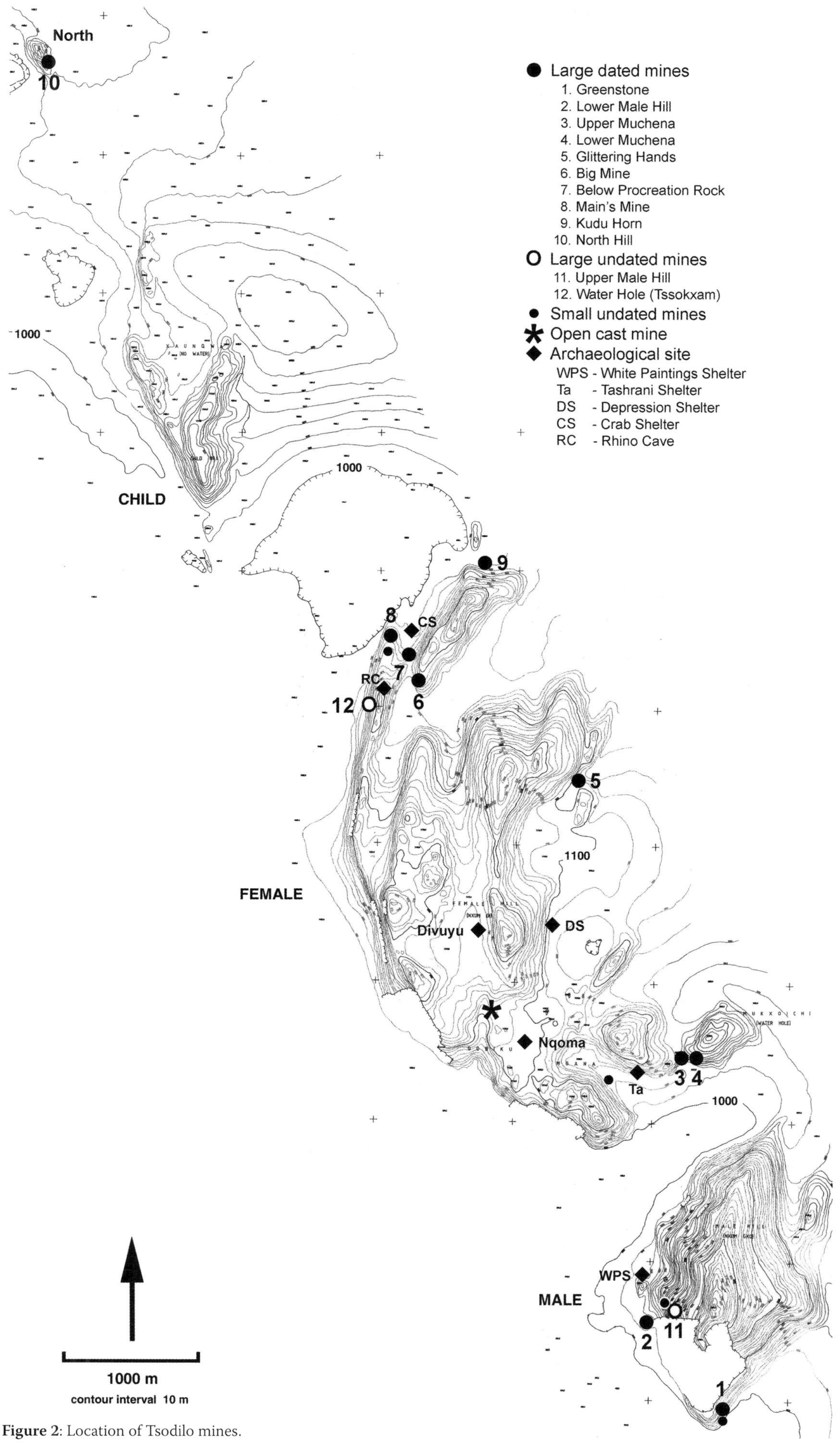

Figure 2: Location of Tsodilo mines.

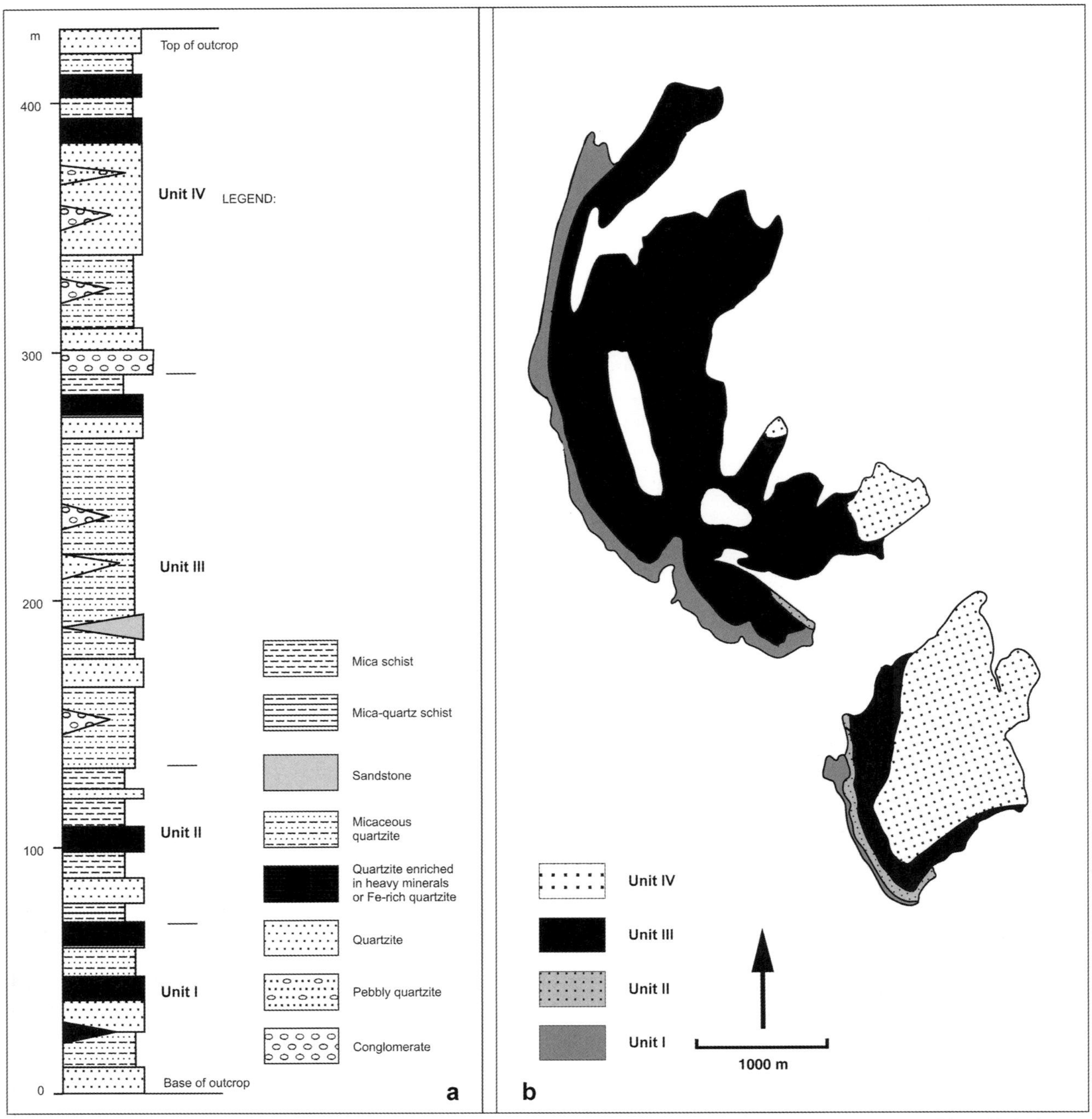

Figure 3: Lithostratigraphy of Tsodilo Hills: a) lithostratigraphic column, south end Male Hill; b) schematic lithological map of the Hills (redrawn from Wendorff, 2005, Figs. 3 and 4).

provenance of archaeological samples throughout the region, thus providing valuable information on past trade routes and patterns.

Tsodilo lithostratigraphy

The strata outcropping in the Tsodilo Hills belong to a metasedimentary succession in the Damara Sequence which was deposited between approximately 1000 to 490 Ma, during the Neoproterozoic to Early Palaeozoic ages (Singletary, 2003). There are four hills in the group: at 395 metres above the sand plain, the 'Male' hill is the highest and the 'Female' hill is the largest in area. The 'Child' and 'North' hills are smaller, separated from the two main hills and each other by roughly 2 kilometres (Fig. 2).

The lithostratigraphic succession of the strata is described by Wendorff (2005), who subdivides it into four units (Fig. 3a). Unit I is exposed only as narrow strips at the base of the west side of the Female Hill and the southern third of the Male Hill, where it is 65 metres thick at the southernmost tip, gently dipping downward to the northeast and east (Fig. 3b). It is characterised by fine grained ferruginous and micaceous quartzites that are often a hard, glassy green; they range from slightly to strongly micaceous, and may be very fine or coarse grained. Unit II is 80 metres thick and, dipping to the east and northeast, is exposed mainly in a narrow band above Unit I on Male Hill; it is composed of mica schist and mica-quartz schist with intercalated quartzite, often with a lower content of silica cement. Unit III is a 150 metre thick complex of quartzite containing intercalations of sandstone; micaceous quartzites occur often in this unit and are often ferruginous in the upper strata. The quartzites range from coarse to fine grained and their colour varies, from grey to green to red. The exposed strata of Female Hill are predominantly Unit III rocks and a moderately wide belt, which dips to the east-northeast, is exposed on Male Hill. Unit IV can be distinguished from underlying units by the presence of quartzites with subordinate intercalations of micaceous quartzite and pebbly conglomerate. Quartzites are often ferruginous and fine to coarse grained, some brownish with green banding, some dark brown, micaceous and rich in specularite. Mica schists appear as subordinate interbeds. Unit IV rocks form most of the upper mantle and surface of the Male Hill and a smaller mantle caps the south-eastern prominence of Female Hill (Fig. 3b).

The micaceous strata of all units often contain abundant metamorphosed muscovite; the angle of dip ranges from 15° to 25° (Wendorff, 2005, pp.19-21). The rock is often sheared and several major shear zones contain thick quartz veins: laminated platelets of steel grey-to-black metallic haematite are encased in the quartz, with kyanite blades often rimming the veins.

The Tsodilo mines

Since 1991 at least twenty mines have been located at Tsodilo as well as numerous small holes less than a metre deep, which indicate where rocks have been struck away from the parent rock (Murphy, et al., 2010). Most of the mines have been hollowed from quartz-mica schist or muscovite schist encasing specularite. Varieties of this specularite appear to have been the principal materials sought at this time (Fig. 4), although middens saturated with pulverised mica at Nqoma suggest that that mineral was also valued.

Of the twelve large mines investigated eight are on Female Hill, in lithostratigraphic Unit III, with the exception of Main's Mine (marked 8 on Fig. 2) which is in Unit I. The Greenstone Mine at the base of Male Hill (1 on Fig. 2) was found to contain a relatively rare green muscovite. The Upper Male Hill mine (11 on Fig. 2) and an adjacent, unnamed mine near the top of this hill, had massive crystalline quartz veins which contained dark grey metallic haematite platelets, up to 2cm long, and kyanite blade-like crystals aproximately 3cm in length.

Many of the mine adits are approximately horizontal and follow specularite seams (Fig. 5c). When the seams dip, in concert with their lithographic strata, the adits also incline

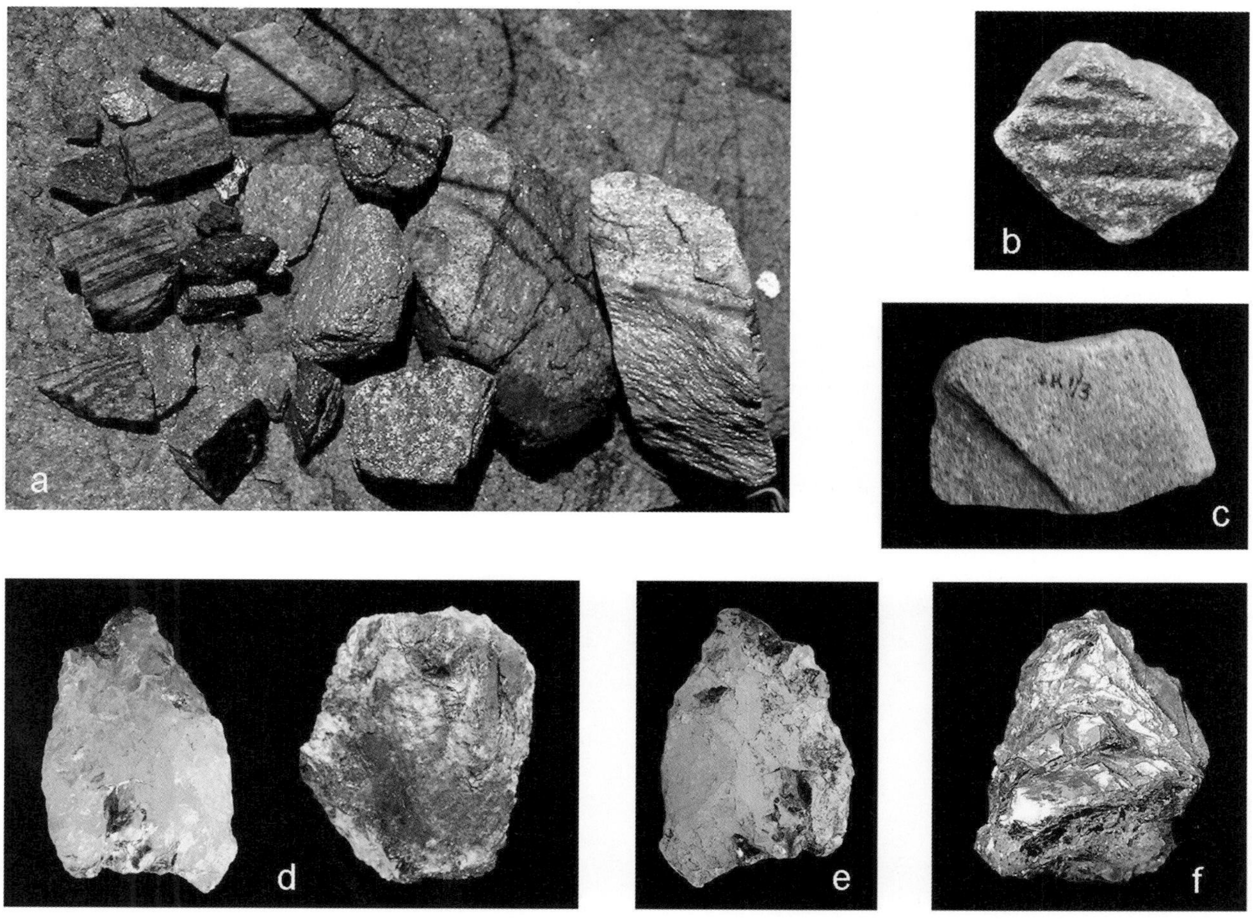

photos: a, A. Campbell; b-e, E. Wilmsen; f, E. Kose.

Figure 4: Specularite ore: micaceous quartzite specularite from a) Greenstone Mine, b) Nqoma, c) Serondela; platy quartz vein specularite from d) Nqoma, e) Matlapaneng, f) Ruuga.

at angles up to 35°. While some mines resemble shallow caves, or shelters with large mouths, entrance areas tend to be constricted and interior chambers are often domed. As a result of rockfall and of miners throwing tailings from the working face (rock debris was generally thrown aside or down the hill at the entrance of the mine) into previously exploited areas, many of the adits within the larger mines are merely crawlways (Fig. 5b). No vertical shafts or evidence for the use of timbers for shoring have been found. The rock was artificially spalled by setting fires, evidenced by black carbon soot on ceilings and walls (Fig. 6c) and large pieces of charcoal found in the fill on the floors of most mines. Miners would use stone hammers and wedges to break up the rocks, and these heavily battered artefacts have been found during excavations inside the mines and in mine tailings.

The mines vary in length from about 4.5 metres to an estimated 43 metres in 'Big Mine' (6 on Fig. 2), which is situated above a valley at the north end of Female Hill and has two conjoined entrances, (Fig. 5a) between which are two curved adits about 43 metres in length. It is not possible to measure the entire length directly as the adits are partially blocked with rubble for several metres near the centre. A consensus exists that the two entrances were initiated separately and became connected in the course of mining. The Mine's northwest corner was filled with excavated rubble, leading to a conservative estimate that at least 500 tons of rock had been removed at some point. Its width ranges from about 2 to 18 metres and maximum ceiling heights vary from approximately 1 to 2.5 metres.

Greenstone Mine (1 on Fig. 2) is named after the greenish muscovite it contains. Located not far above the base of Male Hill in lithostratigraphic Unit I it is another large mine, with four interconnected chambers extending over an approximate 40 metre area and seams of specularite visible in one chamber. Robbins et al. (1998) estimated that at least 1000 tons of rock were moved from this large complex of chambers. The Lower Male Hill Mine (2 on Fig. 2) is also in lithostratigraphic Unit I.

The Upper Male Hill Mine (11 on Fig. 2) and a smaller mine on the same level, some 40 metres to the west, appear to be on a major shear zone between lithostratigraphic Units III and IV (see Wendorff, 2005, Fig. 5). Consequently both have thick quartz veins encasing platy crystalline specularite in addition to micaceous quartzite. Evidence suggests that the smaller mine to the west was formed through cudgelling – 'bashing' - of the cliff face to leave shallow cavities. Upper Male Hill Mine has two parts (Fig. 6a): an upper sub-round chamber penetrates the cliff for about four metres, exposing both granular and crystalline ore, and a lower section, formed by bashing intended to expose granular and platy specularite. A larger instance of surface mining is the only open cast mine so far identified in the area on a ridge above the southwestern corner of Nqoma (* on Fig. 2, Fig. 6b).

A highly consistent cluster of nineteen radiocarbon dates, obtained from ten different mines, is one of the most comprehensive dating records of prehistoric mining available for a single area in southern Africa (Table 1). These dates show clearly that most of the mines were active during the period ca. 750-1050 cal. AD. Closer inspection shows that at least nine mines were being worked within a period of about

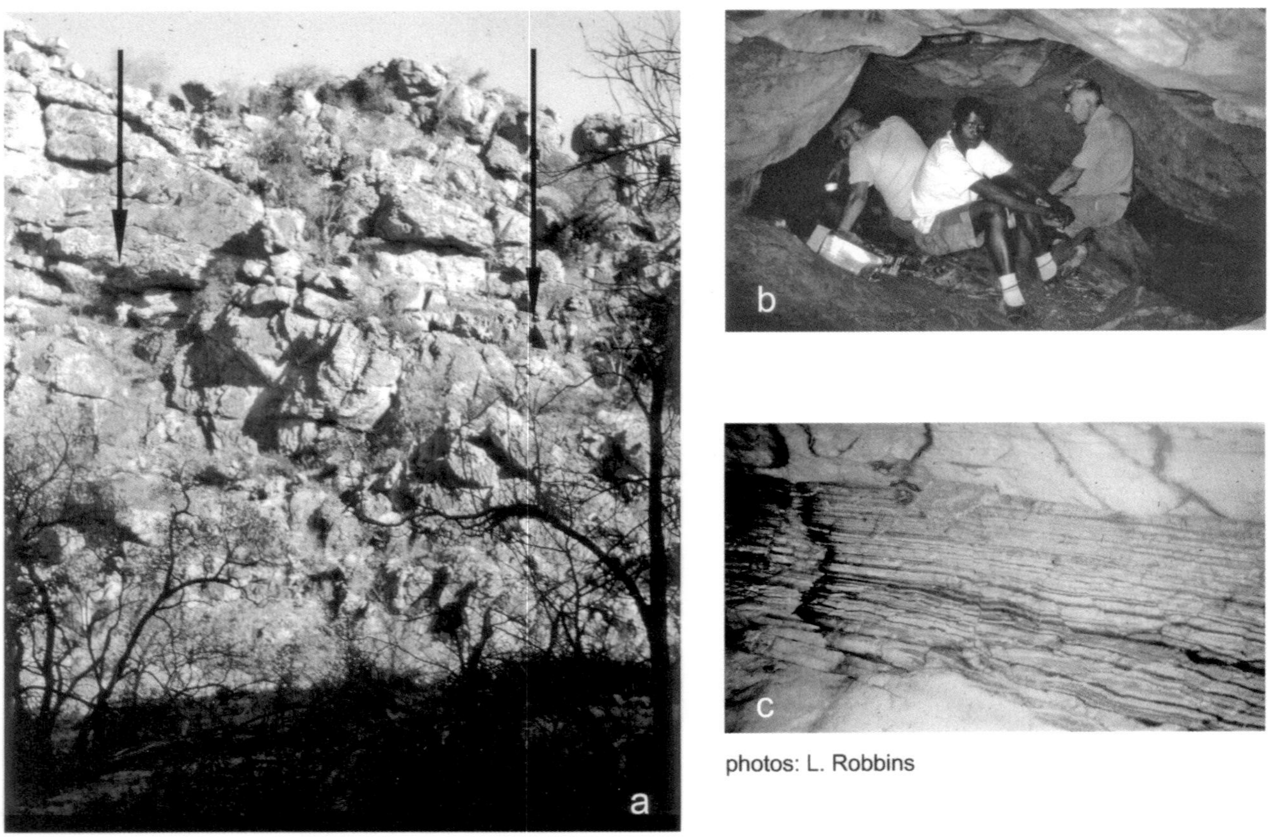

Figure 5: Tsodilo Big Mine: a) two entrances; b) interior; c) specularite seams inside the mine.

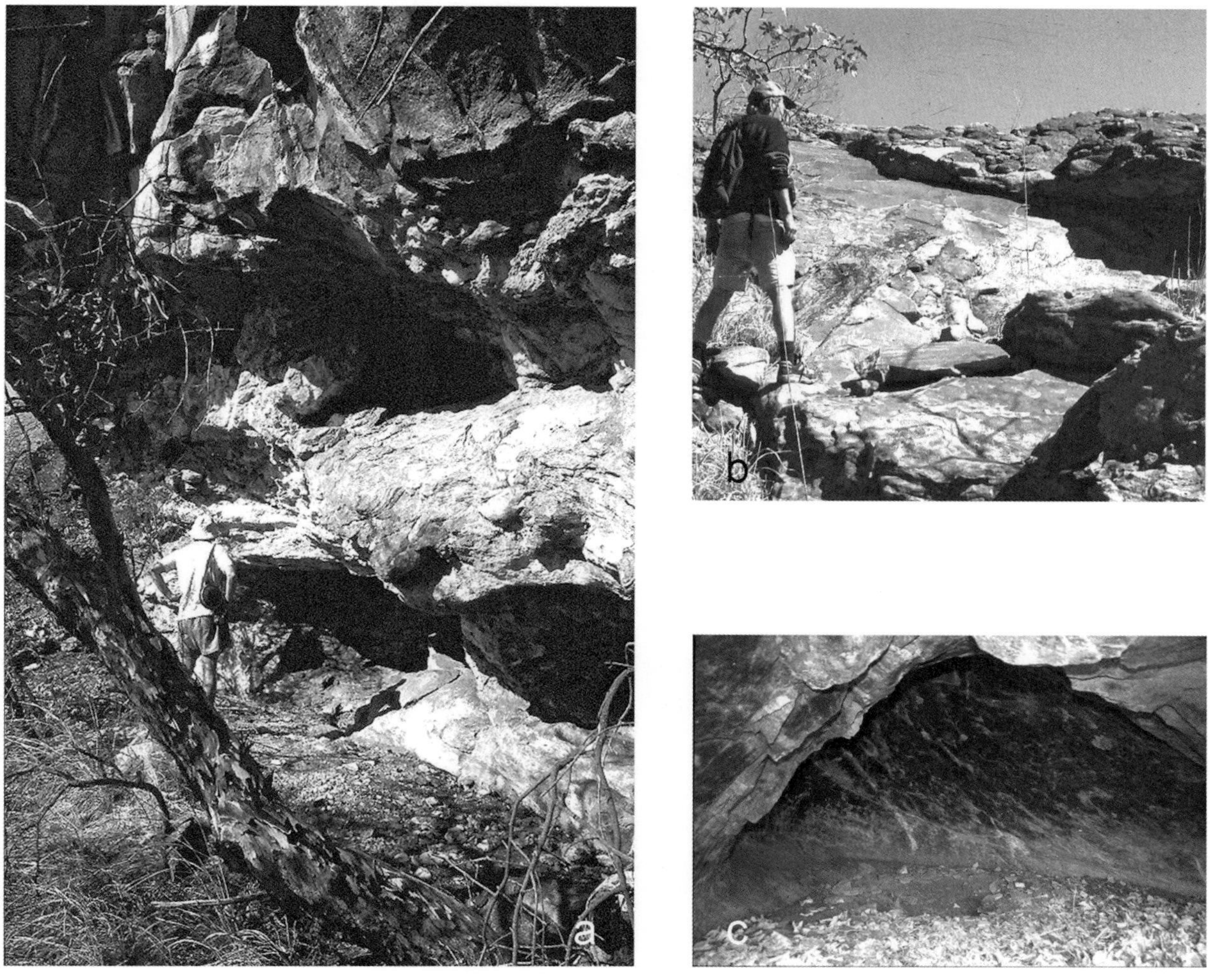

Figure 6: Tsodilo mine varieties: a) Upper Male Hill Mine; b) open cast mine; c) soot on walls and roof of a small unnamed mine. Photos: A. Campbell

a century and, moreover, that three mines (North Hill, Kudu Horn, and Upper Male Hill) have identical radiocarbon ages.

This is remarkable due to their distinct locations: one exists in a relatively inaccessible area high on Male Hill; another is at the base of Female Hill; and the third is a considerable distance away, on the slope of North Hill. Two additional mines (Glittering Hands and Lower Male Hill) have identical dates which are very close to those of the three mines mentioned above; these are also located on different hills, although in different lithostratigraphic units.

Five and three dates, obtained from spatially separated areas within both Greenstone Mine and Big Mine respectively, suggest that the former was worked for about 200 years and the latter for about 125 years.

Fingerprinting the mines

Although archaeological occurrences and early historical accounts indicate that, in southern Africa, specularite was exploited and highly valued as a cosmetic up until the 19th century; very little has been done to provenance these archaeological finds. In establishing the source of archaeological specularite a prerequisite is the analysis of geological sources, and a study by Brook and his students (Kiehn, et al., 2007) marked the first attempt to geochemically characterise and fingerprint specularite deposits from mines in Botswana.

As part of the analysis, samples of specularite ore were collected (in 1995 and 2005) from five separate mines in the Tsodilo Hills and from individual mines at Sebilong and Dikgatlampi (Tsodilo is in the far northwest corner of the country, while Sebilong and Dikgatlampi are located in the southeast, approximately 40km west-southwest and 60km north-northwest of Gaborone respectively). The specularite at both southern mines is located in and near quartz veins in the red quartz sandstones of the Mannyelanong Formation of the Waterberg Group (Campbell and Main, 2003). As at Tsodilo, fire-spalling techniques were used in these mines to extract the ore. Pottery and various other evidence indicate that intensive mining activity took place in the 14th century (although mining earlier than this cannot be ruled out) with historical accounts suggesting that, at both sites, this had ended by the 19th century. As the Tsodilo mines are 750 kilometres as the crow-flies from Sebilong and Dikgatlampi, ores taken from these mines provide a test of how well specularite can be sourced at a regional scale.

Preliminary analyses and previous literature (Smith and Frankhauser, 1996) indicate that transition metals (TM) and rare earth elements (REE) have unique signatures associated with genetic and metamorphic processes. To test how well

Table 1. Tsodilo mine (boldface) and site (thinface) radiocarbon dates.

Site/Mine	^{14}C BP	cal. 1 sigma CE	lab number
Nqoma Periphery	860±60	1150-1250	Beta-13263
Below Procreation Rock	*960±60*	*1015-1070*	*Beta-84717*
Nqoma OP2	970±50	1010-1060	Wits-836
Nqoma Outpost	970±50	1010-1060	Beta-13261
Nqoma Central	970±70	1000-1060	Beta-13262
Nqoma 40w.00	980±60	1000-1160	Beta-13255
Greenstone (lower)	*1000±60*	*995-1115*	*Beta-96235*
Greenstone (right)	*1000±50*	*1000-1040*	*Beta-84711*
Nqoma Blacksmith	1000±60	990-1040	Beta-13260
Greenstone (left)	*1050±60*	*970-1025*	*Beta-84712*
Greenstone (lowest)	*1060±60*	*960-1020*	*Beta-96236*
Upper Male Hill	*1080±50*	*905-1010*	*Beta-65194*
Kudu Horn	*1080±50*	*905-1010*	*Beta-84713*
North Hill	*1080±50*	*905-1010*	*Beta-84718*
Lower Male Hill	*1090±60*	*890-1010*	*Beta-84716*
Glittering Hands	*1090±60*	*890-1010*	*Beta-84710*
Big Mine (upper)	*1100±60*	*885-1005*	*Beta-84707*
Nqoma Society	1100±80	880-1010	Iso-11411
Big Mine (centre)	*1110±60*	*885-1000*	*Beta-84708*
Big Mine (alcove)	*1120±60*	*880-995*	*Bata-84709*
Greenstone (lower)	*1120±70*	*875-1000*	*Beta-96234*
Main's Mine	*1140±60*	*865-985*	*Beat-96237*
Upper Muchena	*1150±50*	*865-975*	*Beta-47863*
Lower Muchena	*1160±60*	*800-975*	*Beta-84705*
Divuyu Baobab 7	1190±70	770-910	Beta-13266
Lower Male Hill, tailings	*1210±60*	*770-890*	*Beta-84715*
Divuyu Baobab 2c	1220±70	700-890	Beta-13267
Nqoma Central	1220±70	700-890	Beta-13256
Lower Muchena, tailings	*1230±60*	*705-885*	*Beta-84706*
Nqoma Central	1290±60	670-780	Beta-13257
Divuyu Baobab 1x	1330±60	650-720	Beta-13265
Divuyu Acacia	1330±60	650-720	Beta-13264
Divuyu Baobab 2c	1370±60	640-690	Beta-13269
Divuyu Baobab 2c	1400±70	610-680	Beta-13268

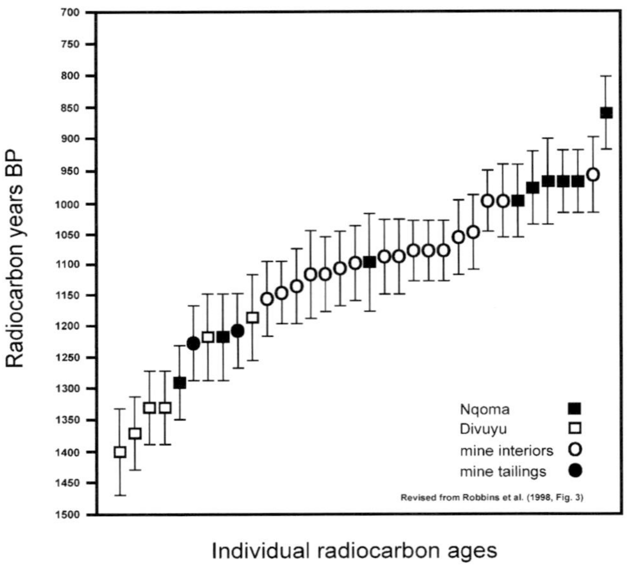

Note: all dates were calibrated by Beta Analytic with southern hemisphere correction.

these two groups of elements can source specularite from the Botswana mines, Kiehn, et al. (2007) applied stepwise multivariate discriminant analysis to the chemical compositions of mine sources determined by instrumental neutron activation analysis (INAA) and compared the results for both a TM model and a TM+REE model. Hand samples, ranging from about 50 to 300 grams each, were ground to a monomineralic grain size and the heavy mineral fraction (density >2.9g/cm^3) of 60 samples, consisting largely of specularite, was subjected to INAA analysis at the University of Missouri Research Reactor.

The INAA data were subjected to discriminant function analysis, using SPSS 13.0. The results presented were obtained by entering all variables together and using the within groups covariance matrix options. Stepwise entering of variables allows the exclusion of variables if they do not significantly contribute to the model's variance and allows the evaluation of each element's utility for fingerprinting haematite at a given scale. Additive log-ratio transform was used to unconstrain the data while preserving the elemental ratios. The small degree of analytical error associated with most INAA analyses does not significantly affect the log-ratios in this study because most elements in the subcomposition occur at levels several orders of magnitude lower than Fe.

Preliminary analyses suggested that the most likely and reliable elements for fingerprinting specularite sources would be transition metals and other ion species that could easily substitute for Fe^{3+} in the crystal lattice. The ratios of these elements would be affected by initial differences in composition and subsequent petrogenic and metamorphic processes specific to each sample location. These differences were expected to leave a unique fingerprint for each source. A data set of 11 elements (Fe, Sc, Ti, V, Cr, Mn, Co, Sb, La, Sm, Th) was transformed into 10 log-ratios for 60 non-replicate samples for discriminant analysis. The TM+REE model included all 10 log-ratios, while the TM model (without REE) excluded the log-ratios of Sb, La, Sm, and Th (Kiehn, et al., 2007, p.468).

Kiehn et al. (2007, p.478) have shown that multivariate statistical analyses of the compositional data can distinguish specularite from specific mines or groups of mines which are separated by tens to hundreds of kilometres. The three mining areas were successfully separated at the regional scale (Fig. 7) with all 60 samples correctly classified by the initial functions – only one outlier from the Tsodilo Hills group was misclassified with the Dikgatlampi samples during cross-validation. The statistical analysis also showed that when REEs were added they contributed to the TM+REE model, although the resulting model was not significantly better than the TM model.

Results are mixed however, at the local Tsodilo mine-to-mine scale. A discriminant analysis was conducted in the same manner as in the regional analysis, on twenty-seven samples from individual Tsodilo prehistoric mines: Main's (n = 2), Big (n = 3), Lower Male Hill (n = 5), Below Procreation Rock (n = 6), and Upper Male Hill (n = 11). The small sample sizes from two of the mines raise concerns about the accuracy and reliability of the method at this scale, concerns which will be rectified in future work. Nonetheless, in the TM+REE model (Fig. 8a),

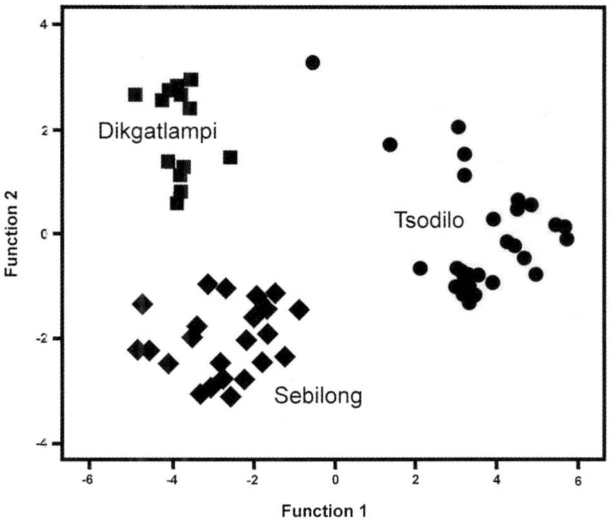

Fig. 7a. Regional discriminant function plot of TM+REE model.

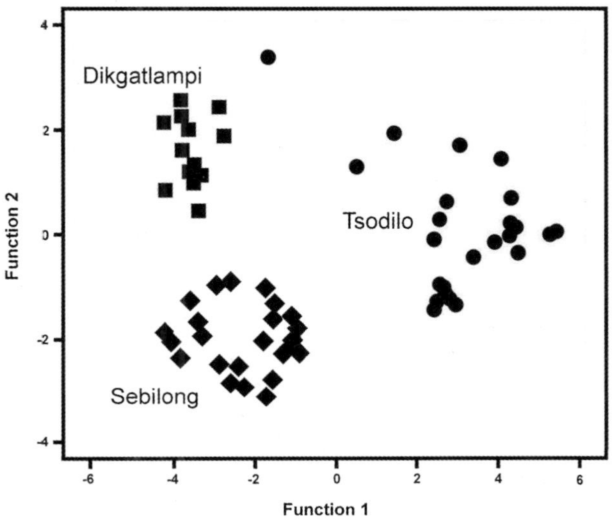

Fig. 7b. Regional discriminant function plot of TM model.

Figure 7: Regional discriminant function plots.

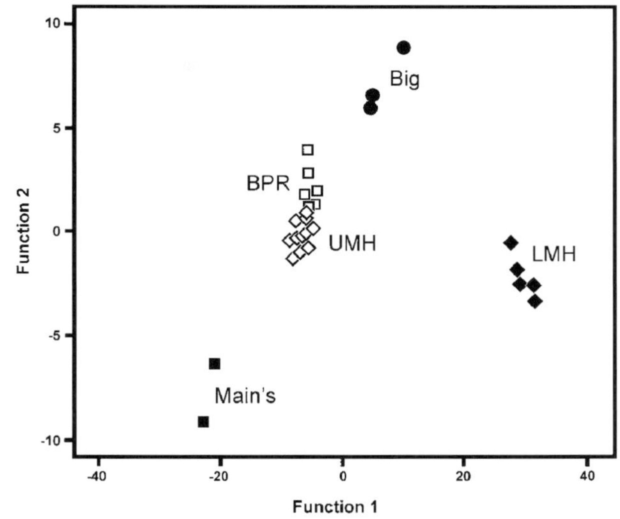

Fig. 8a. Tsodilo mines discriminant function plot of TM+REE model.

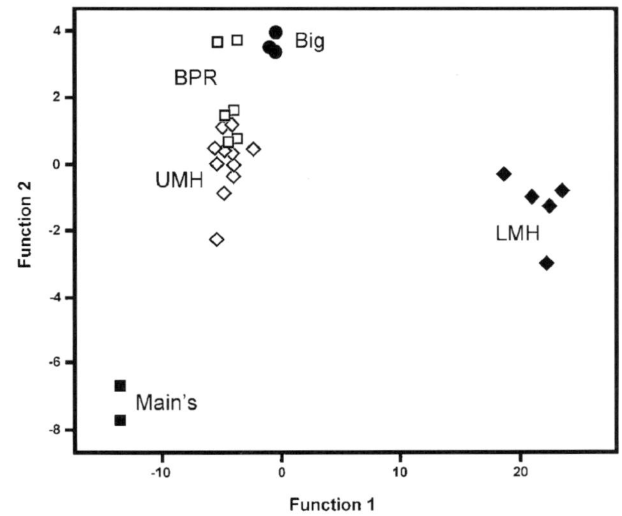

Fig. 8b. Tsodilo mines discriminant function plot of TM model.

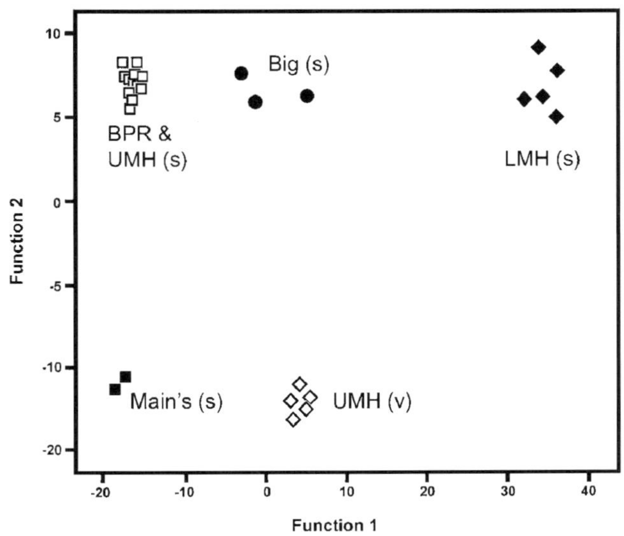

Fig. 8c. Tsodilo mines discriminant function plot of TM+REE model after parent schist (s) and vein (v) rocks subdivided.
Figures 7 & 8 redrawn from Kiehn et al. (2007, Figs. 3-7)

Figure 8: Tsodilo mine discriminant function plots.

three (Big, Main's, Lower Male Hill (LMH)) of the five mines are set apart. The other two (Below Procreation Rock (BPR) and Upper Male Hill (UMH)) are indistinguishable from each other. In the TM model (Fig. 8b) however, Big Mine no longer plots separately from Below Procreation Rock and Upper Male Hill. This indicates that REEs can play an important role in distinguishing some mines from others; however, classification accuracy remains unchanged. Thus it appears that REEs are useful for fingerprinting mines, although their contribution to the overall model is statistically insignificant. Indeed much of the overlap can be said to be due more to a greater similarity among similar rock types at different mines, than among different rock types found within an individual mine.

The Tsodilo results are further complicated by geologic differences and similarities between and within the mines, all of which have low metamorphic grade schist and quartzite rocks and some – notably Upper Male Hill – have hydrothermal quartz vein deposits adjacent to the schist and quartzite. All of these formations contain high quality specularite. Discriminant analysis of the TM+REE model, after subdividing Upper Male Hill specularite according to its parent rock, schist (s), and quartz vein (v), plots the vein specularite discretely from all other groups, but the schist continues to plot closely with that from Below Procreation Rock (Fig. 8c). Specularite found in schist in Upper Male Hill Mine is more similar to that in schist in Below Procreation Rock than it is to specularite from a hydrothermal vein deposit in the same Upper Male Hill Mine. This similarity may be due to the situation of both mines in Unit III of the Tsodilo lithostratigraphy (Fig. 3) although Big Mine, also in Unit III, plots more erratically. On the other hand, Main's and Lower Male Hill are consistently segregated from the others (and from each other), possibly due to their situation in the lower lithostratigraphic Unit I.

Thus the Tsodilo data exposes problems in distinguishing specularite from mines in a constricted geographic/geologic locale. Nevertheless the overlap effect due to rock types does not appear to be a problem at the larger regional scale, as the differences among rock types and formations between the mining regions examined easily dominate the geochemical signatures. Kiehn et al. (2007), having established the feasibility of geochemically fingerprinting specularite sources, have made it possible to confidently associate archaeological materials with geological sources.

Archaeological sites and specularite

We have excavated a total of fourteen sites in the Tsodilo Hills, with seven particularly significant for our present purpose: Nqoma on a low and Divuyu on a high plateau of Female Hill; the small rock shelters at the base of this hill, Depression and Tashrani; Rhino Cave and Crab Shelter, high on its flanks; and White Paintings, a large rock shelter at the base of Male Hill (Fig. 2).

A series of nine radiocarbon dates for Nqoma intercalate the entire range of mine dates, while four of six Divuyu dates predate or just coincide with the earliest mine dates. The remaining two are the same as these earliest dates (Table 1). The 5000 year history of specularite exploitation indicated at Rhino Cave has been noted. At Tashrani, a stratum of probable similar age has platy specularite, substantial quartz rubble, and a single small slab of silvery specularite; the lowest stratum in this shelter (275-290cm below surface) has fragments of silvery-black platy specularite with quartz rubble probably dating to 10,000-12,000 years ago (Wilmsen, forthcoming). Similar levels at White Paintings have the same platy specularite associated with masses of angular quartz debris and large numbers of LSA stone tools (Robbins, et al., 2010) – as does the smaller Depression shelter, in the same time range. Thus, several independent occurrences confirm the use of specularite at sites in the Tsodilo Hills from a very early date. Archaeological evidence suggests that this material was obtained by bashing exposed ore from the parent rock, an action possibly responsible for the creation of the shallow mining holes noted in the Hills.

The lowest levels of the Nqoma site, which remain undated, yielded substantial numbers of stone artefacts, a few shards of which suggest that occupation of this area began in the later Late Stone Age (LSA). Massive quantities of crushed quartz with a little micaceous quartzite constitute evidence for the extensive processing of platy specularite at estimated dates – perhaps as early as 200 AD and certainly before the inception of the Divuyu occupation on the upper plateau, which took place around 650 AD. Almost all of this debris is assumed to be from hydrothermal vein quartz: adherent and loose fragments of platy specularite indicate that this was the most desired mineral at this time (Fig. 4d). Pottery at Tashrani places the 15-25cm level in the same age bracket as both granular and platy vein specularite occur here. The Crab Shelter is a very small natural dolmen which, judging from pottery style, is also contemporary with a rich inventory of granular and platy specularite and items suggesting ritual use of the shelter (Wilmsen, forthcoming). These levels of the three sites predate the dated mines, but the volume of quartz debris suggests that mining, rather than simple rock bashing, must have been undertaken. Possibly the undated Upper Male Hill Mine or its unnamed small neighbour, with hydrothermal veins known to have been mined, was the source.

The later ca. 700-1200 cal. AD, midden soils at Nqoma are saturated with pulverised granular specularite and mica although very little solid ore was recovered from them (Wilmsen and Denbow, 2010). Large quantities of granular micaceous quartzite must have been processed during these centuries, although not on the plateau. It appears that only the finished product was brought to the site while the few ore chunks that have been found typically show signs of grinding or rasping (Fig. 4b). This suggests a possible shift in emphasis from platy quartz vein to granular micaceous quartzite specularite at approximately the time when the dated mining activity was most intense. Divuyu poses something of a dilemma: there is little evidence for the processing of ore having taken place at the site, the midden does not contain visible pulverised specularite and only two small lumps of an ochreous material were recovered (Denbow, 2011). The dates of the site, ca. 650-800 cal. AD, indicate that it was abandoned before or shortly after the beginning of hard rock mining. This suggests that Divuyu people either obtained small amounts of specularite using pre-mining, bashing techniques and were consequently frugal with what they obtained; or that (although unlikely) they were not especially concerned with the type of adornment yielded by specularite

Specularite has also been found at sites in the Okavango-Chobe waterveld region (Fig. 1). Ruuga, on the Okavango River 250km to the west of Tsodilo and Xaro, 45km away on the Delta Panhandle, have yielded a similar, thin-walled pottery dated respectively as the 5th-7th centuries cal. AD (Kose, 2009) and contemporary Matlapaneng, 250km to the southeast, was occupied ca. 670-1030 cal. AD; finely comb-stamped wares as well as Gokomere and Kumadzulo related wares were made in the earlier part of this period, while Nqoma-related pendant triangle wares are more common after 900 AD (Denbow and Wilmsen, 1986). The specularite at these three sites is all of the silvery-grey platy variety in quartz (Figs. 4e and 4f). All three sites exist in deep-sand, waterveld environments, hundreds of kilometres from exposed rocks (except fluvial silcretes near Matlapaneng) and with no known sources of specularite nearer than Tsodilo. Serondela on the Chobe River, 400km east of Tsodilo, has a longer time range of 690-1280 cal. AD and a similar pottery sequence of Gokomere, Kumadzulo, and some Nqoma-related wares is present (Denbow, 1982). A single slab of granular micaceous specularite (Fig. 4c) was found in a level containing Nqoma-style shards, associated with a date of 1180-1280 cal. AD. The predominant basalts and subordinate granites of the Chobe region are unlikely sources of the specularite and, once again, there are no known metasedimentary sources closer than Tsodilo.

Optical petrography of shards from Nqoma at Tsodilo and Serondela on the Chobe has established a strong interactive connection among these sites from at least the Early Iron Age (Wilmsen, et al., 2009). The most likely route by which Zhizo glass beads and pottery wares from the East Africa coast trade reached Tsodilo would have passed through Serondela, along the Chobe-Okavango waterway to Xaro, before crossing a short stretch of sand to the Hills; and we know specularite to have been a valuable commodity in the internal links in this trade.

Work in progress should allow us to determine if specularite from Tsodilo mines was carried along the same routes established for pottery and other goods. Confirmation of this routing by specularite fingerprinting would add potent support from this independent archaeometric technique to the conclusions of the optical petrography study and, ultimately, open significant new horizons for Botswana archaeology.

Acknowledgements

We thank David Killick, Duncan Miller, Dana Drake Rosenstein, and Anne Griffiths for substantial comments. The pottery analysis was supported by Wenner-Gren Foundation Grant 7509.

References

Baines, T., 1864. *Explorations in South-West Africa*. London: Longman, Green.

Beaumont, P., 1973. The ancient pigment mines of Southern Africa. *South African Journal of Science*, 69, pp.140-6.

Beaumont, P. and Boshier, A.K., 1974. Report on test excavations in a prehistoric pigment mine near Postmasberg, Northern Cape. *South African Archaeological Bulletin*, 29, pp.49-59.

Burchell, W., 1967 [1822-1824]. *Travels in the interior of southern Africa*. Cape Town: Struik.

Campbell, A. and Main, M., 2003. *Guide to Greater Gaborone*. Gaborone: Botswana Society.

Campbell, J., 1835. *Journal of travels in South Africa*. London: Religious Tract Society.

Denbow, J., 1982. Archaeology in Botswana: exciting discoveries in Ngamiland. *The Zebra's Voice*, 9, pp.14-21.

Denbow, J., 2011. Excavations at Divuyu, Tsodilo Hills. In: E. Wilmsen ed. *Retrieving Botswana's past: essays in honour of Alec Campbell*. Botswana Notes and Records 43. Gaborone: Botswana Society, pp.76-94.

Denbow, J. and Wilmsen, E., 1986. Advent and course of pastoralism in the Kalahari. *Science*, 234, pp.1509-14.

Humphreys, A., 1974. The occurrence of ostrich eggshells filled with specularite in the Northern Cape. *South African Archaeological Bulletin*, 70, p.48.

Jacobson, L., 1977. A pottery cache from the Bethanie District, South West Africa. *Cimbebasia*, B2, pp.227-33.

Kiehn, A., Brook, G., Glascock, M., Dake, J., Robbins, L., Campbell, A. and Murphy, M., 2007. Fingerprinting specular haematite from mines in Botswana, southern Africa. *American Chemical Society Symposium Series*, 968, pp.460-79.

Kose, E., 2009. New light on the Iron Age along the middle Kavango in northern Namibia. *Southern African Archaeological Bulletin*, 64:130-147.

Metheun, H.H., 1846. *Life in the wilderness*. London: Richard Bentley.

Murphy, M., Robbins, L. and Campbell, A., 2010. The prehistoric mining of specularite. In: A. Campbell, L. Robbins and M. Taylor eds. *Tsodilo Hills - Copper Bracelet of the Kalahari*. East Lansing: Michigan State University Press, pp.82-93.

Robbins, L., Murphy, M., Campbell, A. and Brook, G., 1996. Excavations at the Tsodilo Hills Rhino Cave. *Botswana Notes and Records*, 28, pp.23-45.

Robbins, L., Murphy, M., Campbell, A. and Brook, G., 1998. Intensive mining of specular haematite in the Kalahari, ca. AD 800-1000. *Current Anthropology*, 39, pp.144-50.

Robbins, L., Murphy, M., Campbell, A., 2010. Windows into the past: excavating Stone Age shelters. In: A. Campbell, L. Robbins and M. Taylor eds. *Tsodilo Hills - Copper Bracelet of the Kalahari*. East Lansing: Michigan State University Press, pp.50-63.

Singletary, S.J., 2003. Geochronology of basement rocks in the Kalahari Desert, Botswana, and implications for regional Proterozoic techtonics. *Precambrian Research*, 121, pp.27-71.

Smith, M. and B. Frankhauser, 1996. *An archaeological perspective on the geochemistry of Australian red ochre deposits*. Canberra: Australian National University.

Thackeray, A.I., Thackeray, J.F. and Beaumont, P., 1983. Excavations at the Blinkklipkop specularite mine near Postmasburg, Northern Cape. *South African Archaeological Bulletin*, 38, pp.17-25.

Wendorff, M., 2005. Outline of lithostratigraphy, sedimentation, and tectonics of the Tsodilo Hills Group, a Neoproterozoic – Lower Palaeozoic siliciclastic succession in NW Botswana. *Annales Societatis Geologorum Poloniae*, 75, pp.17-25.

Wilmsen, E., 2011. Nqoma: an abridged review. In: E. Wilmsen ed. *Retrieving Botswana's past: essays in honour of Alec Campbell*. Botswana Notes and Records 43. Gaborone: Botswana Society, pp.95-114.

Wilmsen, E., forthcoming. *Mines and myths: the archaeology of the Tsodilo Hills*.

Wilmsen, E. and Denbow, J., 2010. Early villages at Tsodilo: the introduction of livestock, crops, and metalworking. In: A. Campbell, L. Robbins and M. Taylor eds. *Tsodilo Hills - Copper Bracelet of the Kalahari*. East Lansing: Michigan State University Press, pp.72-81.

Wilmsen, E., Killick, D., Drake Rosenstein, D., Thebe, P. and Denbow, J., 2009. The social geography of pottery in Botswana as reconstructed by optical petrography. *Journal of African Archaeology*, 7, pp.3-39.

Authors' addresses

1,2,3 Edwin N. Wilmsen, School of Geography, Archaeology, and Environmental Sciences, University of the Witwatersrand. Centre of African Studies, Edinburgh University. Department of Anthropology, University of Texas-Austin, anaw630@uts.cc.utexas.edu.

4 Alec C. Campbell, Box 306, Gaborone, Botswana, acpeba@mega.bw.

5 George A. Brook, Department of Geography, University of Georgia, Athens, gabrook@uga.edu.

6 Lawrence H. Robbins, Department of Anthropology, Michigan State University, lrobbins@msu.edu.

7 Michael Murphy, Kalamazoo Valley Community College, michaelmlmurphy@aol.com. Correspondence to ENW.

Late Iron Age technology of Mt. Kenya region: the case studies of the Kangai and Kanyua archaeological sites in Mbeere District

M'Mbogori Freda Nkirote

ABSTRACT: Much archaeological research has concentrated on charting the initial spread of iron technology into East, Central and Southern Africa associated with Bantu speakers. Such research on the Early Iron Age however, has been mostly concerned with pointing out the routes and presence of Bantu speakers and pays little or no attention to the technological aspects of ironworking. For example, specific pottery decorations and forms (together with the remains of iron production) were used as indicators of both Early and Late Iron Age sites of Bantu speakers (Soper, 1967, 1976, 1982; Odner, 1971; Siiriainen, 1971; Chami, 1994; Haaland and Msuya, 2000; Phillipson, 2005). However, ethnographic research (Gosselain, 1998, 2000; Smith, 2000; and Roux, 2007) has shown that any identification of social boundaries based solely on these two archaeological attributes may be inaccurate.

This paper presents the archaeology of iron technologies within the Mbeere area of the Mt. Kenya region and compares it with the ethnographies of ironworking technology of the same area; with a view to offering a different perspective on Iron Age research in Kenya. Where major differences are noted, the paper also points out ethnographic ironworking affiliations in order to prompt new questions into the iron producers of Mbeere region during the Late Iron Age.

Keywords: Bantu, Late Iron Age, ethnographic, tuyères, furnaces, smithies, Mt. Kenya region.

Background

Early Iron Age research in Africa during the mid-20th century focused on the origins of iron technology in Africa and its relationship with Bantu migrations (Karega-Munene, 1993). The original hypothesis of such research focused on the concept of a north-south spread of Bantu speakers who carried the knowledge of iron technology (Forslund, 2003, p.11). According to Phillipson (1977), early Bantu speakers spread south from Cameroon about 3000-4000 years ago, finally reaching the Mt. Kenya region approximately 1000 years ago, during the Late Iron Age. Urewe, Lelesu, and Kwale pottery wares were used in East Africa as indicators of the Early Iron Age sites of Bantu speakers; this method of identifying sites associated with Bantu speakers persisted into evidence gathered from the Late Iron Age (Soper, 1967, 1982; Odner, 1971; Siiriainen, 1971; Chami, 1994; Msuya and Haaland, 2000). Within the Mt. Kenya region Siiriainen (1971) excavated the Gatung'ang'a Late Iron Age site, which was found to contain both Early Iron Age pottery (Kwale ware) and Late Iron age pottery (Gatung'ang'a ware) which he attributed to Bantu speakers. Other sites within the Mt. Kenya and Central Kenya regions, with Late Iron Age materials similar to those found at Gatung'ang'a, were excavated in the areas around Mbeere, Chuka, Tana Grand Falls, Ithanga, and Thika (Gramly, 1971; Cummings, 1978; Mhalstedt and Diblasi, 1979). The technological aspects of ironworking in these areas, however, were not investigated.

The Iron Age dates suggested by archaeological evidence for the presence of Bantu speakers around the Mt. Kenya region are contradicted by historical evidence, which claims later dates: from the 15th and 16th centuries AD (Fadiman, 1973, 1993; Mwaniki, 1973; Muriuki, 1974, 1978). The lack of consensus between historical and archaeological dates may be attributable to the archaeological research methods employed. For example, as significant as pottery decorations and forms may be, research has shown that, long before these areas were occupied by Bantu speakers, there were other different linguistic groups inhabiting them. For instance, the Cushitic-speaking population may have preceded the Nilotic and Bantu speakers by approximately 2000 years (Ehret, 1974; Ambrose, 1982). It would be speculative therefore to attribute all archaeological artefacts in the Mt. Kenya region and dating from the Early Iron and Later Iron Age to Bantu speakers due to the fact that Bantu speakers may have encountered and mixed with autochthonous populations (Vansina, 1995). These interactions render the peculiarity of specific pottery forms or decorations to a specific linguistic group unlikely. Furthermore, recent ethnographic research by Gosselain, 1998, 2000; Smith, 2000; and Roux, 2007 has shown that pottery decorations and forms are salient features but they can be easily copied – in which case, identification of social boundaries based solely on these two attributes may be inaccurate. In order to make more informed assumptions regarding social boundaries, a complete *chaîne opératoires* of pottery production should be considered.

This paper wishes to highlight some of the ironworking technologies of Mt. Kenya region using the sites of Kangai and Kanyua as case studies, with a view to offering new considerations into the Late Iron Age in this region. The archaeological evidence associated with ironworking technology will be compared with ethnographic data concerning ironworking technology (recorded in 1974 by Brown (1995)). Similarities and differences between the technologies of these two periods will be emphasised and possible affiliations to other regional ironworking technologies proposed.

These theories have been made possible by a personal excavation of two Iron Age sites in the Mbeere district of Mt. Kenya region in June 2006. The sites are located in the Kangai and Kanyua villages, from which their names are derived. The Kangai site yielded a small quantity of pottery, tuyère fragments, plenty of iron slag, and an in situ furnace, while the Kanyua site produced only tuyères, which were buried together.

Geography of Mbeere Land

Mbeere is a low altitude area of approximately 1014km^2 except for two hills: Kiang'ombe (1804m) and Kianjiru (1495m) [Mwaniki (1973)]. The land slopes from 1219m to just under 61m above sea level to the east and northeast as it joins the Tana River. About one third of the area lies to the left of the bank of Tana, at an altitude between 610 and 914 metres; the rest is 91-122 metres above sea level. Mwaniki (1973) notes that the Mbeere land is too far from Mt. Kenya to benefit from it's precipitation. As a result rainfall is inadequate and erratic throughout the year and mostly dependent on relief. As a consequence the lowlands which are today occupied by the Mbeere people receive between 500mm and 750mm of rainfall annually, unlike the high altitude areas which can receive up to 2250mm (Muriuki, 1974). Soils are sandy, grey or reddish-brown in colour, with productive soils found only around riverbeds. In general, the land is stony, with scrubs of thorny acacia trees barely 6m high.

People of Mt. Kenya region

The Mt. Kenya region is currently inhabited by Bantu speakers. Phillipson (1977) argues that Bantu speakers may have moved to this region around 1000AD (Fig. 1), while Nurse (1982) argues that the Bantu language of Mt. Kenya region is conservative and has not undergone the sound shifts which may have occurred in most East African languages a millennium or longer ago. Linguistic sources have classified all the Bantu speakers of Mt. Kenya region as Thagicu: a group of dialects that are spoken to the east and south of Mt. Kenya (Bennett, 1967, cited in Muriuki, 1974).

The Thagicu group of dialects includes Kikuyu, Kamba, Mbeere, Meru and Chuka, and Segeju and Sonjo of Tanzania. The Embu, Mbeere and Chuka peoples are included in the Kikuyu history of migration. Mbeere split from the Embu after their settlement in the Embu region (Mwaniki, 1973; Muriuki, 1974). Some historians suggest the 15th century as the date of arrival of the Mbeere people to their present land (Muriuki, 1974)

Sites, settings and excavations

Defunct iron smelting sites are widespread in the Mbeere land and the selection of the Kangai and Kanyua sites was based on the surface finds encountered at these locations. Each site yielded unique materials (discussed in the text) that qualified it for further investigations. Charcoal samples from the Kangai site were collected from an in situ furnace while for the Kanyua site they were collected at a depth of 30cm, before both sets of samples were taken for dating at the Beta analytic radiocarbon dating laboratory in Florida, United States. The results (Table 1) show that these sites were occupied around the same period, an aspect which bears upon the contemporeanity of the excavated archaeological materials. The distance between Kangai and Kanyua is 8.9km and the soils and vegetation of both areas are the same. They are both located on the river banks, on ground where iron ore is easily accessible.

Kangai site

It is possible to deduce, from the extent of surface debris and mud mounds, that the Kangai site occupies an area of approximately 500m^2 . It lies at S 00.465° latitude and E 037.789° longitude and at an elevation of 857m. The site presents evidence of a prehistoric settlement, with seven mud mounds and an area for ironworking situated on the right side of a dry riverbed.

The mounds are suspected to be the remains of dwellings, as attested by the circular formation of the collapsed mud features [Fig. 2, marked 1 to 7] - the substance of which is compact soil consisting of clay and sand. In between these mounds is an open ironworking area (marked 8) bordered by collapsed mud walls. It is at the centre of this ironworking area that the remains of furnace have been found in situ (discussed below). Other Late Iron Age materials e.g. pottery, iron slag and pieces of tuyères were found scattered all over the site. Excavations were carried out at the ironworking area marked 8 and a test pit was dug on mound 7 (Fig. 2). The test pit produced iron slag, pottery, and fragments of tuyères.

Kanyua site

This site is situated 9.4km southwest of Kangai. It lies at S 00.442° latitude and E 037.867° longitude, at an elevation of 749m. It is located on the right side of the road leading to Muthanthara, about 30 metres away from the Kang'ote riverbed and 200m away from the bridge (Fig. 3). The ground

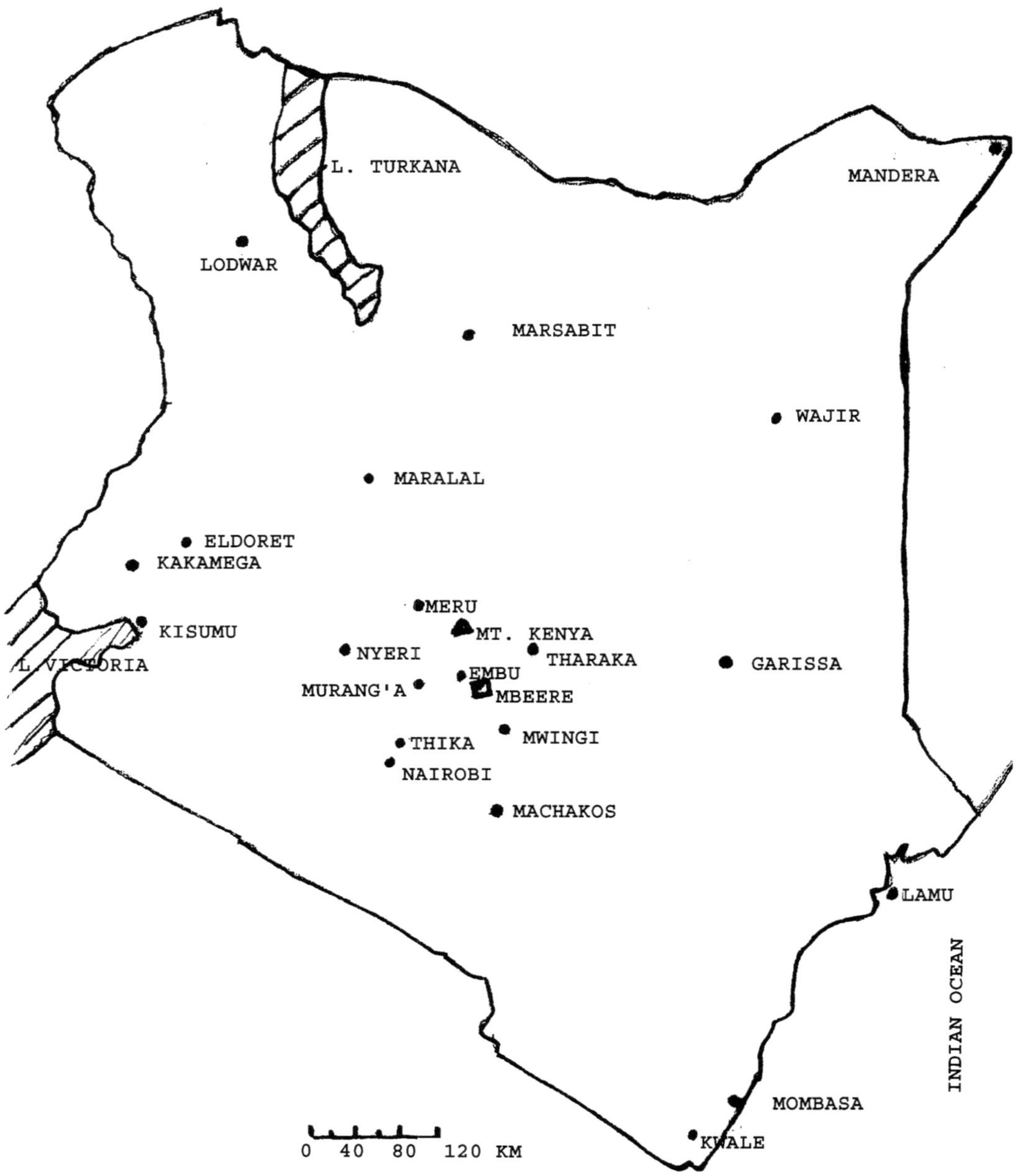

Figure 1: Bantu speakers of Mt Kenya region by F.N. M'Mbogori.

Table 1: Radiocarbon dates for the Kangai and Kanyua archaeological sites.

Sample data	Measured radiocarbon age	13C/12C ratio	Conventional radiocarbon age
Beta-221270	230 ± 40 BP	-24.8	230 ± 40 BP
Sample - Kangai Site			
Analysis: Radiometric - Standard delivery			
Material/pre-treatment: (charred material) acid/alkali/acid			
2 sigma calibration: Cal AD 1530 to 1550 (Cal BP 420 to 400) AND Cal AD 1630 to 1680 (Cal BP 320 to 270)			
Cal Ad 1740 to 1810 (Cal BP 210 to 140) And Cal AD 1930 to 1950 (Cal BP 20 to 0)			
Beta 221271	220 ± 40 BP	-27.7	180 ± 40
Sample - Kanyua Site			
Analysis: Radiometric - Standard delivery			
Material/pre-treatment: (charred material) acid/alkali/acid			
2 sigma calibration: Cal AD 1650 to 1710 (Cal BP 300 to 240) AND Cal AD 1720 to 1880 (Cal BP 230 to 70)			
Cal AD 1910 to 1950 (Cal BP 40 to 0)			

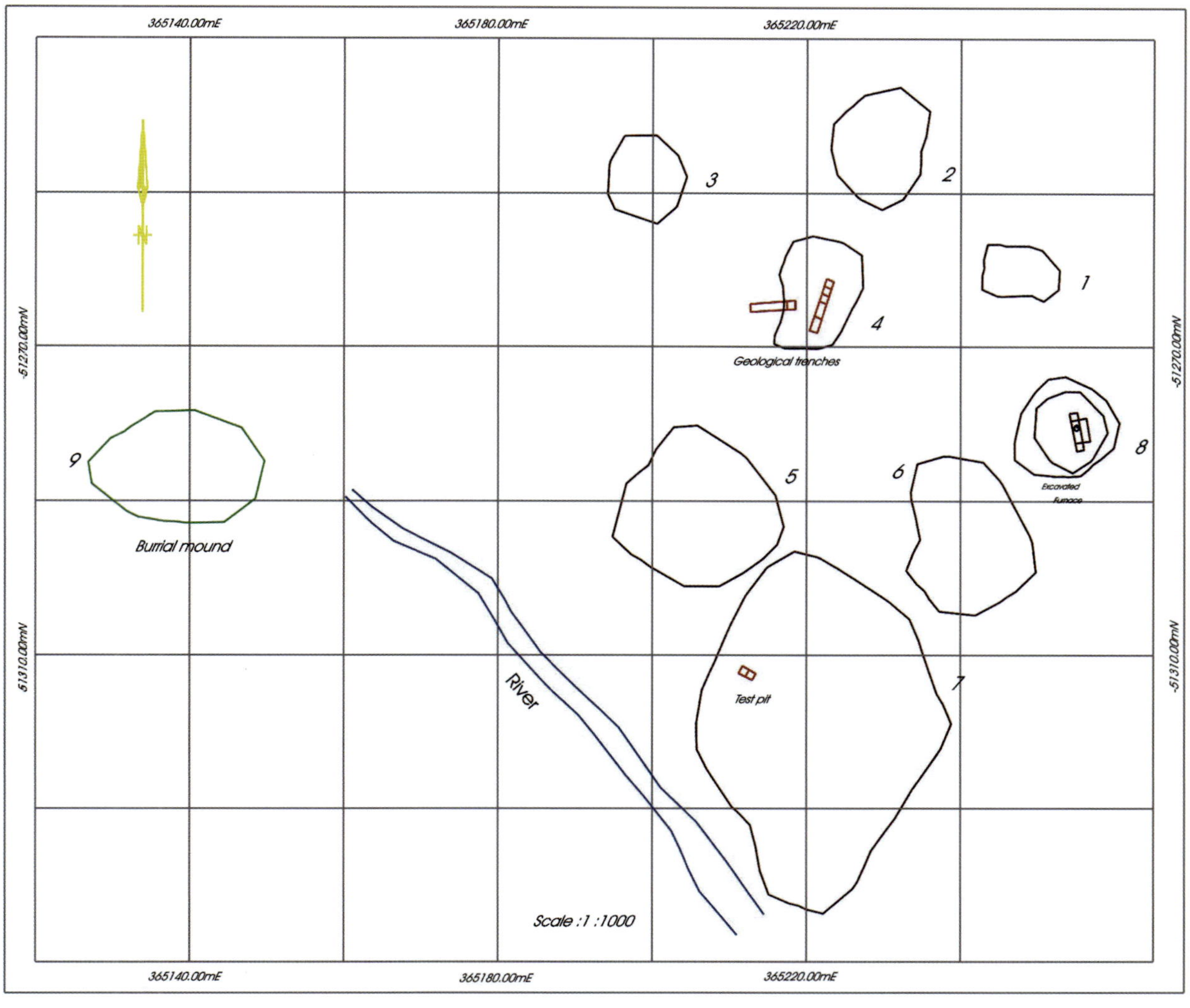

Figure 2: Topographical map of the Kangai site by Stephen Manoa (BIEA).

surface is typically rocky with loose, sandy soil. The Kanyua site yielded a concentration of tuyère fragments, which were buried together. Although none of the tuyères was recovered complete, the vitrification identified on the ends of some of the tuyère fragments exhibited clear evidence of use in iron production activities.

The deposit within which the tuyères were excavated was approximately 20cm deep and contained darkish silt; below which was a sterile red sandy silt deposit that did not yield any artefacts. The area chosen for excavation showed a concentration of tuyère fragments on the surface and only two square test pits (1 x 1m) were excavated: both reaching the bedrock at a depth of 50cm.

As there were no other cultural materials observed or recovered during the excavation or on the surrounding slope, it is possible that the site was used for deliberate burial of damaged tuyères (Fig. 4.). Adjacent to the excavated test pits more tuyères were visible, although no further excavations were carried out.

Iron technology

Tuyères

Tuyères were recovered from both the Kangai and Kanyua sites. Fragmentation or vitrification meant that it was not possible to measure their exact lengths, although the thickness of their walls ranged from 11 to 35mm. Mouths were funnel shaped and flared, with flared lips 32 to 35mm thick and 16 to 22cm in circumference. Both sites yielded two distinctive sizes of tuyères; the average inner diameter of the larger tuyères was 4cm and the outer 7cm; while the smaller tuyères had an average inner diameter of 3cm and an outer diameter of 5cm. At their flared ends, some tuyères exhibited finger impressions and horizontally incised lines which may have been left by the finishing techniques used during the final stages of tuyère forming.

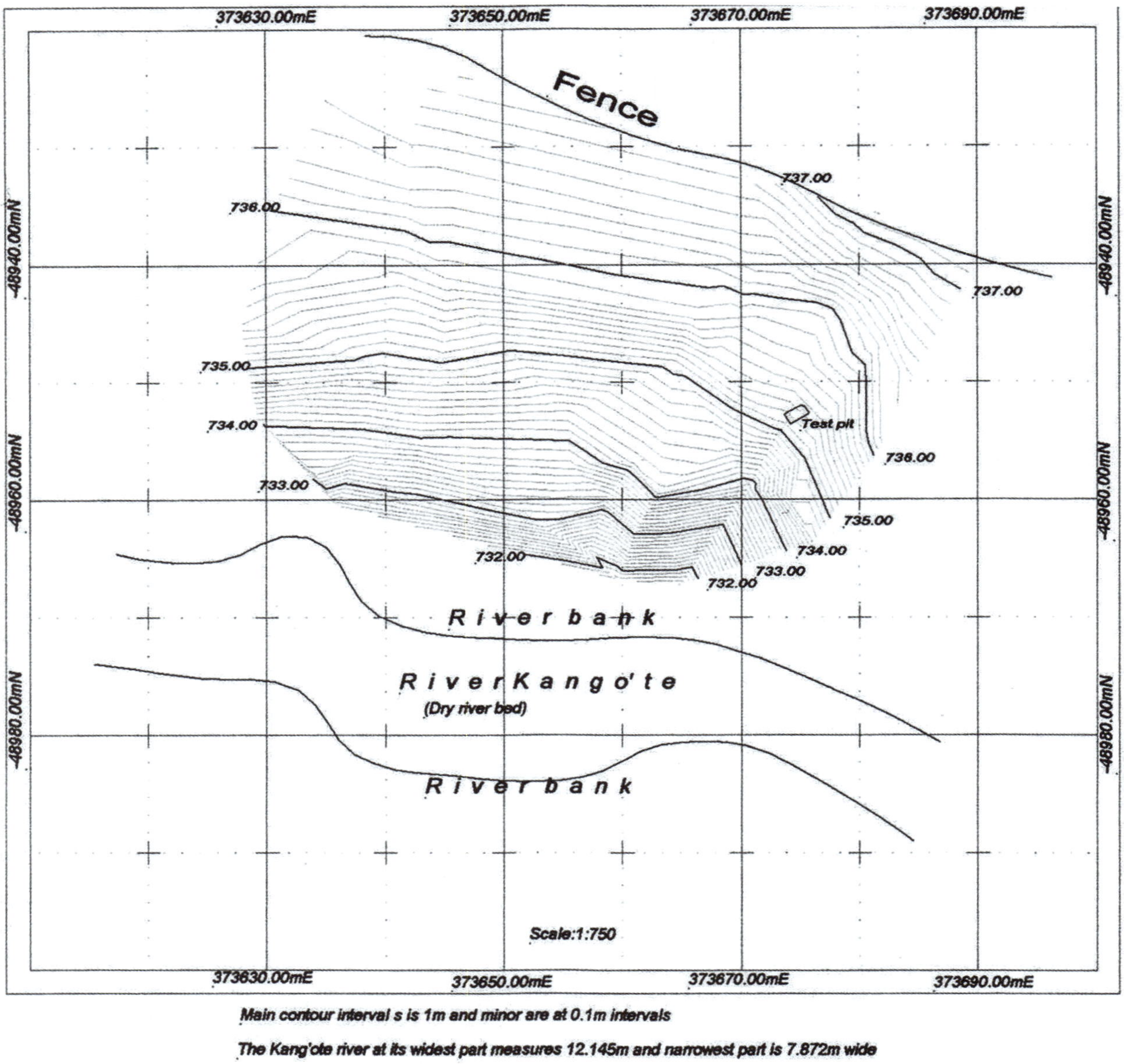

Figure 3: Topographical map of the Kanyua site by Stephen Manoa (BIEA).

Furnace

The excavated furnace appeared to be a low shaft, built with bricks of clay. Most of the bricks that made up the upper parts of the furnace had already been eroded, leaving only two layers protruding on the surface, and the joints between the bricks were clearly visible. All the furnace bricks had slag and soot attached to them, unlike the bricks below or at the base. The bricks were cemented with clay, the ones at the bottom of the furnace being more closely attached than the ones above the ground level. They were again plastered all round with more clay (an outside layer).The sixteen bricks that formed the protruding layers had an average width of 14cm and were between 4 and 6cm thick. The furnace base diameter was 34cm and its depth was 30cm below the surface (46cm below datum) and 20cm above the surface (Fig. 5). The above-ground circumference was 117cm. Three pieces of tuyères and small amounts of slag were recovered from inside the furnace.

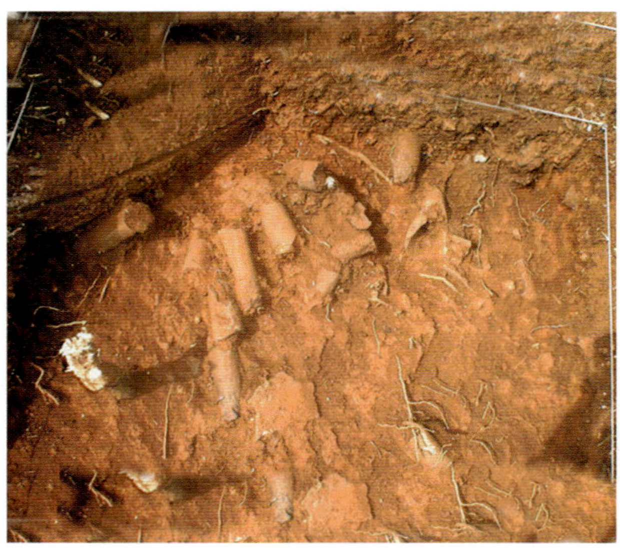

Figure 4: In situ tuyère fragments from the Kanyua site by F.N. M'Mbogori.

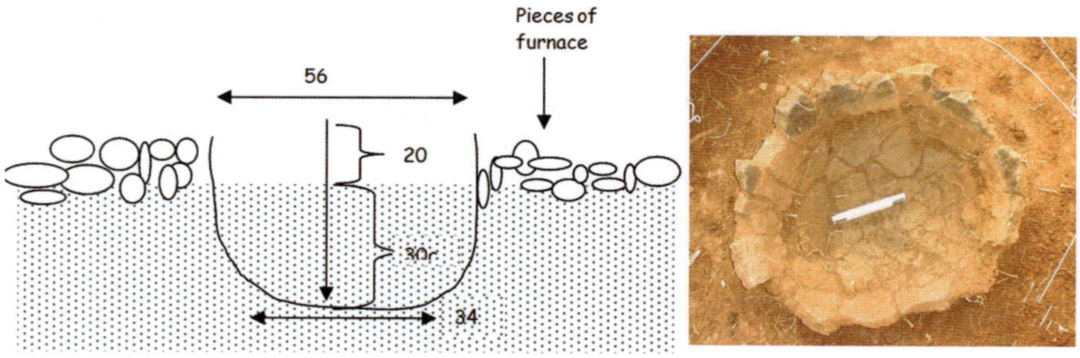

Figure 5: Sketch of the cross-section of the furnace by F.N. M'Mbogori., and photograph of furnace after excavation.

After studying the furnace in situ, a cross-section was cut and further excavations were carried out underneath, as investigation into whether there were indications of ritual materials, as reported in some ethnographic practices (Schmidt 1997). No evidence suggesting ritual materials was exposed; only a termite hole (Fig. 6).

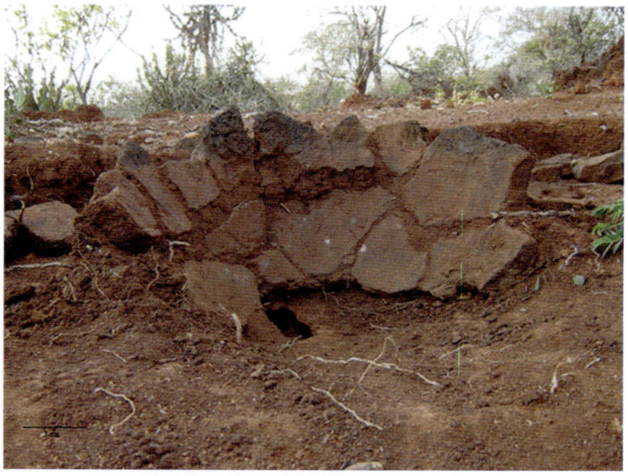

Figure 6: Termite hole beneath the furnace by S. Trygve.

Slag

The slag which was observed and collected from the sites exhibited varying characteristics (Fig. 7). One type of slag was smooth with flow marks while the other type was spongy and porous. One large piece which was recovered weighed over 15kg and is thought to have originated from a slag pit. This slag 'cake' was dark, rounded and without metal lustre. The majority of the slag were blue-grey and showed shiny metallic colour on the top face.

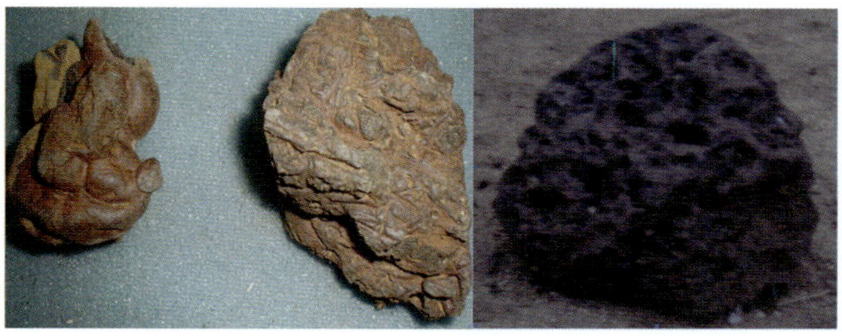

Figure 7: Two types of slag from the Kangai site.

Anthropological significance of Mbeere Late Iron Age technology

Earlier anthropological research conducted by Brown (1995) recorded ironworking technologies in all regions of Kenya. Brown's purpose was to classify the smithies, hearths, furnaces, smith tool kits and products into types to see if they fell into any distinct groupings or assemblages. An additional aim was to categorise the techniques, rituals, taboos, and social status of smiths in order to assess to what extent they were related, thus to relate these to the background culture of the peoples with whom they lived and worked. Brown further divided ironworking into distinct industries and streams, correlating them with tribes or groups of tribes; linguistic groups; means of livelihood; social organisation or environment. In an attempt to suggest the origins of present day ironworking industries in Kenya, Brown examined archaeological, linguistic, and oral traditions for evidence of related ironworking cultures in adjoining territories and historical movements of peoples. To consider evidence of ironworking in Mt. Kenya and the central regions of the country, she used archaeological materials recovered from the Gatunga'ng'a site by Siiriainen (1971), the Ithanga hills by Mahlstedt and DiBlasi (1978), the Grand Falls by Cummings (1978), and Embu, Mbeere and Chuka by Soper (1979).

The results of this research led Brown to conclude that the ethnographic ironworking technology in Kenya can be divided into Eastern and Western streams; the dividing lines of which is the Great Rift Valley. While the streams are geographical, they coincide with a common ironworking technology on each side. Thus to the East of the Rift Valley, bag bellows and bowl furnaces are exclusively used; while to the West bowl bellows and dome furnaces predominate (Fig. 8). According to Brown (1995), ironworking technology within the Mt. Kenya region falls within the Eastern stream.

The ethnographic ironworking research carried out by Brown is the most detailed record available in Kenya, offering important comparative data for archaeological ironworking technologies. It can be used to provide an indication of technological

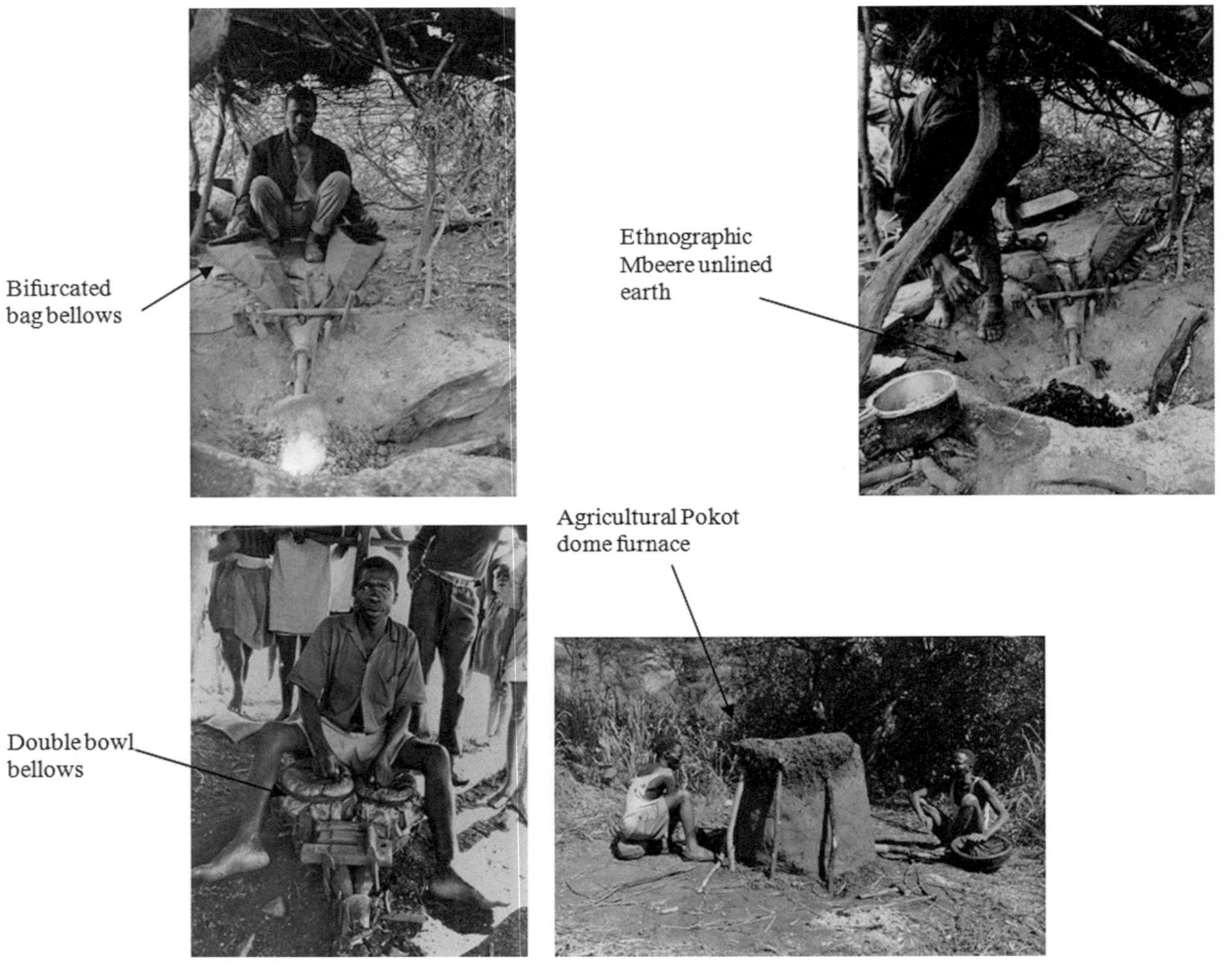

Figure 8: Types of furnaces and bellows used East and West of the Rift Valley (Brown, 1995).

developments or changes in ironworking; of past population interactions or even the past movements of peoples in Kenya.

The Late Iron Age technology findings from Kangai and Kanyua are in significant contrast to the ethnographic ironworking as described by Brown (1995). After excavating and studying an in situ furnace from the Kiburu site in the Mbeere region, Ngari (2004) pointed out that it differed from those described by Brown (1995). It is important to note however that, although the two sites are only approximately 10km away from each other, the Kiburu archaeological furnace also differs from the Kangai one. It is equally possible that these furnaces were used during different archaeological periods, or that they co-existed: the occurrence of dissimilar furnaces within the same archaeological period is not unusual, as shown by archaeological work in Laikipia (northern slopes of Mt. Kenya) by Iles and Martinón-Torres (2009). That particular expedition recovered several furnaces, excavated within the same area and belonging to the same archaeological period, which displayed stylistic variations. This discovery in turn raised questions about the organisation and the identity of the smelters who worked them.

The following section compares the technological variations between ethnographic and archaeological furnaces, tuyères, and smithies of the Mbeere district of the Mt. Kenya regions in an attempt to prompt new questions about the iron workers and ironworking technology of the Late Iron Age in the Mt. Kenya region. Since ironworking in Mt. Kenya and central regions of Kenya is attributed to Bantu speakers, it is not clear as to whether the obvious differences exhibited by the furnace designs and tuyères is a later development, facilitated by different ethnic interactions, or if Late Ironworking in the region was, in fact, carried out by sociolinguistic groups other than Bantu speakers.

Furnaces

Brown describes the ethnographic furnaces in the Mbeere region as simple round holes in the ground, rarely more than 30-35cm in diameter and 20-30cm deep, frequently lined with clay (Brown, 1995, p.45). She goes on to explain that, in the area around Mbeere, the earth is hard and compact enough that it is unnecessary to clay line the holes. Unlike the simple bowl furnaces recorded ethnographically by Brown in 1974 the archaeological furnace excavated at Kangai (M'Mbogori, 2006) appears to have been a shaft furnace measuring 56cm in diameter, with a depth of 30cm below and 20cm above the ground; it stood 50cm high despite several centuries of erosion. On the other hand the furnace excavated by Ngari (2004) was made through moulding inside a pit, and the walls were plastered with a slippery clay. Ngari has maintained (Ngari, 2004, p.17) that this furnace contrasts sharply with the dug-out bowl furnaces described by Brown in 1974. The

Kangai archaeological furnace, although eroded, is comparable with the Western stream furnaces as described by Brown, as the Western stream furnaces are of the shaft and dome types used by Nilotic and Cushitic speakers, and are more reminiscent of the archaeological furnaces recorded here.

Tuyères

Brown describes the ethnographic tuyères used in the Mbeere region as straight pipes, whose maximum length is 26cm and central diameter 5cm (Brown, 1995, p.178) (see Fig. 9). This tuyère technology is similar to that of the Wanga and Pokot of Western Kenya, both of whom fall within Brown's Western stream industry. The Kangai and Kanyua archaeological tuyères were funnel shaped with flared mouths, as shown in Fig. 9. The flared archaeological tuyères are comparable with the ethnographic tuyères of the neighbouring Embu and Kikuyu (Bantu speakers), the Samburu (Nilotic speakers), and the Rendille (Cushitic speakers). This indicates the use of similar tuyère technology within different linguistic groups, making it problematic to assign the Iron Age industry of the Mt. Kenya region exclusively to Bantu speakers, without undertaking further research.

Smithy

Brown's descriptions of and references to the ethnographic smithy indicate that these were used as both smelting and forging areas (Brown, 1995, p.54) although, according to Brown, ethnographic smithy types cannot be classified as belonging to any linguistic or ethnic group: they are erected depending on the environment and means of livelihood of the people, so that a general type is common to a number of different cultures (Brown, 1995, p.1). The ethnographic Mbeere smithy, as witnessed by Brown, was what she subsequently described as being 'brushwood enclosures': mostly found among pastoral people and consisting of a circle of brushwood, heaped against a bush or placed under the shade of a tree (see Fig. 10). There is no evidence for the use of 'brushwood' smithies during the Late Iron Age in the

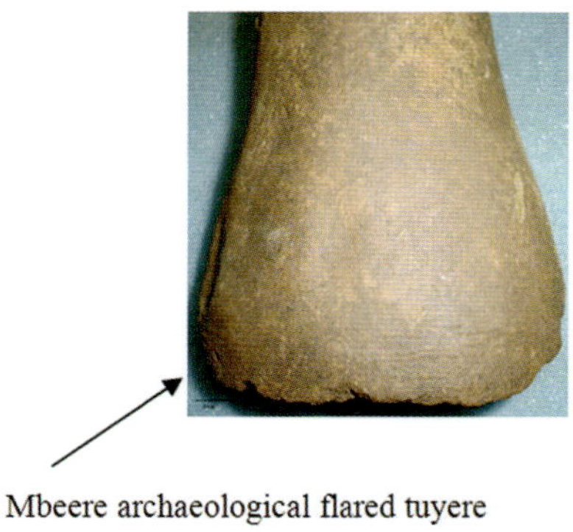

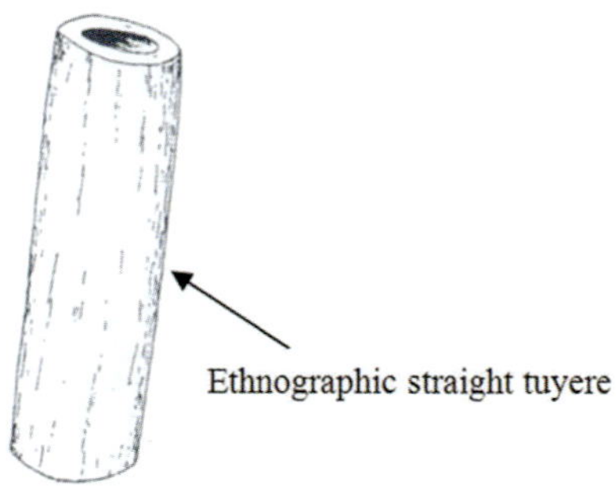

Figure 9: Archaeological flared tuyère (M'Mbogori, 2006) and ethnographic straight tuyère (Brown, 1995).

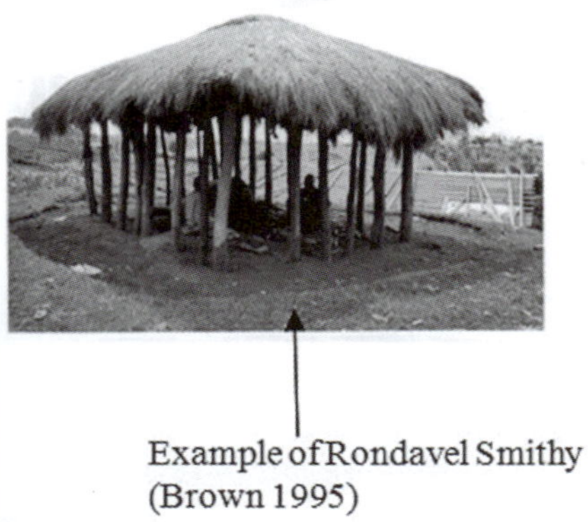

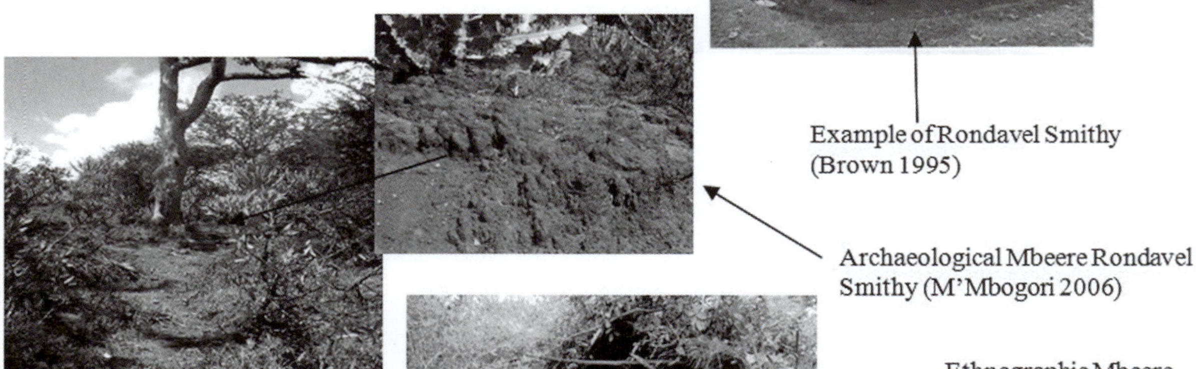

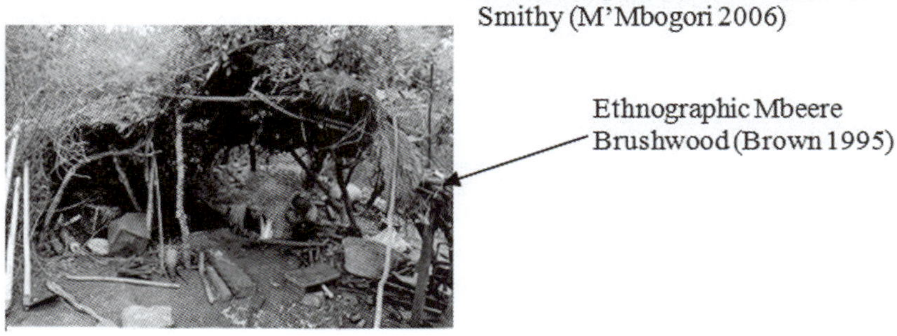

Figure 10: Mbeere archaeological 'rondavel' smithy (M'Mbogori, 2006) and Kikuyu 'rondavel' and ethnographic Mbeere 'brushwood' smithy (Brown, 1995).

Mbeere district; an understandable position since they were solely composed of perishable, organic materials.

Based on Brown's demarcation of smithies, the Kangai site smithy could fall under the 'rondavel' type; round huts with low, grass-thatched roofs and open sides which were widespread amongst agriculturalists and semi-pastoral agriculturalist [Fig. 10].

Their size usually varied between 3 and 6m, depending on the volume of work and number of smiths working there. Earth excavated from the floor was sometimes banked up around the perimeter to prevent water running into the smithy during heavy rain. Although the Kangai archaeological smithy has a compact mud perimeter, it was not possible to tell if it originally had a roof. Measuring 9.6m from east to west and 10.8m from north to south; the remains of only one furnace were found in the middle of this smithy, indicating that probably only one iron smelter used it.

Conclusion and further research

Based on the Mbeere archaeological iron technology, it is possible to make the conclusion that, either the iron technology of the Mbeere people has changed significantly in the last few centuries or, alternatively, that a different population occupied the Kangai and Kanyua sites during the Late Iron Age. In particular, furnaces exhibit forms that are not ethnographically associated with modern Bantu speakers, but which were used by Cushitic and Nilotic speakers in 1974, when Brown's ethnographic research was conducted. Similarly, the Mbeere archaeological tuyères differ from the ethnographic ones yet are similar to those of some other Bantu groups, as well as Cushitic, and Nilotic speakers. Further research on archaeological tuyères may point to their antiquity, which in turn may shed light on peoples' movements, interactions, or the introduction of iron technology.

Due to the fact that the Mt. Kenya region was occupied by Cushitic and Nilotic speakers either before or simultaneously with Bantu speakers, it is possible that any of these linguistic groups may have occupied the Mbeere Iron Age sites. As we have seen, pottery and other archaeological materials occurring in association with ironworking tools unearthed in the Mt. Kenya region are problematic to associate with Bantu speakers (Siiriainen 1971). As such, they cannot be conclusively tied to the Bantu speakers or any other sociolinguistic group until further evidence is obtained. It is necessary therefore to conduct further research at other ironworking sites in the Mbeere region and to compare the results with those obtained from the Kangai and Kanyua sites. Archaeological iron technology research is equally necessary in other regions of Kenya, in order to create and maintain a database which can be used as comparative material.

Questions relating to iron production in Kenya remain poorly understood, owing to the fact that Iron Age research has been, as previously stated, obscured by key issues relating to the migrations of the Bantu speakers and of Stone Age research. As many areas of the Mbeere region and elsewhere in Kenya are currently threatened by agricultural and urban development, urgent archaeological and archaeometallurgical research in this area is needed.

Acknowledgements

I wish to sincerely thank Dr. Jane Humphris and Dr. Xander Veldhuijzen for their editorial comments and for making it possible for me to attend and present this paper at the World of Iron Conference 2009. My sincere gratitude goes to Dr. Stephanie Wynne Jones and Dr. Matthew Davis for accepting to read, edit, and comment on the contents of this paper. To the British Institute in East Africa (BIEA), I am grateful for availing dating funds and help in kind to facilitate the excavations. I thank Dr. Jean Brown very sincerely for allowing me to use the photographs and illustrations from her book and for taking the trouble of scanning and sending me the originals. I must also thank the University of Bergen for providing me with excavation grants and Prof. Randi Haaland for her support and supervision during my master's degree research, when the sites were excavated.

References

Ambrose, S., 1982. Archaeology and Linguistic Reconstructions of History in East Africa. In: C. Ehret and M. Posnansky eds. *The Archaeological and Linguistic Reconstruction of African History.* Berkeley: University of California Press, pp.104-57.

Brown, J., 1995. *Traditional Metalworking in Kenya.* Cambridge Monographs in African Archaeology 38. Oxford: Oxbow Books.

Chami, A.F., 1994. *The Tanzanian Coast in the First Millennium AD: An Archaeology of the Iron-working, Farming Communities.* Uppsala: Societas Archaeologica Upsaliensis.

Cummings, H., 1978. Two Iron Age Sites in the Grand Falls Area, Meru District. *Azania*, 13, pp.173-94.

Ehret, C., 1974. *Ethiopians and East Africans: The Problem of Contacts.* Nairobi: East African Publishing House.

Fadiman, J.A., 1973. Early History of the Meru of Mt. Kenya. *Journal of African History*, 14, pp.9-27.

Fadiman, J.A., 1993. *When We Began, There Were Witchmen: An Oral History from Mount Kenya.* Berkeley: University of California Press.

Forslund, A-L., 2003. *Pottery and East Africa.* C-uppsats i Arkeologi, Uppsalauniversitet. URL:http://www.arkeologi.uu.se/ark/education/CD/Cuppsats/Forslund.pdf

Gosselain, O.P., 1998. Social and Technical Identity in a Clay Crystal Ball. In: T.S. Miriam ed. *The Archaeology of Social Boundaries.* Washington DC: Smithsonian Institution Press, pp.78-106.

Gosselain, O.P., 2000. Materializing Identities: An African Perspective. *Journal of Archaeological Method and Theory*, **7**, pp.187-217.

Gramly, R.M., 1971. *Recent Archaeological Survey in Central Kenya.* Discussion paper 19. Nairobi: University of Nairobi, Institute of African Studies.

Haaland, R. and Msuya, S.C., 2000. Pottery, Iron Working and Trade in the Early Iron Age: The Case of Dakawa, East-Central Tanzania. *Azania*, 25, pp.75-106.

Iles, L. and Martinón-Torres, M., 2009. Pastoralist Iron Production on the Laikipia Plateau, Kenya: Wider Implications for Archaeometallurgical Studies. *Journal of Archaeological Science*, 36, 2314-26.

Karega-Munene, 1993. *Neolithic/Iron Age Subsistence Economies in the Lake Victoria Basin of East Africa*. PhD Thesis, University of Cambridge.

Mahlstedt, T.F. and DiBlasi, M.C., 1978. Archaeological Survey of the Ithanga Hills, Eastern Highlands of Kenya: Preliminary Analysis. *Azania*, 13, pp.192-3.

M'Mbogori, F.N., 2006. *Late Iron Age economies of Mt. Kenya region: A Case Study of Kiburu, Kangai and Kanyua Archaeological Sites*. MA Dissertation, University of Bergen.

Muriuki, G., 1974. *A History of the Kikuyu, 1500-1900*. Nairobi and London: Oxford University Press.

Muriuki, G., 1978. *People around Mount Kenya*. London: Evans Brothers Limited.

Mwaniki, H.S.K., 1973. *The Living History of Embu and Mbeere to 1906*. Nairobi: Kenya Literature Bureau.

Ngari, L., 2004. *Environmental and Ethnoarchaeological Contexts for the Advent of Iron Technology and the Expansion of Bantu Speaking Peoples in the Upper Tana, Central Kenya, circa AD 1000-1950*. PhD Thesis, Kenyatta University, Nairobi.

Nurse, D., 1982. Bantu Expansion in East Africa: Linguistic Evidence. In: C. Ehret and M. Posnansky eds. *The Archaeological and Linguistic Reconstruction of African History*. Berkeley: University of California Press, pp.199-222.

Odner, K., 1971. A Preliminary Report on an Archaeological Survey on the Slopes of Kilimanjaro. *Azania*, 6, pp.91-130.

Phillipson, D.W., 1977. The Later Prehistory of Eastern and Southern Africa. London: Heinemann Educational.

Phillipson, D.W., 2005. *African Archaeology*. Cambridge: Cambridge University Press.

Roux, V., 2007. Ethnoarchaeology: A Non Historical Science of Reference Necessary for Interpreting the Past. *Journal of Archaeological Method and Theory*, 14, pp.153-78.

Schmidt, P.R. 1997. *Iron Technology in E. Africa: Symbolism, Science, and Archaeology*. Indiana, Indiana University Press.

Siiriainen, A., 1971. The Iron Age Site at Gatung'ang'a, Central Kenya. Contributions to the Gumba problem. *Azania*, 6, pp.199-232.

Smith, A.L., 2000. Processing Clay for Pottery in Northern Cameroon: Social and Technical Requirements. *Archaeometry*, 42, pp.21-4.

Soper, R.C., 1967. Kwale: An Early Iron Age Site in South-Eastern Kenya. *Azania*, 2, pp.1-18.

Soper, R.C., 1979. Iron Age Archaeology and Traditional History in Embu, Mbeere and Chuka Areas of Central Kenya. *Azania*, 14, pp.31-58.

Soper, R.C., 1982. Bantu Expansion in Eastern Africa. In: C. Ehret and M. Posnansky eds. *The Archaeological and Linguistic Reconstruction of African History*. Berkeley: University of California Press, pp.223-38.

Vansina, J., 1995. New Linguistic evidence and ' Bantu Expansion. The Journal of African History 36, (2) pp. 173-195.

Author's address

- Nkirote M'Mbogori Freda
  Archaeology Section, Earth Sciences
  National Museums of Kenya
  Email: nkirotef@yahoo.com, fnkirote@museums.or.ke

Pre-colonial iron production in Great Lakes Africa: recent research at UCL Institute of Archaeology

Jane Humphris and Louise Iles

ABSTRACT: Since 2002 the authors have been investigating the iron production industries of several pre-colonial polities of Great Lakes Africa: Buganda and Bunyoro-Kitara in present-day Uganda, and Nyiginya in Rwanda. Combining the results of survey, excavation, archaeobotanical analysis and archaeometallurgical analysis, as well as an ethno-experimental reconstruction and interviews in the field, the intrinsic nature of these technologies within the wider political and social histories of the Great Lakes is gradually being revealed.

Iron was a fundamentally important material in Great Lakes Africa and as production grew to meet increasing demand for tools and weapons during later periods, it is presumed that the iron production industries came to have a significant impact on the environment (e.g. see Taylor and Marchant, 1994/1995), trade networks, and economic and socio-political structures of the region.

This paper presents a synthesis of our recent research, in which we have explored the broader issues surrounding technological change and variation in Great Lakes Africa and examined the evidence with an awareness of the movement of people and ideas amongst these political entities. We suggest that the evidence collected allows us to interpret iron metallurgy as a process which responded to and created a stimulus for larger socio-political forces.

Introduction

Some of the earliest recorded dates for the production of iron have been noted in the eastern and central regions of Africa (for example see Van Grunderbeek, et al. 1983, Van Grunderbeek 1992; Van Noten, 1983; Schmidt, 1996). In addition, the extent of the effect that iron production had on the later histories of numerous Great Lakes polities is now also becoming clear (e.g. Herbert, 1993; Schoenbrun, 1999; Chrétien, 2003; see Fig. 1 for polities and sites mentioned in the text). The production of iron grew to be a particularly powerful driving force for political, social, and economic change in the second millennium AD, providing a catalyst (alongside cattle and, in some cases, bananas) for the shaping of the region's environmental landscape and social organisation. Strong, durable iron was particularly well suited for the manufacture of agricultural tools and weaponry. The production of iron hoes and machetes is thought to have stimulated greater levels of forest clearance and a subsequent increase in agriculture, while the production of iron weapons gave increased might to the territorial battles and disputes common in the region. These valuable roles established iron's integration within society and its frequent association, both practical and symbolic, with leadership and power.

The archaeological context of our research

Iron metallurgy in Africa has long been an area of archaeological interest (Childs and Killick, 1993) yet it has only relatively recently emerged from beginnings dogged by preconceptions of the 'backward' and 'pre-industrial' nature of African technologies (e.g. Cline, 1937; Tylecote, 1965). Today, the discipline carries great academic value, not only due to the recognition of the sophistication and complexity of indigenous, pre-colonial iron production techniques, but also as a result of the new perspectives that African case-studies can offer to the general field of world archaeometallurgy and theory of technology (e.g. Haaland, 1985; Childs, 1991; Gordon and Killick, 1993; Haaland, et al., 2002; Haaland, 2004; Humphris, et al., 2009; Iles and Martinón-Torres, 2009).

It is now understood that the most effective research combines social, ideological, and cognitive considerations of metallurgy (Childs and Killick, 1993, p.317) with modern scientific and technical analysis. Together, such approaches can provide a more thorough appreciation of the interrelated human aspects of past technologies (Killick, 2004). In addition to these investigative methods, detailed ethnographic and experimental research has highlighted the importance of understanding the social contexts of production (e.g. Haaland, 1985; David, et al., 1989; Reid and MacLean, 1995; Schmidt, 1997; Childs, 2000 amongst many others). This understanding presents an indispensable opportunity to understand the full *chaînes opératoires* of iron production in an African context.

Early records of the iron production technologies of the Great Lakes region are attributed to the accounts and diaries of early travellers such as Richard Burton and Emin Pasha within the region (Grant, 1864; Schweinfurth, et al., 1888; Burton, 1860). These accounts were followed, in the early 20th century, by those of missionaries and other westerners dedicated to documenting the daily life of the African populations that they encountered (e.g. Fisher, 1911; Roscoe, 1911, 1915, 1923). These resources, alongside later detailed ethnographic studies (e.g. Célis and Nzikobanyanka, 1976; Célis, 1987; Childs, 1998, 1999, 2000), provide important records of historical iron production techniques. They have proved useful in the interpretation of archaeological remains and offer significant contribution to our understanding of the role of iron within pre-colonial African society.

Research approaches

With regards to the systems of iron production operating within the period and region in question, there are a number of extant research questions that the authors are personally interested in addressing. These approaches aim to deepen our understanding of ferrous technologies on both local and regional scales, with specific reference to the form and mutation of particular technological styles and to how the technologies and the people who undertook them were organised. Other questions of particular interest relate to how resources, both human and natural, were managed and controlled, through what mechanisms technical knowledge was protected and/or communicated, and to the impact of iron production on both local society and regional political structures.

Over a number of years the authors have collected data in order to address these questions with regards to the regions of Rwanda and western and southern Uganda (Humphris, 2004, 2008, 2010; Humphris, et al., 2009; Iles, 2004, 2011; see also Reid, 2003; Reid and Young, 2003). During this time a variety of research techniques have been employed to examine these technologies in more detail, which range from survey and excavation (Humphris, 2008; Iles, 2009a) to archaeobotanical (Thompson and Young, 1999; Iles, 2004, 2009b) and archaeometallurgical analysis (Humphris, 2004; Rehren, et al., 2007; Humphris, et al., 2009). Ethno-experimental

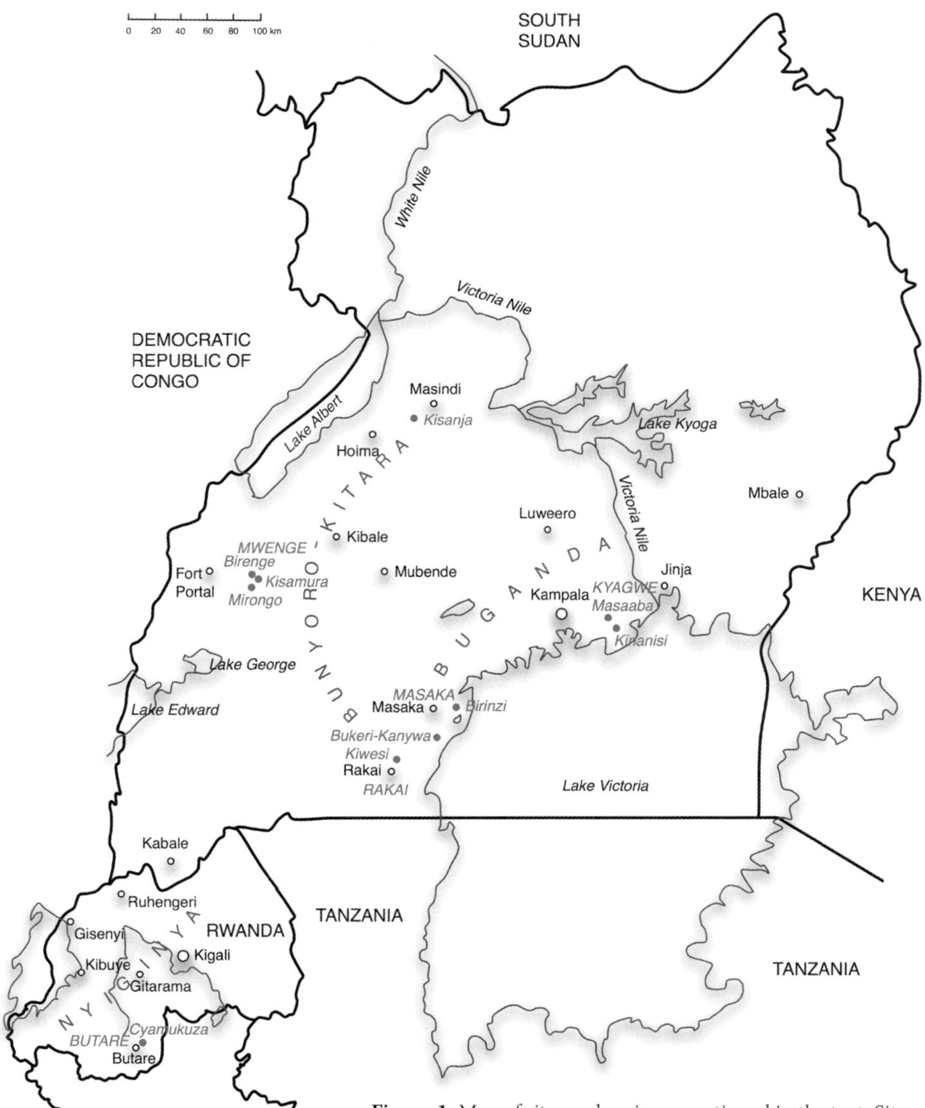

Figure 1: Map of sites and regions mentioned in the text. Sites and towns not to scale.

reconstruction and interviews with local community members who remember recent smelting activity have also been integral to the research (Humphris, 2002, 2010; Iles, in press, and this volume). Fig. 1 shows the locations of the particular areas and sites of interest mentioned in this paper.

In 2000 and 2001, survey teams working within a territory relating to the former Buganda kingdom in Uganda, noted several extensive iron production sites which were dominated by large slag blocks (Reid, 2003). As a result a three-pronged fieldwork strategy was implemented, in order to generate a better understanding of these seemingly industrial-scale technologies.

Firstly, in 2002, an extensive programme of interviews with elderly men who remembered iron production was carried out across the southern areas of the former kingdom of Buganda by JH. In 2003 excavations were carried out at four smelting sites – Masaaba and Kinanisi in Kyagwe, and Bukeri-Kanywa and Birinzi in Masaka (cf. Fig. 1). JH undertook an archaeometallurgical study to assess the technical components of smelting at these sites (Humphris, 2004; Humphris, et al., 2009), which was complemented by a parallel archaeobotanical project which examined the use of plants within these technologies (Iles, 2004, 2009b) and was led by LI.

Both authors continued their investigations into the iron production technologies of the Great Lakes region as the basis for their PhD research. Both utilised archaeometallurgical techniques in the laboratory – optical microscopy and scanning electron microscopy in conjunction with chemical analysis (SEM-EDS and ED-XRF) – to reconstruct the technical processes of iron production in these areas, complementing the data generated in the field.

The combined results of these investigations have significantly increased the body of data available to researchers in the field of iron production and have allowed a wider consideration of the process in the Great Lakes region. We intend to continue to build and clarify the picture of iron production in the region which has been initiated by researchers such as Childs (1998, 1999, 2000) in Uganda, Van Noten (1983) and Van Grunderbeek and co-workers in Rwanda (Van Grunderbeek, et al., 1983), and by Reid and MacLean (1995) and Schmidt in north-western Tanzania (1997).

Early evidence for iron production

The research of Van Noten (1972, 1982, 1983), Van Grunderbeek (1988, 2007), Vignati-Pagis (1995), and Schmidt (1996, 1997) suggests that iron smelting in the south-western corner of the Great Lakes region occurred from very early times. Metallurgists made use of a distinctive stylistic approach to iron production, using furnaces constructed in part from decorated clay rolls or bricks (Hiernaux and Maquet, 1960; Nenquin, 1971, pp.183-187; Schmidt and Childs, 1985, 1996). Although a number of dates for iron production in the region before 1000 BC are now considered more than likely to be erroneous (Schmidt and Childs, 1996: 205-206; Van Grunderbeek, et al., 2001, p.276; Killick, 2004, pp.100-101) the sheer number of these early dates, when synthesised, does suggest that future research may corroborate evidence for the existence of established iron production technologies during the early first millennium BC – and possibly even before (de Maret, 1985, pp.138-139; for radiocarbon dates see Clist, 1987; Van Grunderbeek, 1992; Van Grunderbeek, et al., 2001; Wotzka, 2006). It seems however that this early iron production may have occurred on a relatively small scale and have been confined to particular areas (see Van Grunderbeek 1992; Van Grunderbeek, et al., 1983 for summaries of early iron production sites in the Great Lakes region). Our recent survey, for example, has yielded comparatively few iron production sites associated with distinctive early pottery (in this region, Urewe ceramics) across both Rwanda and Uganda, especially in the areas associated with Bunyoro-Kitara. It should be noted however that, in Rwanda in particular, extensive farming and erosion may have distorted the archaeological and archaeometallurgical records of these early periods.

With regard to the northern area of the Great Lakes region (western Uganda in particular), there appears to have been "scant agricultural settlement" prior to 1000 AD (Robertshaw, 1999, p.124; also cf. Robertshaw, 1994; Robertshaw and Taylor 2000). Although lake cores from Kabata Swamp (cf. Fig. 1) indicate a marked increase in forest clearance and degraded soils from around 500 BC (cf. Taylor, et al., 2000) intensive archaeological survey has been unable to locate any distinctive Urewe pottery (e.g. Connah, 1990; Reid, 1991; Robertshaw, 1994; Iles, 2009a), although occasional fragments have been recorded upon excavation. When present, the sporadic examples of Urewe ceramics in western Uganda have tended to be found in association with Late Iron Age (LIA), roulette-decorated ceramics. As such, it is plausible that there was a lower population density in this area prior to 1000 AD. Indeed, based on the archaeological evidence presently available (or rather, the lack of it), it appears that western Uganda – unlike many other regions of the Great Lakes at this time – was a "sparsely populated agricultural frontier at the beginning of the second millennium AD" which, when viewed on a macro-regional level, was a largely 'peripheral' area (Robertshaw, 1999, p.131, 2003). No early iron production remains have been found in this area to date.

Questions remain about how or why the metallurgists of the Early Iron Age (EIA) in the southern Great Lakes region built the superstructures (or parts of the superstructures) with clay bricks or rolls decorated with C-shaped, S-shaped, or lined ridges (Schmidt and Childs, 1985, pp.82-88; Raymaekers and Van Noten, 1986). Evidence suggests however that this specific early technological approach (Fig. 2) was applied continuously until the mid to late first millennium AD across Rwanda and Burundi (Van Grunderbeek, 1992) and at sites in nearby northern Tanzania (Schmidt and Childs, 1985).

Later iron production industries

By the middle of the second millennium AD the political organisation of the Great Lakes region was changing. New and developing socio-political configurations began to emerge

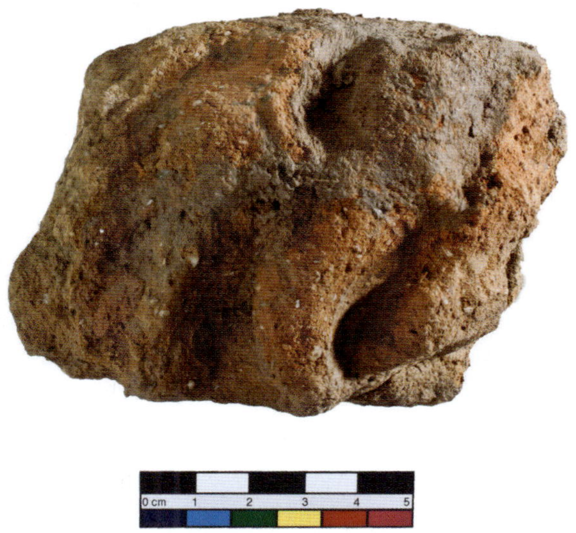

Figure 2: Example of a decorated clay brick excavated from within an EIA furnace pit in Southern Rwanda.

which in 17th century Rwanda, for example, ultimately formed the Nyiginya kingdom (Chrétien, 2003, p.107; Vansina, 2004). Further to the north in present-day Uganda, a patchwork of incipient kingdoms, including Bunyoro-Kitara and Buganda, also began to dominate the socio-political landscapes. This period saw a culmination in societal and environmental transformation as populations grew, technology and food production intensified, and the distribution of power became centralised.

Linguistic evidence for this transformation suggests a flourishing of cattle herding and a rapid increase in banana cultivation sometime between the 12th and 15th centuries (Schoenbrun, 1993, 1998). Terminologies for power and authority also began to develop from the 13th century, indicating a shift from hereditary to appointed power structures (Schoenbrun, 1998): "this semantic shift thus clearly demonstrates the increasing importance of economic assets as a socio-political tool at this time, as individual wealth forged a new path to authority, regardless of clan or lineage" (Ashley, 2010, p.143; see also Robertshaw, 2010). The Entebbe and Luzira figurines suggest that around Lake Victoria a similar shift had occurred, with power beginning to focus on important individuals (Ashley and Reid, 2008); parallels can be posited for the emergence of the 'Big Men' that Vansina recognises in regions further to the west (Vansina, 1990).

From this point onward it is likely that demand for iron grew; a likelihood which corresponds with increasing numbers of smelting sites in the archaeological record. By the 18th century, for example, the Nyiginya army was well established and professional – "from the time of Rujugira onward [1770] the country was almost continually in a state of war" (Vansina, 2004, p.75); and iron weapons were particularly important during territorial disputes and in the drive towards expansion. Similarly, in the 19th century Buganda was described as being "the terror of their neighbours" (Kiwanuka, 1971, p.144). Territorial disputes had become commonplace over the entire region (Chrétien, 2003, pp.139-165), with iron weapons providing material might for these conflicts. The economic value of iron was also important. In the late 19th century, a prince of Bunyoro promised the king of Buganda that he would pay a yearly tribute including 600 hoes in exchange for his help to win the Nyoro throne (Wainwright, 1954). The acquisition of iron by the state through the collection of tribute was a common practice (Humphris, 2002; Vansina, 2004). Large commercial centres became well established, trading in iron tools as well as agricultural products and salt (Tosh, 1970; Uzoigwe, 1976; Connah, 1991).

Iron was often conceptually integrated within the organising structures of these states, with iron symbolism frequently incorporated within iconography, mythology, systems of tribute payment and so on. To provide a few examples, iron is intertwined with early mythology in Bunyoro-Kitara, where the heavens were visualised as being held aloft by two tree trunks and a bar of iron (Fisher, 1911, p.70; Dunbar, 1965, p.10). Within Buganda, one of the iron production sites that the authors worked at was called *Kinanisi*, the Luganda name for the iron bracelets and anklets that were made there to be supplied to the Kabaka's (king's) court as tribute (Humphris, 2002). Power, fertility, and iron are also combined within the concept of the smith-king, which has been well documented across the region (de Maret, 1985; Reid and MacLean, 1995; Chrétien, 2003, pp. 85-137; Herbert, 1993, pp. 131-163). In Rwanda, according to the accounts of Célis (1987, pp. 17-55), sacred iron hammers and the role of the king as forger of iron are symbolically incorporated into the words and symbols of numerous royal rituals, for example:

When the king has become another,
The descendant of Muhinda is sent.
He goes to Nganzo-lez-Mushongi
And hits the iron to make a hoe of
unity...
He hits the iron for four axes
He also hits the iron for two bells
One which rings
and the other which is silent.
He also hits the iron for four hoes.
The ritual follows:
A forge is installed
And the Tsoobe action the bellows
The King holds the arm of the Tsoobe and actions the bellows four times ...
They tell of the hoe of unity
The descendant of Muhinda comes to present Nyarushara to the King,
Saying: here is the hammer that your father bequeathed.
May it defeat Burundi,
May it defeat Bunvabungo,
May it defeat all countries ...
(Célis, 1987, pp.44-46, translated from the original French)

It is clear from the summary above that we know significantly more about iron production and its social context during the later periods as compared to the earlier periods, and that particular trends are apparent across certain areas. So, how did these trends manifest themselves in the archaeological record? Below we introduce the individual research areas:

Buganda

The joint research carried out in Buganda investigated the iron producers who worked within the Buganda kingdom, and combined archaeometallurgical and archaeobotanical approaches. Through these means, a high level of technological variation across the study area was revealed, indicating that distinct, contemporaneous smelting traditions were in operation throughout the different sub-regions of Buganda.

Excavations in the east of Buganda at Kinanisi (Kyagwe district) revealed that local artisans were smelting iron in large pit furnaces packed with grasses and reeds, with each smelting episode producing a single slag block weighing in some cases over 100kg per smelt. The Kinanisi slags seem indicative of the large-scale industrial production of iron in this area, with the refuse from many smelts accumulating as numerous large clusters of slag blocks across the landscape (Fig. 3). A similar pattern of slag block clusters is repeated across the Buganda study area, as well as in Bunyoro-Kitara and Rwanda, which begs questions as to their significance. Do they represent the smelting of specific groups of smelters controlled by a master smelter, as suggested by Reid and Young (2003), potentially family or clan based? Do they represent specific time periods of smelting activity, accumulations of one-off events, or successive seasons of smelting? Alternatively, do they represent later episodes of field clearance, where slag blocks were moved from the land in order to allow agriculture to take place, resulting in clusters of slag blocks around field boundaries? In all study areas instances were encountered which highlight the mobile nature of iron slag, such as blocks being used in road construction, or for the foundation of houses, or as grave markers, so this possibility should not be discounted.

Less than 10km from Kinanisi, the site of Masaaba was investigated. No furnaces were found in this area, the land being either densely cultivated or overgrown, although four metric tonnes of slag was collected (Humphris, 2004). The Masaaba slag was very fragmented, and particularly prominent was the presence of distinct slag flows, which immediately raised the possibility later confirmed through microscopic analysis (Fig. 4) that these slags resulted from the use of a

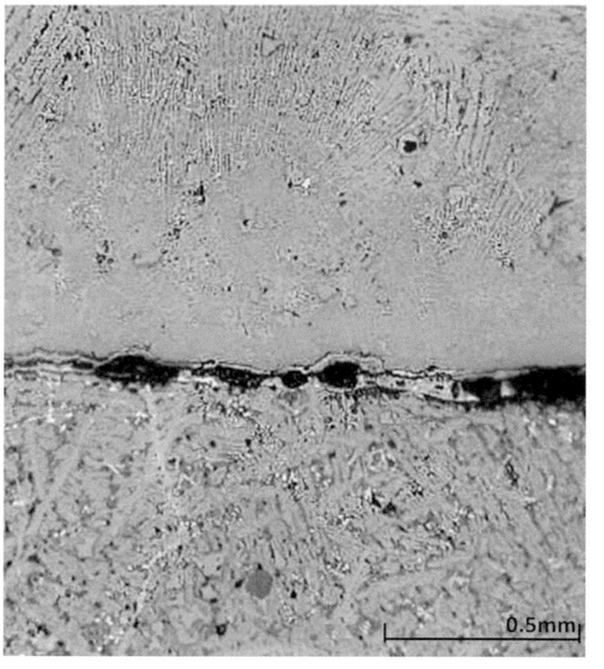

Figure 4: Photomicrograph of a slag sample from the site of Masaaba, illustrating classic tap flows and a distinctive magnetite skin which has formed as the first flow of slag oxidises on contact with the environment outside the furnace.

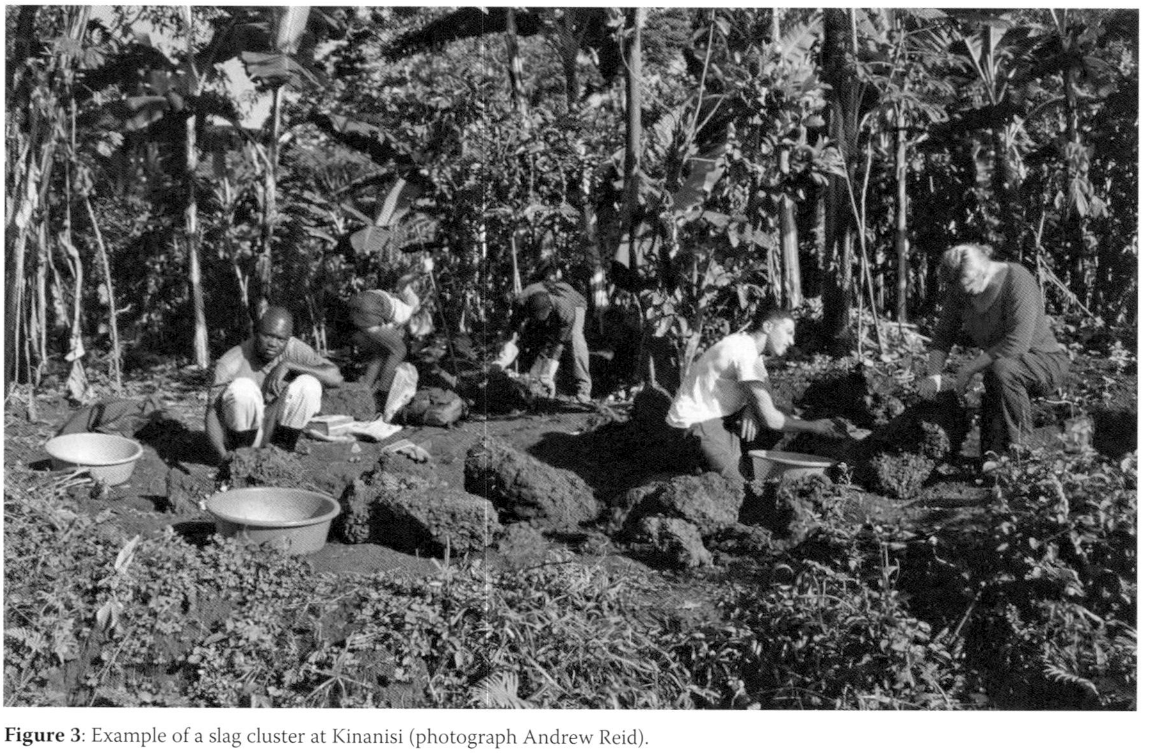

Figure 3: Example of a slag cluster at Kinanisi (photograph Andrew Reid).

horizontal slag-tapping technology: a form of bloomery smelting that is particularly unusual in this region. Moreover, the archaeobotanical evidence from this site identified the additional use of banana pseudostems within these smelting operations (Iles, 2004, 2009b, in press), which had also been noted in slag collected from nearby Nyemerwa (Reid and Young, 2003). It is not known for what purpose banana plants were used in these smelts, but it has been suggested that they may have acted as a protective barrier between the smelters and the heat of the furnace, or as a symbolic ingredient for the smelt (cf. Iles, 2009b).

Just over 100km to the west of Kinanisi, two iron production sites were excavated in the Masaka district – Bukeri-Kanywa and Birinzi. This area lay in what was once the territory of Bunyoro-Kitara, but which was deliberately annexed by Buganda in the late 1800s, partly due to its iron ore deposits and its iron workers. Again, the sites excavated here suggest that the smelters were using large pit furnaces (Humphris, 2004), although here the furnace pit packing generally comprised papyrus exploited from local sources (Iles, 2004), and the smelters were utilising kaolin to produce particularly refractory tuyères: the technological approach was similar to that documented in Kinanisi in terms of furnace design, while making use of locally-available resources.

A striking feature of our Buganda research was the realisation that at least two very distinct technologies (themselves witnessing internal variations) were being carried out within the kingdom of Buganda in the latter half of the second millennium AD. The pit-smelting technology evident at Kinanisi was very distinct from the horizontal tapping technology being used at Masaaba, although the sites were geographically close. How did such different techniques come into use in such a small area? Does this pattern correlate with known different population groups? It is well documented that Buganda invited smelters from outside its territories to work and live in the kingdom (Reid, 2002, pp.76-86): "it was Mawanda who received Kongonge of the Banyoro, whose occupation was that of smelter. He gave him an estate in Kyagwe country [where Masaaba and Kinanisi are located], and employed him as a smelter ... When Djundju annexed Budu territory [where the Masaka sites are located], he imported and employed several qualified smiths" (Kagwa, 1934, p.160). As such, there is a possibility that what we are seeing in the archaeological record is a representation of the differing smelting traditions of communities brought into the area specifically to produce iron for the kingdom.

Bunyoro-Kitara

Contrasting with earlier periods, for which we have little evidence for early iron production, Bunyoro-Kitara is thought to have developed into one of the biggest producers and traders of iron in the region in these later periods, controlling large areas of land containing all the necessary ingredients for successful, large-scale iron production (Roscoe, 1923; Tosh, 1970; Connah, 1996). This is reflected in the large number of iron smelting sites that were identified during a survey in 2007: over 200 sites were located, including mining sites for the procurement of iron ore and iron production sites, indicated by residual slag and tuyère and furnace remains (Iles 2009a).

Following the evidence suggesting varied iron production technologies in Buganda it was considered likely that a comparably high level of variation would also be encountered in Bunyoro-Kitara. Fieldwork revealed that this was the case on both a local and general scale, as distinctive furnace types were identified in the three areas examined during the research. The majority of the fieldwork was carried out in the central research zone of Mwenge, where nine furnaces were excavated over eight distinct smelting sites. In this region, the typical pit-furnaces that were excavated were of moderate size, although there was a reasonable amount of small-scale variation still apparent in their size and shape. At three of these smelting sites (Birenge, Kisamura and Mirongo) impressions of banana stems were again apparent on the slag blocks (Iles, 2009b), in contrast to other sites studied within the kingdom.

Two further regions of Bunyoro-Kitara were examined to supplement the main body of data deriving from Mwenge. In the north near Masindi, two furnaces were excavated. These were both very small, shallow furnace bowls, one of which was very similar to those described by Roscoe when he described the royal Banyoro smelters (Roscoe, 1923), with remains suggestive of a furnace with a 'lipped' superstructure rather than a tall 'chimney' (Fig. 5). In the south, around Rakai – an area that later came under the control of Buganda – a single pit-furnace was excavated at the modern village of Kiwesi (which in itself literally means 'place of the smiths', and whose inhabitants claimed they originated from central Bunyoro to the north-west). The whole village was extensively littered with smelting debris, organised into clusters of activity similar to those seen in the Buganda study, and the furnace that was excavated was over one metre deep. It will be interesting to see from the ongoing analysis whether there are any appreciable

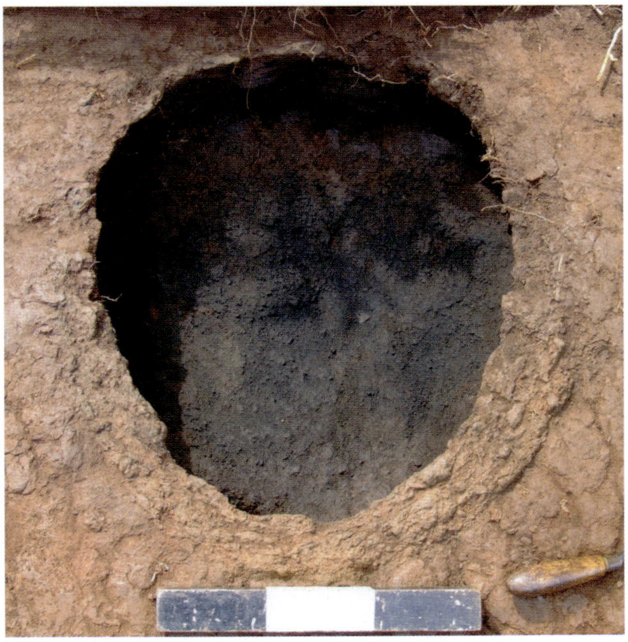

Figure 5: Lipped furnace from Masindi, similar to that seen in Roscoe (1923).

technological links to the Buganda smelting technologies. As mentioned previously, the kingdom had strong links with Buganda, both in terms of oral histories and its trading relationship, as well as being a competitor for territory and control of resources, and it is possible that these connections influenced the development of technology in this oft-disputed frontier area.

Nyiginya

Of particular interest during the research around the town of Cyamukuza in Southern Rwanda were iron production sites dated to the post-EIA period, but before the emergence of the Nyigina Kingdom (see Van Grunderbeek, et al., 1983 for the first mention of sites in this area). The archaeological remains excavated around the southern town of Butare indicate that at some point between the 8th and 10th century AD a technological transition occurred. The approach to smelting moved away from the well-established use of (sometimes decorated) bricks to construct the superstructures of furnaces. Instead it would appear that smelters began to build their furnace shafts on top of less cylindrical pits and, if a superstructure was constructed, this was made from material that leaves no apparent archaeological trace (as opposed to the typical, durable decorated clay rolls found at Early Iron Age smelting locations). It is tempting to suggest that the absence of evidence for furnace walls in the later furnaces could indicate a more roughly assembled, plain clay structure – perhaps more prone to weathering into less recognisable remains –, or possibly one built using old slag blocks, as seen in later times (Célis, 1987, p.123). The alternative is that superstructures stopped being used at all – a possibility again illustrated in some ethnographic examples (Célis, 1987, p.63). This technological shift seems to entail a move away from artistic or decorative expressions, and can therefore be broadly considered in relation to the ceramic record of the region. In pottery technologies we see the transition (at a similar time) from Urewe, a "well-made and highly crafted ceramic ... with a wide palette of horizontal banding, cross-hatching, hanging or pendant motifs" (Ashley, 2010, p.144), eventually to roulette decorated pottery styles which seem to demonstrate "a declining concern with quality and an unwillingness to invest in complex ceramics" (Ashley, 2010, p.156).

Combined with the technological transition visible on the macroscopic level was a new technical approach to the reduction of iron ore. Within these new furnace styles, rather than producing smaller fragmented slag in droplets and flows (Schmidt and Childs, 1996, pp.208-209; Van Grunderbeek, et al., 2001, p.288), the smelters began to create large continuous slag blocks weighing on occasion up to 200kg. The technological change representing such a shift most likely incorporated the use of a more significant amount of material per smelt and also, possibly, the use of different types of ores.

As mentioned above, some of the most significant findings from the work in Southern Rwanda are at least two extensive iron production locations which date to the 10th and 13th centuries AD respectively, hence within the time period preceding the Nyiginya Kingdom. The importance of these sites lies with the fact that little is known so far about the socio-economic and political setting of this period, which lies between the Early Iron Age and the Nyiginya period. While the broader implications of this evidence will be explored in detail in future publications, a few initial conclusions can be suggested here.

The 10th century site displayed relatively extensive production and represented the most standardised methodological approach to smelting out of all of the six sites studied in the area. The significant macroscopic, microscopic, and chemical similarities revealed through the archaeometallurgical analysis both within and between the slag blocks analysed at this site indicate the utilisation of an extraordinarily uniform methodology and control of ingredients used within each individual smelt and from smelt to smelt. The 13th century site, however, was more extensive, and although the general smelting principles such as furnace pit size, the vegetation packing used in the furnace pit, and the arrangement of the tuyères remained similar from smelt to smelt, chemically and microscopically there was a significant level of variation between smelting episodes, according to the archaeometallurgical analysis carried out. If such intensive smelting occurred at this site over a relatively short time, such variation between smelts could suggest that a greater work force was used, and this resulted in less control over smelting episodes by a master smelter. Alternatively, it could be suggested that in the drive to maximise iron output, less control over specific operating parameters or particular raw materials was exercised. Whatever the case, the broader ramifications of major iron industries dating to this period, are significant. Considering the control and access to natural resources and labour that these production locations would have required, as well as the clients for whom the iron was being produced, leads to a necessary reconsideration of the socio-political environment of this period, and how this contributed to the Nyiginya polity of later times.

Combining the evidence – relationships and diversity

Research carried out at UCL in recent years has demonstrated that a key characteristic of iron production in Great Lakes Africa is that of variation. What this research has failed to identify, however, is which factors have influenced this variation. So far, the evidence is still too limited to draw firm conclusions, but it has prompted us to consider one possibility: the dynamic and innovative, yet ultimately interconnected nature of the smelting communities that we are investigating. The last two thousand years of African history, and specifically eastern Africa in this case, can be viewed as a complex mosaic of interactions between communities, bound by "friendship and clientship, alliances, knowledge" and operating within defined arenas of "kinship, status, space and demography ... conflict and competition" (Kusimba and Kusimba, 2005, p.393).

The historical links between Buganda and Bunyoro-Kitara are strong and well-documented, particularly when connections between clans are considered (cf. Buchanan, 1974, 1978, 1979; Tantala, 1989). Add to this the territorial disputes and the movement of Banyoro populations into Buganda and Baganda populations into Bunyoro-Kitara, and a complex picture is painted of human interrelationships. Specifically with regards to iron production, again several shared features of the technology are known to exist from ethno-historical sources: Roscoe describes the genderisation of the two different types of ore smelted at the beginning of the 20th century in both Buganda and Bunyoro (Roscoe, 1911, 1923; see also Buchanan, 1974; Childs 1998), whilst clans of both kingdoms often share histories, including those clans that were the traditional undertakers of iron smelting (Roscoe, 1911; Buchanan, 1974). As such, it is not implausible that the technologies that we unearth in Buganda and Bunyoro-Kitara will have been influenced and shaped by these connections, resulting in stylistic and technical similarities (and differences) that are accessible through the archaeometallurgical record.

If both ideas *and* populations groups can move across the landscape (Rowland and Atmore, 2001), it is unsurprising that contacts and links, including those based on intermarriage and clan lineages, existed in a complex web of interdependence for millennia. Trade certainly played a growing role as the kingdoms of the region grew from the 17th century, with, for example, Buganda trading to the south of the lake and over to the east African coast, with imports and exports moved by caravans both on land and in canoes (Oliver and Atmore, 2001, pp.146-148). The far-reaching trade networks within which Rwanda was involved from early times are also beginning to be better understood (Giblin, et al., 2010). The problem for archaeologists, however, is identifying, with confidence, technological evidence of these fluid interactions.

Nevertheless, considering all of the above, the diversity seen within the iron production industries of the Great Lakes region illustrated in this paper, as well as the technological parallels obvious across the region since at least the Early Iron Age, should not come as too much of a surprise. Diversity and parallels represent the traditions of various groups in combination with the independent innovations of individuals who were competing to supply iron to the king and court. While some ethnographic accounts emphasise the secret nature of smelting, with groups guarding their own smelting techniques (e.g. Reid and MacLean, 1995), other bodies of evidence suggest that members from across communities were invited to participate in smelting as a labour force, from collecting the ore to powering the bellows, in return for a share of the product, or simply for some goat to eat and some banana beer (Iles, this volume).

It is hoped that the above outline of our ongoing work has served to demonstrate the importance we see in considering the interrelationships of smelting communities, in addition to the impact of the movements of these smelters as they responded to political and environmental pressures, situations that become manifest both in the high levels of technological variation and the striking parallels that we have seen throughout the region. Considering the complexities of the political, social, and economic situations that prevailed in the region over the past two thousand years, it is perhaps to be expected that these variations and parallels can be seen in the archaeological record. The key to understanding what is revealed in the studies of iron production is to study the archaeometallurgy of the region as part of a broader context rather than as a technological entity outside the realms of society.

Acknowledgements

The Arts and Humanities Research Council (AHRC) funded the PhD research discussed in this paper. Both authors also benefitted from addition financial support from the UCL Institute of Archaeology, the UCL Graduate School and the Institute for Archaeometallurgical Studies (UCL). We would like to thank our supervisors, Dr Andrew Reid and Dr Marcos Martinón-Torres, and Professor Thilo Rehren, for their guidance and support throughout this research.

References

Ashley, C. 2010. Towards a Socialised Archaeology of Ceramics in Great Lakes Africa. *African Archaeological Review* 27: 135-163.

Ashley, C. and Reid, D.A.M. 2008. A reconsideration of the figures from Luzira. *Azania* 43: 95-123.

Buchanan, C. 1974. *The Kitara complex: the historical tradition of Western Uganda to the 16th century*. Unpublished PhD thesis. Bloomington: Indiana University.

Buchanan, C. 1978. Perceptions of ethnic interaction in the east African interior: the Kitara complex. *The International Journal of African Historical Studies* 11: 410-428.

Buchanan, C. 1979. Courts, clans and chronology in the Kitara complex In: J. Webster (ed.) *Chronology, Migration and Drought in Interlacustrine Africa*. 87-124. London: Longman Press.

Burton, R. 1860. *The Lake Regions of Central Africa: a picture of exploration* (2 Vols.). London: Longman, Green, Longman and Roberts.

Célis, G. 1987. *Introduction a la Metallurgie Traditionelle au Rwanda: techniques et croyances*. Rwanda: Institute National de Recherche Scientifique.

Célis, G. and Nzikobanyanka, E. 1976. *La Metallurgie traditionelle au Burundi (techniques et croyances)*. Tervuren: Musee Royal de l'Afrique Centrale.

Childs, S.T. 1991. Style, technology and iron-smelting furnaces in Bantu-speaking Africa. *Journal of Anthropological Archaeology* 10: 332-59.

Childs, S.T. 1998. 'Find the ekijunjumira': iron mine discovery, ownership and power among the Toro of Uganda. In: A. Knapp, V. Pigott and E. Herbert (eds.) *Social Approaches to an Industrial Past: the archaeology and anthropology of mining*. 123-137. London: Routledge.

Childs, S.T. 1999. 'After all, a hoe bought a wife': the social dimensions of ironworking among the Toro of East Africa. In: M. Dobres and C. Hoffman (eds.) *The Social Dynamics of Technology: practice, politics and world views*. 23-45. London: Smithsonian Institution Press.

Childs, S.T. 2000. Traditional iron working: a narrated ethnoarchaeological example. In: M. Bisson, S.T. Childs, P. de Barros and A. Holl (eds.) *Ancient African Metallurgy: the socio-cultural context*. 199-253. Walnut Creek: Altamira.

Childs, S.T. and Killick, D. 1993. Indigenous African metallurgy: nature and culture. *Annual Review of Anthropology* 22: 317-37.

Chrétien, J.-P. 2003. *The Great Lakes of Africa: a thousand years of history*. New York: Urzone.

Cline, W. 1937. *Mining and Metallurgy in Negro Africa*. Wisconsin: George Banta

Clist, B. 1987. A Critical Reappraisal of the Chronological Framework of Early Urewe Iron Age Industry. *Muntu* 6: 35-62.

Connah, G. 1990. Archaeology in Western Uganda. *Nyame Akuma* 34: 38-45.

Connah, G. 1991. The salt of Bunyoro: seeking the origins of an African kingdom *Antiquity* 65: 479-94.

Connah, G. 1996. *Kibiro: the salt of Bunyoro, past and present*. London: BIEA

David, N., Heinmann, R., Killick, D. and Wayman, M. 1989. Between bloomery and blast furnace: Mafa iron smelting technology in North Cameroon. *African Archaeological Review* 7: 183-208.

de Maret, P. 1985. The Smith's myth and the origins of leadership in central Africa. In: R. Haaland and P. Shinnie (eds.) *African Iron Working: Ancient and Traditional.* 73-87. Oslo: Norwegian University Press.

Dunbar, A. 1965. *A History of Bunyoro-Kitara.* Nairobi: Oxford University Press.

Fisher, A. 1911. *Twilight Tales of the Black Baganda.* London: Marshall Brothers.

Giblin, J., Clement, A. and Humphris, J. 2010). An Urewe burial in Rwanda: exchange, health, wealth and violence c. AD 400. *Azania: Archaeological Research in Africa* 45:3, 276-297.

Gordon, R. and Killick, D. 1993. Adaptation of Technology to Culture and Environment: bloomery smelting in America and Africa. *Technology and Culture* 34: 243-270.

Grant, J. 1864. *A Walk Across Africa: or domestic scenes from my Nile journal.* London: William Blackwood and Sons.

Haaland, R. 1985. Iron production, its socio-cultural context and ecological implications. In: R. Haaland and P. Shinnie (eds.) *African Iron Working – Ancient and Traditional.* Oslo: Norwegian University Press, pp.50-72.

Haaland, R., Haaland, G. and Rijal, S. 2002. Social Life of Iron; A Cross-Cultural Study of Technological, Cognitive and Political Aspects of Iron Symbolism. *Arthropods* 97: 35-54.

Haaland, R. 2004. Iron smelting a vanishing tradition: ethnographic study of this craft in South West Ethiopia. *Journal of African Archaeology* 2/1: 1-15.

Herbert, E. 1993. *Iron, Gender and Power: rituals of transformation in African societies.* Indianapolis: Indiana University Press.

Hiernaux, J. and Maquet, E. 1960. Cultures préhistoriques de l'âge des métaux aux Ruanda-Urundi et au Kivu (Congo Belge). *Bulletin des Sciences de L'Academie Royals des Sciences Coloniale.*

Humphris, J. 2002. *An Ethno-Historic Investigation of Iron Smelting in the Buganda Kingdom.* Unpublished BA dissertation. Manchester: Manchester University.

Humphris, J. 2004. *Reconstructing a Forgotten Industry: an investigation of iron smelting in Buganda.* Unpublished MA dissertation. London: UCL.

Humphris, J. 2008. Iron production in southern Rwanda: a summary of recent research. *Nyame Akuma* 70: 2-10.

Humphris, J. 2010. *An Arcaheometallurgical Investigation of Iron Smelting Traditions in Southern Rwanda.* Unpublished PhD thesis. London: UCL.

Humphris J., Martinón-Torres, M., Rehren, Th. and Reid, A., 2009. Variability in single smelting episodes – a pilot study using iron slag from Uganda. *Journal of Archaeological Science* 36: 359-369.

Iles, L. 2004. *Supporting the Smelt: an archaeological investigation into the selection and use of plants within the Buganda iron smelting tradition.* Unpublished BSc dissertation. London: UCL.

Iles, L. 2009a. Pre-Colonial Iron Production in Western Uganda: recent survey and excavation. *Nyame Akuma* 71: 35-45.

Iles, L. 2009b. Impressions of banana pseudostem in iron slag from eastern Africa. *Ethnobotany Research and Applications* 7: 283-291.

Iles, L. 2011. *Reconstructing the Iron Production Technologies of Western Uganda: reconciling archaeometallurgical and ethnoarchaeological approaches.* Unpublished PhD dissertation. London: UCL.

Iles, L. in press. Iron production in Uganda: memories of a near-forgotten industry. In: H. Orange (ed.) *Reanimating Industrial Spaces.* Walnut Creek: Left Coast Press.

Iles, L. this volume. Applying ethnographic presents to archaeological pasts: the relevance of memories of iron production in western Uganda. In: J. Humphris and H. Veldhuijzen (eds.) *Proceedings of the World of Iron Conference, London 2009.*

Iles, L. and Martinón-Torres, M. 2009. Pastoralist iron production on the Laikipia Plateau, Kenya: wider implications for archaeometallurgical studies. *Journal of Archaeological Science* 36: 2314-2326.

Kagwa, A. 1934. *The Customs of the Baganda.* New York: AMS Press.

Killick, D. 2004. Review Essay: What Do We Know About African Iron Working? *Journal of African Archaeology.* Vol 2 (1): 135–152.

Kiwanuka, M.S.M.S: 1971. *A History of Buganda From the Foundation of the Kingdom to the Present Day.* London: Longman Group Limited.

Kusimba, C. and Kusimba, S. 2005. Mosaics and Interactions: East Africa, 2000 bp to the Present. In: A. Stahl (ed.) *African Archaeology.* 392- 419. Oxford: Blackwell.

Nenquin, J., 1971. The Congo, Rwanda, and Burundi. In: P.L. Shinnie (ed.) *The African Iron Age.* Wotton-under-Edge: Clarendon Press.

Oliver, R. And Atmore, A. 2001. *Medieval Africa, 1250-1800.* Cambridge: Cambridge University Press.

Raymaekers, J. and Van Noten, F., 1986. Early Iron Furnaces with Bricks' in Rwanda: Complimentary evidence from Mutwarubona. *Azania* XXI, 65-84.

Rehren, Th., Charlton, M., Chirikure, S., Humphris, J., Ige, A. and Veldhuijzen, H. 2007. Decisions set in slag – the human factor in African iron smelting. In: S. LaNiece, D. Hook and P. Craddock (eds.) *Metals and mines – studies in archaeometallurgy.* 211-218. London: Archetype.

Reid, D.A.M. 1991. *The Role of Cattle in the Later Iron Age Communities of Southern Uganda.* Unpublished PhD Thesis. Cambridge: University of Cambridge.

Reid, D.A.M. 2003. Recent research on the archaeology of Buganda. In: P. Mitchell, A. Haour and J. Hobart (eds.) *Researching Africa's Past.* 110-117. Oxford: Oxbow.

Reid, D.A.M. and MacLean, R. 1995. Symbolism and the social context of iron production in Karagwe. *World Archaeology* 27: 144-161.

Reid, D.A.M. and Young, R. 2003. Iron-smelting and bananas in Buganda. In: P. Mitchell, A. Haour and J. Hobart (eds.) *Researching Africa's Past.* 118-123 Oxford: Oxbow.

Reid, R. 2002. *Political Power in Pre-colonial Buganda: economy, society and welfare in the nineteenth century.* London: James Currey.

Robertshaw, P. 1994. Archaeological survey, ceramic analysis, and state formation in western Uganda. *African Archaeological Review* 12: 105-131.

Robertshaw, P. 1999. Seeking and keeping power in Bunyoro-Kitara, Uganda. In: S. McIntosh (ed.) *Beyond Chiefdoms: pathways to complexity in Africa.* 136-150. Cambridge: Cambridge University Press.

Robertshaw, P. 2003. The origins of the state in east Africa. In: C. Kusimba and S. Kusimiba (eds.) *East African Archaeology: Foragers, Potters, Smiths and Traders.* Philadelphia: University of Pennsylvania.

Robertshaw, P. 2010. Beyond the Segmentary State: creative and instrumental power in western Uganda. *Journal of World Prehistory* 23: 255-269.

Robertshaw, P. and Taylor, D. 2000. Climate change and the rise of political complexity in western Uganda. *Journal of African History* 41: 1-28.

Roscoe, J. 1911. *The Baganda.* London: Macmillan and Co.

Roscoe, J. 1915. *The Northern Bantu: an account of some central African tribes of the Uganda Protectorate.* Cambridge: Cambridge University Press.

Roscoe, J. 1923. *The Bakitara or Banyoro.* Cambridge: Cambridge University Press.

Schmidt, P. 1996. Reconfiguring the Barongo: reproductive symbolism and reproduction among a work association of iron smelters. In: P. Schmidt (ed.) *The Culture and Technology of African Iron Production.* 74–127. Gainesville: University of Florida Press.

Schmidt, P. 1997. *Iron Technology in East Africa: symbolism, science and archaeology*. Indiana: James Currey.

Schmidt, P. and Childs, S. T. 1985. Innovation and industry during the Early Iron Age in East Africa: KM2 and KM3 sites in Northwest Tanzania. *African Archaeological Review* 3: 53-94.

Schmidt, P. and Childs, S.T. 1996. Actualistic Models for Interprtation of Two Early Iron Age Industrial Sites in Northwestern Tanzania. In: P. Schmidt (ed.) *The Culture and Technology of African Iron Production*. 186-233. Gainesville: University of Florida Press.

Schoenbrun, D. 1993. Cattle herds and banana gardens: the historical geography of the western Great Lakes region, *c.*AD 800-1500. *African Archaeological Review* 11: 39-72.

Schoenbrun, D. 1998. *A Green Place, A Good Place: Agrarian Change, Gender, and Social Identity in the great Lakes Region to the Fifteenth Century*. Oxford: James Currey.

Schoenbrun, D. 1999. The (in)visible roots of Bunyoro-Kitara and Buganda in the Lakes region: AD 800-1300. In: S. McIntosh (ed.) *Beyond Chiefdoms: pathways to complexity in Africa*. 136-150. Cambridge: Cambridge University Press.

Schweinfurth, G., Ratzel, F., Felkin, R. and Hartlaub, G. (eds.) 1888 [trans. Mrs. R.W. Felkin] *Emin Pasha in Central Africa: being a collection of his letters and journals*. London: George Philip and Son.

Tantala, R. 1989. *The Early History of Kitara in Western Uganda*. Unpublished PhD dissertation. Madison: University of Wisconsin-Madison.

Taylor, D. and Marchant, R. 1994/1995. Human impact in the Interlacustrine region: long-term pollen records from the Rukiga highlands. *Azania* 29/30: 283-295.

Taylor, D., Robertshaw, P. and Marchant, R. 2000. Environmental change and political-economic upheaval in precolonial western Uganda. *The Holocene* 10: 527-536.

Thompson, G. and Young, R. 1999. Fuels for the Furnace: Recent and Prehistoric Ironworking in Uganda and Beyond. In: M. van der Veen (ed.) *The Exploitation of plant resources in Ancient Africa*. 221-239. New York: Kluwer

Tosh, J. 1970. The Northern Interlacustrine Region. In: R. Gray and D. Birmingham (eds.) *Pre-colonial African Trade: essays on trade in Central and Eastern Africa before 1900*. 102-118. London: Oxford University Press.

Tylecote, R.F. 1965. Iron-smelting in pre-industrial communities. *Journal of the Iron and Steel Institute* 203: 340-348.

Uzoigwe, G. 1976. Precolonial markets in Bunyoro-Kitara. In: B. Ogot (ed.) *Economic and Social History of East Africa*. 24-66. Nairobi: Kenya Literature Bureau.

Van Grunderbeek, M.-C. 1988. Essai d'étude typologique de céramique urewe dans la région des collines au Burundi et Rwanda. *Azania* 23: 11-55.

Van Grunderbeek, M.-C. 1992. Essai de delimitation chronologie de L'Age du Fer Ancien au Burundi, au Rwanda et dans la region des Grand Lacs. *Azania* 27: 53-80.

Van Grunderbeek, M-C. and Roche, E., 2007. Multidisciplinary Evidence of Mixed Farming during the early Iron Age in Rwanda and Burundi. In: T. Denham, J. Iriate and L. Vrydaghs (eds.), *Rethinking Agriculture: archaeological and ethnoarchaeological perspectives*. Walnut Creek: Left Coast Press, 299-319.

Van Grunderbeek, M-C., Roche, E. and Doutrelepont, H. 1983. *Le Premier Age du Fer au Rwanda et au Burundi, Archaeologie et Environment*. Brussels: Institut National de Recherché Scientifique.

Van Grunderbeek, M-C., Roche, E. and Doutrelepont, H. 2001. Un Type de Font de Fer Associe a la Culture Urewe (Age de Fer Ancien) au Rwanda et au Burundi. *Mediterranean Archaeology* 14: 271-297.

van Noten, F. 1972. *Les tombes du roi Cyirima Rujugira et de la reine-mère Nyirayuhi Kanjogera: Description archéologique*. Tervuren: Musée Royal de l'Afrique Centrale.

Van Noten, F. 1982. *The Archaeology of Central Africa*. Gräz: Akademische Druck-U Verlagsanstalt.

Van Noten, F. 1983. (ed.) *Histoire Archeologique du Rwanda*. Tervuren: Musee Royal de l'Afrique Centrale.

Vansina, J. 1990. *Paths in the Rainforest: Towards a History of Political Tradition in Equatorial Africa*. London: James Currey.

Vansina, J. 2004. *Antecedents to Modern Rwanda: the Nyiginya Kingdom*. Oxford: James Currey.

Vignati-Pagis, E., 1995. *Du Fourneau a la Fosse: Changements Techniques dans la Métallurgie du Fer au Burundi au Cours des Duex Derniers Millenaires?* Unpublished PhD thesis. Paris: Université de Paris.

Wainwright, G. 1954. The diffusion of "uma" as a name for iron. *Uganda Journal* 18: 113-136.

Wotzka, H-P. 2006. Records of activity: radiocarbon and the structure of Iron Age settlement in Central Africa. In: *Grundlegungen. Beiträge zur europäischen afrikanischen Archäologie für Manfred K. H. Eggert*. Tübingen: Francke pp.271-289.

Authors' addresses

- Jane Humphris
  UCL Qatar
  Georgetown Building
  Hamad bin Khalifa University
  Doha, Qatar
  j.humphris@ucl.ac.uk
- Louise Iles
  UCL Institute of Archaeology
  31-34 Gordon Square
  London WC1H 0PY
  louise.iles@gmail.com

Bricolage, ritual performance, and habitus [forgotten] in Barongo iron smelting

Peter R. Schmidt

ABSTRACT: When Victor Turner (1967, p.19) defined ritual as "prescribed formal behavior for occasions not given over to technological routine...", he did not consider that ritual process in a technological setting may be dynamic and contextually sensitive – hardly to be associated with routine and often spontaneously drawing on core rituals performed in other settings, such as medicinal healing. This paper examines how ritual bricolage is linked to technological habitus and, moreover, how ritual provides an experimental template upon which technological success may be realised. Iron technology among the Barongo (southwest of Lake Victoria) must be seen as an improvisational *chaîne opératoire* – responding to and incorporating a rapidly changing and improvisational ritual performance that is never identical from smelt to smelt. These vital and dynamic ritual and technological performances reflect changing social and economic conditions, provide the means by which the tensions of new social and economic relations are resolved, and illustrate how the agents controlling iron technology accommodate challenges when confronting new physical resources.

Keywords: Iron smelting; Africa; bricolage; *chaîne opératoire*; tropes; ritual; Barongo; Tanzania.

Foregrounding the problem

> I believe that the question is not so much to determine where function stops and symbol (or style) begins, but to be aware of their remarkable intricacy. Indeed, making pottery and 'making sense' are two compatible, entangled, and above all, complementary processes (Gosselain, 1999, p.221).

This paper is an attempt to reclaim and recuperate, along with other scholars such as Lemonnier (1992, 2004) and Gosselain (1999), the social domain in *chaîne opératoire* studies focusing on technological production. The social dimension that Pierre Lemonier (2004) has patiently refined is perhaps most poignantly expressed in his treatment of the interplay of myth, the spirit world, and drum making in Papua New Guinea, where the weave of social, bodily, and technological performances provides rich insights into the cultural reasons why various choices in a technological sequence are made. This socially-informed perspective, however, seems increasingly threatened in the archaeological world by what has become, for lack of a better phrase, an obsessive fixation to document lithic reduction series (e.g., Grace, 1997; Close, 2004). It is Lemonnier's goal to show how the social world, including the ancestors, punctuates and guides the daily lives of those who undertake technological activities. In this construction of technology, the technical is inseparable from the social: the ancestral spirits talk and wail, punctuating the gestures and technological motions of the craftspeople, be they drum makers in Papua New Guinea or iron smelters in Tanzania.

The one key contrast that arises between these two regions rests on the unique qualities of the multiple ancestral lines represented in Barongo iron smelting. The Barongo are not answering to the ancestors of one lineage or larger social group with a common progenitor. Rather, their multi-ethnic makeup compels a different approach – addressing the needs and sensibilities of multiple lines of ancestors, a condition that creates a competitive bricolage. It only takes a short look inside a Barongo iron smelting furnace to understand that much of the *chaîne opératoire* is ritually constructed, with ritual choices often changing from smelt to smelt while the improvisational bricolage of the Barongo intertwines ritual solutions with technological solutions. This issue has been succinctly summarised by David (2001, p.62): "The fact that technical and ritual processes are... 'richly interwoven' is not in dispute; the vital question for materials scientists and anthropologists alike is whether they inform upon each other...".

This key question about how we view ritual and technology builds on Gosselain's (1999) question when discussing pottery, "one may wonder whether potters' behavior could in turn be influenced by metaphors, with steps of the *chaîne opératoire* becoming the locus of symbolic discourse" (1999, p.206). I will argue here – in parallel with David's observation – that the technological behaviour of Barongo iron smelters is deeply influenced by the results or what they assess as the effectiveness of ritual interventions (when tropes are enacted) – sometimes electing to engage in different ritual/symbolic solutions in an on-going assessment process. Their behaviour is also profoundly influenced, if not directed, by the spiritual agency of ancestors with whom they conduct an ongoing, complex, and enormously varied tropic discourse – the variety of which is linked to and determined by the multiple ancestral lines represented among the smelters. Arising out of this discourse are ritual processes that engage iron smelters as agents who enact tropic cures and intercessions, thereby affecting the substance, tempo, and configuration of iron smelting operations that affect the final product.

Here I take a perspective that draws on the transformative properties of tropes through the agency of subjects – iron smelters – who through their application of techniques of medicinal curing, magical protections, and ancestral appeasement render themselves as protectors and curative agents vis-à-vis the iron smelting furnace. Warnier (2001) and Rowlands (2005) help to unwrap the praxis engaged in such ritual processes in which relations between different categories of persons in the Grassfields of Cameroon "are conceived metonymically through the bodily processes…, that is, digestion, regurgitation, breathing, spitting, and ejaculation" (Rowlands, 2005, p.78), illustrated by the Fon's (king) spitting raphia wine – tropically constructed as an ancestral substance – over the backs of his subjects and particularly women who desire to ensure conception. Though the Grassfields rituals pertain to different categories of persons, the same principles apply to persons and smelting furnaces, transformed through multiple rituals into persons, thus collapsing the subject/object dichotomies that often prevail when we examine smelting furnaces.

I consider tropes – metaphors and metonyms – under conditions of highly focused social agency during ritual to understand how iron smelting furnaces are embodied as fecund brides. Kirmayer (1993) and Van Wolputte (2004) argue that tropes belong to the domain of praxis (not cognition) because they "suggest instead of conceive, elicit instead of define, or provoke instead of prescribe. At the same time, they are rooted in bodily (sensuous) experience" (Van Wolputte, 2004, p.257), a view shared by Warnier (2001) and Rowlands (2005). Importantly for our treatment, tropes address both social and bodily experiences (Jackson, 1983) – a kind of efficacy that brings together the two domains, imposing them upon inert material (clay, slag, iron, medicines, parts of the furnace), creating conditions of contiguity and association to transform them into social and embodied experience. Meanings arising out of tropes are not just representational but are also embodied, for example, when the metonym for menses – a red bleeding piece of bark – is placed in the innermost reaches of a furnace; or, when potent medicines to appease the ancestors are placed contiguously to the 'bleeding' bark. Such social agency resonates with practices, for example, among the Himba of Namibia when ancestors are embodied (ingested) by particular diets associated with them or when their names are uttered (sound and speech motions) daily (Van Wolputte, 2004). Other examples may be referenced from the Grassfields of Cameroon, where "the principal ancestral substances are saliva, raphia wine, breath, seamen, food, and palm oil; some of these are contained physically in the body, others in bags, bowls, and calabashes that are considered as extensions of that physicality" (Rowlands, 2005, p.78).

As open containers, Barongo iron smelting furnaces (as well as numerous other African examples) receive medicinal tropes (figures of speech) that signify the bodily fluids of both female and male social agents – transformations that occur under conditions of ritual performance. The furnace is technologically viable only if it serves the higher purposes of human reproduction, maintaining good relations with the ancestors, and protection against evil social forces. Anthropologists of Africa have long known that African iron smelting furnaces sculpted with female characteristics are more than static, material phenomena that signify only the reproductive qualities of females (Childs and Killick, 1993; Herbert, 1993; Schmidt, 1997). Rather, they fit into a larger tropic ritual field in which males play key roles as creators of the female images, as sexual partners and impregnators, as well as liminal roles as mid-wives, as healers, and as ritual authorities guiding various dangerous rites of passage. The tropes attached to iron smelting furnaces are multilayered, yet their exegesis has much to tell us about how meaning is conferred through rituals of embodiment, especially rituals that pertain to the inside of smelting furnaces, rituals that embrace healing, protection from angry ancestors, and apotropaia against evil perpetrated against the furnace. Such a perspective moves away from a focus on the surface materiality (clay breasts, painting), also informed by ritual enactment of tropes, which has been virtually the sole focus of studies about transformations of smelting furnaces into female bodies.

Meanings that are infused within the internal female organs of iron smelting furnaces – beyond those that are applied to the surface – usually unfold in a complex play of tropes before, during, and after the smelting activity. The female reproductive cycle is central, for in Barongo iron smelting the female body becomes technological process and vice-versa, a metonymic transformation of the human body into technological process and economic prosperity. Economy and reproduction become inseparable, as do healing and prosperity. If we consider these as interrelated phenomena belonging to the same conceptual field, then an understanding of the play of tropes in African iron technology opens more readily. My study of the Barongo iron smelters southwest of Lake Victoria shows that different ritual processes are drawn extensively from the medicinal world, from apotropaic magic (protective treatments to ward off evil, e.g., those practicing sorcery or black magic), from rituals to appease ancestors, and from sympathetic magic to ensure functional success (Schmidt, 1996).

The Barongo and ritual performance

Heretofore anthropological treatments of African iron smelting furnaces have oversimplified, if not misrepresented, reproductive tropes by focusing on female parturition and by treating menstruation as strictly taboo and as a period of sterility. This facile treatment of menses, often emphasising pollution, isolation, and avoidance – overlooks the biological reality that without menses there is no female reproductive cycle. It is also an uncritical acceptance of normative menses taboos, an artificial academic posture arising out of an absence of prolonged ethnography of the full panoply of beliefs and ritual practices that unfold over multiple smelting activities. Also, the ingestion of specialised drugs during ritual processes (for other contexts see Hamilakis, 1999; Wilkie, 2000) immediately refocuses our attention on iron smelting

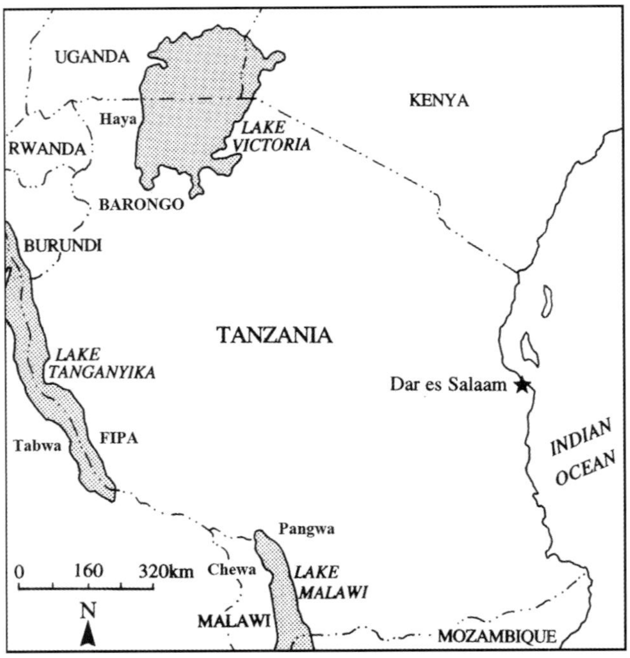

Figure 1: Map of Tanzania showing the location of the Barongo, a multi-ethnic iron smelting group southwest of Lake Victoria.

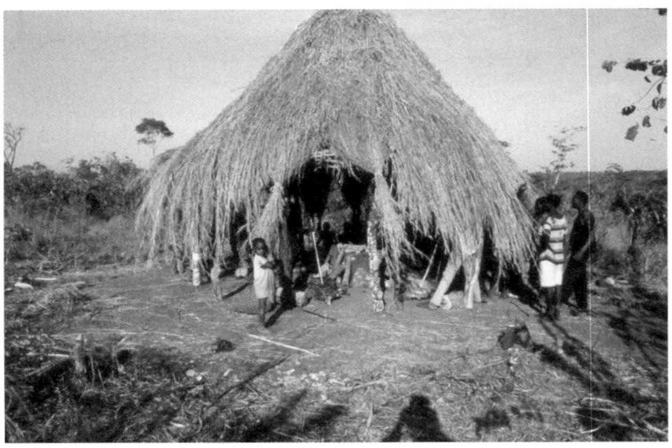

Figure 2: Barongo iron smelting house, eastern Biharamulo District, 1979.

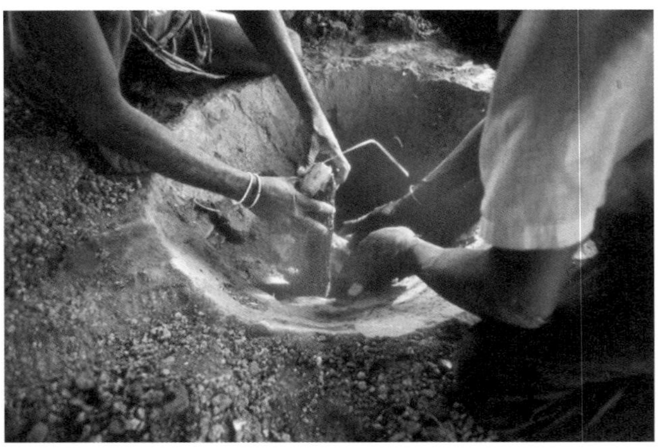

Figure 3: Placing ritual objects within the *nyaruwengi*, the womb of the furnace. The head smelter is holding a piece of *kukuba* wood with red, bleeding bark.

rituals, where the application of potent drugs to the inside of furnaces and to furnace tools placed inside the furnace help us to unlock deeper knowledge about internal embodiments effected by the ritual healers/smelters. A focus on Barongo iron smelting sheds new light on what previously appeared to be simply 'technological ritual', a category that privileges ritual as a mnemonic sequencing of technical steps and thus obscures a deep well-spring of healing knowledge applied to human frailties.

The bricolage of Barongo ritual applied to iron smelting likely arose during the 19th and 20th centuries, when work groups of multi-ethnic makeup arose out of the depredations of Arab slave raiders in central and western Tanzania, driving settled craftspeople into places of refuge, usually forested environments not easily penetrated (Schmidt, 1996). At the same time, the raids mounted by the Buganda state (northern shores of Lake Victoria) into the lands south of the lake also contributed to social upheaval and the mixing of ethnic groups who were thrust together in multi-ethnic configurations (Fig. 1). Practicing their craft secretly, these multi-ethnic work groups of Barongo – speaking their own technical language (Kirongo) – were constantly involved in meeting challenges such as mining new clays for tuyères, obtaining new iron ore resources, and dealing with unfamiliar wood species for charcoal. As they moved from locale to locale, either to avoid attacks from marauding slave traders in the 19th century or the sanctions of colonial officials during the 20th century, they constantly confronted social and physical adversities. Their responses to new political, social, and physical conditions can be read in their ritual repertoire, with extensive borrowings from medicinal cures. Their multiple ancestral lines also compel a plethora of ritual intercessions with ancestral spirits not witnessed in most technologies.

Thus bricolage, the fabrication of new ritual solutions that draw on past experience and spontaneous innovation, is an important part of the technology. Several examples stand out from the first smelt conducted during 1979 (Fig. 2). The ritual events of that day and later smelts focus our attention on sensory processes of embodiment, with an emphasis on sensory stimulations related to sexual intercourse and to the sexual reproductive cycle of a woman – contradicting the normative treatments of menstruation as polluting (Gottlieb, 1982; Gausset, 2002). The intense sensory stimulations activated during sexual intercourse – as with eating – are "stored into the body, generating bodily memory, which can be recalled and re-activated by similar experiences..." (Hamilakis, 2002, p.124). Such bodily memory is invoked when the iron smelters initiate the furnace, much like the sexual initiation of a bride.

After the construction of a shelter and the excavation of a pit within the shelter, the first activity is the placement of key ritual items in the bottom of the furnace pit – the metonymical womb (*nyaruwengi*) of the furnace. A piece of *kukuba* wood – thick with its red sap and a trope for menstrual blood (a local emic explanation) is placed in the *nyaruwengi* first, a sequencing that captures the female reproductive cycle – the womb being cleansed before it is fertile and ready to reproduce (Fig. 3). Importantly, the leaves of the *muhingura* plant are placed beneath the bleeding *kukuba* offering. The *muhingura* tuber, known for its medicinal properties in curing infertility, is also an offering, but what captures our attention in this embodiment process is that the *muhingura* leaves are saturated by the symbolic menses, uniting menses – the restoration of fertility – with a potent device that restores fertility. These *internal* bodily processes, the most intimate domain of female human sexuality, are embodied in the ritual actions and movements of the two smelting officials as they carefully add these ingredients to the furnace womb, wetting the protective leaves in a manner consistent with the biological flow of menses upon a protective apron. These particular rituals did not appear spontaneously in the ritual repertoire but rather were a part of smelting ritual habitus, yet only practiced upon the initiation of a new furnace. Such a ritual would be repeated only until such a time that the potency of the devices had expired.

The next more sensual ritual phase involves the embodiment of the furnace by both the head smelter and the head ritual authority. Both remove their clothes, sit on the ground opposite each other across the furnace pit – back to back, and in a rhythmic manner, using their voices and their bodies, they push a circular mound of earth (earlier excavated from the pit to accommodate the various medicines) into the pit (Fig. 4). This sexual intercourse with the furnace – with direct genital contact – is a profoundly sensory experience in which the embodied sounds and motions of intercourse culminate in the impregnation of the furnace, a fecund bride. A metonymic act of intercourse – because of the contiguity of human genitalia with the earth of the furnace pit (semantically separated domains of flesh and earth) – the play of tropes in this embodiment of the furnace goes much deeper. It also signifies, because of the concurrent display of anuses, a strong contempt for bad and dangerous things. This practice directly neutralises and overwhelms any sorcery possibly aimed at the furnace. Parallel to the treatment of menses, such a display does not pollute; rather, it protects and ensures success.

After the earth had been returned to the furnace pit, the head smelter sprinkled two containers of *rukago* medicine over the new furnace floor, an apotropaic device to make people fear the place and prevent anyone from interfering with the ritual devices that were intended to "bring a heavy child." Notably, multiple tropes are also at work in the *rukago* ritual, which also seals the furnace against the interference of unhappy ancestors. It is also of great interest that the use of a white substance (either salt or ash or a mixture of both) occurs in the *rukago* ritual, applied in the shape of a cross to seal out displeased ancestors. Some will recognise this application as a ritual cosmogram, enacted in similar ways among the BaKango or other peoples of central and western Africa. This potent tropic enactment, now known to be vital in western Tanzania, affirms that the meanings associated with such ritual performances are not only widespread but also trans-domain in their application, in this instance being applied within a technological setting. These two ritual cycles – the sexual impregnation and the protections against ancestors – occur only when a new furnace is being initiated. A furnace pit may be used repeatedly without renewal of the magical and medicinal devices, even over a year hiatus (e.g., between 1979 and 1980, but not between 1980 and 1984, when renewal was required although the same area was used). When the smelting process resumed the next day, the head smelter dug a small trench contiguous to the southern side of the furnace pit and buried the leaves of the *muhingura* tuber (ensures fertility) as well as a bleeding piece of *kakuba* tree (a trope for menses) – both previously buried in the furnace pit – a process that reaffirmed and amplified the earlier treatments.

Once the furnace superstructure (hewed blocks of termite earth with broken tuyères wedged between the blocks) was in place (Fig. 5), the smelters inserted their clay tuyères *inside* the

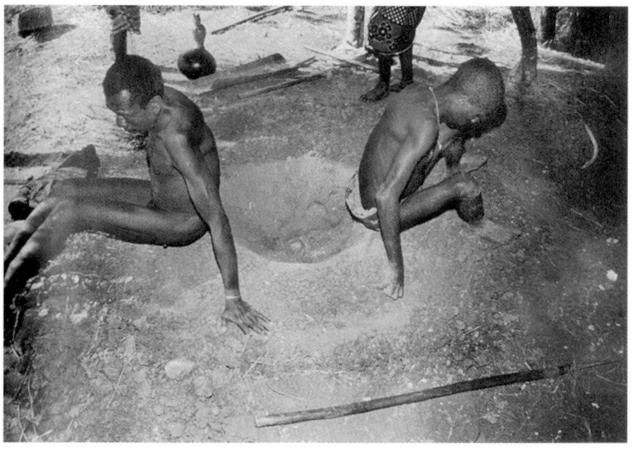

Figure 4: Barongo smelters pushing earth into a furnace pit to seal ritual devices at the bottom. Both their genitals and anuses are in contact with the soil – simultaneously suggesting 1) sexual intercourse with the furnace, and 2) suppression of evil spells and displeased ancestors.

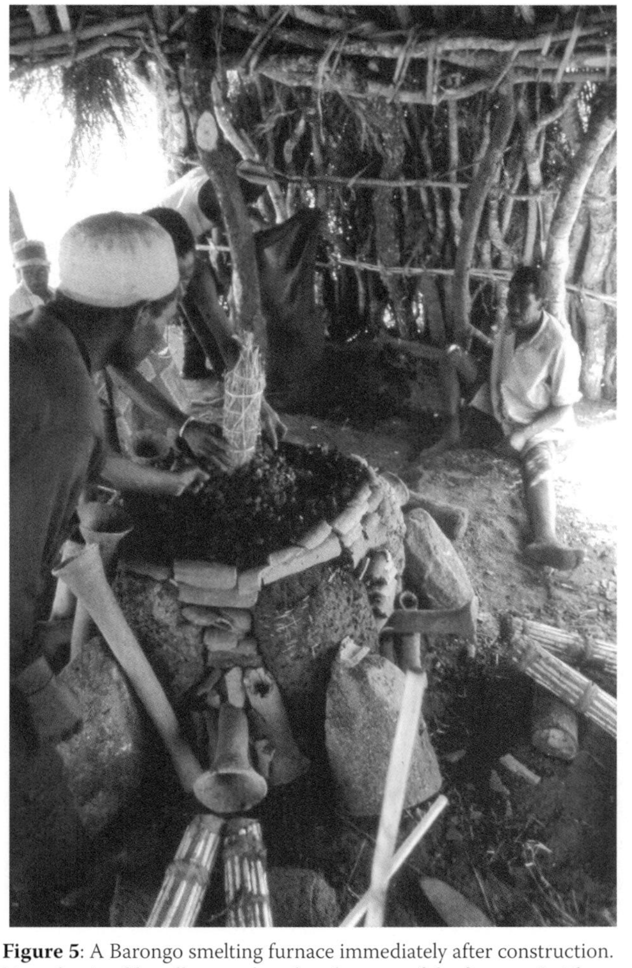

Figure 5: A Barongo smelting furnace immediately after construction. Note the ritual bundle, nsamba, placed on top of the furnace. Inside is an iron needle, used by healers of Bazinza ethnicity to exorcise malevolent spirits, including ancestral spirits. The furnace is built from blocks of termite earth, between which are wedged horizontal, broken pieces of tuyères.

furnace after rubbing their ends with *Maputo* bark, a reddish substance said to strengthen the tuyères. The value of this practice lies in the 'medicine' stiffening the tuyère, a phallus, so it will withstand the heat and not droop, a metonymical Viagra for flaccid, drooping blowpipes. It is both an external application of the ritual cure and an internal one, for the blowpipes are then slipped inside the hot vaginal-like cavity. The play of tropes that integrate menses as lived experience and that treat the menstrual condition of women as normal is amplified during two subsequent rites of passage that initiate the furnace before smelting commences. The first rite – not always practiced according to the same formula (and thus another example of bricolage) – took place just after the fire broke through the top of the charge of sinter and ore stacked on top of the charcoal, at which time the head smelter's wife entered the smelting house with a pot of beer.

This event is marked as distinctive in ritual repertoires of African iron smelting by the inclusion of a reproductive woman (for other exceptions see Redinha, 1953; Herbert, 1993). Each smelter in turn sipped beer through a straw and then spit it on the ground and, again, on the furnace; this ritual mimics a widely practiced rite of passage in the region that ensures fertility when a bride and groom spit beer or milk on one another. Equally important, it simultaneously addresses concern about the ancestors, in which ancestral agency is expressed by the smelters' purposeful propitiation of ancestors through beer spitting.[1] This ritual was repeated five years later, yet quite differently – again showing the play of different interests and the deeply embedded affinity toward bricolage: the wife of the head smelter was joined by another smelter and together they sipped beer from the woven pot and then spit it upon the furnace, followed by the other senior smelters who also spit beer upon the furnace – more closely replicating the ritual as it occurs in nuptial settings, with a living female engaged in the embodiment of the fertile condition. This ritual brings a fecund female (*without* concern for her menstrual condition, as her embodiment of this condition may indeed be critical to the process) into intimate contact with the furnace, a process similar to that practiced by several other groups, such as the Chokwe in Angola (Redinha, 1953) and Chewa in Malawi (van der Merwe and Avery, 1987). It contradicts the taboos that forbid women in menses from contact with the furnace, either because of their polluting condition or because they are too 'hot' – the latter an acknowledgement that menstruation is associated with hotness in females, an attribute also widely associated with sexual intercourse.

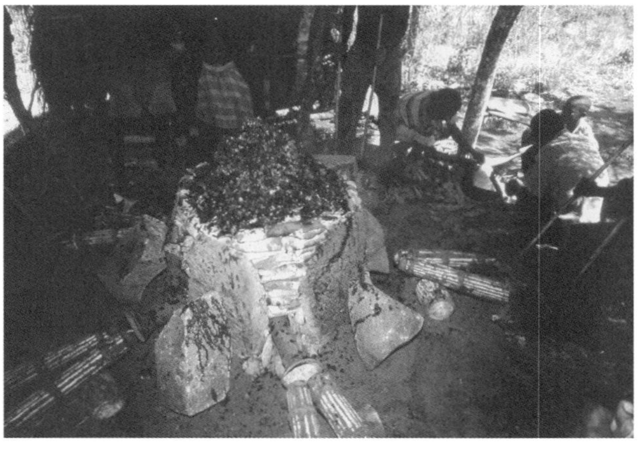

Figure 6: Sacrificial blood splashed on the flow slags, ritual fertility medicine, furnace, and bellows (and smelters) before a Barongo smelt.

The second rite is a blood sacrifice that saturates the furnace and the bellows, as well as the charge of sinter and ore on top (including a medicinal bundle, *nsamba*, that acts as an aphrodisiac). It occurred immediately before smelting commenced and immediately after the beer spitting ritual (Fig. 6), but only on two occasions – once *after* two smelts had already been conducted. The tropic power of blood sacrifice through the social agency of male smelters in this case rests very deeply upon the potency of blood sprayed upon the furnace charge, made up of partly sintered iron and other 'slags' that are referenced as 'giving off water', that is, as *flowing* through the furnace. Covered with purifying sacrificial blood, the metonymical menstruation of the furnace is manifest. Indispensable for conception yet simultaneously holding properties of sterility, purifying while having the potential to pollute, these dualities of menses parallel the sacrificial blood that represents both death and life (Herbert, 1993). Luc de Heusch opens important insights into this process with his exegesis of the Dagon myth, in which "Blood from throat slitting, called 'menses of the earth,' contains the life of the world", a condition in which "Sacrifice is...a...cosmic menstruation (de Heusch, 1985, pp.137-138). The Barongo ritual performers – themselves splattered by the sacrificial blood and tropically transformed by the blood into women – embody smelting furnaces as females in menses.

This second cycle of menstrual cleansing – the first having occurred during the cleansing and impregnation of the furnace pit – was soon followed by the injection of salt into the tuyères, a metonym for impregnation and a ritual performance that also draws from a healing repertoire. During one period of particularly troubling tuyère failure, the head smelter brought a tray with salt and *omunsesene* root to the furnace and at each failing tuyère he took a little salt and root, spit on them, and shoved the concoction to the end of the tuyère (deep inside the furnace), later explaining that this would improve the yield, for it had 'fed' or propitiated the ancestors (Fig. 7). He shortly returned with more salt and pieces of the *musengati* root, which he wet in his mouth and dipped in salt before also shoving them down each tuyère to open them for better flow, a metonymical ritual creating sperm flow to increase reproductive success, a process bearing some resemblance to a Nyakuysa (southern Tanzania) ritual after first menstruation, when a young girl has a pungent tasting root pushed into her mouth along with salt through a funnel (Wilson, 1957, p.87).

It is important to note that salt and white clay are known metonyms for semen and sexual union (for the latter see Roberts, 1991). When the Barongo ritual specialist and head smelter used salt in this manner – quite unpredictably, by their bodily gestures and incantations they were shooting (quickly shoving) their seeds into a 'hot' female. They also alter the tempo of bellowing and the focus of the technological process, weaving together technological solutions with ritual tropes that are materially performed and that have material implications.

There are additional treatments involving ancestors that occur irregularly – sometimes only once – ranging from the

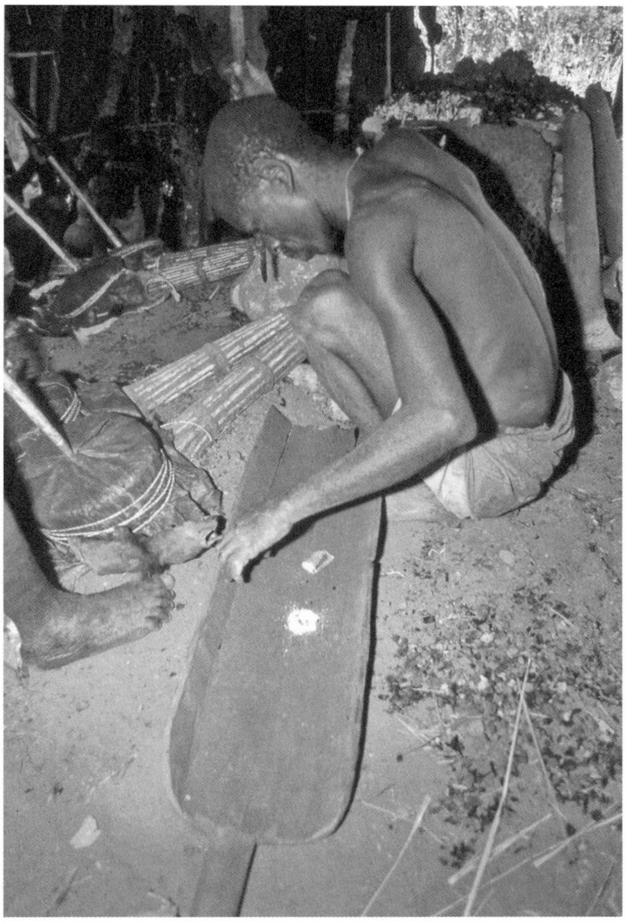

Figure 7: A wooden tray with a piece of *omunsesene* root and salt, soon to be mixed and shoved down the tuyères – addressing concerns over fertility as well as appeasing the ancestors.

use of a ritual tuber (*kalinza*) placed on the top of the charge and also used as part of the internal medicinal package deep within the furnace. As it disappears into the furnace when the flames burst out of the top of the charge, '*kasinge*' or, thank you, is chanted to the ancestors. At other times ritual solutions varied according to the needs and actions of the smelters, when other ancestral concerns came to light. One such ritual treatment parallels the *kalinza* tuber – the *nsamba* package, before the flames break through the charge: the components of the package (made of banana bast) were once displayed (but not explained), including an iron needle (Fig. 5). Bjerke's (1981) study of Bazinza (an identity of some smelters) healing practices tells us that needles are used in exorcism rituals, again affirming the idea that unfriendly spirits must be expunged at multiple stages of the smelt. As the smelt progressed, ancestral agents again guided ritual behaviour, when, for example, the head smelter stopped the bellows from being pumped so he could insert a tuber called *mususemi* into the tuyère. Taken from a shrub growing on land belonging to their Boyovu ancestors some 25 miles away, the root was stripped of its bark, then mixed with salt and placed within the tuyère with the head smelter chanting "*Nakurungira tata*", or, "He has fed you well." Enacting multiple tropes much like the beer spitting, this ritual pertains both to fertility and to the appeasement of direct ancestor(s) who are being 'fed'.

There are yet other ritual processes that irregularly appear as part of the Barongo bricolage, among them a prohibition that appeared during the second smelt of 1979, when the head smelter announced that he did not wash since making charcoal and would continue in an unwashed state until smelting was complete. This resembles a puberty rite of passage among the Nyakusa, where girls reaching menstruation are secluded and not allowed to wash for a month, again indicating that a complete female reproductive cycle is the structuring principle.

Conclusion

Ritual performance compels us to understand as much as we possibly can about the inseparable domains of ritual practice and iron technology. Social agency in iron smelting is now understood to be a much more inclusive practice than just the individual male ritual agent. As the ritual healer tends to his patient, he only does so by involving his cohorts in a collective action in which all attend to the patient's needs, as well as respond to the spiritual agency of the ancestors. Often the remaining healing and apotropaic devices have an enduring materiality, sometimes being preserved for thousands of years (Schmidt and Mapunda, 1997). The inclusion of females as transformative ritual agents and the integration of ritual metonyms pertaining to menses show us that gender dichotomies *do not* always prevail, that in fact the ritual practices of iron smelters and their female companions/wives are social declarations illustrating a gender unity manifest in a host of ritual treatments, ranging from salty tuyères to blood-soaked flowing slag. When both genders are embodied though sensual gestures and sounds, then both are united within the same domain, a relationship of combination and contiguity – key ingredients of metonymy – resulting in an integrative identity (Ricoeur, 1977) – a collapse of Cartesian gendered dichotomies that have too long informed interpretations of African iron smelting. Tropes do not act alone. They require the active intervention and application of focused social agents – the iron smelters who seek a productive outcome from their labours, and, I hasten to add, the ancestors who desire specific behaviours. With this dynamic perspective arising out of Barongo social agency (and that of their ancestors), the dynamic, reflexive, and improvisational characteristics of ritual process are revealed in such a way as to reassess significantly Turner's notion of ritual as prescribed formal behaviour.

Note

1. Space does not allow a complete discussion of all details; however, one important ritual act must be mentioned – both the head smelter's wife and her male companion wear a *kizingo*, the hat a spirit medium wears while communicating with the ancestors.

References

Bjerke, S., 1981. *Religion and misfortune: The Bacwezi complex and other spirit cults of the Zinza of northwestern Tanzania.* Oslo: Universitetsforlaget.

Childs, S.T. and Killick, D., 1993. Indigenous African metallurgy: nature and culture. *Annual Review of Anthropology*, 22, pp.317-37.

Close, A., 2004. *Chaîne-Opératoire analysis of a Northwest Coast lithic assemblage.* Paper presented at the 69th Annual Meeting of the Society for American Archaeology, Montreal.

David, N., 2001. Lost in the third hermeneutic? Theory and methodology, objects and representations in the ethnoarchaeology of African metallurgy. *Mediterranean Archaeology*, 14, pp. 49-72.

De Heusch, L., 1985. *Sacrifice in Africa: a structuralist approach.* Bloomington: Indiana University Press.

Gausset, Q., 2002. The cognitive rationality of taboos on production and reproduction in sub-Saharan Africa. *Africa*, 72, pp.628-54.

Gosselain, O.P., 1999. In pots we trust – the processing of clay and symbols in sub-Saharan Africa. *Journal of Material Culture*, 4, pp.205-30.

Gottlieb, A., 1982. Sex, fertility and menstruation among the Beng of the Ivory Coast: a symbolic analysis. *Africa*, 52, pp.34-47.

Grace, R., 1997. The 'chaîne opératoire' approach to lithic analysis. *Internet Archaeology*, 2, http://intarch.ac.uk/journal/issue2/index.html and http://www.hf.uio.no/iakh/forskning/sarc/iakh/lithic/opchainpaper.html

Accessed June 6, 2009.

Hamilakis, Y., 1999. Food technologies/technologies of the body: the social context of wine and oil production and consumption in Bronze Age Crete. *World Archaeology*, 31, pp.38-54.

Hamilakis, Y., 2002. The past as oral history: towards an archaeology of the senses. In Y. Hamilakis, M. Pluciennik and S. Tarlow eds. *Thinking through the body: archaeologies of corporeality.* New York: Kluwer and Academic/Plenum Publishers, pp.121-36.

Herbert, E., 1993. *Iron, gender, and power: rituals of transformation in African societies.* Bloomington: Indiana University Press.

Jackson, M., 1983. Thinking through the body: an essay on understanding metaphor. *Social Analysis*, 14, pp.127-49.

Kirmayer, L., 1993. Healing and the invention of metaphor: the effectiveness of symbols revisited. *Culture, Medicine and Psychiatry*, 17, pp.161-95.

Lemonnier, P., 1992. *Elements for an anthropology of technology.* Anthropology Papers, Museum of Anthropology, University of Michigan, No. 18. Ann Arbor: University of Michigan.

Lemonnier, P., 2004. Mythiques chaîne opèratoire. *Techniques & Culture*, 43-44, pp.25-43.

Ricoeur, P., 1977. *The rule of metaphor: multi-disciplinary studies of the creation of meaning in* language. Toronto and Buffalo: University of Toronto Press.

Redinha, J., 1953. *Campanha etnografica ao Tchiboco*, No. 1. Lisbon: Companhia de Diamantes de Angola, Serviços culturais.

Roberts, A.F., 1991. Where the king is coming from. In: A. Jacobsen-Wilding, ed. *Body and space.* Acta Univesitatis Uppsalalensis 16. Uppsala: Uppsala Studies in Cultural Anthropology.

Rowlands, M., 2005. A materialist approach to materiality. In: D. Miller, ed. *Materiality.* Durham: Duke University Press, pp.72-87.

Schmidt, P.R., 1996. Reconfiguring the Barongo: reproductive symbolism and reproduction among a work association of iron smelters. In: P.R. Schmidt ed. *The culture and technology of African iron production.* Gainesville: University Press of Florida, pp.74-127.

Schmidt, P.R., 1997. *Iron technology in East Africa: symbolism, science, and archaeology.* Bloomington: Indiana University Press.

Schmidt, P.R. and Mapunda, B.B., 1997. Reading ideology in the African archaeological record. *Journal of Anthropological Archaeology*, 16, pp.73-102.

Turner, V., 1967. *The forest of symbols: aspects of Ndembu ritual.* Ithaca: Cornell University Press.

van der Merwe, N.J. and Avery, D.H., 1987. Science and magic in African technology: traditional iron smelting in Malawi. *Africa*, 57, pp.143-72.

Van Wolputte, S., 2004. Hang on to yourself: of bodies, embodiment, and selves. *Annual Review of Anthropology*, 33, pp.251-69.

Warnier, J-P., 2001. A praxeological approach to subjectivation in a material world. *Journal of Material Culture*, 6, pp.5-24.

Wilkie, L., 2000. Magical passions: sexuality and African-American archaeology. In: R.A. Schmidt and B.L. Voss eds. *Archaeologies of sexuality.* London: Routledge, pp.129-42.

Wilson, M., 1957. *Rituals of kinship among the Nyakyusa.* London: Oxford University Press.

Author's address

- Professor of Anthropology and African Studies; 415 Grinter Hall, University of Florida, Gainesville, Florida, 32611, USA; Telephone: (352) 468-3244; Email: schmidtp@ufl.edu

Refining narratives: transformations of iron-working traditions in Ghana – 19th to 21st centuries

Len Pole

ABSTRACT: This paper examines the changing purpose of iron-working traditions among the Mawu (Akpafu) and Lolobi, populations living in the Togo hills of the central Volta Region of Ghana. It compares present-day oral traditions and other descriptions collected from published sources with the evidence from the author's ethnographic fieldwork conducted in the 1970s, which investigated the processes of mining, smelting, and tool manufacture. The record of the Mawu iron-workers, gleaned from their own elders, from evidence on the ground, and from written reports is considered in the context of archaeological and historical evidence from this and other parts of southern Ghana. The paper delineates and accounts for transformations that have taken place in the content of the tradition in the past 100 years.

Keywords: Iron-smelting, Ghana, oral tradition, technology, reconstructions.

Background

The Mawu (Akpafu) and Lolobi (estimated current combined populations, 13,000-18,000 (Dingemanse 2011)), and to a lesser extent the Santrokofi, have long been famous for iron-making within the southern regions of present-day Ghana and Togo (Fig. 1). 'Akpafu' is the name given to the Mawu by their Ewe neighbours, explained by some as an imitation of the sound of bellows being blown, formerly an everyday sound throughout Mawu and Lolobi villages (see Dingemanse, 2007). The languages spoken by these groups are distinct from those of the surrounding groups, the Twi-speaking Akan, the Guang-speaking Nkonya, and the Ewe, but are closely related to each other. They are known collectively as Ghana-Togo mountain languages (Siwu is the language of Mawu, and also of the Lolobi, with some dialectal variation; Sɛlɛɛ is that of the Santrokofi. For more details, see Dingemanse, 2007; Blench, 2009).

These communities have not made iron since about 1910; more surprisingly, the forges have also been silent since the 1940s. This may be related to the replacement of the cutlass (*'lewanfa'*), the smiths' principal product, by an imported equivalent (see Pole, 2008). Nevertheless, the traditions of how the smelters and smiths worked are often recounted, remains of furnaces dot the forested hillsides, and one or two disused forges provide shelter for goats. The notion of having made a significant contribution to an essential economic activity for hundreds of years is highly significant for the Mawu today, as recent contributors to the Ideophone website, responding to a post 'A visit to Akpafu by David Asante, 1887' have testified.

The first descriptions of these iron-working traditions published in Europe were those of German and British missionaries and colonial officers, starting in the 1860s (most importantly: Asante, 1889; Hupfeld, 1899; Rattray, 1916, Robertson 1921). In the 1960s, oral traditions began to be more carefully collected and studied (see Höftmann and Ayitevi, 1968; Darkoh, 1970). In 1971, researchers from the University of Ghana at Legon taped interviews with people at Lolobi (Van Hussen and Iddah, 1971); among other topics, these also covered iron-working. More recently, much attention has been devoted to the special linguistic features of the northern Volta Region (e.g. Kropp-Dakubu, 2006; Blench, 2009; Dingemanse, 2011). The published archaeological evidence for iron smelting in southern Ghana suggests that the techniques may be understood as a number of variations within a more widespread tradition. The furnaces excavated at Cape Coast in 1964 by Penfold possess internal diameters within the range found in the Togo Hills (Penfold, 1971). The furnace excavated in 1938 at Achimota by Shaw is also commensurate with them, as far as can be ascertained (Shaw, 1969, pp.48-54). It contained a large piece of slag with embedded tuyères and charcoal. The tuyère bears a resemblance to the one found at Tsito, and recognised as an air-inlet pipe by Mawu informants at Todzi. Similar pipes were found by Penfold at Cape Coast, in a context which appears to suggest the existence of a shaft furnace there (Pole, 1973, pp.263-266).

These disparate accounts provide a more rounded narrative of all the aspects of iron-making in this region than any single report could offer. Between 1971 and 1973, I investigated iron-working processes as part of my work at the Ghana National Museum. The fieldwork was carried out in December 1971, May-August 1972, and September 1973; this work included interviews with several old men who claimed to have special knowledge of iron-working practices in Mawu, Lolobi, and Santrokofi villages. (A synthesis of these and other accounts

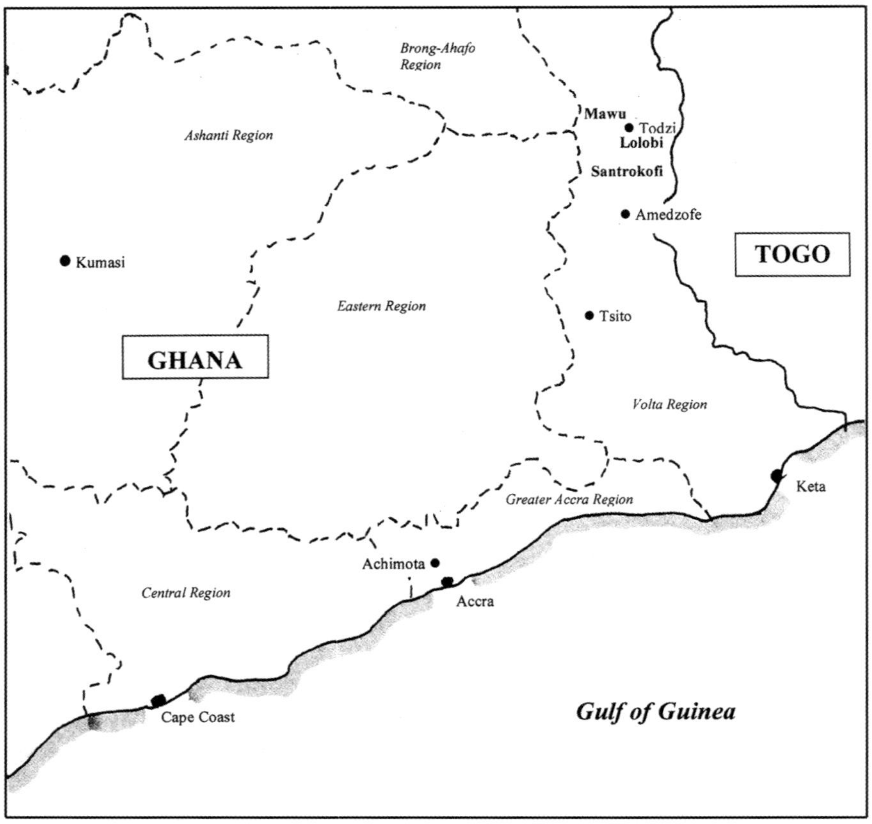

Figure 1: Map of Southern Ghana, showing sites and places mentioned in the text.

forms the basis of a longer paper, "The Hammers of Mawu" (Pole, 2010), from which some of the data quoted here has been taken.) Comparison of these descriptions draws attention to the discrepancies between the technology itself and oral traditions about the technology. This could be attributed to the imperfections of memory, but could also be due to a more deliberate change in purpose for these descriptions. The inappropriateness of the oral tradition as a comprehensive recipe for smelting iron was demonstrated in 1973, when a reconstruction of the process failed to produce the metal in the expected quantity.

The oral tradition in the 1970s

The iron industry in the Togo Hills was already in decline when the first reports were being published in the late 19th century. There were some individuals still alive in the early 1970s who could remember being involved in smelting and smithing activities in their youth. For example, Kwesi Nyarko (interviewed 26/09/1973), who said he was 90 years old, referred to a bank of six furnaces in operation behind the pastor's house to the north of the village when he was about nine years old. This was most likely between 1890 and 1900. Hupfeld reported that only one old man was practising the craft of smelting in Akpafu during his visit in 1898. In Lolobi, one furnace was still preserved but no longer in use (Hupfeld, 1899, p.188). In 1915, Rattray suggested that the furnace he saw "at the foot of the conical mountain at the base of which Akpafu lies" (Rattray, 1916, p.433) was seldom used at that time. It is quite likely that he never saw a smelting operation. In 1918, Robertson inspected a bank of six furnaces located near the mine shafts he visited north of Akpafu. According to his description, very little was left of these; an old man told him he remembered them being used when he was young, i.e., 1860-1870 (Robertson, 1921).

Nana Adibo II, the paramount chief of Mawu in the 1970s, stated that when the Germans arrived, they brought their iron tools with them (interview 19/12/1971), as a result of which the indigenous methods almost immediately became a thing of the past. Other informants reiterated this: the smelting process ended when the Germans came (Van Hussen and Iddah, 1971, tape III, p.11; see also Höftmann and Ayitevi, 1968, p.236). It is impossible to pin down the inception of German influence on technology to a year, but it would have started in the 1880s (Pole, 2010, p.73). This process was also taking place, at varying rates, in many parts of the continent at this time. It has even been the subject of an Akan folk-tale (Appiah, 1977, p.35). In Mawu, the Germans explained that much of the iron in the ore was being wasted, ending up in the slag. They may have been the first to introduce cheaper iron tools of European manufacture into the area, but English steel was also being imported into the area from Keta (Hupfeld, 1899, p.188). Nyarko also mentioned the increasing imposition of European institutions in what became a German colony, particularly in the field of education. His younger brother went from Todzi to the Bremen Mission School in Amedzofe in the early 1890s.

There was an additional reason for giving up iron smelting at this time: the cash attraction of growing cocoa. This crop had been introduced into the neighbouring Gold Coast Protectorate some years earlier, and was already being referred to as "that remunerative industry" in 1888; it reached the German colony by the early 1890s (Dickson, 1969, p.166).

The reconstruction of a smelting operation in 1973

Furnaces were usually built on land owned communally by the entire village, but the building operation was the responsibility of one clan, who held the rights to use the furnace. Individuals within the clan could use the furnace, and paid for the help they received in operating it with food and palm-wine (interview 28/09/1973). It appears that each of the six Mawu clans had their own furnace; Robertson reported seeing a bank of six furnaces, and six furnaces were also mentioned by Kwesi Nyarko.

In 1971-1973, I was researching the furnace-making and smelting processes. At the instigation of Nana Adibo II, a furnace was built by some of the villagers of Akpafu Todzi in 1972 (Fig. 2), at the edge of the village. It was used for a smelting re-enactment in September 1973, after a suitable drying-out time had elapsed. My description of this process was enriched by verbal accounts obtained from Lolobi Kumasi, and also from Todzi and Santrokofi (Pole, 2010, pp.55, 62).

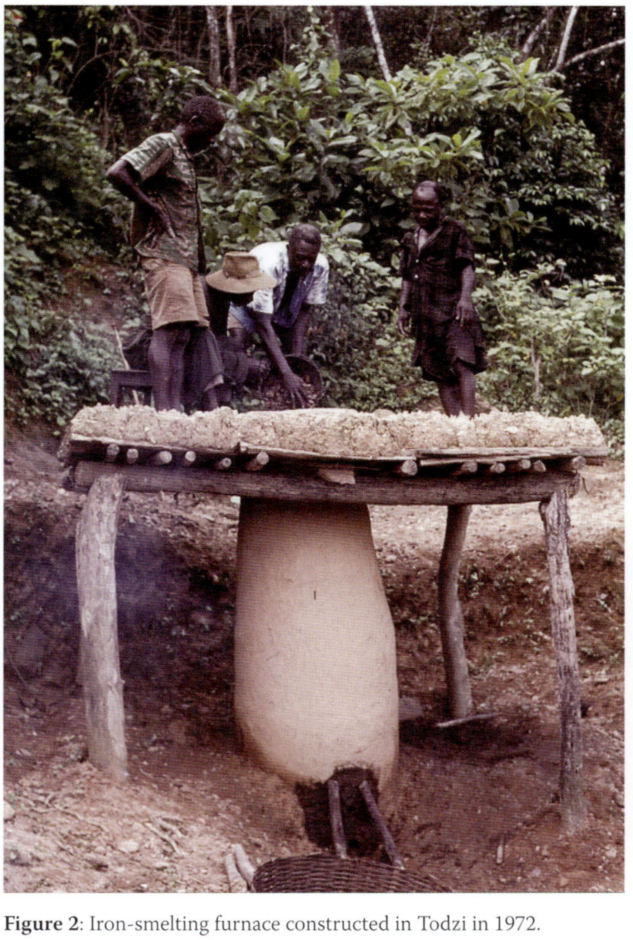

Figure 2: Iron-smelting furnace constructed in Todzi in 1972.

The smelting reconstruction began on 26 September 1973 with a libation to the ancestors asking for their help in the work of the day. It was poured by Benaze Asemoah, a knowledgeable elder who had been instructive in advising on the furnace construction. A large hole in the lower front of the shaft (*'irɛ'* in Siwu, the language of Mawu) was plugged with a mixture of sand and clay taken from an old furnace. Two holes were made in this plug; a lower one for the outlet of slag, and an upper one for the inlet of air (Fig. 3). There were also three other inlet holes spaced equally around the lower part of the shaft.

The furnace was charged with four sackfuls of charcoal. No fire was lit underneath the charcoal. A basketful of ore was placed on top of this, weighing 15.5 kilograms, followed by another sackful of charcoal. The charcoal weighed 92 kilograms in all. Then, a small amount of lit charcoal was put on top of the charge to start the fire. In his report, Rattray says that the shaft was filled with charcoal and the fire lit at the *'irɛ'*, which was only then blocked up with clay, except for a hole left for the slag to run out (Rattray, 1916, p.433). Agbodza at Santrokofi says (interview 04/04/1972) that "half-burned wood was put in first, then charcoal up to a foot [30cm] below the top of the shaft", which was topped with ore in egg-sized pieces. There can be little doubt in comparing all these versions that the workers at Todzi in September 1973 were operationally at fault for not putting fire at the base of the shaft.

By 10.00 am, the workers decided that no more need be done once the fire was well alight. The following morning, after 22 hours of smelting during which time the workers had inspected the progress of the smelt from time to time, the plug of clay at the *'irɛ'* was knocked away; it revealed a thin wall of black slag. Behind this was more slag-like material which the workers called *'ibora'*: iron. The *'ibora'* weighed 3.1 kilograms, but contained negligible amounts of metallic iron when closely examined. (A piece produced at the Iron Festival in 2008 was described as iron, made during the 1973 demonstration (Dingemanse, personal communication, 24 September 2008). It is too light to contain much iron.) A pyrometer and thermocouple had been used to measure temperatures inside the furnace at the level of the air inlets. The maximum temperatures reached were in the range 810-910°C between 4.00pm and 6.00pm, 6.5-8.5 hours after the smelt started. These temperatures are not sufficient to smelt iron.

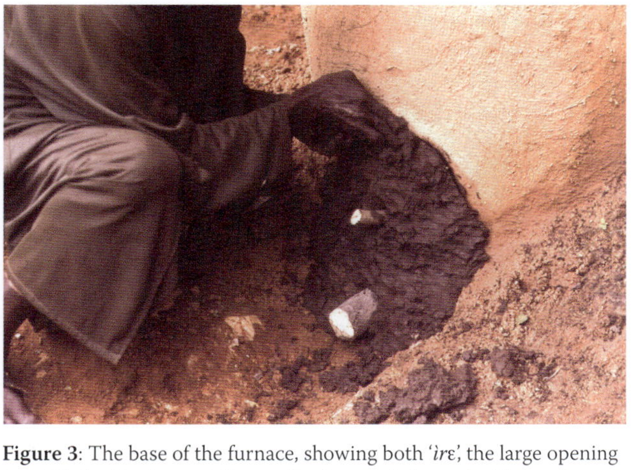

Figure 3: The base of the furnace, showing both *'irɛ'*, the large opening here filled with clay, and the smaller *'kanomi'* plugged with wood.

The reasons given by the elders for the failure of the smelt did not focus on the technology itself. Benaze Asemoah said the furnace had been incorrectly placed, since it could be seen by everybody in the village. At the time when smelting was a normal part of the life of the village, no women would have been allowed to be associated with smelting, or even to have looked upon the furnace. This was one of the reasons why the furnaces were sited away from the village, as was mentioned to me on more than one occasion. Benaze said that since this precaution had not been taken in 1972, the smelt had failed. Part of the reason for Nana Adibo suggesting that the furnace be sited close to the village was so that it could contribute to his project of encouraging tourism to the area, with a focus on the uniqueness of their iron-working traditions. From a modern scientific perspective, the failure of the reconstruction to produce meaningful amounts of metallic iron in 1973 was due to a number of factors: initially, the inadequate preparation of ore and charcoal, a lack of awareness of the need to control the proportion of fuel to ore in the charge, and, above all, an incomplete understanding of the combustion process. It is hard to avoid the conclusion that this last crucial factor would not have made its fatal contribution had the iron-workers still been practising blacksmiths at this time.

The evolution of tradition

It is instructive to compare some elements of the procedure undertaken in 1971-1973 with evidence from earlier descriptions. One difference that becomes immediately apparent regards the preparation of the raw materials. The quality of the ore was inferior to that which was used formerly. The iron content of the ore analysed by Hupfeld at the end of the 19th century was about 55%, whereas a number of samples examined in the early 1970s had an iron content in the region of 40%. Also, Hupfeld describes the pieces of prepared ore as being 'nut-sized'. The pieces I saw in 1973 ranged in size from small hen's eggs to a few millimetres across. Comparing the way the shaft was filled with ore and fuel in 1973 to Hupfeld's description of the same activity in 1899 (Hupfeld, 1899, p.185), there would appear to have been gaps in the elders' knowledge of the need to carefully control this aspect of the work. This also applied most crucially to the failure to light the fuel at the base of the charge, which prevented the necessary temperatures from being reached in the right place in the shaft, or for being sustained for a sufficient length of time. The workers may have remembered their blacksmith ancestors igniting the gases at the top of the shaft, but were not aware of the necessity of putting fire in the shaft before charging it with fuel and ore.

Igniting the gases at the top of the shaft (if they did not ignite spontaneously, as in Hupfeld's account) is an essential action for two reasons. First, it increases the draught created by the hot gases, therefore allowing the temperatures to attain their optimum level. Second, it burns off the excess of the highly poisonous carbon monoxide. This gas is the essential reducing agent of the iron smelting process. There was also a lack of care in making sure that the air inlet holes were kept clear of slag obstructions; they were not regularly inspected to assess the temperature inside the furnace, to ascertain that it remained within appropriate limits. Paradoxically, there appears to be a contrast between these operational lapses and the detailed attention applied to constructing the shaft. The initial preparation of the base, including the incorporation of a twisted vine, said to impart strength to the structure; burying white palm nuts under each ventilation hole; building the shaft in layers, each left for a fortnight to dry out, so that the construction took place over a period of three months; all of these are examples of such attention to detail. I will return to this apparent paradox below.

During the building process in 1972, there were several points at which disagreements arose concerning the structure of the furnace. The first concerned the shape and size of the '*ìrɛ*', the hole at the base of the front of the shaft. Some said there was one hole, others that there were two smaller holes. As has been indicated, both descriptions were correct at different times in the smelting operation. Before the smelting operation begins there is one hole, but during the smelt itself there are two holes, both referred to as '*kanomi*'. At the end of the smelt, when the bloom is to be removed, there is one larger hole again. It appears to have been this change in configuration over time which was responsible for the disagreement, one informant having a memory of the furnace at one stage, others at another.

Another point of contention was the thickness of the walls. One faction among the elders maintained that the shaft wall, as it was being built, should have been reinforced with an outer layer of clay. Benaze Asemoah said that there were occasions when that was done, but, if so, the outer layer was built at the same time as the inner one. At the stage when the height of the shaft wall was about 80cm, an outer wall was begun at the suggestion of the first faction (Fig. 4).

This was later removed, because it was agreed that the outer layer would crack as it dried, and shrink against the already dry surface of the inner wall. Photographic evidence indicates that the double thickness type of furnace existed in the Mawu area (Hupfeld, 1899, p.187; Fisch, 1911, p.29; Rattray, 1916, pp.21, 24). The photograph from the Bremen archives (Fig. 5) and the one in Fisch (1911) appear to be of the same furnace.

Evidence from the inspection of the furnace remains at Lolobi shows that the single thickness furnace was also used (see also a drawing in Hupfeld, 1899, p.185). Here, once again, we are led to the conclusion that apparently conflicting versions of a tradition can both be true, depending not just on the exact locus of the tradition, but on the content of individual experience upon which the informant draws for his expression of this tradition. (It is an exclusively male occupation in Mawu, though in northern Ghana elements of the iron-working procedures were more accurately recalled in some places by women who had been involved in them; see Herbert, 1993, p.29.)

Other accounts of furnace building illustrate these points of difference. In December 1971, a description of the structure of the furnace was given to me by Agyeman Bedu and Moses Adum, from Todzi. The internal chamber was said to have consisted of two parts: an upper one with holes in it for ventilation purposes, then a constriction,

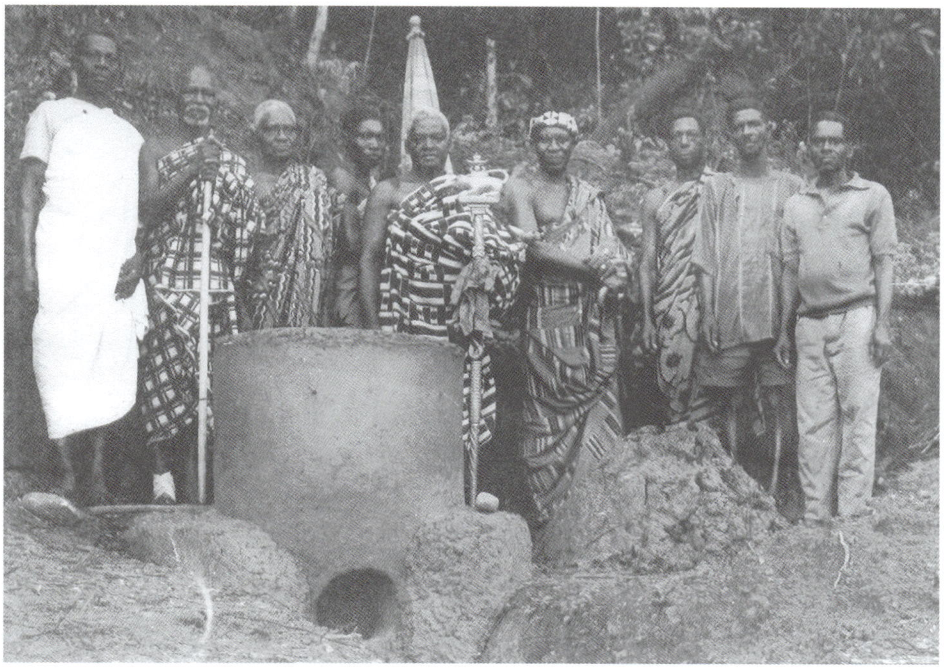

Figure 4: The outer wall temporarily in place. The individuals in this photo, from left to right, are: 1) Joseph Asema; 2) Ebenezer Asamoah; 3) Agyeman Bedu; 4) name not recorded; 5) Ofori (the linguist of Nana Adibo); 6) Nana Adibo; 7) Nana Akuamoah (full name probably Emanuel Kwabena Akuamoah); 8) Gottfried Kojo Buame; 9) Komla Akuamoah. I am grateful to Mark Dingemanse for his help in identifying some of these men.

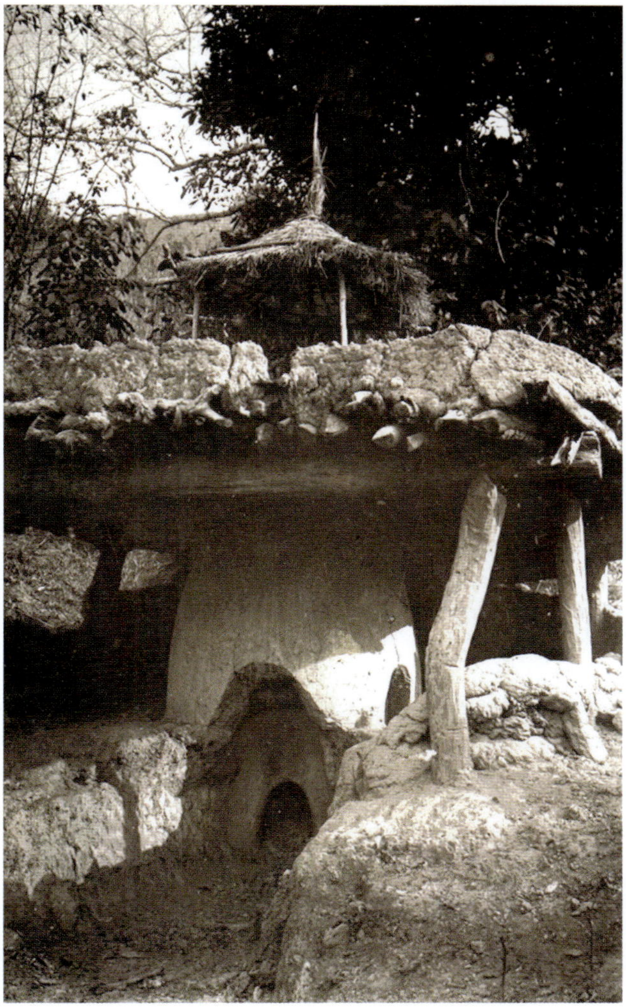

Figure 5: A Mawu furnace in 1910, image courtesy of the Bremen Mission archives.

below which was a lower chamber. This account had the appearance of being a description given by men who had seen the furnace from afar, but had never participated in building one or in smelting. They may have assumed that the charging platform extended into the shaft as a constriction, with a chamber above and below. However, this account also serves to underline points of identity with other descriptions, for instance in their common description of the use of white palm nuts as an element placed below the inlets at the start of the construction process.

The present-day role of Mawu iron-working traditions

Re-evaluating the role of reconstructions after an interval of 35 years, I have a rather more sceptical attitude to this form of experimental research than I had in the early 1970s. (I have explored this in Pole 2012). These reconstructions were useful as an investigative tool in helping to fill in gaps in the operational account of the processes; details that had not been recorded at the time or had faded from the memory of those striving to recall details of the process after a period of over 60 years. They also served to confirm elements in the descriptions remembered by the erstwhile specialists. However, most reconstructions, either in a point of detail or in their entire organisation, paid insufficient attention to the stresses that such activities were likely to create in the corporate stability of the communities into which they were inserted. Peter Schmidt has described this in some detail as he experienced it in Buhaya, north-west Tanzania (Schmidt, 1997).

What is of greater relevance to today's Mawu is the overall vision provided by the oral narrative, rather than a necessarily pedantic process of reconstruction, which inevitably and rightly concentrates on operational details. I think it is illuminating in this context to look at the situation from the old iron-workers' viewpoint: having been shown one or two things about smelting, but never seen or experienced the work over any length of time, how impertinent must it have seemed to be interrogated over 70 years later, and then told to supervise all the details of the process as part of the practical reconstruction, even bits that one could never have witnessed, nor were allowed to know about? How could one be aware after such a long time which parts of what one remembered were crucial and which superfluous? Some elements, like shaft construction, may be clearly recalled, perhaps because they relate to other activities still performed until quite recently, such as compound house or granary building; however, other details are very vague. Moreover, how easy would it be to admit to areas of vagueness when one is supposed to be the repository of ancient knowledge? Perhaps a detailed reconstruction of such a complex technology unperformed for such a length of time is not what men in their 80s should be asked to do, even if they, with aged enthusiasm to relive the skills of their youth, needed little encouragement. Of course they did know many details of the processes, but not all: besides, the keepers of tradition are seldom required to regurgitate it in the form of a practical performance; they are simply required to know better than others in the community what had been done.

Nevertheless, reconstructions of iron smelting have had some worth, both as revivals and as re-enactments. The former occurred as direct contributions to the supply of iron when other sources dried up, as in the region of Sukur in the Mandara mountains of northern Cameroon in the 1950s and early 1960s, following the reduction in availability of imported iron due to World War II (Sassoon, 1964). The latter have contributed, albeit peripherally, to the development of tourism and the heritage industry in more recent times, as in the example described by Gardi (1969, p.22), also from northern Cameroon, and potentially in Mawu itself, as I have mentioned above.

It is hardly necessary to emphasise that oral tradition, by its very nature, is selective in the elements it picks out. It is a re-interpretation of these events to suit the needs of the present community. For the Mawu today, it is neither practically possible nor economically viable to actually make iron – although the remaining ore may be exploitable in these days of increasing raw material values. What is important to the elders of the present community is to be able to demonstrate its independence and viability as a cultural unit. This is in the face of massive influence from the Asante confederacy of nations over the past three centuries, as well as an increasing Ewe presence moving up from the south in the 19th and 20th centuries, more recently encouraged by the emphasis on regional government in the past few decades. As far as the function of the stories about iron-working in the maintenance of this sense of independence is concerned, the description of technological details does not have to be comprehensive or complete in every detail. The esoteric character of the knowledge and its identification with the ancestors are the essential points in supporting the tradition, not necessarily its technological accuracy. This is one reason why, for instance, the details about the ring of vine stems being laid below the base of the shaft and the white palm nuts being inserted below the inlet holes are retained.

This aspect may also go some way to explaining why other equally esoteric details, for instance about the preparation of the ore and the size of the pieces of fuel, have disappeared from the record. There is less in the way of paradox here than might at first appear. The latter elements are of more purely operational significance, less connected to an active ritual belief system. The important thing to keep in mind is that this is not simply a matter of what the present-day elders had themselves seen. It is highly unlikely that even the oldest of them actually witnessed a furnace being built; if there were only half a dozen left in the 1890s, none would have been built after about 1880, even though they may have been used later. It is more likely that they would have seen smelting operations, rather than furnace building, in their youth. Details about the size of particles and the importance of lighting the exhaust gases have no relevance in any other context than smelting, but information about the placement of white palm nuts, stones, and vine stems can be understood by reference to other contexts, in which such items retain a meaning. They are still part of the community's ritual vocabulary. They are more important to retain in the oral tradition than the more technical and utilitarian parts of the process because they emphasise the connection with supernatural agency.

Furthermore, demonstrating in detail how the production of iron was achieved in order to develop a tourist attraction, as was suggested by Nana Adibo in the 1970s, has now been implemented with the establishment of a *Katere Iyi* (Iron) festival in February 2008 (Fig. 6). Apart from its potential in generating income, this festival has become a regular focus for civic pride for all Mawu (the latest festival was held in February 2011). It is also a forum for re-invigorating their cultural identity in the face of the increasingly dominant presence of Akan and Ewe communities in the region.

However, it is crucial for those organising and promoting these re-enactments to realise that it is not sufficient to rely on the hearsay memories of members of former iron-working families, because that is not what the retention of their knowledge is for: it has been transformed into a narrative to illustrate a past tradition, not a blueprint for rebuilding it.

It is enlightening to know that, at the present time, the reconstruction undertaken in 1973 is regarded as a success by the Mawu who are aware of it (Dingemanse, personal communication, 8 February 2011). It may be that part of the reason for this belief is the regard for the product of the smelting operation, the '*ibora*': this is an artefact which does contain metallic iron, even if not in the quantity that would have been expected when smelting was a regular activity. However, it seems likely to me that the more salient and appropriate legacy of the reconstruction, embodied in the memory of it as successful, is the respect which is paid to the past and present Mawu individuals who took part in it. The event took place at a time when many aspects of the process could be recalled and formed the basis of the reconstruction, and in this sense it was clearly a success. Had it been delayed for even a few years, it is doubtful whether sufficient knowledge

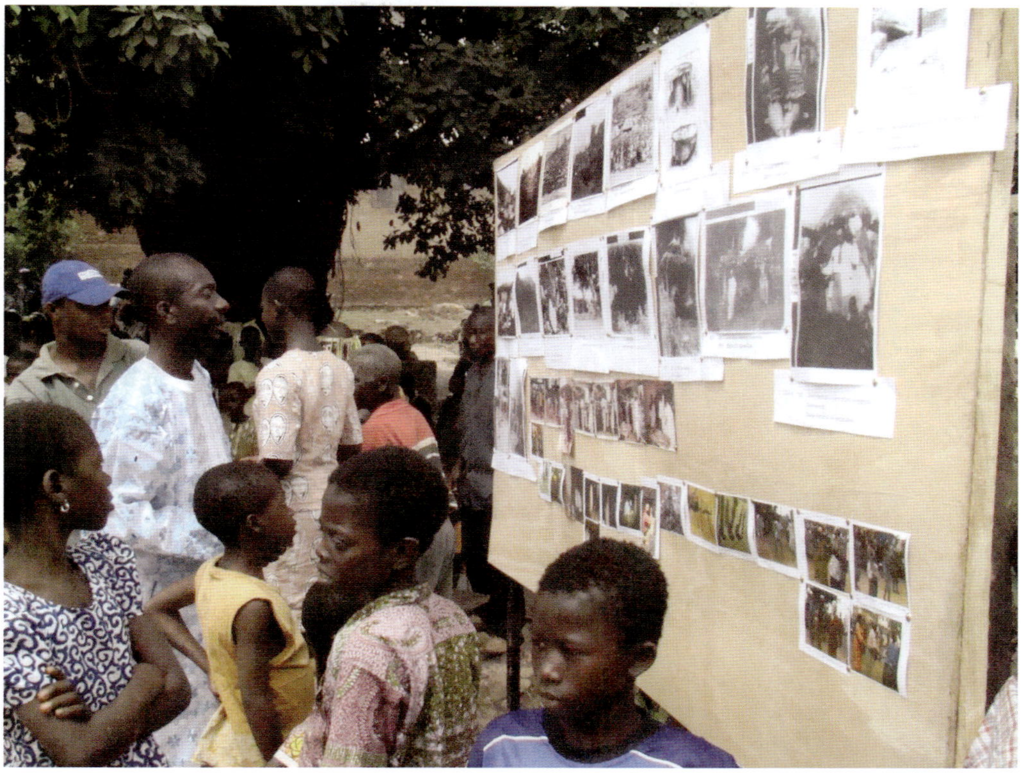

Figure 6: The iron festival held in Todzi in 2008, image courtesy of Mark Dingemanse.

and expertise would have been available to make re-enacting any part of the smelting procedure viable. As it is, the 1973 reconstruction stands as a testament to the longevity of that knowledge in the face of fundamentally altered economic circumstances covering a period of at least 60 years.

Conclusion

It is clear that the traditional oral narrative of iron-working in Mawu does not itself serve as an archive, let alone a detailed technical account of the processes. The narrative has been reworked to serve new purposes: to confirm the specialists within the community as custodians of an esoteric body of knowledge about iron working, and, faced with the presence of other, more dominant communities within the region, to give the Mawu and Lolobi the means to continue to present themselves as distinctive, viable, and independent cultural units.

Although iron is no longer produced by the Mawu or Lolobi, the notion of being important for having done so is significant for them today. Expertise in iron-working is not required to be practically effective in producing iron tools; but what the tradition must still affirm is an account of the processes involved which is technically coherent and compelling for its audience. Specialist expertise inherent in being an iron-worker must continue to generate respect, if not iron itself, for any future re-enactments to be effective in serving the needs of these 21st century Ghanaian communities. The strength of the oral tradition has persisted sufficiently to provide the basis for a revival of interest in recent years, culminating in the Iron Festivals taking place in successive years.

It is now the responsibility of all of these individuals, not only the keepers of the tradition, but also the historians, archaeologists, and ethnographers, to gather their records together with the intention of forming a more comprehensive archive. All the communities will be then in a position to benefit from knowing about this ancient and crucial process of making iron; not just the families of those who used to operate and control the technology, but all those who benefited from its products, throughout what is now southern Ghana and beyond.

Acknowledgements

In the need to take account of work published in recent years, I have benefited from discussions with Mark Dingemanse and Kodzo Gavua. I have also found the Ideophone website a constant source for updating information and contact with Mawu (http://ideophone.org).

References

Appiah, P., 1977. *'Why the Hyena Does Not Care for Fish' and Other Tales from Ashanti Gold Weights*. London: Deutsch.

Asante, D., 1889. Eine Reise in den Hinterländern von Togo. *Geographische Gesellschaft für Thüringen zu Jena*, 7/8, pp.106-33. (An English translation of this is now available on http://ideophone.org)

Blench, R., 2009, Ghana-Togo Mountain Languages Homepage. at <http://www.rogerblench.info/Language%20data/Niger-Congo/GTML%20Website/GTMLopeningpage.htm

Darkoh, M.B.K., 1970. Note on the peopling of the Forest Hills of Volta Region of Ghana. *Ghana Notes and Queries*, 11, pp.8-13.

Dickson, K.B., 1969. *A Historical Geography of Ghana*. Cambridge: Cambridge University Press.

Dingemanse, M., 2007. Gender-based folk etymologies for the name Akpafu. *Ideophone website* <http://ideophone.org/two-folk-etymologies-for-the-name-akpafu>.

Dingemanse, M., 2011. *The Meaning and Use of Ideophones in Siwu*. PhD dissertation.

Nijmegen: Max Planck Institute for Psycholinguistics/Radboud University.

Fisch, R., 1911. *Nord Togo und seine westliche Nachbarschaft*. Basel: Basel Mission.

Gardi, R., 1969. The "Iron-cookers". In *African Crafts and Craftsmen*. pp.15-45. New York: Van Nostrand Reinhold.

Herbert, E.W., 1993. *Iron, Gender, and Power: Rituals of Transformation in African Societies*. Bloomington: Indiana University Press.

Höftmann, H. and Ayitevi, J.K., 1968. Oral traditions of some of the so-called remnant people in the Central Volta Region of Ghana. *Mitteilungen des Institut für Orientsforschung*, 12, pp.199-219.

Hupfeld, F., 1899. Die Eisenindustrie in Togo. *Mitteilungen von Forschungsreisenden und Gelehrten aus den Deutsche Schutsgebieten*, 12, pp.175-94.

Kropp-Dakubu, M.E., 2006. Pushing back linguistic time in the Trans-Volta: movement, assimilation and loss. Paper presented at the *Workshop on oral traditions, Ho, Ghana, July 24-26 2006.*

Penfold, D.A., 1971, Excavation of an iron-smelting site at Cape Coast. *Transactions of the Historical Society of Ghana*, 12, pp.1-15.

Pole, L.M., 1973. Excavation of an iron-smelting site at Cape Coast: Two Comments. *Transactions of the Historical Society of Ghana*, 14, pp.263-6.

Pole, L.M., 2008. Steeling Themselves: Effects of External Trade on West African Ironworkers. *Journal of Museum Ethnography*, 20, pp.17-32.

Pole, L.M., 2010. The Hammers of Mawu: Iron-working Traditions in the Togo Hills. *African Archaeological Review*, 27, pp.43-78.

Pole, L.M., 2012. Errors in Translation: the uses of reconstructions in ethnographic fieldwork, in Dudley, S. H. et al (eds.) *Narrating objects, collecting stories. Essays in honour of Professor Susan M. Pearce*, pp. 203-220. London, Routledge.

Rattray, R.S., 1916. The Iron workers of Akpafu. *Journal of the Royal Anthropological Institute*, 46, pp.431-5.

Robertson, T., 1921. *Report on the Geology of Western Togoland*. Accra: Gold Coast Geological Survey.

Sassoon, H., 1964. Iron-smelting in the Hill Village of Sukur, North-Eastern Nigeria. *Man*, 64(215), pp.174-8.

Schmidt, P.R., 1997. *Iron Technology in East Africa*. Bloomington: Indiana University Press.

Shaw, C.T., 1969. An iron-smelting furnace excavated in Ghana. *Bulletin of the Historical Metallurgy Group*, 3, pp.48-54.

Van Hussen, J. and Iddah, R., 1971. *Iron smelting and smithing in Lolobi*. Legon, Ghana: University of Ghana (typed transcript, dated February 1974, of taped interviews undertaken on and before 5th April 1971)

Author's address

- Independent curatorial consultant 24 Beeches Close, Saffron Walden, Essex, CB11 4BT UK
  www.lenpole.com

Indian sub-continent

Indian iron and steel, with special reference to southern India

Sharada Srinivasan

The Indian subcontinent has had a vibrant and longstanding tradition of iron and steel working. The famed Delhi iron pillar, with a Gupta era inscription (late 4th century to early 5th century AD), stands testimony to the skills of early Indian iron smiths and is the earliest surviving large-scale iron forged work in the world, aptly described as the 'rustless wonder' due to its resistance to corrosion (Anantharaman 1997). It remains one of the best known cultural emblems of the Indian subcontinent, as well as being one of the earliest ironworks to have been studied from an archaeometallurgical or scientific perspective: Robert Hadfield's initial 1912 examination indicated it to be high-purity iron (one of the earliest identifications of high carbon steel was reported by Marshall in 1951, of three sword blades with 1.2-1.7% C from Taxila, in modern day Pakistan (ca 5th c. BC-1st c. AD)).

Another celebrated object of study in the region has been the Indian wootz steel: a high-grade, high-carbon steel known from travellers' accounts to have been made in the southern part of India from the 16th-17th centuries onwards, and widely traded to make the fabled Damascus sword blades. This particular material has been described as one of the four outstanding metallurgical achievements of antiquity Smith (1963). As pointed out by Smith (1960), the Oriental Damascus steel blade, believed to have been forged of Indian wootz steel of a high carbon content of 1.5-2%, with a typical 'watered silk' pattern, was one of the first to have been studied under the microscope. The properties of wootz, or Damascus steel, were enthusiastically investigated and debated during the 19th and 20th centuries by several European scientists, including Faraday. These studies led to certain significant innovations and discoveries such as the development of alloy steels (Srinivasan and Ranganathan 2004).

There has been a rich body of writing by European travellers related to the observance of steel-making practices by crucible processes. For example, Voysey (1832) documented steel producers in parts of Andhra Pradesh, falling under the Nizam's dominions and corresponding to the districts of Telangana while Buchanan (1807), travelling in the provinces of Mysore after the fall of Tipu Sultan, also noted steel production by local communities. Ananda Coomarawamy (1956) was able to observe the making of crucible steel in the Kandyan region of Sri Lanka as recently as the 1950's.

The Indian subcontinent provides a thriving ethnographic record of blacksmithing practices and surviving evidence of iron smelting activities; primarily via the Agarias communities, although these communities are increasing marginalized.

In the past century new insights have emerged concerning the diverse contributions of the Indian subcontinent to metallurgy. For example, active programs of archaeometallurgical research in Sri Lanka have pointed to the unusual use of wind-powered linear furnaces for iron smelting (Juleff 1996). Although more systematic archaeometallurgical and archaeological research is warranted in many parts of the Indian subcontinent in order to achieve a clearer picture of the development of iron and steel, recent finds and studies have thrown promising new light on several aspects.

In the iron smelting 'bloomery process' a solid state iron bloom was produced, whereby the reduction of ore to iron metal took place at a temperature sufficient to reduce the ore, below the melting point of iron. Bloomery iron has a low carbon content, below about 0.04%, which produces wrought iron after the smithing process. However, in the blast furnace technology, which the Chinese mastered during the latter part of the first millennium BC (Wagner 1999) higher temperatures and more reducing conditions were reached, resulting in the formation of cast iron with higher carbon contents (up to 2-4%) – although this is a rather brittle product. Crucible steel has an intermediary composition between that of wrought iron and cast iron, of about 1-2% carbon. This characterises it as an ultra-high-carbon steel with properties of high ductility for forging and, in comparison to cast iron, high impact strength and reduced brittleness, making it highly suitable for weapons.

An interesting aspect of the research relating to iron and steel from the Indian subcontinent has been the question of the reported production of wootz, or higher carbon crucible steel, particularly from parts of southern India. Also crucial to the overall understanding of the development of ferrous metallurgy in South Asia is the evidence for a diversity of production mechanisms; best exemplified by studies from Sri Lanka establishing the presence of a novel technique of wind powered iron smelting using linear furnaces, which is datable from around the mid to late first millennium BCE (Juleff 1998). Experimental reconstructions of such wind-powered techniques demonstrated that such furnaces were able to produce iron of different grades, i.e. from lower to higher carbon steels. As noted by Juleff at the 2009 *World of Iron* conference, this invites analogies to the production of iron and steel by the Tatara process of Japan and, possibly, the movement of technologies from South Asia and Sri Lanka further east (a theory perhaps concurrent with the spread of Buddhism). It may be noted however that in comparison to India, where there are a growing number of reports of early finds from scattered contexts for higher carbon iron alloys, there seem to be fewer such reports coming out of Sri Lanka.

One of the issues with archaeometallurgy in this region that is not clearly resolved concerns the dates of the emergence of iron in South Asia and whether it is possible to attribute this emergence to independent origins or to the diffusion of iron metallurgy. The Iron Age is generally thought to have occurred in different parts of the subcontinent towards the end of the Late Harappan or Chalcolithic phases, which are represented by copper-bronze finds. Conventionally speaking, the main Iron Age cultures have included – in northern India – the PGW culture (Painted Grey Ware) (c. 1100 to 350 BCE) and the NBPW culture (Northern Black Polished Ware) (700-200 BCE). There has however been growing interest in re-examining the early development of iron metallurgy in south Asia. For example, Chakrabarti (1992) has explored the case for the indigenous development of iron metallurgy in geographical regions including northern and southern India. There have been more recent finds from numerous parts of the subcontinent which indicate that iron production is dated to 1300-1200 BCE or earlier. Tewari (2010), citing dates from iron producing sites such as Malhar in Uttar Pradesh and Dadupur in Sai Valley of Lucknow district of Uttar Pradesh, argues that the dates for the emergence of iron in the Gangetic Valley and the north-western Indian subcontinent could be situated in the earlier part of the second millennium BCE.

Tripathi (this volume) argues the case for placing initial iron working in the Indian subcontinent to the first half of the second millennium BCE, owing to a growing number of carbon dates from sites dating to this time. There are, however, often problems with the reliability of Indian radiocarbon dates due to the occasional limited availability of dates for each site (as pointed out by Glover in discussions during the Indian subcontinent session of the World Archaeology Conference). Nevertheless, the growing body of early dates from individual sites have increased significantly and point to the potential of undertaking a system of dating which could throw more conclusive light on the matter.

Tripathi (this volume) highlights the potential that the Middle Ganga region and eastern Vindhyas, with proximity to the mineral rich Vindhya-Kaimur hills, represent a promising region for exploring the early development of iron technology. The paper also distinguishes between the early iron, middle iron and late iron ages; pointing out that in the Middle Ganga Plain further east, iron appears first in a largely Chalcolithic milieu associated with pottery of the Black and Red Ware (BRW). The most intriguing sites are Malhar, Raja Nal-Ka-Tila and Koldihwa, which have yielded iron production dates as early as 1600-1500 BCE. The occurrence of iron in a microlithic setting is also pointed to in parts of the Vindhyas such as the districts of Mirzapur, Chandauli and Sonbhadra, at sites such as Morahana Pahar and Baghair Khoh: the region where traditional iron smelting communities such as the Asurs and the Agarias have thrived for generations. Other significant early sites touched upon by Tripathi include Noh in eastern Rajasthan and Atanjikhera in Uttar Pradesh, with associated painted grey ware pottery.

The Gangetic Valley area has yielded some early evidence for carburization, as mentioned by Balasubramaniam (2008: 247). In studies by Ghosh and Chattopadhyay, an iron sickle found at Barudih in Bihar was dated to 810 BC with 0.35% C, 0.2% Si, 0.2% S, 0.5% P and 0.04% Mn; showing the presence of 40% pearlite phase with some tempered martensite at the sharp edge.

In southern India too, there has been debate (notably Chakrabarti, 1992) regarding the indigenous origins of iron metallurgy. Calibrated radiocarbon dates for two iron specimens from Hallur in Karnataka in southern India were dated to 1385-1050 BCE and 1125-825 BE (Possehl and Rissman 1992). AMS dates from Hallur by Fuller et al. (2007) also support a chronology of 1200-1100 BCE. The notion of the existence of a long tradition of smithing and forging in southern India is also highlighted by Possehl and Gullapalli (1992). Studies by Agrawal (1985) indicated the early use of lamination of low carbon iron with higher carbon steels in artefacts from Tadakanahalli and Kumaranahalli (c 1200-1100 BCE). As Gullapalli (2009) makes clear, however: though it seems that there is more evidence of when iron appears in the archaeological record, the process of experimentation leading to that stage is yet to be clearly understood.

In the Vidarbha region of the Deccan in Maharashtra, excavations of locations including burial grounds and settlements yielded hundreds of iron artefacts – including an iron smelting furnace with tuyere at Naikund which has been dated to at least 700 BCE (Deo 1982). Micro-structural studies conducted by Deshpande et al. (2010 and presented as a poster at the *World of Iron* conference) on some of the artefacts revealed the presence of tempered martensite and pearlitic colonies (pearlitic steel is a technical term used to describe a steel that is fully pearlitic throughout the structure); suggesting that there had been some processing of the material which involved quenching, hardening and steeling.

In light of the debate regarding the dates of the South Indian Iron Age, it is significant to mention ongoing studies conducted on iron artefacts from the reasonably well-dated site of Kadebakele in Karnataka. This site, located on the northern bank of the Tungabhadra River in Koppal District of northern Karnataka, has been excavated by a collaborative team from the Karnataka Department of Archaeology, the University of Michigan and the University of Chicago; led by Morrison, Sinopoli and Gopal. The site, which covers more than 60 hectares, was occupied from at least the early first millennium BC into the early centuries AD. Carbon dates from the Iron Age or 'megalithic' period span from c. 800-400 BCE. Research at Kadebakele was carried out within a larger project, titled 'The Early Historic Landscapes of the Tungabhadra Corridor' (EHLTC), which sought to examine settlement and economy during this sequence through a program of excavation both in Kadebakele and of contemporary smaller settlements and mortuary sites, with a regional survey in a c. 35 square kilometer area straddling the Tungabhadra River.

Research in 2003 and 2005 focused on Kadebakele's Iron Age sequence, with excavations of habitation and ritual mortuary areas in the "upper terrace" – a well-defended zone on top of a high outcropping hill that rises some 110 metres above the river flood plain. Along with architectural remains and a variety of features, excavations recovered enormous quantities of ceramics and other artifacts, as well as faunal and botanical materials from well-documented stratified contexts – on whose study the author of this paper has been engaged, in collaboration with Ranganathan and the IISc and of which some preliminary findings have been reported in Srinivasan et al. (2009) and at the *World of Iron* conference. Of particular interest was the find of an a small iron ring (acc. no. 900, 22E-28 N, Level 7) discovered above a plaster surface, with charcoal from this level radiocarbon dated to 800–440 BCE.

The microstructure of the ring from Kadebakele (Fig. 1) consisted almost entirely of a pearlitic structure with a fine lamellar network, suggesting a composition close to the eutectoid composition of 0.8% carbon. The fact that this seemed to be pearlitic steel could be significant in terms of its possible identification as crucible steel or cast steel, rather than wrought iron carburized in the solid state by case hardening of forging alone. Given the confirmed dates of the layer from which the ring was excavated (800-440 BCE), this is potentially an early example of pearlitic, higher carbon steel. It is intriguing that Sasisekaran (2002) reported a ferrous artefact with a micro-structure akin to cast iron, uncovered from an Iron Age furnace from Guttur in Tamil Nadu.

Further studies must be undertaken before a connection can be made between sites of primary production of iron or steel. Studies on slag finds from the Iron Age smithing and forging sites of Bukkasagara and Ramapuram in the neighbouring area have been made by Johansen (2008), which may provide some evidence of differentiation between activities such as smithing and smelting.

As reported in this volume by Behera and Chattopadhyay there are other examples: the early historic Iron Age site of Badmal in Eastern India (see also Chattopadhyay and Behera 2007), demonstrated that, although on-site iron smelting activity was not clearly identified, there is compelling evidence for the use of medium carbon steel – demonstrated, for example, in a specimen reported to have 0.4-0.5% carbon and 0.5% phosphorus which was analysed by PIXE. A clear aspect of the quantitative analysis of early iron artefacts is that a notable number of them contain a significant amount of phosphorus – of which the best known example is the Delhi iron pillar, which has also been discussed in terms of 'phosphoric iron'.

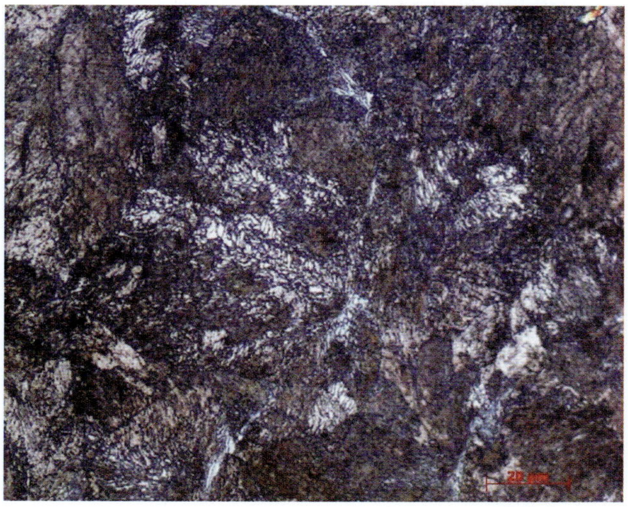

Figure 1. Micro-structure of ring fragment from Iron Age site of Kadebakele, Karnataka, c 800-440 BCE (given to Srinivasan for analysis by Morrison and Sinopoli and Karnataka State Department of Archaeology).

The Delhi iron pillar in the Qutb Minar complex, standing approximately 7.375 m high and 41.6 m in diameter and dated to the 4th century AD, is believed to have been made by cementing and forge-welding 'cakes' of wrought iron. Hadfield's report (1982) pointed to the presence of approximately 0.1% phosphorus. More recently, the Delhi pillar has been studied in some detail by Balasubramiam (2008), in relation to understanding the mechanisms behind its fabled non-corrosion. From his studies related to rust characterization, Balasubramiam postulated that the corrosion resistance was partly due to the higher phosphorus content and the formation of a protective passive film on the surface, and to the formation of an amorphous oxyhydroxide layer next to the metal-rust surface of the phosphoric iron. Another iron pillar, situated in Dhar in the Gangetic Valley, was also found to have a composition described as 'phosphoric iron'.

Whether the phosphoric nature of the pillars' composition was an intentional or an intrinsic aspect, linked to the nature of the iron smelting technology practiced, remains an area for further investigation. A poster presented in the World of Iron conference by Srinivasan and Balasubramaniam also pointed to the general presence of phosphorus in a range of ferrous metallurgical contexts, including in remnants of high carbon steel prills in crucibles for making wootz steel from Mel-siruvalur (Srinivasan, 1994).

A recent study by Raj et al. (2005) using in situ metallography on various parts of the Delhi iron pillar, indicated an heterogeneous structure which could be expected due to the forging of a ferrous cake showing micro-structures more clearly associated with a forged structure. Curiously however, the top of the pillar's ornamental capital and platform displayed characteristics approaching Widmanstatten structures, suggesting faster cooling rates and even some dendritic distribution akin to cast steel. This remains an aspect for further exploration. Fig. 2 shows the micro-structure from the middle portion of the iron pillar undertaken by Raj et al.

As previously mentioned, Marshall (1951: 534) reportedly discovered three blades with high carbon contents between 1.2-1.7% C from the Bhir mound of Taxila which were datable to the latter part of the first millennium BCE. Meanwhile historical texts (accounts of the late 5th century BC) indicate that the swords of Indian steel presented to the King of Persia were praised by the Greek physician Ctesia (Schoff 1915, Bronson 1986: 18). Craddock (1995: 279) pointed out that Greek alchemist Zosimos in the 3rd century AD mentioned that the Indians made steel for high quality swords by melting soft iron in crucibles. The term *wootz* itself seems to derive from the word for steel, *ukku*, which may be linked to the Tamil word, *uruku* (i.e. boiling). Such terms are prevalent in the writings of classic Tamil Sangam poetry of the Christian era (Srinivasan 1994).

From investigations made on steel-making crucibles conducted by the author from a mound in Mel-sirvular, in Tamil Nadu; there is strong archaeometallurgical evidence to suggest that steel was produced by a process of carburising wrought iron to hypereutectoid high-carbon steel, as described by Srinivasan (1994, 2007). Terracotta jars which resembled the broken legs of megalithic sarcofagi were identified in a canal next to the mound with crucible steel debris (ibid.), while megalithic activity around Mel-siruvalur was independently corroborated by Sasisekaran (2002). Crucibles were also found close to trenches with iron smelting slag and debris. The prills of metallic remnants trapped in the vitrified matrix of the crucible displayed a fairly uniform structure of ultra-high carbon steel of about 1.2% carbon, as seen in the pearlitic colonies in prior austenite grains and surrounded by cementite needles (Fig. 3). The process of carburisation of wrought iron to steel has also been described as the Mysore process at southern Indian sites such as Gatihosahalli (noted by Rao (1989), Craddock (1998) and described by Buchanan). Crucibles from undated surface sites identified by Srinivasan in Karnataka at Machnur and Tintini had a similar vitrified vessel fabric, of carbonaceous material packed with a quartz rich exterior (Srinivasan and Griffiths 1997)

Recent evidence suggests correlations between the literary and historical accounts that have pointed to the specialty

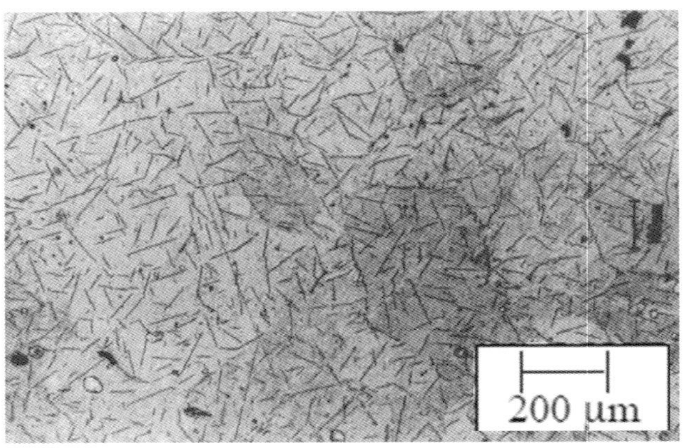

Figure 2: In situ metallograph from the lower middle portion of the Delhi iron pillar, undertaken by IGCAR (photo courtesy Baldev Raj).

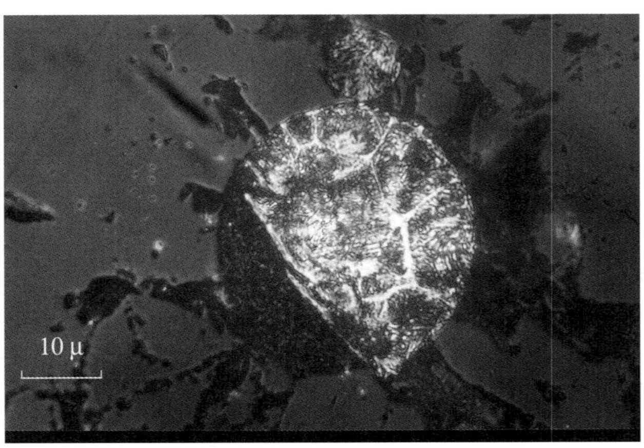

Figure 3: Micro-structure of high-carbon wootz steel prill from crucibles uncovered from Mel-siruvalur in Tamil Nadu by Srinivasan.

of steel from southern India and certain archaeological or archaeometallurgical evidence. Pliny's 'Natural History' makes prominent mention of the import of iron from the 'Seres', which some scholars have identified with the ancient Southern Indian kingdom of the Cheras (Schoff 1915). Investigations on the iron-rich megalithic and early historic sites of Tamil Nadu and the Malabar (mid first millenium BC to early centuries AD) are throwing new light on the early development of iron and steel.

Excavations by the Kerala Council for Historical Research (Cherian et al. 2007), near the village of Pattanam in Kerala on the south bank of the river Periyar, have yielded evidence identifying this area with the Indo-Roman trading port of the Muziris, or Muciri, on the Malabar coast; described in Roman accounts as the 'Periplus of the Erythrean Sea'. As reported by Srinivasan (2007) and as indicated in Fig. 4, microscopic studies made on an iron nail from Pattanam reveal a microstructure typically associated with ultra-high carbon *wootz* steel containing about 1.5% carbon. Such nails originated in the Sangam Chera dynasty and consequently may be identifiable with the Roman accounts of iron from the Seres (or *Chera*) kingdom. The bardic poems of the Tamil poetess Auvaiyar, of the Sangam era (3rd c. BCE-3rd c. AD), richly evoke the skirmishes between rival chieftains of that era. Excavations reported by Rajan (1990) at an iron age megalithic site at Kodumanal, Tamil Nadu (3rd c. BC) – close to Karur, the capital of the Chera kingdom of the Sangam era (3rd c. BCE-3rd c. AD) – have revealed furnaces stacked with vitrified crucibles and separate from an abundance of iron slag. A vitrified crucible fragment (Fig. 5) from the site was analysed (Srinivasan and Griffiths 1997, Srinivasan 2007) and found to have a similar morphology to the steel-making crucibles from Mel-siruvalur, with iron-rich areas in the matrix suggesting that it could have been used for ferrous metal processing.

Fig. 6 is an elemental distribution map for iron in the crucible fragment from Kodumanal, obtained from an x-ray area scan from a wavelength dispersive x-ray spectrometer (EPMA-WDS, using JOEL Superprobe JXA 8600 at 20 kv), indicating iron-rich areas. Sasisekaran (2004: 33) reported a micro-structure pertaining to a chisel from Kodumanal with an edge of higher carbon steel (about 1% carbon).

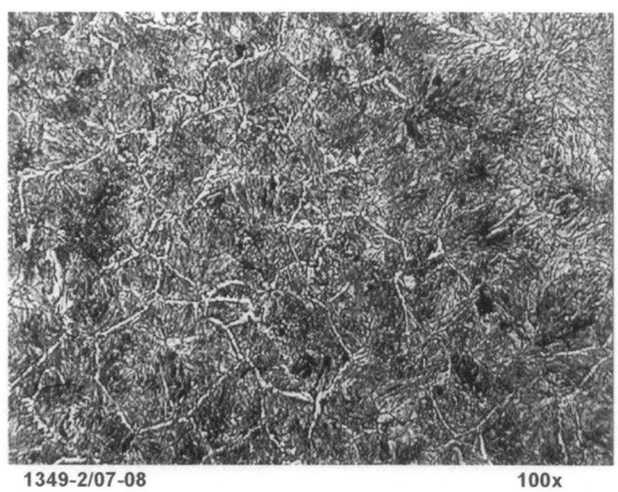

Figure 4: Micro-structure of nail from early historic Pattanam, Kerala showing structure of wootz steel (nail fragment given to Srinivasan by KCHR for analysis and investigated through GML).

While the above examples relate to the carburisation process of wrought iron, it seems that an alternative method of creating crucible steel was applied in southern India, as reported from parts of Andhra Pradesh such as Nizamabad. Here, steel is thought to have been made by the fusion of cast and wrought iron to achieve a steel of intermediate composition, as described by Voysey (1832). This is also known as the Deccani process, as mentioned by Lowe (1990). The fusing approach to making crucible steel has also been reported as a result of evidence gathered from the medieval site of Merv in Turkemenistan (Merkel et al. 1995, Rehren and Papachristou 2003).

Lowe's surveys in the region of Andhra Pradesh in the late 1980's resulted in the identification of at least fifteen sites of crucible steel production in the area of Konasamudram and Nizamabad, in modern day Andhra Pradesh. Although these crucible steel sites had not been dated through archaeological excavation, Lowe (1990) has pointed out that in the area, ironworkers traditionally smelted lateritic and magnetite ores in an iron smelting industry which dated back to the first millennium BC. Intriguing findings made by Lowe and Thomas indicated that the historical Deccani crucible fabric could be characterised in modern terms as a high-performance refractory known as a mullite fiber-reinforced ceramic composite (Lowe et al. 1990).

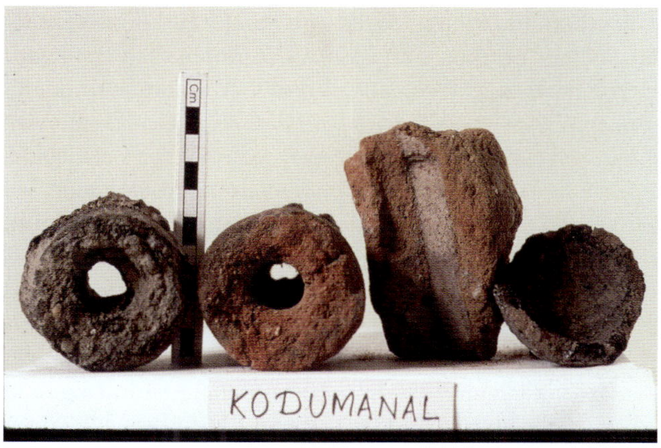

Figure 5: Vitrified crucible fragment found with iron smelting debris from Kodumanal, c 3rd century BCE (given to Srinivasan by K. Rajan, Tamil University Tanjavur).

The present day region of Telangana, which covers the Konasamudram and Nizamabad sites, represents one of the most dramatic landscapes in terms of the vast extent of debris related both to iron smelting and, significantly, to *wootz* production. The area constitutes part of the former provinces of Golconda, of which writers such as Tavernier have alluded to in their vivid descriptions of the trade of *wootz* ingots from this region route to West Asia and Persia.

The ethnographic record of Telengana bears witness to its long tradition of skilled metal processing. In recent years sites related to pre-modern iron and steel making in the region have been extensively documented by Jaikishan (2009), who was able to identify some eleven hundred iron smelting sites and approximately one hundred and eighty crucible steel production sites in the districts of Adilabad, Nizamabad and Karimnagar. Several aspects of ethnographic interest have been brought to light by Jaikishan's documentation, such as the correlation of site and village names with slag heaps or ferrous metalworking evidence, with further insights emerging through a 2010 survey in the region (undertaken through the aegis of UKIERI as a collaborative, funded project between NIAS and the University of Exeter, led by Juleff, Srinivasan and Ranganathan and with invited investigators Jaikishan and Gilmour). The survey involved an attempt to build up a database of sites using geospatial information and GSP recordings and the systematic recording and sampling of the different types of archaeometallurgical debris at different locations. One of the significant preliminary findings from the analysis, using EPMA-WDS methods at Cambourne

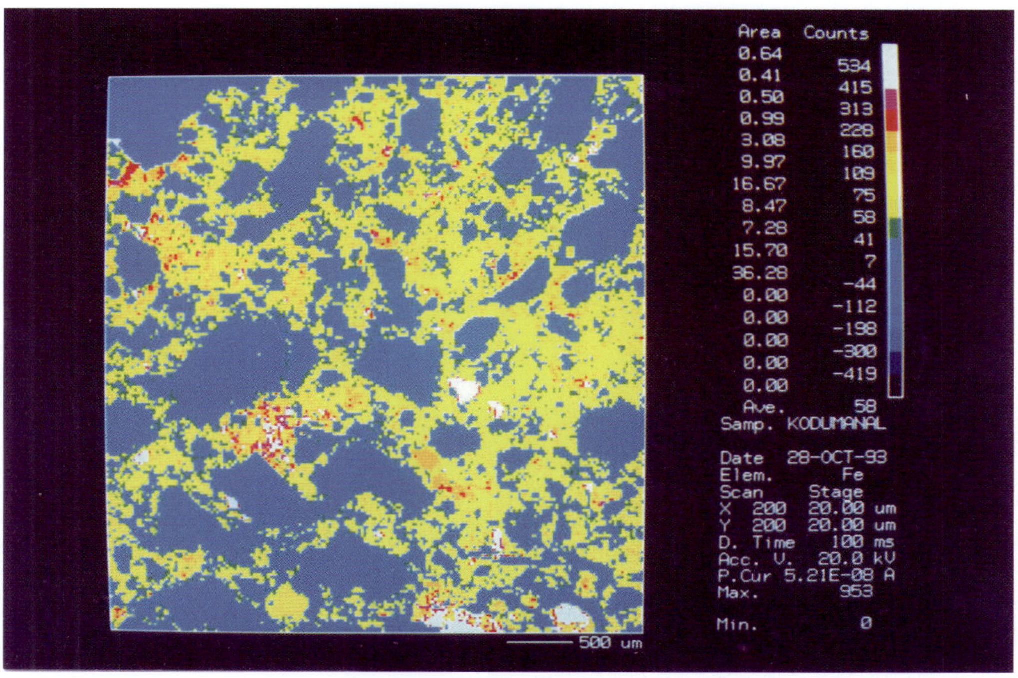

Figure 6: Elemental distribution map for iron in the crucible fragment from Kodumanal through EPMA-WDS analysis (given to Srinivasan by K. Rajan, Tamil University Tanjavur and analysed through Institute of Archaeology, London).

School of Mines was that cast iron prills were detected. This lends weight to the notion that a co-fusion process was very likely followed by the manufacture of high-carbon steel by the decarburization of cast iron (Srinivasan et al. 2011).

Although numerous blacksmithing traditions still survive in the region, it is clear that the specialised skills involved in working *wootz* steel have almost disappeared with the steel itself. There is a certain poignance to Jaikishan's observations concerning the elderly blacksmith, Mattela Gangaram, one of the few who is said to have practiced working with *wootz* and to the cultural myths, perpetuated by Reaumur in 1722, which said that the Indian cakes of steel from India could not be forged by Parisian smiths.

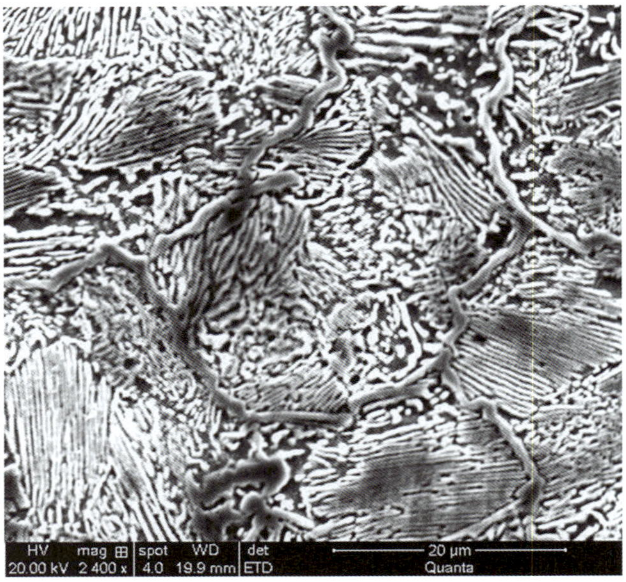

Figure 7: SEM micro-graph of toddy tapper's knife from Telangana showing structure of high-carbon wootz steel (given to S. Ranganathan by S. Jaikishan and analysed through IISc and Prof Satyam Suwas).

Modern experiments in forging blades of a similar composition to *wootz* carried out by Veorhoeven and Pendray have pointed to certain complexities: they tend to brittleness beyond red heat (Verhoeven and Pendray 1998, Srinivasan and Ranganathan 2004: 94-8). However, ultra-high carbon steels have the property of super-plasticity or high workability at certain temperatures which the traditional Oriental blacksmiths are likely to have empirically exploited (Wadsworth and Sherby 1983, Srinivasan and Ranganathan 2004: 89-92). Fig. 7 is an SEM photomicrograph of a toddy tapper's knife from the Telangana region (given by Jaikishan to Ranganathan and investigated in a series of collaborative studies at IISc by Suwas). The knives used for tapping toddy or the sap of the palmyra palm had to be very sharp as the bark of this tree is very tough. The microstructure of the knife shows a typical wootz structure of pearlitic regions within cementite networks.

Elsewhere too, traditions of blacksmithing are dwindling. In 2005 a study of sword blades from Kerala, used in the martial art tradition of Kalaripayattu, the author found that an analysed example of the flexible *urumi* blade (Fig. 8) used in combat was made of spring steel with about 0.8% carbon (Fig. 9). These findings were also presented at the conference 'Metallurgy; a touchstone for cross-cultural interaction' at the British Museum in 2005. Although from an unconnected context, it points to the potential for further studies into the diverse kind of artefacts that the higher carbon steels produced in the region may have been used to make.

To conclude, this overview attempts to point out that while there is a long way to go in terms of producing a comprehensive or systematic understanding of the developments of iron and steel metallurgy in the Indian subcontinent, there is growing evidence pointing to diverse and interesting metallurgical histories that make this a particularly promising region for ongoing archaeometallurgical research.

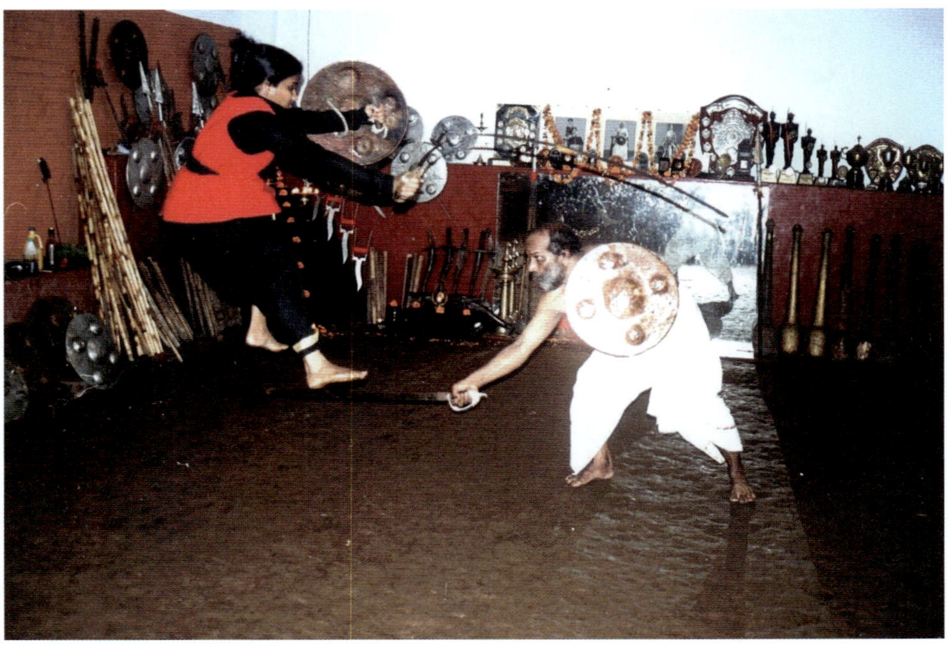

Figure 8: Flexible Urumi blades used in martial art tradition of Kalaripayattu, Calicut, demonstrated by Mr. Vijayan and his daughter.

Figure 9: Analysis of urumi blade collected by Srinivasan from Mr. Vijayan in Calicut showing microstructure of spring steel with 0.8% carbon.

References

Agrawal, O.P., Narain, H., and Joshi, G.P. 1985, 'Iron objects from south Indian megaliths (Karnataka)- A Technological Study and Significance'.

Anantharaman, T.R., 1997, *The Rustless Wonder- A Study of the Iron Pillar at Delhi*, New Delhi: Vigyan Prasar.

Balasubramaniam, 2008, *'Marvels of Indian iron through the ages*, New Delhi: Rupa & Co.

Baldev, R., Kalyanasundaram, Jayakumar, T., Babu Rao, C. and Venkaataraman, B, Kamatchi Mudali, U., Joseph, A. Kumar, A., and Buchanan-Hamilton, Frances, 1807 *"A journey from Madras through the countries of Mysore, Canara and Malabar"*, London. Wood, C. 1893, "Discussion of Turner's paper on production of iron in small furnaces in India", *Journal of the Iron and Steel Institute.* 44 (2): 177-180.

Chattopadhyay, P.K., P.K.Behera and P.K.Datta. 2007. From wrought iron to steel: Beginning of steel making in Eastern India. *Indian Journal of History of Science*, 42(3): 387-02.

Rajkumar, K.V., 'Non-destructive evaluation of Delhi iron pillar', Current Science, 88 (2005), pp.1948-56.

Chakrabarti, D.K. 1992, The early use of iron in India, *Oxford University Press.*

Coomaraswamy, A.K. 1956. *Mediaeval Sinhalese Art,* New York.

Cherian, P.J., Shajan, K.P., Selvakumar, V., (2007) Pattanam Excavation - 2007, Interim Report, KCHR Publication, Trivandrum, pp.107-290.

Deo, S.B and Jamkhedkar, A.P in Naikund, 1982, Excavations-1978-80 (eds) Department of Archaeology and Museums, Govt of Maharashtra and Deccan College, Pune, pp.52.

Deshpande, P.P., Mohanty, R.K., and Shinde, V.S., 2010, Metallographic studies of a steel chisel found at Mahurjhari, Vidarbha, Maharashtra, *Current Science*, 99 (5): 636-639).

Fuller, D.Q. Boivin, N. and Korisettar, R., 2007, Dating the Neolithic of South India: New radiometric evidence for key economic, social and ritual transformation, *Antiquity,* 81: 755-778.

Ghosh, A.K., and Chattopadhyay, P.K. 1982, An Early Steel Implement from Barudih, Bihar Province, India: Metallurgical Studies. *MASCA* 2(2): 63-4.

Gullapalli, P. 2009, 'Early metal in South India: Copper and iron in Megalithic contexts', *Journal of World Prehistory*, 22: 439-59.

Hadfield, 1912. Sinhalese iron and steel of ancient origin, *Journal of the Iron and Steel Institute*, 85: 134-74.

Jaikishan, S., 2009, Iron and wootz steel industry in northern Telangana: a historical perspective, Telangana Jagriti.

Johansen, P.G. (2008). A political economy of space: Social organization and the production of an Iron Age settlement landscape in northern Karnataka, PhD Dissertation submitted to Department of Anthropology, University of Chicago, Chicago.

Juleff, G. (1996), An ancient wind-powered iron smelting technology in Sri Lanka, Letters to Nature, 6560, pp.60-62.

Juleff, G. 1998) Early Iron and Steel in Sri Lanka: A Study of the Samanalwewa Area. AVA-Materialien Bd 54, von Zabern, Mainz.

Lowe, T.L., 1990. 'Refractories in high carbon iron processing: A preliminary study of Deccani wootz making crucibles', in Kingery, W.D. (ed), *Ceramics and Civilization,* The American Ceramic Society, Pittsburgh, pp.237-250.

Lowe T., Nima M., and Thomas G., 1990, An Historical Mullite Fiber Reinforced Ceramic Composite: Characterization of the 'Wootz' Crucible Refractory. MRS Proceedings, 185, 627 doi:10.1557/PROC-185-627.

Marshall, J. 1951. *Taxila.* Cambridge: University Press, pp.534.

Merkel, J., Feuerbach, A. and Griffiths, D., June 1995, Analytical investigation of crucible steel production at Merv, Turkmenistan, IAMS Newsletter, 19: 12-14. London: Institute of Archaeometallurgical Studies.

Possehl, Gregory L., and Praveena Gullapalli, 1999. The Early Iron Age in South Asia. pp. 153–175 in: Pigott, Vincent C. (ed.), The archaeometallurgy of the Asian Old World. (MASCA Research Papers in Science and Archaeology, University Museum Monograph, volume 16.) Philadelphia: The University Museum, University of Pennsylvania.

Possehl, Gregory L., and Paul C. Rissman, 1992. The chronology of prehistoric India: From earliest times to the Iron Age. Vol. II, Ehrich, Robert W. (ed.), *Chronologies in Old World archaeology*, 3rd ed. Chicago: The University of Chicago Press, pp.465-90.

Rajan, K., 1990, "New light on the megalithic cultures of the Kongu region, Tamil Nadu", *Man & Environment*, 16(1): 93-102 Rehren, Th. and Papachristou, O. (2003): Similar like white and black: a comparison of steel-making crucibles from Central Asia and the Indian subcontinent. In: Th. Stoellner, G. Körlin, G. Steffens & J. Cierny (eds), *Man and Mining,* (=Der Anschnitt, Beiheft 16, Bochum), 393-404.

Sasisekaran, B. (2002). Metallurgy and metal industry in ancient Tamil Nadu – An archaeological study, *Indian Journal of History of Science,* 37(1), 17–29.

Sasisekaran, B. 2004. *Iron Industry and Metallurgy: A Study of Ancient Technology.* Chennai: New Era Publications.

Schoff, W.H. 1915, The eastern iron trade of the Roman empire, *Journal of the American Oriental Society,* 35: 224-239.

Smith, C.S., 1960, *"A History of Metallography. The Development of Ideas on the Structure of Metals before 1890",* Chicago: University of Chicago Press, pp.14-6.

Smith, C. S. 1963, Four Outstanding Researches in Metallurgical History, ASTM, Philadelphia, pp.17-23.

Smith, C. S. 1981, *"A Search for Structure. Selected Essays on Science, Art and History,"* Cambridge, Massachusetts: MIT Press, pp.67-72.

Srinivasan, S. 1994. Wootz crucible steel: a newly discovered production site in South India. *Papers from the Institute of Archaeology,* 5: 49-61. London: University College London.

Srinivasan, S. 2007, On higher carbon and crucible steels in southern India: further insights from Mel-siruvalur, megalithic Kodumanal and early historic Pattinam, *Indian Journal of History of Science,* 42.4: 673-95.

Srinivasan, S. and Griffiths, D. 1997. "Crucible steel in South India: Preliminary investigations on crucibles from some newly identified sites", in *Material Issues in Art and Archaeology-V.* Ed. P. Vandivar, J, Druzik, J. Merkel, and J. Stewart, pp.111-27. Warrendale: Materials Research Society.

Srinivasan, S. and Ranganathan, S. 2004, India's legendary wootz steel, National Institute of Advanced Studies with Tata Steel.

Srinivasan, S., Ranganathan, S., Anderson, J. and Suwas, S. 2011, 'From the macroscopic to the microscopic: some scientific insights', *'Pioneering metallurgy: origins of iron and steel making in the southern Indian subcontinent. Telengana Field Survey, Interim Report 2011'*, National Institute of Advanced Studies and University of Exeter, pp.29-32.

Srinivasan, S., Sinopoli, C., Morrison, K., Gopal, R, Ranganathan, S. 2009 'South Indian Iron Age iron and high carbon steel: with reference to Kadebakele and comparative insights from Mel-siruvalur'. *Metallurgy and Civilisation: Eurasia and Beyond: Proceedings of the 6th International Conference on the Beginning of the Use of Metals and Alloys (BUMA VI)*, London: Archetype Books, pp.116-122.

Tewari, R. 2010, Updates on the antiquity of iron in South Asia, *Man and Environment* XXXV (2): 81-97.

Verhoeven, J.D. and Pendray, A., 1998, Muse, Vol. 2, No. 2, pp.35-43.

Voysey, H. W. 1832, "Description of the manufacture of steel in Southern India, *Journal of Asiatic Society of Bengal*", 1: 245-47.

Wagner, D.B. 1999, 'The earliest use of iron in China', In Metals in Antiquity, ed Young, S., Pollard, M, Budd, P., and Ixer, R., (BAR International Series 792), Oxford, Archaeopress, pp.1-9.

Wadsworth, J. and Sherby, O.D., 1983, Damasus steel-making, *Science*, 216, pp.328-30.

Author's address

• National Institute of Advanced Studies
Indian Institute of Science Campus
Bangalore 560 012
India

Recent archaeometallurgical discoveries of iron in Southeast Rajasthan, India

Lalit Pandey

ABSTRACT: Evidence from the Ahar-Banas Chalcolithic culture of the third and second millennia BC has previously defined southeast Rajasthan as an archaeologically distinct region. Until recently, however, no attempt has been made by archaeologists to understand the nature of physical occupation in the region. This paper aims to provide the first such insight into both the archaeometallurgical activities and the occupation within this region; presenting evidence which indicates that ancient iron production played a vital role in the socio-economic, political and cultural formation of southeast Rajasthan.

Keywords: Archaeometallurgy, tuyères, furnace, forging furnace.

Introduction

For the purposes of this paper and, in order to understand the archaeometallurgical activities related to iron production in southeast Rajasthan, cultural development in India will be divided into three periods: the first settled communities; the emergence of iron technology; and the emergence of urban societies.

During the period of the first settled communities, i.e., the third and the second millennia BC, southeast Rajasthan represented a distinct cultural entity. During this early period the people of the Ahar-Banas culture interacted with other contemporary cultures located to the east and southeast of this area. After the second millennium BC, however, there appears to have been an interruption in the occupational sequence in the region, for which there may be two possible reasons. Firstly, there has not yet been sufficient archaeological research to understand the nature of occupation during these later times. Secondly in later periods, occupation developed elsewhere (on the Gangetic plain, for example). Until recently, insufficient archaeological research into southeast Rajasthan during the first millennium BC led to the supposition that the Gangetic plain formed the focus for occupation in this period. The latest archaeological research, however, has provided evidence to indicate that the cultural sequence did in fact continue and that the archaeological landscape of southeast Rajasthan was significantly different from that of the Gangetic plains during this time (Gullapalli, 2005, pp.69-100).

The modern state of Rajasthan is situated in the north-western region of India. Its south eastern area forms a transitional region between the vast Indo-Gangetic plain and the Deccan Plateau, where monsoon rains in the south and southeast sustain the regions' rivers, lakes and springs. As a result, south and south-eastern Rajasthan have historically provided congenial conditions for human settlement (Rosin, 2001, pp.31-33). Due to its geographical proximity to the Indus Valley, the Ganges Valley, the Malwa Plateau and Gujarat, southeast Rajasthan has been intrinsically connected to the cultural development in all of these regions.

As the dominant geographical feature of the region, the Aravalli mountain range has, since ancient times, played a significant role in the formation of culture in Rajasthan.

Cultural development in relation to the process of iron making can be seen here, as a favourable climactic environment and abundant natural resources meant that the region was favourable to settlers, with Ahar providing the earliest evidence of both human settlement and of Chalcolithic culture within the region (this favourable combination has caused Sankalia to remark that southeast Rajasthan has been subject to early attempts at urbanisation at least five times during the last 5000 years (Sankalia, et al., 1969, pp.l-2)).

As the present paper is concerned with south-east Rajasthan, the southern Aravallis will be discussed here. The eminent geographer Bhattacharya describes one of the key features of this area as follows:

> "The highest amongst these ridges is the Jarga Parbat, referred endearingly as Jarga Ji. It has been anthropomorphized, as well as deified, by the people of Mewar. Topographical evidence, availability of perennial water from natural springs, abundance of forest cover, at least in the earlier days, isolation and solitude have all combined to provide it '*genius loci*,' so to say, for the establishment of a series of temples round which an annual fair is held in the month of February." (Bhattacharya 2000, pp.9-10)

Although iron ore deposits are distributed throughout the state of Rajasthan, with noteworthy deposits in the Alwar, Jaipur, Udaipur, and Ajmer districts (the Department of Mines and Geology of Rajasthan has reported iron ore deposits at Nathara-ki-Pal, Dangio-ki-Pancholi and Iswal), the Aravallis

– in addition to their socio-cultural importance to the region – are the site of the most significant mineral belt in Rajasthan. Iron, copper, lead and zinc deposits are found in the south of the range, around Dariba, and in north Rajasthan. There are also copper deposits at Khetri-Singhana Chouhan, 1996, p.75).

When studying the relationships between iron production and socio-economic and political developments in the region it is important to note – as Chakrabarti has remarked – that: "historically, the production and the economic organization of metals and metal crafts in some metal rich areas of the subcontinent (cf. Rajasthan) before the coming of the British should tell us a lot about the traditional way of organizing metal industry" (Chakrabarti, 1999, p.330).

Iron production in Southeast Rajasthan

Ancient metallurgical activities in the region were supported by the richness of readily-available forest products suitable for iron production. Southeast Rajasthan had a long tradition of extraction of charcoal (Bhattacharya, 2000, p.35), which was the primary forest product used in the process, as well as timber, timber poles and fire wood. Evidence suggests that the activities of the majority of the *Bhil* people of the area were non-agricultural (Gahlot and Dhar, 1989, p.221); indicating that, as communities of miners, they were involved in supplying charcoal for metallurgical activities.

Evidence of iron production using the natural materials mentioned above dates from the first millennium BC, as attested at the site of Noh (Ghosh, 1967, pp.28-29, 1969, pp.34-35). In addition to this, sites like Ahar (Sankalia, et al., 1969), Sunari (Mittra, 1983, pp.55-56), Bairat (Ghosh, 1965, p.31, Sahni 1999), Rairh (Puri, no date) and Gilund (Indian Archaeology – A Review 2001, pp.223-228) also provide evidence of the origins of iron technology in Rajasthan. Among them, the site of Balathal was the first to be systematically excavated for the purposes of understanding the complex cultural sequence present. It was observed in the course of the excavations at Balathal that the Iron Age occupants had established their settlement at the top of the settlement in the 4th century BC, a discovery which yielded evidence of Chalcolithic and Iron Age cultures and indicating that the mound was a multicultural site (Menon, 2002, pp.98-99).

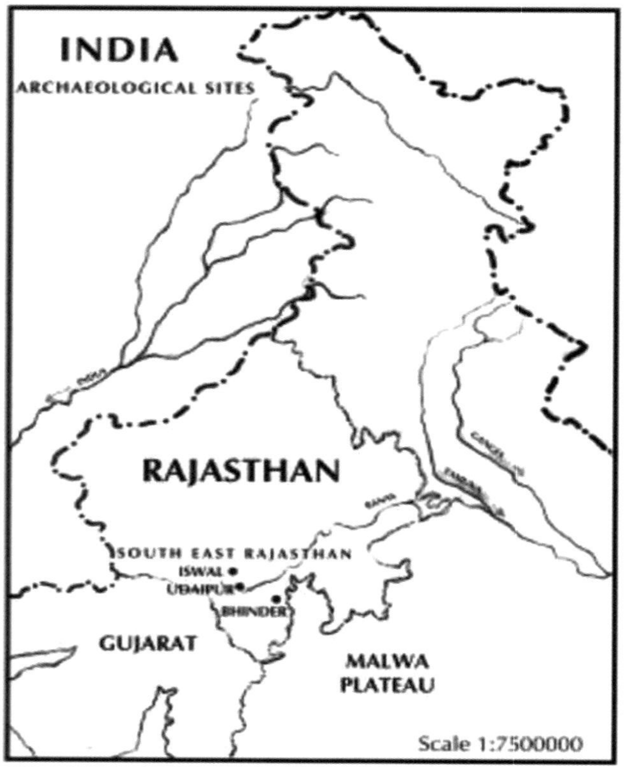

Figure 1: Archaeological map of the region.

A large number of iron objects – comprising points, dagger-blades, adzes, rods, nails, drills, hooks, clamps and hoes – were found at Balathal, as well as a number of iron smelting furnaces. This evidence is indicative of the existence of a complex metal technology developed by communities during the early historic period at Balathal (Menon, 2002, pp.98-99).

Following these findings at Balathal and other early evidence for iron production in the area, at sites such as Ahar (Thakur, 1998, p.363), the author conducted an extensive investigation of the surrounding region (Fig. 1), the main objective of which was to locate and investigate Iron Age sites in southeast Rajasthan with a view to establishing a connection between iron production and cultural development.

Site excavation

In the course of the investigation a number of Iron Age sites were identified at Iswal, Bhoion Ki Pancholi, Bhinder and Nathara Ki Pal (all in Udaipur District). These were selected on the basis of the availability of iron slag deposits on the surface and in the sections (excavation profiles) of the sites. The radiocarbon dates obtained from these locations at the Physical Research Laboratory in Ahemdabad has yielded evidence to suggest that the Iron Age originated in the 4th century BC. The details of the charcoal sample indicating this fact are shown in Table 1.

Table 1. Charcoal sample dated from Udaipur District

S. No.	PRL No.	Trench/layer	Depth	Radiocarbon date half life = 5730
1	1849	3	90 cm	2290 ± 70

Iswal (73°50' E, 24°30' N) was selected for excavation as a result of the large amount of iron deposits within the section (excavation profile) of the site and for its close proximity to two iron ore mines, located to the north-east at a distance of 5km and 1.5km respectively. The site is characterised by two mounds, separated by a deep valley, the first of which is located beyond the medieval Vishnu temple on the eastern side and was excavated during the first season of the investigation in order to better understand the nature of the site as a whole. During the course of excavations, it was found that a trench (trench 10) was filled with ash and large amounts of iron slag, suggesting that the site was a major iron working area during the 12th century AD – also known to be the date when the temple came into existence. The calibrated

^{14}C date of the charcoal, which was recovered at a depth of 12m, is 930 BP (Birbal Sahni Institute of Paleobotany, Lucknow, India).

During five seasons of excavation at Iswal (Fig. 2), the remains of four smelting and forging hearths were recovered, all of whose bases were beneath the ground. The site also provided evidence of early settlement, which had begun about 2973 years BP. The date is based on the charcoal samples collected in the course of the excavation, the details of which are as follows: charcoal, trench HB 5, depth 17.3m, calibrated age 2973 years BP (Birbal Sahni Institute of Palaeobotany, Lucknow, India).

At Iswal, Trench HG5 was excavated down to sterile layers because a number of stone walls were exposed in the trenches, indicating the existence of various habitation structures. In addition to this, trenches HA1, HA2, HA3, HB1, HB2, HB3, HC1, HC2, and HC3 also provided abundant evidence of structural activities.

This structural feature indicates that smelters lived in the vicinity of the industrial area. In the SE quadrant of HG 5, a stone wall 1.20m high was recovered at a depth of 17m, on top of virgin soil (Fig. 3) which was pale-yellow in colour, with sand granules and sand stones.

A 5cm thick slag deposit with a significant quantity of ash was also exposed during the excavation. In the same quadrant, a large amount of iron ore (Fig. 4), slag and ash was also recovered at a depth of 16.75m. Evidence suggests that the iron was natural.

Taken together, this material evidence is sufficient to make a valid proposition that early settlers of the site were engaged in metallurgical activities. The iron ore, slag and one singular tool recovered from Iswal were also analysed, by the Department of Geology at Mohan Lal Sukhadia University in Udaipur. On the basis of the analyses it was verified that the sample of ore is goethite: mainly an anhydrous oxide of

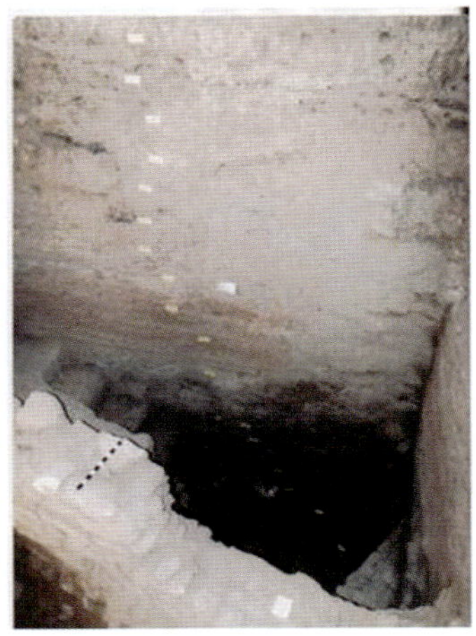

Figure 3: Stone wall at Iswal.

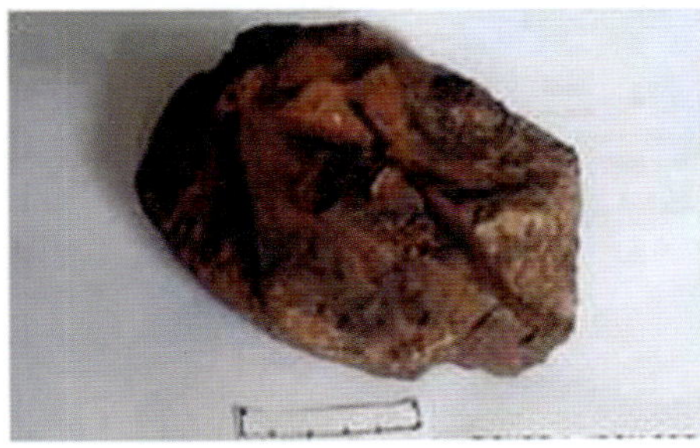

Figure 4: Iron ore found during excavations.

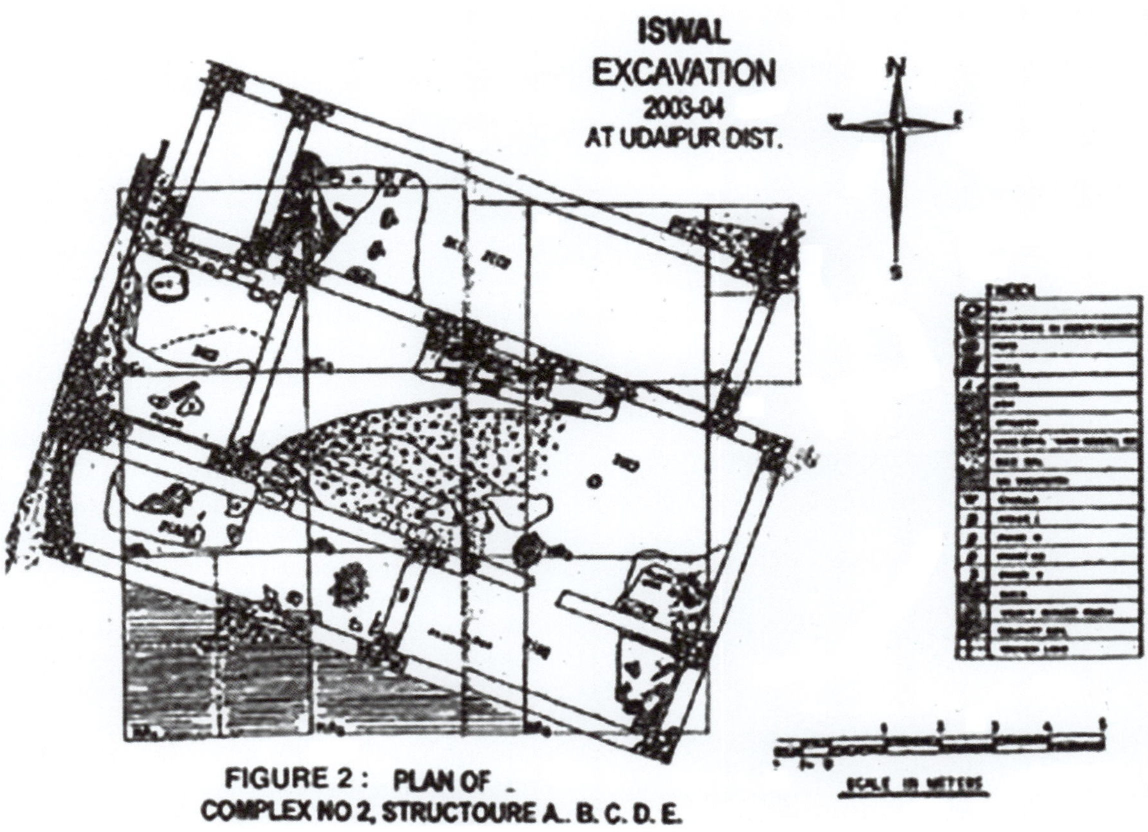

Figure 2: Details of the 2003-2004 excavations at Iswal.

iron with the chemical composition Fe_2O_4 (H_2O) x; brown to blackish dark brown in appearance and showing a polished section which is opaque to translucent. According to the analysis it is mainly a weathered product of haematite. In addition, the quality of the slag indicated that the craft of smelting iron ore was at that point well developed in the area. Analysis also revealed that one of the samples was a fragment of a rusted knife. Although rusting had diminished the pristine character of the ore, it was indicative of the fact that the object was made at the same place where the iron was smelted (Pandey, et al., 2006, p.139). Likewise, the sample of a rusted knife-like object also indicates that it was made at the same place where it was smelted (Pandey, et al., 2006, pp.138-139).

A semi-circular hearth was found in the SE and NE quadrants of HG 4, in the close proximity of a structural complex, at a depth of 10.82m (Fig. 5).

The hearth was made of stones and rectangular-shaped terracotta bricks, with a floor made of fine clay and slag material. A large amount of fine ash was also recovered within it. Its total depth was approximately 52cm and its measurements were 34cm (NS) and 48cm (EW). Evidence points to the fact that the hearth was used for forging. Unlike the smelting furnaces, however, the depth and the lateral walls of the hearth do not show any evidence of iron smelting. A rotary quern was also found at a depth of 11.42m in trench HD 4; it had two transverse slots at each side for the insertion of a handle. This type of rotary quern was also found at Nevasa (Maharashtra), and Sankalia has dated this particular type to the Indo-Roman period (Ghosh, 1956, p.11). The calibrated ^{14}C date of the charcoal sample that was collected from Iswal at a depth of 13.80m, is between 1730 and 1420 BP. All circumstantial evidence, including the related pottery found nearby, suggests that the forging furnace belongs to a period between the 2nd century AD and the 4th century AD.

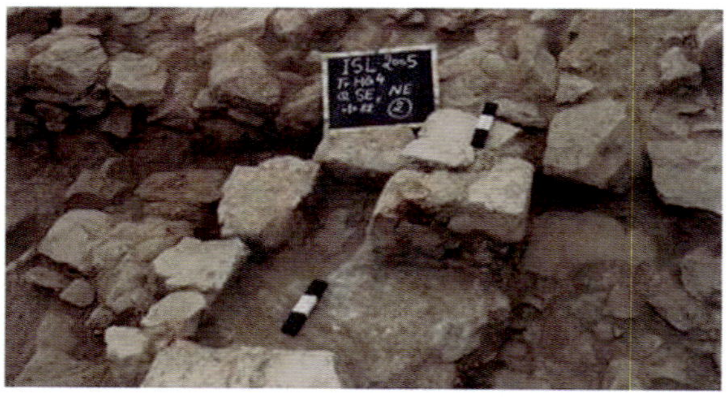

Figure 5: Semi-circular forging hearth.

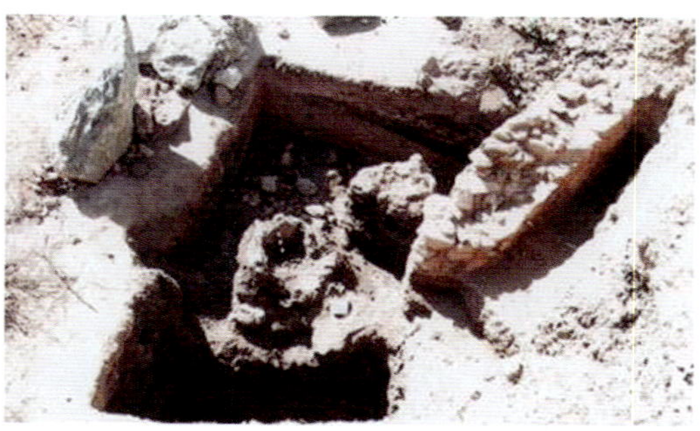

Figure 6: The remains of two furnaces.

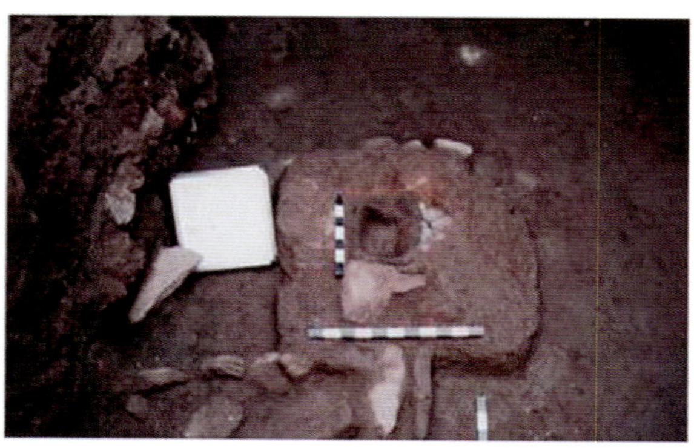

Figure 7: Square-like structure possibly representing the remains of forging furnace.

Two other furnaces (Fig. 6) were exposed in the SE quadrant of trench HAB5 (where excavation was undertaken for purposes of salvage archaeology, as the construction of a state highway had begun on the western corner of the mound on which HAB5 is located), at a distance of 70cm apart.

The first measured 80cm in height and 60cm in width and had a roughly triangular shape. Three broken tuyères were observed in the upper part of the furnace. These tuyères were 12cm in length, with a width of 8cm and a hole of 1cm in diameter. The second furnace was roughly oval in shape, measuring 90cm high and 70cm wide. The walls of both furnaces were about 2.4cm thick and the furnaces contained slag and ash. The SW quadrant was dug to a sterile level in order to establish the beginning of the settlement. During the cleaning of the surface of the quadrant of trench HAB5, a cluster of broken tuyères was observed in the western section of the SW quadrant – after which it was decided to expose the quadrant to collect the tuyères. A total of 32 broken tuyères were found, ranging in length from 15 to 22cm. The quadrant had nine alternative slag levels which were separated by rammed floors of fine clay. The average thickness of each greenish brown level was 10cm. At a depth of 11m, in the NE quadrant of HAB5, a greenish-brown floor 35cm thick was observed, from which pottery evidence was totally absent. In the northern section of the NE quadrant, a stone wall 1.30m long (N-S), with a height of 80cm and which rested on the natural hard rock was also exposed; as was a square-like structure in the SE quadrant of HB 18, at a depth of 12.50m (Fig. 7).

The structure was made of compact mud and spread in the E-W and N-S directions. Its total measurement was 75cm E-W and 60cm N-S and it covered an area of 16 x 30cm. The central part of the structure was filled with loose soil. A stone of 25 x 21cm in size was close to the North-South SE quadrant of HB 18 wall of the structure. In addition, all of its corners had been strengthened with bricks and pebbles, causing the complete structure to cover an area of 82 x 53cm. The upper surface of the entire area was covered with a layer of thick red soil, indicating large scale burning. A significant amount of ash and charcoal was recovered from within the area of the square-like structure. The calibrated ^{14}C date of the charcoal sample that was collected from the square area of HB 18 (10.52m) was 1071 BP; on the basis of which it may be presumed that the forging hearth is of an earlier date, as the difference in depth between the location of the forging hearth and that of the charcoal sample is 1.98m.

The close examination of the sections of trench 10 revealed a number of working levels, characterised by well plastered and rammed layers of soil that were yellowish-brown in colour. The remains of two furnaces were also recovered from the trench. The first furnace was roughly pear-shaped and sealed by a well plastered and rammed layer that covered both the top and the base. The height of its surviving lateral walls was 75cm on the outer surface and 65cm on the inner surface. The width of its base was about 83cm, with 12cm thickness. The other furnace, which was exposed in the same trench, was rectangular in shape with rounded corners and 95 cm in height – although as it was located in the section, its width could not be measured. Both these furnaces were built in situ over a stone foundation. The walls were made of clay, burnt red due to constant use. A huge amount of slag which had been spread in and around the furnaces, constitutes evidence of their constant use. During the course of excavations broken tuyères were collected from the surrounding area of the furnaces, which were round in shape and with an average thickness of about 4cm. The central holes of these tuyères were about 1cm in diameter. The longest broken retort that was recovered from the surrounding area was 40cm long.

Evidence of iron working recovered at the site of Iswal was spread over an area of 75 hectares, with the archaeometallurgical remains identified during these excavations indicating that Iswal was a major industrial iron smelting centre. A contributing factor in this conclusion is the fact that, over five seasons, only 36 iron implements were recovered from the site. This leads to the suggestion that the inhabitants of Iswal were exporting and trading iron ingots or finished products to neighbouring regions.

The available radiocarbon dates of the different levels of the site at Iswal are shown in Table 2:

Table 2. Radiocarbon dates from Iswal.

Depth	Calibrated age
4.7 m	930 BP
5.3 m	1070 BP
10.52 m	1071 BP
13.88 m	1580 BP
17.31 m	2973 BP

(Birbal Sahni Institute of Palaeobotany, Lucknow, India)

On the basis of the ^{14}C dates, it appears that settlement at the site had begun at 2973 years BP and it continued until 930 years BP, spanning various periods. The available material culture evidence reflects the continuity of settlement at the site.

It is my observation that, before the 2nd century AD, habitation at the site was still very much in the initial stages, as the material culture evidence associated with this phase is of poor quality and provides only relative dating. As a result of this it is not actually possibly to provide a precise dating of settlement activities.

Various ancient inscriptions (such as the Kundagram inscription of Aparajita of 661 AD, the Saraneshwar inscription of 953 AD and the Natha Prasasti of 971 AD, as well as the Ahar fragmentary inscription and a marble slab inscription (which both date to the 10th century AD)) corroborate the fact that the Kundagram site was a major centre of political and socio-economic activity. The Kundagram inscription, which is located 8km from Iswal, provides evidence of temple construction by the king Aparajita (Burgess, 1979, pp.29-32); while the Atapur or Ahar inscription of 977 AD (*Epigraphia Indica, Srinivasan, P.R.(Editor)* Vol. XXXIX, Part III, 1971, Published in 1983) and the Saraeshwar Temple Inscription of 953 AD both provide evidence of a number of commercial taxes levied on the merchants of various regions, such as Karnataka, Southern Gujarat and Northern Konkan.

Another important feature of the site is the recovery of glass bangles. In the course of five seasons of excavations, a total of 249 broken bangles were recovered. A variety of production techniques were used for their making, producing monochrome, polychrome, translucent and transparent bangles.

In addition to iron and glass working, a single piece of translucent bead made of beryl was also recovered from Iswal. One pendant of lapis lazuli inlaid with gold was also recovered, indicating the existence of some kind of trade relations with Central Asia.

The sites of Balathal and Iswal provided evidence for iron working in two different contexts: Balathal is associated with iron production for domestic purposes, while Iswal is associated with an intensive production of iron that may have served industrial purposes. As a result both sites have yielded evidence of different types of tuyères. The tuyères found at Balathal have a series of quartz granules on the inner side, which were probably added to increase their refractory properties. The tuyères found at Iswal do not display this feature.

Thus, it can be concluded that southeast Rajasthan was a very prominent centre of metallurgy in the past. The erstwhile inhabitants of this region had developed certain types of guilds of smiths who were successfully running this activity on an industrial scale. In addition to the commercial activity, Iswal was also a place of cultural activities. The area near Iswal has a series of temples (the Someshwar, Charbhuja, Parshuram, Ranakpur, Kundagram temples of Vishnu, and the Badavasan Devi and Vishnu) on either side of the Aravallis (Bhattacharya, 2000, p.11). The temple of Vishnu at Kundgram was built by Yasomati, wife of Aparajita's Commander-in-Chief Maharaja Varahasimha, in the 7th century AD.

"Anthropogeographically, temples have been of great significance in India for several years and provide some of the most tangible historical evidence in terms of their location, the period of their construction, the building material used, the style of architecture, mode of their occurrence (say in clusters or otherwise), and manner of the co-existence of places of worship of different faiths" (Bhattacharya, 2000, p.46).

In general, the evidence offered above appears sufficient to prove that these ancient metallurgical sites played a vital role in the socio-economic, political and cultural formation of southeast Rajasthan.

References

Epigraphia Indica, Srinivasan, P.R.(Editor) Vol. XXXIX, Part III, 1971, Published in 1983.

Bhattacharya, A.N., 2000. *Human Geography of Mewar*. Udaipur: Himanshu Publications.

Burgess, J. ed., 1979. *Epigraphia Indica, Vol. 4*. 2nd edition. New Delhi: Archaeological Survey of India.

Chakarbarti, Dilip. K., 1999. *India, an Archaeological History: Palaeolithic Beginnings to Early Historic Foundations*. New Delhi: Oxford University Press.

Chouhan, T.C., 1996. *Encyclopaedia of Rajasthan*. Jodhpur: Vigyan Prakashan.

Epigraphia Indica, Srinivasan, P.R.(Editor) Vol. XXXIX, Part III, 1971, Published in 1983.

Gahlot, S.S. and Dhar, B., 1989. *Castes and Tribes of Rajasthan*. Jodhpur: Jain Brothers.

Ghosh, A. ed., 1956. *Indian Archaeology: A Review, Vol. 3 (1955-1956)*. New Delhi: Archaeological Survey of India.

Ghosh, A. ed., 1960. *Indian Archaeology: A Review, Vol. 7 (1959-1960)*. New Delhi: Archaeological Survey of India.

Ghosh, A. ed., 1965. *Indian Archaeology: A Review, Vol. 10 (1962-1963)*. New Delhi: Archaeological Survey of India.

Ghosh, A. ed., 1967. *Indian Archaeology: A Review, Vol. 11 (1963-1964)*. New Delhi: Archaeological Survey of India.

Ghosh, A. ed., 1969. *Indian Archaeology: A Review, Vol. 12 (1964-1965)*. New Delhi: Archaeological Survey of India.

Gullapalli, P., 2005. *Smelting and Smithing: The Organization of Iron Production in Early Historic Southeastern Rajasthan*. PhD Thesis, University of Pennsylvania, Philadelphia.

Menon, K.G. ed., 2002. *Indian Archaeology: A Review, Vol. 44 (1996-1997)*. New Delhi: Archaeological Survey of India.

Mittra, D. ed., 1983. *Indian Archaeology: A Review, Vol. 28 (1980-1981)*. New Delhi: Archaeological Survey of India.

Pandey, L. 2006. A Report on Excavations at Iswal 2005-06. *Shodh Patrika*, 57, pp.123-133, Udaipur, Institute of Rajasthan Studies.

Puri, K.N.(no year) Excavation at Rairh, Jaipur.

Rosin, R.T., 2001. Locality and Frontier: Securing Livelihood in the Aravalli Zone of Central Rajasthan, p. 30-64. In: K. Schomer, L. Joan, E. Deryck, O. Lodrick and L. Rudolph eds. *The Idea of Rajasthan*. New Delhi: Manohar.

Sankalia, H.D., Deo, S.B. and Ansari, Z.D., 1969. *Excavations at Ahar (Tambavati)*. Poona: Deccan College Postgraduate and Research Institute.

Thakur, Vijay Kumar, Social Implications of Technology: AStudy of Iron in pre-C.200 B.C. India, p.363. In Archaeometallurgy in India ed. Vibha Tripathi, 1998, Sharda Publishing House, Delhi.

Author's address

- Director of the Institute of Rajasthan Studies, JRN Rajasthan Vidyapeeth, Udaipur (Rajasthan)-313001, INDIA

Smelting: a sacred process. Observations of iron smelting in Madhya Pradesh, India

Jake Keen

ABSTRACT: This paper presents a series of observations made in November 2006 and January 2007 of one family making iron in a remote, rural part of India. The family belong to the Agaria people, who, according to Elwin (1942, pp. 89, 101), believed their ancestors were born out of fire, ate iron, and drank liquid slag. The author had previously attended an event in 1995, at which six groups from different Agaria villages demonstrated their traditional skills of smelting iron, outside the modern blast furnace at Bhilai. Inspired, he spent the next decade learning to smelt in England. He returned to the area in 2006, but despite managing to organise the events described below, was unable to find an interpreter. Furnace building, ore collecting, charcoal making, and the smelting process were carefully observed. Finally, he came back three months later with a friend of the villagers, who was fluent in English, Hindi, and the local dialect, and witnessed two further smelts taking place.

Keywords: Ceremony, kodon chaff, Lohasur, virgin iron.

Introduction

The Agaria people occupy a large territory, mainly in Madhya Pradesh and Chattisgarth. They share this land with several other tribes. To enter Chokh Asur's village, the visitor leaves the country mostly inhabited by the Baiga people, and crosses the river either by a ford upstream of the waterfalls, or by stepping across logs spanning a deep gorge. In either case, before crossing, resin and joss sticks are burned, a coconut is offered, and prayers given. This expression of religious devotion also characterises each significant stage of the iron smelting production that follows.

A small rectangular thatched hut sits beside a dirt road above the river. Next to it, in a tiny plot of kodon millet, is an open-sided shelter containing a broken clay furnace, a small smithing hearth on the ground, and, in between the two, an anvil cut from a length of railway line. Managed forests of sal trees cover the hills for miles around, while paddy fields and pastures grazed by a few cattle and goats occupy the land near the river.

Chokh Asur (C), in his fifties, is in charge, but his brother (H), two of his sons (N) and (K), a nephew (S), an uncle (M), and his wife (J), his brother's wife (W), and daughter in law (B) all share the work over the next eight days. By and large the men perform the rituals, do the heavier work, and anything involving trips away from the smithy, while the women mainly prepare the raw materials, build the furnace, and pump the bellows. There is however no strict demarcation of roles. C's everyday work is as a blacksmith. During the days of furnace building described below, a succession of local men brought axes and sickles to be re-edged or repaired. Next to his hearth, C keeps an axe made from bloomery iron. The locals prefer such tools and say they last longer, but as they are much more expensive to make, little smelting gets done these days. Most smithing is carried out using recycled scrap steel.

Preparations

Day One: Outside the shelter, beside the path, a cow dung fire and incense sticks are lit. C and H mutter prayers and appear to ask for blessings while they (i) crumble sal resin onto the fire, (ii) peel a coconut and feed its husk to the flames, (iii) smash the coconut on a stone and burn its flesh little by little, (iv) pour locally-made rum onto their hands and then in a circle on the ground, (v) make a hole using an iron bar in the circle of alcohol. This ceremony, which takes about twenty five minutes, seems heartfelt and far from perfunctory, as it might have appeared if the belief system had been no longer current. At the end, everyone present, including various children, is given a portion of coconut to eat and asked to bless the furnace.

The hole is then enlarged and deepened to 225mm. A wooden post (90mm diameter) is inserted into the ground and stabilised with tightly packed stones. Three long bundles of grass are carefully and evenly tied round the post to make a tapering mould, the profile of which represents the shape of the inside of the furnace.

Meanwhile two of the women, W and B, have been gathering and mixing the ingredients for the daub. River clay, cow dung, straw and water brought in a brass pot from the river are mixed and remixed for a couple of hours and then left overnight.

Day Two. C and the three women begin to cover the grass mould with the daub. The work is slow and careful. Each handful is kneaded into its neighbour with the thickness gradually tapering from 60mm at the base to 35mm at the top. W and B continue until this job is finished. The work is mundane, but the care and precision employed to make the furnace shaft give it beauty (Fig. 1).

Day Three. With a knife, C cuts an arch in the drying daub, at the point where the bloom will eventually be extracted. Little by little the grass wrapped round the post is pulled out through the new opening (Fig. 2). Then the post is wiggled until loose and lifted out from the top. A smaller opening, the slag tapping arch, is cut about 135 degrees anticlockwise round from the first. All sharp edges made by the knife are rounded off. The inside of the furnace, which is the height of C's arm, is smoothed all over with a wetted hand. Next, a second, much thinner (20mm) layer of daub is applied to the outside of the furnace, working upwards from the base.

A shallow basket load of cow dung is collected by B, broken up, and fed into the furnace until it is full. Embers from the smithy are used by W to light the dung through the bloom arch. When the fire begins to burn too fast, a metal plate is used as a lid to control the rate of burn.

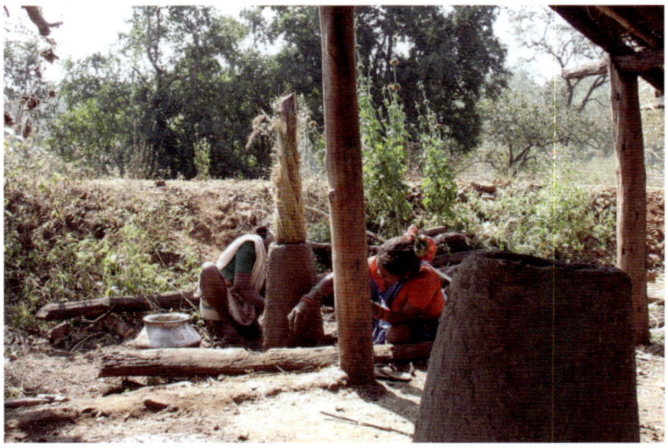

Figure 1: Constructing the furnace shaft around the grass mould.

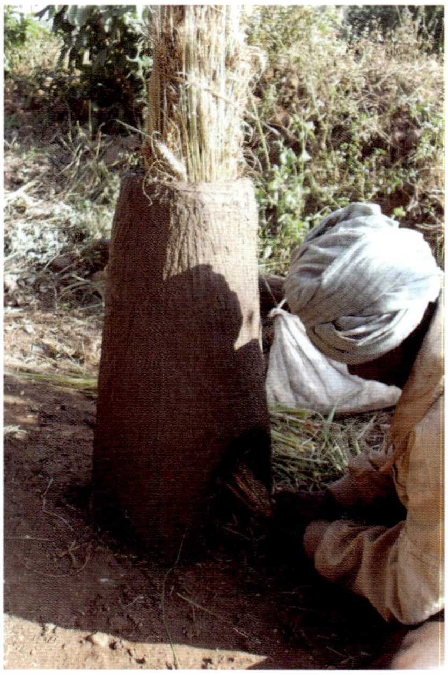

Figure 2: Pulling the grass from the newly constructed opening.

Day Four. By exaggerated mime C and H seem to be explaining the difficulty, distance, and danger of the journey into the hills to collect the ore. Perhaps they consider the journey may be too arduous for the European with a notebook in their midst. The ancient mines, shared by all the Agaria families in the region, are at the summit of a steep ridge, densely covered in thorn bushes, about five kilometres from the village. The overgrown shafts are about 700mm wide at the top and between two and three metres deep. The sides have collapsed, and it appears the shafts were once much deeper, with interconnecting tunnels lower down. They are identical in form to ore pits observed by the author in Sri Lanka and Nigeria, both near village bloomeries. The width of the shaft is presumably determined by the need of the miner, if all else fails, to be able to climb out by bracing his legs against the sides. Because the mines are now unsafe, ore is collected from the ground near the old mine shafts and further down the slope where rock is exposed. (This source may have a bearing on subsequent events.)

On the hilltop, C, H and N collect a basket of iron rich stones which are seen as being of two types: some are purple to crimson in colour, the others scarlet, orange, and yellow. The men gather cross-legged next to the ore, which they have separated into two piles sorted by colour. In front of the best ore (purple-crimson), two small fires of cow dung are lit, surrounded by a ring of joss sticks planted in the ground. Each fire in turn receives a succession of offerings, each accompanied by a mantra lasting two to three minutes; first sal resin, then coconut, next alcohol, and finally raisins (the author's survival rations). Again, as with the earlier ceremony on day one, the ritual words seem to be spoken with great sincerity. Elwin, who lived with the Agaria people for nearly a decade, translated the words he heard recited at the ore pits seventy five years ago. The following sample (Elwin, 1942, p.174) may resemble the words intoned with passion at ore pits in 2006: "Lohasur Baba, we take refuge in thee. O Mother that givest increase we seek refuge in thee. From little may there be much. From shortage may there be double. Twelve dozen times twelve dozen, generation after generation, depend on thee....We dig the earth, may it bear riches." (Lohasur is the god of the furnace, embodied in the form of iron. Twelve brothers were the mythical founding ancestors who lived when the world was being created out of iron.)

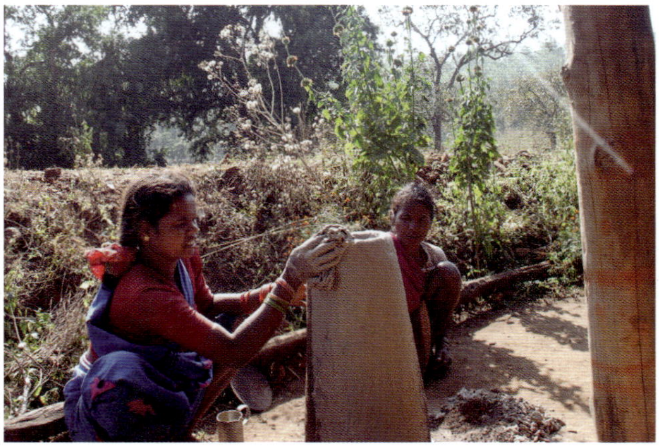

Figure 3: Applying slip with a rag.

Day Five. A basket of light, crumbly, white clay is dug from the river bank by B. It is mixed into a slip and applied to the furnace, inside and out, using a rag (Fig. 3). Later, the furnace is filled with pieces of resinous bark and lit. Presumably this will burn hotter than the dung fire used to dry the furnace on day three.

The two types of ore are separately piled next to the furnace.

Day Six. As C breaks up more of the white crumbly clay, he indicates that this is the same material he throws onto a tool when he is hammer welding at the forge. (See Aftermath.) J, who has recently become blind, mixes the powdered clay with water, kneading out all the lumps, and then working in handfuls of rice husks until it has the consistency of dough. From this mixture three tuyères are made by J, B, and an uncle, M, who joins the group. He brings a rod of polished sal wood (25mm diameter) and having shaped a 'dough' cylinder 80mm wide by 100mm long, he makes a hole, inserts the rod, and then works the clay along it. The end result is about 140mm long tapers from 70mm inside diameter at the thick end to 25mm at the thin end (Fig. 4). Much merriment accompanies the suggestive shaping of the tuyeres.

The old broken furnace in the shelter is demolished and the floor and a metre-wide strip outside are cleared and then swept. N and K, supervised by their father, gently pick up the shaft of the new furnace and carry it inside to the place where the old one used to stand (Fig. 5). It is then tilted so that the side over the bloom arch is vertical (perpendicular to ground level) and the opposite side sloping accordingly.

The gap at the base caused by tilting the furnace is filled with some of the original daub mix, saved for the purpose and kept moist for five days. Then the base is further reinforced to ensure no cracks develop where the furnace joins the hut floor (Fig. 6). Meanwhile the floor inside the furnace is broken up using an iron bar and is then reshaped to leave a sloping channel that funnels downwards to the tapping arch. Additionally, the inside wall of the furnace immediately above the bloom arch, i.e. above where the tuyère will sit, is shaved back, presumably to reduce the chances of wall-melt at the hottest part of the furnace.

The floor around the furnace is smoothed with white clay slip once all these adjustments have taken place. Meanwhile a new pile of daub has been mixed, and S has been busy cutting and debarking two long straight poles and a further pair with forked ends which have been planted securely in the ground. C positions the straight poles on the forks to form a ramp sloping down to the top of the furnace. Across these poles split bamboo lathes are lashed using a rope twisted from a vine, debarked and split for the purpose by S (Fig. 7).

Finally the new daub is smoothed all over the ramp, which is designed to carry the charge of charcoal and ore so that, if necessary, the furnace can be recharged using a long hooked stick by someone who is simultaneously operating the foot bellows (Fig. 8). The ramp has a raised edge (a split bamboo is tied to each side and covered in daub), which extends round the top of the furnace. This way none of the charge can be lost over the edge.

Day Seven. Now follows the most extraordinary part of the process so far. No charcoal has yet been made. In the author's experience in England, this takes as much time as all the other preparations put together.

H, K and S set out for the sal forest three kilometres away well after midday, carrying axes and two sacks. C is fasting from now until the smelt and does not come. The permit granted by the forest authorities allows the Agaria of this community to use any dead wood, standing or already fallen.

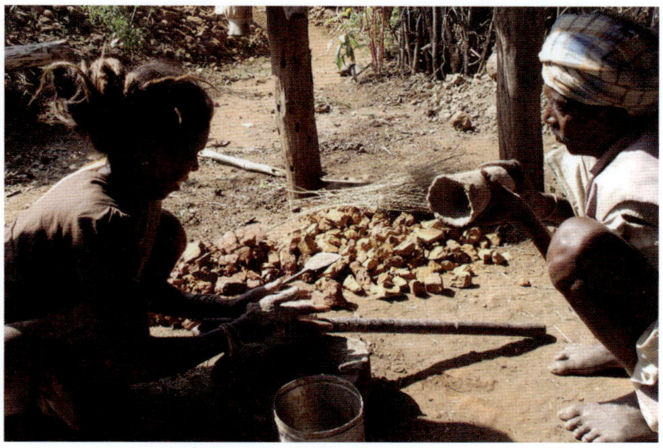

Figure 4: Making the tuyeres.

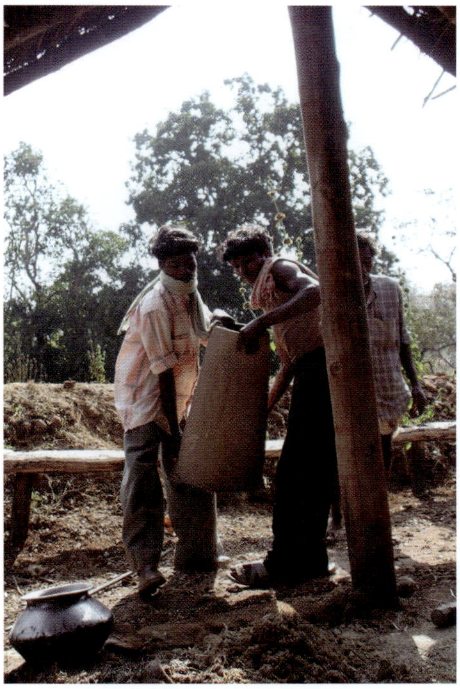

Figure 5: Moving the furnace shaft.

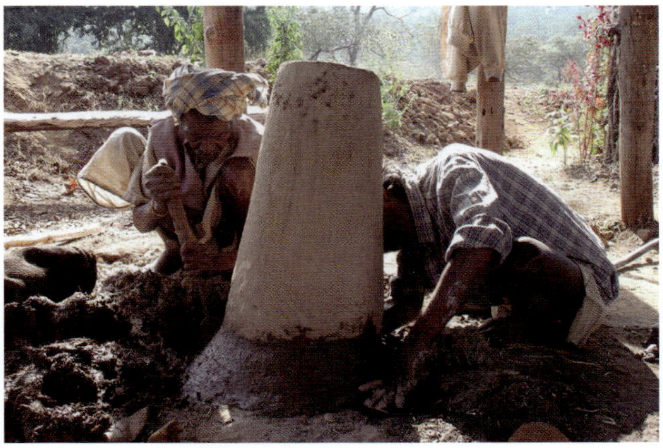

Figure 6: Reinforcing the base of the furnace.

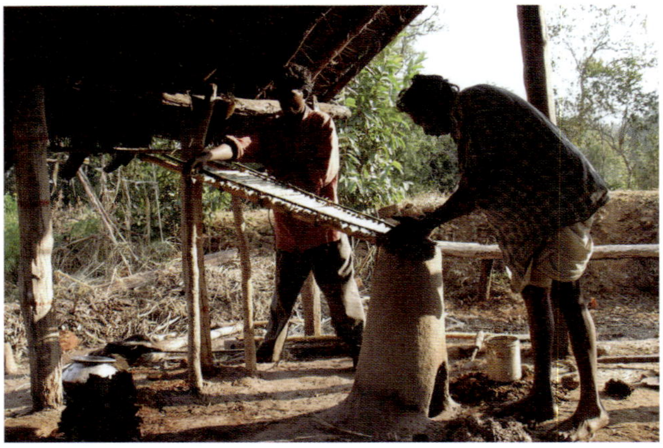

Figure 7: Construction of the charging ramp.

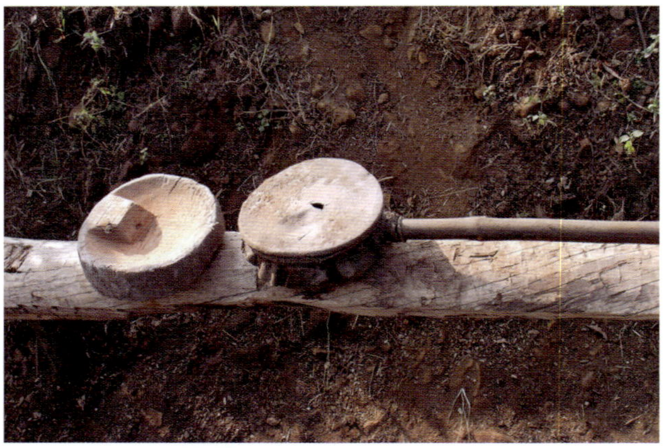

Figure 8: Foot bellows (with skin removed on left).

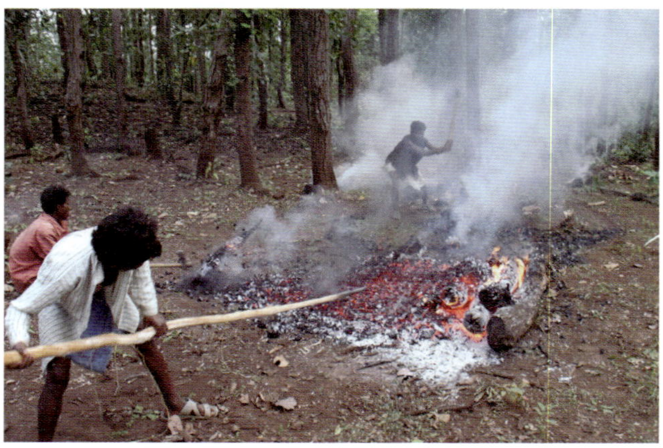

Figure 9: Making charcoal.

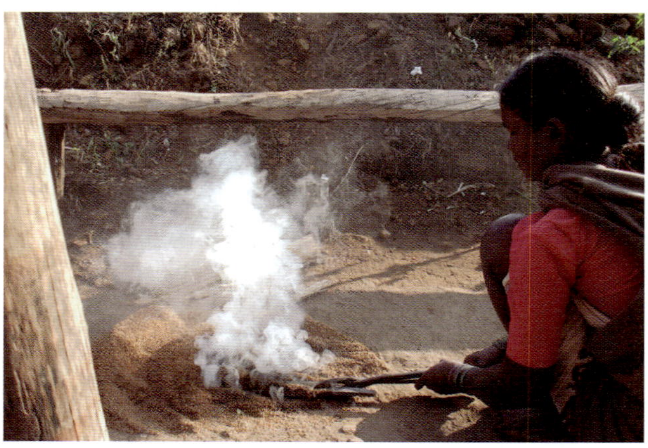

Figure 10: Charring the kodon chaff.

For about thirty minutes each man chops into the hard, reddish timber, producing logs about 1.5m long and 100 to 200mm in diameter. These are then dragged into a 1m high pile and lit with brash. Thereafter the charcoal burners lie on their backs or wander off amongst the trees while the fire burns. After a little over two hours, S sets off to cut and trim three 4m poles. When he returns, these are used to beat the loose charcoal off what remains of the logs (Fig. 9). Any of the larger logs not reduced to charcoal are hauled out of the way by striking an axe deeply into one end and dragging. Then, using the long poles, the charcoal is raked back and forth on the forest floor until no piece is left covering another. Amazingly, after only ten minutes of such work, the charcoal is gathered by hand into the sacks. No water or any kind of cover has been needed to extinguish the fire. The party arrive back at dusk with 25kg of exceptional smelting fuel.

Day Eight. On the morning of the final day the entire floor of the shelter and a 1m strip surrounding it are purified with fresh cow dung and water applied with a rag. (Each homestead is renewed each morning in this way. The pristine floor dries in a few minutes.) B begins meticulously charring a pile of small round seed husks (saved from threshing and winnowing the millet earlier) using a red-hot iron bar (Fig. 10) while another is being heated in the forge. She continues at this for about an hour, passing alternate lengths of red-hot iron back and forth through the fine kodon chaff, until a bucket load has been thoroughly charred. Whatever the purpose of this activity might be, the meticulous attention devoted to it suggests it is crucial.

A good deal of time is spent repairing and tuning a pair of foot bellows. These are 80mm deep, 280mm wide disks of sal wood, each hollowed out into a bowl 60mm deep, except where a small platform has been left to support the ball of a small foot. Into each platform a 30mm hole has been bored from the outside to fit a 500mm length of bamboo with the nodes drilled out. This allows air from each bellows to be blown alternately into the wide end of the tuyère. The inner end of the foot platform has a 13mm hole allowing air to escape into the tube. A groove on the outside of each disk permits a cow-skin membrane to be tied on. Each membrane has a small slot cut where the heel sits. This acts as a valve, closed by the heel on the down stroke, opened to let air in as the foot rises. Each membrane is raised in turn by a pair of springy sticks firmly anchored into the ground, at an angle of 45 degrees, with the aid of boulders from the river. A cord, tied to the top end of each stick, is attached to its respective membrane by a toggle slipped inside the valve slot. This way, if a fault develops in one or other of the bellows, it can be easily detached from its springy pole.

By late afternoon most of the family are busy with the final preparations. B and W are crushing the unroasted ore on two basalt slabs using large pebbles. The ore is reduced to pea-size or less. Eventually three heaps, purple-red, orange-yellow, and dull brown (ore from a different source remaining from an earlier smelt), each approximately three to four kilos, are placed by the furnace.

S strips the woody core from a vine and twists the remainder to make string to bind the leather skins to the bellows. H splits bamboo to make a clamp to hold the end of the tubes supplying the tuyère. C uses an adze to prepare a third bellows disk in case one of the others fails.

Expectation rises as people gather. By dark there are at least fifteen new faces, mainly men in cloaks and turbans, squatting round the forge or standing outside. A big bonfire is lit on the road and a pile of logs and rice straw provided for onlookers to sit on.

C, M, J, B, and W are all absorbed in a task that seems to be demanding particular care. B winnows ash and clay dust saved from the old furnace. This is mixed with the kodon-millet chaff prepared earlier in the day. W puts the hands of J, who is blind, onto the mixture. She feels it between her fingers and gives her approval. M puts handfuls of it through the bloom hole and tamps it down, little by little, shaping a bed for the bloom. (Elwin (1942, pp.108, 187) remarked on the importance the Agaria people gave to this detail of the smelting process.) A little water is added to the remainder of the chaff mixture to make a black, malleable paste. C uses some of this to close the bloom arch, just leaving a hole for the tuyère, which is then fitted with the base horizontal, the top sloping down. The inner end projects marginally into the furnace, the outer end outwards by 60mm. C and M spend much time discussing and then fine tuning the exact positioning of the tuyère. Eventually they raise it about 5mm, by packing bamboo splints and clay under it.

The furnace is filled to the top with charcoal, which S has been beating with a stick to break into smaller pieces. Only now does C use the rest of the black mixture to seal the tuyère into the bloom arch, which he smoothes off into a rounded profile.

The bamboo tubes from the bellows are fixed in position, pointing into the tuyère, and weighed down with a heavy stone. The ramp is piled with charcoal spread evenly to a depth of 80mm. C takes a scoop of the yellow ore and spreads it onto the charcoal at the front of the ramp.

The tension, after so many days of accelerating preparation, is growing palpable. The night is very cold. The crowd continues to grow. C's youngest daughter trips over a sack which had been carried in on the end of a pole at dusk. A piercing squeal confirms the rumour that a piglet, as well as a young chicken, is to be offered to Lohasur, god of the furnace.

C and H, his right hand man, squat for a final smoke round the forge.

The smelt

H begins the supplication, dripping libations of rum near the piglet. Tumeric in oil and peppers are dribbled and touched onto the end of the tuyère and the bellow pipes. Incense sticks are lit. C, H and M clasp hands and touch them to their foreheads. They recite a prayer together in unison.

S detaches the piglet from the pole. It shrieks in panic. It is offered rice and immediately calms down. More prayers are followed by more rice. It is happy.

C and H pray as they hold the chubby black piglet. S cuts off one of its ears and puts it in a brass bowl with other offerings. Then holding its snout he saws off its head with the blade of an axe. The head is leaned against the furnace.

A chick is now offered rice, while being talked to gently. It feeds, cheeping loudly. Its head is removed. C prays deep and long. A coconut is offered, pointing the sharp end to the furnace. M prays. The husk is peeled off. Each man places pieces of coconut husk on the ground in front of the furnace between the tuyère and the tap arch next to the brass bowl. An ember is placed on the coconut husks which catch fire. The coconut is smashed and its meat presented. Sal resin is crumbled and added to the fire as the prayers continue. Now alcohol is tipped into C's palm. The others hold his hand while it is poured onto the pile of gifts. Another offering of rum follows.

The pig's head is put into a basket. All the offerings, including the pig's blood, are raked up and placed in a sal leaf platter and put on the roof above the furnace. The whole area is cleaned up with a cloth and water as though nothing had had happened.

9.31pm. C puts an ember into the end of the tuyère. B starts pumping the bellows. Smoke issues from the furnace. C puts in more embers. The furnace starts to roar.

B is pumping at 160 strokes a minute, 80 each side. This is an easy rate to maintain as the bellows need little effort to operate. It's a bit like jogging without taking one's toes off the ground.

9.38. Carbon monoxide is lit on top of the furnace. An intense blue flame illuminates men, women and children huddling round the forge fire (Fig. 11).

9.50. C spreads the ore along the ramp on top of the charcoal but what system, if any, is used to blend the different types, is not clear. As she pumps, B guides the charge into the top of the furnace with her stick. (Three full ramp-loads of charcoal and ore will complete the smelt.)

10.25. The rate has been more or less consistent for nearly an hour. S took a turn pumping for a while. Usually each person works the bellows for 20-30 minutes before changing.

10.53. The second sack of charcoal is being broken up. The first ramp-load of ore has been used.

10.55. C pokes through the tuyère. M opens the slag tap hole. Flames shoot out. C tries to draw slag for three minutes. He pokes the tuyère again.

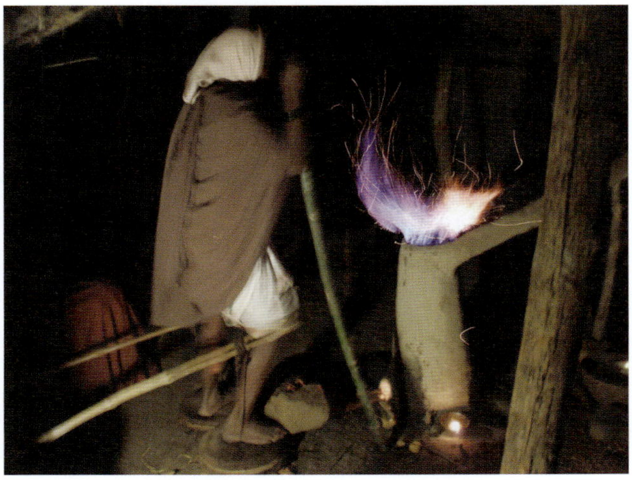

Figure 11: The smelt

11.03. Again C attempts to tap slag by probing with a green stick. M takes over bellows.

11.05. C probes and prods hard through the tuyère as though drips of slag are interfering with the airflow. Again he tries to tap slag without success. He packs the hot charcoal that has escaped from the tap hole back in, but lets the flame continue to squirt out until the temperature has picked up.

11.11. Another tapping attempt. This time a solidified lump of slag comes out. Now the tuyère is blocked again.

During the next one hundred minutes there are a further eight attempts to tap slag, at first with a green stick, later, more frantically, with an iron rod. At the same time there is clearly a problem with slag forming at the inner end of the tuyère and this needs frequent prodding to free the airway. C takes longer and longer breaks from the furnace. No one seems to be in charge of loading ore onto the ramp. Whoever is nearest takes responsibility. The order and care which typified eight days of preparation seems to have disintegrated.

C has been away for a long time. Where is he? The tiny dwelling hut next to the shelter has its door shut. Inside, C and most of the others are eating the sacrificial piglet. Has he lost interest or is the roast meat too great a distraction after fasting for thirty six hours?

M attempts to tap slag with no success. One of the bellows is not working well. C reappears and gets the flames going again, but by now he seems preoccupied elsewhere. All this time the bellows have been maintained at 160 strokes per minute but by now the bloom is obstructing the airflow through the furnace.

1.38. The bellows are removed. The black chaff mixture round the tuyère is removed. The bloom is pulled out, attached to the broken end of the tuyère, and dropped near the bellows. C does not even look at it. He examines the furnace floor instead. Is this a sign of disgust, indifference or supreme confidence?

The aftermath

As first light shines across the river illuminating cloaked bodies sleeping by the forge, the author is able to examine the bloom. It is spherical, silvery grey, and not very heavy. Its bottom half is a perfect cast of the bowl shaped in the kodon chaff.

The huge, inescapable impediment of not being able to question the Agaria about their actions during the last week is a frustration too great to leave unresolved. It is agreed by mime and sign language that the author will take away the bloom to study and will return in three months, this time with an interpreter. When cut at a laboratory in Bangalore the bloom is found to consist almost entirely of slag, with a few, poorly consolidated, iron rich areas near the top.

Two further smelts are conducted early in 2007, this time with a different ore and greater success. A visit to the abandoned village from which C's great-great-grandfather moved in the 19th century reveals an untouched slag heap 35m long by 5m wide. On it are several domed slag casts very similar to that taken to Bangalore.

The two new smelts are done back to back throughout the night. Once again, perhaps under the influence of rum and pork, offered to Lohasur but consumed by C and his male helpers, the smelting operation seems chaotic. Again the slag fails to run; again there is a problem with the tuyère blocking. During the second smelt frozen slag obstructs the air flowing to the back of the furnace, which grows prematurely cold. Even with the help of a skilled interpreter, who has a good relationship with the family, it is difficult to understand what is happening. The third smelt continues through the bitterly cold hours before dawn, but does not reveal any further insights.

By 7am the third bloom is out and smithing commences. The second bloom, although fragmentary, has good iron in it. Two of the better pieces are placed, one on top of the other, in the forge. As they are heated to welding temperature, pinches of white clay crushed from a broken tuyère, are thrown onto them in the bed of charcoal. This 'helps to join the iron.' Repeated reheating and hammering with pinches of clay added from time to time, results in a homogenous bar, hammered out into a blank suitable for a knife.

Finally, C tackles the bloom from the November smelt. He does not see it as a failure and sets out to retrieve the iron. He is able to make a small bar which he draws out into a number of rings. The quantity is very small, less than 100 grams, but the iron, once consolidated, is homogenous and malleable. Moreover it bears a significance (see below) not evident from all the hours of careful observation.

Discussion A: Expectations

It is tempting, when reconstructing the past using research tools available to the archaeometallurgist, to make certain assumptions about the context in which smelting may have occurred. For example, it is easy to assume that economic and environmental factors predominated, or that productivity was paramount. This is probably a consequence of focusing on quantifiable data. It is difficult to consider what cannot be measured. Even more so to accommodate behaviour that leaves no evidence.

The smelts described above did not conform to expectations. It is hoped that the reappraisal this prompted will provide useful insights, even though there may be no cultural parallels with other studies presented in this publication.

The expectations were based on:

i. Proven smelting competence of C and his family. (During the 1995 smelting demonstration at Bhilai the author had watched them produce an excellent bloom consisting of nearly 3kg of well-consolidated soft iron.)
ii. Impressive skill and knowledge demonstrated over eight days of preparation during which it was clear that all family members were familiar with every aspect of the technology.
iii. The author's long experience of bloomery smelting which confirmed that this community had retained its full understanding of the process, even if smelting today is not normally economically viable.

The smelts failed to conform because:

i. C, the 'iron master', was absent at crucial stages of the smelt.
ii. The process seemed haphazard.

iii. The iron produced seemed less important than the event itself.

The following thoughts are offered, despite the belief that they may throw more light on the author's ignorance than the Agaria's knowledge.

i. In the author's view, the smelts produced less iron than they could have done because the inner end of the tuyère was too close to the bed formed in the kodon chaff. Either the bed should have been lower, or the tuyère raised. Once the slag froze in its bowl it could not be tapped and thereafter further slag build up obstructed the flow of air.
ii. Research driven smelting focuses on optimising the results. With the Agaria, however, the outcome may be considered by the smelters to be in the hands of Lohasur, a divine being to whom the many offerings and supplications are made. This may result in what appears to be a more *laissez-faire* approach, once the rituals have been carried out, because ultimately Lohasur, not C, controls what iron is created in a smelt.
iii. Invisible factors unknown to the observer might have intervened. For example, if either the piglet or chick had refused to eat the rice offered, the smelt would have been aborted. In smelt two the piglet was very suspicious and would not eat even when its snout was prodded into the food. Fortunately, H saved the day by bringing some warm cooked rice from next door. This was gobbled up and the sacrifice, followed by the smelt, went ahead. If the animals brought to be sacrificed had refused to take the rice, this would have been seen as Lohasur's unwillingness to proceed.
iv. More complicated explanations, beyond the scope of this brief study, may have been behind the ambivalence displayed by C during the smelt. The first smelt in a new furnace produces 'virgin iron', highly significant in Agaria culture (Elwin, 1942, pp.90-94, 158-161). It protects against lightning and illness and promotes fertility. In a sense it is the *pièce de résistance* of a society whose entire mythology is based on iron production. It is possible that granting so much, so soon, to a stranger caused difficulties.
v. The passion for iron that sometimes enflames archaeometallurgists and experimenters alike is of a different order from the combination of religious belief and enduring myth that drives the Agaria. Neither archaeology nor scientific examination of bloomery products throws much light on the belief systems of early iron producers in Europe or elsewhere. But the Agaria show us that such considerations can be relevant.
vi. The bloom, carried away to Bangalore to be cut in half, seemed to be a failure. But it was shown later to be a failure only in terms of the quantity of iron it produced. C fashioned from it rings of excellent virgin iron, of a value probably immeasurable in economic terms, because such items would not be traded. There is a danger in assuming economic criteria are the only ones influencing the production of bloomery iron.

Discussion B: Kodon chaff

Elwin's monograph on the Agaria (Elwin, 1942, p.148) includes a myth which seemed to refer to the technological requirement for an insulated furnace bottom. After years of suffering frozen slag in experimental smelts, the author began using a bed of charcoal fines, shaped to receive the bloom. This helped keep the slag molten and enabled it to be tapped when required. It seems that when Lohasur is happy in his bed of kodon chaff (or charcoal fines), iron will be found there: "The twelve Agaria brothers put iron stones in the furnace, and for eight days and nine nights they kept up their fire, but the iron did not come. The iron used to go out secretly to a pile of kodon chaff and play with it. Then came an old woman, and said, 'What are you doing my sons? The iron is over there. It is playing with the kodon chaff.' They laughed at her, but she said again and again, 'No, it is over there, in the pile of chaff.' So at last one of the brothers went and saw that Lohasur really was there. That brother picked up Lohasur, but his hands broke; then all the brothers came, and somehow they put Lohasur back into the furnace, but he again crept out and went to play with the kodon chaff. Then Lohasur came in a dream and said, 'Make a place of kodon chaff within the furnace where I can sit.' When they did that, Lohasur remained. Lohasur wanted to marry the virgin kodon chaff. For love of her he used to go to play with her. And to this day, if the kodon chaff is not put in the furnace, the iron does not settle properly."

Conclusion

Can looking at a contemporary survival of traditional bloomery iron smelting in India provide any insights relevant to early iron production elsewhere in the world? Archaeology and history are limited in the information they can provide. The Agaria smelt suggests that: (i) ritual/religion can be a major preoccupation influencing all aspects of the process, (ii) social aspects, e.g. feasting, can overwhelm its purely economic foundations, (iii) mythology may be a way in which technological information can be transmitted, (iv) belief in the efficacy of certain types of iron, e.g. virgin iron, may make sense of the exploitation of otherwise poor quality material.

It is recognised that such a brief study, without an adequate translation of the Agaria's own thoughts, or a quantification or analysis of the materials used, will be of limited academic value. However, the technological traditions observed above may have a very long history (see Tewari, 2003). At present these unique smelting skills are sustained by their community's belief in the divine nature of iron production, as much as they are by economic demand.

References

Elwin, H.V.H., 1942. *The Agaria*. London: Indian Branch, Oxford University Press.

Tewari, R., 2003. The Origins of Iron Working in India: New evidence from the Central Ganga Plain and the Eastern Vindhyas. *Antiquity*, 77, pp.536-44.

Author's address

• jakekeen@hotmail.com

An ethno-archaeological survey of iron working in India

Vibha Tripathi

ABSTRACT: Owing to new radiocarbon dates, the origins of iron production in India are now potentially dated to first half of second millennium BCE (Tewari 2003:536-545; Tripathi 2001, 2008). Excavations across India have revealed furnaces and other archaeometallurgical remains relating to iron production in this era, which will be highlighted within this paper. Reconstructions from the archaeometallurgical record of both furnace design and the method of these furnaces' operation are provided below, using a multidisciplinary approach including literary, archaeological and ethnographic material. The paper will also examine how iron metallurgy in India appears to have evolved from the production of low grade bloomery iron to steely iron, including the ***wootz*** steel that was famous and in great demand until 18th-19th century AD.

Keywords: Indigenous, ethno-archaeological, bloomery iron, metallurgy, typology, shaft furnace, Agaria, Asur.

INTRODUCTION

The present paper proposes to examine the available evidence on ancient Indian iron over the course of millennia. While examining ancient Indian iron technology a number of questions arise, of which key issues include the status of iron and iron producers, and whether ethnographic records can help in reconstructing archaeological examples. In addition, it is interesting to question whether evolutionary stages within the growth of metallurgy are visible within the archaeological record, and what the properties of Indian iron and steel actually were.

Within this paper the origins and development of iron through the ages will be examined and the typology of objects, as well as their role during different chronological and cultural periods, will be highlighted. Ethnographic evidence of iron production in India is drawn upon to consider past metallurgical processes; including the written accounts of travellers from the Middle East, the records of the Geological Survey of India and European correspondence from the 18th-19th century A.D. A diverse body of evidence will be used to provide valuable insights into the status of iron and steel in India, as the following examples demonstrate.

In the 19th century Capt. Presgrave of Sagar (MP) mint analysed the iron produced by the Agarias of Tendukhera (near Hoshangabad), commenting, "----bar iron...of most excellent quality, possessing all the desirable properties of malleability, ductility at different temperatures and of tenacity for all of which I think it cannot be surpassed by the best Swedish iron; ... the Agaria piece when brought to the bend it showed itself possessed of the power of elongating and stood the bend better than the general run of English iron purchased in the Bazar" (cited by Dharmpal 1971 p.289). In 1878 Sir George Braidwood (1878, cited by Krishnan 1954 p.70) noted in the notebook of the British Indian section: "Indian steel was with such properties celebrated from the earliest antiquity and the blades of Damascus which maintain their pre-eminence even after the blades of Toledo became celebrated, were in fact made of Indian iron... The ***Ondanique*** of Marco Polo's travels refers originally, as Col. Yule has shown, to Indian steel, the word being a corruption of the Persian ***Hindwany*** i.e. Indian steel. The swords of Kirman were eagerly sought after in the 15th and 16th centuries AD by the Turks who gave great prices for them. Arrian mentions Indian steel '***Sideros indicos***' (that) was imported into Abyssinian ports".

After the Great Indian Mutiny of 1857, the British Government confiscated all the sharp edged weapons such as swords, daggers and knives which were kept by the people. The weapons of the 'Moplas of Malabar', made with native iron in the Indian blast furnaces, are described by Charles Wood (1894 p.179) :"---and wonderful material they were. To break them was impossible, so a pair of strong hand-shears was made to cut them up. But the remarkable point was this, that if put into the shears with the thin cutting edge first, they could not be cut at all, but notched the shear blades immediately." Likewise, La Touché (1918) observed "...its (iron's) superiority is so marked, that at the time when the Britannia Tubular Bridge across the Menai Straits was under construction preference was given to use of iron produced in India".

These statements are testimonies by European experts pertaining to the excellence of the Indian iron and steel that was being exported to different parts of the world until 19th century AD. This paper proposes to re-examine the salient features of ancient Indian iron.

Archaeological evidence

The issues related to the antiquity of iron in India have been studied in recent decades by a number of scholars, including

Sahi (1979, 1994), Chakrabarti (1992, 1994), Prakash (1991, 1997), Tewari (2003pp.536-544) and Tripathi (2001, 2008). Recent radiocarbon dates going back to the first half of the 2nd millennium BCE from the sites located in the Middle Ganga Plain and the Eastern Vindhyas (Tables 1.1-1.4) have revolutionized the antiquity of iron technology in India (Chakrabarti 1992; Chakrabarti, D.K. and N. Lahiri, 1994:12-23; Tewari, Rakesh, R.K. Shrivastava, K.S. Sarswat, K.K. Singh 2000: 69-98; Tewari, Rakesh, R.K. Shrivastava, K.K. Singh 2002:99-116, 2003; 536-544, 2010: 81-97; Agrawal 2000; Balsubramaniam 2008; Srinivasan and Ranganathan 2004; Magee 2004:37-44; Jayaswal 2009: 141-146).

Table 1.1: Radiocarbon Dates from iron age level (Raja Nal- Ka -Tila).

SI no.	Laboratory no.	Layer/depth	Radiocarbon dates in years BP/BCE 5568 ± 110BP	Calibrated dates based on half life 5530 ± 40 years	Calibrated
1	BS-1378 1996-7 Trench no. U-19	6. 1.95-2.00#m with iron	2550 ± 110 BP 600 ± 110 BCE	2626 ± 110 BP 676 ± 110 BCE	882 (773) 486 BCE
2	PRL-2047 1996-97 Trench no. U-20	6. 2.08-2.10#m with iron	2890 ± 90 BP 940 ± 90 BCE	2980 ± 90 BP 1030 ± 90 BCE	1196 BC - 1188 BCE 1164 BC - 1143 BCE 1132 BC - 976 BCE 970 BC - 930 BCE
3	BS-1299 1995-96 Trench no. A-1	Pit sealed by layer 6. with iron	2830 ± 100 BP 880 ± 100 BCE	2914 ± 100 BP 960 ± 100 BCE	1118 (963) 859 BCE
4	BS-1300 1995-96 Trench no. A-1	6. 2.00#m with iron	3060 ± 110 BP 1110 ± 110 BCE	3150 ± 110 BP 1200 ± 110 BCE	1423 (1307) 1144 BCE
5	PRL-2049 1996-97 Trench no. T-19	6. 2.00#m with iron	3050 ± 90 BP 1100 ± 90 BCE	3150 ± 90 BP 1200 ± 90 BCE	1406 BC - 1198 BCE 1186 BC - 1164 BCE 1143 BC - 1132 BCE

Table 1.2: Radiocarbon Dates from Iron Age level (Malhar).

SI no.	Laboratory no.	Radiocarbon dates in BP/BCE on the basis of half life 5568 ± 30 years 5730 ± 40 years		Cal BCE Stuiver et al. One sigma
1	BS-1623 MLR II Trench no. XA1, layer no. 3. Depth 0.55cm	3450 ± 90 BP 1500 ± 90 BCE	3550 ± 90 BP 1600 ± 90 BCE	1882 (1743) 1639 BCE
2	BS-1614 MLR II Trench no. A1, layer no. 3. Depth 61-63cm	6380 ± 110 BP 4330 ± 110 BCE	6570 ± 110 BP 4620 ± 110 BCE	5475 (5358, 5351, 5340, 5329, 5323) 5262 BCE
3	BS-1593 MLR II Trench no. A1, layer no. 3. Depth 90-100cm	3540 ± 90 BP 1590 ± 90 BCE	3650 ± 90 BP 1700 ± 90 BCE	2012 (1882, 1836, 34) 1742 BCE
4	BS-1590 MLR II Trench no. XA1, layer no. 4. Depth 80cm	3740 ± 80 BP 1790 ± 80 BCE	3850 ± 80 BP 1900 ± 80 BCE	2283 (2141) 1984 BCE

Table 1.3: Radiocarbon Dates from iron age level (Jhunsi).

SI no.		Trench details	Radiocarbon yrs BP			Date in BCE/AD	Calibrated dates in BCE/AD
				5568 ± 30	5730 ± 40		
AU/JHS/9	207 5	C-15 (46) 1210	Pre NBP with iron	2650 ± 90	2730 ± 90	780 ± 90 BCE	897 (806) 789 BCE
AU/JHS/12	207 7	C-15 (49) 1240	Pre NBP with iron	2820 ± 90	2900 ± 90	950 ± 90 BCE	1107 (973, 956, 941) 844 BCE
AU/JHS/16	208 1	C-15 (53) 1325	Pre iron (Chalcolithic)	2700 ± 90	2780 ± 90	830 ± 90 BCE	966 (830) 799 BCE
AU/JHS/18	280 3	C-15 (62) 1520	Pre iron (Chalcolithic)	3200 ± 90	3290 ± 90	1340 ± 90 BCE	1597 (1490, 480, 1450) 1400 BCE

Table 1.4: PGW SITES. All BP dates are based on 5568 yr half-life of radiocarbon.

Hastinapur	TF-91	2450 ± 120	752, 709, 530 cal. BC	Late Painted Grey Ware, Period II
Hulas	PRL- HLS- TL-1	None	None	Period II, Painted Grey Ware, late phase
Khalaua	PRL-67	2450 ± 155	752, 709, 530 cal. BCE	Painted Grey Ware
Khalaua	TF-1228	2420 ± 95	484, 437, 424 cal. BCE	Painted Grey Ware
Khalaua	PRL-68	2370 ± 170	403 cal. BCE	Painted Grey Ware
Noh	TF-1144	2370 ± 85	403 cal. BCE	Painted Grey Ware
Noh	TF-994	2560 ± 100	794 BCE	Painted Grey Ware
Noh	UCLA-703A	2480 ± 250	760, 684, 656, 638, 592, 586, 550 cal. BCE	Painted Grey Ware
Noh	UCLA-703B	2690 ± 220	833 cal. BCE	Painted Grey Ware
Noh	TF-993	2600 ± 145	801 cal. BCE	Painted Grey Ware
Sonkh	1-6277	2570 ± 85	795 cal. BCE	Painted Grey Ware

Sahi highlighted evidence of the occurrence of iron at Ahar in a context that was reported to be Chalcolithic in nature. Two 14C dates of carbon samples (TF 31 and TF 32) had yielded the date of 1270 BCE and 1550 BCE from the above layers. On that basis Sahi stated, "... as indicated by the position of iron objects in the deposits, the introduction of iron at the site may date back to the beginning of 16th century BC, if not earlier". On this reckoning he proposed, "The date of beginning of iron smelting in India may well be placed as early as the sixteenth century BC..." (Sahi, 1979:366), although this assumption remained controversial on the grounds of the uncertain stratigraphical position of the site of Ahar. Since then, however, new evidence of the occurrence of iron has been forthcoming, for example with radiocarbon dates from sites like Gufkral (Sharma 1992: 64,67) from a megalithic context dated to 1550-1300 – not calibrated. A number of other sites like Komaranhalli in Karnataka also yield early dates (T.L. dates ranging from 12th to 15th century BCE); Veerapuram, Ramapuram, Andhra Pradesh (16th-11th century cal. BCE; Raja Nal-ka-Tila (1423-1144 cal. BCE, Tewari, 2000:93), Malhar (1882-1639 cal. BCE Tewari et al., 2000 :88), Dadupur (1882-1465 cal. BCE (Tewari et al., 2002:111). The carbon samples from sites like Lahuradeva, Jhusi and Atranjikhera have also yielded 14C dates ranging from 1200 BCE to 1000 BCE (see Tables 1.1-4). Tewari has recently reviewed in detail the radiometric dates available from Iron Age sites in India (Tewari 2010:79-97) and arrived at similar conclusion that iron started in South Asia in the earlier half of the second millennium BCE. Proceeding with caution we may try and procure more dates, preferably AMS dates from the strata containing iron. As the evidence stands today, however, the beginning of iron in India appears to date to the first half of the second millennium BCE, both on the basis of archaeological and radiocarbon determinations.

These dates are older than those reported from the countries bordering the Indian sub-continent, suggesting an earlier and possibly indigenous beginning to iron in India. Without indulging in further details on beginnings of iron in India we proceed to discuss the status of iron metallurgy in northern India. It is possible to view the developmental stages of iron metallurgy from simple wrought iron to steely iron and, in view of the long period of iron production in ancient India, it is reasonable to divide the early Iron Age into three stages of development that is, Early, Middle and Late Iron Age.

Early Iron Age (1700/1600 – 700/600 BCE)

It is seen that around the first half of the 2nd millennium BCE iron came into use, albeit in small quantities, in several of

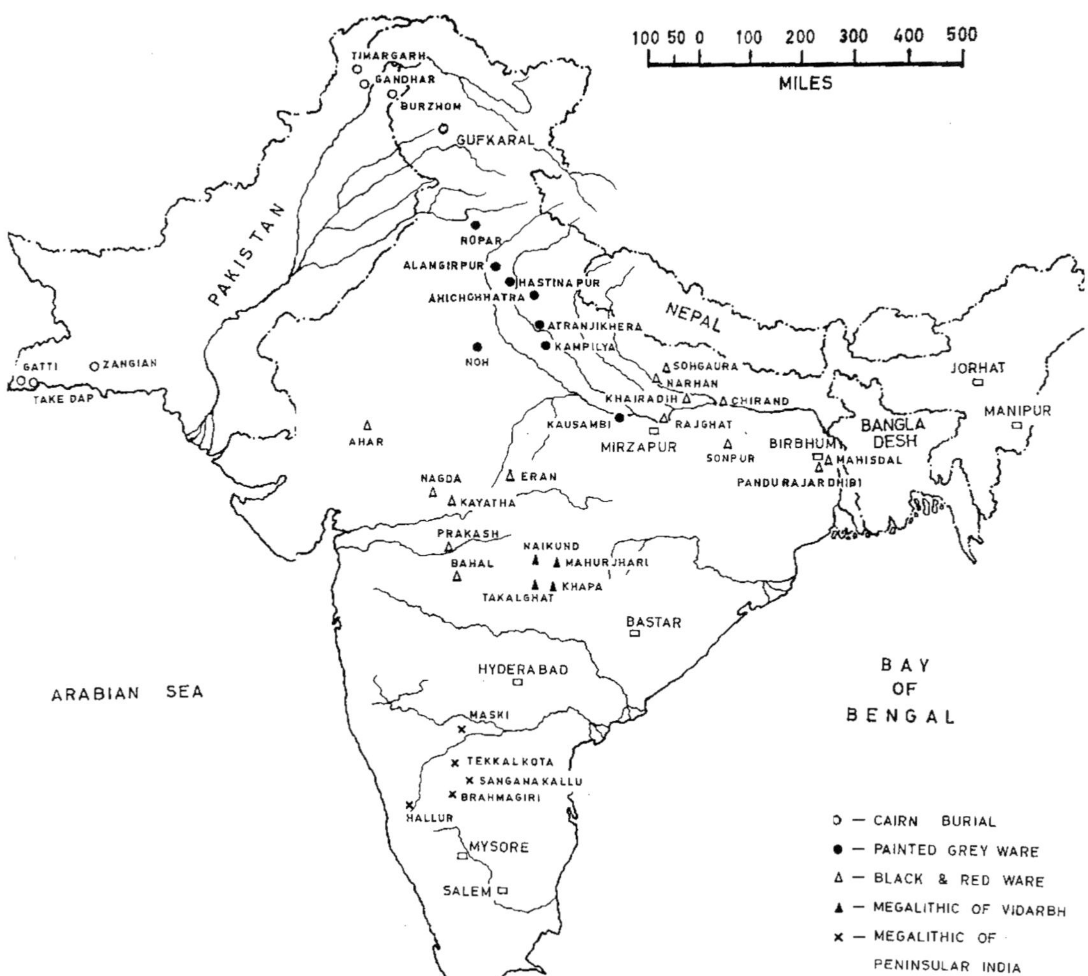

Figure 1: Map showing location of Iron Age sites.

the eco- zones in India (Fig. 1). The findings are summarised below.

In the Upper Ganga Plain, prior to the regular use of iron in India, shapeless tiny pieces of iron were noted in Black-and-Red Ware (BRW, henceforth) phase at sites such as Noh, district Bharatpur in Rajasthan and Jakhera, district Etah; and Dadupur in Lucknow, Uttar Pradesh. Jakhera has notably yielded an indeterminate object and an arrowhead with a long tang from the BRW Phase assignable to the first half of the second millennium BCE. From the succeeding Period III, 19 iron objects including hoes, sickles and arrowheads, along with a lump of iron (slag?) and fragments of slag were found. A furnace base was also recovered in this proto-Painted Grey Ware (PGW, henceforth) phase itself. This phase is followed by Period III B – the PGW culture. Iron objects found during this Period include hoes, sickles, spearheads, arrow-heads, daggers, choppers, chisels, axes, nails, rods etc. (Sahi 1994 pp.142, 148, Figs. 14, 15). Nearby, the site of Atranjikhera displays a similar cultural sequence yet displays no evidence of iron present during this period – only the succeeding PGW period yields iron. This 2.50m thick deposit is divided into three sub-periods. The lowermost phase has yielded only seven indeterminate pieces of iron. Regular tools appear from mid-PGW phase (46 objects). The upper phase yields 81 objects. This indicates a gradual but consistent rise in iron objects between BCE 1100/1000 and 600 at most of the sites mentioned here (Fig. 2). The evidence is repeated uniformly at all of the sites, with the number of objects increasing gradually in the succeeding periods. Therefore, it is reasonable to expect that the state of preservation will be the same in all the periods. Significantly, all of these sites are located near the Agra-Gwalior iron ore deposit.

At most other sites (Hastinapur, Alamgirpur, Allahpur etc) iron came to be used for the first time with PGW. Spear-heads, a barbed arrow-head, nails and pins, clamps, chisels, rods etc, were the main objects found at these sites. Closer examination of PGW culture reveals that western Rajasthan and Haryana had no iron present at earlier phases. Iron technology therefore seems to have been developed, or learnt somewhere, during an eastward movement of PGW culture from Rajasthan to the Upper Gangetic Plain (Tripathi 1997). As the PGW culture moves further east towards the Gangetic plain, iron starts to

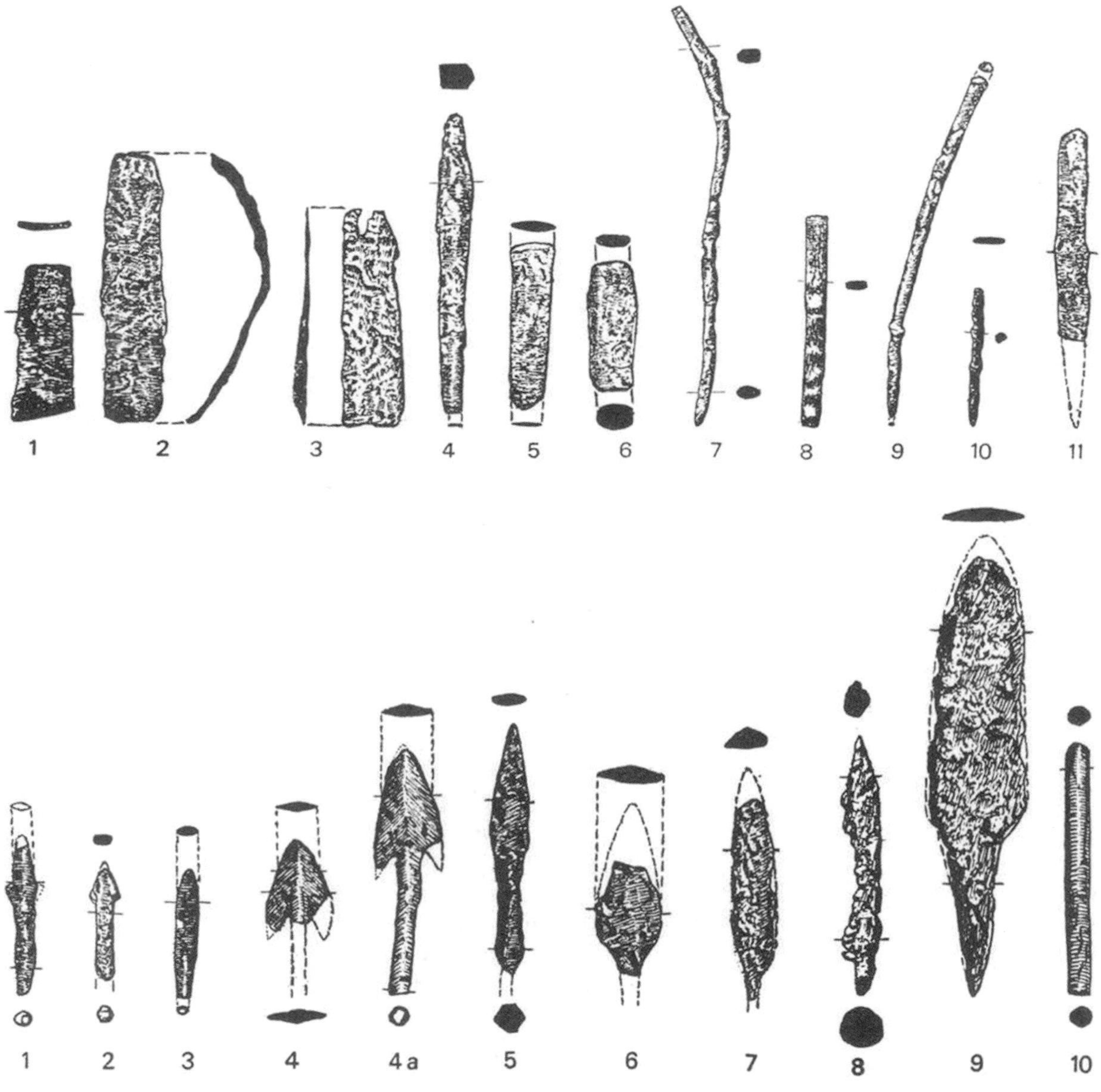

Figure 2: Iron Objects of PGW (Stage I, Atranjikhera).

occur right from the earliest levels. In the latter region iron occurs right from the earliest levels of the culture.

Further east, in the Middle Ganga Plain, iron makes its first appearance with the BRW pottery tradition in an otherwise Chalcolithic milieu. Jhusi and Kausambi near Allahabad yielded iron from the BRW phase (1300-1100 BCE). The number of objects as well as their quality seems to improve with time. A pattern of typological evolution may be discerned at other Early Iron Age sites as well. Mention may be made here of sites such as Chirand in Bihar and Hatigra, Pandu-Rajar-Dhibi, Mangalkot etc, in Bengal. The latter has yielded eight iron objects from the earliest phases of occupation, including an arrow-head, spearhead, nail and rod. A furnace, along with 16 iron objects, was found in a trench of 6m x 6m at Mangalkot, datable from the lowest level to 1200 BCE. The succeeding period yielded more evolved shapes such as sickles, chisels, pegs and a knife or sickle blade (Ray 1990: 131-140; Datta 1992: 293-308).

The most revealing archaeological evidence has come from the sites of the Vidhya-Kaimur hills bordering the Ganga Plain. Radiocarbon dates from Malhar, Raja Nal-Ka-Tila and Koldihwa range between 14th to 17th centuries BCE. Dadupur near Lucknow yields an earlier date of 18th century BCE. Nearby, in the Vindhyan region itself – and especially in the districts of Mirzapur, Chandauli and Sonbhadra – we come across another set of sites such as at Morahana Pahar and Baghair Khoh near Bhaisaur, wherein iron occurs along with microliths. This shows either an early beginning of iron or a late survival of microlithic tradition. Importantly, the traditional Indian iron smelting communities such as the Asurs and the Agarias still live in the area, and traditional iron working still survives.

During the early period iron was produced in bloomery furnaces and can therefore be viewed as relatively elementary. The analysed objects of Hatigra dated to 1100-1000 BCE displayed *widmanstaaten* structure due to prolonged exposure at a temperature of about 1200°C followed by a slow cooling. It is said to be a 'low carbon hypoeutectoid steel' (Ghosh and Chattopadhyay 1987: 88). The concentration of carbon on the edges visible in the metallographic picture indicates carburization. It appears that the furnaces at this stage were not efficient enough to generate sufficiently high temperature. The specimens from Hatigra, Pandurajar Dhibi and Mangalkot contain a high percentage of silica (De and Chattopadhyay 1989, Fig. 4A-C; Datta 1992: 303).

Middle Iron Age (7th-8th Century BCE to 1st to 2nd AD)

The Northern Black Polished ware succeeds PGW culture in North India by 800/700 BCE (Fig. 1), contemporaneous with the late phase of Painted Grey Ware culture. This was a period of consolidation of iron technology with traces of steeling, case hardening and carburization. A relative increase is recorded in the number and types of iron objects (Table 2). There is a qualitative and quantitative improvement in iron objects: more sophisticated weapons including javelins, lances, daggers, blades and elephant goads, along with the earlier types which become more prolific. Agricultural implements, rarely reported from Early Iron Age sites (with a few exceptions like the site of Jakhera), appear to become relatively frequent during this period.

Six iron samples from Rajghat (Varanasi) belonging to 600-400 BCE level were analysed. All of them are found to be of wrought iron with slag inclusions (Bhardwaj: 2000, Figs. 55-58). Evidence of carburization has been attested to. Sample No. 6 has 1.1% carbon, although it is difficult to ascertain whether carburization was deliberate (Bharadwaj 1979:153).

Metallographs of iron objects are illustrated here (De and Chattopadhyay 1989: 37, see Fig. 4A-C). An iron sickle of Pandurajar Dhibi (Period III, NBP phase) analysed by De and Chattopdhyay (1989: 37 Fig. 4C) shows a non-uniform structure with martensite. It also exhibits retained acicularity at certain places, especially around large areas of ferrite. "Electron micrograph obtained at a magnification of 100 x clearly represent its tempered martensitic structure" (De and Chattopadhyay 1989: 37). It may be said that the iron used to make the sickle blade had been forged at a significantly high temperature to extricate the slag particles, giving the metal a more homogenized structure. Carburization was done during manufacture of the tool by subsequent heating and forging.

"Inside the core, the carbon content that is retained is only 0.22%. But the high level of corrosion that took place over the time must have caused depletion of carbon. There is also an uneven distribution of carbon concentration. It indicates that carbon was more than 0.4% initially." (De and Chattopadhyay

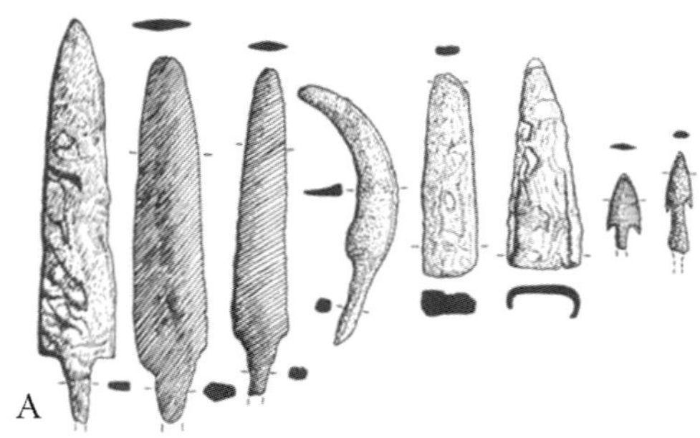

Figure 3A: Iron objects of NBPW (Stage II, Atranjikhera).

Table 2: Iron objects at different cultural stages

clamps	*	*	*
chisels	x	*	*
pipes	x	*	0
sockets	x	*	0
plumb bobs	x	*	0
chains	x	*	*
door hooks	x	*	*
door handles	x	x	*
hinges	x	x	*
spikes	x	x	*
tweezers	x	x	*
anvils	x	x	*
hammers	x	x	*
scissors	x	x	*
saws	x	x	*

KEY

Definite existence	*
Confirmed date not available	0
non-existence	x

1989: 37). Chattopadhyay also finds indications of quenching and tempering of this specimen.

Late Iron Age: (2nd AD to 5th -6th Century AD)

In the opening centuries of the Christian era there was not only a proliferation in tool typology (Figs. 3B, C and Table 2) but iron metallurgy attained greater heights of technical sophistication, with techniques such as lamination and quenching seen within the iron objects (Fig. 4E). Taxila has yielded a rich iron tool repertoire including some armour grade weapons in the opening centuries of the Christian era (Fig. 3B). Hadfield (1913-14) found Taxila objects to be of high carbon steel, the type of which compares closely with iron production by ethnic smelters. Sisupalgarh in Orissa (datable to 5th-6th AD) yielded a caltrop, a weapon to be used in the battlefield to curb elephants. It is well attested that the ancient Indian smiths at this stage had a thorough knowledge of the importance of carbon alloying, case hardening and tempering. However, these techniques seem to have been used judiciously wherever necessary. A nail, for example, differs in composition from a knife or a dagger, showing a

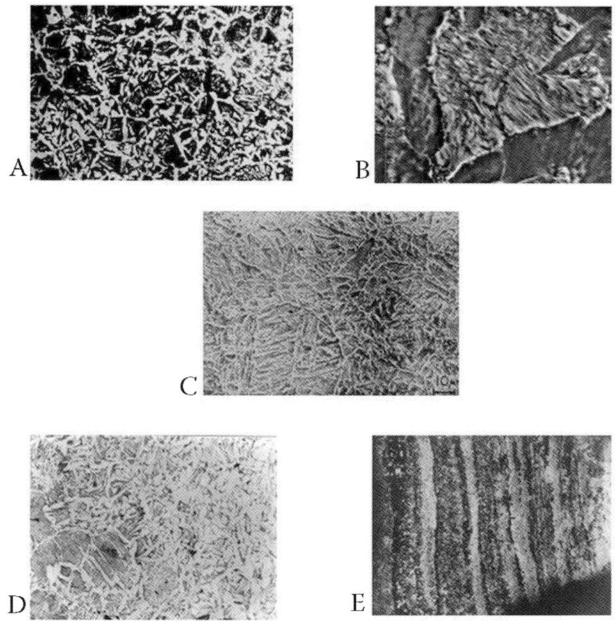

Figure 4 : Metallographs of Iron Objects. A: Photomicrograph, iron implement showing carburization Widmanstaaten pattern, Hatigra (500X); B: Electron Photomicrograph, iron implement, Hatigra (3000X); C: Electron micrograph of a sickle, 1000X, 15KV revealing tempered martensitic structure, Pandurajardhibi; D: Microstructure of a kitchen spoon, Sringverpur; E: Microstructure of a bent knife, Sringverpur

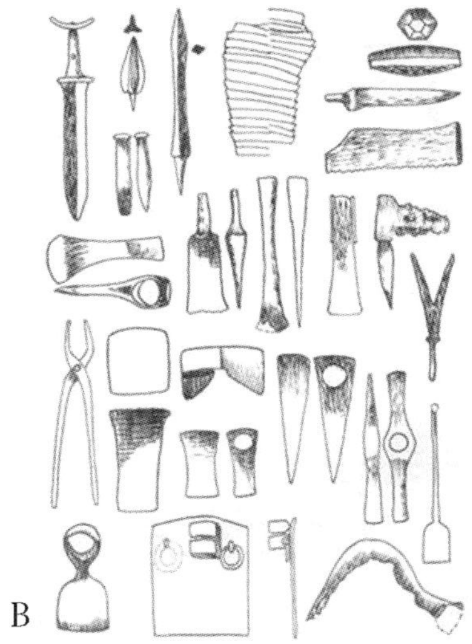

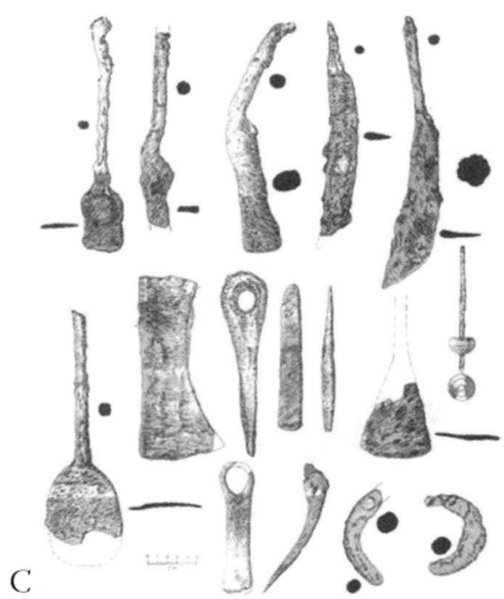

Figure 3B: Iron objects (Stage III, Taxila).
Figure 3C: Iron objects (Stage III, Khairadih, UP).

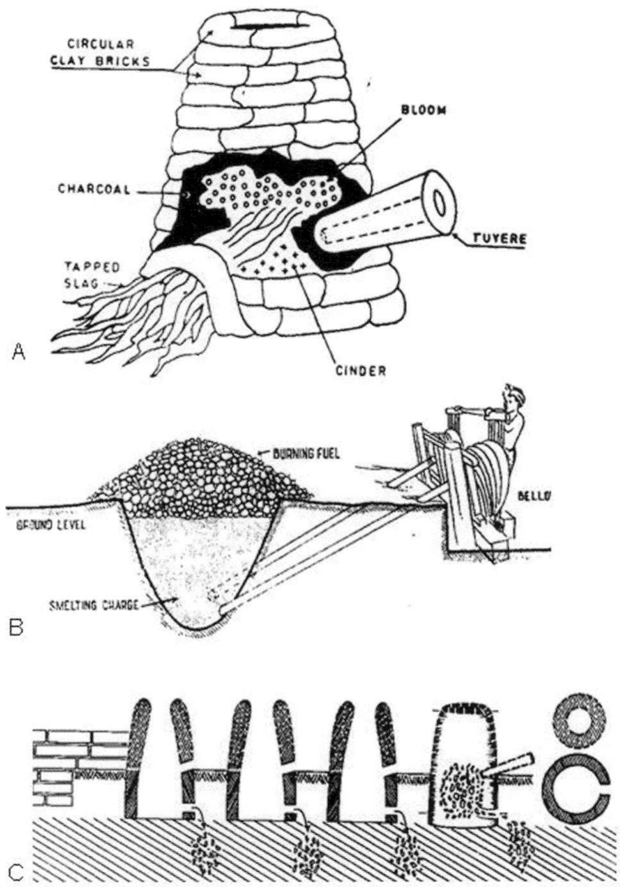

Figure 5: Reconstructed ancient iron smelting furnaces. A: Megalithic Naikund; B: Dhatwa; C: Khairadih

selective use of carburization or quenching by the artisans. These technical skills must necessarily have been acquired over a long period of time, before the craftsmen ventured to take up more challenging jobs – such as the manufacture of a 7-8 ton iron pillar, whose corrosion resistance must have required a skill of high order.

The colossal Delhi Iron Pillar (Fig. 10A), produced with approximately 8000 kg of iron, is an example of the art of mass production by 5th-6th AD. The iron making furnace of Nagpur (17th century AD) could produce about 40 kg of iron per operation (Prakash and Tripathi, 1986), hence the means by which the Delhi Pillar was produced is unknown. It is thought that at least 200 furnaces of that size would have to have been operated simultaneously, or that the same furnace was operated repeatedly, to produce enough iron of consistent quality to forge the pillar (Balasubramaniam, 2008: 48-58).

It appears that over time, artisans mastered the technique of forge-welding to manufacture colossal structures such as pillars and beams. An examination of the fractured surface of the beams used in the Sun Temple at Konark in Orissa (Fig. 10B) clearly indicates that it was manufactured by forge-welding square rods. Jena (cited by Prakash 1997) found traces of lead between rods where the forging joint was not perfect. It is believed that a large lead bath was used for the uniform heating of a bundle of wrought iron bars to the forging temperature, at which point they were forged together. Since the iron surface is non-wettable by lead, normally the lead will flow out when the wrought iron bundle is removed, although some molten lead might get trapped in the crevices. Incidentally, some lead has been noted in the Delhi Iron Pillar (Balasubramaniam 2008:53).

Wootz, or Damascus steel was famous and was much in demand in the marketplaces of the ancient world. This type of steely iron was generally produced in the southern part of India, although it seems that there were several sword making centres in other parts of India in the early centuries of Christian era (Tripathi, 2007 pp.403-426). A 5th century text supposedly composed by Varahmihir, the famous mathematician and astrologer, devoted a section (the *Khadga Lakshanam)* to the techniques of producing sharp edged weapons, notably sword blades.

Reconstructing the Ancient Indian iron technologies

Excavations generally do not yield intact furnaces. Instead we find refractory material, tuyeres, slag pieces, ore etc. Some sites yield ore crushing tools (Fig. 7). On the basis of available finds some attempts have been made to reconstruct iron technology as practiced by ancient iron smelters. The ethnological data relating to these is discussed below.

Furnace remains

The furnaces used by the ancient Indian iron workers are rarely exposed in a good state of preservation because of (a) the destructive nature of the iron working itself and (b) the problem of archaeological preservation. In most cases, furnaces are represented within the archaeological record as shallow pits full of ash. Mention may be made here of a smith's workshop with broken furnace, iron tools including a pair of tongs found in PGW Period (1100/1000 BCE) at Atranjikhera (Dist. Etah, Uttar Pradesh: Gaur 1983 p.127). PGW Period contexts at Jodhpura, (Dist. Jaipur in Rajasthan), have also yielded a forge and a "furnace with a side hole into which the nozzle or bellows could have been inserted. The site yielded iron objects like axes with broad cutting edge, daggers, spearhead, arrowhead with leaf shaped point, socketed tang, nails, tongs etc" (Agarwala and Vijaikumar 1976 pp. 241-244).

Clearer evidence in this context comes from the PGW level from Jakhera in Dist. Etah (mentioned above). Excavations revealed a semi-circular furnace base containing slag and burnt charcoal and appeared similar to an ancient furnace excavated by the author. This bowl-shaped slag lump with cinder and charcoal found in situ appears to be the lower portion of a slag pit furnace (similar in appearance to those observed at several sites in the ancient world). Some other sites, for example Naikund, Dhatwa and Khairadih have yielded sufficient remains to reconstruct the furnaces (Fig. 5A-C). At the latter site the author was able to reconstruct a series of three furnaces. During the excavations at Malhar, Dist. Chandauli a furnace belonging to Period II was exposed at the mound known as Lohsan (Tewari, 1998). This cylindrical clay furnace measured 1.30m in height, its dimensions being 0.80m (inside) and 1.00m on the outside, with an extant depth of 35cm (Fig. 6 B,C). The inside of the furnace was burnt, with slag adhering to it. Tuyeres, slag, iron axe and potsherds (dish, bowl, vases of grey ware, black slipped ware, red ware and a few coarse varieties of BRW pieces) were found associated with it. Incidentally, it compares well with the Agaria furnace commonly used in some parts of India (Fig. 6A).

In the eastern part of the Middle Ganga Plain, similar shallow pits have been excavated at several Chalcolithic sites. At Pandu Rajar Dhibi dated by ^{14}C from 1045±55 BCE

Table 3: Distribution of iron objects at PGW level (Atranjikhera)

Sl no.	Object	Lower phases	Middle phases	Upper phases	Total
1	arrowhead	-	7	14	21
2	spearhead	-	3	5	8
3	shaft	2	5	3	10
4	tong	-	-	1	1
5	clamp	-	10	11	21
6	nail	2	4	14	20
7	bar/rod	-	2	5	7
8	hook	-	2	5	7
9	borer	1	2	3	6
10	chisel	-	4	2	6
11	needle	-	-	1	1
12	axe	-	-	1	1
13	knife	-	1	2	3
14	bangle	-	1	1	2
15	lump (slag)	1	2	4	7
16	indeterminate	2	3	9	14
	Total	8	46	81	135

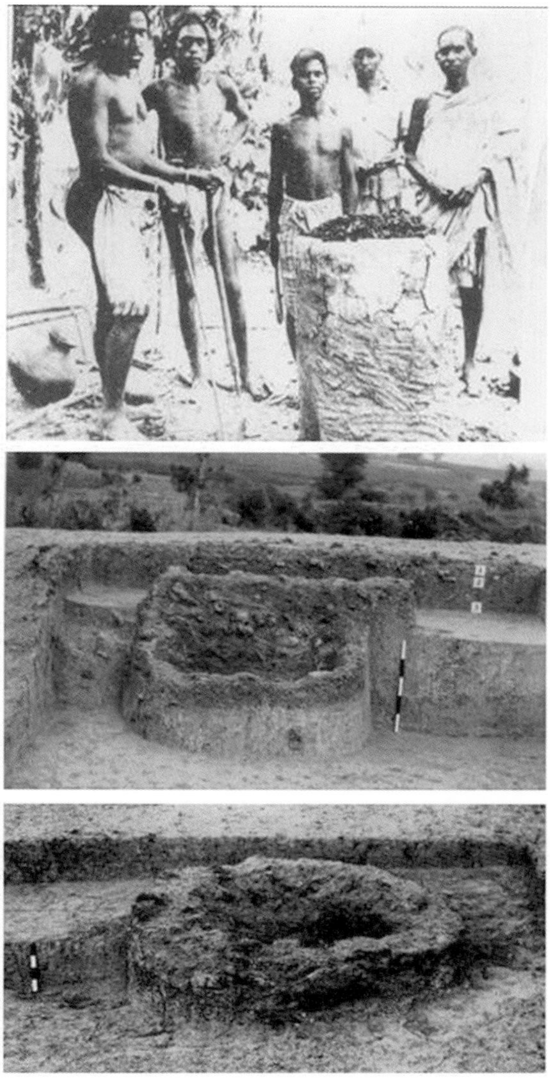

Figure 6: Ancient Furnaces. A: Agaria Workers with their furnace, Ranchi; B: Ancient Furnace at Lohsan, Malhar Pd. II; C: Ancient Furnace at Lohsan, Malhar Pd. II.

to 920±50 BCE iron objects have been found in a small number. The microscopic and chemical analysis of slag led to the conclusion that smelting was carried out at a rather low temperatures (Fig. 4C). The hearth was filled with layers of charcoal and ore and there is no clear evidence to suggest the use of tuyeres. Some of the samples, particularly the one belonging to Pd. III (i.e. post-Chalcolithic) contain fayalite (iron silicate) in high quantities. This proves, according to De and Chattopadhyay (1989), an elementary knowledge of iron smelting, with the presence of significant amounts of iron silicate making the iron quite brittle.

Naikund (Gogte, 1982), a megalithic site 42km North-East of Nagpur in the Vidarbha region in Maharashtra on the left bank of river Pench has yielded valuable evidence of iron smelting. An iron smelting furnace of 30cm diameter and 25cm height was constructed with circular clay bricks arranged one above the other. The upper surface of the lower brick was convex and the lower surface of the upper brick was concave, thus they fitted well into each other (Deo and Jamkhedkar 1982, Fig. 5A). The inner side of the bricks had turned black due to firing in reducing conditions. The outer side of the bricks remained brownish red. A hole was provided at the bottom for tapping slag. A few of the bricks were found fused with slag and cinder. Two heavily vitrified tuyeres were recovered during excavations, made with clay and a heavy mixture of quartz grains, and measuring a length of 16 cm and a diameter of 2.5cm-3.6cm. The trench also yielded pieces of ore which has been described by Gogte as 'crystalline to microcrystalline micaceous hematite quartzite which is generally found with manganese ores' (1982 p.54). The site is located only 3km from the Parshivni open cast manganese mine. A fully-fledged workshop area was exposed at Khairadih in District Ballia (Fig. 5C), where three shaft furnaces in a row were found and dated to the early historic-historic period context (1st AD). Slag, ore, crucible, ore crushers, hammers with short handles, four legged oblong quern (Fig. 7), mica and silica have all been found in a small trench. This appears to be some kind of a composite workshop area as glass tiles and glass beads were also found in close proximity.

This evidence creates an understanding of iron smelting furnaces that were being used in different parts of India at different periods. At the earliest level, it has been suggested that iron was being smelted in small bowl shaped furnaces like the ones used in copper smelting. The bloomery iron produced could easily be forged and beaten to desirable shapes and, as production efficiency was low and the number of iron objects produced was small at the beginnings of the introduction of iron, the limited number of iron objects relating to the early stages of its introduction can be explained. With time, the technology appears to have been perfected. To aid in the reconstruction of ancient technological practice, archaeological evidence can be substantiated with the ethnological models. Since there is a long chronological gap between the archaeological and ethnological evidence, we can only use the latter as a model for understanding the ancient practices. However the practices that survive even today in certain remote parts of India among ethnic communities can provide a good insight into past practices.

Ethnographic evidence

A variety of furnaces seem to have been in use in different parts of the subcontinent in the past (Tripathi and Tripathi, 1994) and it is logical to think that these furnaces have been designed locally, to suit the requirements and skills of the artisans of the region. In the early eighteenth century, furnaces with 40kg to 250kg capacity were being operated at various locations such as Nagpur and in the Malabar area. The furnace at Nagpur was made of red clay mixed with sand and the walls were supported by a wooden frame. The furnace in Malabar region (in erstwhile Madras Presidency) could produce 250kg of iron bloom every 48 to 60 hours. The bloom could be retrieved by breaking the lower part of the furnace (Holland, 1892). Generally, however, ethnographic studies show that small furnaces were the most common across India. The ethnographic evidence of these small furnaces will be examined below.

Several ethnographic studies on iron working have been conducted in different parts of India. These include Elwin

A B C

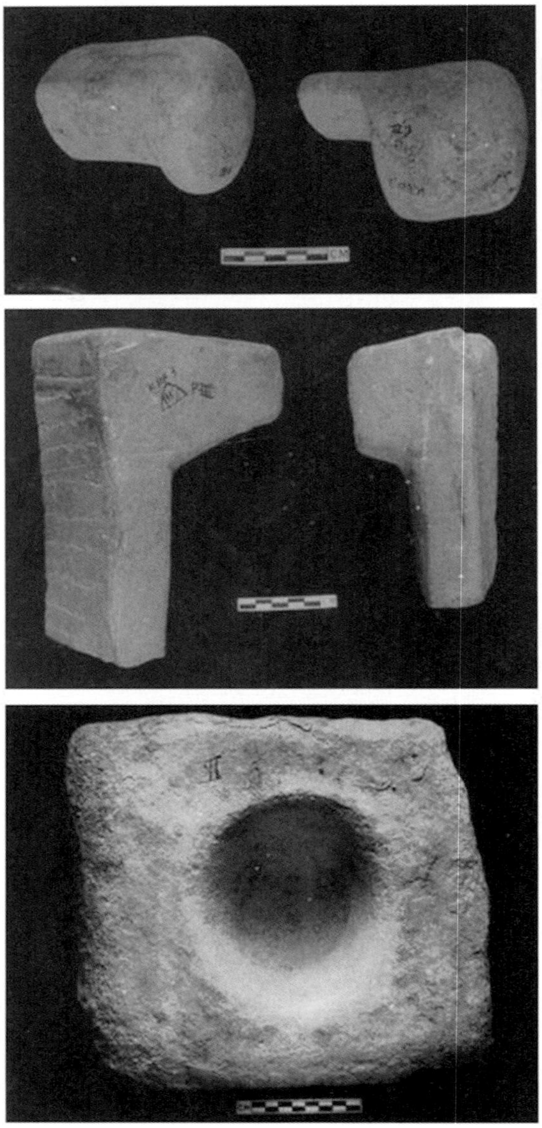

Figure 7: Ore crushing implements. A: Short handled pestles from a furnace complex, Khairadih; B: Legged querns, Khairadih; C: Quern, Chalcolithic Senuwar.

(1942), Leuva (1963) Prakash and Igaki (1984), Sharma (1998) and Tripathi (2001). It has been observed by scholars that the iron workers of Jharkhand, Chhattisgarh, Sonbhadra and Sidhi start the smelting process by performing rituals such as the sacrificing of a fowl or pigeon.

Traditional Indian iron working: a brief survey

Traditional Indian iron workers have a good feel for minerals. Even when a variety of ores is available in the vicinity, tradition dictates that they know to opt for the type of ore suitable for their specific purpose. For instance, the Agarias of Bihar (as reported by Elwin, 1942) prefer limonite or goethite; the Agarias of Wadruffnagar also select similar low-grade ores. The Agarias of Chhattisgarh have exploited the locally available hematite although for specific purposes they choose to work with magnetite. The Kamis (Nepali ironworkers of Darjeeling area) opted for magnetite, a black-coloured and

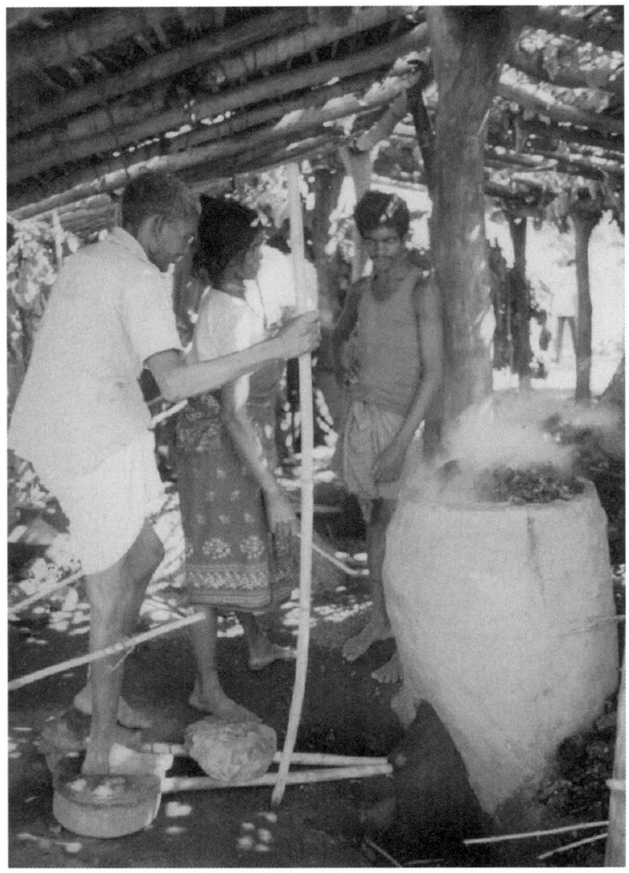

Figure 8: Iron working by Asur Birjia at Bishunpur, Dist. Gumla (Jharkhand).

difficult to smelt ore, as it was more suitable for the production of sharp-edged weapons like the *khukri* (dagger) or the *ban* (arrow). Significantly, both the magnetite and the hematite ore exist side by side in several of these mining belts.

Each group appears to have evolved its own working style and methodology, as is borne out by the fact that the Agarias of Wadruffnagar and Pratappur Block, although they live in close proximity, work in their own individualistic style (Fig. 8-9).

In Sarguja district of Madhya Pradesh there is a group of smelters known as Mahuli Agarias of Parsa group who use magnetic river sand (noted above), available in the local rivers, for iron smelting. They produce white iron, known locally as *Charkā Lohā*. This particular iron has so far not been analysed but its description suggests that it must have been high carbon steel and this product is in demand even today for the local manufacture of gun barrels. The Mahuli Agarias represent another group of ironworkers who have evolved an efficient skill. The furnaces and forges used by some of these artisan groups are illustrated below (Figs. 8-9), as is an analysis of the method of iron extraction used by some of the traditional iron working communities.

The Smelting Process

The process begins with religious rites and offerings to a deity, praying for successful conclusion of the operation.

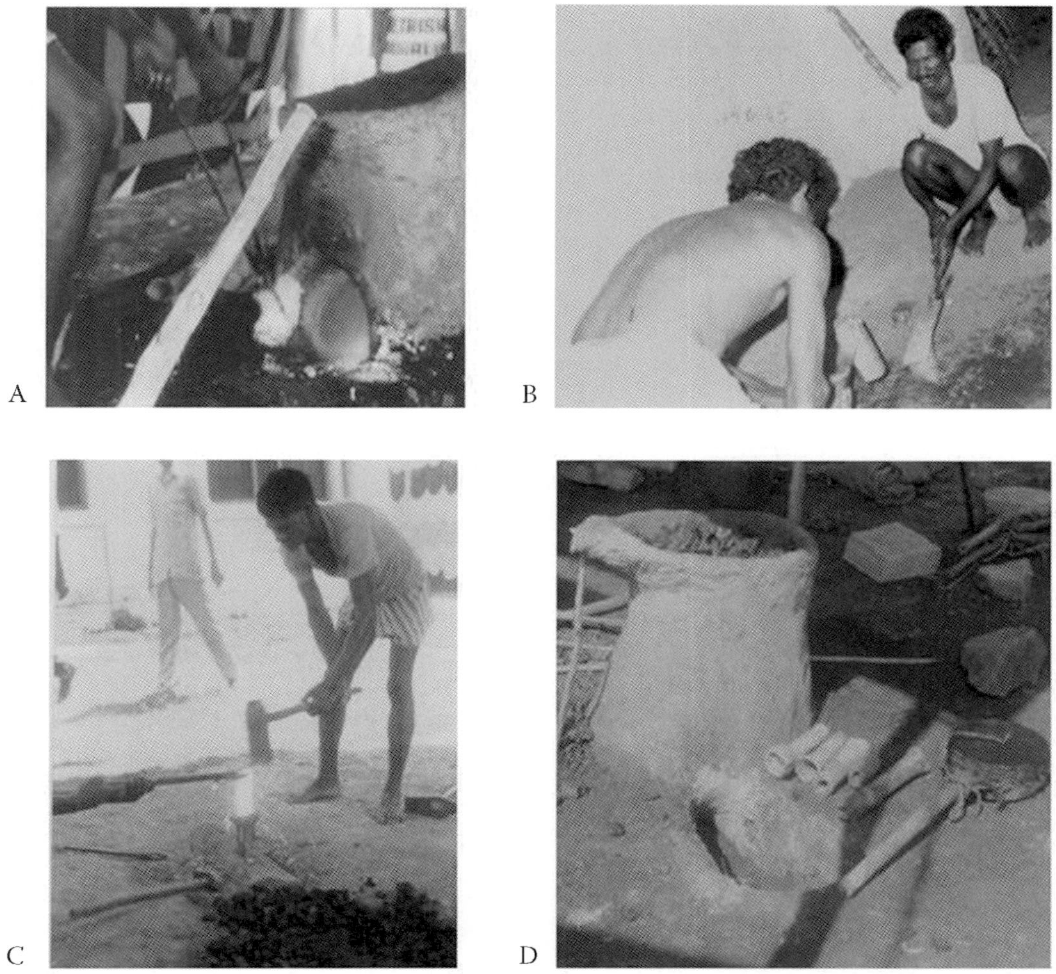

Figure 9: Iron working by Agarias, Wadruffnagar. A: Bloom being taken out; B: Forging in progress; C: Forging in operation; D: The furnace after smelting bellows, tuyeres etc.

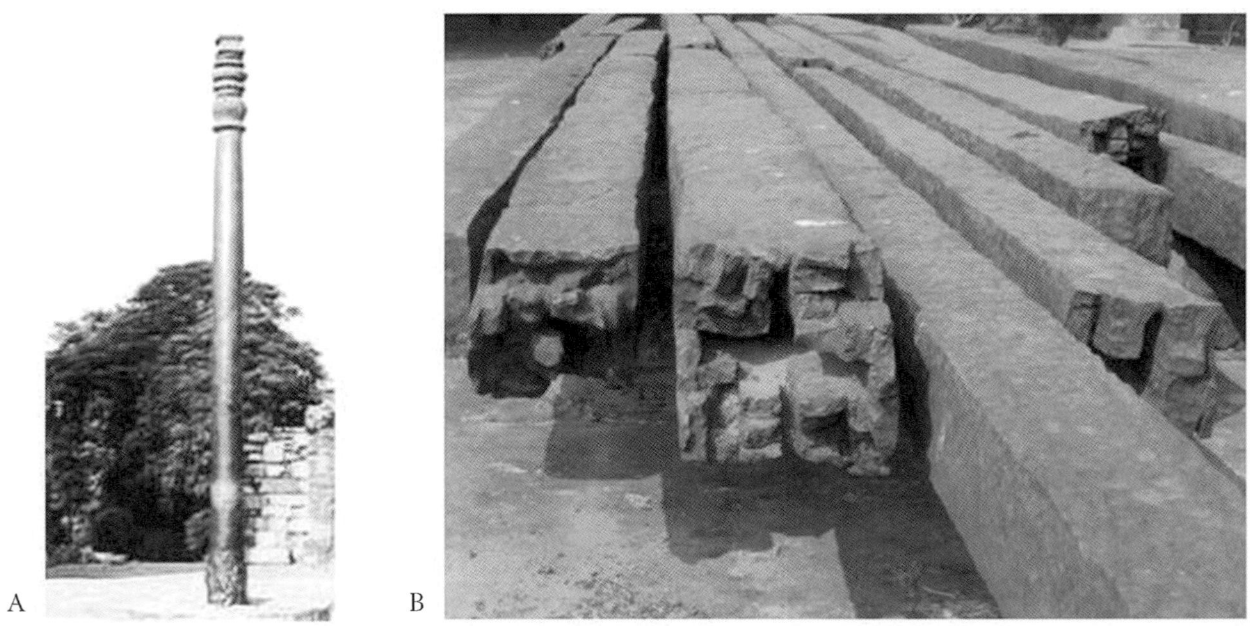

Figure 10: Colossal structures of iron. A: An old picture of the Delhi Iron Pillar; B: Iron beams of Sun Temple (Konark).

The furnace is charged with alternate layers of charcoal and ore. After the furnace has been ignited and the bellows start blowing air with full blast additional charge of the ore, alternated with charcoal, is piled on the top of the platform or *argha* to be later pushed into the furnace. This process, we assume, pre-heats and removes moisture from the charge. Nearly 15-20kg of ore and 30-40kg of charcoal are used in one operation lasting approximately 3-4 hours. This proportion varies as a result of the quality of raw material. For example at Pratappur, the amount of ore and coal that is consumed is lesser i.e., 15-16kg of ore and 25-30kg of charcoal. At intervals, the slag is tapped with a thin stick through the channel at the side of the furnace, wherein the molten slag runs out of this hole in a red-hot condition. The opening of this channel is sealed in the same way as the aperture for the bellows. A hole may be made into this loosely sealed opening by simply poking a thin rod and raking the charge, so that the liquid slag flows out of the furnace. The bellows are operated at a faster and quicker pace at the peak, gradually slowing the pace towards the end. The master artisan makes an assessment at this stage about whether the iron is ready. The bloom is then removed with a pair of tongs through the opening by removing the bellows, tuyeres and the temporary seal of loose clay and ash etc (Fig. 8B; 9A). Generally, approximately 3-4kg of iron is produced depending on the quality of the ore and the success of the operation (Tripathi and Mishra 1997: 59-66; Personal observation).

A study of the heat and mass balance during the process has been undertaken (Prakash and Tripathi 1986). The small circular bellows worked at a frequency of 120 strokes per minute at its peak and were capable of supplying sufficient air to burn charcoal and to produce a flame temperature as high as 1238°C – hence the maintenance of furnace temperature was not a serious problem.

A comparison of the furnaces used by the Agarias and the ancient furnaces exposed during excavations shows some kind of similarity in the tradition; the furnaces exposed at Lohsan in Malhar show a close resemblance with an Agaria furnace that was in use until recent decades (Fig. 6). Mc William (1920) observed continuity in the iron making process from ancient to pre-industrial times. He had examined the iron production at Mirjati in District Manbhum in Chhotanagpur Plateau. He commented,"---the analysis of the Mirjati iron and of Delhi iron pillar are very much alike. The iron produced there contained about 0.9% carbon" (Mc William, 1920: 159-170).

Hadfield (1913-14) had analysed the iron of Delhi and had corroborated this observation of similarity in composition of the two. Likewise, the ancient iron axe at Prakash (district Dhulia), Mahurjhari (district Nagpur) belonging to 700/ 600 BCE also contained 0.9% carbon. Such comparisons have been made by several others who studied traditional Indian iron made by ethnic communities in different parts of India (Tripathi 2001, Table 16). We have come across the earliest evidence of iron technology dating back to first half of 2nd millennium BCE on the one hand and, on the other hand, the presence of Agaria iron smelting in the heartland of India.

There is little doubt that, until recently, the legacy of iron technology has been carried out by ethnic communities against all odds. Iron produced by the Agarias and Asurs is still preferred by villagers. This heritage in metallurgy calls for in-depth study involving archaeologists, ethnographers, and metallurgists alike for better and more fruitful results.

Conclusion

At the present reckoning the chronology of iron in India appears to go back to 1800 BCE. This is earlier than the dates from bordering countries suggesting an indigenous origin of iron. An evolution of iron working may be traced from wrought iron to steely iron and from small iron objects to colossal structures. The furnaces too, show gradual evolution in design. Interestingly, it is possible to trace continuity in furnace design brought to light during excavations, and those used by ethnic iron smelters till recently.

Acknowledgements

The author is thankful to Indian National Science Academy for funding a project on Ethno-archaeological study of iron. I express my gratitude to the organizers of WIC and the Indian National Academy of Engineering for extending financial assistance.

References

Agrawal, D.P. 2002. *Ancient Metal Technology and Archaeology of South Asia.* New Delhi, Aryan Books International.

Agrawala, R.C. and Vijay Kumar, 1976. The Problem of PGW and Iron in north-eastern Rajasthan. *Mahabharat Myth and Reality, Differing Views* S.P. Gupta and K.S. Ramchandran (Eds.) Delhi: Agam Prakashan, pp.241-244.

Balasubramaniam, R., 2008. *Marvels of Indian Iron through the Ages.* New Delhi: Rupa & Co. and Infinity Foundation Series.

Bhardwaj, H.C., 1979. *Aspects of Ancient Indian Technology.* Varanasi: Motilal Banarasidass.

Bhardwaj, H.C., 2000. *Metallurgy in Indian Archaeology.* Varanasi: Tara Printing Works.

Chakrabarti, D.K., 1992. *The Early Use of Iron in India.* Delhi: Oxford University Press.

Chakrabarti, D.K. and N. Lahiri, 1994. The Iron Age in India: the beginning and consequence. *Puratattva.* 24: pp.12-23.

De, S. and P.K. Chattopadhyay, 1989. Iron objects from Pandurajar Dhibi: Archaeometallurgical Studies. *Steel India,* A Biannual Technical Journal for the Iron and Steel Industry, 2(1): pp.33-41.

Deo, S.B. and A.P. Jamkhedkar, 1982. *Excavations at Naikund.* Bombay: Department of Archaeology and Museums. Government of Maharashtra.

Dharmpal. 1971. *Indian Science and Technology in the 18th century.* Hyderabad: Academy of Gandhian Studies.

Datta, A., 1992. Early stages of iron technology and the development of regional pattern in India. In *Man and his Culture: A Resurgence* (Ed.) Peter S. Bellwood, New Delhi: Books & Books: pp. 293-308.

Elwin, V., 1942. *The Agaria.* London: Oxford University Press.

Gaur, R.C., 1983. *Excavations at Atranjikhera.* Delhi: Munshiram Manoharlal.

Ghosh, M.K., 1964. 'Primitive Iron Making in Contemporary India,' *Technical Journal,* Jamshedpur, Tata Iron and Steel Co. No. 2(3).

Ghosh, N.C. and P.K. Chattopadhyay, 1987-88. The archaeological background and an iron sample from Hatigra. *Puratattva* 18: pp.21-27.

Gogte, V.G., 1982. Report on iron-working and furnaces. In *Naikund Excavations-1978-79.* S.B. Deo and A.P. Jamkhedkar, Pune: Deccan College Pune: pp.52-59.

Hadfield, R. (1913-14). Archaeological Survey Report 1913-14, quoted in *Taxila* Sir John Marshall, Reprint 1975, New Delhi: Motilal Banarasi Dass: pp.534-35.

Holland T.H., 1892. Preliminary report on the iron ores and iron industries of the Salem district. *Records of the Geological Survey of India* 25: p.137.

Jayaswal, V., 2009. Antiquity of Varanasi as Revealed by Excavation at Aktha (2008-09), *Puratattva* 39: pp.141-146.

Krishnan, M.S., 1954. *Iron Ore, Iron and Steel.* Bulletins of the Geological Survey of India No. 9. Series A - Economic Geology, Delhi: Geological Survey of India.

Lal, B.B., 1949 *Sisupalgarh 1948: An Early Historical Fort in Eastern India.* Ancient India, Bulletins of the Archaeological Survey of India No.5: pp.62-105.

La Touche, T.H.D., 1918. *A Bibliography of Indian Geology and Physical Geography with an Annoted Index of Minerals of Economic Value. Part, I-B.* Calcutta: Geological Survey of India.

Leuva K.K., 1963. *The Asur.* New Delhi: Bhartiya Adimjati Sevak Sangh.

Magee, Peter, 2004. Mind the Gap: The Chronology of Painted Grey Ware and the Prelude to Early Historic Urbanism in Northern South Asia, *South Asian Studies* 20: pp.37-44.

Mc William A. 1920 Indian Iron Working at Mirjati, Chhotanagpur, JISI (CII) 159-170.

Prakash, B. and V. Tripathi, 1986. Iron technology in ancient India. *Metals and Materials.* (Sept. Issue): pp.568-579.

Prakash, B., 1991. Metallurgy of iron and steel making and black smithy in ancient India. *Indian Journal of History of Science* 26(4): p.351.

Prakash, B., 1997. Metals and metallurgy. *History of Science and Technology.* Vol. I. (Ed.) A.K. Bag. New Delhi: INSA: pp.80-174.

Prakash, B. and Igaki, K., 1984. Ancient iron making in Bastar district. *Indian Journal of History of Science* 19(2): pp.172-185.

Ray, A. 1996. Archaeology of Mangalkot, *Historical Archaeology of India – A dialogue between Archaeologists and Historians*; A. Ray and S.Mukherjee (eds.) New Delhi: Books and Books: pp.131-140.

Sahi, M.D.N., 1979. 'Iron at Ahar', *Essays in Indian Protohistoriy,* D.P. Agrawal and D.K. Chakrabarti (eds.), New Delhi: B.R. Publishing Co.: pp.365-68.

Sahi, M.D.N., 1994. *Aspects of Indian Archaeology,* Publication Scheme, Jaipur.

Sharma, A.K. 2000. *Early Man in Jammu Kashmir and Ladakh.* Delhi: Agam Kala Prakashan.

Sharma, M. 1998. Report on Rural Iron Making Practice at Bishunpur, SAIL, Ranchi.

Srinivasan, S. and S. Ranganathan 2004. India's Legendary Wootz Steel: An Advanced Material of the Ancient World. Bangalore: National Institute of Advanced Studies and Indian Institute of Science.

Tewari, Rakesh 1998. 'In Quest of Early Iron Age Sites in Karmanasa Valley', *Pragdhara* Journal of U.P. State Archaeology Department, 8: pp.57-61.

Tewari, Rakesh, R.K. Shrivastava, K.S. Saraswat, K.K. Singh 2000. Excavations at Malhar 1998-99: District Chandauli (U.P.), *Pragdhara* 10: pp.69-98.

Tewari, Rakesh, R.K. Shrivastava, K.K. Singh 2002. Excavations at Dadupur,: District Lucknow 1999-2000, *Pragdhara* 12: pp.99-116.

Tewari, Rakesh 2003. 'The Origin of Iron Working in India: New Evidence from the Central Ganga Plain and the Eastern Vindhyas' *Antiquity* 77: pp.536-544.

Tewari, Rakesh 2010. Updates on the Antiquity of Iron in South Asia, *Man and Environment,* 35 (2): pp.81-97.

Tripathi, V. and A. Tripathi, 1994. Iron working in ancient India: an ethnoarchaeological study. In *From Sumer to Meluhha.* Wisconsin Archaeological Report, Vol. 3, (Ed.) Wisconsin. Madison: J.M. Kenoyer: pp.241-251.

Tripathi, V., 1997. 'Stages of the Painted Grey Ware Culture' *Indian Prehistory*-1980, V.D. Misra and J.N. Pal (eds.), Allahabad: University of Allahabad: pp.318-322.

Tripathi, V. and A.K. Mishra, 1997. Understanding Iron Technology - An Ethnographic Model, *Man and Environment,* 25 (1): pp.59-67.

Tripathi, V., 2001. *The Age of Iron in South Asia: Legacy and Tradition.* New Delhi: Aryan Books International.

Tripathi, V., 2007. Towards the Wootz: Iron and Steel Technology in India, *Indian Journal of History of Science.* 42 (3): pp.403-426.

Tripathi, V., 2008. *History of Iron Technology in India (From Beginning to Pre-modern Time).* New Delhi: Rupa & Co. and Infinity Foundation Series.

Wood, C., 1894. In a Discussion in *Journal of the Iron and Steel Institute.* X LIV:179.

Author's address

- Department of Ancient Indian History, Culture and Archaeology, Banaras Hindu University
  Varanasi – 221 005, India
  E-mail: vibhahist@gmail.com

New insights on the *wootz* steel of Telangana

Jaikishan Sriperumbudur

ABSTRACT: New insights into the production of wootz steel in Telangana, based on physical and material evidence from the ancient to early modern period, have been obtained and are presented below. A detailed survey of iron and wootz steel production sites was conducted in four districts of northern Telangana, as well as the identification and exploration of a large number of previously unknown sites in order to determine the extent of archaeometallurgical industry in the region. Findings which will be highlighted from this survey include the interesting aspect of names of villages which co-relate to several aspects of iron and wootz steel production, as well as the exposition of important wootz steel centers in Telangana and the fact that large heaps of slag and crucible can be found today in most of those villages which produced metal. Relevant details from the travelogues of visitors will be briefly touched upon in order to obtain firsthand accounts of the state of the wootz steel industry in this area. The survey also helped to understand the existence of wootz culture and to identify a few artisans who were still familiar with wootz steel working, whose knowledge of processing wootz steel cakes was obtained through interviews. It was discovered that the thriving nature of the wootz steel industry in the past can be confirmed by the survey of available implements and objects related to wootz steel. The need for careful study and cataloguing of these traditional tools of the blacksmiths of the region will be highlighted.

Keywords: Wootz, *hinduwani, inmu, ukku, kammari*, furnace, tuyere, crucible, steel.

Introduction

Telangana is one of three geographical divisions within the Andhra Pradesh State in South India, along with Rayalaseema and Coastal Andhra. The region of Telangana is the largest of the three at 114,800 km^2 and is divided into ten districts, including the state capital of Hyderabad. The geographical and cultural characteristics of Telangana are unique and significant quantities of iron ore exist in the region, although it is currently difficult to say when iron production began in this area.

The main purpose of this paper is to present the brisk wootz steel metal industrial activity in the ancient to early modern period in the Telangana region. "The best steel was produced at 'Konasamudram' near Nirmal in Nanded District" (Tavernier, 2001, I, p.127). I intend to put forth that the brisk iron industrial activity has not seen the light of modern archaeometallurgy and industrial archaeology. This paper presents the results of a detailed survey of iron and wootz steel production sites that was conducted in four districts in northern Telangana, and includes a large number of previously unknown sites. Over 1800 villages were surveyed and ample evidence was found regarding iron smelting, wootz steel processing and the forging of different kinds of iron and wootz steel implements. More than 1100 villages were identified as iron and wootz steel manufacturing centres and it is clear that a unique wootz steel usage was and still is prevalent in this region.

Iron implements have been found in megalithic burials in this region. A megalithic burial site in Dacharam village, Illenthakunta Mandal in Karimnagar district, was illegally dug out and a few iron implements were found in burial. An iron smelting site was also discovered, located very near to this burial site. Iron and wootz steel implements were excavated in early Shatavahana sites dating back to the pre-Christian era (Department of Archaeology and Museums, 2008). Although no absolute dating has been carried out in this region, the megalithic iron smelting site of Naikund in Maharashtra is considered to be one of the oldest (Deo and Jamkhedkar, 1982). The same kind of smelting remains are found at the Ranamkota site, near Buggaram Village, Dharmapuri mandal in Karimnagar district (see Figs. 1 and 2). In recent studies, thermo luminescence (TL) dating carried out on pottery has produced dates of 1995 and 2505 BC at two excavated megalith burials in the Hyderabad University campus in Gachibowli, Hyderabad (Thomas, et al., 2008, p.787; Rao, 2010, p.111).

It is interesting to note that most of the village names are etymologically identified in the regional language (Telugu) as terms related to iron ore, iron slag, iron smiths, iron furnaces and iron manufacturing. For example, in Telugu '*Inumu*' means

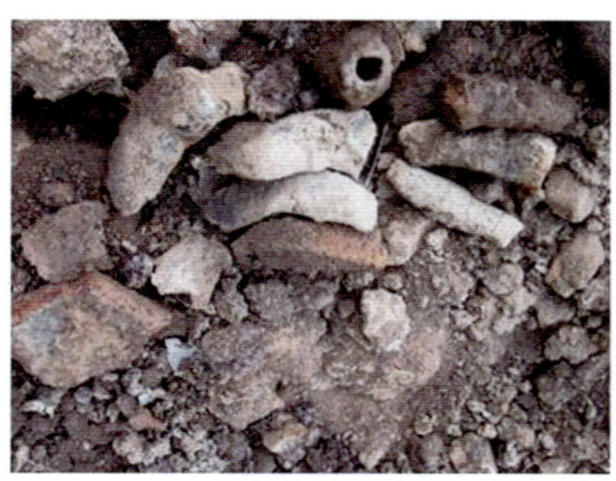

Figure 1: Slag site at Ranamkota, Buggaram.

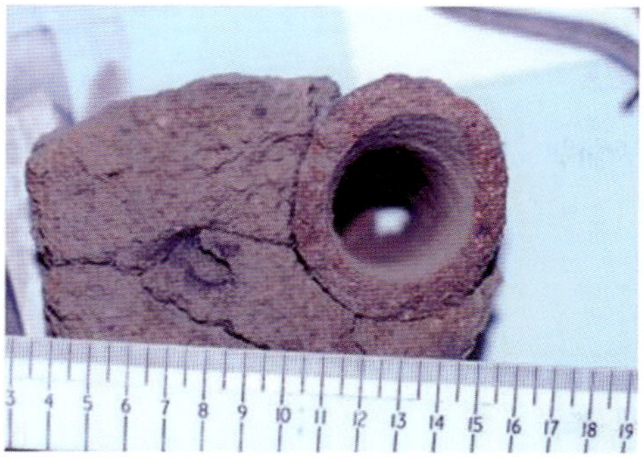

Figure 2: Furnace wall with tuyère, Ranamkota, Buggaram. Dharmapuri Mandal, Karimnagar district.

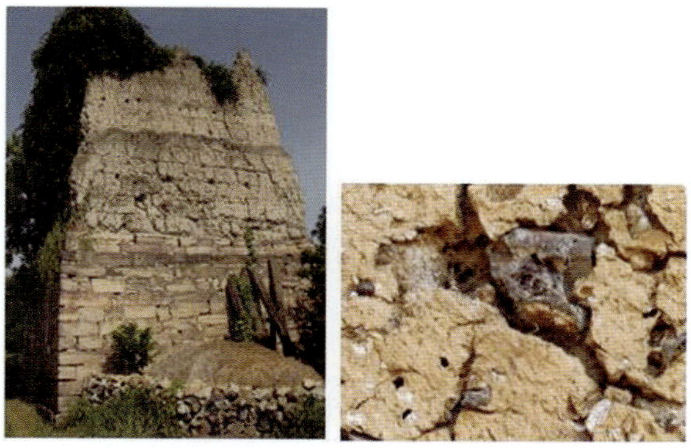

Figure 3: Slag pieces can be seen within the mud wall fortalices.

iron, and *'Dindurthi', 'Inukurthi', 'Induoor',* and *'Indurthi'* are all villages named after iron; *'Chityamu'* means slag or scum of melted metal, and villages named after this include *'Chityala', 'Chittapuram', 'Chittlapenta',* and *'Chttemula'*; *'Kammari'* means blacksmith: *'Kammaripeta', 'Kammaripally', 'Kammarashala',* and *'Kammarala'* are all villages presumably linked with this profession; finally, *'Kolimi'* means furnace, and *'Kolimkunta', 'Kolimiyala',* and *'Kodimayala'* are village names related to the word furnace. The nomenclature of the village names is also related to continuously burning furnaces, dried cakes of slag or the soft stone of the iron ore, etc.

"Ancient mining activity for iron is noticed at several places in Armour taluk of Nizamabad district, the Peddapalli and Jagtial taluks in Karimnagar district, the Jangoan taluk in Warangal district and Siddipet in Medak district" (Shastri, 1983, p.108). Shastri further states that: "Iron ore extensively found in the ferruginous quartzite in Armur in Nizamabad District; Nirmal and Luxettipet taluks in Adilabad district and broad stretches of Chikiyala bed in Sirpur taluk of Adilabad district constitute an extensive source of iron ore" (Shastri, 1983, p.108). Iron ore was available everywhere in this region on the surface of the earth, as well as in cap type deposits on the top of the hills and hillocks. Thelma Lowe stated that: "As determined in the field, the upper surface of all outliers with iron-rich lateritic hard-caps are above 600m" (Lowe, 1990, p.8). A Jean de Thevenot, a French traveller who visited India in the 17th century, stated that: "the mines are mere holes at a depth of a man" (Surendranath, 1949, p.112). Iron ore stones were collected and pulverised to powder and washed with water to clean the clay and dust particles. Sometimes the iron ore was collected from the stream beds during the rainy season. Cleaned iron ore was sold in every market and traders would sell it in the streets.

Interviews and observations from this recent research revealed a number of circumstances relating to iron production sites dating to the early or later medieval period. For example:

1. Slag sites that were originally located on the outskirts of villages, where agriculture is now being practiced, are beyond recognition because of modern agricultural disturbance.
2. Slag sites which are currently located in the centre of villages are difficult to identify because many houses are constructed on the remains of the slag.
3. Within most of the iron producing villages, the remains of slag pieces can be seen within the mud walls of the houses and in other structures, such as fortalices, that were constructed in the early medieval period, i.e. 12th-15th centuries AD (Fig. 3).
4. Houses are constructed around the slag sites.

A huge number of sites were identified in the deep forests of Adilabad and Karimnagar. The slag site remains in Ranamkota mentioned above and Parusurampalli, in Ghanapur Mandal in Warangal district, are similar to those of Naikund in Maharashtra.

At this early stage, the scientific techniques of mining, smelting and forging iron were not fully developed. However, extraordinary craftsmanship was applied to the manufacturing of various kinds of iron objects, for example the iron pillar in Delhi, the iron beams at the Konark Sun Temple and the massive, forge-welded iron cannons in Deccan forts, which all stand testimony to the skill of the artisans.

Wootz steel

The invention of wootz steel is a subject of debate, generating much discussion among metallurgists, archaeometallurgists and archaeologists. Various hypotheses have been suggested regarding the origin and development of high carbon steel (i.e., crucible steel). The carburisation process employed to harden the iron is attributed to the ancient world cultures of Greece, Persia and China. The method adopted by these cultures, however, was crude when compared to the wootz process.

The wootz steel industry flourished in northern parts of Telangana from time unknown. The specific term for steel in local Telugu language is *'Wokku'* or *'ukku'*; with the same word, *'ukku',* still used in modern times. The meaning of the term *'ukku'* in Telugu is 'strength'. The literal meaning of *'ukku'* is *'ayassaramu',* or 'essence of iron'. The term *'teilga ukku'* is used for the same metal, which was a very tough type of iron, known as *'teliga ukku'* or 'tiling iron' (Brown, 1903). Lowe states that "Archaeometallurgical field work on known production sites can address the problem of history of 'wootz' technology by providing new information based in known province" (Lowe, 1990, p.3). Lowe identified 74 ore smelting sites, of which 14 were crucible production sites, in the Telangana region (Lowe, 1990, p.3). In the field survey in question in this paper, 1100 iron smelting sites and nearly

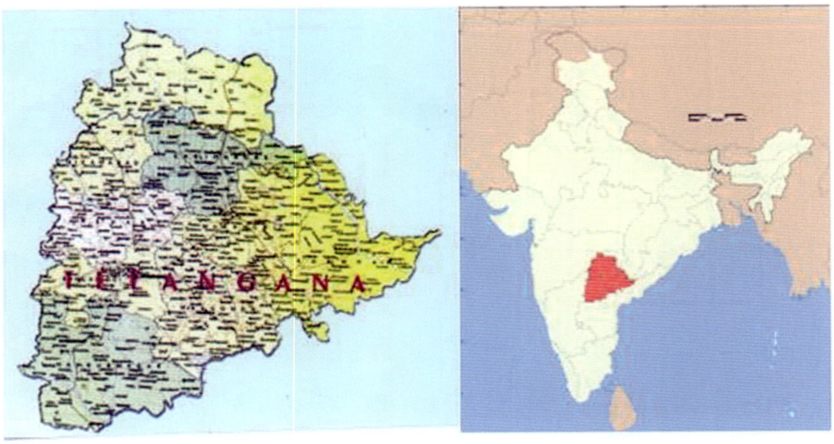

Figure 4: Geographical location of Telangana.

180 crucible steel production sites have been identified in only four Telangana districts: Karimnagar, Adilabad, Nizamabad, and Karimnagar. The region of Telangana is shown on the map below (Fig. 4).

"In several Arab accounts we hear of the terms *hindvi* and *teling* swords, which were highly prized, and were manufactured using Deccani wootz. The word *teling* could be a reference to Telungu (a Tamil phonetic variant for the language Telugu) and '*teling*' to a sword obtained from Telugu-speaking smiths" (Feuerbach, et al., 2007, p.380). "All the Dravidian languages own a word for steel; most of them (*uruku, ukku, karugu, urku, ukku*, etc.) are based on the meaning: 'melt, dissolve' describing the wootz process" (Toussaint, 2002, p.45). Most scientists, archaeometallurgists and historians discuss the Deccani wootz without mentioning the region from where it originated. There is ample evidence of archaeometallurgical remains in hundreds of villages in this region – not only the remains of archaeomaterials but also evidence that the usage of wootz metal still exists. Wootz material usage in the day to day life of the people of the region has enabled the identification of a specific wootz culture.

To produce wootz steel, the iron was heated at high temperatures, with herbs, in a clay crucible. Krishna Shastri stated that, "it was the perhaps the ores of Karimnagar and Nizamabad regions that were supplied to the ancient iron smelters of Konasamudram, ... iron for the Damascus steel is said to have been exported. During their voyages this region was visited by the earlier Roman and Persian traders for the steel, as this Indian 'Wootz' was well known throughout the Ancient East" (Shastri, 1983, p.5).

"Wootz was also known as '*hinduwani*' in Arab countries, who reportedly claimed that it had no parallel in the world. In several Arab accounts, we hear of the terms '*hindvi*' and '*teling*' swords, discussed above, which were highly priced and were manufactured with Deccani wootz. The word *teling* could be a reference to Telugu (a Tamil phonetic variant for the language Telugu) and "*teling*" to a sword obtained from Telugu speaking smiths. (Feuerbach, et al., 2007, p.380)

Al-Beruni describes the manufacturing process of crucible steel which is called '*pulad*' as having great demand in Arab countries. The Arabic accounts mentioned the '*hindvi*' or '*hinduwani*' steel; signifying the geographical location of India as it was known as Hindustan in Central Asia. "The European term for '*hinduwani*' was mentioned as '*ondanique*' by Marco Polo, fame of the region of Telangana in central Deccan" (Marco Polo, 2001, p.39). "Arab accounts also point to the fame of '*teling*' steel which can be taken to refer to the region of Telangana in Andhra Pradesh in South Western India. Indeed, the Golconda region of Andhra Pradesh was clearly a nodal centre for the export of wootz steel to West Asia" (Srinivasan and Ranganathan, 2004, p.54).

A translation of Tavernier, given by Le Coze in his article 'On the Question of Possible Transfer of Steel Technology from India to Europe through the Muslim Middle East', states that: "the Persians know perfectly well the damascening with vitriol for instance of swords, knives and the like, but the nature of the steel which they use has a great importance because they could not obtain the same result with their own nor with ours. This steel is brought from Golconda, and is the only sort which can be well damasked. It is different from ours, for when it is put in the fire for quenching it needs no more than small red like the colour of a cherry, and instead of quenching it in the water as we do, it must only be wrapped in a moist linen cloth, because if it was given the same heat as ours, it would become so hard that when you would like to handle it, it would break like glass" (Le Coze, 2007, p.340).

Craddock stated that "The most famous region of steel production by co-fusion is in the Nizamabad District of Andhra Pradesh, which was previously part of the state of Hyderabad in the Deccan of central India. One of the most important sites was Konasamudram which was visited in 1820s by the English traveller, H.W. Voysey" (Craddock, 2007, p.598).

Results of the study in question in this paper have revealed that the village of Konasamudram may have been the nucleus of wootz steel production. Moreover, although the villages surrounding Konasumudram were producing huge quantities of wootz, these are not mentioned in any traveller's accounts or in contemporary specialist literature. The village name of Konasumudram, however, was mentioned by Tavernier in his "Travels in India" (Tavernier, 2001, I, p.127); with Thevenot's account stating that "The best steel was produced at 'Konasamudram' near Nirmal in Nanded District" (Surendranath, 1949, p.112).

Evidence suggests that most of the wootz marketing was centralised in the village of Konasamudram, where a dilapidated trader's bungalow remains, belonging to the

Bhandri family of Bania community. Records state that Haji Hassan, a Persian trader, came to Konasamudram and stayed in the same house when he failed to get the required amount of wootz ingots at Indur (the big marketing centre was Indur, whose original name may have been Inumoor, or 'iron village'. It was located in Nanded Pargana, the modern town of Nizamabad, district headquarters of the Andhra Pradesh state) (Voysey, 1832, p.246).

Wootz steel in academic literature

Prof. R. Balasubramaniam's article provides details of the ancient Sanskrit text *"Rasaratna Samuchahaya"*. This text mentions three types of iron: *'munda'* (cast iron), *'kanta'* (tool iron), and *'tikshna'* (hard iron or steel). He says "Two common properties mentioned for the medium carbon steel in the 'Rasaratna Samuchahaya' – verses 74-80 are, first, they were lustrous (…) and secondly, they possessed lines on the surface (…). The lustrous nature of the broken surface is clearly due to the presence of cementiate in the wootz steel ingot." Thus, "The best quality wootz steel (normally *'vajra tiksna loha'*) is said to possess 'minute lines' which implies that an aligned arrangement of finer carbide particles in the original wootz steel ingot resulted in good cutting properties in the finished product" (Balasubramaniam, 2007, p.437).

The 18th century Telugu poetic work *"Khadga Lakshana Shiromani"* also provides valuable information regarding the swords made with wootz steel. According to this text, the quality of the sword blade was determined by the imparted tint and shining shades on it. It describes the best wootz steel as having a tint and lustre that resembled castor oil seeds, and that the tint or lines on the surface of the blade would resemble script or the beautiful hair of a female (Jaikishan, 2007).

The term wootz first appears in European accounts in 1795, to describe ingots of Eastern and, more particularly, Indian crucible steel, when Robert Pearson reported that specimens of wootz had been sent from Bombay for examination at the Royal Society: "Doctor Scott, of Bombay, in a letter to the President, acquaints him that he has sent over specimens of a substance known by the name of wootz: which is considered to be a kind of steel, and is in high esteem among the Indians" (Pearson, 1795, p.322-23).

An account given by the Venetian traveller Marco Polo stated that: "The iron at Hyderabad, at a very early period, obtained a wide renown, being in fact the material which, when made into steel, afforded the source of supply for the manufacturer of Damascus blades, the raw material having been exported to Persia and the Punjab for that purpose" (Biswas, 2001, p.137). It is in some doubt whether Hyderabad city had been founded by the time of Marco Polo's visit in the 13th century. Instead, the city referred to here might be Warangal (the capital city of the Kakatiya Kingdom) or the famous sea port of Machilipatnam.

Another account noted by Briggs states that, "During 16th and 17th centuries mining and manufacturing work was very brisk. Iron mines were located at Indur, Nirmal, Nagalwancha, Machilipatnam, Ramayapatnam, Bhimunipatnam, Mootapalli and Singarayakonda" (Briggs, 1981, III, p.272).

On his way to Golconda, Thevenot visited Indur and Indalvai, describing that: "We past next by the town *Indalwai*, of which nothing is to be said in particular, but that a great many Swords, Daggers, and Lances are made there, which are vended all over the *Indies* and that Iron is taken out a Mine near the Town, in the Mountain of *Calagatch*" (Kalaghat) (Surendranath, 1949, p.112).

Travelling around the same period, Tavernier claimed that, "The iron at Hyderabad, at a very early period, obtained a wide renown, being in fact the material which, when made in to steel, afforded the source of supply for the manufacturer of Damascus blades, the raw material having been exported to Persia and the Punjab for that purpose" (Tavernier, 2001, I, p.127). Tavernier praised the quality of the products and stated: "The barrels of their muskets are stronger than ours, and the iron is better and purer; this makes them not liable to burst" (Tavernier, 2001, I, p.127).

Some years later, in 1832,Voysey said that the industry "seems to have started in South India by 300 BC. This steel is known by the name of wootz in 'telinga'. Small pieces of iron were placed in refractory crucibles and flux added was a piece of wood and green leaves in it. The cylindrical shape crucible, at bottom made with a small concavity and reduced towards lid. The top of the crucible made with lump of clay in a shape of pinnacle was fixed to it" (Voysey, 1832, p.247).

In the early years of the 20th century, Moreland noted that, "Mining and metallurgy were well developed in the Southern part of the country, especially in Hyderabad (Golconda), Mysore (Karnataka)" (Moreland, 1923, p.92), while Abul Fazal said that the "Nirmal and Indur possessed iron ore mines, and other metals also arms production centres" (Fazal, 1962 II, p.230).

Lord Egerton of Tatton stated in "A Description of Indian and Oriental Armour" that: "Hindus excel in the manufacture of iron, and the preparation of those ingredients along with which it is usually styled Indian steel. They also have workshops wherein are forged the most famous sabres in the world. It is impossible to find anything to surpass the edge you get from Indian steel" (Egerton, 2001, p.56).

He went on to claim that: "The East has always been famous for its blades. Damascus has obtained a reputation, which it can no longer claim for the manufacture of the finest steel weapons, but in all probability, the metal from which the ancient blades of Damascus made that was brought from India. Wootz or Indian steel was exported from Cutch to the ports on the Persian Gulf." He further mentioned that at "Elgundel there is a manufacture of swords, daggers, and spearheads from the steel obtained there. At Lingampilly, in the same district, barrels for pistols and matchlocks are prepared in the following manner: bits of old iron are formed into rods the thickness of a man's finger, which are then twisted; three or four of these are joined lengthways, another band of iron of the same breadth and one third of an inch thick is welded to this, and both are formed in to a band which is twisted after words beaten in to a solid cylinder, which is bored by a hard steel chisel" (Egerton, 2001, p.57). The two places mentioned in the text above are located in the present day Karimnagar district of Telangana.

In a summary of the history of wootz in the world, Arun Kumar Biswas stated, "The wonder material of the Orient – wootz steel – supplied to far corners of India and the world from this region, even at the beginning of the industrial revolution in the Europe" (Biswas, 2001, p.300).

Manufacturing wootz steel

The course of this survey discovered that crucible making is critical to the process of manufacture. The preparation of the clay crucible for contemporary use was described by two senior smiths: Mattela Pedda Gangaram, from Ibrahimpatnam in the Karimnagar district; and Mandalogi Gangaram from Konapuram, in Nizamabad, as follows:

The clay is collected from a pond near the village. For collecting the soft clay for crucibles, a ditch was dug and filled with a mixture of clay and water. After the water is absorbed, the next day sediment clay is collected from the pond. This soft clay is mixed with rice husk and a small quantity of castor oil is added to it. Then the smiths made the bottom of the crucible, two to three inches high according to the required size of the ingot, and kept it for greening. After drying, these refractory crucibles were filled with the required quantity of bloomery iron pieces, one wood piece of '*Thengedu*' and the green leaves of '*pulyeilauku*', grinded into paste, were placed in the crucibles, before a small quantity of Borax was added. This was closed with a lump of clay with a conical shape. The crucibles thus filled in were kept for greening until they were totally soaked in the shade. The soaked crucibles, three to four dozen depending on their size, were kept within the furnace, which was constructed about four to five feet high. Tuyères made of the same red clay were fixed on four sides in opposite directions around the furnace, which was constructed in a round shape using locally made bricks. At first the bed of the furnace was filled to about a foot high with small pieces of charcoal, before the crucibles were placed on top of the bed of charcoal in a circular pattern. Subsequently, the furnace was filled with charcoal. The furnace was fired continuously for about 24 hours. While blasting was in progress the crucibles were lifted and rotated with the help of long pincers (Fig. 5) and charcoal was continually added to the furnace.

Egerton offers a similar description of the process presented above: "The smelted iron was hammered and made into small pieces and those were placed in refractory crucibles and flux added was a piece of wood and green leaves in it. The round concave shaped bottom of the crucible was gradually reduced from bottom to top, the lid in pinnacle shape" (Egerton, 2001, p.57).

The base of the crucible was slightly concave and the lids were in the shape of a pine cone (as shown in Fig. 5): consequently these were called '*Konam pavulu*' because of their shape; or '*Wukku pavulu*, meaning wootz crucibles, in the local language. The village names Konasamudram and Konapuramu etymologically denote the pine cone-shaped cover of the wootz steel crucible. There are a number of villages in the region named after 'Konapuramu', due to their production of wootz steel and iron.

Coal was not mined or used during this period: instead, iron and wootz steel were extracted using charcoal made with *Sandrugu* (mimosa Sandra), which provides a hard charcoal. Other preferred wood species used for charcoal were bamboo, rose wood (from the *Gittregi* tree) and mahua wood or *Bassia latifolia* (*Ippa* tree).

The wootz ingots were produced in different sizes and weights as per the requirement of the weapons made out of the metal.

Production requirements also changed from time to time. Observance of the crucible sites at Parsurammpalli yielded evidence of very small crucibles, which could have produced not more than 50 grams of iron (See Fig. 6).

Three ingots collected for study weigh 400 g, 1500 g and 2000 g respectively (Figs. 7, 8 and 9), demonstrate the significant difference in the dimensions of the wootz steel ingots. The quantities produced ranged from 100 grams to twenty kilograms in weight (Fig. 10). The ingots are round shape.

There was much demand for wootz steel within the region from local buyers, echoed by demand from external traders from West Asia and Europe. Moreover, the use of wootz steel for artisans' implements was widespread and remains fashionable in the region today.

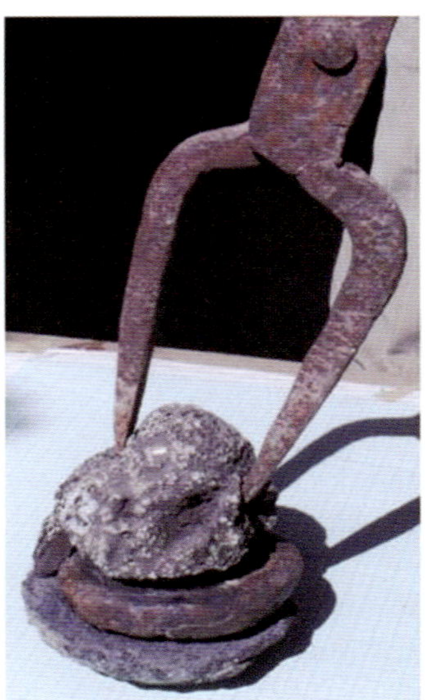

Figure 5: Image showing the handling of a crucible.

Modern accounts of local wootz steel production

Following his account of the clay crucible, in January 2006 the author was directed to the village of Ibrahimpatnam by Mattela Gangaram, the most senior smith (see Fig. 11),

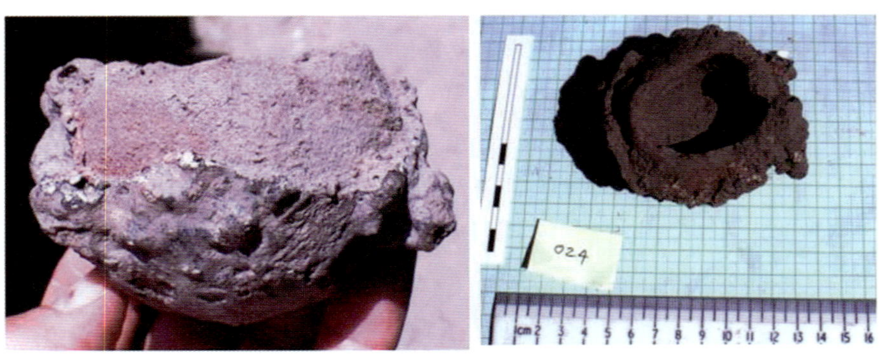

Figure 6: Small crucibles collected from Parsurammpalli, Ghanpur Mandal, in Warangal district.

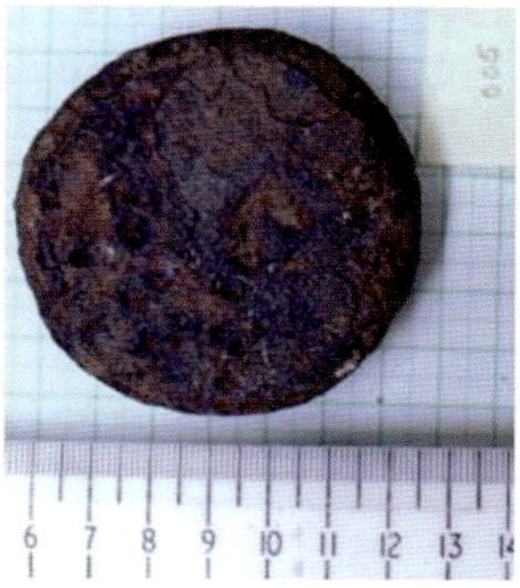

Figure 7: 400 grams wootz ingot.

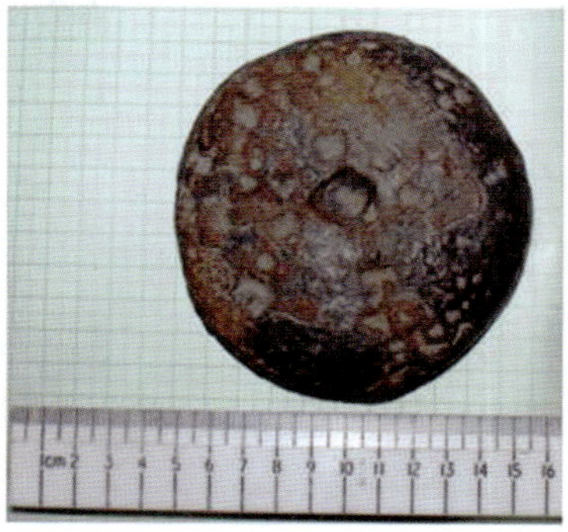

Figure 8: 1500 grams wootz ingot.

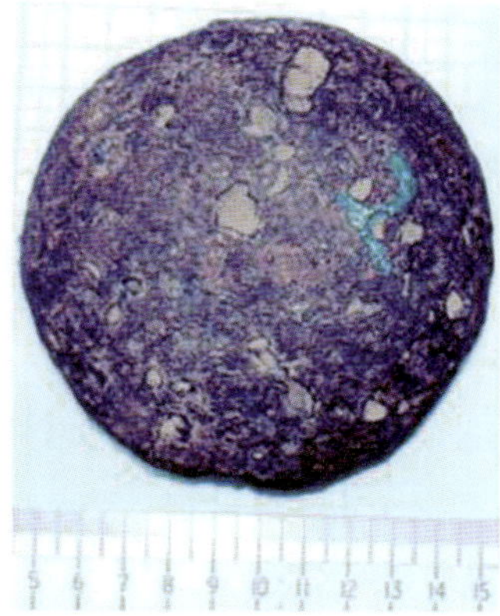

Figure 9: 2000 grams wootz ingot.

where he pointed out the heaps of crucibles and slag before explaining the production process, which is described below:

Holes in the road are filled with the remains of crucible and slag debris which are now reduced to ground level. This place is highly respected in the village and presently two big protected water supply tanks are constructed on the heap of crucibles . A community hall and a temple of the '*Mammayee*' Goddess belonging to the '*Vishwavakarma*' – the smithing families – are beside the water tank. There are about 30 families of smiths in this community. Mattela Gangaram guided the author to an '*inupagutta*' (iron hillock) near the village that was the source of the ore used for producing the iron. A heap of slag was spread in this area. Gangaram said that the iron lumps produced were cut into pieces to then be transformed into wootz. Red gravel clay was used for making the soft clay employed for making clay crucibles or '*matti pavulu*'.

Another firsthand account came from Mandalogi Gangaram, a senior blacksmith from Konapuram village, Kammerpally Mandal in Nizamabad district, who possessed special skills in the working of wootz steel. He revealed the details of wootz making as he had heard them from his father in his childhood. He told me that the clay used for the crucibles was brought from a specific place called the '*Vandubonda*' clay pond (Fig. 12). This place is located in the forest, about one and half kilometres from the village. The green crucibles were dried up in the shade and used for making wootz. Each crucible held four to five small pieces of iron, as well as one piece of '*thangedu*' wood and was then filled with the leaves locally available and known as '*pulyeelauku*' (Fig. 13), adding a small quantity of Borax before closing the crucible. The crucible was then closed with a cover made of a lump of clay. Mandologi Gangaram and I went to the forest, visited the '*Vandubonda*' clay pond, and collected the leaves.

The furnace, or '*bhatti*', was constructed to be big enough for three to four dozen crucibles to be placed in it at one time. For the construction, locally produced clay bricks were used, and the furnace itself was filled with charcoal. Four tuyères were arranged in the north-south and east-west directions, and the air blast was produced with bellows that were made of buffalo skin. The blasting process continued for twenty-four hours. Then the same temperature was maintained for eight to ten hours, during a process called '*theepulu*', or annealing. The furnace was then allowed to cool down naturally. On the third day flat, round-shaped wootz ingots were removed by breaking open the crucibles.

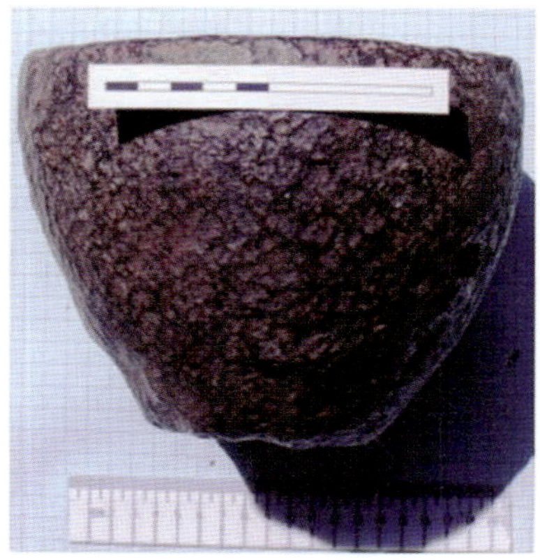

Figure 10: 19 kilograms wootz ingot.

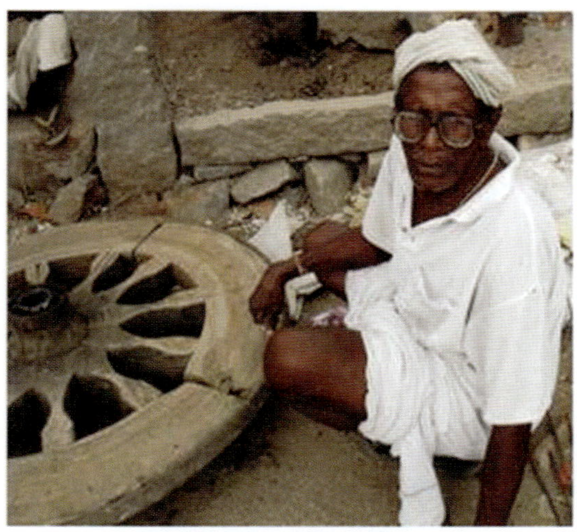

Figure 11: Smith Mattela Gangaram-Ibrahimpatnam.

Three of these ingots, of different weights, were collected from Konapuram and Ibrahimpatnam, before being tested and identified by ironsmiths from Habsipuram and Ibrahimpatnam, who easily tested the metal with a small hammer.

Evidently, not all of the wootz ingots are of the same quality. Some crucibles in the furnace may be exposed to a temperature that is too low, while others may be subject temperatures that are too high. Therefore, some of the crucibles and lumps of wootz were discarded (Fig. 14). The ingots were then categorised according to their quality by the experienced smiths. The best quality ingots, which demonstrate a fine finish and maintain their sharp edges for a long time, would be kept for sword making; the ingots of slightly lesser quality would be preferred by the local artisans who use sharp items in their daily routine.

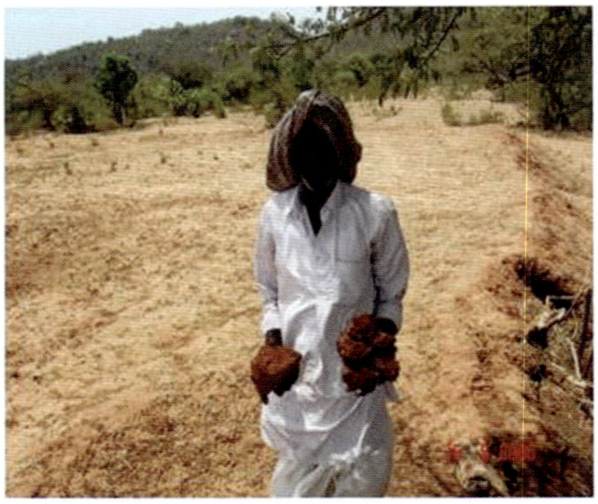

Figure 12: *Vandubonda* at Konapuram village in Kammerpally Mandal, Nizamabad District.

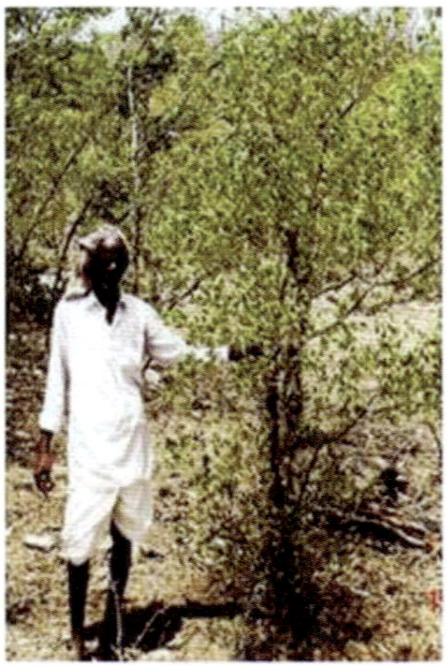

Figure 13: '*Pulayelaku*' tree at Konapuram village in Kammerpally Mandal, Nizamabad District.

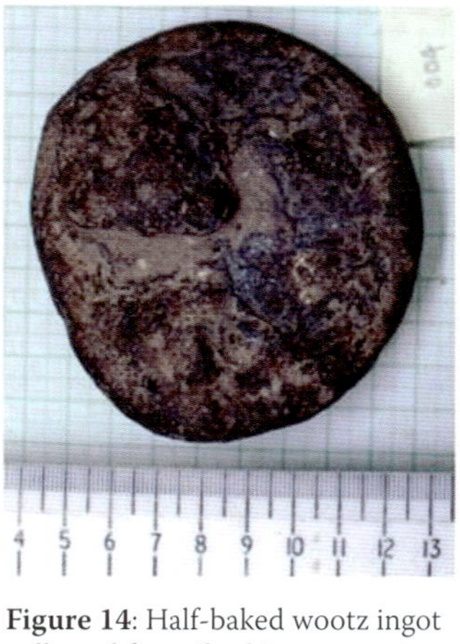

Figure 14: Half-baked wootz ingot collected from Ibrahimpatnam in Karimnagar district.

Mandalogi Gangaram is the only living smith in the surrounding area who is able to work with wootz metal to make implements. Wootz is difficult to handle because it cannot resist high heat: if it is subjected to very high temperatures the metal can break up, or the objects produced will be very brittle. An example of this is a wootz steel axe that lost its edge when handled by an amateur smith, who sharpened the edge of the axe and quenched it in water. The smith claimed that the edge had broken because the axe was made of inferior iron, although expert opinion confirmed that it was made from wootz steel (Fig. 15).

"The leaves used to cover the iron and the wood is those of the Asclepiads gigantean, or when that is not to be procured those of the *Convolvulus laurifolia*. As soon as the clay used to stop the mouths of the crucibles is dry, they built up in the form of an arch with their bottoms inwards, in a small furnace urged by two goatskin bellows. Charcoal heaped over them, and the blast kept up without intermission for about two hours and a half, when it stopped and the process considered was complete. The furnace contains from 20 to 24 crucibles" (Egerton, 2001, p.57). A number of aspects of the production process of wootz steel can be variable. For example, in some areas of India, the quality of wootz steel is said to be strengthened with herbs and shrubs placed in the crucible.

Summary

It is clear that the making and processing of wootz steel was a significant industry in the villages of Northern Telangana districts until the end of the 19th century, and that the region was famous for making wootz steel tools, weapons and armour. Several thousand people were employed in this production industry. Poodur and Kodimayal were the best production centres for cutlery and arms. Elgandal, Jagtial, Konasamudram and Kathalapur were famous for sword production and generally for the arms industry in the medieval period. Kodimayal produced forge-welded cannons, as well as the best tailor's scissors and bronze ware products. It is encouraging to note that the people in this region excelled in craftsmanship and in handling wootz steel implements until relatively recently, compared to other regions in India. Further historical case studies and archaeometallurgical investigations are needed to probe the quality of wootz steel and its production process in this

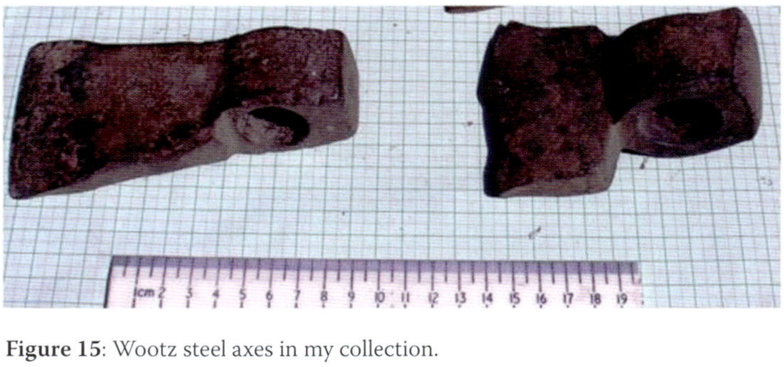

Figure 15: Wootz steel axes in my collection.

region. At present it appears that Telangana was the best local artisan base for making the world famous wootz steel implements in India.

Conclusion

During survey the author located the remains of several unexcavated furnaces in the Adilabad forests, and it is clear that there are more wootz production sites still to be discovered. Further excavation and analysis (including dating) is needed before the region's archaeometallurgical history can be fully appreciated and, in addition, before the origin and development of the wootz steel industry in general can be revealed.

I presume that the wootz processing technology might have originated in the Telangana region and then spread to other places. Although I carried out an extensive and exhaustive survey of this region, I feel there are number of sites which remain to be discovered. Even the last time I visited the region on 14 November 2008, I visited the Gopulapur village in Checode Mandal of Medak district, where some old temples can be seen in dilapidated condition. There I found ample iron slag and after some time I also saw, to my surprise, crucible fragments. I asked a villager, named Rajaiah, about wootz. He told me that he knew of wootz implements and he was using them, but he did not know about the crucibles and slag.

Once flourishing iron and wootz steel industrial centres, employing a considerable number of workers and artisans, these villages were responsible for the economic prosperity in this region. As with any time honoured crafts that prospered in this region, Telangana shows a continuity in cottage industry that stand as testimony to the traditional lifestyle of the people. The globalisation policies of the Indian government have put an end to this living tradition of Telangana.

Acknowledgements

I am expressing a deep sense of gratitude to Late Professor. R. Balasubramaniam, Head of the Department of Materials and Metallurgical Engineering, IIT Kanpur, for support, encouragement and help in this study.

References

Balasubramaniam, R., 2007. New insights on classification of iron-carbon alloys, especially high carbon steels. *Indian Journal of History of Science*, 42, pp.427-44.

Biswas, A.K., 2001. Minerals and metals in medieval India. In: A. Rahman ed. *History of Indian science, technology and culture, AD 1000-1800*. New Delhi: Oxford University Press, pp.275-313.

Briggs, J. ed., 1981. *History of the rise of the Mahomedan power in India till the year AD 1612*. New Delhi: Oriental Books Reprint Corporation.

Brown, C.P., 1903. *Telugu-English Dictionary*. Madras: Society for Promoting Christian Knowledge.

Craddock, P.T., 2007. Cast iron: the elusive feedstock of crucible steel. *Indian Journal of History of Science*, 42, pp.593-607.

Deo, S.B. and Jamkhedkar, A.P., 1982. *Excavations at Naikund, 1978-80*. Bombay: Department of Archaeology and Museums, Government of Maharashtra.

Kotilingala: A Report on Excavations (1979-1983), Published by the Department of Archaeology and Museums, Government of Andhra Pradesh, 2006.

Egerton, Lord W., 2001. *A Description of Indian and Oriental Armour*. New Delhi: Asian Educational Services.

Fazal, A., 1962. *Ain-i-Akbari*. New Delhi: Oriental Books Reprint Corporation.

Feuerbach, A., Balasubramaniam, R. and Kalyanaraman, S., 2007. On the origin of the terms wootz, Hinduwani and Pulad. *Indian Journal of History of Science*, 42, pp.377-86.

Jaikishan, S., 2007. Khadgalaksanasiromani of Navanappa. *Indian Journal of History of Science*, 42, pp.i-vi and T1-49.

Krishna Shastri, V.V., 1983. *The proto and early historical cultures of Andhra Pradesh*. Hyderabad: Department of Archaeology, Government of Andhra Pradesh.

Le Coze, J., 2007. On the question of possible transfer of steel technology from India to Europe through the Muslim Middle East. *Indian Journal of History of Science*, 42, pp.336-77.

Lowe, T.L., 1990. "Indian iron ore and the technology of Deccani 'wootz' production." Paper presented at Iron paleo metallurgy and cultures International symposium, University de Technologie de-Compiegne a Savannas, 90010, Belfort.

Marco Polo, 2001. *The travels of Marco Polo*. New Delhi: Asian Educational Services.

Moreland, W.H., 1923. *From Akbar to Aurangzeb: a study in Indian economic history*. London: Macmillan.

Pearson, G., 1795. Experiments and observations to investigate the nature of a kind of steel, manufactured at Bombay, and there called wootz: with remarks on the properties and composition of the different states of iron. *Philosophical Transactions of the Royal Society of London*, 85, pp.322-46.

Rao, K.P., 2010. Excavations at Gachibowli: A preliminary report. *Puratattva*, Bulletin of the Indian Archaeological Society, No.40, 2010. pp.102-111.

Srinivasan, S. and Ranganathan, S., 2004. *India's legendary wootz steel, an advanced material of the ancient world*. Bangalore: National Institute of Advanced Studies and Indian Institute of Science.

Thevenot, J. de. 1949, *Indian travels of Thevenot and Careri*, S. N. Sen (Ed), National Archives of India, New Delhi.

Tavernier, J.-B., 2001. *Travels in India*. Second English edition by V. Ball. New Delhi: Munshiram Manoharlal Publishers.

Thomas, P.J., Nagabhushanam, P. and Reddy, D.V., 2008. OSL - Optically stimulated luminescence dating of heated materials using single-aliquot regenerative-dose procedure: a feasibility study using archaeological artefacts from India. *Journal of Archaeological Science*, 35, pp.781-90.

Toussaint, F., 2002. Iron and steel in ancient India. In: G. Pande and J. AF Geijerstam eds. *Tradition and innovation in the history of iron making*. Nainital: Pahar, pp.39-46.

Voysey, H.W., 1832. *Description of the native manufacture of steel in southern India. Journal of Asiatic Society of Bengal*, 1, pp.245-7.

Author's address

- Reader in History
  SLNSA Oriental Degree College
  DHARMAPURI- Karimnagar district, AP., India, PIN – 505425
  dr.sjai@gmail.com

Iron Age – Early Historic period in Eastern India. A study in material culture and technology: evidence from the site of Badmal-Asurgarh, District Sambalpur, Orissa, India

Pradeep K. Behera and Pranab K. Chattopadhyay

ABSTRACT: Badmal-Asurgarh is a fortified settlement of the Iron Age-Early Historic period (c. early 1st millennium BCE to 2nd century BCE) located in the middle Mahanadi valley region of Orissa highlands in eastern India. Situated on the left bank of the *Harihar* stream, which is a minor tributary of the river Mahanadi, the site spreads over an area of about four hectares and is bounded by a lozenge-shaped earth rampart-and-moat with three passage ways. The available radiocarbon dates, stratigraphic contexts and a careful study of changes in the ceramic and other cultural assemblages, revealed two broad periods of human habitations, viz. Iron Age and Early Historic. The earliest settlers, whose subsistence economy was mainly based on limited farming, animal husbandry and hunting-gathering, had definite knowledge of iron technology, though its use was restricted to a few hunting weapons. After a brief period of settlement desertion, the site was reoccupied during the Early Historic period and rose to prominence as a bead manufacturing centre in this part of the Orissa highlands. Domestic trade was perhaps one of the important factors in this process of development. Our study has revealed that, in addition to locally available gem resources for manufacturing beads for regional and extra-regional trade or exchange, the bead workers of Badmal also imported *cowry* shells and finished items of onyx and coral from more distant places. Iron played a catalytic role in the development of the specialised bead industry at the site, as attested to by the iron tools that were employed to craft the beads. The present study appears to suggest that, in exchange for finished beads, the settlers were importing iron objects of improved quality, as well as other raw materials, such as banded-hematite-red-jasper, agate, etc. from multiple sources for facilitating large-scale bead production. However, it is much too early to arrive at any firm conclusions on the basis of this preliminary study. More intensive investigations from multidisciplinary perspectives are required to understand the formation processes at Badmal and its spatio-temporal roles in the middle Mahanadi valley region and beyond.

Keywords: Badmal, middle Mahanadi Valley, iron technology, beads, PIXE, trade, micro-alloy.

Introduction

The second half of the first millennium BCE witnessed a resurgence of urbanisation across most of the Indian sub-continent, with much of this developing out of the preceding regional iron-using cultures. Until recently it was believed that iron technology appeared in the Indian sub-continent by the beginning of the first millennium BCE, but recent radiocarbon dates from northern India have pushed this date back to the early second millennium BCE (Tewari, et al., 2002).

In Eastern India, Iron Age-Early Historic settlements are widely distributed and scattered across the country in varied environmental settings. The vast mineral resources of the highland areas; the alluvial tracts of the Gangetic and the Mahanadi deltas which favoured agricultural intensification and expansion; the long coast line that accommodated maritime trade; as well as contemporary socio-cultural forces; must have played catalytic roles in the emergence of several urban centres in this part of the sub-continent. Chandraketugarh, Tamluk, Mangalkot and Bangarh in West Bengal, and Sisupalgarh, Jaugarh, Radhanagar, Manikpatana, Kharligarh and Narla Asurgarh in Orissa (Ghosh, 1989; Behera, 1994; Allchin, 1995; Ray, 1995; Smith, 1995; Ray and Salles, 1996; Sengupta and Panja, 2002; Yule, et al., 2005) are but a few of the many notable Early Historic urban centres of this region frequently referred to in the literature.

Archaeological investigations conducted during the last few decades in various parts of Orissa have brought to light several Iron Age-Early Historic settlements, some of which are fortified (Lal, 1949; Das, 1990; Basa and Mohanty, 2000; Yule, 2006; Patra, 2007). From the available literature, it appears that by the 5th-4th century BCE there was a proliferation of urban centres of various sizes and styles in Orissa (Fig. 1). These have been reported from three environmental zones: the hilly terrain of the Eastern Ghats and Chhottanagpur Plateau; the coastal plains comprising the alluvial stretch of the Mahanadi delta and the Rusikulya river system; and the littoral areas. Thus, the region of Orissa included distinct environments to which the Early Historic settlers adapted. The distribution pattern of these settlements appears to suggest that, in addition to the latter two environmental zones, highlands were also preferred, obviously due to the availability of rich and varied mineral resources and a wide spectrum of subsistence-related forest products. Our survey in the middle Mahanadi Valley alone identified thirty-four Early Historic sites associated with black-and-red ware pottery. Most of these also contain Iron Age deposits (Behera, 2002-2003).

The site and culture sequence

The site of Badmal Asurgarh (21° 07' 84" N and 84° 04' 42" E) is located in the middle Mahanadi Valley (Singh, 1971, p.773) of the Orissa highland region, and is situated on the bank of a minor stream called *Harihar Jhor,* which is a tributary of the river Mahanadi (Fig. 2). The area is flanked on the west, north and east by a series of hill ranges of varying heights. The general elevation of the terrain floor varies between 155 and 170m above mean sea level. The valley, as well as the hilly areas, is covered with thick vegetation growth of a dry, deciduous, mixed type of forest. Although the area receives an average annual precipitation of about 1500mm, the present population largely depends upon seasonal rains, besides lift irrigation for crop cultivation. In addition to limited agriculture, local populations depend upon the vast natural resources available in the area to meet their day-to-day requirements, such as large and small game animals, aquatic resources and wood from the forests, etc. (Fig. 3).

The geological formations represented in the Sambalpur district are shown in Table 1.

The oldest rock formation of the area belongs to the Archaean series, followed by those of the Dharwarian and Cuddapah series. Overlying Cuddapah, the sediments of the Gondwana formations represented by Talchirs, Damuda, etc. were deposited on a Precambrian platform, almost separating the Eastern Ghats and the North Orissa craton (Mahalik, 1994, pp.41-51). The Quaternary formations are represented by thin to thick deposits of secondary laterite, colluviums and alluvium of variable extent.

The most noteworthy geological feature of this belt is its rich gem stone deposits. Some 200 pegmatite bodies, varying in size from 2 to 15m in width and about 1 km in length, have been identified in the area (Mishra and Mohanty, 1995). Most of the gemstones in this area occur in secondary contexts where they have been derived from weathering of

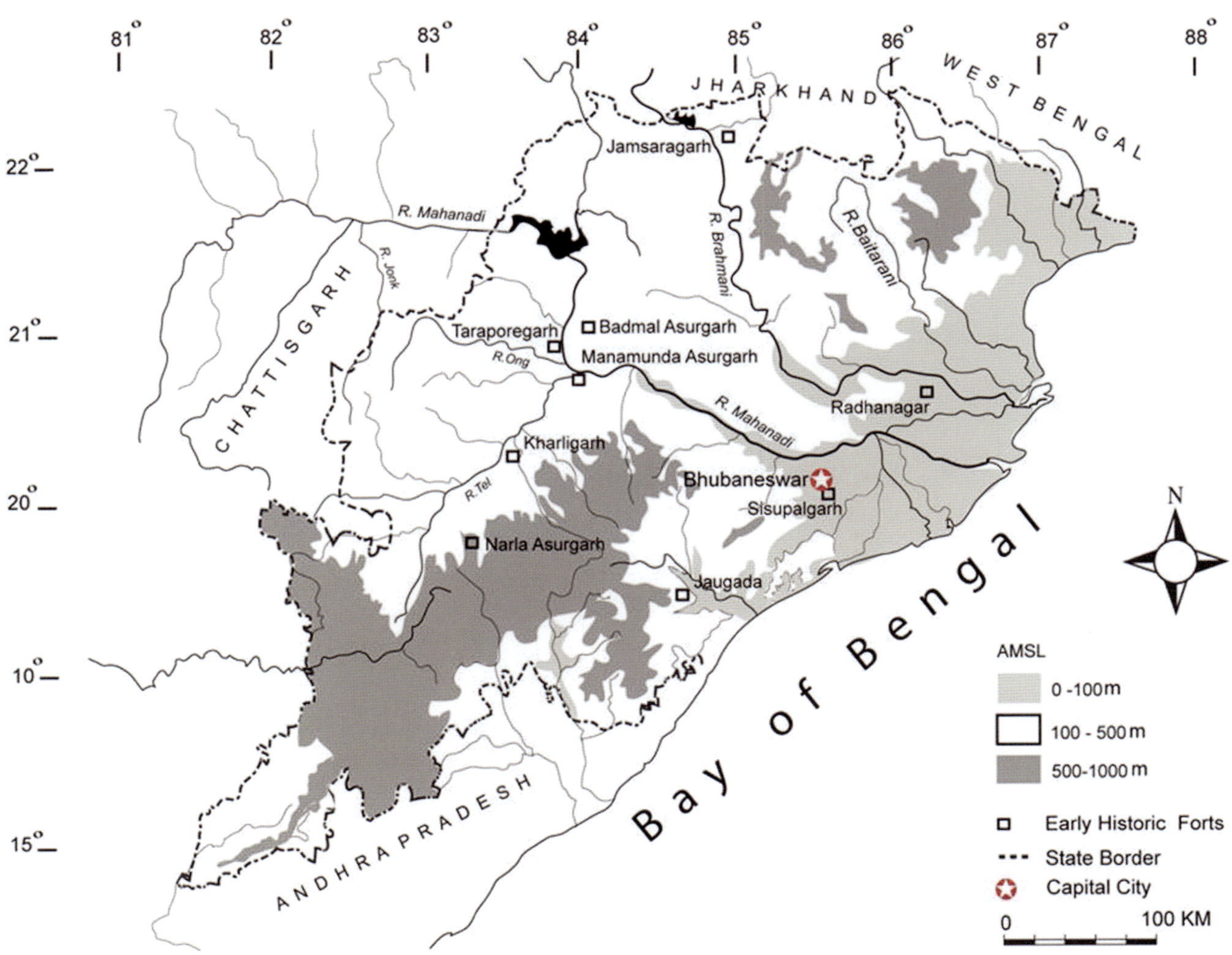

Figure 1: Distribution of major early historic sites in Orissa.

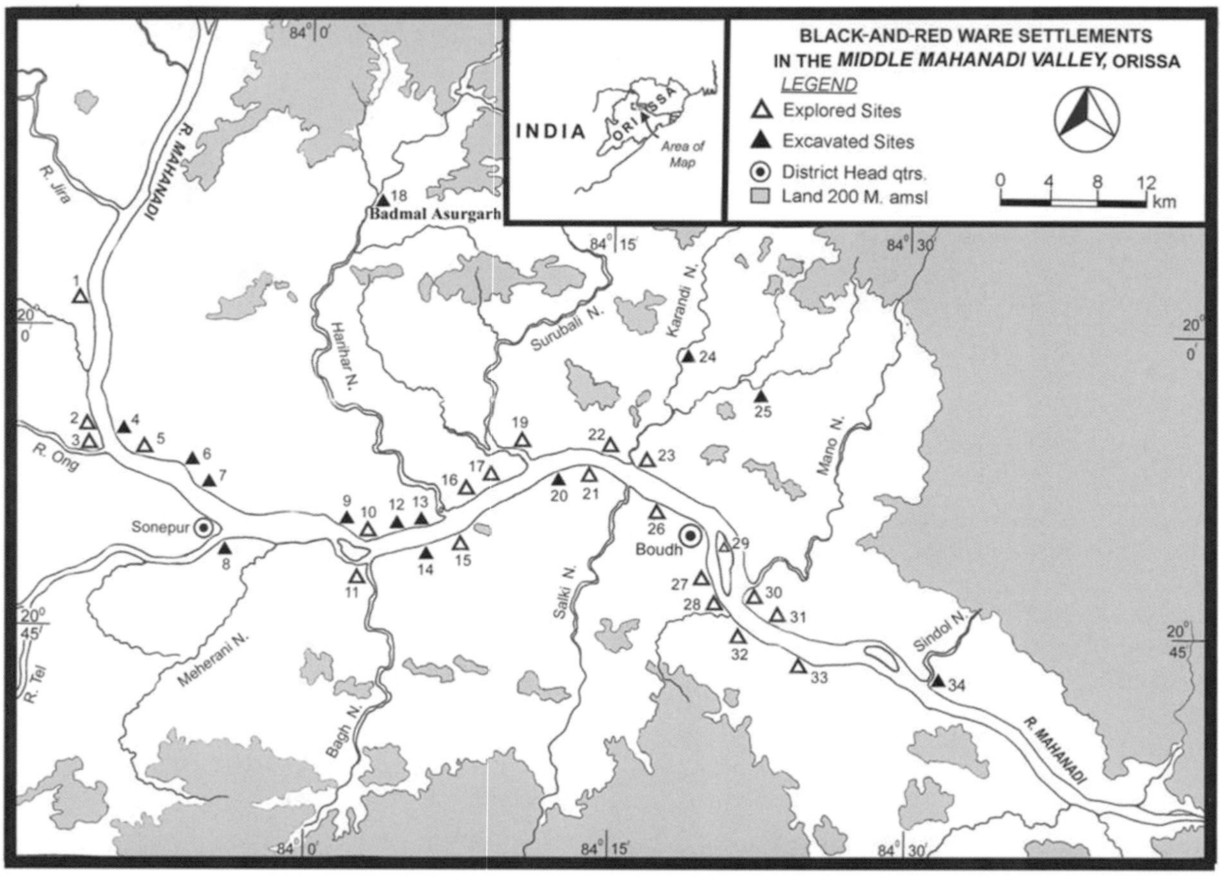

Figure 2: Location of the Badmal Asurgarh site in relation to other Iron Age-Early Historic settlements associated with black-and-red ware pottery in the middle Mahanadi Valley of Orissa (based on Survey of India topographic maps and archaeological site information).

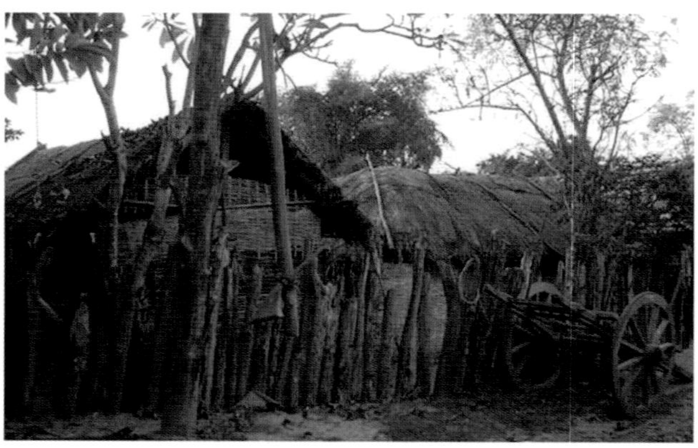

Figure 3: Use of natural resources by the modern settlers of Badmal village.

pegmatite bodies with subsequent transfer of the minerals into the fluvial network active in the region. These secondary deposits constitute an easily amenable source for good quality gems. Gemstones to be found within this area include beryl, aquamarine, topaz, garnet, heliodor, ruby, sapphire, tourmaline and amethyst. For manufacturing beads and pendants, the Early Historic settlers of Badmal Asurgarh exploited some of these gemstones, as well as other semi-precious stones such as chalcedony, agate, banded-hematite-red-jasper (BHRJ) and chert.

The fortified site (Fig. 4) is roughly lozenge-shaped and spreads over an area of about four hectares (180 x 220m). The site is protected on all sides by a massive earth rampart, which rises to a height of about six to seven metres above

Table 1: Geological formations represented in the Sambalpur district, Orissa, in order of increasing antiquity (after Senapati and Mahanti, 1971, pp.15-16).

Age	*Rock formations*	
Recent	.......Soil and alluvium.	
Tertiary	.......Laterites and lateritic iron ore – usually manganiferous.	
Lower Gondwana	Damuda series.	.... Kamthi stage.
	Talchir series	Barakar stage, needle shales and sandstone. Boulder bed.
	------UNCONFORMITY------	
Cuddapah	.... Shales and sandstones, quartzites and limestones.	
	Newer dolerites, pegmatites, quartz veins and amphibolites.	
Dharwars	.... Gondites, mica schist, dolomite-marble, phyllites.	
Archaeans	.... Gneisses, granites, charnockites, mica schist, quartzites and khondalites, etc.	

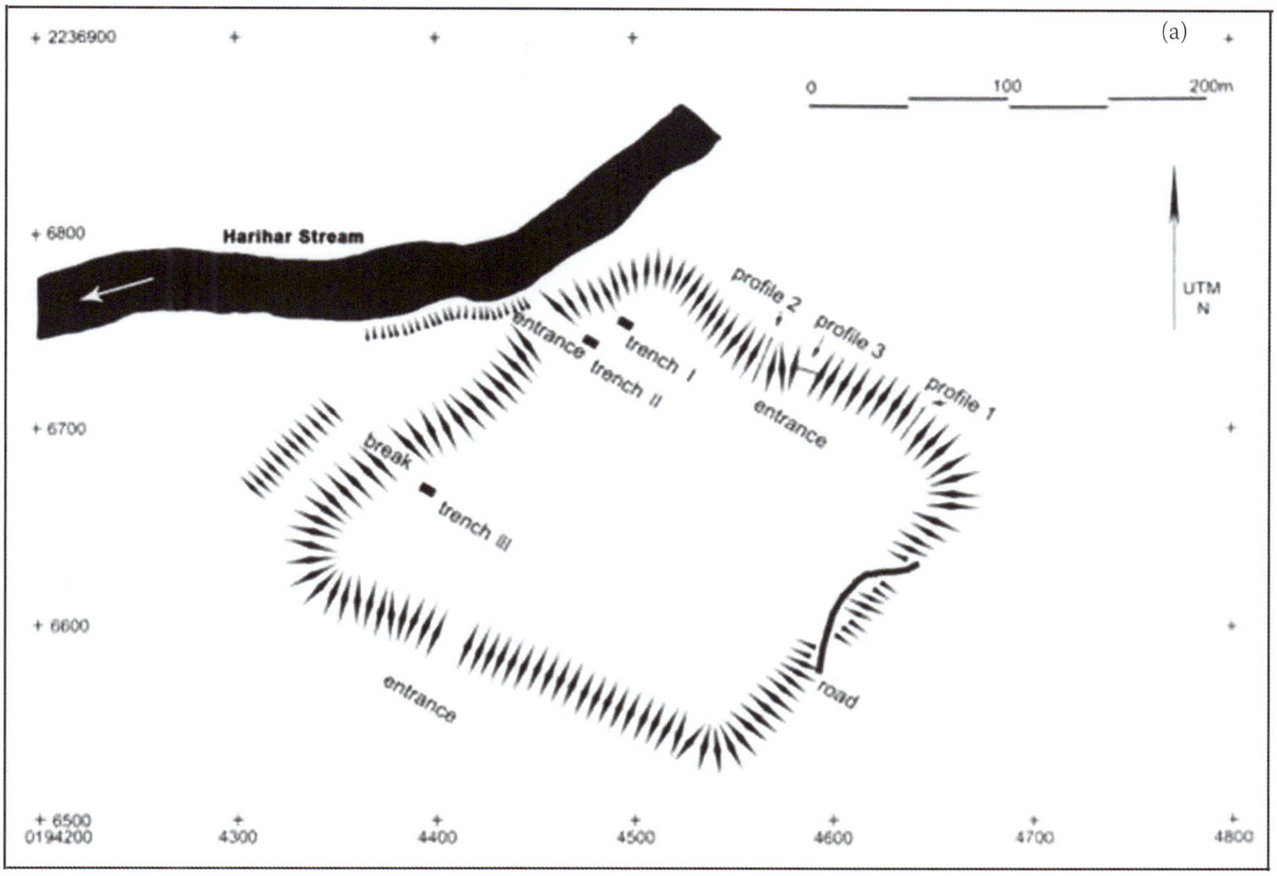

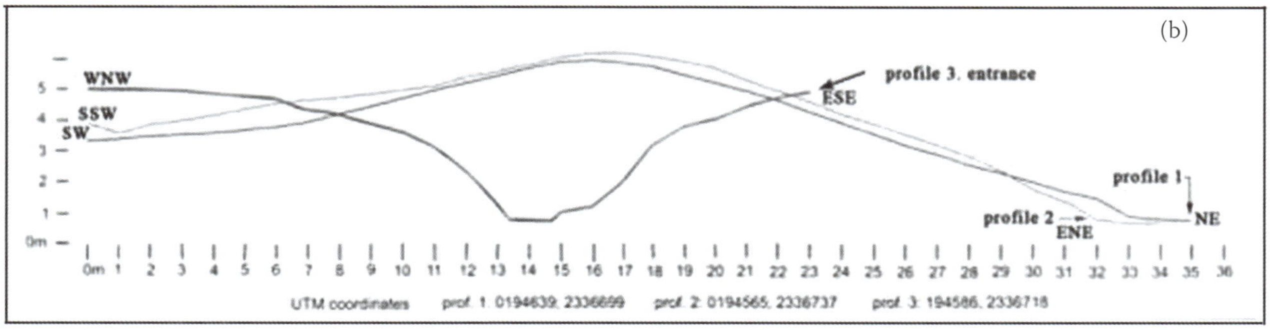

Figure 4: a) Plan of the Iron Age-Early Historic fort at Badmal Asurgarh; b) profile drawings of the fortification wall (after Yule, et al., 2005, p.311).

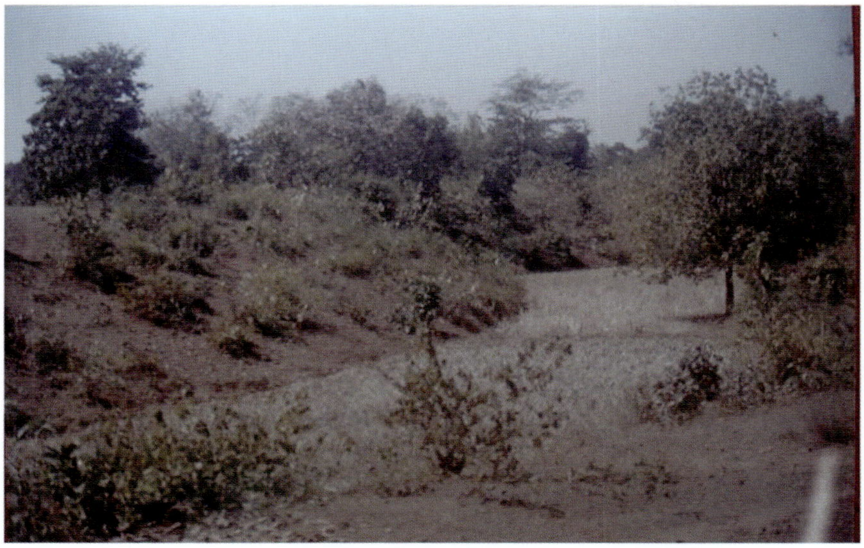

Figure 5: Flanking remains of moat visible outside the north-eastern defence wall.

the surrounding plains and is approximately 20-25 metres wide at the base. In addition to the main fortification, two additional small earth ramparts are located on the north-western side of the site towards the stream, presumably to protect the settlement from seasonal flooding. At least three entrances to the settlement have been identified, located on the south-western, north-western, and north-eastern sides. While the north-eastern and north-western prominences of the rampart wall are to some extent better preserved, other parts have already been destroyed by the intensive surface agricultural activities carried out by the present villagers. The rampart was quite possibly originally encircled by a 15-20m wide moat, which has now been converted into agricultural land (Fig. 5). Another important factor which has seriously impaired the preservation of the site is the frequent illegal mining operations undertaken by the present villagers inside the fortified area, particularly in the southern sector.

Trial excavations were conducted in 2002 to understand the nature and succession of cultures at the site. On the basis of available radiocarbon dates, stratigraphic contexts and a careful study of changes in the ceramic and other cultural assemblages, two broad periods of human occupation, corresponding to the Iron Age and the Early Historic period, have been identified. The Iron Age is further divided into three sub-Periods (IA, IB and IC), mainly on the basis of changes in the ceramic assemblages, as well as radiocarbon dates (Table 2).

In Trench-I (Fig. 6), excavated in the northern sector of the site, an occupational break represented by a 30-35cm thick culturally sterile deposit was noticed between the two major periods. Most probably due to frequent inundation from the Harihar stream, the site was deserted by the Iron Age people and was reoccupied subsequently during Early Historic times.

Chronology

Several radiocarbon dates were obtained from various levels of the habitation deposit corresponding to the Iron Age period occupation of the site (Behera, et al., 2007). No date could be retrieved from Period IA of the Iron Age occupation and the succeeding Early Historic levels. The available dates are indicated in Table 2. In the absence of any date from the earliest level of Period IA, but on the basis of available dates from Periods IB and IC, the beginning of habitation at the site may be traced back to the early first millennium BCE. All of the available dates from the site are consistent (except one isolated date from sub-Period IC) and fall within the time range of the Early Iron Age period in India (Tripathi, 2001). Most probably due to the frequent inundations from the *Harihar* stream, the site was deserted by the Iron Age people by about the 5th century BCE and,

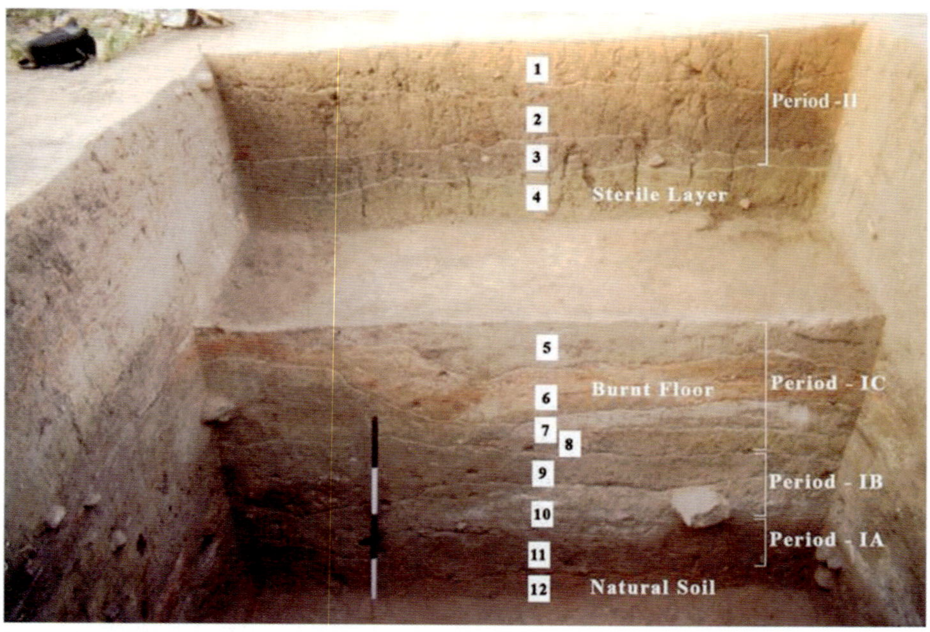

Figure 6: Trench-I, excavated in the northern sector, showing habitational layers and sequence of cultures.

Table 2: Radiocarbon dates from the Iron Age deposit of Badmal Asurgarh.

Sl. No.	Trench No./ Depth (cm)	Cultural Affiliation & Periods	Lab. Code	Calendar Age (BCE)
1	Trench-I/120	Iron Age (IC)	IP-275	468±76
2	Trench-I/160	Iron Age (IB)	IP-274	838±86
3	Trench-II/45	Iron Age (IB)	IP-285	803±44
4	Trench-II/50	Iron Age (IB)	KIA-20155	617±27
5	Trench-II/115	Iron Age (IB)	KIA-20154	785±12
6	Trench-II/125	Iron Age (IB)	KIA-20153	786±11

after a short occupational break, it was reoccupied during Early Historic times, probably by about the early 4th century BCE. The date of the construction of the rampart could be considered as falling within the Early Historic period based on circumstantial evidence.

Period I (Iron Age)

During this period the settlement was confined to a limited area, situated towards the highest north-western part of the mound, and lacked any fortification wall. The inhabitants lived in wattle-and-daub houses. They used a variety of materials, including iron spears and arrow-heads and copper bangles, bone and antler tools, sling balls of stone, pottery discs and a variety of earthen wares. Although they were exposed to locally available precious and semi-precious stones, none of these materials were found during the trial excavations. The early settlers mainly subsisted on limited agriculture and domestic as well as wild animals. The deposit of this period yielded a large quantity of charred and un-charred skeletal remains of wild and domesticated animals bearing distinct butchery marks (Fig. 7). The ceramic assemblages, which demonstrate close affinity with those from contemporary Iron Age settlements of the middle Mahanadi Valley, are represented by black-and-red ware pottery of fine to medium fabric, black slipped ware, red slipped ware and plain red ware pottery. Important shapes in the black-and-red ware include flat and round-based convex-sided bowls, dishes and medium-sized vases (Fig. 8). While vases of different sizes and forms predominate in the red slipped ware group, large-sized storage vessels of coarse fabric are rare and confined to the plain red ware.

Period II (Early Historic)

After a brief habitational break, the site was again occupied by the people of the Early Historic period. During this period, the settlement expanded to its fullest extent and was protected on all four sides with a massive earth rampart with at least three entrance passages. While some of the cultural features of the preceding period continued to exist, the site now transformed into a stone bead manufacturing centre. The low-lying southern part of the mound was exclusively used for manufacturing precious and semi-precious stone beads.

For manufacturing beads, the craftsmen of this period not only exploited the locally available beryl of greenish-blue, golden-yellow and pale blue to sea-green colour, aquamarine, smoky quartz, amethyst, corundum, and tourmaline, but also imported materials such as banded-hematite-red-jasper, agate, chalcedony and amygdaloidal basalt obtained from regional and extra-regional contexts by way of trade or exchange. In addition, beads and pendants of coral and onyx, which were directly imported from outside the region, have also been found.

A small trench (2 x 2.5m) excavated in the southern sector of the mound yielded more than 3000 bead wastes, 347 bead roughouts and blanks, 61 semi-finished beads and only 33 finished beads (Figs. 9, 10 and 11), besides craft tools including bead polishers, small-sized pestles with sub-triangular cross-section, anvils, hammers, mulers and two-legged querns (Fig. 12). In addition to beads and bead wastes indicative of different stages of the manufacturing process, the trench also yielded a sizeable number of iron objects including drill bits, pottery discs of various sizes and weights and a large quantity of pottery, which also included black-and-red ware.

Despite the technological advances associated with the emergence of specialised craft industry, there was no marked change in the general lifestyle of the people of this period. The Early Historic people continued to live in wattle-and-daub houses.

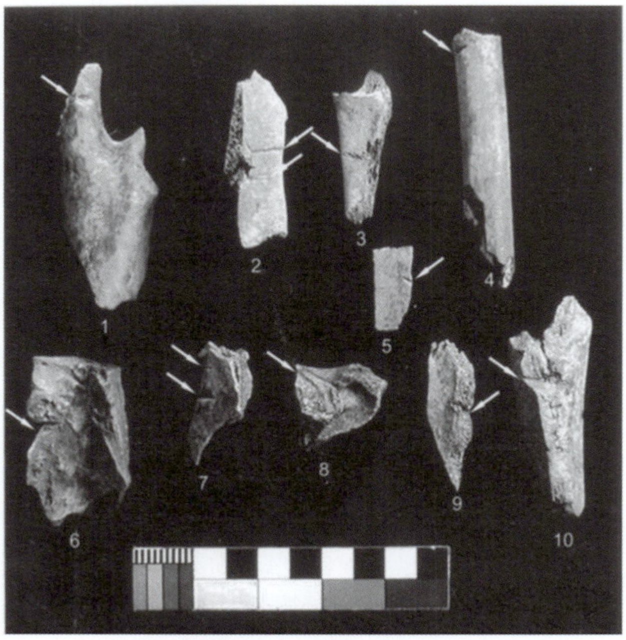

Figure 7: Faunal remains from Period IB, showing cut marks.

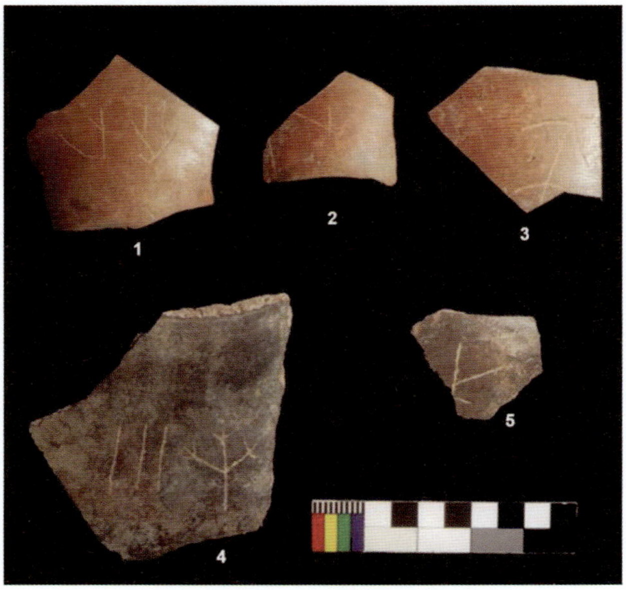

Figure 8: Fragments of black-and-red ware bowls with graffiti marks from the Iron Age deposit at Badmal.

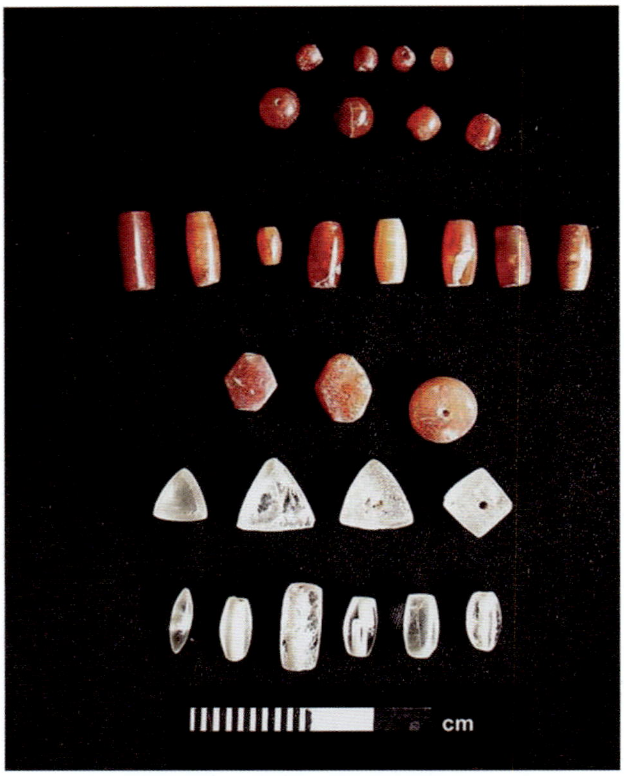

Figure 9: Finished beads of carnelian and crystal quartz from Trench-III.

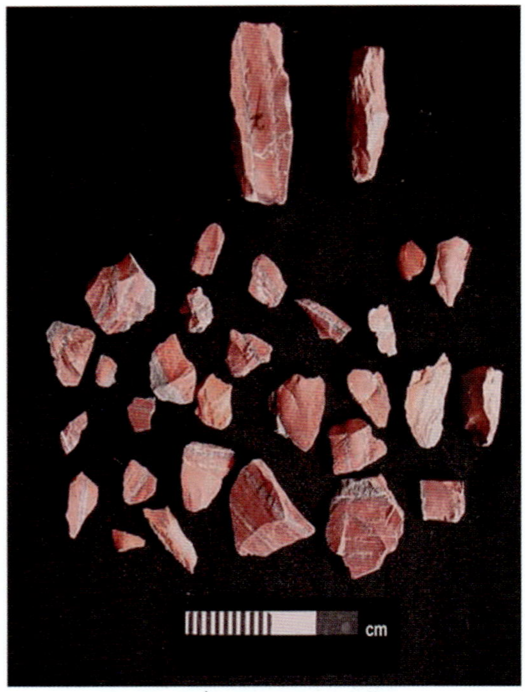

Figure 10: Bead roughouts and wastes of banded-hematite-red-jasper from Trench-III.

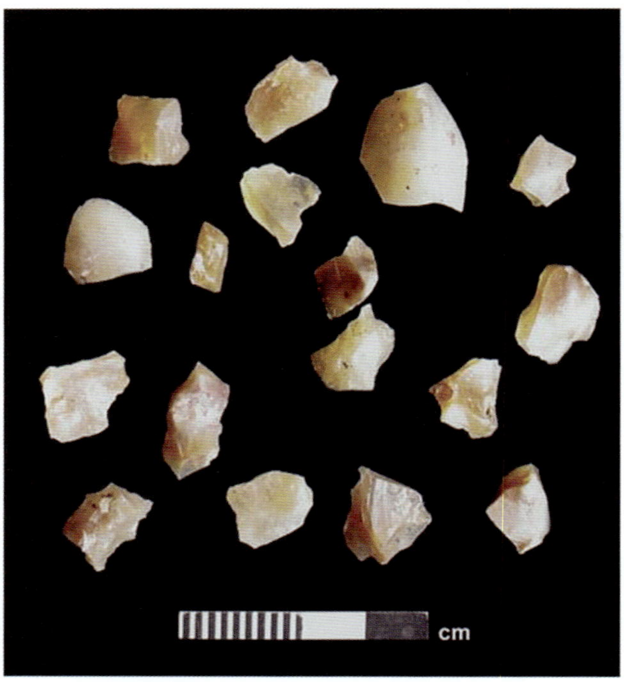

Figure 11: Bead blanks of agate from Trench-III.

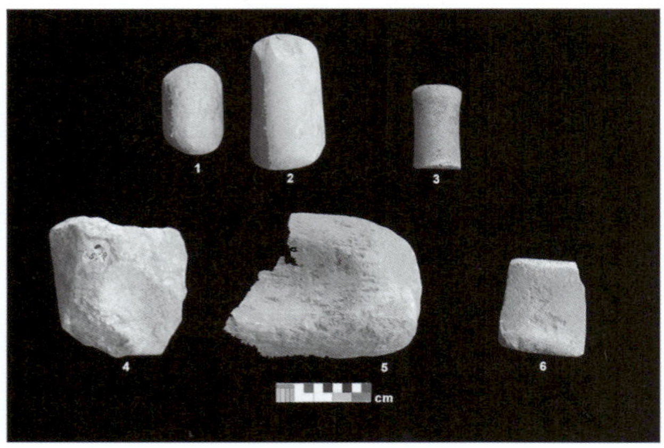

Figure 12: Stone tools used in manufacturing stone beads at Badmal: 1-2) pestles with sub-triangular cross-section; 3) muler (vertical pestle); 4-5) bead polisher; 6) two-legged quern.

Iron technology at Badmal

Although iron objects have been found associated with the earliest levels of human occupation at Badmal, there is no direct evidence for on-site production of this metal. The period-wise distribution of iron objects, as indicated in Table 3, clearly reveals proliferation of iron technology during the Early Historic period. It can be suggested that the emergence of large-scale craft production of specialised items, such as stone beads, led to the diversification in the typology of iron objects (Fig. 13). Unlike the preceding period, in which only a few hunting weapons are represented, it seems that the emphasis at this time was placed on procuring and using craft-related tools, including drill bits, saws, nails, etc.

Iron procurement strategy

In Orissa, iron ore deposits are found east of the Brahmani river valley, in the northern and north-eastern highland zones, comprising the Bonai-Keonjhar, Gandhamardan, Tomka-Daitari and Gorumahisani-Badampahar regions (Fig. 14). The parent rocks of these deposits are represented by Banded

Table 3: Distribution of the excavated iron objects from Periods I and II.

Sl. No.	OBJECT TYPE	Pd. IA	Pd. IB	Pd. IC	Pd. II
1	Spear head	1	1	1	1
2	Arrow head	-	1	-	-
3	Drill bit	-	-	-	7
4	Nail parer (?)	-	-	-	3
5	Flattened ring	-	-	-	3
6	Toothless saw (?)	-	-	-	5
7	Weight with a hole / Net sinker (?)	-	-	-	1
8	Short pipe-like clamp	-	-	-	1
9	Nail	-	-	-	5
TOTAL		**1**	**2**	**1**	**26**

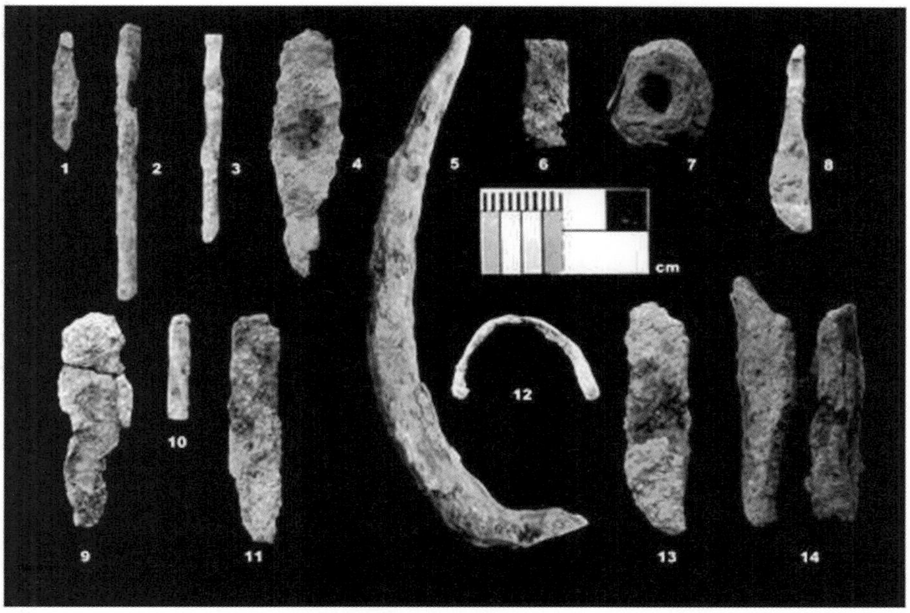

Figure 13: Iron objects from the Early Historic period at Badmal.

Iron Formation (BIF) and ferruginous shales. The iron bands are also represented by magnetite, marmite, goethite and magnetite, although hematite constitutes the predominant band in the layered BIF (Mohanty, 1995).

Excavations at Badmal did not yield any evidence pertaining to on-site iron smelting activities. However, during the excavations, a lump of chromate ore was retrieved from Trench-III, located in the southern sector of the mound. In order to determine the source of this iron and the technology involved in the manufacture of the iron objects found at the site, four iron objects, one from Period IA and the rest from Period II, were subjected to External Particle Induced X-ray Emission (PIXE) and metallographic examinations (Behera and Chattopadhyay, 2004-2005; Chattopadhyay, et al., 2007).

The PIXE analysis was conducted at the Institute of Physics, Bhubaneswar, India, which uses a 3mA tandem type Pellet Ron Accelerator with proton beam energy of 3 Me (Vijay, et al., 2003). This method is extremely useful for analysing archaeological samples, since it is not only fast, sensitive and capable of simultaneous multi-elemental analysis, but it also ensures that samples of any size can be quantitatively analysed without causing physical damage to the artefacts (Johansson and Campbell, 1988; Campbell, et al., 1995a, 1995b; Demortier, 1997, 2000; Hajivaliei, et al., 1999; Govil, 2001). However, in PIXE experiments, all the calculations are based on the hypothesis that the material is homogeneous in all the depth of the analysed material (5-10μm). The analysed depth of the irradiated artefact is less than 10μm, even for particles crossing 25 to 30μm in the material, due to the decrease of X-ray production with decreasing energy of the projectiles. Thus, archaeological applications are generally restricted to non-corroded materials or to drillings involving partial destruction of the sample (Mando, 1994). In the present analysis, the target areas (1cm^2 surface) of the iron samples were thoroughly cleaned by carefully removing the outer corrosive layer and keeping the original shape intact before the artefacts were irradiated. The results of the analysis are shown in the Table 4.

Table 4 indicates that the iron objects were possibly produced from multiple ore sources and may well therefore have been procured from different locations. However, the most interesting result of the analysis is the presence of vanadium and chromium in three of the four samples analysed. Vanadiferous magnetite deposits are found in small

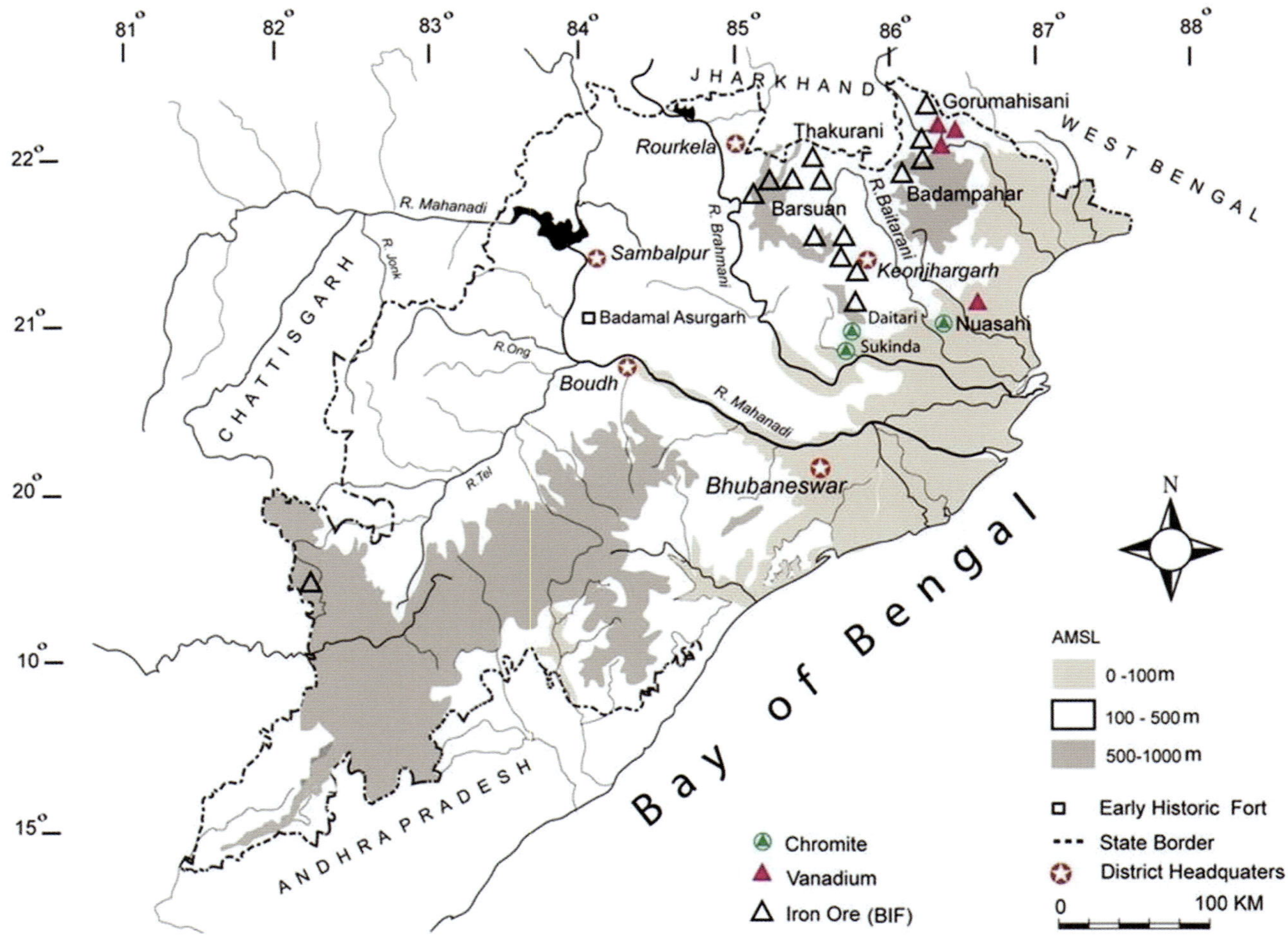

Figure 14: Distribution of iron, vanadium and chromate deposits in Orissa (after Mohanty, et al., 1995).

Table 4: Elements detected through external PIXE in weight percent.

Sample No.	Period	Object Type	P	S	Ca	Ti	V	Cr	Mn	Fe	Ni	Cu
BDM-18	IB	Spear head	0.15	0.13	-	-	0.13	-	0.70	95.15	0.15	-
BDM-7	III	Drill bit	0.49	0.41	0.41	0.29	0.076	-	0.58	95.76	0.29	0.12
BDM-8	III	Drill bit	0.58	0.40	0.23	-	-	-	0.44	95.71	0.27	-
BDM-16	III	Toothless saw	0.53	0.43	0.12	-	-	0.06	0.56	95.08	0.43	-

pockets near the Nuasahi-Boula area on the eastern boarder of Keonjhar, in the Baripada-Rairangpur belt in Mayurbhanj district, and in the Betei-Rangamatia area in the district of Balasore (Nayak and Das, 1995: Fig. 14). It is known that the addition of even less than 0.1% of vanadium to steel or cast iron can significantly increase strength, toughness and ductility. Whether this characteristic feature of vanadium was known to the early settlers of Badmal cannot be determined with any amount of certainty based on the presently available evidence. Similarly, one of the iron samples also contains chromium, sources of which are located in the Tomka-Daitari belt of Keonjhar district (Sahoo, 1995). The source lies some 180 km east of the Badmal Asurgarh site. As mentioned above, a lump of chromite ore has also been recovered from the Early Historic deposit at the site. In addition, the bead workers of Badmal extensively used banded-hematite-red-jasper raw material for bead production. The nearest source for this lies about 140km north-east of Badmal, in the Bonaigarh area of Sundargarh district. The possibility that the early settlers of Badmal imported iron objects from this area cannot therefore be ruled out. Thus, circumstantial as well as scientific evidence appear to suggest that during the Iron Age and Early Historic periods, iron was procured either as finished goods or in the form of ore by way of trade or exchange from locations at least 140-200km away from the site of Badmal. However, only future investigations using geochemical and other scientific techniques will solve the question of the provenance of iron at Badmal.

Metallographic studies

For metallographic analysis, small samples were taken from each of the four objects (Table 3) and mounted on perspex using a Buhler Simplimate 2 mounting press. Subsequently, the specimens were observed in polished and etched conditions. Observations were made using a Leica DMLM microscope in different areas, with different magnifications, ranging from 50 to 1000x. Each of the specimens was heterogeneous in grain size, from large coarse grains of ASTM grain size 1 to very fine ones of ASTM grain size 8. Each sample was also subjected to 2-3 micro-hardness tests using the Leica VMHT at 300gm load for 15 seconds. The average Vickers Hardness Values (HV) are shown in Table 5. Slag inclusions were commonly observed in the analysed samples.

Table 5: The HV Values of the samples obtained by micro-hardness tests.

Specimen No.	HV at ferrite region	HV at other region
BDM-18	113.7	-
BDM-07	99.0	-
BDM-08	134.8	441.6
BDM-16	110.0	195.6

Sample BDM-18, a spearhead, is chronologically the earliest of the sample range. The microstructure indicates the presence of oxide scale and silica-rich slag inclusions. When etched, the heterogeneous nature of the grains was revealed, with ASTM grain sizes between 3 and 8. No evidence for carburisation was visible and the average micro-hardness obtained was 113.7 HV within a ferrite region.

Sample BDM-8 is a drill bit. At 500x, a heterogeneous microstructure was revealed, including elongated slag inclusions, massive ferrite and ferrite grains with Neumann bands. The latter structure indicates that the tool was forged at low temperatures (below 500°C), which caused a shock to the metal. No evidence for annealing was visible, and the micro-hardness value obtained was 134.8 HV within a ferrite region.

Sample BDM-7, also a drill bit, was identified as low carbon steel. The microstructure was again heterogeneous, with evidence of partial spheroidisation of the carbides, and Widmanstatten side plates had formed in hypoeutectoid regions.

Finally, sample BDM-16, identified as a toothless saw, included a fine-grained corrosion layer with ferrite matrix. In the central region, both coarse and fine grain matrices could be observed, along with pearlite grains. A few grains with Neumann bands were also noticed, along with elongated slag inclusions. The micro-hardness value obtained in a ferrite region was 110 HV, while that in a pearlite region was 195.6 HV.

Concluding remarks

The foregoing discussion on the Badmal Asurgarh site clearly reveals that the rich bio-habitat and the vast mineral resources of the region have attracted human occupation right from the beginning of the first millennium BCE. The settlement developed out of the existing Iron Age cultures of the middle Mahanadi Valley, and rose to prominence during the Early Historic period as a bead manufacturing centre in this part of the Orissa highlands. Domestic trade was perhaps one of the important factors in this process of development. This study had revealed that, in addition to locally available gem resources for manufacturing beads for regional and extra-regional trade or exchange, the Early Historic bead workers of Badmal also imported cowry shells and finished items of onyx and coral from more distant places. Iron definitely played a catalytic role in the development of the specialised bead industry at the site, as attested by the iron tools that were employed to craft the beads. Presumably in exchange for finished beads, the settlers were importing iron objects, as well as other raw materials, such as banded-hematite-red-jasper, agate, etc. for large-scale bead production.

The chemical analysis clearly indicates a non-meteoric origin of the analysed iron samples. While the uniform distribution of slag inclusions is indicative of wrought iron, one of the specimens may be identified as low carbon steel. The earliest iron object, sample BDM-18, is definitely primitive in terms of the technology employed for its making. In Table 4, while the composition of each iron object is shown, the values of aluminium contents, which are considered as essential constituent of the entrapped slag particles, are not shown. The total slag contents, therefore, may be identified as 3.59, 1.57, 2.37, and 2.79 per cent, respectively. This is clearly indicative of technological improvements regarding the elimination of slag particles between the Iron Age and Early Historic periods. Not only did we find an increase in craft-related tool types, but the introduction of low carbon steel making is also visible during the post-Iron Age period. It is often mentioned that the Iron Age smiths of the Indian sub-continent applied 'layering and forging techniques'.

One of the recent advancements in making modern steel and alloy steels has been the use of micro-alloy. The presence of vanadium and chromium in three of the four analysed samples of Badmal is attested in the results of the PIXE analysis. Thus one may conclude that the smiths realised the characteristic capability of the ores was the potential of producing tough steels, which they used to manufacture drill bits (Chattopadhyay et al., 2011)

Future investigations will definitely shed more light on the various aspects of the state of iron technology in the pre-Christian era middle Mahanadi Valley of Orissa highlands.

Acknowledgements

The authors are thankful to the Archaeological Survey of India for permission to investigate the site of Badmal Asurgarh, and to the Ion Beam Laboratory, Institute of Physics, Bhubaneswar, Orissa, for analysing the iron samples by PIXE. The authors are grateful to the reviewers for their thorough critiques of the original manuscript. They are also thankful to Mr. Tosabanta Padhan for preparing some of the maps used in this paper.

References

Allchin, F.R. ed., 1995. *The Archaeology of Early Historic South Asia: The Emergence of City and States.* Cambridge: Cambridge University Press.

Basa, K.K. and Mohanty, P. eds., 2000. *Archaeology of Orissa.* New Delhi: Pratibha Prakashan.

Behera, K.S., 1994. Maritime Contacts of Orissa: Literary and Archaeological Evidence. *Utkal Historical Research Journal,* 5, pp.55-70.

Behera, P.K., 2002-2003. Excavations at Kumersingha and Kurumpadar – The Iron Age Settlements in the Middle Mahanadi Valley, Orissa: Results of the First Season Work. *Pragdhara,* 13, pp.87-103.

Behera, P.K. and Chattopadhyay, P.K., 2004-2005. Iron Objects from the Iron Age-Early Historic Level at Badmal (Dist. Sambalpur, Orissa): Archaeometallurgical Studies. *Puratattva,* 35, pp.118-25.

Behera, P.K., Ray, D.K., Dutta, K., Ravi Prasad, G.V., Routray, T.R. and Choudhury, R.K., 2007. Radiocarbon Dates from the Middle Mahanadi Valley, Orissa. *Man and Environment,* 32, pp.41-6.

Campbell, J.L., Johansson, S.A.E. and Malmqvist, K.G., 1995a. *Particle Induced X-Ray Emission Spectrometry (PIXE).* New York: John Wiley & Sons.

Campbell, J.L., Teesdale, W.J. and Halden, N.M., 1995b. Theory, Practice and Application of PIXE Microanalysis and Element-Distribution Maps. *The Canadian Mineralogist,* 33, pp.279-92.

Chattopadhyay, P.K., Behera, P.K. and Datta, P.K., 2007. From Wrought Iron to Steel: Beginning of Steel Making in Eastern India. *Indian Journal of History of Science,* 42, pp.387-402.

Chattopadhyay, P.K. and Sengupta, G. 2011. *History of Metals in Eastern India and Bangladesh,* New Delhi: Pentagon Press and Infinity Foundation, pp.73, 86, 89.

Das, H.C., 1990. Urban Centers in Ancient Orissa. In A. Ray and S. Mukherjee eds. *Historical Archaeology of India.* New Delhi: Books and Books, pp.175-93.

Demortier, G., 1997. IBA Application to Ancient Metallic Items. In M.A. Respaldiza and J. Gómez-Camacho eds. *Applications of Ion Beam Analysis Techniques to Arts and Archaeometry.* Seville: Secretariado de Publicaciones, Universidad de Sevilla, 91-114.

Demortier, G., 2000. Essentials of PIXE and RBS for Archaeological Purposes. In G. Demortier and A. Adriaens eds. *Ion Beam Study of Art and Archaeological Objects.* Luxembourg: Office for Official Publication of the European Communities, pp.125-36.

Ghosh, A. ed., 1989. *An Encyclopaedia of Indian archaeology.* New Delhi: Munshiram Manoharlal.

Govil, I.M., 2001. Proton Induced X-ray Emission – A Tool for Non-Destructive Trace Element Analysis. *Current Science,* 80, pp.1542-49.

Hajivaliei, M., Garg, M.L., Handa, D.K., Govil, K.L., Kakavand, T., Vijayan, V., Singh, K.P. and Govil, I.M., 1999. PIXE Analysis of Ancient Indian Coins, *Nuclear Instruments and Methods in Physics Research B,* 150, pp.645-50.

Johansson, S.A.E. and Campbell, J.L., 1988. *PIXE: A Novel Technique for Elemental Analysis.* Chichester: John Wiley & Sons.

Lal, B.B., 1949. Sisupalgarh 1948: An Early Historic Fort in Eastern India. *Ancient India,* 5, pp.62-105.

Mahalik, N.K. 1994. Geology of the Contact between the Eastern Ghats Belt and North Orissa Craton, India. *Journal of the Geological Society of India,* 44 (1), pp. 41-51.

Mando, P.A., 1994. Advantages and Limitations of External Beams in Applications to Arts and Archaeology, Geology and Environmental Problems. *Nuclear Instruments and Methods in Physics Research B,* 85, pp.815-23.

Mishra, B.P. and Mohanty, B.K., 1995. Gemstones. In B.K. Mohanty, N.K. Mahalik and R.N. Mishra eds. *Geology and Mineral Resources of Orissa.* Bhubaneswar: Society of Geoscientists and Allied Technologists, pp.185-202.

Mohanty, B.K., 1995. Iron Ore. In B.K. Mohanty, N.K. Mahalik and R.N. Mishra eds. *Geology and Mineral Resources of Orissa.* Bhubaneswar: Society of Geoscientists and Allied Technologists, pp.229-51.

Nayak, J.C. and Das, J.N., 1995. Vanadium Ore. In B.K. Mohanty, N.K. Mahalik and R.N. Mishra eds. *Geology and Mineral Resources of Orissa.* Bhubaneswar: Society of Geoscientists and Allied Technologists, pp.288-97.

Patra, B., 2007. Early Historic Urban Centers of Western Orissa: An Archaeological Study. *Man and Environment,* 32, pp.27-36.

Ray, H.P., 1995. Maritime Archaeology and Early Historic Port Sites: The Evidence from the East Coast. In F.R. Allchin and B. Allchin eds. *South Asian Archaeology, Vol. 1.* New Delhi: Oxford and IBH, pp.393-402.

Ray, H.P. and Salles, J.F. eds., 1996. *Tradition and Archaeology: Early Maritime Contacts in the Indian Ocean.* New Delhi: Manohar.

Sahoo, R.K., 1995. Chromite and Nickel Ore. In B.K. Mohanty, N.K. Mahalik and R.N. Mishra eds. *Geology and Mineral Resources of Orissa.* Bhubaneswar: Society of Geoscientists and Allied Technologists, pp.108-44.

Senapati, N. and Mahanti, B. eds., 1971. *Orissa District Gazetteer: Sambalpur.* Cuttack: Government of Orissa Press.

Sengupta, G. and Panja, S. eds., 2002. *Archaeology of Eastern India: New Perspectives.* Kolkata: Center for Archaeological Studies and Training, Eastern India.

Singh, H.H., 1971. Orissa Highland Region. In R.L. Singh ed. *India: A Regional Geography.* Varanasi: National Geographical Society of India, pp.754-75.

Smith, M.L., 2005. Archaeological Research at Sisupalgarh, An Early Historic City in Eastern India. In U. Franke-Vogt and H.-J. Weisshaar eds. *South Asian Archaeology 2003.* Aachen: Linden Soft Verlag, pp.297-306.

Tewari, R., Srivastava, R.K. and Singh, K.K., 2002. Excavation at Lahuradewa, District Sant Kabir Nagar, Uttar Pradesh. *Puratattva,* 32, pp.54-62.

Tripathi, V., 2001. *The Age of Iron in South Asia.* New Delhi: Aryan Books International.

Vijjayan, V., Choudhury, R.K., Mallick, B., Sahu, S., Choudhury, S.K., Lenka, H.P., Rautray, T.R. and Nayak, P.K., 2003. External Particle-Induced X-ray Emission. *Current Science,* 85, 772-7.

Yule, P., Behera, P.K., Brandtner, M., Modarressi-Tehrani, D. and Rath, B.K., 2005. Preliminary Report of the Third Field Season, 2002: Contexting Early Historic Western Orissa. In U. Franke-Vogt and H.-J. Weisshaar eds. *South Asian Archaeology 2003.* Aachen: Linden Soft Verlag, pp.307-18.

Yule, P., 2006. *Early Historic Sites in Orissa.* New Delhi: Pragun Publication.

Authors' addresses

- Pradeep K. Behera
  Post-Graduate Department of History
  Sambalpur University
  ORISSA-768 019
  INDIA
  mailto:pkbehra@rediffmail.com
- Pranab K. Chattopadhyay
  CASTEI
  KOLKATA-700 016
  INDIA
  pranab.chattopadhyay@gmail.com

Invention, innovation and inspiration

Invention, innovation and inspiration: optimisation and resolving technological change in the Sri Lankan archaeological record

Gillian Juleff

Dirty, heavy, intractable, dependent on the muscle and tenacity of people most often situated on the physical and cultural fringes of society, iron has nonetheless gravitated towards the centre of human development since its first known appearance over three millennia ago.

The aim of this session of World of Iron was to examine evidence in the archaeological record that allows us to explore the interface between technology and culture dynamics and the circumstances of technological change. What emerged was a disparate collection of perspectives, each insightful in their own right but which, when combined, are demonstrative of the infancy in our collective study of the drivers of innovation in this field, whether individual, societal, technological or evolutionary. While other sessions of World of Iron showcased the wealth and maturity of our understanding of the nature of iron as a material, its production through the smelting of ore minerals and its optimisation through alloying and heat treatment, far less apparent was our understanding of the nature of the relationships between the people and societies who contextualised its use.

Over some five decades or more of research the discipline of archaeometallurgy has progressed from a focus on the material itself in the form of artefacts. Detailed, laboratory-based metallographic studies of individual and assemblages of objects have brought to light the range of iron-carbon alloys and heat treatments utilised. The place of a metallurgical tradition within the development of iron was largely based on a linear scale of increasing technological repertoire and competency.

The need to understand how and where the raw materials of artefacts were produced soon followed and, with this, field studies of smelting and refining sites and processes gained importance. By their nature such studies are location specific, leading to claims evidenced by major concentrations of technological waste (slag) to 'centres of production'. This 'dots on maps' approach can have the effect of isolating the evidence from both its physical (topographical and environmental) and human (cultural) landscapes. Technological characterisation, through the analysis of site-level data and micro-structural and compositional evidence from collected samples has become the core task of the archaeometallurgist, where standardisation of projected production figures through the use of mass-balance calculations is the measure of technological success.

As data has accumulated in more recent years, however, there has been a growing appreciation that – given the ubiquitous use of iron within most past societies – a more continuous landscape of production, contoured by the physical topography of raw material distribution and the cultural topography of human settlement and economy, would be a more realistic model for understanding the development of iron in both spatial and temporal dimensions. With a broader canvas to view and data drawn from the field as well as the laboratory, detecting where and when changes happen and the pathways by which they are transmitted should become more visible. Importantly, it also gives the opportunity to recognise the impetus behind change and to consider whether that change is technologically or culturally determined, or indeed driven by evolutionary processes. Conversely, looking at broad socio-cultural landscapes may also help explain why technological change may be slow to occur or, when it does happen, what impact it has on local or regional cultural dynamics.

Thus, to address an expanded set of objectives that incorporates cultural archaeology as a vehicle for explaining invention, innovation and inspiration in metallurgy, it seems that, as a discipline, there is a need to share trajectories with archaeology, in all its forms, with material culture and with the history of science and technology. In summarising the papers presented in this session of the conference I frivolously commented how refreshing it was to hear so much about iron and see so few phase diagrams. Instead of the expected quick in-breath of opprobrium, there were mutterings of approval, suggesting the possibility that we have all now arrived at this realisation.

Invention, innovation and inspiration, in the context of archaeology in general and the development of iron technology in particular, are terms that have invited important debate in recent years (Fitzhugh, 2001, Shennan, 2002, Killick, 2004, Rehren et al, 2007, Charlton et al, 2012) and their definition offers considerable potential for understanding the nature of technology. However, rather than attempt to distinguish between them at a theoretical level or in the archaeological record, a more constructive approach with practical application may be to simplify the task by examining evidence of technological change to determine what drives or motivates change and from that define what processes are in action – innovation, invention, adoption, adaptation, optimisation – if this indeed is a useful conclusion to draw. More useful may be what can be inferred from the rate or pace of change. Archaeology abounds with examples of technological change. What constrains our chances of determining with any precision the drivers of change will be the resolution of available data. Here I will consider instances from the Samanalawewa Archaeological Project in Sri Lanka, now perhaps better known as Monsoon Steel (Juleff 1996, 1998).

Tracing technological optimisation

Setting

Samanalawewa lies in the southern foothills of the Central Highlands of Sri Lanka (Fig. 1) in an area traditionally regarded as of low archaeological interest. During the first millennium AD the broad lowland plains of the north and southeast of Lanka were densely settled with affluent agricultural communities, based on irrigated rice cultivation and two strong, stable, regional kingdoms, each closely aligned with the established Buddhist priesthood (De Silva 1981). The Samanalawewa area occupied a physiographic and culturally marginal position, remote from the heartland of the polity. It saw only small scattered valley-bottom settlements in otherwise wild, jungle-clad hilly terrain. The ethnic make-up of the local population is uncertain but traditions and early chronicles suggest that they were in part *pulindas*, a semi-vedda (aboriginal) tribal group (Perera 1959).

Climate

Sri Lanka's climate is dominated by the tropical monsoonal system and the seasonal weather pattern of Samanalawewa

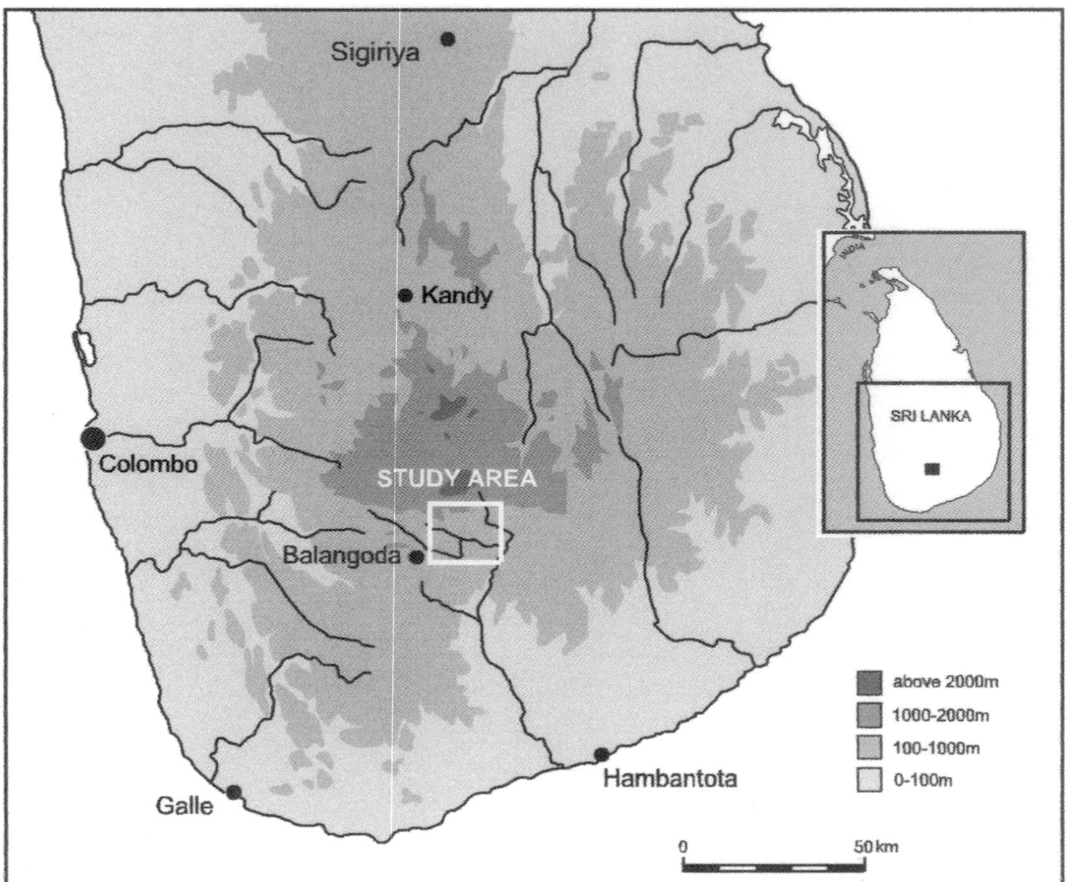

Figure 1: Map of Sri Lanka marking the Samanalawewa project area.

is transitional between the south-western Wet Zone and the northern and eastern Dry Zone, with constant, desiccating winds emerging from the Wet Zone to the west, where the burden of rain is dropped during the monsoon months of June to September. Orographic uplift, caused by the highlands, increases wind velocities in the area to an average of ~35km/h, rising frequently to 50+km/h and occasionally above 100km/h (Somasekaram et al, 1988, Juleff, 1998).

Survey

In the 1980's Samanalawewa became the focus of a major hydro-electric scheme that necessitated environmental and cultural impact studies, including an archaeological survey. Despite the lack of settlement during the first millennium Early and Middle Historic periods of the Island (for periodisation see Deraniyagala 1992), the area was closely associated with iron smelting and blacksmithing traditions and, significantly, crucible steel-making during the Late Historic Kandyan and colonial periods. Eye-witness accounts of smelting and steel-making by Ananda Coomaraswamy (1956) in the first decade of the twentieth century located these processes within specific named villages (Ondaatje, 1854, Wayman and Juleff, 1999, Juleff et al 2009). Hence, the initial priority for the Samanalawewa archaeological survey was to identify field evidence relating to these processes. From the outset of fieldwork, however, a previously unknown group of smelting sites dominated the archaeological record. These sites quickly became known as *west-facing* sites because of the consistent location of slag waste deposits on the west-facing slopes at the summits of exposed hilltops with uninterrupted aspects in that same direction. Correlation between smelting and strong directional, seasonal winds was clearly apparent from the first recognition of these sites. Slag morphologies across the seventy-seven sites of this type, recorded during the first survey (a number that continues to rise with every return visit to the area), demonstrated a similar high level of consistency, with forms that suggested a large furnace of unusual design and shape.

Excavation

On the strength of the data from field survey, in 1991 the Archaeological Survey of Sri Lanka initiated the excavation of one west-facing smelting site on a ridge known as Galewalahinna (site code SM88) (Fig. 2). The strategy was to open a series of 10x10m excavation areas to reveal the lateral extent of the site and to excavate two long, metre-wide trenches through the full depth of the deposits, to quantify output and examine the development of the site over time. In addition, attention focussed on the detailed investigation of individual furnace structures. Despite six months of digging, quantitative sampling of 17 tonnes of technological debris and exposure of 41 furnace structures it was estimated that only 20% of the site had been excavated and that the potential number of furnaces on this site alone was in the region of two hundred (Malim et al, 1995, Juleff, 1998).

Furnace design

The furnace design comprises a semi-permanent elongated C-shaped clay structure, reaching no more than ~0.5m in height, terraced into the leading westerly edge of the ridge. Evidence shows that the wall was terminated at each end with a large stone set in the ground and that a straight clay wall extended across the open front of the furnace, between the stones – although the evidence suggests that this was a single-smelt construction, deliberately broken down as the final act of each smelt. Embedded in the front wall were a line of pre-fired clay tuyérès, for the purpose of channelling air into the furnace (Figs. 4 and 5).

Furnace function

The manner in which the furnaces functioned was deduced from careful study of the archaeological evidence and, most valuably, the experimental reconstruction of the process, combined with the application of CFD (computational fluid dynamics), and has been described in a number of papers (Juleff, 1996, 1998, Tabor et al, 2005). Low-level incident wind, travelling up the contour of the hill-slope in front of the furnace, reaches its maximum velocity at the leading edge of the summit where it encounters the front wall of the furnace. It is forced upwards off the outer lip of the top of the wall, causing boundary layer separation and creating an area of extreme low pressure along the top of the wall. The pressure difference between the top of the wall and the external mouth of the tuyérè then generates a powerful air flow through the tuyérès embedded in the front wall (Tabor et al, 2005).

Phasing and chronology

Analysis and interpretation of the excavation record resolved 32 stratigraphic groups, each comprising multiple contexts, which in turn formed a sequence of five major phases in the development and use of the site (Juleff 1998). Of these, Phase 1 refers to circumstantial evidence for prehistoric occupation unrelated to the smelting site. Phase 2 sees the initial use of

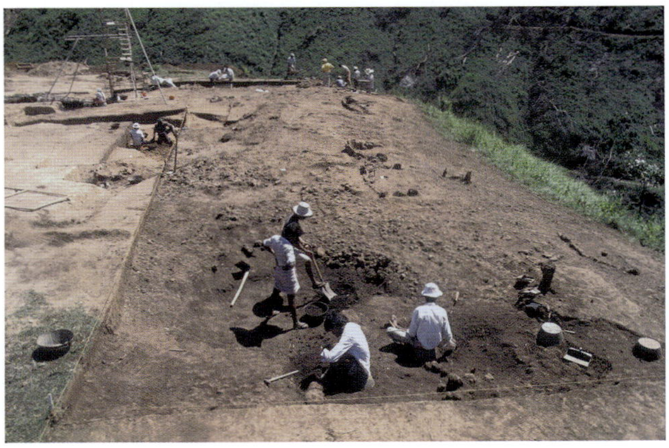

Figure 2: View of excavation area from north after removal of topsoil. The upper levels of structures at furnace locations L and J, in foreground, and G, in background, are visible.

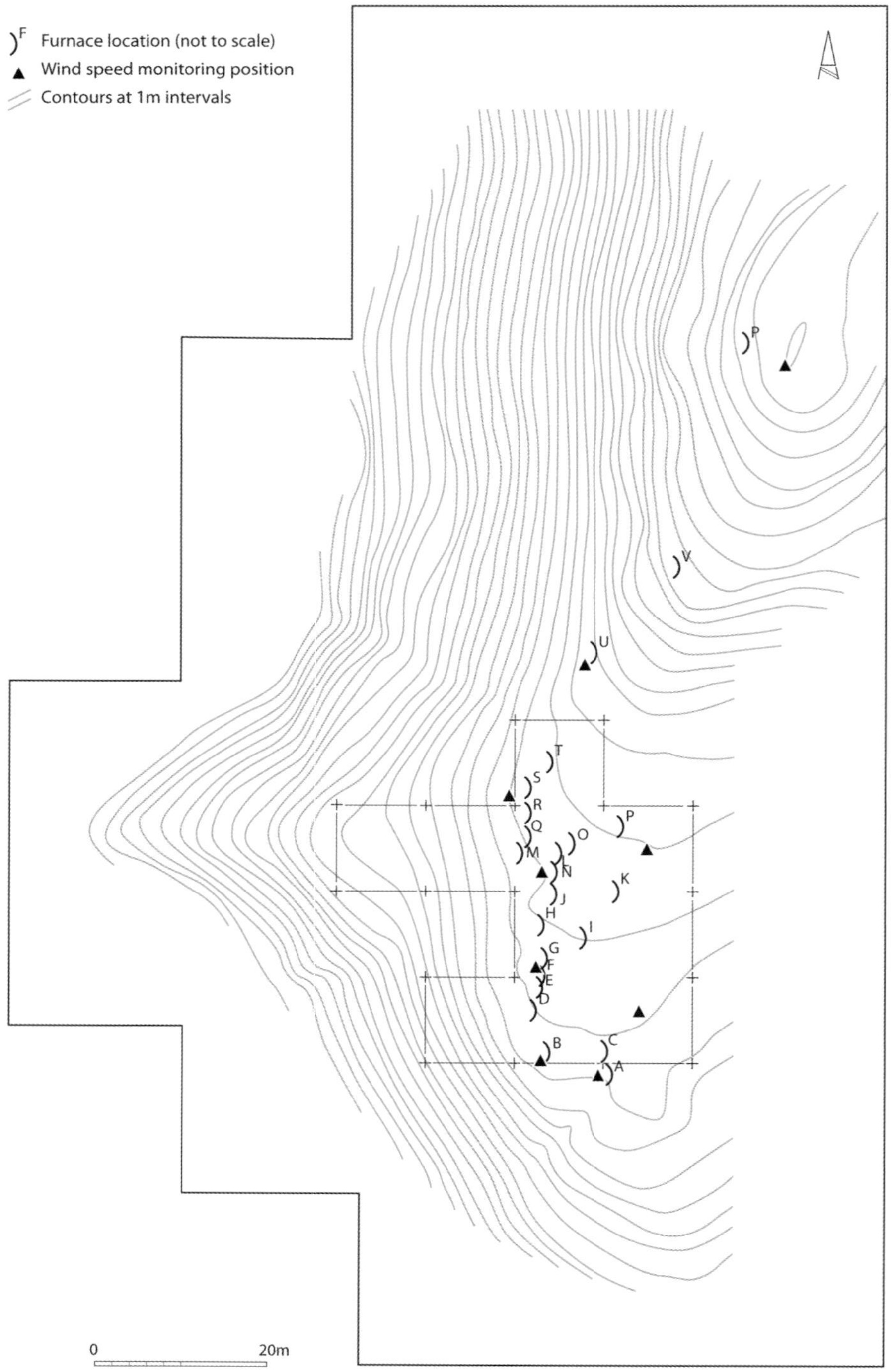

Figure 3: Site plan showing excavation areas and trenches, furnace locations and wind monitoring positions.

the site for smelting and a series of radiocarbon dates indicates activity from a start in the mid-7th century AD through to the early 9th century. Phase 3 represents a break in the smelting sequence and site-wide colluvial deposition of soil, suggesting abandonment. The duration of this abandonment is uncertain but a single radiocarbon date places it in the 9th century. Phase 4 sees a return to full scale smelting activity from the mid-9th to the early 11th century. Two radiocarbon dates from the first evaluation excavation at the site record the termination of smelting activity at the end of this phase. From then until the present the site remained abandoned and Phase 5 is assigned to the build up of topsoil over time. Reinforcing the series of radiocarbon dates, 350 rim sherds were recovered from the site, conforming to known Middle Historic types and dating broadly to the period 300 to 1100AD. Pottery finds from a number of west-facing sites, recorded during field survey, comprised the same dominant Middle Historic storage jar forms, suggesting that the dating of SM88 holds true across Samanalawewa. Using a combination of radiocarbon, pottery finds and consistency of morphologies, there is no evidence at present to extend the dating of what is clearly an industry earlier or later than the mid-7th to early 11th century AD (Juleff 1998).

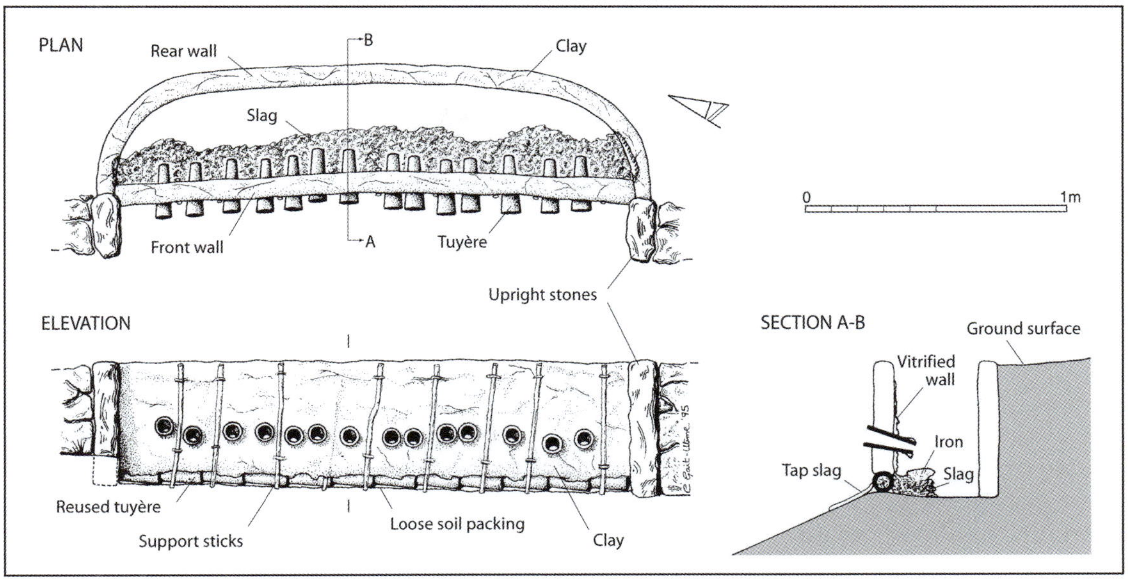

Figure 4: Reconstruction drawing of west-facing, wind-powered furnace with front wall in position.

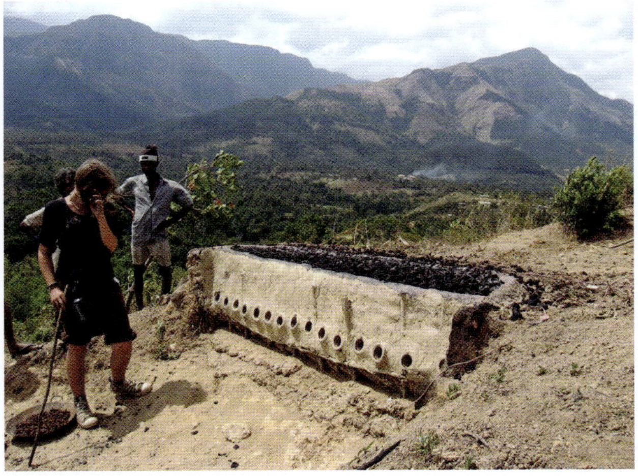

Figure 5: Experimental smelting in reconstructed furnace, Samanalawewa 2007.

Synchronic and diachronic variations

Against this general background we are now able to examine in more detail the technological changes that took place during the working life of the site and to begin to tease out the character of technological change.

Evidence from the excavation identified three distinct smelting campaigns during phase 2, alternating with brief episodes of localised disuse. The number of furnaces in use during any one campaign varies from three to seven, and in total fourteen furnaces at ten locations operated during the phase (within the area excavated). It is important here to distinguish between furnaces and furnace locations. Furnace indicates a single structure in which a series of smelts took place while a furnace location records a position on the site where a succession of furnaces were constructed, one on top of another. For example, furnace location B on the southern edge of the excavated area (Fig. 3) saw a succession of at least seven furnaces (labelled B1-B7).

The resumption of smelting in phase 4 saw at least six smelting campaigns, with 28 furnaces at fifteen locations. Again the smelting campaigns appear to alternate with localised episodes of disuse. The data allows us to chart the movement of activity across the site in terms of combinations of furnace locations going in and out of use and re-use but no distinct pattern of usage was discernible. For example, furnace location B is used during four of the campaigns but not in succession and in random combination with other locations.

Correlation of wind and furnace

The major changes from Phase 2 to Phase 4 are shown in Figs. 6 and 7, with the most distinct change being in furnace position. During Phase 2 furnaces are constructed both on the leading edge of the ridge and on the marginally flatter summit of the ridge, less than 10m from the leading edge. Their positioning also places some furnaces directly behind others, as in those at locations J and K and B and C. Our understanding of fluid dynamics and CFD analysis enables us to surmise that boundary layer separation off the front wall of the furnace, on the leading edge, would create a lee-eddy effect for some meters behind the furnace (Juleff, 1998, Tabor et al, 2005). With a gap of no more than 5m between J and K, B and C, the airflow over furnaces at K and C would be significantly reduced and turbulent. We know from stratigraphic evidence that furnaces at B and C did operate during the same smelting campaign and we can only imagine that smelting in C would have been sub-optimal at the least.

Whether shielded by another furnace or not, we also know that the winds incident on the flatter area along the summit of the ridge are substantially lower than those at the leading edge. During the monsoon season of 1990, a week-long wind monitoring exercise was conducted on the site and ~450 half-hourly velocity and direction readings were collected from positions along the ridge (see Fig. 3, positions marked with

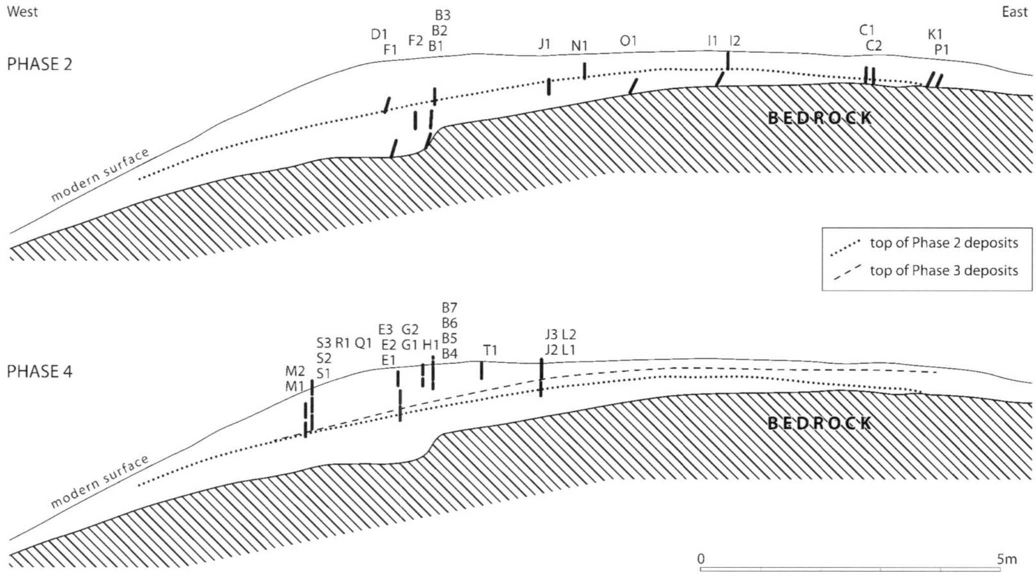

Figure 6: Schematic view of site phasing in section with all furnace wall positions.

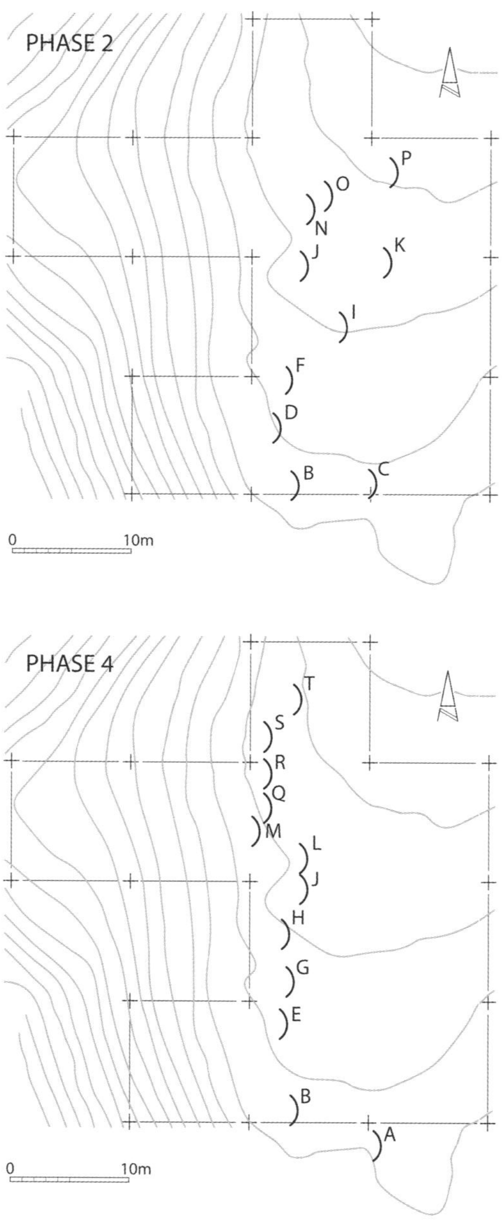

Figure 7: Schematic view of site phasing in plan showing main excavation area and furnace locations.

triangles). At the time the wind data was collected excavations had not taken place and furnace positions were unknown. Thus any correlation between wind records and furnace position only became apparent later, when data sets were compared. The records clearly show that winds recorded at the two positions to the east of the smelting debris on the flat of the ridge have average velocities of 10 and 12km/h, while positions along the leading edge average between 26 and 35km/h (Juleff, 1998). In consequence the furnaces of Phase 2 would have varied in their performance, with those at the leading edge probably being noticeably hotter and more stable.

The response to these observations in Phase 4 is dramatic, with furnaces lining up along the leading edge with no overlap or shielding of positions (Figs. 6 and 7). This principle is adhered to regardless of which of the locations are used during any one campaign. Only location A does not strictly line up with other Phase 4 furnaces within the main excavation area, although its position at the southernmost tip of the ridge (Fig. 3) suggests it would have experienced uninterrupted winds. Three other furnace locations are shown in Fig. 3 outside of the main excavation area, at U, V and W. Interestingly, the 1990 wind data shows that location U recorded the highest average wind speeds at 35.3km/h while at location W, the highest part of the ridge, wind speeds averaged only 26.3km/h. Overall the wind data from SM88 and other sites recorded in the course of the project, demonstrates a high degree of localised velocity variation, undoubtedly determined by local topographies. What is striking is that these phenomena were empirically observed, learned and exploited in the technological development of the site. The micro-level correlation between wind conditions and furnace location is seen at location B, which coincides with one of the 1990 wind monitoring positions. Winds at this position were the second highest of the dataset, with an average of 34.4km/h. Excavation data shows that the first furnace constructed at B was at the outset of smelting on the site in the first campaign of Phase 2. A succession of at least six more furnaces were built at this position spanning both Phase 2 and 4, each furnace being constructed on top of the previous as the surrounding slag and debris accumulated.

The fine-tuning of furnace with the wind can be discerned in a further change from Phase 2 to 4, with a gradual and almost imperceptible shift in orientation from west-facing to west-northwest (Fig. 8). Again, the close correlation between wind and furnace was not recognised until the data sets from two strands of research, wind monitoring and excavation, were finally compared later on in the project analysis. It was mentioned above that the 1990 wind-monitoring exercise included recording wind direction. Of 215 readings, the major component (64.5%) records the wind direction as WNW, with W making up most of the remainder at 33%, N comprising only 1.3% and WSW just 0.9% (Juleff 1998). It was also noticeable that the most sustained periods of velocities above the averages coincide with the wind veering to WNW. The subtlety of this observation, recorded through relatively sophisticated instrumentation, appeared unimportant until the footprint plans of the most complete furnaces were seen together (Fig. 8) and the shift in orientation of the Phase 4 furnaces towards WNW became apparent. The act of re-aligning the furnace structure towards WNW would ensure that the straight front wall of the furnace was presented broadside to the strongest and most constant winds, thereby optimising furnace performance.

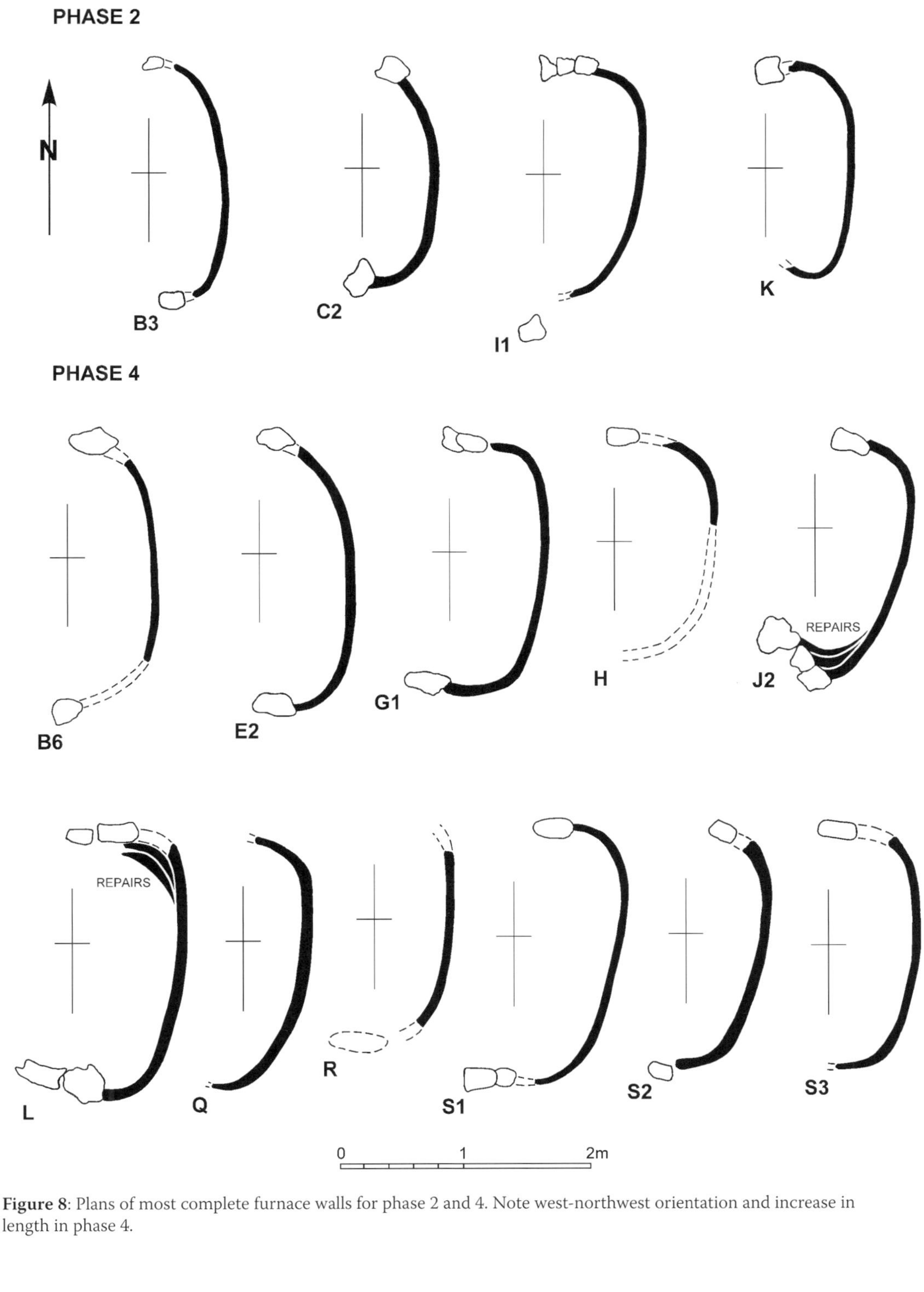

Figure 8: Plans of most complete furnace walls for phase 2 and 4. Note west-northwest orientation and increase in length in phase 4.

Furnace size and construction

Changes are also seen in furnace size and construction between the two phases. Fig. 8 also shows the increase in furnace length from Phase 2 to Phase 4 for the most complete examples. Furnace length is the internal N-S measurement and seven examples were captured for phase 2, ranging between 1.35m and 1.76m, with the mean falling at 1.60m. For phase 4 the range starts at 1.60m and extends to the longest examples at 2.10m, with the mean falling at 1.85m. While furnace length increases over time, depth-in-plan, the distance from the centre of the back wall to the line of the straight front wall, although problematic to record accurately, shows little variation between the two phases. It remains within a range from 0.3m to 0.5m with 0.4m being the mean, and with only one anomalous outlier at 0.8m (Malim et al, 1995, Juleff, 1998). Using both length and depth-in-plan to calculate furnace volume (on the basis that the two curved ends form 90° quadrants of radius 0.4m, leaving a rectangular central section), the overall increase in furnace length from 1.35m to 2.10m represents a 39% increase in furnace capacity.

In the early stages of Phase 2 furnaces were constructed with finesse and careful attention to detail. The clay used was of uniform texture, with few visible inclusions. Back walls were carefully modelled to angle backwards (to the east) and extended into a thin but equally carefully-modelled base to take on a scoop-like form. The angled walls of B1, D1, F1, O1, I1, K1 and P1 are shown in Fig. 6. The wall of B1 is set into a purpose-dug scoop, cut into the bedrock, as was the base of F1. By Phase 4 furnace building is stripped of any unnecessary effort and reduced to uniform functionality, with back walls being universally vertical and crudely constructed from coarse clay, loaded with massive inclusions of readily available garnetiferous granulite bedrock. The walls are poorly finished, displaying uneven plastering marks, and the idea of a clay furnace base is entirely dispensed with. Despite the apparent lack of attention to detail, the furnace walls of Phase 4 are clearly functionally robust and survive as well as those of Phase 2.

Characterising change

The technological developments between the two phases of smelting at SM88 all point to processes of optimisation, with changes made to maximise the interaction between smelting furnace and the most constant and powerful wind conditions. These changes would thereby improve furnace performance and increase output, coupled with changes to eliminate the unnecessary expenditure of energy. The changes observed are tightly proscribed by geographical, environmental, chronological and archaeological means and allow a high-resolution view of events. What is powerfully conveyed is the conscious nature of the changes enacted between the inception of smelting in the 7th century and the abandonment of the site in the early 11th century. This is not a process of random copying-error or diffusion of innovation as described in recent studies of technological evolution, but conforms more closely to conscious and directed invention as creative human behaviour (Fitzhugh, 2002, and see also Rehren et al, 2007, Charlton et al, 2010). The subtle re-orientation of the furnace from west to west-northwest could only have resulted from empirical and scientific observance of regional and local wind and weather patterns, combined with micro-level observation of the furnace in action.

When weighing the case for invention versus innovation, the rate or pace of change may provide useful insight. The changes seen at SM88 took place over a period of 300 to 400 years, perhaps the span of ten or twelve generations. Given the nature of radiocarbon date ranges and the lack of additional precisely datable material, this is the best chronological resolution we can expect. Within this span, the archaeology suggests a steady rate of seasonal site use punctuated by fallow episodes, perhaps when charcoal fuel trees were allowed to regenerate and smelting continued at other nearby west-facing sites. With the changes to furnace design, size, location and orientation comes an increase in activity, with more furnaces constructed and more debris generated (with corresponding increase in metal produced) (Juleff, 1998). Experimental reconstructions of the smelting process have demonstrated the possibility of achieving consistent furnace temperatures of 1450°C and above and of smelting directly to slag-free, high-carbon steels. The rate of optimisation conjures a scenario of cumulative knowledge consciously transmitted from one generation to the next, as well as a kinship between generations that requires custodial preservation of knowledge, as might be expected in craft/caste specialisation.

Interestingly, the dynamic picture of optimisation and increasing focus on maximising efficiency argues strongly for an underlying economic motivation. However, the archaeology provides us with insights that suggest otherwise. As mentioned, Samanalawewa lay at the fringe of mainstream society during the Middle Historic period and there are strong hints that its inhabitants were ethnically distinct from the core Sinhalese society of the lowland plains. There is no evidence in the archaeological record of any material or social gain, in the form of built structures, religious endowments, increased settlement size or material culture, from the development of the highly productive and sophisticated iron and steel industry. The likely interpretation, that the high-carbon steel produced in the wind-driven furnaces of Samanalwewa was one and the same as the *sarandibi* steel described by the Early Islamic writer, al-Kindi, as highly prized for the making of swords in the Islamic world, has been hypothesized elsewhere (Juleff, 1996, 1998). For the valuable products of the west-facing furnaces to reach end-users in the Near East, a trade network connecting the remote hills of Samanalawewa with coastal ports via the lowland plains must be envisaged, with state officials or entrepreneurs to maintain the link. The lack of evidence for local economic or material gain from this trade suggests that smelting and the production of steel was carried out as tribute to the state or as a contractual obligation. In which case, it becomes more difficult to identify the driving force behind technological change and the process of optimisation. Perhaps what we see here is ultimately innovation and invention inspired by curiosity, natural learning combined with ownership of, and identification with, the technology.

Discussion

The arguments for conscious processes of optimisation based on empirical learning embedded within a specialist craft society developed here, appear to run counter to the overview presented in a recent paper (Juleff, 2009) that examined the development of the west-facing furnace type from its earliest appearance in the 4th century BC in the Samanalawewa area, through other manifestations of long, low furnaces both within Sri Lanka and further afield, across Southeast and East Asia. In this broad sweep the linear configuration of the furnace was traced as a distinct, long-range evolutionary development independent of the means of air supply used to drive the furnaces, e.g. monsoonal wind or bellows, and processes of memetic inheritance were postulated for a divergence in technological evolution which distinguished circular plan furnaces from an Asian tradition of linear furnaces. The data discussed here, however, describes localised, meso- and micro-level processes of adaptation, innovation and optimisation. In conclusion, the potential for detecting and characterising processes of technological change and development exist at a variety of levels. Evidence from field survey and excavation can prove a valuable source of insight into the cultural dynamics of technology.

Acknowledgements

The original fieldwork and excavations of the Samanalawewa Archaeological Project were carried out in collaboration with the Archaeological Survey Department of Sri Lanka. The project was supported by the British Government (Overseas Development Agency) and the Japanese (Kumagi-Hazima-Kagima) and British (Sir Alexander Gibb and Partners, and Balfour Beatty) developers of the Samanalawewa Hydro-Electric scheme. Logistical and financial support were also given by the British Council and the British High Commission, Colombo. I wish to acknowledge the professional archaeological expertise of Tim Malim and Drew Shotliff who contributed to both the excavation and post-excavation stages of the project and Caroline Malim and Seán Goddard for preparation of site drawings and illustrations. I would also like to thank the organisers of World of Iron, especially Jane Humphris for her exceptional patience.

References

Charlton, M.F., Crew, P., Rehren, Th., Shennan, S.J. 2012. Explaining the evolution of ironmaking recipes – An example from northwest Wales. *Journal of Anthropological Archaeology*, 29, 352-367.

Coomaraswamy, A.K. 1956. *Medieval Sinhalese Art,* (2nd edn), New York, Pantheon Books.

Deraniyagala, S.U. 1992. *The prehistory of Sri Lanka: an ecological perspective,* Memoir 8, Archaeological Survey Department, Colombo.

De Silva, K.M. 1981. *A history of Sri Lanka,* London, Hurst and Co.

Fitzhugh, B. 2001. Risk and invention in human technological evolution. Journal of *Anthropological Archaeology,* 20, 125-167.

Juleff, G. 1996. An ancient wind-powered iron smelting technology in Sri Lanka, *Nature,* 379(6560): 60-63.

Juleff, G. 1998. *Early iron and steel in Sri Lanka: a study of the Samanalawewa area,* Mainz, Von Zabern.

Juleff, G. 2009. Technology and evolution: a root and branch view of Asian iron from first millennium BC Sri Lanka to Japanese steel. *World Archaeology* 41, 557-577.

Juleff, G., Craddock, P.T. and Malim, T.J.P. 2009. In the footsteps of Ananda Coomaraswamy: Veralugasmankada and the archaeology and oral history of traditional iron smelting in Sri Lanka. *Historical Metallurgy,* 43(2), 109-134.

Killick, D. 2004. Social constructionist approaches to the study of technology. *World Archaeology 36: debates in World Archaeology,* 571-8.

Malim, T.J.P., Shotliff, A. and Juleff, G. 1995. Excavating iron smelting furnaces at Samanalawewa, Sri Lanka: a methodological approach, in F.R. and B. Allchin (eds) *South Asian Archaeology* (New Delhi), 911-924.

Ondaatje, W.C. 1854. The Kandyan mode of manufacturing steel, *Ceylon Almanack, appendix,* 40.

Perera, L.S., 1959. The early kings of Ceylon up to Mutasiva. In Paranavitana, S. (ed.) *University of Ceylon history of Ceylon,* pt 1, 98-111, University of Ceylon Press, Colombo.

Rehren, Th., Charlton, M.F., Chirikure, S., Humphris, J., Ige, A. Veldhuizen, H.A. 2007. Decision set in slag – the human factor in African iron smelting. In La Niece, S., Hook, D., Craddock, P. (eds), *Metals and Mines: Studies in Archaeometallurgy,* Archetype Publications, London, 211-218.

Shennan, S.J. 2002. *Genes, memes and human history, Darwinian archaeology and cultural evolution.* London, Thames and Hudson.

Somasekaram, T. et al (eds), 1988. *National atlas for Sri Lanka,* Survey Department, Colombo.

Tabor, G.R., Molinari, D. and Juleff, G. 2005. Computational simulation of air flows through a Sri Lanka wind-driven furnace, *Journal of Archaeological Science,* 32: 753-766.

Wayman M.L. and Juleff G. 1999, Crucible steelmaking in Sri Lanka, *Journal, Historical Metallurgy,* 33(1), 26-42.

Author's address

- University of Exeter
  g.juleff@ex.ac.uk

Iron working in an Indian Ocean context

Randi Haaland

ABSTRACT: The Kushite Kingdom was a powerful state from the 8th century BC to the 4th century AD in what is now modern-day Sudan. Its early phase, when the seat of power was in the north at Napata, was clearly influenced by Egyptian culture. When the capital moved south to Meroe in the 4th century BC however, Egyptian influence declined, with indigenous African and distant Indian cultures appearing to become more prominent in Meroitic culture. This paper explores features of both the iron technologies and the religious ideology which are indicative of this Indian connection. It will also propose an exploration of the Meroitic Kingdom which considers the sources of power that were the basis for its longevity.

Introduction

Texts on early world systems tend to focus on Eurasian, Indian, Chinese and Southeast Asian trade connections, while African connections within such systems are given less consideration, although Edens and Kohl (1993) refer to Sudan and the Arabian Peninsula as being included in the West Asian Bronze Age world system. During Pharaonic times, Sudan's position in the world system was largely related to the flow of goods such as ivory, gold and slaves which were channelled to Egypt. Later trade between the Roman Empire and South Asia, however, altered this flow dramatically, as Sudan occupied a strategic position on this route: "Regular trade started in the reign of Augustus (27 BC-AD 15), when envoys from the various states of India visited the Roman emperor" (Warmington, 1974 p.36). The beginning of frequent direct exchanges between Rome and India stemmed from the discovery of the monsoon on the Arabian Sea by Hippalus in the first century AD (Liu, 1988, p.7). From the ports of Barbaricon and Barygaza (north-western India), Indian products (e.g. ivory, cloth, indigo, myrrh and pepper) were acquired by both Roman (Pliny xxxvii, 23; Peryplus: 39, 48, 49; Watt, 1966, p.891; Warmington, 1974, p.200) and Indian traders (Ray, 2003, p.245), who brought the goods to Red Sea ports such as Adulis and Berenike. These ports in the "hinterland of Ethiopia", in turn "provided African ivory and gold" which was sent back across the sea (Thapar, 1966, p.114).

The importance of sea trade increased as conflicts between the Roman and Persian states restricted overland trade – an increase which is significant when considering sea trade during the first centuries of the present era. Today, as in the past, trade and politics are essentially intertwined, with events in some parts of the world system causing repercussions – such as the restructuring of local production and international distribution and the growth and decline of states – in widely disparate regions. An important but frequently overlooked aspect of this is that world system connections are not only a matter of the flow of goods and political influence but also the diffusion of ideas and cultural artefacts, brought about by migration of occupational specialists whose skills are in demand in distant communities.

At some point in the cycle of crises and hegemonic changes in the Eurasian-African world system, a particular civilisation emerged in the Sudan Nile valley. The archaeological material left by what is now called the Meroitic civilisation is enigmatic in several respects:

1. Evidence clearly indicates that the civilisation held close economic, political and symbolic relationships to Pharaonic Egypt. Its written scripts, however, show that the Meroitic language was not Egyptian.
2. Large slag heaps indicate extensive iron production, but few iron products have been recovered from local sites.
3. Although evidence suggests that the agricultural resource base of the civilisation was very small, the remains of the ancient city of Meroe indicate a complex, stratified society.
4. Iconography expressing the ideological legitimisation of rulership demonstrates certain features that are similar to Egyptian styles, while other elements are so different that their iconographic inspiration is likely to have been stimulated by "non-Egyptian iconographic tradition".

I shall discuss some of these enigmas by exploring the interplay between local iron production, division of labour, political centralisation, symbolic legitimisation of power and diffusion of symbolic elements mediated via world system participation.

The Meroitic context

The Kushite Kingdom of Meroe represents a later development of the Kushite Kingdom of Napata, located in the fourth cataract area of the Nile. Egyptian written sources as well as inscribed statues show that during the 25th dynasty, in the 8th century BC, Egypt was ruled by Kushite kings from Napata (Fig. 1). Statues of these rulers, as well as iconography found on the funerary and temple walls of this dynasty, are modelled with clearly Negroid features but display evidence of being heavily influenced by Egyptian stylistic ideals. The

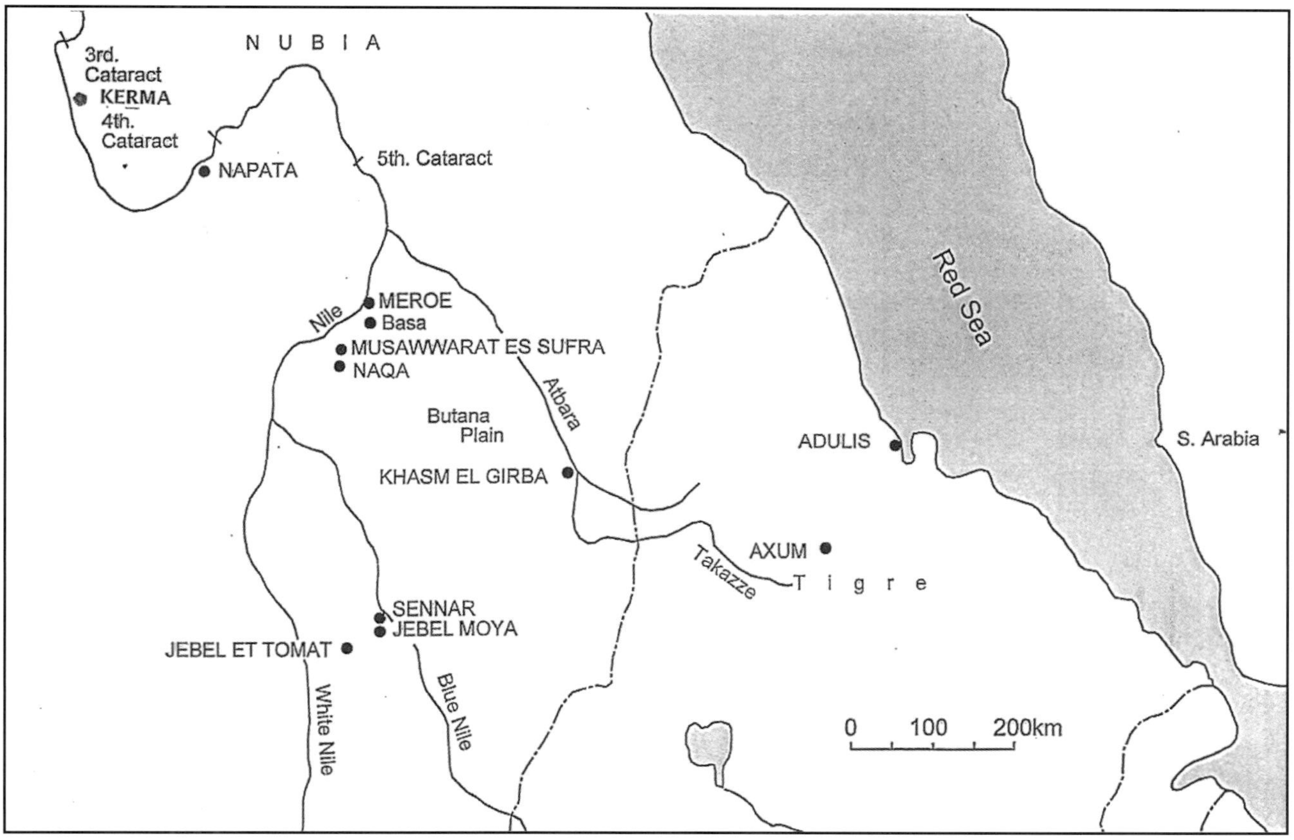

Figure 1: Map of major Meroitic sites.

Meroitic language, however, differed from Egyptian and did not belong to the same Afro-Asian language family. Rilly (2009) has recently demonstrated that Meroitic belonged to the Eastern Sudanic Branch of the Nilo-Saharan language family. It is reasonable to assume that the Meroitic elite were highly influenced by Egyptian culture, while African cultural traditions were dominant among the lower strata of society (as visible in remains of wattle and daub buildings and household pottery).

From written sources we know that the capital of the Kushite Kingdom was moved during the 4th century BC from Napata southward to Meroe, near the fifth cataract (Shinnie, 1967; Adams, 1981; Welsby, 1996; Edwards, 2004). At this time Egyptian influence remained important but new influences began to appear, for example from eastern cultures across the Indian Ocean, the Roman and the Hellenistic worlds and from more local cultural traditions.

The present paper proposes an exploration of the Meroitic Kingdom which considers the sources of power that were the basis for its longevity. The local presence of ferrocrete sandstone provided the basis for centralised control of iron production, which played a vital role in the kingdom's power as iron was primarily used for the production of weapons (see below). Without this centralised control the meagre agricultural potential afforded by the narrow Nile belt and the surrounding savannah could not have provided a basis for long-term state power. Control of iron production, however, cannot be seen as a sufficient condition for this power and must be viewed in conjunction with two further conditions that both played a crucial role: trade on a global scale; and a legitimising ideology as expressed in different symbols, particularly those dedicated to the war god *Apedemak*, whose association with rulers as well as with iron production rendered this a compelling force (Haaland and Haaland, 2007).

The distribution of trade items uncovered at Meroe show that the kingdom participated in a network of material exchanges that connected it with the Mediterranean world as well as with the Horn of Africa, lands across the Red Sea and even to India (Arkell, 1961, 1966; Shinnie, 1967; Welsby, 1996; Phillips, 1997; Edwards, 2004). The Meroitic civilisation was clearly influenced by the Egyptian and Mediterranean cultures to the north, although trade and cultural influence did not solely traverse a north-south axis. There is strong evidential indication of influence from across the Indian Ocean, in addition to local influences, which became more prominent after the capital was moved south to Meroe.

Iron working at Meroe

Archaeological material in the form of enormous slag heaps located just outside the ancient city (Fig. 2) has shown that iron smelting was a large-scale activity at Meroe, although very few iron objects from the Meroitic period have been found .

During the 1990s, Rehren (2001) made an estimate of the amount of iron production that had taken place at Meroe, based on the quantities of slag in the vicinity of the city. Rehren surmised that this indicated some 5000 tons of iron or 2500

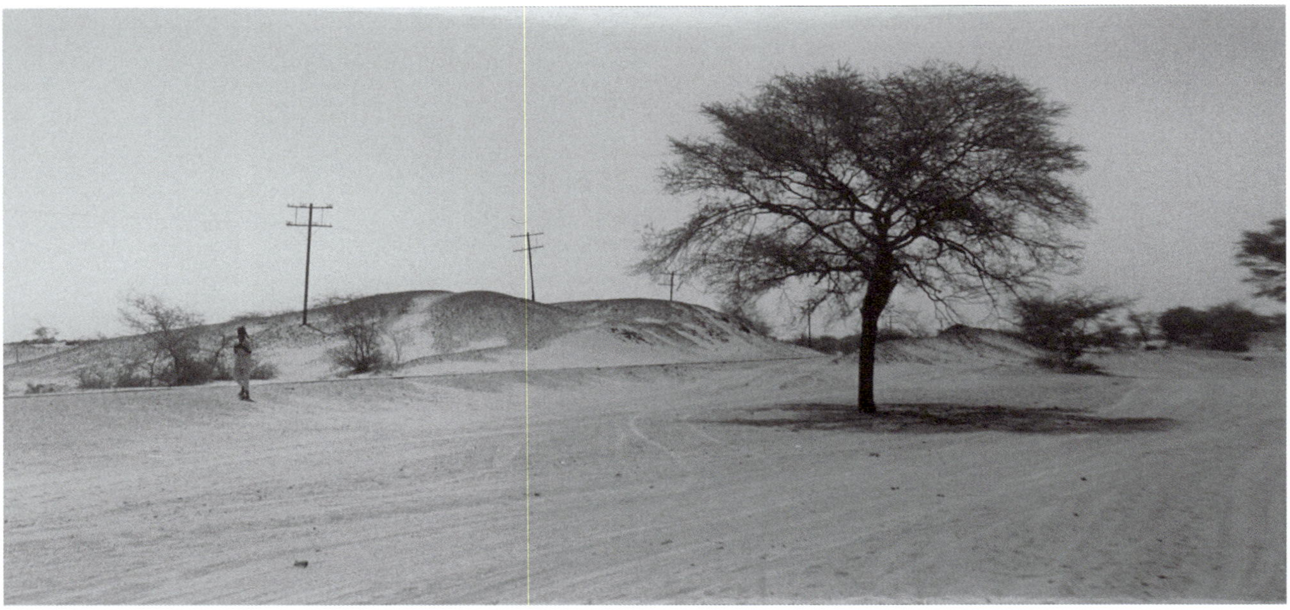

Figure 2: Slag mounds east of the ancient city of Meroe (Photo: Randi Haaland).

tons of finished metal iron objects, after smithing, which would in turn indicate an annual production of perhaps five to twenty tons of iron objects over a period of 500 years. Since the archaeological remains of iron products recovered in Meroe are very few compared to the large amount of iron that must have been produced, it appears plausible that a substantial amount of iron was exported, probably to Egypt (Haaland and Haaland, 2007).

An early survey of iron products was published by Wainwright (1945) and by Trigger (1969) and, recently, more general works have been written by Welsby (1996) and Edwards (2004). In a more detailed study of the iron objects in the National Museum at Khartoum, Beshir (2006) has shown that most of the objects recovered during excavations at Meroe were small items, such as light weapons and artefacts used for personal decoration; only rarely were tools for agriculture or other forms of production found. Objects from the late Meroitic period such as iron-tipped spears and arrows have been found in burials spread over a wide area. Arkell stated in 1966 that of the 1,500 graves at Napata which preceded Harsiotef (4th century BC), only eighteen contained iron objects: small items such as arrowheads and tweezers. At other Meroitic cemeteries from the period between the 2nd century BC and the 3rd century AD, single barbed arrowheads are the most prominent artefact, particularly at those sites dating from the 2nd century AD (Beshir, 2006).

At this point, few absolute dates have been confirmed that can tell us much about the chronology of Meroitic iron working. However, dates from furnaces indicate that the main period of iron smelting occurred rather late in the Meroitic period (Shinnie and Kense, 1982). At this time ore was found in the hills to the east of the ancient city, where it occurred as a black crust on the surface of the sandstone which covers a large part of the Northern Sudan (Trigger, 1969). The hardwood acacia trees needed to produce charcoal for iron smelting would have been available in the surrounding area as two thousand five hundred years ago, the vegetation around Meroe was probably low woodland savannah (Edwards, 2004). The current poor, semi-arid vegetation found in the area is most likely a consequence of human activities leading to over-exploitation of trees for charcoal production for iron smelting in the past, and overgrazing by domestic animals today (Haaland, 1985).

While providing little substantive evidence of the earliest furnaces, the discovery of bowl-like pits could indicate either the remains of slag pits from non-slag tapping furnaces or from forging activities (Tylecote, 1970; van der Merwe, 1980). The increased importance of iron from the 1st century BC onward is consistent with the use of a more efficient slag tapping technique, as practised in the Roman period. The Meroitic slag-tapping furnace (Fig. 3) was dome-shaped and made of fired bricks. It was surrounded by six pot-bellows,

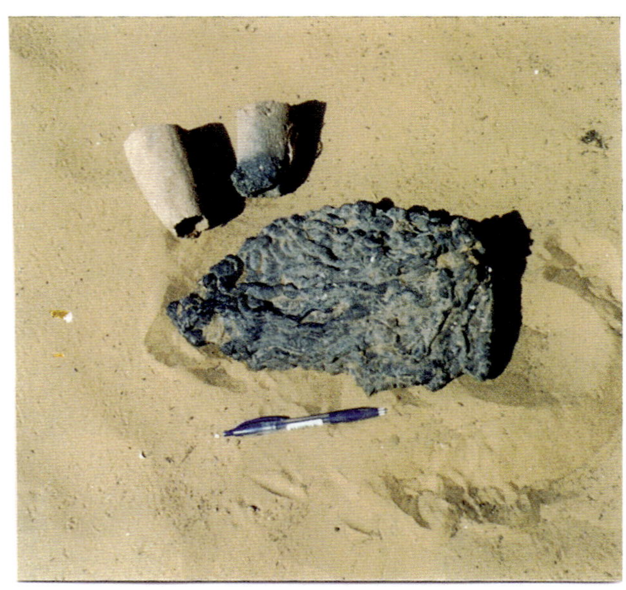

Figure 3: Flow slag from a slag-tapping furnace and remains of tuyères from the eastern part of ancient Meroe (Photo: Randi Haaland).

with two holes made for fastening two tuyères to the furnace. The pot-bellows and tuyères were made of the same type of clay used for making household pottery (Shinnie, 1985).

While direct evidence is scarce, it can be hypothesised that the slag tapping technique probably spread southward along the Nile valley to Meroe during the 1st century AD, at a time when there was an active Roman military presence in Nubia (Shinnie, 1967, 1985; Torok, 1975; Tylecote, 1982; Welsby, 1996; Edwards, 2004). Prior to this iron producers in the region may have used slag pits which may have been introduced as early as the 6th century BC either from Egypt (Arkell, 1961; van der Merwe, 1980; Kense, 1985; Miller and van der Merwe, 1994) or possibly from the Horn of Africa, via the Arabian Peninsula (van der Merwe, 1980; Kense, 1985; Miller and van der Merwe, 1994; Mapunda, 1997).

It does not seem that the technique of slag tapping continued as a mainstream method in iron working, nor did it experience a limited spread to the rest of Africa until almost 1000 years later (Chirikure, 2005). As stated by Miller and van der Merwe (1994, p.9): "The iron smelting technique at Meroe relates to the Egyptian and Roman technology and appears to have no bearing on the development of metallurgy in West or East Africa". All other recorded indications of iron smelting in the Sudan and adjacent areas of Ethiopia today are based on slag pits (van der Merwe, 1980; Todd, 1985; Haaland, 1985, 2004 a, b).

The remains of large, well-used, brick-built furnaces in the same area of the ancient city of Meroe indicate a well-organised, specialist industry, with society divided into areas of activity (Shinnie and Kense, 1982). Iron smelting appears to have taken place within these constricted areas, as did pottery making (Shinnie and Kense, 1982). Tylecote (1970) analysed a number of samples taken from iron production remains, concluding that the ore was roasted and broken up into smaller pieces before being smelted. The smelting furnaces and finds associated with roasting of the ore are enclosed by brick structures in the constricted iron smelting area of the city.

Abdu and Gordon (2004) have conducted a recent study of Meroitic iron production, analysing objects from the early to the classic period (300 BC-250 AD). Their studies have led them to suggest that these objects represent a distinctive iron making style which is not found in the eastern Mediterranean or in sub-Saharan Africa but is rather indicative of a transmission of knowledge to Meroe from the east, from Arabia or perhaps India, by way of the Horn of Africa. They emphasise the deliberate use of iron-arsenic alloys and piled metal (probably to achieve distinctly decorated surfaces: the Meroitic smiths displayed a high level of skill in the preparation of the metal and in the shaping of small, complex objects), arguing that this material suggests that knowledge regarding the manufacture of iron objects might have been introduced via trade across the Indian Ocean.

Other cultural features have demonstrated Indian influence on Meroitic culture, leading Vercoutter to suggest that Meroitic art was, *tout aussi indianisant qu' égyptisant* – equally influenced by Indian and Egyptian culture (cited in Shinnie, 1967, p.114).

Division of labour and centralised redistribution

Excavations have shown that iron smelting activities within the walled city of Meroe were located close to the palace, indicating the significance of state control over the iron industry. The unpredictability of attacks and raids from enemy groups such as the nomads of the savannah were a serious political threat, perhaps contributing to the location of the activities within the royal quarters.

In my earlier work (Haaland, 1980, 1985; Haaland and Haaland, 2007) I argued that iron production at Meroe was based on a caste-like division of labour, organised by the Meroitic state. There is little evidence to indicate that control over iron production was of any strategic importance with respect to improving means of production (agricultural tools are virtually absent in the archaeological record). Rather, the focus appears to have primarily been the improvement of the means of destruction.

War and conflict were important features of the Meroitic socio-political setting and the success of the Kushite army is demonstrated by the survival of their state for over a thousand years. It seems practical to suggest that this was made possible not only as a result of the large amount of iron which was produced and made into weaponry, but that it was directly related to the centralised control of iron production – an important factor in the formation of the Meroitic state. State formation, however, was also affected by Meroe's position in the world system, from the first centuries BC to the first centuries AD. In this respect, it was the impact of those world system processes which led to the decline of centralised power in Egypt and the growth of sea travel in the Roman period, which in turn affected the strategic position of Meroe, economically as well as politically.

The role of trade in the political economy of Meroe

Trigger (1969) has argued that trade between Egypt and Sudan continued to flourish during Ptolemaic rule and probably extended into the early period of Roman rule. However, when Egypt's control of political developments in Sudan was weakened, this increased the Meroitic state's ability to increase revenue via control over Egyptian trade. On the Meroitic side there is abundant evidence, most notably found within burial excavations, pointing to long-distance exchange of goods, including metal vessels (especially vessels for serving and display), jewellery, worked stone, glassware faience and a wide range of ceramic wine and oil containers (Torok, 1989; Edwards, 2004, pp.167-168). The distribution of these imports suggests that most foreign artefacts entering the Meroitic culture were distributed through an elite network, which was likely to be a royal monopoly. During the previous Napatan period this seems to have taken the form of 'Embassy Trade', a form of elite gift exchange widely encountered in the Mediterranean and Near Eastern World (Edwards, 1999) which appears to have continued during the Meroitic period and which linked the Meroitic kings with successive Ptolemaic and Roman rulers. In such redistributive systems, kings acted as pivots, mediating the flow of goods and services internally between different occupational specialists, including cultivators, as well as externally between different kingdoms.

The control over iron production which constituted a fundamental pillar in the military power of the Meroitic state (Goody, 1971) also contributed to the strategic importance of iron in the field of trade, especially with Egypt. Iron was a scarce commodity in Egypt, due both to the rarity of charcoal for smelting and the limited access to ore (Arkell, 1966; Shinnie and Kense, 1982). The Meroitic state's power was also greatly affected by its control over Red Sea ports during the period of increased sea faring trade between India and the Roman Empire.

Most of our information on this early trading network comes from a single source, the *Periplus Maris Erythraei* (the southern Red Sea, together with the Indian Ocean, was then known as the Erythraean Sea). A Greek merchant trader wrote this account in Egypt during the 1st century AD, at a time when Roman power in the region was at its height. It provides the earliest evidence of valuable information for the benefit of traders on the route between Egypt, India and East Africa (Casson, 1989; Ray, 2003, p.170). Drawing on information from *Periplus*, Thapar (1966, p.144) argues that the African hinterland provided ivory and gold as well as a market for Indian cloth (especially muslin), with the Meroitic elite most likely being the consumers.

Goods which came from India included carnelian beads as well as others which are difficult to detect archaeologically such as spices – especially pepper – cotton and indigo dye. Pepper and cotton are both trade items which are mentioned in the *Periplus* (Casson, 1989, pp.194-195). A large cache of pepper from the Roman period which indicates the spice's importance, has been discovered at Berenike, as well as numerous pieces of cotton which are known to be Indian-made due to their Z-spun characteristics, the usual twist for Indian cotton thread (India was also the main producer of textiles in the Indian Ocean region) (Barnes, 1997, p.49; Ray, 2003, p.218).

Textiles and especially indigo dye were important markers of social identity and it is likely that these products, imported through the port of Adulis, were important to the Meroitic elite.

The Ptolemies and later the Romans carried out trading expeditions from Egyptian ports in the North, stopping at the Red Sea port of Adulis where they collected objects of trade from the African hinterland. From the 2nd to the 1st century BC, the number of ships sailing to India increased from twenty to more than a hundred annually, and Adulis became a trade centre between India and Egypt (Phillips, 1997). The Meroitic state was heavily involved in providing the goods needed for this trade, which probably came from the African savannah in the west as well as from southern Sudan and which included ivory, ebony, rhinoceros horn, leopard and other skins, ostrich feathers, slaves, gold – and probably iron. Although knowledge of the two major trading corridors of the Red Sea and the Nile is quite extensive, the overland trade network is not well documented. It is difficult to assess how far inland trade goods went based on presently available archaeological data, although in 1950 Arkell proposed that the connections went westward across the savannah to Darfur and as far as southern Sudan (Arkell, 1961); an area which was rich in gold, as documented by the Greek writer Kosmas Indikopleutes during the 6th century AD. It is thought that the Meroitic kingdom was undermined by conflicts with nomadic groups on the overland trade routes between the Red Sea and the Nile, a factor that may have weakened it in competition with the growing power of the Axum Kingdom in Ethiopia. Consequently, it is probable that the fall of Meroe was mainly caused by a decline in trade, primarily resulting from increasing Axumite control over southern trade goods such as ivory and slaves (Welsby, 1996; Edwards, 2004).

Symbolic legitimation of power

The stability of centralised states does not only depend on physical and administrative control over both the means of destruction (weapons) and of production but also on the means of legitimising rulership, as expressed in ritual events and material symbolic constructs. There are two relevant factors to discuss with reference to this: firstly, the role of temples in maintaining centralised control of production and thereby the system of redistribution; and, secondly, the ritual role of temples in production.

Situated on a slag mound in the south-eastern part of Meroe (dated to 246-266 AD) was a temple dedicated to the war god *Apedemak* (Haaland and Haaland, 2007). This can be taken as an indication of the importance of the temple in the ritual and political aspects of iron working (Fig. 4).

The temples of *Apedemak* are all located in the Butana region, close to Meroe at Naqa (a trading city with remains of temples and palaces) and Musawwarat es Sufra (a ritual centre just north of Naqa). The *Apedemak* temples at Musawwarat es Sufra were built during the period between the 3rd century BC to 2nd century AD when Meroe was at the height of its political power, while the temple at Naqa was constructed slightly later. Torok (1986) has argued that when the seat of power shifted from Napata to Meroe in the south, *Apedemak* became one of the principal gods, probably because of the higher incidence of conflict in the area, and became closely associated with the ideology of kingship. One of the main scenes on the outer wall of the temple at Musawwarat es Sufra, for example, shows the king and the prince facing *Apedemak* in a posture of adoration. The temple was also important in the investiture of any king.

Apedemak

With regards to the centralised political control of iron working, temples dedicated to the god *Apedemak* can be interpreted as having both ideological and ritual importance in terms of legitimating this power. Ethnographic evidence has found that items associated with iron working were employed to a large extent as symbolic constructions which expressed social concerns such as those in rites of passage or ritual. Similarly, the extensive slag remains outside the city of Meroe are evidence of large-scale iron working taking place in connection with the palaces and temples. The special activity areas related to iron production (and pottery production) indicate that such activities were restricted to people with a caste-like identity.

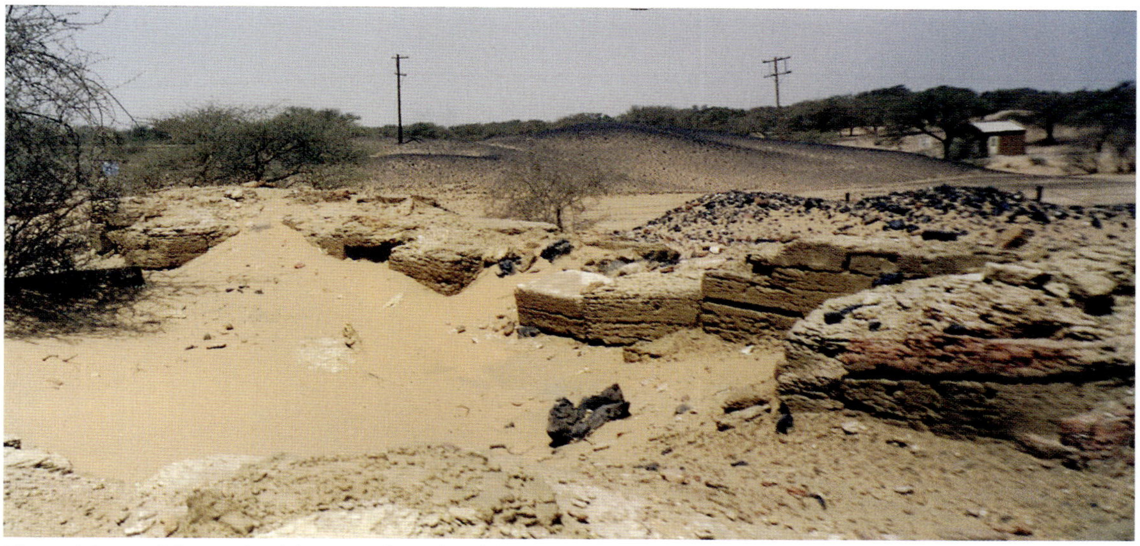

Figure 4: Ruins of the temple of the war god Apedemak built on top of a slag mound (Photo: Randi Haaland).

The emergence of a new symbol of power (added to the Egyptian-inspired symbols), namely the god *Apedemak* (Fig. 5), manifested the importance of political control over iron working and it seems plausible that the image of *Apedemak* served to legitimise royal rule, with reference to the rulers (possibly divine kings) as both the source of the country's fertility and protectors of its people (Zabkar, 1975). However, in the material imagery, it is symbols which express the king's military power as derived from *Apedemak* that are emphasised.

This material connection between the war god *Apedemak*, the royal centre of Meroe and the residue of iron production (indicated by the location of the *Apedemak* temple on a slag mound) suggests a close symbolic association between the legitimising god, the political ruler and iron as the most important source of the ruler's power.

The *Apedemak* temples and iconography are unique to the late Meroitic state and probably served not only to represent the identity of the ruler but also metonymically; to express the identity of the subjects of the Meroitic state versus subjects of other states. By examining the political and economic contexts in which the Meroitic state was positioned we may throw some light on enigmas found in the inventory of components associated with *Apedemak*.

I consider the following political aspects as salient: that the population of the Meroitic and the Egyptian states spoke unrelated languages and most likely differed with respect to other cultural traditions, as well that Meroe and Egypt were connected in relations of competition as well as symbiosis (trade); that legitimisation of power (Meroitic) favoured the development of symbolic expressions to communicate similarity as well as contrast to rivalling powers (Egypt), as well to serve as differentiation between rulers and commoners; that the figure of *Apedemak* was demonstrative of elements associated with iron production and symbolically placed political rulership in a mediating role between earthly practical activities and cosmological powers; and that such expressions require craftsmen skilled not only in monumental work but also in the creation of contrasting iconographic images. I suggest that an exploration of the

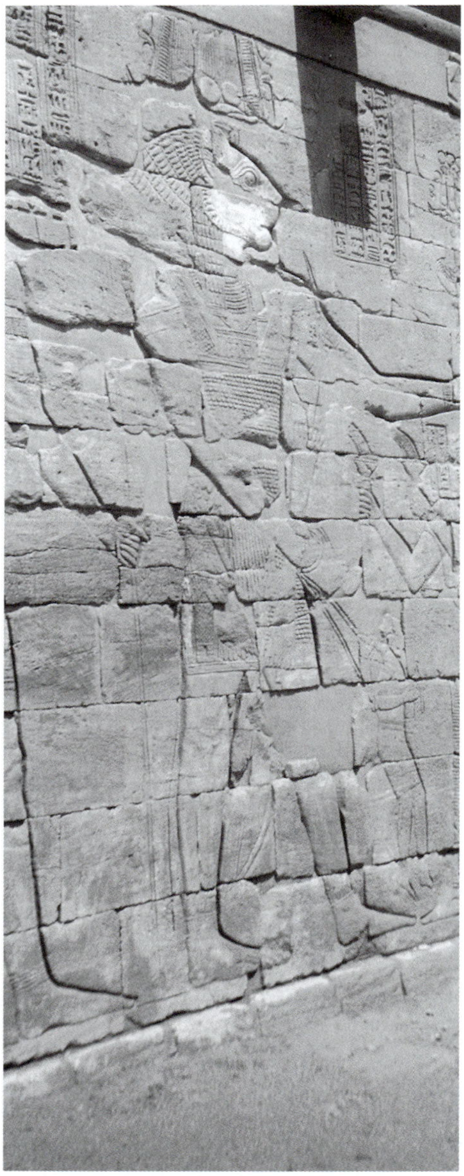

Figure 5: Apedemak holding a bow and arrow in one hand and a cord of prisoners in the other, from Musawwarat es Sufra (Photo: Randi Haaland).

Indian connection may throw some light on the *Apedemak* enigmas.

The region in which representations of *Apedemak* have been found held a strategic position on the trade routes going east toward the Red Sea. The most important centres in this area were the ancient cities of Naqa and Musawwarat es Sufra.

Naqa is located 50km east of the Nile, along the wadi Awateib, which has flowing water during good seasons of rain and where cultivation can consequently be practiced. Westwards the site is connected to Wad ben Naqa, which is situated closer to the Nile and where evidence of large, vaulted storerooms with stocks of ivory and exotic goods have been recovered. The remains of cemeteries and several temples have been discovered at Naqa, the best preserved of which is the Lion Temple, dated to the 1st century AD, which was dedicated to *Apedemak* manifest as a lion god (Shinnie, 1967). The Lion Temple provides at least five different representations of *Apedemak*: emerging from a lotus flower in the shape of the body of a snake (Fig. 6), and *Apedemak* as a three-headed god with four arms, holding his hands and fingers in positions somewhat similar to those that can be found in Indian imagery (Fig. 7).

At Musawwarat es Sufra, Apedemark is also represented: here with bow and arrow (the attributes of a war god) and holding a cord of prisoners in his hand.

At the base of a temple column at Musawwarat, the depiction of gods in an unusual high relief and seated in a yoga-like posture was noted by Shinnie, who remarks that the engravings are similar to Indian reliefs (Shinnie, 1967, p.22). Elements such as the lotus flower, snake, yoga-like posture,

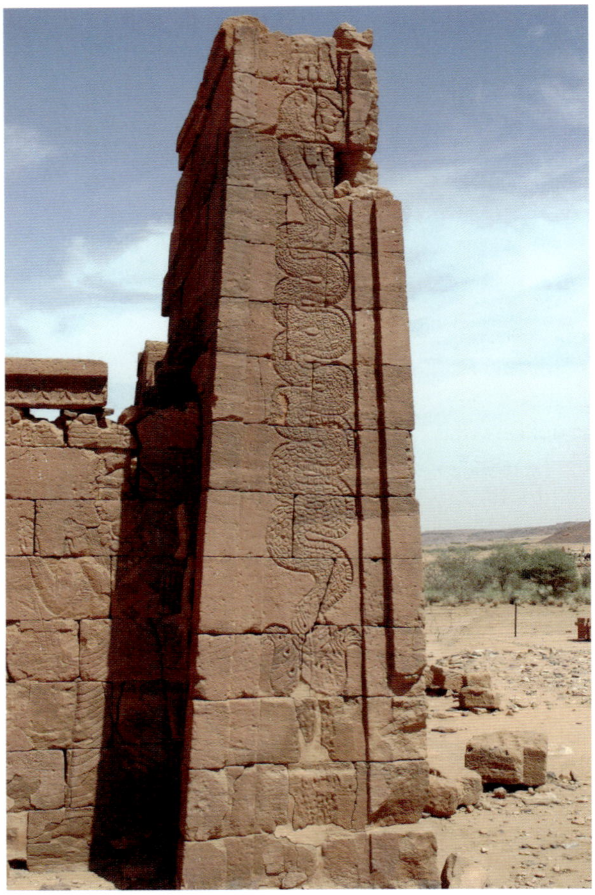

Figure 6: Apedemak depicted as a lion headed snake emerging from a lotus, from the Lion Temple at Naqa (Photo: Idea Heierland).

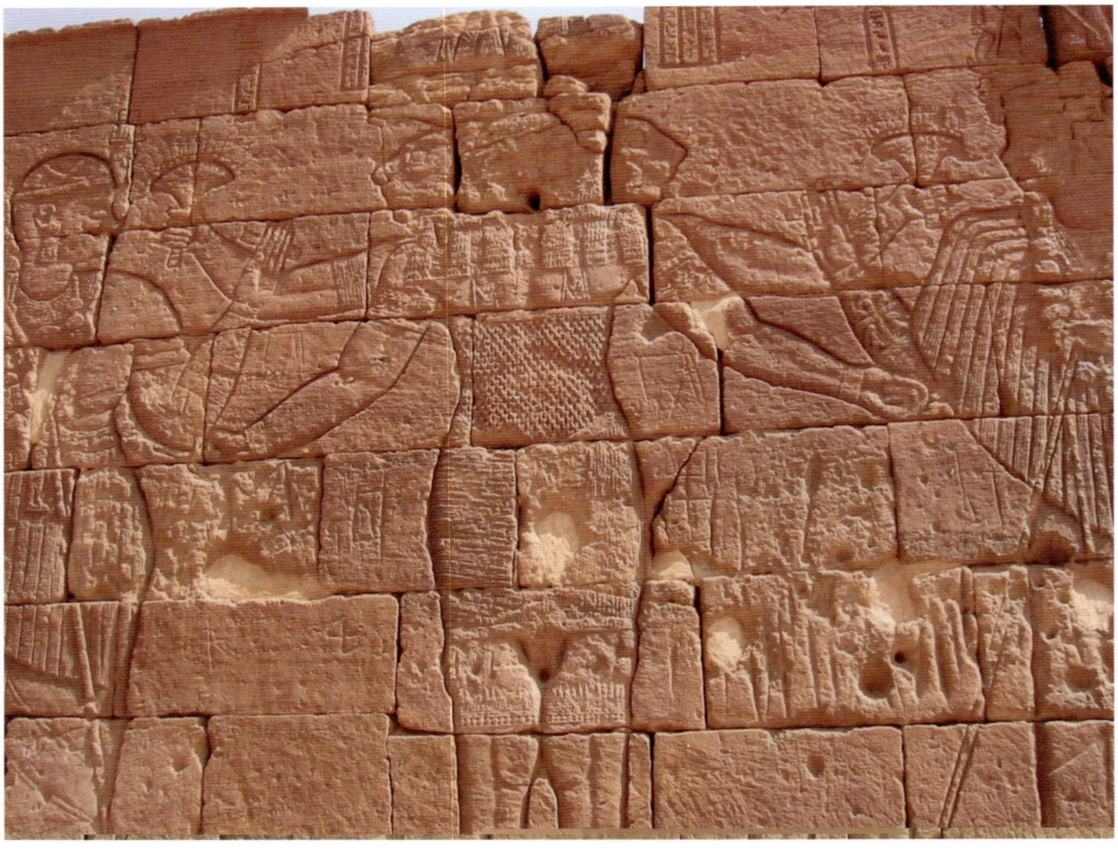

Figure 7: A relief of Apedemak at the Lion Temple at Naqa. Apedemak depicted with three heads and four arms (Photo: Randi Haaland).

finger positions, three-headed representations, four armed gods and the high relief are all design features that display similarities to those found in comparable Indian contexts. Questions necessarily arising from these similarities relate to the identity of those who commissioned the creation of symbolic *Apedemak*-linked representations. Why did they choose designs with Indian attributes? Who were the specialists with the knowledge and skills to create symbolic constructions that blended iconography from different civilisations?

Concerning the decision-makers, a plausible argument is that the rulers of the Meroitic state played a major role in promoting the creation of representations of *Apedemak*, as it is unlikely that a priesthood inspired by Egyptian religion played a major role.

As to the choice of representations displaying attributes linked to both Indian and Egyptian civilisations, these have the potential to communicate the message that the represented state (the Meroitic) is on the same level as its competitor (in this case, Egypt), while simultaneously stating its difference.

The critical question that remains is how craftsmen skilled in the Indian traditions of symbolic constructions in stone were brought into the Meroitic state.

It is important that this aspect is viewed in the context of the Indian trade, taking into consideration the possibility that this may have not only involved an exchange of objects but that it may also have involved the import of people: for example, *mahouts* (Indian elephant instructors) or other skilled craftsmen who migrated between states, based on a similar caste-like division of labour as that seen at Meroe (Haaland and Haaland, 2007).

During the time of the Kushan Empire in north-west India there was a boom in Buddhist-inspired art and architecture related to guild/caste specialisation and monasteries (Liu, 1988, p.5). The art produced was a kind of hybrid Indian and Greek style associated with the so-called Gandhahara tradition and was influenced by the Graeco-Roman style of Alexandria (Thapar, 1966, p.119). Craftsmen from Kushan would thus be familiar with the creation of hybrid expressions such as those we see in the *Apedemak* iconography. The migration of craftsmen between regionally separate states was an important mechanism in the diffusion of cultural elements across civilisations.

Cultural features which may be indicative of the Indian influence of craftsmen are the so-called royal Royal Baths at Meroe, influenced by the idea of ritual baths in India which goes back to the Indus civilisation (Possehl, 1999). Similarly, the great enclosure at Musawwarat es Sufra is thought to have possibly been an area for elephant training, as well as probably being used for religious processions and warfare (Shinnie, 1967; Welsby, 1996).The use of elephants in Indian warfare influenced the Meroitic state to adopt the same practice, which in turn inspired the Ptolemies to do the same. The Ptolemies and Meroites were dependant on Indian trainers, the so-called *mahouts*, to train their African elephants (Kistler, 2007, p.72). The *mahouts* may also have introduced ideas relating to the symbolic and ritual use of elephants in India. Some commentators see the elephant as a divine animal, viewed in the same way as the lion god *Apedemak* (Shinnie, 1967, p.146), although this interpretation has created heated discussion (see Hofmann, 1975; Wenig, 1976). Hofmann argues that the elephant itself was not regarded as divine but that it was the carrier of gods, who were divine. One example of this is an engraving found on the temple wall at Musawwarat es Sufra of an elephant carrying the divine king with the royal religious insignia of a winged python on its back (Shinnie, 1967, p.95).

Conclusion

In conclusion, the power of the state of Meroe was to a significant degree based on military technology in which iron products played an essential role.

The political control of iron products was thus very important. Recovered iconography and burial finds demonstrate the significance of iron objects in relation to the ruling elite. Iron was also of ritual significance, and iconographic representations in temples, pyramid chapels and graves emphasise iron objects. These are seen in the context of transformations from life to death and most importantly in the investiture of rulers.

The war god *Apedemak* was a local Kushite god and was primarily associated with the Meroitic heartland of central Sudan. There do not seem to be any temples or shrines built in his honour outside this region. The temples dedicated to *Apedemak* were built during the later phase of the Meroitic period. The oldest was built around 200 BC, while most of the others were built in the 1st century AD.

The iconography of the god *Apedemak* is related to iron, weapons and royalty. In addition to being the local god, he also exhibits features that indicate international contacts and influences, especially from India. The power and scale of the centralised Meroitic state may thus have been significantly related to import-export linkages of a global scale. The contacts with the north are well documented and discussed. However, to date there has been far less focus on the eastward contacts across the Indian Ocean.

An additional feature to consider has been posited by Helms (1993), whereby metal workers may have moved over long distances between culturally different communities, leading to both the diffusion of artefacts produced by the metal workers and the diffusion of cultural ideas associated with these objects. Furthermore, as suggested by Appadurai, the craft objects themselves create social relationships on a global scale, i.e. across the Indian Ocean in this case (Appadurai, 1986; Ray, 2003, p.214). I suggest that this is what is demonstrated by the Meroitic iron objects and the iconography of the god of war *Apedemak*.

Acknowledgements

I would like to thank the organisers; Jane Humphris, Xander Veldhuijzen, and Thilo Rehren for inviting me to participate at the conference and for asking me to be on the scientific committee. It turned out to be one of the most interesting conferences I have attended. I also want to thank the Norwegian Research Council through the Global Moments project for the financial support. My thanks to Dorian Fuller and Pawel Wolf for helpful comments after my presentation at the Nubian conference in London, August 2010. Most of all my deepest gratitude goes to Jane for her important comments.

References

Abdu, B. and Gordon, R., 2004. Iron artefacts from the land of Kush. *Journal of Archaeological Science*, 31, pp.979-98.

Adams, W.W.Y., 1981. Ecology and Economy in the Empire of Kush. *Zeitschrift für Aegyptische Sprache und Alterumskunde*, 108, pp.1-11.

Appadurai, A., 1986. *The Social Life of Things: Commodities in Cultural Perspective*. Cambridge: Cambridge University Press.

Arkell, A.J., 1961. *A history of Sudan – from the earliest times to 1821*. London: Athlone Press.

Arkell, A.J., 1966. The Iron Age in the Sudan. *Current Anthropology*, 7, pp.451-2.

Barnes, R., 1997. *Indian block printed textiles in Egypt*. Oxford: Oxford University Press.

Beshir, M.S., 2006. *Iron Objects from Meroitic and Post-Meroitic Periods. Discussion on Chronology, Contexts, and Ritual*. MA Dissertation, University of Bergen.

Casson, L., 1989. *The Periplus Maris Erythraei: Text with Introduction, Translation and Commentary*. Princeton: Princeton University Press.

Childs, S.T. and Killick, D., 1993. Indigenous African Metallurgy: Nature and Culture. *Cultural Review of Anthropology*, 22, pp.317-37.

Chirikure, S., 2005. *Iron production in Iron Age Zimbabwe: stagnation or innovation*. PhD Thesis, University College London.

Edens, C.M. and Kohl, P.L., 1993. Trade and World Systems in Early Bronze Age Western Asia. In: C. Scarre and F. Healy eds. *Trade and Exchange in Prehistoric Europe*. Oxbow Monographs 33. Oxford: Oxbow Books, pp.17-34.

Edwards, D.N., 1999. Meroe in the Savannah – Meroe as a Sudanic Kingdom? *Meroitica*, 15, pp.312-20.

Edwards, D.N., 2004. *The Nubian Past. An Archaeology of the Sudan*. London: Routledge.

Gills, B.K. and Frank, A.G., 2002. A structural theory of the five thousand year world system. In: S.C. Chew and J.D. Knottnerus eds. *Structure, culture and history: recent issues in social theory*. Lanham, MD: Rowman & Littlefield Publishers.

Goody, J., 1971. *Technology, Tradition and the State in Africa*. Oxford: Oxford University Press.

Haaland, R., 1980. Man´s role in the changing habitat of Mema during the old Kingdom of Ghana. *Norwegian Archaeological Review*, 13, pp.31-46.

Haaland, R., 1985. Iron production, its socio-cultural context and ecological implications. In: R. Haaland and P.L. Shinnies eds. *African iron working: ancient and traditional*. Oslo: Norwegian University Press, pp.50-72.

Haaland, R., 2004a. Iron smelting – A Vanishing Tradition: Ethnographic Study of this Craft in South-West Ethiopia. *Journal of African Archaeology*, 2, pp.1-13.

Haaland, R. 2004b. Technology, Transformation and Symbolism: Ethnographic Perspectives on European Iron Working. Norwgian Archaeological Review, 1. vol. 37, pp. 1-19.

Haaland, R. and Shinnie, P.L. eds., 1985. *African iron working: ancient and traditional*. Oslo: Norwegian University Press.

Haaland, G., Haaland, R. and Rijal, S., 2002. The social life of iron. A cross-cultural study of technological, cognitive, and political aspects of iron symbolism. *Anthropos*, 97, pp.35-54.

Haaland, R. and R. Haaland 2007.God of war, worldly ruler, and craft specialists in the Meroitic Kingdom of Sudan 7 pp.373-392

Helms, M., 1993. *Craft and Kingly Ideal. Art, Trade and Power*. Austin: University of Texas Press.

Herbert, E., 1993. *Iron, Gender and Power. Rituals of Transformations in African Societies*. Bloomington: Indiana University Press.

Hofmann, I., 1975. *Wege und Möglichkeiten eines indischen Einflusses auf die meroitische Kultur*. Bonn: Verlag Des Anthropos-Institus St. Augustin.

Kense, F., 1985. The Initial Diffusion of Iron to Africa. In: R. Haaland and P.L. Shinnies eds. *African iron working: ancient and traditional*. Oslo: Norwegian University Press, pp.11-27.

Kistler, J.M., 2007. *War Elephants*. Lincoln and London: University of Nebraska Press.

Liu, X., 1988. *Ancient India and Ancient China. Trade and religious exchanges AD 1-600*. New Delhi: Oxford University Press.

Mapunda, B.B.B., 1997. Patching up the evidence for iron working in the Horn. *African Archaeological Review*, 14, pp.107-19.

de Maret, P., 1985. The smith's myth and the origin of leadership in Central Africa. In: R. Haaland and P.L. Shinnies eds. *African iron working: ancient and traditional*. Oslo: Norwegian University Press, pp.73-87.

Miller, D. and van der Merwe, N.J., 1994. Early metal working in sub-Saharan Africa: a review of recent research. *Journal of African History*, 35, pp.1-36.

Phillips, J., 1997. Punt and Aksum; Egypt and the Horn of Africa. *Journal of African History*, 38, pp.423-57.

Possehl, G.L., 1999. *Indus Age: The Beginnings*. Philadephia: University of Pennsylvania Press.

Ray, P.H., 2003. *The Archaeology of Seafaring in Ancient South Asia*. Cambridge: Cambridge University Press.

Rehren, Th., 2001. Meroe, iron and Africa. *Mitteilungen der Sudanarchäologischen Gesellschaft*, 12, pp.102-9.

Rilly, C., 2009. *Le Méroitique et sa famillie linguistique*. Louvain: Peeters.

Sassoon, H., 1983. Kings, Cattle and Blacksmiths; Royal Insignia and Religious Symbolism in the Interlacustrine States. *Azania*, 18, pp.93-107.

Schmidt, P., 1997. *Iron Technology in East Africa: Symbolism, Science and Archaeology*. Oxford: James Curry.

Schmidt, P. and Mapunda, B.B.B., 1997. Ideology and the archaeological record in Africa: interpreting symbolism in iron smelting technology. *Journal of Anthropological Archaeology*, 16, pp.73-102.

Schoff, W.T., 1912. *The Periplus of the Erythreaen Sea*. New York: Longmans Green and Co.

Shinnie, P.L., 1967. *Meroe: A civilization of the Sudan*. London: Thames and Hudson.

Shinnie, P.L., 1985. Iron working at Meroe. In: R. Haaland and P.L. Shinnies eds. *African iron working: ancient and traditional*. Oslo: Norwegian University Press, pp.28-35.

Shinnie, P.L. and Kense, F., 1982. *Meroitic iron working*. Meroitica Studies 6. Berlin: Akademie-Verlag.

Thapar, R., 1966. *A history of India*. London: Penguin Books.

Todd, J., 1985. Iron production by Dimi of Ethiopia. In: R. Haaland and P.L. Shinnies eds. *African iron working: ancient and traditional*. Oslo: Norwegian University Press, pp.88-101.

Trigger, B.G., 1969. The Myth of Meroe and the African Iron Age. *African Historical Studies*, 11, pp.23-50.

Trigger, B.G., 1985. Land and Trade as Patterns in Sudanese History. In: *Studi di paleontologia in Honore di Salvatore M. Puglisi*. Rome, Origini, pp.465-75.

Torok, L., 1975. Economic offices and officials in the Empire of Kush. A review of the written evidence. *Zeitschrift für Aegyptische Sprache und Alterumskunde*, 1, pp.45-69.

Torok, L 1986. Der Meroitische Staat 1. Untersuchungen und Urkunden zur Geschischte des Sudan im Altertum. Meroitica, 9. Berlin.

Torok, L., 1989. Kush and the external world. In: S. Donadoni and S. Wenig eds. *Studia Meroitica*. Berlin: Akademie Verlag.

Tylecote, R.F., 1970. Iron working at Meroe, Sudan. *Bulletin of the Historical Metallurgy Group*, 2, pp.23-50.

Tylecote, R.F., 1982. Metal working at Meroe, Sudan. *Meroitica*, 6, pp.29-42.

van der Merwe, N.J., 1980. The Advent of Iron in Africa. In: T.A. Wertime and J.D. Muhly eds. *The Coming of the Age of Iron*. New Haven: Yale University Press, pp.463-506.

Wainwright, G.A., 1945. Iron in the Napatan and Meroitic Ages. *Sudan Notes and Records*, 26, pp.23-50.

Warmington, E.H., 1974. *The commerce between the Roman Empire and India*. 2nd ed. London: Curzon Press Ltd.

Watt, G., 1966. *The commercial products of India*. London: Today and Tomorrow's Printer and Publishers.

Welsby, D., 1996. *The Kingdom of Kush*. London: The British Museum Press.

Wenig, S., 1976. *Review: Inge Hofmann: Wege und Möglichkeiten eines indischen Einflusses auf die meroitische Kultur*. London: The Egypt Exploration Society.

Wildung, D., 1998. Naga Project (Sudan). Egyptian Museum Berlin Preliminary Report 1995-1996 Seasons 1 and 2. *Archéologie du Nil Moyen*, 8, pp.183-90.

Wildung, D., 1999. *Die Stadt in der Steppe – Grabungen des Ägyptischen Museums Berlin in Naga, Sudan*. Berlin: Stiftung Preussischer Kulturbesitz.

Wildung, D., 2004. Kushite religion: Aspects of the Berlin excavations at Naga. In: D.A. Welsby and J. Anderson eds. *Sudan: Ancient Treasures*. London: British Museum Press, pp.174-85.

Wolf, P., 1997. Recent fieldwork at Musawwarat es Sufra. *Sudan & Nubia*, 1, pp.20-9.

Zabkar, L.V. 1975. *Apedemak Iron God of Meroe*. Warmeister: Ariel and Phillips.

Author's address

- Oysteinsgt. 3, University Bergen 5020. Norway
  e-mail randi.haland@ahkr.uib.no

The world's earliest iron smelting? Its inception, evolution and impact in northern Nigeria

Patrick Darling

ABSTRACT: Around lateritic mesas and quartzite ridges in northern Nigeria lies one of the world's most extensive concentrations of very early iron smelting sites (Darling, 1986, 1989a). Similar very early dates from nearby Niger have subsequently strengthened the case for seriously considering the role of West Africa close to, or even at, the inception of global iron smelting (Alpern, 2005). Intriguingly, these areas lie at one end of an ancient trans-Saharan trade route, the other end of which is just a short sail across the Mediterranean to Anatolia.

The earliest smelters often used bluffs in gullies and along mesa scarps to create vertical slag-tapping furnaces with drip-pits, and there is possible evidence of experimentation in the varied drip-pit typologies, but seasonal destruction of previous furnace superstructures precludes similar observations for furnace evolution. There is an urgent need for further survey work in this area, especially since cattle are in the process of destroying many sites. More samples, to cross-check previous very early dates, have been submitted for thermoluminescence (TL) dating.

The transition to horizontal slag-tapping furnaces in this area took place around the mid-first millennium AD. The occurrences of ellipsoidal furnaces and furnaces with vertical tuyères indicate that, for whatever reasons, experimentation was on-going, and major elements analyses indicate that there may have been an improvement in smelting efficiency – some of which may be attributed to the discovery of higher yielding ores. Fuel, not ore, was the major locating factor for these later sites.

Interviews with the last generation of people with boyhood memories of iron smelting revealed a much wider range of smelting techniques than those observable in other accounts of this area or in the archaeological record, and most historical analyses of major trade routes have failed to incorporate the trading role of this dry-season *cinrani* activity in inter-regional commerce. Using recent survey information, some of the wider impacts of iron production on state formation and maintenance are discussed.

Introduction

The region in and around present-day Nigeria includes some of the highest concentrations of ancient iron smelting sites in Africa (Goucher, 1981; Darling, 1986; Okafor, 1993; Quéchon, 2000; Aremu 2004; Eze-Uzemaka 2009). This paper will focus on archaeometallurgical sites in today's Hausaland in northern Nigeria, where Durham thermoluminescence (TL) and a Harwell radiocarbon date provide an overall framework for major developments of iron smelting practices (Fig. 1) (Darling, 1983a, 1983b, 1984, 1985, 1986, 1989a, 1989b).

Earliest phase

Between the second and third millennia BC, vertical slag-tapping furnaces with drip-pits began operating around the hard laterite-capped mesas (level low plateaux) of northern Nigeria (Darling 1983a, 1989a). These lateritic or ironstone caps were a complex mix of iron oxides and clays from some of the world's earliest surviving land surfaces. Thousands of circular black-coloured slag blocks (or 'furnace-bottom' slags), some up to 1.25m in diameter and edged with 1 to 2cm thick 'walls' of baked clay, fringed the mesas in semi-continuous clusters (Fig. 2).

There was little evidence of any tuyères, sherds, slag heaps or additional slag fragments. Almost all the slag found was in the intact drip-pits – there were only a few visible fragments of slag from upper furnace structures or broken drip-pits (as Fulani cattle now trample the area). Single drip-pits (Fig. 3a), drip-pits divided by an internal baulk (Fig. 3b) and multiple mini drip-pits (Fig. 3c) exhibit structural variations that would seem to indicate a degree of technical experimentation, possibly in response to material or cultural constraints.

At the time of survey, 'furnace-bottoms' were widely thought to be remains of 'bowl furnaces' (Sutton, 1976). However, an ephemeral gully-side exposure near Sokoto (Fig. 4) showed that 'furnace-bottoms' themselves may be no more than thin baked clay-edged drip-pits underneath, offset from slightly thicker clay-lined furnace superstructures. In this example, the furnace superstructure was too weak to exist without being embedded in the gully-side and its collapse by the time of the next visit a few months later helps to explain

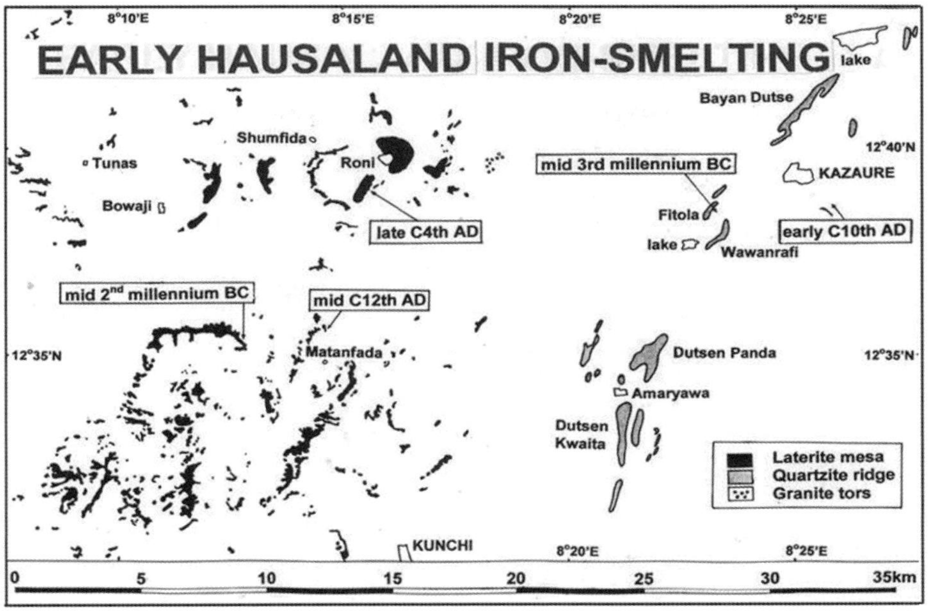

Figure 1: Early Hausaland iron smelting.

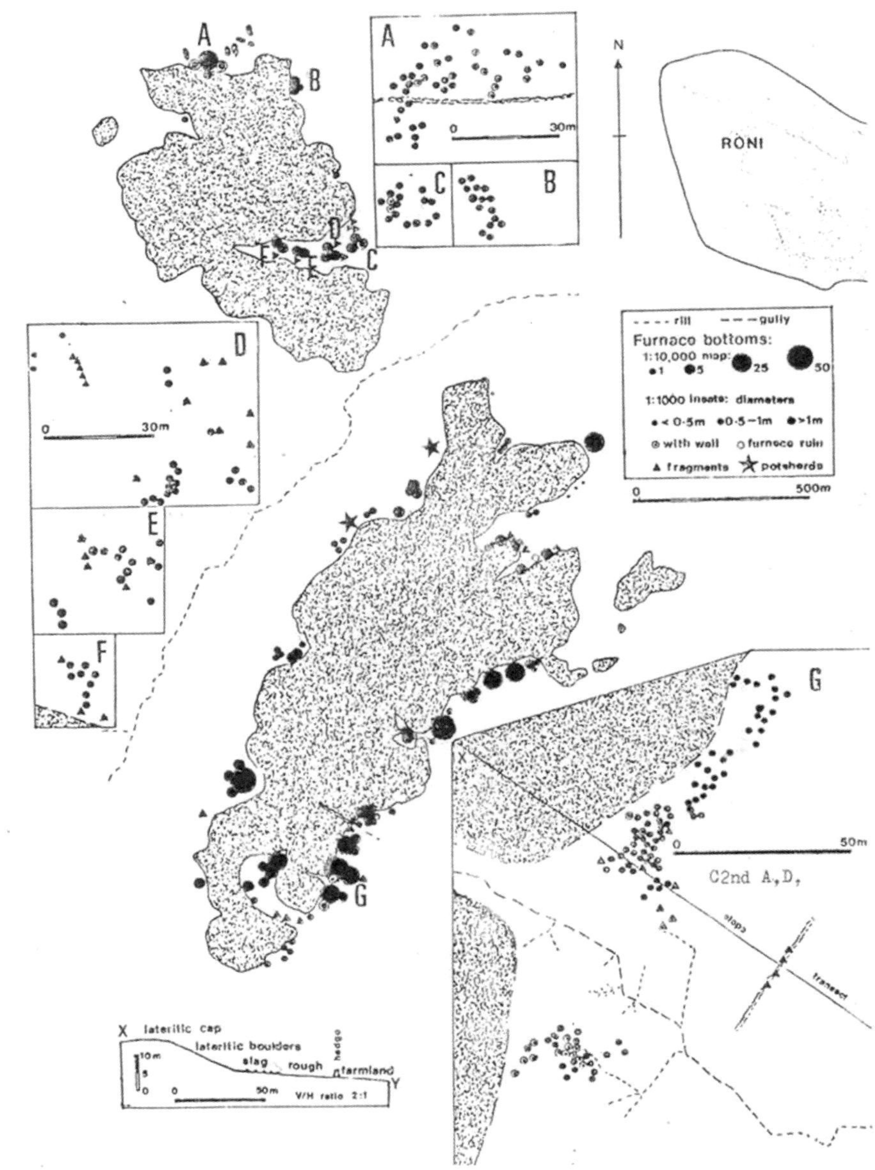

Figure 2: Furnace-bottoms around laterite mesas near Roni.

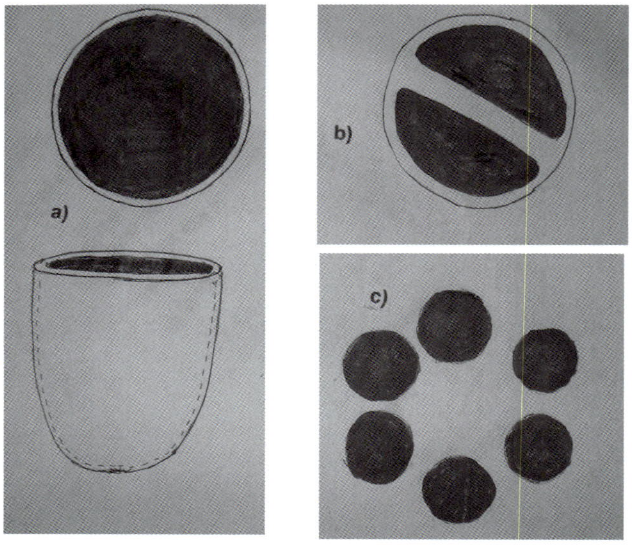

Figure 3: Variations in early phase drip-pit typology.

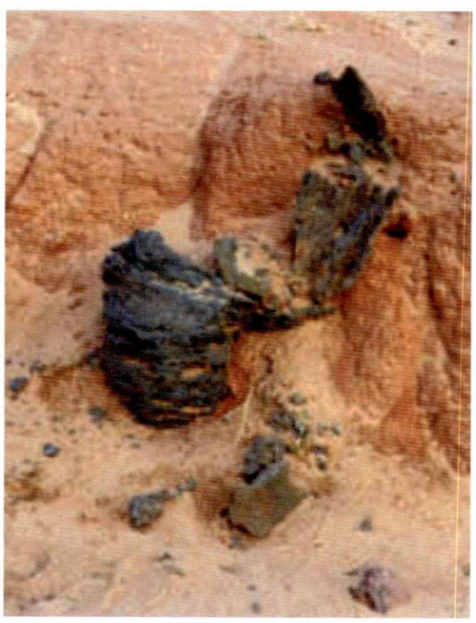

Figure 4: Vertical slag-tapping furnace superstructure over drip-pit, Sokoto gully.

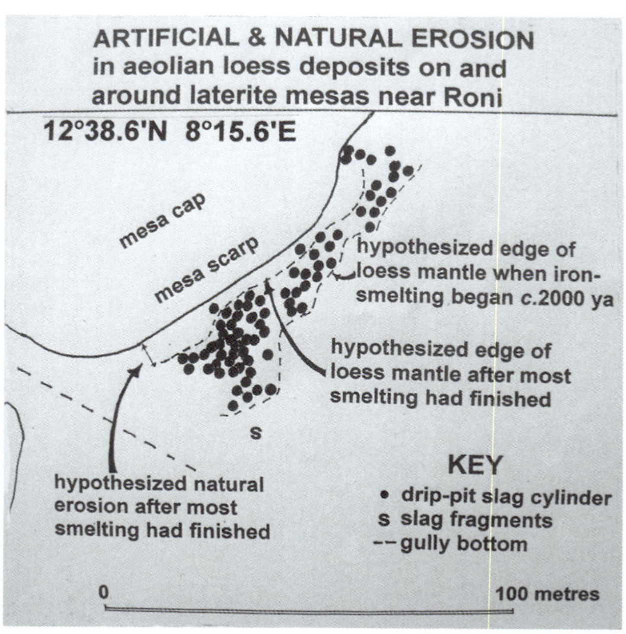

Figure 5: Artificial and natural erosion.

why no other furnace superstructures have been identified from this early phase – bloom removal being probably the initial cause of furnace superstructure destruction.

Smelting location determinants

This particular structural weakness of the Sokoto furnace also revealed the need for vertical bluffs in early vertical slag-tapping furnace construction with its underlying drip-pit; and in this region's peneplains, natural bluffs suitable for early smelting occurred only along mesa scarp faces and hillside gullies, where past loess dune mantles may have provided a homogenous and cohesive medium for drip-pit and upper furnace construction – the excavated holes apparently being lined with clay. Each year, it is hypothesised, the bluff was worked back, with the past season's superstructures broken during the bloom extraction process and the ruins removed to create the next season's batch of furnaces: only the underground drip-pits (furnace bottoms) were left intact. This left the older drip-pits further from today's mesa scarp faces than later furnace bottoms. On this assumption, at one site near Roni, artificial scarp erosion during the smelting era averaged 11m, nearly twice the mean natural backward erosion over the subsequent two millennia (Fig. 5).

Firewood and charcoal were too bulky to transport across large distances, and consequently were another important location determinant for iron smelting sites. It can be hypothesised that trees used to produce charcoal were more abundant than today, and, furthermore, that the preferred *kirya* species (*Prosopis africana*) was largely obliterated from the surrounding area by intense past smelting activity, along with the last vestiges of loess soils atop the mesas – as indicated by a cluster of 130 furnace-bottoms on a mesa near Roni.

Early phase dating

Three thermoluminescence dates obtained from Durham University for slag-pit furnaces were all based on sub-surface fired-clay samples and their adjacent background soil:

- 2,400 BC ± 1100 (Dur TL 57-2AS) from the lowest section of a half-buried tuyère fragment; sample and adjacent sub-soil collected at Fitola, Kazaure Hills;
- 1,400 BC ± 850 (Dur TL 57-3AS) from the clay lining and adjacent soil underneath a drip-pit near Matanfada;
- 380 ± 320 AD (Dur TL 45-2AS) from the clay side lining of a drip-pit and adjacent sub-surface soil, south of Roni.

At the time of dating, the collection criteria for thermoluminescence dating cited sub-surface sampling but no depth specification and initial publications of the results (Darling, 1984 *et seq* – *op.cit.*) did not include sample depths. As sample depth may now prove to be an issue, fieldwork notes and retrospective memory estimates qualify those results with the following depths and provenances:

- Dur TL 57-2AS. Both the fired clay and adjacent soil samples were collected from c. 25cm below the soil surface: the tuyère protruded vertically above the surface, so the sample was collected from the lowest sub-surface point of the tuyère fragment.

- Dur TL 57-3AS. Both the baked clay and soil samples were collected from under an undisturbed massive slag disc some 40cm thick and 90cm in diameter.
- Dur TL 45-2AS. Both baked clay and soil samples were collected c. 45cm below the surface after digging down nearly one cutlass blade length. Care was taken to collect undisturbed soil on one side, avoiding loose soil at the bottom of the hole.

The 2,400 BC date from Fitola could be seen as a challenge to the world's earliest evidence for iron smelting in Anatolia (Turkey), in about 2,300 BC. In the mid-1980s, it stood out as an embarrassing and isolated anomaly, explicable only in terms of a weak internal dose rate or an unreliable background count – even though a Durham laboratory check of this result did not discover any anomaly. The reliability of TL dating of near-surface material, the high standard error, and the highly unusual presence of tuyères at Fitola underlie other queries, which are being addressed by a new set of collected material. In the wider region, the second to third millennia BC dates later obtained for fired structures in eastern Niger by Grébénart have been dismissed as the result of 'old' charcoal or copper smelting and iron smelting was only supposed to have reached the Sahel in around 500 BC (Killick, et al., 1988). However, iron slag noted at all levels at a Birnin Kudu rock-shelter excavation (Fagg, personal communication, 1986) and a Nok radiocarbon date of 925± 70 BC (Rustad, 1980) led to a reconsideration of the arguments put forth by Killick and co-authors.

Recent, controversial dates from eastern Niger – 2,600 ± 300 BC and 2,098 ± 423 BC (Quéchon, 2000) – are based on sherds having a proximal association with iron smelting remains. The issue of the earliest dates for iron-production in this region and possibly the world (Aremu, 2004; Alpern, 2005) failed to discover the previously published corroboratory dates from Fitola and Matafada (Darling, 1983a, 1983b, 1984, 1985, 1986, 1987a, 1987b, 1989a) in their literature search; thus, the issue merits further pursuit – particularly in the collection and dating of further samples. While far from unassailable, this combined Nigeria-Niger evidence strengthens claims that some of the world's earliest iron smelting occurred in what is now a Hausa-speaking region (Phillips, 1989). Thus the focus on trans-Saharan links between West Africa and Roman North Africa in the first millennium BC and the first millennium AD (Killick, 2009) may need to be shifted further east and more than a millennium back in time, while, perhaps, reconstructing a link to Anatolia.

Late phase

Horizontal slag-tapping furnaces with their characteristic slag-heaps, tuyère fragments and much thicker (5-15cm) shaft furnace walls (Fig. 6) made their appearance in this area sometime between the 380 ± 320 AD vertical slag-tapping drip-pit furnace near Roni (noted above) and the mid-late first millennium AD slag-tapping furnace dates reported for Dala Hill in Kano, Samaru West near Zaria, the Kabba-Okene area and Igboland (Fagg A., 1972a, 1972b; Sutton, 1976; Obayemi, Personal Communication,1993; Darling, 1983a, 1983b; Okafor, 1993).

The chimney shafts were removable in order to allow multiple smelts and consequently their incidence was correspondingly lower than for the vertical slag-tapping drip-pit furnaces. During this period, smelting shifted to the nearby Kazaure quartzite ridges (Fig. 7) to use charcoal made from the hard red wood of the *kirya* (*Prosopis africana*): fuel, not bluffs or ore, was now the locating factor (Darling, 1986).

Late phase dating

The vast majority of horizontal slag-tapping furnaces were slightly tapering shaft furnaces with radial tuyère(s) (Fig. 8), such as the one near the Kazaure reservoir.

Their dates were as follows:

Radiocarbon date:

- 580 AD ± 70 (HAR 6258): 814 to 986 AD on Clark's calibration curve; based on charcoal sample from c. 90cm below the surface inside and at the base of a breaking furnace exposed in the wall of a recent Kazaure reservoir borrow-pit.

Thermoluminescence dates:

- 720 AD ± 250 (DurTL 57-6AS): ellipsoidal shaft furnace on laterite mesa near Waira: samples from outer clay wall and adjacent subsoil c. 35cm deep.
- 830 AD ± 230 (DurTL 57-7AS): straight cylindrical quartzite ridge, Gabasani, near Bichi; samples from outer clay wall and adjacent subsoil c. 30cm deep.
- 1140 AD ± 170 (DurTL 57-4AS) vertical tuyère in furnace structure, Matanfada mesa: samples from outer wall near base of furnace and adjacent sub-soil c. 40cm deep.

This variety of furnace typologies indicates a degree of experimentation during the late first millennium AD. It is not yet known, however, whether this reflects the practices of different smelting groups evolving different cultural typologies over time, or whether it was a response to different types of ore or some other technical factor. At present, interpretation leans towards the first explanation as the slag products are very similar regardless of the type of furnace used (Rehren, et al., 2007, p.212). Furnaces with single, large, vertical tuyères incorporated into the furnace structure (Fig. 9) were rare. Similar down-draught furnaces reproduced experimentally at Mafa in North Cameroon yielded cast iron, steel, and low-carbon iron (David et al., 1989).

At the time of field survey in the 1980s, it was thought that some tuyères encased in slag and penetrating the basal furnace wall resulted from the collapse of such vertical tuyères in failed smelts. However, this phenomenon is better explained by Schmidt and Mapuda's (1997, pp.98-99) observations in Tanzania of tuyères full of kaolin being placed vertically at the bottom of furnaces for ritual purposes before the smelt. Not only does this explain the tuyères being encased in slag and their apparent neat penetration of the furnace wall, but

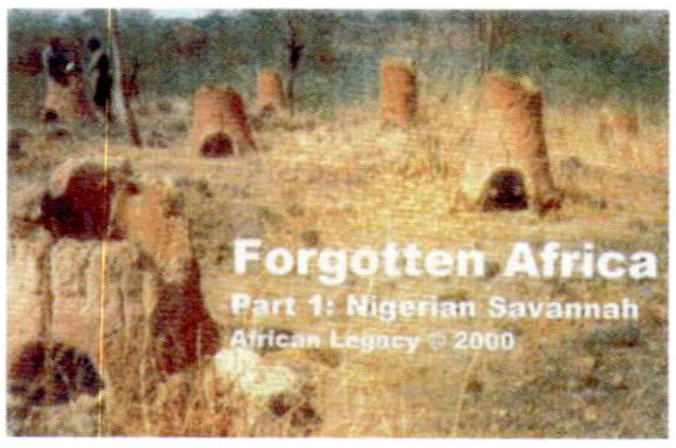

Figure 6: Well-preserved shaft furnaces (Yankari Game Reserve).

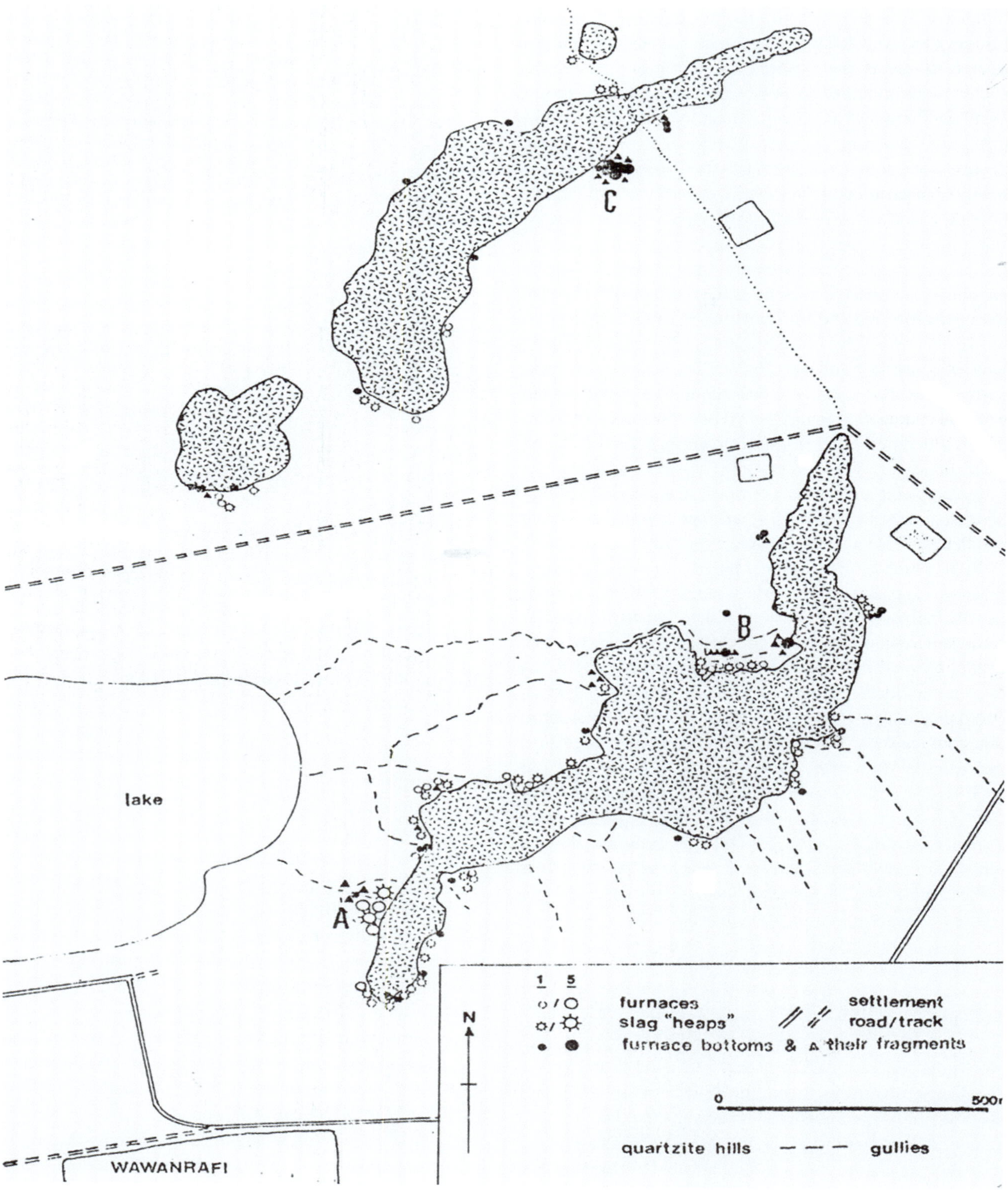

Figure 7: Horizontal slag-tapping ruins around some of the Kazaure Hills.

it reinforces the notion of a single source for the spread of iron smelting and associated practices over the rest of Africa.

XRF analyses

The difference in determining smelting efficiency between vertical slag-tapping furnaces with drip pits and horizontal slag-tapping furnaces depended on finding past ore sources; something with which local informants helped, by identifying recent ore sources in various parts of Nigeria. XRF (X-Ray Fluorescence) analyses, based on pulverised ores set in resin and analysed at Southampton University, produced a series of graphs with pronounced peaks easily identifying: a) magnetite ores in Yorubaland; b) hard grey haematite ores (soft brown where weathered) for the 12th century AD natural-draught smelting furnaces using numerous short narrow tuyères located at Itakpe, near Lokoja; c) most ores having a limonite component; and d) a friable golden brown goethite ore from Hausaland.

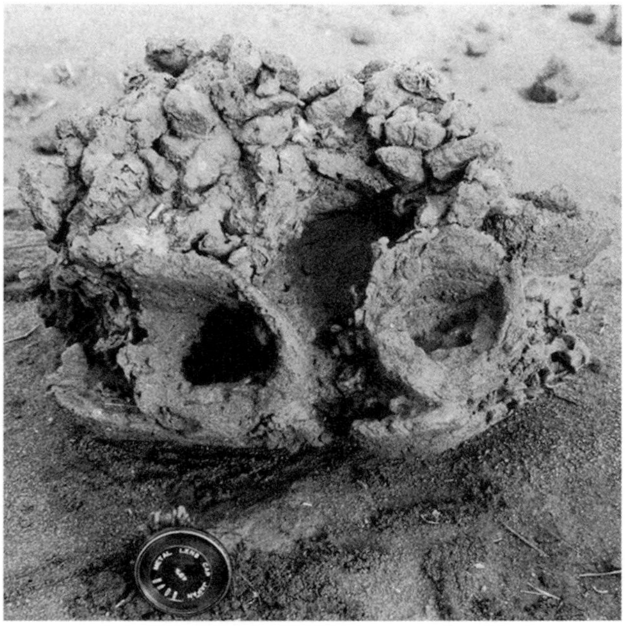

Figure 8: Radial tuyères from furnace ruins near Sokoto.

Underground survey of one iron ore source

Local informants claimed that this last goethite ore had been mined for centuries in the Makurdi hills near Bakori (Keen, personal communication, 2006), creating well over 200 shaft mines. The shafts, each around 90cm in diameter and spaced about 5m apart, lay between two parallel quartzite ridges running approximately north-south, roughly 15m apart. Weighted tape measurements indicated a past undulating enriched layer of ore 8 to 14m below ground level (Fig. 10).

In lowering a lighted oil-lamp no scorpions, snakes, collapse, carbon dioxide or methane could be detected, although disturbed bats flew out of an adjacent shaft. A perilous belay down one shaft to collect an ore sample avoided touching the shaft walls and, at the base, horizontal mining extended to create a dangerously large underground cavern linked to two other shafts (thus explaining the disturbed bats): the sample of friable, golden-brown ore was collected here. On

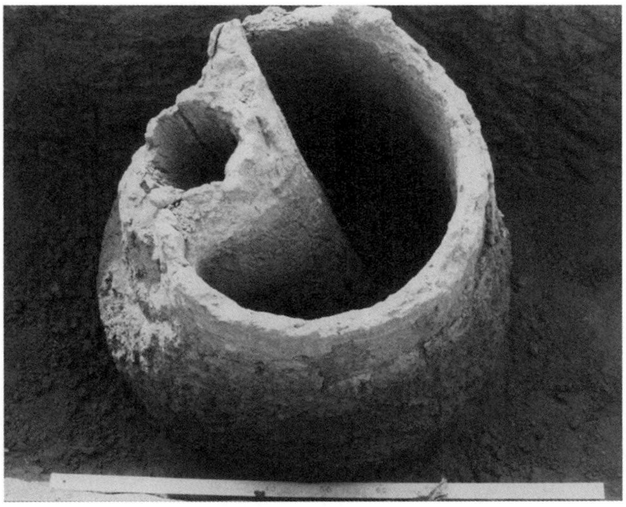

Figure 9: Vertical tuyère incorporated in furnace structure.

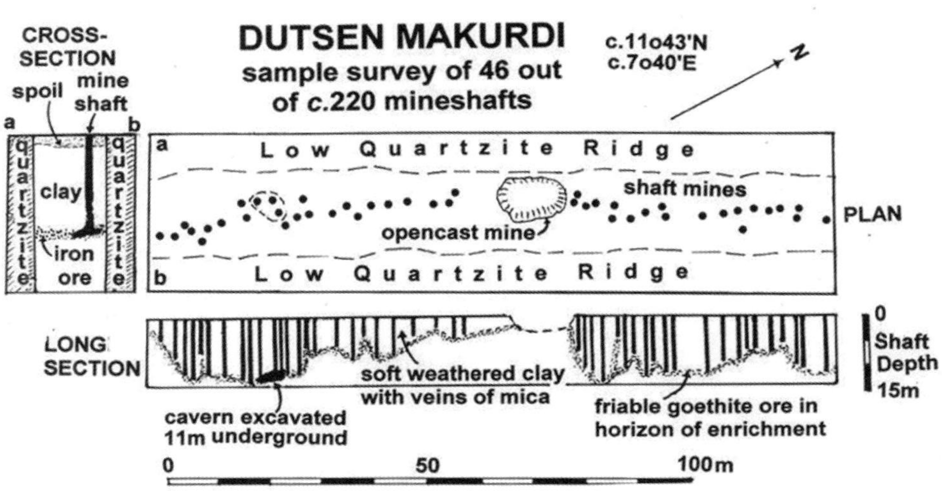

Figure 10: Makurdi mine shafts and ore depths.

the ascent back up, it was necessary to use the old handholds and footholds, causing sections of the shaft wall to give way with loud thumps as they hit the bottom. The resultant sudden shortening of one's life expectation encouraged a very healthy respect for the miners of long ago.

Other mine shafts

Other shaft mines were observed near Tureta, reputed to have been a steel sword-making centre in the 19th century and possibly much earlier (Duffil, personal communication, 1984); and, in Nigeria, there are present-day shafts dug by gold-diggers near Birnin Gwari and near Ilesha. The only other similar past shafts observed by the author in Nigeria were tens (or even hundreds) of thousands of similarly spaced mines up to 20m deep dug through basal Palaeocene Kerri Kerri Deposits lying unconformably on the Upper Cretaceous white Gombe Sandstone cliffs along the northern rim of the Gongola River flood-plain, a few hundred kilometres further east (Darling, 2011, forthcoming). These shafts, however, had been dug by women in the quest for yellow clay. This clay was prepared entirely by women, who baked the clay pats red: women were the main retailers and purchasers as well as the main users for these artefacts. Some of the clay products acted as an ochre for decorating wedding calabash interiors or as a major component of red slip baked on pottery. Its major use, however, was as an edible clay consumed by pregnant women to quell morning sickness nausea, to satisfy pica cravings for trace minerals and to absorb plants and other toxins at a time when their immunity levels were lower, to guard against miscarriage (Darling, forthcoming). This ancient practice of geophagy antedates iron smelting by a matter of forty thousand years and may even go back to *Homo habilis* (Abrahams, 2003, p.31). Accordingly, there are new questions raised about the relationship between ancient edible clay baking (by women), later pottery firing (by women) and the even later discovery of smelting. It twists the gender issues raised by Herbert (1993), Berns (1993) and Maclean (1998) and it takes their arguments a step further to speculate that women may well have been better placed than men to have obtained iron ore and to have accidentally discovered iron smelting.

Major elements analyses

In the northern Hausaland region, the quest to discover early phase iron ore sources was challenging: major elements analyses of uncorroded slags and potential ores indicated that the mesa laterites probably used for large-scale smelting had lower iron oxide contents than the local slag. The ancient ore has, therefore, not been precisely located. Nevertheless, using all ten major elements resulting from analyses undertaken at Southampton University, Roni slag and laterite were closely correlated, indicating that the ore still had a local origin; although the high level of phosphorous in Roni slag and laterites might have seriously impaired product quality in terms of brittleness. The iron content percentages calculated from the major elements analyses were as follows:

Main technique:	Site	Laterite/ Ore?	Slag	Iron extracted
Vertical slag-tapping furnaces with drip-pits	Bichi	29.8%	48.6%	ore not found
	Roni north	49.2% *	-	-
	Roni south	47% *	48.6%*	ore not found
Horizontal slag-tapping furnaces	Makurdi	55.7%	39%	16.7%
	Wawanrafi	43.2%*	42.7%	ore not found

* denotes phosphorous content of about 1%, possibly impairing product quality.

The lack of ancient ores and the small sample size inhibit full comparison of the conversion efficiency of vertical and horizontal slag-tapping furnaces, but they do not preclude it. The mean iron content percentages of drip-pit slags (from early vertical slag-tapping furnaces) were nearly 50%, whereas tapped slag (from later horizontal slag-tapping furnaces) contained about 40% iron. Assuming, *pro tem*, that similar ores were used, this difference would represent an appreciable increase in the extraction of iron. One interpretation of this could be that Early Iron Age vertical slag-tapping furnaces with drip-pits were less efficient than Late Iron Age horizontal slag-tapping furnaces. Indeed, Okafor (1993) attributes similar changes in Igboland smelting by-products to this technical shift.

Yet the numerous mine shafts indicate that the Makurdi ores were clearly regarded as being high grade – this involved perceptions concerned with self-fluxing properties and rates of smelting success rather than with the percentage of iron extracted or phosphorus content. Over time, the discovery of improved ore sources in terms of smelting success (which is what the rituals sought to enhance) may provide a better explanation for any productivity increase than technology *per se*. Without many more samples being analysed, however, one cannot press this small data set too far for other viable interpretations. Whatever the exact cause, it is possible that a surge in Late Iron Age productivity was one key factor in the changes in wealth, trade and state-formation involved in the creation of, *inter-alia*, Old Ghana, Kanem-Borno and early Hausaland.

Twentieth century interviews

It was impossible to examine directly the subtleties of 'migration' and 'diffusion' in the advent of Late Iron Age techniques taking place over a thousand years ago. Even the insights provided by ethnoarchaeology or participatory observation were unobtainable because iron smelting in these regions had died out about sixty years ago; on-site interviews were similarly impractical apart from in a few cases. The best proxy solution was to interview old farmer-blacksmiths with boyhood memories of iron smelting, even though they lived and farmed (now, as then) far from where they used to smelt. During their youth in the early 20th century, they recalled that iron smelting was a dry-season *cinrani* activity – a seasonal coping mechanism inducing a consumption time-lag to conserve vital food stores for the prolonged 'hungry season' energy demands by both farmers and smiths (Darling, 1987b). The key to *cinrani* calculations lay in the sealing of granaries (Fig. 11), leaving a generous *mudu* of grain per person per day for those remaining (Mortimore, 1989).

By the early 20th century, farmer-blacksmiths of the fertile Kano plains were journeying mainly east and west to whichever was the nearest smelting location (Fig. 12), to create numerous, great *matamaci* (seasonal encampments) of one to five hundred *bukka* (temporary houses), around each dry season's smelting activities.

The site of each *matamaci* shifted each year to exploit new trees for charcoal-making (Darling, 1987b), whereas the best sites for the exploitation of iron ore lasted for decades or centuries (Keen, personal communication, 2006). As the *matamaci* were located in the no-man's lands between powerful political entities they acted as large seasonal markets, enabling the farmer-blacksmiths to generate extra wealth from the sale of smelted blooms and to meet current expenses incurred for smithing (Fig. 13). The Maguzawa, Siri, Kare-Kare and other non-Muslim farmer/smelters tended to reside more permanently in these no-man's lands (Last, 1993; Darling, 2004).

In this process of exchange between smelters from different regions, it is reasonable to assume that the merits and demerits of different local ores, charcoal and smelting practices were discussed, thus facilitating the diffusion of a wide variety of experimental smelting techniques. Eye-witnesses in the 1980s recollected a whole range of early 20th century techniques over and above the single standard Hausa technique recorded in the available literature (Jaggar, 1973a, 1973b; Pole, 1975; Sutton, 1976a). Some of them showed smelting sites where they had worked – a salutary experience, for underground firewood pits and the use of crucible techniques provided no remains for the visible archaeological record. However, as a range of similar variations has been reported elsewhere (Killick, et al., 1988), it is clear that the techniques were being diffused, and this implies that knowledge about them was being shared. This is at odds with traditions maintaining that smelting secrets were retained strictly within each smelting family and passed down the generations, as this would seriously constrain any technical diffusion.

The role of magic

A curious spiked bowl (Fig. 14), probably once used for magical potions (Schmidt and Mapunda, 1997, pp.87-90, 95) may solve this apparent enigma. Smelting was a risky business, so special herbs used in night-time rituals before a smelt and often before every stage of the process (Herbert, 1993; Keen, Personal Communication 2006) were perceived to make the crucial difference between success and failure. It may well be that these herbs and magical rituals were what each family kept secret – not the practical experimentation that was part of everyday pooled knowledge (Darling, 2004, pp.106-107). These magical secrets, playing with fire and fire-walking (Madauci et al., 1968), became part of the mystique and

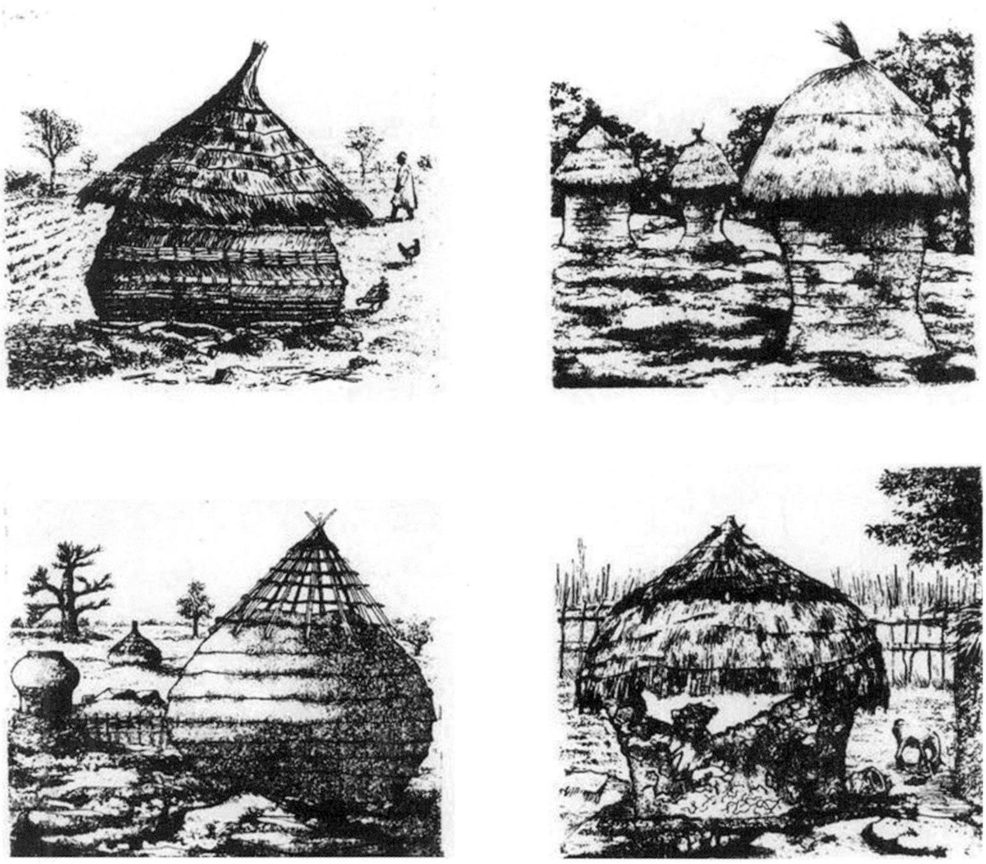

Figure 11: Granaries (*rumfa*) in Kano farmland settlements.

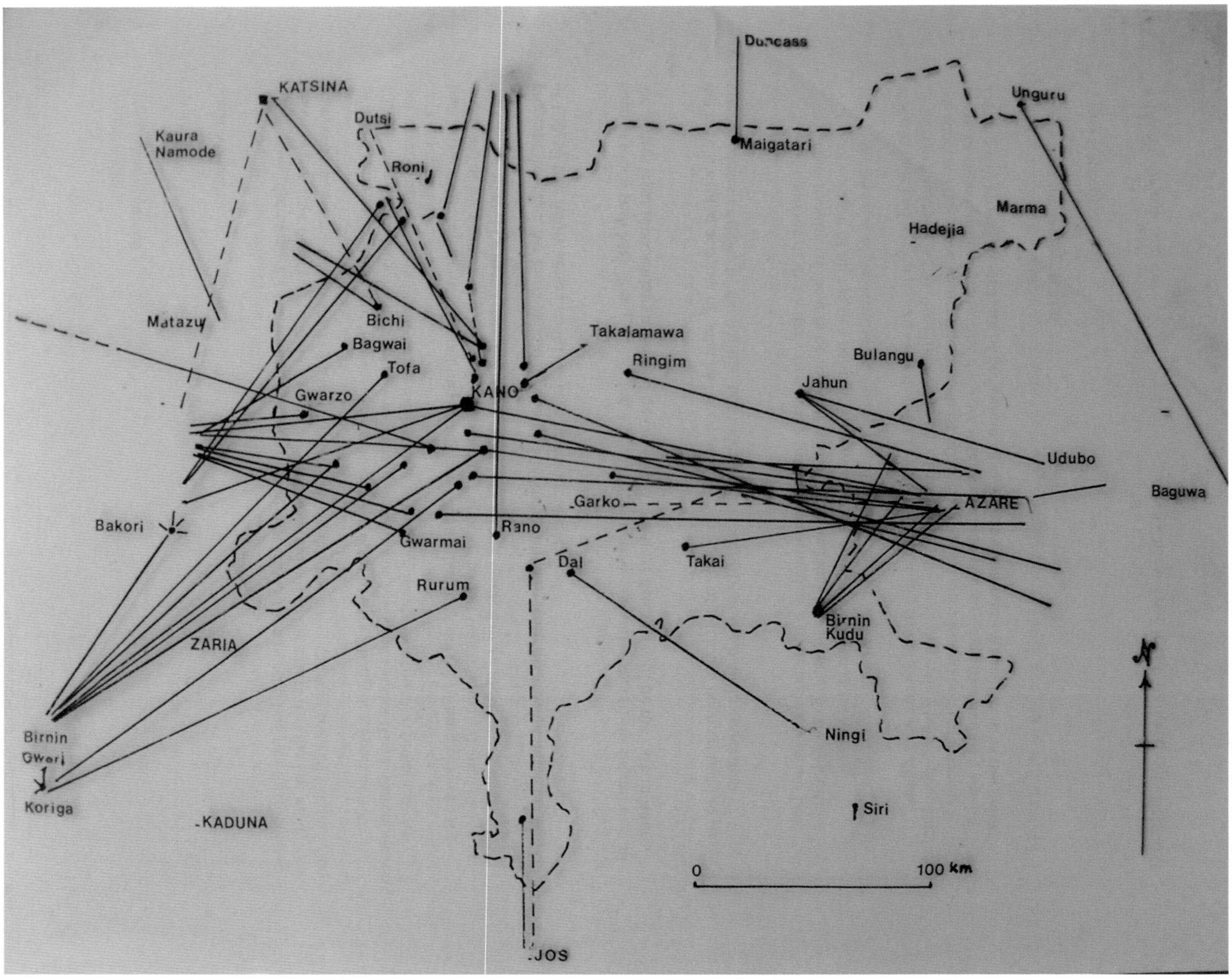

Figure 12: Early 20th century seasonal migrations of Kano farmer/blacksmiths from farmlands out to smelting sites.

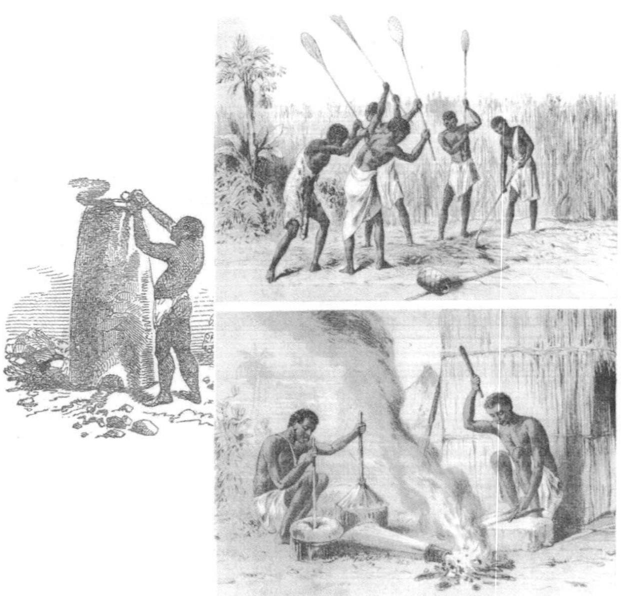

Figure 13: Farmer-blacksmiths at work in the 19th century.

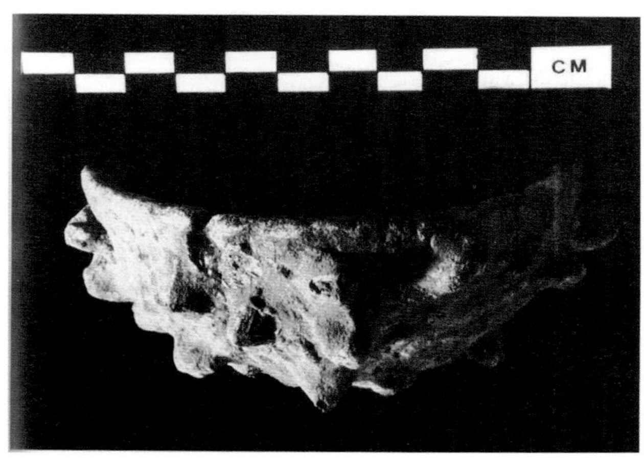

Figure 14: Spiked crucible-like bowl, Wawanrafi.

socio-political identity of the non-Muslim Maguzawa. Their 'identity rejection' of growing Hausa religious and political dictates was permitted by early Muslim traders because their ancient practices meant that they could be considered 'Magians' – not pagan *arna*. Therefore, the Maguzawa counted as *ahl al-dhimma*, with whom trade was legitimate without conversion to Islam or sharing in Islamic literacy, mathematics or science (Last, 1993).

Secrecy

At Siri, a ban on iron smelting during the 19th century Jihad led to a different kind of secrecy. Donkeys were used on secret night-time journeys to collect iron ore along the Bunga River. This was then smelted at night inside hill-top houses where the furnace faced away from the door lest its glow betray the forbidden construction of tools and weapons for the renegade Ningi (Figs. 15a, 15b and 15c).

Secrecy is a strong expression of identity resistance; and Siri identity was partly forged as being 'not those Hausa people I am hiding from' (Darling, 2009). Similarly, the secrets held by all blacksmiths/smelters were to give them a special identity, often independent from the rest of society (De Barros, 2000, p.175).

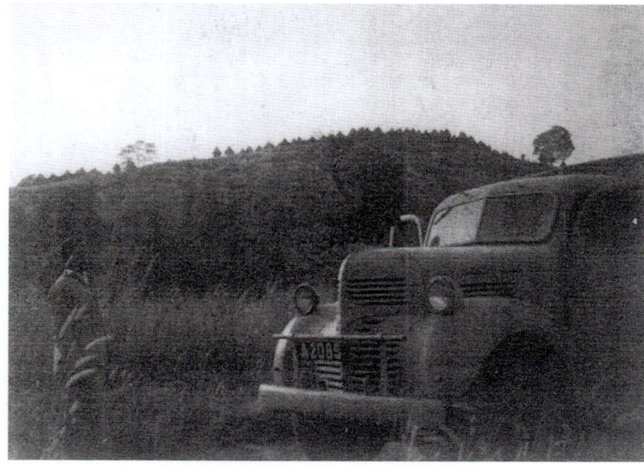

Figure 15a: Siri houses on a hilltop skyline in the 1930s.

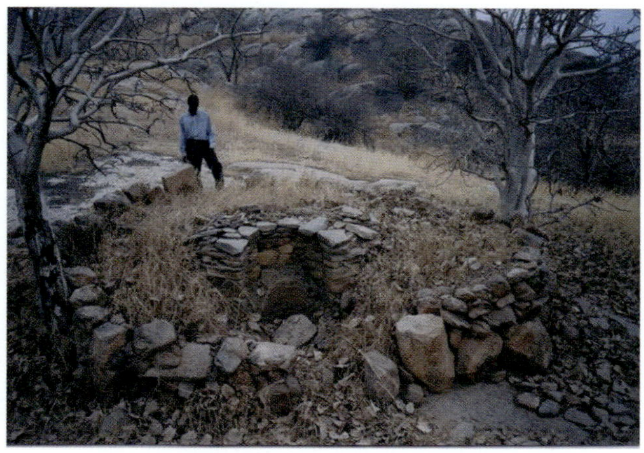

Figure 15b: Secret smelting in furnace facing away from doorway, Siri.

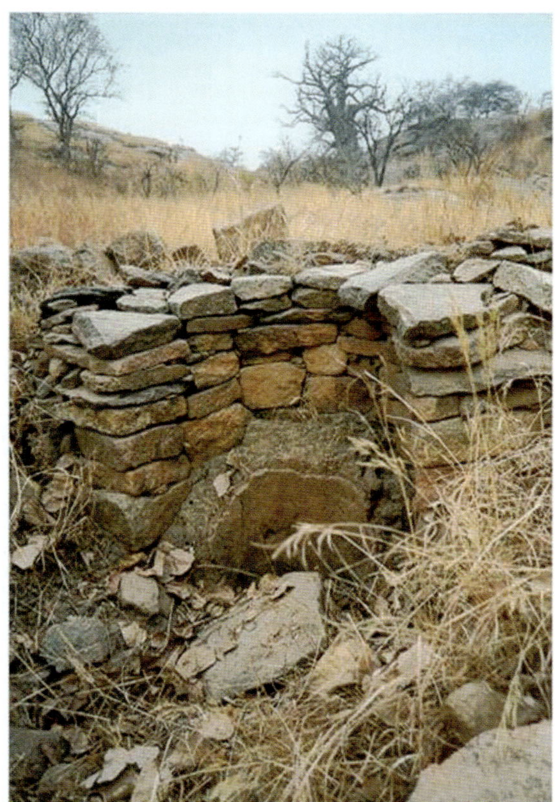

Figure 15c: Close-up of secret furnace hidden by wall from far side doorway.

Impact

The main tools forged were the triangular *fartagnya* and wide-tined *galma/garma* – the different types of hoes used respectively for strong clayey and loose sandy soils; and the dyadic role of farmer-blacksmith led to an efficient, appropriate development of this agricultural technology. Sorghum grains dated to the 4th-6th centuries AD and the emergence of huge settlement mounds in the Lake Chad *firki* (Connah, 1985; Magnavita, 2001, 2002) are indicative of a massive agricultural intensification process in north-east Nigeria. When these changes reached Hausaland, the resulting increase in population density on the fertile loess plains caused shifting agricultural settlements to become sedentary as they took up rotational bush/fallow land management.

It should of course be remembered that gender, ritual and symbolism were vital facets of past African iron production (Saltman, 1986, p.8; Herbert, 1993; Schmidt and Mapunda, 1997; de Barros, 2000; Katz-Hyman and Rice, 2011). However, it was the technical efficiency of sharp, durable iron tools that actually achieved change. Land clearance, cultivation, house construction, well-digging, firewood collection and a host of other activities encouraged increasingly sophisticated levels of intensive expertise, later able to support permanent agriculture using *taki* (manure) in a densely populated 'close-settled zone', the largest trading emporium in West Africa at Kano City and the largest concentration of walled urban settlements in Africa (Mortimore and Wilson, 1965). As soon as the increased productivity of the Late Iron Age farmer-blacksmiths became apparent, élite religious, political, military and commercial power groups vied with one another to control or to tax iron supplies and to expropriate the surplus rural wealth into more urban centres (Darling, 1987b). Here, the same blacksmiths who beat out the hoes generating the agricultural prosperity of Hausaland now also beat out the swords, spear heads and fetters that underpinned the state control needed to create and sustain its increasingly urban identity (Jaggar, 1973a, 1973b).

Demise of traditional smelting

Omoweh (2005, p.57) argues that iron smelting in Nigeria collapsed because "the colonial state deliberately prohibited any form of manufacturing from taking place in Nigeria"; thus enabling companies like the United Africa Company to import hoes, steel pipes, rods and coils with little competition (Omoweh, 2005, pp.58-59). Many historians would probably argue that taxation played a greater role than 'prohibition'. However, it is the iron smelters themselves (those farmer-blacksmiths interviewed in the 1980s) who most strongly contradict Omoweh's contention. They all claimed that the use of broken lorry springs became their new, more accessible source of steel and that this was the reason why they gave up smelting. Just as European settlers unintentionally introduced new pathogens causing widespread mortality in South America (Acemoghu, et al., 2003, p.402), so, perhaps, the introduction of a new mode of transport into Nigeria had the unintended knock-on effect of rapidly collapsing Nigeria's traditional iron smelting industry. Whatever the exact causes, this collapse in traditional smelting signalled the end of an era; for farmer-blacksmiths terminated their engagement with ritual and practical activities that had been closely related to those developed by the first smelters, whose experimental playing with fire here over three-and-a-half millennia ago had ushered in a major new technology to the rest of the world.

References

Abrahams, P.W., 2003, Human Geophagy: A Review of Its Distribution, Causes and Implications. In: H.C.W. Skinner and A.R. Berger eds. *Geology and Health: Closing the Gap*. New York: Oxford University Press.

Acemoghu, D., Robinson, J. and Johnson, S., 2003. Disease and development in Historical Perspective. *Journal of the European Economic Association*. 1, pp.377-405.

Alpern, S.B., 2005. Did They or Didn't They Invent It? Iron in Sub-Saharan Africa. *History in Africa*, 32, pp.41-94.

Aremu, D., 2004. Iron Roots in Africa. In Bacoum (ed) *The Origins of Iron Metallurgy in Africa – new lights on its antiquity.* UNESCO.

Barth, H., 1890. *Travels and Discoveries in North and Central Africa*. London: Ward Locke & Co.

Berns, M.C., 1993. Art, history and gender: women and clay in West Africa. *The African Archaeological Review*, 11, pp.129-48.

Connah, G.E.C., 1985. Agricultural intensification and sedentism in the *firki* of NE Nigeria. In: S. Farrington ed. *Prehistoric intensive agriculture in the tropics, Part II*. Oxford: British Archaeological Reports International Series 232, pp.765-85.

Darling, P.J., 1983a. Iron-smelting in the Kazaure Hills. *Zaria Archaeological Papers*, 5, pp.40-1.

Darling, P.J., 1983b. *Iron-smelting in Hausaland, northern Nigeria*. Pan-African Congress on Pre-History and Related Studies, Jos, Nigeria.

Darling, P.J., 1984. Report on Archaeological Activities at Bayero University, Kano. *Zaria Archaeological Papers*, 6, pp.24-7.

Darling, P.J., 1985. The Coming of Iron to Kano. *The Triumph* newspaper, Kano, 07/05/1985.

Darling, P.J., 1986. Fieldwork Surveys in and around Kano State, Nigeria. *Nyame Akuma*, 27, pp.39-41.

Darling, P.J., 1987a. *The Splendours and Miseries of Technical Change in Agriculture* (set title). Winegarten Essay Prize (winning entry), National Farmers' Union/Wye College.

Darling, P.J., 1987b. *Seasonal food strategies and coping mechanisms of blacksmiths and smelters in Hausaland, northern Nigeria: some implications for development studies*. Unpublished paper for Conference on The Changing Sahara, Wye College.

Darling, P.J., 1989a. Archaeology and the Dating of Historical Events in Kano. In: B.M. Barkindo ed. *Kano and Some of Her Neighbours*. Zaria: Ahmadu Bello University Press for the Department of History, Bayero University, Kano, pp.31-8.

Darling, P.J., 1989b. How cattle, iron, sorghum and irrigation influenced social and agricultural development in Hausaland. *Nigerian Field*, 54, pp.11-8.

Darling, P.J., 2004. Migrants, millenialists and mayhem: Kari-Kari, Ningi, Gwari, Kebbi, Kwatarkwashi and Zuru 'enclaves' in northern Nigeria. In: L. Smith, P. Rose, S. Wahida and G. Wahida eds. *Fifty years in the archaeology of Africa. Themes in archaeology and practice. Papers in honour of John Alexander*. Azania 39. Nairobi: British Institute in Eastern Africa, pp.90-109.

Darling, P.J., 2009. *The Visible Archaeology of Hausaland: sherds, stones, slag, shrines, words, walls and baobab tree dating*. Power-point presentation and paper for Conference on Hausa Identity, UEA and Liverpool.

Darling, P.J., 2011. August-September 2011 Fieldwork Notes. *African Legacy*.

Darling, P.J., forthcoming. *Exploring Nigeria's past: its visible archaeology, cultural landscapes and local lore*.

David, N., 1988. The Ethnoarchaeology and Field Archaeology of Grinding at Sukur, Adamawa State, Nigeria. *African Archaeology Review*, 15, pp.13-63.

David, N., Heimann, R., Killick, D. and Wayman, M., 1989. Between bloomery and blast furnace: Mafa iron-smelting technology in North Cameroon. *African Archaeological Review*, 7, pp.183-208.

De Barros, P., 2000. Iron Metallurgy: Sociocultural Context. In: M.S. Bisson, S.T. Childs, P. De Barros and A.F.C. Holl eds. 2000. *Ancient African Metallurgy: The Sociocultural Context*. Walnut Creek, CA: AltaMira Press, pp.147-98.

Eze-Uzomaka, P., 2009. *Iron and its influence on the prehistoric site of Lejja*. World of Iron Conference, London.

Fagg, A., 1972a. Pottery from the Rop Rock Shelter Excavations of 1944 and 1964. *West African Journal of Archaeology*, 2, pp.20-38.

Fagg, A., 1972b. A preliminary report on an occupation site in the Nok valley, Nigeria. *West African Journal of Archaeology*, 2, pp.75-9.

Goucher, C., 1981. Iron is iron 'til it is rust: trade and ecology in the decline of West African iron-smelting. *Journal of African History*, 22, pp.178-89.

Herbert, E.W., 1993. *Iron, Gender and Power – Rituals of Transformation in African Societies*. Bloomington: Indiana University Press.

Jaggar, P., 1973a. Kano City Blacksmiths: Precolonial Distribution, Structure and Organization. *Review of African Political Economy*, 13, pp.21-36.

Jaggar, P., 1973b. A Kano blacksmith's vocabulary. *Kano Studies*, 1, pp.99-110.

Katz-Hyman, M.B. and Rice, K.S., 2011. *World of a Slave: encyclopaedia of the material life of slaves in the United States*. Santa Barbara, CA: Greenwood.

Killick, D., Van der Merwe, N.J., Gordon, R.B. and Grebenart, D., 1988. Reassessment of the evidence for early metallurgy in Niger, West Africa. *Journal of Archaeological Science*, 15, pp.367-94.

Last, M., 1993. History as religion: de-constructing the Magians ("Maguzawa") of Nigerian Hausaland. In: J.-P. Chrétien ed. *L'invention religieuse en Afrique: historie et religion en Afrique noire*. Paris: Karthala, pp.267-96.

Maclean, R., 1998. Gendered Technologies and Activities in the Interlacustrine Early Iron Age. In: S. Kent ed. *Gender in African Prehistory*. Walnut Creek, CA: AltaMira Press, pp.163-77.

Madaucl, I., Isa, Y. and Daura, B., 1968. *Hausa Customs*. Zaria: Northern Nigerian Publishing Company.

Magnavita, C., 2002. Recent finds of domesticated *Sorghum bicolour* in the Lake Chad region. *Nyame Akuma*, 57, pp.14-20.

Magnavita, C. and Magnavita, S., 2001. New evidence of proto-urban settlements in the Lake Chad area. *Nyame Akuma*, 55, pp.46-50.

Meek, C.K., 1925. *The Northern Tribes of Nigeria*. London: Oxford University Press.

Mortimore, M., 1989. *Adapting to drought: Farmers, famines & desertification in West Africa*. Cambridge: Cambridge University Press.

Mortimore, M. and Wilson, J., 1965. *Land and people in the Kano close-settled zone: a survey of some aspects of rural economyin the Ungogo District, Kano Province*. Occasional Paper 1. Zaria: Department of Geography, Ahmadu Bello University.

Okafor, E.E., 1993. New Evidence on Early Iron-smelting from Southeastern Nigeria. In: T. Shaw, P. Sinclair, B. Andah and A. Okpoko eds. *The Archaeology of Africa: Foods, Metals and Towns*. London: Routledge, pp.432-48.

Omoweh, D.A., 2005. *Political Economy of Steel Development in Nigeria: Lessons from South Korea*. Trenton, NJ: AfricaWorld Press.

Patton, A. Jr., 1975. *The Ningi Chiefdom and the African Frontier: Mountaineers and Resistance to the Sokoto Caliphate, ca. 1800-1908*. PhD Thesis, University of Wisconsin.

Pole, L.M., 1975. Iron-working Apparatus and Techniques: Upper Regions of Ghana. *West African Journal of Archaeology*, 5, pp.11-39.

Phillips, J.E., 1989. A History of the Hausa Language. In: B.M. Barkindo ed. *Kano and Some of Her Neighbours*. Zaria: Ahmadu Bello University Press for the Department of History, Bayero University, Kano.

Quéchon, G., 2000. Les datations de la métallurgie du fer à Termit (Niger): leur fiabilité, leur signification. In: H. Bacoum ed. *Aux origins de la métallurgie du fer en Afrique: une ancienneté méconnue*. Paris: UNESCO Publishing.

Rehren, Th., Charlton, M., Chirikure, Sh., Humphris, J., Ige, A. and Veldhuijzen, H.A., 2007. Decisions set in slag; the human factor in African iron-smelting. In: S. LaNiece, D. Hook and P. Craddock eds. *Metals and Mines – Studies in Archaeometallurgy*. London: Archetype, pp.211-8.

Rustad, J.A., 1980. The Emergence of Iron Technology in West Africa, with Special Emphasis on the Nok Culture of Nigeria. In: B.K. Swartz and R.E. Dumett eds. *West African Culture Dynamics: Archaeological and Historical Perspectives*. The Hague: Mouton, pp.227-46.

Saltman, C., 1986. *The Blooms of Banjeli: Technology and Gender in West Africa*. Documentary Education Resources (Film and Study Guide).

Schmidt, P.R. and Mapuda, B.B., 1997. Ideology and the Archaeological Record in Africa: Interpreting Symbolism in Iron Smelting Technology. *Journal of Anthropological Archaeology*, 16, pp.75-102.

Shaw, T., 1978. *The Archaeology of Nigeria*. London: Thames & Hudson.

Sutton, J.E.G., 1976. Iron working around Zaria. *Zaria Archaeological Papers* 8.

Tylecote, R., 1975. The origins of iron-smelting in Africa. *West African Journal of Archaeology*, 5, pp.1-10.

Author's address

- African Legacy

Casting iron in ancient China

Han Rubin and Chen Jianli

ABSTRACT: Since at least the 5th century BC, cast iron was used widely in China in agriculture, warfare and craft production. Technological inovations in iron production led to societal developments such as increased food production, which in turn stimulated population growth. This paper presents the metallographic analyses performed on iron and steel artefacts from this period. The aim of this analysis was to understand the broader context and impact of iron casting technology, which ultimately helped to shape the development path of Chinese civilisation.

Introduction

The casting of iron in ancient China was a significant invention that occurred during the 5th century BC or earlier (Han 2000). Due to the absence of historical records regarding iron technology in this early period, scholarship concerning early iron production must rely on the archaeological record. Iron artefacts have been metallographically studied and the results of these investigations, including SEM-EDX analyses, are presented below.

According to current research, the earliest worked iron was meteoric iron, which was made into blades during the middle of the Shang Dynasty period, around the 14th century BC. The use of meteoric iron lasted more than 500 years, until the 9th to 8th century BC (Han, Jiang et al., 1999). The cooling and transformation of meteorites in space can take 4×10^9 years. The slow cooling rate (about 1-10°C per million years) creates a special Widmanstätten microstructure, in which nickel and cobalt are interspersed (Li, 1975). The iron-nickel alloy is easy to forge and quite tough, making it perfect for use as a sharp blade. Even within badly corroded iron artefacts, the meteoric microstructures can still be determined by metallographic analysis and electron microscopy. So far, seven meteoritic iron artefacts have been identified in China. The results indicate that ancient craftspeople were aware of the differences between iron and bronze, and familiar with the heat treatment of iron to enhance its performance (Han, Jiang et al. 1999). All seven meteoric iron objects were buried in high-status tombs, indicating that they had symbolic value for their owners.

Six iron weapons and tools excavated from the Guo State tombs dated to the 9th century BC in Sanmenxia, Henan Province, demand particular attention. Three of the objects are made from smelted iron, while the other three are made from meteoric iron, suggesting that the ancient craftspeople used both sources of iron simultaneously (Han, Jiang et al., 1999). Similar situations have been recorded in other parts of the world, such as in the ancient civilisations of Mesopotamia, Egypt, and Anatolia (Waldbaum 1980). These results have also pushed back the date of the earliest use of man-made iron in China to the 9th century BC or even earlier. (The earliest man-made bloomery iron objects dating to about the 14th century BC were unearthed from the Mogou site in Gansu Province in 2009; the results are forthcoming). The earliest smelted iron was made by the bloomery process, although archaeometallurgical remains of bloomery furnaces dating before the 2nd century BC are yet to be found in China. Shortly after the use of bloomery iron was established, cast iron was invented. The relationship between the use of meteoric iron and the invention of iron smelting technology is still not very clear.

Iron production before the 5th century BC

Iron artefacts dating to this early period have been excavated from various tombs. Prior to 1998, approximately 150 objects had been excavated, and 40 of these were studied using metallographic methods. From the analysis of these iron artefacts, the early iron production techniques can be summarised as follows.

Early iron artefacts were small and thin in size and simple in form and design. Many of the objects have handles of gold, jade, and bronze; some of them are inlaid with turquoise, gold, and jade, suggesting that they served as prestige items. Using such materials to decorate iron was one of the signature features of these early objects. The iron objects shared similar designs with the bronze wares found in the same tombs and dating to the same period. Agricultural tools such as shovels, hoes, sickles, and adzes also started to be produced in iron at this time.

The study of the 40 iron artefacts showed that 15 of them were produced from bloomery iron or bloomery steel and 25 were produced from cast iron or decarburised iron (Han 1998), as shown in Table 1.

Most of the early iron objects were badly corroded. Through careful identification and observation, however, smaller pieces of metallic iron with a cementite structure and minute inclusions could be found in the microstructure. In the

Table 1: Results obtained from 40 iron artefacts by metallographic examination.

Location	Object	Material
Changsha, Hunan	Sword 1	Bloomery steel, 0.5% C
Liuhe, Jiansu	Rod 1	Bloomery iron
Wuxian, Jiangsu	Shovel 1	Bloomery steel, 0.2% C
Linyi, Shandong	Knife 1	Bloomery iron
Changqing, Shandong	Dagger-axe 1	Bloomery iron
Sanmenxia, Henan	Sword 1, dagger-axe 1, spear 1	Bloomery steel or bloomery iron, weapons with bronze handles
Hancheng, Shaanxi	Dagger-axe 1, knife 1	Bloomery steel
Lingtai, Gansu	Sword with bronze handle 1	Bloomery steel
Guyan, Ningxia	Sword 2	Bloomery steel
Xiji, Ningxia	Sword 1	Bloomery steel
Pengyang, Ningxia	Sword 1	Bloomery steel
Tianma-Qucun, Shanxi	Iron fragment 2	White cast iron
Changzhi, Shanxi	Shovel l	Decarburised cast iron
Liuhe, Jiangsu	Bola 1	White cast iron
Changsha, Hunan	Ding 1	White cast iron
Jiangling, Hubei	Axe 1	Decarburised cast iron
Laohekou, Hubei	Shovel 1, adze 3, sickle 1, iron fragment 3	White iron 3, Decarburised cast iron 3, Solid state decarburising steel 2
Daye, Hubei	Axe 1	Decarburised cast iron
Dengfeng, Henan	Adze 5, hoe 1	Decarburised cast iron
Luoyang, Henan	Adze 1	Decarburised cast iron
	Shovel 1	Malleable iron
Xinzheng, Henan	Thin plate 1	White cast iron
Lixian, Gansu	Ding 1	White cast iron

bloomery iron very low levels of carbon (< 0.06%) were found, indicating a ferritic iron structure. Within the inclusions of ferrous oxide – an iron-olivine eutectic mixture – there were several areas of phosphorus, sulphur, manganese, silicon and other volatile elements from the original ore. During the forging of bloomery iron, the metal came into contact with charcoal. The carbon diffused into the iron, thereby changing it into steel, with increased strength and durability. Bloomery steel was used for sharp weapons and tools because of its higher hardness than bronze tools. The development of bloomery steel techniques played a critical role in the history of iron and steel technology.

From the data obtained so far, the earliest cast iron pieces were found in Tianma – Qucun of Shanxi Province, dating to around the 9th to 8th century BC. One sample's microstructure is shown in Fig. 1 (Han 2000). However, the original shape of these cast iron pieces was indistinguishable because they were too small. In addition, they were excavated together with one piece of bloomery iron. Therefore, it is not clear if these two cast pieces were intentionally made or rather produced accidentally during the iron smelting process. Cast iron and bloomery iron artefacts were also excavated from the tomb of Liuhe, Jiangsu Province, dating to the middle of the 8th century BC (Jiangsu 1965; Nanjing 1974). One excavated adze from Laohekou cemetery (Hubei Province) and one thin plate from Xinzheng (Henan Province) were identified as having been made during the Warring States Period, around the 5th century BC, and contained white iron micro-structures as shown in Figs. 2 and 3. Artefacts associated with the iron production

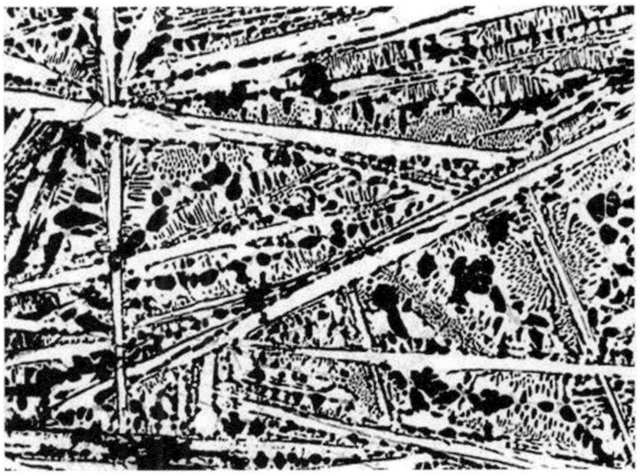

Figure 1: The microstructure of the earliest known cast iron fragment dating to the 9th-8th century BC and found at the Tian-ma Qucun sites in Shanxi Province.

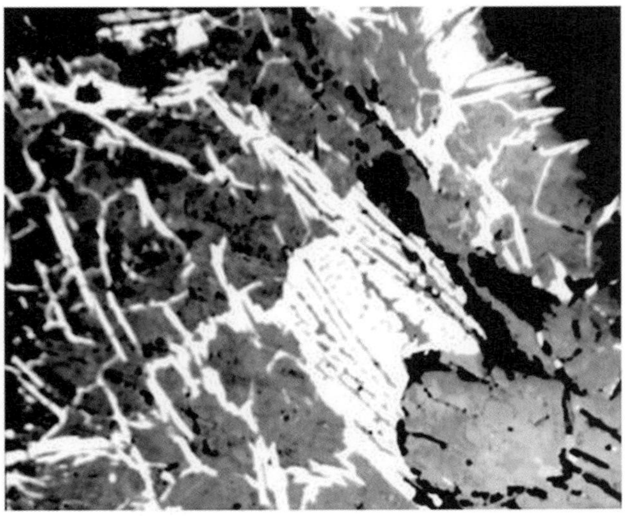

Figure 2: Microstructure of an iron artefact of Eastern Zhou period, excavated from Liuhe City in Jiangsu Province (mid-8th century BC).

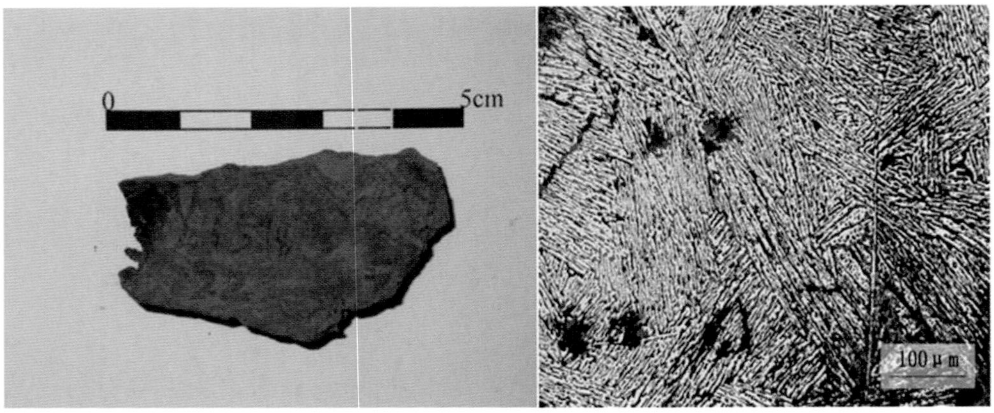

Figure 3: An adze unearthed from Laohekou, Hubei Province, and its microstructure (5th century BC).

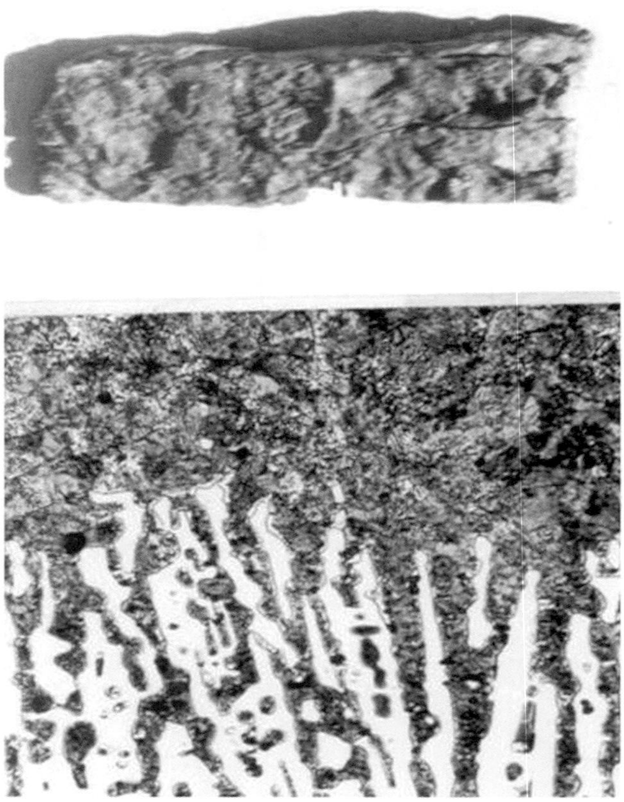

Figure 4: Adze unearthed from Luoyang, Henan Province, and its microstructure (5th century BC).

process were mainly found within the areas of the Jin and Chu states, where advanced iron-making technology had been known from an early time. The same raw materials and fuel were used to produce bloomery iron and cast iron; the only differences between the two technologies were the smelting temperature and the furnace style.

Ancient Chinese craftspeople invented the technique of cast iron and the use of the annealing process to decrease the brittleness of cast iron. Cast iron objects subsequently contributed significantly to the development of society. Decarburisation was used in many tools and weapons. For example, one of the excavated adzes found in Luoyang, Henan Province, was dated to around the 5th century BC, and comprised of white iron in the core of the object with a decarburised layer of steel on the surface, as shown in Fig. 4.

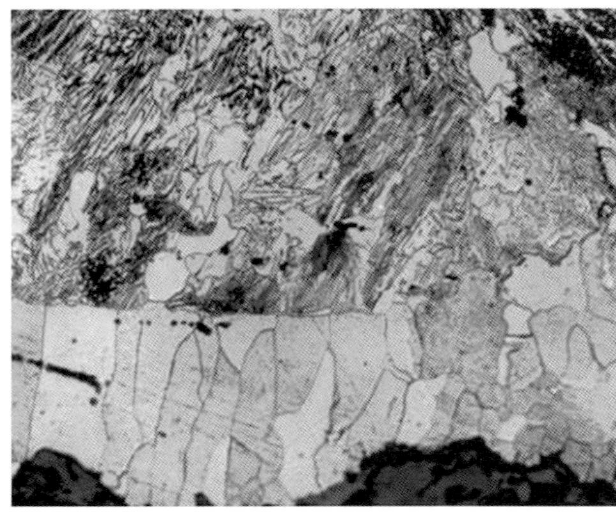

Figure 5: Microstructure of a hoe (T3H32:1) excavated from the Yangcheng smelting site, Henan Province (5th century BC).

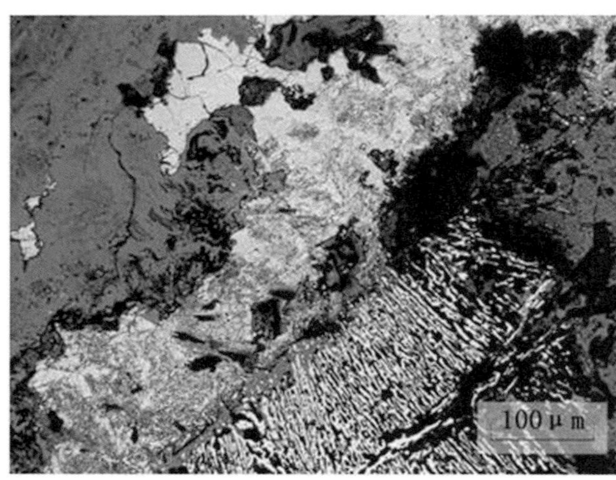

Figure 6: Microstructure of a sickle (43007) unearthed from Laohekou, Hubei Province (5th century BC).

A decarburised steel structure was formed by heat treatment at a temperature around 900°C and a short annealing time. Iron items including a hoe from Yangcheng, Henan Province and a spade from Laohekou of Hubei Province all displayed similarly characteristic indicators of decarburised cast iron, as shown in Figs. 5 and 6.

It is worth mentioning that three major factors were involved in the technological shift from bloomery to cast iron: furnace structure, the use of bellows to produce a forced air blast through the furnace and bronze casting technology. The same raw materials, e.g. ore and fuel, are used to obtain bloomery iron and cast iron; the main differences are structure of the furnace and smelting temperature. Higher temperatures can be reached in a blast furnace to produce cast iron. Furnace structure and blasting facility were well developed in the copper smelting and casting technologies of the Shang and Zhou dynasties as many very complicated bronze vessels were cast with molten bronze rather than shaping and forging solid metal. The Chinese metallurgical tradition favoured casting over deformation-based shaping techniques and it is thus not surprising that the Chinese invented iron casting technology. The blasting facility is the other important factor for obtaining appropriate thermal conditions inside a furnace. Smelting temperature can be calculated based on the compositions of iron objects and slag. The silica content is lower in early iron casting technology than in later periods, due to the limits of blast and the lower smelting temperature.

Iron production during the 5th-3rd century BC

Excavations of ten burial and smelting sites have yielded approximately 2,000 iron pieces, out of which 100 pieces have been systematically analysed. The results demonstrate that as early as the 4th century BC, cast iron and steel had already played an important role in the development of agriculture, craft industries and the production of weapons and tools. The materials used for the iron items include decarburised iron, white iron, grey iron, malleable iron and malleable iron with nodular graphite. During this period, ancient Chinese craftspeople used cast iron to produce farming tools and developed an annealing process to decrease the brittleness of cast iron.

The number of iron objects excavated from 20 provinces and dating to the 3rd century BC is 56 times that of the preceding period, as shown in Fig. 7 (Duan 2001).

The quantity and variety of iron farm tools increased rapidly during this period following the developments in agriculture during the Warring States period, as shown in Table 2.

The quantity and variety of iron farm tools increased rapidly during the Warring States period. Based on re-investigations of ten tombs and three iron smelting sites of the Warring States and Han periods in Shanxi Province, sixty iron artefacts have been examined by metallographic methods. The results are shown in Table 3. This investigation was the first study of the development of iron making process in the Jin state before the 3rd century BC. 90% of the sixty iron artefacts were made of cast iron that had been annealed. High quality weapons made of steel gradually replaced bronze tools.

In 1977, the Cultural Heritage and Archaeology Institute of Henan Province discovered an iron casting site dating to the 4th-2nd century BC in Yangcheng, Dengfeng City. Pieces of furnace wall, kiln remains, a large number of broken pieces of pottery, pieces of models, and a variety of iron items were found. 90% of the excavated iron items were farming tools and weapons including pickaxes, hoes, chisels, adzes, knives and arrowheads, as well as several plates of iron. The total number of iron objects was 1158, together weighing 110 kilograms (Henan and Zhongguo 1992). The metallurgical study of thirty-four iron items selected from this group indicated that thirty-three of them had gone through surface heat treatment (Han 1992). The microstructures indicated that craftsmen were able to choose heat treatment conditions in order to attain various desired levels of hardness on the iron surfaces.

Table 2: The number of iron farm tools dating before the 3rd century BC.

Location	Number of iron objects	Number of iron farm tools
Lianhuabao, Fushun, Liaoning	More than 80	68
Changzhi, Shanxi	36	21
Xinglong, Hebei	87	58
Huixian, Henan	93	58
Changsha Chu tombs, Hunan	More than 70	17
Yinshanling, Pingle, Guangxi	184	91

Table 3: Sampled iron objects excavated from several tombs in Shanxi Province.

Material	Houma (24)	Changzhi (27)	Yuci (7)	Yongji (2)	Total (60)
White cast iron	Adze 3, shovel 2, belt hook 9, container 1, ironware 1, jar 1	Belt hook 11	Adze 2, pickaxe 1	Container 1	32
Decarburised cast iron	Shovel 1, belt hook 1, knife with round hoop 1, ploughshare 1	Shovel 2, belt hook 3		Adze 1	10
Solid state decarburised steel	Sickle 1, belt hook 1	Knife with round hoop 3, sword 2, dagger 1, iron ring 1	Sword 2, knife with round hoop 1		12
Bloomery or steel		Knife 1, nail 1			2
Steel	Belt hook1	Belt hook 1, ironware 1 (quenched)	Belt hook 1		4

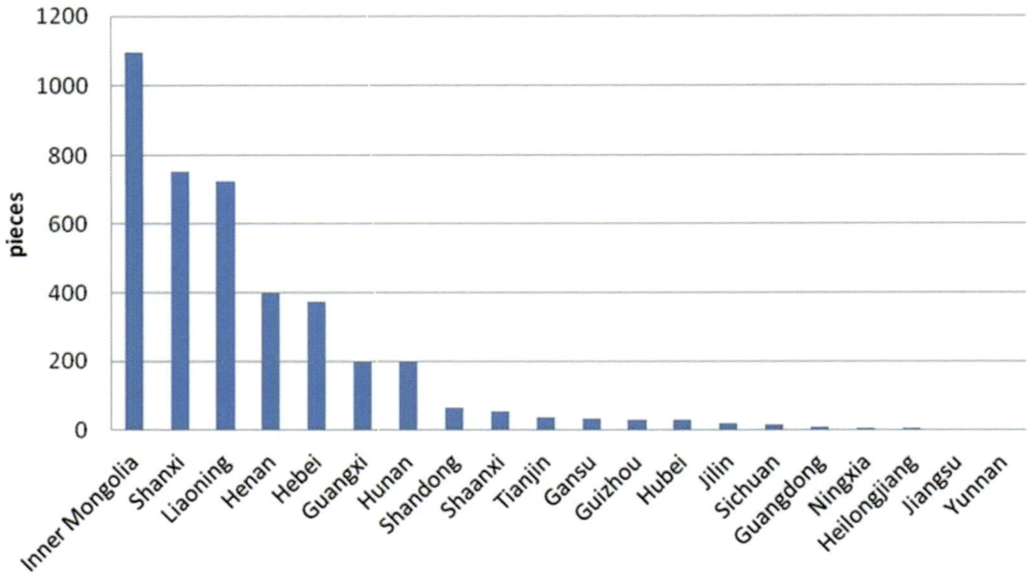

Figure 7: The number of iron objects excavated from 20 provinces.

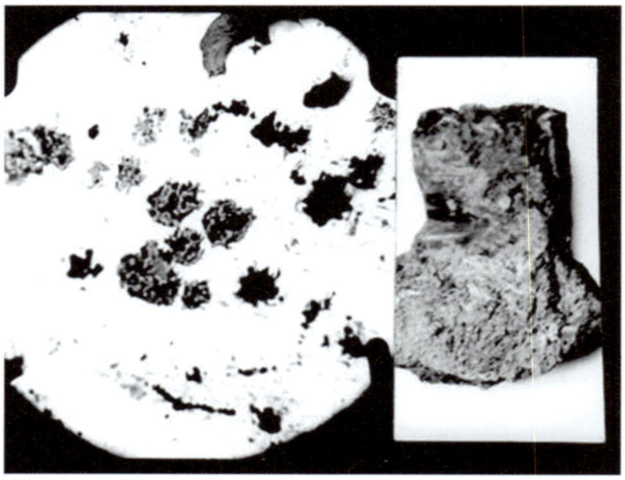

Figure 8: A shovel unearthed from Luoyang, Henan Province, and its microstructure.

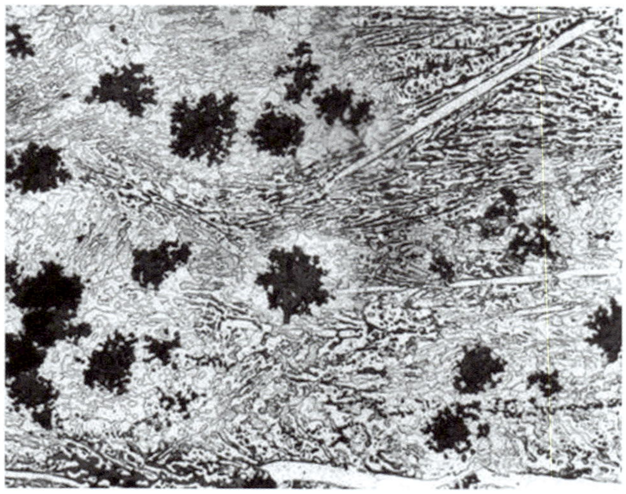

Figure 9: The microstructure of the Han shovel (16) 4109 excavated from the iron casting site at Yangcheng, in Dengfeng, Henan Province.

Although only a hundred objects dating to the period between the 5th and 3rd centuries BC have been metallographically analysed and subsequently published, and the sample size is thus limited, the results can be summarised as follows:

The materials are mainly white iron, decarburised cast iron, malleable iron, and solid state decarburised steel, but bloomery steel was still being used and the production techniques were continually being improved.

Craftspeople used annealing treatment to improve the performance of cast iron. Annealed cast iron with a structure corresponding to malleable iron was also found in a spade excavated from Luoyang, Henan Province (Zhongguo 1978), as shown in Fig. 8.

Several farming tools excavated from other smelting sites and tombs also demonstrated a decarburised layer on the surface, or a malleable structure, indicating the invention of a process used to render the iron malleable. When the annealing temperature of cast iron surpassed 900°C and the iron was subjected to a longer annealing time, the cementite structure in white iron would decompose into the graphite structure shown in Fig. 9. After the graphite gathered into floccus, the structure would enhance the performance of cast iron by improving its ductile strength. Depending on annealing temperature, atmosphere and the various annealing times, the properties of cast iron can differ significantly.

The microstructures of some farming tools showed matrices consisting of ferrite, ferrite and pearlite, pearlite and flocculation, or graphite blocks, as shown in Figs. 10, 11, and 12. Spherical graphite structures were also observed in excavated iron objects. The earliest known iron artefact with a spherical graphite structure is an iron hoe of the late Warring States period which was unearthed from the Yangcheng site in Dengfeng county, Henan Province (Fig. 13). More than ten iron objects excavated in Henan Province during the 1980s had spherical graphite structures and were dated to a period from the Warring States to the Han Dynasty, between the 4th century BC and the 1st century AD (Qiu 1981). Some examples of the finds from

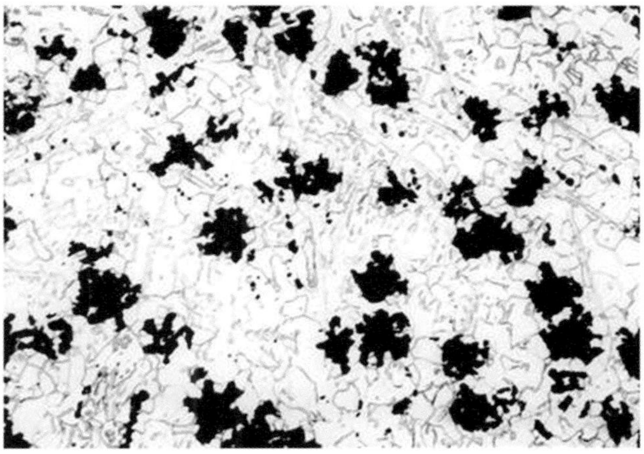

Figure 10: The microstructure of a Han shovel T12:8 (excavated from Tieshenggou, Henan Province, and dating to the Han Dynasty) consists of a ferrite matrix with graphite blocks.

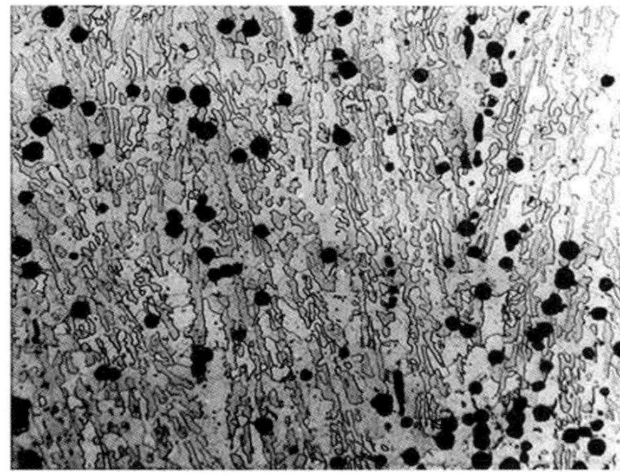

Figure 13: The spherical graphite structure of a hoe excavated from Yangcheng, at Dengfeng, Henan Province.

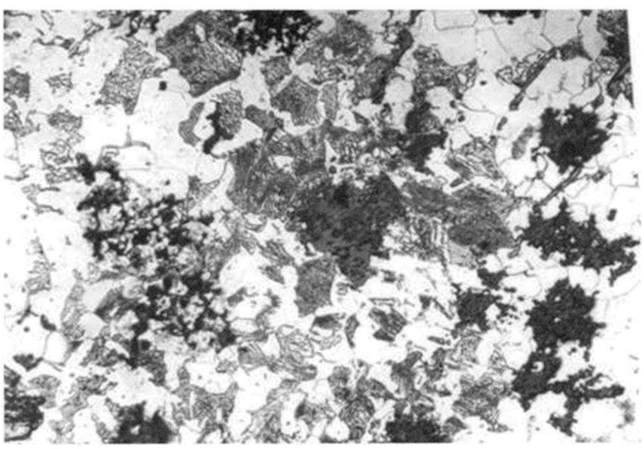

Figure 11: The microstructure of an iron shovel consists of a ferrite and pearlite matrix with graphite blocks.

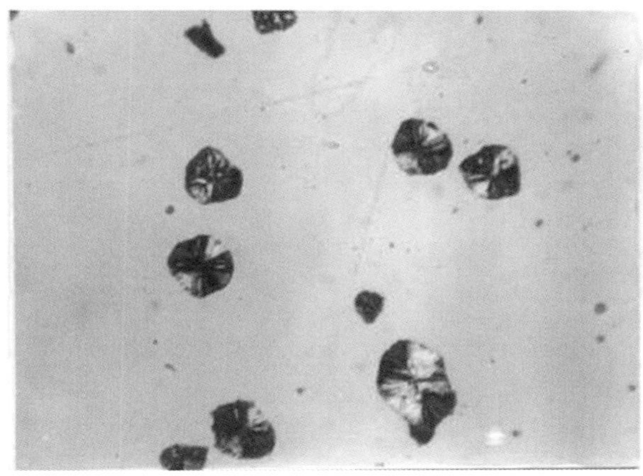

Figure14: An iron plate with spherical graphite in malleable iron unearthed from Xinzheng, Henan province and its polarised photograph.

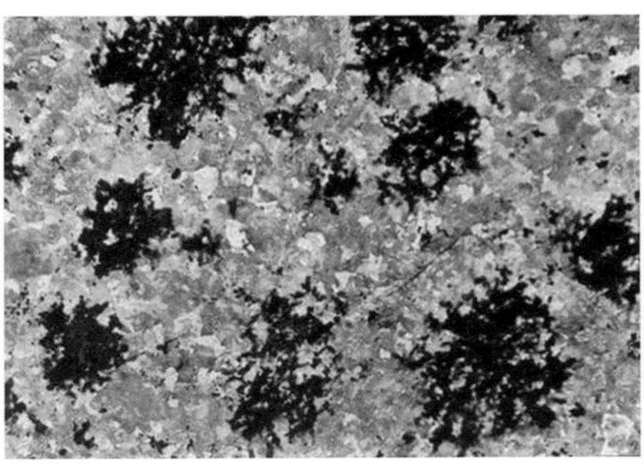

Figure 12: The microstructure of a Han iron object unearthed from a cellar at Mixian, Henan Province, consists of pearlite matrixes with graphite flocculation.

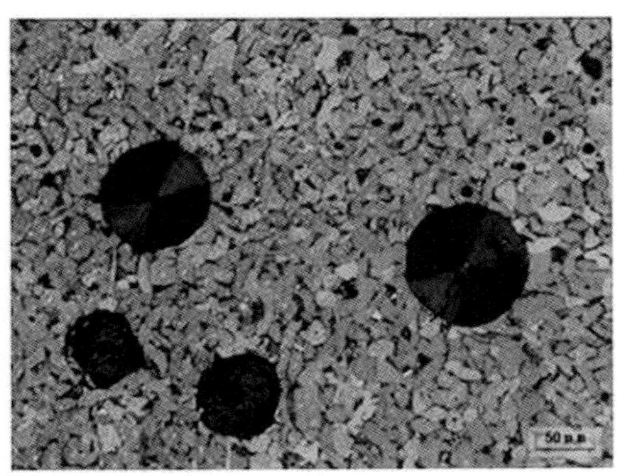

Figure 15: An iron plate with spherical graphite in steel unearthed from Xinzheng, Henan Province.

this group are shown below: an iron plate and strip from the Warring States Period unearthed in Xinzheng (Figs. 14 and 15) and iron tools from the Han Dynasty unearthed in Zhengzhou. By studying the microstructures of cast iron, it was concluded that the spherical graphite in the excavated iron artefacts was formed by annealing white iron. This study also concluded that the spherical graphite could have been formed during either the ancient charcoal smelting process or the modern coke smelting process for white iron, as long as the annealing temperature reached 920°C.

The invention and the earliest use of malleable iron in China took place no later than the 5th century BC. This discovery means that, based on the data currently available, China was the first country in the world to achieve this technique.

By studying the excavated iron tools it is clear that malleable iron was widely used in the late Warring States period and down through the Han, Wei (4th century BC to 4th century AD), and Northern and Southern Dynasties, and to the Tang Dynasty (5th-9th centuries AD). The use of malleable iron therefore has a history of 1300 years in China.

Another important contribution of ancient Chinese craftspeople was the invention of decarburisation of solid iron into steel during the Warring States period (ca. 4th century BC). The ancient craftspeople developed a unique decarburising method to change iron into steel through annealing treatments. Some pieces of broken casting moulds and strips of iron were found at the Yangcheng iron casting site at Dengfeng, Henan Province, which was dated to the 4th century BC; examples are shown in Fig. 16 (Qiu 1981). The microstructure of the excavated iron tools from this site consisted of a homogeneous ferritic structure with fewer inclusions, as shown in Fig. 17. In addition there was one iron object that still displayed surface adhesion traces, indicating a failed annealing process. After studying these items, the process used by the ancient Chinese craftspeople can be summarised as follows: first they made various iron tools using cast iron with 3-4% carbon. Then, after the iron pieces cooled down, they were decarburised via oxidation in an annealing furnace. When the cast iron became steel after decarburisation, it could be forged or heat-treated again, depending on the desired object. Observations of the iron objects show that flash lines are sometimes visible. The carbon concentration on the surface of iron objects was lower and there was very little to no graphite precipitation. It was found that overall the majority of the structure had become steel.

At the ancient cast iron production site excavated at Guxing Zhengzhou, Henan Province, dozens of kilograms of trapezoidal iron plates were found. These plates were about 19cm long, 7-10cm wide and 0.4cm thick. The analysis indicated that the iron plates had undergone decarburisation and had a steel composition of 0.1-0.2% carbon (Zhengzhou 1978). Other similar discoveries were made at Nanyang, Henan Province. As shown in Fig. 18, the iron items had various shapes, for example 74 pieces of flat iron, 44 pieces of cubic iron, and 36 pieces of round iron (Li and Chen 1995). A number of strips made of decarburised solid state steel and a bar-shaped iron object actually made from a rolled up iron sheet were also excavated.

Another important piece of evidence that China was the first country to discover steel making technologies was the finding of a decarburising furnace at the casting site of Gong Xian – T12 furnace 11, shown in Fig. 19 (Zhao, Li et al., 1985).

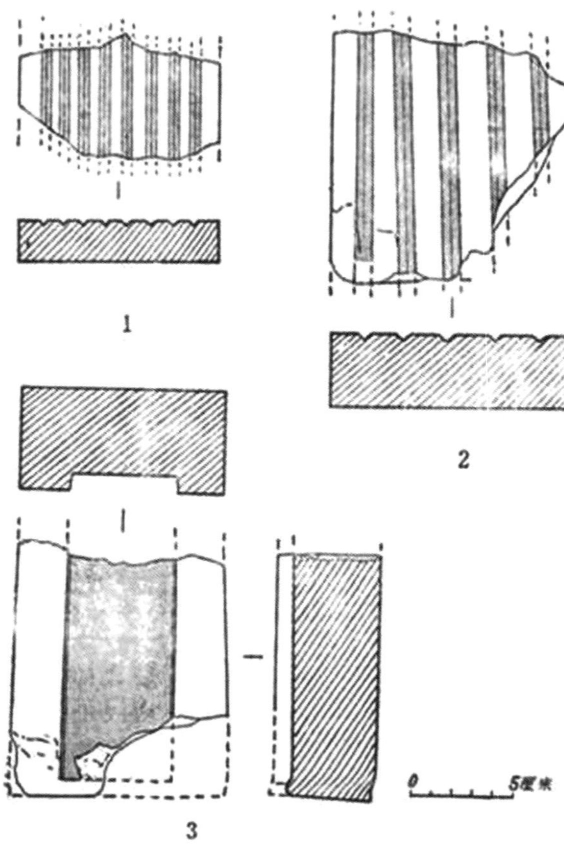

Figure 16: Clay moulds unearthed from the iron-casting site of Yangcheng, Henan Province, dating to the early Warring States period (Henan and China, 1992).

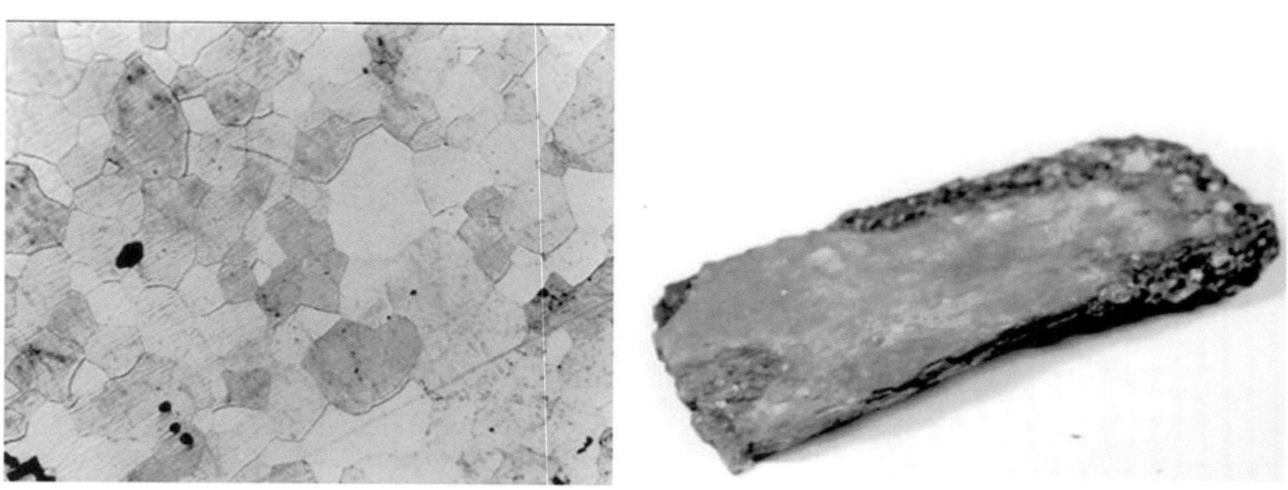

Figure 17: A decarburised iron plate excavated from the Guxing ironworks site in Zhengzhou, Henan Province.

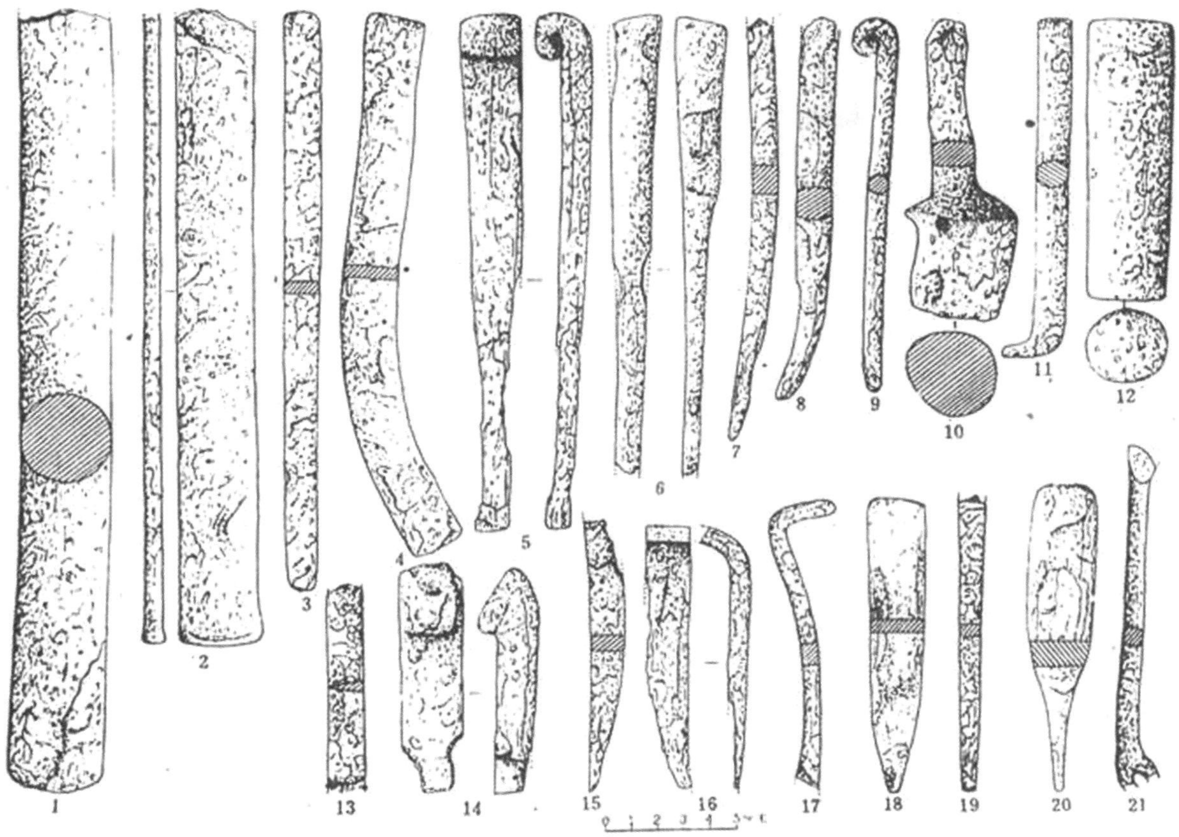

Figure 18: Iron strips unearthed from the Wafangzhuang iron-casting site, Nanyang City, Henan Province (Li and Chen, 1995).

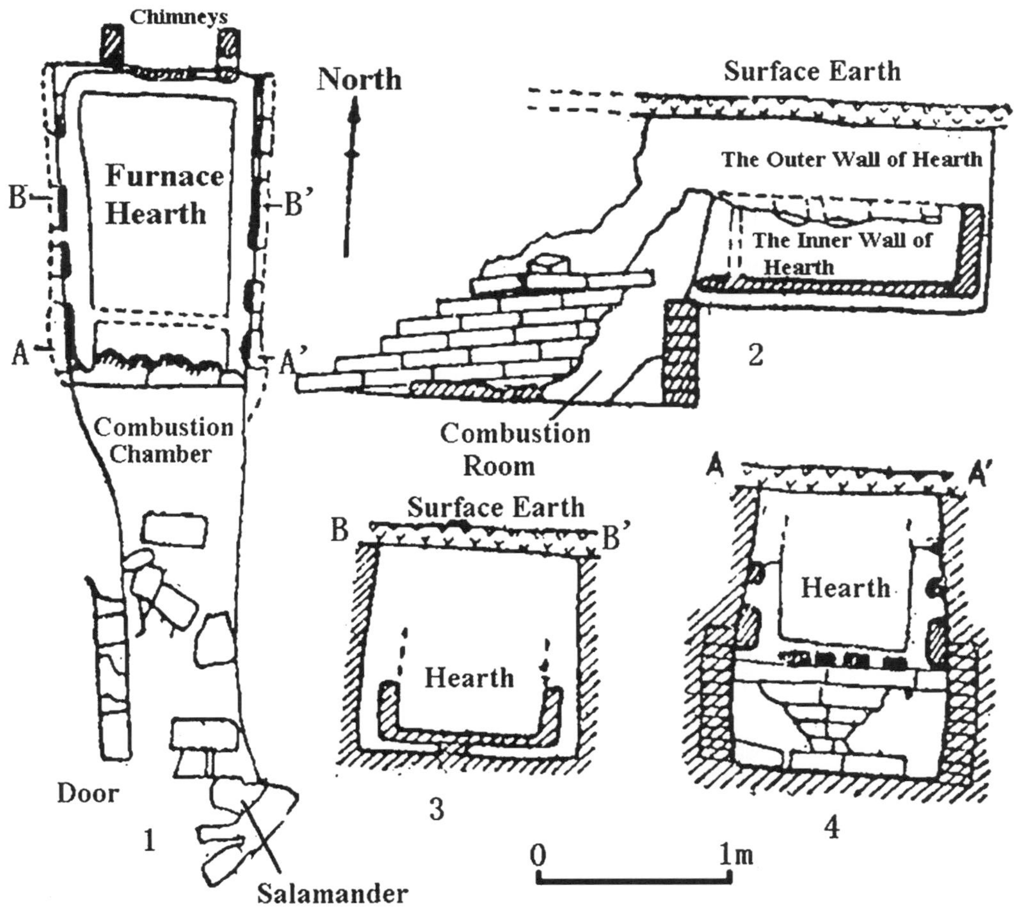

Figure 19: The plan and profile of decarburising furnace T12:11 at the casting site of Tieshenggou, in Gong Xian, Henan Province (Zhao, et al., 1985).

By the late 4th-3rd century BC, permanent moulds for casting iron items had been developed (Zheng 1956), greatly increasing productivity, the precision of casting and the capacity to replicate objects such as ploughshare tips. This can be considered a primitive model of mass production and standardisation. Standard mould for making casting moulds were issued by the Iron Official of the Han Dynasty to various foundries and the permanent moulds subsequently made were then used to cast iron and steel objects. This shows that cast iron and steel production played a very important part in the advancement of agriculture and water irrigation and was a significant technological development promoting the formation of a unified China.

It should be pointed out that although bloomery iron objects have not been found after the 3rd century AD in central China, this technology did not disappear. For example, one bloomery iron smelting site dated to the 2nd century BC – 2nd century AD was found in Pingnan County, Guangxi Zhuang Autonomous Region of Southern China (Huang and Li, 2011) and several bloomery objects dated to the 9th century AD were unearthed in Xinjiang (Chen, 2010), while many cast iron objects were also found in these areas. The social, economic and technical level of development may have been among the factors influencing the choice of bloomery or (decarburised) cast iron under certain circumstances. The production efficiency of cast iron is higher than that of bloomery iron, and this may be the real reason behind such choices.

Comparative studies of iron technologies between central China and outlying areas show that the earliest use of cast iron took place in central China. Cast iron objects found in northwest, northeast, south and southwest China, the Korean Peninsula, Japan and Vietnam indicate that cast iron objects or technologies had been adopted in these areas no later than the 2nd century BC, though the dates of the earliest use of iron and smelting technology in these outlying areas are not the same. It is very likely that the earliest iron objects excavated in these areas came from central China. The use of raw, iron-like ingots to make implements was then adopted by local people, before the eventual development of indigenous iron smelting (Chen and Han, 2007). Iron casting technology contributed to the progress of civilisation in these areas. Recently, more iron casting evidence dating from the Han Dynasty has been found in Xinjiang, Hainan, Yunnan and other frontier areas; we will discuss these materials further in future publications.

Conclusion

The invention of cast iron and the use of iron tools contributed significantly to the development of agriculture in ancient China. Archaeological excavations show that iron tools were widely used during the Warring States period. Evidence from the latter stages of this period indicates that iron agricultural tools occupied a dominant position, which led to an unprecedented veritable revolution in farming methods. The development of agriculture in turn promoted the progress of ancient culture, science, and technology and consequently led to changes in social structure, after which feudalist relations of production were established.

The extensive use of iron promoted the development of agriculture, handicrafts and the exchange of commodities, which enhanced economic vitality and allowed cities to prosper. In turn, the development of cities also generated demand for more iron objects, promoting the prosperity of iron smelters. Linzi, the capital of the Qi state during the Warring States period, was well-known for its iron industry and was described in several historical records as a crowded, busy, and bustling city with 70,000 households.

High-quality iron weapons were an important factor in the development of the Qin state into the first unified empire in Chinese history and the large-scale iron industry can be considered as one of the most important foundations of the political and economic might of the Han Dynasty, the powerful empire following that of Qin. Bloomery iron and carburised steel weapons of superior quality to bronze weapons, gradually replaced bronze weapons during the Warring States period. Iron weapons played an important role in warfare, especially in the wars among the Qin and other states, like Han and the Huns. At the beginning of the Western Han Dynasty, the empire established 49 local iron officials to manage iron casting works. To date, about 30 iron smelting and melting sites of the Han Dynasty have been identified in China, some of them occupying an area of 1,000,000 m^2 or more. Every iron workshop had one or more furnaces. For example, eight blast furnaces and one puddled steel furnace were found within the '*He San*' (河三) site in Gongyi County, Henan Province. Iron working sites and unearthed iron objects dating to the Han Dynasty indicate that iron had become the main raw metal material that was used in agriculture, military, handicrafts, transportation etc.

In short, by the 5th century BC or even earlier, cast iron had been invented and was widely used in agriculture, warfare, production, etc., representing a discovery significant to the development in Chinese society for the next 2500 years. Cast iron and steel-making technologies made significant contributions to the progress of civilisation in China and the world.

Acknowledgments

The research was supported by NSFC (No.51074010) and the State Administration of Cultural Heritage of China. We wish to thank Prof. Thilo Rehen, Prof. Mei Jianjun and Dr. Duan Hongmeifor giving us support and help; we especially thank Dr. Wang Dongning for the English translation of this paper.

References

Bian, X. 边. (1972). "Shanxi Changzhi Fenshuiling 126Hao Mu Fajue Jianbao山西长治分水岭126号墓发掘简报 (Excavation of Warring States tomb no.126 at Fenshuiling in Changzhi City, Shanxi). ." Wenwu文物 (*Cultural Relics*)(4): 38-46.

Chen, J. 陈. (2010). "Zhongguo Yetie Jishu Qiyuan yu Fazhan de Xin Tansuo 中国冶铁技术起源与发展的新探索 (New findings and research on the beginning and development of iron technology)." Wenwu Keji Yanjiu文物科技研究(*Scientific and technological research on cultural heritage*) **7**: 118-131.

Chen, J. 陈. and Han, R. 韩. (2007). Hanjin Zhongyuan Ji Beifang Diqu Gangtie Jishu Yanjiu 汉晋中原及北方地区钢铁技术研究 (Iron and Steel technology in Central Plain and Northern Region during the Han and Jin Dynasties). Beijing 北京, Beijing Daxue Chubanshe 北京大学出版社.

Duan, H. 段. (2001). Shanxi Houma Diqu Chutu Zhanguo Tieqi de Jinxiangxue Yanjiu 山西侯马地区出土战国铁器的金相学研究 (*Metallographic Study of the Iron Artefacts of the Warring States Period Unearthed from the Houma Area, Shanxi*). Yejin yu Cailiao shi Yanjiusuo冶金与材料史研究所（Insititute of Historical Metallurgy and Materials）. Beijing 北京, University of Science and Technology Beijing 北京科技大学. PhD

Guangxi, Z. Z. W. G. 广. (1978). "Pingle Yinshanling Zhanguo Mu平乐银山岭战国墓 (The Warring States tombs at Yinshanling county in Pingle city)." Kaogu Xuebao 考古学报 (*Acta Archaeologica Sinica*)(2): 211-250.

Han, R. 韩. (1992). Yangcheng Zhutie Yizhi Tieqi de Jinxiang Jianding阳城铸铁遗址铁器的金相鉴定 (The Metallographic Examination of Iron Artefacts Unearthed from the Iron Casting Site at Yangcheng). Dengfeng Wangchenggang yu Yangcheng 登封王城崗與陽城 [Wangchenggang site at Dengfeng and the ancient cities of Yangcheng]. Z. L. B. K. 中. Henan & Zhongguo (Henan Sheng wenwu Yanjiusuo 河南省文物研究所. Beijing 北京, Wenwu chubanshe 文物出版社(Cultural Relics Press): 329-336.

Han, R. 韩. (1998). "Zhongguo Zhaoqi Tieqi (Gongyuan 5 shiji yiqian) de Jinxiangxue yanjiu 中国早期铁器(公元前5世纪以前)的金相学研究 (A Metallographic study on Early Iron Objects of China)." Wenwu 文物 1998(2): 87-96.

Han, R. 韩. (2000). Tianma-Qucun Yizhi Chutu Tieqi de Jianding 天马-曲村遗址出土铁器的鉴定 (The Examination of Iron Artefacts Unearthed from Tianma Qucun, Shaxi). *Tianma-qucun, 1980-1989* 天马--曲村1980-1989. Z. H. 邹衡. Beijing 北京, Kexue Chubanshi 科学出版社 (Science Press): 1178-1180.

Han, R. 韩., Jiang,T. 姜. et al. (1999). Guoguo Mudi Chutu Tieren Tongqi de Jianding yu Yanjiu虢国墓出土铁刃铜器的鉴定与研究 (The Study and Examination of Copper Artefacts with Iron Edge Excavated from Graves of Guo State). Sanmenxiao Guoguo Mudi 三门峡虢国墓地 (*Graves of the Guo State in Sanmenxia*). H. S. W. K. Y. 河. H. I. o. C. R. a. Archaeology). Beijing 北京, Wenwu Chubanshi 文物出版社 (Beijing: Cultural Relics Press): 559-573.

Henan, S. w. Y. 河. and Zhongguo, L. B. K. 中. (1992). Dengfeng Wangchenggang yu Yangcheng 登封王城崗與陽城 [Wangchenggang site at Dengfeng and the ancient cities of Yangcheng]. Beijing 北京, Wenwu chubanshe 文物出版社.

Huang, Q. 黄. and Li, Y. 李. (2011). "Guangxi Pingnan Liuchen Pozui Yizhi Yelian Jishu Yanjiu广西平南六陈坡嘴遗址冶炼技术研究 (Study on the Pozui Iron Smelting Site in Pingnan County of Guangxi)." Youse Jinshu有色金属 (*Non-ferrous Metals*) 63(1): 140-146.

Huixian, 辉. (1956). Huixian Fajue Baogao辉县发掘报告 (*Excavations in Huixian County, Henan*). 北京, Kexue Chubanshe 科学出版社 (Beijing: Science Press).

Hunan, P. C. R. W. L. (1956). "Changshan, Hengyang Chutu Zhanguo Tieqi长沙、衡阳出土战国时代的铁器 (Excavation of iron artefacts of the Warring States Period at Changsha City and Hengyang City, Hunan)." Kaogu Tongxue 考古通讯 (*Archaeological Bulletin*)(1): 77-79.

Jiangsu, P. C. R. O., Nanjing Museum (江苏省文物管理委员会、南京博物院) (1965). " Jiangsu Liuhe Chengqiao Dongzhou Mu江苏六合程桥东周墓 (Excavated from the Eastern Zhou Dynasty Tomb at Liuhe, Jiangsu)." Kaogu 考古 (*Archaeology*)(3): 113.

Li, C. 李. (1975). "Zhongguo Fengjian Shehui Qianqi Gangtie Yelian Jishu Fazhan de Tantao 中国封建社会前期钢铁冶炼技术发展的探讨 (The Development of Iron and Steel Technology in Ancient China)." Kaogu Xuebao 考古学报（*Acta Archaeologica Sinica*） (2): 1-21.

Li, J. 李. and Chen, C. 陈. (1995). Nanyang Handai Yetie 南阳汉代冶铁 (*The smelting iron of Han dynasty at Nanyang City, Henan Province*). : . Zhengzhou 郑州, Zhongzhou Guji Chubanshe 中州古籍出版社 (Zhongzhou Ancient Books Press).

Nanjing, M. 南. (1974). "Jiangsu Liuhe Chengqiao Erhao Dongzhou Mu江苏六合程桥二号东周墓 (Excavated from the No.2 Tomb of the Eastern Zhou Dynasty at Liuhe, Jiangsu)
" Kaogu 考古 (*Archaeology*)(2): 116-120.

Qiu, L. 丘. (1981). Gudai Zhanxing Zhutie zhong de Shimo 古代展性铸铁中的石墨 (Spherules Graphite in Ancient Malleable Iron) *The First International Conference on the Beginning of the Use of Metals and Alloys* Beijing, Beijing Gangtie Xueyuan 北京钢铁学院.

Shanxi, P. C. R. O. 山. (1957). "Shanxi Changzhi Fenshuiling Gumu de Qingli 山西长治市分水岭古墓的清理 (Cleaning up of Warring States tombs at Fenshuiling in Changzhi City, Shanxi). ." Kaogu Xuebao 考古学报 (*Acta Archaeologica Sinica*)(1): 103-118.

Shanxi, P. C. R. O. 山. (1964). "Shanxi Changzhi Fenshuilling Zhanguo Mu Dierci Fajue山西长治分水岭战国墓第二次发掘 (The second excavation of Warring States tombs at Fenshuiling in Changzhi City, Shanxi Province)*Sinica*)." Kaogu Xuebao考古学报 (*Acta Archaeologica Sinica)*(2): 132-133,137.

Waldbaum, J. C. (1980). The First Archaeological Appearance of Iron and the Transition to the Iron Age. *The coming of the age of iron*. T. A. Wertime and J. D. Muhly. New Haven, Yale Universtiy Press**:** 69-98.

Wang, Z. 王. (1964). "Liaoning Fushun Lianhuabao Yizhi Fajue Jianbao 辽宁抚顺市莲花堡遗址发掘简报 (Excavation of sites at Lianhuaao in Fushun City, Liaoning)." Kaogu 考古 (*Archaeology*)(6): 286-293.

Zhao, Q. 赵., J. 李. Li, et al. (1985). "Gongxian Tieshenggou Handai Yetie Yizhi 巩县铁生沟汉代冶铁遗址再探讨 (A re-investigation of the remains of the iron and stell works of the Han Dynasty at Tieshenggou, Henan)." Kao Xuebao 考古学报（*Acta Archaeologica Sinica*）(2): 157-183.

Zheng, S. 郑. (1956). "Rehe Xinglong Faxian de Zhanguo Shengchan Gongjv Zhufan 热河兴隆发现的战国生产工具铸范 (Moulds and casting production implements discovered at a Warring States period site at Gudonggou in Xinglong County, Hebei)." Kaogu Tongxun 考古通讯 (*Archaeological Bulletin*)(1): 29-35.

Zhengzhou, C. M. 郑. (1978). "Zhengzhou Guxingzhen Handai Yetie Yizhi Fajue Jianbao 郑州古荥镇汉代冶铁遗址发掘简报 (Excavation of a Han period iron smelting site at Guxing town in Zhengzhou City, Henan)." Wenwu 文物 (*Cultural Relics*)(2): 28-43.

Zhongguo, Y. B. 中. (1978). "Cong Guxing Yizhi Kan Handai Sheng Yellian Jishu 从古荥遗址看汉代生铁冶炼技术 (Han period cast iron metallurgy in light of the ironworks site at Guxingzhen in Zhengzhou, Henan)." Wenwu 文物 (*Cultural Relics*)(2): 44-47.

Authors' addresses

- Han Rubin
  Institute of Historical Metallurgy and Materials
  University of Science and Technology Beijing
  Xueyuan Road No. 30, Beijing, 100083
- Chen Jianli
  School of Archaeology and Museology
  Peking University
  Yiheyuan Road, Beijing, 100871

Questioning research on early iron in the Mediterranean

Martina Renzi[1], Salvador Rovira[2], M. Carme Rovira Hortalà[3], and Ignacio Montero Ruiz[1]

ABSTRACT: Iron technology reached the Western and Central Mediterranean at the beginning of the first millennium BC, when intense commercial contacts between Phoenicians and local indigenous communities were established. It is assumed that during this period iron technology was fully developed in the Phoenician motherland (i.e. the use of bowl and shaft furnaces and slag tapping), but this advanced metallurgy does not seem to have affected the colonised areas. To date, archaeological research has not identified tap slags in the Iberian Peninsula which predate the Roman presence.

This article will confront related common assumptions about early iron production in the Western Mediterranean by asking if we really have archaeological evidence for the use of bloomery furnaces during this period, and, if so, why this developed technology was not used from the beginning of siderurgical production. In light of these issues certain assumptions about early iron production in that region should be critically reviewed.

Keywords: Iron; siderurgy; slags; Phoenicians; Iberian Peninsula; Iron Age; SEM-EDX.

Introduction

It is generally assumed that iron technology reached the Western and Central Mediterranean coasts at the beginning of the first millennium BC, when intense commercial contacts between Phoenician colonists and local indigenous communities were established (Tylecote, 1979, p.40; Pleiner, 2000, pp.29-31). It is also assumed that during this period iron technology was fully developed in the Phoenician motherland. It is reasonable to assume, then, that metallurgical knowledge reaching the Western world would have been at a similar technological level as that of the Eastern Mediterranean, in the use of advanced reduction furnaces and the production of iron debris such as tap slags.

In spite of this, current archaeological research shows a rather different situation both in the Near East and in the Western world. Few ancient Near Eastern settlements have provided evidence of iron production and this evidence, when it occurs, is very scarce and any interpretation of it confusing. Due to the lack of an exhaustive analytical study of these archaeometallurgical materials, their identification as sub-products of siderurgical processes (i.e. smelting or smithing slags) is also sometimes questioned or rejected. Late Bronze Age iron production debris from Tel Yin'am (Israel) (Liebowitz and Folk, 1984) or Kamid el-Loz (Lebanon) (Frisch, et al., 1985), for example, which dates from the 13th century BC and is identified as reduction slags, is traditionally considered by the authors to be the oldest examples of iron metallurgy in the entire region – although this interpretation is not accepted by many specialists in this field. Veldhuijzen and Rehren (2007, pp.189-191), after reviewing the archaeological and archaeometric evidence, have argued convincingly against these conclusions. According to them, Kamid el-Loz and other sites "are most likely related to copper metallurgy and the Israeli site to pigment production" (Veldhuijzen and Rehren 2007, p.191).

Evidence of iron pyrotechnology generally accepted as reduction debris has been collected in Greece and Cyprus but its chronology cannot be placed before the mid-first millennium BC (Pleiner, 2000). Thus, the most ancient Near Eastern debris related to iron processing is represented by the findings from Tell Hammeh (Jordan); an in-depth study and exhaustive analytical investigation of this material has been carried out by Veldhuijzen and Rehren (2006). This archaeometallurgical debris can be dated to a phase that spans the period from approximately 900 BC to 750 BC and includes almost 750kg of slaggy materials of different types and morphology.

These findings are generally not regarded as evidence of a new technology but seem to be indicative of a well-established production process that includes both primary mineral reduction and bloom-purifying procedures. More than half of the slags found at the site are tap slags while tuyères, partially reduced nodules of iron ores and some possible furnace linings were also found. Evidence of pyrometallurgical structures have been identified as well and some of those furnaces are of an impressive size: reaching 2-2.5m in their external diameter and 0.60-0.80m in their preserved upper part, having probably been bigger originally. Slags, ashes, mud bricks with evident fire marks and hematite nodules have been found in association with these furnaces. As suggested by the presence of tap slags, these structures seem to be indicative of an advanced type of bloomery furnace (Veldhuijzen and Rehren, 2006).

Evidence of ancient iron production from other settlements in Israel, Jordan and Syria – dated to the 8th century BC – has been mainly classified as smithing debris. This would also be the case for the ferric materials from Tel Beth-Shemesh, which have been clearly attributed to secondary smithing activities (Veldhuijzen and Rehren, 2007).

Early iron objects in the Central and Western Mediterranean

In the Western world a number of iron objects known to be from archaeological contexts of the Iberian Peninsula can be confidently dated to the first quarter of the first millennium BC (1000-750 cal BC). Details of these items, mainly from Portugal, have been published in the comprehensive study of Vilaça (2006) and indicate an early introduction and circulation of iron in the Late Bronze Age societies of the Western Iberian Peninsula.

Twenty eight iron objects have been collected from seven Portuguese sites, which cover an extended area that includes the Beiras, Estremadura and Alentejo: Monte do Frade (Penamacor); Moreirinha (Idanha-a-Nova); Monte do Trigo (Idanha-a-Nova); Castro de Nossa Senhora da Guia (Baiões, S. Pedro do Sul, Viseu); Outeiro dos Castelos de Beijós (Carregal do Sal); Quinta do Marcelo (Almada) and Rocha do Vigio 2 (Requengos de Monsaraz). Other iron items have been collected at Cachouça (Idanha-a-Nova) but these objects should be ascribed to a slightly later chronology, to the beginnings of the Iron Age.

All of these are sites where Mediterranean imports have been documented in the archaeological record and all, aside from the site of Quinta do Marcelo which lies on the estuary of the Tajo River, are located in the hinterland, far from the coast. Most of these settlements have been dated by ^{14}C and, according to the results obtained, can be safely dated to a chronological period that precedes the 9th century cal BC. These items are not associated with any kind of iron production residues but, especially in the case of the Beira Interior, they appear in contexts where copper-based metallurgical debris has been detected. They are also associated with the presence of prestige goods such as amber beads or faience items, imported ware and bronze objects of Mediterranean and Atlantic origin, etc. In some cases, bi-metallic objects have also been found, for example Cu-Fe or Au-Fe items (Vilaça, 2006).

Metallographic examination of a number of these Portuguese iron objects revealed that they were poorly made. As shown in Fig. 1a, abundant slag inclusions are detectable within a sample of a knife from Moreirinha. Its microstructure presents polygonal ferrite grains, leading to the conclusion that this was a soft iron knife (due to the reduced thickness of the sample, it was not possible to carry out a Vickers microhardness test on the metal).

Fig. 1b shows another example of an iron knife from Moreirinha: in this case, the sample contains a slightly higher quality metal and is constituted by two steel sheets welded by piling. The upper steel sheet has a carbon content of approximately 0.4% C while the lower is soft steel with a carbon content of about 0.2% C. The welding line is clearly visible. Vickers microhardness testing revealed a value of 250 HV for the steeled sheet and approximately 100 HV for the softer one.

The last image (Fig. 1c) shows a sample taken from a knife blade found at Cachouça: made of soft iron, it displays abundant slag inclusions. The grains are polygonal in section and show a slight orientation which is an effect of the forging process. The microhardness test confirmed the low quality of the iron, with a mean value of 93.4 HV.

Overall, the hardness level of these items is relatively low and, in fact, inferior to contemporary bronze materials.

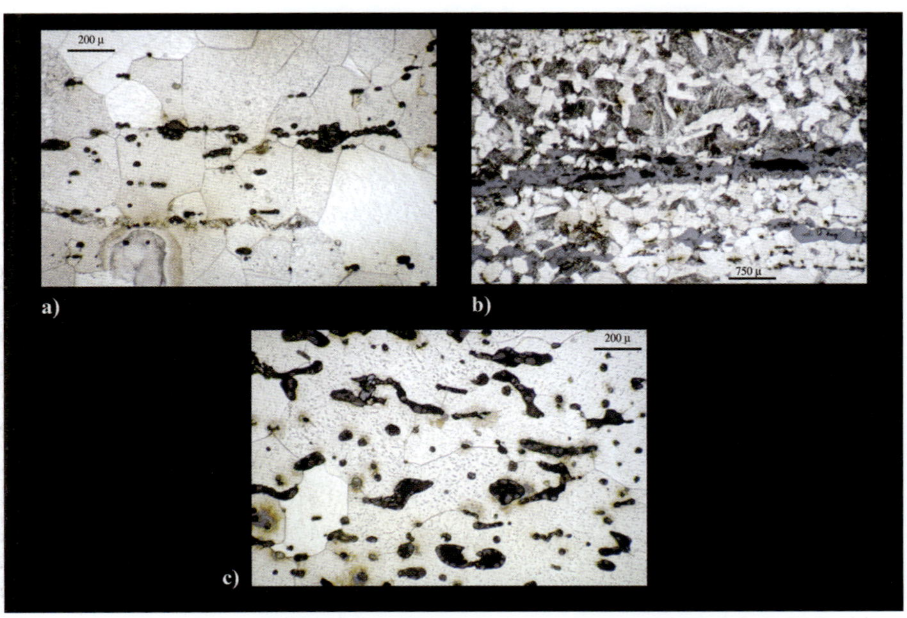

Figure 1: Metallographic examinations of some Portuguese iron objects. a) knife 1, cutting edge, Moreirinha (Idanha-a-Nova); b) knife 2, blade, Moreirinha (Idanha-a-Nova); c) knife, cutting edge, Cachouça (Idanha-a-Nova). Optical microscope images; etched with Nital.

Furthermore, some of these artefacts appear to replicate local bronze prototypes, suggesting that the iron objects could have been made locally. Raw iron bars or ingots would have been traded and artefacts subsequently manufactured at the indigenous sites (Vilaça, 2006).

Iron objects have also been collected in Late Bronze Age contexts in Spain, at El Berrueco (Salamanca), Huelva, Campotéjar (Granada), Peña Negra (Alicante), El Castellar de Librilla (Murcia), El Castillico de Moratalla (Murcia) and Villena (Alicante) (Almagro Gorbea, 1993). Some necropolises – Palomar de Pintado (Toledo) and Arroyo Culebro (Madrid) – have also yielded iron items, providing more evidence for the presence of iron artefacts in the west-central part of the Iberian Peninsula (Pereira Sieso, et al., 2003). As in Portugal, the Spanish sites are situated mainly inland. Furthermore, these findings are not related to any kind of iron production residues but appear in association with imported elements, especially of Mediterranean origin. Due to the lack of ^{14}C analyses, however, the chronology of these Spanish sites cannot be established as precisely as in the previous Portuguese cases described above. A similar situation also occurs in Italy, where some iron artefacts have been found within Late Bronze Age deposits, for example at Gabbro (Livorno), San Martino (Elba) and Casalecchio (Rimini). Sardinia has also yielded some iron objects that can be dated to the 12th century BC: their typology generally suggests a relation with Cypriot trade (Almagro Gorbea, 1993).

Both the recovery contexts – votive deposits, habitats and necropolises – and the typology of these items are quite varied, suggesting different uses and functions of the new metal. However, all these findings precede a stable Phoenician presence in the Iberian Peninsula. In fact, even if recent radiocarbon dating pushed back the traditional chronology for the first Phoenician colonies in the Iberian Peninsula – dating them to the 9th century cal BC (Torres Ortiz, 1998) – some of these findings are old enough to be considered 'pre-colonial' and therefore not related to the foundation of the colonies. Furthermore, at present, the lack of any debris that could be related to metallurgical processes for obtaining iron in this chronological period seems to confirm the hypothesis of a first introduction of iron as raw metal. Thus, iron ingots may have been part of the trade route across the Mediterranean before the colonies were established and local communities developed their own ways to work and shape this metal. This would be particularly true for iron objects with a simple structure, reproducing local types. Other more structurally complex objects, such as blades built with piled sheets, are likely to have been imported.

This new data encourages a review of the penetration routes for prestigious goods in the Western world and especially in the Iberian Peninsula (Almagro Gorbea, 1993; Vilaça, 2006).

Iron production in the Iberian Peninsula

There is little data for the beginning of iron production in the Iberian Peninsula. In fact, studies carried out at various Early Iron Age settlements that have yielded evidence of siderurgic activities show a situation similar to the Near Eastern region. For example, data on iron production obtained in some Phoenician colonies – such as Toscanos (Keesmann, et al., 1989), Morro de Mezquitilla (Schubart, 1999), Cerro del Villar (Rovira Hortalà, 2005) in the Malaga province, Adra in Almeria (Aguayo de Hoyos, et al., 1989), sa Caleta at Sant Josep de Talaia in Ibiza (Torres, 2007) or La Fonteta in the Alicante area (González Prats and Ruiz Segura, 1999; Renzi and Rovira, 2007), among others – are not clear yet.

These archaeometallurgical findings are attributed mainly to the first half of the first millennium BC, probably between the end of the 9th and the beginning of the 8th century cal BC, according to recent radiocarbon dating. Metallurgical evidence has also been identified at some Orientalising period sites. During this period Iberian indigenous sites were strongly influenced by the Phoenicians or by their culture. This would be the case with Los Castillejos de Alcorrín (Manilva), close to the Malaga coast (Marzoli, et al., 2009); Carmona in the Seville area (Belén Deamos, et al., 2000); and the aforementioned El Castellar de Librilla in the province of Murcia (Ros Sala, 1993; Fig. 2). These locations have all provided iron slags dating from the beginning of their occupation, in the 8th century BC.

Some Italian contexts have also provided early evidence of iron technology. For example, slags and other production debris were collected at Elba, Livorno (Coretti and Benvenuti, 2001) and at Ischia (Pithekoussai) and Napoli (Buchner, 1969). Their recovery contexts and interpretation, however, are not always clear and their chronology cannot be precisely established.

The Phoenician colonies of Morro de Mezquitilla and Toscanos, both in Malaga province, have been the first sites in the Iberian Peninsula to yield residues related to iron pyrotechnology. At Morro de Mezquitilla, various pyrometallurgical structures were detected. Four of them, discovered in Sector 21, were not very well preserved and it is impossible to reconstruct their original shape. The archaeological evidence consisted of simple, irregularly circular hearths, approximately 0.40-0.50m in diameter. In one case, the hearth was surrounded by stones and mud, resulting in an external diameter of about 0.65m. Two more hearths were identified but their interpretation as metallurgical pits is dubious (Schubart, 1999, pp.243-248).

A better preserved pyrometallurgical structure was discovered in Sector 23. Part of it could not be excavated as it was located exactly in the section of the excavation, but its size seemed to be relevant. The furnace had a preserved height of 0.55m and its external diameter could be about 1.40m. Two openings were identified in the furnace walls, which were probably used to insert the tuyères or for natural ventilation. These structures were associated with slags and various fragments of prismatic tuyères with a double perforation (Schubart, 1999, pp.246-248).

Another hearth used for siderurgic activities was discovered at Cerro del Peñón, at the west side of Toscanos, next to the Vélez river (Keesmann, et al., 1989). This hearth is quite poorly preserved and was excavated directly into the rock. A furrow with a width of about 0.30-0.40m was located next to the hearth, forming three sloped steps or levels ending in a wider flat step. This furrow was filled with charcoal and abundant slag fragments. The combustion chamber was

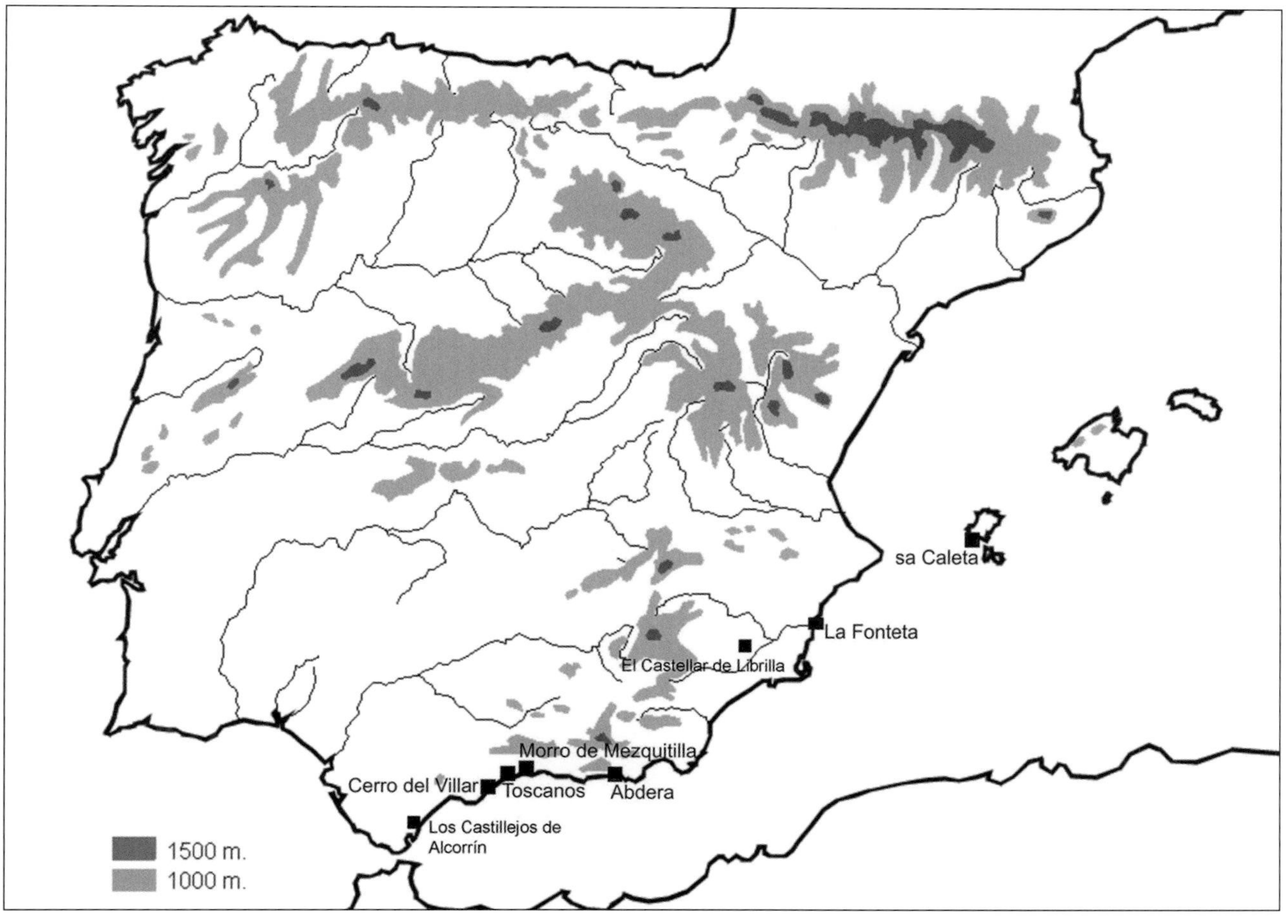

Figure 2: Phoenician sites in the Iberian Peninsula presenting siderurgical evidence.

supposedly placed on the upper part of the furnace slope; it was just a simple oval hearth. The existence of a possible dome covering this hearth has also been suggested. Slag cakes, a sand layer, charcoal, tuyères similar to those found at Morro de Mezquitilla and furnace linings, have been collected in association with this structure.

Siderurgic materials from both Morro de Mezquitilla and Toscanos have been published as evidence for smithing activities (Keesmann, et al., 1989; Schubart, 1999). However, the interpretation of the slag morphology and microstructure is not always clear. Furthermore, the absence of mineral fragments at these sites, together with their relative distance from the sources, invites the authors to point at the production of objects, excluding the possibility of interpreting part of these materials as remains related to iron ore reduction.

At Cerro del Villar (Rovira Hortalà, 2005) and La Fonteta (González Prats and Ruiz Segura, 1999), both located on river estuaries of the Spanish Mediterranean coast, the situation is similar.

Parts of residential, productive and commercial areas were excavated during the archaeological campaigns carried out at Cerro del Villar. Some of these structures were used as workshops for iron production and they are located in two different areas of the settlement. At La Fonteta, abundant archaeometallurgical material and several fire structures were found for all occupational phases, especially the earliest one (Renzi and Rovira, 2009).

At both sites, various pyrometallurgical hearths were excavated but their poor preservation did not allow a clear interpretation of their original shape. Slags, tuyères and charcoal have been detected; moreover, hammerscales were identified in association with some of these structures, documenting their use for post-reduction processes, including refining of the bloom and secondary smithing activities.

Siderurgic debris from the Phoenician sites of sa Caleta, Adra and Carmona are still in the process of being studied. Interestingly, in all of these Phoenician colonies, copper-based metallurgy has also been documented.

To sum up, a great part of the aforementioned materials have been interpreted as the result of smithing activities. Even if in some cases their relation to the production of iron objects is unquestionable, in others we cannot exclude their use in some other stage of the *chaîne opératoire* for processing iron. This general interpretation is mainly due to problems in distinguishing primary siderurgic materials from refining or smithing debris. In fact, it is true that some of the pyrometallurgical debris found at these sites could be easily interpreted as evidence of smithing activities, but there is also a fair amount that – due to their morphological, structural, and chemical characteristics – cannot be safely included in this typology.

Discussion and analysis of some iron slags

Analytical study of some slags from La Fonteta, Cerro del Villar and Los Castillejos de Alcorrín has been carried out. A first insight into these materials was gained through optical microscopy and, subsequently, the samples were analysed using a scanning electron microscope combined with an X-ray microanalyser (SEM-EDX). The SEM analyses have been performed by Marta Furió and Laura Tormo (National Museum of Natural Science of Madrid, CSIC and by Esperanza Salvador (SIdI, *Universidad Autónoma de Madrid*).

Different kinds of slags – nut-sized, amorphous specimens, and plano-convex slags – were collected. Their size was also different, varying from fragments to small nodules and bigger slag cakes (Fig. 3).

Most of the analysed specimens have a similar heterogeneous microstructure, often with abundant formations of iron oxides, especially wüstite and/or magnetite. Continuous bands of secondary iron oxides were also identified in samples from both settlements (Fig. 4).

The sample matrix is usually composed of melted material; fayalite crystals are present in some slags from Cerro del Villar and at Los Castillejos de Alcorrín, often with traces of calcium and magnesium oxides (Fig. 5a). Crystallisations of complex iron silicates with high contents of calcium were detected in La Fonteta samples and in one specimen from Los Castillejos de

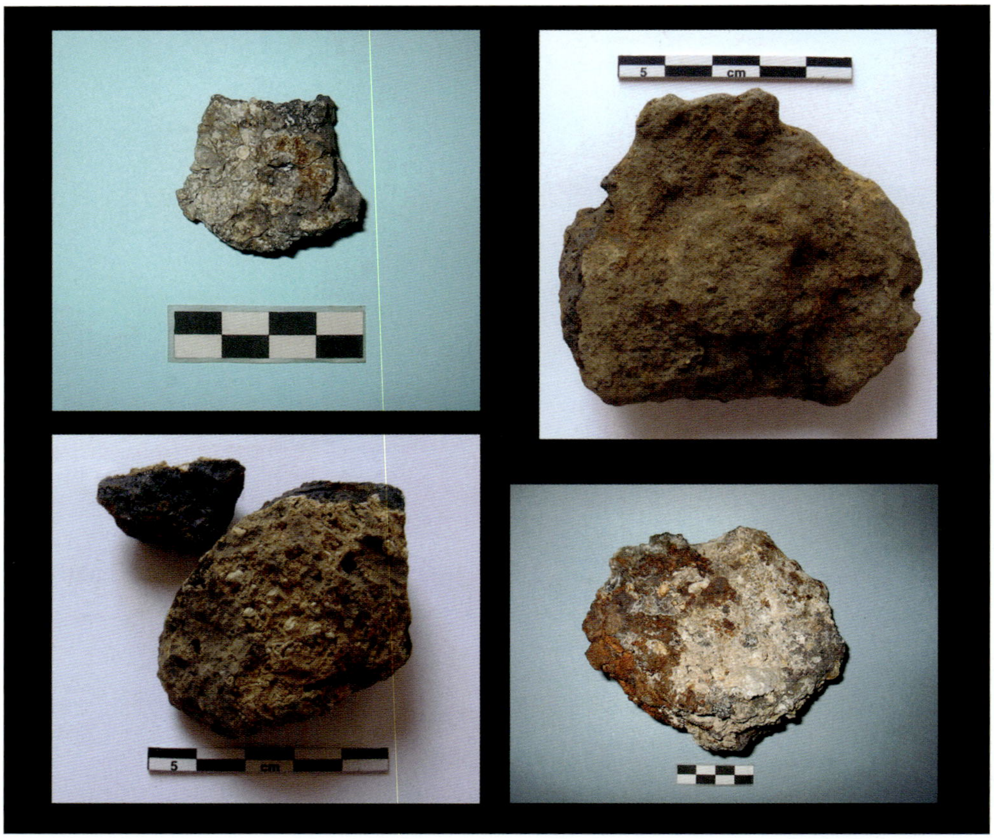

Figure 3: Different types of iron slags from Cerro del Villar (Malaga, Spain; above, right and down, left) and La Fonteta (Alicante, Spain; above, left and down, right).

Table 1: SEM-EDX results of the bulk analyses of metal inclusions in some slags (wt%; nd: not detected).

Sample	*Si*	*S*	*Ca*	*Fe*	*Ni*	*Cu*	*As*	*Sn*	*Sb*
Cerro del Villar (CV9A-140)	nd	nd	nd	98	2	nd	nd	nd	nd
La Fonteta (F41536-3)	2.22	nd	1.33	31.41	3	26.52	35.6	nd	nd
La Fonteta (F41536-3)	1.8	nd	1	32.32	3.61	25.7	35.6	nd	nd
La Fonteta (F41536-3)	2.05	nd	1.27	6.05	nd	90	1.05	nd	nd
La Fonteta (F41536-3)	14.04	0.63	15	35	1.3	7	18.11	nd	10.04
La Fonteta (F41536-3)	5	nd	2.06	34.54	11.7	12.5	34.2	nd	nd
La Fonteta (F41711-1)	1.32	nd	1.04	10.5	0.63	77	0.7	8.8	nd
La Fonteta (F41711-1)	nd	17.34	nd	2.14	0.3	79	0.9	0.7	nd
La Fonteta (F41711-1)	nd	nd	nd	12.2	nd	87.8	nd	nd	nd
Los Castillejos de Alcorrín (13S1-2)	nd	nd	nd	22.54	34.37	18.54	24.55	nd	nd
Los Castillejos de Alcorrín (13S1-2)	nd	nd	nd	26	29.15	18	26.4	nd	nd
Los Castillejos de Alcorrín (13S1-2)	nd	nd	nd	92.44	6.85	0.72	nd	nd	nd
Los Castillejos de Alcorrín (13S1-2)	nd	nd	nd	99	nd	1	nd	nd	nd
Los Castillejos de Alcorrín (13S1-2)	nd	nd	nd	84.4	14.15	0.2	nd	nd	nd

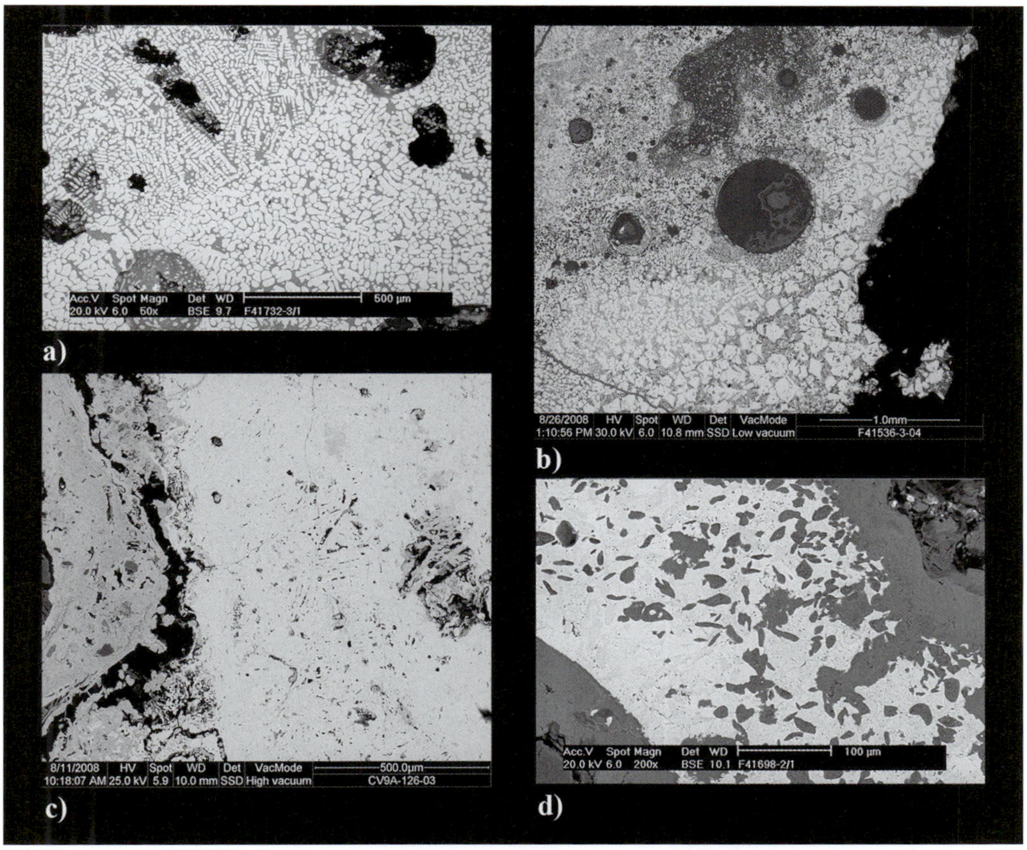

Figure 4: Microstructure of some iron slags from Cerro del Villar (Málaga, Spain) and La Fonteta (Alicante, Spain): a) wüstite-rich slag; b) magnetite-rich slag; c-d) bands of iron oxides (SEM backscattered electron images).

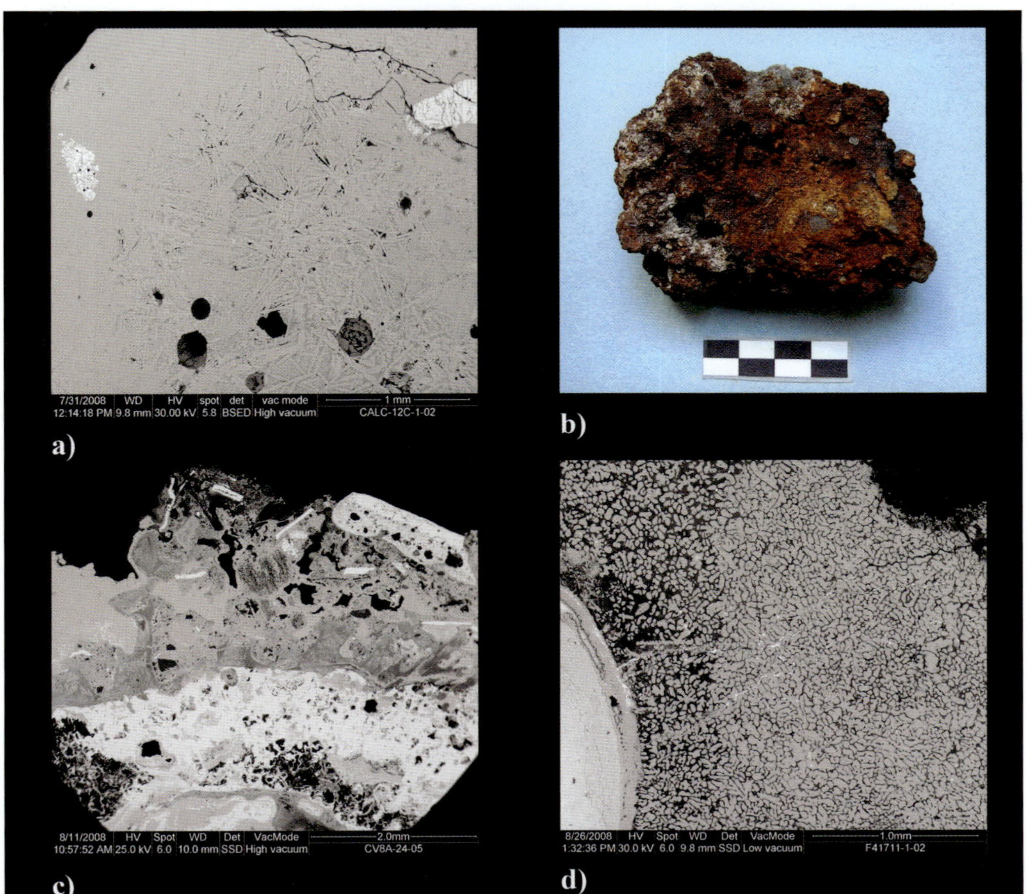

Figure 5: a) Microstructure of an iron slag from Los Castillejos de Alcorrín (Malaga, Spain). Crystals of fayalite; b) a smithing slag from La Fonteta (Alicante, Spain); c) microstructure of a smithing slag from Cerro del Villar (Malaga, Spain); d) microstructure of sample F41711-1 from La Fonteta (Alicante, Spain). SEM backscattered electron images (a, c, d).

Alcorrín. In some cases, an absolute predominance of wüstite with scarce silicate material in the interstitial filling has been detected in samples from the three sites. For example, some slags present a structure almost entirely made up of formations of globular and dendritic wüstite. This characteristic, together with the presence of ferric hammerscales stacked on the surface of the slag or embedded into the sample matrix, allowed us to interpret them quite safely as smithing slags (Fig. 5b-c).

Other samples present abundant metal inclusions with complex compositions including, for example, Fe, Cu, Ni, Sn, As and Sb. Iron prills with variable nickel content (2-3%) were detected in one slag from Cerro del Villar. Various iron inclusions with relevant quantities of Cu, As and Ni – up to approximately 20% Cu, 30% As and 4% Ni – were identified in another sample from La Fonteta. Los Castillejos de Alcorrín also provided a sample in which metal prills with different contents of Fe, Ni, Cu and As have been detected (Fig. 6).

The atypical composition of these inclusions – sometimes embedded into continuous iron oxide bands that could correspond to mineral relics – seems to suggest that we are dealing with materials related to mineral processing. Some examples of bulk compositions of these metal inclusions are shown in Table 1.

The interpretation of these slags is still problematic. In some cases as, for example, sample F41711-1 from La Fonteta (Table 1), its relation to iron metallurgy is not very clear, although its microstructure is very similar to that of ferric slags (Fig. 5d). We must also emphasise the complexity of the iron (or copper) minerals exploited at these sites, exemplified by the composition of the metal prills trapped in the slags investigated. In the Malaga area, for example, where several coastal Phoenician colonies were established, abundant mineral resources are known (García Alonso, 2007). These ores show a strong polymetallic composition, in some cases including noble metals and also Fe-Cu-Ni sulphides, as in the ores from the Serrania de Ronda area (Torres-Ruiz, et al., 1991). Despite the atypical compositions of metal inclusions in some of these slags, we have no doubt that they are iron slags. The main argument supporting this interpretation is that both their chemical composition and microstructure are completely different from copper and copper-tin slags found at the same sites. Unfortunately the analytical studies of these copper-based slags are mostly unpublished, with the exception of a few materials from La Fonteta that have been the object of a preliminary report (Renzi and Rovira, 2007).

If iron was obtained from minerals containing arsenic, nickel and other elements (as these slags suggest), it is very likely that their presence should be detected in the finished objects manufactured from the metal in question. This will represent the next step in our research, although it has to be said in advance that the few iron objects recovered so far are extremely corroded and do not seem to maintain a metallic core.

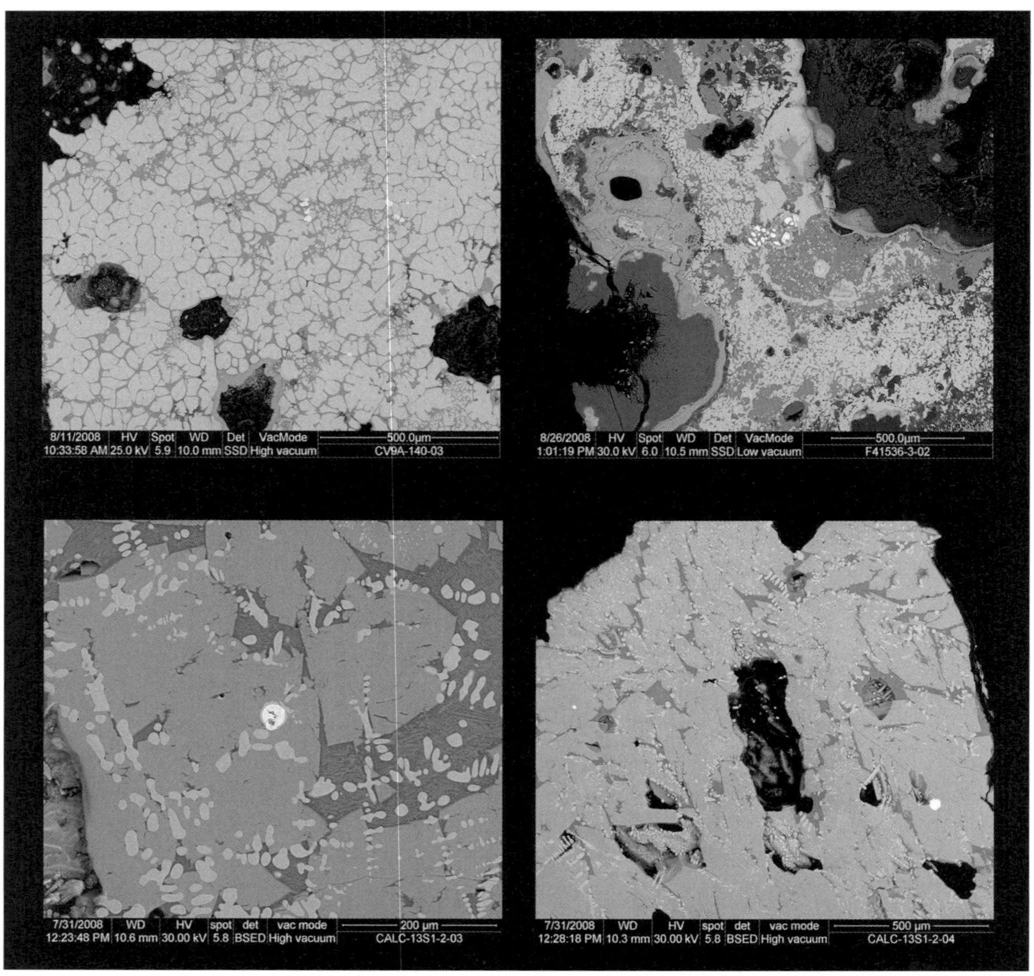

Figure 6: Examples of metal inclusions embedded in the matrix of the slags (SEM backscattered electron images).

It is only through more detailed analyses and broader sampling that it will be possible to shed more light on many aspects of Phoenician iron metallurgy. In fact, it is possible that this type of slag could be more widespread than initially thought. For example, in the analytical study of the archaeometallurgical materials from Toscanos and Morro de Mezquitilla, the presence of metallic iron inclusions with variable contents of copper was observed (Schubart, 1983; Keesmann, et al., 1989), although no other elements – such as Ni, As, Sb, etc. – were detected. However, more than twenty years have passed since those slags were studied: it could be worthwhile to re-analyse them in view of recent analytical results obtained during the study of the metallurgical debris from the other Phoenician colonies mentioned above.

Conclusions

We consider that, despite the fact that the metallurgical debris analysed in this paper cannot be easily explained as the result of secondary smithing processes, it could be interpreted as sub-products of bloom refining activities or as evidence of primary metallurgy to obtain iron, probably by exploiting mixed ores.

The smelting method employed could be a direct reduction of iron-rich ores, performed through an unintentional slagging process, without the addition of flux agent. If that was so, we can expect the production of small amounts of slag and, of course, such slag would not be comparable with the commonly known fayalite slag. The presence in the slag matrix of fayalite-like materials or the formation of complex iron silicates, whose composition is similar to the one of the kirschteinite group, can be easily explained as the reaction of the silicate gangue minerals with the ashes of the charcoal used as fuel, in addition to the contamination from some furnace linings.

So far, when comparing iron pyrotechnology during the Early Iron Age in the Western world with the siderurgical evidence documented in the ancient Near East – at Tell Hammeh (Veldhuijzen and Rehren, 2006, 2007), for example – the absence of furnaces and tap slags at the Phoenician settlements of the Iberian Peninsula seems to suggest a less advanced technological level for processing iron or, at least, a different kind of metallurgical process or furnace used. To date, however, no evidence of pyrometallurgical structures employed to carry out this kind of smelting processes has been clearly identified. The only archaeological evidence for fire structures are simple hearths with irregularly circular shapes and diameters varying between 0.40 and 0.80m, such as the ones excavated at La Fonteta (Renzi and Rovira, 2009). Furthermore, tap slags have not been archaeologically documented in this area until Roman times.

In conclusion, these findings pose the following question: could this evidence be the result of direct reduction activities of iron-rich minerals, in which only a limited amount of slag is produced? Our iron smelting experiment at the Celto-Iberian site of Segeda (Mara, Zaragoza, Spain), involving the reduction of iron-rich hematite in a small cylindrical furnace, points to this hypothesis (Rovira and Burillo, 2007). The experiment was carried out in the framework of the project I+D+I BHA2001-2439 "Social and economic processes in the formation and development of the city-state of Segeda" (*Procesos sociales y económicos en la formación y desarrollo de la ciudad-estado de Segeda*), directed by Dr. Francisco Burillo Mozota.

The experiment was performed in a slag-pit furnace. Fig. 7a shows the pit around which a structure of approximately 0.50m in height was built. At the soil level, openings for two tuyères were provided; these were situated one in front of the other and were operated with the help of manual bellows (Fig. 7b). About 10kg of hematite were crushed to 2-3mm sized fragments and charged when the temperature inside

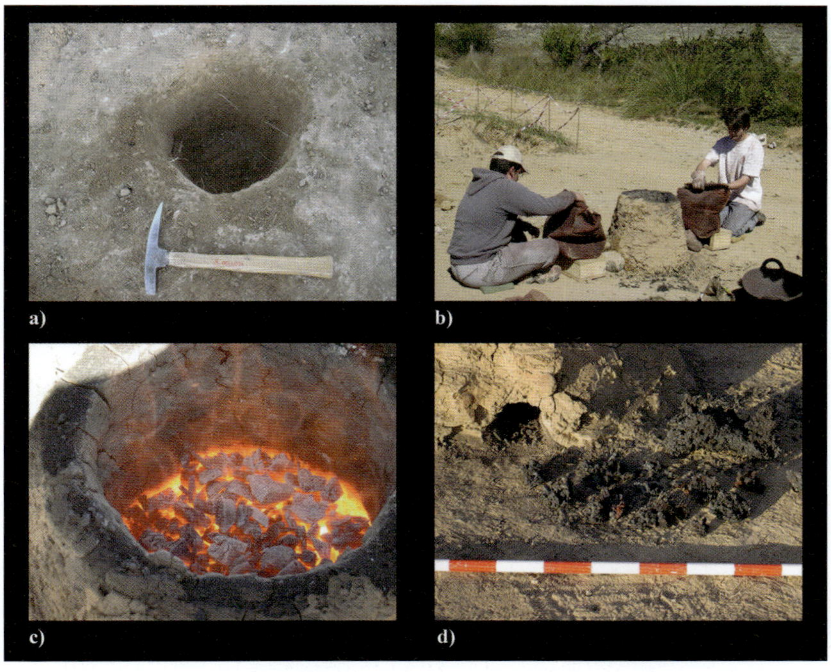

Figure 7: Iron smelting experiment carried out at the Celto-Iberian site of Segeda (Mara, Zaragoza, Spain).

the furnace was about 1000ºC (light cherry red), in alternating layers of ore and charcoal. The furnace reached high interior temperatures, ranging from 1100ºC to 1200ºC (orange-yellow to yellow) (Fig. 7c). After eight hours, a big and compact slaggy lump and several smaller ones made of partially reduced iron ore and metal, filled by melted silicates, were obtained at the bottom of the furnace (Fig. 7d). SEM analyses showed a slag which was generally rich in wüstite-magnetite (Fig. 8a). Furthermore, accumulations of metallic iron have been detected in samples of the big slaggy cake (Fig. 8b). The experimental debris is similar in aspect and composition to some of the archaeological finds.

Should we really assume the use of bloomery furnaces during this chronological period in the Western world? And, if a more advanced technological level is documented in the Near East, why was it not used since the first introduction of siderurgy, at least in the Phoenician settlements of the Western Mediterranean?

Presently, many questions concerning these issues are being asked, but very few have been answered convincingly. For this reason, considering the information obtained from these recent archaeometallurgical findings, some of the assumptions about early iron production, and not only in the Western Mediterranean, should be subjected to new critical revision.

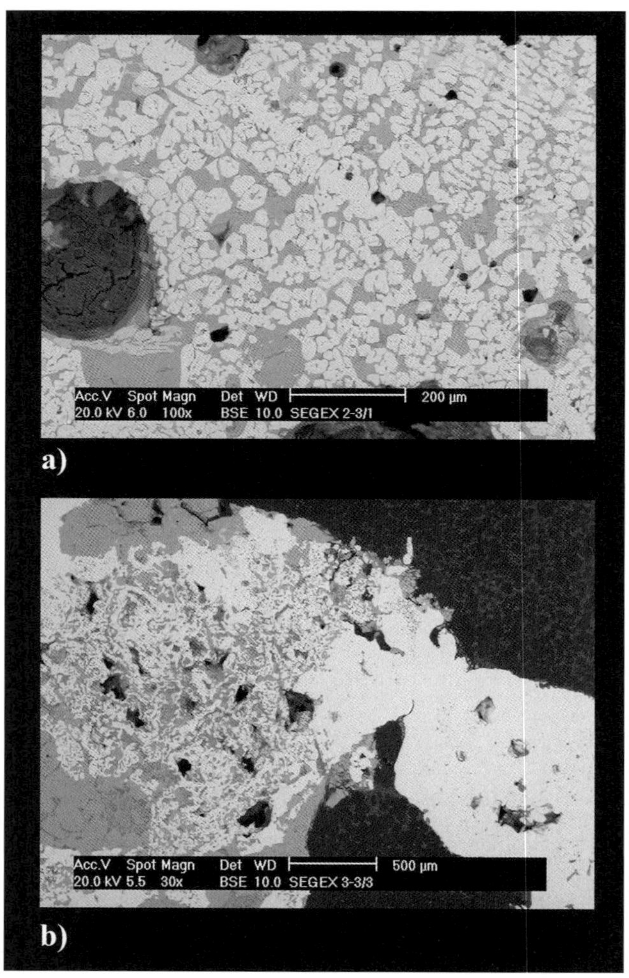

Figure 8: Samples from the iron smelting experiment at the Celto-Iberian site of Segeda (Mara, Zaragoza, Spain): a) magnetite-wüstite-rich nodule; b) iron-rich lump. SEM backscattered electron images.

Acknowledgements

This research has been carried out within the framework of the programmes Consolider-Ingenio 2010 (CSD2007-00058) – "Technologies for the conservation and valorisation of cultural heritage" (*Programa Consolider de Investigación en Tecnologías para la valoración y conservación del Patrimonio Cultural – TCP*) and "Relation between local raw materials and metallurgical production: Southern Cataluña as a comparative model" (HAR2010-21105-C02-02) (*Relación entre materias primas locales y producción metalúrgica: Cataluña meridional como modelo de contraste*) of the Spanish Ministry of Science and Innovation.

The authors would like to thank: Prof. Dr. Alfredo González Prats, director of the archaeological excavations at La Fonteta (Alicante); Prof. Dr. M. Eugenia Aubet, director of the archaeological excavations at Cerro del Villar (Málaga) and Prof. Dr. Dirce Marzoli, director of the German Archaeological Institute of Madrid and of the archaeological excavations at Los Castillejos de Alcorrín (Manilva, Malaga) for giving us the possibility of studying the archaeometallurgical materials from these sites.

References

Aguayo de Hoyos, P., Suárez Márquez, A., San Martín Montilla, C., López Castro, J.L. and Carrilero Millán, M., 1989. Abdera: una colonia fenicia en el Sureste de la Península Ibérica. *Madrider Mitteilungen*, 30, pp.135-50.

Almagro Gorbea, M., 1993. La introducción del hierro en la Península Ibérica. Contactos precoloniales en el Período Protoorientalizante. *Complutum*, 4, pp.81-94.

Belén Deamos, M., Anglada Curado, R., Jiménez, A., Rodríguez, L., Escacena, J.L. and Lineros Romero, R., 2000. Presencia e influencia fenicia en Carmona (Sevilla). In: *Actas del IV Congreso Internacional de Estudios Fenicios y Púnicos, Vol. 4.* Cádiz: Servicio de Publicaciones, Universidad de Cádiz, pp.1747-61.

Buchner, G., 1969. Mostra degli scavi di Pithecusa. *Dialoghi di Archeologia*, 3, pp.85-101.

Corretti, A. and Benvenuti, M., 2001. The beginning of iron metallurgy in Tuscany with special reference to Etruria Mineraria. *Mediterranean Archaeology*, 14, pp.127-45.

Frisch, B., Mansfeld, G. and Thiele, W.-R., 1985. *Kamid El-Loz 6, Die Werkstäten Der Spätbronzezeitlichen Paläste.* Bonn: Habelt.

García Alonso, E., 2007. *En la orilla de Tartessos. Indígenas y fenicios en las tierras malagueñas. Siglos XI-VI a.C.* Málaga: Fundación Málaga.

González Prats, A. and Ruiz Segura, E., 1999. Una zona metalúrgica de la primera mitad del siglo VII en la ciudad fenicia de La Fonteta (Guardamar, Alicante). In: *XXIV Congreso Nacional de Arqueología, Vol. 3.* Murcia: Instituto de Patrimonio Histórico, pp.355-7.

Keesmann, I., Niemeyer, H.G., Briese, C., Golschani, F. and Schulz-Dobrick, B., 1989. Un centro primitivo de la elaboración de Hierro en la factoría fenicia de Toscanos. In: C. Domergue ed. *Minería y Metalurgia en las antiguas civilizaciones Mediterráneas y Europeas, Vol. 1.* Madrid: Ministerio de Cultura, Direccion General de Bellas Artes y Archivos, Instituto de Conservación y Restauración de Bienes Culturales, pp.99-108.

Liebowitz, H. and Folk, L.R., 1984. The dawn of iron smelting in Palestine: the Late Bronze Age smelter at Tel Tin'am, Galilee, Israel. *Journal of Field Archaeology*, 11, pp.265-80.

Marzoli, D., González Wagner, C., Suárez Padilla, J., López Prado, F., Mielke, D.P., Leon Martín, C., Thiemeyer, H. and Torres Ortiz, M., 2009. Vorbericht zu den deutsch-spanischen Ausgrabungen in der endbronzezeitlichen Siedlung von Los Castillejos de Alcorrín, Manilva (Prov. Málaga) 2006 und 2007. *Madrider Mtteilungen*, 50, pp.118-48.

Pereira Sieso, J., Ruiz Taboada, A. and Carrobles Santos, J., 2003. Aportaciones del C-14 al mundo funerario carpetano: la necrópolis de Palomar de Pintado. *Trabajos de Prehistoria*, 60, pp.153-68.

Pleiner, R., 2000. *Iron in Archaeology. The European Bloomery Smelters*. Praha: Archeologický ústav AVČR.

Renzi, M. and Rovira, S., 2007. Escorias metalúrgicas del yacimiento fenicio de La Fonteta (Alicante). Estudio preliminar. In: J. Molera, J. Farjas, P. Roura and T. Pradell eds. *Avances en Arqueometría 2005. Actas del VI Congreso Ibérico de Arqueometría*. Girona: Universitat de Girona, pp.163-71.

Renzi, M. and Rovira, S., 2009. Slags of Early Iron Production in the Iberian Peninsula. In: J.-F. Moreau ed. *Proceedings: ISA 2006: 36th International Symposium on Archaeometry, 2-6 May 2006, Quebec City, Canada*. Cahiers d'archéologie du CELAT. Série archéométrie 7. Quebec: CELAT, Université Laval, pp.399-406.

Ros Sala, M.M., 1993. El trabajo del hierro en el poblado protohistórico de El Castellar (Murcia). I: análisis arqueológico. In: S.F. Ramallo Asensio, A.M. Muñoz Amilibia, M.M. Ros Sala and R. Arana Castillo eds. *Metalurgia en la Península Ibérica durante el primer milenio a.C.: estado actual de la investigación*. Murcia: Universidad de Murcia, pp.71-110.

Rovira, S. and Burillo, F., 2007. Experimentos de fundición de minerales de hierro en la ciudad-estado celtibérica de Segeda (Mara, Zaragoza). In: J. Molera, J. Farjas, P. Roura and T. Pradell eds. *Avances en Arqueometría 2005. Actas del VI Congreso Ibérico de Arqueometría*. Girona: Universitat de Girona, pp.137-43.

Rovira Hortalà, M.C., 2005. Los talleres metalúrgicos fenicios del Cerro del Villar (Guadalhorce - Málaga). In: A. Spanò Giammellaro ed. *Atti del V Congresso Internazionale di Studi Fenici e Punici, Marsala-Palermo, 2-8 ottobre 2000, Vol. 3*. Palermo: Università degli Studi di Palermo, Facoltà di Lettere e Filosofia, pp.1262-70.

Schubart, H., 1983. Morro de Mezquitilla: vorbericht über die Grabungskampagne 1982 auf dem Siedlungshügel an der Algarrobo-Mündung. *Madrider Mitteilungen*, 24, pp.104-31.

Schubart, H., 1999. La forja fenicia del hierro en el Morro de Mezquitilla. In: A. González Prats ed. *La cerámica fenicia en Occidente: centros de producción y áreas de comercio. Actas del I Seminario Internacional sobre Temas Fenicios, Guardamar del Segura 21-24 noviembre de 1997*. Valencia: Direcció General d'Ensenyaments Universitaris i Investigació, Consellería de Cultura, Educació i Ciència; Alicante: Instituto de Cultura "Juan Gil-Albert", pp.241-56.

Torres, J.R., 2007. *Excavaciones arqueológicas en el asentamiento fenicio de "sa Caleta" (Ibiza)*. Cuadernos de Arqueología Mediterránea 16. Barcelona: Edicions Bellaterra.

Torres Ortiz, M., 1998. La cronología absoluta europea y el inicio de la colonización fenicia en Occidente. Implicaciones cronológicas en Chipre y el Próximo Oriente. *Complutum*, 9, pp.49-60.

Torres-Ruiz, J., Gervilla, F. and Leblanc, M., 1991. Mineralogía y comportamiento geoquímico del oro en los macizos lherzoliticos bético-rifeños (España y Marruecos). *Estudios Geológicos*, 47, pp.281-93.

Tylecote, R.F., 1979. *A History of Metallurgy*. London: Metals Society.

Veldhuijzen, H.A. and Rehren, Th., 2006. Iron smelting slag formation at Tell Hammeh (az Zarqa), Jordan. In: J. Perez-Arantegui ed. *Proceedings of the 34th International Symposium on Archaeometry*. Zaragoza: Institución "Fernando el Católico", pp.245-50.

Veldhuijzen, H.A. and Rehren, Th., 2007. Slags and the City. Early Iron Production at Tell Hammeh, Jordan, and Tel Beth-Shemesh, Israel. In: S. La Niece, D.R. Hook and P.T. Craddock eds. *Metals and Mines – Studies in Archaeometallurgy*. London: Archetype, pp.189-201.

Vilaça, R., 2006. Artefactos de ferro em contextos do Bronze Final do território português: Novos contributos e reavaliação dos dados. *Complutum*, 17, pp.81-101.

Authors' addresses

1 CCHS-CSIC. Albasanz, 26-28. 28037 Madrid (Spain), martina.renzi@cchs.csic.es; ignacio.montero@cchs.csic.es

2 Museo Arqueológico Nacional (MAN). Serrano, 13. 28001 Madrid (Spain), s_rovirallorens@hotmail.com

3 Museu d'Arqueologia de Catalunya (MAC). Passeig de Santa Madrona, 39-41, Parc de Montjuïc. 08038 Barcelona (Spain), crovirah@gencat.cat; carmerovirah@gmail.com

Iron, myth and magic: perceptions of smiths and metal workers through cultural evidence from Britain and beyond

Peter Halkon

Introduction

"Lady sits at her front door straight as the willow wand
When by there comes a coal blacksmith with his hammer in his hand
Well may you sit you lady fair all in your coat of red
But by tomorrow at this same hour I'll have your maidenhead
Chorus: *Bide lady bide for there's nowhere you can hide for the lusty smith will be your love and that will lay your pride*
Away, away you coal blacksmith you have done me great wrong
You never shall have my maidenhead that I have kept so long
I rather I was dead and cold and my body lain in the grave
Than a fusty dusty coal black smith my maidenhead shall have
So she became a turtle dove and flew all in the air
And he became a cock pigeon and they flew bare and bare
She became a brisk young hare running all on the plain
And he became a greyhound dog and fetched her back again
She became a brisk young mare dark as the night was black
And he became a golden saddle and clung onto her back
She became a little yew sheep grazing upon the common
He became a shaggy old ram and he swiftly fell upon"

Adapted from: "The Two Magicians" (Child Ballad no. 44). *"English and Scottish popular ballads"* by F.J. Child (Boston, 1857-1858).

This song, popularised by modern folk groups such as Steeleye Span, highlights the magical and transformative powers and sexual potency attributed to blacksmiths in past societies. With the inclusion of shape-shifting – an attribute well suited to a transformer of metals – into hares, greyhounds and birds, the song has clear roots in a more distant past. The creatures included are associated with the birth of Taliesin, the bard of Welsh mythology (Ford, 1999), and make frequent appearances in the art and ritual of Europe in the 'Celtic' Iron Age. According to Julius Caesar's account of late Iron Age Britain (The Conquest of Gaul, V.12, Hanford and Gardner, 1982, p.111), hare was a taboo meat. This animal is also depicted, hotly pursued by hounds, on the decorated pottery vessels of the Romano-British production centre of the Nene valley (Howe, et al., 1981).

The blacksmith is referred to in many other British folk songs. For example, the index of the Vaughan-Williams Memorial Library of the English Folk Dance and Song Society at Cecil Sharp House, London, contains 171 references to this trade, some of which are perhaps relics of rituals relating to iron production in later prehistoric and early medieval Britain.

In many societies around the world the role of the iron smelter and blacksmith has been perceived as special or magical in some way. In certain regions of Africa, for example, those involved in the production of iron have been seen as marginal; on the mysterious edges of society or, in contrast, central to it (Shinnie and Haaland, 1986; Herbert, 1993). Iron production has been related to religious cults and traditions and, through past and present anthropological observation and oral tradition, we know in some detail the relationship between the iron workers and the societies in which they lived. In the modern western world this process is far more difficult to determine and, apart from brief references by Classical writers, 'Celtic' tales and Scandinavian sagas, we have to rely on archaeology to provide evidence for the socio-cultural context of this fundamental technology.

Iron objects appeared in burial and votive contexts by the 8th century BC, soon after the adoption of iron in Europe, implying that the manufacture, use and deposition of the metal had already accrued special associations. This article assesses the extent to which it is possible to reconstruct the way blacksmiths and other metal workers were perceived where living traditions have been largely forgotten with reference to north-western Europe, and particularly Britain and Scandinavia, from the Iron Age to the Medieval period.

The Iron Age

The importance of iron production in Iron Age Britain and beyond has been somewhat understated and it is only in recent years that the scale of effort needed to smelt iron has been realised (Crew and Crew, 1995; Milcent, 2007; Halkon 2012). Greater attention has been paid to stylistic attributes and other aspects of the manufactured objects themselves (Harding, 2007; Garrow, et al., 2008). It is clear that the smiths of Iron Age Europe were highly accomplished and seemed to have enjoyed a special status, attested through the writing of ancient classical commentators and reconstructed from Irish epic poetry, the latter providing a glimpse into a pre-Christian Iron Age world (Scott, 1991; Green, 1995). Olc Aiche, described as a druid-smith (*druígoba*) in *Scéla Eogain* (Scott, 1991, p.186) placed protective magic circles around Cormac mac Art, one of the greatest legendary High Kings of Ireland, at his birth. In contrast the *Faed Fiada*, a prayer attributed to St Patrick, Ireland's patron saint, includes this plea for protection:

> "I summon today all these virtues ... against spells of women, smiths, and druids" (Scott, 1991, p.186).

It is interesting to consider to what extent the literary evidence for the status of iron workers is supported by archaeology. There is evidence of burials containing blacksmithing tools dating from Iron Age Britain. At Whitcombe in Dorset, one of these burials contained a spearhead, a sword, a file and an iron hammerhead (Stead, 1991a; Fell, 1998; Cunliffe, 2004). At Rudston (Stead, 1991b), a pair of blacksmith's tongs and a hammer were found in Grave 154, a male aged around twenty (Fig. 1).

The tips of two spears had been jammed between the arms of the tongs and a short sword was also found in the tomb. As at Whitcombe, the status of the buried male is ambiguous and the extent to which the deposition of these tools reflected his role in life as a smith or warrior, or was simply an expression of his status by those responsible for his disposal, is unclear. The purely hierarchical significance of these burials has recently been challenged by Giles (2007), who sees the symbolic transformative abilities of the blacksmith as being at least of equal importance.

On the other hand, it has been convincingly argued that those interred with both tools (such as hammers and tongs) and weapons, in a large area of Europe north of the Alps, were of relatively high status in society (Henning, 1991). Of the twenty or so Iron Age burials discussed in Henning's study, eight are in the area of present-day Poland, with the remainder spread evenly across northern Europe from Gotland to France (Henning, 1991, Fig. 1, p.67). The riverine distribution of many of these smiths' graves and other geographical factors are not considered in any depth in Henning's article but nevertheless are aspects worthy of further investigation.

In Britain there have been few attempts to elicit meaning or even quantify iron objects in Iron Age burials. A recent survey of published material noted that iron objects were recorded in 21% of a total of 856 burials in Iron Age East Yorkshire (Walker, 2007; Halkon, 2007, 2012). The total of 418 iron artefacts was subdivided into various classes: personal adornment (37%), vehicle/harness fittings (37%), weapons (24%) and tools and objects of uncertain function (2%). Unsurprisingly, the cemeteries with chariot burials such as Wetwang (Dent, 1985) included the most iron artefacts and from experimental work (Crew, 1991 and forthcoming; Halkon, 2007, 2012) it can be calculated that the iron objects alone contained in such a grave would have involved over 200 person days of labour. The chariot burial at nearby Kirkburn (Fig. 2) contained a tunic of mail, one of the earliest known in Europe (Stead, 1991a).

The most frequently occurring objects were brooches (30%) and spearheads (16%). Weapons were almost exclusively found in male burials and objects associated with personal adornment were most common in the graves of females. By comparison, a catalogue of Iron Age burials in Britain (Whimster, 1981) recorded only around a hundred out of 1,952 burials (excluding those from Yorkshire) that contained 189 iron objects: 29% of these were weapons and 34% items of personal adornment, including brooches.

The most bizarre burials in East Yorkshire are those where spears had been deliberately thrown at the corpse during interment, in what has been described as a "killing ceremony"

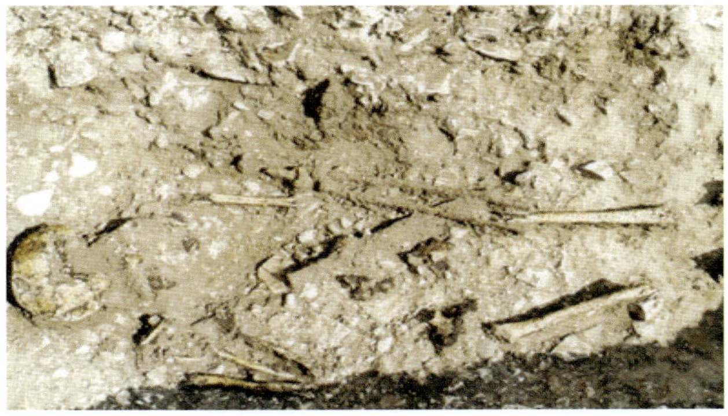

Figure 1: Rudston Grave 154, the Iron Age burial of a male, aged around 20, with a sword, two spearheads, a pair of blacksmith's tongs and a hammer. (Photo: courtesy I. Stead and J. Joy British Museum).

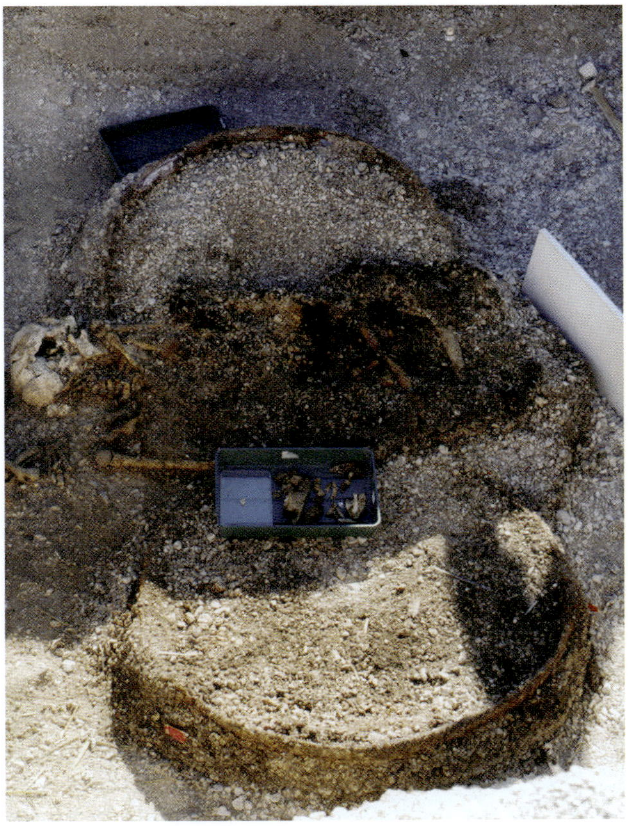

Figure 2: The chariot burial at Kirkburn. The iron tyres are clearly visible and the remains of the mail tunic can be seen on the torso of the skeleton. The tunic had been placed in the grave over the corpse. (Photo: P. Halkon).

(Stead, 1991a, p.75). At Garton Station (burial GS10), for example, the tightly crouched burial of a male aged 25-35 with a sword at his back contained 14 iron spearheads.

In the later Iron Age around the 1st century BC, Welwyn-type burials such as Baldock, resembling those of the near European continent, contained iron objects: particularly fire-dogs (stands used to support logs over an open fire on which to rest spits and other culinary devices) and other items relating to feasting (Fitzpatrick, 2007).

Fire dogs are amongst the single items and hoards of iron objects found particularly in northern and western Europe in non-cemetery contexts, such as bogs, streams and rivers, which are now usually interpreted as part of a pattern of structured deposition (Hill, 1995; Bataille and Guillamet, 2006) rather than simply caches of scrap metal. One of the most spectacular fire dogs found at Capel Garmon in Snowdonia (Fig. 3), which was embellished at both ends by horned cattle heads with manes, may have taken the equivalent of three person years to manufacture (Green, 1998, p.141).

The choice of such beasts was not accidental and may have ritual and mythological associations, as well as being an appropriate decoration for an instrument associated with the spit-roasting of large joints of beef.

In Britain tools relating to blacksmithing, such as hammers (Fell, 1998), have been discovered mainly in the ditches and pits of hill forts, as have bars of trade iron, commonly known as 'currency bars' (Crew, 1991, 1995; Hingley, 1990; Haselgrove and Hingley, 2006). Many iron objects from this period have been found in 'watery places', such as the famous type site of La Tène, where over one hundred and sixty-six swords were found (De Navarro, 1972). This phenomenon extends across Europe, with iron swords being deposited in rivers and blacksmith's hammers and tongs being notable amongst the larger items disposed of in this way (Dumont, et al., 2006). At Fiskerton in Lincolnshire, hammers and other tools were amongst the objects associated with a timber causeway (Field and Parker-Pearson, 2002).

To whom were such votive deposits dedicated? Was it a water deity or one relating to iron? According to Welsh and Irish mythology the smith god was variously known as *Gofannon* or *Goibniu*, meaning 'Great Smith', viz. a version of

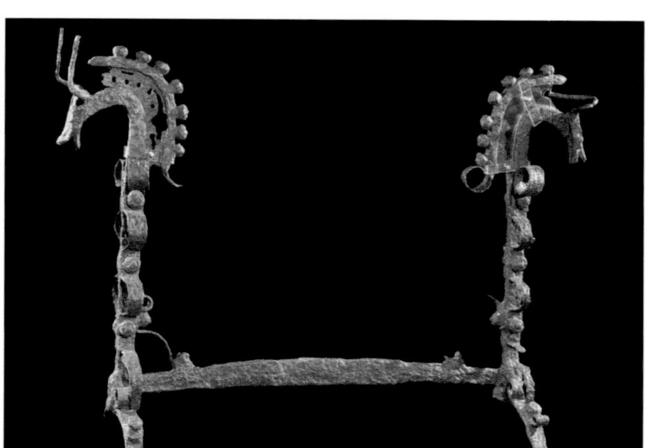

Figure 3: The Iron Age fire dog found at Capel Garmon, Snowdonia, North Wales. National Museum of Wales.

the name, "*Cobannus*", is known from an inscription found at Fontenay-près-Vézelay in France, as well as on a metal plaque originating from Bern in Switzerland (Fellmann, 1991, 2000). It is clear that 'Celtic' smith deities soon became associated with the Roman god *Vulcan* and it is to the Romans that we now turn.

The Roman period

Vulcan (the Greek equivalent of Hephaestus), god of blacksmiths and other craftsmen, was the son of Jupiter and Juno, and husband of Maia (Goddess of Spring) and Venus (Goddess of Love and beauty). His smithy was believed to be situated underneath Mount Etna in Sicily, hence 'Volcano'. His feast, the *Vulcanalia*, was celebrated on 23rd August, when fish and small animals were thrown onto a sacrificial fire (Varro, *De Lingua Latina* VI.20, "On the Latin Language" translated by Kent 1951, 193). In Roman art, Vulcan is depicted with a withered leg, which by legend was caused by his being hurled from Olympus by Jupiter in a rage or, in an alternative version, by Juno because of his ugliness. He is usually shown as bearded, wearing a brimmed hat and a tunic which leaves his right shoulder uncovered for action, and holding a hammer (Henig, 1995).

Small bronze statuettes and figurines which might have been contained in a *lararium* or household shrine have been found quite widely across Roman Britain, amongst the most well-known being those from Southbroom in Wiltshire, now in the British Museum (Henig, 1995; Potter, 1997). There is a better quality bronze figure from North Bradley, also in Wiltshire (Henig, 1995), and from Richborough in Kent, a superbly modelled statuette which is almost certainly imported, of a seated and naked Vulcan about to strike with a now missing hammer (Henig, 1995, p.81 and Fig. 50). Another votive item in Roman Britain were 'leaves' of silver, perhaps best known in a Christian context in the Water Newton Treasure. At Barkway in Hertfordshire, Vulcan is depicted on two of these (Henig, 1995; Crerar, 2006, p.3).

In Roman Britain, sculpture and pottery associated with Vulcan and his 'Celtic' equivalent have a marked northern pattern of distribution. At Maryport in Cumbria, on the Northern Frontier, the prefect of the Roman fort, Helstrius Novellus, dedicated an altar to Vulcan (*RIB* 846); while the inhabitants of the vicus of Old Carlisle set up a temple to both the smith god and Jupiter (*RIB* 899). The *vicani* of Vindolanda dedicated an altar to the smith god (*RIB* 1700) and, at Corbridge, a mould for making plaques depicting him in typical pose confirms the importance of this deity in the region (Webster, 1989).

At Dringhouses near York (Ottaway, 2004) a fine stone relief, once thought to be the tombstone of a blacksmith, is now considered more likely to be Vulcan due to his off-the-shoulder tunic and withered leg, both attributes of the deity (Fig. 4).

The York relief is similar, though a little cruder in style, to a bronze figure of Vulcan found at Catterick in North Yorkshire (Wilson, 2002). Amongst the products of the pottery kilns of Malton and Norton, also in North Yorkshire, were pots

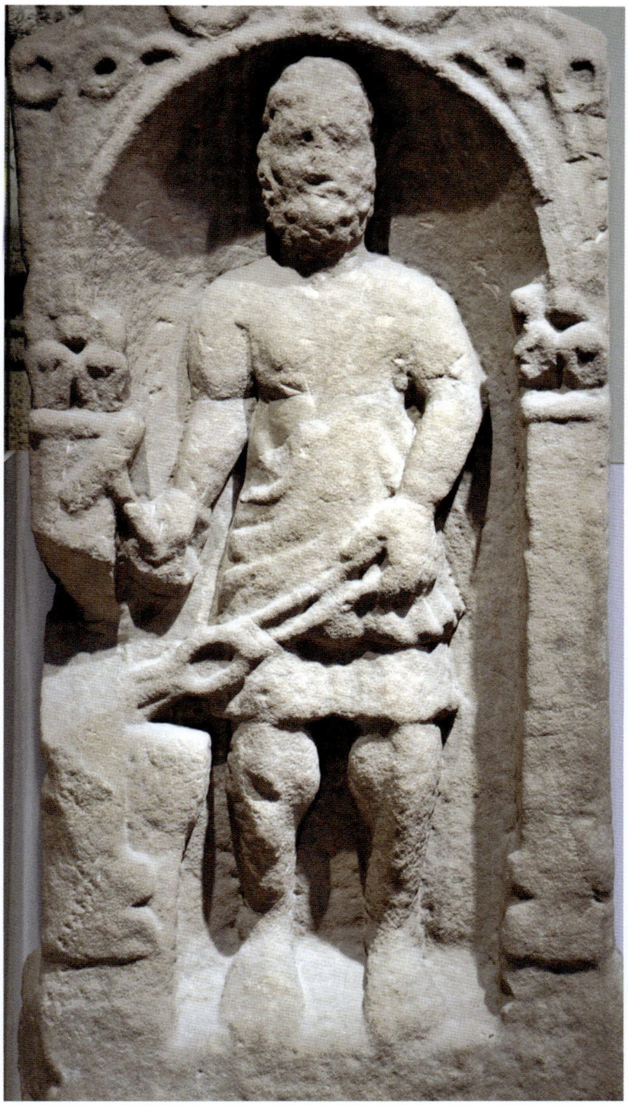

Figure 4: Long thought to be the tombstone of a blacksmith, this Roman sculpture from Dringhouses, near York, is more likely to represent Vulcan (Photo: M. Park, courtesy Yorkshire Museum, York).

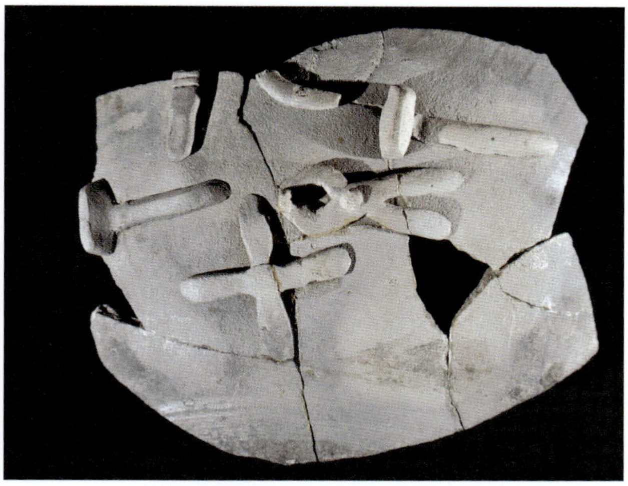

Figure 5: Part of the body of a Roman greyware jar from Norton, North Yorkshire decorated with applied tools. (Photo: M. Park, courtesy Malton Museum).

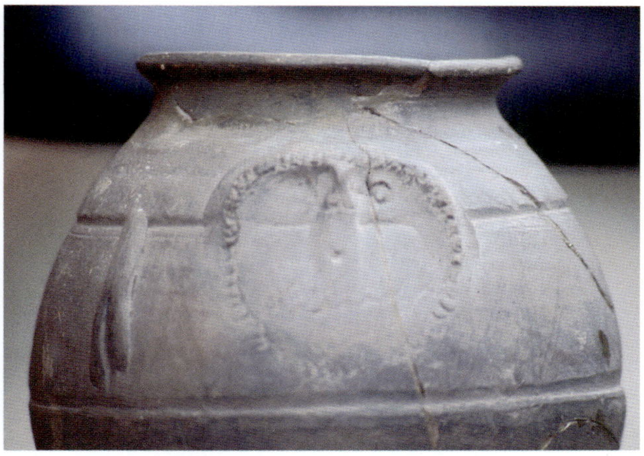

Figure 6: Detail of a complete Roman greyware jar from Shiptonthorpe, East Yorkshire decorated with a bearded face and blacksmith's tongs, which represent Vulcan. (Photo: P. Halkon).

decorated with blacksmith's tools (Fig. 5) and faces of Vulcan, which have also been identified elsewhere in eastern Yorkshire (Halkon, 1992, 2006).

At Shiptonthorpe, for example, a complete vessel decorated with a bearded face and smith's tongs is almost certainly a depiction of Vulcan. (Fig. 6). On the same site a sherd displaying an applied bearded face and a brimmed hat was found close to the main Roman road between York and Brough-on-Humber, on the site of a water hole found during later excavation (Millett, 2006).

A further smith's tool pot sherd was found at Bielby (Didsbury, 1984), on a Roman site where some iron smelting is also known to have taken place.

At nearby Hayton miniature axes and a hammer, usually associated with Vulcan or his Celtic counterpart (Green, 1983) have been identified amongst metal detector collections (Stratton, 2005) on a roadside settlement (Halkon, et al., 2000; Halkon, 2008). A miniature anvil, (Fig. 7) now in Hull and East Riding Museum, was found just outside the gateway of the walled enclosure at Brough.

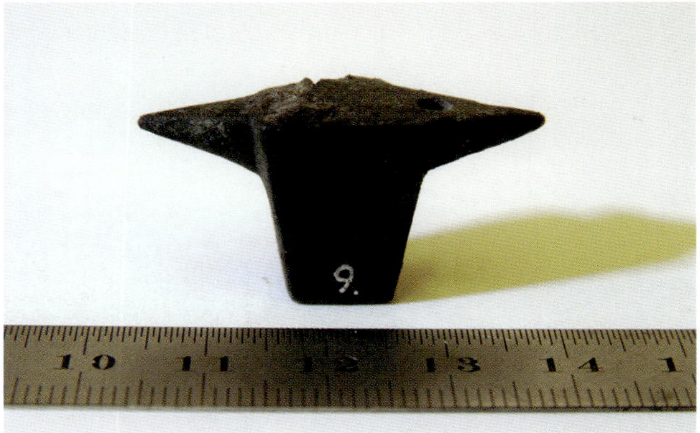

Figure 7: A miniature copper alloy anvil, now in Hull and East Riding Museum, Hull, found just outside the gateway of the Roman walled enclosure at Brough. The hole suggests that it was suspended as an item of adornment. (Photo: M. Park, courtesy Kingston-upon-Hull City Museums).

It is clear from the distribution of metal votive objects, combined with the smith's tool pots and face pots described above, that that there was a regional cult in East Yorkshire associated with Vulcan; appropriate in a landscape where the furnace-based industries of iron smelting and potting flourished in the Iron Age and Roman periods respectively (Fig. 8).

Objects associated with Vulcan have not been confined to East Yorkshire however, and other items associated with the smith god have been discovered on Romano-British sites elsewhere (Fig. 9).

Across the river Humber in Lincolnshire, finds associated with Vulcan include a limestone relief from Keisby (Henig, 1995, p.113 and Fig. 74) and a gold bezel, still in its setting in a silver ring, from Brant Braughton (Johns, 1991, pp.61-62, no. 18). Just over the modern county border at Newark, Nottinghamshire, a further Roman gold finger-ring setting showing Vulcan was located by a metal detectorist in 2005 (Hobbs and Henig, 2006, p.62). It is not possible to determine where these items belonged to smiths or whether Vulcan was the favoured deity of their original owner.

With the exception of the figurines mentioned above, dedications to Vulcan are much rarer in southern England. At Dun's Tew in Oxfordshire, however, a worn relief, now in the Ashmolean Museum, was found and there are smith's tool pots or pots depicting Vulcan from the Roman cities of Colchester and Wroxeter. A smattering of Vulcan pots in the East Midlands, including Water Newton and Chatteris, is thought to be associated with the Nene Valley pottery industry (Webster, 1989) and the iron production of the Northamptonshire region (Jackson, et al., 1978; Jackson and Tylecote, 1988; Crew, 1998).

Votive deposits of iron objects, some containing blacksmith's tools, are thought to have continued into the Roman period demonstrated by objects such as those found at Waltham Abbey in the Lea Valley north of London, though the objects themselves could range in date from 100BC to 100 AD (Merrifield, 1987, Fig. 8, p.31). It is noticeable that the handle of the tongs has been deliberately bent before deposition, like those from Llyn Cerrig Bach on Anglesey (Fox, 1945). It has been claimed recently that this hoard may have been lost from a ship (Roberts, 2007) rather than ritually deposited in a lake. That votive deposition took place elsewhere in the Roman period is beyond reasonable doubt, as hammers and various other items related to smithing, such as a field anvil, were contained within hoards from Carlingwark Loch, Blackburn Mill and Eckford in Scotland (Piggott, 1953). The quantity of iron objects in these hoards is small by comparison with the huge Roman period votive deposits from Denmark (Jorgensen, et al., 2003), some of which contain hundreds of iron items, largely weapons.

An extensive search of the relevant literature failed to locate any Romano-British burials containing full-sized blacksmith's tools. North of the Alps only three of the forty-one Roman period *Schmiedegräber* (smith's graves) were inside the Empire, with the largest number in present-day Poland (seven) and Denmark (fourteen) (Henning, 1991, Table 1, p.71 and Fig. 4, p.69). The absence of smith's graves within the Roman Empire is significant and may relate to the greater control of mineral resources by the Roman state and the application of the Augustan ***Lex Iulia De Vi Publica***; which forbade the carrying of weapons except for hunting and journeys. If burial with smith's tools and weapons did serve

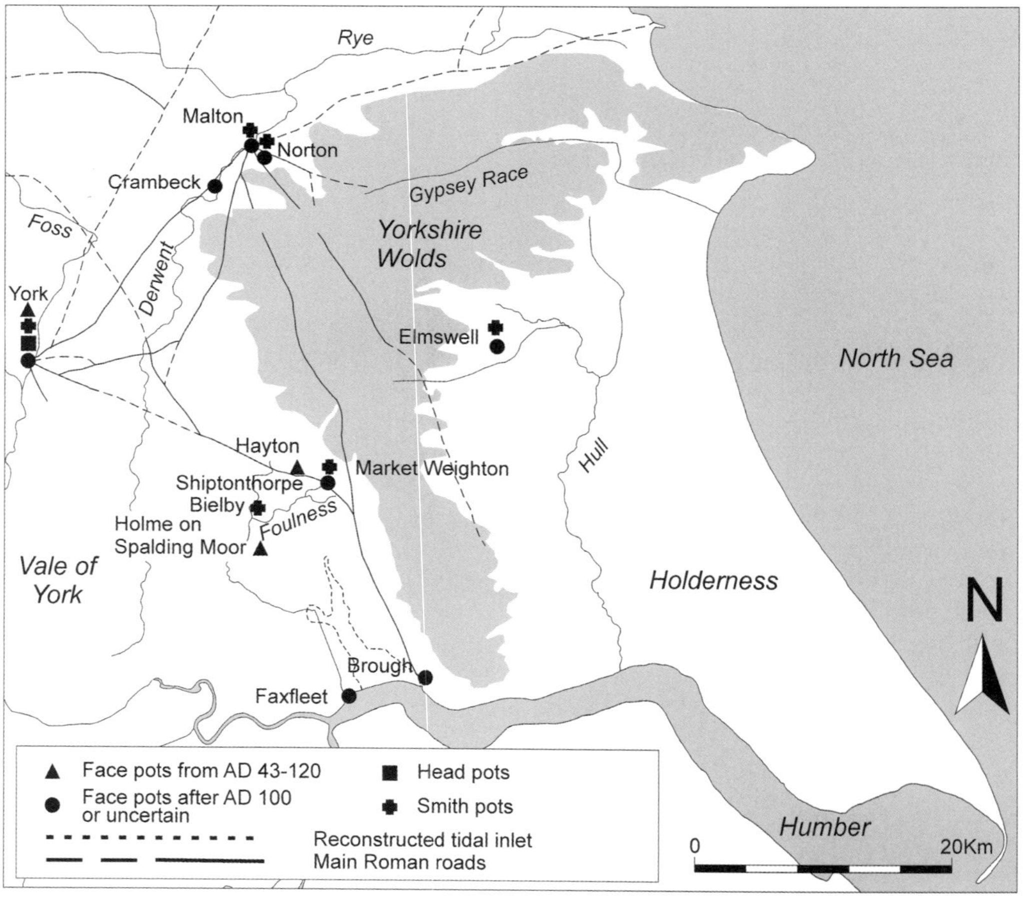

Figure 8: A map of Roman East Yorkshire showing the distribution of face pots, (some of Vulcan), head pots, and smith's tools pots. There are Roman pottery kilns at Crambeck, Norton, and Holme-on-Spalding Moor and evidence for iron production at Holme-on-Spalding Moor, Malton, Bielby and Elmswell. (Map: H. Woodhouse, after K. Dennett).

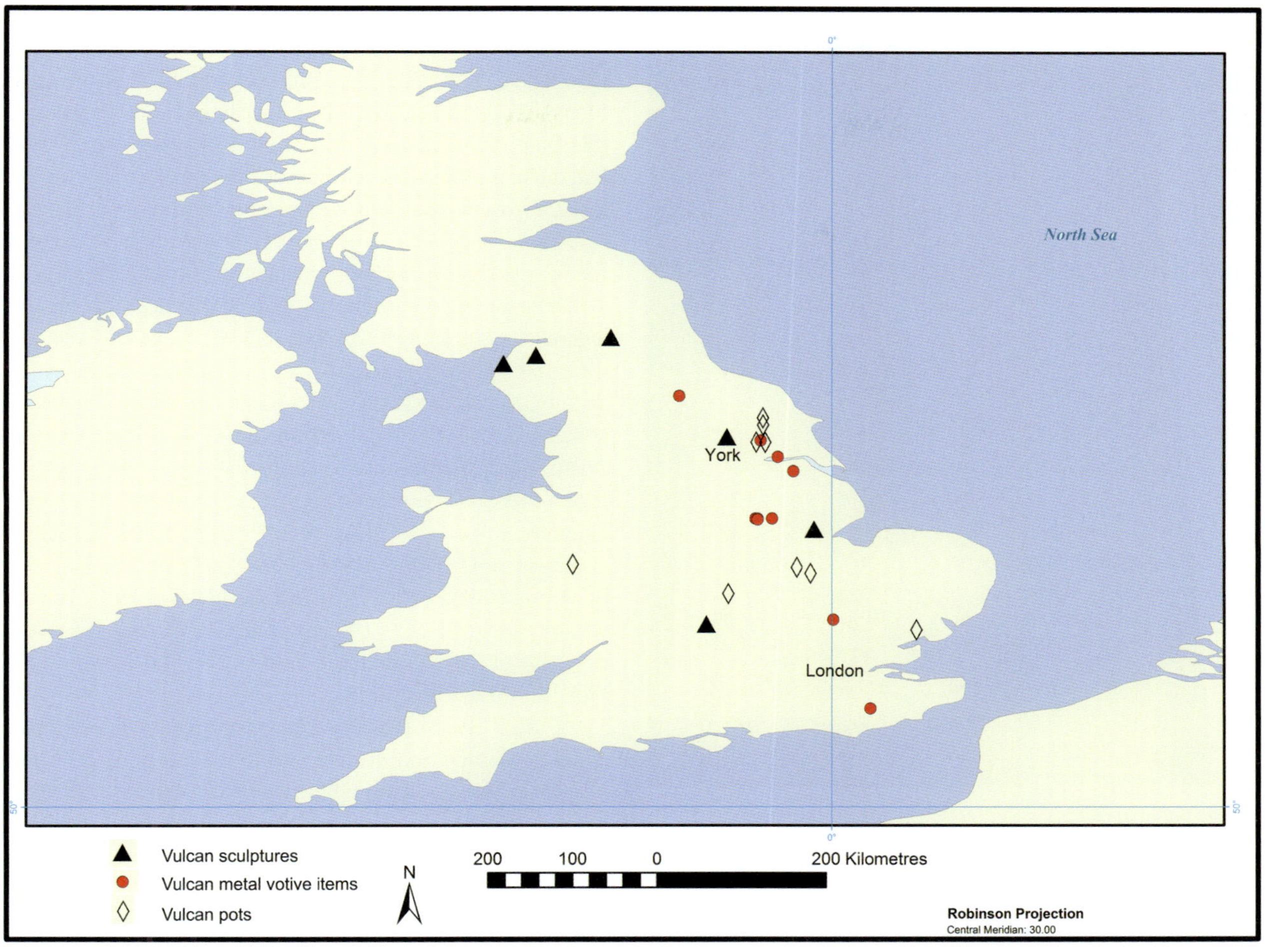

Figure 9: A map of Roman Britain showing the distribution of artefacts associated with Vulcan. (Map: P. Halkon).

to represent important individuals in Iron Age society it is unlikely that the Roman Empire, in its wisdom, would have encouraged such a practice: such people would have been the obvious foci of resistance.

A set of miniature silver blacksmith's tongs, with slightly bowed jaws and extended tips, was, however, found in an unusual cemetery containing decapitated inhumations, some of which may have been gladiators, at Driffield Terrace in York (Hunter-Mann, 2006; Rogers, 2006). So far there is nothing in Roman Britain to compare with the collection of stone funerary *stelae* from a large Gallo-Roman cemetery excavated at Autun in Burgundy, which portrays tradesmen with their tools, including a number of blacksmiths holding their hammer and tongs. Unfortunately it was not possible to equate the skeletons and *stelae*, as the latter were found in secondary positions (Y. Labaune, personal communication; Labaune, 2009; Venault, et al., 2009).

The post-Roman period

It is clear that iron remained of fundamental importance in north-western Europe during the post-Roman era and that this was far from a 'Dark Age' as far as smiths were concerned: some of the finest examples of the blacksmithing craft date from this period. It could be argued that blacksmiths remained important during this era exemplified by the recovered burial of a male with weapons including a spear, sword and shield, from Herouvillette in Normandy (Fig. 10) (Decaens, 1971).

What distinguishes this particular discovery from the many burials of the 6th century AD which contained weapons is the large quantity of blacksmithing tools, originally contained in a wooden chest, at the foot of the body. Even if the deceased was not solely a blacksmith, as some of the tools related to gold working and carpentry, the craft was regarded as being of sufficient importance for smithing tools to be included.

The Herouvillette burial is one of twenty-eight *Schmiedegräber* (smith graves) listed for the Migration and Merovingian periods in Europe north of the Alps (Henning, 1991, p.71). There is a contrast between the distribution of these and Roman period burials. As in the Iron Age, no smith's graves are recorded in Denmark, but 57% of the burials were found in present-day Germany. The vast majority of burials lie close to major rivers.

In some respects, the discovery of a 7th century AD interment found at Tattershall Thorpe in Lincolnshire (Hinton, 2000; Leahy, 2003), which contained two bundles of metalworking tools and other items, including tongs, hammers, a small anvil and files fits into this category. The burial also included iron, copper alloy, sliver and lead scrap, a small piece of gold, glass fragments, small garnets and Roman coins, but no weapons. The study of Anglo-Saxon codes of law tells us that, to some extent, a smith's status in society was

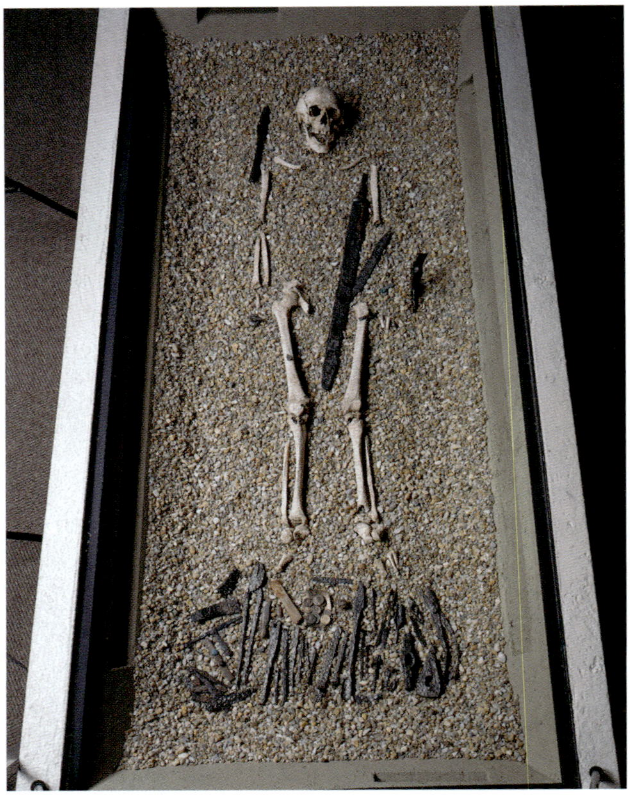

Figure 10: Burial of a male with tools, including those of the smith and including a spear, sword and shield from Herouvillette, Calvados, Normandy. (Photo: Copyright, Musée de Normandie, Caen).

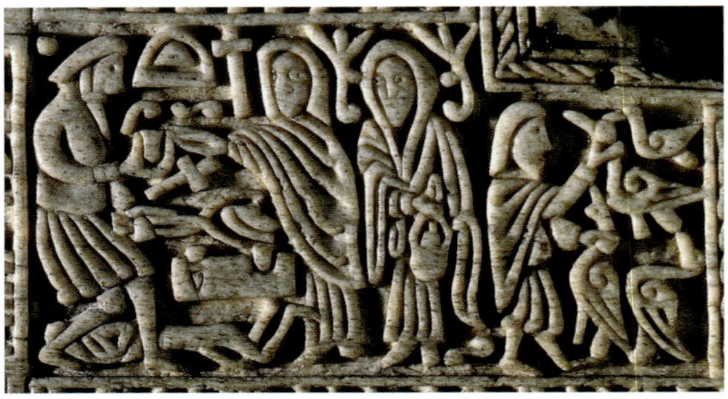

Figure 11: Detail of the panel of the whalebone Anglo-Saxon Frank's Casket showing Wayland in his smithy. (Photo: Copyright Trustees of the British Museum).

determined by the material with which he worked (Henning, 1991; Hamerow, 2002). Not all toolkits were as extensive as that from Tattershall Thorpe. For example, only a set of pincers or tongs was found in the grave of a male with a spear, shield and knife in the antiquarian excavation of a burial mound at Silbertswold in Kent (burial 115) (Roach Smith, 1856).

In the Anglo-Saxon lands on both sides of the North Sea, the religious and magical associations of the blacksmith continued. Welland or Wayland the Smith in England, or Volundr/Voelund in the Germanic/Scandinavian regions, is an important character in myth and legend; the depiction of whose exploits are found on many objects. His legend, which appears in the *Poetic Edda* (Larrington, 1999), is as follows:

Wayland and his brothers married Swan Maidens. After seven years their wives returned to their duties as Valkyries, leaving their husbands desolate. While Wayland's brothers went in search of their wives, Wayland stayed at home. He was then captured by a Swedish king called Nithad (Nidud), who robbed him of his sword and ring, cut his hamstrings and forced Wayland to work for him. Wayland took exception to this, gaining his revenge by secretly killing Nithad's two sons and making goblets and jewellery for the royal family from their skulls, eyes and teeth. He then either seduced or raped Beaduhild or Böðvildr (Nithad's daughter) and flew from the crime scene with the aid of the mechanical wings that he had forged!

Once again we are provided with the example of a legendary blacksmith attributed with transformative and inventive powers regarded as an outsider by the rest of society. It has also been noted (Leahy, 2003, p.172) that the Wayland story tells us something of the relationship between aristocratic or royal patrons and craftsmen (though in this case a rather dysfunctional one) and the ability of this smith, albeit legendary, to work in several materials. The best known portrayal of his story is evidenced in the Anglo Saxon carvings of Frank's Casket (Webster and Backhouse, 1991) a lidded whalebone box thought to have been made sometime in the first half of the 8th century AD, probably in the kingdom of Northumbria, England (Fig. 11).

At first glance the left side of the casket's front panel appears to depict an ordinary smithy, until one notices that a human head is clamped between Wayland's tongs and a torso lies by his feet. This grisly carving clearly portrays the moment when the body parts of Nithad's sons are transformed into the precious objects described above. A similar scene appears on a standing stone at Gotland in Sweden (Ardre VIII), which may date from the 8th to 9th century AD (Staecker, 2006). Inside the forge are two sets of tongs and two hammers, the headless bodies of Nithad's sons lie to the right of the forge and Wayland's flying machine, his eventual method of escape, to the left.

By AD 955 a Neolithic long barrow in what was the county of Berkshire (now Oxfordshire), with a prominent stone facade and associated chambers of the Cotswold-Severn type, was named in an Anglo-Saxon charter as Wayland's Smithy (Birch, 1885-93).

Antiquarian accounts, including the following, show that various superstitions associated with the site prevailed:

> "At this place lived formerly an Invisible Smith, and if a traveller's Horse had lost a Shoe upon the road, he had no more to do than to bring the Horse to this place with a piece of money, and leaving both there for some little time, he might come again and find the money gone, but the Horse new shod"
>
> (Francis Wise, 1738 quoted in Ellis Davidson, 1958).

For many years it was thought that two bars of iron found at the site during excavations in 1919-20 were currency bars. Subsequent examination, however, showed them to be the hinges of doors (Grinsell, 1991).

The symbolic importance of blacksmithing tools to the pagan Anglo-Saxons is demonstrated by the great ceremonial whetstone from the Sutton Hoo ship burial which, though clearly skeumorphic, may be a sceptre (Carver, 1988; Enright, 2006). A further large whetstone was found in an upright position, close to a burial containing weapons, during the excavation of a 'Final Phase' Anglo-Saxon cemetery which was dug into a large prehistoric burial mound at Uncleby, in the East Riding of Yorkshire (Smith, 1912; Williams, 1998).

In Norse mythology/religion the role of the blacksmith remains important, especially in the stories associated with Sigurd who killed the dragon Fafnir with a magical sword named Gram, forged by Regin, the evil blacksmith who raised him. Here the smith is a villain rather than a hero. Their story is portrayed in a series of stone sculptures in Viking Age York and north-eastern Yorkshire (Lang, 1976; Bailey, 1980; Thompson, 2004). At Nunburnholme (Pattison, 1973; Lang, 1991) in East Yorkshire, Sigurd and Regin are shown eating dragon rings (Fig. 12).

A similar scene appeared on the portal of the Stave church at Hylestad, in Setesdal in Norway, which dates from the 12th century AD and is now in the Kulturhistorisk Museum, University of Oslo (Fig. 13).

The choice of this theme for the decoration of the church was not coincidental, as the Setesdal region was important for iron production in the Viking Age. Regin and Sigurd are also shown in a cartoon-like style in a carving at Ramsund in Sweden, where Regin's corpse, minus its head, lies surrounded by smithing tools echoing the earlier Wayland sculpture from Ardre. The distribution of Wayland and Sigurd sculptures is shown in Fig. 14.

As in earlier periods a number of Viking Age burials furnished with the tools of the blacksmith have been discovered, with one of the most impressive at Byggland in Morgedal, Norway (Graham Campbell and Kidd, 1980). This burial also contained weapons, including four swords, richly-inlaid spears, hammers and tongs of different sizes and an anvil. The spears have been interpreted as the products of the deceased's own hand (Martens, 2004). Weapons are also included in a further rich burial, believed to be that of a blacksmith, at By in Hedemark, Norway, which is dated to the 11th century AD (Martens, 2004). In the Norse colony

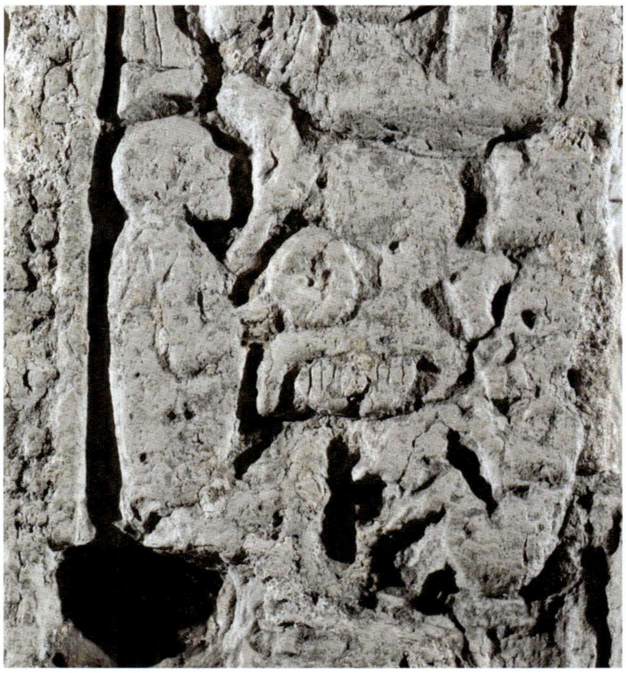

Figure 12: A detail of the Nunburnholme Cross, East Yorkshire, showing Sigurd and Regin eating dragon rings. (Photo: M. Park).

(a)

(b)

(c)

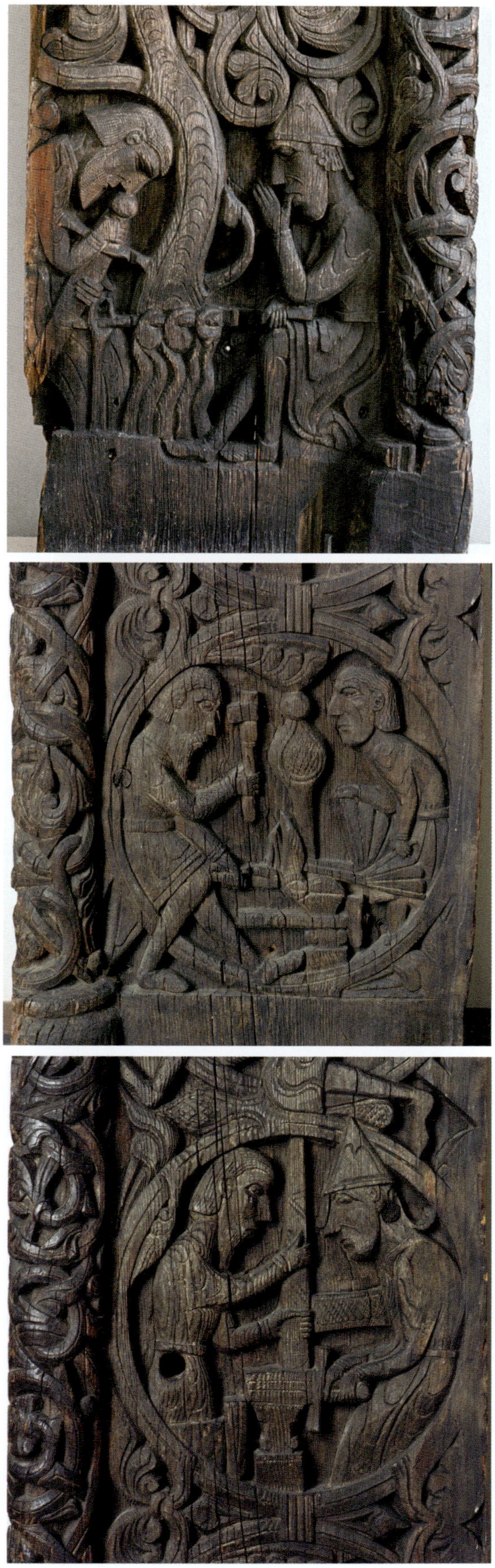

Figure 13: Details from the Hylestad Stave church, Setesdal, Norway. **a)** Sigurd sucks his thumb, scalded by the blood of the dragon Fafnir. The same scene appears on the Nunburnholme Cross (Fig. 12). b) Regin and Sigurd forge the sword Gram. c) Sigurd (helmeted) testing the sword on the anvil. (Photos: Copyright: Kulturhistorisk Museum, University of Oslo).

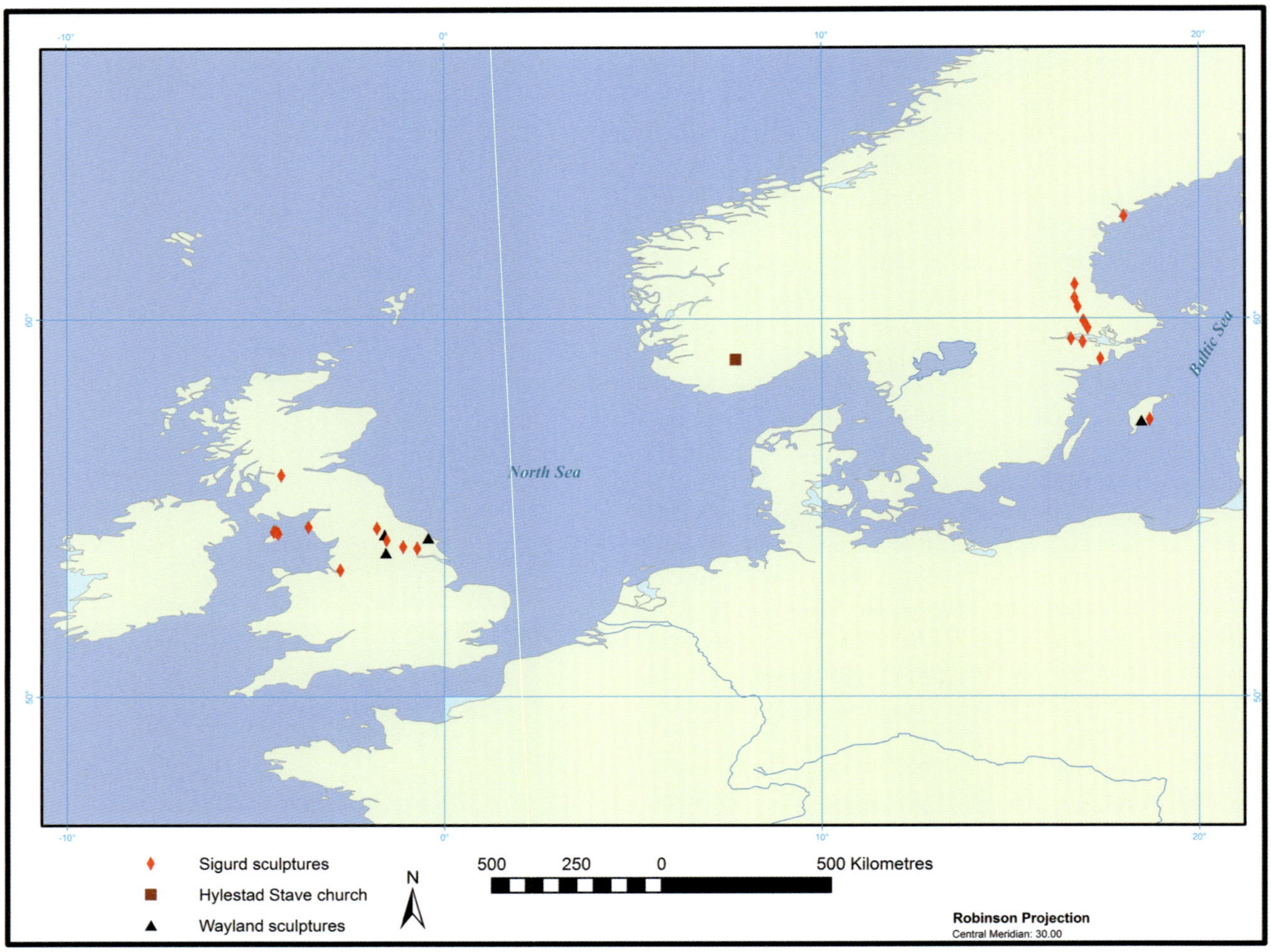

Figure 14: Map showing the distribution of Wayland and Sigurd carvings. (Map: P. Halkon).

of Man, a combination of weapons and blacksmithing tools was deposited in the boat burial excavated at Knoc y doonee (Kermode, 1927) and, within the cemetery at Kilmainham-Islandbridge near Dublin, blacksmith's tools are reported to be among the finds from the Viking cemetery (O'Brien, 1995, 1998). At Ballinaby on the island of Islay (Anderson, 1879), one of a pair of richly furnished burials unearthed in 1877 included a sword and scabbard, parts of a shield, a spear, two axes, an adze, and fragments of a drinking horn and cauldron, as well as a hammer and a pair of tongs (Fig. 15). According to Graham-Campbell and Batey (1998, p.122):

> "The man's social status is confirmed by the presence of blacksmith's tools, for in the context of pagan Norse graves these are not necessarily to be interpreted as workaday equipment of an artisan but as belonging to a skilled weapon smith."

As in the Iron Age, blacksmiths' tools have also been found in hoards in the north western areas of post-Roman Europe, although it is unlikely – especially in the Christian period – that they were votive, as most of the Iron Age examples probably had been.

A significant example of an English hoard was found at Flixborough in North Lincolnshire; comprising tools of the type used for heavy timber work, contained within lead vats dating from the 8th or 9th century AD (Leahy, 2003).

Perhaps the most famous tool hoard from this era, however, was uncovered at Mastermyr in Sweden, which included tools for heavy iron smithing and working, as well as carpentry, found in a well preserved wooden chest. Such was the scale of the find that it is unlikely to have been an accidental loss (Graham-Campbell and Kidd, 1980; Ardwisson and Berg, 1983) and may represent a later continuation of the tradition of votive deposition, as in the pre-Roman Iron Age.

Conclusion

It is clear that blacksmiths and the craft that they pursued were of considerable significance, from the first use of iron tools and weapons into the medieval period and beyond. From 'Celtic' Ireland to Viking Scandinavia, blacksmiths are associated with magic and superstition. Deities and legendary characters

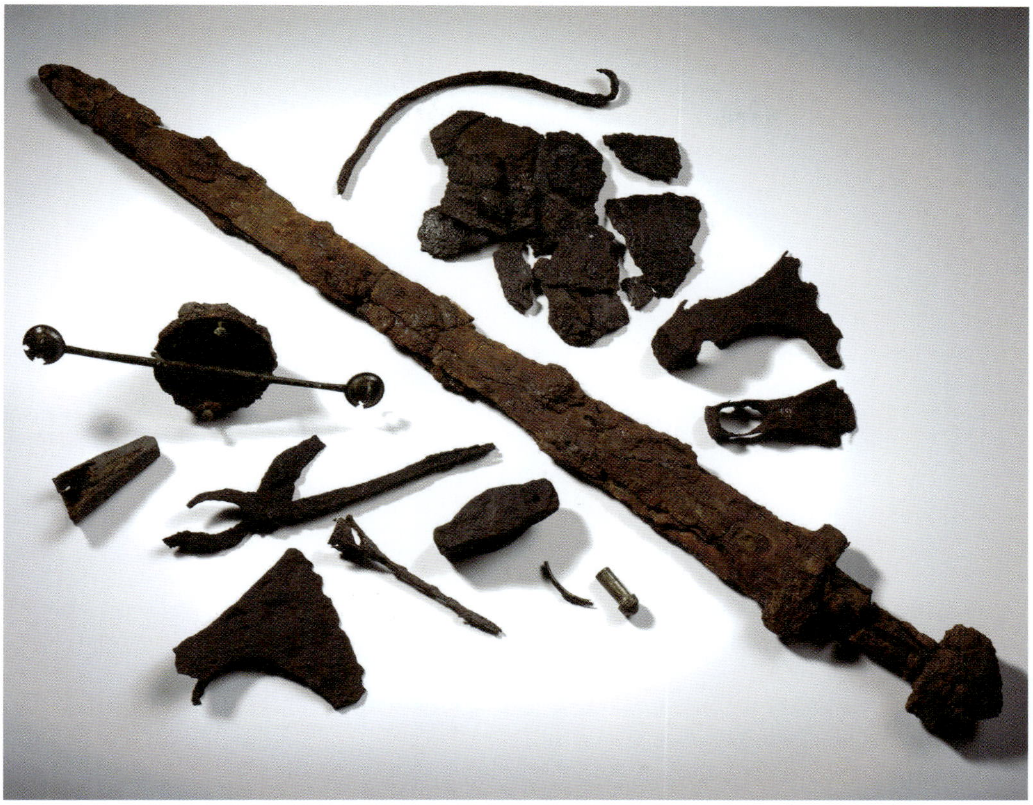

Figure 15: Objects from a richly furnished burial at Ballinaby on the island of Islay found in 1877, including a sword, scabbard, parts of a shield and spear, two axes, an adze, parts of a drinking horn and cauldron, and a hammer and a pair of tongs. (Photo: Copyright National Museum Scotland, Edinburgh).

associated with iron working are portrayed on artefacts in a way that no other craft is. Iron objects, especially swords, are some of the most frequently deposited votive items. It remains difficult, however, to be sure that all those individuals interred with the tools of the trade were actually blacksmiths; although whenever a grave with a hammer or tongs appears in the archaeological record, it is usually presumed that they were. Only 25% of the 90 graves listed in Europe north of the Alps were without weapons (Henning, 1991, p.71). Apart from the burial at Tattershall Thorpe which contained only tools, all the others British examples from Rudston and Whitcombe in Iron Age Britain to the Viking burial at Ballinaby, contained weapons, sometimes a complete set, often laid out in the correct position, as though the interred had been a warrior in life. Henning (1991) argues that many of the burials may represent "warrior-smiths", and that the person in the grave was of high status.

There is still much research to be done on the so-called smith burials. Where skeletons are well preserved, it may be pertinent to examine the physique of the deceased and to analyse the bones for larger than normal proportions of trace elements associated with metalworking in order to obtain more concrete evidence in a similar way to analyses undertaken on skeletons found at the ancient copper mining and smelting centres in the Jordanian desert (Grattan, et al., 2005).

Further analysis of tools buried in the graves is also necessary, in order to ascertain their precise function. Although the basic shape of tools remains broadly the same, subtle typological differences have been noted in some such as tongs (Henning, 1991). Although evidence indicates that *Schmiedegräber* persisted only on the fringes of the Roman Empire, the hammers found in these locations, even in unconquered territories, were often of Roman type (Henning, 1991). It is often presumed that grave goods were included to enable the deceased to continue activities in the next life; however, the handles on the tongs found in some graves, such as those found at Dessau-Grosskühnau, Germany (Henning, 1991, Fig. 7, p.72) – like those from Llyn Cerrig Bach and Waltham Abbey – appear to have been deliberately bent to put them beyond use. This resembles the earlier Bronze Age acts of ritual deposition in which copper-alloy swords were bent and thrown into watery places.

Although the deposition of tools and weapons continues from the Bronze Age into the Iron Age and beyond, workers in copper-alloy do not seem to have accrued the same status as blacksmiths. If there was a mythology associated exclusively with copper working, it has not survived in the same way as that associated with iron, at least in the geographical area covered here.

A key reason for this may lie in the means of production and attendance at a number of recent experimental archaeology conferences, in which both copper alloys and iron were being produced, enabled a simultaneous comparison. It was noticeable that bloomery shaft furnaces attracted a much bigger audience than the hearths used for copper production. Most copper-alloy objects are made by the smelting of copper ores followed by the pouring of liquid metals into moulds, which is a relatively quiet process: iron production is much more theatrical. The constant rhythmical blast of larger bellows than those usually needed for working copper alloys; the flames shooting from the top of the furnace; the tapping of the slag (or demolition of the furnace itself for the non-tapping, slag-pit variety), the grasping of the bloom with tongs and its smithing into a semi-product and, finally, the clash of hammer on the object and anvil during blacksmithing all combine to create a spectacle. The making of an iron object rather than a copper-alloy one is intrinsically more complex and, perhaps as a result, appeared more miraculous to spectators. In the mythology explored here, metalworkers are notably portrayed as being 'different' and occupying liminal positions within society and, relatedly, it is interesting to note that most of the Iron Age iron production sites researched by the author in the Foulness Valley in East Yorkshire (Halkon and Millett, 1999; Halkon, 2008) were similarly marginal in comparison to denser settlement concentrations, whereas smithing slags were more common in 'domestic' contexts.

The author's first interest in ancient iron arose from finds of slag and bog ore on his father's farm at Hasholme in East Yorkshire, which eventually led to a large scale landscape archaeology project, revealing one of the largest prehistoric iron industries in Britain. With the aim of popularising this research a website, www.ironmasters.hull.ac.uk, provides an explanation of the process and the importance of iron working. A story of Iron Age iron production was constructed around an invented character – Bryn the Ironmaster – and based on an assortment of archaeological and palaeoenvironmental evidence, in an attempt to recreate the type of myths and legends relating to iron and its production still surviving in other parts of the world and which are referred to in this volume, with a view to introducing the magic of iron to a new generation.

Acknowledgements

I wish to thank the individuals and institutions who kindly supplied illustrations, which are acknowledged in the appropriate captions. I am particularly grateful to Dr John Pearce for pointing out several useful sources, particularly the Autun sculptures, and to Yannick Labaune, who sent me further information about them. Thanks are due to Professor Paul Busch for his help in translating J. Henning's important article. Finally, I am most grateful for the helpful comments of anonymous referees and the patience of the editors.

References

Anderson, J., 1879. Notes on the Contents of Two Viking Graves in Islay, Discovered by William Campbell, Esq., of Ballinaby; With Notices of the Burial Customs of the Norse Sea-Kings, as Recorded in the Sagas and Illustrated by their Gravemounds in Norway and in Scotland. *Proceedings of the Society of Antiquaries for Scotland*, 2, pp.51-94.

Ardwisson, G. and Berg, G., 1983. *The Mastermyr Find: A Viking Age Tool Chest from Gotland*. Stockholm: Kungl. Vitterhets Historie och Antikvitets Akademien: Almqvist & Wiksell International.

Bailey, R.N., 1980. *Viking Age Sculpture in Northern England*. London: Collins Archaeology.

Bataille, G. and Guillaumét, J-P. eds., 2006. *Les dépôts métalliques au second âge du Fer en Europe tempérée*. Glux-en-Glenne: Bibracte, Centre Archéologique Européen.

Birch, W. de Gray, 1885-93. *Cartularium saxonicum: A Collection of Charters Relating to Anglo-Saxon History*. London: Whiting & Co.

Carver, M., 1998. *Sutton Hoo: Burial Ground of Kings?* Philadelphia: University of Pennsylvania Press.

Crerar, B., 2006. Votive Leaves from Roman Britain. In M. Henig ed. *Roman Art, Religion and Society. New Studies from the Roman Art Seminar*. British Archaeological Reports International Series 1577. Oxford: Archaeopress, pp.71-89.

Crew, P., 1991. The Experimental Production of Prehistoric Bar Iron. *Historical Metallurgy*, 25, pp.21-36.

Crew, P. and Crew, S. eds., 1995. *Iron for Archaeologists: A Review of Recent Work on the Archaeology of Early Ironworking Sites in Europe*. Plas Tan y Bwlch Occasional Paper No. 2. Maentwrog : Plas Tan y Bwlch.

Crew, P., 1998. Laxton Revisited: A First Report on the 1998 Excavations. *Historical Metallurgy*, 32, pp.49-53.

Crew, P., forthcoming, Twenty five years of bloomery experiments: perspectives and prospects. In D. Dungworth, and R. Doonan, eds., *Accidental and Experimental Archaeometallurgy. Proceedings of the 2010 Conference*. London: Historical Metallurgy Society.

Cunliffe, B., 2004. *Iron Age Communities in Britain: An Account of England, Scotland and Wales from the Seventh Century BC until the Roman Conquest*. London: Routledge.

De Navarro, J.A.L., 1972. *The Finds from the Site of La Tène. Vol. 1: Scabbards and the Swords Found in Them*. London: Oxford University Press.

Decaens, J., 1971. Un nouveau cimetière du haut Moyen Age en Normandie, Hérouvilette (Calvados). *Archéologie Médiévale*, 1, pp.1-126.

Dent, J.S., 1985. Three Chariot Burials from Wetwang, Yorkshire. *Antiquity*, 59, pp.85-92.

Didsbury, P., 1984. A Grey-ware Cult Shard from Bielby. *Britannia*, 15, p.239.

Dumont, A., Gaspari, A. and Wirth, S., 2006. Les objets métalliques des âges du Fer découverts en contexte fluvial. Les exemples de la Saône (France), du Danube (Allemagne) et de la Ljublanica (Slovénie). In G. Bataille & J.-P. Guillaumet eds. *Les dépôts métalliques au second âge du Fer en Europe tempérée*. Glux-en-Glenne: Bibracte, Centre Archéologique Européen, pp.257-77.

Ellis Davidson, H.R., 1958. Weland the Smith. *Folklore*, 69, pp.145-59.

Enright, M.J., 2006. *The Sutton Hoo Sceptre and the Roots of Celtic Kingship Theory*. Dublin: Four Courts Press.

Fell, V., 1998. Iron Age Ferrous Hammerheads from Britain. *Oxford Journal of Archaeology*, 17, pp.207-25.

Fellmann, R., 1991. Die Zinktafel von Bern - Thormebodenwald und ihre Inschrift. *Archäologie Schweiz*, 14, pp.270-3.

Fellman, R., 2000. Une divinité gallo-romaine inconnue, son précurseur Gaulois et un dépôt de statuettes trouvé en France et actuellement aux États-Unis. In H. Walter ed. *La sculpture d'epoque romaine dans le nord et l'est des Gaules et dans les regions avoisinantes: Acquis et problematiques actuelles*. Paris: Collection Annales Litteraires, pp.163-77.

Field, N. and Parker-Pearson, M., 2002. *Fiskerton: An Iron Age Timber Causeway with Iron Age and Roman Votive Offerings*. Oxford: Oxbow Books.

Fitzpatrick, A.P., 2007. The Fire, the Feast and the Funeral: Late Iron Age Mortuary Practices in South-Eastern England. In V. Kruta and G. Leman-Delerive eds. *Feux des morts, foyers des vivants - Les rites et symboles du feu dans les tombes de l'Age du fer et de l'époque romaine*. Lille: Revue de Nord, Hors série, 11, pp.123-45.

Ford, P.K., 1999. *The Celtic Poets: Songs and Tales from Early Ireland and Wales*. Belmont, Mass: Ford & Bailie.

Fox, C., 1945. *A Find of the Early Iron Age from Llyn Cerrig Bach, Anglesey: Interim Report*. Cardiff: National Museum of Wales.

Garrow, D., Gosden, C. and Hill, J.D. eds., 2008. *Rethinking Celtic Art.* Oxford: Oxbow Books.

Giles, M., 2007. Making Metal and Forging Relations: Ironworking in the British Iron Age. *Oxford Journal of Archaeology*, 26, pp.395-413.

Graham-Campbell, J.A. and Batey, C.E., 1998. *Viking-Age Scotland. An Archaeological Survey.* Edinburgh: Edinburgh University Press.

Graham-Campbell, J.A. and Kidd, D., 1980. *The Vikings.* London: British Museum Publications.

Grattan, J., Abu Karaki, L., Hine, D., Toland, H., Gilbertson, D., Al-Saad, Z. and Pyatt, B., 2005. Analyses of Patterns of Copper and Lead Mineralization in Human Skeletons Excavated from an Ancient Mining and Smelting Centre in the Jordanian Desert: A Reconnaissance Study. *Mineralogical Magazine*, 69, pp.653-66.

Green, M., 1983. *The Gods of Roman Britain.* Princes Risborough: Shire Archaeology.

Green, M., 1995. *The Celtic World.* London: Routledge.

Green, M., 1998. *Animals in Celtic Life and Myth.* London: Routledge.

Grinsell, L.V., 1991. Wayland the Smith and His Relatives: A Legend and Its Topography. *Folklore*, 102, pp.235-6.

Halkon, P., 1992. Romano-British Facepots from Holme-on-Spalding Moor and Shiptonthorpe, East Yorkshire. *Britannia*, 23, pp.222-8.

Halkon, P., 2006. Facepots from Shiptonthorpe. In M. Millett ed. *Shiptonthorpe East Yorkshire: Archaeological Studies of a Romano-British Roadside Settlement.* Yorkshire Archaeological Report 5. Leeds: Yorkshire Archaeological Society, Roman Antiquities Section and East Riding Archaeological Society, pp.29-37.

Halkon, P., 2007. *Iron in the Arras Culture and Later Iron Age East Yorkshire.* Paper presented at Early Iron Working in Europe Conference, Plas Tan y Bwlch, September 2007.

Halkon, P., 2008. *Archaeology and Environment in a Changing East Yorkshire Landscape: The Foulness Valley c. 800 BC to c. AD 400.* British Archaeological Reports 472. Oxford: Archaeopress.

Archaeological Report 4, Yorkshire Archaeological Society (Leeds)

Halkon, P., 2012. Iron, landscape and power in Iron Age East Yorkshire. *The Archaeological Journal*168 pp. 133-165.

Halkon, P. and Millett, M. eds., 1999. *Rural Settlement and Industry: Studies in the Iron Age and Roman Archaeology of Lowland East Yorkshire.* Yorkshire Archaeological Report 4. Leeds: Yorkshire Archaeological Society, Roman Antiquities Section and East Riding Archaeological Society.

Halkon, P., Millett, M., Easthaugh, E., Taylor, J. and Freeman, P., 2000. *The Landscape Archaeology of Hayton, East Yorkshire.* Hull: University of Hull.

Hamerow, H., 2002. *Early Medieval Settlements: The Archaeology of Rural Communities in Northwest Europe, AD 400-900.* Oxford: Oxford University Press.

Hanford, A. (translator) and Gardner, J. (reviser), 1982. Julius Caesar, *The Conquest of Gaul.* London: Penguin Books.

Harding, D.W., 2007. *The Archaeology of Celtic Art.* Abingdon & New York: Routledge.

Haselgrove, C. and Hingley, R., 2006. Iron Deposition and Its Significance in Pre-Roman Britain. In G. Bataille and J-P. Guillaumet eds. *Les dépôts métalliques au second âge du Fer en Europe tempérée.* Glux-en-Glenne: Bibracte, Centre Archéologique Européen, pp.148-63.

Henig, M., 1995. *The Art of Roman Britain.* London: Batsford.

Henning, J., 1991. Schmiedegräber nördlich der Alpen. *Saalburg Jahrbuch*, 46, pp.65-82.

Herbert, E.W., 1993. *Iron, Gender, and Power: Rituals of Transformation in African Societies (African Systems of Thought).* Bloomington: Indiana University Press.

Hill, J.D., 1995. *Ritual and Rubbish in the Iron Age of Wessex.* Oxford: Tempus Reparatum.

Hingley, R., 1990. Iron Age 'Currency Bars': The Archaeological and Social Context. *Archaeological Journal*, 147, pp.91-117.

Hinton, D.A., 2000. *A Smith in Lindsey: The Anglo-Saxon Grave at Tattershall Thorpe.* Lincolnshire Society for Medieval Archaeology Monograph 16. London: Society for Medieval Archaeology.

Hobbs, R. and Henig, M., 2006. Newark area, Nottinghamshire: Roman gold finger-ring setting. *Treasure Annual Report2005/6.* London: British Museum.

Howe, M.D., Perrin, J.R. and Mackreth, D.F., 1981. *Roman Pottery from the Nene Valley: A Guide.* Peterborough City Museum Occasional Paper 2. Peterborough: City Museum and Art Gallery.

Hunter-Mann, K., 2006. *Romans Lose Their Heads. An Unusual Cemetery at The Mount, York.* The Archaeology of York (Web Series 6, AYW6). York Archaeological Trust. http://www.iadb.co.uk/driffield6/index.php

Jackson, D.A. and Tylecote, R.F., 1988. Two New Romano-British Iron-working Sites in Northamptonshire – A New Type of Furnace. *Brittania*, 19, pp.275-98.

Jackson, D.A., Ambrose, T.M., Pacitto, A.L., Woods, P.J., Tylecote, R.F., Biek, L., Hartley, B.R., Hartley, K., Jenkins, F., Goodburn, R.A., Esmonde Cleary, S., Butcher, S.A., Bayley, J., Charlesworth, D., Anderson, F.W., Cook, A., Jones, R., Evans, J.G. and Arthur, J.R.B., 1978. Excavations at Wakerley, Northants, 1972-75. *Britannia*, 9, pp.115-242.

Johns, C.M., 1991. Some Unpublished Jewellery from Roman Britain. *Jewellery Studies*, 5, pp.55-64.

Jorgensen, L., Storgaard, B., Thomsen, L.G., 2003. *The Spoils of Victory – The North in the Shadow of the Roman Empire.* Copenhagen: Nationalmuseet.

Kermode, P.M.C., 1927. Knoc-y-doonee. *Proceedings of the Isle of Man Natural History and Antiquarian Society*, 10, pp.241-6.

Labaune, Y., 2009. La topographie funéraire antique d'Autun Bilan et nouvelles propositions à la lumière des découvertes récentes. *Bulletin Archéologique Antiquité Archéologie Classique*, 35, pp.97-128.

Lang, J.T., 1976. Sigurd and Weyland in Pre-Conquest Carving from the North of England. *Yorkshire Archaeological Journal*, 48, pp.83-94.

Lang, J.T., 1991. *Corpus of Anglo-Saxon Sculpture, III: York and Eastern Yorkshire.* Oxford: Published for the British Academy by Oxford University Press.

Larrington, C. trans., 1999. *The Poetic Edda.* Oxford: Oxford World's Classics.

Leahy, K., 2003. *Anglo-Saxon Crafts.* Stroud: Tempus.

Martens, I., 2004. Indigenous and Imported Viking Age Weapons in Norway – A Problem with European Implications. *Journal of Nordic Archaeological Science*, 14, pp.125-37.

Merrifield, R., 1987. *The Archaeology of Ritual and Magic.* London: Batsford.

Milcent, P.-Y. ed., 2007. *L'économie du fer protohistorique: de la production à la consommation du metal.* XXVIIIe colloque de l'AFEAF, Toulouse, 2004: (Thème specialise). Pessac: Fédération Aquitania Supplément 14/2.

Millett, M. ed., 2006. *Shiptonthorpe East Yorkshire: Archaeological Studies of a Romano-British Roadside Settlement.* Yorkshire Archaeological Report 5. Leeds: Yorkshire Archaeological Society, Roman Antiquities Section and East Riding Archaeological Society.

O'Brien, E., 1995. A Tale of Two Cemeteries. *Archaeology Ireland*, 9, No. 3. pp.13-15.

O'Brien, E., 1998. The Location and Context of Viking Burials at Kilmainham and Islandbridge, Dublin. In H.B. Clarke, M. Ní Mhaonaigh and R.Ó. Floinn eds. *Ireland and Scandinavia in the Early Viking Age.* Dublin: Four Courts Press, pp.203-21.

Ottaway, P., 2004. *Roman York.* Stroud: Tempus.

Pattison, I.R., 1973. The Nunburnholme Cross and Anglo-Danish Sculpture in York. *Archaeologia*, 104, pp.209-34.

Piggott, S., 1953. Three Metalwork Hoards of the Roman Period from Southern Scotland. *Proceedings of the Society of Antiquaries of Scotland*, 87, pp.1-53.

Potter, T.W., 1997. *Roman Britain*. London: British Museum Press.

RIB, 1965, *The Roman Inscriptions of Britain*. Volume 1, edited by R.G. Collingwood and R.P. Wright. Oxford: Clarendon Press.

Roach Smith, C., 1856. *Inventorium sepulchrale: An Account of Some Antiquities Dug Up at Gilton, Kingston, Sibertswold, Barfriston, Beaksbourne, Chartham, and Crundale in the County of Kent, from A.D. 1757 to A.D. 1773 Edited from the Original Manuscript in the Possession of Joseph Meyer*. London: Printed for the Subscribers Only.

Roberts, O., 2007. Accident not Intention: Llyn Cerrig Bach, Isle of Anglesey, Wales – Site of an Iron Age Shipwreck. *International Journal of Nautical Archaeology*, 31, pp.25-38.

Rogers, N., 2006. The Small Finds. In K. Hunter-Mann, *Romans Lose Their Heads. An Unusual Cemetery at The Mount, York*. The Archaeology of York (Web Series 6, AYW6). York Archaeological Trust. http://www.iadb.co.uk/driffield6/index.php

Scott, B.G., 1991. *Early Irish Ironworking*. Belfast: Ulster Museum Publication 266.

Shinnie, P.L and Haaland, R., 1986. *African Iron Working: Ancient and Traditional*. Oslo: Scandinavian University Press.

Smith, R., 1912. On the Excavation by Canon Greenwell FSA in 1868 of an Anglo-Saxon Cemetery at Uncleby, East Riding of Yorkshire. *Proceedings of the Society of Antiquaries*, 24, pp.146-58.

Staecker, J., 2006. Heroes, Kings and Gods: Discovering Sagas on Gotlandic Picture Stones. In A. Andrén, K. Jennbert and C. Raudvere eds. *Old Norse Religion in Long-Term Perspectives: Origins, Changes and Interactions*. Lund: Nordic Academic Press, pp.363-9.

Stead, I.M., 1991a. The Grave Goods. In G.M. Aitken and G.N. Aitken eds. Excavations at Whitcombe, 1965–1967. *Proceedings of the Dorset Natural History and Archaeology Society*, 112, pp.57-94.

Stead, I.M., 1991b. *Iron Age Cemeteries in East Yorkshire*. London: English Heritage.

Stratton, S., 2005. *Analysis of Roman Metal Detected Artefacts from Hayton*. BA Hons Dissertation, University of Hull.

Thompson, V. 2004. *Dying and death in Later Anglo-Saxon England*. Woodbridge: Boydell Press.

Venault, S., Deyts, S., Le Bohec, Y. and Labaune,Y., 2009. Les stèles funéraires de la nécropole de Pont-l'Evêque à Autun: contextes de découverte et étude du corpus. *Bulletin Archéologique: Antiquité Archéologie Classique*, 35, pp.129-204.

Walker, M., 2007. *Iron Objects in Iron Age Burials of East Yorkshire*. BA Dissertation, University of Hull.

Webster, G., 1989. Deities and Religious Scenes on Romano-British Pottery. *Journal of Roman Pottery Studies*, 2, pp.1-19.

Webster, L. and Backhouse, J., 1991. *The Making of England. Anglo-Saxon Art and Culture AD 600-900*. London: British Museum.

Whimster, R., 1991. *Burial Practices in Iron Age Britain*. Oxford: British Archaeological Reports, British Series 90.

Williams, H., 1998. Monuments and the Past in Early Anglo-Saxon England.

The Past in the Past: The Reuse of Ancient Monuments. *World Archaeology*, 30, pp.90-108.

Wilson, P., 2002. *Cataractonium, Roman Catterick and its Hinterland, Excavations and Research 1958-1997*. York: Council for British Archaeology.

Author's address

- Dr Peter Halkon BA MA FSA
  Lecturer in Archaeology, University of Hull

West and Central Asia

Early Iron in the Near East, a review of evidence

H. Alexander Veldhuijzen

ABSTRACT: More than a decade after the first finds of metallurgical debris at Tell Hammeh, Jordan, and Tel Beth-Shemesh, Israel, it is surprising how little new material has come to light. The 'coming of the age of iron' (Wertime and Muhly 1980) near the end of the Second Millennium BC, an event of major historical importance which gave rise to the subsequent period being called the Iron Age, therefore remains surrounded by questions rather than answers. How did iron production start and how did it evolve? How was it embedded in society? Drawing on the author's work in recent years, this article reviews the scarce evidence for iron smelting and smithing in the Near East, and tries to answer some of the questions on early iron using the answers suggested by material found at Tell Hammeh, Jordan, Tel Beth-Shemesh and Tell es-Safi/Gath, Israel.

Introduction

This article explores the impact and role of iron in society at the end of the Second Millennium BC (Fig. 1), an event often referred to as the 'coming of the age of iron' (Waldbaum 1999, Wertime and Muhly 1980). Reconstructing what can be known about this period draws on textual evidence and a relatively small number of iron artefacts. It is neither intended to discuss each source or find, nor to make sweeping statements about heavily debated subjects such as origin or spread of iron technology.

Where little has been archaeologically attested about the appearance of iron in the Second Millennium beyond a scattering of artefacts, perhaps even less is known

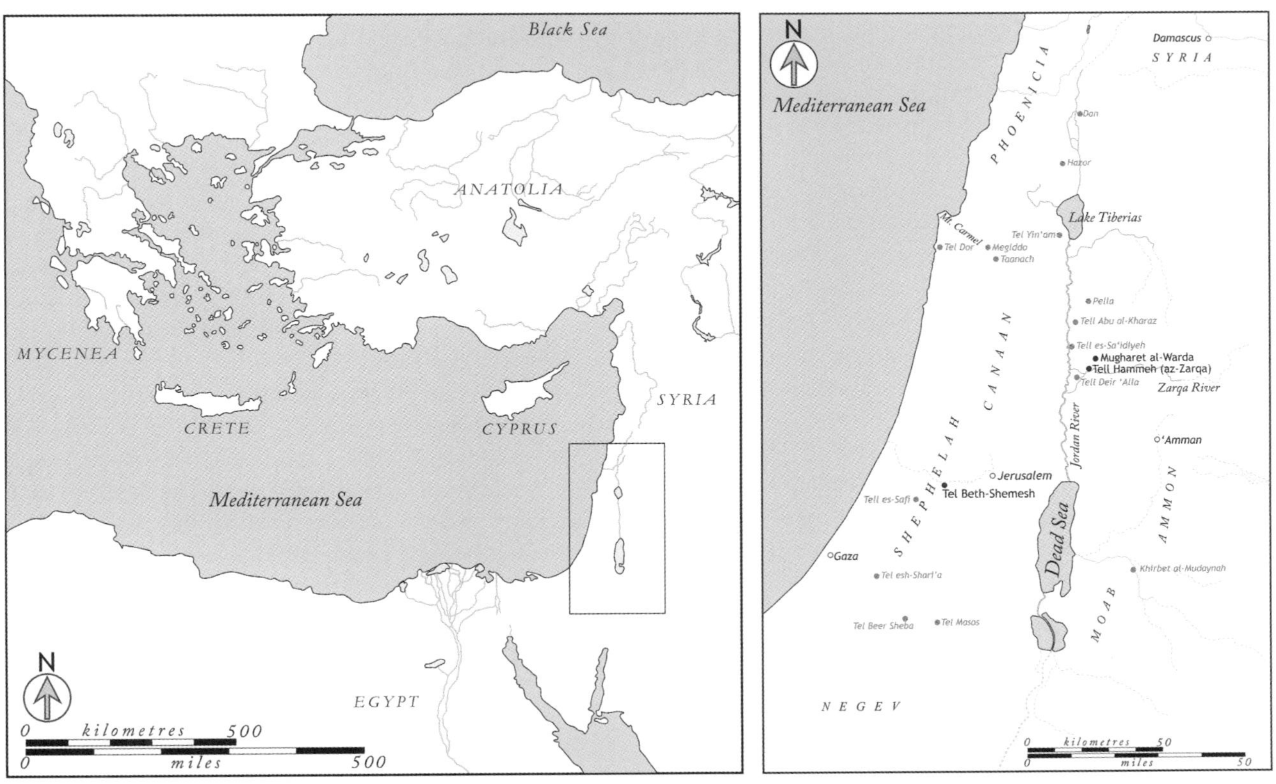

Figure 1: Map of the Eastern Mediterranean. Inset: Map of the Southern Levant, indicating the location of Tell Hammeh, Tel Beth-Shemesh, Tell es-Safi/Gath, Mugharet al Warda, and several of the major sites in the Jordan valley.

about developments and technology at the turn of the First Millennium BC. This is, of course, a problematic archaeological lacuna, as it is exactly at this time that quantities of iron artefacts increase significantly and iron starts to make a real and lasting impact on the Mediterranean world. The marked increase in the adaptation of iron coincides with society itself witnessing significant changes due to the withdrawal or collapse of the Hittite, Mycenaean, and Egyptian political and economic powers. It is through the unique finds of early iron smelting evidence at Tell Hammeh in Jordan (930/910 ± 40 cal BC), and of the contemporaneous iron smithy at Tel Beth-Shemesh, Israel (c. 900 cal BC), that an attempt can be made to interpret aspects of early iron technology and its place in the society and economy of the Mediterranean and Near East at the transition from the second to the First Millennium BC.

The innovation and adaptation of iron

Intentionally smelted iron – as opposed to processed chance finds of meteoritic or native iron – first appears in the Second Millennium BC. The artefacts produced are generally carbon poor or carbon-free, restricting their use to a predominantly ornamental function. Most of these artefacts are of a high status appearance, often bi-metallic, where the iron is quite frequently the secondary and decorative component.

The origin and early history of iron are still widely debated topics. Despite a few new discoveries of smithing workshops in recent years, it still proves problematic to reconstruct the innovation and spread of this new technology today. This is certainly aggravated by the scarceness of iron artefacts, the commonly bad state of their preservation, and especially the absence of evidence for iron production proper. Several reasons have been proposed for the demise of bronze and rise of iron: collapse of the tin trade, the greater ease of iron ore procurement over copper and tin, the withdrawal of major powers from the region and collapse of international trade, and improving smithing techniques. In all likelihood, it is a combination of most or all of these factors that lies behind the ascendency of iron (see discussions in: Curtis et al. 1979, Muhly 1983, 1977, 1973, Muhly et al. 1977, Pleiner 2000, 7-22, Snodgrass 1980, Veldhuijzen and Rehren 2007, Waldbaum 1999, 1989, 1980, 1978, Wertime and Muhly 1980).

Iron artefacts that were, in all likelihood, intentionally smelted do appear in quite early contexts in the Near East. Evidence for their actual *smelting* (direct reduction of ore to produce an iron bloom), *primary smithing* (consolidation of the bloom to bar or billet; *bloomsmithing*) and *secondary smithing* (working of consolidated iron to artefacts in a smithy; *forging*) prior to the middle of the 1st millennium BC has hardly been attested, with the exception of a few sites with evidence of secondary smithing (Craddock 1995, 259, Eliyahu-Behar *et al.* 2008, Eliyahu-Behar *et al.* 2012, Pleiner 2000, 7-8, Rothenberg and Tylecote 1991, Veldhuijzen 2012, 2009c, Veldhuijzen and Rehren 2007, 189-191, Waldbaum 1978, 65).

Literary and artefactual evidence

Literary evidence

The literary evidence on iron particularly concerns the supposed monopoly of the Hittites in producing and using iron, and predominantly consists of Hittite tax lists and palace inventories. When interpretating these texts it should be taken into account that they may serve propaganda purposes as well as administrative ones. Nevertheless, from these texts a picture emerges of an initially rare and valuable metal used for ornamental purposes, that gradually becomes more common and applied in daily life. The literary evidence abruptly dissapears with the beginning of the 'dark ages' of the 11th and 10th century BC. With the often violent collapse of the Mycenaean kingdoms, the Hittite Empire in Anatolia and Syria, and the Egyptian Empire in Syria and Canaan, existing economic structures, long distance international trade, and (supposed) control over technologies and their products are extinguished.

It appears that the absence of literary evidence due to the Late Bronze Age collapse of the economic and social structures in the region coincides with the crucial period when adaptation of iron in daily life undergoes a rapid acceleration. The end of the 'Dark Ages' is characterised by the emergence in the 10th century BC of scores of small (semi-)independent kingdoms and fiefdoms around the Levant, together with the rise of Neo-Hittite and Aramean kingdoms in upper Mesopotamia, followed in the 9th and 8th century by the Neo-Assyrian Empire.

In the 9th and especially the 8th century BC, a range of Assyrian and Urartean texts are available that mention the quantities of iron that the various kings plunder and capture, and that this happens even in small remote places. These texts show that iron is exchanged between kings in large quantities, and strongly suggest that by the end of the 8th century BC, iron is widely used and available in all layers of society and daily life throughout the region. From the fact that the Assyrian kings procure their iron by the ransacking and plunder of small remote settlements, it appears that the industry and especially production (mining, smelting, and smithing) was not under (complete) state control, but rather organised on a far more localised basis.

Besides the important role of the Assyrians in this period, it is interesting to note that the biblical texts suggest a monopoly on iron technology held by the Philistines on the coast of Palestine. Such a Philistine 'monopoly' clearly refers to the suppression of local iron-workers rather than a monopoly on the technology itself. One could even argue that there is an *argumentum ex silentio* here that there is widespread local smithing going on throughout Cis-Jordan, which is of concern to the Philistines trying to maintain their control over the territory they occupy. The idea of there being no smith in the land of Israel becomes especially poignant when considering the finds at Tel Beth-Shemesh (see below).

One can only speculate how, during the undocumented 'Dark Ages', a fully state controlled industry producing small quantities of an elite metal in the second millenium developed into what appears to be a widespread locally organised 'democratic' enterprise with a much higher quantitative output (albeit in small batches) of a metal for daily use and warfare. Completely lacking are sources hinting at where the iron itself may originate (for a detailed review of literary evidence see Veldhuijzen 2012, 238-239 and references therein).

Artefactual evidence

By and large, the appearance of iron begins in Central and East Anatolia (Fig. 1) at the beginning of the 2nd millennium, as attested by iron artefact fragments that generally stem from funerary or ceremonial contexts. From there, iron 'spreads' to the east into the Assyrian realm and the north-east with finds along the Black Sea coast of modern-day Georgia. Towards the end of the 2nd millennium the main concentration of iron artefacts shifts to the south-east of Anatolia and Northern Syria, from there spreading through the Levant.

From c. 1200 BC iron artefacts become increasingly utilitarian and less rare. For many years modern day Georgia has been regarded as the potential origin of iron technology in the region, however, recent (re-) analysis on previously and newly excavated material show that no early evidence exists for actual smelting of iron in this area, and that iron more likely arrives here from Anatolia than vice-versa (Nieling 2009).

Between c. 1000 BC to 800 BC, iron gradually becomes the primary utilitarian metal in society. However, one should notice that copper/bronze neither disappears from the archaeological record, nor is reduced to a minor or strictly ornamental metal. Quantities of iron artefacts now start to increase significantly. Their presence in practically any settlement of this period clearly confirms the impression from the textual sources that iron is widely available and a utilitarian metal, not an exclusively high-status or ceremonial one.

From circa 11th century BC, larger concentrations of iron artefacts are found in Syria and Palestine than in Anatolia or Mesopotamia, although this is an archaeological reality, and one in constant flux, rather than a historical truth. Compared to the Second Millennium, the Early First Millennium BC is characterised by a dearth of literary sources together with more extensive but scattered and physically poor artefactual evidence (for a detailed review of artefactual evidence see Veldhuijzen 2012, 239-240 and references therein). Through the finds of early smelting activity at Tell Hammeh in Jordan, and quite early smithing activities attested at Tel Beth-Shemesh, Israel and at a smaller scale also at Tell es-Safi/Gath, Israel, interpretation of the role of iron in this period is getting one small step closer (Fig. 1).

Production evidence

Iron artefacts of the Eastern Mediterranean are notoriously very heavily corroded (Fig. 2), and – if released for analysis at all – usually do not reveal much more than the smithing techniques applied, the presence or absence of carbon, and stylistic preferences. Slag and related production debris however can reveal a host of information about the technology used, from information on ore and technical ceramics involved to data on furnace temperature, redox conditions, potential metal yield, scale of operations, and embedding in local economic structures.

Another factor that makes production debris an important source is that slag and technical ceramics are far less spatially mobile than for example artefacts, and thus clearly relate the information derived from them to the location where the debris was recovered. Obviously, the fact that iron smelting often takes place away from settlements, or at the very least is located near resources rather than consumers, does not improve our chances of discovering it. It is therefore logical that more evidence has been found for smithing, located near the consumer, than smelting (see e.g. Rehren et al. 2007, Veldhuijzen and Rehren 2007).

The discovery of very early smelting operations at Tell Hammeh in Jordan, and that of one of the earliest known smithing workshops at Tel Beth-Shemesh, Israel, provides an unprecedented opportunity to learn more about the nature and role of iron in society of the Early First Millennium BC (the review of Hammeh and Beth-Shemesh below appeared earlier in Veldhuijzen 2012). In the last few years, a few more instances of iron metallurgy have been attested and analysed (see Veldhuijzen and Rehren 2007, 191) with, most recently, a secondary pit-deposit of smithing remains at Tel Dor (Eliyahu-Behar et al. 2008), and evidence for iron working in parallel with copper/bronze processing at Tell es-Safi/Gath (Eliyahu-Behar et al. 2012)

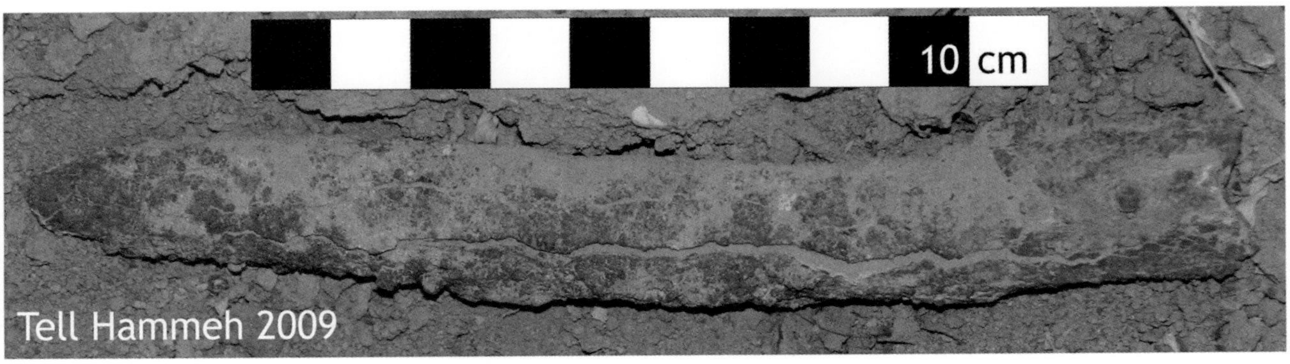

Figure 2: Tell Hammeh 2009: one of the rare iron artefacts found in the iron smelting strata, a long dagger or knife blade. On the right, where the blade is broken, a rivet can be seen, as well as remains of the wooden handle, preserved in the corrosion of the iron metal.

Early Smelting at Tell Hammeh

Tell Hammeh was excavated by teams from the University of Leiden, the Netherlands and Yarmouk University, Jordan, in four seasons, in 1996, 1997, 2000 and 2009 (Fig. 3). The last two seasons, directed by the author, specifically targeted the iron production remains (Veldhuijzen 2012, Veldhuijzen and Rehren 2007). Dedicated metallurgical excavation methods (Figs. 4, 5, 6) were developed and applied here that were further developed during the excavation of the Beth-Shemesh smithy and the Hammeh season of 2009 (Veldhuijzen 2009b).

The material excavated at Hammeh consists of close to one tonne of slags, hundreds of square-section tuyères (Figs. 7, 8), molten and burnt technical ceramics, charcoal and ash, the remains of several collapsed or destroyed furnace structures (Fig. 9), and very few iron artefacts (Fig. 9). Short-lived olivewood charcoal from the furnace structures dates the active production of iron at 930/910 Cal BC (±40 years; 1 sigma ranges of 1000-900 and 940-850 Cal BC; AMS analysis with ^{13}C-^{12}C correction). With due caution concerning ^{14}C dating (van Strydonck *et al.* 1999), these dates make Hammeh the earliest if not the only known find of iron smelting in the entire Near East (see among others Nieling 2009, Pleiner 2000, Waldbaum 1999).

The site itself is a relatively small tell near the central Jordan Valley (Fig. 2), from where one has easy access to the natural resources desirable in iron smelting: water, clay, and above all the only iron ore deposit of the wider region at Mugharet al-Warda (Abu-Ajamieh et al. 1988, al-Amri and Hauptmann 2008, Bender 1975, 1974, Bender 1968, 149-151, van den Boom and Lahloub 1962, Pigott 1983, Pigott et al. 1982). The tell was occupied intermittently from the Chalcolithic up to Medieval times, and importantly appears settled and uninterrupted during the transition from the Late Bronze Age

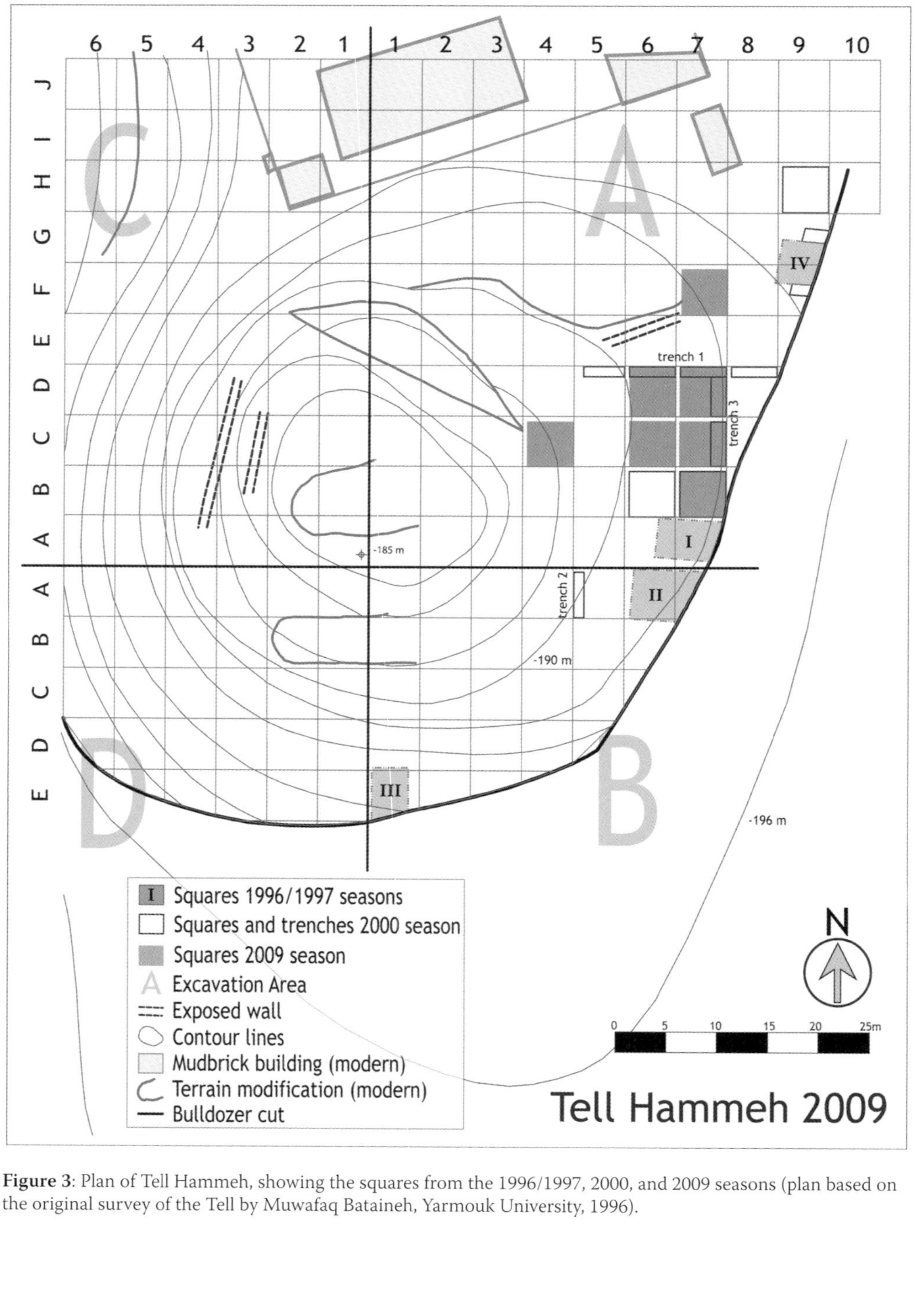

Figure 3: Plan of Tell Hammeh, showing the squares from the 1996/1997, 2000, and 2009 seasons (plan based on the original survey of the Tell by Muwafaq Bataineh, Yarmouk University, 1996).

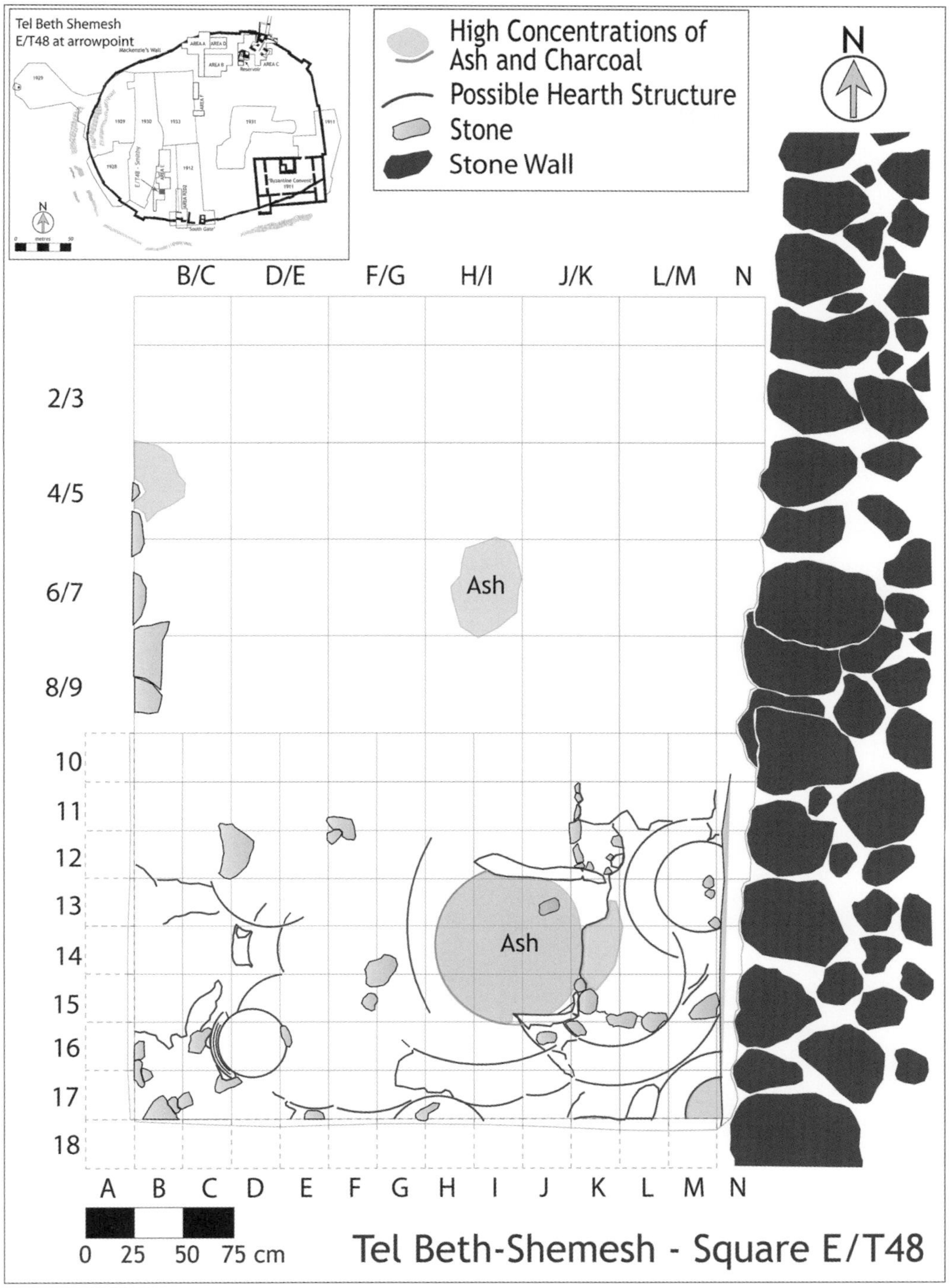

Figure 4: Plan of square E/T48 at Tell Beth-Shemesh, showing the 25x25cm grid system, various partial potential hearth structures, concentrations of ash, and the large stone wall (Veldhuijzen 2009c). Inset: Plan of Tel Beth-Shemesh, after Zvi Lederman).

(c. 1600-1150 BC) to the Iron Age I (c. 1150-1000 BC) (van der Steen 2004). Then, domestic settlement disappears and, after a hiatus, is replaced with a stratigraphically well defined phase of iron production.

Taking the ^{14}C dates as *terminus post quem* and combining this with the non-metallurgical occupation attested at Hammeh immediately following the metallurgical strata and dating from c. 800 BC onwards, iron production at Hammeh takes place for c. 100 to 150 years (Veldhuijzen 2005a, 2005b, Veldhuijzen and Rehren 2006, 2007, van der Kooij, personal communication).

Iterative mass balance calculations have shown that 100kg of average Mugharet al-Warda ore produces c. 58kg of Hammeh slag and c. 47kg of iron metal, but require an addition of almost 20kg of tuyère and furnace wall ceramic as a flux (for the interesting relationship between ore, slag, and technical ceramics in the Hammeh smelting process, see Veldhuijzen 2005a, 2005b, Veldhuijzen and Rehren 2006, 2007, van der Kooij, personal communication). Extrapolating the excavated slag quantities at Hammeh over the assumed total production area as well as the assumed time span smelting took place, a total production of iron metal between c. 6

to 12 tonnes may have taken place. Such quantities show that Hammeh represents a well-established and substantial operation, as opposed to an early attempt or experimentation with a new technology, but at quantities more in keeping with a seasonal activity than a full-time specialisation (Veldhuijzen 2009a, 2005a, Veldhuijzen and Rehren 2006, 2007).

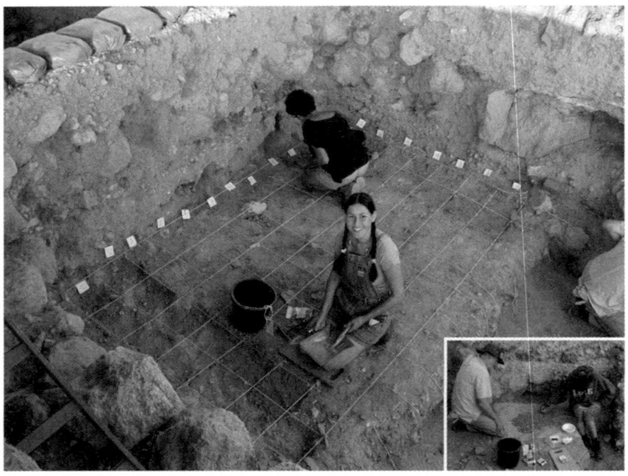

Figure 5: Excavation of the 25x25cm grid system in square E/T48 at Tel Beth Shemesh in 2003. Soil from each grid is recovered in separate units. The inset shows the dragging of a magnet above the soil retrieved from one unit to recover micro-magnetic material (see Veldhuijzen 2009b).

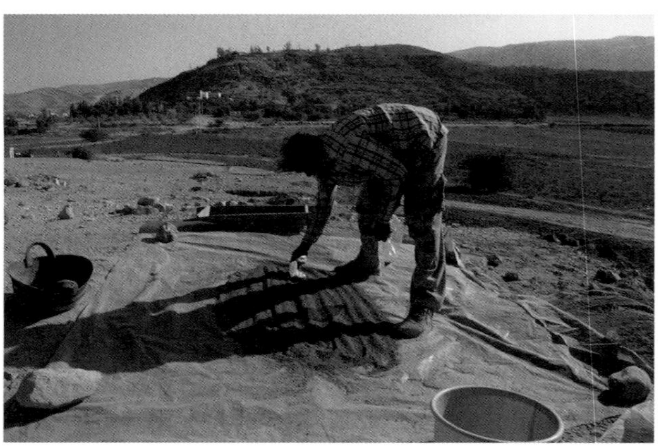

Figure 6: Tell Hammeh 2009: Magnet-dragging on soil from a unit from square A/D6 to recover micro-magnetic material (see Veldhuijzen 2009b).

It is very likely that Hammeh represents a seasonal smelting operation, taking place close to the available ore and other necessary resources, and in synchronization with the equally seasonal pruning of olive trees to obtain fuel for the furnace.

That the metalworkers at Hammeh were able to perform and sustain such an operation repeatedly over a considerable period of time must indicate they had access to and control over the various resources necessary in the process, particularly the ore. This seems unlikely without a larger socio-economic structure in which the iron production is embedded, but the political situation for this region at this period is not well enough understood to say more about the nature of this structure.

In summary, the material found at Tell Hammeh and its archaeological context create a strong impression of a well-established and dedicated iron smelting operation that forms part of a wider web of activities. The use of *sacrificial tuyères* (Veldhuijzen 2005b, Veldhuijzen and Rehren 2006) (Fig. 8) and the tapping of slag indicate a clear understanding of the process by the smelters, using and adapting to the particularities of the locally available materials, and embedding itself in seasonal cycles of the wider economy. Iron was produced away from regular settlement in quantities that clearly exceed individual consumption, but fall short of constituting a large-scale regional industrial centre.

Early Smithing at Beth-Shemesh

Tel Beth-Shemesh is located in the north-eastern Shephelah, Israel, approximately 20km west of Jerusalem, and c. 75km southwest of Tell Hammeh (Fig. 1). It lies at what once formed the border area between the Philistine territory of the lower Shephelah and coastal plain, and the Judean hill country. Previously excavated in the 1910s (MacKenzie 1914) and 1930s (Grant and Wright 1939), Shlomo Bunimovitz and Zvi Lederman from Tel Aviv University began a third series of excavations in the 1990s (Bunimovitz and Lederman 2009, 2003) (Inset Fig. 3).

In 2001, work started in Area E, a narrow strip between previously excavated areas in the south-western part of the site (Fig. 4). Several phases of industrial and commercial activity were excavated until evidence for metallurgical activity began

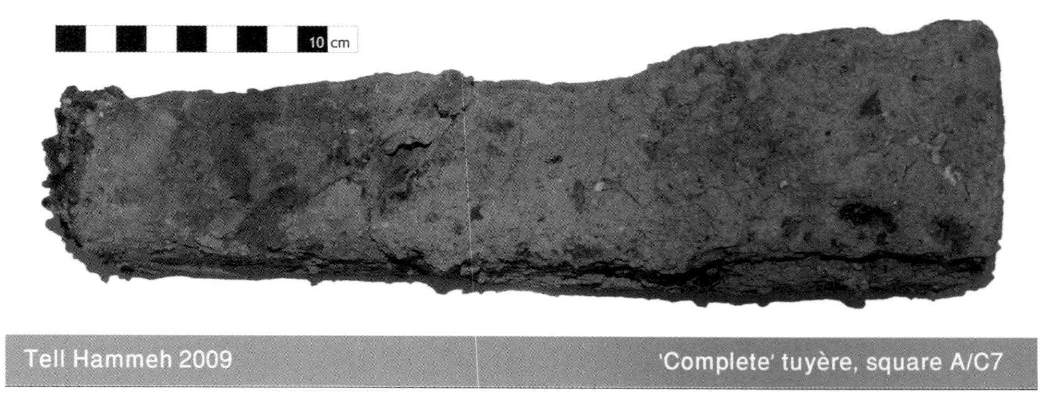

Figure 7: Tell Hammeh 2009: The longest complete tuyère found to date, where complete refers to the fact that the c. 28cm long tuyère is retained from the molten nozzle which extended into the furnace to the widened rear end, where the bore is enlarged for the insertion of bellows.

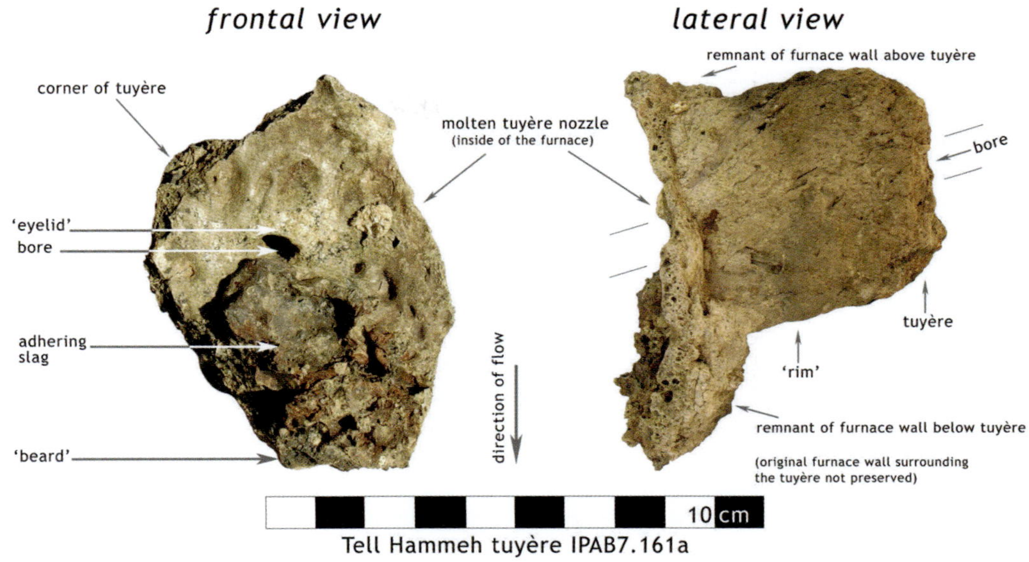

Figure 8: Frontal and lateral view of tuyère IPAB7.161a, showing the molten nozzle of a square section tuyère that is partially fused with the technical ceramic of the furnace wall. Remnants of unvitrified furnace wall are also preserved, both below and above the tuyère. An 'eyelid', formed by molten technical ceramic flowing down and pushed up and forward by the airflow, is visible above the bore. A large 'beard' of molten ceramic material with adhering slag can also be seen. In the lateral view, some 3 cm behind the molten nozzle, can be seen the 'rim', a greyish band indicating the extent to which reducing conditions in the furnace affected the ceramic, simultaneously suggesting a downward angle for the tuyères (Veldhuijzen 2009a, 2005, Veldhuijzen and Rehren 2007, 2006).

to appear here, comprising a single type of morphologically homogeneous slag: concavo-convex 'smithing hearth bottom' (SHB) slag (Fig. 10), remains of hearth structures (Fig. 4), molten and vitrified technical ceramics, square-section tuyères, large quantities of iron artefacts and fragments, charcoal and ash, and large quantities of very fine magnetic material cm (i.e. hammerscale; on smithing technology, see Serneels and Perret 2003). In 2003, 2006, and 2009, the author directed the excavation of this suspected smithy, applying and further developing dedicated metallurgical excavation and processing methods (Veldhuijzen 2009b, 2009c, 2005a) (Figs. 4, 5).

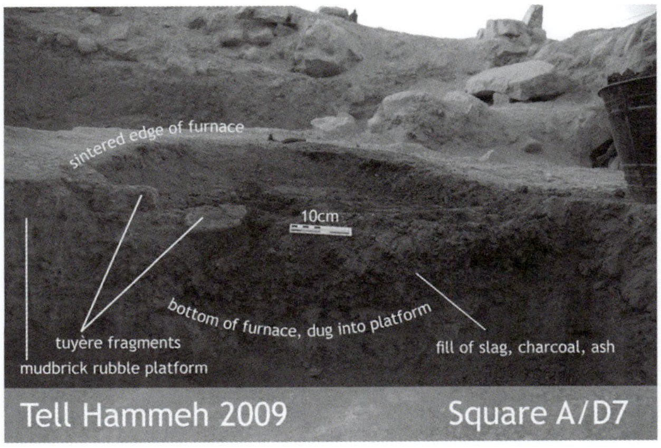

Figure 9: Tell Hammeh 2009: bottom of an iron smelting furnace during excavation. The hollow forming the bottom of the furnace was apparently dug into a specially constructed platform consisting of compact mudbrick rubble. Visible here are the sintered edge of the inside of the furnace, and a fill of charcoal, ash, and some tuyère fragment. No superstructure was preserved.

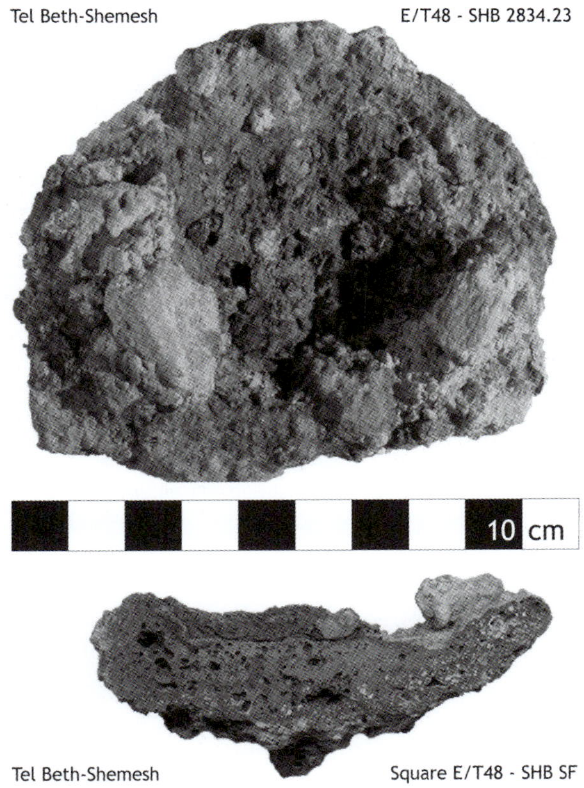

Figure 10: Above: view from the top of SHB slag E/T48 2824.23, showing a depression at the centre, indicating where the airflow from the tuyère hit the collecting liquid slag. The roughly straight side at the bottom of the image may be the place where the SHB was in contact with the hearth wall, indicating that the tuyère was located on this side of the SHB (see Serneels and Perret 2003). Below: view at the section of smithing hearth bottom slag E/T48 SF. The SHB reveals a porous slag core of concavo-convex shape, with rust adhering to the top and soil embedded at the bottom.

Intriguingly, the Beth-Shemesh tuyères are virtually identical to the Hammeh ones in all macroscopic aspects, from size, colour, feel, and temper, to shape. Less abundant here than at Hammeh, the tuyères at both sites are square and approximately 5x5cm in section, with a bore of c. 10mm in diameter. ^{14}C Radiocarbon analyses of three burned olive pits from the smithy resulted in a date of c. 900 CalBC (±45 years; AMS analysis with ^{13}C-^{12}C correction).

The finds at Beth-Shemesh clearly represent a *secondary smithing* operation, as opposed to iron *smelting* and/or *primary (bloom-) smithing*. This was further confirmed by the dedicated recovery and interpretation of the micro-magnetic material using a grid system of 25x25cm (Veldhuijzen 2009c, Veldhuijzen and Rehren 2007) (Figs. 4, 5). Based on the micro-magnetic material, which at Beth-Shemesh consists almost exclusively of hammerscale, use of space in the smithy could be plotted, likely indicating the presence and location of the last hearth in use in the workshop (Veldhuijzen and Rehren 2007).

Smithing slags are often very similar to bloomery smelting slags in general chemical and micro-structural composition (Pleiner 2000, 255). The morphology, size, and uniformity of the Beth-Shemesh slag all point towards *secondary smithing*. Their morphological features of a mostly round, concavo-convex shape with a diameter of up to 10cm, with rust adhering to the top and soil to the bottom form a classic example of a 'smithing hearth bottom' (SHB) or 'plano-convex bottom' (PCB) (Fig. 10).

Chemically, the Beth-Shemesh slags compare quite well to other smithing slags discussed in the literature (e.g. Kronz 1998, 225), with the notable exception of a higher lime content, and a lower iron oxide content than 'regular' smithing slags. In that respect, the Beth-Shemesh slags are quite similar in composition to the Hammeh tapped smelting and primary smithing slags, suggesting the use of similar materials and methods (Veldhuijzen 2005a, Veldhuijzen and Rehren 2007).

Such a connection between the Beth-Shemesh smithing operation and the Hammeh smelting is strongly suggested by the shared tuyère design as well as the compositional parallels between the slags at both sites. An attempt was made to show this suspected connection through the analysis of (smelting) slag inclusions in iron artefacts from the Beth-Shemesh smithy. Whereas several different iron smelting systems appear to be represented in the Beth-Shemesh metal, it can't be concluded that Hammeh is one of those (Blakelock et al. 2009).

The quantity of SHBs at Beth-Shemesh, where each specimen probably represents a separate smithing operation, cycle, or workday (Serneels and Perret 2003, 473), indicates that the smithing work must have been a regular operation aiming at local consumption. The choice of location, near substantial public buildings and the outer perimeter of the city, reflects a desired closeness to the consumer (Veldhuijzen 2009c, Veldhuijzen and Rehren 2007).

In summary, the material found at Tel Beth-Shemesh and its archaeological context evokes an image of a smithing workshop that is operated regularly and at considerable (local) scale within the confines of one of the larger cities in the Iron Age Levant, catering to the needs and wants of the inhabitants of that settlement. As an aside, the presence of smithing activity at Beth-Shemesh, situated right on the border of the Philistine territory of the lower Shephelah and coastal plain, gains extra poignancy as it clearly contradicts the biblical sources that suggest a Philistine monopoly on iron.

Parallels at Tell es-Safi/Gath

It is extremely interesting to see that the latest iron metallurgical find in the region, that at Tell es-Safi/Gath (Eliyahu-Behar et al., 2012), shares many elements with the material at Beth-Shemesh, and indirectly also Hammeh. At Tell es-Safi, Iron Age IIa (contemporaneous or slightly younger than Beth-Shemesh) fragmented iron slags were found, that are quite similar to the Beth-Shemesh SHBs, in chemical and mineralogical terms as well as in their macroscopic appearance (Eliyahu-Behar et al., 2012, 257, Fig. 3). Hammerscale and slag prills were also recovered, as well as hammering stones (the last conspicuously absent at Beth-Shemesh).

Especially interesting is the fact that part of the iron working material at Tell es-Safi consists of square section tuyères that are almost identical in appearance (size and shape) to the Beth-Shemesh and Hammeh ones (Eliyahu-Behar et al. 2012, 262, Fig. 6). The square section of tuyères is not a technical requirement, and as such forms a *technological choice*, i.e. a choice not guided by technological constraints, but one made by the person performing the technological activity based on, for example, social or cultural considerations, perceived requirements, or local traditions (see discussions in, among others: Costin 2001, Costin and Wright 1998, Dobres and Robb 2000, Killick 2004, Lechtman 1999, 1977, Lemonnier 1992, Pfaffenberger 1992, Veldhuijzen 2009a). This factor may well represent an ethnic, cultural, or traditional identifier that will help to elucidate how iron production is embedded in the local society and economy. With these clear parallels, a range of possible future comparisons of chemical and mineralogical data as well as technical and archaeological data from all three sites becomes possible, which may considerably enhance our knowledge of iron production in this region and time.

Conclusions

It is clear that 30 years after Ted Wertime and Jim Muhly's seminal book on the 'Coming of the Age of Iron' (Wertime and Muhly 1980) and more than a decade after Jane Waldbaum's last extensive review of the artefactual evidence (Waldbaum 1999), not much has changed with regard to our knowledge of iron throughout the Eastern Mediterranean (see also earlier presentation of these ideas in Veldhuijzen 2012). We still largely refer to the same literary evidence we did then, and whilst the number of known artefacts has increased, archaeometallurgical analysis of artefacts is still a rare occurrence. Most crucially, there is still a dire need to find and study more production evidence. Although it is remarkable how so little actual production has been attested in a region that is quite widely and intensively studied archaeologically, it is still possible to create a picture of the role of iron in the region using the information we do possess.

The available literary sources and the nature of artefacts let us deduce that iron plays an increasingly important role in society during the Second Millennium. Iron first appears as a rare and valuable government controlled luxury commodity, almost exclusively used for palatial and temple related status objects – likely going hand in hand with (a lack of) metallurgical properties that make it impractical for use in daily life. From the 13th century BC, iron starts to become more commonplace, and increasingly appears in utilitarian and weaponry contexts.

When looking at the apparent widespread small-scale working of iron in the Early First Millennium BC, when iron really starts to impact on society, it seems quite unlikely that either the Hittites or later the Philistines had a proper monopoly on the knowledge of this metal, but they likely exercised some level of control over procurement and/or production or trade of iron. The role of Iron in the Early First Millennium BC is less clear, hampered as our knowledge is by the absence of literary sources for this period. It is often suggested that the political fragmentation of the Iron Age I and early Iron Age II stimulated local industries to exploit locally available raw materials see (Stech-Wheeler et al., 1981, van der Steen 2004). This is exactly the picture that emerges from the smelting finds at Tell Hammeh, where local people use and adapt their technology to local materials and seasonal cycles in the local economy.

The lack of contemporary habitation on the tell raises interesting questions about ownership, organisation, and control of space, know-how and resources in the local region, but no answers can be given as yet. What is noticeable is a marked absence of evidence for the presence of major powers during the iron smelting strata. This suggests local control over and organisation of the production, clearly different from the situation of the 8th century BC, when iron processing seems to be, or come, under control of the invading Assyrians (e.g. Rothenberg and Tylecote 1991).

The smithing activities at Beth-Shemesh and Tell as-Safi/Gath show a quite different choice of location, as it takes place within the confines of a large city, which corresponds with other finds identified as smithing in the region (Veldhuijzen 2009c, Veldhuijzen and Rehren 2007). Here the choice of location likely reflects a desire to be close to the consumer, rather than being close to resources.

The identical design characteristics of Beth-Shemesh, Hammeh, and Tell es-Safi/Gath tuyères is likely indicative of cross-cultural contacts, shared technological characteristics, or even a possible socio-ethnic link between the people conducting two consecutive processes at three different sites (see earlier presentation of these ideas in Veldhuijzen 2005a, 2005b, 2009a, 2012, Veldhuijzen and Rehren, 2007). The square cross-section of the tuyères is certainly not a technological requirement for their use, and may therefore represent a *technological choice*, which significantly widens the scope of possible interpretation of the remains at these sites.

Neither the Beth-Shemesh smithing nor the Hammeh smelting shows any sign of innovation or development of the technology practised. Both sites represent well-established technological processes of a considerable scale and with indications of standardisation. This clearly suggests that iron production and working were known and practised in the region prior to the start of iron smelting at Hammeh no later than 930 BC, and the smithing at Beth-Shemesh in c. 900 BC.

Bringing all sources of information together, one can put forward that the demise of empires and the rise of iron at the end of the Late Bronze Age do go hand in hand. Major changes in the political and economic situation allow the accelerated adaptation of a 'new', previously elite metal, where production shifts from centrally organised production centres to independent local workshops that operate within the wider web of local economic and social activities.

Acknowledgments

The author wishes to gratefully acknowledge the Institute for Archaeo-Metallugical Studies (IAMS), the Leverhulme Trust, and the UCL Institute of Archaeology for their invaluable support and assistance towards the Hammeh and Beth-Shemesh research over the last years. I am indebted to many colleagues around the world, and especially the Scientific Directors of the Deir 'Alla Regional Project, Dr Zeidan Kafafi and Dr Gerrit van der Kooij, and the Scientific Directors of the Tel Beth-Shemesh Excavations, Dr Shlomo Bunimovitz and Dr Zvi Lederman.

A huge debt is owed to Professors Thilo Rehren and Vince Pigott, for their tireless efforts to instil in me the principles of archaeometallurgy and its interpretation. It is because of them all that so much analysis of the iron production and processing debris from Hammeh and Beth-Shemesh became possible.

Last but not very certainly not least I have to thank Jane Hatton-Humphris, my partner in crime in organising WIC2009; together we got iron to be the prime and only material at a conference, and my sincere admiration goes to her for getting these wonderful proceedings of WIC together.

References

Abu-Ajamieh, M.M., Bender, F.K. and Eicher, R.N. (1988) *Natural Resources in Jordan, Inventory-Evaluation-Development Program*, Amman, Natural Resources Authority.

al-Amri, Y. a. E.S. and Hauptmann, A. (2008) The Iron Ore Mine of Mugharat el-Wardeh/Jordan in Southern Bilad al-Sham, Excavation and New Dating. In: Yalçın, Ü., Özbal, H. and Paşamehmetoğlu, A. G. (Eds.) *Ancient Mining in Turkey and the Eastern Mediterranean*. Ankara Atılım University.

Bender, F.K. (1975) *Geology of the Arabian Peninsula: Jordan*, USGS Prof. Paper 560-I.

Bender, F. K. (1974) *Geology of Jordan*, Berlin, Gebruder Borntraeger Press.

Bender, F.K. (1968) *Geologie von Jordanien (Beiträge zur Regionalen Geologie der Erde, Band 7)*, Berlin, Gebrüder Borntraeger Verlagsbuchhandlung.

Blakelock, E., Martinón-Torres, M., Veldhuijzen, H.A. and Young, T. (2009) Slag Inclusions in Iron Objects and the Quest for Provenance: an Experiment and a Case Study. *Journal of Archaeological Science* 36, 1745-1757.

van den Boom, G. and Lahloub, M. (1962) *The Iron-Ore Deposit 'Warda' in the Southern Ajlun District*, Hannover/Amman, Unpublished report, Federal Institute of Geoscience and Natural Resources, Hannover (Unveröffentlichte Bericht Deutsche Geologische Mission Jordanien, Archiv Bundesanstalt für Bodenforschung)/National Resources Authority, Amman.

Bunimovitz, S. and Lederman, Z. (2009) The Archaeology of Border Communities: Renewed Excavations at Tel Beth-Shemesh, Part 1: The Iron Age. *Near Eastern Archaeology* 72 (3), 114-142.

Bunimovitz, S. and Lederman, Z. (2003) Tel Beth-Shemesh, 2001-2003. *Israel Exploration Journal* 53 (2), 233-237.

Costin, C.L. (2001) Craft Production Systems. In: Feinman, G.M. and Price, T.D. (Eds.) *Archaeology at the Millennium: A Sourcebook*, 273-327. New York, Kluwer Academic/Plenum Publishers.

Costin, C.L. and Wright, R.P. (1998) *Craft and Social Identity*, Arlington, VA, American Anthropological Association.

Craddock, P.T. (1995) *Early Metal Mining and Production*, Edinburgh, Edinburgh University Press.

Curtis, J.E., Stech-Wheeler, T., Muhly, J.D. and Maddin, R. (1979) Neo-Assyrian Ironworking Technology. *Proceedings of the American Philosophical Society* 123 (6), 369-390.

Dobres, M.-A. and Robb, J. (2000) Agency in Archaeology: Paradigm or Platitude? In: Robb, J. (Ed.) *Agency in Archaeology*, 3-17. New York and London, Routledge.

Eliyahu-Behar, A., Shilstein, S., Raban-Gerstel, N., Goren, Y., Sharon, I., Gilboa, A. and Weiner, S. (2008) An Integrated Approach to Reconstructing Primary Activitiesfrom Pit Deposits: Iron Smithing and Other Activities at Tel-Dor (Israel) Underneo-Assyrian Domination. *Journal of Archaeological Science* 35 (11), 2895-2908.

Eliyahu-Behar, A., Yahalom-Mack, N., Shilstein, S., Zukerman, A., Shafer-Elliott, C., Maeir, A.M., Finkelstein, I. and Weiner, S. (2012) Iron and Bronze Production in Iron Age Iia Philistia: New Evidence from Tell Es-Safi/Gath, Israel. *Journal of Archaeological Science* 39 (2), 255-267.

Grant, E. and Wright, G.E. (1939) *Ain Shems Excavations (Palestine): Part V (Text)*, Haverford, Pennsylvania, Haverford College.

Killick, D.J. (2004) Social Constructionist Approaches to the Study of Technology. *World Archaeology* 36 (4), 571-578.

Kronz, A. (1998) *Phasenbeziehungen und Kristallisationsmechanismen in Fayalitischen Schmelzsystemen. Untersuchungen an Eisen- und Buntmetallschlacken*, Bielefeld, Friedland.

Lechtman, H. (1999) Afterword. In: Dobres, M.-A. and Hoffman, C.R. (Eds.) *The Social Dynamics of Technology: Practice, Politics, and World Views*, 223-232. Washington D.C., Smithsonian Institution Press.

Lechtman, H. (1977) Style in Technology -- Some Early Thoughts. In: Lechtman, H. and Merrill, R. S. (Eds.) *Material Culture: Styles , Organization, and Dynamics*, 3-20. St Paul, American Ethnological Society/West Publishing.

Lemonnier, P. (1992) *Elements for an Anthropology of Technology*, Ann Arbor, Michigan, University of Michigan; Anthropological Paper 88.

Mackenzie, D. (1914) *Excavations at Ain Shems (Beth-Shemesh)*, London, Harrison and Sons.

Muhly, J.D. (1983) Early Iron-Working in Anatolia and the Question of the Hittite Iron Monopoly. *American Journal of Archaeology* 87 (2), 246-247.

Muhly, J.D. (1977) The Copper Ox-Hide Ingots and the Bronze Age Metals Trade. *Iraq* 39, 73-82.

Muhly, J.D. (1973) *Copper and Tin. The Distribution of Mineral Resources and the Nature of the Metals Trade in the Bronze Age.*, Hamden, CT, Archon Books.

Muhly, J.D., Stech-Wheeler, T. and Maddin, R. (1977) The Cape Gelidonya Shipwreck and the Bronze Age Metals Trade in the Eastern Mediterranean. *Journal of Field Archaeology* 4 (3), 353-362.

Nieling, J. (2009) *Die Einführung der Eisentechnologie in Südkaukasien und Ostanatolien wärend der Spätbronze- und Früheisenzeit*, Aarhus University Press, Black Sea Studies, 10.

Pfaffenberger, B. (1992) Social Anthropology of Technology. *Annual Review of Anthropology* 21, 491-516.

Pigott, V.C. (1983) The Innovation of Iron (as Steel) in Palestine. *American Journal of Archaeology* 87 (2), 252.

Pigott, V.C., McGovern, P.E. and Notis, M.R. (1982) The Earliest Steel from Transjordan. *Journal of the Museum Applied Science Center for Archaeology* 2 (2), 35-39.

Pleiner, R. (2000) *Iron in Archaeology. The European Bloomery Smelters*, Praha, Archeologický Ústav Av Cr.

Rehren, T., Charlton, M., Chirikure, S., Humphris, J., Ige, A. and Veldhuijzen, H.A. (2007) Decisions Set in Slag—the Human Factor in African Iron Smelting. In: La Niece, S., Hook, D.R. and Craddock, P.T. (Eds.) *Metals and Mines – Studies in Archaeometallurgy*, 211-218. London, Archetype, British Museum.

Rothenberg, B. and Tylecote, R.F. (1991) A Unique Assyrian Iron Smithy in the Northern Negev (Israel). *Institute for Archaeo-Metallurgical Studies (IAMS) Journal* 17, 11-14.

Serneels, V. and Perret, S. (2003) Quantification of Smithing Activities Based on the Investigation of Slag and Other Material Remains. *Proceedings of the International Conference Archaeometallurgy in Europe*, 469-479. Milano, Associazione Italiana di Metallurgia.

Snodgrass, A.M. (1980) Iron and Early Metallurgy in the Mediterranean. In: Wertime, T. A. and Muhly, J. D. (Eds.) *The Coming of the Age of Iron*, 335-374. New Haven; London, Yale University Press.

Stech-Wheeler, T., Muhly, J.D., Maxwell-Hyslop, K.R. and Maddin, R. (1981) Iron at Taanach and Early Iron Metallurgy in the Eastern Mediterranean. *American Journal of Archaeology* 85 (3), 245-268.

van der Steen, E.J. (2004) *Tribes and Territories in Transition. The Central East Jordan Valley in the Late Bronze and Early Iron Ages: A Study of the Sources*, Leuven, Peeters.

van Strydonck, M., Nelson, D.E., Crombé, P., Bronk Ramsey, C., Scott, E.M., van der Plicht, J. and Hedges, R.E.M. (1999) What's in a ^{14}C Date. *^{14}C et Archéologie. Actes du 3e Congrès International (Lyon 6-10 Avril 1998)*, 433-440. Lyon, Mémoires de la Société Préhistorique Française, XXVI, 1999 et Supplément 1999 de la Revue d'Archéométrie.

Veldhuijzen, H.A. (2012) Just a Few Rusty Bits: The Innovation of Iron in the Eastern Mediterranean in the 2nd and 1st Millennia BC. In: Kassianidou, V. and Papasavvas, G. (Eds.) *Eastern Mediterranean Metallurgy in the Second Millennium BC.* Oxford, Oxbow Books.

Veldhuijzen, H.A. (2009a) Hip to Be Square. How Tuyères Shaped the Hammeh Technology. In: Gebel, H.G., Kafafi, Z. and Ghul, O. (Eds.) *Modesty and Patience. Studies and Memories in Honour of Nabil Qadi (Abu Salim).* Berlin / Irbid, Ex Oriente / Yarmouk University.

Veldhuijzen, H.A. (2009b) Of Slag and Scales, Micro-Stratigraphy and Micro-Magnetic Material at Metallurgical Excavations. In: Kaptijn, E. and Petit, L.P. (Eds.) *A Timeless Vale, Archaeology and Related Studies of the Jordan Valley*, 163-174. Leiden, Leiden University Press.

Veldhuijzen, H.A. (2009c) Red Hot: The Smithy at Tel Beth-Shemesh. *Near Eastern Archaeology* 72 (3), 129-131.

Veldhuijzen, H.A. (2005a) *Early Iron Production in the Levant. Smelting and Smithing at Early 1st Millennium BC Tell Hammeh, Jordan, and Tel Beth-Shemesh, Israel*, London, Unpublished PhD Thesis, Institute of Archaeology, University College London.

Veldhuijzen, H.A. (2005b) Technical Ceramics in Early Iron Smelting. The Role of Ceramics in the Early First Millennium BC Iron Production at Tell Hammeh (Az-Zarqa), Jordan. In: Prudêncio, I., Dias, I. and Waerenborgh, J.C. (Eds.) *Understanding People through Their Pottery; Proceedings of the 7th European Meeting on Ancient Ceramics (Emac '03)*, 295-302. Lisboa, Instituto Português de Arqueologia (IPA).

Veldhuijzen, H.A. and Rehren, T. (2007) Slags and the City. Early Iron Production at Tell Hammeh, Jordan, and Tel Beth-Shemesh, Israel. In: La Niece, S., Hook, D.R. and Craddock, P.T. (Eds.) *Metals and Mines - Studies in Archaeometallurgy*, 189-201. London, Archetype, British Museum.

Veldhuijzen, H.A. and Rehren, T. (2006) Iron Smelting Slag Formation at Tell Hammeh (Az-Zarqa), Jordan. In: Pérez-Arantegui, J.

(Ed.) *Proceedings of the 34th International Symposium on Archaeometry, Zaragoza, 3-7 May 2004*, 245-250. Zaragoza, Institución «Fernando el Católico» (C.S.I.C.) Excma. Diputación de Zaragoza.
Waldbaum, J.C. (1999) The Coming of Iron in the Eastern Mediterranean. Thirty Years of Archaeological and Technological Research. In: Pigott, V.C. (Ed.) *The Archaeometallurgy of the Asian Old World*, 27-57. Pennsylvania, MASCA Research Papers in Science and Archaeology 16, University of Pennsylvania Museum.
Waldbaum, J.C. (1989) Copper, Iron, Tin, Wood, the Start of the Iron Age in the Eastern Mediterranean. *Archeomaterials* 3, 111-122.
Waldbaum, J.C. (1980) The First Archaeological Appearance of Iron and the Transition to the Iron Age. In: Wertime, T. A. and Muhly, J. D. (Eds.) *The Coming of the Age of Iron*, 69-98. New Haven; London, Yale University Press.
Waldbaum, J.C. (1978) *From Bronze to Iron, the Transition from the Bronze Age to the Iron Age in the Eastern Mediterranean*, Goteborg, Paul Astroms Forlag.
Wertime, T.A. and Muhly, J. D. (1980) *The Coming of the Age of Iron*, New Haven; London, Yale University Press.

Author's address

• H. Alexander Veldhuijzen
Honorary Research Fellow, UCL Institute of Archaeology

The iron ore mine of Mugharet al-Wardeh, Jordan: excavations and new dating

Yosha Alamri and Andreas Hauptmann

ABSTRACT: Archaeological investigation and survey at the iron ore mine of Mugharet al-Wardeh (MaW) in the Ajlun Mountains uncovered intensive Ayyubid-Mamluki (1169-1516 AD) exploitation and smelting activities. In addition, some Iron Age remains were found, as well as indications for Roman-Byzantine activity related to iron production. Mineralogical, chemical and lead isotope analyses of Chalcolithic period (4500-3500 BC) mace heads and other finds from archaeological sites in the region were consistent with compositional and isotopic patterns of the hematitic iron ore from MaW. This aspect points strongly to a prehistoric utilisation of this ore body which, so far, is not visible at the mine itself. In this paper we present, with a high degree of certainty, an indirect new dating for the use of iron ores from this very important deposit.

Keywords: Iron, iron ore mining, iron smelting, Iron Age, Mugharet al-Wardeh, Ajloun, Jordan, Southern Bilad al-Sham, Ayyubid-Mamluki Period.

Site description

The iron ore mine and settlement of Mugharet al-Wardeh (MaW) is located about 70km northwest of Amman, at coordinates 32° 13' 20.57" N, 35° 42' 42.09" E, within Ajlun province in Jordan. It sits at an elevation of ca. 650 m above sea level. The site lies about 15km east of the archaeological site of Dier A'lla, on a mountainous area covered with oak trees and recently planted olive trees.

Four collapsed adits (entrances to the underground mine) are located on the south-western slope of Aljabal Alakhdar (The Green Mountain) and south of the road which leads down to the Jordan Valley. Two of these adits are blocked with huge limestone blocks. According to many locals, these adits were blocked during the levelling and reconstructing of the old road. At least one of these adits leads to a wide underground gallery in front of the main ancient iron mine of MaW.

The Zarqa River runs approximately 4km to the south of the site. In addition to travelling the long, bending asphalted roads, it is also possible to reach the Zarqa River from MaW on foot or using domesticated animals (donkeys) along one of the many dusty unlevelled tracks. The Zarqa River is the main source of irrigation water in the region. Rain water and some seasonal springs (which were also used for drinking water until relatively recently) are also used for irrigation purposes.

Today, approaching the site of MaW is possible either from the eastern or the western side, the new asphalted road might have been placed over the ancient one used and described by earlier travellers (e.g., Glueck, 1935). This road would have connected MaW along the natural trade routes and linked it to population centres in the Jordan Valley with the eastern heights (Fig. 1). This iron ore deposit was located along two of the key ancient natural trading routes. The first one connects the Red Sea and Aqaba in the south with Syria and Anatolia in the north through the Jordan Valley and the Wadi Arabah. This north-south route is crossed by another trading route along the Zarqa River which connects the eastern Jordanian Plateau and Baqa'h Valley with the Palestinian coast in the west. The area of MaW is currently witnessing a sharp increase in the extent of domesticated land, with new olive trees plantations and many houses constructed in recent decades.

Geological background

Despite the limited reserves (561 000t), the MaW hematitic iron ore deposit is the largest iron ore deposit known in the region of southern Bilad al-Sham (sBaS). These ores have a high concentration of iron, with the average content of iron oxide (Fe_2O_3) at about 68 wt% (Bender, 1968; Abu Ajamieh, et al., 1988; Saffarini, 1989; Alomari, 1991). Other iron ore occurrences have been reported in this area; however, these represent even smaller reserves which exhibit very low iron content.

The Jabal Ajloun (Ajloun Mountains) area is part of the eastern Jordanian heights, with an elevation of ca. 1600m above sea level at its highest point. This series of mountains separate the eastern plateau from the Jordan Rift Valley. It has been subjected to continuous erosion that demolished the so-called Balqa Group of soft limestone, leaving the Ajloun limestone (Quennell, 1952; Bender, 1974). The oldest rock unit that crops out in the area is the Triassic Abu Ruweis Gypsum Formation (Zarqa Main Group), which occurs in

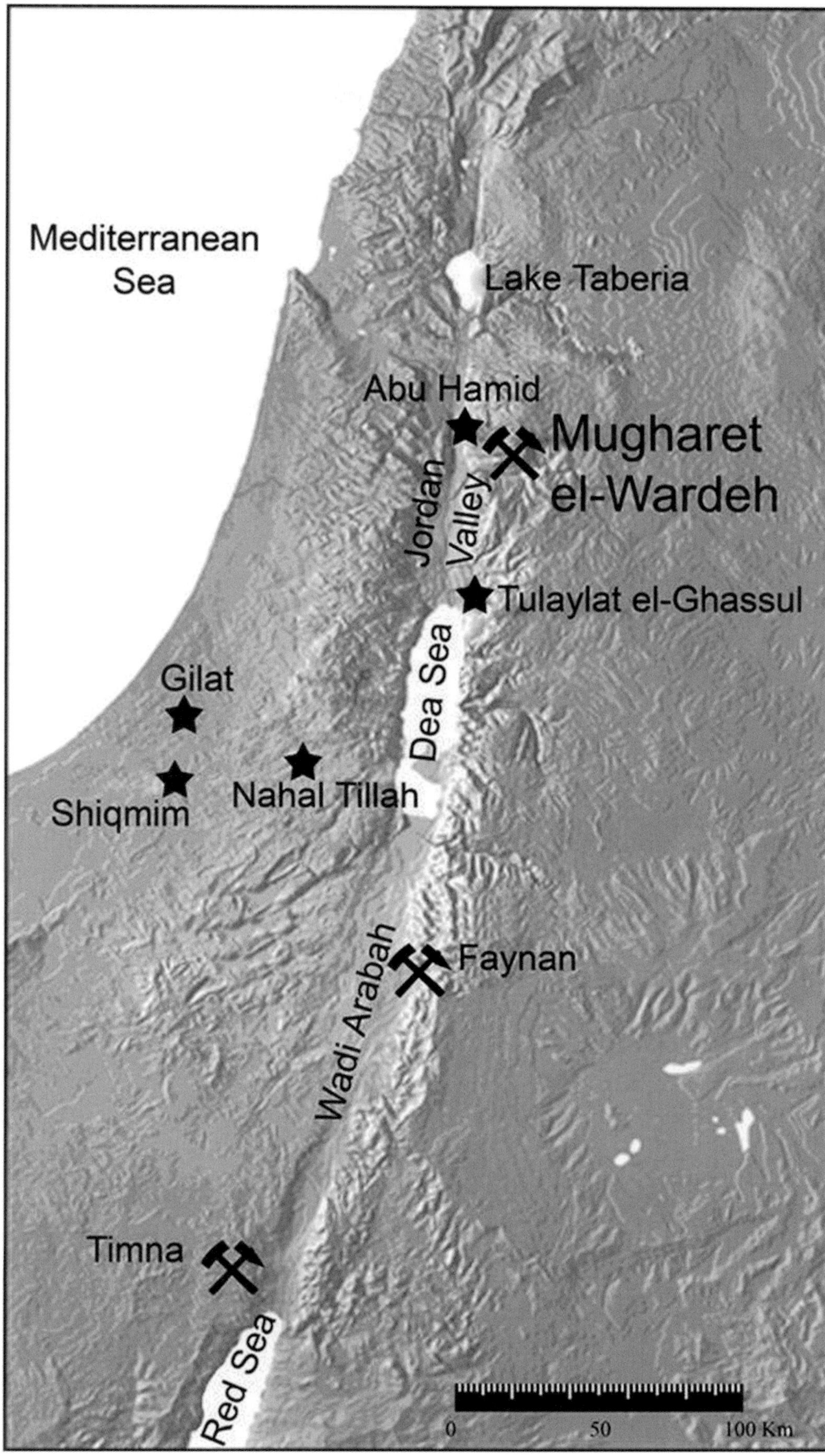

Figure 1: The region of southern Bilad el Sham, showing the distribution of some sites mentioned in the text. Note the ancient iron ore mine of Mugharet al-Wardeh in the Jordan Valley, between Lake Taberia and the Dead Sea. The ancient copper mining districts of Faynan and Timna are also shown in this map. Their importance in the history of early iron in southern Bilad el-Sham is discussed in the text.

the vicinity of Zarqa River and comprises of mainly massive gypsum and thin, bituminous, silty mudstone. Most of the Jurassic succession of Jordan (Azab Group) is exposed in the area as a narrow belt extending from Wadi Bassat el-Faras in the south to the Zarqa River in the north. The main rock types that comprise this unit are: claystone, (silty) mudstones, limestone, dolomite, sandstone, and many transitions between these rocks. The depositional environment of these rocks was coastal with a continuously shifting shoreline. The Jurassic calcareous rocks are suggested to have been deposited on a carbonate platform where marine regimes prevailed (Moh'd and Muneizel, 1998; Abed, 2000).

The Lower Cretaceous Kurnub Sandstone Group contains a thickness of 200-300m of various coloured sandstone (Fig. 2a). In the vicinity of Zarqa River, the base of the group overlies a paleosoil horizon at the top of the Jurassic sequence. The Kurnub Group is comprised of conglomerates, sandstone, quartzarenites, siltstone, and shale. These rocks indicate fluvial deposition interrupted by short periods of marine transgression (Bender, 1974; Moh'd and Muneizel, 1998).

The iron ore body of MaW has a lenticular or vein-like shape; it is up to 10m thick, 300m long, and about 200m wide (Fig. 2b). This ore deposit has a 68 wt% concentration of iron oxide (Fe_2O_3), consistent with a high quality of iron ore (Bender, 1968; Abu Ajamieh, et al., 1988; Saffarini, 1989). This iron deposit consisted of hematite as a major component, along with minor percentages of limonite, goethite, calcite, quartz, and chalcedony. It was formed by hydrothermal, late magmatic, epithermal metasomatism (Bender, 1968; Saffarini, 1989). Hydrothermal solutions containing a high percentage of iron replaced the limestone in the massive limestone member. These hydrothermal solutions have moved upwards through faults and fractures resulting from tectonic activity and were precipitated to form the iron ore body (Abu Ajamieh, et al., 1988).

According to Basha (1968), the mineralised layers can be divided into two parts; the upper part which is dark-reddish brown with massive cellular and fractured ore, interspersed with ferrous and dolomitic limestone, and the lower part which is dark brown and rusty yellow. Cellular and dense Cenomanian-Turanian Age limestone layers are distributed over and around the iron ore deposit itself (see Fig. 2). and occasionally limestone layers are integrated with the hematite massive itself, which is another indication of the hydrothermal origin of this ore.

The excavation at MaW

In 2005, an extended archaeological excavation was conducted at MaW by Y. Alamri (from 12.03.2005 until 31.05.2005). The site was divided into two areas depending on the nature and the expected ancient use of the site. The two areas (Fig. 3) are separated now by the new asphalted road:

1. The iron ore mine area. This area lies on the southern side of the road and includes the ancient iron ore mine. Two squares of 4x4m^2 were dug in this area in addition to excavations within one gallery.
2. The settlement area. It is located on the northern side of the same road and on top of a small hill at the foot of Aljabal Elakhdar. Metallurgical remains and potsherds were observed in this area during the first inspection of the site. Two 4x4m^2 squares were also dug in this area.

The underground mine (Area A)

Local people mentioned that the entrance to this main cave had been damaged during the work of the Natural Resources Authority (NRA) to explore and estimate the reserve during the beginning of the second half of the last century, and today dynamite holes and drilling marks are still visible on some of the walls of the mine. The still existing adit might have once functioned as part of a larger entrance which led into two different sized but parallel galleries:

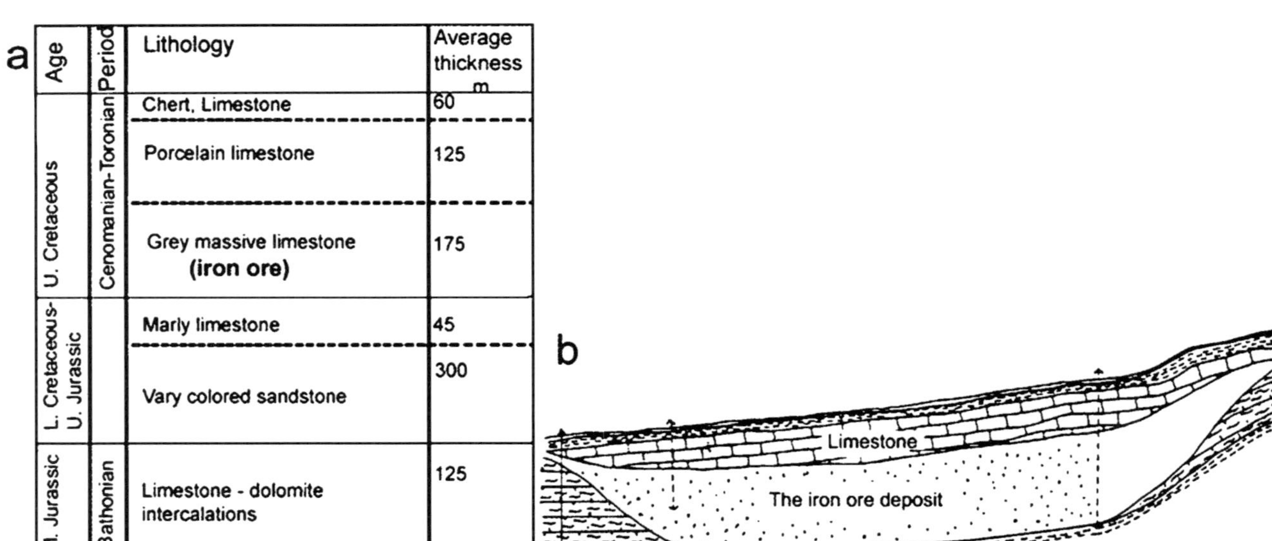

Figure 2: Thermal solutions have penetrated into the limestone layers, replaced it, and occur as an iron ore body. A 10m thick lenticular 'layer' of hematite ore is embedded in the Upper Cretaceous Limestone layer. a) The geological sequence of the region of Mugharet al-Wardeh. b) Iron ore deposit in the limestone layer (after Saffarini 1989; Alomari 1991).

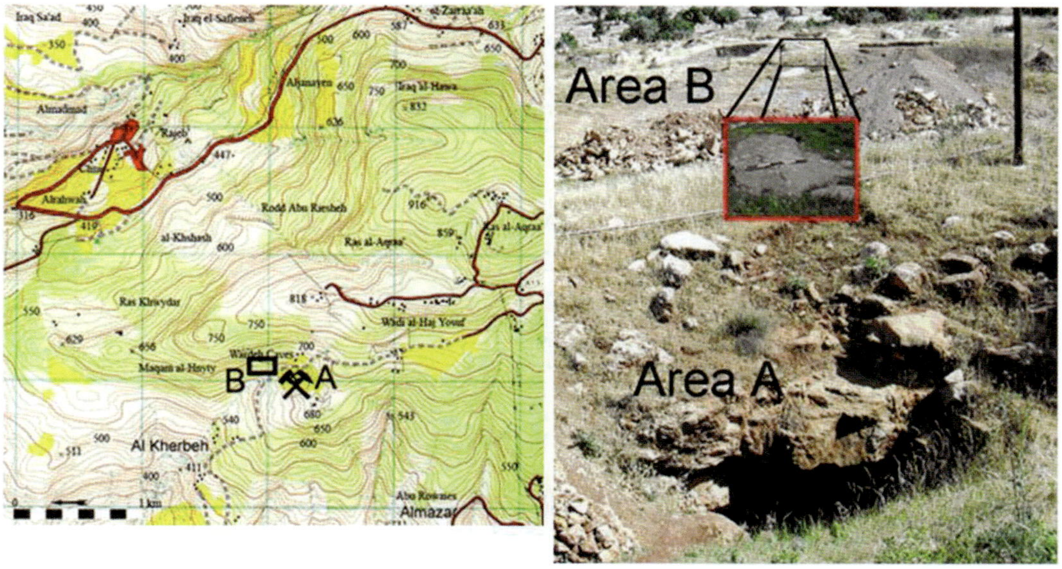

Figure 3: Part of the contour map of Ajloun area. It shows the location of the iron ore mine of Mugharet al-Wardeh (Area A: the settlement and Area B: the mine). The road separates the mine to the north and the settlement to the south of the road. Zarqa River runs less than 4km to the south of the site. In the red frame are two large slabs of slag and charcoal, which still lying on the top of the iron ore mine of Mugharet al-Wardeh. These represent clear evidence of previous metallurgical activities at the site.

- The first gallery extends on both sides of the main adit, and the left part extends about 10m before it is blocked by soil and pebbles washed in by rain over a long period. Cultivation of the nearby fields by the inhabitants of the village has increased the amounts of debris brought in by rain-wash. It is obvious that this gallery continues further to the west, but it is not possible to reach its end. A $3x2m^2$ trench (Fig. 4) was dug in this side of the gallery to uncover the depth of the accumulated debris and to reveal any archaeological remains that might indicate the periods of use of the mine.
- The second gallery (Fig. 4): A low and narrow passage connects the first gallery with the second one. The second gallery extends to the right and left of this low passage. It is much larger than the first one, with a higher and wider ceiling. The left part (Fig. 4) extends to approximately 15m, and it is 5-8m wide. It is impossible to proceed further due to the accumulation of rain-wash debris. All attempts to explore the exact extension and direction have failed, although it is presumed that this wing is much deeper.

The right part extends about 40m and is between 2-12m wide. Its northern side is partially collapsed (having been destroyed by the NRA), although part of this side is preserved as it may have been in ancient times. In the middle of the northern side of this wing is a small adit, which is almost blocked with rain-wash debris.

A $4x4m^2$ trench was excavated at the eastern end of the right wing to reveal the original shape of the gallery and to look for *in situ* archaeological remains or installations left by the ancient miners. A gallery (5x2m long) was revealed by further excavations. The structure of this feature suggests that the miners were following a natural crack in the rock from which iron ore (hematite) was being mined, widening this crack up to 2m in some places.

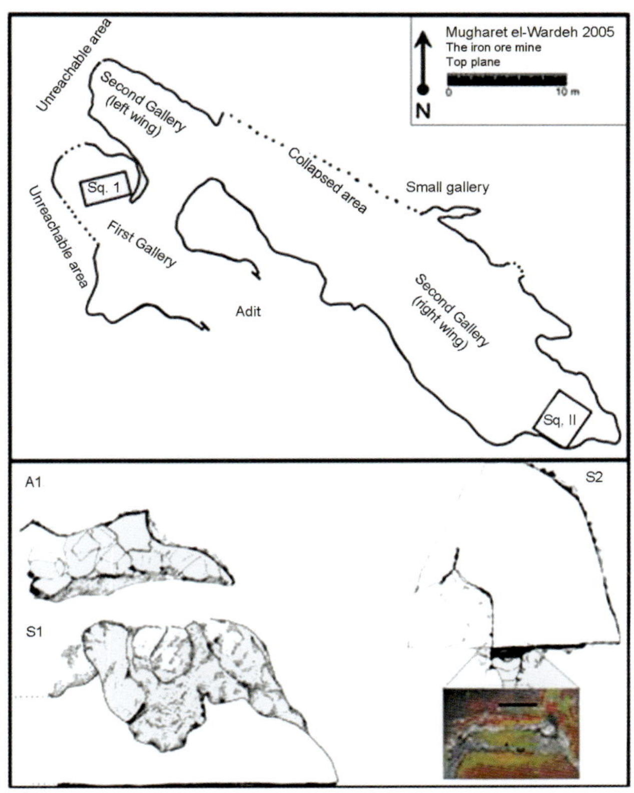

Figure 4: Top plan of the iron ore mine of Mugharet al-Wardeh in area A. It shows the excavated squares (I and II) and the small gallery. Limited pottery sherds and charcoal samples were obtained from these excavated locations. Many parts of the mine were inaccessible due to the accumulated fluids and the collapsed parts of the mine. Many parts of the mine are inaccessible or very dangerous due to the possible collapse of these parts.

The first trench

A 3x2m² trench was excavated across the width of the western wing of the first gallery, close to the entrance connecting the two galleries (Fig. 4). The main aim was to uncover any mining tools or other archaeological remains left by the ancient miners. This trench was also intended to test the depth of the rain-wash debris, and provide dateable material for the purposes of clarifying the period during which this mine was used.

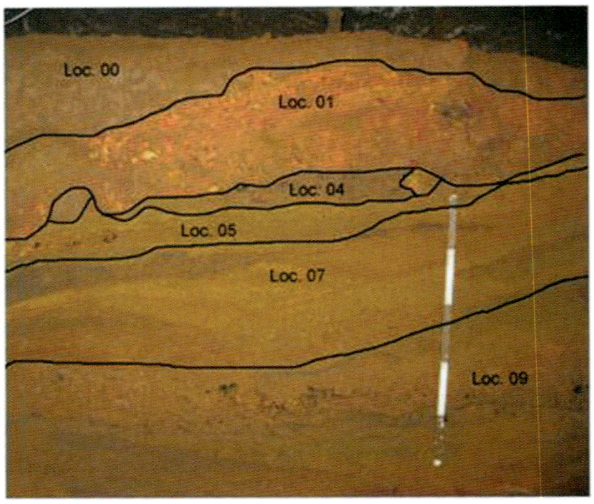

Figure 5: Southern section of the second trench at the iron ore mine of Mugharet al-Wardeh. Apparently this mine has passed through several periods of use and some periods of abundance. ^{14}C samples and a pottery lamp have dated Loc. 09 of this trench to the Ayyubid-Mamluki period.

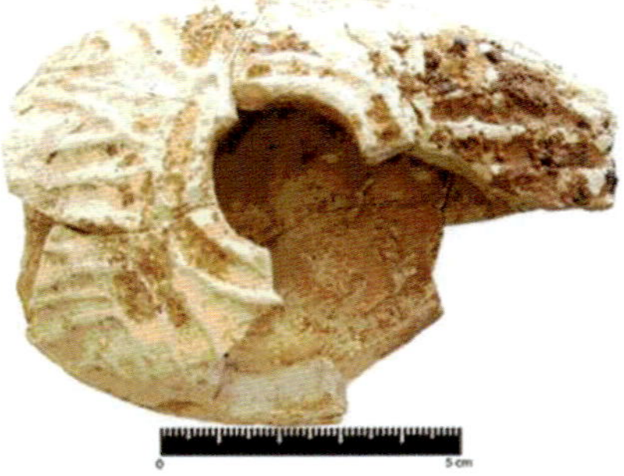

Figure 6: The Ayyubid-Mamluki partly restored pottery lamp. It was excavated from a 'sand-like' muddy layer above the fire pit.

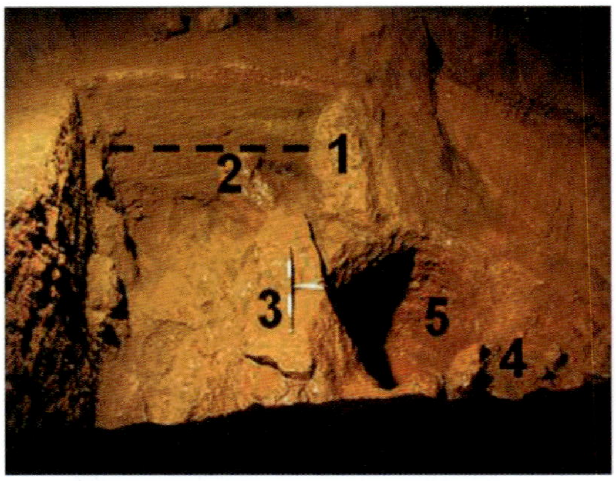

Figure 7: The second trench in the mine yielded clear Ayyubid-Mamluki layers of occupation. A pottery lamp of this period was uncovered in the layer above the line. (1) and (2) a circle of stones "fire pit" (3) is a stone slab in front of an ore exploring or working shaft (5), and (4) is a layer of stones of different sizes that accumulated in the corner behind the shaft.

Fragments of a pottery lamp of the Ayyubid-Mamluki period (1188-1516 AD) were uncovered in Loc. 07 (Fig. 5); some of these fragments were successfully reconstructed. No other diagnostic potsherds or any other material was found in this trench. A charcoal sample was extracted from a charcoal layer Loc. 06. In general, the trench consisted of an accumulation of layers of red-brownish soil interspersed with sandy layers mixed with lime and hematite stones of different sizes.

The second trench

At the end of the right wing of the second gallery, a 4x4m² trench was excavated. The aim of this trench was to uncover what might have been the last stage of the use of this hall. If any installations or tools were being used at the time when this mine was abandoned, they could possibly still be *in situ*.

The upper layer (Loc. 00) in (Fig. 5) formed a levelled surface, which was possibly the result of rain-wash from the last period of abundant rainfall. It consisted of dark, reddish-brown soil very similar to the soil outside the mine. Below this, was topsoil Loc. 01 consisting of fragments of hematite of different sizes. It seems that this layer accumulated because of the collapsing ceiling of this mine. These collapsed hematite-rich fragments had never been gathered and remained *in situ*.

Loc. 04 and 05 were composed of reddish-brown, mixed soft and dense areas of deposit, in addition to pieces of charcoal. These layers contained no potsherds.

Below Loc. 05, a layer of very compact soil of different colours accumulated, and this has been interpreted as further rain-wash debris indicative of a longer period of abundant rainfall. The lowest layer (Loc. 09) within this trench was just above the bedrock of the mine and was formed of a mixture of different kinds of stone fragments, charcoal, and soil. An Ayyubid-Mamluki (1169-1516 AD) pottery lamp (Fig. 6) was uncovered in this Locus about 12 cm above the floor. The style of this lamp suggests it could be contemporary to the ^{14}C dated (1480 AD) samples from the small gallery.

Within Loc. 09, just under the lamp, a pavement of unshaped stones surrounded a hole in the earth, approximately 80cm in diameter and 120cm deep (Fig. 7). A large stone slab (c. 150cm long by c. 20-50cm wide) with the remains of wooden beams on the eastern and western ends formed part of the surround of this hole (Fig. 7:5). Two possible interpretations of this feature can be currently offered:

1. These remains might have been left after an unsuccessful attempt to dig out a shaft hole in order to prospect for any possible new layers of iron ore, but was abandoned when miners found only lime layers that contained no traces of iron ore.
2. This feature could represent an ore beneficiation installation: a miner would have stood in the hole behind this stone slab, hammering the ore on the stone to remove the best ore and thus to avoid the efforts of transporting poor quality ore from the mine (Thomas Stöllner, personal communication). Tylecote (1987) certainly stated that in some cases in Europe, mining and ore processing were conducted simultaneously within a mining location. In this case, such an installation may have been the result of a combination of the two possibilities above; an unsuccessful ore prospecting shaft that was converted into an ore-dressing workshop situated within the mine.

Archaeological and metallurgical remains (Area B)

On top of the hill, above the mine, slag pieces can be seen scattered over the area, as well as two large mixed formations of slag and ash (each one approximately 1.5m in diameter), situated one on top of the other. A few deteriorated potsherds are scattered across the hill, in addition to some small tuyère fragments. Two 4x4m^2 trenches were dug in this area to test for metallurgical activity or any past settlement evidence, and to obtain secure datable material. Other features possibly associated with the ancient mine are also visible in the vicinity.

Square T.1

Within the 4x4m^2 trench, a 1m thick layer of mixed ash and crushed slag (from 1cm up to 15cm in length) was uncovered, indicative of an intensive period of iron smelting activities.

Very limited quantities of glazed and red decorated potsherds were also obtained from this layer. The Ayyubid-Mamluki period is the most probable date for this layer (D. Vieweger, personal communication).

Below this mixed ash-slag layer, a layer of soft light brown soil with scattered stones of different sizes was revealed. No pottery sherds, metallurgical remains, or any distinguishable artefacts were found associated to this layer, which is situated directly over the bedrock of the mountain.

Square T.3

The second trench was excavated 5m to the west of the first, with the aim of locating more metallurgical remains and evidence of any earlier activities. Again, an ashy-slag mixed layer over 1 metre thick, it slopes toward the northern side of the hill.

Different diagnostic pottery sherds were uncovered from this locus. They were of the same glazed and red decorated pottery type of the Ayyubid-Mamluki (1169-1516 AD) period

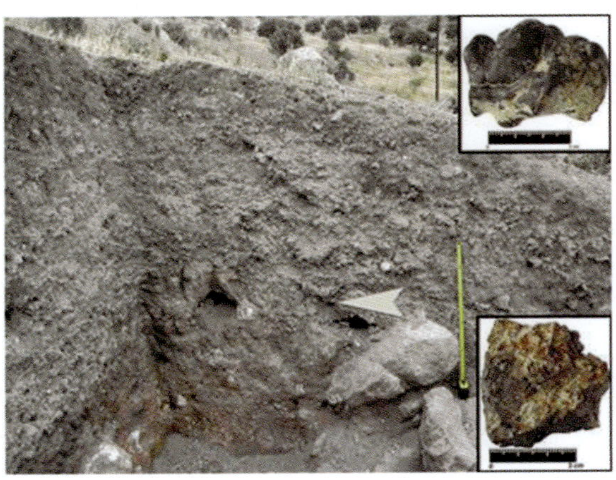

Figure 8: A thick layer of ash mixed with metallurgical remains (tuyères and slag) has been uncovered during the 2005 excavation at MaW. This ashy layer has been dated back to the Ayyubid-Mamluki period according to the associated pottery sherds. In the upper right corner is a sample of tab slag and in the lower one a sample of broken slag. The arrow shows the north direction.

uncovered in square T.1. In this layer, as in the ashy one of the first trench, scattered fragments of tuyères were uncovered in addition to some broken and other almost complete ones. These tuyères are another indication of the metallurgical activity that took place at this location. In most of the cases, 'tears' of molten slag were still attached to the air output of these tuyères.

Loc. 06, lying directly underneath the above mentioned ashy layer (Loc. 03), consisted of light-brownish, compact soil with 1-2 rows of unshaped stones of different sizes. It is probable that these stones are part of a collapsed construction that extends into the northern and eastern sections of this square.

Limited amounts of potsherds excavated from Loc. 06 were typologically dated to the Iron Age, the Late Roman, Byzantine and Early Islamic periods. Nevertheless, more excavations and dating are necessary to determine the exact dates of the early use of the mines and the iron smelting events at this site.

Direct dating of the settlement and the mine

Dating of the smelting location on the hill and the mine was obtained by morphological studies of the excavated potsherds, as well as by ^{14}C dating. Both methods suggest an intensive exploitation of this mine during the Ayyubid-Mamluki period (1169-1516 AD). Potsherds excavated from the mine gave the same date as the analysed charcoal samples. It appears that this mine reached its maximum size during this period. Any earlier remains would have been demolished during this later period. The excavated remains of metallurgical activity were also dated to the Ayyubid-Mamluki period.

Indirect dating of the mine

Since direct dating of pottery and charcoal samples at the site provided limited results, an indirect approach has been used to clarify the periods during which this mine was in use. This indirect method involves tracing artefacts produced from the raw materials originating from the mine that were exported to other sites. In using this method, it was hoped that any earlier exploitation of the mine would be detected.

Chalcolithic period (4500-3500 BC) hematitic mace heads were sampled from the archaeological sites of Gilat (Levy and Alon, 1994, 1996), Nahal Tillah (Levy, et al., 1997), Shiqmim (Levy and Alon, 1983, 1987), Abu Hamid (Dollfus and Kafafi, 2001) and Teleilat el-Ghassul (Malon, et al., 1934). They were analysed for their trace elements and lead isotopes in order to establish the provenance these finds.

Fig. 9 shows a comparison of the concentrations of six different trace elements (Bi-Cr-As and Sr-Pb-Ba) between the analysed samples from the iron ore deposit at MaW and the Chalcolithic period (4500-3500 BC) hematitic artefacts (including the mace heads). Chemical analyses were carried out by ICP at the laboratories of the Deutsches-Bergbau Museum in Bochum, Germany.

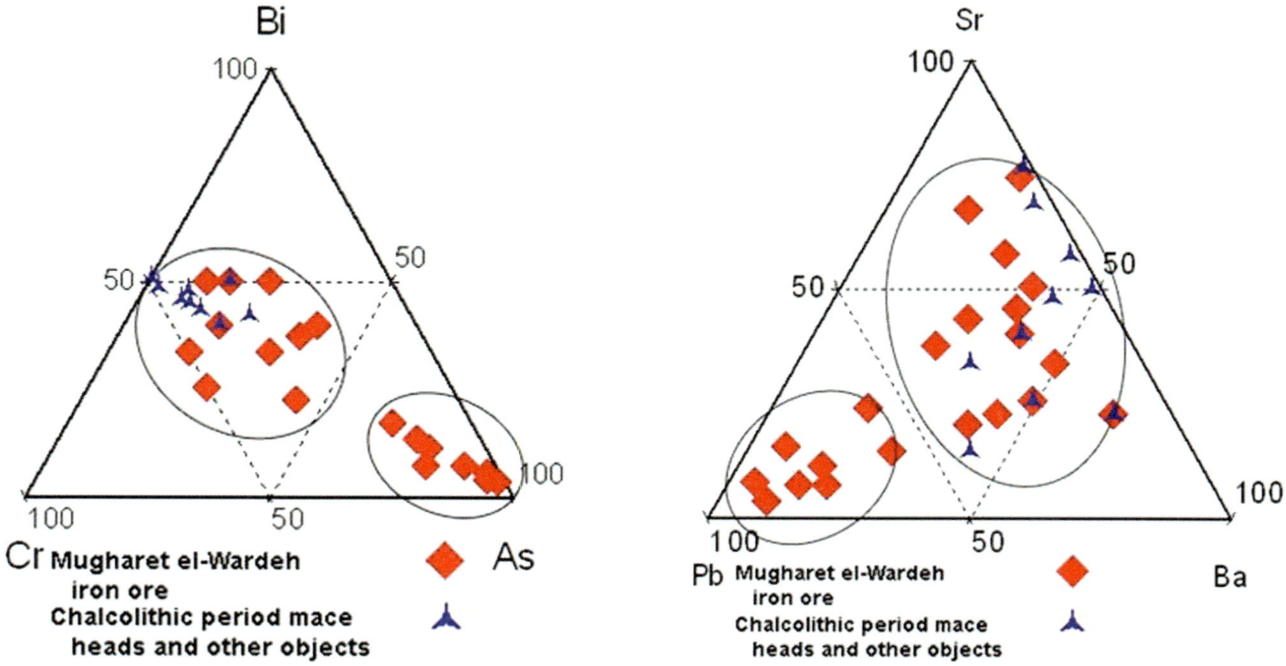

Figure 9: Two plotted ternary diagrams showing the concentrations of six different trace elements (Bi-Cr-As and Sr-Pb-Ba) in the iron ore deposit of Mugharet al-Wardeh and Chalcolithic period (4500-3500 BC) hematitic artefacts. It is very clear that two different groups can be distinguished within the ore deposit. It is most probable that the hematitic artefacts were obtained from one of the two groups of the iron ore of Mugharet al-Wardeh during the Chalcolithic period (4500-3500 BC).

It is clear that the iron ore samples can be divided into two groups according to the concentrations of their trace elements (Bi, Cr, As, Sr, Pb, and Ba). This would be due to the original formation of this iron ore deposit and its composition.

The Chalcolithic hematitic artefacts cluster around/within the same group of ore samples, with at least four of these hematitic artefacts overlapping with some of the ore samples in this group. Based on the distribution of the ore samples from MaW and the hematitic artefacts, it is clear that the artefacts were obtained intentionally from the deposit of this hematitic iron ore at MaW. Therefore, it is possible that this iron ore deposit was the source of these Chalcolithic period (4500-3500 BC) artefacts from sites in the greater region of sBaS.

Lead isotope results

Fig. 10 shows the distribution of lead isotopes in the samples analysed in this study. It includes the Chalcolithic period (4500-3500 BC) mace heads and hematite artefacts and the Iron Age/Classical period iron slag samples, as compared with the iron ore samples from MaW. Five iron metal artefacts were employed in this study. Among them are two artefacts excavated from the archaeological site of Safut, one Ayyubid-Mamluki knife excavated from Mahis, and two other unidentified iron artefacts excavated from the ash layer in Square T.3 at MaW. The last two objects were found mixed in with the ash and other metallurgical remains.

A two-dimensional, mirror-image diagram shows that the mace heads and the hematitic artefacts overlap with the iron ore deposit. Only one mace head sample (S-S374) lies outside the clustering samples of iron ore, but still very close to the lead isotopic profile of the MaW iron slag samples. There might be two explanations for this: some layers of the iron ore deposit do not exist anymore and may have been completely consumed in antiquity during the intensive smelting activities of the Ayyubid-Mamluki period (1169-1516 AD). Alternatively, this might be due to the limited number of samples collected for lead isotope analysis.

Moreover, the lead isotopes of the Iron Age/Classical period iron slag samples from Tulul edh Dhahab also cluster very clearly around/over the iron ore deposit samples and the iron slag samples from MaW.

It is probable that all the analysed samples of mace heads, hematite artefacts, and Iron Age/ Classical Periods iron slag had been extracted from the iron ore deposit of MaW. Therefore, it is possible to re-date the use of the MaW hematitic iron ore mine; as such, the Chalcolithic period (4500-3500 BC) would be the earliest period of exploitation of this ore deposit, instead of the Ayyubid-Mamluki period (1169-1516 AD). Exploiting this iron ore deposit may well have started by using this ore for colouring and manufacturing of objects of raw hematite. In later periods, i.e. Iron Age/Classical Periods, charges of iron ore obtained from this deposit were reduced in smelting furnaces to produce metallic iron.

Fig. 10 shows that the lead isotope profile of only one iron artefact overlaps with the lead isotope profile from MaW. Therefore, it could be at least probable that the iron knife excavated from Mahis was made of iron ore from the MaW deposit.

Lead isotope profiles of the two iron artefacts from MaW are located outside the cluster of the lead isotopes of the MaW iron ore and also outside of the range of distribution of this profile. This leaves no doubt that these two artefacts, as well

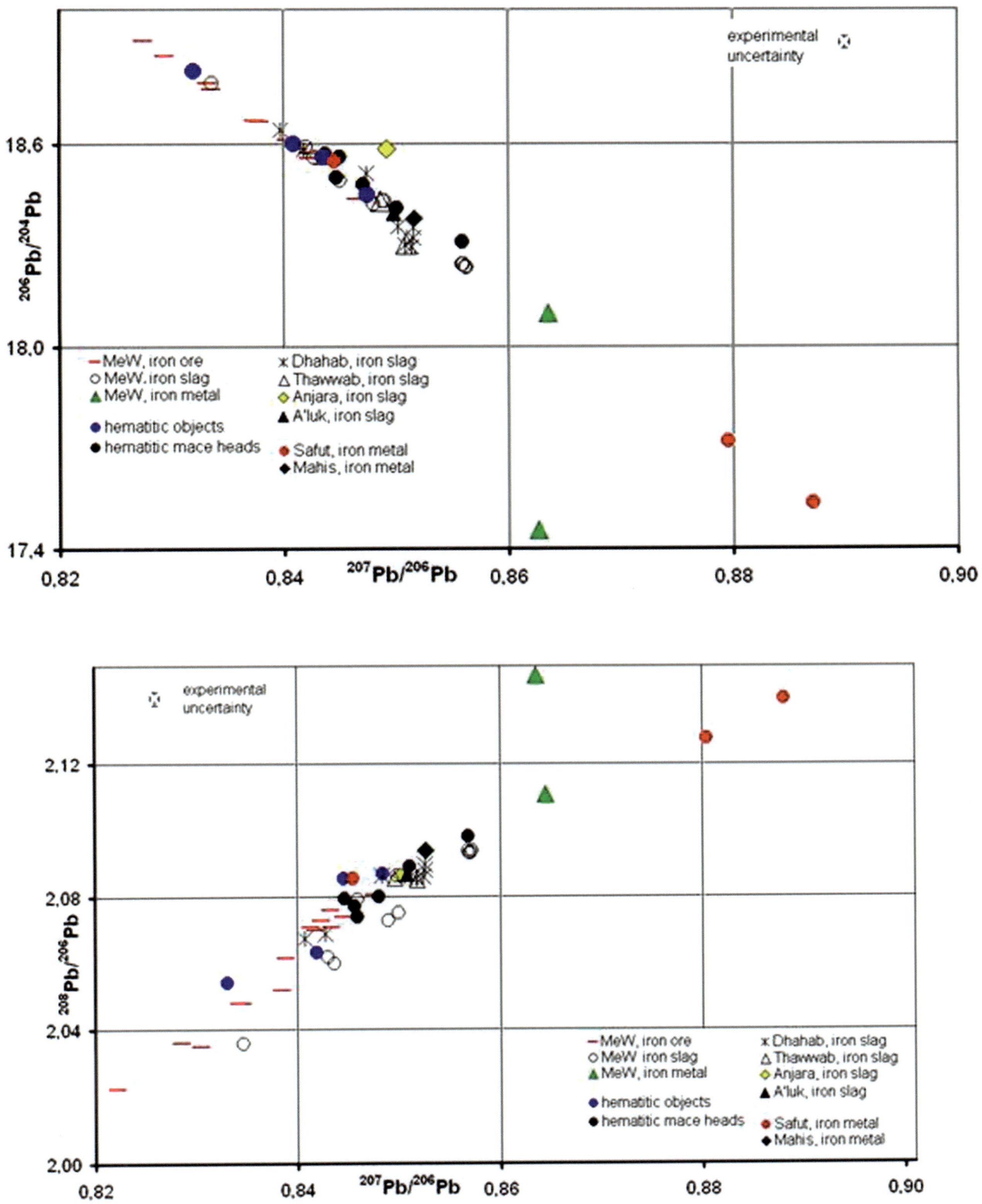

Figure 10: Two dimensional mirror-image lead isotopes diagrams of the iron ore from the mine of MaW and other hematitic, slag, and iron metal samples. It is very possible that all the hematitic artefacts, slag samples, and at least one of the metal artefacts from Safut have been produced with raw material from this iron ore deposit.

as the other two iron artefacts from Safut, do not belong to this group of lead isotopes. These artefacts would have been produced from iron ores other than those originating from the MaW iron ore deposits.

Conclusion

Direct dating of the archaeological site of MaW relates the exploitation of this mine to the Ayyubid-Mamluki Period (1169-1516 AD) based on the typological dating of the archaeological remains (mainly pottery sherds). Nevertheless, scientific analyses of hematitic artefacts (of different periods) excavated from the sBaS proved that most of the analysed objects were produced with material obtained from the iron ore source of MaW. This iron deposit is most likely to be the main source of hematite in this region throughout the ages. Hematite was used for producing red colours and hematitic artefacts since the Chalcolithic period (4500-3500 BC), long before it was used in the production of iron metal and manufacturing of iron objects after the 13th century.

More analyses are needed and will be applied on more artefacts, either hematitic or iron objects from different periods, to both estimate the distribution of these artefacts and the early connection between archaeological sites and the iron ore deposit of MaW, and to reach a better understanding of the development of the ore exploitation and ore smelting techniques at this site.

References

Abed, A., 2000. *Geology, Environment and Water of Jordan*. Amman: Jordan Geologists' Association, Amman (in Arabic).

Abu Ajamieh, M., Bender, F., Eicher R., 1988. Natural Resources in Jordan, Natural Resources Authority, Amman

Alomari, Q., 1991. Ancient Mining in the Hashemite Kingdom of Jordan. *The Jordanian Engineer*, 11, pp.7-10 (in Arabic).

Basha, S.H., 1968. *A Report on the Stratigraphy, Occurrence, Origin and Prospection of the Iron Ore Deposit at Warda-South Ajloun District*. Amman: Natural Resources Authority.

Bender, F., 1968. *Geologie von Jordanien*. Beiträge zur Regionalen Geologie der Erde, Band 7. Berlin and Stuttgart: Borntraeger.

Bender, F., 1974. *Geology of Jordan*. Berlin and Stuttgart: Gebrueder Borntraeger.

Dollfus, G. and Kafafi, Z., 2001. Jordan in the Fourth Millennium. In: *Studies in the History and Archaeology of Jordan VII*. Amman: Alrai Commercial Press, pp.437-49.

Glueck, N., 1935. *Exploration in Eastern Palestine, II*. Annual of the American Schools of Oriental Research 15. New Haven: Jane Dows Nies Publication Fund.

Levy, Th. and Alon, D., 1983. Shiqmim, 1983. *Israel Exploration Journal*, 33, pp.272-4.

Levy, Th. and Alon, D., 1987. *Shiqmim I: Studies Concerning Chalcolithic Societies in the Northern Negev Desert, Israel (1982–1984)*. Oxford: British Archaeological Reports International Series 356.

Levy, Th. and Alon, D., 1994. Gilat – 1991. *Excavations and Surveys in Israel*, 12, pp.91-4.

Levy, Th. and Alon, D., 1996. Gilat – 1992. *Excavations and Surveys in Israel*, 14, pp.88-93.

Levy, Th., Alon, D., Smith, P., Yekutieli, Y., Rowan, Y., Goldberg, P., Porat, N., Brink, E., Witten, A., Golden, J., Grigson, C., Kansa, E., Dawson, L., Holl, A., Moreno, J. and Kersel, M., 1997. Egyptian-Canaanite Interaction at Nahal Tillah, Israel (4500-300 B.C.E.): An Interim Report on the 1994-1995 Excavations, *Bulletin of the American Schools of Oriental Research*, 307, pp.1-51.

Mallon, A., Koeppel, R. and Neuville, R., 1934. *Teleilat Ghassul I*. Scripta Pontificii Instituti Biblici. Rome: L'Institut.

Moh'd, B. and Muneizel, S., 1998. *The Geology of Salt Map*. Map Sheet No. 3154 III, Natural Resources Authority (NRA) 41, Amman.

Quennell, A.M., 1952. The Geology and Mineral Resources of (former) Trans-Jordan, in Colonial Geology and Mineral Resources. *The Quarterly Bulletin of the Colonial Geological Surveys*, 2, pp.85-115.

Saffarini, G., 1989. Geochemical Characterization of a Carbonate-hosted Hydrothermal Iron Ore: The Wardeh Iron Ore Deposit/ Ajlun-Jordan, *Dirasat*, 16, pp. 46-63.

Tylecote, R., 1987. *The Early History of Metallurgy in Europe*. London: Longman.

Authors' addresses

- Yosha Alamri
  The Jordan Museum, Amman Jordan
- Andreas Hauptmann
  Deutsches Bergbau-Museum Bochum, Germany

Forging activities in Jaffa during Ottoman Palestine

Sylvain Bauvais

ABSTRACT: Within the framework of research on networks of iron exchange in the Mediterranean area, this article presents an analysis of forging wastes from two workshops of the Palestinian city of Jaffa during the Ottoman period (end of the 18th / beginning of the 19th century). The results of a metallographic study are presented here and used to suggest that the metal worked in these two workshops came from both the import of metal from indirect reduction and, in a larger proportion, of metal from recycling. This hypothesis comes to confirm the situation of a supply dependent on European traders, in particular the French. This pressure on resources would have urged the craftsmen of Jaffa to develop innovative techniques to recover the maximum metal from the metallic waste, including the fragments of cannonballs made of cast iron.

Keywords: Ottoman period, Palestine, Jaffa, steel-making, slag, iron scrap, metallography, provenance.

Context and methodology

Research context

In this article a case study with a new geographical and historical research focus is presented. While my previous work concerned the Iron Age in the North of France, here forging activities in Jaffa during Ottoman Palestine are discussed, which to some may seem a daring divergence. However, the research problems are the same: those regarding the reconstruction of iron- and steel-making techniques, the definition of the 'qualities' of raw materials in trade and the regional organisation of the production. Furthermore, the region has important features in common with the North of France.

Ore deposits are few and are found distributed around the periphery of historic Palestine (Beirut, Damascus, Ajlun, and Sinai) (Al-amri, 2007; Bauvais, 2008). In such conditions, exchange networks were essential for the development of local production. Throughout history, the southern Levant area suffered constant conquests and occupations, and was subject to a variety of forms of political powers and controls. The production and trade of iron depended on the metallurgists being able to meet the changing conditions and demands which resulted from the new geopolitical and economic affinities.

This work therefore considers the city of Jaffa during the Ottoman period in relation to diachronic fluctuations within the organisation of local iron industries, and aims to understand the role of this material and its importance for the societies of the Levant (Bauvais, 2008). This article presents an analysis of forging wastes from two workshops, both of which date to a period between the end of the 18th century and the beginning of the 19th century (around the time of the conquest of Jaffa by the Napoleonic troops). The present study is part of a more global research programme on the entire city, the "Jaffa Cultural Heritage Project" (JCHP), introduced in 2007 by Martin Peilstöcker (from the IAA) and Prof. Aaron Burke (from UCLA) (Peilstöcker, 2006, 2007; Burke and Burke, 2008).

Initially, there were three key aspects of the research that were expected to yield particularly good data:

First, it was assumed that waste from more recent periods would be better preserved; second, that the analysis of a greater number of elements would yield a better understanding of the technical practices employed; finally, it was believed that access to historical texts and local traditions would enrich the results of this research. In fact, none of these assumptions proved to be true.

The city of Jaffa sits beside the sea on sandy substratum, leading to particularly intensive corrosion of both iron objects and production waste dating from all historical periods. The workshops are located in an urban archaeological context, which means that the evacuation of waste outside the spaces inhabited during ancient times does not yield archaeometallurgical material. Finally there is almost a complete absence of historical data concerning metallurgical techniques dating to this period. As such, the situation is comparable on many levels with that of pre-Roman France.

Iron metallurgy in Jaffa

Excavations carried out since the 1950s in Jaffa revealed eight metallurgical zones, either workshops or areas of waste dumps, dating from the Persian period to the Ottoman period (Fig. 1). The fact that Jaffa is a port and, more especially, the port of Jerusalem put this city in an important position within the maritime iron trade networks in the Western Mediterranean

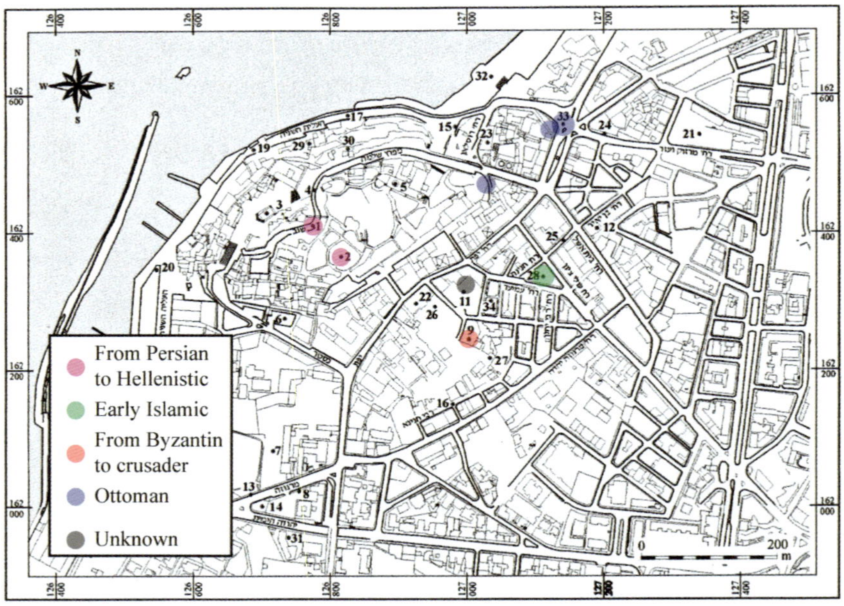

from Peilstöcker 2007 modified

Period		Jaffa - Kaplan	Jaffa - Bet Eshel Street (A)	Jaffa - Ganor comp. D, H	Jaffa - Clock-Tower Square	Jaffa - Police comp.	Jaffa - Rehov HaTsorfim
1517-1917	Ottoman						
1250-1517	Mamluk						
1099-1250	Crusader						
638-1099	E. Islamic						
330-638	Byzantine						
63-330	Roman						
323-63	Hellenistic	2					
539-323	Persian						

Figure 1: Localization of forging workshops in Jaffa and their dating.

area and explains the large number of craftsmen who seem to have settled in its quarters.

The first analysis presented below concerns steel-making wastes from the old city in the Police Compound (A 5037) and HaTsorfim Street (A 5378), both dated to the end of the 18th and the beginning of the 19th century.

During the 18th century, international trade dominated by the French involved South Syrian ports such as Saïda, Haifa, Acre, and Jaffa (Panzac, 1990). Annual records drawn up by the French consuls in the Levant concerning the trade under their jurisdiction contain relatively little information about traded metallic products. Between 1756 and 1787, only manufactured goods such as nails are quoted as iron artefacts coming from France, while where metallic raw materials are concerned, iron bars shipped by the French from Marseille seem to have come from Northern European countries. It is clear therefore that archaeological data resulting from new excavations in Jaffa have the potential to improve our understanding of production techniques and of the raw material worked along the coasts of the South Levant.

In terms of production techniques, the 14th and 15th centuries in Europe witnessed the emergence and the development of indirect ferrous metallurgy in the blast furnace (Arribet-Deroin, 2001; Fluzin, 2002; Dillmann et al., 2007a, 2007b). Very little is known about the introduction of this technique in the Middle East, and it seems that only provinces of the Northern Ottoman Empire produced iron, steel and cast iron on a large scale (Danisman, 2007). Here, production was mainly steered towards the imperial cannon foundries subject to imperial decrees and scrupulously controlled by the Ottoman administration. In these early periods, very little of the produce of this industry reached the borders of the empire, and even less civilian use.

Meanwhile, in Europe, production increased enormously and European markets had the capacity to meet the demands of the Middle East.

Many questions therefore arise when one considers the introduction of the metal produced by the indirect reduction method to the Levantine coasts during this period. What proportion of the metal was European and how much was produced in Anatolia via this method? How much iron was still produced using the bloomery method after the introduction of the indirect method? Answers to such questions may prove crucial for understanding trade and production during this period.

A particularly interesting account comes from a study carried out on the L. Stein Factory in Jaffa (Avistar, 1990). This account concerns a Polish Jewish smith who settled in Jaffa in 1887, founded the first repairing forge, then went on to became one of the most important metal producers of this time in Palestine. The author (Shmuel Avitsur) presents the know-how and skills of Palestinian craftsmen, while simultaneously revealing his own prejudices, as follows:

> "The methods, machinery and tools of metal-working in Europe were unknown to blacksmiths in Palestine in the late nineteenth century. A handful of German (Templar) blacksmiths did carry out some repairs, mainly agricultural tools which they had brought with them from Europe. (...) Their use became widespread with the Jewish settlement drive of 1882 (...). Since they were not suited to local conditions, these tools broke down frequently, necessitating repairs by skilled craftsmen in modern equipped workshops, neither of which were available in the country at this time."
> (Avistar 1990 p.159.)

The two sites analysed for the purposes of this paper present a revised version of metallurgy during this period in Jaffa. Before the present research, only limited information was available about the city of Jaffa and its steel making activities, including its commercial networks, organisation, and the function of the steel items produced. However, this harbour city seems to have been at the centre of a large scale trade network encompassing Europe, Anatolia, Syria and Egypt. The results presented here therefore finally allow for an understanding of this essential metallurgical activity in relation to the key position of Jaffa during ancient, medieval, and modern times.

Methodology

The methodology used in this study is based on a metallographic approach to slags and iron scraps coming from workshop contexts in order to define the main lines of the *chaîne opératoire*. Slags are sorted according to a typology defined by Philippe Fluzin based on their morphology – slag-cake, fragment of slag-cake, and formless slag (Fluzin, 2000; Bauvais and Fluzin, 2005) – and by Vincent Serneels based on the first macrographic interpretation of their composition – rusty ferrous slag, dense grey slag, and clay-sandy slag (Anderson et al., 2003; Serneels and Perret, 2003). After this first step, slag-cakes are sorted depending on their morphology and composition in order to define 'families'. Then, samples are selected from these families to undergo a full metallographic analysis under the optical microscope, considering them as representative of their family. After a photographic cover, samples are cut in half and mounted in epoxy resin. They are pre-polished using abrasive paper (from grade 80 to 1200) and final polishing is performed using diamond polishing (medium 6, 3 and 1μm). During the analysis, crystalline structures and their proportions are characterised according to the possible stratigraphy (silicate phases, iron oxide, metal, adding, etc.). Finally, samples containing metallic iron undergo a metallic etching using 2% natal in order to define their composition.

For the metallic set of samples, the sampling depends on the quantity of material available. Artefacts directly stemming from shaping activities (semi-products, scraps, rough shapes, etc.), and artefacts identified with certainty as resulting from the production of the workshop are chosen first and foremost. These objects undergo the same protocol of preparation. During the metallographic analysis, the internal structures are described (degree of compactness, welding lines, types of slag inclusions, etc.) and after metallic etching their composition and their potential thermo-mechanical and thermo-chemical treatments are characterised.

From crude masses to elaboration in the Police Compound

The set of artefacts originating from the Police Compound is only composed of slags and no iron scrap is attested. These slags were discovered in several pits and floors and they seem to be linked to dumps located outside the workshop. No archaeological structures directly linked to forging activities were identified.

Macrographic analysis

The first macrographic approach to the set of artefacts shows that all the wastes stemming from this group of samples are related to a post-reduction activity, i.e. posterior to the iron ore reduction. Thus, the iron arrived in this workshop in its metallic shape in order to be transformed into objects or intermediary products (semi-products).

The group of slags consists of 128 pieces weighing a total of 10.4kg, which represents a very limited group for the post-medieval and modern periods. It can be divided up into a majority of whole slag-cakes (57%) and slag-cake fragments (28%). The rest of the samples belong to the category of indefinite fragments and formless slags (15%) (Fig. 2a).

This waste material is mostly made up of rusty ferrous slag (51%), clay-sandy slag (33%), and dense grey slag (16%) (Fig. 2b). Metallic losses seem to have been significant. The clay-sandy slag-cakes appear very homogeneous, suggesting an important practice of siliceous addition during certain working phases.

Metallographic analysis

According to their morphology and composition, three families of slag-cakes can be distinguished. The sampling was only carried out on the whole slag-cakes because, in contrast to other types of slags and slag fragments, they theoretically represent complete working sequences. Because slag-cakes belonging to family 2 are particularly homogeneous (5 slag-cakes), we selected only one of them (sample INV 007), whereas for family 1 (3 slag-cakes), two samples were analysed to confirm their belonging to the same working phase (samples INV 002 and INV 008). The only slag-cake from family 3 (sample INV 003) was also analysed. As such, we are going to present the metallographic study of four samples, representative for the set of slags (Fig. 3).

Family 1 consists of very heavy slag-cakes (from 1kg to 1.5kg), reacting to the magnet test. They are strongly oxidised and granular. Both of the analysed slag-cakes belonging to family 1 present a strong resemblance and

	Weight	Number	Average	Number of SC
Slag-cakes	5,958	8	0,74	8
Frgts of slag-cakes	2,865	46	0,06	4
Formless slags	1,549	69	/	/
Total	10,372	123	/	12

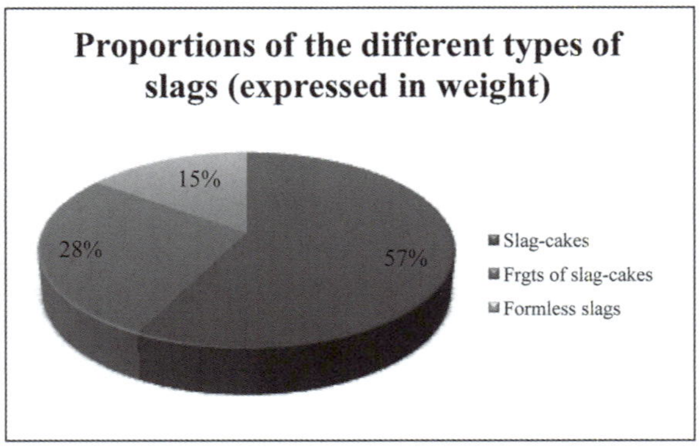

a - Proportion of the different categories of slags.

	Weight	Number
Clayey-sandy	3,374	72
Dense grey slag	1,703	39
Rusty ferrous slag	5,295	12

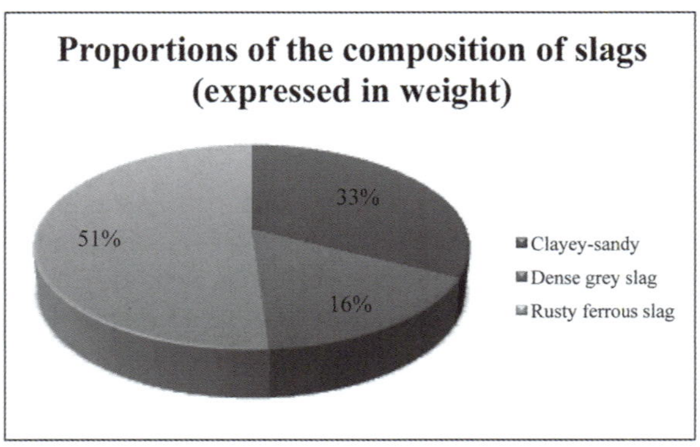

b - Proportion of the differents compositions of slags.

Figure 2: Presentation of the set of slags from Police Compound.

are linked to the same working sequence. Sections reveal a composition of 15-20% and 45% of metal and metal forms stratifications, respectively. At the bottom, the metal consists of a juxtaposition of metallic grains or metallic surface separated by slag. Towards the surface the metal becomes compact and forms homogeneous masses (Figs. 4a-b). The silicated matrix of the samples is fayalitic in nature, with very small slats. The absence of visible iron oxide in the matrix is the most striking feature of these slags – there is no wüstite present. When the metal becomes denser, it is characterised by porosities and by numerous foldings. The metal is completely composed of very well formed and equiaxed ferrite grains (0.02% C).

Family 2 is characterised by its clay-sandy composition. The average mass of slag-cakes is reduced (going from 300 to 500g) for a volume equivalent to other families.

The section reveals a completely clay-sandy composition with a very significant proportion of porosity (around 70%). In some places fragments of limestone are visible and we can even recognise a fragment of shell on the right-hand side of the section (Fig. 4c).

In some other places located in lower zones, in contrast to the clay-sandy zones located at the top of the section, more homogeneous areas of amorphous silicate and of crystallised fayalite with very small cobbles reveal a locally higher temperature that has allowed partial liquefaction of the clay-sandy material. These areas contain small concentrations of metallic prills, which show more important losses at the beginning of the activity (lower part) and then a decrease of the losses related to the use of additives (higher part). No traces of wüstite or other iron oxides are detectable.

The slag-cake making up family 3 is the only one that differs from the others morphologically. It is comparatively light (584g) and reacts to the magnet.

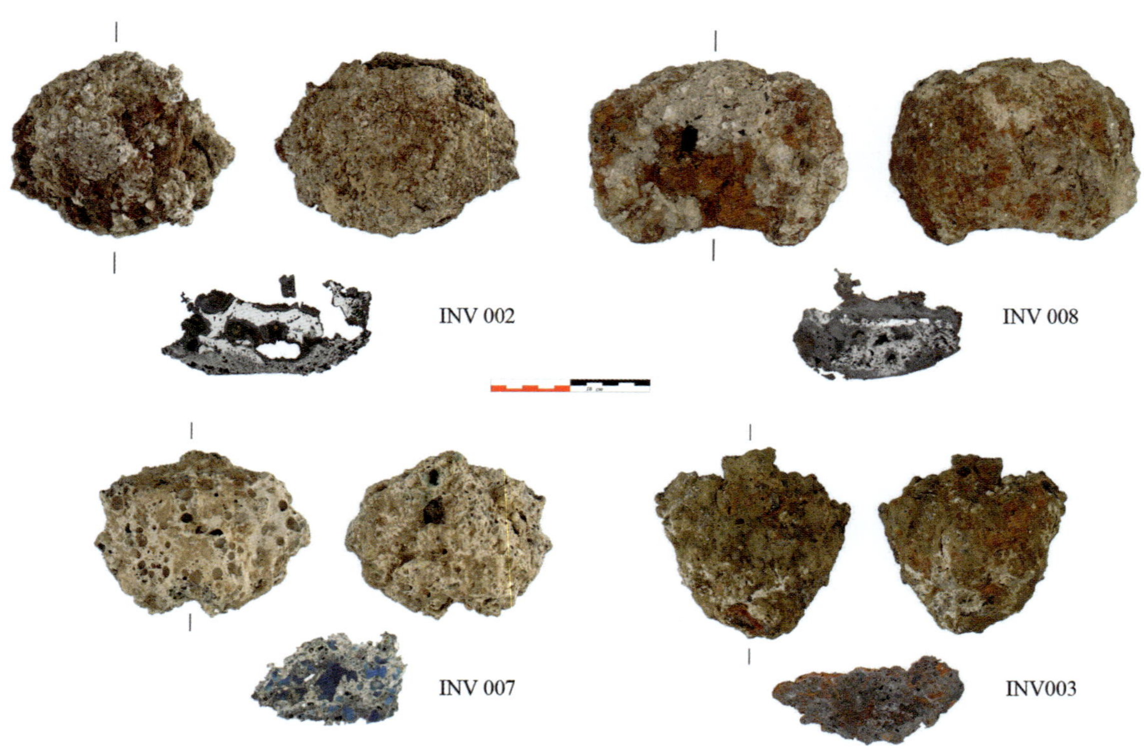

Figure 3: Presentation of the four slag-cakes sampled from Police Compound.

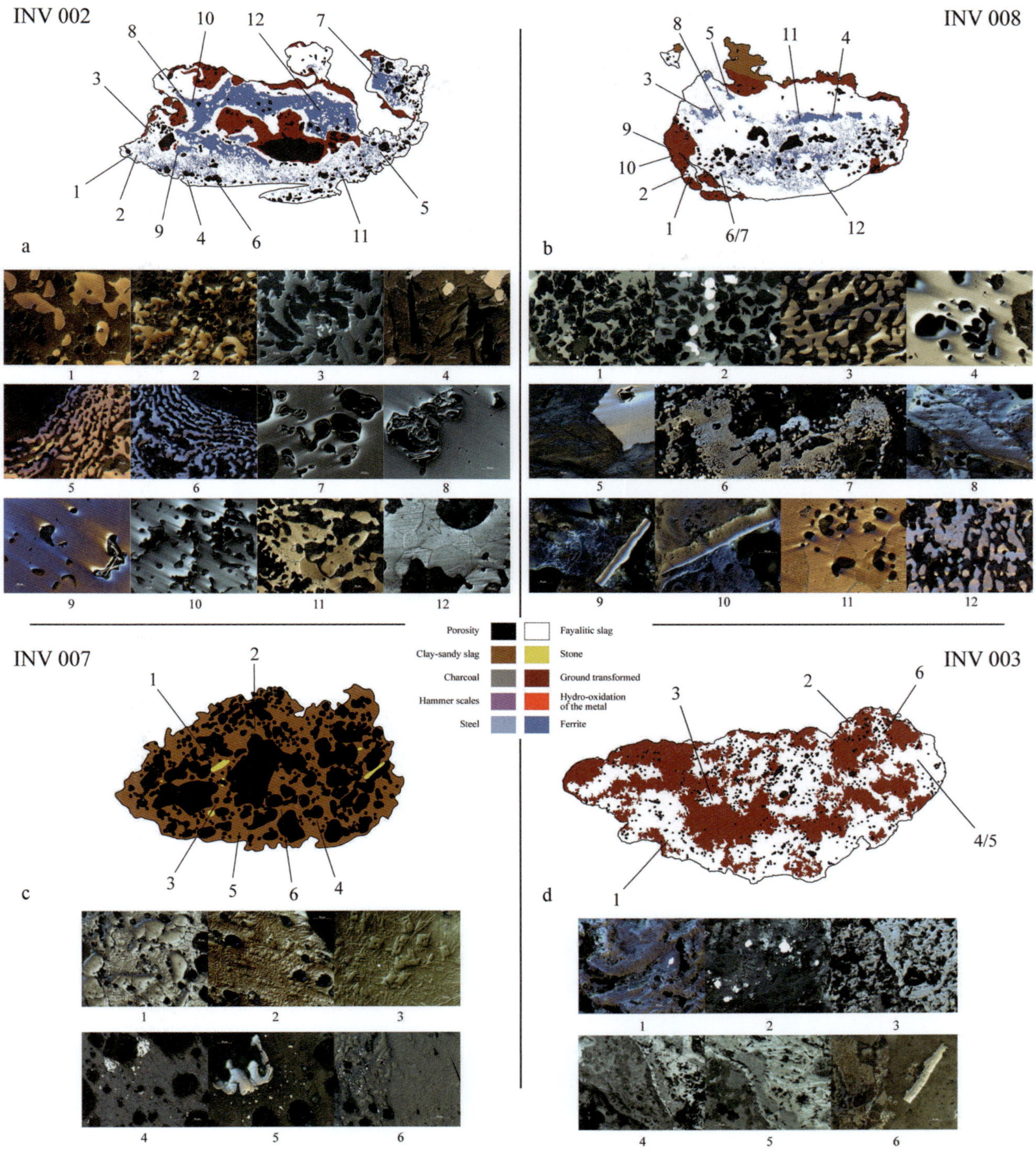

Figure 4: Analyses results for the four slag-cakes samples from the Police Compound.

The section presents a completely corroded surface. Areas which could correspond to metallic fragments adjoin corroded zones, as do areas which appear to have been former zones of porosity (Fig. 4d).

Due to the level of corrosion, chemical and microscopic analyses of this slag-cake are problematic. However, flat hammerscale is present in amorphous areas protected from the corrosion.

Only few metallic zones have been protected in the centre of initial fragments. In some places, at the periphery of what seems to be a former fragment of metal, a layer of wüstite seems to indicate a peripheral hot oxidation. A zone also seems to show a ghost-structure of steel (Widmanstätten structure – 0.4% C).

Due to the high levels of corrosion, it is difficult to extrapolate much information from this slag. However, the peripheral hot oxidation of the metallic fragments seems to indicate a forging phase in an oxidising atmosphere. This impression is confirmed by the presence of flat hammerscales in the matrix.

The most significant aspect within this set of slags is the complete absence of iron oxide losses in family 1, while a lot of metallic iron is present in the slag. For this reason, it is difficult to link these wastes to a specific type of refining activity (whether from direct reduction or from iron casting).

Another possibility is that family 1 represents a recycling activity in a reducing atmosphere at high temperature (1300-1400°C), leading to the agglomeration of fragments of metal in a forging hearth covered with charcoal in order to preserve a reducing atmosphere. The product of this technique is close to a crude mass of iron, which afterwards will need to be hammered (refining).

At the moment, and in the absence of comparative data, be they archaeological or experimental, it is difficult to confirm this interpretation.

Whatever the case, it seems that we are in presence of a complete *chaîne opératoire*, from the production of masses (family 1) to finishing (family 2) and elaborations (family 3).

Between importations and recycling in HaTsorfim Street

This excavation took place in a very narrow area crossing a forging workshop. Soils are conserved and a great quantity of material was discovered. The forging hearth and the location of the anvil were identified, and two complete hammers were discovered (Figs. 5t-u).

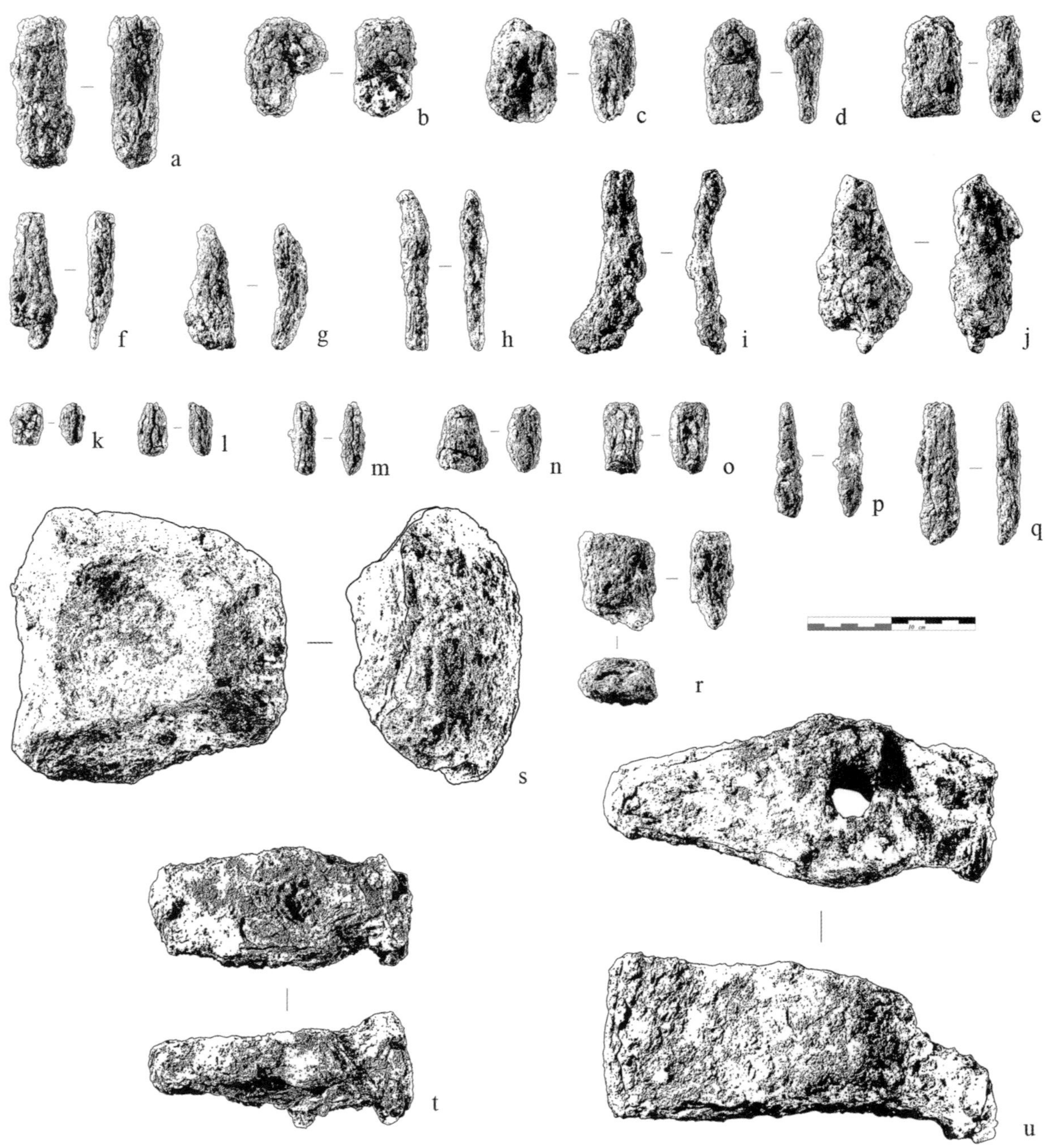

Figure 5: Presentation of a selection of metal artefacts.

Macrographic analysis

Metal elements from this site include waste from the forging activities (Figs. 5a-r), as well as many fragmented objects. These elements weigh a total of 15kg and a large majority come from the same locus, adjacent to the hearth. Fragments of cannonballs dated to Napoleon's conquest of Jaffa were found in this accumulation of iron (Fig. 5s).

	Weight (kg)	Number	Average	Number of SC
Slag-cakes	187,30	175	1,07	175
Frgts of slag-cakes	46,10	237	0,19	43
Formless slags	4,05	/	/	/
Total	237,45	412	/	218

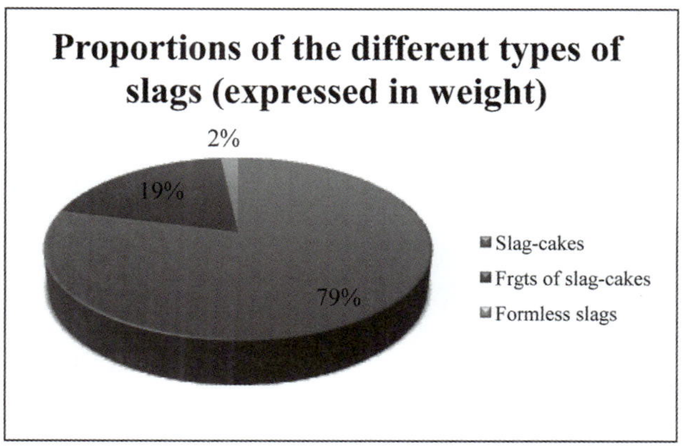

Figure 6: Presentation of the set of slags from HaTsorfim Street.

The set of slags weighs a total of 237kg, representing a large quantity as compared to the materials resulting from the excavation surface. The great majority of slags are whole slag-cakes (79%) and slag-cake fragments (19%). Only 2% of the material are formless slags (Fig. 6).

The 175 slag-cakes are rusty-ferrous slags, with only 4 slag-cakes having a clay-sandy composition.

Metallographic analysis

The homogeneity of these slags allows for the definition of two morphological types on the basis of their composition, one being composed of rusty-ferrous slag (family 1), and the other of clay-sandy slag (family 2). Two samples were analysed from the first family (samples INV 10106/1 and INV 10090/1) and one from the second (sample INV 1148/1) (Fig. 7).

For the metal, four forging scraps were chosen to reveal information about the quality and the composition of the worked metal; two iron bars reveal evidence of hot-cutting (samples INV 1027 bar 1 and 2) and two are irregular and quadrangular masses (samples INV 1027 mass 1 and 2) (Fig. 8).

Slags

Family 1 represents 171 slag-cakes with an average weight of 1.1kg. These slags are strongly reactive to the magnet and their texture is highly corroded and granular. They appear similar to the slag-cakes belonging to family 1 from the Police

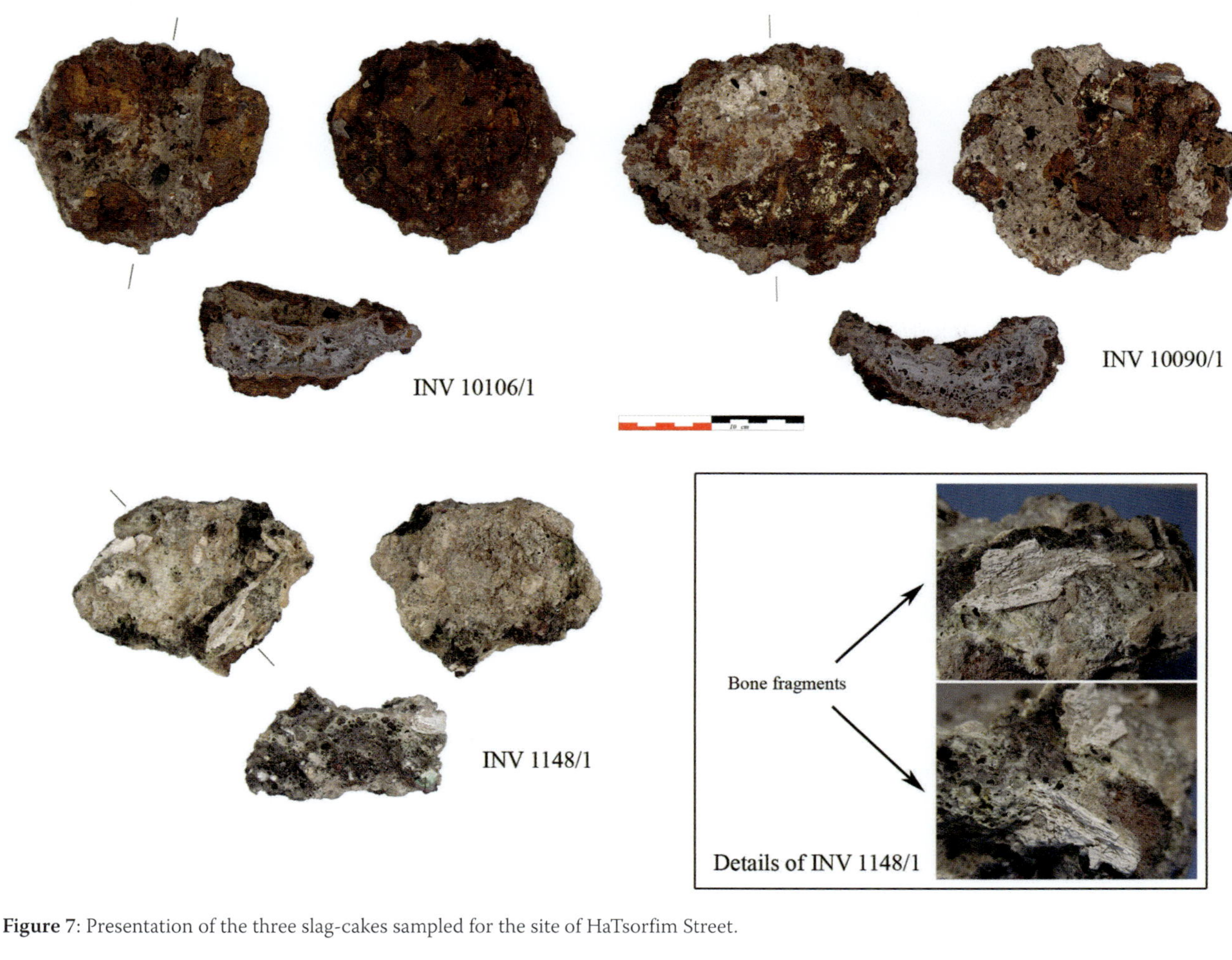

Figure 7: Presentation of the three slag-cakes sampled for the site of HaTsorfim Street.

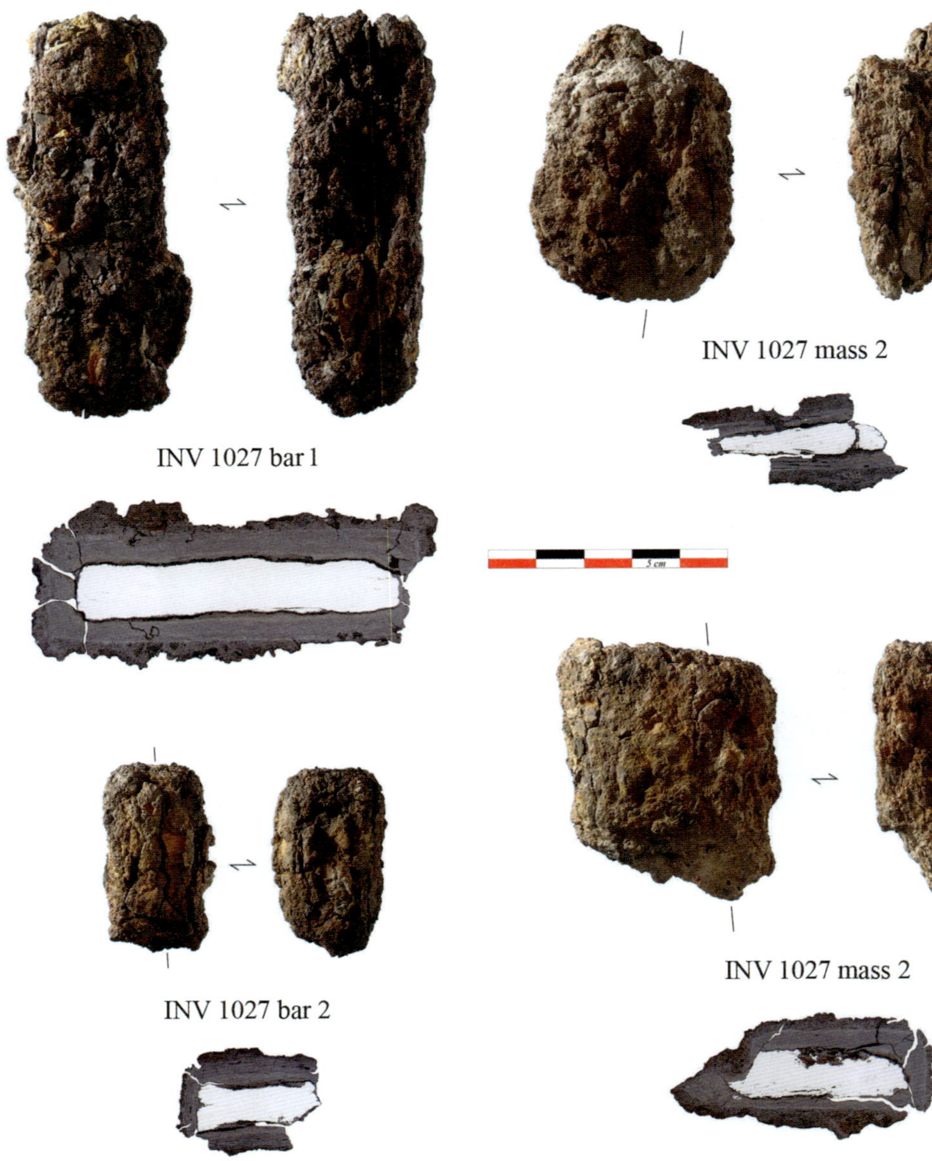

Figure 8: Presentation of the four metal-scraps sampled for the site of HaTsorfim Street.

Compound, because in the section a strongly corroded surface with occasional metal areas is revealed.

The matrix is composed of fayalite in long slats, as well as of two-phase vitreous structures containing globules of wüstite (Figs. 9a-b). The location of the iron oxide remains and the presence of ghost structures of cementite needles within the corroded areas allow for the definition of the metallic proportion. In addition, in the non-corroded matrix, prills and small fragments of iron metal can be seen. Metal represents at least 30% of the section. Hot oxidation can also be seen at the periphery of certain metal masses.

The presence of cementite reveals the work of hypereutectoid steel, but the quantity of losses is not compatible with a simple shaping phase.

Once again, the interpretation becomes difficult here because of structural particularities which do not fit with usual scenarios.

The four slag-cakes belonging to family 2 are composed of clay-sandy slag. They are of a very weak density (from 300 to 500g), and their main peculiarity is the presence of bones inside the matrix (Fig. 7). The section of the analysed sample shows a great number of porosities, large in the periphery and smaller in the middle. The matrix is composed of silica grains with round angles and sometimes of vitrified structures which contain iron fragments, prills and hammerscales (Fig. 9c). One of the metal fragments is composed of hypereutectoid steel, while the others are ferritic.

Most of the time, the use of clay or sand is linked to finishing work, in order to protect the metal from oxidation. The fact that bones are found in this material can be linked to a case-hardening technique, the clay playing the role of an isolating container.

Iron scraps

Among the four analysed samples, two types of metal can be distinguished. One of the bars (INV 1027 bar 1) and one of the masses (INV 1027 mass 2) are of a very good compaction quality, with less than 5% of inclusions (Figs. 10a-b). The bar is composed of eutectoid steel with one quenched extremity (Fig. 10a). The mass is composed of ferritic iron with traces

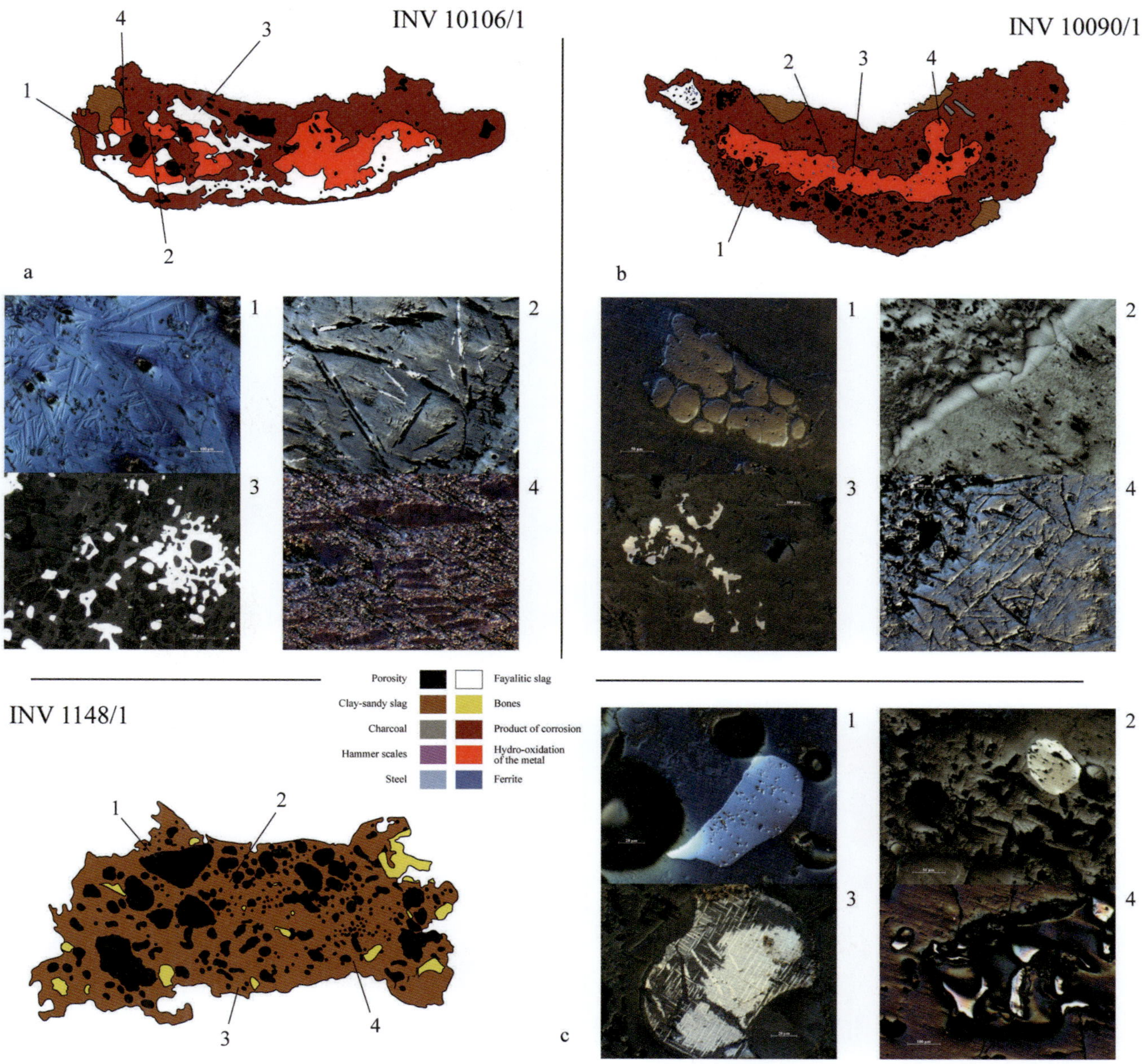

Figure 9: Analyses results of the three slag-cakes sampled for HaTsorfim Street.

of steel at the periphery, but most of it has been completely corroded. It seems to be the remains of a cementation (Fig. 10b).

The other bar and the other mass are less compacted. They contain 5 to 15% of inclusions (Figs. 10c-d) and display a heterogeneous composition. Both thin and slightly larger ferrite layers, and layers of steel with 0.2% of carbon (Fig. 10c) were found within the bar. A juxtaposition of ferritic layers and steel layers with 0.2/0.3% of carbon was also found within the mass. A welding line runs through the section and confirms the heterogeneous nature of the artefact (Fig. 10d).

Thus it seems that at least two types of metal were worked in this workshop, associated with different production techniques. While further investigation is necessary to confirm this hypothesis, the first two samples seem to be linked to an indirect process according to their homogeneous composition, while the other two samples seem to come from a direct process or from a recycling operation according to their heterogeneous composition.

The presence of an accumulation of iron scrap may suggest that a recycling activity was performed in this workshop (as described above). Recycling is also suggested by the fragment of a worked object seen in one of the two slags from family 1 (Fig. 9a4). Ghost structures of cementite reveal hypereutectoid steel, meaning that at least some of the recycled metal was highly carburised.

Furthermore, the presence of cannonballs made from pig iron may not be accidental. The recycling could be linked with refining (here decarburisation) this pig iron, and this would also explain the presence of hot oxidation in the slag-cakes. The hypereutectoid steel remaining in slag-cakes would therefore represent the part of the metal not completely decarburised and blocked under the mass of metal.

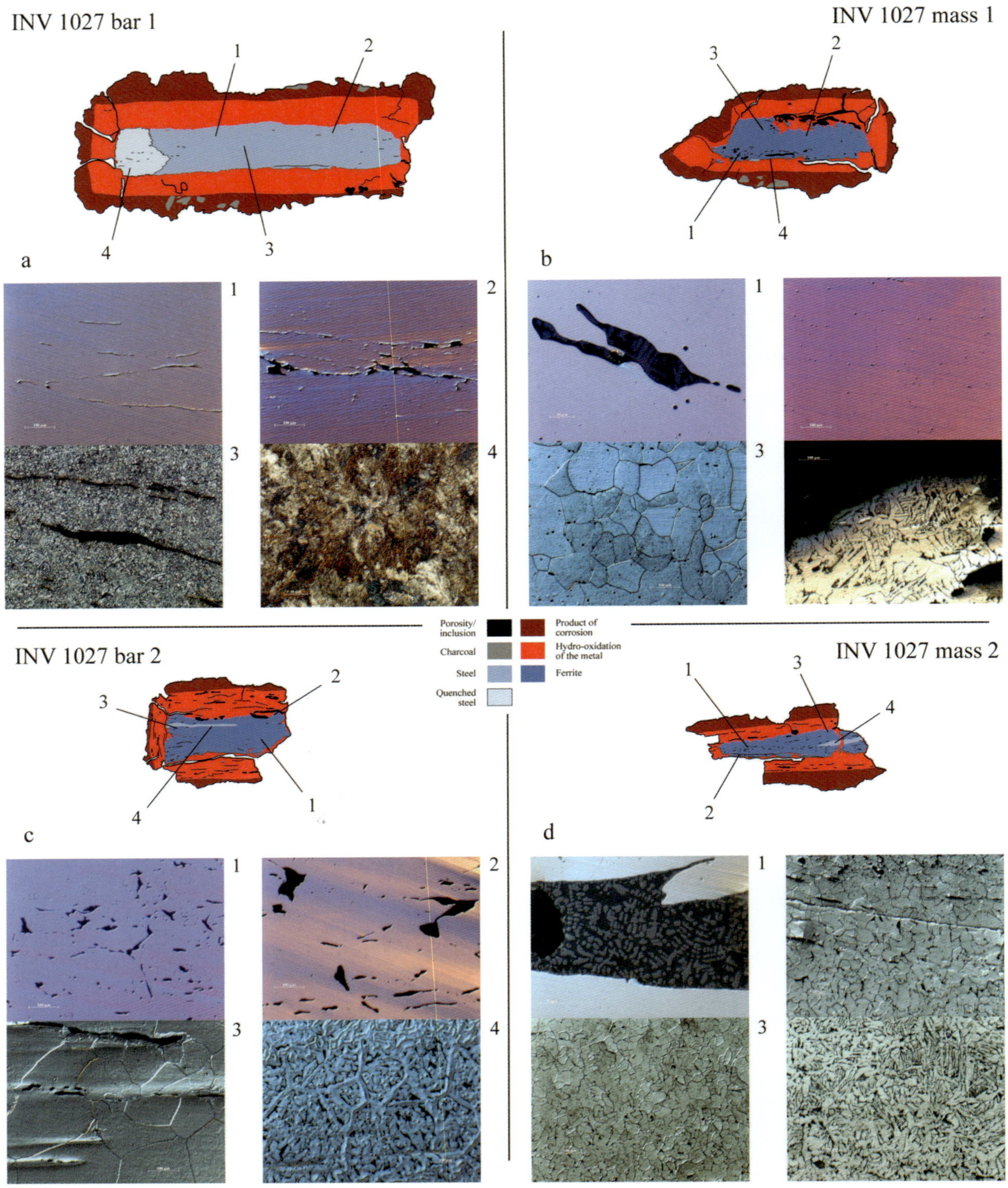

Figure 10: Analyses results of the four metal-scraps sampled for HaTsorfim Street.

It would certainly appear that a large range of techniques were carried out in this workshop: cementation, quenching, decarburisation, welding, and recycling.

Recycling or the pressure of dependence

The two analysed sites described in this paper seem to be complementary and lead to an interpretation that seems to indicate recycling, in addition to typical forging activities using imported metal.

While additional analyses are required, in particular slag inclusion analyses (Dillmann and L'Héritier, 2007), it seems that some of the worked metal is derived from an indirect process making use of iron scrap. This may confirm both the dominance of European imports in Jaffa during this period, as well as the lack of iron in this region.

Results illustrate the great competence of the craftsmen at Jaffa, who were skilled at complex processes such as welding, cementation, and decarburisation. Therefore, at least at the beginning of the 19th century, Arab metal workers from Jaffa were not only able to repair tools, but they also had the skills to produce them at a standard comparable to those seen in Europe. Furthermore, the situation of shortage (or of limited supply coupled with intense demand) seems to have urged the craftsmen to develop innovative techniques of recycling.

Further analyses of the HaTsorfim site will enrich our understanding of the two (or three) techniques used here.

References

Al-Amri, Y.A.S., 2007. *The Role of the Iron Ore Deposit of Mugharet el-Wardeh/Jordan in the Development of the Use of Iron in Southern Bilad el-Sham.* PhD Thesis, Ruhr-Universtät Bochum.

Anderson, T.J., Augustoni, C., Duvauchelle, A., Serneels, V. and Castella, D., 2003. *Des artisans à la campagne. Carrière de meules, forge et voie romaine à Châbles (FR).* Fribourg: Academic Press Fribourg.

Arribet-Deroin, D., 2001. *Fondre le fer en gueuses au XVIe siècle. Le haut fourneau de Glinet en pays de Bray (Normandie).* PhD Thesis, Université de Paris I, Panthéon-Sorbonne, Paris.

Avistar, S., 1990. The Stein Factory: early attempts to establish a modern industry. In G.G. Gilbar ed. *Ottoman Palestine, 1800-1914.* Leiden: Brill, pp.159-72.

Bauvais, S., 2008. Prolégomènes à une histoire de la métallurgie du fer au Levant Sud. *Bulletin du Centre de Recherce Français à Jérusalem*, 19, pp.1-14.

Bauvais, S. and Fluzin, P., 2005. Analyses métallographiques des déchets de forge du site de Bazoches-sur-Vesle "les Chantraines": confrontation et perspective régionale (Aisne). In G. Auxiette, F. Malrain, S. Desenne and B. Robert eds. *Hommage à Claudine Pommepuy.* Amiens: Revue Archéologique de Picardie, pp.115-30.

Burke, A.A. and Burke, K.S., 2008. Investigating a Forgotten Port: The Jaffa Heritage Project. *Backdirt*, 2008, pp.70-5.

Danisman, H.H.G., 2007. Ottoman Mining and Metal Working in the Balkans: Its Impact on Fire-Arms Technology of Southeast Europe (15th-17th centuries). *Foundation of Science Technology and Civilisation*, 2007, pp.1-11.

Dillmann, P., Arribet-Deroin, D., Vega, E. and Benoit, P., 2007a. Early modern cast iron and iron at Glinet. In I.-M. Petterson Jensen and U. Ohman eds. *Norberg-Nora, 700 years of ironmaking.* Norberg: Jernkontorets Bergshistorika Utskott.

Dillmann, P., Desaulty, A.-M., Leroy, S., Arranda, R., Lechevallier, G., Arribet-Deroin, D., Neff, D., Bauvais, S., Vega, E., Bellot-Gurlet, L. and Tereygeol, F., 2007b. Early Walloon refining process: first experimental restitution and metallographic analyses. In *Archaeometallurgy in Europe: 2nd international conference, 17-21 June 2007, Aquileia, Italy: proceedings.* Milano: Associazione Italiana di Metallurgia (cd-rom).

Dillmann, P. and L'Héritier, M., 2007. Slag Inclusion Analyses for Studying Ferrous Alloys Employed in French Medieval Buildings: Supply of Materials and Diffusion of Smelting Processes. *Journal of Archaeological Science*, 34, pp.1810-23.

Fluzin, P., 2000. Les ateliers métallurgiques: archéométrie. Examens métallographiques des déchets d'ateliers. In M. Mangin, P. Fluzin, J.-L. Courtadon, and M.-J. Fontaine eds. *Forgerons et paysans des campagnes d'Alésia (Haut-Auxois, Côte-d'Or). La terre, le Fer, la Route en pays Mandubien, Haut Auxois (Côte-d'Or), Ier siècle avant-VIIIe siècle après J.-C.* Paris: CNRS, pp.185-265.

Fluzin, P., 2002. La chaîne opératoire en sidérurgie: matériaux archéologiques et procédés. Apport des études métallographiques. In H. Bocoum ed. *Aux origines de la métallurgie du fer en Afrique. Une ancienneté méconnue.* Paris: UNESCO, pp.59-91.

Panzac, D., 1990. Commerce et commerçants des ports du Liban et de Palestine (1756-1787). *Revue des mondes musulmans et de la Méditerranée*, 55, pp.75-93.

Peilstöcker, M., 2006. La ville franque de Jaffa à la lumière des fouilles récentes. *Bulletin Monumental*, 164, pp.99-104.

Peilstöcker, M., 2007. Urban Archaeology in Yafo (Jaffa): Preliminary Planning for Excavations and Research of a Mediterranean Port City. *Palestine Exploration Quarterly*, 139, pp.149-65.

Serneels, V. and Perret, S., 2003. Quantification of smithing activities based on the investigation of slag and other material remains. *Archaeometallurgy in Europe. Proceedings of the International Conference, 1, 24-26 September 2003, Milano.* Milano: Associazione Italiana di Metallurgia, pp.469-78.

Author's address

- CNRS researcher, LMC – IRAMAT – UMR 5060 – CNRS and LAPA – SIS2M – UMR 3299 – CNRS/CEA, CEA Saclay, 91191 Gif-sur-Yvette Cedex.
  Email: sylvain.bauvais@utbm.fr

Characterisation of iron technology at Medieval Talgar in Kazakhstan

Jang-Sik Park and Dmitriy Voyakin

ABSTRACT: Examination of iron artefacts from medieval Talgar (900-1300 AD) in southeast Kazakhstan showed that both low carbon iron and cast iron were used as key materials. A variety of iron objects were forged from ingots weighing 1 to 1.5 kg each. Microstructures observed within the ingots and in a fragment of slag found at the same site suggest that they originated from a bloomery process. The use of cast iron was limited by comparison, but the material seemed readily available. Cast iron artefacts varied significantly in their carbon and sulphur levels and also in their microstructures, revealing the technical diversity applied in smelting, casting, and heat-treatments. The accelerator mass spectrometry analyses on the carbon samples extracted from two cast iron cauldrons were fairly consistent and placed their radiocarbon age between the late 10th and early 12th century AD, before the Mongol invasion.

Keywords: Talgar in Kazakhstan; medieval period; iron technology; low carbon iron and cast iron; microstructure; radiocarbon age.

Introduction

Talgar is located approximately 25 km to the east of Almaty (Fig. 1), in the Republic of Kazakhstan. The site has been a major settlement for centuries, and is situated in southeast Kazakhstan, in a region called Semirechye (Seven Rivers), which is bounded by the Zailiisky Alatau Mountains to the south, the Dzunghar Alatau Mountains to the northeast and the Balkhash Desert to the west. Located at the crossroads between the desert-oasis region of Central Asia proper and the semi-arid and desert areas of Mongolia and western China, this region played an important role in the establishment and

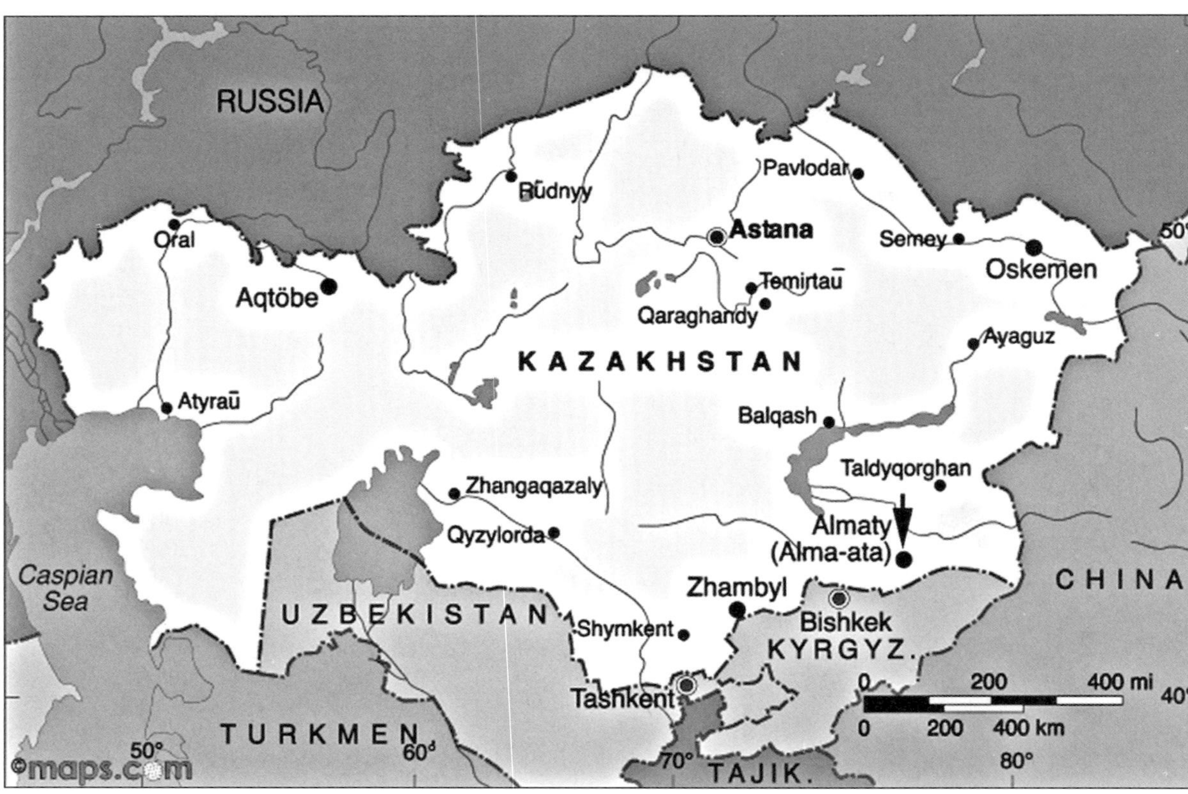

Figure 1: Political map of the Republic of Kazakhstan. The arrow indicates the medieval site of Talgar, from which the present objects were excavated.

operation of a branch of the Great Silk Road from around 130 BC through to the 14th-15th century AD (Frachetti, 2008). Medieval Talgar was one of the steppe towns and cities that had developed along this Silk Road in Semirechye from the 8th to the 13th century AD (Chang, et al., 2002). It faced the Mongol invasion at the beginning of the 13th century AD.

The recent excavation of the medieval site at Talgar has recovered numerous iron objects, along with a variety of other cultural remains (Chang, et al., 2002). Iron artefacts from Talgar have a special historical significance in view of the site's unique geographical location connecting the East and West, and of the fundamental difference in the early iron industry between the two worlds. The difference arose primarily from the early production of cast iron in the East around the mid-first millennium BC, in comparison to the West where no large-scale cast iron production was carried out until around the 14th century AD (Needham, 1980; Rostoker and Bronson, 1990; Tylecote, 1992; Maddin, 1996; Wagner, 1996). Iron artefacts from Talgar therefore provide a unique opportunity to understand the production techniques established under strong influences coming from both the East and West. Currently, aside from this study, little material evidence is known from this region dating to this important period. This paper presents the results of visual and metallographic examinations carried out on a variety of iron objects, both forged and cast, and offers a characterisation of the local iron industry in terms of raw materials and major manufacturing processes. The radiocarbon ages of cast iron objects, measured using accelerator mass spectrometry (AMS), are also presented, along with data showing significant variation in their carbon and sulphur levels.

Comments on artefacts

The majority of the iron artefacts excavated from Talgar were forged objects. Cast iron was apparently less important, but its use can nevertheless frequently be encountered in objects such as cauldrons and ploughs.

Figs. 2a to 2d present a series of iron objects at various stages of the manufacturing process. Those in Fig. 2a, roughly in the form of rectangular bars approximately 4cm thick and weighing 1 to 1.5kg each, were ingots of negligible carbon content containing profuse levels of slag inclusions. Their appearance suggests that these objects are representative of the raw iron brought to the workshop to be worked. The objects in Fig. 2b, including bars, axes, a plough, an anvil, and a pick are items that could simply have been forged directly from these ingots. The slender bars in the upper left corner are apparently an intermediate stock that could be shaped into diverse final products, such as the axes in Fig. 2b, ploughs in Fig. 2c, and sickles, swords, and a variety of other tools as shown in Fig. 2d. The two ploughs shown in Fig. 2c and marked by the arrows were not forged but cast.

The objects illustrated in Figs. 3a to 3h are cast iron objects that were examined as part of this study. Figs. 3a to 3e present fragments from five broken cauldrons. The object in Fig. 3f

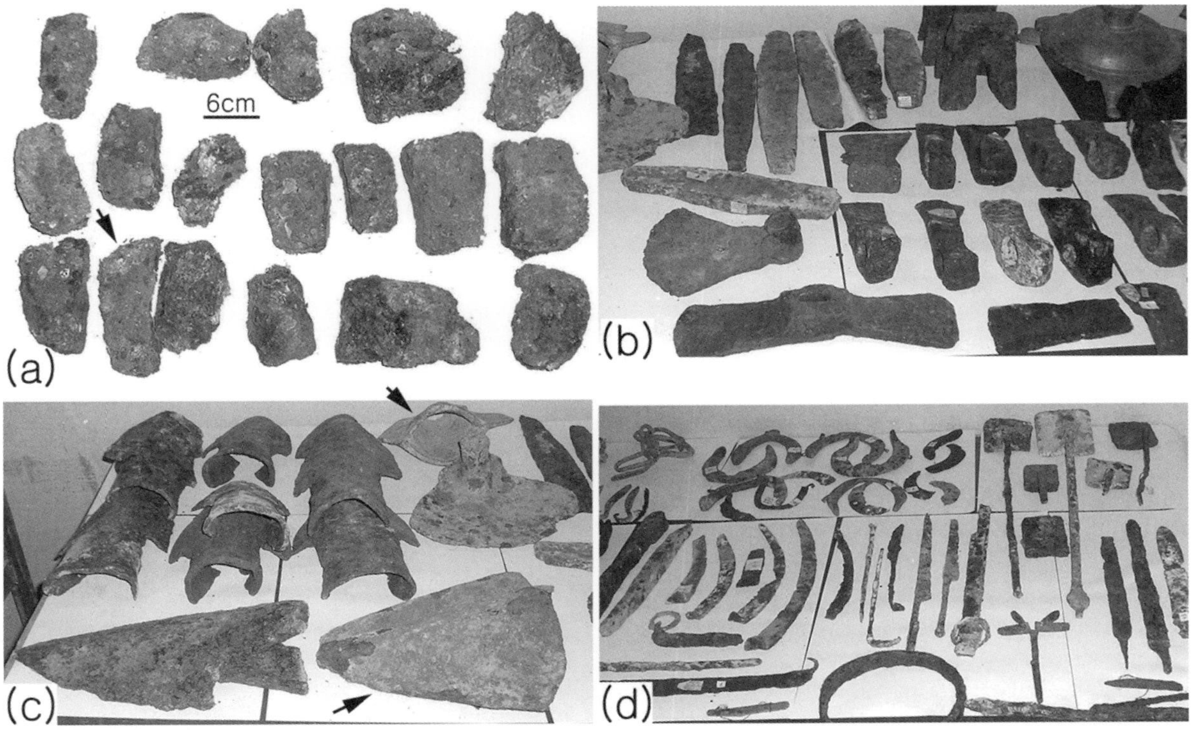

Figure 2: Iron objects recovered from the medieval site of Talgar in Kazakhstan shown in Figure 1. (a) Iron ingots – the arrow indicates the ingot chosen for structure examination; (b) bars, axes, a plough, an anvil, and a pick; the artefact in the upper right corner is a bronze object; (c) ploughs; (d) varieties of finished iron objects. The objects were all forged to shape, except for the two cast iron ploughs marked by the arrows in (c).

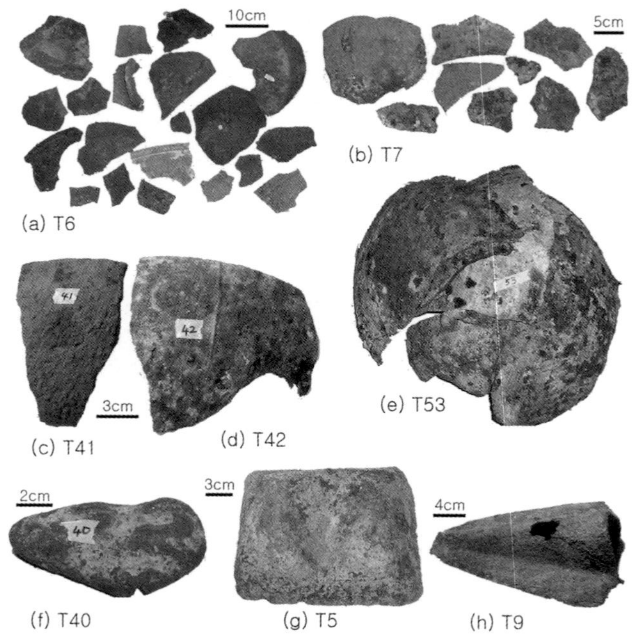

Figure 3: Cast iron objects recovered from the medieval site of Talgar in Kazakhstan shown in Figure 1. (a)-(e) Fragments from five cast iron cauldrons; (f) indeterminate cast iron chunk; (g) cast iron block with two parallel ridges on top and a flat bottom; (h) cast iron plough.

does not indicate a particular functional object. Fig. 3g is a flat-bottomed square block with two parallel round ridges on its top, and Fig. 3h shows a plough.

Microstructure examination

For the examination of the microstructure, small specimens taken from the artefacts were mounted and polished following standard metallographic procedures and then etched with a solution of 2ml nitric acid in 98ml methanol. They were examined using the optical microscope and scanning electron microscope (SEM) equipped with an energy dispersive spectrometer (EDS). Chemical information was obtained by analysing the EDS spectra. Quantitative chemical analysis on carbon and sulphur was performed using the C/S Determinator. All chemical compositions presented in this article are given as weight fractions.

Fig. 4a is a magnified view of the ingot marked in Fig. 2a, one of the four that were selected for examination of the microstructure. The results were all qualitatively similar and are displayed in Fig. 4b, an optical micrograph showing the structure of the ingot in Fig. 4a, which consists of the bright α phase and dark slag inclusions. In the latter, the light dendrites correspond to the wüstite phase (FeO) and the dark background to a glassy material. Fig. 4c, a secondary electron image magnifying the area marked by the arrow in Fig. 4b, was chemically analysed using the SEM-EDS. Figs. 4d, 4e and 4f are the EDS spectra taken from arrows 1, 2 and 3, respectively, in Fig. 4c. Fig. 4d shows that the wüstite phase contains a noticeable amount of manganese (Mn). Fig. 4e shows that the glassy background contains a measurable amount of sulphur (S). The bright particles included in the slag area of Fig. 4b are shown in Fig. 4f to correspond to iron (Fe). These particles were similar to the iron phase in the bulk and consisted of ferrite only. The peak at carbon (C) in these spectra and those that follow is due to a thin carbon layer coated for the SEM-EDS examination, and is irrelevant to the specimens. No manganese was detected in the iron phase, but only in slag inclusions. No sign of plastic flow was observed either in the ferrite area or in the shape of the slag inclusions, which indicates that the microstructures still maintain most features determined during smelting. It seems that the ingots in Fig. 2a were prepared with minimal work following smelting.

Fig. 5a illustrates a finished, forged object (in this case a sickle), where arrows 1 and 2 locate spots in the blade and the back, respectively, from which specimens were taken for examination. The structures at arrows 1 and 2 are illustrated in the optical micrographs, Figs. 5b and 5c, respectively. The structure of the blade is seen in Fig. 5b to contain martensite in the dark areas, and the Vickers hardness inferred from the size of the dark rhombuses was approximately 450 at the lower left corner and 320 at the right. This high hardness cannot be obtained by merely controlling the carbon content, if not accompanied by a quenching treatment. In contrast, the structure of the back of the object is shown in Fig. 5c to consist mostly of ferrite grains, and the dark indentation marks correspond to a Vickers hardness of approximately 160. It is evident, therefore, that this object was forged from an ingot such as those in Fig. 2a, with a carburisation and then a quenching treatment applied subsequently to the blade part.

Fig. 6a shows a bottom (left) and a top view (right) of a chunk of slag recovered from the same site along with other iron objects. It has flow marks on the surface, suggesting that it was accumulated by a slow flow of molten slag during smelting. Fig. 6b, an optical micrograph taken from the slag specimen, illustrates the typically observed microstructure. Arrows 1-4 locate areas of the four different phases that were identified: wüstite, fayalite, glassy material, and ferrite, respectively. Fig. 6c, an SEM micrograph magnifying the area labelled 'A' near the left edge of Fig. 6b, provides a better view of the first three major constituents at arrows 1-3, in the same order. The basic chemical makeup of each phase is illustrated in the EDS spectra from arrows 1 through 4, respectively, as seen in Figs. 6d to 6g. This specimen is distinguished from the slag inclusions in Fig. 4c in at least three aspects: the greatly increased fraction of fayalite at the expense of wüstite and the glassy background, wüstite containing no manganese, and the absence of sulphur. The coarse fayalite structure suggests that the slag probably cooled at a slow rate.

Fig. 7a, an optical micrograph from the cast iron object in Fig. 3a, shows dark dendrites spread out across the light background of white cast iron eutectic. These morphological characteristics, typical of white cast iron, are mostly established during solidification in casting. The presence of dendrites places the average carbon content somewhat below the eutectic composition of 4.3%. The EDS analysis showed that iron and carbon were the only elements detected, the detection limit being in the order of a few tenths of a percent. Microstructures of the other four cauldrons in Fig. 3b to 3e were almost identical and similar to that in Fig. 7b, an optical micrograph taken from the cauldron in Fig. 3b. This

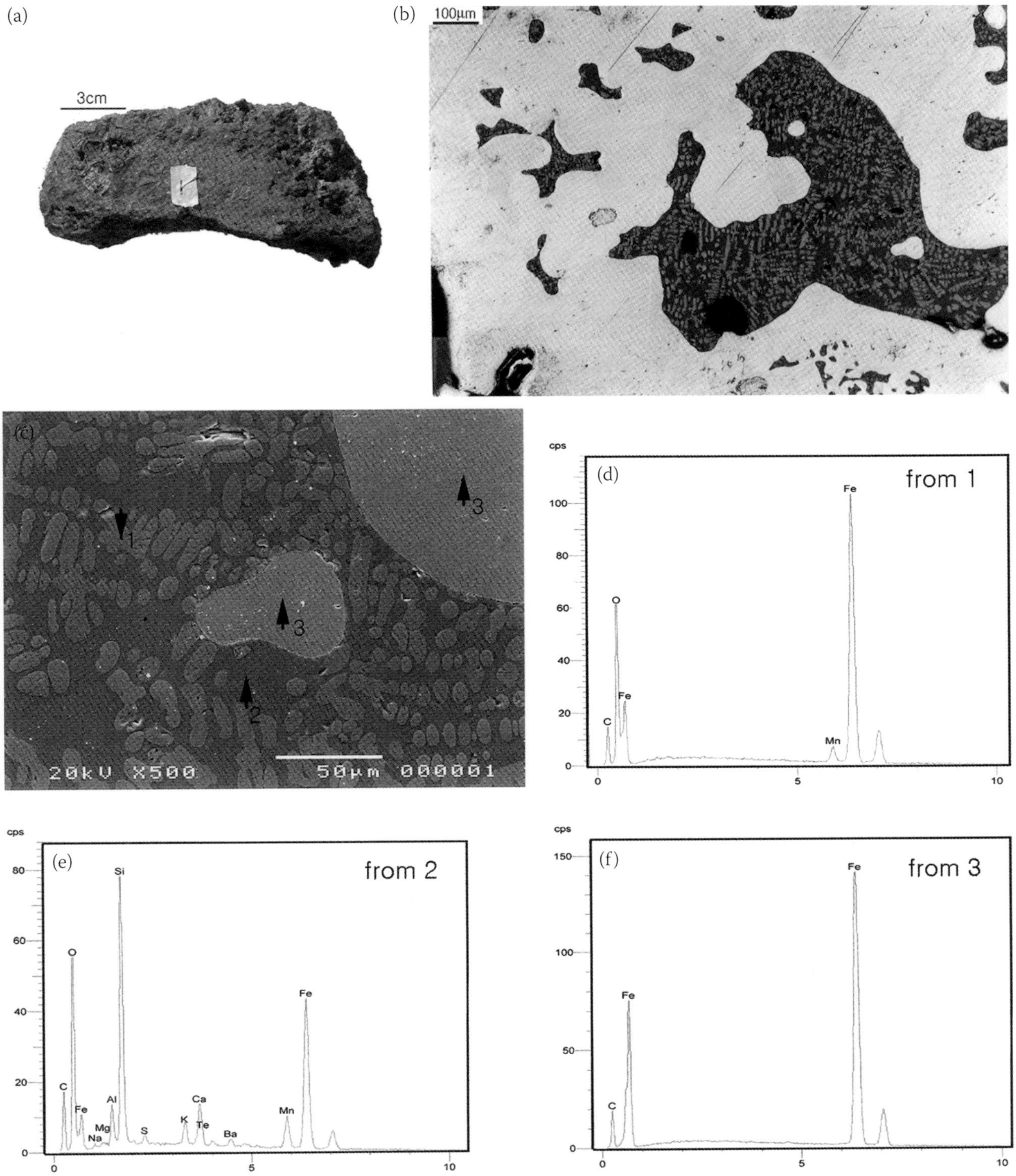

Figure 4: Iron ingot, its microstructure and chemical composition. (a) The ingot marked by the arrow in Figure 2a; (b) optical micrograph showing the structure of the specimen taken from (a); (c) SEM micrograph magnifying the area marked by the arrow in (b); (d), (e), (f) EDS spectra from arrows 1, 2 and 3 in (c), respectively.

microstructure, mostly resulting from solidification, is clearly distinguished from that in Fig. 7a in at least two aspects, i.e., the substantial increase in the fraction of dendrites, and the absence of such fine eutectic as observed in Fig. 7a. Both differences are associated with the reduced carbon content and have significant practical implications, as lower carbon means higher melting points and requires higher temperatures in smelting and casting. Fig. 7c, an optical micrograph, and Fig. 7d, an SEM micrograph, both magnifying the area marked by the white arrow in Fig. 7b, show the presence of a second phase particle near the centre. It was only one of the many that were observed in the cauldrons presented in Figs. 3b through 3e. No such particles, however, were found in the cauldron in Fig. 3a. Fig. 7e, an EDS spectrum from the white arrow in

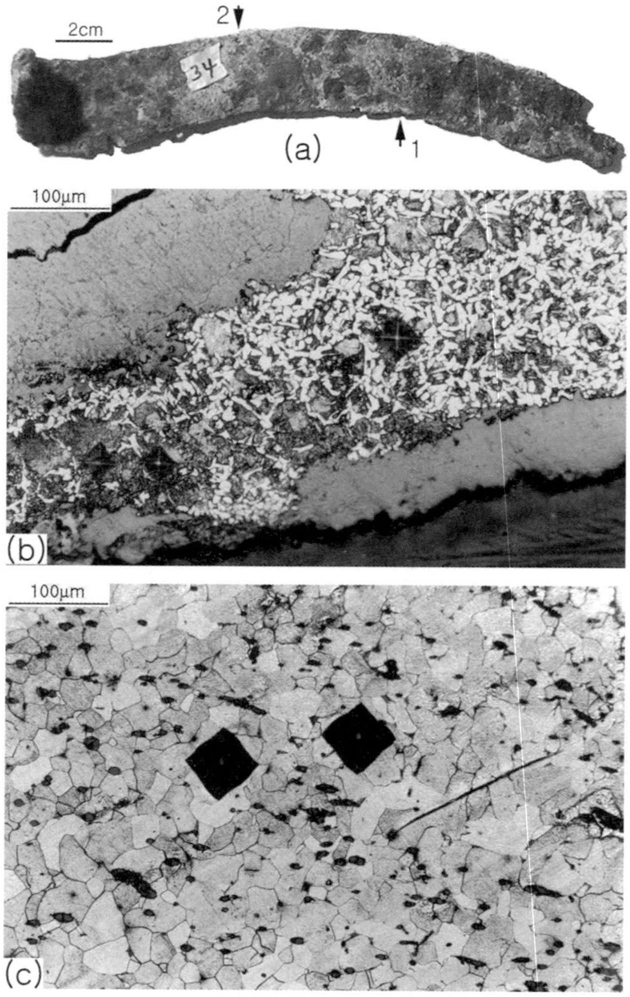

Figure 5: Iron sickle and its microstructure. (a) Sickle; (b) optical micrograph showing the structure of the blade marked by arrow 1 in (a); (b) optical micrograph showing the structure of the back marked by arrow 2 in (a). The dark rhombuses resulted from the Vickers hardness measurement.

Fig. 7d, shows two major peaks at sulphur and manganese in addition to the peaks at iron and carbon. The peaks at gold (Au) are due to the gold coating for the SEM observation and have no relevance to the specimen. The particle is evidently a sulphide of iron and manganese. Fig. 7f, a spectrum from another such particle, shows major peaks at sulphur and iron, with the peak at manganese being greatly reduced.

Table 1 summarises the results of the quantitative chemical analysis carried out on the carbon and sulphur levels of the cauldrons, where each value is the average of two or three

Table 1: Carbon and sulphur contents of the cast iron cauldrons excavated from the medieval site of Talgar in the Republic of Kazakhstan.

Artifact	Carbon [mass%]	Sulfur [mass%]
T6 in Figure 3a	4	0.015
T7 in Figure 3b	2.88	0.27
T41 in Figure 3c	3.03	0.22
T42 in Figure 3d	3.03	0.22
T53 in Figure 3e	2.87	0.21

Measurements were made using the C/S determinator.
All the data are the average of two or three separate measurements.

measurements. The cauldron in Fig. 3a contains 4.00% C and 0.015% S, whereas the other four contain 2.95% C and 0.230% S on average. It is seen that the sulphur level is substantially higher in the lower carbon cauldrons.

Figs. 8a, 8b and 8c present optical micrographs taken from the objects in Figs. 3f, 3g and 3h, respectively. They all consist of two different areas, white and dark. The white areas correspond to white cast iron eutectic; however, the structure in the dark areas requires some explanation. The presence of flake type graphite is seen in Fig. 8a. Similar flakes are also visible in Fig. 8c and in Fig. 8d, an optical micrograph magnifying the near central area of Fig. 8b. These flakes are typically observed in grey cast iron, which forms in place of white cast iron when the solidification reaction during casting occurs at extremely slow rates. Its formation is promoted in the presence of silicon, but silicon is not detected in these objects. The morphological characteristics of the dark regions, however, are very similar to those observed in malleable cast iron, particularly in Figs. 8b and 8c, which is produced from treating white cast iron at elevated temperatures.

Accelerator mass spectrometric dating

Carbon samples for AMS measurements were extracted from the high carbon-low sulphur cauldron, T6, in Fig. 3a and one of the low carbon-high sulphur caldrons, T7, in Fig. 3b. The method of carbon extraction from cast iron is detailed in a separate publication (Park, et al., 2010). The samples were then treated and prepared as graphite targets for AMS measurements at the Christian-Albrechts-University's Leibniz Labor AMS Facility in Germany. The results are summarised in Table 2, where the calibrated or calendar ages were calculated using Calib Rev 5.0.1 (Stuiver and Reimer, 1993; Stuiver, et al., 2005) and the data set IntCal04.14c (Reimer, et al., 2004).

Table 2: Results of the AMS radiocarbon measurements on two cast iron cauldrons excavated from the medieval site of Talgar in the Republic of Kazakhstan. The measurements were performed at the Christian-Albrechts-University's Leibniz Labor AMS Facility in Germany.

Artifact	$\delta^{13}C$ (‰)	1σ ^{14}C age (yr BP)	95.4% (2σ) cal age ranges (AD)	Lab Code (Comments)
T6 in Figure 3a	-23.8±0.09	1013±23	982-1041 (98.4%) 1108-1116 (1.6%)	KIA28679 (high C and low S)
T7 in Figure 3b	-7.93±0.18	1002±25	986-1047 (85.1%) 1089-1122 (12.1%) 1139-1149 (2.8%)	KIA28680 (high $\delta^{13}C$ value, low C and high S)

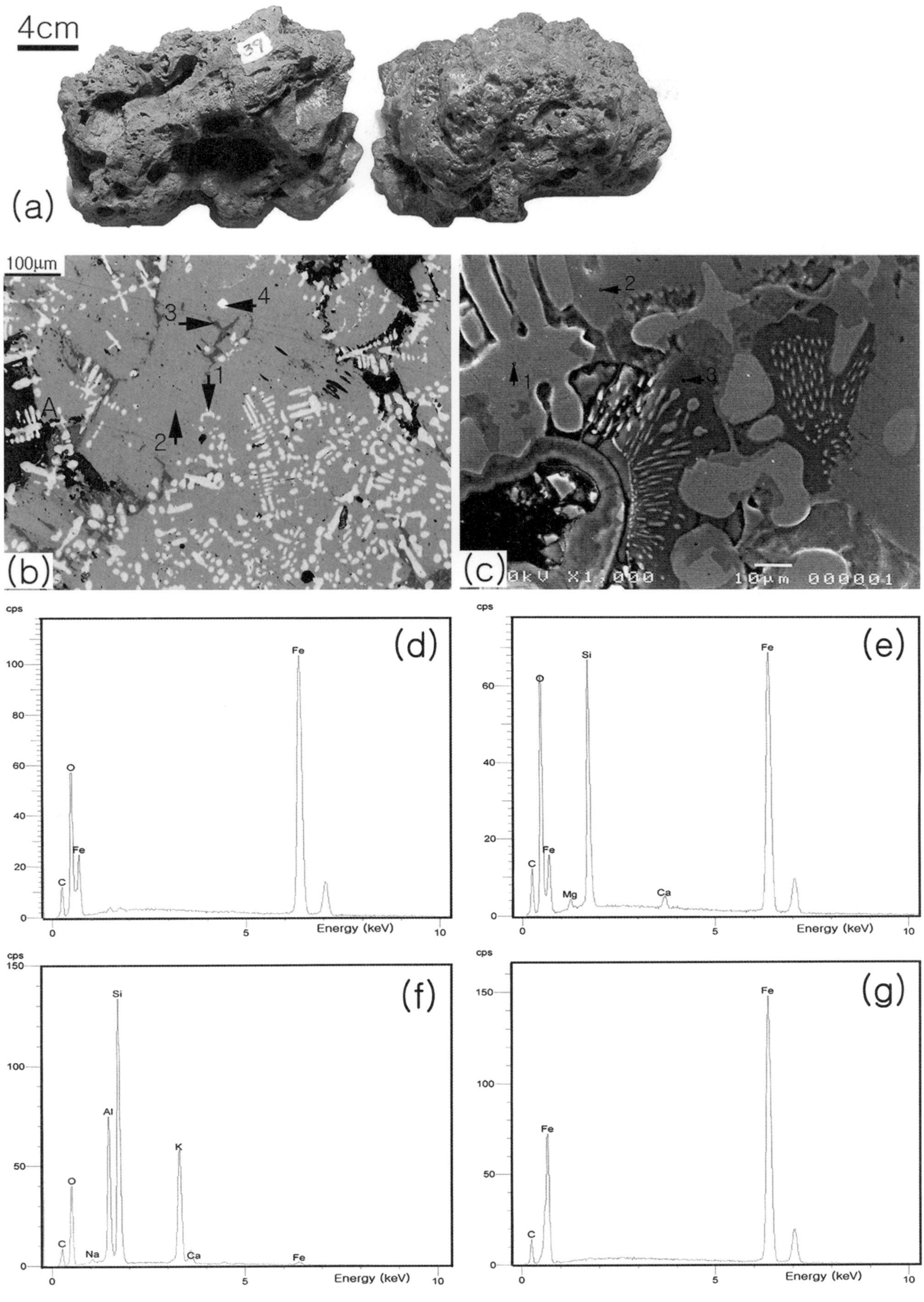

Figure 6: A piece of slag, its microstructure and chemical composition. (a) Bottom (left) and top (right) view of the slag; (b) optical micrograph showing the structure of the slag in (a); (c) SEM micrograph magnifying the area at A near the left edge of (b); (d)-(g) EDS spectra from arrows 1-4 in (b).

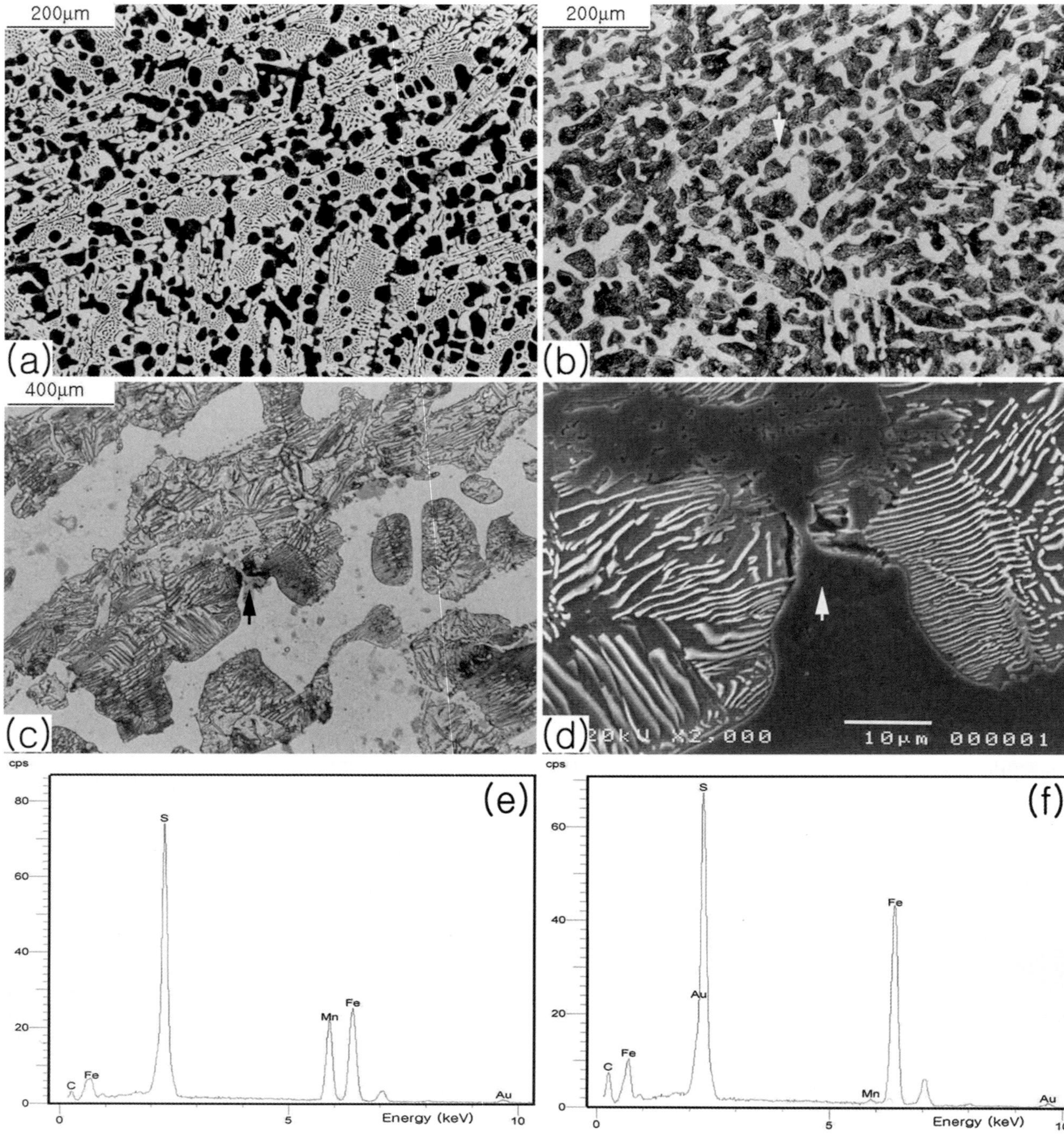

Figure 7: Microstructures and chemical composition of cast iron objects. (a), (b) Optical micrographs showing the structures of the cauldrons in Figures 3a and 3b, respectively; (c) optical micrograph magnifying the area marked by the white arrow in (b); (d) SEM micrograph magnifying the area marked by the dark arrow in (c); (e), (f) EDS spectra from the white arrow in (d) and another spot not shown in (d).

The 1σ ^{14}C ages of T6 and T7, 1013±23 and 1002±25 years before present (yr BP) as of 1950, are nearly equivalent within a 1σ probability range. These results show that the carbon samples originated from charcoal and, when calibrated within a 2σ probability range, consistently place the radiocarbon date between the late 10th and early 12th century AD. There is a substantial difference, however, in the δ^{13} values between the two samples, implying that the material used in making their charcoal came from two different sources (Bowman, 1990).

Discussion

The collection of objects shown in Figs. 2a to 2d clearly indicates that iron ingots such as those in Fig. 2a served as a major raw material in the iron industry of medieval Talgar. The ingots were made of almost pure iron and contain abundant slag inclusions high in iron content. These traits suggest that they came from a bloomery process, although it is not absolutely impossible for them to have been produced from treating cast iron. The slag in Fig. 6a, consisting primarily

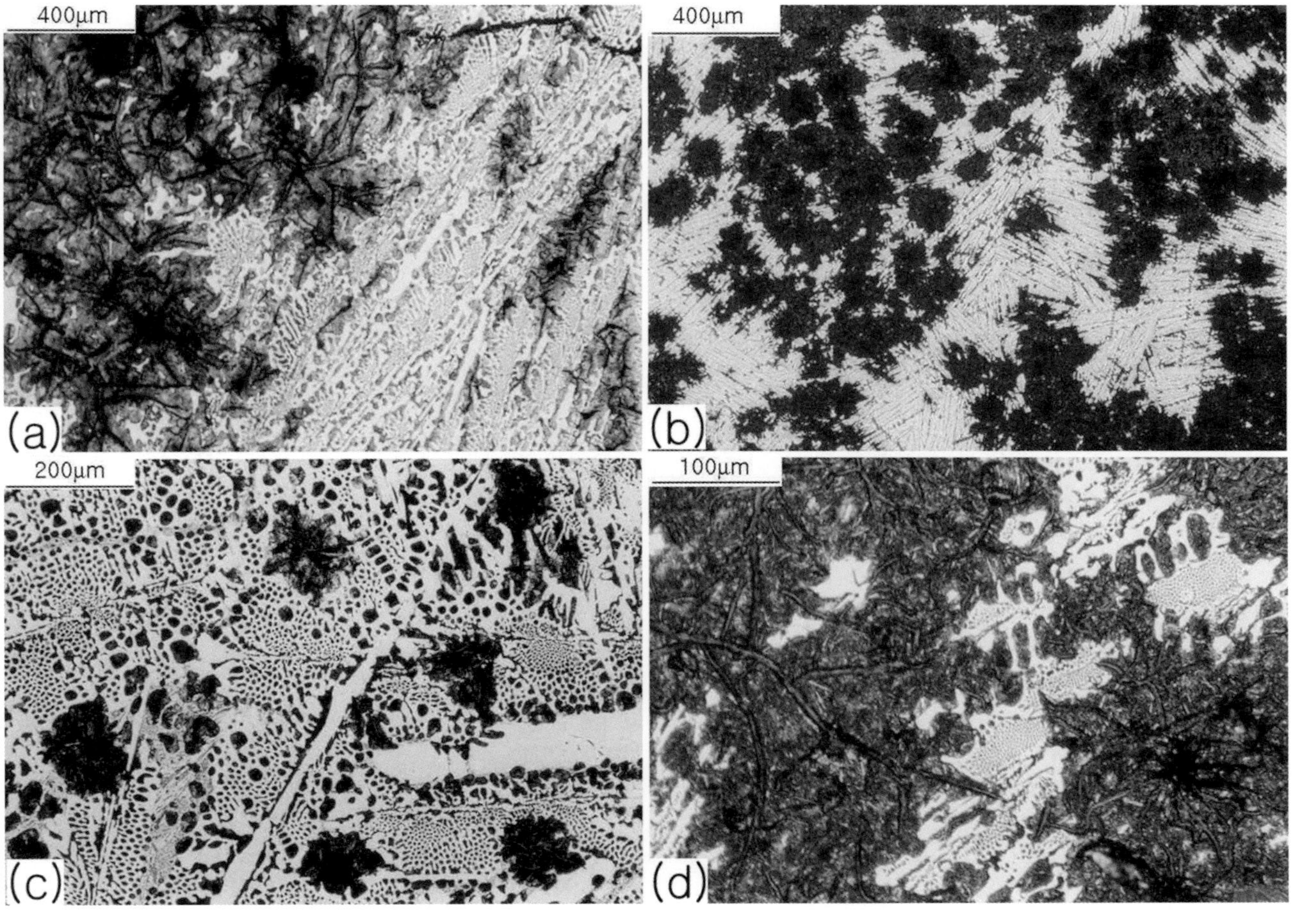

Figure 8: Microstructures of cast iron objects. (a), (b), (c) Optical micrographs showing the structures of the objects in Figures 3f, 3g and 3h, respectively; (d) optical micrograph magnifying the structure in (b).

of a fayalite phase with some dendrites of wüstite, is typical of slag produced in bloomery smelting. Although their microstructures are not presented, of the bars in Fig. 2b and the ploughs in Fig. 2c, those that were examined were all made of almost pure iron. It is clear that they were simply forged out of ingots such as those in Fig. 2a. The carbon content of finished products – such as axes, sickles, swords, arrowheads, and smithery tools – was also very low, and high carbon structures were located only at the blades of some edged objects. It is of significance that some of the axes examined were made of almost pure iron, even at the blade, while some of them had the carbon content increased at the blade. This difference suggests that ingots or bars such as those in Figs. 2a and 2b were first forged to shape and then given a treatment for hardening if needed. The hardening was generally achieved by raising the carbon content and then applying a quenching treatment, as is observed in the sickle in Fig. 5a. The best method for carbon control in this case may be carburisation applied in solid state, aiming at specific parts to be hardened. Evidence was also found that some of the finished products were forged from ingots or bars that had their carbon content partially increased.

It remains uncertain whether cast iron was smelted at the site or imported from the neighbouring regions under the influence of China, where the iron casting technology had long been established. By medieval times cast iron was smelted on a large scale in Mongolia (Gelegdorj et al., 2007; Park et al., 2008), and there is no reason to believe that the cast iron objects used in medieval Talgar were all imported. It is apparent, however, that the use of cast iron was quite limited in comparison to that of low carbon iron. Nevertheless, cast iron seemed readily available so that it could be selected for objects, like cauldrons, that might best be shaped by casting. The availability of cast iron is also reflected in the fact that implements with a similar function, such as ploughs, were either forged or cast. However, the number of cast ploughs does not compare with that of forged ones, in spite of the advantage of casting for mass production. The conspicuous difference in both carbon and sulphur contents between the cauldron T6 and the others must have resulted from the different fuels used in smelting. This may represent a significant technological variation to be studied in the future, which may hold a clue allowing the origin, either regional or temporal, of the cauldrons to be inferred. The microstructures observed in the cast iron objects in Figs. 3f-3h are not normally obtained without special thermal treatments, either during solidification or in solid state. Such treatments could not have been tried without a complete understanding of the mechanical properties of the cast iron objects that could be obtained as a result. It may therefore be concluded that most technological aspects relating to cast iron were well understood in medieval Talgar, although its use was restricted.

Conclusion

The iron technology in medieval Talgar, in south-eastern Kazakhstan, is characterised by the use of low carbon iron as a major material and the use of forging as a major manufacturing technique. The use of cast iron was limited, in spite of the evidence showing both the technological awareness and the ready availability of cast iron. The preferential selection of low carbon iron and forging seems to have resulted from infrastructures that had long been established, and not from the lack of technological knowledge of cast iron. This premise is supported by the unique microstructures suggesting the bloomery process as the most probable source of low carbon iron and the carburisation in solid state as the major method of carbon control. The AMS measurements carried out on carbon samples extracted from two cast iron cauldrons placed their radiocarbon date between the late 10th and early 12th century AD, before the Mongol invasion. In view of its availability and the associated technological awareness, however, cast iron may have made its first appearance at Talgar much earlier. It seems that the delayed propagation of cast iron from the East to the West resulted not from the slow diffusion of technological ideas, but from the slow transformation of traditional infrastructures.

Acknowledgements

The encouragement and support of Drs. K.M. Baipakov and T.V. Savelieva at the Institute of Archaeology of the Republic of Kazakhstan are gratefully acknowledged. The time and effort the editor spent on improving the English expression is gratefully acknowledged. This work was supported by the Korea National Research Foundation (NRF-2011-0029808).

References

Bowman, S., 1990. *Radiocarbon dating*. Berkeley and Los Angeles: University of California Press.

Chang, C., Tourtellotte, P.A., Baipakov, K.M. and Grigoriev, F.P., 2002. *The Evolution of Steppe Communities from the Bronze Age through Medieval Periods in Southeastern Kazakhstan (Zhetysu)*. The Kazakh-American Talgar Project 1994-2001. Sweet Briar and Almaty: Sweet Briar College, The Ministry of Education and Science of the Republic of Kazakhstan, A.Kh. Margulan Institute of Archaeology.

Frachetti, M.D., 2008. *Pastoralist landscapes and social interaction in Bronze Age Eurasia*. Berkeley and London: University of California Press.

Gelegdorj, E., Chunag, A., Grodon, R.B., Park, J.S., 2007. Transitions in cast iron technology of the nomads in Mongolia. Journal of Archaeological Science 34, 1187-1196.

Maddin, R., 1996. The History of the Evolution and Development of Metal. In: R. Maddin ed. *Proceedings of the Forum for the Fourth International Conference on the Beginning of the Use of Metals and Alloys (BUMA-IV)*. Shimane: The Japan Institute of Metals, pp.1-7.

Needham, J., 1980. The Evolution of Iron and Steel Technology in East and Southeast Asia. In: T.A. Wertime and J.D. Muhly eds. *The Coming of the Age of Iron*. New Haven: Yale University Press, pp.507-41.

Park, J.S., Burr, G.S. and Jull, A.J.T., 2010. A thermal and acid treatment for carbon extraction from cast iron and its application to AMS dating of cast iron objects from ancient Korea. *Radiocarbon*, 52, 1312-21.

Park, J.S., Chunag, A., Gelegdorj, E., 2008. A technological transitions in Mongolia evident in microstructure, chemical composition and radiocarbon age of cast iron artifacts. Journal of Archaeological Science 35, 2465-2470.

Reimer, P.J., Baillie, M.G.L., Bard, E., Beck, J.W., Bertrand, C.J.H., Blackwell, P.G., Buck, C.E., Burr, G.S., Cutler, K.B., Damon, P.E., Edwards, R.L., Fairbanks, R.G., Friedrich, M., Guilderson, T.P., Hogg, A.G., Hughen, K.A., Kromer, B., McCormack, F.G., Manning, S.W., Ramsey, C.B., Reimer, R.W., Remmele, S., Southon, J.R., Stuiver, M., Talamo, S., Taylor, F.W., van der Plicht, J. and Weyhenmeyer, C.E., 2004. IntCal04 Terrestrial radiocarbon age calibration, 26-0 ka BP. *Radiocarbon*, 46, 1029-58.

Rostoker, W. and Bronson, B., 1990. Pre-industrial iron: its technology and ethnology. Philadelphia: Archaeomaterials Monograph 1.

Stuiver, M. and Reimer, P.J., 1993. Extended ^{14}C database and revised CALIB radiocarbon calibration program. *Radiocarbon*, 35, pp.215-30.

Stuiver, M., Reimer, P.J. and Reimer, R.W., 2005. *CALIB 5.0*. [WWW program and documentation- http://calib.qub.ac.uk/calib/].

Tylecote, R.F., 1992. *A History of Metallurgy*. London: The Institute of Materials.

Wagner, D.B., 1996. *Iron and Steel in Ancient China*. 2nd edition. Leiden: E.J. Brill.

Authors' addresses

- Jang Sik Park
  Corresponding author; Department of Metallurgical Engineering, Hongik University, Jochiwon Sejong, 339-701, Korea
  e-mail jskpark@hongik.ac.kr
- Dmitriy Voyakin
  Department of Documentation and Archaeological Conservation, Institute of Archaeology, 44, Dostyk Ave., 480100, Almaty, Republic of Kazakhstan
  e-mail d_voyakin@hotmail.com

New aspects of the 9th and 10th century Islamic period metallurgical workshop excavated at Merv, Turkmenistan: reassessing the slag compositions

John Merkel

ABSTRACT: This report publishes additional compositional data for the green, glassy slag associated with crucible-steel production from part of the archaeometallurgical remains from an Early Islamic Period (9th-10th Century) workshop in Area 4 within the Gyaur Kala enclosure at Merv, Turkmenistan. Similarities with blast furnace slag compositions from the production of cast iron are discussed. Laboratory co-fusion experiments using wrought iron, cast iron and calcium-rich slag compositions in modern refractory crucibles readily produced steel at temperatures of 1400°C. The steel product is comparable to a description of 'Second Quality' steel by the medieval (12th century) Islamic writer Al-Birūnī. The aim is to assess the evidence for slag compositions used in crucible steelmaking at Merv in a wider technical and historical context.

Introduction

In the later first millennium, steelmaking was a complex process with many options which could all produce a high-carbon steel. In the archaeometallurgical evidence for crucible steel production across Asia, one should expect some corresponding variability for steelmaking from diverse contexts. The technical aspects of steelmaking techniques can serve to identify regional practices and traditions. Inevitably, there will be some similarities as well as differences in the technical aspects. These data are still being collected as new sites are discovered. The academic debate can be presented as a simple table on which technical details are listed (Rehren and Papachristou, 2003; Table 1). Likewise, the possible origins or region of 'most frequent occurrence' for specific technical details are also relevant. Most research on ancient and medieval steelmaking is focused on the carburization of iron (direct production) or the decarburization of cast iron (indirect production), but there are other options, such as co-fusion which represents a combination of the desired products. (Needham 1964: 13). Mixing and melting variable proportions of cast iron with malleable wrought iron is referred to as co-fusion and there are a number of relevant technical aspects for this option which are distinctive. However, a full set of technical parameters are not agreed for the identification of co-fusion (Papakhristu and Rehren, 2002).

Slag compositions associated with steelmaking can provide one more insight into the possible attributions and geographic region. Microanalysis of slag inclusions in ancient and historic iron/steel objects (Craddock and Lang, 1993; Williams, 1990) utilize concentrations of Ca, Al, Fe, Si, Mn, and P in attempt to identify links to indirect production using a blast furnace to produce cast iron for the finery process. Slag compositions for blast furnace operation are not optimised; there is a range of acceptable concentrations with low viscosity at high temperatures (White, 1980). A range of operational slag compositions connected with steelmaking options should be expected, with or without other options such as crucibles. Slag compositions represent primary evidence, documented in the laboratory, through the metallurgical investigation of an archaeological site. Therefore, slag compositions associated with crucible steel production are increasingly relevant to the debate.

The evidence of crucible steel remains from an Early Islamic Period (9th-10th century) workshop in Area 4 within the Gyaur Kala enclosure at Merv, Turkmenistan (Hermann, 1998; Simpson 2001: 14-15) has been interpreted as co-fusion by Merkel et al. (1995:12-14) and Gilmour (2000: 55), but the evidence has also been re-interpreted and published as carburizing bloomery iron, also known as "cementation", by Feuerbach et al. (1998: 37-44); Feuerbach (2002); Feuerbach et al. (2003: 258-266) and Rehren and Papachristou (2003: 397). The objective now is to reassess the evidence and focus specifically on the calcium-rich slag associated with the crucibles from steelmaking at Area 4. The definitions of terms and classification of techniques have been discussed (Feuerbach, 2006). A new programme of sample selection and analysis of the crucible slag from Area 4 with CaO, FeO, SiO_2, Al_2O_3 and other major slag-forming oxides confirms the compositional ranges published Feuerbach et al. (2003: 258-266). Two new laboratory experiments with co-fusion under a calcium-rich slag composition are presented. The interpretation of the observed calcium-rich slag adds a new aspect to the continuing research.

Crucible steel production is one of several activities recognised from the metallurgical remains in Area 4 (Fig. 1); other foundry work includes casting and use of non-ferrous alloys. Area 4 is the more recent designation for the same location within the industrial quarter in Gyaur-Kala at Merv, previously identified with the site code MGK 7.F.II (Merkel et al., 1995:12-14). Fundamentally, the obvious mounds of crucible fragments with adhering slag, concentrated within one specific location, Area 4, within the enclosure walls of Gyaur Kala enclosure at Merv, clearly indicate a high-temperature metallurgical process. The broken edges of the crucible fragments reveal a remarkable, white-coloured core composition. Several colour plates are published by Feuerbach et al. (2003, Fig. 17.2). The crucible fragments appeared to be quite refractory, with very few fragments displaying any distortion attributable to heating. Initial refining experiments within a modern electric furnace were used to confirm a refractory character of the crucible fragments to above 1500°C. This is much higher than required for other non-ferrous foundry activities. Most of the recognizable fragments of crucible walls also have a characteristic green, glassy slag adhering to the inner surface. No complete crucibles with fully intact contents have been discovered, so the process and crucibles must be reconstructed from fragmentary remains. Crucible dimensions are based upon approximate positioning of recognised fragments of bases, walls and lids collected from the mounds of metallurgical debris in Area 4 (Fig. 2). The thin lids are sealed to the nearly vertical crucible walls with a layer of additional refractory clay. There is also, however, a characteristic hole or perforation in the lid, presumably allowing for some internal pressure release, but not enough for significant oxidation of the charge. Initial microscopy of the green, glassy slag adhering to some crucible fragments revealed numerous uncorroded, entrapped metallic steel prills up to an observed diameter of about 0.3mm along with fewer prills with cast iron microstructures (Merkel, et al., 1995; Feuerbach et al., 1998: 37-44; Feuerbach et al., 2003: Fig. 17.4). Microhardness values measured 140-320 for steel prills in the crucible slag and the range of carbon concentrations were estimated from <0.1% to 0.8% (Merkel et al., 1995: 13). The discovery of furnaces (Fig. 1) associated with the mounds of crucible fragments (Simpson, 2001), provided the conclusive evidence for high-temperature process operating with crucibles to produce steel. However, the presence of prills of cast iron alongside prills of high-carbon steel in the crucible slag is not conclusive proof of co-fusion. The construction and mode of operation have been discussed by Feuerbach et al. (2003: 259). Again, these metallurgical remains for steelmaking in crucibles are dated securely, upon numismatic and ceramic evidence, to the 9th -10th C ((Hermann, 1998; Simpson 2001).

In his book *The Development of Iron and Steel Technology in China*, Needham (1964: 13) presented a diagram illustrating several different techniques for making steel; crucibles are not connected in the diagram for co-fusion in first millennium China. The diagram is still worthwhile because it depicts a schematic flow chart for steelmaking with various options in a single figure. Further details outlining the various techniques are published by Rostoker and Bronson (1990), Wagner (2008) and others. The evidence for Area 4 at Merv differs most significantly from the schematic diagram, published by Needham (1964: 13), in that the reconstructed technique is based upon melting in a closed, refractory crucible under a layer of slag. The high-temperature melting and chemical reactions between wrought iron and cast iron are separated from the fuel in the furnaces. All of the necessary components of the charge must be placed in the crucible before it is sealed. Use of a refractory crucible represents a clear contrast to the evidence for co-fusion schemes in China (Needham 1964: 13). Quality and quantity of the steel products should be controlled better in a crucible. Some aspects of the crucible-steel production in

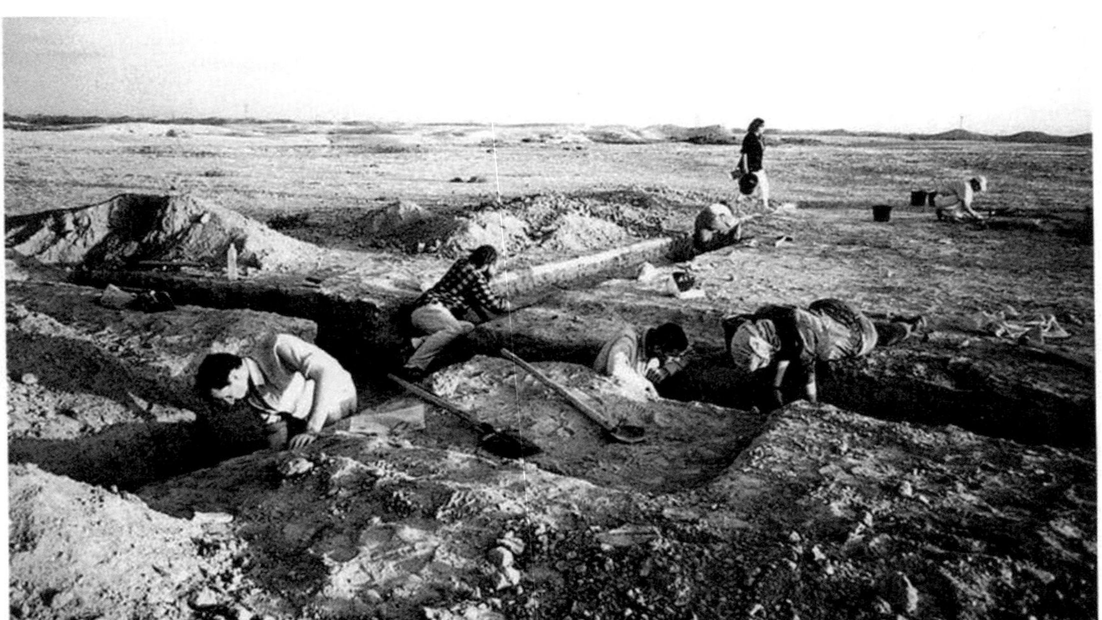

Figure 1: Within the enclosure at Gyaur Kala, excavations in Area 4 reveal a section through a surface dump of crucible fragments associated with the remains of high-temperature furnaces. Four furnaces in the Area 4 industrial quarter at Merv were identified as crucible furnaces. A magnetometer survey failed to identify any further furnaces, so metallurgical activity is limited to one area within the enclosure (Simpson, 2001).

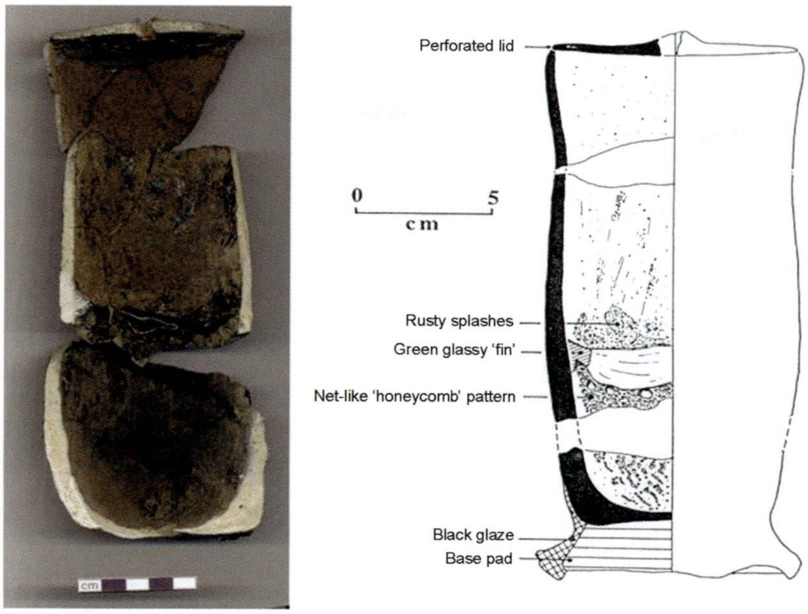

Figure 2a: Three fragments (presumably from different crucibles) of a base, wall and top with a perforated lid are oriented to provide a representative example of a complete crucible. Fragments were assessed on the wall thickness tapering toward the top as well as matching interior surface textures. Note the thin layer of green, glassy slag in the crucible at a level approximately one-third of the distance from the bottom indicates a working volume of about 0.25 litre. Perhaps due to settling of the charge, the whole volume of the crucible was not utilised effectively. (Photograph: J. Merkel; Illustration: A. Feuerbach).

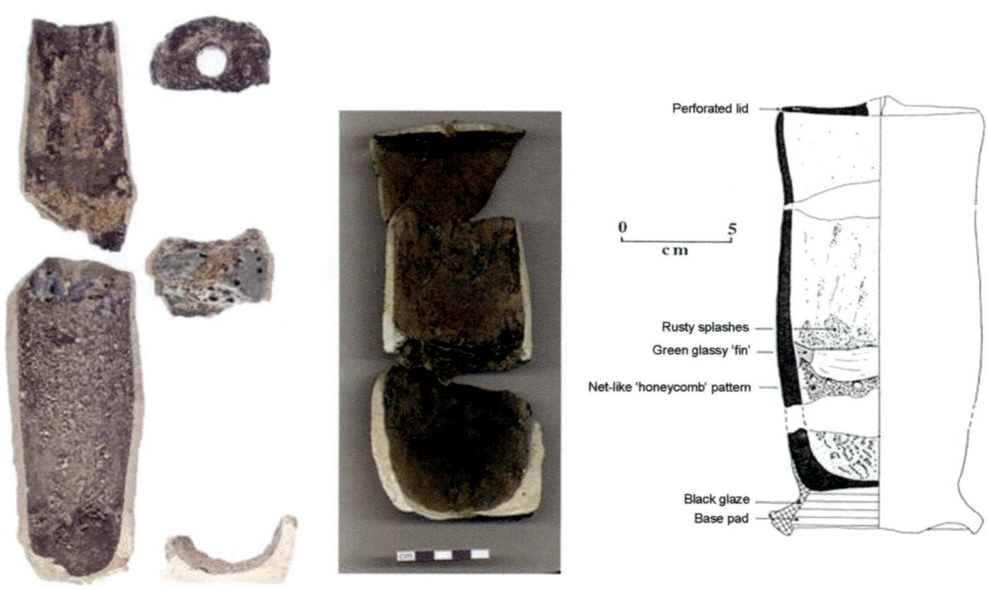

Figure 2b: Image of crucibles from Akhsiket, Uzbekistan from Bayley and Rehren, 2007, 52 alongside the typical crucible from steel production of Merv from Merkel et al., 1995. Note the different levels for the remaining slag on the crucible interiors and the characteristic, thick layer of slag from the Akhsiket crucibles. The Akhsiket crucibles are reported to be approximately 300mm tall and 80mm wide internally. Slag composition data from the Akhsiket crucibles is unpublished. The crucibles from Merv have a reconstructed height of about 200mm and similar diameter. Provisionally, based on crucible dimensions (Fig. 2a) and a characteristic layer of slag approximately 1cm thick, an estimate of the mass of slag across single crucible would range between 100-150g total.

Area 4 at Merv are similar to those at other early sites in Central Asia (Feuerbach et al., 2003: 265; Rehren and Papachristou, 2003; Papachristou and Rehren 2008), while other aspects are clearly different; a theme already noted for evidence from crucible-steel production sites. The key differences are the crucible slag quantity, crucible size and the lid dimensions (Rehren and Papachristou, 2003). Compositional data for the crucible slags from Akhsiket and Pap have not been published (Rehren, personal communication).

The author's participation started in 1993 with a visit at the invitation of Georgina Herrmann (Co-Director with K. Kurbansakhatov) to Merv to collect further metallurgical samples for technical investigation. The first publication of the preliminary results was in the *IAMS Newsletter* (Merkel et al., 1995) with a second publication by Feuerbach et al. (1998). Excavations of the metallurgical remains were started in 1994 and discovery of four furnaces for crucible-steel production were published by Simpson (2001) and assessed by Griffiths and Feuerbach (1999). The archaeometallurgical investigation was co-ordinated under the International Merv Project at the Institute of Archaeology, University College London, as well as the YuTake, South Turkmenistan Multi-Disciplinary Archaeological expedition, the Turkmen Academy of Sciences, Ashgabat and the Institute for the History of Material Culture, St. Petersburg. Selected samples from the Area 4 excavations at Merv were also passed to Brian Gilmour for further investigation and published in 2000. Most of the technical studies were completed as supervised post-graduate student research in the Wolfson Archaeological Science Laboratory at the Institute of Archaeology, University College London. Ann Feuerbach completed her Ph.D. on the investigation of crucible-steel at Merv at the Institute in 2002. Results have been published more recently by Feuerbach et al. (2003: 258-266). Other relevant publications on research at the site of Merv are in the Journal *Iran* (Herrmann, et al., 1995, 1997).

This report will describe several experiments into the technique of co-fusion based upon technical investigations of the archaeometallurgical remains and highlight several aspects of the historical context for the metallurgical remains at Merv. An experimental approach serves as a means to assess further the complexity and control of the co-fusion technique. It was hoped that a number of fundamental questions would be readily answered during the experiments: Which specific aspects are unique to co-fusion? How complicated it is in practical terms? And what are the controlling technical parameters? The results of technical studies of the experimental products and associated remains can be compared to the archaeometallurgical remains and few relevant textual sources to gain confidence in the interpretation. The emphasis of these experiments will be placed upon crucible slag compositions and straightforward observations of the steel products as well as effects of carburization and decarburization during subsequent forging. Most published experimental investigation has focused on the production of Damascus steel (e.g. Wadsworth and Sherby 1980; Verhoeven and Jones, 1987). These publications are readily available and will not be reviewed here. The classification by Feuerbach (2003) is useful to reassess the experimental evidence. Experiments by Wraith (1978) are relevant for furnace operation in Area 4 at Merv.

Fragments reveal a high porosity, so the measurement of slag density is variable. It is significant to note that there is limited reaction of the slag against the crucible wall. In separate melting experiments, fragments of the crucible slag were observed to melt and become free-running at a temperature between 1150 and 1250°C. This temperature is below the temperature of cast steel.

Technical evidence: crucible slag compositions and crucible steel products

According to the description of co-fusion production of steel by Biringuccio, (ca. 1540) in Pyrotechnia, the technique does not use a crucible, but rather in a shallow hearth or cupel (Smith and Gnudi, 1942: 68, 69). The text describes immersing wrought iron in molten cast iron. Most relevant, however, is the description of the slag layer produced over the molten metal,

> '... Then when the masters see that all is afire and well heated, especially the receptacle, they begin to work the bellows more and to add some of that iron in small pieces mixed with saline marble, crushed slag, or other fusible and nonearthy stones. Melting it with such a composition they fill up the receptacle little by little as far as desired.
>
> Having previously made under the forge hammer three or four blooms weighing thirty to forty pounds each of the same iron, they put these while hot into this bath of molten iron. ... for four to six hours, often stirring it up with a stick ...'

The term translated as saline marble is *marmo saligno* (Smith and Gnudi 1942: 68). Therefore, the description provides details for the slag composition used in the additional crucible steel experiments by co-fusion. Williams (2003), however, has questioned aspects of the interpretation. Needham (1964: 13) connects the Biringuccio description to co-fusion; immersion of wrought iron in molten cast iron. This description most importantly connects calcium-rich slag with co-fusion. According to Killick and Gordon (1987) other slag compositions for bloomery slags, finery slag and puddling slag consist of magnetite, fayalite and glass with low CaO. Tylecote (1976: 88) collected similar compositions for low CaO in finery slags, but also illustrates the addition of scrap with cast iron pigs in the process. Nevertheless, Pipping and Barraclough (1988) note the use of 'crushed green glass' with lime powder as a reported constituent in the crucible melting of steel in the Huntsman process in 1767. A calcium-rich slag is used to exclude air and cover the surface of molten steel. Flux additions to form a slag can seem quite confusing due to the range of acceptable compositions to simply cover the underlying molten metal.

Most simplistically, the 'ingredients' of the charge in each crucible in Area 4 at Merv would be expected only to be cast iron and wrought iron in varying proportions which were also tested in the new experiments in open crucibles. However, (unspecified) slag compositions are also noted in some traditional 'recipes' for crucible-steel production (Craddock, 1995: 276-279), presumably to provide some protection from oxidation of the underlying molten metal. In other words, a slag layer would effectively help decrease loss of carbon from the molten metal and provide more control of the mixing. A relatively neutral atmosphere in the crucibles (Feuerbach et al., 2003: 266) would not affect the molten metal under the slag and would not contribute to a suggested carbon 'boil' and splashing. There may also be related aspects with the use of slag for refining and improvements in the quality of the steel products. Provisionally, based on crucible dimensions (Fig. 2) and a characteristic layer of slag approximately 1cm thick, an estimate of the mass of slag across a single crucible would range between 100-150g. A deliberate addition of flux is required to produce the observed slag layer in the crucibles. There is no obvious reason for such a characteristic slag layer of some 100-150g to develop in a closed crucible, even with

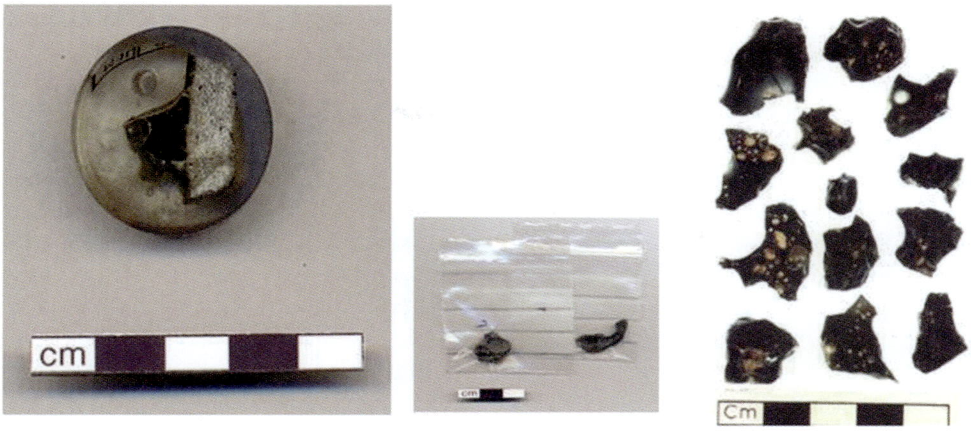

Figure 3a: The image on the left is typical of the adhering layer of the green, glassy slag in a metallographic section. Metallic prills of steel are present in the crucible slag. The collection of crucible slag fragments (note the glass-like fracture patterns) on the right, collected from the dump of crucibles in Area 4, represents about 20g in weight.

a small perforation in the lid, from charges of wrought iron and cast iron alone. A final working volume of each crucible, below the slag line, is estimated at approximately 0.25 litres. The lid was not entirely necessary with the accompanying use of a calcium-rich slag over the molten steel lower in the crucible, except to keep out additional fuel ash.

Slag compositions for crucible-steel production in Area 4 at Merv were first determined in the Ph.D. research by Feuerbach (2002). Based upon SEM/EDS, elemental analysis of the green, glassy slag adhering to the crucible wall fragments as well as loose fragments could be assigned into two groups, noting differences in the average manganese concentrations: High Manganese low K, low Ca Group (7 samples) and Low Manganese Group (14 samples).

For further characterisation, compositions of a new selection of ten slag fragments were determined with SEM/EDS. These ten fragments were selected at random from available samples. It is uncertain if any are from the same or different crucibles. The data are presented in Fig. 4. Altogether, the analytical results could be grouped by manganese concentrations, but the CaO compositions seem most significant and also

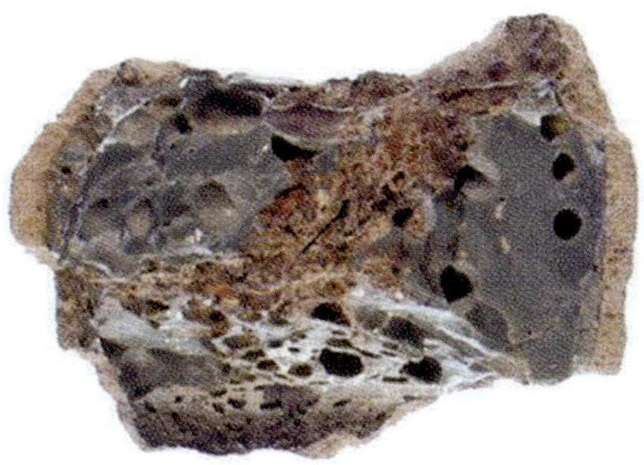

Figure 3b: Enlarged image (left) for the published slag from Akhsiket crucibles (Bayley and Rehren, 2007). Note, it extends from wall to wall across the crucible interior. Also it is not as glassy, but much thicker than the Merv slag fragments (Fig. 3a).

	Na2O	MgO	Al2O3	SiO2	SO3	K2O	CaO	TiO2	MnO	FeO	Total
A1	0.22	1.81	17.73	46.91	0.33	1.53	11.61	1.33	16.64	1.89	100.00
A2	0.27	1.85	17.37	47.21	0.18	1.53	11.82	1.39	16.63	1.74	100.00
B1	0.43	4.50	17.87	47.86	0.09	2.34	24.35	0.47	0.95	1.13	100.00
B2	0.34	4.44	17.86	47.95	0.00	2.41	24.27	0.51	1.03	1.19	100.00
C1	0.70	4.58	17.17	51.45	0.12	4.47	18.82	0.55	1.02	1.14	100.00
C2	0.63	4.03	18.33	52.99	0.13	4.11	16.29	0.64	1.27	1.58	100.00
D1	0.50	4.97	15.78	48.65	0.05	2.89	16.87	0.53	8.12	1.65	100.00
D2	0.50	5.09	15.75	48.39	0.07	2.88	16.88	0.54	8.14	1.76	100.00
E1	0.32	2.90	18.98	46.25	0.17	1.72	24.46	0.68	3.73	0.79	100.00
E2	0.21	2.87	18.83	45.45	0.23	1.72	25.18	0.74	3.87	0.91	100.00
F1	0.29	3.65	21.92	52.10	0.13	2.28	14.97	0.58	2.53	1.56	100.00
F2	0.33	3.55	22.11	52.07	0.04	2.25	14.73	0.57	2.56	1.78	100.00
G1	0.49	4.12	17.80	50.92	0.00	3.23	18.68	0.58	2.84	1.34	100.00
G2	0.58	4.21	18.11	50.52	0.11	3.13	18.59	0.56	2.87	1.32	100.00
H1	0.41	5.02	15.98	48.76	0.14	2.80	16.57	0.48	8.10	1.73	100.00
H2	0.53	4.91	15.72	48.58	0.15	2.88	16.89	0.52	8.06	1.75	100.00
I1	0.53	2.12	25.05	52.81	0.24	3.93	10.00	0.39	0.55	4.39	100.00
I2	0.61	1.94	25.39	52.91	0.27	3.86	8.59	0.42	0.41	5.61	100.00
J1	0.31	2.13	16.02	44.72	0.34	1.72	14.52	1.27	16.55	2.43	100.00
J2	0.33	2.19	15.97	44.55	0.25	1.63	14.71	1.37	16.59	2.43	100.00

Figure 4: Compositions for major slag forming oxides were determined by SEM/EDS on ten more mounted samples, labelled A to J. Two new point analyses were made per sample by selected area, avoiding iron/steel prills. These data are normalized to 100% for comparison using assumed oxides and standard calibration procedures.

could be interpreted to represent a single group. The CaO concentrations have an observed range between about 9% and 22%. (Feuerbach et al., 2003: 261); several more samples extend the upper range to about 25%. It would seem that CaO would represent a more dominant factor, rather than MnO concentrations, in actually determining desired slag properties that were readily apparent and obviously different in colour and composition from fayalite-type slags (Bachmann, 1982) from direct production of wrought iron. Except for two samples from a total of 31, all of the other FeO concentrations are below 2%.

It is also noted that the observed green, glassy slag compositions on the crucible fragments are not too dissimilar from historic and modern blast furnace slag compositions. Examples of selected charcoal blast furnace slag compositions are published by R.F. Tylecote (1986: 217), Rostoker and Bronson (1990: 105) and White (1980). Compositions are variable within a range, but the characteristic feature is the calcium, rather CaO, concentrations of the blast furnace slags. Too few data are available representing ancient or early historic blast furnace slag compositions. Examples of blast furnace slag compositions with 25% CaO are reported by Wagner (2008: 237) during the Han State Monopoly production of cast iron. Blast furnace slag compositions, while under-represented in relevant archaeometallurgical publications, were well-known at this time in China; a technical knowledge necessarily associated with the widespread production of cast iron. Cast iron production and blast furnace slag compositions are unknown at this time throughout Central Asia. However, some of the first examples of cast iron objects with Early Islamic contexts in Central Asia have now been reported at this conference.

The consistent control of CaO concentrations in the slag is a critical property for blast furnace operation. Fundamentally, CaO replaces FeO losses in the slag and governs slag viscosity. Blast furnace slag compositions are characteristically green and glassy. Crew (2000) used mass balances to assess fuel ash contributions of CaO to bloomery slag. Merkel (2007) discusses the practical effects of CaO in iron-rich slag compositions used in smelting experiments. There is a conceptual continuum or range of CaO concentrations from iron-rich slag compositions to calcium-rich slag compositions used for cast iron production, but other physical, chemical and mineralogical properties significantly change across the range.

A simple mass balance of CaO can be estimated to assess better the practical effects of CaO in the crucible slag compositions from crucible-steel production in Area 4 at Merv. The CaO concentrations in the crucible slag reveal several aspects related to crucible charges as well as guide the undertaking of simulation experiments. With an estimate of some 100-150g slag in each crucible (Fig. 2) and some 10-25% CaO in the slag (Fig. 4), a weight of up to about 38g CaO would have been added to the crucible with other documented slag-forming oxides before the lid was closed and sealed. The equivalent weight of limestone as $CaCO_3$ would represent about 70g as a flux.

Along with limestone or marble, further additions of sand or further reaction with the silica-rich crucible walls would be required to produce the slag compositions observed in each crucible. Based on the slag compositions and an estimated weight of 150g, the charges to make the crucible slag would include appropriately 75g SiO_2, 15g Al_2O_3 and 15g MnO. Other concentrations such as TiO_2 and MgO and K_2O would seem incidental in the charge. Thus, a total of about 200g of mixed limestone/marble and sand in variable proportions would produce a range of slag compositions depending upon quality, selection and processing of the flux. A small amount of iron oxide in the crucible slag would derive from the cast iron and wrought iron charges. However, because the refractory crucibles are closed with the charge inside to react upon melting, fuel ash contributions, sulphur, and crucible wall contributions would be minimal. These calculated inputs of slag forming 'ingredients' could produce the observed blast furnace slag compositions within the crucibles. Gilmour (2000: 55) favours this variation utilising the addition of slag forming minerals to produce the observed crucible slag compositions. As a further indicator of complexity, Wagner (2008: 236) presented a material balance for the CaO from the ore and limestone flux to produce observed blast furnace slag compositions for production of cast iron.

Alternatively, a more simplified approach would be the addition of some 100-150g crushed fragments of blast furnace slag, the waste by-product from production of cast iron, to produce the observed crucible slag compositions. This approach was also tested experimentally. Slag additions to crucible-steel charges might only be infrequently mentioned in the relevant texts (Craddock, 1995: 276-279), because its significance was not understood clearly by various writers. Nevertheless, there is still no other published evidence yet for knowledge of blast furnace slag compositions, except in the crucible-steel production remains at Merv, during the Early Islamic period of Central Asia.

A vital clue towards understanding the Merv evidence was the discovery of a corroded egg-like crucible-steel ingot (Fig. 5) from near the steelmaking workshop at Merv. In the excavation of Area 4, everything that looked like iron slag or

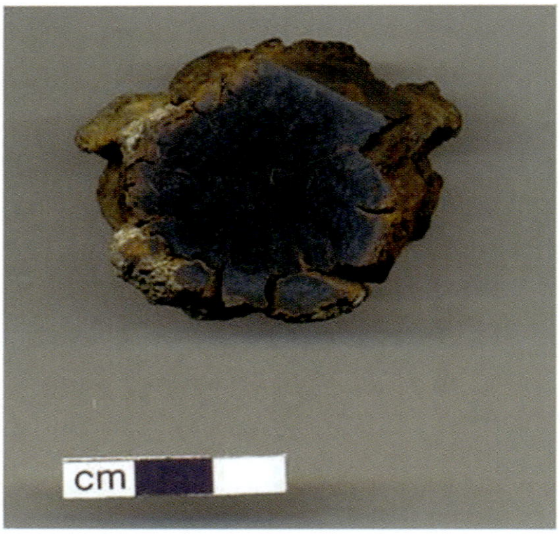

Figure 5. The corroded steel ingot measures about 5 cm across and after sectioning now weights approximately 150g. From relative molar volumes for Fe_3O_4 and FeOOH, the original weight as a steel metal is estimated at about 300-400g, rather 2-3 times the original weight. After the section was cut for metallography by B. Gilmour, the specimen was polished as illustrated. Although almost completely corroded, the centre 'ghost' structures suggest that this represents a once, rather homogeneous mass of steel.

rusty iron was sectioned and examined. This critical specimen was studied by B. Gilmour and published in 2000. Technical details of the ingot are published; hypereutectoid steel with 1.2-1.4% C with a grain boundary network of cementite in a pearlitic matrix. Referring to the Merv steel ingot, Gilmour (2000: 55) states,

> 'This process seems almost certain to have involved a crucible charge of which the main constituents were plain (bloomery) iron and cast-iron pieces, perhaps imported from more than one source, probably mixed with slag forming minerals and vegetable matter. This procedure was described by Al-Birūnī as that used...'

From the often-quoted, medieval (12th century) Islamic writer Al-Birūnī, several points may assist the interpretation of this steel ingot from Area 4 within the Gyaur Kala enclosure at Merv. The discussion of specific technical terms is more extensive in Allan (1979: 75). Selectively quoting from the passage by Al-Birūnī published by Feuerbach et al. (2003: 259), two qualities of steel could be produced:

The first quality was the result of melting components 'equally so that they become united in the mixing operation and no component can be differentiated or seen independently ... such steel is suitable for files or similar tools'.

The second quality was obtained 'if the degree of melting of the wrought and cast iron was different for each substance and thus the intermixing between both components is not complete, and their parts are shifted so that each of the two colors can be seen by the naked eye ...'.

Therefore, this seemingly homogeneous steel ingot from Area 4 could be more plausibly interpreted as 'First Quality' steel. The steel ingot examined by B. Gilmour would not be sufficiently heterogeneous to produce any watered pattern upon forging. It seems unlikely that it was a deliberate discard (cf. Gilmour, 2000: 55). Alternatively, what would the expected properties be for a 'Second Quality' steel ingot?

Experiments

An experimental approach was taken to help resolve several aspects of the interpretation of the archaeometallurgical remains from Merv. Low temperature co-fusion experiments were undertaken in 1955 by Mr. P. Whitaker of Stewarts and Lloyds, Ltd., Corby, and the following details are published by Wagner (2008: 257):

> 'Strips of wrought iron, 1mm thick, were piled up and wired together with varying amounts of crushed white cast iron between layers. This faggot was heated in a non-oxidising atmosphere to a temperature between 950° and 1000°C, then immediately withdrawn from the furnace and hand-forged, welding it into a single bar. At the original welded surfaces in this bar were highly carburised zones as well as some unchanged white cast iron. A homogenising anneal of 8 hours at 900°C resulted in a uniform carbon content of 0.8 percent throughout the bar.'

These experiments by Whitaker were used as a general guide as a technological context for the present series of co-fusion simulation experiments. However, at such temperatures, under 1000°C, and with unequal proportions of wrought iron and cast iron, the observed results would seem more like straightforward carburisation of the wrought iron strips.

Because calcium-rich slag compositions were found in the crucible-steel remains in Area 4 at Merv, a most significant change for the new experiments included the addition of blast furnace slag as a deliberate 'ingredient' in the charge of each crucible. Molten blast furnace slag was intended to prevent oxidation of the cast iron and wrought iron charge. Even if it seems uncertain if the actual intent in the historic past was also to prevent oxidation, the effect of a slag layer was nevertheless quite successful. A lid or other covering of the crucible was not used in the new experiments to assess

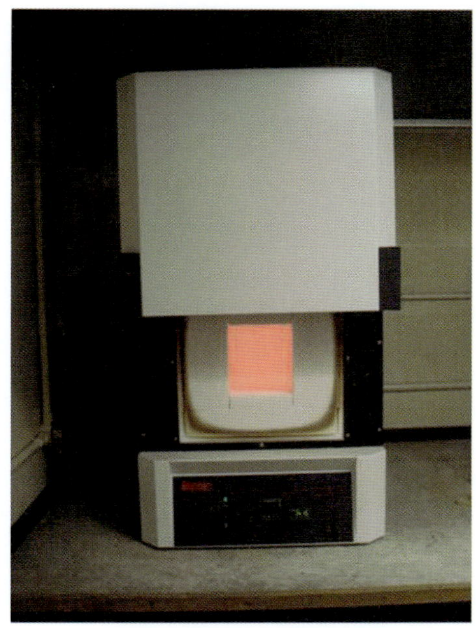

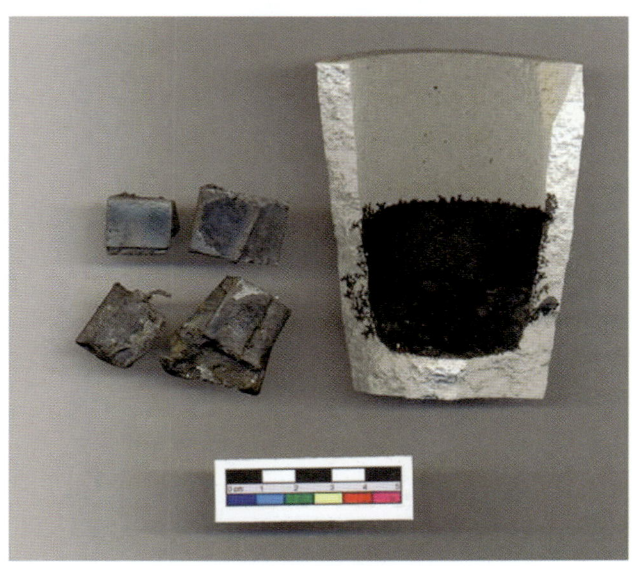

Figure 6. Experiment A parameters:
- 11g cast iron and 45g folded wrought iron coupons with a cover of 50g crushed blast furnace slag
- Two hours to reach 1300°C, then held for one hour
- Held at 900°C for four hours (annealing and carburization)
- Cooled overnight in the muffle furnace
- Crucible broken open, slag discarded
- 40g 'ingot' of fused coupons and slag
- The iron 'ingot' was forged with difficulty to produce a piled structure with clear weld lines on the surface

the effectiveness of the calcium-rich slag layer to prevent oxidation. The observed compositions (approximately 70% SiO_2 and 25% Al_2O_3) and mineralogy (including mullite and cristobalite) of the Merv crucibles and lids determined by XRF, SEM/EDS, XRD and petrography (Feuerbach et al., 2003: 261) were difficult to match from a commercial supplier. For the laboratory simulation experiments, sillimanite crucibles to approximate Merv crucible compositions were obtained from DYSON Thermal Technologies (UK). The DYSON small pouring crucible, Quality SGP, selected for use was reported to contain 41% SiO_2, 56% Al_2O_3, with 0.7% Fe_2O_3 and 0.05% TiO_2. This was judged a good substitution, the key parameters are that the crucible was sufficiently refractory and sufficiently unreactive with blast furnace slag compositions. Other changes from the Whitaker series of experiments include higher temperatures (1300° and 1400°C) and longer, overnight cooling.

A ratio of about 20% cast iron to 80% wrought iron were selected to produce an anticipated 0.8% carbon steel (Wagner, 2008: 258). The cast iron selected for the experiments was architectural scrap, grey cast iron with an observed melting temperature of about 1200°C. The wrought iron was commercial sheet, cut to coupons 1mm thick. The wrought iron was confirmed metallographically to be hot-rolled ferrite. The blast furnace slag was selected from the teaching collection at the Institute of Archaeology. XRF determined 46% SiO_2, 24% CaO, 13% Al_2O_3, 4% MgO and 3% MnO for the slag used in Experiment A. It had an observed melting temperature of 1200°C. In Experiment B, the blast furnace slag composition was 53% SiO_2, 17% CaO, 13% Al_2O_3, 3% MgO and 3% MnO with an observed melting temperature of about 1250°C. Because the crucibles in Area 4 were closed with an attached, sealed lid, there could be no addition of fuel ash into the slag compositions. Another, further campaign of simulation experiments would be necessary to evaluate better the calculated proportions of limestone and sand necessary to produce the observed quantity of blast furnace slag within the crucibles for co-fusion production of steel.

At 1400°C from a Iron-Iron Carbide Diagram, the slag and cast iron contents of the crucible should have been fully molten at equilibrium. However, obviously the wrought iron coupons did not fully react or melt with the quantity of cast iron and slag (Fig. 7). Insufficient time, temperature and perhaps low proportions of cast iron to wrought iron would have been responsible. Higher temperatures, near 1500°C would have eventually melted the charge. Nevertheless, under the parameters for Experiment B at 1400°C, a less than ideal product resulted. Fusion of the cast iron and wrought iron were incomplete, so the product may be representative of a 'Second Quality' steel ingot; lower temperature co-fusion product. Based upon carburization penetration diagrams (Rostoker and Bronson, 1990: 123), two to four hours should have been sufficient for the selected wrought iron coupon thickness. Metallographic examination confirmed the production of steel, but the structure was highly variable with gradients and banding of ferrite and pearlite (Fig. 8). Subsequent forging by a blacksmith under reducing conditions would contribute further toward developing overlying patterns of co-fusion, piling, case hardening and distortion as well as probably some decarburization of exposed surfaces. T. Jones at the Slade School of Fine Art forged the steel product into a compact, dense mass, but the conditions were too oxidising. Thus,

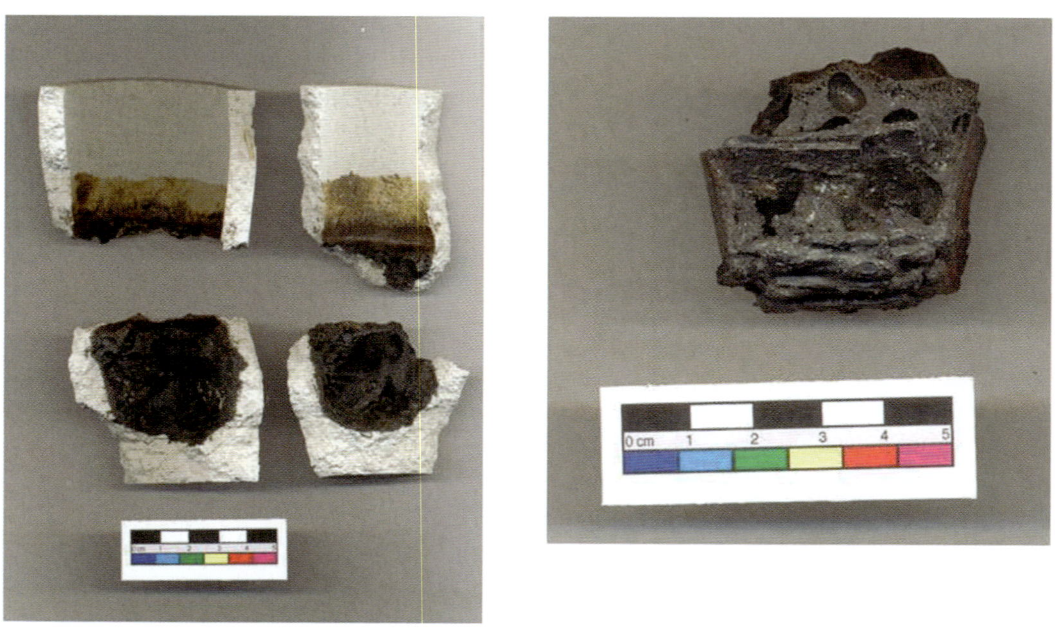

Figure 7: Experiment B parameters:
- 31g cast iron and 128g folded wrought iron coupons with a cover of 100g crushed blast furnace slag
- Two hours to reach 1400°C, then held for one hour
- Held at >1100°C for four hours (annealing and carburazation)
- Cooled overnight in the muffle furnace
- Crucible broken open, 185g mass collected (note: the greater reaction of the blast furnace slag with the crucible and 1cm thick layer of slag with large entrapped 'bubbles'), slag discarded
- 120g 'ingot' of fused coupons and adhering slag
- The steel 'ingot' was forged and folded repeatedly to produce a plied structure with clear weld lines on the surface

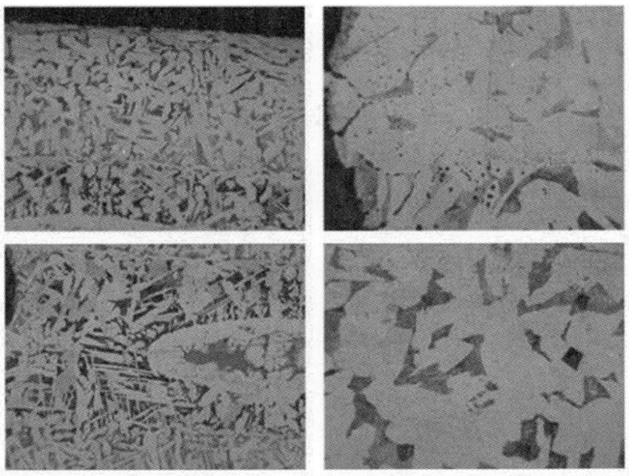

Variable ferrite and pearlite microstructures (magnification x50-100; etch 2% nital; microhardness of 180-220 Hv 100)

Figure 8: Experiment B produced steel microstructures with variable proportions of ferrite and pearlite (magnifications x50 and x100; etchant 2% Nital; microhardness values between 180-220Hv100). Thus, the 'Second Quality' steel product obviously retained a piled structure as well as variable carbon gradients. Higher temperatures would be expected to produce a more homogenized microstructure and a more elliptical shape to the steel ingot.

it becomes most difficult to unravel each distinct aspect of production in a finished object. The problem seems to be that crucible steel made by co-fusion would also be to the effects of piling and cementation in a forge as well as some oxidation in the processing of forging an object. In this respect, the microstructure may seem most complicated, but relatively straightforward when considered in a stepwise process with competing effects.

Historical context on the silk route from China

The parameters for these co-fusion experimental simulations were based upon the archaeometallurgical evidence from Early Islamic Gyaur Kala at Merv. However, it is uncertain exactly how or where this technology developed. Certainly, the co-fusion technique seems to appear at Gyaur Kala in a fully developed state using blast furnace slag compositions in crucibles in an Early Islamic context. What other foreign influences might be inferred from the technological parameters observed in the workshop of Area 4 within the Gyaur Kala enclosure at Merv? The origin of the blast furnace in Europe is somehow attributed to the early diffusion of the indirect process from China (Wertime, 1962; Tylecote 1986), but the evidence is not conclusive. The presence of calcium-rich slag associated with crucible steel production at Merv does suggest at least one line of investigation to follow; a complex system of trade and transfer of some metallurgical techniques along the Silk Route from China. However, the technology can readily change and be adapted. Excavated examples of calcium-rich slag as well as examples of cast iron should be more actively and widely sought in Central Asia. There are many unresolved issues for technology transfer along the Silk Route.

D. Wagner (2008: 264) states, 'The co-fusion of cast and wrought iron is of course a Chinese innovation – it was practised in China long before cast iron was known anywhere else.' Blast furnace slag compositions are then also a Chinese innovation. Whether the blast furnace slag compositions in the crucibles for steel-production are made from charges of raw materials, such as limestone, or 'recycled' blast furnace slag cannot be determined conclusively, but the knowledge and utilisation of such slag compositions must be linked with the Chinese origin of cast iron.

Some other aspects of the technology, such as the use of high-quality refractory crucibles in steel production, seem most obscure. Co-fusion in China is not a crucible based technology (Needham, 1964); other variations were used and reproduced in the experiments by P. Whitaker (Wagner, 2008: 257). No crucibles similar in composition or function to those at Merv have yet been found in China. Alternatively, white crucibles for steel-production have been published in Korea (Jung Maun Kim, 1994), but the earliest dates still seem somewhat uncertain. The published colour plates (ibid.) clearly illustrate similar refractories, but with different shapes. Technical studies of these crucibles are not yet available. Crucibles used for production of steel by direct smelting had been published by Papachristou and Rehren (2008). These crucibles and technical parameters have some similarities, but many differences, too. The differences for the evidence from Area 4 at Gyaur Kala include smaller crucible dimensions and working volumes, thinner refractory lids, more consistent calcium-rich blast furnace slag compositions and variations in steel prill microstructures and distributions (Feuerbach et al., 2003: 265). Crucible smelting in Manchuria (Zenshiro Hara, 1992) also documents crucibles for steel production, but from later dates and much larger furnaces. The foreign influences for the crucibles, and indeed, the circular furnace plan discovered in Early Islamic Gyaur Kala at Merv reflect more diverse influences and may ultimately be shown to derive from various sources.

Most interesting, there is an historical text documenting the presence of a Chinese prisoner at Merv during this chronological period. Herrmann (1996: 3, 11) highlights a little-known Chinese account by Du Huan. In 765, after his return from ten years in captivity at Merv, he described the enclosure walls, buildings, canals, irrigation and two Buddhist temples. The relevant points are that there is apparently a thriving Buddhist community as well as Chinese captives within the city enclosure. Other associations establish links between the Chinese military and iron production (Wagner, 2008: 250). These few associations remain rather too tenuous. Notwithstanding, the crucible-steel production at Gyaur Kala seems fully-developed and so far, without local precedent, when it first appears in the Early Islamic Period (9th-10th Century) at Merv. The knowledge of the process is apparent in the workshop remains of steel production. Details such as the use of calcium-rich slag in the crucibles are distinctive and unparalleled at the time. It is significant that the calcium-rich slag compositions are documented in the Merv workshops. It is not a local process, so the Silk Route is a factor in technology transfer. The technique apparently did not end here either because similar slag composition for steelmaking are described centuries later by Biringuccio (1540) in Siena and Florence compositions and finally documented (on a larger scale) in the Renaissance. The description of the working zone of a crucible furnace for glass making by Biringuccio (Smith and

Gnudi, 1942) is not dissimilar to the Merv type furnace also operated with bellows and a central tuyere.

Conclusions

The simulation crucible-steel experiments using modern furnaces have provided further background towards understanding product qualities described by the Islamic writer Al-Birūnī, (Allan 1979: 75; Feuerbach et al., 2003: 259). The new experiments help illustrate the low temperature co-fusion technique discussed by Wagner (2008: 257-258). The use of calcium-rich slag had advantages in the experiments, as described by Biringicco (1540), preventing oxidation of the molten metal over several hours and decreasing loss of carbon. The mass balance of cast iron to wrought iron 'scrap' in each crucible is the key, but a range of proportions produce steel. Therefore, prediction of carbon concentrations in the steel product could be relatively straightforward because of the decreased loss of carbon under slag in a sealed crucible. There is no need for additional charges of organic material within crucibles to maintain a reducing atmosphere (cf. Gilmour, 2000: 55). Calcium-rich slag compositions could be consistently produced in each crucible with either calculated additions of limestone and alumina-rich quartz sand or additions of existing calcium-rich slag. As a result, a charge of about 20% cast iron by weight in ratio to the wrought iron was adequate to provide sufficient carbon necessary to produce steel (Fig. 8). Increasing the proportion of cast iron to wrought iron in the crucible would result systematically in higher carbon steel. Observational control of the steel product was based upon visual inspection as noted by Al-Birūnī, (Allan 1979: 75) and subsequent forging. A ratio with too much cast iron would simply result in a brittle, cast iron-rich product. With a modern furnace at higher temperatures nearer 1500°C, the use of crucibles allow for complete homogenization of the wrought iron and cast iron charge. The essential variable was, however, temperature. Across a furnace filled with crucibles, a consistent temperature nearer 1500°C would have been most difficult to achieve and maintain. For an approximation of carbon diffusion from liquid cast iron at such high temperatures, several hours would seem sufficient for 1mm thick wrought iron sheet (Rostoker and Bronson, 1990: 123). The observation during fieldwork that very few examples of the thousands of crucible fragments showed any distortion attributable to heating is most significant. Calcium-rich slag effectively covered the mixed charge within each crucible and the crucibles were placed into the furnace near the highest expected temperatures. The refractory properties of the crucible compositions serve as a pyrometer; temperatures were not over 1500°C in the furnace. Furthermore, the furnace remains do not show the expected effects of such high temperatures. With temperature as the essential parameter, sorting through the steel ingots from a numerous crucibles in the furnace, only two qualities could be differentiated as described by Al-Birūnī. The two new laboratory experiments illustrate a "Second Quality" product. Experiment B was approximately half-scale, input weights should be doubled to match better the estimated weights and related volumes for the crucibles from Area 4 at Merv. Further experiments could, of course, examine the ratios for cast iron to wrought iron as well as homogeneity of the steel product at higher temperatures. Each closed crucible allows control over the charges and excludes fuel ash, but the limitations of furnace design results in a range of temperatures which introduce variability in the steel products. Griffiths and Feuerbach (1999) also consider impurity effects influencing the steel product quality, such as manganese, vanadium or molybdenum. From among the steel ingots recovered from each crucible, the quality would be assessed on observed physical properties, and then distributed, as described by Al-Birūnī, for various tools.

The technical details for steelmaking at Merv are significantly different from other sites. The composition and microstructure of the steel ingot from Area 4 at Merv, investigated by Gilmour (2000), are unique for an ingot from an excavated context with crucible fragments. The ingot represents a 'First Quality' intermediate product ready for further forging. Calcium-rich slag in the crucible effectively covered the mixed charge and the crucible was set in the furnace near the highest temperatures.

Diffusion of technical knowledge perhaps can be inferred in this case when there is no evidence of any relevant development within the region or at Merv. Crucible-steel production represents a new metallurgical technique requiring skilled use and access to advanced materials and sequences of actions. The earliest evidence for crucible-steel production in Area 4 in Gyaur Kala at Merv represents a fully developed and sophisticated technique relying of technology not present elsewhere yet in the larger area. The refractory crucibles used for steel-making are quite distinct from the lower-grade sand-rich clay crucibles used for casting of brass and bronze in the metallurgical zone in Area 4 at Gyaur Kala. The use of cast iron and calcium-rich slag compositions are also seemingly out-of-place at this time in Central Asia. Industrial activities at Gyaur Kala were presumably under close administrative control within the enclosure, so the discovery of crucible-steel production here using a co-fusion technique may even represent a unique combination of historical and technological influences at this time. Economic factors such as major, long-distance trade routes across Central Asia, and even religious factors, such as the spread of Buddhism, may be correlated. Additional military factors may also seem relevant for the diffusion of metallurgical technology along the frontiers crossed by the Silk Route. So far, there are only tantalizing clues suggesting transmission of crucible-steel technology using crucibles and a calcium-rich slag composition with a chronological sequence at major sites along the Silk Route and further west.

References

Allan, J.W. 1979 *Persian Metal Technology, 700-1300 AD.* London: Ithica Press.

Bachmann, H-G. 1982 The identification of slags from archaeological sites. *Institute of Archaeology, Occasional Papers,* No. 6.

Bayley, J. and Th. Rehren 2007 Towards a functional and typological classification of crucibles. In *Metals and Mines.* London: Archetype and British Museum, 46-55.

Craddock, P.T. and Lang, J. 1993 Gizah iron revisited. *Historical Metallurgy* 27,2: 57-59.

Craddock, P.T. 1995 *Early Metal Mining and Production*. Edinburgh: Edinburgh University Press. 276-279.

Crew, P. 2000 The influence of clay and charcoal ash on bloomery slags. In *Iron in the Alps: deposits, mines and metallurgy from antiquity to the XVI century. Proceedings of the Conference*, 38-48.

DYSON Thermal Technologies (UK). Online Technical Reports.

Feuerbach, A. Merkel, J. and Griffiths, D. 1998 An examination of crucible steel in the manufacture of Damascus steel, including evidence from Merv, Turkmenistan. In *Metallurgica Antiqua*, Der Anschnitt, Beighft 8, eds. Th. Rehren, A. Hauptmann, and J.D. Muhly. Bochum, 37-44.

Feuerbach, A. 2002 *Crucible steel in Central Asia: Production, Use, and Origins*. Ph.D. Thesis, University of London.

Feuerbach, A. Griffiths, D. and Merkel, J.F. 2003 Early Islamic Crucible Steel Production at Merv, Turkmenistan. In *Mining and Metal Production through the Ages*, Chapter 17, eds. P. Craddock and J. Lang. London: British Museum, 258-266.

Feuerbach, A. 2006. The Production and Identification of Crucible Steel. In *Make All Sure: The Conservation and Restoration of Arms and Armour*. Smith, R.D. ed., Basiliscone Press, UK. 12-17.

Gilmour, B. 2000 The development of iron and steel technology in southern Asia. In *Persian Steel, The Tanavoli Collection*. eds. J. Allan and B. Gilmour. Oxford Studies in Islamic Art XV. 41-79.

Herrmann, G. Kurbansakhatov, K. and Simpson, St J. 1995 International Merv Project. Preliminary Report on the Third Season, *Iran* 33: 42-49.

Herrmann, G. 1996 Early and Medieval Merv: A Tale of Three Cities. *Proceedings of the British Academy*, 94-1-43.

Herrmann, G., Kurbansakhatov. K., and Simpson, St J. 1997 International Merv Project. Preliminary Report on the Fifth Season, *Iran* 35: 10-13.

Hermann, G. 1998 Central Asian city on the Silk Road, ancient and medieval Merv, *Archaeology International*, 1: 32-36.

Jung Maun Kim 1994 Discovered crucibles from Puyo Province, Korea (In Korean). *Kogohakchi* Vol. 6: 109-122 (Translation).

Killick D.J. and Gordon, R.B. 1987 Microstructure of puddling slags from Fontley, Roxbury, Connecticut, USA. *Historical Metallurgy*, 21,1: 28ff.

Merkel, J. Feuerbach, A. and Griffiths, D. 1995 Analytical Investigation of Crucible Steel Production at Merv, Turkmenistan. Institute for Archaeo-Metallurgical Studies, *IAMS*, No. 19: 12-14.

Merkel, J. 2007 Imperial Roman production of lead and silver in Upper Moesia (Mt. Kosmaj Area). *Journal of the Serbian Archaeology Society,* 23: 39-78.

Needham, J. 1964 *The Development of Iron and Steel Technology in China*. Newcomen Society, 13.

Papakhristu, O.A. and Rehren, Th. 2002 Techniques and technology of ceramic vessel manufacture crucibles for Wootz smelting in Central Asia. In *Modern Trends in Scientific Studies on Ancient Ceramics*. V. Kilikoglou, A. Hein, and Y. Maniatis, BAR International Series 1011, 69-74.

Papachristou, O. and Rehren, Th. 2008 A tentative comparison of steel-making crucibles from Central Asia and the Indian Subcontinent. In: *Proceedings of the 4th Symposium of the Hellenic Society for Archaeometry,* Y. Facorellis, N. Zacharias and K. Polikreti, eds, BAR IS 1746, Oxford, 519-528.

Pipping, G. and Barraclough, K.C. 1988 An important contribution to the history of the crucible steelmaking process. *Historical Metallurgy*, 22,2: 96-101.

Rehren, Th. and Papachristou, O. 2003 Similar like white and black: a comparison of steel-making crucibles from Central Asia and the Indian subcontinent. In: *Man and Mining*, Th. Stoellner, G. Körlin, G. Steffens & J. Cierny (eds), (=Der Anschnitt, Beiheft 16, Bochum), 393-404.

Rostoker, W. and Bronson, B. 1990 Pre-Industrial Iron, Its Technology and Ethnology. Archaeomaterials Monograph 1. Philadelphia.

Smith, C.S. and Gnudi, M.T. 1942 Translated and edited *Pyrotechnia of Vannoccio Biringuccio 1540.* MIT Press.

Simpson St. J. 2001 The Early Islamic crucible steel industry at Merv. Institute for Archaeo-Metallurgical Studies, *IAMS*, 21: 14-15.

Tylecote, R.F. 1976 *A History of Metallurgy*. London: Metals Society.

Tylecote, R.F. 1986 The early history of the iron blast furnace in Europe: a case for east-west contact? In *Medieval iron in society*. Hook, E. and Palsson, R. eds., Stockholm.

Verhoeven, J.D. and Jones, L.L. 1987 Damascus Steel, Part II: Origin of the Damask Pattern. *Metallography* 20: 153-180.

Wadsworth, J. and Sherby, O.D. 1980 On the Bulat-Damascus steels revisited. *Progress in Materials Science*, Vol. 25: 35-68.

Wagner, D.B. 2008 Chemistry and Chemical technology, Vol. 5, Part 11: Ferrous Metallurgy. In *Joseph Needham Science and Civilisation in China.* Cambridge: Cambridge University Press.

Wertime, T.A. 1962 *The Coming of the Age of Steel*. Chicago.

White, J.R. 1980 Historic blast-furnace slags: archaeological and metallurgical analysis. *Historical Metallurgy* 14,2: 55-64.

Williams, A.R. 1990 Slag inclusions in armour. *Historical Metallurgy* 24,2: 69-80.

Williams, AR. 2003 Biringuccio and the metallurgy of Italian armour. *Archaeometallurgy in Europe*, Proceedings, Vol. 1, 143-151.

Wraith, A. E. 1978 The performance of an experimental charcoal fired crucible furnace. *Historical Metallurgy*, 12,1: 44-47.

Zenshiro Hara 1992 Crucible smelting in Manchuria. *Archaeomaterials*, 6: 131-139.

Author's address

- Institute of Archaeology, University College London

Theoretical approaches to technology

Variability in iron smelting practices: assessment of technical, cultural and economic criteria to explain the metallurgical diversity in the Dogon area (Mali)

Caroline Robion-Brunner, Vincent Serneels and Sébastien Perret

Abstract: Since 2002, our research team has been investigating the development of iron metallurgy in the Dogon Country (Mali), from its origins to the present. To understand this activity at a regional scale and through time, a multidisciplinary approach was developed. At present, more than one hundred bloomery sites have been inventoried and mapped in the Dogon Country. The establishment of a typology based on technological, cultural and economic criteria permitted demonstration of the existence of six metallurgical traditions. Such diversity in a geographically restricted area is unusual. In this article, we present the technological and cultural diversity of iron metallurgical activity, and discuss the classification criteria selected and the impact of the level of production on variability in the socio-economic systems established in the Dogon Country.

Keywords: History, iron, metallurgy, classification, production regime, socio-economic system, Dogon, Africa

Introduction

Since 2002, within the framework of an international and interdisciplinary project entitled 'Human populations and paleoclimatic evolution in West Africa', directed by Eric Huysecom, the paleometallurgical objective has been to examine the development of iron metallurgy in the Dogon Country (Mali) from its origins to the present (Huysecom, 2002; Serneels, 2005; Serneels, et al., 2006; Huysecom, et al., 2007; Robion-Brunner, 2010). The Dogon Country, a place of multiple influences, possesses a cultural heritage remarkable for its uniqueness, its diversity, and its degree of preservation. Such a favourable context is auspicious for the study of the social, economic and historical organisation of the iron industry, and of the relationships linking metal, population movements, social and political change, and technological development. The principal goals of this project are to describe and explain the metallurgical techniques established in the Dogon Country and their impacts on society and environment.

To do so, a multidisciplinary approach was developed. The ethnohistorical approach has made it possible to collect oral traditions related to metallurgy and information about the history of producers and consumers of metal concerning the past two or three centuries. These surveys contribute to understanding the historical, social and economic aspects of metallurgy that are impossible to identify through the material record alone. They also inform on the spiritual and symbolic world in which iron production and working are integrated. Finally, the practical knowledge of modern craftspersons is of help in understanding and reconstructing early practices.

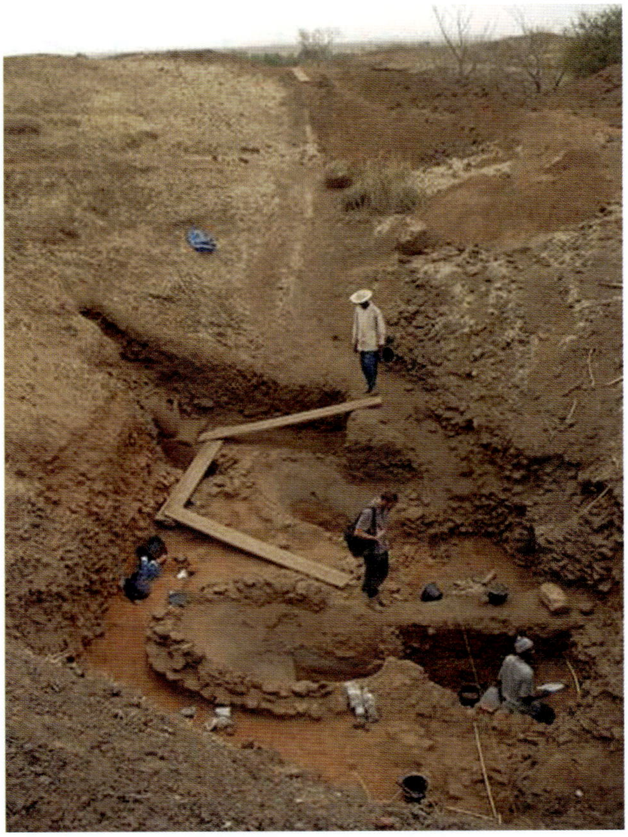

Figure 1: Fiko bloomery site: excavation of the northern crater of heap 3, with a pair of bloomery furnaces in the centre (2005 season).

The archaeological approach focuses on the inventory, description and explanation of the material evidence of iron metallurgy: early ore extraction pits, iron smelting furnaces, and smithies at which objects were made. These sites have been located, visited and documented through detailed description, photographs, and topographic mapping. Characteristic sites were selected and studied in greater detail through limited or extensive excavations.

Such fieldwork has exposed bloomery furnaces, making it possible to study their operation, to collect charcoal samples in stratigraphic context to establish the chronology of the sites and the species of wood used as fuel, and to sample metallurgical residues (ore, slag, technical ceramics). These samples were analysed in the laboratory to better understand the technological aspects of the furnaces, including different modes of production and their development through time. Detailed topographic maps were created in order to examine the spatial organisation and to estimate the quantity of waste, and thus the corresponding levels of production.

During our eight field seasons in the Dogon Country, a zone of 100 by 150km was systematically surveyed. This includes the Bandiagara Plateau and Escarpment between the Inland Niger Delta and the Seno Plain. In November 2008, more than one hundred bloomery and mining sites have been inventoried, mapped and studied for the first time.

Iron production in this region of Africa (and across the continent) has progressively decreased since the 20th century, replaced by the use of scrap iron. A series of fourteen AMS dates and oral traditions confirm that the iron production started at least in the beginning of the second millennium AD (Robion-Brunner, 2008, 2010; Perret and Serneels, 2009). Data are currently insufficient to assess the beginnings and fluctuations in iron production over this long period, but significant variability in metallurgical remains has been observed within this relatively limited restrained geographic region. In order to explain such diversity, we have attempted to construct a classification based principally on technological criteria. Criteria such as morphology, size and mode of construction of structures further refine this classification and reflect cultural choices of the metallurgists. Historical and economic contexts are examined through analysis of geographic and linguistic data, as well as by the quantification of metallurgical waste.

In this article, we present the technological and cultural diversity of the iron metallurgical activities, discuss the classification criteria selected, and show possible relations between the level of production and variability in the socio-economic systems identified in the Dogon Country.

Comparative analysis of metallurgical production: selection of classification criteria

Since the pioneering work of W. Cline (1937) in Africa, many researchers have attempted to explain the astonishing variability in metallurgical practices by creating different typological systems. The first is based on furnace morphology (Sutton, 1985; Martinelli, 1993; McNaughton, 1993) and considers the size and morphology of the superstructure, dimensions and form of the furnace pit, number of openings, construction materials, etc. The second is based on technological criteria: presence or absence of a superstructure, mode of ventilation, and types of slag, in order to identify iron smelting practices (Kense, 1985; Pole, 1985; Tylecote, 1987; Killick, 1991). The third classifies metallurgical workshops based on social, political and economic organisation (de Maret, 1980, 1985; Childs, 1991; Martinelli, 2002; Langlois, 2005-2006), taking into consideration the social identity of the metallurgists. Such work has influenced our choice of classification criteria. However, we have attempted to develop an approach that is not only technological and cultural, but also economic, in order to place archaeological material in relation to socio-economic systems.

The following criteria were used to describe metallurgical activities in the Dogon Country (Table 1, Robion-Brunner, 2008, 2010):

1) Metallurgical residues constitute a significant source of information. Including them permits inclusion of sites where furnaces are in poor condition or remain unexcavated. Residue classification shows differences between the broad families of furnaces. Archaeometric investigations of these materials help us understand the smelting processes (thermodynamic conditions, efficiency, output).
2) The morphology of bloomery furnaces is extremely variable (cylindrical, tapered or domed shafts). The poor state of preservation of the superstructures often makes their reconstruction difficult. The architecture (dimensions, form) and construction materials (nature, modes of use) provide essential elements for understanding their operation. From data on the architecture and construction materials, we can identify the dimensions and forms of furnaces, and the type, nature and mode of use of different elements. It should be noted that the use of specific building material varies according to local or regional resources (rock, clay, sand, etc.) and the choices of the metallurgists (re-use of materials, maintenance of the structure). Furnace morphology is strongly influenced by cultural choices and as such can reflect the identity of the producers.
3) Reconstruction of the spatial organisation yields information on the management of the workplace and the volume of production. The bloomery furnace is only one part of the smelting workshop. It generates a long series of technological processes and auxiliary infrastructures: beneficiation, crushing and sorting of the ore, charcoaling, storage of raw materials (ores, charcoal, clay, stone, etc.), bloomsmithing, etc. These activities, sometimes observable by the presence of typical waste products (ore fragments, charcoal, wood, metallic residues), are used to reconstruct the operational process, management of space, and the life of a workshop. The size of a site varies greatly, depending on duration of use, the number of active furnaces, and the frequency of the smelts. Quantitative criteria for all of these elements contribute to estimate a site's production level. In our study area, the volume of metallurgical waste material relates essentially to the quantity of ore treated and iron produced; in a

controlled chronological framework, it can be used to determine the mode of production (sporadic, regular, intensive) and helps us understand the impact of this technology on the way of life of the population (agricultural output, growth of political and military power) and on the environment (management of wood resources).

4) Geographic and linguistic data enable us to precisely locate the bloomery sites and to analyse the spatial distribution of metallurgical traditions. Sites are defined in terms of whether they are located in a restricted or extended perimeter, and whether these assemblages overlap or are found in specific regions. Based on these observations, we attempt to evaluate whether there is a relationship between metallurgical traditions and geolinguistic zones and to identify historical reasons for these concordances.

5) Historical data include the dates of establishment and abandonment of a bloomery site, the identity of the metallurgists (social origin, patronyms, village of residence), the organisation of the operational process, the need for a labour force, specific operating parameters of the bloomery furnaces, and the ultimate destination(s) of the iron produced. Obtained through interviews or archaeological excavation, these data allow us to place the bloomery sites into their historical and economic contexts, and must be compared to the paleometallurgical data.

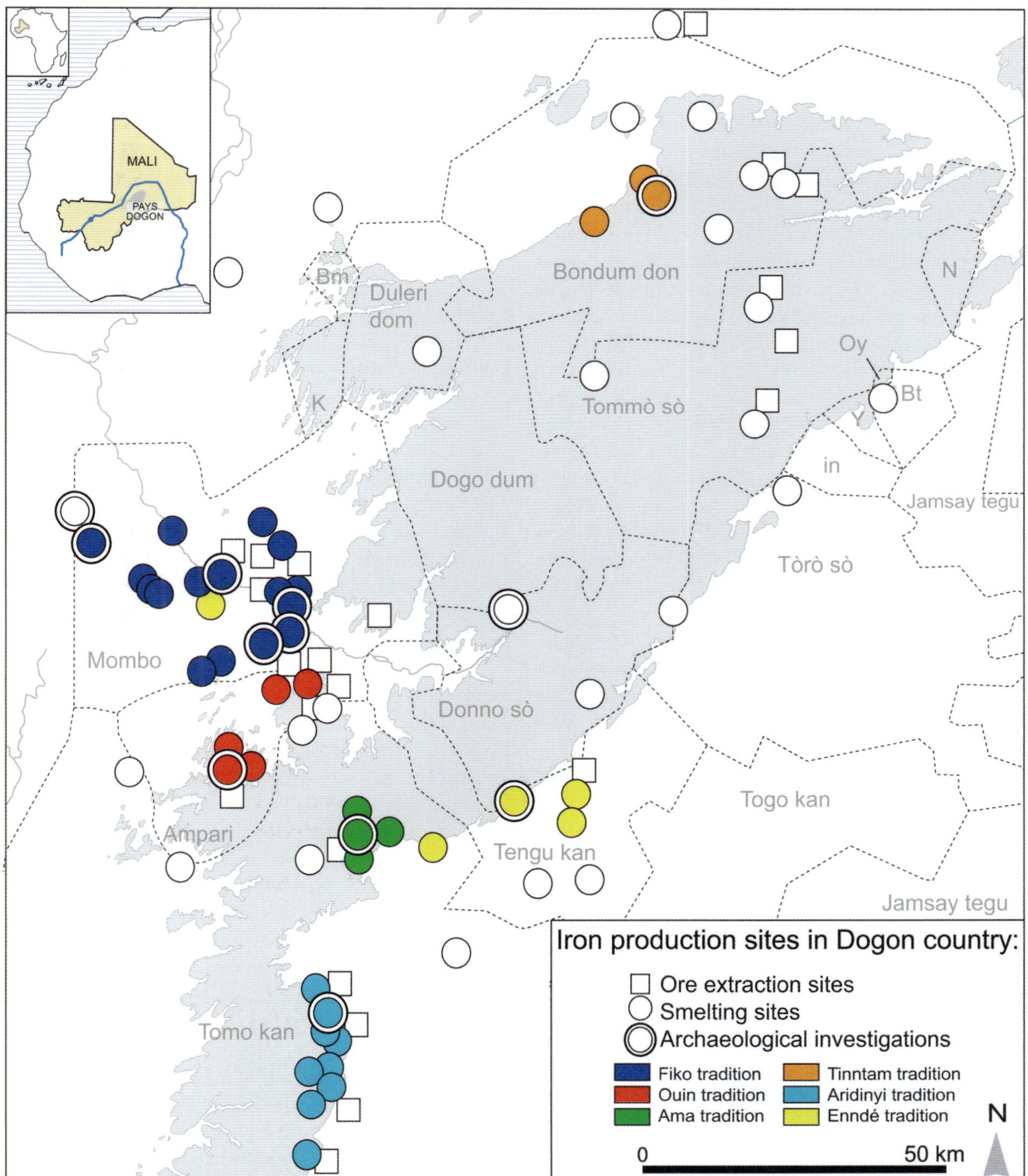

Figure 2: Location of extraction and bloomery sites, and the six metallurgical traditions in the Dogon Country.

The six metallurgical traditions identified in the Dogon Country

Simultaneously taking into account technological, cultural and economic criteria, we have now identified six distinct metallurgical traditions in the Dogon Country (Fig. 2), including a total of 41 bloomery sites (Robion-Brunner, 2008, 2010).

Located on the western part of the Bandiagara Plateau, along the lower valley of the Yamé and in the Mombo linguistic zone, the *Fiko tradition* is differentiated by very large dumps of metallurgical waste. They form irregular or closed craters generally several metres high, building up around single or paired bloomery furnaces.

The working areas around the furnaces are kept clear by supporting walls maintaining the slag and furnace residues outside the perimeter. The furnaces consist of a large elongated pit buried in the soil or embedded in the slag, and a horseshoe-shaped shaft. The tuyères were inserted just above the pit, through about twenty large openings; in front of the furnace, a large door extended to the bottom of the furnaces and may have contained other tuyères and the tapholes. The superstructure was built with sandy slag, re-used tuyères, and clay. The internal volume is impressive and reaches between 3 and 4m^3. The metallurgical assemblage is primarily composed of tapped slag (70%), associated with silica-rich platter shaped slag cakes (20%), and internal furnace slag blocks (10%).

The *Ouin tradition* is found in the Pignari-bana, a region southwest of the Bandiagara Plateau where the inhabitants speak Ampari. Metallurgical waste forms one or more ring-shaped heaps of small size (50cm to 2m high), in the centre of which a single bloomery furnace is located. The furnace has a circular shaft with tapered walls pierced at the base by eight openings and a door. Bloomery furnaces belonging to the Ouin tradition can be recognised by the systematic use of lozenge-shaped bricks to build the base of the superstructure, while the rest of the chimney is constructed using silica-rich slag and clay

The metallurgical assemblage is fairly similar to that of the Fiko tradition, although with smaller slag and tuyère remains.

The *Ama tradition* is located along the edge of the Bandiagara Escarpment, west of Kani-Bonzon, in the Tomo Kan linguistic zone. The spatial organisation of the waste zone is similar to that found within Ouin tradition sites. The bloomery furnaces consist of a circular pit pierced by five openings and a door. The walls are built with different kinds of material: slag, re-used tuyères and sandstone blocks. The tuyères of this tradition frequently have an extremity highly altered by heat or obstructed by a mass of slag. The metallurgical assemblage is composed of tapped slag (80%), silica-rich slag cakes (15%) and rare, large blocks of furnace slag (<5%). The morphology of tapped slag is unique because the fragments tend to be hollow slabs. This indicates the formation of a large gas bubble during the cooling phase.

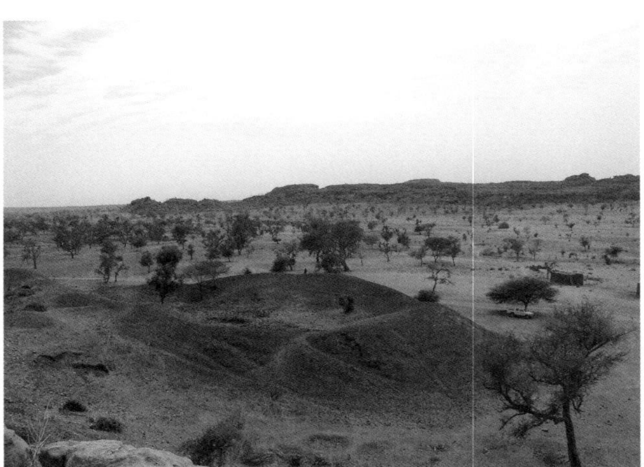

Figure 3: A metallurgical waste heap at the Kowa bloomery site.

The *Tinntam tradition* is found along the north-eastern border of the Bandiagara Plateau, in the Bondum Don linguistic zone. The spatial organisation of the sites shows a high degree of diversity. They can have one or several adjacent crater-shaped mounds often several metres high, in the centre of which one or two bloomery furnaces are located. The volume of waste is also quite variable. The distinguishing characteristics of the bloomery furnaces are to be found in the construction of the shaft walls, which are built using three different materials: an internal coating of clay mixed with broken tuyères, a central layer made with clayey silica-rich slag and clay, and an external pavement made of rectangular sandstone blocks. The waste is largely dominated by tapped slag (95%) with a slightly triangular section, giving the impression that the material flowed out of the furnace in a channel.

Bloomery sites of the *Aridyini tradition* are located along the escarpment of the southern plateau, in the Tomo Kan linguistic zone. The metallurgists discarded the waste around the narrowly spaced bloomery furnaces, forming extended but shallow dumps (1-3m high); sometimes, a network of small alleys linked the different working areas together. The bloomery furnaces are massive, with very thick walls built of slag and clay. Tapered, they stand on six feet between which there are five arch-shaped openings and a door. The metallurgical waste of the Aridinyi tradition is characterised by large furnace slag blocks that formed in the furnace pit. They can weigh up to 80kg.

To our present knowledge, the *Enndé tradition* covers three linguistic zones: Tomo Kan, Tengu Kan, and Mombo, situated respectively at the base of the escarpment, the northeast sector of the Seno Plain, and along the western margin of the plateau. These workshops are distinguished by their spatial arrangement, which consists of thinly spread isolated bloomery furnaces individually associated to very small slag heaps. The bloomery furnaces include a circular pit, a slightly conical shaft made of clay containing disjointed slag, a door and five openings for the tuyères. A pit is dug in front of the door because these bloomery furnaces are partially embedded in the ground. Slag of the Enndé tradition is composed of two equally present types: vertically flowing, imbricated internal furnace slag and silica-rich slag with irregular morphology.

While all metallurgical traditions of this area belong to a low-temperature, natural draft bloomery iron smelting process, the individual technologies show more or less substantial differences in the furnace morphologies, the slag separation process, and the type and nature of the waste products.

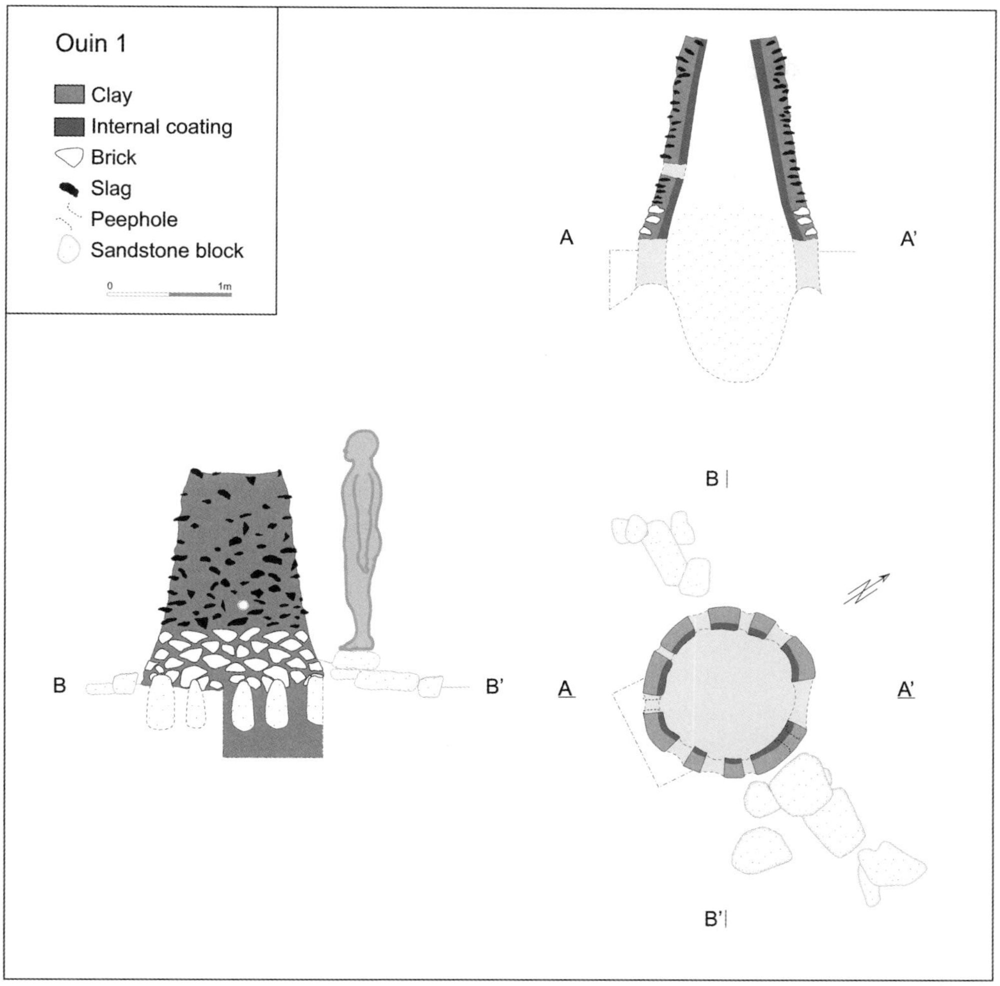

Figure 4: Plan and sections of the bloomery furnace at the Ouin 1 site.

The Fiko, Ouin, Ama and Tinntam traditions are characterised by a slag-tapping process; a substantial part of the slags from these sites was tapped and shows more or less specific morphologies. The Aridinyi smelters used slag-pit furnaces, and the slag formed large furnace slag blocks in the pit. The Enndé furnaces did not have tap holes, and the slag flowed into the lower part of the furnace, where it cooled down with characteristic, intricate vertical flow structures.

From a morphological point of view (see Table 1), the furnaces of the different traditions demonstrate a high degree of architectural diversity. In a geographically limited area, the Dogon metallurgists showed creativity in the construction of bloomery structures. All of the bloomery furnaces have a pit, mostly of circular section, and a door to remove the slag and raw iron. However, the number of openings, the presence or absence of stairs or a peephole, and the nature of the construction materials differ. In the end, each tradition has its own characteristic bloomery structures; of course, this does not exclude a degree of variability within a single tradition, in particular for those which persisted over long periods.

From an economic point of view (Table 1), the spatial organisation and the volume of slag heaps indicate high diversity in the intensity and level of production. The Fiko tradition is distinguished by enormous craters of waste, with fifteen sites, each containing a slag volume of more than 10,000m^3. Sites in the Tinntam and Aridinyi traditions generally have volumes of slag between 1,000 and 10,000m^3. The spatial organisation and waste volume for the Ouin, Ama and Enndé traditions reflect smaller scale metallurgies, with slag volumes in the order of 10 to 1,000m^3. The Enndé tradition is characterised by sites composed of very small scattered slag heaps of no more than a few m^3, in most cases far less.

Levels of metallurgical production and socio-economic systems

Obviously, these differences reflect very different economic realities. What were the markets for the products? Was the level of production meant to meet the needs of the metallurgist's own villages or the broader region, or to produce iron to be traded across more distant networks?

Considering the huge quantities of metallurgical waste, the sites of the Fiko tradition form a genuine metallurgical district in which iron was intensively produced. This near-industrial activity seems to exceed local or even regional consumption needs, and must probably be understood in the context of large-scale and long-distance trade, or of the West African empires. Located on the western margins of the Bandiagara

Table 1: Summary table of the principal characteristics of the different metallurgical traditions

	1- Technological criteria							2- Morphological criteria									3- Economic criteria											
	Method		Slag				Shaft section		Opening								Construction materials				Spatial organization					Volume of waste dumps per site (order of magnitude)		
			External			Internal				Tuyere holes			Stairs		Peep-hole						Waste dump morphology			Number of furnace/dump				
Metallurgical traditions	Superstructure	Natural draft	Normal degassing	Large gas bubble	Large furnace bottom slag	Vertical flow structure	Circular	Elliptical	Door	~ 20	8	5	No	Yes	No	Yes	Slag	Clay brick	Re-used tuyere	Sandstone block	Open dump	Closed dump	Scattered small dumps	1 furnace/dump	> 1 furnace/dump	> 10,000 m³	1,000 to 10,000 m³	10 to 1,000 m³
Fiko	X	X	X					X	X	X			X		X		X		X		X	X			X	X		
Ouin	X	X	X				X		X		X			X		X	X	X			X			X				X
Ama	X	X		X			X		X			X		X	?		X	X	X	X	X			X				X
Tinntam	X	X	X				X		X		X		X		X		X		X	X	X	X		X	X	X	X	X
Aridinyi	X	X			X		X		X			X		X	X		X				X			X		X	X	X
Enndé	X	X				X	X		X			X	?		?		X						X	X				X

Plateau, we believe that this production was destined for a market outside the Dogon Country, most presumably the Inland Niger Delta.

The Tinntam and Aridinyi traditions sites display different levels of output and duration of exploitation. Most sites seem to have been aimed at sustaining the local demand. A few sites show an intensive production for a regional market. Sites in the Arou region (Aridinyi, Yélé, Ségué) and the village of Tinntam developed rapidly in the 19th century, and, up to the 1970s, they provided raw iron and iron objects for regions located far away from the Dogon Country. For forty years, their production supplied the villages of the plateau, the escarpment, and the northern part of the Seno Plain, complementing the recycled iron coming from Europe.

The metallurgical sites of the Ama and Ouin traditions are characterised by a small-scale but regular production. This most certainly seasonal activity appears to have provided iron only to nearby villages. Production is thus oriented towards a purely local market, and is strongly tied to a village and a metallurgist family.

The spatial organisation of sites of the Enndé tradition gives the impression of a sporadic and temporary activity. Metallurgists did not settle to smelt iron ore. After having carried out one or a few smelting operations in a bloomery furnace, they abandoned it to build another one later when they needed more iron.

In order to explain such different situations, one needs to examine the impact of the production levels on the organisation of the iron metallurgy in the Dogon Country. Who controls iron production? Who are the metallurgists?

Few researchers have attempted to precisely identify the people involved in the different phases of the operational sequence of the production of iron. In most ethnographic research, it is the social and symbolic dimension of iron and the blacksmith that have been studied. Researchers have favoured the analysis of myths related to iron and smiths over technological activities. Moreover, they have often extrapolated what has been observed at the smithy to be valid for the extraction and smelting of iron ore. Thus, the blacksmith is often described as the sole actor of iron production, although he is not necessarily the only social category to master the production chain of the extractive metallurgy of iron. The confusion between blacksmith and metallurgist (smelter) leads to a poor perception of the political, economic and social organisation that governs this activity. Bruno Martinelli (1993, 2002) was among the first researchers in West Africa to be interested in the social organisation of metallurgical production. Based on his observations and fieldwork in Togo, Burkina Faso and Mali, he has identified three different techno-economic systems.

This schema (Table 2), simultaneously including technological, economic, social and political concepts, makes

Table 2: The three socio-economic systems of iron metallurgy

TECHNO-ECONOMIC SYSTEM	PRODUCTION CHAIN		
	Extraction	Smelting	Forging
Unitarian	Smith	Smith	Smith
Mixed	Smith and farmer	Smith and farmer	Smith
Dualist	Farmer	Farmer	Smith

it possible to consider iron production in a more general context, and can be applied to the Dogon traditions.

In order to determine the identity of the metallurgists, we have taken into account the status of the actors (landholders, possessors of magical and technological knowledge, labour force) for each of the principal stages of the operational sequence (extraction, smelting, smithing), and the level of production. The available ethnohistorical data yields valuable information concerning the situation in the later periods of the traditional iron production in the Dogon Country (18th century to the beginning of the 20th century).

In all metallurgical traditions, the holder of the land from which the iron ore was extracted belongs to the social category of the cultivators. Indeed, the mines are always situated within the respective village's territories, which are under the responsibility of the land masters (descendants of the village founders). The technically more complex blacksmithing activities are a domain exclusively executed and controlled by the blacksmiths, and gives them their status of endogamous craftspersons separate from the cultivators.

In the Ouin, Ama and Tinntam traditions, the unitarian system prevails.

Blacksmiths direct and execute all stages of the operational sequence of iron production. Cultivators may approach the primary production sites, but only as part of accessory labour force. This techno-economic system seems to have developed in the context of small to middle scale local or regional productions, in which the diffusion of the raw product was restricted.

The Aridinyi tradition can be assigned to the category of mixed systems (see Table 3). In effect, the Jèmè-yélin blacksmiths, founders of the workshops of this tradition, permitted cultivators and Jèmè-irin blacksmiths to work alongside them. The latter could build their own bloomery furnaces and thus smelt iron ore in exchange for a part of the bloom produced. The mixed model is included in a double trade network, both local and interregional. The distribution zone was more widespread, but does not appear to have exceeded the Dogon Country.

The Fiko tradition belongs to the dualist techno-economic system, in which cultivators direct the ore extraction operations and the iron smelting, while the blacksmiths treat the raw iron and transform it into semi-finished or finished products (see Table 3). Our ethnohistorical data suggest that dualist systems are related to higher production levels, and possibly developed from unitarian or mixed systems as the production intensified. The abundance of ore, or perhaps a geographic situation favourable for peripheral trade, may be reasons contributing to an increase in iron production. The principle of a symbiotic relationship between the producing region and the importing region leads to economic specialisation in the region where iron is produced, and increases the dissociation between the production and transformation phases of iron metallurgy.

Conclusions and perspectives

The recent palaeometallurgical research carried out within the framework of the programme 'Human populations and palaeoclimatic evolution in West Africa' has yielded new data on the traditional iron metallurgy in the Dogon Country. Through an extensive and diachronic approach, we now know much more about the level and diversity of this activity in this regional context. More than a hundred bloomery sites have been identified in a territory of around 15,000km^2.

Based on a classification taking into account technological, cultural and economic criteria, six distinct metallurgical traditions have been identified. The diversity of primary iron production in a geographically limited area, and extending over several centuries, is surprising. It suggests that technological differentiation should not be considered from a purely functional perspective, because functional factors do not satisfactorily explain the causes for the observed variability. Cultural and economic factors are essential if we want to study and understand the metallurgical production as a whole.

The diversity of the metallurgical remains is certainly also related to the complex historical context of this region, where the identity of the metallurgists is expressed through material culture. Ethnohistorical data reveal that the smith caste on the Bandiagara Plateau developed out of the local population substrate, but also shows that other persons were incorporated into this caste from other populations, which arrived in this

Table 3: The social categories of the metallurgists involved in the different phases of the production chain of iron metallurgy, and the corresponding socio-economic systems (F = farmer; S = smith)

		OUIN AMA TINNTAM		ARIDINYI		FIKO	
		F	S	F	S	F	S
Ore extraction	Landholder	X		X		X	
	Possessor of magical knowledge		X	X	X	X	
	Possessor of technological knowledge		X	X	X	X	
	Labor	X	X	X	X	X	
Smelting	Landholder		X	?	X	X	
	Possessor of magical knowledge		X	?	X	X	
	Possessor of technological knowledge		X	X	X	X	
	Labor	X	X	X	X	X	
Smithing	Landholder		X		X		X
	Possessor of magical knowledge		X		X		X
	Possessor of technological knowledge		X		X		X
	Labor		X		X		X
		Unitarian		Mixed		Dualist	

area subsequently, mostly from the Inland Niger Delta, but also from the northwest and northeast of the Gondo Plain. These individuals did not originally all belong to the social category of smiths. Following a voluntary or restrained social transformation, local cultivators, of Dogon or external origin, and slaves became iron specialists to meet economic needs.

The information collected at the bloomery sites of the Fiko tradition shows a level of production that goes well beyond local needs, or even those of the entire Dogon Country. The intensification of the production may have been driven by an increasing external demand, which could tentatively be related to the emergence of the West African Empires.

We propose however the hypothesis that this intensification could only develop in a context in which the operational sequence of metallurgy was controlled by the social category of the cultivators. While the blacksmiths owned the smithy and their tools, the manufactured products belonged to the cultivators. Factors such as the availability of labour force (blacksmiths alone cannot sustain large-scale production levels) and the control of the natural resources (land and therefore access to ore and charcoal was controlled by the cultivators) probably played an important role in these developments.

This phenomenon of mass production clearly had a very important economic and social impact, and constitutes a decisive ecological factor. It is a key objective of this research to explain this specific situation. More systematic surveys and complementary excavations are necessary to establish in more detail the history of this production district, and, in particular, the chronology must be refined. At the same time, it is indispensable to initiate research aiming to track the objects produced on the western margins of the Bandiagara Plateau in order to identify the markets to which these exports were destined.

Acknowledgments

We thank the Swiss National Science Foundation (FNS) for the financial support of the field missions during which the data used in this study were obtained. Our fieldwork would not have been possible without the support of the Cultural Mission of Bandiagara, the Institute of Human Sciences of Bamako, the University of Mali, and the Swiss Consulate in Bamako.

We would also like to thank our team in Dimmbal and particularly Ancé Tessougué for his efficient collaboration, as well as all of the Dogon blacksmiths and villagers who had the kindness to share their history.

References

Childs, S.T., 1991. Style, Technology, and Iron Smelting Furnaces in Bantu-Speaking Africa. *Journal of Anthropological Archaeology*, 10, pp.332-59.

Cline, W., 1937. *Mining and Metallurgy in Negro Africa*. General series in Anthropology 5. Menasha: George Banta Publishing.

Huysecom, E., 2002. Palaeoenvironment and Human Population in West Africa: An International Research Project in Mali. *Antiquity*, 292, pp.335-6.

Huysecom, E., Ozainne, S., Robion-Brunner, C., Mayor, A., Ballouche, A., Coulibaly, N., Guindo, N., Kéita, D., Le Drezen, Y., Lespez, L., Neumann, K., Eichhorn, B., Rasse, M., Schaer, K., Selleger, C., Serneels, V., Soriano, S., Terrier, A., Traoré, B.D. and Tribolo, C., 2007. Peuplement humain et paléoenvironnement en Afrique de l'Ouest: résultats de la neuvième année de recherches. *Jahresbericht 2006*. Zürich et Vaduz: Fondation Suisse-Liechtenstein pour les recherches archéologiques à l'étranger, pp.41-122.

Kense, F.J., 1985. The Initial Diffusion of Iron to Africa. In: R. Haaland and P. Shinnie eds. *African Iron Working: Ancient and Traditional*. Oslo: Norwegian University Press, pp.11-27.

Killick, D., 1991. The Relevance of Recent African Iron-smelting Practice to Reconstructions of Prehistoric Smelting Technology. In: P. Glumac ed. *Recent trends in archaeometallurgical research*. MASCA Research Papers in Science and Archaeology 8. Philadelphia: MASCA, pp.47-54.

Langlois, O., 2005-2006. De l'organisation bipartite du travail du fer dans les monts Mandara septentrionaux. *Techniques et cultures*, 46-47, pp.175-209.

de Maret, P., 1980. Ceux qui jouent avec le feu: la place du forgeron en Afrique Centrale. *Africa, Journal of the International African Institute*, 50, pp.263-79.

de Maret, P., 1985. The Smith's Myth and the Origin of Leadership in Central Africa. In: R. Haaland and P. Shinnie eds. *African Iron Working: Ancient and Traditional*. Oslo: Norwegian University Press, pp.73-87.

Martinelli, B., 1993. Fonderies ouest-africaines. Classement comparatif et tendances. *Techniques et cultures*, 21, pp.195-221.

Martinelli, B., 2002. Au seuil de la métallurgie intensive – Le choix de la combustion lente dans la boucle du Niger (Burkina Faso et Mali). In: H. Bocoum ed. *Aux origines de la métallurgie du fer en Afrique – une ancienneté méconnue*. Paris: Editions de l'UNESCO, pp.165-88.

McNaughton, P.R., 1993. *The Mande Blacksmith: Knowledge, Power, and Art in West Africa*. Bloomington and Indianapolis: Indiana University Press.

Perret, S. and Serneels, V., 2009. Technological Characterization and Quantification of a Large Scale Iron Smelting Site in Fiko (Dogon Plateau, Mali). In: J.-F. Moreau, R. Auger, J. Chabot and A. Herzog eds. *Proceedings of the 36th International Symposium on Archaeometry*. Cahiers d'archéologie du CELAT 25. Quebec: University of Laval, CELAT, pp.453-63.

Pole, L.-M., 1985. Furnace Design and the Smelting Operation: A Survey of Written Reports of Irin in West Africa. In: R. Haaland and P. Shinnie eds. *African Iron Working: Ancient and Traditional*. Oslo: Norwegian University Press, pp.142-63.

Robion-Brunner, C., 2008. *Vers une histoire de la production du fer sur le plateau de Bandiagara (pays dogon, Mali) durant les empires précoloniaux: peuplement des forgerons et traditions sidérurgiques*. PhD Thesis, Department of Anthropology and Ecology, University of Geneva.

Robion-Brunner, C., 2010. *Peuplements des forgerons et traditions sidérurgiques: Vers une histoire de la production du fer sur le plateau de Bandiagara (pays dogon, Mali) durant les empires précoloniaux*. Journal of African Archaeology Monograph Series 3, Peuplement humain et paléoenvironnement en Afrique de l'Ouest, Série Monographique 1. Frankfurt: Africa Magna Verlag.

Serneels, V., 2005. An Ongoing Research Project on Iron Production and Use in the Dogon Country, Mali. *Historical Metallurgy Society News*, 60, pp.1-3.

Serneels, V., Robion-Brunner, C. and Perret, S., 2006. La sidérurgie en pays dogon : problématiques, premiers résultats et perspectives. *Études Maliennes*, 65, pp.113-26.

Sutton, J.E.G., 1985. Temporal and Spatial Variability in African Iron Furnaces. In: R. Haaland and P. Shinnie eds. *African Iron Working: Ancient and Traditionnal*. Oslo: Norwegian University Press, pp.164-91.

Tylecote, R.F., 1987. *The Early History of Metallurgy in Europe*. London: Longman.

Authors' addresses

- Caroline Robion-Brunner
  Laboratoire TRACES UMR 5608
  Université de Toulouse 2 le Mirail
  Maison de la recherche - Bât. 26
  5 allée Antonio MACHADO
  31058 Toulouse Cedex 9
  France
  Courriel : caroline.robion@univ-tlse2.fr
- Vincent Serneels and Sébastien Perret
  Department of Geosciences, University of Fribourg,
  Chemin du Musée 6,
  1700 Fribourg
  Switzerland
  vincent.serneels@unifr.ch
  sebastien.perret@unifr.ch

Epistemic communities and epistemological translocations in early modern southern Africa, ca. CE 200-900

Edwin N Wilmsen

ABSTRACT: Philosophers of material value are engaged to outline an archaeoethnography of iron-using peoples in southern Africa during the first millennium CE. I first draw on Locke and Marx-Engels to revisit the proposition that human history and sociality are inextricably bound up with materials. I then turn to Simmel, who found that the value of an object is not an inherent property of that object but reflects a judgment made by persons. Modifying Haas' concept of an epistemic community, I examine the means by which the intellectual property core of such a community could be widely disseminated while its members retained exclusive ownership of that knowledge core. From this perspective it becomes clear that processes of translocation in regimes of knowledge rather than simply movements of people and things were taking place at different levels of socio-political organisation to transform the social formation of Early Iron Age southern Africa.

Keywords: Southern Africa, Iron Age, metallurgy, exchange value.

Of persons and things

At first blush, the term *early modern* in juxtaposition to southern Africa in an Early Iron Age context will seem incongruous. Let me explain. This essay, like its companion (Wilmsen, 2009), is an exercise in epistemology rather than a finished thesis. In it, I engage a social geography in order to map the processes underlying the introduction of iron, and copper, metallurgy into interior southern Africa during the Early Iron Age (ca. CE 200-900 in the chronology of Huffman, 2007) and its subsequent maturation in the Middle Iron Age (ca. CE 900-1300). I borrow the term *early modern* from the Brenner Debate, begun in the 1970s and continuing today, in which historians contemplate the transition of modern Europe from an early modern phase. Although there are bipolar disagreements, all parties to this debate concur that the roots of this transition are to be found in transformations of principally agrarian economies through incipient industrialisation to 19th century monetary capitalism. As Hilton (1985, p.1) wrote of this debate, it is "of intense interest ... to all concerned with the causes behind transitions between successive social formations". The transition from stone-based to iron-based technologies in first millennium southern Africa brought about transformations in the social formation of the subcontinent just as far-reaching as those which occurred in early modern Europe (see Childs and Killick, 1993; Miller and van der Merwe, 1994; and papers in Schmidt, 1996 and Vogel, 2000 for syntheses). Thus, it seems to me that pre-colonial southern Africa has an analogous history of transformation from principally foraging economies, with the introduction of metallurgy serving the role of incipient industrialisation culminating in the Zimbabwean states of the Late Iron Age (ca. CE 1300-1700), and further development to fully modern economies truncated by colonial intervention. The analogy can be stretched too far, but I employ it only to stress that an integral trajectory cuts through the compartments of Early-Middle-Late Iron Age, adding another dimension essential for a comprehensive appreciation of southern African history. This is an orientation focused on social actors, networks, and multiple regulatory constellations that operated not only at a local level, but also at regional and, as we shall see, intercontinental levels.

To trace this trajectory, I focus not on artefacts, things, or even sites as such, but on the social geography in which they were situated. I review ideas of several philosophers of materials and material value to outline an archaeoethnography of iron-using communities in southern Africa during the first millennium CE. In doing so, I heed Gupta and Ferguson's (1997, p.6) critique of spatially territorialised notions of culture and Giddens' (1979, p.202) observation that most forms of social theory fail to take seriously not only the temporality of social conduct, but also its spatial attributes. Annette Weiner, in the early 1990s, turned to the founding text on which the materiality of social life was established; reading John Locke, she found that for him "the boundaries between persons and things are not rigidly separate" (Weiner, 1992, pp.34–37). The basis for this finding is Locke's (1980:114 [1690], original primes) thesis: "yet every man has a 'property' in his own 'person' annexed [through] the 'labour' of his body and the 'work' of his hands." Reading further, we find that Locke enumerated material as well as abstract property that a person annexes as one's own: material property is "whatsoever [a person] removes out of the state that Nature hath provided and left it in, he hath mixed his labour with it". Tragically, Weiner died before elaborating this insight.

This leads, however, to another proposition, one that I elaborate more fully in the essay alluded to above (Wilmsen, 2009). This is that material artefacts have the same ontological status as words, and, marked by distinct intentions of their makers and users, are potentially as comprehensible as verbal documents; thus, they have communicative as well as utilitarian functions (Wilmsen, 1994). This perception derives ultimately from Locke, followed by Marx and Engels (1947 [1846], p.16) who, in *The German Ideology*, wrote:

> The first historical act is thus the production of the means to satisfy [human] needs, the production of material life itself. And indeed this is an historical act, a fundamental condition of all history, which today, as thousands of years ago, must daily and hourly be fulfilled.

Central to Marxian thought is the proposition that human social relations are not primordial but are constructed by human agents through their labour, that human history and sociality are inextricably bound up with materials. Note carefully the convergence with Locke.

Another seminal theorist, non-Marxian, writing at around the turn of the 20th century, articulated the founding premises upon which this proposition could be made operative in the investigation of what Appadurai (1986) somewhat poetically has called 'the social life of things'. Georg Simmel noted in *The Philosophy of Money* (1978 [1907], p.73) that value is not an inherent property of objects, but is a judgment made by subjects. In *The Philosophy of Value* (Simmel, 1900:82), he had previously stressed that in any instance of exchange,

> While the object itself is the thing in controversy – which means that the sacrifice [i.e., value] which it represents is being determined – its significance for both contracting parties appears much more as something outside of these parties and of the self-existent object than if the individual [acquirer or purveyor] thought of it only in its relation to himself.

What Simmel is saying here is that the value each party to an exchange places on an object is mediated by transcendent temporal, cultural, and social caveats; that is, to paraphrase Friedman (1994), it is not things that have social life, social life has things. For Simmel, then, as for Locke and Marx-Engels, the boundaries between persons and things are not rigidly separate (Simmel 1978 [1907], pp.97-98). Simmel was interested primarily in the analysis of money, for him the most complex instrument of exchange; however, it is clear that his observations (particularly in *The Philosophy of Value*, as quoted above) may be applied to any conditions under which objects circulate in regimes of value in space and time, not only those specific to that of monetary capitalism. Together, Locke, Marx and Engels, and Simmel provide a powerful platform from which to address the role of materials in fashioning and fostering social order.

Social order and property

I have presented this extended introduction so that I may fix your attention on metallurgists, rather than on metallurgy as such or on metal itself, for, as Knapp (1998) has remarked, too little attention has been paid to them. Iron (and copper), implements, and jewellery were demonstrably passed from place to place in all phases of the Iron Age, but metallurgists created those objects and would have had at least some control over the initial stages of their distribution. More importantly, it was not metal objects that were the most significant elements in exchange associations, rather it was the dissemination of the knowledge required to reproduce those objects that was of greatest value.

More than a decade ago, Miller (1996, p.91) pointed out that an elucidation of African iron working depends crucially on "an understanding of the social and ideological aspects of indigenous technology". Subsequently, Dietler (1999, p.194) called for archaeologists to develop a firmer grasp of the anthropological principles needed to "consider the intervening practices by which people actually negotiate, reproduce, and contest social order". Recently, Killick (2009, p.181) again urges an expanded archaeological "study of the strategies of individuals, groups and institutions within societies, and of how structure emerges from their complex interplay". Human social order is, of course, the result of multiple outcomes of human agency, an agency that lies at the interstices between systemic conditions, knowledge, and social action; individuals deploy materials as components in strategies to realise some aspect of that order.

To see how this may be applied to Iron Age persons and materials, I adopt the concept of an epistemic community as articulated by Haas (1992) and subsequently applied by Wiber (2005) to questions pertaining to modern Canadian fisheries. An epistemic community is a network of persons with recognised competence in a particular domain, situated in various locations, who share and disseminate a particular form of knowledge; these locations are typically both geographically near and far apart. Haas was interested in the ways in which coteries of academics and others with expertise in global warming, nuclear waste, etc. contribute to policy formation by governments today. While these specific capabilities were not available to Iron Age intellectuals, the concept of an epistemic community is well-suited to examine the role that networks of knowledge-based specialists play in disseminating and activating material innovations, such as metallurgy. According to Haas (1992, p.3), the members of an epistemic community share a set of principles, an episteme, in common: 1. normative beliefs which provide a value-based rationale for social actions; 2. causal beliefs derived from analysis of practices in their domain which serve as the basis for elucidating the multiple linkages between possible actions and desired outcomes; 3. notions of intersubjective, internally defined criteria for validating knowledge in their domain of expertise; and 4. common practices associated with the problems to which their professional competence is directed.

For Early Iron Age (EIA) metallurgists, normative beliefs regarding social actions would have entailed, inter alia, observance of ritual requirements for successful smelts and proper behaviour by smelters and smiths for avoiding pollution; causal beliefs would have centred around their understanding of the physics of heat and ductility. For them, criteria for validating knowledge and associated practices would have stemmed from their understanding of geology and mineralogy.

Today, this set of understandings is thought of as intellectual property. On the premise that control over knowledge and information is an important dimension of power and that the diffusion of ideas can generate transformations in social formations, the means of acquisition and transmission of this intellectual property must be examined.

To do this, we must first place intellectual property in the broader structure of pre-colonial southern African property relations in general. This is important because, as will be made clear, different forms of property were acquired and transmitted differently. With regard to real property, Gluckman (1971, pp.45-46) offers guidance:

> Property law in tribal societies defines not so much rights of persons over things, as obligations owed between persons with respect of things ... The crucial rights of such persons are demands on other persons in virtue of control over land and chattels, not any set of persons, but persons related in specific, longstanding ways ... To understand the holding of property, we must investigate the system of status relationships.

Ownership constrained in this manner cannot be absolute, because property acquires its critical role in a specific nexus of social relationships. Under these circumstances, there can be no definition of ownership in the sense of incontestable control, such as inheres in modern capitalist conceptions of property. Rather, being in a property relationship involves being bound within a set of reciprocal obligations among persons and things; land, and rights to its use, must be subject to a complex of claims arising from this nexus (Schapera 1938, pp.195-238; Gluckman, 1965, p.163). More to the point, it is not land itself that is inherited. What is actually inherited is a set of status positions binding an individual to a network of obligations owed between persons, among which are those with respect to land. It is through this network that persons become associated with geographic space. Other obligations in different polities may include adherence to descent group food avoidance prescriptions, performance of duties to ancestors, conserving corporate (as distinct from personal) herds, and other imperatives for safeguarding group coherence in geographic space. Entitlement seems a better gloss than ownership for such a notion of relations to property, and I have used this term elsewhere (Wilmsen, 1989, pp.158-194). Entitlements of this kind are simultaneously material and ideological and are inalienable. Iron Age metallurgists would have been just as bound to the obligations entailed in their entitlements as any other member of their communities. This must not be confused with status; that metallurgists are known to have been accorded a range of high to low statuses in their respective 19th-20th century communities (Childs and Killick, 1993) obviates neither their entitlements nor their obligations. Even a 'Chief's life is not merely one of privilege. He has many duties to perform for the tribe' (Schapera, 1938, p.68).

The important thing to note is that - despite significant differences in specific kinship-marriage arrangements and in status hierarchies – there is an equally fundamental structural commonality underlying the southern African systems of land tenure considered here (Wilmsen, 1989, pp.158-194): bilateral ascription of entitlements and obligations by birth into a set of corporate, tenure-holding groups; cousin marriage as a mechanism for defining and restricting the extent of these groups; ritualised reciprocal inter-generational exchanges; preferential birth within tenurial bounds of the mother's family; reinforcement of entitlement through marriage or adoption into a group; reciprocal obligations among members; inalienability of entitlements so long as membership of the kin-affine set is maintained; and inalienability of land and associated entitlements. In short, the right possessed natally by anyone to participate in the assets and affairs of any one group is the result of a series of parental natal accumulations and of parental maintenances or losses; continued possession of that right is contingent on the exercise of it. These ascribed and acquired kinship and affinal relations allow tenurial relations to land to remain intact when the geographical space actually occupied is changed. Such changes in occupied space would have been common during the expansion of Early Iron Age societies, and linked institutions for translocations of tenure relations correlating society with the physical environment must have accompanied these moves (cf. Bohannon, 1957, pp.54-55; von Benda-Beckmann, et al., 2009).

Many southern African polities are recorded to have had variants of this form of real property relations: Barotse, from whom Gluckman derived his insights (Gluckman, 1965); Batswana (Schapera, 1938, 1941, 1943; Comaroff and Comaroff, 1981; Comaroff and Roberts, 1981; Krige, 1981; Ramsay, 1991; Griffiths, 1997) and other Sotho-speaking peoples (Kuper, 1975, 1982); Ovaherero (Luttig, 1933; Vedder, 1938) and Ovambanderu (Gibson, 1956); and the Kavango peoples (McGurk and Gibson 1981). These are all Bantu-speaking agro-pastoralists, but Schapera (1943, pp.5-7) speaks of "this system of land tenure characteristic of [Khoisan-speaking foragers]" in exactly this way. Barnard (1988, p.37) highlights the fact that "the deep structure of all the Khoe [Central Khoisan and Nama] kinship systems is essentially the same". This aspect is confirmed by cognate terms for locational place and hereditary overseer in six San languages, including non-Khoe Zhu (Wilmsen, 1989), plus Nama (Hoernlé, 1925; Vedder, 1966; Carstens, 1983), which are derived from common roots referring respectively to a person's or a group's possessed country and to leadership. Thus, it would appear that before direct colonial intervention, many southern African tenurial systems, pastro-foraging Khoisan as well as agro-pastoral Bantu, had as much in common as they now superficially appear to lack. That this should be so is not really surprising, given the millennia-long history of associations of peoples whose descendants are now seen as different tribes. We have the testimony of Tshekedi Khama, Ngwato kgosi in the 1920s-1930s, that in the early 18th century there was little difference between San and Tswana social-political praxis and their internal economic relations, at least as was discernable to the members of those entities at the time (Tagart, 1935). The pervasiveness of these systems along with deep-rooted lexicons associated with their entitlements allows us to see their recorded manifestations as derived from Iron Age prototypes in a manner analogous to the derivation of modern languages from ancestral proto-languages (but for cautions regarding such derivations see Heine and König, 2008).

Personal possessions, on the other hand, are private property. Material possessions – a person's house, tools, adornments, livestock, self-obtained food, and so on – are acquired from various sources, are privately owned, and may be disposed of as the owner may wish. Normally, those of these materials acquired through entitlements, especially food, are shared with or devolve upon lineal or collateral relatives and affines, but an owner may choose to risk conflict if a wider dissemination seems advantageous. Disposal of property acquired through other means, for example trade, bears the risk of only personal dispute.

Crucially for this discussion, esoteric abilities such as divination and curing, as well as learned skills such as woodworking and pottery-making, and in the past metalworking, are also considered to be private property, but in a sense related to what is currently thought of as intellectual property. Among Zhu, for instance, a curer, especially one perceived to be particularly effective, is referred to as 'the owner of medicine', and Willis (1999) found that among Ulundu 'some spirits heal, others only dance', that is, some curers possess the esoteric knowledge needed for curing, others only mimic it. These abilities and skills may be passed on to lineal and collateral juniors, but they may equally be taught to any person who wishes to acquire them. For example, in a recent study of Botswana potters, of all women making traditional hand-built wares it was found that a third had learned their craft from grandmothers and another third from mothers; these women were from families with five known generations of potters (more-or-less the limit of non-recorded memory), but the remaining third learned the craft from unrelated potters, not necessarily from the same village (Thebe, et al., 2009). Again, all rural pastoral Herero and Mbanderu households keep a set of hand-carved wooden pails in which to store the milk from different categories of cows (maternal descent, paternal descant, personal), but very few households make their own pails. Rather, the pails are custom-carved by itinerant Herero and Mbanderu men, each of whom is noted for an individual aesthetic applied in his craft; these men pass their skills on to novices, who may or may not be from their kin cohort.

There is some historic evidence that craft itinerancy has significant time depth, and that metallurgists were among the itinerants. Galton (1971 [1853], pp.178-179) describes in detail one of four yearly Ovambo trading caravans which he met in what is now northern Namibia in 1851:

> their necks were laden with necklaces for sale, and every man carried a long smoothed pole over his shoulder, from either end of which hung a quantity of packages. These were chiefly little baskets holding iron articles of exchange, packets of corn for their own eating, and water bags.

Green (1857) portrays similar Herero caravans, and Hahn (Petermann, 1859) documents the scope of indigenous trade in ores, although in this case copper rather than iron, as well as, as it were, trade secrets:

> At the lowest estimate that I can make, 50 to 60 tons of copper ore [from Tsumeb] must go yearly to Ondonga [to Ovambo]. The Bushmen [Heixum] are so jealous of this trade, that to this day they have not allowed strangers, not even people of Ondonga, to see the places where they dig.

These accounts from the first Europeans to report on the region document the wide-ranging scope of an already mature trade network in the mid-19th century Angola-Botswana-Namibia border region, embracing peoples from several language groups and different economic pursuits. One must assume that these networks had evolved over a very long time. The probable diversity of ore sources from which both iron and copper objects were made at Divuyu, an EIA site, and Nqoma, a stratified EIA to historic period site, in the Tsodilo Hills of Botswana (Miller, 1996, 2002), strongly implies that congruent systems were already in place in the Early Iron Age. Iron Age metallurgists may thus be seen as prototypical ancestors of their 19th century and more recent counterparts.

Communities, communication, and continuities

It is now well established that a bloomery smelting process producing an inhomogeneous iron was typical of African indigenous iron production, although in an astonishing diversity of technologies (Killick, 1991), and that the fabrication sequence followed to form iron articles was essentially the same at locations separated by more than 1000km (Miller, 1996, p.93). Furthermore, iron pendants have been found in EIA contexts at Divuyu and Nqoma that are stylistically comparable to contemporary Kisalian materials in central Zaire, some 1500 km to the north (van Noten, 1982, Fig. 31), and others to Broederstroom in South Africa (Miller, 1996, p.92). That is, a fairly standardised production and design ethos spanned an area of at least 2500km north-south and an as yet undetermined extent east-west during the Early Iron Age.

Ores present a more complex picture; articles at a single site were often made from metals obtained from widely separated sources, as indicated by varying elemental compositions (Miller, 1996, 2002). In some circumstances, this variation would require modifications to furnace conditions in order to produce a uniform product. It is also the case that there is little evidence for smelting at settlement sites or in their near vicinity, except where ore bodies are close by, as at Phalaborwa. Most EIA settlement sites are located tens to hundreds of kilometres away from ore sources. Even when a settlement is near a source - as are the Xaro sites in the Okavango Panhandle, just 10 km from the large Shakawe iron mine, and the Tsodilo sites, 45 km away – there is usually no clear evidence for smelting. While it is unlikely that heavy ores were routinely transported very far from their source, the Tsumeb-to-Ondonga case, spanning a distance of some 200 kilometres, shows that this was sometimes done.

It is more likely that, as Pigott (1998) suggests for Thailand, autonomous producers made many journeys to mines to supply their needs. Similarly, in South Africa people from several language groups and polities are recorded to have independently mined specularite, for cosmetic rather than metallurgical use, at mines such as Doornfontein (Beaumont and Boshier, 1974) and Blinkklipkop (Thackeray, et al., 1983). At Tsodilo, pottery style diversity implies open-access rather than monopolistic mining of specularite beginning before the EIA (Wilmsen, 2011; Wilmsen, et al., this volume). Most settlement sites show at least suggestive evidence for en loco smithing. At Nqoma there is direct evidence in the form of a pair of bell-shaped tuyères preserved in a burned MIA house (Fig. 1); their shape and location in a house rather than a furnace suggests they were made for smithing rather than smelting. This implies that resident smiths were active in these settlements, and each would either make periodic supply trips to mines or engage a colleague to share stock. In either case, opportunities for communication and exchange of ideas would be present. Miller (2002) is of the opinion that based on presently available evidence it is not possible to say whether EIA smelting was or was not carried out in ritual seclusion. Huffman (1993, citing Cline, 1937), on the other hand, argues that because many Bantu-speaking peoples associated smelting with human procreation, ritual prohibitions surrounded smelting and required a site outside settlements, while smithing was associated with kraals inside settlements. In any case, metallurgical knowledge would have been essential to success. Only smiths had this knowledge, and in this they formed an epistemic community, a network of knowledge-based specialists, whose activities were vital to the normal functioning of their society.

This knowledge and the ability to employ it may have entailed other commissions. Along with the tuyères on the floor of the burned Nqoma house were iron and copper jewellery, Zhizo type glass beads, a cowrie shell of Indo-Pacific provenance, ivory ornaments, stone tools and chipping debris, as well as a mass of charred sorghum seeds. Robertshaw and his associates (Robertshaw, et al., 2003; see also Wood, 2011) have shown that Zhizo beads were probably manufactured in South Asia and/or the Middle East from Iranian glass. This compels us to conclude that interior southern Africa was at that time part of an early form of 'globalisation' encompassing minimally Indo-Asia and the Middle East; in such a system, Killick (2009) argues that knowledge is a commodity exchanged as are material goods. EIA metallurgists, notionally members of an epistemic as well as of corporeal communities, may have been agents of transmission in this system, as historic accounts suggest for later times. Optical petrographic analysis of shards from 25 Iron Age sites and 74 geological clay deposits carried out by David Killick, Dana Drake Rosenstein, and me (Wilmsen, et al., 2009) allows us to trace the internal pathways along which basic commodities and exotic goods of sumptuary distinction flowed. At Nqoma during the EIA (ca. CE 200-900), the major links were to the Middle Zambezi region, with suggestions that this region was further connected to the northeast; these connections continued through the MIA (ca. CE 900-1300). This is the probable route by which Zhizo beads and marine shells reached Nqoma from the East African coast some 1500-2000km distant.

Miller (1996, p.90) found the remarkable continuity of metal artefact design intriguing in light of the differences in ceramic style and technology at different sites. The advent of the Iron Age in southern Africa saw conditions of accelerating mobility of people, communication, knowledge, technology, and indeed of the form of capital required to support this system. A set of processes was set in train that generated a multiplicity of linkages and interconnections transcending the regions and societies that made up the Iron Age world. We should, thus, not assume that iron objects, pottery vessels, and sumptuary goods circulated in a purely functional capacity. Rather, their materiality was invested with meaning through social interaction and accumulated histories linking persons and families across class and identity distinctions in space and time. From this perspective, it becomes clear that processes of epistemological translocation in regimes of knowledge rather than simply movements of people and things were taking place at different levels of socio-political organisation in early modern southern Africa. Metallurgical epistemic communities whose members were able to communicate intellectual assets free of prohibitions placed on real property and constraints on entitled private property would seem to have been necessary for the dissemination of smelting and smithing technologies among language clusters and across political borders.

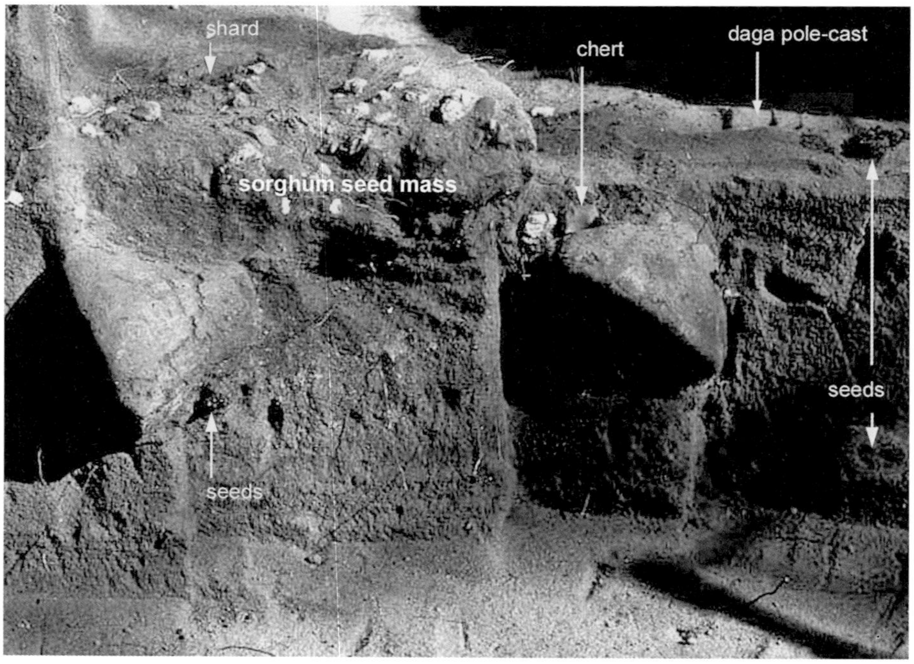

Figure 1: A pair of bell-shaped tuyères preserved in a burned MIA house at Nqoma.

Acknowledgements

I thank Anne Griffiths, David Killick, Duncan Miller, and Dana Drake Rosenstein for substantive comments; I also thank Anne for the notion of translocations. The pottery analysis was supported by Wenner-Gren Foundation Grant 7509.

References

Appadurai, A., 1986. Introduction. In: A. Appadurai ed. *The social life of things: Commodities in cultural perspective*. Cambridge: Cambridge University Press, pp.3-63.

Barnard, A., 1988. Kinship, language, and production: a conjectural history of Khoisan social structure. *Africa*, 58, pp.29-50.

Beaumont, P. and Boshier, A.K., 1974. Report on test excavations in a prehistoric pigment mine near Postmasberg, Northern Cape. *South African Archaeological Bulletin*, 29, pp.49-59.

von Benda-Beckmann, F., von Benda-Beckmann, K. and Griffiths, A., 2009. Space and legal pluralism: an introduction. In F. von Benda-Beckmann, K. von Benda-Beckmann and A. Griffiths eds. *Spatializing law: an anthropological geography of law and society*. Aldershot: Ashgate, pp.1-29.

Bohannon, P., 1957. *Justice and judgement among the Tiv*. Oxford: Oxford University Press.

Carstens, P., 1983. The inheritance of private property among the Nama of southern Africa revisited. *Africa*, 53, pp.58-70.

Childs, T. and Killick, D., 1993. Indigenous African metallurgy: nature and culture. *Annual Review of Anthropology*, 22, pp.317-37.

Cline, W., 1937, *Mining and metallurgy in Negro Africa*. General Series in Anthropology 5. Menasha: George Banta.

Comaroff, J. and Comaroff, J., 1981. The management of marriage in a Tswana context. In: E. Krige and J. Comaroff eds. *Essays on African marriage in southern Africa*. Cape Town: Juta, pp.29-49.

Comaroff, J. and Roberts, S., 1981. *Rules and processes: the cultural logic of dispute in an African context*. Chicago: University of Chicago Press.

Dietler, M., 1999. Review of C. Kramer, Pottery in Rajasthan: ethnoarchaeology in two Indian cities. *American Anthropologist*, 101, pp.194-5.

Friedman, J., 1994. *Cultural identity and global process*. London and Thousand Oaks: Sage.

Galton, F., 1971 [1853]. *The Narrative of an explorer in tropical South Africa*. New York: Johnson Reprint.

Gibson, G., 1956. Double descent and its correlates among the Herero of Ngamiland. *American Anthropologist*, 58, pp.109-39.

Giddens, A., 1979. *Central problems in sociology: action, structure, and contradictions in social analysis*. London: Macmillan.

Gluckman, M., 1965. *The ideas of Barotse jurisprudence*. New Haven: Yale University Press.

Gluckman, M., 1971. *Politics, law, and ritual in tribal society*. New York: Basil Blackwell.

Green, F., 1857. Narratives of an exploration to the northwest of Lake Ngami. *Eastern Province Monthly Magazine*, 1, p.537.

Griffiths, A., 1997. *In the shadow of marriage: gender and justice in an African community*. Chicago: University of Chicago Press.

Gupta, A. and Ferguson, J., 1997. *Culture, power, place: explorations in critical anthropology*. Durham: Duke University Press.

Haas, P., 1992. Introduction: epistemic communities and international policy coordination. *International Organization*, 46, pp.1-35.

Heine, B. and König, C., 2008. What can linguistics tell us about early-Khoekhoe history? *Southern African Humanities*, 20, pp.235-48.

Hilton, R., 1985. Introduction. In: T.H. Aston and C.H.E. Philpin eds. *The Brenner debate: agrarian class structure and economic development in pre-industrial Europe*. Cambridge: Cambridge University Press, pp.1-7.

Hoernlé, W., 1925. The social organization of the Nama Hottentots in Southwest Africa. *American Anthropologist*, 27, pp.1-24.

Huffman, T., 1993. Broederstroom and the Central Cattle Pattern. *South African Journal of Science*, 89, pp.220-6.

Huffman, T., 2007. *Handbook to the Iron Age: the archaeology of pre-colonial farming societies in southern Africa*. Scottsville: University of KwaZulu-Natal Press.

Killick, D., 1991. The relevance of recent African iron-smelting practice to reconstructions of prehistoric smelting technology. In: P. Glumac ed. *Recent trends in archaeometallurgical research*. MASCA Research Papers in Science and Archaeology 8. Philadelphia: MASCA, pp.47-54.

Killick, D., 2009. Agency, dependency and long-distance trade: East Africa and the Islamic world, ca. 700–1500 C.E. In: S. Falconer and C. Redman eds. *Polities and power: archaeological perspectives on the landscapes of early states*. Tucson: University of Arizona Press, pp.179–207.

Knapp, A.B., 1998. Introduction. In A.B. Knapp, V. Pigott and E. Herbert eds. *Social approaches to an industrial past: the archaeology and anthropology of mining*. London: Routledge, pp.1-11.

Krige, E., 1981. A comparative analysis of marriage and social structure among the Southern Bantu. In: E. Krige and J. Comaroff eds. *Essays in African marriage in southern Africa*. Cape Town: Juta, pp.1-28.

Kuper, A., 1975. The social structure of the Sotho-speaking peoples of Southern Africa. *Africa*, 45(1), pp.67-81; 45(2), pp.139-49.

Kuper, A., 1982. *Wives for cattle: bridewealth and marriage in Southern Africa*. London: Routledge and Kegan Paul.

Locke, J., 1980 [1690]. *Second treatise of government*: Chapter 5, *Of property*, section 27, C.B. Macpherson ed. Indianapolis: Hackett.

Luttig, H., 1933. *The religious system and social organization of the Herero*. Utrecht: Kermink.

Marx, K. and Engels, F., 1947 [1846]. *The German ideology*. New York: International Publishers.

McGurk, C. and Gibson, G. eds., 1981. *The Ethnography of southwestern Angola: The Kavango peoples, by Carlos Estermann*. New York: Africana Publishing.

Miller, D., 1996. *The Tsodilo jewellery: metal work from northern Botswana*. Cape Town: UCT Press.

Miller, D., 2002. Smelter and smith: metal fabrication technology in the southern African Early and Late Iron Age. *Journal of Archaeological Science*, 29, pp.1083-131.

Miller, D. and van der Merwe, N., 1994. Early metal working in Sub-Saharan Africa: a review of recent research. *Journal of African History*, 35, pp.1-36.

van Noten, F.L., 1982. *The archaeology of Central Africa*. Graz: Akademischer Druck-u. Verlagsanstalt.

Petermann, A., 1859. Reise der Herrn Hugo Hahn und Rath im südwestlichen Afrika, Mai bis September 1857. *Petermann's Geographische Mitteilungen*, 7, pp.295-303.

Pigott, V., 1998. Prehistoric copper mining in the context of emerging community craft specialization in northeast Thailand. In: A.B. Knapp, V. Pigott and E. Herbert eds. *Social approaches to an industrial past: the archaeology and anthropology of mining*. London: Routledge, pp.205-25.

Ramsay, J., 1991. *The rise and fall of the Bakwena dynasty of south central Botswana*. PhD Thesis, Boston University.

Robertshaw, P., Glascock, M., Wood, M. and Popelka, R., 2003. Chemical analysis of ancient African glass beads: a very preliminary report. *Journal of African Archaeology*, 1, pp.139-46.

Schapera, I., 1938. *Handbook of Tswana law and custom*. Oxford: Oxford University Press.

Schapera, I., 1941. *Married life in an African tribe*. New York: Sheridan House.

Schapera, I., 1943. *Native land tenure in the Bechuanaland Protectorate*. Cape Town: Lovedale Press.

Schmidt, P., 1996. *Culture and technology of African iron production.* Gainesville: University Press of Florida.
Simmel, G., 1900. *Philosophie des Wert.* In: *Gesamtausgabe*, Bd. 6 *Philosophie des Geld (1900-1907).* Frankfurt: Suhrkamp.
Simmel, G., 1978 [1907]. *The philosophy of money*, T. Bottomore trans. London: Routledge and Kegan Paul.
Tagart, E., 1935. *Report on the conditions existing among Masarwa in Bamangwato Reserve of the Bechuanland Protectorate and certain other matters pertaining to the natives living therein.* Pretoria: Government Printer.
Thackeray, A.I., Thackeray, J.F. and Beaumont, P., 1983. Excavations at the Blinkklipkop specularite mine near Postmasberg, Northern Cape. *South African Archaeological Bulletin*, 38, pp.17-25.
Thebe, P., Wilmsen, E., Killick, D. and Drake Rosenstein, D., 2009. Mmopi le mmopa: making pottery in Botswana today and c. 1000 years ago. *Botswana Notes and Records*, 41, pp.25-38.
Vedder, H., 1938. *South West Africa in early times.* London: Frank Cass.
Vedder, H., 1966. The Nama. In: C.H. Hahn, H. Vedder and L. Fourie eds. *The native tribes of Southwest Africa.* London: Frank Cass, pp.107-52.
Vogel, J., 2000. *Ancient African metallurgy: the socio-cultural context.* Walnut Creek: Altamira.
Weiner, A., 1992. *Inalienable possessions: the paradox of keeping while giving.* Los Angeles: University of California Press.
Wiber, M., 2005. Mobile law and globalism: epistemic communities versus community-based innovation in the fisheries sector. In F. von Benda-Beckmann, K. von Benda-Beckmann and A. Griffiths eds. *Mobile people, mobile law: Expanding legal relations in a contracting world.* Aldershot: Ashgate, pp.75-94.
Willis, R., 1999. *Some spirits heal, others only dance: a journey into human selfhood in an African village.* Oxford: Berg.
Wilmsen, E., 1989. *Land filled with flies: A political economy of the Kalahari.* Chicago: University of Chicago Press.
Wilmsen, E., 1994. *Pots and power in the pastoral economy of pre-colonial South Africa.* Research paper, Institute for Advanced Study and Research in the African Humanities, Northwestern University.
Wilmsen, E., 2009. The regulation of commodity exchange in southern Africa during the eighth to fifteenth centuries CE. In: F. von Benda-Beckmann, K. von Benda-Beckmann and A. Griffiths eds. *Spatializing law: an anthropological geography of law and society.* Aldershot: Ashgate, pp.177-94.
Wilmsen, E., 2011. Nqoma: an abridged review. *Botswana Notes and Records*, 43:95-114.
Wilmsen, E., Killick, D., Drake Rosenstein, D., Thebe, P. and Denbow, J., 2009. The social geography of pottery in Botswana as reconstructed by optical petrography. *Journal of African Archaeology*, 7, pp.3-39.
Wood, M., 2011. Interconnections: glass beads and trade in southern and eastern Africa and the Indian Ocean, 7th to 16th centuries A.D. Studies in Global Archaeology 17. Uppsala: Uppsala University.

Author's addresses

- School of Geography, Archaeology, and Environmental Sciences, University of the Witwatersrand
  Centre of African Studies, Edinburgh University, 15a George Square, Edinburgh EH89LD, UK
  and
  Department of Anthropology, University of Texas, Austin TX 78712, USA
  anaw630@uts.cc.utexas.edu

Superstition or ingeniousness: re-thinking symbolism in indigenous African iron technology

Bertram B.B. Mapunda

ABSTRACT: It is common knowledge that every community that produced iron in Africa deployed varied forms of techno-medicines and incorporated various taboos as indispensable ingredients for both smelting and forging processes. With this knowledge, archaeometallurgists and other students of Africa's past have differed greatly on their approach to past technological functions and the significances of these symbolic expressions. There are those who believe that techno-medicines and taboos did not have any technological importance; their deployment simply reflecting the users' propensity for superstition. But there are also others who oppose this viewpoint and argue that these practices had some functional benefits. In this paper the two standpoints are re-examined with the help of ethnoarchaeological evidence from sub-Saharan Africa, particularly Tanzania. That the deployment of techno-medicines and taboos in African indigenous ironworking was carried out for scientific and technological reasons is affirmed. Those who argue otherwise seem to be using criteria alien and prejudicial to the African worldview and system of reasoning.

Introduction

In sub-Saharan Africa, ironworking is generally accepted to have started about 3-2.5 millennia ago (Alpern, 2005; Pringle, 2009; Clist, this volume). Since then, ironworking expanded rapidly both in space and technological complexity, so that by the mid-nineteenth century AD, when European imperialists colonised Africa, almost every community in the region knew about iron and a good number of them had mastered its production (Childs and Herbert, 2005; Chirikure, 2005; Mapunda, 2010). Reports from the early European visitors, including missionaries, traders and explorers, also show that in some places the product was of impressive quality, sometimes better than what was being produced within their respective countries in Europe (Kjekshus, 1996; Chirikure, 2005). This drew the attention of the visitors, who viewed it as a spectacle (Fagan, 1961), considering the fact that they had come with pre-conceived prejudices that Africa was a dark and backward continent. Hence, they had not expected to find ironworking regularly practiced in Africa, a place of the savages, given that this technology was, at that time, considered to symbolise civilisation (Morgan, 1877). That is to say, sub-Saharan Africans were, in the view of Europeans, savages, yet the level of technological sophistication and the quality of its end products did not belong to savagery but rather to the civilised societies of Europe and Asia. On account of this there were people who believed that sub-Saharan Africa was the cradle of iron technology, as Wainwright once noted: "A theory [had] been strongly held in some quarters that the art of iron-working originated in Central Africa and spread thence over the world" (Wainwright, 1942, p.103). This echoed Robert Forbes' remark made earlier on, that "the native African negro, far from being the stark savage of common imagination, is the possessor of varied though crude cultures to which in some respects civilized people are deeply indebted" (Forbes, 1933, p.230). One of these cultures expounded in his paper is iron technology (see Alpern, 2005 for additional comments on this subject). This paradox was itself an attraction calling for close observation.

The basic question was 'how did these savages, some of them living deep in the heart of the continent, learn about the technology of the civilised people?' At that time culture change across space was almost exclusively explained in terms of migration and diffusion, hence the answer to the question of origins was sought through the identification of the nearest place of ironworking in the civilised world from which sub-Saharan Africans might have borrowed. The source that seemed obvious to many of them was Mesopotamia. This area had already been established by then as the cradle of not only ironworking but also metallurgy in general, and of other markers of civilisation such as plant and animal domestication, and centralised political organisation (Childe, 1951). However, two issues remained unresolved. First, tracing the route through which the technology could have passed; and second, establishing circumstances around which the transfer took place long before Europeans penetrated into the interior of the continent. In other words, they wanted to know who had visited the land before the arrival of colonialists and when. This again was a cause for many speculations. But one such speculation that gained popularity and has continued to do so involves the Phoenicians. These were believed to have picked up the technology from the eastern Mediterranean and transplanted it in the south-central Mediterranean, and from there Berber traders took it to West Africa across the Sahara desert (Kense, 1985; Miller and van der Merwe, 1994; Alpern, 2005).

The early visitors noted yet another irony imbedded in the metallurgical sophistication of sub-Saharan Africa; this was its entanglement with symbolism, in the form of rituals, medicines, and taboos. The killing of sacrificial animals prior to and during smelting; the deployment of ritual ceremonies, some of which involved nudity; secrecy; sexual discrimination; and so on; all were considered to contradict the apparent cultural advancement evinced by iron technology. These practices were simply labelled as 'superstition', which in turn was used to prove the savageness of Africans, hence disqualifying all the good qualities demonstrated by African metallurgists.

Unfortunately, this viewpoint was adopted by the early students of African metallurgy – historians, archaeologists, and anthropologists (Cline, 1937; Wainwright, 1942). Since then, metallurgical symbolism has by and large continued to be considered so, in a way contributing subjectivism in the search for the origins of metallurgy in sub-Saharan Africa (Chirikure, 2010; Mapunda, 2010; Zangato and Holl, 2010).

Using an historical approach, this paper provides a contextual and holistic explanation of metallurgical symbolism as practiced by ironworkers in sub-Saharan Africa prior to colonial invasion in the region. The paper is premised upon the thesis that metallurgical symbolism or any other symbolic expression for that matter is an integral part of the socio-cultural environment where it belongs, and its worth or worthlessness can best be understood only when assessed within that context.

Symbolism in production and reproduction

Symbolism in this paper is conceptualised with a metallurgical bias. As such, it entails four main elements: a) techno-medicines, such as charms, talismans, preventives, and curatives; b) taboos, which involve restrictions against some material objects or actions before, during and after iron working in order to acquire purification, power, or energy; c) rituals, as demonstrated through performances, costume, procedural activities or rhetoric expressions; and d) sacrifices, which in principle were gifts offered to spiritual powers (gods or ancestral spirits) for appeasement or simply as a gesture of commitment to the activity being undertaken. Habitually, sacrificial offerings were objects of high value or of the best quality, so as to express the reverence one would have to the recipient. Often these would be products of the toil of the person(s) making the offering. For example, a metallurgist would offer bloom or products thereof; a potter would offer pots; a farmer would give out crops; a fisherman, fish; and sometimes human beings were sacrificial objects – a parent or a king would offer a child or a subject, respectively (Mapunda, 2002).

We need also to note that the *modi operandi* for symbolic expressions were multiple and varied, as noted in the type of objects used for each activity, the profile of the person(s) ideally acceptable for performing the ritual or handling sacrificial objects and the status of the targeted recipient. For example, at one place a cow would be ideal, while at another a bull would be so; at one place a sheep, at another a goat; at one place the chief smelter himself handles the medicines, at another an 'innocent' juvenile does that; in some cases ironworking had its own god, whereas in others the same god/goddess controlled ironworking and other means of livelihoods, and so forth (for further examples see Wainwright, 1942; van der Merwe and Avery, 1987; Herbert, 1993; Rowlands and Warnier, 1996; Schmidt and Mapunda, 1997; Barndon, 2001, 2004; Mapunda, 2010).

Although the concern here is metallurgy, one needs to note that symbolism of the type discussed above was not confined to metallurgy; it spread across a wide range of technologies and means of subsistence, including hunting, honey collecting, fishing, herding, and farming, as well as socio-cultural activities such as sports, games, and traditional dances (Wainwright, 1942; Herbert, 1993; Mapunda, 2002; Haaland, 2006). For example, Fipa hunters were required to strictly observe sexual abstinence the night before departure and during the entire hunting period. Failure to observe that would result in either a fruitless or disastrous mission, which might involve serious injuries or death from attacks by wild animals (Fabian Makanta, retired professional elephant hunter, personal communication, 1997). The same applied to warriors. On the eve of an offensive mission (or planned defence), a soldier was supposed to observe sexual abstinence, failure of which could result in his death, or, if he was a commander of the army, his 'misbehaviour' could lead to a disastrous defeat of his army. G. Gwasa noted the strict observance of such a taboo by local warriors during the Maji Maji War fought in southern Tanzania against Germans in 1905-1907 (Gwasa, 1973).

One should note that although symbolism characterised all production processes and crafts, in metallurgy, and especially in iron production, it was more deeply involved than in the others. This was not only because iron production was a more complex process than the rest, involving transformation of materials across different forms – solid-liquid-solid – but also because ironworking produced tools and weapons upon which the other production processes, such as agriculture, hunting, farming, fishing, basketry, weaving, woodwork, as well as warfare, depended.

Profiling symbolism in iron production

Given its pivotal role in contributing towards societal survival, iron production was almost always highly respected and hence closely related with human reproduction. As such, it abounded in symbolism along its entire process, from ore procurement to tool production. For example, some people (e.g., the Pangwa of southern Tanzania) engendered their ores, limonite and magnetite, as female and male respectively, and mixed them during smelting. This was equated to subjecting the ore to sexual intercourse, the end product of which would be bloom (Barndon, 2001, 2004). The smelting furnace was in many communities considered as a woman, or more specifically a bride, who was ready to be impregnated by the bridegroom, the smelter. The anthropomorphising and engendering of the furnace was expressed in various ways, including the performance of wedding ceremonies no less seriously and respectfully carried out than those done for a human bride. Such practices could be found among the Fipa of southern Tanzania (Barndon, 2001, 2004; Mapunda, 1995, 2002, 2010); the Grassfields societies in Cameroon (Rowlands and Warnier, 1996); and in Dawro, Ethiopia (Haaland, et al., 2004), to name but a few. Some people went further, and constructed the furnaces with feminine anatomical features, such as breasts, thighs, hips, and birth track, as was the case with the Shona of Zimbabwe, and Bena and Pangwa of southern Tanzania (Fig. 1). There were also cases of cultural feminine features, such as scarification and ornamentation (Childs, 1991; Herbert, 1993).

Another anthropomorphic expression observed across societies in sub-Saharan Africa and beyond is its association with human reproduction. Most often ritual practices applied in ironworking bear a direct or derivative relationship with those practiced in human reproduction (Herbert, 1993; Rowlands and Warnier, 1996; Schmidt and Mapunda, 1997; Barndon, 2001, 2004; Haaland, et al., 2002; Haaland, 2007; Mapunda, 2010). For example, across the region a smelter was equated to a husband whose wife was the furnace with which he was producing iron, and the bloom which formed inside the furnace and was finally raked out of it was equated to a newborn baby. Hence, rituals and taboos applied to human reproduction applied also to ironworking. For instance, within many societies which considered infertility to be a punishment from God, the same meaning was applied when a smelt failed to yield a bloom. Furthermore, adultery, even when committed under deep cover, was believed in many societies to have the potency of revealing itself in the form of mishaps, either on the doer or his or her spouse. These would include miscarriage, ill health, or death. Similarly, a smelter who was the 'husband' to the furnace was not supposed to have sexual relationships with any other person, including his biological wife, during the time he was engaged with the 'metallurgical wife', the furnace. Failure to adhere to this was considered no less adulterous than cheating on a biological wife and the resultant symptoms and punishments would be the same.

Furthermore, in the event two blooms formed in a single smelt, they were considered to be twin babies and were treated as such, as Rowlands and Warnier (1996, p.61) note:

> Washman Chi, an old We man, said that every now and then, the bloom would come out in two separate lumps. It would then be considered that the furnace had delivered twins, and the smelters would perform the same birth celebration as for twins, with twin dances and songs.

A wide range of symbolic expressions, however, fall under medicinal applications. Again, these were applied at every stage of the process, from ore exploration to sales of forged metal products. The Fipa, for example, placed a heavy emphasis on medicine during furnace construction, perhaps because their furnaces were reused many times. They used herbal medicines that were buried underneath the furnace floor, and then covered by a clay pot followed by a layer of mud which formed the floor of the furnace (Figs. 2a and 2b).

This practice has also been documented along the coast of Tanzania, where it dates back over 2000 years (Mapunda, 2007). As they were believed to have everlasting powers, these medicines were permanently placed underneath the furnace. When the furnace superstructure collapsed, the

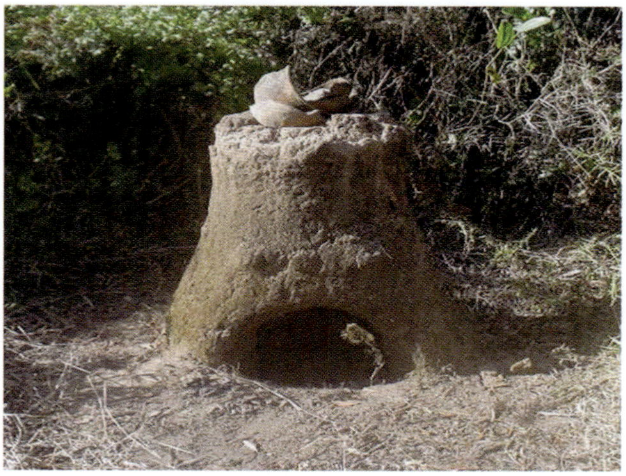

Figure 1: Anthropomorphised furnace of the Bena, southern Tanzania – note the breasts just above the rake hole. (Photograph by the author).

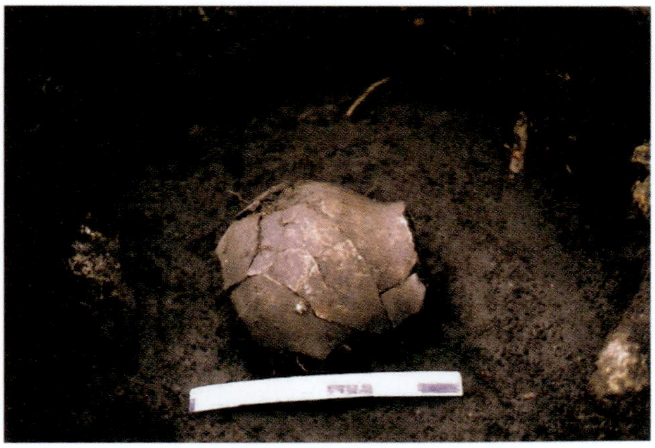

Figure 2a: A half-pot covering herbal medicines buried at the furnace's centre in Ufipa. (Photograph by the author).

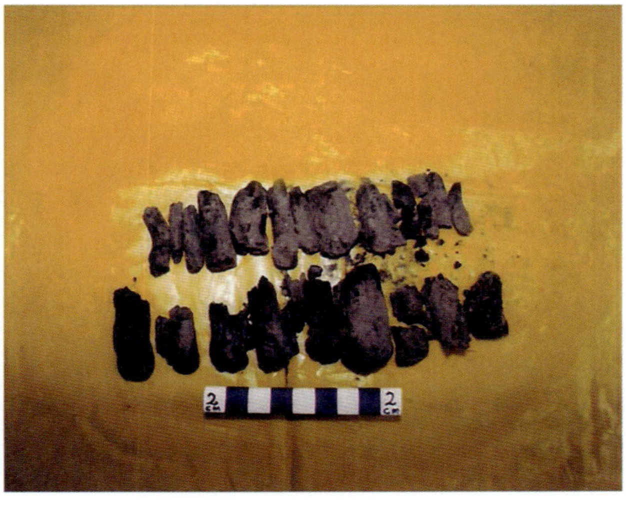

Figure 2b: Charred medicinal herbs retrieved from beneath a pot in Ufipa. (Photograph by the author).

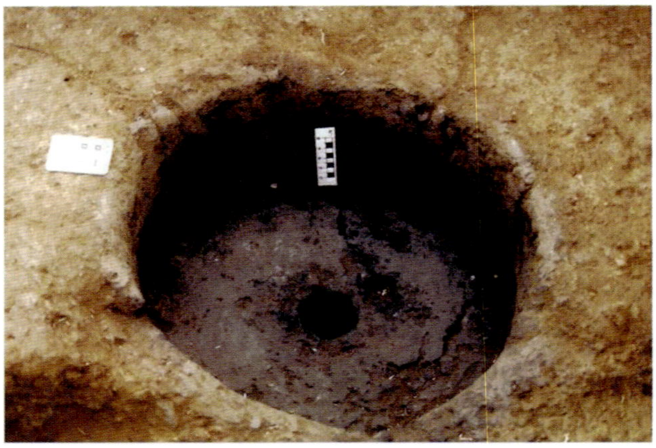

Figure 3: Ritual hole at the centre of a bowl furnace of the Rangi, central Tanzania. (Photograph by the author).

smelters often constructed a successive one at almost the exact location of the old one, so that the two share the same centre, and hence the same medicine (Mapunda, 1995, 2010). Those who used their furnaces only once often buried their medicine in a hole dug at the centre of the furnace base (Fig. 3).

Offering of sacrifices was another common form of rituals practiced by iron smelters. These ranged from animals to crops and brews. An animal (cow, goat, or chicken) would be killed and its blood sprinkled in the furnace foundation prior to construction, as it was the case among the Fipa (Barndon, 1996) and the Dawro of south-western Ethiopia (Haaland, et al., 2004), or on the furnace wall and/or other accessories, as was the case with the Mawu of Ghana and Togo, West Africa (Pole, 2010). The Fur of southern Sudan and the Phoka of northern Malawi used millet flour (van der Merwe and Avery, 1987; Haaland, et al., 2002;); whereas the Chewa of southern Malawi sprinkled maize flour (van der Merwe and Avery, 1987), and the Mawu applied rice flour and palm wine in the hole prepared at the furnace base (Pole, 2010, p.51).

Finally, symbolism was expressed in the form of taboos. While sexual abstinence and avoidance of menstrual blood were cross-cutting, there were other taboos that varied between and within communities. For example, while at the smelting camp, the Pangwa smelters (southern Tanzania) were not allowed to wash their bodies, cut their hair or get angry or violent, because any of these actions was believed to have the power to change the balance of hot and cold in bodies (Barndon, 2001). Observance of peace and reconciliation prior to and during ironworking is also reported among the We of Cameroon (Rowlands and Warnier, 1996) and the Mawu of Togo and Ghana (Pole, 2010). Pole, for example, observes that when the Mawu forged smithing hammers, they had to settle all quarrels before the work could begin; this was in addition to abjuring sexual intercourse before the appointed day (Pole, 2010).

In short, iron smelting was given the same status as human reproduction. This was in recognition of its contribution to subsistence and the existence of the human race.

What was all this for?

Metallurgical symbolism has been interpreted variably. However, when the various versions are closely examined, one finds that they fall under three broad categories: a) Symbolism as idiocy (ingenuous superstition); b) Symbolism as inadvertent science; and c) Symbolism as ingenious metaphors.

a. **Ingenuous superstition:** This interpretation views metallurgical symbolism as plain superstition. We encounter this interpretation in most of the reports of European missionaries and travellers in the 19th century. Although the viewpoint anchored much upon Edward Tylor's and Lewis Morgan's socio-cultural theories and racial prejudice, both of which dominated the intellectual thinking of the time. Religious factors also contributed, especially to the thinking and argument of the missionaries. For them, symbolism, especially the one involving sacrificial offering to ancestors and African deities, contradicted the teaching of one God, the biblical one.

   At the same time, according to Tylor's and Morgan's socio-cultural schemes, Africans were lacking science and therefore used superstition as an alternative means of explaining nature and other phenomena around them. Hence, symbolism was an expression of ignorance. As stated earlier on, and as Schmidt and Mapunda (1997) argue, this perspective has continued to dominate the thinking of many people, including professionals, until recently.

b. **Symbolism as inadvertent science:** According to this viewpoint, symbolism, though executed in a superstitious manner, had some logical, technological, or scientific values. This line of thinking has received support from a number of 'materialist' archaeometallurgists, including Herbert (1993), Schmidt and Mapunda (1997), Mapunda (2002), and Haaland (2006). The following case can help to illustrate this viewpoint. The Rangi iron smelters of central Tanzania had a taboo against eating vegetal relish such as vegetables, beans, peas, and the like during the entire period of smelting; they had to eat meat. This meat had to come from a juvenile bull with horns approximately 10-15cm long. The bull would be slaughtered immediately after the furnace had been dug out (they used bowl furnaces – see Fig. 3). They would take one of the horns, stuff it with techno-medicines and then insert it in the hole dug at the centre of the furnace floor and cover it with soil; the other horn would be reserved in case a second furnace was needed. The meat was consumed by the smelters as food during the work. But they were supposed to eat steak only, not bones; all bones were thrown into the furnace, becoming part of the raw materials in the process. The Rangi believed that eating vegetal relish or gnawing bones during iron smelting would result in an abortive operation, i.e., the bloom failing to form (Shaaban Kideso, personal communication, 1999).

   Although on the surface this narrative may sound as superstitious as any other, or as one participant in my presentation commented: "This was just one of those pretexts African men used to cover up for their greediness", a serious consideration can discern the logic and science hidden behind the taboo. Being rich in calcium, the bones must have acted as a flux in the smelting process; and to ensure a constant and sufficient supply of the flux (calcium), the Rangi made it mandatory for smelters to consume meat at the workshop.

A materialistic explanation can be detected in most if not all taboos. For example, the reason why most smelters placed their workshops away from residential areas must be materialistic. Although most informants tend to explain it in terms of symbolism, such as to hide from people with evil intents or menstruating women (Schmidt, 1997; Barndon, 2001), we cannot ignore the fact that smelting iron away from residential areas removed the risk of setting their houses ablaze, as most of them were constructed with wattle and grass.

c. **Ingenious metaphors:** This interpretation views symbolism as reality according to the worldview of those who applied it, i.e., the iron smelters and their respective communities. They treated metallurgical symbolism and other forms of symbolic expressions as an integral part of "the natural process of the transformation of materials from one state into another" (Rowlands and Warnier, 1996, p.52). The only problem is that in the pre-colonial African epistemological process, knowledge, including the logical justifications of symbolic actions, was not literarily recorded as was the case in the Western World, but rather preserved in the intellectuals' or experts' own mind and memory. As a result, transmission of knowledge from an expert to a disciple depended not only on the pedagogical skills of the resource person and the intellectual absorption parameter of the recipient, but also on the will of the knowledge provider. As there was no written syllabus for any given expertise allowing the disciple to crosscheck the accuracy or content of the knowledge received, omissions and deviations were inevitably high. Although on the one hand such variations were a blessing as they enriched the subject in question, on the other they were a reason for missing logical explanations of certain phenomena, some of which would be found bizarre and illogical in subsequent circumstances. This is the situation we find ourselves in today, especially with regards to taboos and rituals.

However, when viewed objectively, especially from the viewpoint of the users, and while believing that there was intelligence, let alone ingeniousness, among the metallurgists, most of the so-called superstitious beliefs are found to be no less scientific and logical than modern science and logic. In the remaining part of the paper, we shall examine four such logics. Whereas there is direct ethnographic support for some of them, others are indirectly and syllogistically inferred.

i. **Symbolism for respect and value:** We have already established that, on account of its contribution to the well-being of society, ironworking was, in many societies, highly respected; and so were ironworkers. It was for this reason that people equated it with procreation, the most valued human engagement. As such, ironworking was given all the rights and respects bestowed upon procreation. This was reflected in the socio-cultural treatment of metallurgical materials – furnaces, bellows, tuyères, bloom and slag – as well as the types of medicines, rituals, and care granted to them, as Rowlands and Warnier inform:

> every now and then, the bloom would come out in two separate lumps. It would then be considered that the furnace had delivered twins, and the smelters would perform the same birth celebration as for twins, with twin dances and songs. They would slaughter a goat and the whole ward would rejoice
> (Rowlands and Warnier, 1996, p.61).

They further report that the herbs used in 'curing' a spoiled smithy are the same as those used in treating women who have suffered from miscarriage (Rowlands and Warnier, 1996, p.63).

In short, by equating ironworking with human reproduction, along with all symbolic expressions involved therein, the ironworkers wanted to portray the value of ironworking to the community.

There is still another equally important point regarding respect, and that is the need to indicate that ironworking is a no-nonsense business. In achieving this, ironworkers devised a host of taboos and rituals, just as most traditional healers and shamans do today. Upon receiving a client, healers do not simply record her or his history and attend or treat her or him, but subject the patient to a complex ritual procedure, under the name of 'diagnosis'. During the process, some exotic or extra-traditional attires, utensils, tools, songs, and even incomprehensible language would be employed. All these are intended to mesmerise the client and enhance her or his compliance to both the diagnosis and the resultant curative procedures. Ironworkers also had this impressionistic goal in mind when deploying the various forms of symbolism in their craft. The underlying goal was to demonstrate the importance of their business (Stephano Malimbo, personal communication, 1997).

ii. **Symbolism for medicinal purposes (cure, protection and charm):** Some rituals and techno-medicines were applied for healing ailments 'suffered' by the various technological accessories or the metallurgical process as a whole. Rowlands and Warnier note:

> When the smelt does not produce bloom, or when the blooms are light and spongy, or when a dispute occurs in the workshop, or when blood is spilled in it, then the workshop is spoiled and has to be cured.

Failure to do so would end in permanent 'barrenness' of the process or the workers might be affected with barrenness, illness, leprosy, and death (Rowlands and Warnier, 1996, p.65).

Curing the ailing workshop required attracting the blessings of divinity to act through medicines (Rowlands and Warnier, 1996). It is at this juncture that sacrifices often involving either blood or beer came in so as to appease the gods. The Mawu of Togo and Ghana, West Africa, for example, sprinkled rice floor and poured a libation of palm wine (*nda*) in the furnace base before construction started in order to invoke the help of the grandfathers in the furnace building work. In addition, a length of twisted vine (*kana*) was laid in the hole; this was said to be indicative of strength (Pole, 2010).

The important point to note here is that the medicinal herbs used in treating metallurgical 'ailments' would be the same ones used in healing humans, especially in facilitating childbirth, alleviating pain, and promoting fertility and peace (Rowlands and Warnier, 1996; Mapunda, 2010). In so doing, ironworking was conceived as a living and sentient process that shared with human beings both somatic and socio-cultural characteristics.

iii. **Symbolism for patent right:** This type of symbolism is premised upon the fact that during the time in question there were no international laws for safeguarding people's inventions as we have them today. Yet the history of ironworking, especially in sub-Saharan Africa, is full of experimentations which led to new discoveries and, hence, to a multitude of technological variations. At the same time, we learn that ironworking was jealously guarded by both the experts and the communities in which the experts operated. For example, not everyone could become a smelter or a blacksmith; there were specific clans designated by the community for the task, based on specific historical or socio-cultural factors. However, in most cases there were no formal means put in place by the given community to safeguard an individual smelter's technological discoveries within the ironworking clan. That was left at the discretion of each craftsman.

One simple method that the experts employed in protecting their discoveries against plagiarism was symbolism – in the form of rituals, taboos, techno-medicines, and sacrifices. This way symbolism also served as the identity of one's invention, in the sense that in addition to the cross-cutting rules such as sexual abstinence and avoidance of menstrual blood, each master smelter used to have detailed symbolic practices that were specific to him which he would not share, not even with his immediate assistants (Mapunda, 2010). And because symbolism was a reality for them (see subsections c and i above), aspirants of the craft, especially those from outside the designated clans, knew straight away that they could not smelt iron on their own because they did not have the knowledge of techno-medicines. That is why when asked to conduct iron smelting today, e.g., for experimental purposes, most elders (former smelters and their descendants) become concerned about symbolism; they would say that they have forgotten the techno-medicinal menu, or that morals have eroded and the young generation cannot observe rituals (Ibrahim Mruke, informant, Kondoa, Tanzania, personal communication, 2001). Similarly, when experiments fail, the blame is always centred on symbolism (Herbert, 1993; Schmidt, 1997; Barndon, 2001).

Another tool metallurgists used for protecting their discoveries was witchcraft. "In some societies like the Ba Ila of the Kafue Flats", Msambachime notes, "iron workers were simultaneously praised for their workmanship and despised for their alleged involvement in sorcery" (Msambachime, n.d., p.174). The belief that iron smelters were also masters of magic and witchcraft was widespread across Sub-Saharan Africa (van der Merwe and Avery, 1987; Herbert, 1993; Mapunda, 2010; Msambachime, n.d.). The association of iron smelting with witchcraft and magic seems to have emanated from the 'miraculous' ability the smelters had of turning rocks (ore) into iron. Interestingly, the allegation seems not to have bothered ironworkers, who took pride in it. As one prominent iron smelter confessed during an interview, "we enjoyed being feared by other people. You know, it was this [witchcraft] which scared away aspirants from other clans from doing what we were doing [smelting iron]. And so there was no competition (Stephano Malimbo, personal communication, 1997, Fig. 4).

We further note that some societies protected their inventions by establishing guilds, each with their own passwords or codification systems that helped them to protect their discoveries against "poachers" (Msambachime, n.d., p.177).

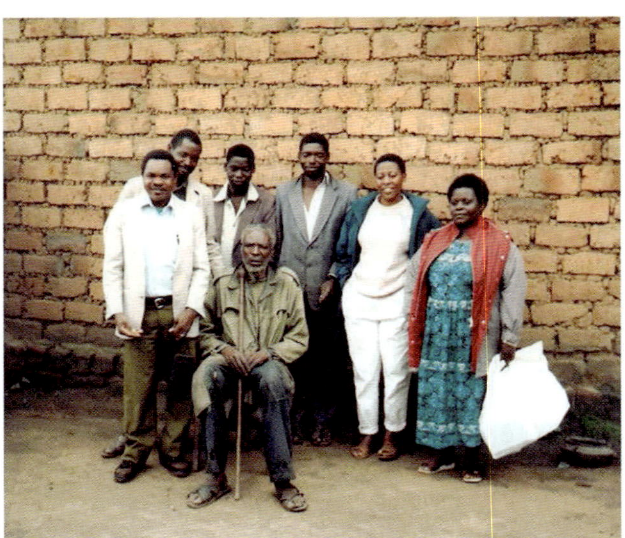

Figure 4: A souvenir photo with Stephano Malimbo (seated) after an interview with him at his residence, Katumba Azimia Village, Ufipa, in 1997. Standing to the extreme left is the author, followed by three of Malimbo's disciples who are running a smithy at the compound; the remaining two women are research assistants, then undergraduate students at the University of Dar es Salaam.

iv. **Symbolism for peace of mind:** This applied mainly to taboos that involved sexual abstinence and seclusion. Iron production was an arduous work that required concentration, tranquillity, and peace of mind. To fulfil these conditions, smelters placed their workshops away from villages. As noted above, this also protected their houses from accidental fires.

Stephano Malimbo (Plate 4), an iron smelter who survived until recently (died in 2007), justified sexual abstinence in terms of "peace of mind". This taboo applied to a husband and his spouse(s). It was believed that if a smelter's wife back home acted in a distrustful manner, the husband would know, because some mishaps would occur in the metallurgical process. When that happened, a diagnostic test was undertaken to determine the cause of the problem. The test could simply involve psychological determination by an experienced elder or the master smelter, or, if complicated, would involve a diviner (van der Merwe and Avery, 1987; Schmidt, 1997). The latter often involved inviting the crew to drink an herbal concoction which would lead to headache, stomach ache, intoxication, or even death for the wrongdoer. If the smelter or any of his assistants was found guilty, he was punished in accordance with the severity of the offence, depending on such factors as the length of time from the action to the smelting time or the type of partner one was involved with. The weight of punishment increased respectively from "committing adultery" with one's own wife to an unmarried woman or one married to another man. The latter case could lead to the death penalty. In the event that none of the members at the smelting site was found guilty, then their spouses back home were 'tested'. If they also turned out to be 'clean', then the mishap would be attributed to evil eye or the anger of spirits, and hence the matter would be handled differently.

With this invisible security system in place, both the smelters and their spouses could have peace of mind during the entire time of separation, which sometimes took up to four months. This symbolic practice, together with the restriction against women, especially sexually active ones, coming close to the workshop or in direct contact with the crew members, enabled smelters to concentrate on their work with minimal jealous feelings about their spouses, less thoughts about sex, and at the same time allowed their spouses back home to take care of the normal family responsibilities with a similar peace of mind.

In addition to sexual abstinence and seclusion, avoidance of blood, including menstrual blood, is also attributed to peace and tranquillity. Justifying the relationship, Rowlands and Warnier write:

> ...iron production is explicitly expected to be a nonviolent process. No blood should be spilled in the workshop, after the blood of a fowl has been sprinkled on the medicines and on the furnaces once and for all at the time of their erection. Should a smith wound himself, he must run out of the workshop, pressing the wound with his hand so as to contain the blood. No menstruating woman is admitted in the workshop. (Rowlands and Warnier, 1996, p.63).

This means that iron production was supposed to be carried out in a non-violent context; otherwise, it would cause the destruction of the community (Rowlands and Warnier, 1996, p.64). This probably explains why smiths and smelters in some societies, e.g., the Zoa of Grassfields Cameroon (Rowlands and Warnier, 1996) or the Bena of the southern highlands of Tanzania (Nyagava, 1999), were not allowed to go to war – in addition to the fact that they were needed to produce iron for weaponry. This interpretation of taboo against menstrual blood varies from the popular view of menstrual blood as a symbol of infertility, hence its avoidance (Herbert, 1993; Schmidt, 1997).

Conclusion

This paper has argued that the interpretation of African metallurgical symbolism started from a false premise; one which was based on religious, racial, and ethnocentric grounds. Unfortunately, most attempts to rectify the trend so far have not incorporated the African worldview seriously. As a result, metallurgical symbolism has by and large continued to be viewed as superstitious. In this paper weight has been given to the thinking of an African metallurgist. Through this perspective, metallurgical symbolism is found to have been founded upon both materialistic (technological) and socio-cultural functions and values of the respective societies in the region. Previous interpretations were therefore based on incorrect premises, and did not take into account the African epistemological and cultural contexts.

References

Alpern, S.B., 2005. Did They or Didn't They Invent It? Iron in Sub-Saharan Africa. *History in Africa*, 32, pp.41-94.

Barndon, R., 1996. Fipa Ironworking and Its Technological Style. In: P. Schmidt ed. *The Culture and Technology of African Iron Production*. Gainesville: University Press of Florida, pp.58-73.

Barndon, R., 2001. *Masters of Metallurgy - Masters of Metaphors: Iron Working among the Fipa and the Pangwa of SW-Tanzania*. Bergen: University of Bergen.

Barndon, R., 2004. A Discussion of Magic and Medicines in East African Iron Working: Actors and Artefacts in Technology. *Norwegian Archaeological Review*, 37, pp.21-40.

Childe, V.G., 1951. *Social Evolution*. London: Watts and Company.

Childs, S.T., 1991. Style, Technology, and Iron Smelting Furnaces in Bantu-Speaking Africa. *Journal of Anthropological Archaeology*, 10, pp.332-59.

Childs, S.T. and Herbert, E., 2005. Metallurgy and its Consequences. In: A.B. Stahl ed. *African Archaeology*. Oxford: Blackwell Publishing, pp.276-300.

Chirikure, S., 2005. *Iron Production in Iron Age Zimbabwe: Stagnation or Innovation?* PhD Thesis, Institute of Archaeology, University College London.

Chirikure, S., 2010. On Evidence, Ideas and Fantasy: The Origins of Iron in Sub-Saharan Africa. Thoughts on E. Zangato and A.F.C. Holl's 'On the Iron Front'. *Journal of African Archaeology*, 8, pp.25-8.

Cline, W., 1937. *Mining and Metallurgy in Negro Africa*. Wisconsin: George Bantu Publishing Company.

Fagan, B., 1961. Pre-European Ironworking in Central Africa with Special Reference to Northern Rhodesia. *Journal of African History*, 2, pp.199-210.

Forbes, R.H., 1933. The Black Man's Industries. *The Geographical Review*, 23, pp.230-47.

Gwasa, G.C., 1973. *The Outbreak and Development of the Maji Maji War 1905-1907*. PhD Thesis, University of Dar es Salaam, Tanzania.

Haaland, R., 2006. Iron in the Making – Technology and Symbolism: Ethnographic Perspective on European Iron Working. In: A. Andrén, K. Jennbert and C. Raudvere eds. *Old Norse Religion in Long-Term Perspectives: Origins, Changes and Interactions*. Lund: Nordic Academic Press, pp.79-85.

Haaland, R., 2007. Say it in Iron: Symbols of Transformation and Reproduction in the European Iron Age. *Current Swedish Archaeology*, 15, pp.1-20.

Haaland, G., Haaland, R. and Rijal, S., 2002. Social Life of Iron: A Cross-Cultural Study of Technological, Symbolic and Social Aspects of Iron Making. *Anthropos*, 97, pp.35-54.

Haaland, G., Haaland, R. and Dea, D., 2004. Smelting Iron: Caste and Its Symbolism in South-Western Ethiopia. In: T. Insholl ed. *Belief in the Past*. Oxford: Oxbow Books, pp.75-86.

Herbert, A., 1993. *Iron, Gender, and Power: Rituals of Transformation in African Societies*. Bloomington: Indiana University Press.

Kense, F., 1985. The Initial Diffusion of Iron to Africa. In: R. Haaland and P. Shinnie eds. *African Iron Working: Ancient and Traditional.* New York: Norwegian University Press, pp.11-27.

Kjekshus, H., 1996. *Ecology, Control and Economic Development in East African History: The Case of Tanganyika 1850-1950* (second impression). Dar es Salaam: Mkuki na Nyota.

Mapunda, B., 1995. *An Archaeological View of the History and Variation of Iron Working in Southwestern Tanzania.* PhD Thesis, University of Florida.

Mapunda, B., 2002. *Ufundichuma Asilia Afrika Mashariki: Chimbuko, Kukua na Kukoma Kwake* (Indigenous Ironworking in East Africa: Origins, Development, and Termination). Peramiho: Peramiho Printing Press.

Mapunda, B., 2007. *Re-examining Limbo in Regional and Temporal Contexts.* Unpublished Field Report submitted to the Antiquities Division, Dar es Salaam.

Mapunda, B., 2010. *Contemplating the Fipa Iron Working.* Kampala: Fountain Publishers.

Miller, D. and van der Merwe, N., 1994. Early Metal Working in Sub-Saharan Africa: A Review of Recent Research. *Journal of African History*, 35, pp.1-36.

Morgan, L.H., 1877. *Ancient Society.* New York: H. Holt and Company.

Musambachime, M.C., n.d. *Mining and Salt Making in Zambia to 1920: A Study of the "Dying" Indigenous Industries.* Unpublished manuscript in possession of the author, Lusaka, Zambia.

Nyagava, S., 1999. A *History of the Bena to 1914.* Peramiho: Peramiho Printing Press.

Pole, L.M., 2010. The Hammer of Mawu: Ironworking Traditions in the Togo Hills, Ghana. *African Archaeological Review*, 27, pp.43-78.

Pringle, H., 2009. Seeking Africa's First Iron Men. *Science*, 323, pp.200-2.

Rowlands, M. and Warnier, J.P., 1996. Magical Iron Technology in the Cameroon Grassland. In: M.J. Arnoldi, C.M. Geary and K.L. Hardin eds. *African Material Culture.* Bloomington: Indiana University Press, pp.51-72.

Schmidt, P.R., 1997. *Iron Technology in East Africa.* Oxford: James Currey.

Schmidt, P. and Mapunda, B., 1997. Ideology and the Archaeological Record in Africa: Interpreting Symbolism in Iron Smelting Technology. *Journal of Anthropological Archaeology*, 16, pp.73-102.

van der Merwe, N.J. and Avery, D., 1987. Science and Magic in African Technology: Traditional Iron-Smelting in Malawi. *Africa*, 57, pp.143-72.

Wainwright, G.A., 1942. The Coming of Iron to Some African People. *Man*, 42, pp.103-8.

Zangato, E. and Holl, A.F.C., 2010. On the Iron Front: New Evidence from North-Central Africa. *Journal of African Archaeology*, 8, pp.7-23.

Author's address

• Archaeology Unit
Department of History
University of Dar es Salaam, Tanzania

Applying ethnographic presents to archaeological pasts: preliminary thoughts on memories of iron production from western Uganda

Louise Iles

ABSTRACT: The question of how to make ethnographic models relevant to archaeological interpretations is a pressing concern in understanding the social mechanisms of past metal production. This paper presents some preliminary results of a recent archaeometallurgical survey and excavation in western Uganda, and discusses these in relation to the simultaneous, but more informal collection of locally-held memories of the iron industries of the more recent past. Together, these highlight the potential for innovation and change on the level of the individual, challenging the tendency of archaeologists and ethnographers to present homogenised ethnographic metallurgical 'traditions', and broadening our expectations of the diversity of past technologies.

Keywords: Africa, Uganda, ethnography, iron, smelting, gender.

Introduction

Ethnographic examples of African iron production recorded over the last hundred years or so have radically transformed our understanding of the social contexts of iron technologies (e.g. Cline, 1937; Celis and Nzikobanyanka, 1976; van der Merwe and Avery, 1987; David, et al., 1989; Childs and Killick, 1993; Brown, 1995; Reid and MacLean, 1995; Kusimba, 1996; Schmidt, 1996; Childs, 2000; Barndon, 2004, among many others). Studies such as these have brilliantly demonstrated to a global audience the intricacies and organisation of the various technologies that, when brought together, transform iron ore into iron. The value of these accounts has been much emphasised (e.g. Killick, 1991), and they have, amongst other things, acted to remind researchers of the complex ideologies and concepts that surround and shape these crafts, many of which would not be immediately perceptible from material remains generated by archaeological excavation and interpreted through archaeometallurgical analysis alone.

The broad body of such ethnographic research has in recent years underlined the variation that is apparent in modern-day (or recent) metalworking techniques across sub-Saharan Africa, from the small bowl furnaces of the Marachi of western Kenya (Brown, 1995) to the large single-tuyèred furnaces of northern Cameroon (David, et al., 1989). Nevertheless, when addressing matters of the more ideological aspects of iron production, there has been a tendency to highlight unifying themes, not just within the continent, but also on a global scale (cf. Herbert, 1993; Schmidt and Mapunda, 1997; Haaland, 2004, 2006; Blakely, 2006; Giles, 2007). In particular, conceptual links between reproduction, fertility, and iron production, often associated with the exclusion of women from smelting activities, have become almost stereotypical characteristics readily associated with a generic idea of iron production in sub-Saharan Africa:

> *"In the composite image of the African smelt, young men celebrate their sexual potency around a furnace that is the metonymic bride, wife, and mother."*
> (Blakely, 2006, p.9)

As Peter Schmidt attests, as archaeologists we are "engaged in constructing generalising stories ... that emphasise technological commonalities rather than the differences that prevail at local scales" (Schmidt, 2001, p.219): a necessity if we are to construct coherent and accessible stories of the past. However, there is also a need to ensure that an awareness of these commonalities does not become a preoccupation; the consideration of alternative concepts of and approaches to technology and its role within society should not be precluded by overexposure to a popularised yet simplified version of the past. It is these insights which, together with a consideration of technological variation in general, can open a window into the organisation and operation of past societies – the ultimate aim of all archaeological research.

This paper will examine the importance of considering variation, not just in technical practice, but also in the (often) intangible behaviour that surrounds such technologies. Our expectations of metal working in the past (and therefore our research strategies and interpretations) are unavoidably formed with reference to our experience of metalworking in the present. As such, this paper emphasises the need to keep the discussion open about the deviations from the norm made by modern metalworkers if we are to fully understand the organisation of metalworking in the past. It is through an examination of small-scale variation – within technologies that are undoubtedly multifaceted and complex production systems – that the finer elements of technological behaviour, change, and innovation can be identified, and the broader questions of social organisation can be addressed.

Research background and methodology

Ideas for this paper were developed during six months of archaeological fieldwork carried out in western Uganda in 2007, although they evolved from a more general and longstanding interest in the presentation and interpretation of ethnographic data. Within Great Lakes Africa, iron is known to have been of great relevance to the economic and political structure of the region, as a material used to make both weaponry and agricultural implements. Some of the earliest evidence for iron production in sub-Saharan Africa comes from this region (van Grunderbeek, et al., 1983; van Noten, 1983; van Grunderbeek, 1992; Schmidt, 1996), and the exploitation and control of iron went on to contribute towards the formation and development of several Great Lakes kingdoms in the second millennium AD, playing a key role in the acquisition and maintenance of power and authority.

The overall aim of the 2007 fieldwork was to provide archaeometallurgical data for the author's doctoral thesis (Iles, 2011; see also Humphris and Iles, this volume), which comprises an examination of the role and organisation of iron production within the pre-colonial kingdom of Bunyoro, one of a network of late second millennium polities within the borders of present-day Uganda (Fig. 1). Historical and archaeological sources indicate that iron production, alongside salt production, played a significant role in Bunyoro's diverse economy, and made a considerable contribution to both local and long-distance trading systems, and to the later development of the kingdom (e.g. Uzoigwe, 1976; Connah, 1996). The strategic importance of iron is clearly indicated on a local scale by the territorial expansion of neighbouring Buganda into parts of Bunyoro in the early eighteenth century, an expansion led by a desire to gain access to, and control of, the iron smelting resources available there – both natural and human (Reid, 2002). Through the study of these consequential production industries it is possible to provide a more refined picture of the internal mechanisms of these polities, as well as of their interrelationships across the region.

The fieldwork

The 2007 fieldwork was based primarily in Mwenge, one of two counties in the district of Kyenjojo in western Uganda (Fig. 2). Smaller, complementary studies were also undertaken in Rakai to the south and Masindi to the north, although these areas will not be discussed in the present paper. Mwenge was chosen as the principal study area as it was well known to have been plentiful in the raw materials necessary for iron production, and oral histories speak of its rich background of smelting and smithing. Terry Childs had also undertaken extensive ethnographic work regarding iron production in the local area in the 1990s (Childs, 1998a, 1998b, 1999, 2000), building upon the earlier ethno-historical work of Buchanan (1974, 1979) and John Roscoe's documentation of metalworking practices in the early twentieth century (Roscoe, 1923).

Brought together, these resources have been instrumental in forming a picture of recent iron production and use within the kingdom, portraying in-depth explorations of all stages of iron production within the context of their social and ritual associations. Alongside technical descriptions, these accounts all emphasise the ritualised nature of the iron production episodes that they depict, illustrating the social rules and taboos that governed smelting activities. The symbolism surrounding these recorded rituals tended to be placed within a highly sexualised framework, with frequent references to reproduction, menstruation, and marriage, as well as spirit mediums and colour. These technologies were seen to be governed in such a way that they remained "as exclusive as any medieval guild system" (Ingham, 1975, p.16).

The first task of the fieldwork was to undertake a combination of systematic and informant-led surveys in order to locate new iron production sites, utilising extensive walking surveys along roads and exposed surface areas, and following leads provided by local community members. Over the course of several weeks, 120 previously unrecorded sites across a 25km by 25km area were located and recorded within Mwenge (for a full report of the survey and excavation results, see Iles, 2009). Eight of these sites were selected for excavation (now known to range in date from the fourteenth to the twentieth centuries), and recovered slag and technical ceramics were sampled and shipped back to London for further archaeometallurgical analysis. From initial macroscopic and microscopic investigations of the slag and furnace morphologies, there already appeared to be a reasonable extent of technical and technological variation in the smelting techniques of pre-colonial Mwenge (cf. Iles, 2011).

However, this paper will focus mainly on the glimpses into local memories of past iron production that were encountered as the fieldwork progressed, and how these insights can contribute to wider discussions regarding the organisation of iron production within such communities. As the most active part of the fieldwork, the walking survey provided an ideal opportunity to talk to many older residents of the Mwenge area who had either taken part in or remembered witnessing smelting activities taking place in their childhoods. Some were from families with a long tradition of iron smelting, who were often eager to recall stories they had been told about their fathers' or grandfathers' smelting, in addition to what they could remember themselves from their own first-hand experiences. As such, informal interviews were carried out with all of those who retained such knowledge, and these interviews (which were conducted in the local languages of Rutoro and Lunyoro by Elijah Kisembo) covered topics ranging from the technical 'recipes' of smelting to more socially oriented aspects of the production processes. The next section outlines some of the major points of interest determined by several of these interviews.

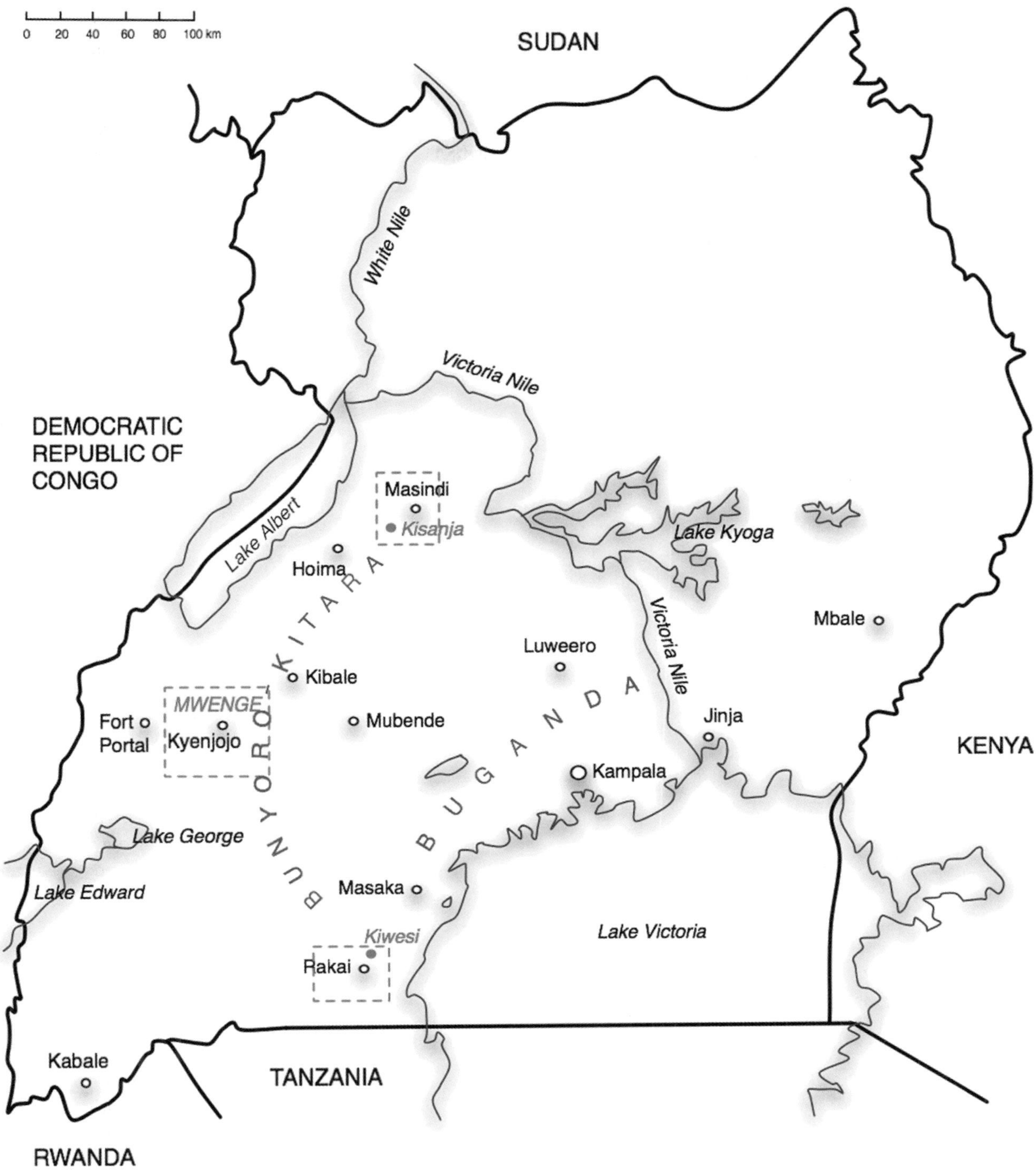

Figure 1: Map of Uganda, showing survey zones and the kingdoms of Buganda and Bunyoro-Kitara.

Witnesses to smelting technologies in Mwenge

In general, the information that was received from informants was in agreement with Childs' earlier work, correlating with prevailing ideas surrounding the complex relationship between women of child-bearing age and smelting, the concerted efforts that were made to retain knowledge within clans, and so on – themes that are in some form, and as mentioned previously, common to many of the recent iron producing communities of eastern Africa (e.g. Celis and Nzikobanyanka, 1976; Herbert, 1993; Reid and MacLean, 1995). As an example, an old man named Ezekiel Kaheru (for interview location, see Fig. 3) clearly asserted that the presence of women at a smelt, as well as sexual interaction around the time of a smelt, were strict taboos, and if not observed would compromise the successful production of iron. He also stated that only the members of four local clans could be smelters, in order for these clans to protect knowledge of their smelting expertise, thereby retaining their monopoly over the process.

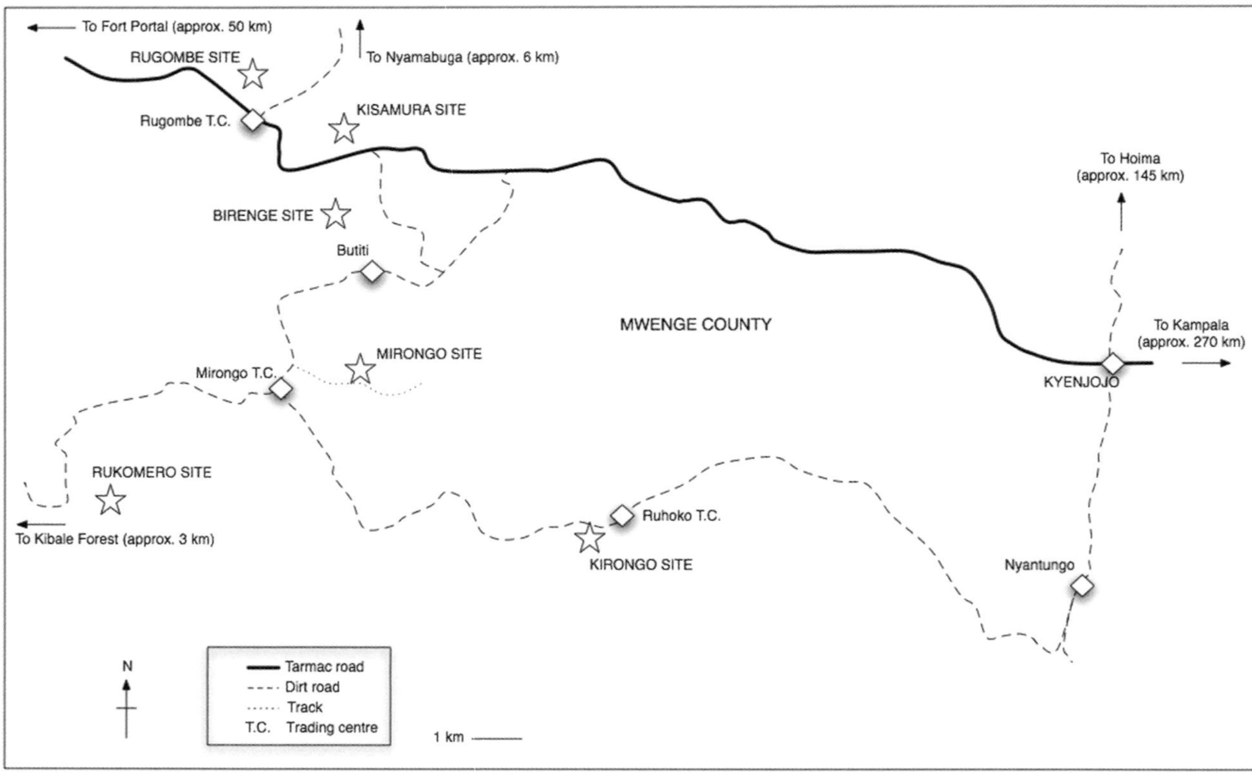

Figure 2: Map of the Mwenge survey zone, with excavated sites marked with a star.

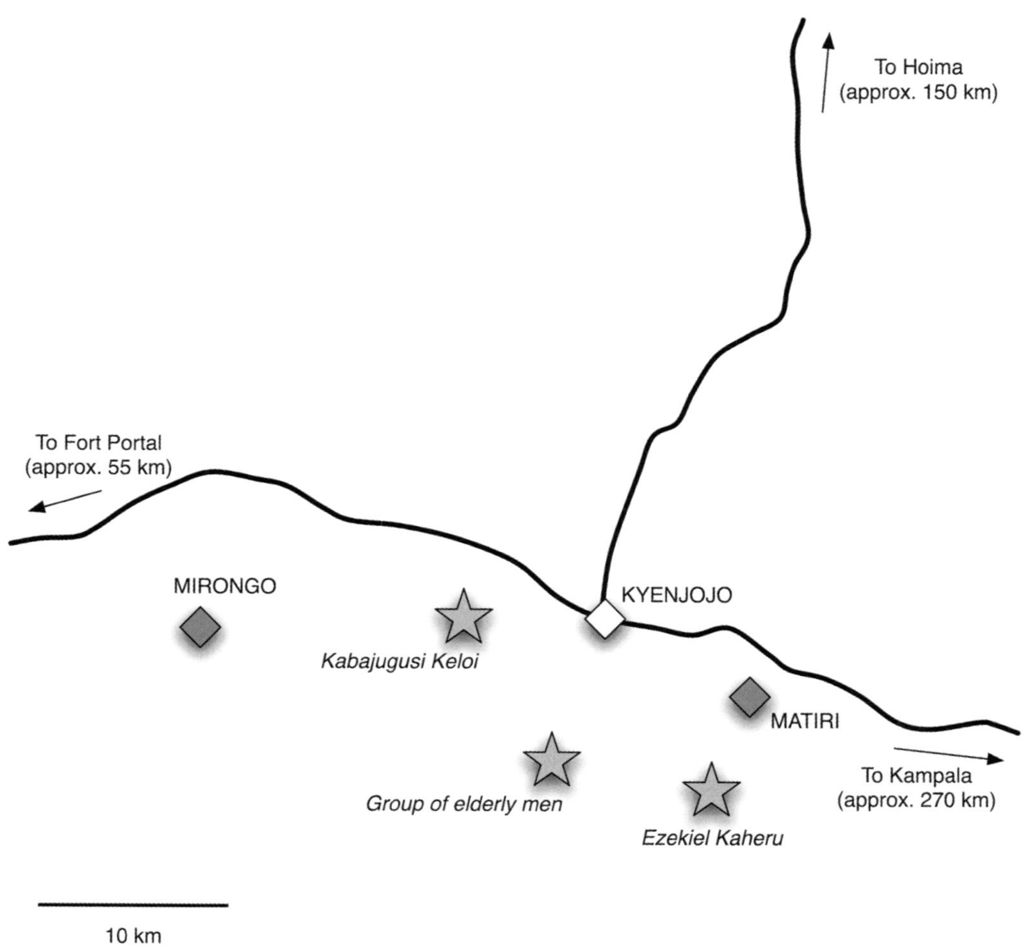

Figure 3: Map showing location of interviews mentioned in the text.

Ezekiel's interview correlated strongly with the prevailing presentation of ethnographic smelting norms in western Uganda; however, other interviewees' stories did not appear to fit so neatly with what is considered 'traditional' smelting practice in the region. One of the most interesting informants we spoke to was a woman named Kabajugusi Keloi, whom we interviewed at the location of her father's compound, approximately 19 kilometres to the north west of the location of our interview with Ezekiel (cf. Fig. 3).

Kabajugusi was born to a smelter father and assisted him in his work when she was a young girl; her name is derived from the Lunyoro word for 'smelter(s)' (*bajugusi*) (Roscoe, 1923, p.217), itself related to the Lunyoro verb 'to bellow' (*jugusi*). She was able to tell us about her personal experiences of smelting, as she remembered it from her youth. Unlike many studies that document the exclusion of girls and women from iron production activities (e.g. Childs, 1998a, p.113; Reid and MacLean, 1995), Kabajugusi told us that in her father's work there were no such restrictions, and women often took part in carrying and sorting the ore and pumping the bellows for the furnace. Kabajugusi herself was frequently put in charge of preparing the ore, and she also often joined in with the bellowing. Further to this, in contrast to accounts that stress the seasonality of smelting, her father carried out his iron production throughout the year, using stored charcoal and iron ore, and only stopping when supplies of these materials ran out. This would trigger journeys back to the forests and hills to collect and prepare more of the raw materials needed to allow the work to continue.

Kabajugusi also spoke of the sounds of smelting – a sensory aspect of iron production that is often overlooked, yet one that appears to have played a significant role in the social organisation of this particular production system. The noise of the pounding bellows echoed around the surrounding hills, alerting people that a furnace had been fired. Those who wanted to buy iron would come from their homes and join in with the arduous work of bellowing, and in return for this assistance Kabajugusi's family would supply them with some of the meat that was kept continually roasting within the compound. There was often a line of people queuing up to take over the bellowing as soon as the person in front grew tired. No direct payment was made to those who came to help with the smelt, but they were often smiths themselves who were going to buy the resulting iron.

Kabajugusi's memories regarding the inclusion of women in smelting technologies did not represent an isolated case in our survey zone. A group of elderly men interviewed approximately 10 km to the south-east of Kabajugusi's compound (cf. Fig. 3) shared this opinion of the inclusion of women in smelting activities. The general agreement amongst them was that although women tended not to take part in smelting activities as often as men, there were no formalised restrictions on this, and they believed it to stem from a tendency of women to take less interest in smelting than men, which meant that they did not acquire smelting knowledge (see also Kusimba, 1996, p.398). They also asserted that clan affiliation was not linked to participation in smelting.

Discussion

These informal yet suggestive interviews provided an excellent opportunity to contribute towards the existing body of ethnohistorical knowledge of the region, as well as the opportunity to question the generalisations that are often made regarding these aspects of smelting technologies. It is evident from the above that although many of the people we interviewed agreed with the current view of recent smelting practices in the Mwenge region, there are some examples that diverge from these patterns. It was especially surprising to hear the diverse opinions regarding the inclusion of women in smelting technologies within such a small geographical area. If this study can be seen to highlight that there is a much more complex relationship between gender, work roles, and iron production than is commonly presented, there are several possibilities that might be instrumental in accounting for these discrepancies.

A first suggestion could be that there may have been a change in the accepted work/gender roles in the more recent past, which meant that attitudes towards smelting (and potentially work in general) may have been changing during the period that the informants were recalling (namely the early twentieth century; although in some cases, stories stretched back into the end of the nineteenth century, with dates established using political and historical markers). Did the falling populations and resultant demographic pressures within western Uganda during this time (cf. Doyle, 2006) contribute towards a shift in 'traditional' work patterns? Would the observation of fewer rituals and taboos be seen as beneficial in terms of making tasks less timely, costly, and labour specific, and allowing more room for process experimentation? Or would deviating from established work patterns, widely considered to have grown "as a response to procedural failures" (Blakely, 2006, p.11), make the process intolerably risky and unpredictable?

Herbert, in her work on the relationships between iron production, gender and power in sub-Saharan Africa, discusses how 'traditional' tasks were continually contested by "competing social groups or by men and women", forcing shifts in work roles and boundaries (Herbert, 1993, p.222). Although she states that "unless or until it became impossible ... change was accommodated to pre-existing belief rather than undermining it", she goes on to illustrate how significant change has been unavoidable in several recent cases, using as examples the introduction of non-native crops and the advent of cash-cropping, both of which triggered adjustments in the sexual division of labour. Extrapolating from this, if changes in gender roles in relation to agriculture were seen locally to prove beneficial, perhaps (as various stigmas were removed or diminished) this would dispel reluctance to apply similar changes to other economic endeavours, including that of iron production. Similarly, might the movements of different population groups throughout the region have instigated changes in ritual and social practice as well as technical practice (cf. Schmidt, 2009, p.271)?

In the case of smelting in Mwenge, it is now possible to suggest that there is variation within aspects of iron production relating to cultural taboos and restrictions that might previously have been thought of as quite resilient to change. Judging from the evidence presented above, it seems prudent to be receptive (and explicit) about the possibility of such variation in archaeological interpretations and discussions.

It is also relevant to consider an additional possibility: that of a potential bias in the survival or reporting of highly ritualised industries in particular. It is not inconceivable that those technologies that stress the connection between reproduction and smelting have been emphasised by researchers, who see them as perhaps more fascinating and exotic (more publishable?), leaving more 'mundane' examples of technology unreported. Alternatively, but resulting in the same outcome, it may be that highly ritualised technologies are more widely remembered in contemporary Ugandan culture, surviving more robustly in the collective memory. Importantly, these may include the technologies that functioned as 'royal' industries, which were documented in the early twentieth century (i.e. Roscoe, 1923), and which therefore were more strongly incorporated into the ethno-historical records.

There is one additional possibility that may be relevant to this paper – that of research design. The methodology employed during this fieldwork was designed not to generate interview material, but so as to effectively locate new archaeological sites. In contrast to common ethnoarchaeological approaches within a new study area, where a local community leader (often male) would generally be approached in order to gain introductions to informants, this fieldwork methodology guaranteed exposure to a greater cross-section of rural society. This may be the key in generating maximum diversity in future ethnoarchaeological studies, not only within eastern Africa.

Conclusions

In contrast to the prominent emphasis on highlighting variation within the physical and technical elements of the recent iron production technologies of sub-Saharan Africa, the diversity of social approaches to craft organisation and metalworking has not yet been fully elucidated. This is despite a growing number of studies that identify, for example, the complex and changeable scenarios regarding the involvement of women in several smelting and smithing technologies (e.g. Sheridan, 2001; Chirikure, 2007; Schmidt, 2009, p.274). As a consequence of this, discussions have remained dominated by broad generalisations, thereby further precluding the identification of small-scale variation and giving disproportionate weight and value to a single series of observations.

With regards to the cultural and ritual aspects of smelting, there has been a tendency to focus on the unifying similarities between smelting 'traditions' rather than recognising the potential for individual innovation and invention on more than just a technical level (cf. Fowler, 1990). If it is accepted that groups or individuals would experiment with different technical variants, then experimentation with social and ritual rules may be equally plausible in the search for more effective working methodologies. As such, these interview opportunities have served as a reminder of just how much care should be taken when discussing the organisation of production in the recent past, let alone the more distant past. It is vital that we are as critical as possible of our expectations in order to remind ourselves of the very high potential for change and variation in technology and social norms, as we have seen here even within a small time scale and across a small geographical area. By looking for such differences, and balancing them against the need to generalise, we can develop a greater understanding of why these patterns and diversities exist.

Acknowledgments

This paper is based on fieldwork that would not have been possible without the continual support of Elijah Kisembo, Dismas Ongwen, and the National Museum of Uganda, as well as all the students who worked on the project in Uganda. Sponsorship for the fieldwork was provided by the Arts and Humanities Research Council, the University of London and University College London, and the British Institute in Eastern Africa. Thanks also to Jane Humphris and Xander Veldhuijzen for organising the World of Iron conference, as well as offering advice and encouragement, and to an anonymous referee for their valuable and insightful comments.

References

Barndon, R., 2004. *An Ethnoarchaeological Study of Iron-smelting Practices among the Pangwa and Fipa in Tanzania*. British Archaeological Reports International Series 1308. Oxford: Archaeopress.

Blakely, S., 2006. *Myth, Ritual, and Metallurgy in Ancient Greece and Recent Africa*. Cambridge: Cambridge University Press.

Brown, J., 1995. *Traditional Metalworking in Kenya*. Oxford: Oxbow Books.

Buchanan, C., 1974. *The Kitara Complex: The Historical Tradition of Western Uganda to the 16th Century*. PhD Thesis, Indiana University, Bloomington.

Buchanan, C., 1979. Courts, Clans and Chronology in the Kitara Complex. In: J. Webster ed. *Chronology, Migration and Drought in Interlacustrine Africa*. London: Longman, pp.87-124.

Celis, G. and Nzikobanyanka, E., 1976. *La Metallurgie Traditionnelle au Burundi*. Tervuren: Musée Royal de l'Afrique Centrale, Archives d'Anthropologie.

Childs, S.T., 1998a. Social Identity and Craft Specialization among Toro Iron Workers in Western Uganda. *Archaeological Papers of the American Anthropological Association*, 8, pp.109-21.

Childs, S.T., 1998b. 'Find the ekijunjumira': Iron mine discovery, ownership and power among the Toro of Uganda. In: A.B. Knapp, V.C. Pigott and E.W. Herbert eds. *Social Approaches to an Industrial Past: The Archaeology and Anthropology of Mining*. Routledge: London, pp.123-37.

Childs, S.T., 1999. "After all, a hoe bought a wife": the social dimensions of ironworking among the Toro of East Africa. In: M. Dobres and C. Hoffman eds. *The Social Dynamics of Technology: Practice, Politics and World Views*. Washington DC and London: Smithsonian Institution Press, pp.23-45.

Childs, S.T., 2000. Traditional Iron Working: A Narrated Ethnoarchaeological Example. In: M. Bisson, S.T. Childs, P. de Barros and A. Holl eds. *Ancient African Metallurgy: The Socio-Cultural Context*. Walnut Creek: Altamira, pp.199-254.

Childs, S.T. and Killick, D., 1993. Indigenous African metallurgy: nature and culture. *Annual Review of Anthropology*, 22, pp.317-37.

Chirikure, S., 2007. Metals in society: iron production and its position in Iron Age communities of southern Africa. *Journal of Social Anthropology*, 7, pp.72-100.

Cline, W., 1937. *Mining and Metallurgy in Negro Africa*. Wisconsin: Menasha.

Connah, G., 1996. *Kibiro: the salt of Bunyoro, past and present*. London: The British Institute in Eastern Africa.

David, N., Heinmann, R., Killick, D. and Wayman, M., 1989. Between bloomery and blast furnace: Mafa iron smelting technology in North Cameroon. *African Archaeological Review*, 7, pp.183-208.

Doyle, S., 2006. *Crisis and Decline in Bunyoro: Population and Environment in Western Uganda 1860-1955*. Oxford: James Currey.

Fowler, I., 1990. *Babungo: A Study of Iron Production, Trade and Power in Nineteenth Century Ndop Plain Chiefdom (Cameroon)*. PhD Thesis, University College London.

Giles, M., 2007. Making metal and forging relations: ironworking in the British Iron Age. *Oxford Journal of Archaeology*, 26, pp.395-413.

van Grunderbeek, M.-C., 1992. Essai de delimitation chronologie de L'Age du Fer Ancien au Burundi, au Rwanda et dans la region des Grand Lacs. *Azania*, 27, pp.53-80.

van Grunderbeek, M.-C., Roche, E. and Doutrelepont, H., 1983. *Le Premier Age du Fer au Rwanda et au Burundi, Archaeologie et Environment*. Brussels: Institut National de Recherché Scientifique.

Haaland, R., 2004. Iron smelting – a vanishing tradition: ethnographic study of this craft in south-west Ethiopia. *Journal of African Archaeology*, 2, pp.65-79.

Haaland, R., 2006. Iron in the making – technology and symbolism: ethnographic perspectives on European iron working. In: A. Andrén, K. Jennbert and C. Raudvere eds. *Old Norse religion in long-term perspectives: origins, changes, and interactions*. Lund: Nordic Academic Press, pp.79-85.

Herbert, E., 1993. *Iron, Gender and Power: Rituals of Transformation in African Societies*. Bloomington: Indiana University Press.

Iles, L., 2009. Pre-colonial iron production in western Uganda: recent survey and excavations. *Nyame Akuma*, 71, pp.35-45.

Iles, L., 2011. *Reconstructing the Iron Production Technologies of Western Uganda: Reconciling Archaeometallurgical and Ethnoarchaeological Approaches*. PhD Thesis, University College London.

Ingham, K., 1975. *The Kingdom of Toro in Uganda*. London: Methuen.

Killick, D., 1991. The relevance of recent African iron-smelting practice to reconstructions of prehistoric smelting technology. In: P. Glumac ed. *Recent Trends in Archaeometallurgical Research*. MASCA Research Papers in Science and Archaeology 8. Philadelphia: MASCA, pp.47-54.

Kusimba, C., 1996. The social context of iron forging on the Kenya coast. *Africa*, 66, pp.386-410.

van der Merwe, N.J. and Avery, D.H., 1987. Science and magic in African technology: traditional iron smelting in Malawi. *Africa*, 57, pp.143-72.

van Noten, F. ed., 1983. *Histoire Archeologique du Rwanda*. Tervuren: Musee Royal de l'Afrique Centrale.

Reid, A. and MacLean, R., 1995. Symbolism and the Social Contexts of Iron Production in Karagwe. *World Archaeology*, 27, pp.144-61.

Reid, R., 2002. *Political Power in Pre-colonial Buganda: Economy, Society and Welfare in the Nineteenth Century*. London: James Currey.

Roscoe, J., 1923. *The Bakitara or Banyoro*. Cambridge: Cambridge University Press.

Schmidt, P. ed., 1996. *The Culture and Technology of African Iron Production*. Gainesville: University of Florida Press.

Schmidt, P., 2001. Resisting homogenization and recovering variation and innovation in African iron smelting. *Mediterranean Archaeology*, 14, pp.219-27.

Schmidt, P., 2009. Tropes, Materiality, and Ritual Embodiment of African Iron Smelting Furnaces as Human Figures. *Journal of Archaeological Method and Theory*, 16, pp.262-82.

Schmidt, P. and Mapunda, B., 1997. Ideology and the Archaeological Record in Africa: Interpreting Symbolism in Iron Smelting Technology. *Journal of Anthropological Archaeology*, 16, pp.73-102.

Sheridan, M., 2001. Cooling the Land: The Political Ecology of the North Pare Mountains, Tanzania. PhD Thesis, Boston University Graduate School of Arts and Sciences.

Uzoigwe, G., 1976. Precolonial markets in Bunyoro-Kitara. In: B. Ogot ed. *Economic and Social History of East Africa*. Nairobi: Kenya Literature Bureau, pp.24-66.

Author's address

- UCL Institute of Archaeology, 31-34 Gordon Square, London WC1H 0PY
  louise.iles@gmail.com

Evolutionary analysis of ironmaking slag

Michael F. Charlton, Stephen J. Shennan, Thilo Rehren, and Peter Crew

ABSTRACT: The remnants of pre-industrial ironmaking record a diverse array of forms and activities across space and time. The most commonly encountered explanation for this diversity stipulates that variations in ironmaking tools and techniques arise as local adaptations to social and economic norms. Despite this assertion, there is remarkably little evidence to demonstrate how and why new ironmaking variants were introduced, adopted, or transmitted. Neo-Darwinian evolutionary theory is specifically crafted to generate explanations for diversity in any system involving the transfer of variable information between individuals. The application of neo-Darwinian theory to bloomery iron production can add significant insights into the structure and trajectories of technological lineages. Two models are presented for building and explaining ironmaking lineages from slag chemistry. Their application to slag from Llwyn Du, a late medieval bloomery in northwest Wales, highlights the sometimes unexpected results obtained from evolutionary analysis and the new questions it stimulates.

Keywords: Evolutionary theory, slag chemistry, bloomery, lineage, selection.

Introduction

The means by which humans have extracted, crafted and used iron metal are incredibly diverse. Explaining how this diversity arose remains one of the most challenging topics in archaeometallurgy (e.g., Pigott, 1982; van der Merwe and Avery, 1982; Gordon and Killick, 1993; Charlton, 2009; Juleff 2009). While ironmaking variants are routinely interpreted as adaptations to available resources, economic networks and societal values, there remains little analytical investment in building explanations for how and why the technology developed along specific paths. Full explanations must ultimately state how variation in chemically constrained ironmaking systems arises and is shaped by cultural transmission processes, psychological biases, behavioural strategies, and fluctuating environments. The application of Neo-Darwinian evolutionary theory to the remnants of past ironmaking activity can provide a means to build and test hypotheses about the causes of the technology's diversification and change.

Neo-Darwinian evolutionary theory, not to be confused with progressive theories of social and cultural change, is a collection of principles that relate historical phenomena – those which display variation, heritability, and the differential persistence of variants across space and time – to natural processes that lead to 'descent with modification'. Models derived from evolutionary theory generate explanations for patterns of variation in terms of constrained and unconstrained transmission, behavioural ecology, drift and selection (Shennan, 2002). Once applied to the empirical record, the final products include testable narrative histories and probabilistic predictions of the kinds of patterns expected under specified conditions (O'Brien and Lyman, 2000).

Ironmaking technologies are historical phenomena that exhibit high levels of both spatial and temporal diversity (e.g., Killick, 1991; Pleiner, 2000). Knowledge of ironmaking technologies is transmitted from master to apprentice (e.g., Kusimba, 1996) or professor to student through varied mechanisms of social teaching and learning. Iron production technology also changes through time, with new variants emerging in the course of history and competing with others for adoption and replication (e.g., Schubert, 1957). Neo-Darwinian evolutionary theory is therefore applicable to the archaeological study of ironmaking and can provide new insights into processes of innovation and change, as well as into the relationships of the technology to other aspects of culture, society, and economy.

The basic tenets of neo-Darwinian theory are introduced below. Two models are then presented for describing the evolution of smelting recipes as approximated by slag chemistry. The first model identifies lineages of ironmaking recipes, establishing or falsifying evidence of technological heritability. The second model predicts the optimal smelting strategy in terms of metallic yield for specified economic environments.

Both models are applied to the chemistry of ironmaking slag excavated from Llwyn Du, a small to medium scale bloomery in northwest Wales (Crew and Crew, 2001; Crew and Charlton, 2007). Part of a cluster of bloomeries in Coed y Brenin, Merioneth (Crew and Crew, 1995; Smith, 1995), Llwyn Du produced iron in three distinct non-continuous phases between 1355 and 1419 AD (Crew and Mighall, this volume). These phases are roughly recorded in five main deposits of slag (A → B1-B2 → B3 → C1 → C2) that form, along with other ironmaking residues, the site's prominent slag heap. The slag was tapped from a bellows-driven low-shaft clay furnace with an internal diameter of about 40cm and wall thicknesses ranging between 20 and 40cm. The furnace was charged with two types of locally obtained bog ore (Charlton, 2007) – a red ore rich in Fe_2O_3 (≈ 83 wt%) and a black ore rich in MnO

(≈ 54 wt%) – and charcoal dominated by oak and birch species (Crew and Mighall, this volume).

Results show that the recipes at Llwyn Du descended with modification, though not in the direction predicted. Both high and low yielding variants evolved from a more generalised recipe and were used simultaneously. This implies that metallic yield was not the only criterion under the influence of selection and raises important questions about the qualities of iron produced in late medieval Welsh bloomeries. This example highlights the utility of neo-Darwinian theory for shedding new light on problems related to the origin and adaptation of bloomery technologies even when its hypotheses yield negative results.

Neo-Darwinian evolutionary theory

Neo-Darwinian evolutionary theory seeks to generate testable explanations for historical patterns of variation arising from the production and differential transfer of information between entities within a population. In the case of technology, artefacts and documented ideas (usually in the form of notes, plans, or models) constitute entities with variable attributes and attribute-states that are transmitted and replicated through human agency and processes of social learning. Causes of differential transmission are attributed to natural sorting processes, the two most notable being drift and selection. Constraints are also imposed on evolving systems, effectively pushing variation along certain structurally preferred pathways. Indeed, historical phenomena are likely to contain patterns of variation produced by a combination of selection, drift, and constraints.

Drift is nothing more than sampling error in the transmission process and causes variation to be distributed stochastically across spacetime (O'Brien and Lyman, 2000). In other words, drift patterns the frequency distribution of variants at any point in spacetime due to the natural consequences of random transmission and replication events. All things being equal, some variants will be replicated more often than others simply by chance. Some ironmasters, for example, may have more apprentices than others of equal quality through happenstance. This leads to a relative increase in the replication of their smelting recipes in the next generation, as the apprentices become masters. The cumulative effects of drift lead to decreasing variation and trait fixation unless checked by the introduction of novel variants through invention or innovation (e.g., Neiman, 1995).

Selection is a covering term for any sorting process that systematically increases the relative proportion of variants because they possess a specific quality or set of qualities that bias their probability of survival and reproduction (O'Brien and Holland, 1995; O'Brien and Lyman, 2000). The result is a correspondence between specified variants and environmental contexts. Selection operates at different scales and can take a number of different forms. Selective sorting occurs in technological systems when there is variation in engineering performance capabilities, production strategies, training regimes, or alternative ways of making a living. Importantly, selection favours those variants that are optimally suited to a given environment rather than those variants that perform best along some individual performance dimension.

Constraints are biases on the generation and production of variant forms (Maynard Smith, et al., 1985; Gould, 1989; Amundson, 1994; O'Brien and Lyman, 2000). They arise as natural consequences of universal laws and configurational circumstance. Universal laws create natural boundaries for the possibilities of particular variants ever existing. Likewise, unique historical configurations also limit the production of variants. Configurational constraints on culture include such factors as the mode and medium of transmission (Cavalli-Sforza and Feldman, 1981), as well as psychological biases on cultural trait replication (Boyd and Richerson, 1985). Regardless of where constraints originate in any particular case, it is imperative that they be identified, at least at a cursory level, and included amongst the stated initial conditions in any evolutionary analysis.

Drift, selection, and the biasing influence of constraints act continuously on an evolving system, leading to shifting patterns of variation and diversity through the course of history. Variation provides the fuel, transmission of information provides the spark, and sorting mechanisms provide the motion of the evolutionary machine. While it is impossible to foretell the final destination of an evolving system, it is possible to predict the causes of particular patterns of variation and the conduits they take by modelling the action of sorting mechanisms on specific kinds of variants in defined environmental settings. The two models below can be used for identifying evolving iron production systems and hypothesising one possible selective influence on iron production strategies.

Modelling the evolution of bloomery recipes

Investigating problems of evolutionary change requires two steps. First, lineages are hypothesised from models designed to reconstruct transmission pathways. The identification of cultural lineages or traditions ensures that changes in variation are confined to related systems rather than replacement of one system with another. Second, lineage structures are explained using models of evolutionary sorting mechanisms acting in defined contexts. The result is a testable historical narrative, rather than a simple chronicle or listing of disassociated events (O'Hara, 1988).

Constructing technological lineages from slag chemistry

Identifying a lineage requires the inference of transmission from patterns of variation distributed across spacetime. Seriation provides the most well established means of constructing cultural lineages in archaeological contexts (Neiman, 1995; Lipo, et al., 1997; O'Brien and Lyman, 2000; Shennan, 2002). The seriation procedure involves the temporal arrangement of artefact assemblages based on the distribution of artefact classes or according to some rule of development. Assuming that all assemblages under examination represent approximately similar durations of accumulation and derive from the same local area, chronology can be inferred when each class in the arrangement occupies a single segment and overlaps with at least one other class in the same segment. Computer models have demonstrated that this is precisely the pattern of evolution expected from neutral trait transmission (Neiman, 1995; Lipo, et al., 1997). Although most applications of seriation have relied on artefact classes based on decorative attributes, there is no logical reason to assume that compositional data cannot be used to yield similar stochastic patterns.

Seriation-like techniques have been used to examine temporal patterns of variation in slags (Koucky and Steinberg, 1982; Charlton, 2002), but the systematics involved in building classes is somewhat problematic. First, most artefact classes and other taxonomic units are defined by discrete qualitative attributes and similarly discrete qualitative attribute-states. The chemical attributes of slag are all continuous quantitative variables. Continuous attributes are culturally transmitted, but classes constructed from them tend to break up variation along morphological dimensions (morpho-space), in this case chemical dimensions (chem-space), in wholly arbitrary and non-theoretically driven ways. Second, the seriation method hinges on the existence of historical types, i.e., types that have a continuous distribution and occur only once in spacetime (O'Brien and Lyman, 2000). Often this requirement has been worked out on the basis of experience and trial and error. It is unclear whether or not classes constructed from slag chemistry represent single distributions in spacetime. Paleobiologists (e.g., Saunders and Swan, 1984; Reyment, 1985) have noted that even while species morphology is inherited and displays a stochastic distribution in spacetime, the actual location of a specimen in morpho-space may be more a function of local constraints or selection than of heredity. The factors influencing slag chemical composition are analogous.

Principal component analysis (PCA) provides a means of representing multiple continuous quantitative attributes, many of which may be transmitted through social learning. Biologists routinely use principal component (PC) axes as a natural coordinate system for representing the evolution of morphology (Saunders and Swan, 1984; Foote, 1992, 1993, 1994; Maynard Smith, et al., 1985; Reyment, 1985). A PCA approach similar to those employed by biologists should be amenable to representing heritable variation in iron production slag and enabling changes to be tracked across time.

The first step consists of ensuring that the slag assemblages are derived from the same parent materials, with allowances for some natural variation. This procedure prevents geological constraint from being an explanation for observed patterns. Second, all slag assemblages must be assigned to single segments of spacetime, preferably with equal durations. This requires an independent chronological indicator, such as radiocarbon dating or superposition. Third, a PCA is conducted on a data matrix constructed from the entire sample. The chemical variables used in the analysis should be chosen so that the structure of their loadings will have some meaning (see Charlton, et al., this volume). The research presented here uses twelve oxides (Table 1), chosen because of their association with specific parent materials and their relative ease of determination. Fourth, the scores for each specimen are calculated and plotted in PC space, and arranged in temporal order. The fifth and final step assesses the shape of the PC score distribution. While there are several approaches one might adopt (see Charlton, 2009), Kernel Density Estimation (KDE) is employed here. KDEs form data 'islands' with contours or shells that describe the density of points within them. When graphed over time, the distribution of points will overlap between contiguous time steps if they are the product of an evolving technological system. Interrupted distributions are only permissible where sample size is recognisably poor.

This procedure was applied to slag excavated from each stratum of the Llwyn Du slag heap. Fifty-one specimens were chemically analysed with a polarised energy dispersive X-ray fluorescence spectrometer at the Wolfson Archaeological Science Laboratories housed at the UCL Institute of Archaeology. The sample consisted entirely of tap slags assigned to production phases based on their stratigraphic provenance (Charlton, 2007). A PCA was conducted on the sample using the oxides listed in Table 1. These were then classified according to the procedure described above, using the first two PC axes. Fig. 1 provides a graphical representation of the PCA data over time. Differences in point colour are derived from the groups formed by a KDE of the scores projected on to the first and second PC axes.

The pattern of overlapping point distributions in Fig. 1 shows that all slag derives from a single continuous lineage of iron production technology, albeit one that evolved into two distinct recipes. Two noteworthy trends are apparent from an inspection of PC01-PC03. The first two trends are linked to the branching evolution that seems to originate in Phase B3. Both recipes appear to be divergent specialisations that evolved from a more generic ancestor. The slags associated with the right-hand branch (Group 1) exhibit chemistries that are more depleted of FeO than their predecessors. This pattern

Table 1: Variables used to construct lineages and their affiliation to slag parent materials

	MgO	Al_2O_3	SiO_2	P_2O_5	SO_3	K_2O	CaO	TiO_2	MnO	FeO	SrO	BaO
ore	●	●	●	●			●	●	●	●	●	●
clay		●	●			●		●				●
fuel ash	●			●	●	●	●				●	

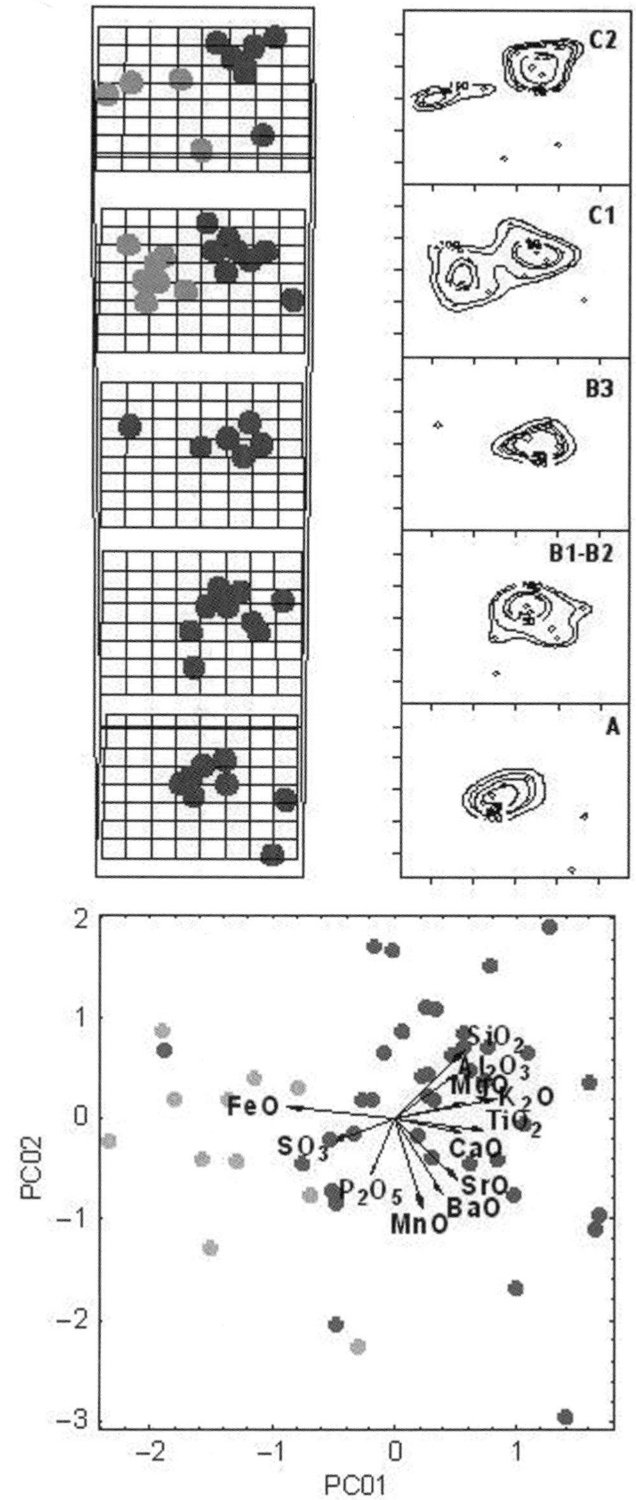

Figure 1: Llwyn Du slag chemistry projected into principal component space and arranged by stratigraphic position. Group designations were determined by two-dimensional KDE. Group 1 slag (black) displays a relatively low proportion of FeO relative to other oxides, while group 2 slag (grey) contains a higher relative abundance of FeO.

is especially clear when rendered as a two-dimensional KDE of PC01 vs. PC02. Slags associated with the left-hand branch (Group 2) have chemistries with higher concentrations of FeO. This pattern of change implies the evolution of two recipes with different yield efficiencies.

Modelling selection from slag chemistry

There are several ways in which selection can manifest itself in iron production systems. Modelling the adaptive landscape (or chem-scape) for metallic yields has the fewest data requirements and is the simplest to consider. Selection will always favour those variants of the bloomery process capable of maximising the value of iron produced relative to the total costs in labour and resources. In so doing, it will favour those bloomery methods whose iron yields are optimal with respect to given market conditions and promote reduction in product variance. Specific changes that affect the yield include altering techniques of ore preparation, furnace construction/repair and furnace operation, as well as shifting choices of ore, clay, and flux resources.

Understanding the relationships between iron yield and the bulk chemistry of ore and slag is facilitated by the use of phase diagrams. Ternary phase diagrams derived from the (CaO + BaO + Alk_2O)–(FeO + MnO + ZnO + MgO)–Al_2O_3–(SiO_2 + P_2O_5) quaternary system are the most commonly used for investigating iron smelting technology (Bachmann, 1982). The most valuable phase diagram for the specific set of materials under scrutiny here is the FeO–SiO_2–Al_2O_3 system (Fig. 2).

Two low-temperature optima exist for fayalitic slag compositions as their chemistry accommodates increasing quantities of Al_2O_3. Although most bloomery furnaces can easily obtain the free-running temperatures for the entire fayalitic portion of the FeO–SiO_2–Al_2O_3 diagram, it is of practical significance that Optimum 1 utilises approximately 20% less iron than Optimum 2. However, each optimum is associated with very different levels of risk and attainability.

Smelting at Optimum 1 results in a higher metallic yield but is also at the base of a very steep temperature precipice and borders a high viscosity glass phase. Producing slag with this chemistry without introducing significant contributions of silica from the furnace wall demands precise control of reducing conditions and temperature. This carries an enormous risk of failure. The reducing conditions necessary to inhibit the formation of Fe_2SiO_4 or break the FeO-SiO_2 bond are significantly greater than that necessary to reduce FeO. They are so great that reduced iron in the system is quickly saturated with carbon, resulting in the production of steel or cast iron.

Smelting at Optimum 2 results in a much lower yield but has a much gentler surrounding temperature gradient for free-running slag with varying chemistry. The phases that it borders also have relatively low viscosity. This increases the range of permissible variation in the smelting recipe while minimising the risk of failure.

The manner by which an individual goes about making a living is critical to his or her fit with a given environment (broadly construed to include social, cultural, and natural components). Iron production is, at its basal level, an economic activity, and the ironmaker's fitness is in large part determined by the resources, prestige, or other honours he or she is able to acquire in return for expended effort. Selection is, therefore, expected to play a strong role in the evolution of iron production strategies. Depending on the nature of the economy, Optima 1 and 2 act as basins of attraction for high and low yielding production strategies. Selection

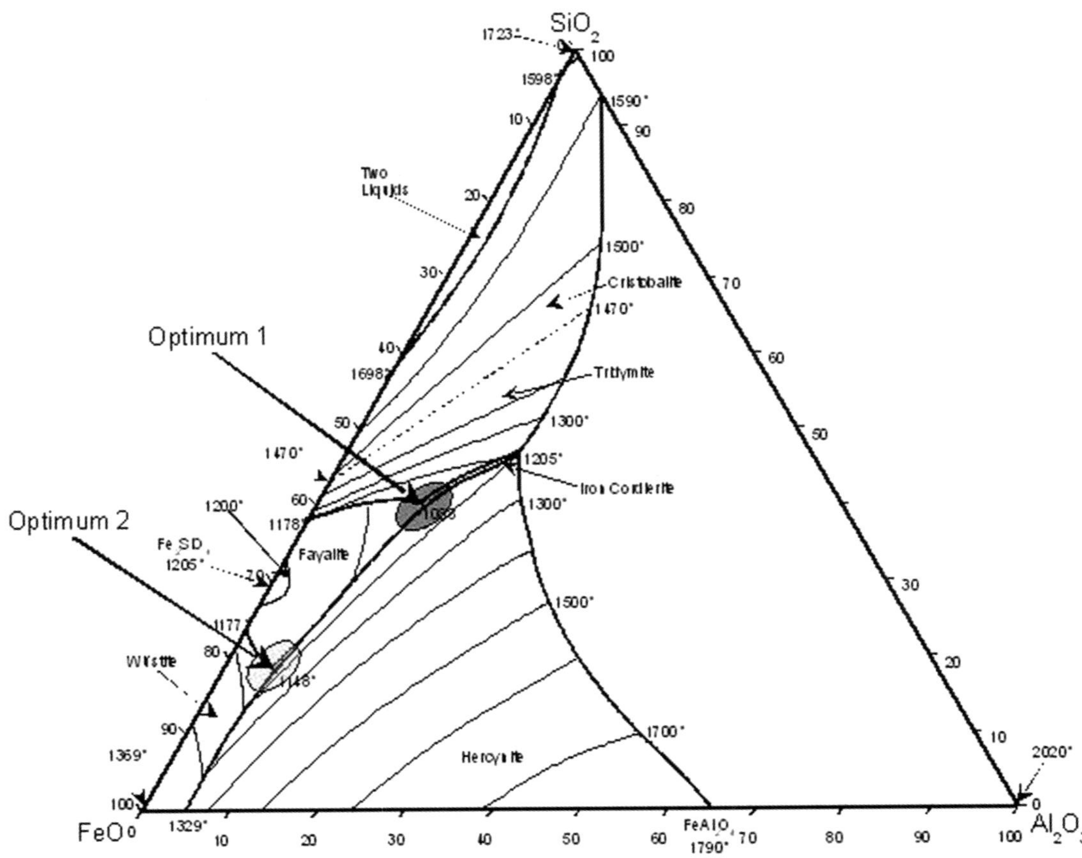

Figure 2: Partial rendering of the $FeO–SiO_2–Al_2O_3$ ternary system (after Levin, et al., 1964, p.241, Fig. 696). Optima 1 and 2 identify the two ternary eutectics that serve as basins of attraction for bloomery slag chemistry under different socioeconomic environments.

favours the movement of slag chemistry towards Optimum 1 when the demand for iron is high and the cost of procuring resources is high relative to the benefits of production, or when both demand and competition are high. Greater energy and more skilled labour are needed to maintain the more exacting reducing conditions. Socioeconomic environments such as this should experience increasing returns to scale. In economic environments where the costs of acquiring resources are low, the benefits of producing iron are high, but demand is low (or competition between ironmakers is low), selection favours slag chemistries that approximate Optimum 2. Less energy and labour are required to maintain appropriate temperatures and reducing conditions. If the economic control of iron is monopolistic or oligopolistic, benefits to the ironmaster might actually decrease with increasing scales of production.

The economy of late medieval Britain was a complex multi-tiered state system in which smelters operated a large variety of furnace types and served markets of varying scales. Ironworkers of all scales had to compete with foreign producers in both general and specialised product markets (Childs, 1981; Threlfall-Holmes, 1999). The price of iron was high and became inflated during the mid-fourteenth century following the Black Death, as labour costs increased. The demand for iron, however, continued to grow.

Given some basic facts about the nature of the British economy in the late middle ages, selection is expected to promote high yielding iron production strategies that generate slag approaching Optimum 1. A projection of Llwyn Du slag onto the $FeO–SiO_2–Al_2O_3$ ternary phase diagram is shown in Fig. 3. Slag from phases A, B1-B2, and B3 are clustered between Optima 1 and 2 and remain stable. Slag from Phases C1 and C2 break into two groups (as noted in the lineage), with one cluster centred closer to Optimum 1 and the other centred on Optimum 2.

The failure to meet modelled expectations implies that the model itself is too simplistic to adequately encompass the complexities of the medieval iron economy. Certainly, increasing metallic yield was not the only factor involved in the ironmaker's production strategy. One possibility is that quality also played an important role. The most likely scenario involves the production of two metallic products – bloomery iron and steel. The strength of the reducing conditions that would be required to force slag chemistry above the pure fayalite line are likely to promote the diffusion of carbon into iron. While steel was expensive and valued for many purposes, its hardness also made it unsuited for some tasks. The Llwyn Du ironmakers may have found that a mixed production strategy allowed them to maximise their returns relative to material demands. Testing this possibility requires more detailed knowledge of the price differentials between iron and steel, as well as accurate assessments of resource costs and rates of consumption. Evolutionary game theory could then be used to construct a model for the optimum production strategy for smelting enterprises at Llwyn Du.

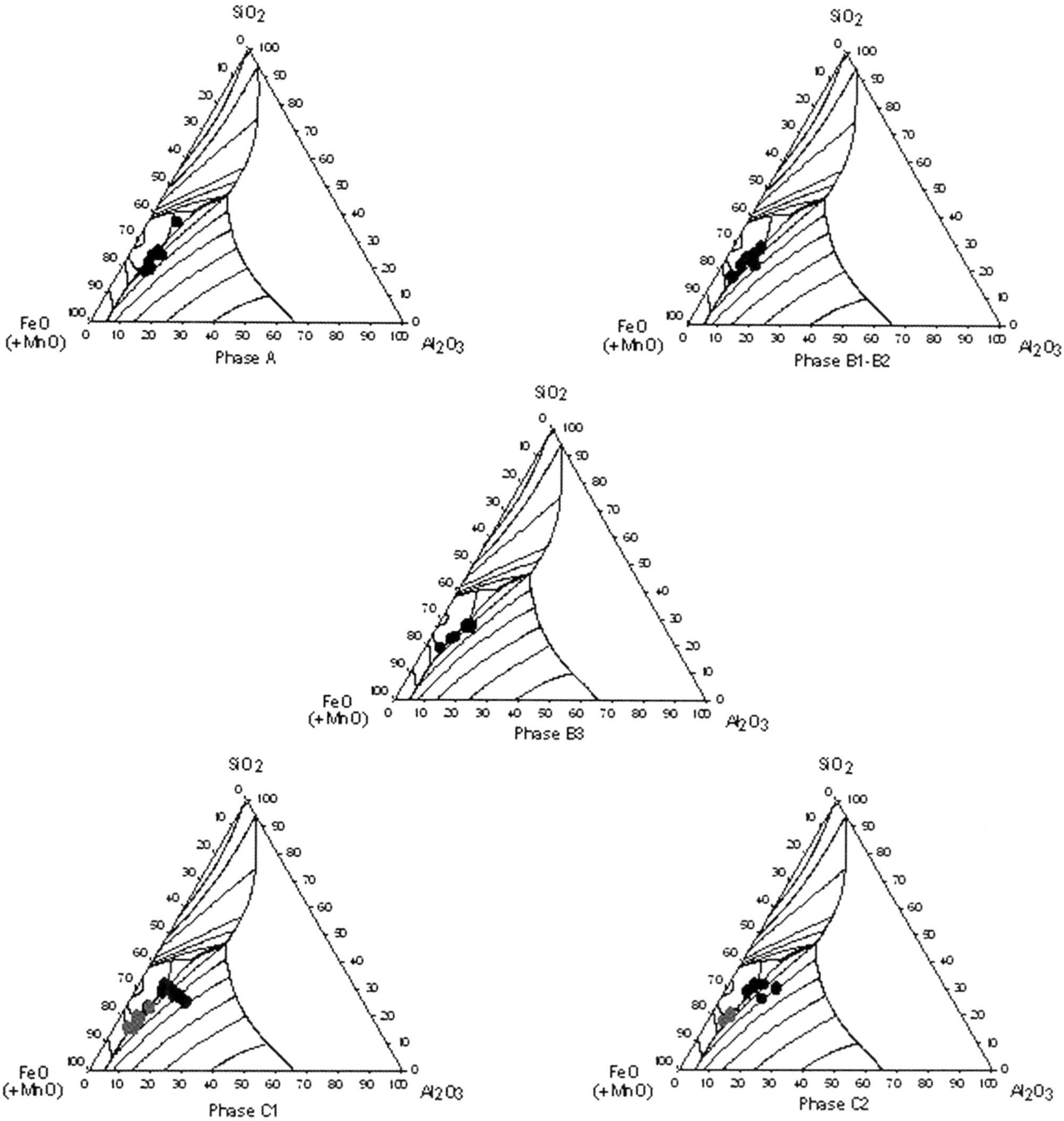

Figure 3: Llwyn Du slag chemistry projected into the $FeO–SiO_2–Al_2O_3$ ternary phase diagram by stratigraphic position. Group 1 slag is plotted in black. Group 2 slag is plotted in grey.

Conclusions

Continuity and change are topics often discussed in relation to iron production technology. While many studies have documented their co-occurrence, little effort has been devoted to explaining the phenomenon in theoretical terms. Neo-Darwinian evolutionary theory provides a powerful scientific framework for explaining the history of iron production. Continuity is identified in terms of technological lineages created through the transmission of knowledge via social learning. Variation and change occurring within and between traditions are explained in terms of random processes, selection, and constraints. These processes can be used to formulate predictions for how variation should be expressed in different environmental circumstances (including the structure of social relationships when observable). Processes that took place in the past cannot be tested directly, but predictions can be compared to patterns of variation that have accumulated in the archaeological record. Even when hypotheses are falsified, knowledge is gained.

The models presented in this paper are early efforts to apply neo-Darwinian evolutionary theory to iron production slag chemistry. When applied to slag from Llwyn Du, these models generated hypotheses about lines of descent and production strategies. Not only did iron smelting recipes change over time, they diverged. Although conflicting results were obtained from applying the yield strategy model to Llwyn Du slag, many more questions can now be raised about the qualities of iron produced in Welsh furnaces and the nature of the medieval iron economy. The preferred hypothesis is that the two recipes produced distinct qualities of iron, and their combined production represented an optimal production strategy for maximising profits relative to capital investments. Testing this hypothesis will require more intensive examinations of both the archaeological and historical records.

Theories of technology are only useful to the archaeology of iron production if the questions they raise lead to greater understanding of the empirical record. Neo-Darwinian evolutionary theory is not perfect and not likely to solve every problem it confronts. However, it will advance our understanding of iron production and other technologies by suggesting plausible processes to explain ultimate level patterns of variation observed in the archaeological record.

Acknowledgements

The authors wish to thank the organisers and sponsors of the World of Iron conference, without whose generous support this paper could not have been presented. An anonymous reviewer provided insightful and constructive criticism. We also thank the many conference participants for their comments and informal discussions. Much of the research presented in this paper was conducted with the help and patience of the staff at the Wolfson Archaeological Science Laboratories housed at the Institute of Archaeology, University College London.

References

Amundson, R., 1994. Two Concepts of Constraint: Adaptationism and the Challenge from Developmental Biology. *Philosophy of Science*, 61, pp.556-78.

Bachmann, H-G., 1982. *The Identification of Slags from Archaeological Sites*. Occasional Publication 6. London: Institute of Archaeology.

Boyd, R. and Richerson, P.J., 1985. *Culture and the Evolutionary Process*. Chicago: University of Chicago Press.

Cavalli-Sforza, L.L. and Feldman, M.W., 1981. *Cultural Transmission and Evolution: A Quantitative Approach*. Princeton: Princeton University Press.

Charlton, M.F., 2002. *Explaining Temporal Variation in Slag Chemistry at the Titelberg Hillfort, Luxembourg*. MA Dissertation, University of Missouri – Columbia.

Charlton, M.F., 2007. *Ironworking in Northwest Wales: An Evolutionary Analysis*. PhD Thesis, University College London.

Charlton, M.F., 2009. Identifying iron production lineages: a case study in northwest Wales. In: S.J. Shennan ed. *Pattern and Process in Cultural Evolution*. Berkeley: University of California Press, pp.133-44.

Childs, W.R., 1981. England's Iron Trade in the Fifteenth Century. *The Economic History Review*, 34, pp.25-47.

Crew, P. and Charlton, M., 2007. The anatomy of a furnace...and some of its ramifications. In: S. La Niece, D. Hook and P. Craddock eds. *Metals and Mines: Studies in Archaeometallurgy*, pp.219-25. London: Archetype Publications.

Crew, P. and Crew, S., 1995. Medieval bloomeries in north-west Wales. In: G. Magnusson ed. *The Importance of Ironmaking: Technical Innovation and Social Change*. Stockholm: Jernkontorets Bergshistoriska Utskott, pp.43-50.

Crew, P. and Crew, S., 2001. Excavations at Llwyn Du, Coed y Brenin, Merioneth, 2001: woodland management and charcoal processing at a late 14th century ironworks. *Archaeology in Wales*, 41, pp.83-7.

Foote, M., 1992. Rarefaction Analysis of Morphological and Taxonomic Diversity. *Paleobiology*, 18, pp.1-16.

Foote, M., 1993. Discordance and Concordance between Morphological and Taxonomic Diversity. *Paleobiology*, 19, pp.185-204.

Foote, M., 1994. Morphological Disparity in Ordovician-Devonian Crinoids and the Early Saturation of Morphological Space. *Paleobiology*, 20, pp.320-44.

Gordon, R. and Killick, D., 1993. Adaptation of Technology to Culture and Environment: Bloomery Iron Smelting in America and Africa. *Technology and Culture*, 34, pp.243-70.

Gould, S.J., 1989. A Developmental Constraint in *Cerion*, with Comments of the Definition and Interpretation of Constraint in Evolution. *Evolution*, 43, pp.516-39.

Juleff, G., 2009. Technology and evolution: a root and branch view of Asian iron from first millennium BC Sri Lanka to Japanese steel. *World Archaeology*, 41, pp.557-77.

Killick, D., 1991. Relevance of Recent African Iron-Smelting Practice to Reconstructions of Prehistoric Smelting Technology. In P.D. Glumac ed. *Recent Trends in Archaeometallurgical Research*. MASCA Research Papers in Science and Archaeology 8. Philadelphia: MASCA, The University Museum of Pennsylvania, pp.47-54.

Koucky, F.L. and Steinberg, A., 1982. The Ancient Slags of Cyprus. In: J.D. Muhly, R. Maddin and V. Karageorghis eds. *Early Metallurgy in Cyprus, 4000-500 B.C.* Cyprus: Plerides Foundation, pp.143-52.

Kusimba, C.M., 1996. The Social Context of Iron Forging on the Kenya Coast. *Africa*, 66, pp.386-410.

Levin, E.M., Robbins, C.R., and McMurdie, H.F., 1964. *Phase Diagrams for Ceramists*. Columbus: The American Ceramic Society.

Lipo, C.P., Madsen, M.E., Dunnell, R.C. and Hunt, T., 1997. Population Structure, Cultural Transmission, and Frequency Seriation. *Journal of Anthropological Archaeology*, 16, pp.301-33.

Maynard Smith, J., Burian, R., Kauffman, S., Alberch, P., Campbell, J., Goodwin, B., Lande, R., Raup, D. and Wolpert, L., 1985. Developmental Constraints and Evolution. *The Quarterly Review of Biology*, 60, pp.265-87.

Neiman, F.D., 1995. Stylistic Variation in Evolutionary Perspective: Inferences from Decorative Diversity and Interassemblage Distance in Illinois Woodland Ceramic Assemblages. *American Antiquity*, 60, pp.7-36.

O'Brien, M.J. and Holland, T.D., 1995. The Nature and Premise of a Selection-Based Archaeology. In: P.A. Teltser ed. *Evolutionary Archaeology: Methodological Issues*. Tucson: The University of Arizona Press, pp.175-200.

O'Brien, M.J. and Lyman, R.L., 2000. *Applying Evolutionary Archaeology: A Systematic Approach*. New York: Kluwer Academic/Plenum Publishers.

O'Hara, R.J., 1988. Homage to Clio, or, Toward an Historical Philosophy for Evolutionary Biology. *Systematic Zoology*, 37, pp.142-55.

Pigott, V.C., 1982. The innovation of iron: cultural dynamics in technological change. *Expedition*, 25, pp.20-5.

Pleiner, R., 2000. *Iron in Archaeology: The European Bloomery Smelters*. Praha: Archeologický ústav AVČR.

Reyment, R., 1985. Phenotypic Evolution in a Lineage of the Eocene Ostracod *Echinocthereis*. *Paleobiology*, 11, pp.174-94.

Saunders, B.W. and Swan, A.R.H., 1984. Morphology and Morphological Diversity of Mid-Carboniferous (Namurian) Ammonoids in Time and Space. *Paleobiology*, 10, pp.195-228.

Schubert, H.R., 1957. *History of the British Iron and Steel Industry from c. 450 B.C. to A.D. 1775*. London: Routledge and Kegan Paul.

Shennan, S.J., 2002. *Genes, Memes and Human History: Darwinian Archaeology and Cultural Evolution*. London: Thames & Hudson Ltd.

Smith, K.E.S., 1995. Iron-working in north-west Wales in the late fourteenth century. *Archaeological Journal*, 152, pp.246-90.

Threlfall-Holmes, M., 1999. Late Medieval Iron Production and Trade in the North-East. *Archaeologia Aeliana*, 27, pp.109-22.

van der Merwe, N.J. and Avery, D.H., 1982. Pathways to steel. *American Scientist*, 70, pp.146-55.

Authors' addresses

- Michael F. Charlton,
  Permanent Address: 5430 127th PL SE, Snohomish, WA 98296 USA
  E-mail: mike.charlton@ucl.ac.uk
- Michael F. Charlton and Stephen J. Shennan,
  I nstitute of Archaeology, University College London, 31-34 Gordon Square, London, WC1H OPY, UK
- Thilo Rehren,
  UCL Qatar, Hamad bin Khalifa University, Georgetown Building, Doha, Qatar
  Email: director.qatar@ucl.ac.uk
- Peter Crew,
  Pen Cefn, Penrhyndeudraeth, Gwynedd LL48 6DR, UK
  Email: crew.peter.susan@gmail.com

Nature's laws & culture's norms. A proposed systematic classification framework for the study of material culture and technology (SOMCAT)

Ann Feuerbach

ABSTRACT: The purpose of this paper is to address problems that have hindered material culture and technology research over the years, and to introduce a simple, flexible, systematic classification framework where data is compatible, comparable, expandable, and inherently allows for different approaches. This framework is designed to facilitate objective analysis by separating the empirical and ideational characteristics of material culture and technology and to create units that can be used with or without existing statistical programs. A case study on crucible Damascus steel will also be presented.

Keywords: Material culture, technology, methodology, classification, Damascus steel, wootz, crucible steel.

Introduction

The lack of a systematic approach in archaeology has been previously discussed primarily by Dunnell (1971, 1980) and O'Brien and Lyman (2000). O'Brien and Lyman (2000, p.189) give three reasons why systematics are generally ignored. The first reason is because it deals with ontological, epistemological, and metadata issues. The second reason is that precise terminology is necessary and this needs to be learned. The third is that it can be tedious to teach. Nevertheless, a systematic approach is invaluable for advancing research. To address these problems, this paper introduces a simple systematic classification framework for the Study of Material Culture and Technology (SOMCAT) which reduces the influence of our own cultural preconceptions and intellectual traditions and inherently allows for different research questions, epistemologies, and methodologies, while producing data that is compatible, comparable, and reusable, within and between different disciplines. This is achieved by separating the observable (empirical) and cognitive (ideational) characteristics of material culture and technology into objective units (data). These units can be used with or without existing statistical programs. The framework is based on the premise that all material culture is fundamentally composed of linked interrelated Variation Units: Attributes, Actions, and Socio-Cultural Variation Units.

While works of art are often coveted for their aesthetic appeal, and for the artistic and cultural information they convey, the technology used to create them is often overlooked. Technology involves taking what nature provides and then transforming these raw materials into artefacts. Both technology and material culture originate from, propagate through, reflect and reinforce the beliefs and behaviours of people. Not only do they provide people with the resources to control their environment, they also help people understand and explain the world in which they live. Thus, all technology and material culture have two inseparable parts. First, there are the observable physical properties and processes. Depending on the discipline, these are referred to as the tangible (Dobres, 2000), hard (Jin, 2002), phenomenological or empirical (Dunnell, 1971; O'Brien and Lyman, 2000) parts. Second, there are the associated mental ideas. These are referred to as the intangible, soft, or ideational parts. To use the definitions put forth by Dunnell (1971, p.199), an empirical (phenomenological) unit "is anything which has an objective existence". These are the physical characteristics of the objects that can be documented with a reasonable amount of objectivity, visually or with the use of analytical equipment. Ideational units, on the other hand, are defined as "anything which does not have an objective existence" (Dunnell, 1971, p.199). These are the intangible aspects that do not survive as physical entities in the archaeological record, yet exist in the minds of people from the past and today. Consequently, it is necessary to consider all the parts in order to fully understand technology and material culture.

There are many disciplines that seek to better understand the relationship between people and the products they choose to make and use. Each discipline tends to concentrate its studies on either the empirical or the ideational parts of the object, but rarely on both. This results in data that is incompatible and incomparable, and therefore cannot be reused by other academics. However, "This separation suggests a dichotomy which is purely artificial and not inherent in the object of study. Unfortunately, the data obtained through these two approaches are often fundamentally incompatible although they clearly concern the very same activities. ... Both approaches are very much complementary, and allow very different aspects of the same operation to be documented, interpreted and possibly even understood. A survey of the literature, however, reveals an astonishing level of an 'either/or' approach, as if they were mutually exclusive" (Rehren, et al., 2007, p.211). In the passage cited above, Rehren and co-authors raise three important points relevant to this paper. First, that "the dichotomy is purely artificial"; second, that "the data obtained is incompatible"; and third, that researchers tend to take an "either/or approach". All three stem from a problem common to multidisciplinary research: each discipline has different intellectual traditions, jargon, focus, epistemologies, ontologies, methodologies, research perspectives and priorities (e.g., Wisseman and Williams, 1994, p.9; Dobres, 2000). Lemonnier points out additional problems; he states "an anthropology of technological systems, too, can only be developed if it has access to precise, detailed, and above all, comparable data. A study of the relations between technology and society must necessarily start from the study of differences, of variation in technological actions, observed in one particular society as well as among many societies, though time and space. Without such studies, the study of technology is just a black box which remains outside of anthropology" (Lemonnier, 1992, p.19). Thus, a systematic classification framework is needed that can accomodate compatable, comparable and reusable data from multiple disciplines.

A systematic approach

The proposed systematic classification framework for the Study of Material Culture and Technology (SOMCAT) is designed to reduce the influence of our own cultural preconceptions and intellectual traditions. It does this by separating the characteristics of material culture and technology into units that can be studied with a greater amount of objectivity. Ramenofsky and Steffen (1998, p.3) define units as "the packages that we create to measure the world". The framework is based on the premise that all material culture is fundamentally composed of interrelated Variation Units:

- Attribute Variation Units
- Socio-Cultural Variation Units
- Action Variation Units
  - Process Variation Units
  - Cognitive Variation Units

Recording the raw data as empirical units without concurrently evaluating or interpreting the data is of the utmost importance because it allows for objective comparison. O'Brien and Lyman (2000) note that failure to keep empirical and non-empirical units separate has crippled archaeology and anthropology. This is because, no matter how objective the scholar tries to be, evaluations are subjective and inherently contain preconceptions based on the scholar's knowledge. Conversely, the raw data can be used to objectively compare archaeological, ethnographic, industrial, and commercial materials (Rice, 1987, p.310). Furthermore, "because the results of these analyses are often expressed at the ordinal or interval level of measurement a variety of statistical manipulation can be employed in their interpretation" and "these in turn allow evaluation of hypotheses about similarities, differences, patterns, or changes...." (Rice, 1987, p.310).

Attribute Variation Units are the object's physical properties that can be determined with a reasonable amount of objectivity. Attributes are usually studied by scientists, archaeologists, and art historians and recorded as units derived "from direct observation, whether macroscopic or microscopic, and the resulting units are correlated with something physical" (Ramenofsky and Steffen, 1998, p.5). Laying aside the changes associated with burial, natural deterioration, and corrosion, it is reasonable to assume that an object's physical attributes are virtually the same today as they were in the past, or at the very least, they can be deduced from evidence, such as relic structure in corrosion products, impressions, staining or mineral replacement. In other words, an iron sword will corrode, but it does not magically turn into a ceramic vessel.

Recording the Attributes as distinct units also allows us to faithfully reconstruct material culture and technological processes, allowing us to 'experience' them. Experimental archaeology offers us the opportunity to share, as best we can, a common experience with people from another culture. We can feel the heat of the furnace, smell the smoke, see the colourful flames indicating temperature and furnace conditions, and hear the unforgettable sound of the 'breathing' and roar of the furnace. Accurate reconstructions allow us to compare the materials' and objects' performance and design with modern ones. Thus, we can suggest how a product or process may have been perceived and how well it performed its intended function (if known) by observing its characteristics, and noting its strengths, weaknesses and limitations.

Socio-Cultural Variation Units are a culture's characteristics, including language, religion, political system, economic system, and kinship structure. These features are usually studied by anthropologists and social scientists who collect data about living people, using methods such as participant observation, field notes, interviews, and surveys. They often endeavour to maximise objectivity by recording data in such a way that it can be used for statistical manipulation. Although Cultural Variation Units are intangible, they influence Actions that can leave markers that can be traced over time. This is significant because archaeologists are also interested in determining Cultural Variation Units about people from the past, but they must deduce their conclusions from the remaining material culture and not directly from the people themselves.

Action Variation Units are physical or mental changes which produce an effect. For example, the physical action of freezing water can be observed and measured. However, the action of blessing water by a priest to make it 'holy' will not change its physical Attributes, but the water will change in the mind of the believers. The difficulty with determining Action Variation Units is that they are fleeting and different Actions can produce the same or similar results. Although they are ephemeral by their very nature, Actions can be recorded orally, textually, and/or within the finished product and by-products.

The 'laws of nature' can dictate why an ingredient or process was necessary, but a specific culture or society will also have an explanation for why it was needed. This is because what we 'know' and how we explain what we 'know' is dictated by our culture (Hench, 1971). Thus, there are two types of linked Action Variation Units: ***Process Variation Units***, which represent measureable physical processes, such as heating a material to a specific temperature for a specified length of time, and ***Cognitive Variation Units***, which are the ideas transferred between people. These ideas produce traditions or "patterned ways of doing things that exist in identifiable form over extended periods of time" (O'Brien, et al., 2010, p.3797).

Actions occur during stages in the object's "life history" (Schiffer and Skibo, 1997). SOMCAT recognises that one or many people and cultures can be involved in an object's life history. It also takes into account that an artefact can mean different things to different people, even at the same time and place. In SOMCAT, the life history of an object is divided into five distinct stages: Conception, Production, Distribution, Consumption, and Elimination. These stages can be further subdivided depending on the discipline's focus or research questions. The term Conception is used here, rather than invention or innovation, because it reduces semantic issues and emphasises the cognitive action and the ideas of birth, beginnings, and origin. It is important to note that these life stages can be repeated an unlimited number of times. For example, the cycle can begin again when an object is reused, repurposed, recycled, or recovered and studied as an artefact (Holtorf, 2002; Bacharach, 2006, 3). Furthermore, archaeologists, conservators, scientists, and scholars are also influencing the artefact's life history. Because of these differing perceptions, it is important to identify, if possible, the people's involvement with, or contribution to, a particular object's life stage.

Dividing material culture into these distinct Variation Units and life stages allows for the construction of linked datasets that can be built upon, combined, analysed, and reused. Creating datasets and analysing them with various software programs is not a new concept; however, developing a systematic classification system that allows us to combine and link these diverse approaches, is a new approach. The proposed framework datasets have the ability to cope with different temporal scales, geographic distributions, and discipline specific data and questions. Using diverse data can, for example, be employed to determine clusters, visualise evolutionary relationships, and identify hierarchies. This is because the datasets are in a form that makes them compatible with each other and with different types of statistical, phylogenetic and other related types of analysis and software programs. Thus, the SOMCAT framework can be used to address a variety of complex questions and can lead to a better understanding of the relationship between technological, cultural, and social change.

Potential

Killick (2004, p.573) asks the following question: "Can our interpretations of their choices ever be more than merely conjectural?" While we will never be completely certain of the reasons for another person's choices, SOMCAT's systematic classification framework reduces our own cultural subjectivity and offers real potential to determine the likelihood (probability) that a certain choice was made, as well as propose the likely reason it was made in that particular culture. This is because choices, or Actions, can be made deliberately, subconsciously, or the choice can be beyond the person's control, but the choice must follow both 'Nature's Laws' and Culture's norms.

It has often been assumed that the reason why a particular choice was made in the past would correspond to a similar motivation today. This is a misleading assumption, because people who lived in the past had a different worldview and set of knowledge as compared to people living today. Certain actions are often preferred over others and the reason for the particular choice is not purely practical but may be due to tradition, economics, available knowledge, or it may have ritualistic or symbolic connotations, regardless of any practical purposes. What is considered to be 'practical', 'logical' or 'sensible' to one person is not necessarily the same to another. Ethnographic and sociological studies offer numerous examples where factors other than "practicality" are the basis for decision making (Lemonnier, 1992; Smedley, 1999). "The human animal has the capacity to come to conclusions and make decisions out of self-interest, out of devotion to some abstract principle, or of his or her perceptions of the larger interests of the group, however that is defined" (Smedley, 1999, p.19). Today, science allows us to understand and predict how materials will behave, but before the advent of modern science, religion and beliefs explained where substances originated from and why materials behaved as they did. Ancient products and processes tend to be evaluated and judged from our own modern perspective. However, knowledge and tastes change over time. It is, therefore, incorrect to presuppose that the Attributes and Actions which are considered desirable and appropriate today are the same as those of the past, because this assumption implies that the person from the past was applying the same rational or explanation for the choice as the person evaluating or judging the choice today. Thus, the specific culture must be considered when trying to determine which Action occurred, and why a particular choice was likely to have been made.

SOMCAT has a very high chance of success because studies using each of the Variation Units have already been successful. For example, Calvert (2011) uses a statistical analysis of the Attributes to identify clusters for her art history dissertation (see the project "Art of Counting: Discovery of patterns, trends, and messages in art throughout history", http://artofcounting.com/about/). Other Attribute studies including typology classifications, cluster analysis of chemical elements (Kato, et al., 2009), and decorative textile characters were used to determine phylogenetic relationships, as well as Consistency and Retention Indexes (Tehrani and Collard, 2002). Action Variation Units were used by Hosfield (2009) to identify clear patterns in knowledge transmission, learning, and cultural patterns. Additionally, O'Brien and co-authors (2010) convincingly demonstrated that cultural traits are valid units of analysis that can be used to study transmission, while also establishing their suitability for phylogenetic analysis. Creating datasets that are compatible with each other and with different types of statistical, phylogenetic, cluster, and other related types of analyses and software programs allows us to link disciplines, approaches, and methodologies. With additional information, the framework may help us answer the ideational questions of *how* and *why* change happens with supporting evidence from multiple disciplines. Thus, by combining features of the above datasets, the newly proposed systematic classification framework SOMCAT has real potential to allow us to ask and answer broader questions, including identifying evolutionary and causal relationships, which factors influence the transmission of ideas, and what causes variation in the archaeological and art historical record.

Crucible Damascus steel: a case study

The most significant hindrance to research on crucible Damascus steel is the preconception that so-called 'wootz' ingots imported from southern India were the primary source for creating Damascus steel blades in Syria, and that the terms 'genuine Damascus steel' and 'wootz' denote one and the same material (e.g., Verhoeven, et al., 1996; Reibold, et al., 2009). It should be noted that the scholars who formulated these assumptions have tended to be from science and engineering academic traditions, and thus the provenance and history of crucible steel were not the primary focus of their research. However, it is argued here that these preconceptions may indeed be indirectly hindering their research, and that of others, because the research is based on assumptions for which there is no firm supporting evidence. Using the systematic classification framework SOMCAT outlined above, the general characteristics (Attribute Variation Units) of crucible Damascus steel were determined. A distinction was then made between the people and places (Socio-Cultural Variation Units) that were involved with various stages of the 'life history' of crucible steel (Action Variation Units). The SOMCAT method helped separate statements for which there is supporting evidence from those assumptions that lack proof, and it revealed areas in need of research, including an order of research priorities.

Defining and determining the Attribute Variation Units of so-called 'Damascus steel' were the first two problems, as there are different types of steel that go by the name 'Damascus'. In his seminal book on 'Damascus steel', Sachse (1994, p.13) points out that 'Damascus steel' is an accepted generic term. It is often indiscriminately used to denote objects with a pattern, usually blades, and the term is so ingrained in literature and popular imagination that it is unlikely to fall out of use. The type of Damascus steel referred to here, also known as watered steel or misleadingly as wootz, is made of crucible steel that, due to its microstructure, often exhibits a pattern if etched. However, not all crucible steel has the necessary microstructure to produce a pattern. Furthermore, there are different types of crucible steel processes, and, as pointed out by McWilliam in 1920, "All steel made in a crucible is crucible steel, but steel made in a crucible and allowed to solidify in the crucible is not crucible cast steel, and in considering world practice and the history of the subject the two methods of manufacture must be carefully differentiated" (McWilliam, 1920, p.166). To clarify the subject of the research, the term crucible damascus steel was coined. The adjective 'damascus' is used because it is such an accepted term, but the lowercase 'd' is meant to break the association with the city of Damascus and to differentiate it from other types of so-called Damascus and crucible steels. This is similar to the way the adjective 'china', with a lowercase 'c', is used to denote a porcelain-like material and not the country. Thus, the present definition for the term crucible damascus steel is a hypoeutectoid (below 0.8% C) or hypereutectoid (above 0.8% C) ferrous product, created using traditional materials and methods, and which may or may not exhibit a pattern if etched.

Next, it was helpful to identify and classify the evidence by determining what stage of the artefact's 'life history' the evidence is associated with. For example, is the evidence associated with the artefact's production, distribution, consumption or elimination? The archaeological evidence published to date indicates that crucible steel was produced in various cities in Central Asia, India, and Sri Lanka. This does not mean that other locations do not exist, as textual evidence by Al-Kindi and Al-Beruni suggest (Hoyland and Gilmour, 2006), but without physical evidence it is not certain if these other locations were primary or secondary production centres. Distinguishing between primary and secondary production centres is important. In the Middle East and Central Asia, there is evidence from Islamic period texts indicating that many people, each with their own specialty, could have been involved in the sword making process (Kennedy, 2001), and they did not necessarily need to be in close proximity to one another. Pinpointing the location of sites that produced ingots is significant because crucible steel is a particular technology requiring a specific set of knowledge and raw materials. The technology was transmitted via socio-cultural dynamics and this can provide information on available knowledge, cultural beliefs, and social relationships, while the raw materials can be used for determining the trade of ingots and/or artefacts. Secondary centres, where forging and decoration occurred, could provide evidence of trade contacts, and locally preferred styles.

Based on the lack of evidence, Elgood has argued that Damascus has never actually been a centre of sword production (Elgood, 1994, pp.103-108). However, large Islamic cities like Damascus did have markets for imported goods and, therefore, Damascus was likely to have been a distribution centre or perhaps a place where imported ingots were forged. Furthermore, Arabic poetry is filled with references to blades with wavy or water-like appearances (Hoyland and Gilmour, 2006), but this does not mean that the Arabs, particularly the nomadic Bedouin Arabs were producers. It is more likely they were distributors and consumers of such artefacts. Nomadic people often forge, decorate and distribute metals, but they typically do not produce it from raw materials. Additionally, crucible steel swords have been found in the Russian Caucasus in nomads' burials (Feuerbach, 2005), and deposited along rivers in Northern European sites associated with Vikings (Williams, 2009). Although there is evidence for the use of crucible steel objects associated with these cultures, there is absolutely no evidence that any of these groups of people produced the steel, but rather it is more likely that they acquired them by trade or booty and disposed of them in ways that allowed the artefacts to become part of the archaeological record and be discovered at a later date.

It is frequently mentioned in the literature that the hypereutectoid crucible steel pattern is the only 'genuine Damascus steel' (Verhoeven, et al., 1993). Not only is this misleading, but it has led to an imbalance in the research. Generally, crucible damascus steel comes in two forms: hypoeutectoid and hypereutectoid. In hypoeutectoid form, the pattern is formed by ferrite-pearlite banding (Samuels, 1980), whereas in the hypereutectoid steels the pattern is formed by the alignment of cementite (Verhoeven, et al., 1996). There are historical examples of both types (see Sachse, 1994) and if we consider the traditional crucible steel process, it is more than likely that they were created, possibly by the same producer, using the same raw materials and even during the same furnace firing. Indeed, cast iron (iron with a carbon content greater than 2%) was also produced (Lowe, 1989b, pp.732 -733) but addressing that issue is far beyond the scope of this paper and calls for consideration in a separate study.

The crucible steel process involves placing an iron rich substance, such as iron bloom or scrap iron, and a carbon rich substance, such as charcoal, plant matter or cast iron, in the crucible and heating it. The carbon diffuses into the iron and creates steel. There is no evidence that the ancient smiths knew about the role of carbon, trace and alloying elements in the production of steel; however, they would have observed the differences in steel with dissimilar carbon contents or compositions. Thus, suggesting that the producers were aiming for a specific carbon content is misleading. It might be more accurate to state that they were attempting to create a product with the properties expected in a certain percentage of carbon steel. Given that ancient materials were not as homogenous as those available today, and that the difference between the two types of steels can be less than a fraction of a percent, it would have been difficult to control the carbon content to such a degree of accuracy and repeatability in antiquity. Therefore, it is highly likely that both types were produced using the exact same materials and methods, and they may have been produced at the same time. Thus, stating that the hypereutectoid steel is 'genuine' and other types are not is inaccurate and misleading. Given that they could have been produced at the same time with the same ingredients, and could exhibit a pattern, both the hypereutectoid and hypoeutectoid varieties should be classed as 'genuine' crucible damascus steels.

It is likely the craftsmen would have noticed the difference in workability; but would the consumer have realised these were different types of steel? The mechanisms that cause the patterns to form are different and they may have required different forging techniques, but this aspect has not been investigated in any detail. No one has explored how forging affects the ferrite-pearlite banded steel pattern. Furthermore, not all crucible damascus steel will exhibit a pattern. From a history of technology point of view this fact does not make the steel any less 'genuine'. Indeed, some crucible damascus steel objects could exhibit a pattern and some could not. Both the early Islamic writer Al-Beruni (Said, 1989, p.218) and the famous sword maker Wilkinson (1837, p.193) in England, said that it is 'accident', not 'industry', that creates the pattern. In the history of crucible damascus steel technology, identifying the range of possible products and appearances, determining what causes these variations, and how these different types of crucible damascus steels might have performed their intended function are important topics in need of research.

Two areas where crucible damascus steel was produced are India and Central Asia. While the processes are related, the materials and methods differ significantly. Some of the differences are the crucible's shape, fabric, and the steel ingot solidification rate. In India, the crucibles are typically made of ordinary clay tempered with rice husks, the crucible bases are rounded or pointy, and the ingot solidified quickly. These materials and technology are most often associated with the term 'wootz'. In Central Asia, the crucibles are made out of refractory clay and tempered with grog or crushed quartz, the bases are flat, and the ingots solidified slowly. In the Middle East, Central Asia and Northern India, the process is called *pulad* or one of its many derivations (Feuerbach, et al., 2007).

There is textual, archaeological and ethnographic evidence of crucible damascus steel production in Central Asia (Papakhristu and Rehren, 2002; Feuerbach, 2006). Much of the written evidence indicates that the steel exhibited a pattern (e.g., Massalski, 1841). There are numerous ethnographic reports of crucible steel production, mostly from the 18th-19th century in India. It is significant to note that Bronson (1986, p.40), who carried out a detailed study of wootz accounts, states that "none of the sources states or implies that the South Indian wootz they saw being made exhibited the classic water patterns of true Damascus when forged in to weapon blades". Perhaps this is the reason why scientists who based their research on the wootz process have had such a difficult time trying to reproduce the pattern. There is no evidence to indicate that South Indian wootz ingots produced steel with the characteristic pattern, and therefore the association of 'wootz' with the characteristics patterned steel is erroneous.

The exception, however, is Hyderabad in Southern India, where the steel, according to Bronson, "is in fact the only type of Indian wootz that we can be fairly sure was regularly found to exhibit a damascene structure when forged into blades" (Bronson, 1986, p.45). The remains from Hyderabad are interesting for a number of reasons. The process appears to be a recombination of characteristics from the South Indian wootz tradition and the Central Asian *pulad* tradition. The shapes of the Hyderabad crucibles are similar to those from Central Asia; both are cylindrical and flat bottomed, except that the Hyderabad crucibles are shorter. Both types of crucibles have gravel or similar debris embedded into the bottom surfaces (Lowe, 1989a, p.247), which indicates that the Central Asia and Hyderabad crucibles sat flat on the floor of the furnace. The temper used in the Hyderabad crucible lids is different from the one used for the body. The lids have grog and quartz as temper (Lowe, 1989a, p.246), similar to that used in Central Asia, whereas the temper of the crucible body is similar to other Indian crucibles in that they both use rice husks. Voysey (1832) also reported that the crucibles were heated for 24 hours and cooled slowly in the furnace. Thus, the materials and methods from Hyderabad appear to contain features of both traditions.

While the crucible damascus steel technology likely originated in India, the variation in the technology which produced the pattern may have originated elsewhere. Bronson (1986, p.47) notes that the origin of the patterned blades seems to be "entirely Middle Eastern" or perhaps more accurately, Central Asian. There is also evidence from the northern regions of the Indian subcontinent, including Pakistan, specifically Gujarat (Abbott, 1847). Bronson (1986, p.44) also notes "that the Hyderabad method has connections of some kind with the larger tradition of Islamic steel making and that analogous processes may legitimately be sought within that tradition". Voysey also stated that in Hyderabad a Persian was in charge of the process. The technology used in Hyderabad and the connection with Central Asian traditions is not surprising if one examines the socio-cultural history of India. Throughout the centuries, Islamic empires from Central Asia invaded and occupied parts of India, particularly the northern regions, but reaching as far south as Hyderabad. It is possible that the technological innovations needed to produce the pattern occurred in Central Asia and were subsequently introduced into India. Although more research is needed, it is possible that the changes in materials and/or methods needed to produce the pattern were brought by the invaders.

To return to the original claim, based on the empirical data, there is no evidence that wootz ingots imported from Southern India were the primary source for creating 'Damascus' steel blades in Syria. From a history of technology perspective, there are many different varieties of crucible damascus steels that could be called 'genuine' (as opposed to 'fake'), of which wootz is one variety. Even when applying the same methods and materials, crucible damascus steels can vary in carbon content, minor and trace elements, and they may or may not exhibit a pattern.

Conclusion

The case study of crucible damascus steel illustrates that it is not necessary to create a database to apply the systematic approach to the study of materials. However, being consciously aware of how indigenous cultural beliefs, in addition to an academic discipline's epistemology and methodology, will influence the way data is collected and viewed, is imperative. Simply mentally separating the evidence into these Variation Units will facilitate objectivity. However, the use of the proposed SOMCAT framework could help answer more complex questions. Thus, the next step will be to begin to develop the framework, input data, and test the validity of the approach. Each Variation Unit has been successfully used to identify patterns and clusters, phylogenetic relationships, and units of cultural transmission. The proposed systematic classification framework SOMCAT offers a flexible approach that can link these datasets. Thus, it is expected that the amount of new and useful information that can potentially be produced will be exponential to that of the current methods.

Acknowledgements

I would like to thank the anonymous reviewer for his/her detailed comments regarding systematics and other issues. The reviewer's comments have greatly improved and focused the SOMCAT systematic classification framework.

References

Abbott, J., 1847. Process of Working the Damascus Blade of Goojarat. *Journal of the Asiatic Society of Bengal*, 16, pp.417-23.

Bacharach, J.L., 2006. *Islamic History through Coins: An Analysis and Catalogue of Tenth-Century Ikhshidid Coinage*. Cairo and New York: American University in Cairo Press.

Bronson, B., 1986. The Making and Selling of Wootz, A Crucible Steel of India. *Archaeomaterials*, 1, pp.13-51.

Calvert, A., 2011. *The Integration of Quantitative and Qualitative Research in a Study of the Regalia of Ramses III*. PhD Thesis, Institute of Fine Arts of New York University.

Dobres, M.-A., 2000. *Technology and Social Agency: Outlining a Practice Framework for Archaeology*. Oxford: Blackwell.

Dunnell, R., 1971. *Systematics in Prehistory*. New York: The Free Press.

Dunnell, R., 1980. Evolutionary Theory and Archaeology. In: M.B. Schiffer ed. *Advances in Archaeological Method and Theory, Volume 3*. New York: Academic Press, pp.35-99.

Elgood, R., 1994. *The Arms and Armour of Arabia in the 18th-19th and 20th Centuries*. Aldershot: Scolar Press.

Feuerbach, A., 2005. Metallographic Investigations of 3rd-13th Century AD Swords, Sabres and Knives from the Russian Caucasus. *Journal of the Institute for Archaeometallurgical Studies*, 25, pp.27-43.

Feuerbach, A., 2006. Crucible Damascus Steel: A Fascination for Almost 2,000 Years. *Journal of the Minerals, Metals and Materials Society*, 58, pp.48-50.

Feuerbach, A., Balasubramaniam, R. and Kalyanaraman, K., 2007. On The Origin of the Terms Wootz, Hinduwani and Pulad. *Thematic Issue: History and Characteristics of Wootz Steel in India and Abroad. Indian Journal of History of Science*, 42, pp.377-86.

Hench, L., 1971. Introduction to the characterization of ceramics. In: L. Hench and R. Gould eds. *Characterization of Ceramics*. New York: Marcel Dekker, pp.1-5.

Holtorf, C., 2002. Notes on the Life History of a Pot Sherd. *Journal of Material Culture*, 7, pp.49-71.

Hosfield, R., 2009. Modes of Transmission and Material Culture Patterns in Craft Skills. In: S. Shennan ed. *Pattern and Process in Cultural Evolution*. Berkeley: University of California Press, pp.45-60.

Hoyland, R. and Gilmour, B., 2006. *Medieval Islamic Swords and Swordmaking: Kindi's Treatise on Swords and Their Kinds*. Warminster: Gibb Memorial Trust.

Jin, Z., 2002. Soft Technology – The Essential of Innovation. *Futures Research Quarterly*, 22, pp.1-24.

Kato, N., Nakai, I. and Shindo, Y., 2009. Change in chemical composition of early Islamic glass excavated in Raya, Sinai Peninsula, Egypt: on-site analyses using a portable X-ray fluorescence spectrometer. *Journal of Archaeological Science*, 36, pp.1698-1707.

Kennedy, H., 2001. *The Armies of the Caliphs: Military and Society in the Early Islamic State*. London: Routledge.

Killick, D., 2004. Social constructionist approaches to the study of technology. *World Archaeology*, 36, pp.571-8.

Lemonnier, P., 1992. *Elements for an Anthropology of Technology*. Anthropological Papers 88. Ann Arbor: Museum of Anthropology, University of Michigan.

Lowe, T.L., 1989a. Refractories in High-Carbon Iron Processing: A Preliminary Study of the Deccani Wootz-Making Crucibles. In: P.E. McGovern ed. *Cross-craft and Cross-cultural Interactions in Ceramics.* Ceramics and Civilization 4. Westerville, OH: American Ceramic Society, pp.237-51.

Lowe, T.L. 1989b. Solidification and the Crucible Processing of Deccani Ancient Steel, in R. Trivedi, J.A. Sekhar and J. Mazumdar (eds.) *Proceedings, Indo-US Conference on the Principles of Solidification and Materials Processing.* Dehli: Oxford and IBH. pp.729-740.

Massalski, Capitaine en second, 1841. Préparation de l'acier damassé en Perse. *Annuaire du Journal des Mines de Russie*, pp.297-308.

McWilliam, A., 1920. Indian Iron Making at Mirjati, Chota Nagpur. *Journal of the Iron and Steel Institute*, 102, pp.159-70.

O'Brien, M.J. and Lyman, R.L., 2000. *Applying evolutionary archaeology: a systematic approach.* New York and London: Kluwer Academic/Plenum.

O'Brien, M.J., Lyman, R.L., Mesoudi A. and VanPool, T.L., 2010. Cultural traits as units of analysis. *Philosophical Transaction of the Royal Society B Biological Sciences*, 365, pp.3797-806.

Papakhristu, O. and Rehren, Th., 2002. Techniques and technology of ceramic vessels manufacture – crucibles for wootz smelting in Central Asia. In: V. Kilikoglou, A. Hein and Y. Maniatis eds. *Modern Trends in Scientific Studies on Ancient Ceramics.* British Archaeological Reports International Series 1011. Oxford: Archaeopress, pp.69-74.

Ramenofsky, A.F. and Steffen, A. 1998. *Unit issues in archaeology: measuring time, space, and material.* Salt Lake City: University of Utah Press.

Rehren, Th., Charlton, M., Chirikure, S., Humphris, J., Ige, A. and Veldhuijzen, H.A., 2007. Decisions set in slag: the human factor in African iron smelting. In: S. LaNiece, D.R. Hook and P.T. Craddock eds. *Metals and Mines – Studies in Archaeometallurgy.* London: Archetype, pp.211-8.

Reibold, M., Paufler, P., Levin, A.A., Kochmann, W., Pätzke, N. and Meyer, D.C., 2009. Discovery of Nanotubes in Ancient Damascus Steel. *Physics and Engineering of New Materials*, 127, pp.305-10.

Rice, P., 1987. *Pottery Analysis: A Sourcebook.* Chicago: University of Chicago Press.

Sachse, M., 1994. *Damascus Steel: Myth, History, Technology, Applications.* Düsseldorf: Stahleisen.

Said, H.M., 1989. *Al-Beruni's Book on Mineralogy: The Book Most Comprehensive in Knowledge on Precious Stones.* Islamabad: Pakistan Hijra Council.

Samuels, L.E., 1980. *Optical Microscopy of Carbon Steels.* Washington: American Society of Metals.

Schiffer, M.B. and Skibo, J., 1997. The Explanation of Artifact Variability. *American Antiquity*, 62, pp.27-50.

Smedley, A., 1999. *Race in North America: Origins and Evolution of a Worldview.* Boulder: Westview Press.

Tehrani, J. and Collard, M., 2002. Investigating cultural evolution through biological phylogenetic analyses of Turkmen textiles. *Journal of Anthropological Archaeology*, 21, pp.443-6.

Verhoeven, J., Pendray, A.H. and Berge, P.M., 1993. Studies of Damascus steel blades: Part II – Destruction and reformation of the patterns. *Materials Characterization*, 30, pp.187-200.

Verhoeven, J., Pendray, A.H. and Gibson, E.D., 1996. Wootz Damascus Steel Blades. *Materials Characterization*, 37, pp.9-22.

Voysey, H.W., 1832. Description of the Native Manufacture of Steel in India. *Journal of the Asiatic Society of Bengal*, 1, pp.245-7.

Wilkinson, H., 1837. On the Cause of the External Pattern or Watering of the Damascus Sword-Blades. *Journal of the Royal Asiatic Society*, 4, pp.187-93.

Williams, A., 2009. A Metallurgical Study of Some Viking Swords. *Gladius*, 29, pp.121-84.

Wisseman, S. and Williams, W., 1994. Why Study Artifacts? In: S. Wisseman and W. Williams eds. *Ancient Technologies and Archaeological Materials.* Amsterdam: Gordon and Breach Science Publishers, pp.3-13.

Author's address

• Hofstra University
New York
USA

Symmetry in the archaeology of technology: nailing down manufacturing processes and networks of innovation in colonial ironworking

Krysta Ryzewski

ABSTRACT: The complex processes of craft technique and production involved in ironworking can be discerned and understood at multiple scales of archaeological visibility – from a single object's internal microstructure to the broader social networks within which iron objects were created and circulated. Seeing and interpreting these scales of visibility requires opening the colonial ironworking industry's black box of technological practices by accounting for material, social, and theoretical relationships from archaeology and beyond. Beginning with microscopic details of iron objects, this discussion builds from various scales of archaeological, historical, and metallurgical evidence to examine networks of innovation and tradition across colonial ironworking sites in New England. In exploring the inseparable material and social components of ironworking through a contextually-based, symmetrical enquiry, the heterogeneous nature of the colonial ironworking process is underscored. The results present socio-material networks and theoretical questions that were heretofore invisible historically and archaeologically.

Keywords: Ironworking; archaeology; nails; technology; symmetry; industry; colonial Rhode Island.

Introduction

This story begins with the excavation of and findings from an archaeological site. The site, the materials recovered from it, and their contextual relationships make visible long forgotten interactions between people, materials, objects, and objects-in-the-making. Where this story diverges from a familiar archaeological case study is in the conception of the site, which in this case is not a single bounded property, landscape, or identifiable geographical space for that matter. It concerns a group of fairly unremarkable iron objects from multiple, historically-related places (Ryzewski, 2012).

The site of enquiry is an assemblage of iron nails; several dozen of them excavated from within archaeological contexts at two Greene family colonial period ironworks in Warwick, Rhode Island: Greene Farm and Potowomut. The nails are accompanied by abundant historical evidence from the family's Coventry Forge, located in the adjacent town of Coventry (Fig. 1).

The nails' *in situ* associations with ore roasting areas, tuyères, bog ore, slag, a furnace area, clay, fuel, and other by-products and landscape features strongly suggest that they were produced during the Greene Farm operations (Fig. 2). No historical records detailing ironworking at Greene Farm exist, but based on the diagnostic ceramics recovered during the excavations of the Greene Farm ironworks, the iron production activities date to the mid-18th century, from between 1730 and 1781, most likely during the 1760s. According to the excavated data, this ironworking operation was small, intermittent, and served the plantation's immediate needs (Ryzewski, 2008).

Operating at the same time, the ironworks at nearby Potowomut (ca. 1690-1820) and Coventry (ca. 1742-1800) were larger scale forges whose production peaked during the mid-late 18th century. The timing and profitability of the Greene family ironworks is especially significant given that the British passed the Iron Act of 1750 to outlaw the production of finished iron objects in the American colonies. This prohibition was a direct response to the shifting trans-Atlantic patterns of iron production and consumption; by the mid-18th century, England consumed more North American-made iron products than it produced (Bridenbaugh, 1950). Such indications of increasing self-sufficiency demonstrated the American colonists' capability to forge a measure of material independence from England, an achievement that rightly concerned English politicians and manufacturers alike. Although historical accounts claim that no ironworks of consequence existed in Rhode Island during the colonial period (Bining, 1933), the archaeological remains of the Greene family ironworks and over a dozen other small scale ventures throughout Rhode Island and nearby south-eastern Massachusetts illustrate the fact that many of these American-based iron production ventures thrived despite political pressures (Ryzewski, 2008).

Archaeological research into historical North American ironworking operations, and the Greene family ironworks in particular, encounters many of the same methodological and theoretical complexities as prehistoric contexts of ironworking.

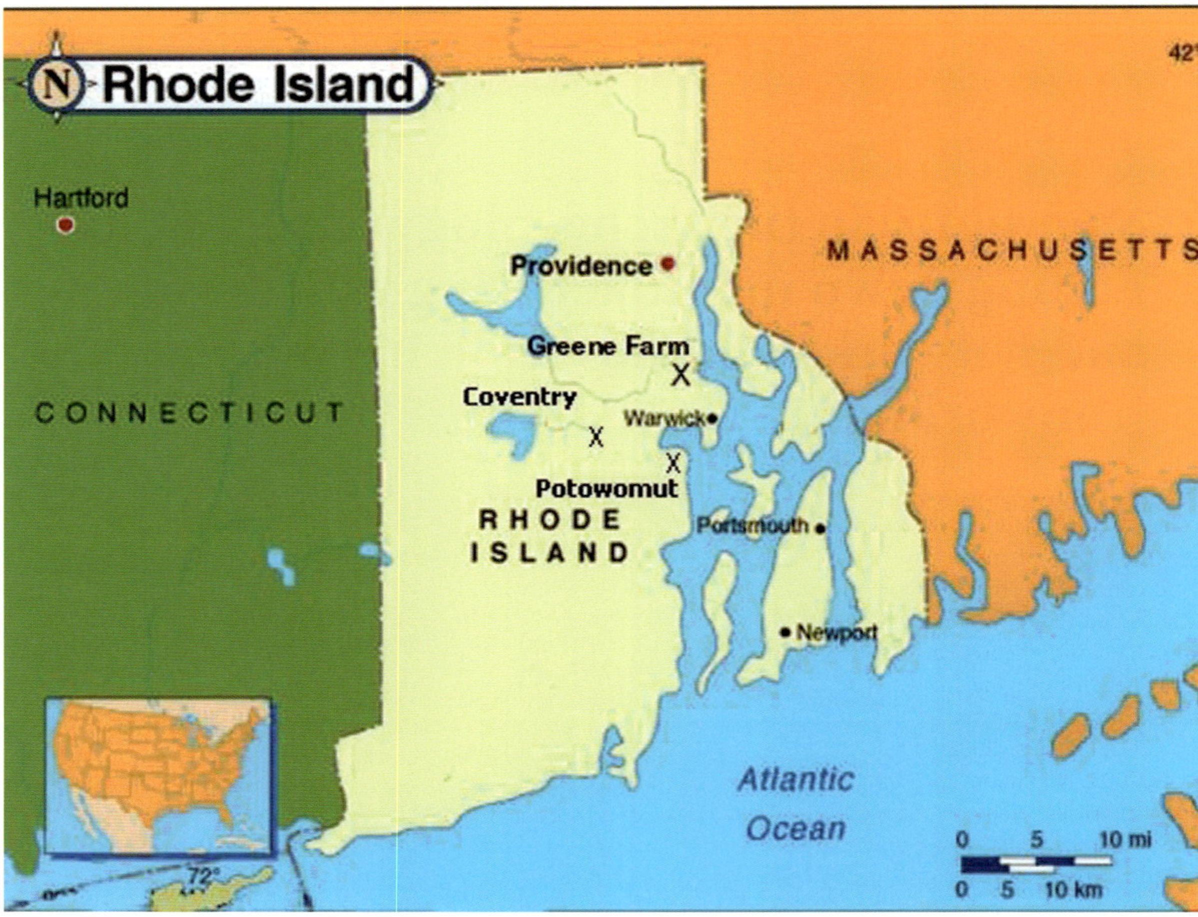

Figure 1: Location of Greene family ironworking operations in Rhode Island (modified by Ryzewski, 2008).

Consequently, the theoretical issues raised here are potentially relevant for archaeological studies of ironworking and technology across time and space. Two particular issues are addressed in the following discussions: (1) how to examine the nature of technological variation or change as expressed in material culture, and (2) the situation of technology in colonial contexts. The challenge of understanding the organisation of production from the archaeological record underlies both of these, and remains an important point of consideration throughout (see Costin, 1991).

The following discussion employs a systemic approach guided by the general principles of symmetry (Law, 1987), to examine the techniques at play in the Greenes' ironworking processes and networks of production. First, an example of nail manufacture illustrates the different perspectives on technology that can arise from close readings of the material culture of ironworking, especially with regard to the techniques of manufacture. The discussion then considers how one might move from the microstructure (the inside of iron objects) outwards. Examining the material and microscale properties of these objects in tandem with other archaeological and historical evidence will elucidate some of the many socio-technical networks that emerged in relation to these iron objects. These networks travel along and between sets of relationships (material, inter-personal, historical, etc.) that, when followed, provide specific information about the dynamics of technology and ironworking techniques in certain situations (on paths, see Gille, 1978; on multi-sited trajectories, see Ryzewski, 2012). The initial results conclude with examples of an end-product stemming from this symmetrical enquiry: biographies that illustrate the Greenes' and their objects' socio-technical networks.

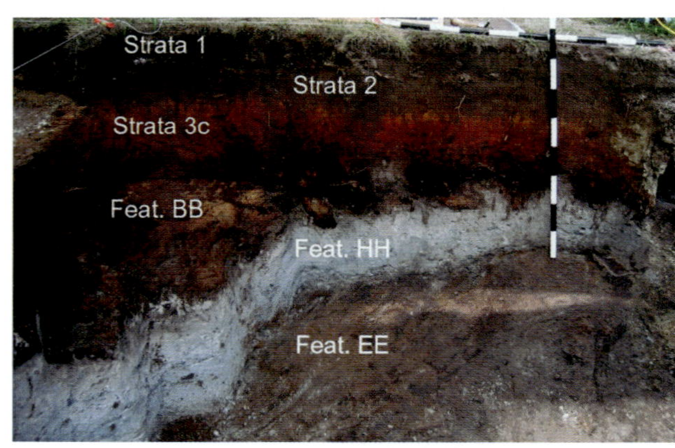

Figure 2: Stratigraphic evidence of ironworking activities at Greene Farm (Ryzewski, 2007).

Symmetry and visibilities

In the 18th century, the Greene family ironworking operations were created and sustained by broad social and material networks. In an attempt to understand these historically invisible socio-technical dynamics of production, it is necessary to view the material and social relationships of ironworking in tandem, paying attention to both social and material evidence. Stemming from the work of scholars in sociology, Science and Technology Studies (STS) and anthropology, such as John Law, Michel Callon, and Bruno Latour, the principle of symmetry does not grant special explanatory status to either social or material elements, but instead encourages the same type of explanation to be used for all elements that go into comprising a heterogeneous network (Law, 1987; Callon, 1987; Latour, 1993). Symmetry is not a theoretical model *per se*, nor does it seek to convey balance between material and social components of a network. Rather, it is a form of enquiry that encourages recognition of both the material and social components of action, in order to articulate how, in networks of production, distribution, and consumption, material properties and human behaviour play overlapping and inseparable roles. Although a long-standing concept in other disciplines, symmetry has only recently gained attention among archaeologists, who have incorporated it into theoretical discussions and have applied it to case studies ranging from media and temporality to material agency (Lechtman, 1977; Callon, 1987; Law, 1987; Latour, 1993; Shanks, 2004; Witmore, 2004, 2007, 2009; Ehrhardt, 2005; Webmoor, 2007; Ramenofsky et al., 2008).

A symmetrical perspective encourages the orientation of method and theory towards a more balanced and integrated approach, perhaps encouraging a greater interdisciplinary approach to anthropological studies of technology. Symmetrical examinations are particularly useful for archaeological interpretations of craft technologies and technical choices. They encourage finding ways to increase observation and visibility through the historical and archaeological information directly at the researchers' disposal (Pfaffenberger, 1992). The archaeological record of Greene ironworking provides an immediate opportunity to work with both materials and the contextual archive to increase these visibilities many times over (Hodder, 1999; Latour, 2005, p.186). Visibilities, or traces of action and interaction, can take many forms, and the process of articulating them can achieve varying degrees of success. Given the variety of site or context-specific information in archaeological examples, the most appropriate starting points for symmetrical enquiries will vary from case to case. Generally, however, enquiries begin by articulating the connections (and disconnections) between people, materials, objects, and objects-in-the-making. The process of identifying various scales of visibilities, and articulating the overlapping social and material relationships that emerge from them, aims to encompass a range of actors and scales, from the microscopic to the global movements of materials. In this case, the archaeology of colonial ironworking is informed by Michel Callon's (1987, p.84) advice, that it is important to recognise heterogeneity and complexity of material and human actors as much as they are present in the process of manufacture as they are in the end product.

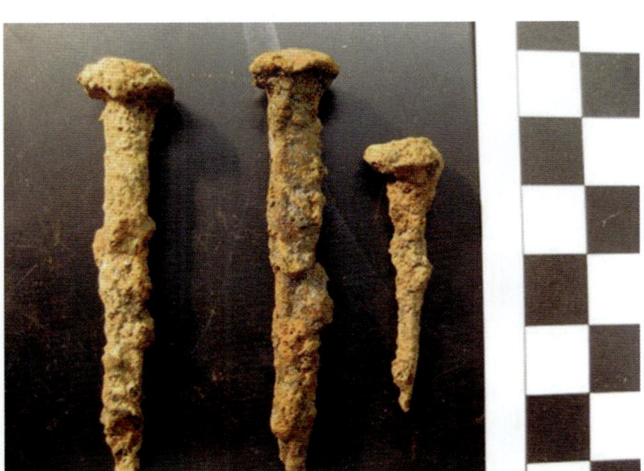

Figure 3: Sample of relatively well-preserved nails excavated from the Greene Farm ironworking site (Ryzewski, 2007).

Objects in the making

These nails are portals into the black box of ironworking technology. As objects, they are crystallisations of activities within a network – their form, structure, and properties embody and document the techniques that created them (Ingold, 2000, p.345). They are also material vehicles that illustrate theory in action. They put past actors and networks of technology in motion, and they call anthropological understandings of technology into deeper consideration.

The problem with nails is that on the surface, it is impossible to know much, if anything, about the processes and complexities of their production, with the exception of general information about form and style (Fig. 3). A problem familiar in studies of technology resurfaces here (see Lemonnier, 1986; Pfaffenberger, 1992; Ingold, 2000). Historically, when dealing with iron objects, archaeologists have relied upon the external, visible, and stylistic features of objects to identify the techniques used in crafting them, to create typologies, and to reconstruct developmental chronologies (Nelson, 1968; cf. Lenik, 1977; Phillips, 1993, 1996; Wells, 1998). From this perspective, examinations of iron or nail production that intend to begin from the object itself can be hampered by views of ironworking technology that are portrayed as following a smooth, progressive technological development from 'simple' to 'advanced', handmade to machine-made processes. But if one attempted to fit a nail produced at Greene Farm into prevailing typologies of nail-making technology, it would fall out of place, somewhere in the middle of the spectrum (Fig. 4).

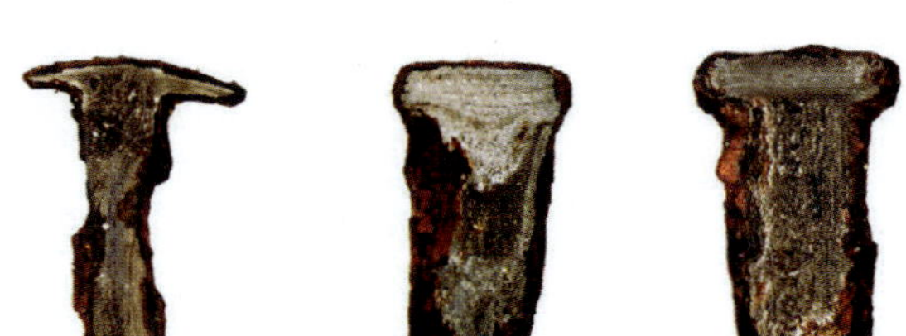

Figure 4: The general chronology of nail types from wrought (left) to machine-made nail (right), with an example (KR-17) of the nails excavated from the Greene Farm ironworking site in between (Ryzewski, 2009).

Nail classifications are intended to act as basic references for identifying general nail types, and for ordering assemblages to tell time based on the objects' morphology and manufacturing techniques (see Lenik, 1977). These typologies often differentiate broadly between hand-wrought and machine-made nails, and, in doing so, order them from hand-made (i.e. least technologically sophisticated) to fully-mechanised (i.e. most sophisticated) production technologies. While nail (and other object) typologies are widely accepted by historical archaeologists as temporal indicators, they can be inappropriate bases for questioning the social dynamics of production (Lyman et al., 1997). The *relations* of nails to one another in these typologies do, however, offer an example of how certain assumptions about technology are packaged and perpetuated. These assumptions have the possible consequence of directing archaeological discussions of ironworking techniques and industrial development towards linear and progressive frameworks at the expense of broader anthropological enquiries into the social complexities, materials, and skills involved in production processes (Killick, 2004).

This impulse to order technological processes into a neat, chronological, and object-centred framework has started to change as researchers examine different scales and visibilities of socio-technical systems. Recent archaeometallurgical research by Ryzewski and Gordon (2008) demonstrates variation and non-linear development in regional metalworking traditions. Based on microstructural analysis, they examine how hand wrought, cut, and machine-made nail-making technologies co-existed at the same sites or in the same local area from the 18th to the 20th century. Other alternatives move away from object-centred research to broader scales of enquiry, such as the work of Neff (1995) and Petréquin (1993), both of whom have turned to production processes to locate evidence for technological variability in production over time.

Opening the black box from the microstructure

Examining nail-making techniques using the analytical and theoretical tools from archaeology and related disciplines opens the black box that shrouds knowledge about colonial ironworking technology and the larger ironworking industries in Rhode Island. Though specific to the examples discussed here, the theoretical orientation of this discussion is meant to provoke and exemplify how a symmetrical and interdisciplinary consideration of manufacturing processes related to iron or other materials could involve similar forays into the black box of a given technological process. If the gathered information is successful in retracing some of the obscured historical connections, then the type and depth of anthropological questions researchers can pose about technology will change. Of course, the task of retracing the rich social, historical, and material connections makes the researchers' tasks more complicated and time-consuming. But opening the black box permits movement beyond the problem that pervades many technological histories and limits anthropological understandings; this is the tendency to trace these technological lineages rather than to work out the complex nature of technologies (Sigaut, 1994, p.435). As David Kingery and others emphasise, the focus on the dynamic interrelationships in the manufacturing processes of material culture is necessary for furthering understandings of technology (Kingery, 1993; Knappett, 2005).

From the exterior, these appear to be nails that were cut and headed mechanically. And they are, *almost*. Visible in the nail's microstructure is evidence of a nail-making technique that does not fit with the traditional and still widely applied chronology of nail production (Fig. 5). The process of making nails in a one-step machine is assumed to be more efficient, quicker, and more reliable than the crafting of wrought nails.

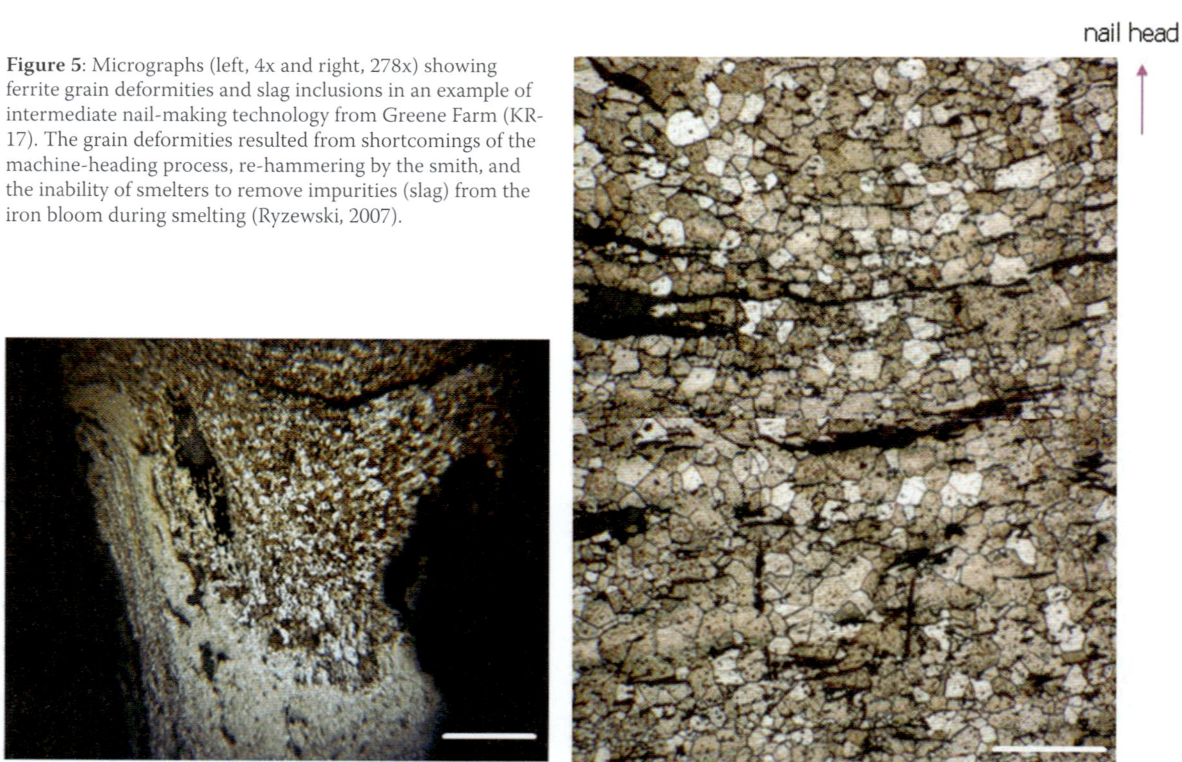

Figure 5: Micrographs (left, 4x and right, 278x) showing ferrite grain deformities and slag inclusions in an example of intermediate nail-making technology from Greene Farm (KR-17). The grain deformities resulted from shortcomings of the machine-heading process, re-hammering by the smith, and the inability of smelters to remove impurities (slag) from the iron bloom during smelting (Ryzewski, 2007).

It is also a technology that has a patented date of 'invention' attached to it, 1790 in this case (U.S. National Archives, Waltham, 1819).

As Fig. 5 illustrates, this particular nail's body was machine-cut, but its head was not successfully attached mechanically. This is also the case for several of the other excavated nails. Such variability raises several questions about the materials and techniques involved in its production processes: Was this failure caused by the continued presence of phosphorous and slag in the material, which compromised the iron's ductility and strength? Was the machine inadequate? Was the ironworker unfamiliar with this technique of nail production? A look inside the nail tells us that these were all possibilities. In addition, microscopic investigation of the nails provides overlapping explanations about the nature of nail-making / ironmaking techniques and technology, and the ways in which these change over time.

These internal material properties indicate the employment of an innovative technique, the (semi)automatic nail machine, which was designed to cut and head nails in one step. The properties also point towards a degree of experimentation, because this technique involved working with warm rather than cold metal. Based on their prior experience and specific knowledge, the ironworkers who used this nail-making technique were innovative; they recognised that the available techniques and machines were neither durable nor strong enough to cut and head cold iron nails, so they worked warm iron in a process that was carried out in part by machine, and in part by hand (Lemonnier, 1992). The microstructural details of this archaeological object provide a record of a technique that was a compromise in response to the failures of the one-step machine's capabilities.

After locating information about techniques of production in particular starting points (in this case the object), discussions can then turn towards the possibilities of multidirectional development, successes or failures of the materials, and questions about evidence for innovation, however subtle. Here, the microstructures of these nails show how the transition from hand to machine in this instance was not smooth, it was in fact problematic, and involved many other material and social relationships in its creation and circulation.

Details revealed through metallurgical analysis about the micro- and macro-structures of these nails illustrate the difficulties people encountered in the process of manufacture. When other nails in similar contexts at the Greene Farm ironworking site are taken into account, the assemblage becomes one of objects that are similar on the surface, but contain variation in internal structure and properties, despite having been crafted with the same nail-making technology (Fig. 6). This variation is an important reminder about the many ways of producing an object, even when using similar mechanical or crafting techniques. Defying any sort of useful classification, the technology of these nails' manufacture illustrates the nonlinearity and unpredictability of production processes (Pinch and Bijker, 1987, p.29; Gould, 2001). Consequently, the structure and properties of these nails draw attention to the necessity of examining how variation, compromise, and failure are integral components of the pace of technological adoption and the nature of production (see Petroski, 1996; Skibbo and Schiffer, 2001).

The black box that the Greene nail-making technology represented is now partly opened, although the contents of the black box did not necessarily require knowledge of its workings to identify it, at least from its exterior (or the surface of the object) (Layton, 1977, p.198). Using tools from inside and outside of archaeology, and considering the converging lines of evidence, it can be concluded that a technique that produced nails mechanically was in use in variable forms, decades before its patented invention in 1790. This is not, however, an accidental date, but one that conveniently coincides with the 'arrival' of the Industrial Revolution in America. This is a situation where historical and archaeological evidence diverge, and this divergence should be taken as a sign of caution against uncritically accepting generalised historical notions, and also as a stimulus for viewing technological development as a longer term, uneven, and gradual process (Latour, 2005; Ryzewski, 2008).

The problem remains, however, that if archaeologists do not look deeper than the surface of these iron objects and beyond the text of surviving historical documents, many possible understandings of the complexities of the actors involved in the production processes and the broader questions about technology will stall at a critical crossroads. As anthropologists

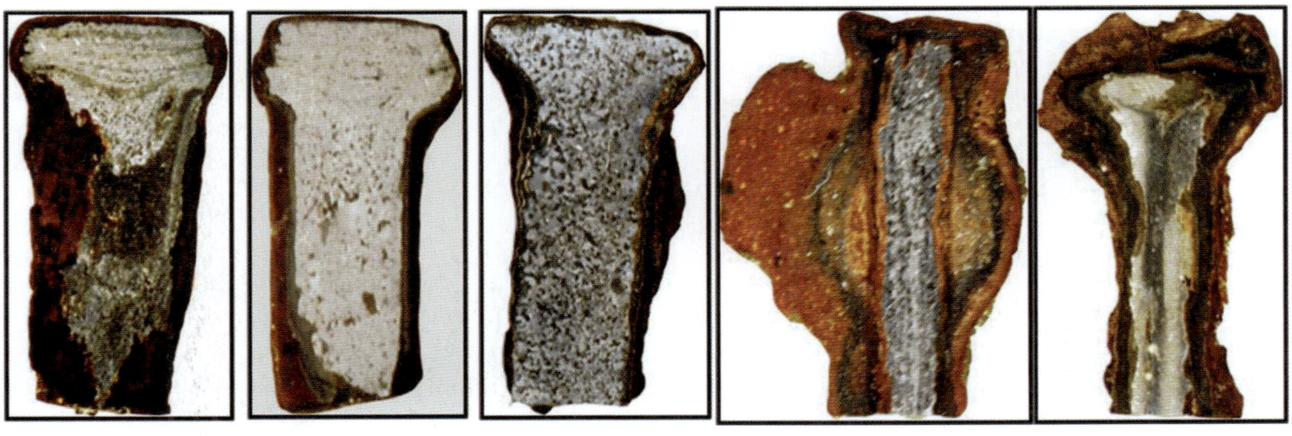

Figure 6: Cross-sections of nails excavated from the Greene Farm ironworking site showing macro- and micro-structural variety despite having been crafted by the same techniques (Ryzewski, 2009).

and archaeologists Michael Schiffer (2001), David Killick (2004), Tim Ingold (2000), Brian Pfaffenberger (1992), and others lament the tendency to stop at this crossroads and this is one reason why understandings of technology remain under theorised, or perhaps oversimplified in archaeology.

Moving through the iron production network

Up to this point the nails have acted mainly as a descriptive point of departure, and it is necessary to push much further to understand the complex dynamics of technological change in colonial contexts. Moving away from the nails, the discussion turns to how interactions between people, materials, and things are expressed in networks, and how archaeologists might build on seeing these networks through traceable associations between material culture and the archaeological record.

If the intention is to recognise techniques and their interrelated components, what processes are involved in reaching this goal, and how far can they take archaeological enquiry? Following the leads found inside the nail, in association with contextual archaeological evidence, discussions turn to the relationships of mining, charcoal production, cross-industry networks, environmental impacts, and the adaptation of ironworking practices to new material constraints (Ryzewski, 2008). Considerations also arise about how these technological practices and their changes over time depended upon both the senses and the intellect of the people involved in the network of production activities.

The recognition of these complexities is propelled by observation, a method that is of paramount importance in anthropological research, and especially in understanding technology (Mason, 1888, cited in Sigaut, 1994, p.515; Sigaut, 1994, p.423). Ethnohistory and ethnoarchaeology offer invaluable sources of information for anthropological studies of technology; much can be learned by simply watching a craftsperson in action, or learning metalworking oneself (Keller, 2001). Even in the case of these observations and experiences, however, there always exists the risk of searching for analogies in the archaeological record, and the danger of injecting and imposing elements of presentism onto interpretations of ancient practices (see Gould, 1986; Wolverton and Lyman, 2000). With this caution in mind, if the principle of symmetry affords a suitable possibility to account for the material and social actors in the production process, then the following question must be addressed: how can these ideas of symmetry play out in more detail in archaeologies of technology and in understandings of the dynamics of colonial industry?

Recognising the heterogeneity of materials is the first step towards tracing archaeologically the complex networks of production, the diversity of actors and techniques involved in production, and the importance of industry on all scales (spatial and temporal). The next part involves tracing actual associations: using whatever tools necessary to do this, whether scanning electron microscopy, oral histories, or trade records. The point is to begin, as Latour notes, with the assemblage (what is already assembled) and work towards accounting for what is not yet in the assemblage by tracing associations along the paths that constitute their network(s) (Latour, 2005, p.12; see also Gille, 1978, p.16, and Sigaut, 1994).

In the case of the Greene Farm nails, the information gained from metallurgical data not only illustrates the nature of technological variation, but also links the material within the nail to other material inputs, infrastructure, by-products of the nail-making process (e.g. slag, ore, tuyères, and fuel), and to the traditions of Greene ironworking (Fig. 7). None of these existed in isolation, and, as Timothy Ingold notes, none of these objects or materials were fixed; they are, instead, processual, relational and transformative in association with their surrounding contexts (Ingold, 2007; Nye, 2006, p.47).

Returning to the example of Greene ironworking during the colonial period, three short vignettes follow that articulate some of the heterogeneous material-human visibilities that emerge through movements from the nail to the paths that circulate around it, and the networks in which they participate. Each vignette illustrates the variety of directions and interrelationships in archaeological conversations about technology.

Adaptation to new environments: the ironworkers and their materials

From the inside of this nail, one can follow the [micro]visibilities of these materials along trajectories that do not necessarily complement prior or traditional archaeological observations of the colonial ironworking industry. In the nail's microstructure its structure and properties are visible. These mechanical and crystalline features document how and why the nail-working technology operated, and the range of materials and conditions that were involved in their manufacture. The resulting relations run contrary to historical accounts; colonial ironworking did, in fact, exist and was a vibrant industry in Rhode Island. Connecting to long-lost and obscure references, one learns that the Greene family was a key player in iron production at the operations at Greene Farm, Potowomut, and Coventry (see Fuller, 1875). Drawing comparisons between materials at these locations, patterns emerge in the material structures of the nails and in other iron objects from Potowomut and Greene Farm. One of these patterns is a presence at both sites of low grade ore that is high in titanium, a quality unique to local Rhode Island outcrops of Cumberlandite (3-6%, in Ryzewski, 2008; Urban et al., 2010), and a characteristic feature that would have adversely affected the heating process. The activities at local ore sources can be connected to the forges, both geochemically and through historical documentation; one learns how important and demanding ore mining and harvesting was; and how this activity increasingly required a scale of labour beyond the family unit. The accounts detail that slaves operated at the mines, some of whom were owned by the local ironworking families, including the Greenes (RIHS MSS 9001, MSS 459, RI Wills).

Records from the Potowomut forge reveal much more about the myriad of social influences at play in Greene ironworking, and, with more extensive analysis, so may the materials (Accounts 1765-66 (Fig. 8). Following the material and social leads, it is possible to trace how, as the operations

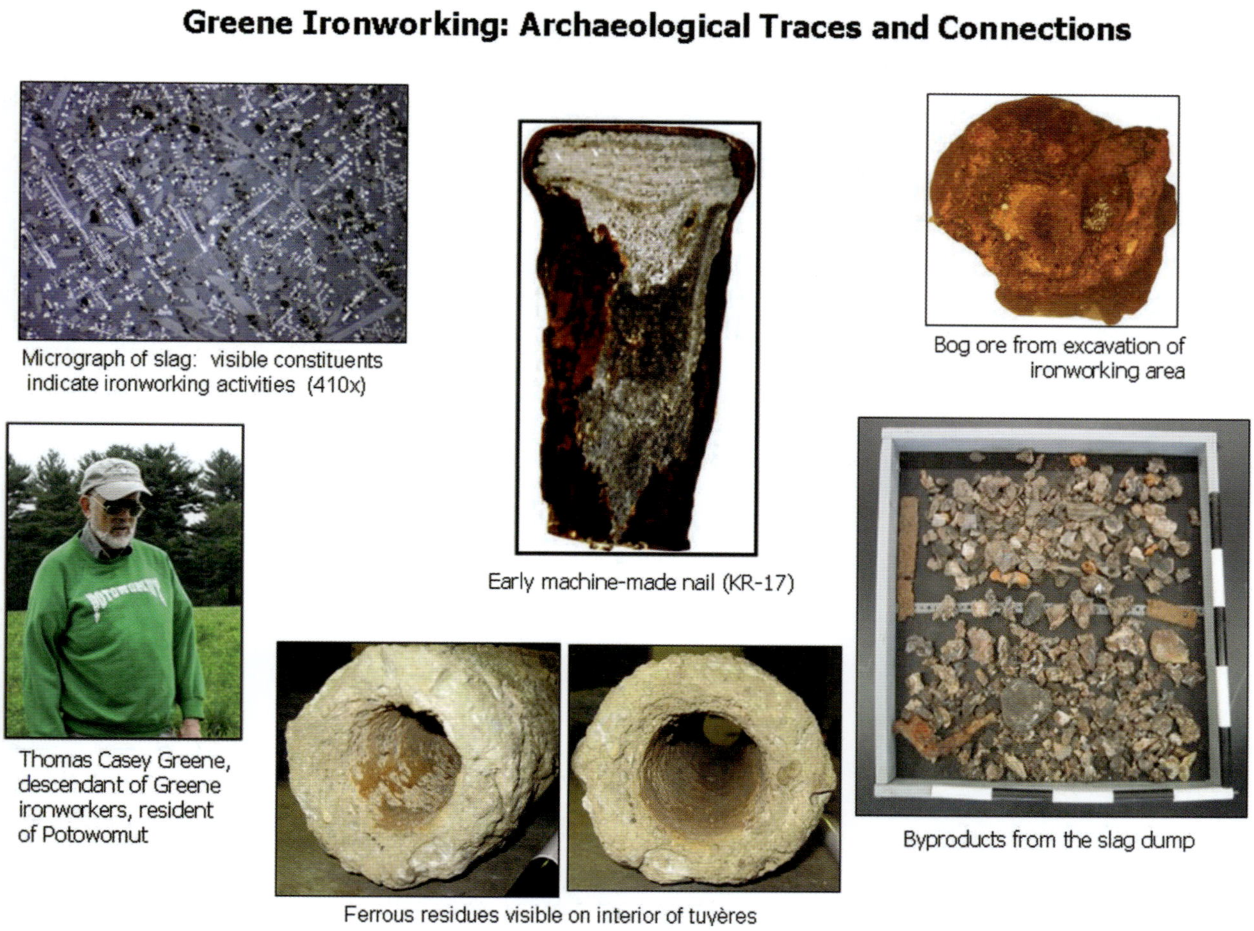

Figure 7: An example of the relationships that can be identified and traced beginning with information from the nail's microstructure and contextual archaeological data (Ryzewski, 2009).

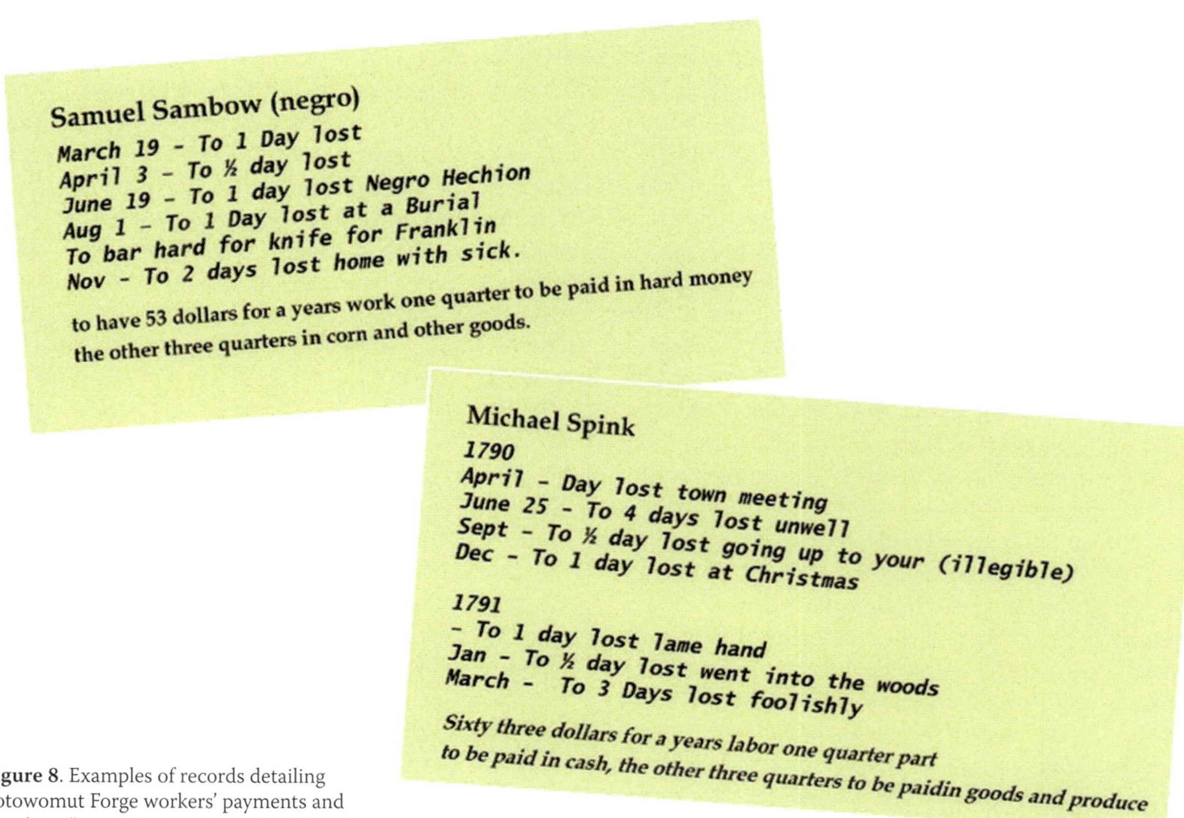

Samuel Sambow (negro)

March 19 - To 1 Day lost
April 3 - To ½ day lost
June 19 - To 1 day lost Negro Hechion
Aug 1 - To 1 Day lost at a Burial
To bar hard for knife for Franklin
Nov - To 2 days lost home with sick.

to have 53 dollars for a years work one quarter to be paid in hard money the other three quarters in corn and other goods.

Michael Spink

1790
April - Day lost town meeting
June 25 - To 4 days lost unwell
Sept - To ½ day lost going up to your (illegible)
Dec - To 1 day lost at Christmas

1791
- To 1 day lost lame hand
Jan - To ½ day lost went into the woods
March - To 3 Days lost foolishly

Sixty three dollars for a years labor one quarter part to be paid in cash, the other three quarters to be paidin goods and produce

Figure 8. Examples of records detailing Potowomut Forge workers' payments and penalties (late 1780s-1791, RIHS MSS 462).

expanded, the Greenes incorporated indentured and enslaved African and Native American workers, and indentured or itinerant European labourers into their ironworking activities; some worked at the forge; many also settled in industrial village housing nearby (RIHS MSS 9001, MSS 459, Wills). As they worked in the mines, farm, and at the forge, they mixed their ironworking craft traditions and specialised technical knowledge into the production processes. From this mixture these archaeologically traceable paths take a more holistic turn as the ironworkers are recognised not as anonymous labourers, but simultaneously as skilled individuals and as participants in multiple social groups (e.g., based on kin, ethnic, or economic ties). This mixture speaks of the many relationships between ironworkers and the various processes of production. The archaeological record of ironworking (both material and documentary) articulates the ironworkers' diversity, knowledge, personal experiences, and technical expertise in relation to the processes of iron production, circulation, and industrial development. It is these interrelationships that inform us about the multicultural composition of colonial industry.

Metals, power, and innovation

In a related circulation, but in a different topical orientation, the nails also detail processes of experimentation, innovation, and networks of power and influence. The residues inside the tuyères at Greene Farm indicate that a magnetic black sand may have been used in smelting iron, an experimental technique written about by contemporaries (Jared Eliot in Connecticut (Eliot, 1762; Oliver, 1762). This is a potential connection to activities documented in the records of the Potowomut forge, which detail that the Greenes imported black sand from nearby Block Island. This path arrives at Nathanael Greene, one of the five Greene brothers who were both ironmasters and managers of the forges, and who secured the supply of sand by marriage into the Littlefield family of Block Island (Clarke, 1875; Fuller, 1875). Nathanael Greene was Major General in the Continental Army during the American Revolution. In the course of his military deployment Greene travelled among networks of people associated with colonial ironworking (including George Washington). Greene's wartime letters to his brothers in Warwick give details about the forges and ironworking technologies he visited across the American seaboard, especially in New Jersey and Pennsylvania. The descriptions occasionally debate technological issues, such as the quality of their bar and pig iron, and include the discussion of whether or not to incorporate a slitting mill into the forge operations (Greene, 1867). While these decisions might be interpreted as examples of the Greenes' technological conservatism or limited technical knowledge, they are more appropriate and useful indicators of the shape and boundaries of the Greenes' ironworking traditions. The material and historical evidence communicates very clearly that the Greenes invested heavily, both socially, economically, politically, and materially, in alternative technical choices. These technical choices are visible in increasing scales of labour, networks of marriage, political office holding, and in the microstructure of their iron objects, the ironworkers' adaptation to local resources (black sand, ore), and landscape transformations, all of which were inseparably material and social considerations related to the rich traditions and techniques of Greene ironworking.

Folding industrial spaces: industries in a long term perspective

The Greene family operated two successful and relatively large scale regional forges during the colonial period and into the 19th century at Potowomut and at Coventry, and a small bloomery at Greene Farm. In all three cases, ironworking was closely related to, overlapped, and laid the foundation for other industrial and social relationships. Archaeological and historical evidence details how the forge at Potowomut was located adjacent to a grist mill, and the Greene Farm ironworking operations were near a large ice house and saw mill. Scars on the landscape at both sites are remnants of intensive mining activities, in which bog ore was harvested for fuelling the iron smelting. The productivity of iron eventually waned at the Potowomut forge, but industry did not disappear. Instead, the Greenes adapted to the demands of the time and converted the forge building and equipment into a textile mill operation. Specialised skill and knowledge, in the case of the Greenes, was not exclusive to a single industrial operation or technology, it was transferable. This transferability was situated among and mobilised by the variety of people involved in the ironworking processes. This notion of transferability is an important, but often overlooked component in archaeological studies of technology, which demands a longer term perspective. At Potowomut and Coventry, over the course of two centuries, the ironworking labour force grew from small family-based operations to include a range of indentured labourers, seasonal workers, and slaves; small cottages were built near the forge to accommodate these workers. At the Coventry forge a small village named Greene housed over 100 iron workers and their families. From this longer term perspective material traces make visible the emergent social structures that laid the foundation for the organisation of factory production, the division of labour between managerial and working classes, and the development of different industrial technologies in America.

Conclusion

The symmetrical view is a mode of enquiry that calls attention to the interrelationships between material and social processes in examining technological change and iron production processes. As an orienting perspective, the concept of symmetry does not necessarily change the types of questions archaeologists can ask, or the theoretical influences one might draw upon to interpret the relations that emerge, but it does lead research towards places or sites that have been rather speculative or inaccessible previously. It refocuses archaeologies of technology: towards incorporating multiple sites, to questions of variability and innovation, to the inside of objects, to identifying networks and relationships between ore, metal, indentureships, nails, marriage records, and Native Americans, and towards incorporating a diversity of interdisciplinary tools into the traditional archaeological repertoire. Visible from these connections is the case study of a technology that is archaeologically-grounded, but is yet still part of a system containing many complex components (see Hughes, 1987, p.51). Nails, in this case, act as mere points of departure to encourage more detailed and daring theoretical approaches to ironworking and other technologies that are complex in nature, variable in pace, longer term in scope, but much more interesting in detail.

Acknowledgements

Special thanks to Xander Veldhuijzen, Jane Humphris, and Thilo Rehren for their efforts in organising WIC 2009 and for inviting me to participate. I am especially grateful to Pierre Lemonnier, Michael Shanks, Chris Witmore, and the anonymous reviewer for their insightful comments and feedback on earlier versions of this paper, both at WIC in London, at TAG Stanford, and during the revision process. Any shortcomings are exclusively my own.

References

Primary Sources

Accounts, Potowomut Forge, 1765-1766. In private collection of Thomas Casey Greene, Potowomut, Rhode Island.
Rhode Island Historical Society Manuscript Collections
Greene Family Papers
Clarke Family Papers
MSS 9001 collection, Elihu & Christopher Greene Papers
MSS 459 collection, James Greene & Co. Papers
Proposed Iron Works at Providence in 1655 (1935, 28:67-70)
Rhode Island Wills, Warwick, RI
Volume I, 1703-1745, Rhode Island Historical Society.
Volume II 1746-1760
Volume III, 1761-1781
U.S.National Archives, 1819. Jesse Reed, Ms, Witnesses for Amesbury in Odiorne v. Amesbury Nail Factory, U.S. Circuit Court, May 1819. U.S. National Archives, Northeast Region, Waltham, Massachusetts.

Secondary Sources

Bining, A.C., 1933. *British Regulation of the Colonial Iron Industry.* Philadelphia: University of Pennsylvania Press.

Bridenbaugh, C., 1950. *The Colonial Craftsman.* New York: New York University Press.

Callon, M., 1987. Society in the Making: The Study of Technology as a Tool for Sociological Analysis. In: W.E. Bijker, T.P. Hughes and T. Pinch eds. *The Social Construction of Technological Systems: New Directions in the Sociology and History of Technology.* Cambridge: MIT Press, pp.83-104.

Clarke, L.B., 1875. *The Greenes of Rhode Island with Historical Records of England Ancestors.* New York: Knickerbocker Press.

Costin, C., 1991. Craft specialization: issues in defining, documenting, and explaining the organization of production. In: M. Schiffer ed. *Archaeological Method and Theory, Volume III.* Tucson: University of Arizona Press, pp.1-55.

Eliot, J., 1762. *An Essay on the Invention, or Art of making very good, if not the best Iron, from black Sea Sand.* New York: New Printing Office.

Ehrhardt, K.L., 2005. *European Metals in Native Hands: Rethinking the Dynamics of Technological Change 1640–1683.* Tuscaloosa: University of Alabama Press.

Fuller, O.P., 1875. *A History of Warwick.* Providence: Angell, Burlingame & Co.

Gille, B., 1978. *Histoire des techniques.* Paris: Gallimard.

Gould, S.J., 1986. Evolution and the triumph of homology, or why history matters. *American Scientist,* 74, pp.60-9.

Gould, R., 2001. From Sail to Steam at Sea in the Late Nineteenth Century. In: M. Schiffer ed. *Anthropological Perspectives on Technology.* Albuquerque: University of New Mexico Press, pp.193-214.

Greene, G.W., 1867. *The Life of Nathanael Greene, Major-General in the Army of the Revolution, Volume 1.* New York: G.P. Putnam and Son.

Hodder, I., 1999. *The Archaeological Process: an Introduction.* Oxford: Blackwell.

Hughes, T.P., 1987. The Evolutions of Large Technological Systems. In: W.E. Bijker, T.P. Hughes and T. Pinch eds. *The Social Construction of Technological Systems: New Directions in the Sociology and History of Technology.* Cambridge: MIT Press, pp.51-82.

Ingold, T., 2000. *The Perception of the Environment. Essays in livelihood, dwelling and skill.* London: Routledge.

Keller, C.M., 2001. Thoughts and production: insights of a practitioner. In: M. Schiffer ed. *Anthropological Perspectives on Technology.* Albuquerque: University of New Mexico Press, pp.33-45.

Killick, D., 2004. Social Constructionist Approaches to the Study of Technology. *World Archaeology,* 36, pp.571-78.

Kingery, D.W., 1993. Technological systems and some implications with regard to continuity and change. In: S. Lubar and D.W. Kingery eds. *History from Things.* Washington, DC: Smithsonian Institution Press, pp.215-30.

Knappett, C., 2005. Thinking through Material Culture: An Interdisciplinary Perspective. Philadelphia: University of Pennsylvania Press.

Latour, B., 1993. *We Have Never Been Modern.* Hemel Hempstead: Harvester Weatsheaf.

Latour, B., 2005. *Reassembling the Social: An Introduction to Actor-Network-Theory.* Oxford: Oxford University Press.

Layton, E., 1977 . Conditions of technological development. In: I. Speigel-Rosing and D. de Solla Price eds. *Science, Technology, and Society: A Cross-Disciplinary Perspective.* London and Beverly Hills: Sage, pp.197-222.

Law, J., 1987. Technology and Heterogeneous Engineering: The Case of Portuguese Expansion. In: W.E. Bijker, T.P. Hughes and T. Pinch eds. *The Social Construction of Technological Systems: New Directions in the Sociology and History of Technology.* Cambridge: MIT Press, pp.111-34.

Lechtman, H., 1977. Style in technology: some early thoughts. In: H. Lechtman and R. Merrill eds. *Material Culture: Styles, Organization, and Dynamics of Technology.* St. Paul, Minnesota: American Ethnological Society, pp.3-20.

Lemonnier, P., 1986. The study of material culture today: Toward an anthropology of technical systems. *Journal of Anthropological Archaeology,* 5, pp.147-86.

Lemonnier, P., 1992. *Elements for an Anthropology of Technology.* Ann Arbor: University of Michigan.

Lenik, E.J., 1977. A Study of Cast Iron Nails. *Historical Archaeology,* 11, pp.45-7.

Lyman, R.L., O'Brien, M.J., and Dunnell, R.C., 1997. *The Rise and Fall of Culture History.* New York: Plenum Press.

Neff, H., 1995. A Role for "Sourcing" in Evolutionary Archaeology. In: P.A. Teltser ed. *Evolutionary Archaeology: Methodological Issues*. Tucson: The University of Arizona Press, pp.69-112.

Nelson, H.L., 1968. Nail Chronology as an Aid to Dating Old Buildings. *History News*, 24, American Association for State and Local History Technical Leaflet 48.

Nye, D., 2006. *Technology matters: questions to live with*. Cambridge: MIT Press.

Oliver, P., 1762. Letter to Jared Eliot. Reprinted in: H. Carman and R. Tugwell, *History of American Agriculture* [1934?], p.249.

Petréquin, P., 1993. North Wind, South Wind: Neolithic Technical Choices in the Jura Mountains, 3700-2400 BC. In: P. Lemonnier ed. *Technological Choices: Transformation in Material Cultures Since the Neolithic*. London: Routledge, pp.36-76.

Petroski, H., 1996. *Invention by design: how engineers get from thought to thing*. Cambridge: Harvard University Press.

Pfaffenberger, B., 1992. Social Anthropology of Technology. *Annual Review of Anthropology*, 21, pp.419-516.

Phillips, M., 1993. Mechanic Geniuses and Duckies: A Revision of New England's Cut Nail Chronology before 1820. *APT Bulletin*, 25, pp.4-16.

Phillips, M., 1996. Mechanic Geniuses and Duckies: Nail Makers and their Machines. *APT Bulletin*, 27, pp.47-56.

Pinch, T.J. and Bijker, W.E., 1987. The Social Construction of Facts and Artefacts: Or How the Sociology of Science and the Sociology of Technology Might Benefit Each Other. In: W.E. Bijker, T.P. Hughes and T. Pinch eds. *The Social Construction of Technological Systems: New Directions in the Sociology and History of Technology*. Cambridge: MIT Press, pp.17-50.

Ramenofsky, A.F., Vaughan, C.D. and Spilde, M.N., 2008. Seventeenth-Century Metal Production at San Marcos Pueblo, North-Central New Mexico. *Historical Archaeology*, 42, pp.105-31.

Ryzewski, K., 2008. *Archaeology of a Colonial Industry: Domestic Ironworking and Industrial Evolution in Rhode Island, 1642-1800*. Ann Arbor: University Microfilms.

Ryzewski, K., 2012. Multiply Situated Strategies? Multi-Sited Ethnography and Archaeology. *Journal of Archaeological Method and Theory*, 19, pp.241-68.

Ryzewski, K. and Gordon, R., 2008. Historical Nail Making Techniques Revealed in Metal Structure, *Historical Metallurgy*, 42, pp.50-64.

Schiffer, M. ed., 2001. *Anthropological Perspectives on Technology*. Albuquerque: University of New Mexico Press.

Shanks, M., 2004. Three rooms. *Journal of Social Archaeology*, 4, pp.147-80.

Sigaut, F., 1994. Technology. In: T. Ingold ed. *Companion Encyclopedia of Anthropology*. London: Routledge, pp.420-59.

Skibbo, J. and Schiffer, M., 2001. Understanding Artifact Variability and Change: a Behavioral Framework. In: M. Schiffer ed. *Anthropological Perspectives on Technology*. Albuquerque: University of New Mexico Press, pp.139-50.

Urban, T., Ryzewski, K. and Hermance, J., 2010. *Cumberlandite: Using Residual Gravity to Delineate an Historically Important Ore Body*. Paper presented at the 75th Annual Meeting of the Society for American Archaeology, St. Louis.

Webmoor, T., 2007. What about 'one more turn after the social' in archaeological reasoning? Taking things seriously. *World Archaeology*, 39, pp.564-78.

Wells, T., 1998. Nail Chronology: The Use of Technologically Derived Features. *Historical Archaeology*, 32, pp.78-99.

Witmore, C.L., 2004. On Multiple Fields. Between the Material World and Media: Two Cases from the Peloponnesus, Greece. *Archaeological Dialogues*, 11, pp.133-64.

Witmore, C.L., 2007. Landscape, Time, Topology: an Archaeological Account of the Southern Argolid, Greece. In: D. Hicks, Laura McAtackney and Graham Fairclough eds. *Envisioning Landscape. Situations and Standpoints in Archaeology and Heritage*. Walnut Creek: Left Coast Press, pp.194-225.

Witmore, C.L., 2009. Prolegomena to Open Pasts: On Archaeological Memory Practices. *Archaeologies*, 5, pp.511-45.

Wolverton, S. and Lyman R.L., 2000. Immanence and Configuration in Analogical Reasoning. *North American Archaeologist*, 21, pp.233-47.

Author's address

- Wayne State University
  Department of Anthropology
  3054 F/AB, 656 W. Kirby
  Detroit, MI 48226, USA

Manufacturing techniques, technological traditions and social choices in ancient Greece

Maria Kostoglou

ABSTRACT: This paper provides a theoretical and methodological model for the archaeological and archaeometallurgical study of the interaction between ancient technology and society by using the iron metallurgy of Aegean Thrace, Northern Greece, as a case study. The paper combines archaeological evidence from recent excavations in the area with data from the metallographic analysis both of wastes from the production of iron and of finished iron artefacts. I aim to demonstrate that a theoretically informed analytical methodology can (a) produce a better archaeological understanding of the long-term interrelation between local/regional technologies and social change; and (b) expose the social dynamics and attitudes of diverse societies within the colonial/imperial setting of the Greco-Roman world.

Keywords: Archaeometallurgy, iron technology, manufacturing techniques, Greece, Thrace, Greco-Roman.

Theoretical background

Soon after the introduction of archaeometry, it was quickly realised that scientific analysis alone was insufficient to answer complicated questions about the past, and the application of interdisciplinary approaches to archaeology began in earnest. Today, within the discipline of archaeological science, the implications and limitations of archaeological methods can confuse scientists, while archaeologists remain sceptical about the application of scientific techniques in archaeology, arguing for more socially oriented focus on science-based archaeology (Muhly, 1998; Knapp, 2000).

Archaeologists have dealt with the study of technology in an indirect way to accomplish other objectives, such as locating sites in time or space, reconstructing past practices, or identifying status differences. In studying technology and its products, they form stylistic groups and they are captured by the form or design of the object. Most archaeological studies have been devoted to the typology of artefacts, which have been used by archaeologists of the culture historical approach to explain behavioural phenomena in the past, such as movements of people, inter-group interaction, or boundary creation and maintenance (for a detailed discussion on the limitations of the typological approach for the study of metals, see Bruno and Kostoglou, in press). However, artefacts constitute only one aspect of the technological process. Technology is a social phenomenon and can only be studied in this framework.

Since the 1970s, anthropologists, ethnologists, sociologists and others have emphasised how techniques can be understood as cultural choices, which depend on the social, economic and ideological setting of the technological act. Furthermore, understanding why one given aspect of material culture, and not another, is used can provide archaeologists with a better insight into certain social relations (Lemmonier, 1992, pp.1-4). Therefore, technologies can only be understood if the analyses consider the wider socio-cultural context in which those technologies are embedded (Lechtman and Steinberg, 1979; Herbert, 1983; Lemmonier, 1992; Dobres 2000 Killick, 2004). From anthropological studies also stems the suggestion that the dynamics of social and technological developments within a colonial/imperial context can be better understood if studied on a micro-scale, time and place specific context (Dobress 2000 Lightfoot, 1995). In this context, the present work deals with iron technology in ancient Greece as a culturally generated activity and tries to explore its diverse role within society.

Iron in Greek society

Iron technology in Greece has been associated with major socio-cultural changes such as population decrease, discontinuity in settlement patterns, and change in burial customs (Snodgrass, 1987). However, our understanding of the roles of iron in ancient Greek society is seriously restricted. Selective documentary evidence, fragmented archaeological finds, and the limited amount of analysed material from archaeological contexts (Pleiner, 1969, 1980; Snodgrass, 1980, 1987; Muhly, 1980; Photos, 1987; Craddock, 1995, pp.259-261; Pigott, 1996) have led to a series of misunderstandings, not only about the spread of iron technology in the Mediterranean, but also about the role of iron in Greek society (Kostoglou, 2010).

From the advent of iron technology within the region and up to the Roman period, technological developments (as indeed cultural) have until now been viewed according to the centre-periphery model (Champion, 1989; Hadrill, 1992). This model explains changes through the prism of economic (supply and demand) or cultural (population displacement) determinism and imposes a series of artificial dichotomies upon ancient societies. Over the past few years, a number of authors have criticised the misleading interpretations of ancient cities as foci of civilisation as compared to their barbaric peripheries (van Andel and Runnel, 1987; Shanks, 1996), as consumer towns versus their supplier country (Rowlands, et al., 1987; Champion, 1989), and overall as a civilised Greco-Roman world versus the indigenous barbaric people (Webster, 1996).

The conventional history and archaeology of Aegean Thrace did not escape the biases of the centre-periphery model. Although we can accept that economic institutions between different communities and social units exist, as do power differentials and access to wealth (Shennan, 1999), defining these units based on ancient sources could be misleading. The centre-periphery history of Aegean Thrace attests that the coastline was colonised intensively in the 8th and 6th century BC by people from Athens, Aegean Islands such as Thasos, Chios, and Samothrace, and Asia Minor, and that the colonies flourished from the 5th to the 3rd centuries BC. The colonists were called 'Greeks' as opposed to the indigenous population of the area, who in the ancient Greek and Latin literature are known under different tribal names and under the collective term 'Thracians' (Kostoglou, 2008a, pp.24-29). Arguments based upon architecture, pottery, sculpture, jewellery and the ancient literature claim that the indigenous population of Aegean Thrace was either assimilated or pushed inland, to the Rhodope mountains and beyond, leaving the coastal area to colonists. Furthermore, cultural, artistic, and technological developments are deemed to have occurred in the centres of the ancient world and to have been passed on from these to the 'barbaric' indigenous population of Aegean Thrace, which had nothing to offer in return but slaves and raw materials.

Today, the explanatory validity of such models is questioned because they are based on ancient written sources (Snodgrass, 1987) and 19th century discourse (Morris, 1994; Shanks, 1996). They imply cultural uniformity (Barrett, 1997), exclude the variability of the process of acculturation (Hingley, 1996), and appear inadequate for accommodating cultural diversity and local developments (Dietler, 1998). The previously described theoretical concerns about a humanistic input into science based archaeology and about the inadequacy of the core-periphery models to explain technological traditions have influenced the creation of the main research hypothesis for this work. If such discontinuities are evident in the archaeological record and if the development of the region is characterised by a gap between the Greco-Roman period and what existed before, then similar discontinuities could be evident in the development of iron technology (and if this is not the case, then an alternative explanation is required). To investigate this possibility I undertook a detailed contextual, diachronic, and strictly regional approach to the material under study for reasons explained earlier in this paper. To achieve these aims, my methodology combines material culture studies with materials science techniques and a solid archaeological case study.

Methodology and analytical work

All the iron related material under study (Table 1) comes from well stratified and dated archaeological contexts (Kostoglou, 2008a), in order to facilitate correlations with the rest of the evidence (pottery, settlement patterns, and burial practices). After the systematic recording, measuring, photographing and drawing of the objects, samples (< 0.015m) were removed from slag and blooms, and (0.005m) also from the iron blades, using a TS 8 / E Trim saw with soluble cutting oil (typical dilution 3-10% in water). All of the samples were set in fast setting resin, with the sample sections placed for analysis. The resin mounts were then wet grinded using three grinding papers (nos. 220, 500, 1200) and polished using the standard metallographic techniques of diamond polishing. A Rotosystem 250 machine was used for both grinding and polishing the samples. The polished sections were then etched in nitric acid in alcohol (Nital with 2% nitric acid and 98% methanol) and examined by both binocular and optical microscopes. The total number of samples is presented in Table 1. Hardness measurements, according to the Vickers system, were made in order to check the phases present and to evaluate the heat-treated blades (Tylecote and Gilmour, 1986, p.7). In order to define the chemical composition of the blooms and slag, as well as of some inclusions in the blades, SEM analyses and examinations

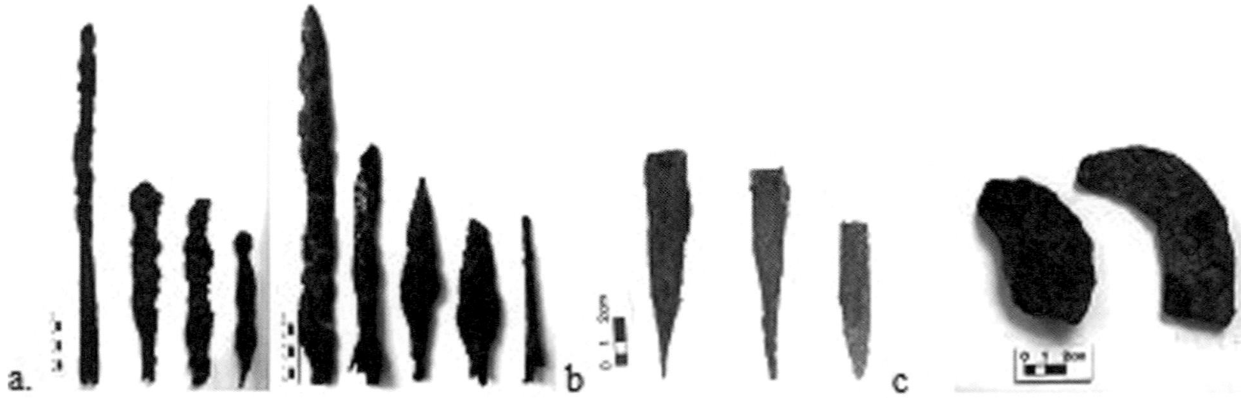

Figure 1: a) Iron spearheads from Messemvria-Zone (6th-4th centuries BC); b) Iron wedges from Avdera (5th century BC); c) Billhooks from Messemvria-Zone (5th century BC).

Table 1: List of objects recorded and sampled for analysis.

TYPE OF FIND	MESSEMVRIA-ZONE		AVDERA		KALYVA-KASTRO	
	Recorded	Sampled	Recorded	Sampled	Recorded	Sampled
KNIVES	15	5	34	8	21	17
SICKLES	2	2	2	2	3	3
SCISSORS			6		2	
SPEARHEADS	11	5	6		1	
ARROWHEADS	12	4	6		3	
AXES			6	3	2	1
SCISSELS				2		
WEDGES			7	2		
SPITS	10	4	9	3		
BARS	1	1				
SAWS			1			
PLOUGHS	1					
HORSE-BITS					1	
STRIGGILS	2		21	2		
KEYS	2		7		7	
PINS			9		2	
RINGS		1		8		
SPATULAE				8		
BELIS				3		1
NAILS	6		145		9	
OTHER (hooks, sheets, chain rings)		3		24		21
TOTAL	66	21	158 (+145 nails)	21 (no nails)	69	30
SLAG	36	16 (all SEM)	72	15 (12 SEM)	91	28 (21 SEM)

were carried out in a "Leo microscopy" Stereoscan S360 SEM instrument with integrated "Linch Analytical" AN10855 energy dispersive EDAX micro analyser.

The metallurgy of iron in Aegean Thrace

From the river Nestos to the river Evros, and from the Rhodope Mountains to the Aegean Sea, between Classical and Roman times, different local technological traditions emerged in particular regions. The following section draws attention to studies carried out on three of these traditions (for a detailed account on other areas of interest see Kostoglou, 2008a, pp.30-35) and presents the results of the study of iron technology from Messemvria-Zone and Avdera (the two sites have parallel lives and the analysed objects cover mainly the 5th century BC) and from Kalyva-Kastro (objects mainly from the 4th century AD).

Messemvria-Zone

Messemvria-Zone is a 'commercial colony' (*emporion*) founded on the Thracian coast in the 7th century BC by the inhabitants of the island of Samothrace. It lies on the coast, at the southern end of the Zonaia Mountains.

The archaeological finds cover mainly the period from the 6th to the 4th century BC, but there are also indications, mainly in the form of coins, that the occupation of the site was continued until the 5th century AD (Tsatsopoulou, 1997). Evidence for the earliest phase of the city comes from excavations in the sanctuary of Apollo (Tsatsopoulou, 1988-1989). A large group of buildings, with the temple of Apollo in the north, are organised around an open paved area (Fig. 2). The temple is 9m wide and 15m long and has a north-south orientation. The complex occupies a total area of 35x45m^2. The buildings are arranged in a Γ-shaped complex around the open area. The architectural complex of the sanctuary of Apollo is the only building dated to the 6th century in the entire excavated area. There is no clear evidence of the size, extension, and the general layout of the city in the 6th century. Most of the finds were discovered both inside and outside the temple of Apollo. There are many marble and clay architectural features, fragments of marble inscriptions, parts of archaic statues (*kouroi*), and many fragments of pottery. They are all dated to the 6th and 5th centuries BC. The pottery is comprised of Attic black-glazed and black-figured vases, of mainly table-drinking wares (*skyphoi, pinakia, cylix*), and local hand-made pottery known as Thracian. Half of the 6,000 pottery sherds (usually Attic black-glazed or black-figured wares) discovered within the temple and its vicinity bear incised inscriptions with the name of the god Apollo, written in a local idiom resembling the Aeolian dialect. One third of these inscriptions are written in the archaic way, from right to left (Tsatsopoulou, 1988-1989). This evidence leads to the safe assumption that the sanctuary was functioning on a local/regional scale. Since Messemvria-Zone is one of the commercial colonies of the

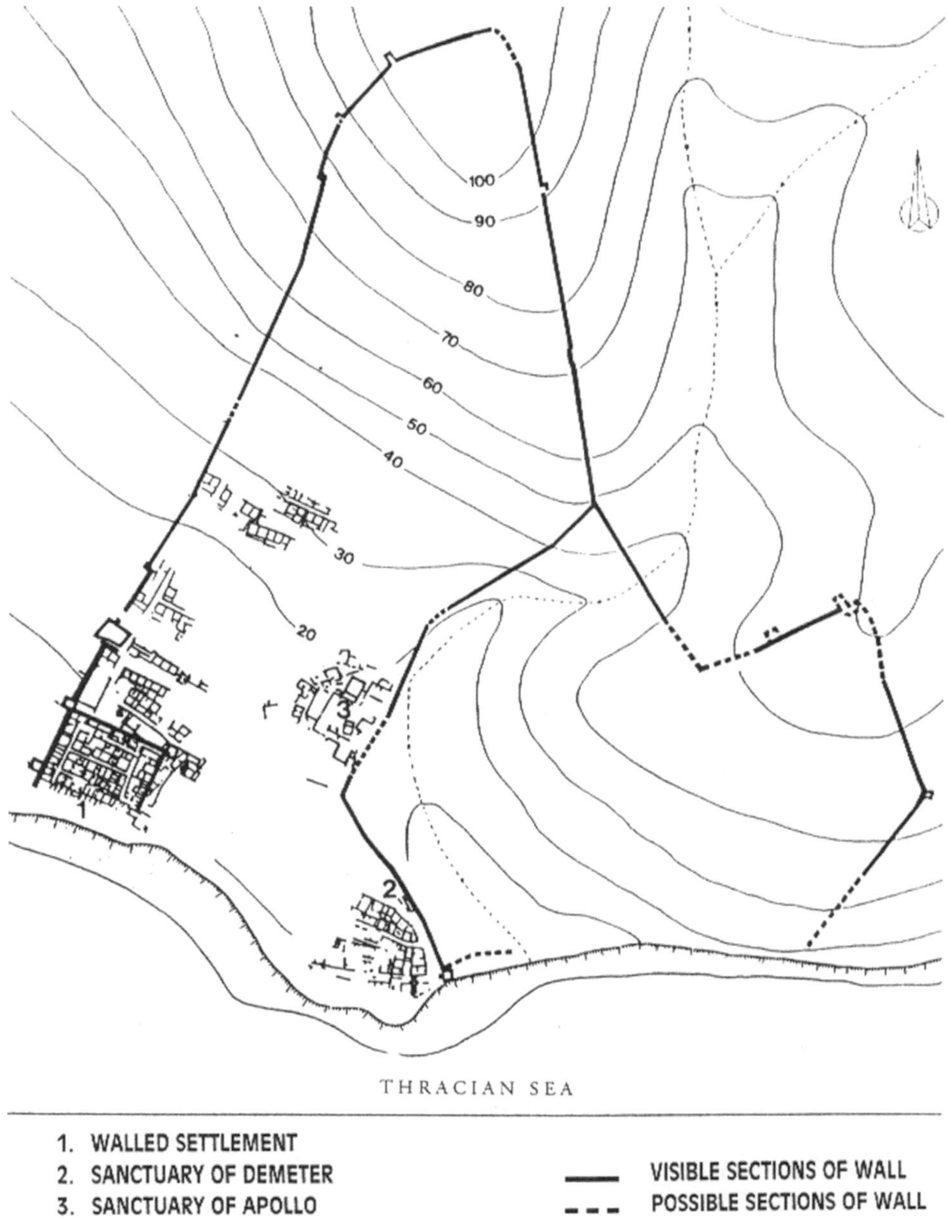

Figure 2: Topographic map of Messemvria-Zone (© Archaeological Museum of Komotini, Greece).

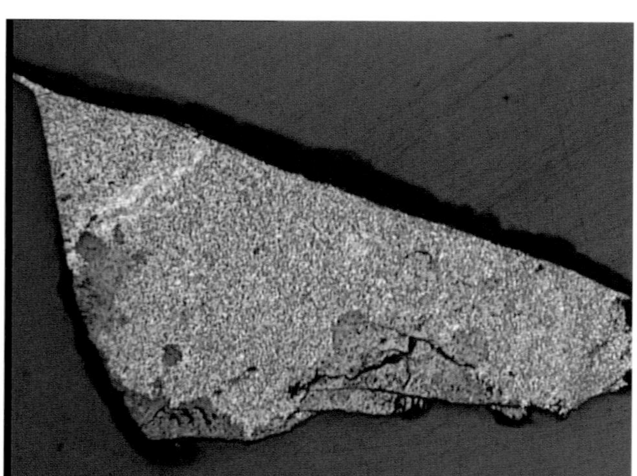

Figure 3: Microphotograph of knife MES9 (66x) showing uniform distribution of high carbon steel structure (Spaeroidised cementite and pearlite, caused by the heat working of a quenched structure, probably tempered martensite).

island of Samothrace, it could be suggested that the sanctuary of Apollo was under the protection of the great temple of Apollo and Kabeiroi in Samothrace.

The Messemvria-Zone slag is characterised by moderate silica content, is high in lime, and rich in magnesia. The unstable lime presence (< 10%) points to insufficient control of smelting or occasional use of lime as flux. The constant presence of manganese in all slag samples, including smithing slag, and in the slag inclusions of artefacts is very characteristic. An unreduced fragment of Mn/Fe ore has been also analysed as part of the same context (Kostoglou, 2008a).

Iron was circulated in small elongated bars forged probably from a single bloom; an elongated bar made from wrought iron and dated to the 6th century BC was discovered among the votive objects in the Apollo sanctuary (Kostoglou, 2003). This bar, along with the hoard of Assyrian type billets found in Delphi and iron ingots from Kommos (Risberg, 1997, p.188) represent the earliest forms of marketable iron found in Greece. Edged tools and weapons were also produced locally and they can be distinguished from the imported examples by the characteristic level of manganese present in the metallic iron and the slag inclusions in the artefacts. All known techniques for manufacturing steel tools and weapons are evident and all were successfully applied, with the use of high carbon steel dominating the record.

Avdera

The city was first founded by Klazomenai (c. 650 BC) and second by Teos (c. 550 BC), in the East Nestos valley, on the cape Baloustra. The geomorphology of the region was influenced by the movement of the bank of the river Nestos and extensive marshlands dominated the area, influencing the development of the colony over the years (Syrides, 1997). Industrial waste and iron finds are described chronologically, with reference to their archaeological context. According to new archaeological data (Koukouli-Chryssanthaki, 1988), the first colonists founded a fortified city (Fig. 4) using the natural protection of the foothills.

The excavation unearthed part of the archaic wall in the north of the city, an archaic temple of Demeter, and burials. The temple of Demeter was in use from the end of the 6th century BC to the end of the 4th century BC, when the Hellenistic cemetery of the city took its place. The first small colony became a city-state during the Classical times, having all the typical characteristics: city walls (5.340m), acropolis, theatre, temples for the city's Gods and Heroes, two harbours, political autonomy, and independent coinage. Imported pottery from Athens, Corinth, and the Ionian coast, silver coins of Avdera found as far away as Egypt and Syria, and the finds from the city's cemetery indicate that the city flourished during the Archaic-Classical times (Isaac, 1986, pp.73-123; Koukouli-Chryssanthaki, 1988). The city walls cover an area of 10 acres and they include areas without buildings in the east, presumably for cultivation or for herding animals (Koukouli-Chryssanthaki, 1988, p.41). If this is compared with the double walls of Messemvria-Zone, then it could be concluded that the land used for agro-pastoral activities was fortified. This evidence suggests that the hinterland (*chora*) of these two colonies was limited. This is a very important aspect for the

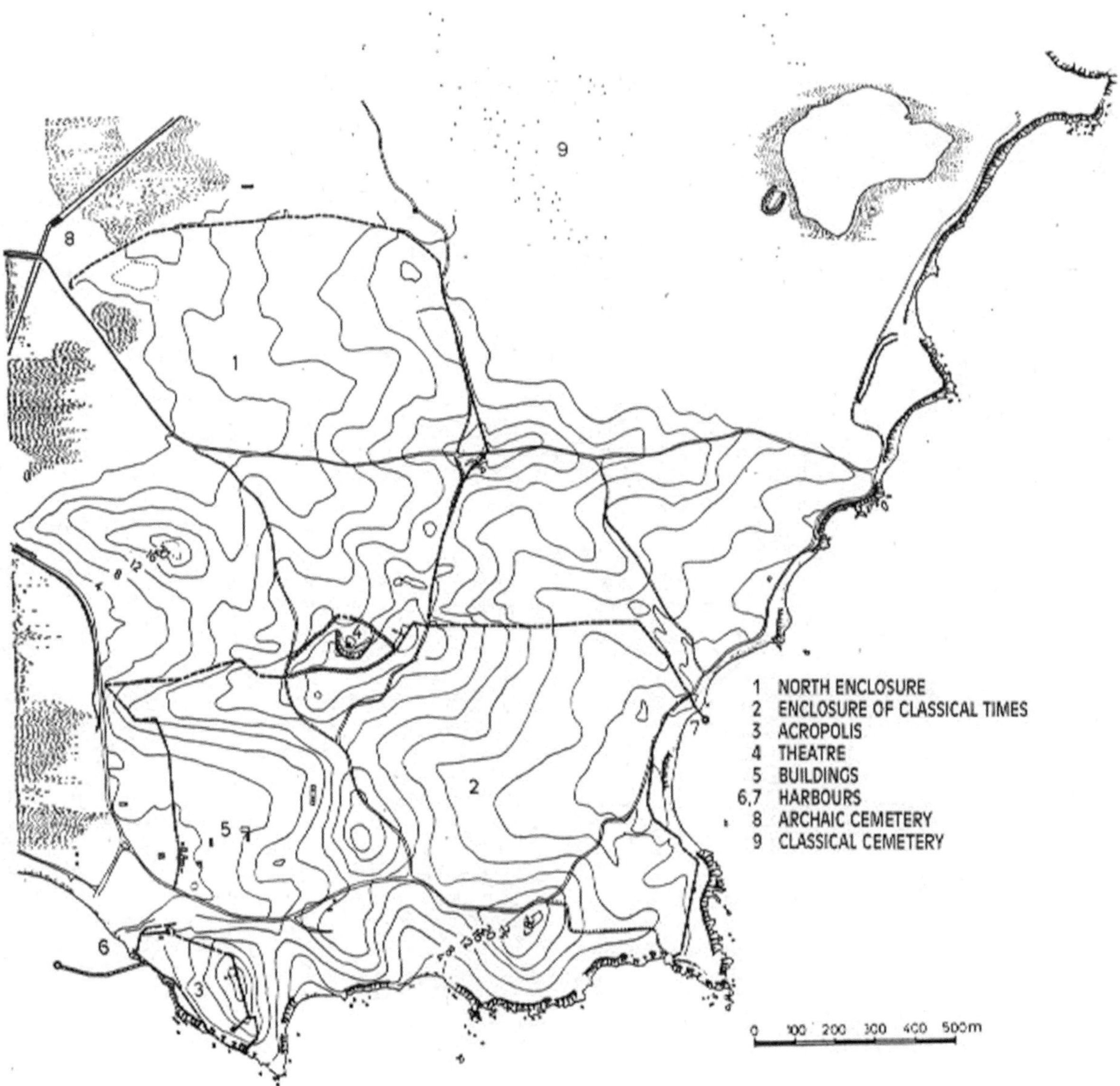

Figure 4: Map of Ancient Avdera (© Archaeological Museum of Avdera, Greece).

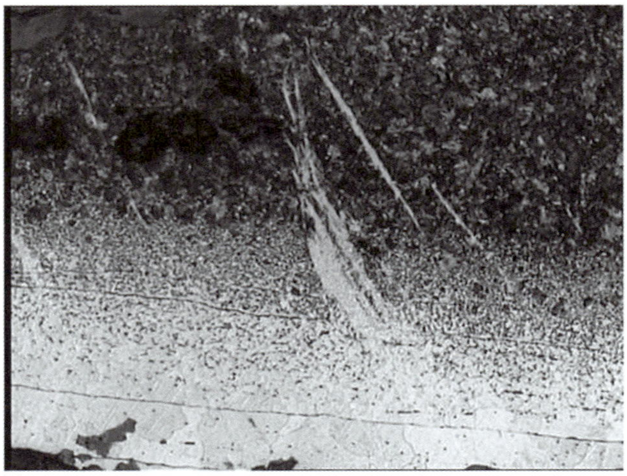

Figure 5: Microphotograph of knife AVD53 (66x) showing welding between wrought iron (lower part) and high medium carbon steel (upper part).

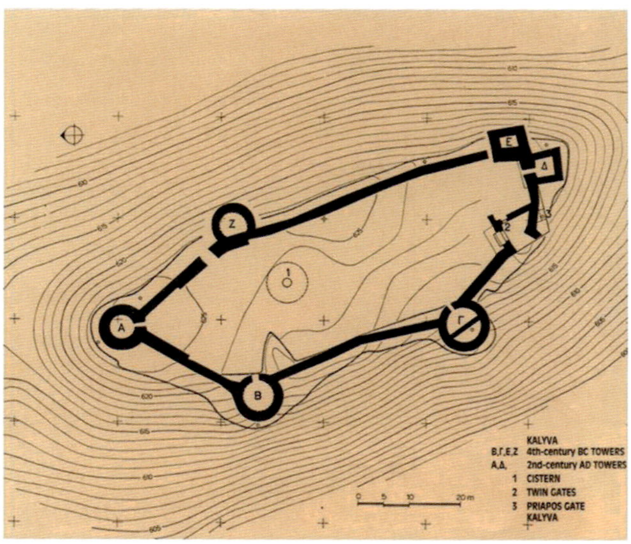

Figure 6: Map of Kalyva-Kastro (© Archaeological Museum of Komotini, Greece).

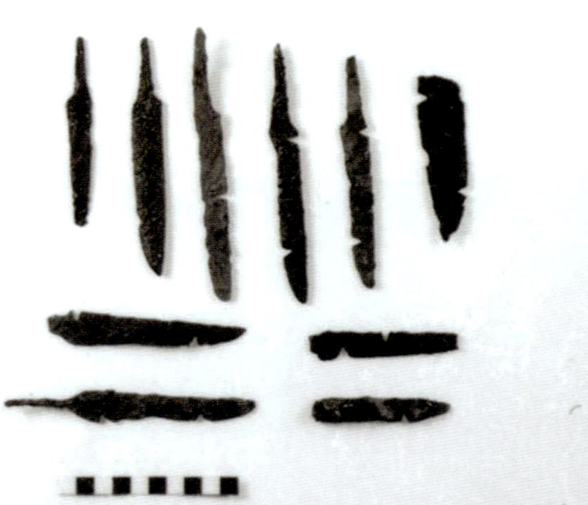

Figure 7: Straight knives from Kalyva-Kastro (3rd century AD) after sampling.

discussion of the impact of Greek colonies on the hinterland, and for the discussion of the Hellenisation process.

No iron slag or iron finds were found in any of the excavations dating to the 7th and 6th centuries BC, indicating that iron was not produced locally and that most of the iron was imported in finished or semi-finished forms. The first iron artefacts were found in the settlement and in the cemetery of the Classical colony. Iron artefacts of the 5th and 4th centuries BC include axes, wedges and spits (Table 1). The first evidence for smelting and smithing iron at the site was found outside the Classical enclosure and dates to the Roman period (2nd and 3rd centuries AD; Kostoglou, 2008a, appendix 107-109). Both ores and fuel deposits were located outside the territory of the city-state of Avdera. This problem was solved only after the re-organisation of the territorial boundaries and under the administrative control of the Romans, as indicated by the majority of artefacts and all slag found in Avdera, which date to Roman times.

The Avdera slag is characterised by moderate silica (8-30%) and lime (2-13%) contents, and is high in alumina (> 10%). The amounts of manganese and magnesium oxides vary (2-4% in the fayalitic phase) and phosphorous is high (2-4% in the glassy phase). The oxides of magnesium, manganese and phosphorous change proportionately, suggesting that the ore consisted of limonite. Most of the industrial waste from Avdera consists of fragments of smithing slag and furnace lining.

Most of the artefacts studied exhibited either surface carburisation or welding of low carbon to high carbon steel (Fig. 5); this appears to be the predominant technique for manufacturing tools and weapons at the site.

Kalyva-Kastro

Philip II built the fort of Kalyva (c. 340 BC) in the Highlands of Xanthi during his second campaign in Thrace (346-342 BC). Philip II founded the fort (Fig. 6) in order to safeguard the strategic route next to the River Nestos from the attacks of the Thracian tribes. Local hand-made pottery dated to the 9th century BC indicates that indigenous Thracian tribes occupied the site before the coming of the Macedonians. Local hand-made pottery of the type often found in Messemvria-Zone is found

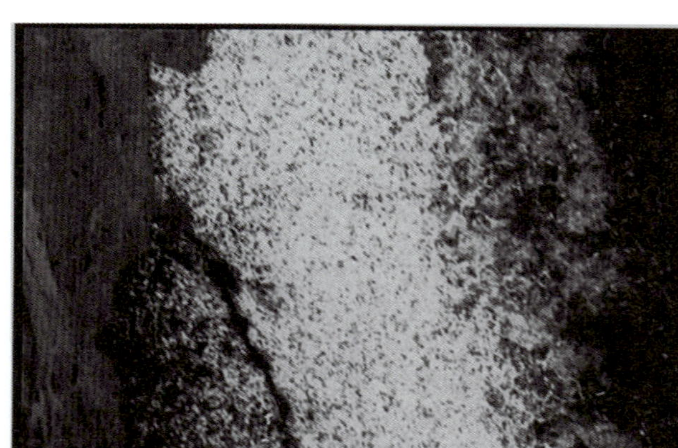

Figure 8: Microphotograph of straight knife KL 20 (50x) displaying carburisation on both edges (Hv on the edge = 383).

at Kalyva-Kastro as late as the 2nd century BC. Macedonians, Thracians and Romans occupied the fort between the 4th century BC and the 3rd century AD (Triantafyllos, 1988).

The fort is situated in one of the richest metalliferous zones of Greece (IGME 1973). The Kalyva-Kastro slag is characterised by high iron loss in the slag and low silica and lime. However, it is high in potash (K_2NA) and soda (N_2O_2), which function to lower the melting point. The metalworkers were efficient, since they were able to produce enough metallic iron from relatively pure iron oxide by using appropriate fluxes (Kostoglou, 2008a). The iron artefacts (for example see Fig. 7) found in this area were probably made locally and mainly include implements used in agro-pastoral activities. Moreover, the techniques used in their manufacture suggest that the smiths were highly skilled: surface carburisation of wrought iron (Fig. 8) or heat treatment of high carbon steel were used to produce steeled tools and weapons. None of the artefacts sampled showed a welded structure.

Interpretation of the analytical results

The archaeological and analytical data suggest that distinctive local technologies have developed in Aegean Thrace from Classical to Roman times. Each local technology was characterised by the use of particular raw materials and by the application of specific manufacturing techniques. Although the study of raw materials and mining is beyond the scope of this paper, it should be emphasised that Aegean Thrace is one of the richest metalliferous zones in Greece: most of the iron minerals occur in close proximity to the archaeological sites under study, meaning that choice and human agency rather than availability of raw materials were responsible for the use or exclusion of the ores used. Over the past twenty years, the systematic study of excavated material from North Greece has established the existence of the following smelting traditions that developed around the use of different ores such as:

- Manganese/ferrous ores in the Symvolon Mountains (Vavelidis, et al., 1996) and in Messemvria-Zone (Kostoglou, 2008a,b).
- Limonites in Aggisron Serres (Chiotis, et al., 1996) and Kalyva-Kastro (Kostoglou, 2008a).
- Iron pyrites in Maroneia (Melfos, 1995).
- Titanium-rich magnetite sands on the island of Thassos (Photos, 1987).

The study of industrial waste and artefacts from excavated contexts made possible a comparative understanding of the specific technological traditions. These traditions and local preferences are more visible in the study of manufacturing techniques. The main conclusion is that manufacturing techniques appear to be very diverse on a regional scale and constant on a local scale. For example:

1. Ferritic knives are absent from the record of the Messemvria-Zone, but they dominate the record in Avdera.
2. High-carbon steel edges are not found at Avdera but they are in preferred in Meseemvria-Zone.
3. Surface carburisation was the dominant technique in Kalyva-Kastro and was applied successfully on the right, left or both edges.
4. Welded structures are not attested at Kalyva-Kastro but they were preferred in Avdera.

We can only assume why one specific technique was preferred to another at a time when choice was available. However, it is safe to conclude that this kind of choice attributes a unique character to every local technology, and that the differentiation is socially and culturally significant.

The evidence presented suggests a pattern characterised by the existence of individual traditions at the local scale, with preferences for the use or exclusion of some techniques and a lack of standardisation on a regional scale. The analytical data combined with the archaeological data supports the conclusion that different local technologies were operating in the region. The individual characteristic rise and fall of each individual technological tradition follows the life of each settlement. Each local tradition is characterised by continuity in terms of the ores, fuel, and manufacturing techniques used. The existing local traditions are variations within the metallurgy of bloomery iron and create a diverse pattern on a regional scale. The emerging picture contradicts the one described in the conventional historical framework of Aegean

Table 2: Manufacturing techniques of knives

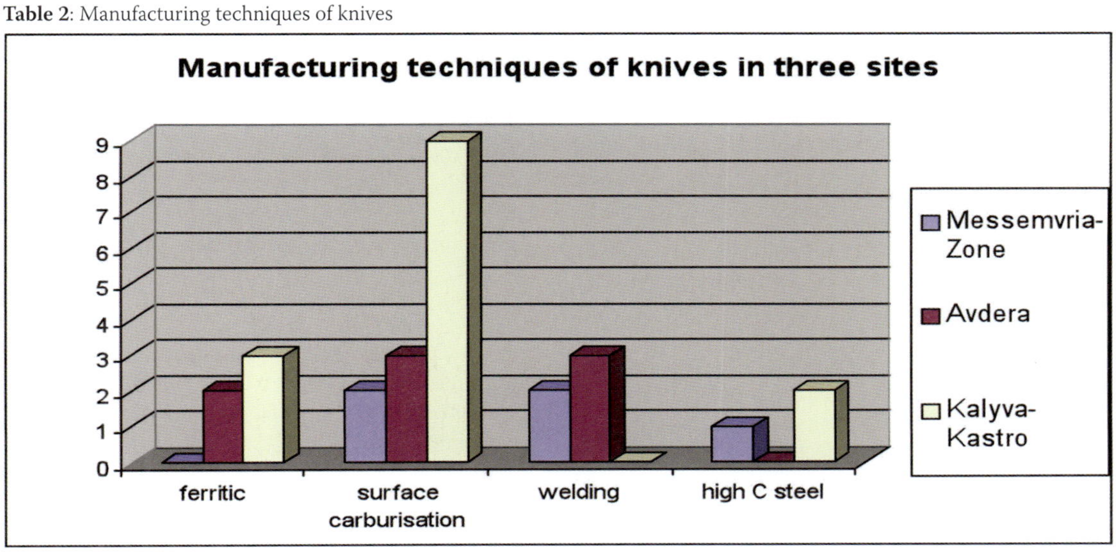

Thrace and suggests that there are no major discontinuities in the local iron production that is associated with the indigenous and not colonial context.

Conclusions

The analytical data suggest that different technological choices were made in the production of iron and steel in the three contemporary sites examined in this work. Each local technology is characterised by the deliberate use of specific raw materials, such as manganese/ferrous ores in Messemvria-Zone or iron oxides in Kalyva-Kastro, along with a preference for certain manufacturing techniques, which in turn indicates a series of technological choices. Skilled metal workers existed in Messemvria-Zone and Kalyva-Kastro. The situation seems to be different in Avdera, where iron was not locally produced or commonly used during the early phases of the colony (7th to 6th centuries BC), but was widely used during the Roman period (2nd and 3rd centuries AD; Kostoglou, 2008a, 2008b). The manufacturing techniques under consideration appear in the archaeological record of Aegean Thrace as early as the end of the 6th century BC. There is no evidence to suggest that new techniques were introduced to the iron technology of the region during the Roman period. No innovation or invention in terms of new ores, fuel, or techniques of manufacture at this time has been noted so far. Furthermore, the results question the assumed uniformity of iron production in ancient Greece and the technological superiority of the dominant Greco-Roman cultures, and they demonstrate that a theoretically informed analytical methodology can expose the complexities and social dynamics of culture contact and technological change.

Acknowledgements

Special thanks are ought to the organisers of the 'World of Iron' conference for inviting my contribution and to all participants for stimulating discussions. I am particularly grateful for the peer review and editorial comments. This work would not be possible without D. Triantafyllos, director, P. Tsatsopoulou and K. Kallintzi of the 19th Department of Antiquities in Greece, who kindly permitted the study of the material and provided all contextual information, and without the generous support of the A.S. Onassis Foundation and the British Academy.

References

Barrett, J.C., 1997. Romanization: a critical comment. In: D.J. Mattingly ed. *Dialogues in Roman Imperialism. Journal of Roman Archaeology Supplement Series*, 23, pp.51-66.

Bruno, A. and Kostoglou, M., in press. Object life cycles and the study of ancient metals: theories, methods and limitations. In: A. Kordas and L. Filippaki eds. *History, Technology and Conservation of Ancient Metals, Glasses and Enamels*. Athens: National Centre of Scientific Research 'Demokritos.

Champion, T.C. ed., 1989. *Centre and periphery: comparative studies in archaeology*. Cambridge: Cambridge University Press.

Chiotis, E., et al. 1996. Old mining and metallurgical activities in Aggistron, Serres. In *Proceedings of the 2nd Symposium of teh Hellenic Archaeometric Society, Archaeometrical and Archaeological Research in Macedonia and Thrace*. Thessaloniki: Aristotle University.

Craddock, P.T., 1995. *Early metal mining and production*. Edinburgh: Edinburgh University Press.

Dietler, M., 1998. Consumption, agency, and cultural entanglement: theoretical implications of a Mediterranean colonial encounter. In: J.G. Cusick ed. *Studies in Culture Contact: Interaction, Culture Change, and Archaeology*. Occasional Paper 25. Carbondale: Center for Archaeological Investigations, Southern Illinois University Press, pp.288-311.

Dobres, M.A., 2000. *Technology and social agency*. Oxford: Blackwell Publishers.

Hadrill, A.W., 1992. City and country in the ancient world: introduction. In: W. Hadrill and J. Rich eds. *City and country in the ancient world*. London: Routledge, pp.ix-xviii.

Herbert, E. 1983. Iron*, Gender and Power: Rituals of Transformation in African Society.* Indiana: Indiana University Press.

Hingley, R., 1996. The 'legacy' of Rome: the rise, decline and fall of the theory of Romanization. In: J. Webster and N. Cooper eds. *Roman Imperialism: Post-Colonial perspectives*. Leicester Archaeology Monographs 3. Leicester: School of Archaeological Studies, University of Leicester, pp.35-48.

IGME 1973. *Geological Map of Greece*. Athens: Institute of Geology and Mineral Exploration.

Killick, D., 2004. Social Constructionist Approaches to the Study of Technology. *World Archaeology*, 36, pp.571-8.

Knapp, A. B., 2000. Archaeology, science –based archaeology and the Mediterranean Bronze Age metals trade. *European Journal of Archaeology*, 3 (1): pp.31-56.

Kostoglou, M., 2003. Iron and steel currency bars in ancient Greece. *Journal of Mediterranean Archaeology and Archaeometry*, 3, pp.63-8.

Kostoglou, M., 2008a. *Iron and steel in Ancient Greece: artefacts, technology and social change in Aegean Thrace from Classical to Roman times*. British Archaeological Reports International Series 1833. Oxford: John and Erica Hedges.

Kostoglou, M., 2008b. Towards an integrated approach for the study of ancient Greek technology. In: Y. Fakorelis, N. Zacharias and K. Polkreti eds. *Proceedings of the 4th International Symposium of the Hellenic Society for Archaeometry, Athens 2003*. British Archaeological Reports International Series 1746. Oxford: Archaeopress, pp.459-67.

Kostoglou, M. 2010. Iron, connectivity and local identities in the Iron Age to Classical Mediterranean. In P. van Dommelen, and A. B. Knapp *Material Connections in the ancient Mediterranean*. London: Routledge: chapter 9, pages 170-189.

Koukouli-Chryssanthaki, C., 1988. Excavation at Avdera. In: *The historical, archaeological and laographical research in Thrace*. Thessaloniki: Institute of Balkan Studies, pp.39-74.

Lechtamn, H. and Steinberg, A., 1979. The history of technology: an anthropological perspective. In: G. Bugliarello and D.B. Doner eds. *History and Philosophy of Technology*. Urbana: University of Illinois Press, pp.135-60.

Lemmonier, P. ed., 1992. *Technological choices: transformations in material culture since the Neolithic*. London: Routledge.

Lightfoot, K.G., 1995. Culture contact studies: redefining the relationship between prehistoric and historic archaeology. *American Antiquity*, 60, pp.199-217.

Mattingly, D., 1996. From one colonialism to another: imperialism and the Maghreb. In: J. Webster and N. Cooper eds. *Roman Imperialism: Post-Colonial perspectives*. Leicester Archaeology Monographs 3. Leicester: School of Archaeological Studies, University of Leicester, pp.49-70.

Melfos, V. I., 1995. Ereuna ton vasikon kai eugenon metallon sthn perirodopiki zeniths Thrakis. Unpublished Ph. D thesis, Department of Geology, Aristotle University. Thessalonika.

Morris, I., 1994. Archaeologies of Greece. In I. Morris (ed) *Classical Greece: ancient histories adn modern archaeologies:* pp: 8-47. Cambridge: Cambridge University Press.

Muhly, J.D., 1998. Metals and Metallurgy: using modern technology to study ancient technology. In *Ancient Greek Technology: Proceedings of the First International Conference on Ancient Greek Technology* (Thessaloniki 4-7 September 1997): pp. 27-33. Thessaloniki: Technology Museum.

Muhly, J.D., 1980. The Bronze Age Setting. In: T.A. Wertime and J.D. Muhly eds. *The Coming of the Age of Iron.* New Haven and London: Yale University Press, pp.25-67.

Photos, E., 1987. *Early extractive metallurgy in north Greece.* PhD Thesis, Institute of Archaeology, University College London.

Pigott, V.C., 1996. Near Eastern archaeometallurgy: modern research and future directions. In: J.S. Cooper and G.M. Schwartz eds. *The study of the ancient Near East in the twenty-first century: The William Foxwell Albright Centennial Conference.* Winona Lake: Eisenbrauns, pp.139-76.

Pleiner, R., 1969. *Iron working in ancient Greece.* Praha: Národní Technické Muzeum v Praze.

Pleiner, R., 1980. Early Iron Metallurgy in Europe. In: T.A. Wertime and J.D. Muhly eds. *The Coming of the Age of Iron.* New Haven and London: Yale University Press, pp.375-416.

Risberg, Ch., 1997. Evidence of metal working in ancient Greek sanctuaries. In: C. Gillis, Ch. Risberg and B. Sjöberg eds. *Trade and production in premonetary Greece: production and the craftsman. Proceedings of the 4th and 5th International Workshops, Athens 1994 and 1995.* Jonsered: Paul Åstrom, pp.185-96.

Rowlands, M. 1987. Centre and Periphery: a review of a concept. In M. Rowlands (ed) *Centre adn Periphery in the Ancient World*: pp.1-11. Cambridge: Cambridge University Press.

Shanks, M., 1996. *Classical Archaeology of Greece. Experiences of the discipline.* London: Routledge.

Shennan, S.J., 1999. Cost, Benefit, and Value in the Organization of Early European Copper Production. *Antiquity*, 73, pp.352-63.

Snodgrass, A., 1980. *The dark ages of Greece.* Cambridge: Cambridge University Press.

Snodgrass, A., 1987. *An archaeology of Greece: the present state and future scope of a discipline.* Berkeley and London: University of California Press.

Syrides, G., 1996. The Holocene stratigraphy and palaeoenvironments in the area of ancient Avdera. In *Proceedings of the 2nd Symposium of the Hellenic Archaeometric Society, Archaeometrical and Archaeological Research in Macedonia and Thrace.* Thessaloniki: Aristotle University.

Triantafyllos, D., 1988. Excavation at Kalyva. In: *The archaeological work in Macedonia and Thrace.* Thessaloniki: Aristotelio Panepistēmio Thessalonikēs.

Tsatsopoulou, P., 1988-1989. *Messemvria-Zone Excavation Diaries.* Komotini: 19th Ephoreia of Prehistoric and Classical Antiquities.

Tsatsopoulou, P., 1997. The excavation at Mesemvria-Zone the last decade (1990-2000). In: *The archaeological work in Macedonia and Thrace.* Thessaloniki: Aristotelio Panepistēmio Thessalonikēs.

Tylecote, R.F. and Gilmour, B.J., 1986. *The metallography of early ferrous edge tools and weapons.* British Archaeology Reports British Series 155. Oxford: British Archaeological Reports.

Van Andel, T.H. and Runnel, C.N., 1987. *Beyond the Acropolis: a rural Greek past.* Stanford: Stanford University Press.

Vavelidis, M., et al. 1996. Skapti Ili: ancient gold mining in Mandra Kari, Palaia Kavala. In *Proceedings of the 2nd Symposium of teh Hellenic Archaeometric Society, Archaeometrical and Archaeological Research in Macedonia and Thrace.* Thessaloniki: Aristotle University.

Webster, J., 1996. Roman imperialism and the 'post-imperial' age. In: J. Webster and N. Cooper eds. *Roman Imperialism: Post-Colonial perspectives.* Leicester Archaeology Monographs 3. Leicester: School of Archaeological Studies, University of Leicester, pp.1-18.

Wertime, T.A. and Muhly, J.D. eds., 1980. *The Coming of the Age of Iron.* New Haven and London: Yale University Press.

Woolf, G., 1990. World System Analysis and the Roman Empire. *Journal of Roman Archaeology*, 12, pp.223-34.

Author's address

- Department of Archaeology, University of Manchester, Oxford Road, M13 9PL, UK
  Kostoglou.maria@gmail.com

The acquisition and circulation of iron during Iron Age II in the northern region of the Paris Basin (France)

Sylvain Bauvais and Philippe Fluzin

ABSTRACT: During Iron Age II (La Tène period – from the beginning of the 5th century to the Roman conquest) in the Northern Paris Basin, although forging activities were well developed across the entire region, the smelting of iron was only practiced as a significant production activity in specific areas. Eighty-eight sites linked to iron metallurgy have been inventoried, and the archaeometric analysis of twenty of these sites enables us to consider the evolution of regional relational relationships between iron production sites and the outlying areas. During the early La Tène period (475-330 BC), the raw material (smelted iron that had not yet been forged) was acquired through a network of limited restricted local exchange systems. As the local consumption of iron objects gradually increased, this situation evolved to incorporate trade over longer distances and a focus on economic centres. By the late La Tène period (150-25 BC) the metal was traded in the form of iron objects on a regional scale, via the sites that held the highest position in the economical hierarchy, the *oppida*. In addition, the increasing importance of individual specialists caused a more intense circulation of material inside the sites themselves. The situation in the Northern Paris Basin during this period suggests that an evolving social and political system with an increasing consumption of metal resulted in the acceleration of the diversification of production, and led to more specialisation within all levels of the production cycle.

Keywords: Iron and steel making; forge; semi-product; trade; economic organisation; archaeometric studies; Northern Paris Basin; Iron Age II.

Context and definition of the scales of analyses

A major technological change caused by the arrival of iron and steel making activities occurred in the North of the Paris Basin at the end of the Iron Age I. This region is characterised by a marked disparity in terms of mineral presence, with zones containing weak mineral deposits juxtaposed to others with propitious geologic formations, allowing for the development of durable iron and steel making activities in this region (Bauvais and Fluzin, 2007) (Fig. 1). In such a context, the organisation of production and the diffusion of raw materials must be organised in a singular way. This paper considers the question of the circulation of raw materials by attempting to characterise its principal manifestations and evolutions during the Iron Age II. This matter is approached through a comparison between the situation prevalent during the Early La Tène Period and during the Late La Tène Period. This study is based on a large corpus of sites, including 88 inventoried sites (Bauvais, 2007), of which 20 underwent archaeometric analyses following the protocols developed by the laboratory "Métallurgies et Cultures" of the French CNRS – UMR 5060 – IRAMAT (Fluzin, 2000, 2002; Bauvais and Fluzin, 2005, 2009; Dillmann, 2006; Dillmann and L'Héritier, 2007). This paper will treat this chronological comparison on three different levels (Fig. 2):

- On the regional level, through an analysis of the North of the Paris Basin as a whole;
- On the local level, through examining the Aisne and Oise valleys;
- On the micro level, through a study of the archaeological site of Condé-sur-Suippe.

By focusing on different levels of social and economic organisation, these three complimentary degrees of analysis provide a more comprehensive picture of the socio-technological changes which occurred in the region during this period.

The Regional Level: the North of the Paris Basin

The development of the organisation and production of iron depended to a large degree on the evolving demand for iron objects. This demand can be qualified by the degree of integration of metal into society (Bauvais, 2007; Bauvais and Fluzin, 2009).

During the Early La Tène Period the archaeological discoveries made during the excavations of necropolises and settlement sites show that the demand for iron was low. At this time iron was a prestige good destined for social elites (Bauvais, 2009). During the Late La Tène Period the need for iron increased. Iron became a more common material and was integrated into all social strata and all activities (including domestic, craft, and agricultural activities). In this context, the forms of organisation and production experienced an intensified evolution. In terms of smelting activities the Early La Tène Period is characterised by small sites composed of one to three slag-pit furnaces (Pleiner, 2000), although whether these furnaces were operated at the same time or whether they are indicative of successive smelting activities is as yet unclear. The production does not seem to have been constant and was undertaken according to needs (Dunikowski, et al., 2007). There appears to have been a simple separation between smelting sites and domestic settlements, as well as between smelting and forging activities. This situation is identical throughout the region, visible both in areas of strong mineral density and in areas with little such potential. Two good examples of the two types of geographic zones are the sites of Les Clérimois (Sarthe – Dunikowski and Cabboi, 1995) and of Travecy (Aisne Valley – Bauvais, 2007). Although situated in very different geological zones, these two sites both contain limited traces of furnaces and a very small quantity of smelting slags dated to the beginning of the La Tène Period. During the Late La Tène Period, disparities between the regions appear. The South-West zone (Pays d'Othe and Sénonais – Dunikowski, et al., 2007), and to a lesser extent the Pays de Bray, reach a much more intense level of production, with sites used over longer periods of time and with more constant smelting occurring in furnaces designed for multiple use. However, all other sectors of the zone maintain the scale of their mode of production or in some cases even show a decrease. In the Aisne Valley only refining activities (compaction of crude masses of metal) carried out in the *oppida* indicate that this type of primary activity (reducing) took place at a short distance if we assume that crude masses are less mobile (Bauvais, 2000, 2007).

During the Early La Tène Period, limited post-smelting activities are visible, and where there is evidence of forging it is usually found linked to smelting activities (Dunikowski, et al., 2007). In the rare instances of forging found in isolation, the remains are so limited that they are likely to indicate itinerant forging (Bauvais, 2007). As with smelting, this situation is the same throughout the region. During the Late La Tène Period the number of forging sites becomes more significant in all areas and, in addition to domestic production, dedicated production centres appear in the *oppida*. The forms of semi-product also seem to diversify, with the circulation of currency bars including a recently identified socket-bar type. These new semi-products represent a regional production (Berranger, et al., 2007).

As a partial conclusion, it can be remarked that that the changes in practices of consumption created differentiations within the network of production and led to the formation of regional specialisations. On the one hand, during the Early La Tène Period, production was local and could meet the needs of the local communities (Fig. 3). On the other hand, during the Late La Tène Period, some areas required the importation of semi-products in order to respond to the demand for manufactured products, while other areas became specialised and produced iron for trade (Fig. 3).

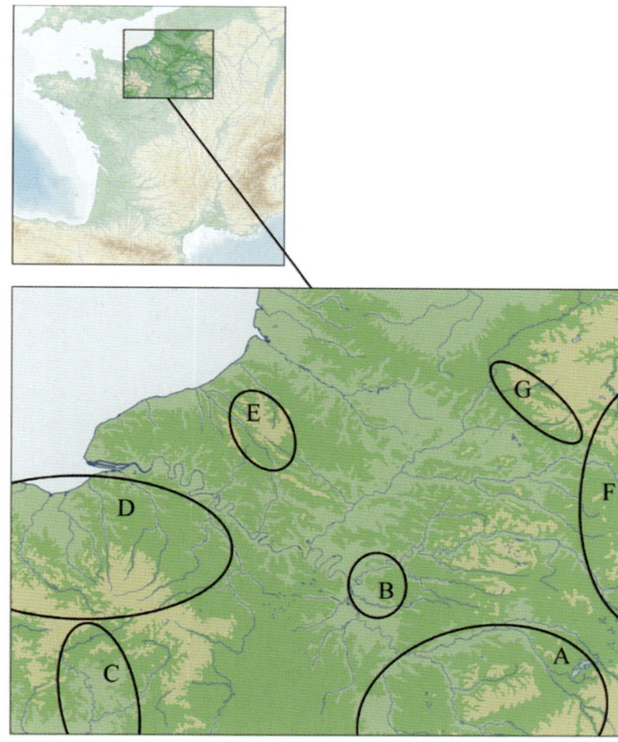

Figure 1: Localisation of the propitious geologic formations for the development of a durable iron and steel-making activity.
A: Sénonais, Pays d'Othe and Puisaye -
B: East of the Brie -
C: The Sarthe -
D: The Basse Normandie -
E: The Pays de Bray -
F: District "Entre Sambre et Meuse" -
G: The Ardennes
The rest of the area is characterized by weak mineral deposits.

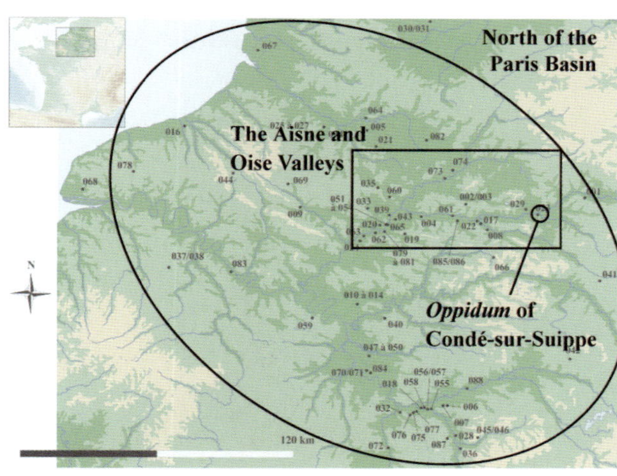

Figure 2: Localisation of the three levels of analysis.

These disparities inside the regional production network had as a principal consequence an increase of the intra-regional exchange of raw materials. The separation between smelting sites and forging sites becomes visible from the Middle La Tène period, and the exchange of materials at the local level also increases.

We can remark on the complexification of the system and a specialization of certain areas.

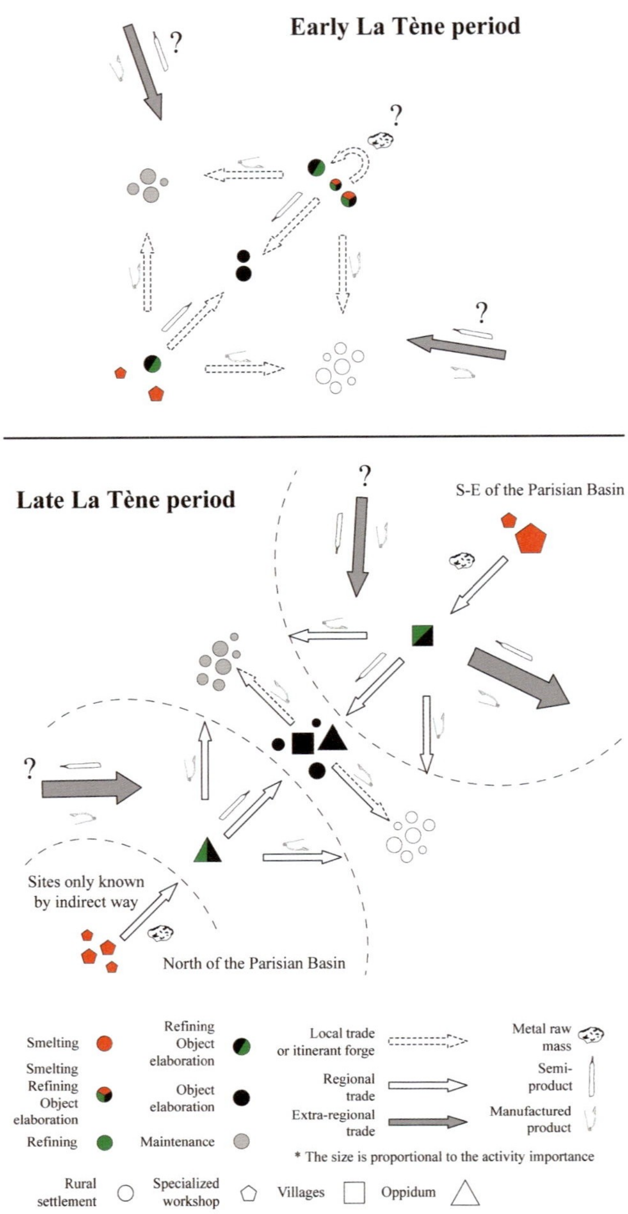

Figure 3: Theoretical diagrams of the organization of iron and steel-making activities during the early La Tène and the late La Tène periods in the north of the Paris Basin.

The local level: the Aisne and Oise Valleys

This area, situated around the two valleys mentioned above, is characterised by limited smelting activities but more extensive forging activities. During the Early La Tène Period the small number of sites and the low intensity of their activity give the impression of a simple organisation for the production and movement of raw materials over short distances. This activity seems to have been carried out by itinerant craftsmen and, in this case, the material must have circulated with the producers (Bauvais, 2007). During the Late La Tène Period the forms of commercialisation of the raw material in the region diversified. The analysis of the waste from these activities has allowed the identification of at least four degrees of integration between the sites, based on the network of acquisition and of the redistribution of the raw materials (Bauvais and Fluzin, 2007; Bauvais, et al., 2007) (Fig. 4). The first degree of integration involved the *oppida* of Villeneuve-Saint-Germain and of Condé-sur-Suippe. They imported crude masses of iron which were, after transformation, distributed as semi-products and manufactured products. These sites, occupying high positions in the economic hierarchy, seem to have monopolised the production of raw products coming from sites belonging to the sectors under their domination (Bauvais, 2007). The second degree of integration was represented by open agglomerations and rural establishments, which acquired coarsely refined products that were transformed into finished products or sometimes into semi-products of better quality (as is the case with socket-bars). The third degree of integration involved sites that acquired good quality semi-products for specialised production (for example socket-bars), and finally the fourth degree of integration included sites that consumed iron objects which were maintained and recycled.

The fact that sites did not acquire the same type of raw materials suggests that the methods of working and the output of the productions were different at different locations. Consequently, this diversification of the raw material is indicative of more specialised production.

The chemical analysis of slag inclusions contained in 19 socket-bars from eight different sites of this sector permitted

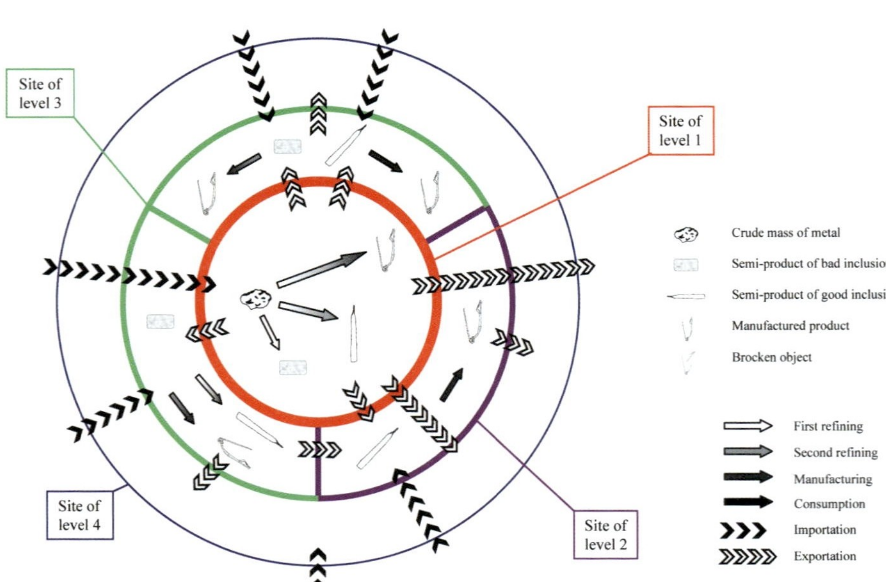

Figure 4: Levels of integration inside the networks of acquisition / redistribution of the metal.
Schematic representation of the possible relations between each of the levels within a micro-region.

a more precise understanding of their circulation. The socket-bars, which were made and used locally, are composed of metals from a variety of origins (Fig. 5), which suggests the major importation of crude iron over long distances (Berranger, et al., 2007). In addition, the site of Villeneuve-Saint-Germain "Les Etomelles" and Braine "La Grange des Moines", which are strictly contemporary, present bars of clearly common origin, thus demonstrating the existence of a true network of exchange and diffusion across the Aisne Valley.

As a partial conclusion, local trade of raw materials during the Early La Tène period in this micro-region seems to have been related to producers. More significant evidence for metallurgical activities can be seen inside the settlement network when the consumption of iron began to increase, during the Middle La Tène (330-150 BC). As diversification of production appears at the sites, these specialisations tend to render the local trade more complex, involving the circulation of several types of semi-product. During the final La Tène, the *oppida* seem to monopolise the local primary production and imports of raw material. This monopoly could have been the consequence of the limited extent of local smelting of crude iron, which is a characteristic of this region. Hence when a material becomes fundamental for the economic and social functioning of a society and the demand relies on trade over average or long distances, it seems (based on the evidence from this region at least) that those in control impose systems of tighter and more drastic control.

The Micro Level: the *Oppidum* of Condé-sur-Suippe

Little data that would enable us to consider the spatial organisation of iron production during the Early La Tène Period exists. However, during the Late La Tène Period, some structured centres of production and of interdependent activities appeared inside the agglomerations. At Condé-sur-Suippe, the spatial analysis of the dispersion of the artefacts and the metallographic analyses carried out on the slag-cakes allow us to define the function of three workshops in the iron and steel making sector (Bauvais, 2000). The relationship between different stages in the *chaîne opératoire* thus allows the reconstruction of the interaction between these workshops. The raw material would arrive at the site of the I2 workshop (Fig. 6), in the shape of either crude masses of iron or fragments of crude masses. This trend is not rare at the *oppida* (see also Entremont and Bibracte – Berranger, 2008). After refining, the raw material would be distributed to workshops devoted to object manufacturing, as shown in Fig. 6. The semi-products would also have as destinations other forging workshops located somewhere else in the *oppidum* or more distant settlements. After a complete or partial working in the two workshops, I1 and O1, the iron elements requiring treatment or particular assembling would have been distributed to the specialised workshops. This separation between manufacturing workshops and specialised workshops is not clear. For example, the presence of copper inclusions in the slag-cakes resulting from iron forging could indicate that copper and iron working was taking place in the same workshop, and even in the same hearth. Only very specialised activities would require particular workshops. For instance, the shaping of wood pieces such as the assembling of buckets, barrels, or of cart wheels seems to have been specialised enough to require particular workshops. Furthermore, practices of maintenance and possible recycling activities are interspersed in this system and contribute to its complexity. The often sparse and partial character of the remains related to iron and steel making activities prevents us

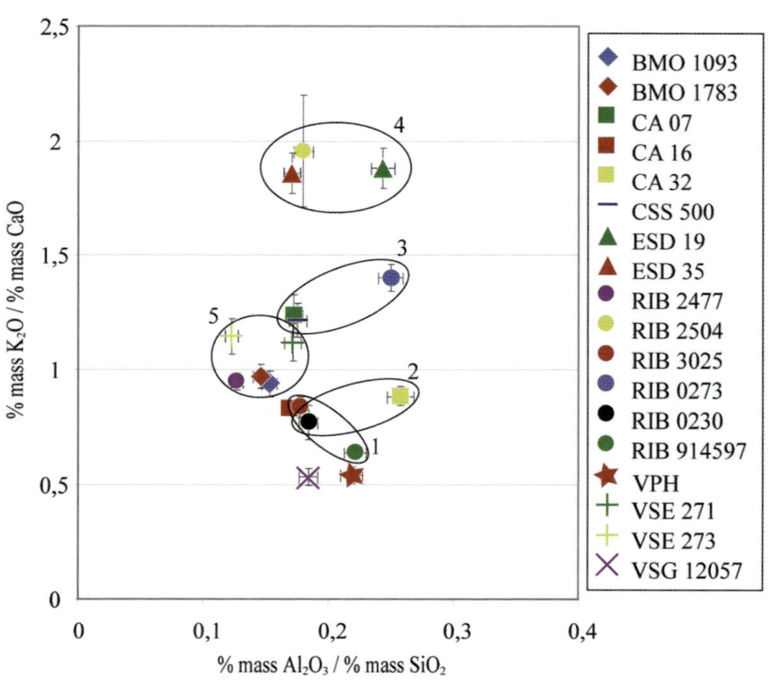

Sample	Group	Composition				
		Al/Si	K/Ca	Mg	Mn	P
RIB 91-45-97	1	0,22	0,64	0,99	0,66	0,62
RIB 3025	1	0,18	0,84	1,08	0,15	1,01
CA 32	2	0,26	0,84	1,74	1,87	4,10
RIB 0230	2	0,18	0,77	1,42	0,94	4,80
RIB 0273	3	0,25	1,42	0,65	1,47	3,54
CA 07	3	0,17	1,23	0,22	1,00	2,19
CSS 500	3	0,18	1,22	0,30	1,41	3,00
RIB 2504	4	0,18	1,95	0,34	0,26	3,58
ESD 19	4	0,24	1,88	0,56	0,61	2,61
ESD 35	4	0,17	1,86	0,24	0,09	0,62
BMO 1093	5	0,15	0,94	0,40	0,15	3,14
BMO 1783	5	0,15	0,97	0,36	0,29	7,41
VSE 271	5	0,17	1,09	0,07	0,20	2,21
VSE 273	5	0,12	1,14	0,30	0,13	4,71
RIB 2477	5	0,13	0,95	0,19	0,17	3,21
VSG 12057	6	0,18	0,54	0,20	0,27	4,33
VPH	7	0,22	0,54	0,69	0,96	0,95
CA 16	8	0,18	0,82	0,55	0,32	0,96

Figure 5: Analysis of the major elements composing the inclusions of 19 "socket bars" from the North of the Paris Basin.
On the left – % mass Al_2O_3/% mass SiO_2 versus % mass K_2O/% mass CaO ;
On the right – ranking by group from average % mass of Al_2O_3, SiO_2, K_2O, CaO, MgO, MnO et P_2O_5.
In grey – bars of the same group which come from the same site.

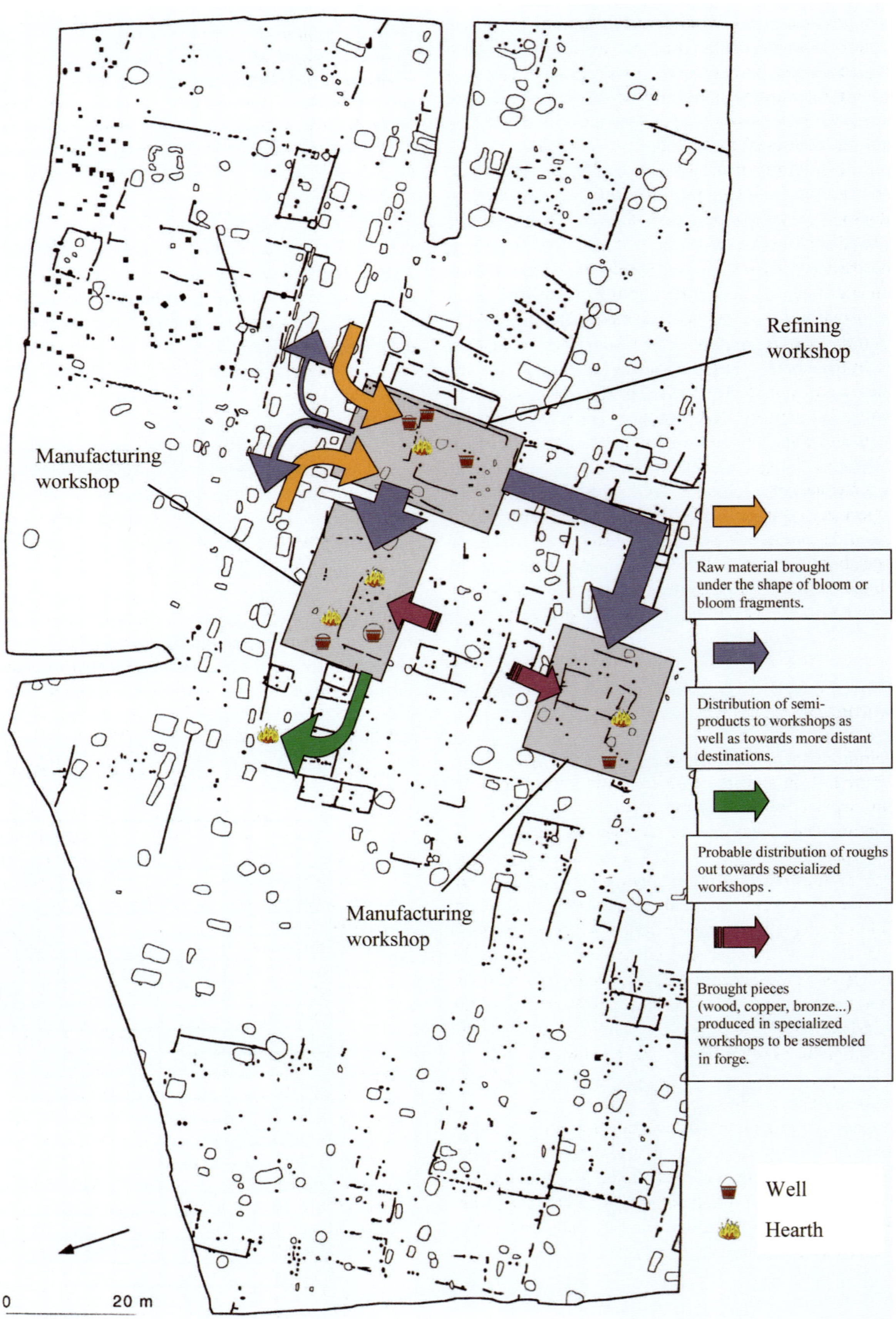

Figure 6: Spatial analysis of the oppidum of Condé-sur-Suippe.
Interactions between workshops

from evaluating the quantitative importance of a production and of a particular method. For the same reasons, it is also impossible for us to define with good precision the exact types of products which could have been made inside these workshops. Likewise, the final destination of the production, i.e. its place of use, can only be surmised. However, and with some reserve, according to the metallic wastes, it would seem that these workshops were directed towards the production of domestic objects (sheets, nails, tools for craft industry or agriculture, furniture of all kinds, etc.), while the theory that copper and iron were being worked in the same workshop could also suggest the production of luxury items intended for a certain elite, as well as of chariot tyres.

As a partial conclusion, the analysis of the *chaîne opératoire* of iron and steel making activities in the craft district of the *oppidum* of Condé-sur-Suippe, excavated in 1987, enables us to determine some aspects related to the type of production, as well as to the organisation of these workshops.

Their productions focused on the final stages of the *chaîne opératoire*, ranging from practices of refining (semi-products, crude masses of iron) to manufacturing (tools, utensils, etc.), or to the assembling of the metal (nails, sheet metals) with other materials (wood, bronze, etc.) within the framework of a specialised craft industry. These activities had a very particular distribution, in specialised sectors to which space was specifically allocated. These specialisations were rather broad, and were directed not towards a particular type of craft industry, but towards a particular type of production.

Conclusion

A multidisciplinary and global approach to the *chaîne opératoire* in iron and steel making permits us to consider certain aspects of the socio-economic organisation of production. The archaeometric analyses carried out allow us to surpass the absence of discoveries of semi-product through characterisations of the raw material derived from activity wastes. These analyses also allow for a more precise definition of the phases of the *chaîne opératoire* and for a reconstruction of the interactions and relationships at the regional level, as well as within the archaeological site. Thus, iron and steel making activities in the northern area of the Paris Basin experienced an increased evolution as the rate of consumption increased. When the demand was too high for small local producers, certain sectors began to specialise in particular products and methods.

Forging activities grew and diversified in all geographic sectors as a result of the increased diffusion of the raw material. A more and more complex social and political system linked to an increscent consumption of metal had the consequence of accelerating the diversification of production, leading to a more specialised production at all levels of analysis.

References

Bauvais, S., 2000. *Analyse de la chaîne opératoire de sidérurgie sur l'oppidum de Condé sur Suippe (Aisne).* MA Dissertation, Université de Paris I, Panthéon Sorbonne.

Bauvais, S., 2007. *Evolution de l'organisation des activités de post-réduction dans le Nord du Bassin Parisien au 2nd Age du fer: études multidisciplinaires de la chaîne opératoire en métallurgie du fer.* PhD Thesis, Université de Technologie de Belfort Montbéliard / Université de Besançon.

Bauvais, S., 2008. From prestige to proto-industry: Iron production evolution during the Iron Age II in the north of the Paris Basin. *The Arkeotek Journal*, Volume 2, N°4, online publication (http://www.irit.fr/thearkeotekjournal).

Bauvais, S. and Fluzin, P., 2005. Analyses métallographiques des déchets de forge du site de Bazoches-sur-Vesle "les Chantraines": confrontation et perspective régionale (Aisne). In: G. Auxiette, F. Malrain, S. Desenne and B. Robert eds. *Hommage à Claudine Pommepuy*. Amiens: Revue Archéologique de Picardie, pp.115-30.

Bauvais, S. and Fluzin, P., 2007. Réflexion sur l'organisation technico-sociale des activités de forge à La Tène finale dans l'Aisne (02). In: *Actes du 16ème colloque du GMPCA, INSTN-Saclay*. Rennes: Presses Universitaires de Rennes, pp.25-43.

Bauvais, S. and Fluzin, P., 2009. Archaeological and Archeometrical Approaches of the Chaîne Opératoire in Iron and Steelmaking: Methodology for a Regional Evolution Study. In: V. Roux and S. Rosen eds. *Techniques and People: anthropological perspectives on technology in the archaeology of the proto-historic and early historic periods in the Southern Levant. Workshop of the CRFJ, Jerusalem*. Paris: De Bocard, pp.157-78.

Bauvais, S., Gaudefroy, S., Gransar, F., Malrain, F. and Fluzin, P., 2007. Premières réflexions sur l'organisation des activités de forge en contexte rural à La Tène finale en Picardie. In: P.-Y. Milcent ed. *L'économie du fer protohistorique: de la production à la consommation du métal. Actes du XXVIIIe colloque international de Toulouse*. Bordeaux: Aquitania, pp.239-61.

Berranger, M., 2007. Organisation de la chaîne opératoire du fer sur l'oppidum d'Entremont (Aix-en-Provence, IIe-Ier siècle av. J.-C.): la circulation du métal. *ArcheoSciences, Revue d'Archéométrie*, 31, pp.7-22

Berranger, M., Bauvais, S. and Fluzin, P., 2007. Socket bars: multidisciplinary results (archaeology and archaeometry) on a specific iron semi-product in north of France. In: *Archaeometallurgy in Europe 2007. Actes de la 2ème conférence internationale, Grado et Aquileia (Italie)*. Milano: Associazione Italiana di Metallurgia (cd-rom).

Dillmann, P., 2006. *Etude des alliages ferreux anciens: élaboration, utilisation, altération. Apport des techniques microfaisceau.* Habilitation à Diriger des Recherches, Université de Technologie de Compiègne.

Dillmann, P. and L'Héritier, M., 2007. Slag inclusion analyses for studying ferrous alloys employed in French medieval buildings: supply of materials and diffusion of smelting processes. *Journal of Archaeological Science*, 34, pp.1810-23.

Dunikowski, C. and Cabboi, S., 1995. *La sidérurgie chez les Sénons: les ateliers celtiques et gallo-romains des Clérimois (Yonne)*. Paris: Maison des Sciences de l'Homme.

Dunikowski, C., Séguier, J.-M. and Cabboi, S., 2007. La production du fer protohistorique au sud-est du Bassin parisien. In: P.-Y. Milcent ed. *L'économie du fer protohistorique: de la production à la consommation du métal. Actes du XXVIIIe colloque international de l'AFEAF, Toulouse*. Bordeaux: Aquitania, pp.279-89.

Fluzin, P., 2000. Les ateliers métallurgiques: archéométrie. Examens métallographiques des déchets d'ateliers. In M. Mangin, P. Fluzin, J.-L. Courtadon, and M.-J. Fontaine eds. *Forgerons et paysans des campagnes d'Alésia (Haut-Auxois, Côte-d'Or). La terre, le Fer, la Route en pays Mandubien, Haut Auxois (Côte-d'Or), Ier siècle avant-VIIIe siècle après J.-C.* Paris: CNRS, pp.185-265.

Fluzin, P., 2002. La chaîne opératoire en sidérurgie: matériaux archéologiques et procédés. Apport des études métallographiques. In H. Bocoum ed. *Aux origines de la métallurgie du fer en Afrique. Une ancienneté méconnue.* Paris: UNESCO, pp.59-91.

Pleiner, R., 2000. *Iron in archaeology: the European bloomery smelters.* Praha: Archeologický ústav AVČR.

Authors' addresses

- Sylvain Bauvais
  CNRS researcher, LMC – IRAMAT – UMR 5060 – CNRS and LAPA – SIS2M – UMR 3299 – CNRS/CEA, CEA Saclay, 91191 Gif-sur-Yvette Cedex.
  Email: sylvain.bauvais@utbm.fr
- Philippe Fluzin
  Research director, LMC – IRAMAT – UMR 5060 – CNRS, Université de Technologie de Belfort-Montbéliard, 90010 Belfort Cedex.
  Email: philippe.fluzin@utbm.fr

East Asia

The East Asia Session

Han Rubin

Based on new archaeological findings, East Asia's ancient iron and steel technologies are now relatively better understood. However, the questions of where, when, and how the invention of iron smelting techniques occurred and how to explain the subsequent wide-spread adoption of iron production (two epoch-making processes in the history of humankind), remain unsolved. The ancient Chinese began to produce bloomery iron in the 9th century BC – or perhaps several hundred years earlier – and from the 8th century BC on they made cast iron. These processes laid the technological foundation for the up-coming Iron Age of East Asia.

China is the earliest country in the world to have invented and utilized cast iron. This technology system is different from the ferrous technologies of ancient West Asia, Europe and Africa. Yet there are many unanswered questions that warrant further research, such as those concerning possible mechanisms that led to changes from bloomery to cast iron. Why did people make the decision to change technologies? What factors enabled them or encouraged them to make this shift?

Papers presented in this session of the World of Iron Conference proceeding provide some answers to such questions using the results of field investigations and lab analysis. Cooperation between archaeometallurgists and archaeologists is the foundation for understanding the history of metallurgy. In recent years, through their excavations, field researchers have developed a better understanding of ancient mining, smelting, casting, smithing and transpiration, as well as the organization of iron production. Murakami Yasuyuki, Huang Quansheng and Li Yanxiang, report about iron production in Han and pre-Han periods in southern China, Sichuan's Chengdu Plain and Guangxi's Pingnan County. Some of the most important information they provide concerns the structures of iron smelting furnaces. Blast furnaces were used to produce cast iron, and the furnace's structure and size were borrowed from earlier copper smelting furnaces, according to evidence from the Yangcheng site in Henan province. The earliest blast furnace, dated to the 4th-1st century BC, was found at the Jiudian site, Xiping county, Henan province. It had an elliptical hearth of 65x106cm, a furnace wall - with adhered slag of 37cm thick - supported by rammed earth, and a residual height of 2.1m. Unearthed furnaces at the Wangchenggang site in Lushan County and the Guxing site in Zhengzhou from the 2nd-1st century AD were bigger than those at Jiudian. The elliptical interior diameter of these furnaces measures about 4.0x2.8m. However, they had been reconstructed with a smaller, oval-shaped hearth (2.0x1.18m) to avoid freezing. Changes in the structure and size of the furnaces, as Murakami Yasuyuki mentions, also indicate that the small blast furnace would have been easier to operate during that time. As Huang Quansheng and Li Yanxiang say, more field work should be done to find evidence of technological changes related to the transition from bloomery iron to cast iron production. The paper by Chen Jianli and Han Rubin is a survey of the development of iron and steel technology of ancient China, based on new archaeological findings and laboratory research. It should be mentioned that the results presented in this paper are not final, and work is continuing on this important topic.

Composition and microstructure of slag can provide much information unobtainable through the study of iron objects, such as details concerning ore, flux, fuel, smelting temperature, etc. Western scholars have published many results on the topic of bloomery slag, and I and my colleagues have also analyzed iron slag from several cast iron works dating to the Han Dynasty in Henan province. Very low iron content and high calcium levels in our samples indicate that the slag studied was produced in a blast furnace and that limestone was used as a flux. This composition is similar to slag produced in Europe during the 18th century. However, a variety of slag compositions were also found, including samples with higher iron contents. More research is needed to characterize these slags and to provide a method for distinguishing between slags produced by various smelting, smithing and refining technologies such as the cast iron, bloomery iron and steel puddling processes. According to the historical literature and to studies of iron objects, coal and coke were used as fuel to produce iron during the Song Dynasty (10th century AD) and the Ming Dynasty (16th century AD) respectively. It is worth asking how to investigate this via the composition and microstructure of iron objects and slag. More papers on the study of slag and shaft furnaces from East Asia are needed in the future.

The issue of interregional technological diffusion is emphasized in the papers of this session. For example, iron and steel technology in Korea is presented by Jang-Sik Park, while relationships and technological approaches across the wider region are discussed by Anna Bennett and by Brian Gilmour *et al.*

The origin of iron smelting technology and the regular use of iron objects may not be isolated incidents. In ancient China, the iron technological system is not the same as that of western Eurasia's bloomery iron technology. While Chinese ferrous metallurgy has its own characteristics, it cannot be denied that the earliest iron objects and smelting technology came from outside of China. By encouraging further investigations and international collaborations to investigate the archaeometallurgy of iron production, it is hoped that the global context of ferrous technology in East Asia, and issues such as the spread of cast iron technology from central China to the Korean Peninsula, the Japanese Islands, Southeast Asia and Central Asia, can finally be understood.

Preliminary studies on Western Han Dynasty iron smelting sites and slag found in Pingnan County, Guangxi Province, China

Huang Quansheng and Li Yanxiang

ABSTRACT: This paper presents a recent study on iron smelting sites and slag dating to the Western Han dynasty (206 BC-24 AD) found in Pingnan County, Guangxi Zhuang Autonomous Region, Southern China. Archaeological fieldwork was conducted in Pingnan County in spring 2005 and winter 2006. The ancient iron smelting sites are distributed in an area of about 100km^2 in the vicinity of the town of Liuchen. Eight iron smelting sites and one iron mining site, all of which were located on hilltops, were investigated. A bowl-shaped furnace has been found, as well as many slag heaps, eroded iron blocks, broken *tuyères*, and Western Han dynasty pottery shards. Preliminary analyses using SEM-EDS of 89 slag samples collected from the sites are presented. The results show that the slag is mainly composed of crystal fayalite as the dominant phase. Analysis of the microstructure of the metallic prills in the slag by metallographic microscopy shows that they were the products of bloomery and cast iron smelting techniques. This is the first time that ancient bloomery iron smelting ruins have been confirmed in China. The typology of the remains of the straight-round-shaped *tuyères* indicates a shared tradition with earlier copper and iron smelting finds in the Lingnan area, closely resembling those from South and Southeast Asia.

Keywords: Lingnan area; Han dynasty; iron smelting; slag; archaeometallurgy.

Introduction

The Guangxi Zhuang Autonomous Region, located in Southern China, an important region connecting Central China with Southeast Asia, was a frontier for cultural contact and exchange from ancient times. Archaeological excavations and preliminary studies indicate that iron artefacts started to be used in the Guangxi region at the latest during the Warring States period (475 BC-221 BC) (Guangxi Cultural Relics Team, 1978; Lan Riyong, 1989). However, these iron artefacts have so far not been examined by scientific methods, and most scholars in China believe that they were imported from Central China (IHMM, 1991; Li Jinghua, 1991). Due in part to the lack of ancient records, the date of the appearance of iron smelting technology in the Guangxi region is still not clear. Research on iron smelting sites, their waste and other contemporary iron artefacts from the area may help to elucidate this problem.

The Western Han dynasty (206 BC-24 AD) iron mining and smelting sites at Liuchen in Pingnan County, Guangxi, discovered during the National Survey of Cultural Relics in the 1980s (Pingnanxian Bowuguan, 1988), cover about 100km^2 and are the earliest iron smelting remains found in the Lingnan area (present-day Guangxi and Guangdong provinces) to date (Zheng Chaoxiong, 1990, 1996). Studies on the remains offer the possibility of discerning the characteristics of early iron smelting technology in the region, as well as its spread and exchange, both locally and to neighbouring areas, such as Central China, Southeast Asia, and the Indian Subcontinent.

This paper presents one component of our research achievements in the Lingnan area since 2004. In addition, materials – such as a *tuyère* and a piece of slag – collected at the Pozui site during the National Survey of Cultural Relics in the 1980s are also presented.

Background to the sites

Pingnan County is located in the south-eastern part of Guangxi Province. At the time of the Qin dynasty invasion of the Lingnan area from the north, in 214 BC, the aboriginal inhabitants of the region were known in Chinese literary sources as the Xiou (西呕) and Luoyue (骆越) (Huang Xianfan, et al., 1988). The Qin Emperor established three administrative regions or Commandries (*jun* 郡) in the Lingnan area, namely the Guilin Commandry, Xiangjun Commandry and Nanhai Commandry; the area of present-day Pingnan County was controlled by the Guilin Commandry at that time (Zhongguo Lishidituji Bianjizu, 1975). During the early Western Han dynasty the region was controlled by the Nanyue Kingdom (203-111 BC) (Zheng Chaoxiong, 1996).

Aims

This research had three major goals: to conduct archaeological fieldwork at the ancient metallurgical sites

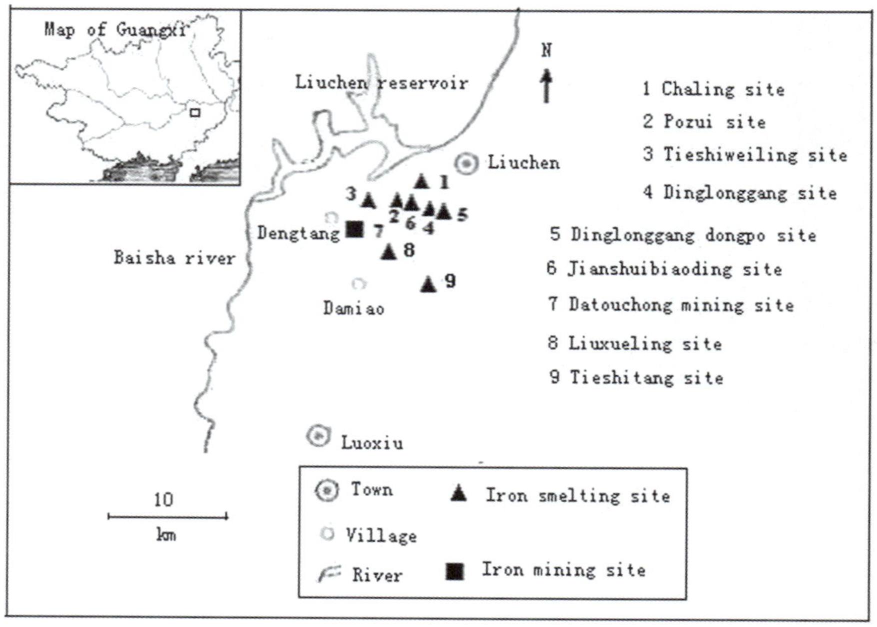

Figure 1: Distribution of iron smelting sites in Pingnan County.

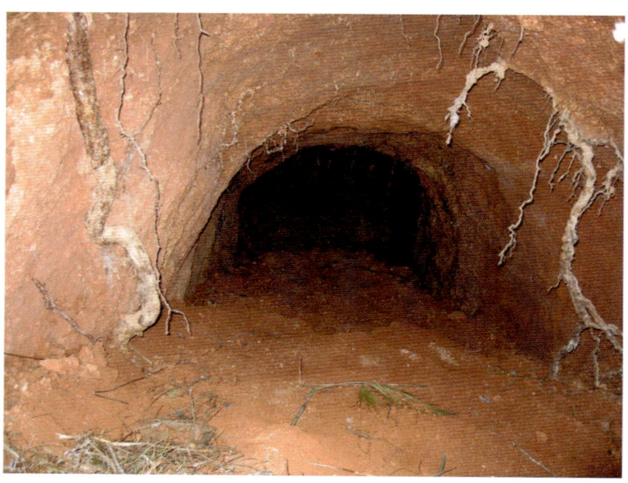

Figure 2: Shaft at the Datouchong mining site.

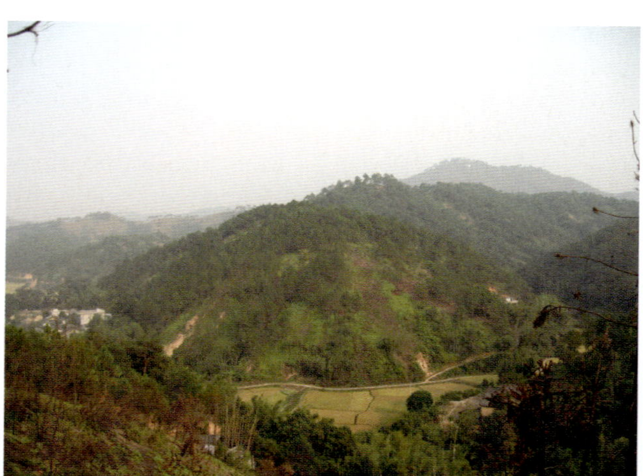

Figure 3: View of the Pozui, Jianshuibiaoding and Dinglonggang smelting sites taken from the southwest at the Tieshiweiling smelting site in early December 2006.

of the region; to reconstruct the history of the indigenous iron smelting technology following scientific analysis of the archaeometallurgical remains; and to consider the wider geographical connections from the point of view of technological and cultural exchanges.

Fieldwork

With the assistance of Professor Zheng Chaoxiong of the Guangxi Provincial Museum, Feng Guichun of the Guigang City Museum, and Gong Hai and Mo Jianhua of the Pingnan County Museum, archaeological fieldwork was conducted in Pingnan County in spring 2005 and autumn 2006. The geographical coordinates of the investigations are between 23° 04'N and 23° 11'N, 110° 17'E and 110° 21'E (GPS measured data).

In the course of the fieldwork we identified nine sites, including eight iron smelting sites and one iron mining site (Fig. 1). The smelting sites included those at Liuxueling, Pozui, Tieshiweiling, Dinglonggang, Chaling and Dinglonggangdongpo at Dengtang village, Tieshitang at Damiao village, and Jianshuibiaoding at Dengtang village. The iron mining site was Tatouchong at Dengtang village, and had six shafts (five of them collapsed) (Fig. 2). All of the ancient remains were situated on hilltops of more than 100m in altitude (Fig. 3), located in the middle reaches of the Zhujiang River (Pearl River). Slag heaps, iron ore, broken fragments of *tuyères* and large quantities of pottery shards, identified by typology and pattern as dating from the Western Han dynasty (Fig. 4), were collected from the surface and from trenches. One almost intact *tuyère* (Fig. 5), which was of a round-straight shape, about 50cm in length, with an inner diameter of about 5-7cm, an outer diameter of about 12-15cm and made of white mud and sand, was collected at the Pozui site during the National Survey of Cultural Relics

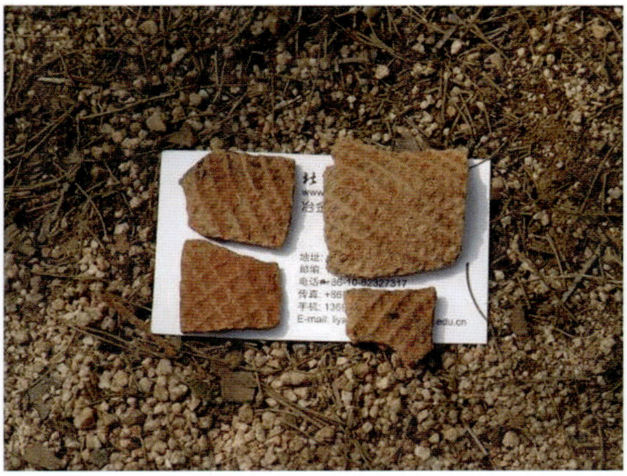

Figure 4: Western Han dynasty pottery shards unearthed from the bottom layer beside the bowl-shaped furnace at the Pozui smelting site.

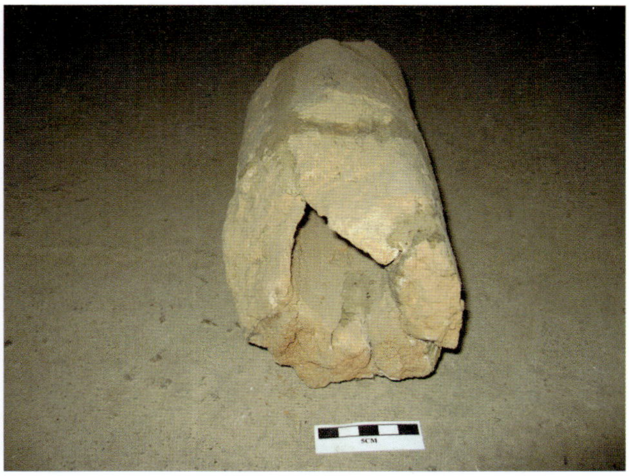

Figure 5: Straight-round shaped *tuyère* from the Pozui smelting site.

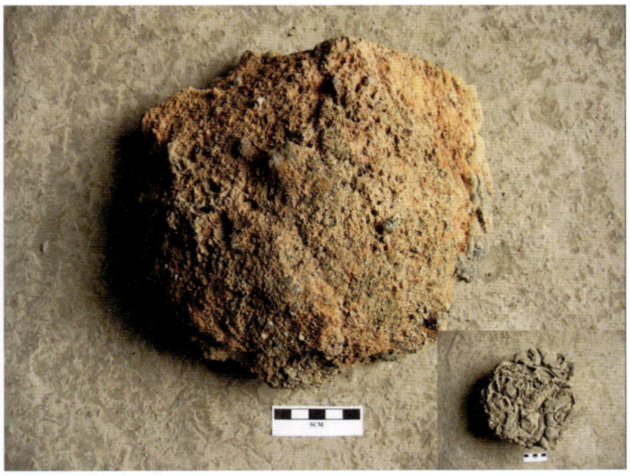

Figure 6: Bowl-shaped (plano-convex) slag at the Pozui smelting site, inset photograph showing the slag surface.

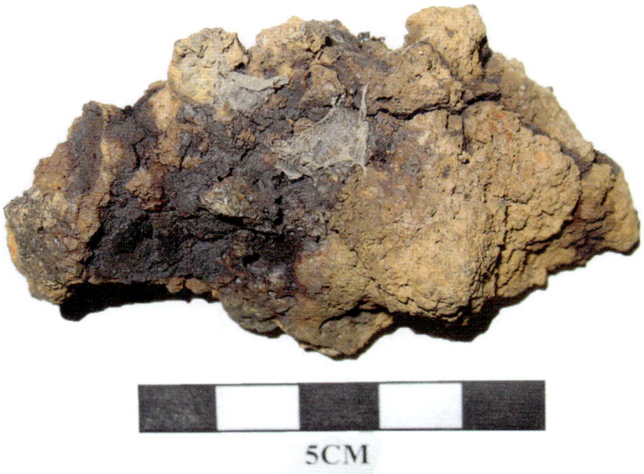

Figure 7: Bar-shaped slag from the Pozui smelting site.

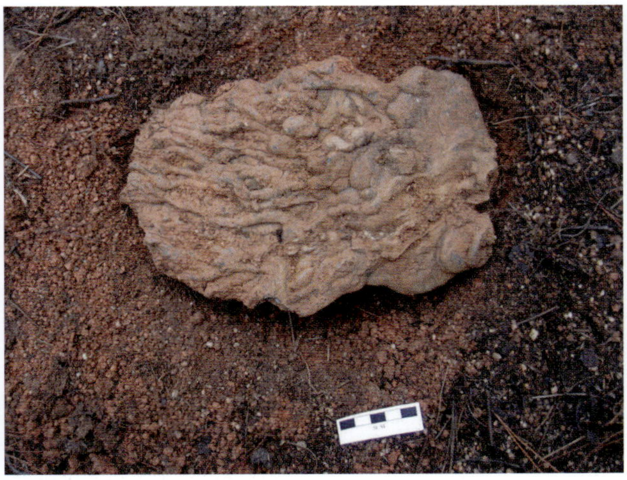

Figure 8: Tapped slag at the Pozui smelting site.

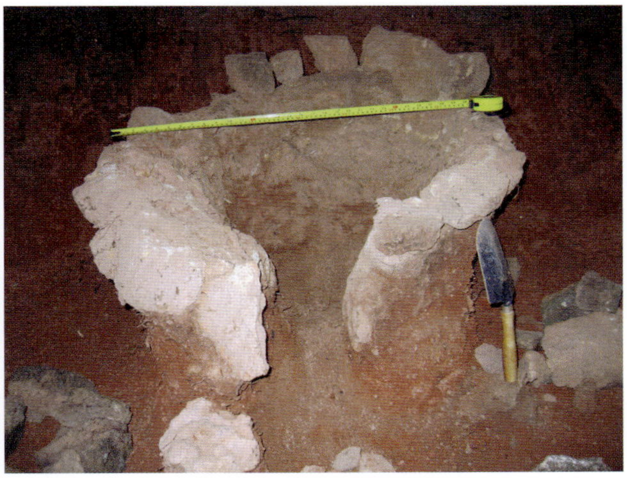

Figure 9: Bowl-shaped furnace at the Pozui smelting site viewed from the southwest, showing a bulk fragment of tapped slag (about 10×14×8cm) with a fragment of iron ore (about 1×3cm) embedded into its surface in front of the furnace, on the left of the opening.

in the 1980s (Pingnanxian Bowuguan, 1988). It is now stored in the Pingnan County Museum.

The slag that was collected can be divided into bowl-shaped (plano-convex) slag (Fig. 6), bar-shaped slag (Fig. 7), and tapped slag (Fig. 8). The bowl-shaped slag was collected at the Pozui site during the National Survey of Cultural Relics in the 1980s (Pingnanxian Bowuguan, 1988). The upper surface diameter of the bowl-shaped slag is about 25cm, and its maximum diameter of the plan section about 20cm.

A unique 'bowl-shaped' furnace (Figs. 9 and 10) never before found in China was also unearthed at the Pozui smelting site. Its remains are 55cm high, the upper rims flaring out with

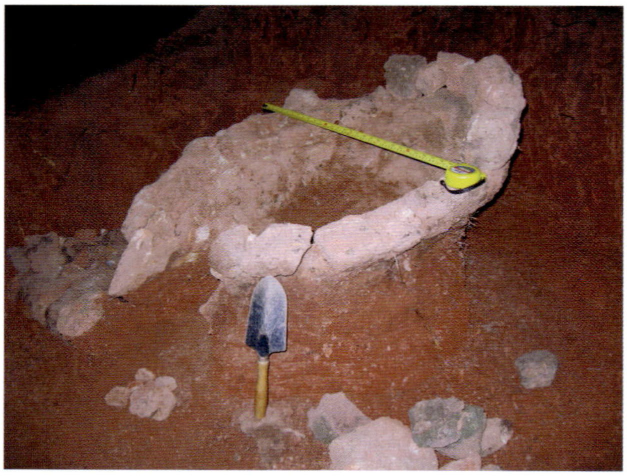

Figure 10: Bowl-shaped furnace at the Pozui smelting site photographed from the southeast.

an outer diameter of about 80cm and an inner diameter of about 60cm. There is a hole for a *tuyère* on the left upper rim. The inner diameter of the middle of the furnace is about 40cm and the inner diameter of the bottom of the furnace about 20cm. No slag was found clinging to the furnace wall. The hearth of the furnace was full of red mud, with one fragment of tapped slag collected within the fill. An opening of about 20cm in width situated 20cm above the bottom of the furnace in its south-western face was also observed. There was also a bulk fragment of tapped slag (about 10×14×8cm) with a fragment of iron ore (about 1×3cm) embedded into its surface in front of the furnace on the west side of the opening. The thickness of the furnace wall was about 10cm. The walls of the furnace were apparently of tamped earth, finished with a lip of mud at the top. It is thought that the furnace is almost complete. In addition, two pieces of iron (about 1×2cm) were found in the lowest layer, about 60cm in depth from the ground, near the opening outside the furnace.

The remains of several furnaces were identified within archaeological deposits at a number of other sites during the investigations, and these will be excavated in the future.

Scientific examination

In order to determine the operating parameters of the smelting technologies employed, samples of slag, ore, iron pieces and furnace pit lining from the nine sites were examined and analysed using scanning electron microscopy with energy-dispersive spectrometry (SEM-EDS).

Microstructural and chemical analyses

The compositional analysis of the slag, iron inclusions, and ore is significant for our understanding of the smelting techniques employed and the sources of the ore used. The microstructure of the iron prills in the slag and iron products is also very important for determining the smelting techniques (Tholander, 1989).

Scanning electron microscopy (SEM) was used to observe the microstructures and compositions of polished sections of the samples. SEM with energy-dispersive spectrometry (EDS) was used to carry out non-invasive quantitative analysis. The analyses were undertaken using a Cambridge S-360 SEM-EDS at the School of Materials, University of Science and Technology Beijing. The excitation voltage was 20 Kv. Since light elements such as carbon and oxygen whose atomic numbers are less than 11 could not be detected, only a qualitative analysis of the corroded objects or only trace elements could be given. Oxidised components could not be determined. To determine the average components, surface scanning was used, with multifaceted scanning on different parts of the samples to discover the precise composition of each sample. Based on previous analyses, the lower confidence limits for this instrument may be established at 0.3 wt%; values below this limit can be considered as indicative only. The analysis results for 89 samples (including 1 bowl-shaped slag, 1 bar-shaped slag, and 87 tapped slags) are shown in Fig. 11. The 87 tapped slag samples were randomly selected.

Metallographic microscopy was used to observe the microstructures of the iron prills in polished sections of samples of the slag and iron products, and this research was undertaken using a German LEICA DM4000M at the Institute of Historical Metallurgy and Materials, University of Science and Technology Beijing, and at the School of Archaeology and Museology, Peking University.

Mineralographic microscopy was used to observe the microstructures of polished sections of samples of the slag and iron ore, using a Japanese OLYMPUS BH2-UMA at the School of Geology, University of Science and Technology Beijing and a German LEICA DM4000M at the Institute of Historical Metallurgy and Materials, University of Science and Technology Beijing.

The SEM-EDS results show that the microstructures and compositions of polished sections of the slag samples can be divided into two types. One type includes one bowl-shaped slag, one bar-shaped slag, and 65 tapped slag samples from eight smelting sites. The phases present include fayalite as the dominant constituent, dendritic wüstite, and a relatively small proportion of glassy matrix (Figs. 12 and 13). We found areas of metallic iron in the bowl-shaped slag and bar-shaped slag samples (Fig. 12), and iron prills in ten of the tapped slag samples. The composition of these slag samples is very similar. Their average composition of FeO (adding oxide) was 77.88 wt%, with SiO_2 about 17.64 wt%, Al_2O_3 about 2.59 wt%, K_2O about 0.49 wt%, CaO about 0.70 wt%, S about 0.22 wt%, MgO about 0.18 wt%, MnO about 0.08 wt%, and P_2O_5 about 0.03 wt%. The other type includes 22 tapped slag samples from four smelting ruins, namely Pozui, Dinglonggang, Dinglonggangdongpo, and Tieshitang. The phases present within this group of slags include fayalite, relatively little glassy matrix, but no wüstite (Fig. 14), and iron prills could be seen in 13 of these tapped slag samples (Fig. 14). Once again, the chemical composition across this sample range was very similar. For the tapped slag samples the average composition of FeO (adding oxide) was 62.60 wt%, with SiO_2 about 30.20 wt%, Al_2O_3 about 3.39 wt%, K_2O about 1.63 wt%,

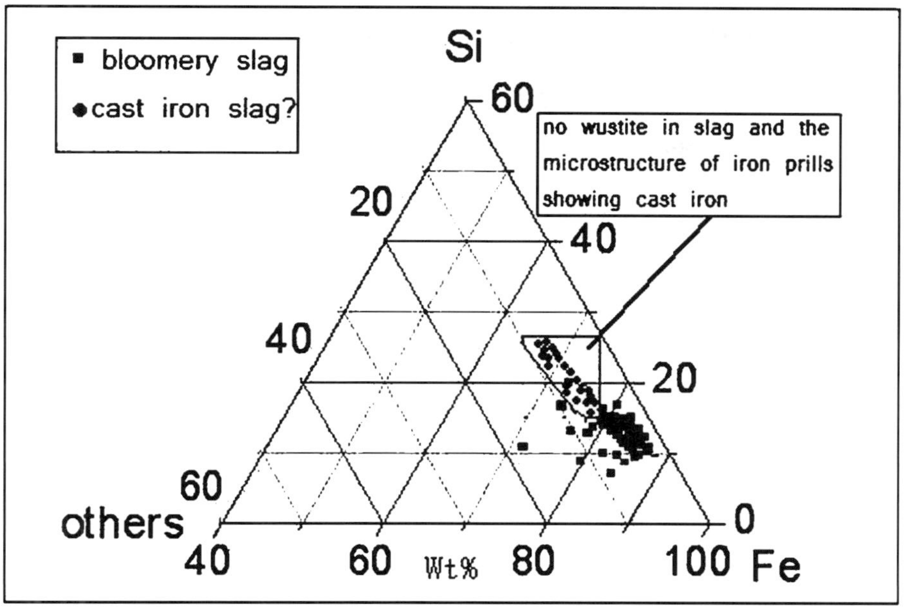

Figure 11: Distribution of the main compositions of Fe and Si for iron smelting slag from Pingnan County.

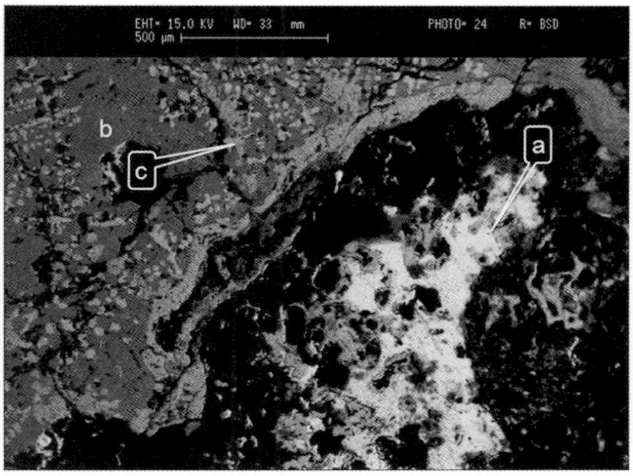

Figure 12: SEM backscattered electron image of the matrix of the bowl-shaped slag sample No. PNPZ09 from the Pozui smelting site (voids are black). The phases present include fayalite (b: black-grey), dendritic wüstite (c: light grey), and metallic iron (a: white).

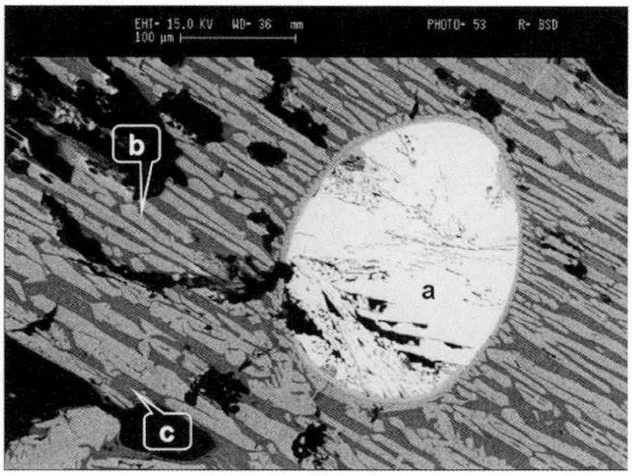

Figure 14: SEM backscattered electron image of the matrix of the tapped slag sample No. PNTS05 from the Tieshitang smelting site. The phases present include metallic iron (a: white), fayalite (b: light grey), and glassy matrix (c: black-grey); voids are black.

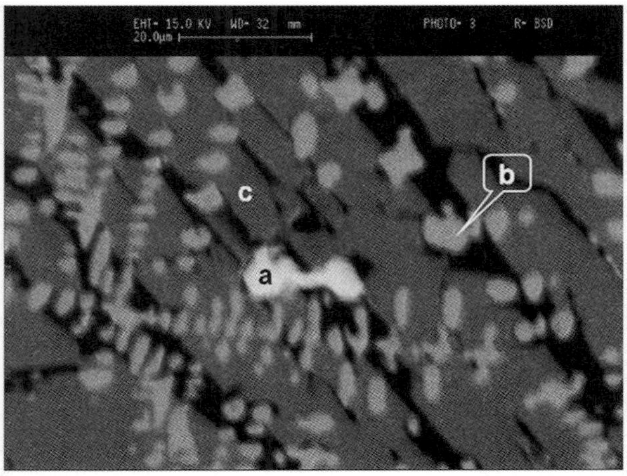

Figure 13: SEM backscattered electron image of the matrix of the tapped slag sample No. PNPZ17 from the Pozui smelting site. The phases present include a metallic iron prill (a: white), dendritic wüstite (b: light grey), and fayalite (c: black-grey); voids are black.

CaO about 1.45 wt%, S about 0.24 wt%, MgO about 0.23 wt%, MnO about 0.07 wt%, and P_2O_5 about 0.03 wt%.

The composition of all the iron ore samples was high in Fe. Their average Fe composition was 95.43 wt%, with Si about 3.69 wt%, Al about 0.28 wt%, K about 0.11 wt%, Ca about 0.13 wt%, S about 0.24 wt%, Mg about 0.27 wt%, Mn about 0.03 wt%, and P about 0.02 wt%.

The bulk composition of the slag inclusions seen within the iron piece sampled from the Pozui site was also FeO-SiO_2. Their average Fe composition was 25.15 wt%, with Si about 63.97 wt%, Al about 3.68 wt%, K about 2.94 wt%, Ca about 2.69 wt%, Mg about 0.29 wt%, and Mn about 1.23 wt%. The elements P and S could not be detected, and the microstructures of the inclusions in the corroded iron piece sample were mono-phase (Fig. 15).

All of the 89 slag samples were well-melted, as observed by mineralographic microscopy. The results show that no other solid residue remained in the slag, indicating a more

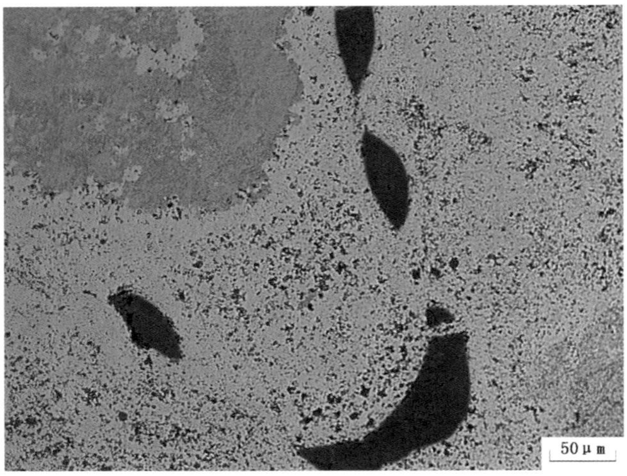

Figure 15: Optical micrograph of the inclusions in the corroded iron piece sample No. PNPZ12 from the Pozui site, mono-phase.

uniform temperature in the process of smelting and that blast technology had reached a certain level of development.

The iron ore samples were oxidised magnetite, as observed by the method of mineralographic microscopy. The results show that the iron ore samples were composed of Fe_2O_3, Fe_3O_4, and small amounts of gangue (Fig. 16). All of them have strong magnetism.

The microstructures of the areas of metallic iron within the bowl-shaped slag and bar-shaped slag samples were ferrite, as observed by metallographic microscopy (Fig. 17). The microstructures of iron prills in the tapped slag samples which contained dendritic wüstite phases could not be observed due to their small size. However, the microstructures of the iron prills visible within the 12 tapped slag samples which contained no wüstite were seen to be white cast iron (Fig. 18). The microstructure of the iron prills seen in the single slag sample that contained no wüstite could not be observed due to their small size, and, finally, the microstructure of the iron debris in the corroded iron pieces showed ferrite-pearlite (Fig. 19).

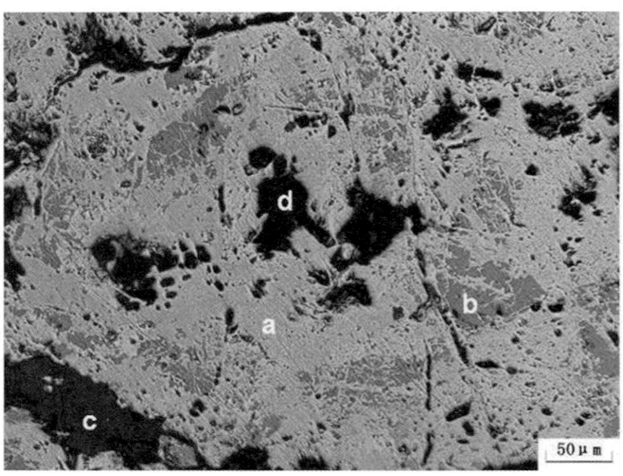

Figure 16: Optical micrograph of iron ore sample No. PNCLK01 from the Chaling site: a: Fe_2O_3 (white), b: Fe_3O_4 (grey-white), c: gangue (black-grey), d: voids (black).

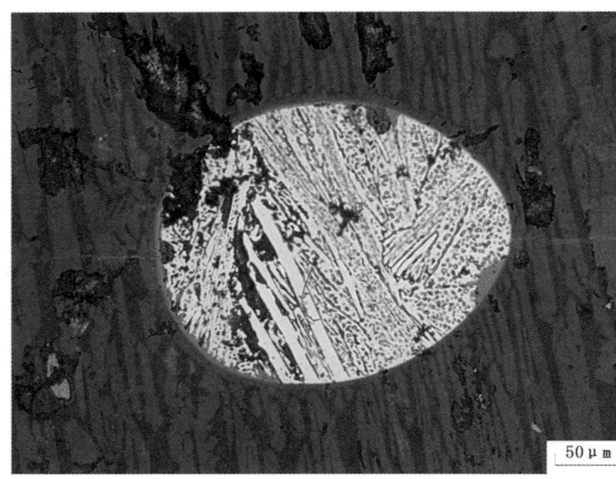

Figure 18: Optical micrograph of the metallic iron prill in tapped slag sample No. PNTS05 from the Tieshitang smelting site. Cast iron, Nital etch.

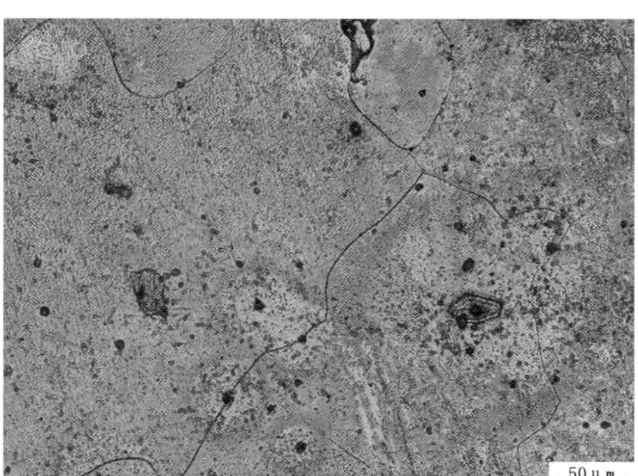

Figure 17: Optical micrograph of the metallic iron from the bottom of the bowl-shaped slag sample No. PNPZ09 from the Pozui smelting site; ferrite, Nital etch.

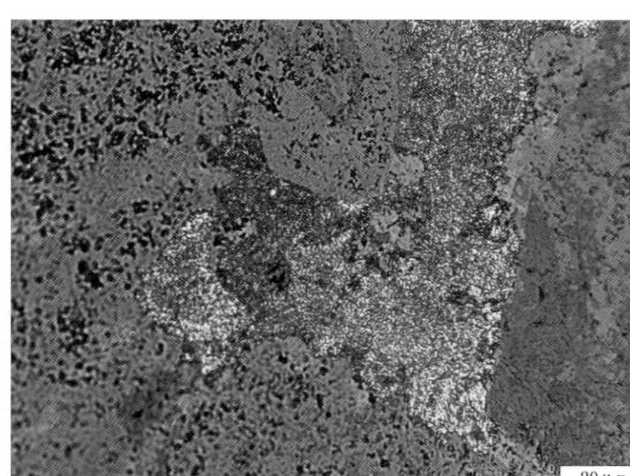

Figure 19: Optical micrograph of iron debris in the corroded iron piece sample No. PNPZ12 from the Pozui site; ferrite-pearlite, Nital etch.

Discussion

Smelting techniques

We have not found Cu, Pb, Zn or Ag or other relevant metal particles during the SEM-EDS analyses of the samples discussed in this paper. Based on current archaeological research on ancient non-ferrous metallurgical slag and iron smelting slag classification criteria (Rothenberg and Blanco-Freijero, 1981, p.312; Craddock, et al., 1985; Li Yanxiang, 1995, p.55), we may exclude the possibility that they are copper smelting slag, lead smelting slag, zinc smelting slag, or silver smelting slag.

The first type of slag, consisting of 67 slag samples from eight smelting sites, including one bowl-shaped slag, one bar-shaped slag, and 65 tapped slag samples. These samples showed phases of fayalite as the dominant constituent, dendritic wüstite, and small areas of glassy matrix, all belonging to the category of slag produced by bloomery iron smelting technology.

The 22 samples of tapped slag which showed phases of fayalite, some areas of glassy matrix, and some evidence of white cast iron within the iron prills, is certainly not slag typical of bloomery iron smelting, and also not typical of cast iron smelting technology. Usually, the typical slag resulting from cast iron smelting contains a flux agent, such as lime, but these slag samples clearly did not display traces that would indicate the addition of a flux agent during the process of smelting. Therefore, it is suggested that there may be two possibilities for this phenomenon: one is that the slag is waste from occasional bloomery iron smelting; the other is that these slag samples represent the waste of cast iron smelting that had no flux agent added during the production process.

Smelting temperature judged from slag compositions

According to the traditional FeO_n-Al_2O_3-SiO_2 phase diagram, the liquid temperature of the bloomery iron smelting slag samples was about 1177°C, judging from the average composition of the slag samples. The liquid temperature of the 22 tapped slag samples which showed evidence of white cast iron was about 1200°C, judging from the average composition of the slag. There are also additional factors to consider when comparing these slags.

Fuels

The local archaeologists excavated significant quantities of charcoal in the furnace hearths at some sites during the National Survey of Cultural Relics in the 1980s (Pingnan County Museum, 1988). However, furnaces with little charcoal present also exist within the archaeological record. Based on the potassium content in the slag samples, it can be considered that the fuel used during the smelting process was charcoal. However, the tree species that were selected for this fuel are still unknown and thus further research is needed to investigate this aspect of the production process.

Furnace

Despite suggestions by an anonymous reviewer that consideration should be given to the possibility that the 'bowl-shaped' furnace is a hearth for fining/puddling (*chaolu*), there may be very good reasons to reject this suggestion. No hearth for fining/puddling (*chaolu*) has been found in China to date, although records of such hearths exist in ancient Chinese literary resources. According to Gordon (1997), finery slag is usually distinguished from bloomery slag by the higher wüstite content. However, the slag examined here has crystal fayalite as the dominant constituent; therefore, it is more likely that it originates from bloomery smelting rather than finery. This means that the 'bowl-shaped' furnace is more likely to have been used for bloomery smelting.

Based on the morphology of the furnaces and slag found at the ruins, we consider that there were at least 2 types of furnace in operation during the period under investigation. One was the bowl-shaped bloomery furnace for bowl-shaped (non-tapped) slag, and the other was the bloomery furnace for tapped slag. We also propose the presence of shaft furnaces.

Ore

Based on the abovementioned analysis of the ore samples, it can be concluded that the local people had the ability to select high-grade iron ore.

Ferrous metal

Based on the analysis presented above, it can also be concluded that steel was made at the sites. However, further investigations into the methods employed to make and forge the steel are necessary before they can be properly understood.

Tuyères

All 89 slag samples were well-melted and relatively uniform, indicative of refractory *tuyères* and a consistent temperature within the furnace. Based on the remains on the 'bowl-shaped' furnace, we can assume that a single *tuyère* was most likely used for a smelt.. The typology of the remains of the straight-round *tuyères* indicates a shared tradition with early copper and iron smelting ruins in the Lingnan area (Guiping Bowuguan, 1988), and resembles closely those from South and Southeast Asian ancient iron smelting sites, such as Ban Di Lung in Thailand (Suchitta, 1983, p.110), Naikund in India (Prakash, 1989), and so on. This aspect has important implications for understanding cultural interactions between Southern China and South and Southeast Asia during the Bronze and Iron Ages.

Further questions

Where did the iron smelting technologies come from?

Cast iron smelting technology had been developed in central/northern China by the Spring and Autumn period at the latest (Han Rubin, 2000). So where did the iron smelting technologies discussed in this paper come from, i.e., from the north or from South or Southeast Asia? The presence of distinctive traditions of metallurgy in South and Southeast Asia (Higham, 1988; Murowchick, 1988; Stech and Maddin, 1988; Li Yanxiang, 1993) means that careful consideration should be devoted to this question through detailed investigation of these metallurgical traditions.

The dates of the finds

Dating the finds and sites discussed above is still a major issue. Based on the patterns of pottery shards unearthed at the sites, some archaeologists consider that they date to the Western Han dynasty at the latest. However, by then the Lingnan area had already been absorbed into the Qin and Han Empires, so one would expect that the tradition of cast iron smelting technology should have already been transmitted there. Why then did bloomery iron smelting technology still flourish in the Lingnan area? According to the comment of an anonymous reviewer, it should not be surprising if bloomery smelting and blast-furnace smelting were used at the same time in the same region. In Europe, the two technologies were used side-by-side for many centuries. This aspect helps us widen our horizon. And did bloomery iron smelting technology exist there before the Qin invasion? The AMS-14C examination of the charcoal samples directly related to the smelting activities identified at the sites should help clarify some of these issues.

Conclusions

The smelting sites in Pingnan County were iron smelting sites. This is the first time that ancient bloomery iron smelting remains have been confirmed in China. The results show that all the slag from the ruins was $FeO\text{-}SiO_2$, composed of crystal fayalite as the dominant constituent. The iron ores used were all of high-grade oxidised magnetite, and the technique of bloomery iron smelting had been mastered. The 'bowl-shaped' furnaces identified were more likely to have been used for bloomery smelting.

This research demonstrates that the Lingnan area had its own iron smelting industry by the time of the Western Han dynasty at the latest, and that the typology of the remains of the straight-round *tuyères* indicates a shared tradition with early copper and iron smelting sites in the Lingnan area, as well as in South and Southeast Asia.

Acknowledgements

The research was supported by the National Science Foundation of China (50874015). We would especially like to thank Professor Tsun Ko, Professor Han Rubin, Professor Sun Shuyun, Professor Mei Jianjun, and Associate Professor Li Xiuhui of the University of Science and Technology Beijing, for their careful supervision, advice and help in the research. Many thanks also to Professor Zheng Chaoxiong of the Guangxi Provincial Museum, Feng Guichun of the Guigang City Museum, Gong Hai and Mo Jianhua of the Pingnan County Museum, and Zhu Lishan, Lu Ying and Zeng Jing of the Guiping City Museum for their help during fieldwork. Dr Qu Yongxin participated in the first investigation in spring 2005. We would especially like to thank Professor Thilo Rehren of University College London, UK, for his encouragement and John Moffett of the Needham Research Institute, Cambridge, UK, for his encouragement and help with the revision of the text. Finally, we would also like to thank the peer reviewer and the editor for their careful comments and useful suggestions.

References

Craddock, P.T., Freestone, I.C., Gale, N.H., Meeks, N.D., Rothenberg, B. and Tite, M.S., 1985, The investigation of a small heap of silver smelting debris from Rio Tinto, Huelva, Spain. In: P.T. Craddock and M.J. Hughes eds. *Furnaces and Smelting Technology in Antiquity*. British Museum Occasional Paper 48. London: British Museum, pp.199-218.

Gordon, R.B., 1997. Process deduced from ironmaking wastes and artefacts. *Journal of Archaeological Science*, 24, pp.9-18.

Guangxi Zhuang Autonomous Region Cultural Relics Team, 1978. Pingle Yinshanling Zhanguo Mu (Pingle silver mountain tomb of the Warring States period). *Kaogu Xuebao*, 2, pp.211-58.

Guiping Bowuguan, 1988, *Guyizhi (1). Guangxi Zhuangzu Zizhiqu Guigangshi Guipingshi, Guiping Bowuguan* (Ancient ruins (1). Guangxi Zhuang Autonomous Region, Guigang City, Guiping Museum). Unpublished Report.

Han Rubin, 2000, Tianma-Qucun yizhi chutu tieqi de jianding (Identification on iron artefacts unearthed at Tianma-Qucun sites). In: Shang Zhou Panel at the Department of Archaeology of Peking University, Shanxi Provincial Institute of Archaeology, and Zouheng eds. *Tianma-Qucun 1980-1989*. Beijing: Science Press, pp.1178-80.

Higham, C., 1988. Prehistoric metallurgy in Southeast Asia: some new information from the excavation of Ban Na Di. In: R. Maddin ed. *The Beginning of the Use of Metals and Alloys. Papers from BUMA-II. Zhengzhou, China*. Cambridge, MA: The MIT Press, pp.130-55.

Huang Xianfan, Huang Zengqing and Zhang Yimin, 1988. *Zhuangzu Tongshi*. (Zhuang History). Nanning: Guangxi Ethnic Publishing House, pp.3-8.

Institute of Historical Metallurgy and Materials (IHMM), University of Science and Technology Beijing, 1991. Xihan Nanyuewangmu Chutu tieqi jianding baogao (The appraisal report on iron artefacts unearthed from the tomb of King Nanyue of the Western Han dynasty). In: Guangzhou City Cultural Relics Management Committee, Institute of Archaeology, Chinese Academy of Social Sciences and Guangdong Provincial Museum eds. *Xihan Nanyuewangmu (shang), Wenwuchubanshe* (The tomb of King Nanyue of the Western Han dynasty (first)). Beijing: Cultural Relics Publishing House, pp.389-96.

Lan Riyong, 1989. Guangxi Zhanguo Tieqi Chutan (A study on Guangxi iron artefacts of the Warring States period). *Kaogu yu Wenwu*, 3, pp.77-82.

Li Jinghua, 1991. Nanyuewangmu Chutu Jinshuqi Zhizuojishu Shixi (An analysis of the production technology of metallic artefacts unearthed from the tomb of King Nanyue of the Western Han dynasty). In: Guangzhou City Cultural Relics Management Committee, Institute of Archaeology, Chinese Academy of Social Sciences and Guangdong Provincial Museum eds. *Xihan Nanyuewangmu (shang), Wenwuchubanshe* (The tomb of King Nanyue of the Western Han dynasty (first)). Beijing: Cultural Relics Publishing House, pp.411-6.

Li Yanxiang, 1993. Dongnanya Yejinkaogu (The archaeological metallurgy in Southeast Asia). *Zhongguo Yejinshiliao*, 2, pp.39-45.

Li Yanxiang, 1995. Tonglushan, Jiuhuashan gudai liantongluzha yanjiu (Studies on ancient copper smelting slag at Tonglushan and Jiuhuashan). PhD Thesis, University of Science and Technology Beijing.

Murowchick, R.E., 1988. The development of early bronze metallurgy in Vietnam and Kampuchea: a re-examination of recent work. In: R. Maddin ed. *The Beginning of the Use of Metals and Alloys. Papers from BUMA-II. Zhengzhou, China*. Cambridge, MA: The MIT Press, pp.182-99.

Pingnanxian Bowuguan, 1988. *Guyizhi. Guangxi Zhuangzu Zizhiqu Guigangshi Pingnanxian, Pingnan Bowuguan* (Ancient ruins. Guangxi Zhuang Autonomous Region, Guigang City, Pingnan County, Pingnan Museum). Unpublished Report.

Prakash, B., 1989. Methods of iron-making in early India. In: R. Pleiner ed. *Archaeometallurgy of Iron: International symposium of the Comité pour la sidérurgie ancienne de l'UISPP: Liblice, 5-9 October 1987.* Prague: Archaeological Institute ČSAV, pp.307-32.

Rothenberg, B. and Blanco-Freijero, A., 1981. *Studies in ancient mining and metallurgy in south-west Spain.* London: Institute for Archaeo-Metallurgical Studies.

Stech, T. and Maddin, R., 1988. Refletions on early metallurgy in Southeast Asia. In: R. Maddin ed. *The Beginning of the Use of Metals and Alloys. Papers from BUMA-II. Zhengzhou, China.* Cambridge, MA: The MIT Press, pp.163-74.

Suchitta, P., 1983. *History and development of iron smelting technology in Thailand.* PhD Thesis, Brown University.

Tholander, E., 1989. Microstructure examination of slags as an instrument for identification of ancient iron-making processes. In: R. Pleiner ed. *Archaeometallurgy of Iron: International symposium of the Comité pour la sidérurgie ancienne de l'UISPP: Liblice, 5-9 October 1987.* Prague: Archaeological Institute ČSAV, pp.35-42.

Zheng Chaoxiong, 1990. Pingnanxian Liuchen Handai Yetieyizhi (Han dynasty iron smelting sites of Liuchen, Pingnan County). In: Chinese Archaeological society ed. *Zhongguo Kaoguxue Nianjian (1989)* (Chinese Archaeology Yearbook 1989). Beijing: Cultural Relics Publishing House, pp.236-7.

Zheng Chaoxiong, 1996. Guanyu Lingnan Yetieye Qiyuan de Ruoganwenti (A number of issues on the origin of iron smelting industry in Lingnan area). *Guangxi Minzu Yanjiu*, 3, pp.50-6.

Zhongguo Lishidituji Bianjizu, 1975. *Zhongguo Lishidituji (2)* (Historical Atlas of China, 2). Beijing: China Map Publishing Society.

Authors' addresses

- Huang Quansheng
  The President Office, Guangxi Radio and TV University, Nanning 530022, China. Email:chainsun@126.com;
- LI Yanxiang
  Institute of Historical Metallurgy and Materials, University of Science and Technology Beijing, Beijing 100083, China. Email: liyanxiang@metall.ustb.edu.cn)

Iron production in the Han Period in the Chengdu Plain, Sichuan, China. Preliminary report on recent fieldwork

Yasuyuki Murakami

ABSTRACT: Collaborative research into ancient iron smelting in China began in the 1950s. However, the number of excavated smelting sites remains limited, and the timing and location of the investigated sites tend to focus on the Eastern Han period and, on the middle of the Yellow River (Hwang he), where the capital cities were situated (Li Jinghua, 2003). Furthermore, discussions concerning stylistic aspects of ancient ferrous metallurgy in China, e.g. the organisation of smelting sites, are rare. It follows that the processes of technological change and the issue of regional variation are aspects of Chinese archaeometallurgy in need of greater exploration.

Keywords: Han Period; ancient iron production; excavation.

Introduction

In the Sichuan Province, named Shu in ancient times, a region located in the upper valley of the Changjiang River, historical records about ancient iron production are well-known. According to Dr Li Jinghua (2003), the leader of research into ancient iron manufacture, the number of iron smelting sites in Sichuan is second only to that in the Henan Province. However, only superficial surveys have been carried out, and no excavations were undertaken before the 21st century.

The Research Centre of Ancient East Asia Iron Culture at Ehime University, Japan began a survey on the Chengdu Plain in Sichuan, in collaboration with the Chengdu Archaeological Institute (Director Wang Yi) and Sichuan University (Prof Li Yinfu), in 2006. This paper presents a report on the progress of this survey to date, including details on the excavations that were carried out and the results that were obtained concerning iron smelting in the Han period.

The Chengdu Plain in Sichuan and ancient iron production

The Sichuan Province is known as a region rich in resources. Archaeological sites have been found here that belong to the Sanxingdui culture (Late Shang to West Chou Period, 1050-770 BC), and to the Shierqiao/Jinsha culture (West Chou period, 770-221 BC). Gold, bronze, and jade artefacts found within the region demonstrate the abundance of geological resources throughout the Chengdu Plain. According to historical texts (Tong Enzheng, 1998), Huiwenwang, the king of Qin (337-311 BC), destroyed the nation of Shu in 316 BC (the Warring States Period) and established the earliest iron officer to control the production and distribution of iron in Chengdu, the capital of Shu, in 311 BC (Huayangguozhi・Shuzhi).

After the First Emperor established the Qin Dynasty (221-206 BC), he forced the famous iron manufacturers to move to the Linjiao prefecture in the Chengdu Plain, and made them produce large quantities of iron, and so they became the richest iron workers of that time in China (Shiji・Huozhiliechuan). Historical texts also show that iron production in the region continued successfully after the establishment of the Han Dynasty (202 BC-220 AD), and records denote place-names as locations of iron manufacture (Huayangguozhi・Shuzhi).

Until recently, historical texts, place names, and ancient iron implements found during surveys (He Ping and Long Tao) have been the only evidence of the iron production that took place at that location in ancient times. All this evidence was utilised to plan the excavations that have recently been carried out.

Selection of sites for excavation: hypothesis and general survey

The purpose of this ongoing research is to clarify the techniques used to smelt iron during the Han period. Before beginning the excavation, a strategy for site selection had to be decided upon. Due to the rarity of surface finds of iron objects that could be used to estimate the dating of the site, it was decided instead that furnace bricks found on the surface would be used to provide an initial chronological marker for the archaeological sites. Li Jinghua (1994) suggested that materials used to construct furnaces changed from clay to actual bricks during the middle of the Warring States Period (furthermore, it is well known that furnaces were generally constructed with stones and clay in the Song Period before the Warring States Period).

Upon further investigation of the furnace bricks, it was noticed that in fact two types of bricks could be identified in the survey material. One type is small, elaborately formed, and burnt at high temperatures, while the second type is slightly larger, less elaborate, and does not appear to have been exposed to high temperatures. The first type was found scattered on the surface together with sherds of pottery dated to the Western Han Period. Based on this information, four chronological periods were identified within the archaeological record:

Period 1: The Western Han Period, when elaborate hardened bricks emerged, as well as other elaborate material culture objects.

Period 2: The Eastern Han Period, when bricks were fired at lower temperatures, similar to other materials of the time.

Period 3: The Three Nation Period to the Sui Period, when bricks were fired at lower temperatures and stones were incorporated into the material.

Period 4: The Tang Period to the Song Period, when clay and stones were used together.

During the research, sites presumed to belong to the first, second, and third period based on this construction typology were chosen for excavation.

Preliminary reports on excavations in the Chengdu Plain

The Linjiao prefecture, as it existed in the Qin and the Han periods, is now divided into two regions; the northern half of the area contains the city of Qionglai, while the southern half contains the Pujiang prefecture. Our archaeological research began in the Pujiang prefecture in 2006. As mentioned above, He Ping and Long Tao (members of the Pujiang Cultural Property), carried out a general survey until the 1980s and created a map of the smelting sites in the region. However, twenty years after their research, it was found that most of the sites had been destroyed due to development and land reform. Our 2006 survey revealed that many smelting sites belonging to the first period had a wide distribution along the Linxi River, running from west to east in the northern area of the Pujiang prefecture. Three particularly well-preserved sites were selected and excavated: Gushishan, Tieniu, and Shazitang.

The site of Gushishan

At Gushishan, dated to the Eastern Han Dynasty based on pottery found at the site (25-220 AD), a well-preserved furnace shaft was discovered. Inside the shaft, a reduction area (upper) and a melting area (lower) could be seen. Based on the level of the top of the melting area, the original height of this furnace was estimated at around 4m. The furnace itself was constructed using bricks that are relatively large and burnt at low temperatures. Slags found around the furnace are light and porous in nature, and consequently it has been suggested that pig iron must have been produced in the furnace. Pottery dated to the Eastern Han Period was excavated within a test pit located in the field around the furnace.

The site of Tieniu

At this site, two locations were investigated: Location 1 is situated at the foot of the hill, while Location 2 is located on the slope of the hill. A large block of pig iron can be seen at the foot of the hill in front of Location 1 (identified as pig iron due to the distinctive colour and gloss). This block of iron is called Tieniu (iron cattle), and is believed to have been produced at this site. At Location 1, large bricks burnt at low temperatures were excavated alongside slags, charcoal, and pottery dated to the Eastern Han Period. At Location 2, a huge pit and a lot of hearths ('chaoganglu') were found. The pit is rectangular in shape, measuring 7x6m, with a depth of over 2m. A large quantity of bricks and charcoal and a small amount of slags were excavated from the inside of the pit. A lot of sherds of pottery dated to the latter half of the Western Han Period were found in the pit. This pit was estimated to have been dug for the disposal of industrial waste. The hearths ('chaoganglu') are round or oval in shape.

The site of Shazitang

At Shazitang the lower half of the furnace shaft was unearthed. It was well-preserved enough to observe the effect of the reducing atmosphere on the clay, and slags were found *in situ* inside the furnace. The furnace was constructed using bricks, and its depth measured about 2m. The bricks are large and coarsely made. Although dating is difficult due to the lack of associated pottery, Shazitang is estimated to date to the same period as the site of Gushishan, based on the observations of the bricks used to construct the furnaces.

Discussion

The excavations of iron production sites in the Chengdu Plain provided considerable information about iron production from the Western Han to the Eastern Han Period. Hypotheses relying on the relative dating of the styles of bricks allowed us to reconstruct a chronological framework for the sites. The bricks seem to change over time from elaborate bricks to coarsely made ones. It is assumed that the complicated construction methods using smaller bricks changed to a more simple method of building furnaces with larger bricks. Further surveys in other regions will allow us to test this hypothesis.

It is also noteworthy that many hearths ('chaoganglu') were found in addition to the smelting furnaces. Although only a few hearths have until recently been known in China, just under ten hearths were discovered in our excavations. It will be possible to obtain new information about the forging technique through the investigation of by-products from these hearths.

Interestingly, at the iron smelting sites surveyed and excavated, no moulds for the production of cast iron implements were found. In the Sichuan Province, cast iron implements with the cast inscription 'Shu jun Chendu' are very famous. It is generally assumed that the public iron commanders of Shu worked to manufacture cast iron implements. Therefore it is possible that the iron manufacturers in the valley of the Linxi River produced pig iron and converted it to steel, which was then supplied to the iron commanders. It is also possible that these iron workers were making iron for private rather than public use. These aspects will be explored further in the future.

Conclusions

Through the investigation of the ancient iron smelting sites presented above, some light has been shed on the smelting techniques in the upper valley of the Changjiang River. Large pits were excavated by the iron workers for the disposal of waste, and the hearths ('chaoganglu') found in a small area at the site of Tienue are the first hearths discovered in China. Although originally research on iron production in ancient China focused on the valley of the Hwang he River, recent research in the Chengdu Plain in the Changjiang Valley has highlighted iron production in the era of earlier centralised states.

Moreover, iron production must have continued in the Chengdu Plain after the Warring States Period, as mentioned above. A more primitive furnace has just been recorded during a recent survey in a peripheral area of the Chengdu Plain. The possibility that iron was produced here before the establishment of centralised states is an issue that will continue to investigated.

References

Tong Enzheng, 1998. *Ancient Bashu.* Zhongqing.

Li Jinghua, 1994. The origin and development of melting furnaces in ancient times. *Studies of metallurgical techniques in the ancient Central Plains*, Zhengzhou.

Li Jinghua, 2003. The investigation and study of iron-smelting ruins in the Warring States Period in Henan Province, China. *Studies of metallurgical techniques in the ancient Central Plains*, vol.2, Zhengzhou.

Author's address

- Research Centre of Ancient East Asia Iron Culture
  Ehime University, Japan

Manufacturing techniques and dates of iron objects found recently at Chinese archaeological sites

Chen Jianli and Han Rubin

ABSTRACT: This paper is a survey of the development of iron and steel technologies in ancient China based upon new archaeological findings and laboratory research. Iron produced by the bloomery process was first used around the 14th century BC in Gansu province, and by the 9th-8th century BC or earlier in the Central Plains and Xinjiang areas. The earliest cast iron fragment, dated to the 8th century BC, was found in the Central Plains. From the 5th century BC cast iron was widely used and a sophisticated cast iron technological system was established during the period between the 3rd century BC and the 2nd century AD. Cast iron technologies were spread widely from Central China into neighbouring areas.

Keywords: Radiocarbon dating; iron; steel; inception.

Introduction

It is generally accepted that the Iron Age began in Asia Minor, where iron-using people occupied the area from about 2000 BC (Tylecote, 1992, p.47). A dagger with copper handle and iron blade, unearthed in a Hittite tomb (2500 BC), was thought to be a man-made iron object. It indicated that iron smelting technology emerged in Anatolia during that period; however, the dagger might just be an accidental discovery. After 1500 BC, man-made iron objects became more and more common in Mesopotamia, Anatolia, and Egypt. The trade of iron objects between Hittites and Assyrians was recorded in documents. Between 1500 and 1000 BC, iron techniques were diffused throughout Europe, Asia, and Northern Africa. Tylecote (1992, pp.47-48) pointed out that ironworking spread to India and China from Iran around 400 BC. Bennet Bronson (1999) also pointed out that the making of iron emerged later in China, than in many other parts of Eurasia.

The origin of iron smelting technology in China has puzzled academics from as early as the 1920s. Significant archaeological fieldwork and scientific investigation on iron and steel has been carried out in China since the 1950s, and there is now a considerable amount of literature on the topic. As a result, scholars have already achieved some preliminary understandings regarding the development, characteristics, and organisation patterns of early Chinese iron production (Yang Kuan, 1955; Chinese Archaeometallurgy Study Group at Beijing University of Iron and Steel Technology, 1978, pp.148-50; Li Jinghua and Chen Changshan, 1995; Hua Jueming, 1999, pp.303-5; Bai Yunxang, 2005; Chen Jianli and Han Rubin, 2007; Han Rubin and Ko Tsun, 2007, pp.357-358,383; 377-378; Wagner, 2008).

However, due to the vague descriptions in historical documents and the lack of archaeological evidence from early periods, the origin of the iron smelting technology in China is still under debate among scholars. Some think it emerged no later than the Erlitou culture, 1500 BC (Zhang Binglin, 1925), while some others believe it began in later periods, such as the Shang Dynasty, 1600-1046 BC (Tong Shuye, 1955; Xia Xiangrong, 1980, pp.212-13), the Western Zhou Dynasty, 1056-771 BC (Guo Moruo, 1976, p.313; Yang Kuan, 1982; Tang Jigen, 1993; Bai Yunxiang, 2005, p.9; Zhao Huacheng, 1997; Hua Jueming, 1999, pp.303-5), or even the Spring and Autumn Period, 771-476 BC (Li Jiannong, 1962, p.42). In addition, whether smelting techniques were introduced from the west or invented indigenously within China is an issue that has not yet been determined. Therefore, systematic studies of early Chinese iron artefacts and smelting technology, as well as their careful dating, are essential for our understanding of the origins of the Chinese Iron Age.

Although the currently available data suggests that the development of iron and steel technologies varies from culture to culture, the technologies of cast iron production and steel made from cast iron in ancient China represent a unique approach that differs considerably from the ferrous technologies of West Asia and Europe (Han Rubin and Ko Tsun, 2007, pp.357-358,383; 377-378.). The distinguishing characteristics which can be used as the criteria for differentiating between puddled steel, decarburised steel, and carburised steel produced in the solid state were previously discussed by Chen Jianli and Han Rubin (2007). Over the past several years, many new iron smelting sites and archaeological iron objects have been excavated in China, providing new information for the study of the development of iron and steel technologies in ancient China. More than 500 samples of slag, iron objects, furnace walls, ores, charcoal, bones, etc. have been collected by the authors from archaeological sites in central, north-east, north-west, and south China, as shown in Fig. 1. These samples have been analysed by metallographic microscopy, SEM-EDS, and AMS-^{14}C dating methods. This paper is a survey of the iron objects and their chronology based on these new findings and our laboratory research results.

The radiocarbon dating of iron objects

Because sometimes iron objects are separated from related archaeological materials that would permit assigning them an absolute date, direct radiocarbon dating of iron materials is a desirable approach to ascertaining their ages. The radiocarbon dating of iron objects was first proposed and carried out by van der Merwe in 1969 (van der Merwe, 1969). Later, AMS-^{14}C dating was used by researchers to obtain dates for iron objects (Creswell, 1991, 1992; Eylon, 2002; Cook, et al., 2003). In order to verify whether ancient Chinese iron can be successfully used for radiocarbon dating, several iron objects with definite dates, including one coin dating to the 6th century AD, 12 coins dating to the 11th-12th century AD, seven iron bridge fragments from the Pujindu site of the 8th century AD (Chen Jianli, et al., 2008), as well as 15 iron objects without definite dates (Chen Jianli, et al., 2008; Chen Jianli and Ma Qinglin, 2009), were examined using metallographic methods, and dated using AMS-^{14}C methods. The results are listed in Table 1.

The 13 coins are made from white cast iron. The calibrated radiocarbon dates obtained for four coins (laboratory codes BA05680, BA05684, BA06516, and BA06518) are approximately one century earlier than the dates indicated by the inscriptions of the coins. The reason is probably that older carbon was introduced from the fuel during the process of iron smelting and casting. This problem has been discussed previously (Craddock, et al., 2002) and we do not believe that there are any reasons to doubt the reliability of the methods of radiocarbon dating of Chinese cast iron objects. Six samples are noted as containing fossil carbon, and these samples also contained FeS slag inclusions. It can therefore be inferred that coal was used in the iron smelting process during the Chongning Period (1102-1106 AD) of the Northern Song Dynasty (960-1127 AD). The radiocarbon dates of iron coins BA06511, BA06521, and BA06522 are 1950±35, 4255±35, and 2260±35BP respectively, much earlier than the dates shown on the coins (1102-1110 AD), which could indicate that coal and charcoal may have been used together during the smelting or melting process. This possibility creates some problems for radiocarbon dating research on iron objects.

The Pujindu is a famous ancient ferry on the Yellow River, located in Yongji county, Shanxi Province. Four iron 'tractors' (each 3.3m long and 1.5m high, weighing about 50 to 70 tons), four iron men (positioned next to each tractor, about 1.9m high and weighing about 3 tons), and other iron objects were found at the site. The records showed they were cast during the Kaiyuan period (about 724 AD) of the Tang Dynasty (618-907 AD). As for the iron samples from the Pujindu site,

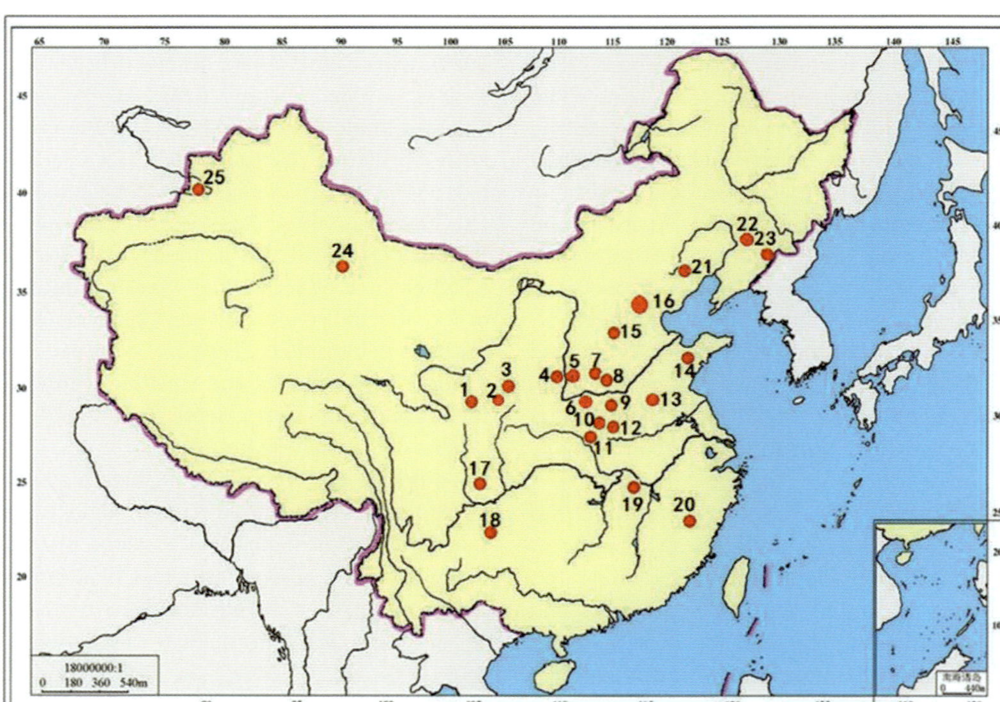

Figure 1: Map of China showing the archaeological sites mentioned in the paper. 1. Mogou site, Lintan county; 2. Li county; 3. Majiayuan site, Zhangjiachuan county; 4. Liangdaicun site, Hancheng; 5. Pujindu site, Yongji county; 6. Sanmenxia; 7. Tianma-Qucun site, Quwo county; 8. Maijiehe and Beiyecun site, Jiaozuo; 9. Zhenghan old city, Xizheng; 10. Wangfangzhuang site, Nanyang; 11. Yangying site, Laohekou county; 12. Five iron workshops in Tongbai county; 13. Shizishan tomb in Xuzhou; 14. Dongpingling site, Zhangqiu city; 15. Dongheishan site, Xushui county; 16. Shuiquangou site, Beijing; 17. Iron workshops in Chengdu; 18. Hezhang; 19. Ezhou; 20. Chengcun site, Wuyishan; 21. Lamadong site, Beipiao county; 22. Erlonghu site, Tongyu county; 23. Goguryeo tombs, Ji'an; 24. Dongheigou site, Balikun county; 25. Tombs in Ili Valley.

Table 1: The AMS-^{14}C dates of iron objects with definite ages*

BA06521	Yuanyou Tongbao	4255±35	2930 BC (75.3%) 2850 BC 2810 BC (17.3%) 2750 BC 2720 BC (2.8%) 2700 BC	Huayin, Shaanxi province, 1086-1094 AD
BA06522	Yuanyou Tongbao	2260±55	410 BC (95.4%) 190 BC	
BA05684	Yuanyou Tongbao	1045±50	880 AD (88.1%) 1050 AD	
			1060 AD (7.3%) 1160 AD	
BA06523	Chongning Tongbao	39985±235	Not calibrated for fossil carbon	Huayin, Shaanxi province, 1102-1106 AD
BA05681	Chongning Tongbao	39130±220		
BA05682	Chongning Tongbao	38140±180		
BA06511	Daguan Tongbao	1950±35	40 BC (95.4%) 130 AD	Yuncheng, Shanxi province, 1107-1110 AD
BA06513	Daguan Tongbao	25720±95	Not calibrated for fossil carbon	
BA06514	Daguan Tongbao	15275±55		
BA06515	Daguan Tongbao	37000±235		
BA06516	Daguan Tongbao	1060±35	890 AD (95.4%) 1030 AD	
IAAA31134	No.1 iron man	1530±40	420 AD (95.4%) 610 AD	Iron bridge at the Pujindu site, Shanxi province, 724 AD
IAAA31135	No.1 iron man	1460±40	530 AD (95.4%) 660 AD	
IAAA31136	No.1 iron man	1430±40	550 AD (95.4%) 670 AD	
IAAA31137	No.1 iron man	1430±40	550 AD (95.4%) 670 AD	
IAAA31138	No.2 iron man	1370±40	590 AD (90.1%) 720 AD	
			740 AD (5.3%) 770 AD	
IAAA31139	Iron stick under the No.3 iron man	1410±40	560 AD (95.4%) 680 AD	
IAAA31140	Iron pillar besides the No.4 iron man	1910±40	0 AD (95.4%) 220 AD	
BA07335	Adze	2320±45	518 BC (73.3%) 349 BC	Yangying site of Laohekou City, Hubei province, about the 5th century BC
			309 BC (22.1%) 208 BC	
BA07336	Sickle	2330±30	507 BC (7.8%) 437 BC	
			420 BC (85.9%) 359 BC	
			273 BC (1.7%) 259 BC	
BA07337	Adze	2300±35	409 BC (66.6%) 351 BC	
			296 BC (28.8%) 209 BC	
BA07338	Fragment	2310±35	415 BC (74.2%) 350 BC	
			299 BC (21.2%) 209 BC	
BA07542	Tripod caldron	2525±35	800 BC (95.4%) 530 BC	Li county of Gansu province, about the 5th century BC
BA07543	Sword	2380±50	760 BC (12.7%) 680 BC	Lingtai county of Gansu province, from 5th century BC to 1st century AD
			670 BC (4.0%) 630 BC	
			600 BC (78.7%) 370 BC	
BA07544	Tripod caldron	2240±40	400 BC (95.4%) 200 BC	
BA07545	Ploughshare	2105±35	350 BC (2.7%) 320 BC	
			210 BC (92.7%) 40 BC	
BA07546	Sword	2315±40	510 BC (73.5%) 340 BC	
			300 BC (21.9%) 200 BC	
BA06507	Iron blade	2150±45	360 BC (95.4%) 50 BC	Chengcun site in Wiuyishan city, Fujian province, about the 3rd-2nd century BC
BA06508	Iron blade	2240±35	400 BC (27.1%) 340 BC	
			330 BC (68.3%) 200 BC	
BA06509	Iron blade	2180±40	380 BC (92.2%) 150 BC	
			140 BC (3.2%) 110 BC	
BA07548	Knife	2220±35	390 BC (95.4%) 200 BC	Suizhou Museum of Hubei province, about the 2nd century BC
BA07549	Knife	2100±35	350 BC (1.3%) 320 BC	
			210 BC (94.1%) 30 BC	
BA07550	Knife	2100±40	350 BC (3.2%) 310 BC	
			210 BC (92.2%) 0AD	

"*Tongbao*" is the name of the coins from Tang through Qing dynasties. "Xining", "Yuanyou", "Daguan", and "Chongning" are the names of the coins referring to periods of reign of various emperors. For example, "Chongning Tongbao" was one such example, cast during the years of rule of Zhao Ji (Huizong) of the Song Dynasty (1102-1106 AD).

the metallographic results show that six of them, the No.1, No.2, and No.4 men (laboratory codes IAAA31134-31138 and IAAA31140) are made of grey cast iron and the iron pillar beside the No.3 iron man (IAAA31139) is made of middle carbon steel. The dates of these six grey cast iron samples are almost the same, while their calibrated dates are also about one century earlier than indicated by the records. However, the date of the iron pillar was older than that of the other six samples. It is possible that older iron materials were used to build the bridge. The archaeologists also pointed out that the pillar is older than the other parts of the iron men. As a result, the radiocarbon dates obtained for the iron men are in agreement with the dates indicated in the historical records.

The radiocarbon dates of iron objects from the Chengcun site of Fujian province, the Li and Lingtai counties of Gansu province, the Yangying site and Suizhou Museum of Hubei province are in good agreement with their archaeologically estimated ages. The results obtained for these archaeological artefacts indicate that the iron objects can be used as materials for ^{14}C dating credibly in some cases, and that rust does not add a significant amount of contaminant carbon. The radiocarbon dates together with the carbon extraction yields from iron objects indicate that much of the original carbon is retained. However, the influence of old carbon used during the smelting, annealing, and forging processes should be considered. When the age of the iron object is approximately known archaeologically, as in the case of the objects discussed in this paper, the contemporary technology by which the iron is likely to have been produced can be considered. Here, the analysis of the microstructures of the iron objects and a consideration of manufacturing techniques provide a better interpretation of the radiocarbon dates. For example, in the case of cast iron and decarburised steel, it may be useful to study their smelting or casting dates. However, when considering carburised steel, because some carbon from different sources may have been introduced to the iron objects, investigating and exploring the dates may prove less reliable. Therefore, the selection of iron objects and the decision as to which part of the object can be radiocarbon dated should be decided based on the microstructure.

Because of the relative lack of data pertaining to the direct radiocarbon dating of iron materials, it can be suggested that it is better to select short-lived plant and bone samples for AMS-^{14}C dating. The radiocarbon dates mentioned in this paper are mainly measured from organic materials rather than iron objects.

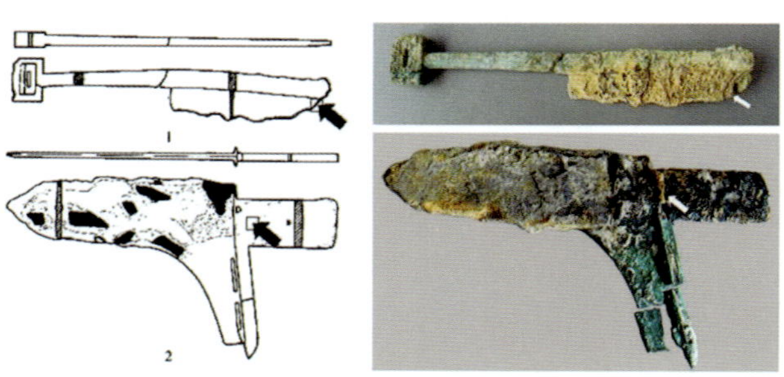

Figure 2: Two copper-iron bimetallic wares excavated from the Liangaicun site, Shaanxi province.

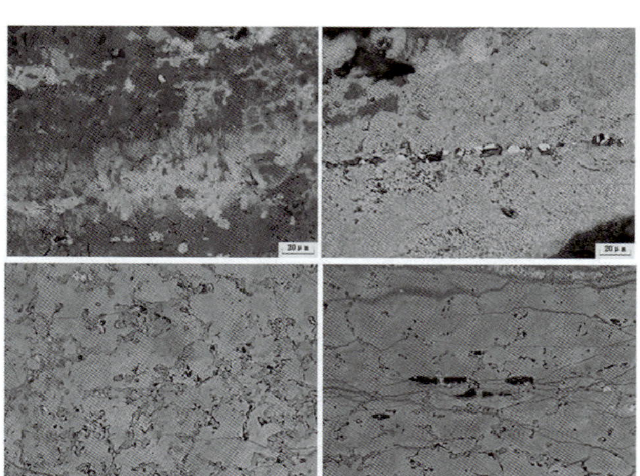

Figure 3: Microstructure of two copper-iron bimetallic wares excavated from the Liangaicun site, Shaanxi province.

The beginning of the use of iron in China

Presently there is no definitive conclusion regarding the time and place of the origin of iron smelting in China. Three meteoritic iron objects and three bloomery iron objects excavated at the Guo State cemetery in Sanmenxia, Henan Province, dated to the 9th century BC, belong to the late Western Zhou Dynasty (1046-771 BC) (Han Rubin, et al., 1999). Two shapeless remnant pig iron fragments and a triangle-shaped iron fragment dating to the middle of the Spring and Autumn period (770-476 BC) were excavated in Qucun village, Quwo county, Shanxi Province (Han Rubin, 2000). They are dated to around 800-600 BC. The fabrication technology of two copper-iron bimetallic wares (a bronze knife with an iron blade and a bronze *ge*戈 with an iron blade, Fig. 2), excavated in tomb M27 of the Liangdaicun site, Hancheng of Shaanxi province, was analysed by metallography, EPMA, and AMS-^{14}C dating methods. The microstructure of the two samples is typical of wrought bloomery iron containing a substantial amount of carbon, i.e. carburised steel made from bloomery iron by cementation in the solid state (Fig. 3).

The objects can be dated back to the early Spring and Autumn period (770-476 BC). Thus, the study of these objects provides new evidence regarding the beginning of iron smelting in Central China. Most of the previously known iron objects dating to the period between the late Western Zhou Dynasty and the early Spring and Autumn period were excavated in the region at the junction of the Henan, Shanxi, and Shaanxi provinces, in the middle reaches of the Yellow River, suggesting that this region might have been one of the earliest centres of iron smelting technology in China and thus deserves further archaeological investigation (Chen Jianli, et al., 2009). In the late Spring and Autumn period, the iron smelting industry in this area developed rapidly. For instance, Han Rubin studied the fabrication technology of approximately 4000 iron objects dating to about the 3rd century BC and earlier in China and found that the area of the Shanxi province was the area where most of the iron objects dating to the period between the 5th and the 3rd centuries BC were excavated. This area became the centre of iron smelting during the Warring States Period (Han Rubin and Duan Hongmei, 2009). The adjacent area of western Henan also yielded a large number of excavated iron objects. The fact that most man-made objects from the western Zhou and early Spring and Autumn periods have been excavated in this area is important evidence in the effort to locate the origins of iron working technology in China. Due to the potential for finding more early Spring and Autumn period iron objects in this area, further archaeological excavations and laboratory research are necessary.

More than 50 iron objects dated to the 5th century BC or earlier were excavated in Baoji, Shaanxi, and the neighbouring Gansu and Ningxia provinces (Han Rubin, 1998). It is worth exploring the relationships between these iron objects and those in the area at the junction of the Henan, Shanxi and Shaanxi provinces. Among the 50 iron objects, nine of them belong to the early Spring and Autumn period and the others belong to the end of the Spring and Autumn period and the early Warring States period (475-221 BC). One carburised steel sword with a gold handle excavated in Cemetery No.2 in Yimencun, Baoji City, was analysed by metallographic methods (Bai Chongbin, 1994). No differences were found in terms of production technology between the early iron objects in this area and those in Sanmenxia, Quwo, or Liangdaicun, located about 300-370km to the east. The dates of the iron objects from these two areas were almost the same, and consequently an investigation of the communication and transportation of the iron objects and smelting technologies in the two areas could provide some very important results.

The early iron objects excavated in Xinjiang have drawn great interest among academics because they provide a clue about the beginning of the use of man-made iron in Central China. Some scholars believe that the iron smelting technology in Central China originated in Xinjiang, but there are still debates about the date of early iron objects excavated in Xinjiang. Iron objects from Yanbulake, Subeixi, Chawuhugoukou, Ili River, and other areas of Xinjiang included small iron knives, swords, sickles, finger rings, arrowheads, thimbles, needles, residual iron blocks, and other small tools and weapons, which may indicate that there was no large-scale iron production before the Han Dynasty. To date, only three iron workshops dating to the time of the Han Dynasty have been found. Details about the sites and tombs where the iron objects were excavated are generally not fully released and a few ^{14}C dates of these objects have been obtained. So far, there are only six ^{14}C dates from samples taken directly from tombs where ironware was found, and the dating error is significant. Therefore, the framework of ^{14}C dates for the early iron objects in Xinjiang needs further research. A total of 32 iron objects and ten radiocarbon dating samples (bones) were collected from the Ili River reaches of Xinjiang for metallographic and AMS-^{14}C dating analysis by Chen Jianli. The results show that objects made of bloomery iron and carburised steel were mainly found in the early period (Fig. 4); decarburised steel made from cast iron (Fig. 5) was first used during the Han Dynasty. The beginning of the use of iron in the Xinjiang area can be dated back to the 9th-8th century BC.

In 2009, an iron bar excavated from tomb M444 at the Mogou site, Lintan county, Gansu province, was analysed

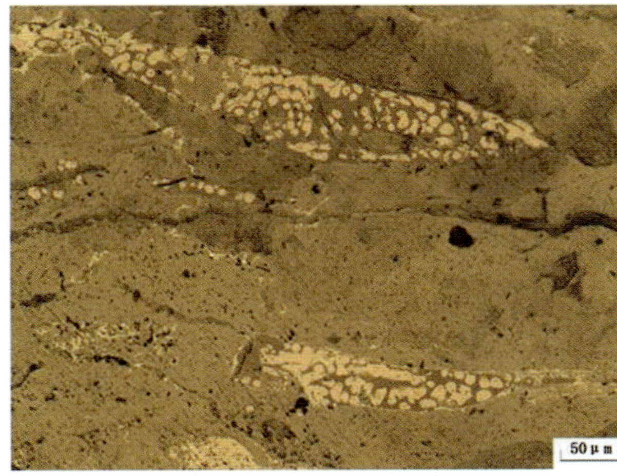

Figure 4: The microstructure of a bloomery iron object excavated from the Ili Valley, Xinjiang

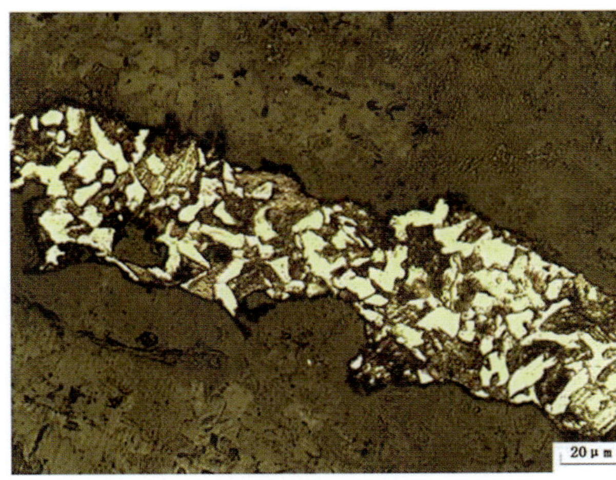

Figure 5: The microstructure of a carburised steel object excavated from the Ili Valley, Xinjiang.

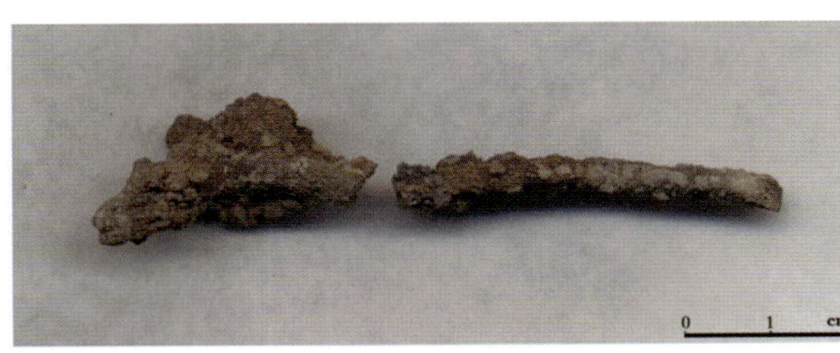

Figure 6: Iron bar excavated from tomb M444 at the Mogou site, Lintan county, Gansu province.

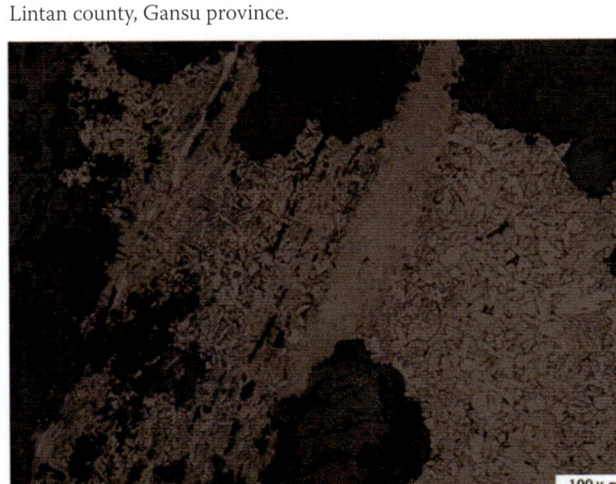

Figure 7: The microstructure of the iron bar excavated from tomb M444 at the Mogou site, Lintan county, Gansu province.

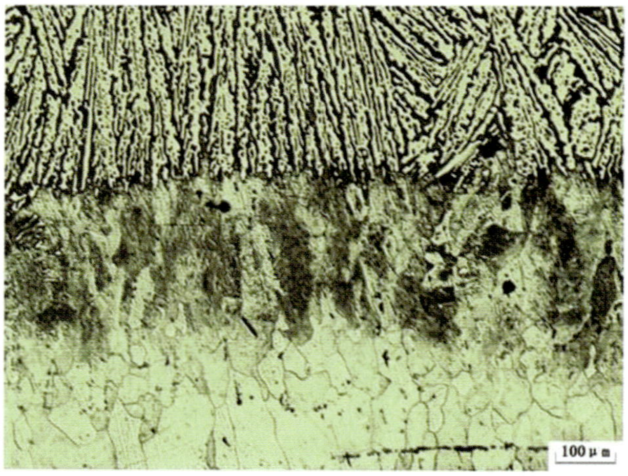

Figure 8: The microstructure of a decarburised iron unearthed from the Yangying site, Hubei province.

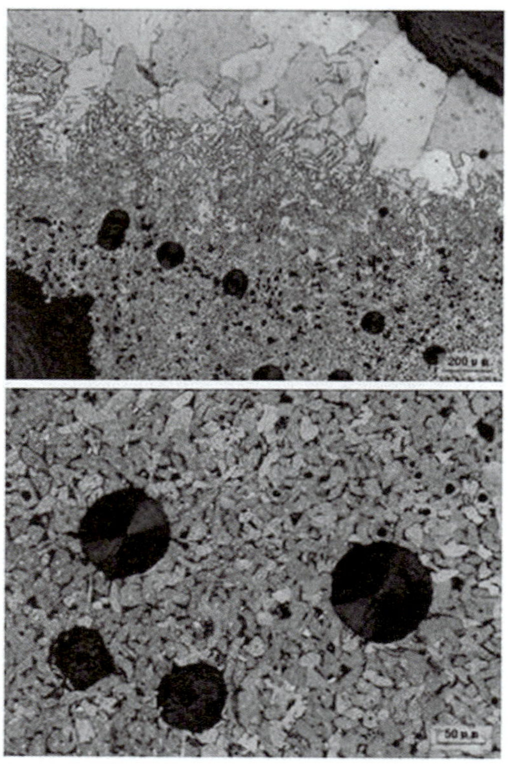

Figure 9: Spherical graphite in a malleable iron object excavated from Xinzheng, Henan province.

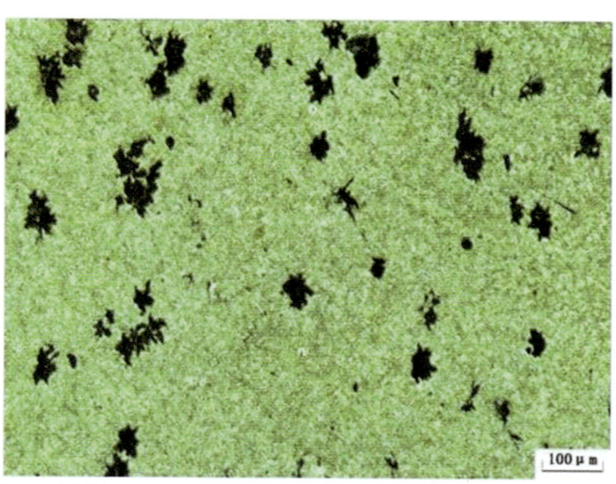

Figure 10: The microstructure of a malleable iron tool excavated from Chengcun, Fujian province.

by metallographic techniques and SEM-EDX by Chen Jianli. The AMS-^{14}C analyses of charcoal and human bones from this tomb and from another tomb containing corroded iron artefacts revealed that both of these tombs can be dated to the 14th century BC. They should thus belong to the local Siwa culture (14th century BC to 11th century BC). The technical study of this artefact showed that it was made with carburised bloomery iron and shaped by forging (Figs. 6 and 7). This new finding may be able to push back the date of the earliest use of anthropogenic iron in China from the 9th century BC to the 14th century BC and provides significant information for the discussion regarding the origin of iron metallurgy in China. Based on the current evidence, the use of smelted iron in the Central Plains of China emerged later than in northwest China, but its technology is unique and may indicate an indigenous invention.

New findings of cast iron and of steel made from cast iron

One of the most significant developments in ancient Chinese metallurgy was the production and use of cast iron, which became the main focus of the iron making technology following the invention of the annealing process for toughening white cast iron and the making of steel from cast iron. The earliest white cast iron fragments in Tianma-Qucun, Shanxi province, were dated to the 8th-7th century BC (Han Rubin, 2000). The annealing process was invented in the early 5th century BC. Several annealing furnaces from the Han Dynasty (Western Han Dynasty 202 BC – 9 AD, Eastern Han Dynasty 25-220 AD) have been excavated at several sites in Henan province, such as Tieshenggou (Archaeology Team of Henan Province, 1962), Guxing (Zhengzhou Museum, 1978) and Wafangzhuang (Henan Provincial Archaeology Institute, 1991).

More than 40 iron objects were found at the Yangying site in Hubei province. Their radiocarbon dates and microstructures show that cast iron, decarburised iron (Fig. 8), and decarburised steel were used at this site. This is one of the earliest pieces of evidence in China that the annealing process was conducted so as to make good use of white cast iron around the 5th-4th century BC. The typical microstructure of decarburised iron consists of ledeburite in the centre of the casting and decarburised layers on the surfaces, i.e. white cast iron encased in steel surface layers. Three iron tools from the Yanying site, Laohekou of Hubei province, and one iron object from Linzi, Shandong province (Chen Jianli, 2007), have this same structure.

Malleable iron can be forged to a certain extent; this aspect was revealed by several analysed objects, which show broken graphite elongated in the direction of working. A shovel excavated at the Luoyang Cement Factory has the typical structure of black-heart malleable iron, and this is the earliest such example found to date, dating to the 5th century BC (Li Zhong, 1975). It should be noted that spherical graphite was produced during annealing in the course of malleable iron production, which also appeared in the 5th-4th century BC in Xinzheng of Henan province (Fig. 9). An arrowhead and an axe excavated from the Lamadong graves of Liaoning province, dated to the 3rd-4th century AD, were made from black-heart malleable iron, with some snowflake graphite embedded in the ferrite matrix visible in the microstructure (Fig. 10). This represents direct evidence that the Xianbei people living in North China had acquired the annealing technology by this time (Chen Jianli, et al., 2003).

The examination of ferrous artefacts showed that a simple, economical process of steel making was in use by the 5th century BC in China; thin or narrow plates or tools were cast and then annealed into steel. This type of steel contains little spheroidised carbide without graphite and inclusions (Han Rubin and Ko Tsun, 2007, pp.357-358,383; 377-378.), as visible in the samples of six pairs of scissors from Zhengzhou, a pickaxe from Dengfeng, a ring handle knife from Changzhi, an arrowhead from Mancheng, and objects from the Wangchenggang iron smelting site (Chen Jianli and Han Rubin, 2007), the Shizishan tomb of Jiangsu province (Chen Jianli and Han Rubin, 1999), the Lamadong tombs of Liaoning Province, Tianshui of Gansu (Chen Jianli and Ma Qinglin, 2009), and the Balikun tombs of Xinjiang. Although they were found at different localities, the objects had similar structure (Fig. 11), and mould joint marks still remained on some of them.

The making of steel from cast iron by decementation was simple, economically viable and easily available in adequately high quality. The plates could be transported, traded, and then forged directly into tools and weapons. Some cast iron plates, clay mould materials and objects forged from steel plates dating to the 3rd century BC and the 2nd century AD have been discovered in ancient iron workshops such as Dengfeng, Guxing, Tieshengou, Nanyang in Henan province (Li Jinghua and Chen Changshan, 1995), Dongpingling in Shandong province, and Taicheng in Shaanxi province. New field excavations are being carried out at the Dongpingling and Taicheng sites, and these will be continued for several years by Peking University and other institutions.

A more advanced steel making process, producing 'puddled steel', first appeared in the 2nd century BC and became very popular by the 2nd century AD. Several furnaces used to produce puddled steel in the Han Dynasty were excavated in Tieshenggou, Wafangzhuang, and Zhaohe of Henan province (Li Jinghua and Chen Changshan, 1995). Five pieces of puddled steel objects excavated from the tomb of Shizishan indicate that ironsmiths in the earlier period of the Western Han Dynasty had produced this type of steel (Chen Jianli and Han Rubin, 1999). They were made from steels with different carbon content, produced by puddling followed by forging until the desired content was achieved. A large number of inclusions elongated in the direction of working are present in the matrix of the steels (Fig. 12).

This was the earliest finding in China that confirmed that the puddling technique had been used by the middle of the 2nd century BC. More than 10 pieces of steel weapons made of puddled steel during the Han Dynasty, excavated from tombs at Suizhou, Hubei province; Hezhang, Guizhou province (Chen Jianli, et al., 2008a), the Chengcun site of Fujian province, (Chen Jianli, et al., 2008c), and other sites in Gansu province indicate that puddled steel technology was widely used during that time. A puddled steel plate originating from the Lamadong graves contains large quantities of inclusions and has a non-uniform structure with different carbon distribution and grain size. It is possible that the raw output of the puddling process was exchanged between different areas (Chen Jianli, et al., 2003). *Hundredfold Refined steel* (*Bailiangang steel*) 百炼钢 was developed after the spread of puddled steel during the 1st-2nd century AD.

Grey cast iron objects appeared in the Warring States period (475-221 BC) at the Yangying site in Hubei province (about the 4th century BC), the Yanxiadu site in Hebei province (about the 3rd century BC) (Li Zhongda, et al., 1996) and the Tonglushan site of Hubei province (about the 3rd century BC) (Ye Jun, 1975). More grey cast iron was produced after the Western Han Dynasty, as indicated by the objects found in the Mancheng tomb (113 BC) of Hebei province (Optical Microscopy Laboratory at Beijing University of Iron and Steel Technology, 1980), the Wangchenggang iron-making site (Fig. 13) (Chen Jianli and Han Rubin, 2007), the Lingtai site of Gansu province (Chen Jianli and Ma Qinglin, 2009), and the Chengcun site of Fujian province (Chen Jianli, et al., 2008c). Many large objects, such as the iron lion of Cangzhou (953 AD) (Song Wei, et al., 2009) and the iron men and oxen of Pujindu (Chen Jianli, et al., 2008b), were made from grey cast iron.

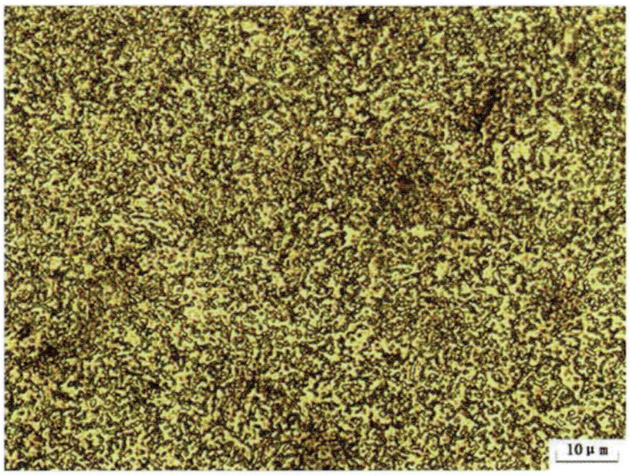

Figure 11: The microstructure of decarburised steel object excavated from Hezhang, Guizhou province.

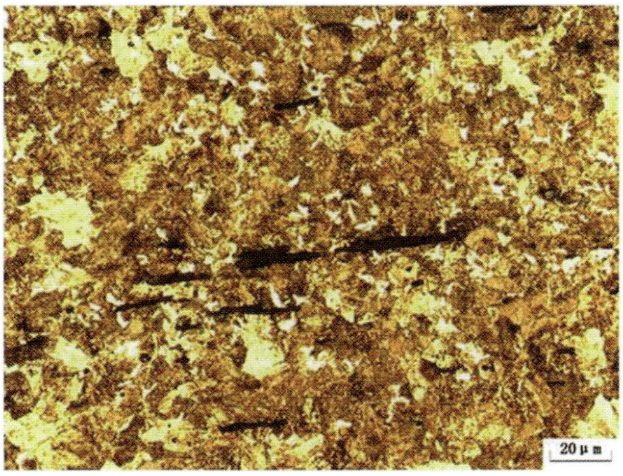

Figure 12: The microstructure of a puddling steel sword excavated from Suizhou, Hubei province.

Based on these results, a scheme of iron- and steel-making techniques in ancient China can be illustrated in the following diagram (Fig. 14).

The history of iron and steel technology in ancient China and its neighbouring countries is of great importance and relevance for the study of the development of the politics, economy, and culture of these countries. In recent years, researchers from many countries have published books and papers that discuss the development of iron and steel technologies in China, employing archaeological, typological, and archaeometallurgical methods. Scientific research on iron and steel artefacts and smelting methods in China, however, has so far focused on objects and sites from Central China, with little attention being paid to the metal findings in outlying areas.

Microstructural analyses of iron and steel artefacts excavated from Laoheshen and Lamadong, two important sites in Northeast China, show that pig iron, decarburised cast iron, malleable iron, puddled steel, *nip steel* 夹钢, and *Guangang steel* 灌钢 were widely used. The *Guangang* process is to seal pieces of cast iron and wrought iron layer by layer and heat them; the carbon within cast iron will diffuse into the wrought iron, and all the parts will become steel. This procedure indicates that these artefacts were made by workers with relatively high-level skills. It is possible that the raw steel materials came from the Central Plains, as one piece of puddled steel bar was found in a tomb. Comparative studies on materials and manufacturing techniques of iron artefacts between northeast China and Central China show that the beginning of the use of man-made iron in northeast China was later than that in Central China, and the techniques and styles of iron objects in Northeast China were strongly influenced by the latter. The appearance of ironware, raw iron and steel materials, and iron smelting did not occur at the same time in northeast China. First, iron objects brought from other areas were used, followed by raw materials for making implements, and finally the skills of iron smelting were mastered. We suggest that war, migration, robbery, and other activities among people living in different areas accelerated the spread of advanced iron technology from Central China to northeast China (Chen Jianli, et al., 2005).

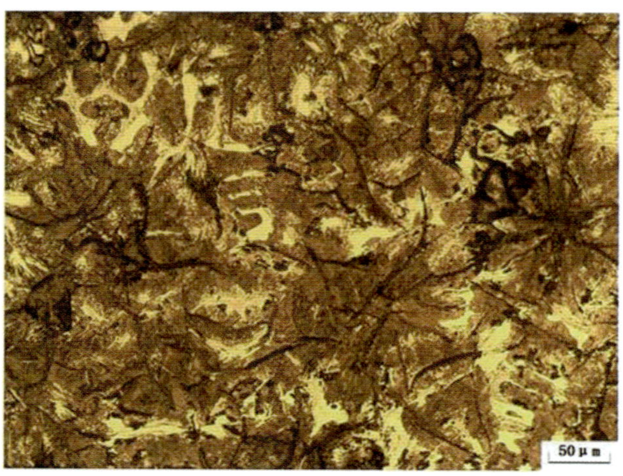

Figure 13: A grey cast iron object excavated from the Chengcun site, Fujian province.

Conclusions

This study provides new evidence for understanding the beginning of iron smelting in China. Most of the previously-known iron objects of the period between the late Western Zhou Dynasty (the Western Zhou Dynasty, 1050-771 BC) and the early Spring and Autumn period were excavated in the region at the junction of the Henan, Shanxi and Shaanxi provinces, in the middle reach of the Yellow River. This suggests that this region might be one of the earliest centres of iron smelting technology in China and deserves further archaeological research. As early iron products were

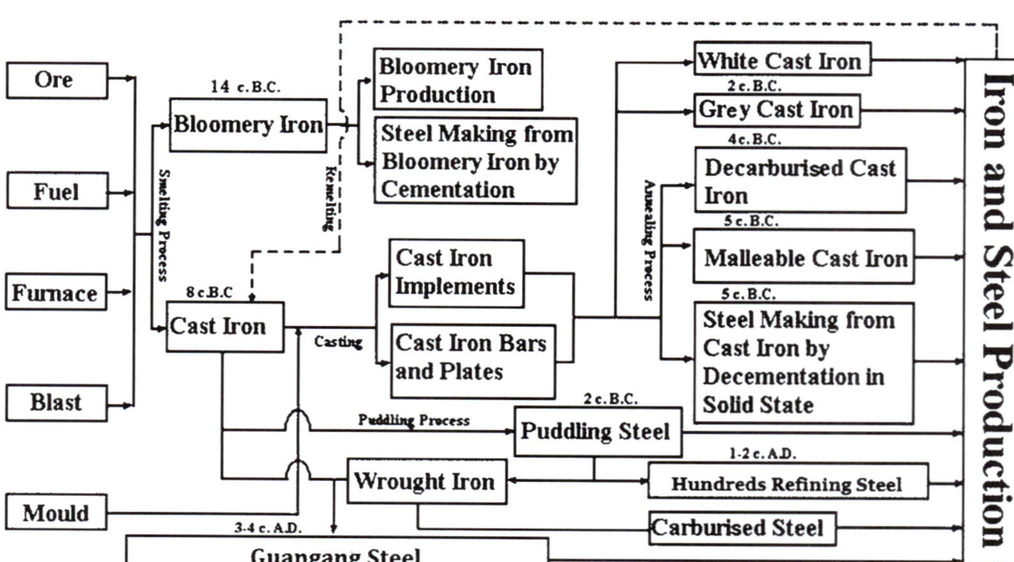

Figure 14: Scheme of iron and steel production techniques in ancient China.

also discovered in the upper reach of the Yellow River and in Xinjiang, appropriate attention should also be paid to the relationship between these two areas in terms of the origin of iron smelting. New findings at the Mogou site may be able to push back the date of the earliest use of artificial iron in China from the 9th century BC to the 14th century BC and provide significant information for the discussion regarding the origin of iron metallurgy in China.

This paper also provides new evidence for understanding the wide use of cast iron and of steel made from cast iron. From the 8th-7th century BC, the Chinese began to use cast iron, and cast iron objects were used to a larger degree from about the 6th century BC onwards. Around this time, the annealing processes were developed to produce decarburised cast iron, malleable cast iron, and decarburised steel. During the 2nd century BC, puddled steel objects appeared in Central China and soon the technology was widely spread into neighbouring areas. Coal was used as fuel for mould baking at the iron foundry sites of the Han Dynasty, but not for iron smelting until the 11th century AD.

Acknowledgements

This research was supported by NSFC (No.51074010) and the State Administration of Cultural Heritage of China. We are especially grateful to Professors Ko Tsun, Sun Shuyun, Li Yanxiang, and Mei Jianjun from the University of Science and Technology Beijing, and Professors Chen Tiemei and Wu Xiaohong from Peking University for their useful supervision, advice and help with the research.

References

Archaeology Team of Henan Province, 1962. *Gongxian Tieshenggou*. Beijing: Cultural Relics Press.

Bai Chongbin, 1994. The examination of iron swords excavated from M2 of the village of Yimen, Baoji City. *Cultural Relics*, 9, pp.82-5.

Bai Yunxiang, 2005. *Archaeological studies on the iron objects of Pre-Qin and two Han Periods*. Beijing: Science Press.

Bronson B., 1999. The Transition to Iron in Ancient China. In: V.C. Pigott eds. *The Archaeometallurgy of the Asian Old World*. Philadelphia: University of Pennsylvania Museum of Archaeology and Anthropology, pp.177-98.

Chen Jianli, 2007. Metallographic Experiment Study on Ironware in the Warring States Period Excavated in Linzi, Shandong. In: Shandong Provincial Institute of Cultural Relics and Archaeology eds., *Linzi Qi Tombs, Volume 1*, Beijing: Cultural Relics Press, pp.117-8 and 489-91.

Chen Jianli and Han Rubin, 1999. Metallographic Study on Iron Artifacts Excavated from Tomb of Chu Precinct of the Western Han Dynasty in Xuzhou. *Cultural Relics*, 7, pp.84-91.

Chen Jianli and Han Rubin, 2007. *The iron and steel technology of Central Plain and Northern China during Han and Jin Periods*. Beijing: Peking University Press.

Chen Jianli and Ma Qinglin, 2009. Metallographic and AMS-14C dating studies on the iron objects excavated from Gansu Province. In: Chinese Academy of cultural heritage eds., *Scientific and technological research on cultural heritage, Volume 6*. Beijing: Science Press, pp.1-13.

Chen Jianli, Han Rubin and Wan Xin, 2003. The Manufacture Technique of Ferrous Artifacts Excavated From Xianbei Gravesite at Lamadong Village of Liaoning Province. In: P. Jett, J.G. Douglas, B. McCarthy and J. Winter eds. *Scientific Research in the Field of Asian Art: Proceedings of the First Forbes Symposium at the Freer Gallery of Art*. London and Washington DC: Archetype Publications, pp.70-8.

Chen Jianli, Han Rubin, Tsutum Saito and Imamura Mineo, 2005. The Development of Iron and Steel Technology of Ancient Northeast China – View from the Metallurgical Studies of Ferrous Artifacts. *Northern Cultural Relics*, 1, pp.17-28.

Chen Jianli, Huang Quansheng, Li Yanxiang and Han Rubin, 2008a. Metallographic Experiment Study on Ironware Excavated in the Ruins at Kele Township, Hezhang County. In: Guizhou Provincial Institute of Cultural Relics and Archaeology eds., *Excavation Report in 2000 in the Ruins of Kele Township, Hezhang County*. Beijing:Cultural Relics Press, pp.195-206.

Chen Jianli, Chen Tiemei, Wang Jianping, Han Rubin, Sakamoto Minou and Imamura Mineo, 2008b. Age Determination of AMS-14C of Ironware Excavated at Pujindu, Yongji, Shanxi. In: Thomas Lawton eds., *New Frontier in Global Archaeology: defining China's Ancient Traditions*. West Palm Bch: American AMS Foundation for the Arts, Science and Humanities, pp.247-262.

Chen Jianli, Yang Cong, Zhang Huanxin and Lin Fande, 2008c. Metallographic Experiment Study on Ironware Excavated in the Ruins of a Han City at Chengcun, Wuyishan, Fujian Province. *Cultural Relics*, 3, pp.89-96.

Chen Jianli, Yang Junchang, Sun Bingjun and Pan Yan, 2009. Manufacture technique of bronze-iron bimetallic objects found in M27 of Liangdaicun Site, Hancheng, Shaanxi. *Science in China Series E – Technological Sciences*, 52, pp.3038-45.

Chinese Archaeometallurgy Study Group at Beijing University of Iron and Steel Technology, 1978. *A Brief History of Metallurgy in Ancient China*. Beijing: Science Press.

Cook, A.C., Wadsworth, J., Southon, J.R. and van der Merwe, N.J., 2003. AMS Radiocarbon Dating of Rusty Iron. *Journal of Archaeological Science*, 30, pp.95-101.

Craddock, P.T., Wayman, M.L., Jull, A.J.T., 2002. The Radiocarbon Dating and Authentication of Iron Artifacts. *Radiocarbon*, 44, pp.717-32.

Creswell, R.G., 1991. The Radiocarbon Dating of Iron Artifacts Using Accelerator Mass Spectrometry. *Journal of the Historical Metallurgy Society*, 25, pp.76-85.

Creswell, R.G., 1992. Radiocarbon Dating of Iron Artifacts. *Radiocarbon*, 34, pp.898-905.

Eylon, D., 2002. On the Radiocarbon Dating of Iron. In: G.-H. Kim, K.-W. Yi and H.-T. Kang eds. *Proceedings of the Fifth International Conference on the Beginnings of the Use of Metals and Alloys*. Seoul: Korean Institute of Metals and Materials, pp.161-8.

Guo Moruo, 1976. *Chinese Historical Manuscript, Volume 1*. Beijing: People Press.

Han Rubin, 1998. A Metallographic Study on Early Iron Objects of China. *Cultural Relics*, 2, pp.87-96.

Han Rubin, 2000. Study on the Iron Objects Excavated from Tianma-Qucun. In: Zou Heng eds. *Tianma-Qucun 1980-1989*. Beijing: Science Press, pp.1178-80.

Han Rubin and Duan Hongmei, 2009. An early iron-using centre in the ancient Jin state region (8th-3rd century BC). In: Jianjun Mei and Thilo Rehren eds. *Metallurgy and Civilisation: Eurasia and Beyond. Proceedings of the Sixth International Conference on the Beginnings of the Use of Metals and Alloys (BUMA VI)*. London: Archetype, pp.99-106.

Han Rubin and Ko Tsun, 2007. *The History of Chinese Science and Technology: Mining and Metallurgy*. Beijing: Science Press.

Han Rubin, Jiang Tao and Wang Baolin, 1999. Study on the iron blade of a bronze object unearthed at tombs of Guo State. In: Henan Provincial Institute of Cultural Relics and Archaeology, Archaeological Team of Shanmenxia City eds. *The Guo State Tombs in Sanmenxia*. Beijing: Cultural Relics Press, pp.559-73.

Henan Provincial Archaeology Institute, 1991. Excavations of the Iron-making Site of the Han Dynasty at Wangfangzhuang, Beiguan, Nanyang. *Huaxia Kaogu (Huaxia Archaeology)*, 1, pp.1-110.

Hua Jueming, 1999. *The Chinese ancient metal – the civilization created by copper and iron*. Zhengzhou: Daxiang Press.

Li Jiannong, 1962. *Historical Economy Manuscript of Pre-Qin and Han*. Beijing: Zhonghua Shuju Press.

Li Jinghua and Chen Changshan, 1995. *Iron Smelting of Han Dynasty at Nayang*. Zhengzhou: Central China Ancient Books Press.

Li Zhong, 1975. The Development of Iron and Steel Technology in Ancient China. *Acta Archaeological Sinica*, 2, pp.1-22.

Li Zhongda, Wang Suying, Su Rongyu, Shi Yongshi, 1996. Metallographic Study on the Iron and Steel Objects Excavated from Yanxiadu. In: Hebei Provincial Institute of Cultural Relics eds. *The Lower Capital of the Yan State*. Beijing: Cultural Relics Press, pp.881-95.

Optical Microscopy Laboratory at Beijing University of Iron and Steel Technology, 1980. Microscopic Examination of Metallic Objects Excavated from a Han Tomb (113 B.C.) at Mancheng. In: Hebei Archaeology Institute of China Social Science Academy, Cultural Relics Management Department of Hebei Province eds. *Han Tombs at Mancheng*. Beijing: Cultural Relics Press, pp.369-76.

Song Wei, Li Xiuhui and Han Rubin, 2009. The research of the relationship between the manufacture technique and the material of the Cangzhou Lion and its corrosion. *Scientific and technological research on cultural heritage, Volume 6*. Beijing: Science Press, pp.14-25.

Tang Jigen, 1993. The origin of iron smelting in China. *Archaeology*, 6, pp.556-65.

Tong Shuye, 1955. Reviews to Hu Shi's Historical Method based on the beginning use of iron in China. *Journal of Literature, History & Philosophy*, 2, pp.1529-32.

Tylecote, R.F., 1992. *A History of Metallurgy*. London: The Metals Society.

van der Merwe, 1969. *The Carbon-14 Dating of Iron*. Chicago: University of Chicago Press.

Wagner, D.B., 2008. *Science and Civilisation in China: Volume 5, Chemistry and Chemical Technology. Part 11, Ferrous Metallurgy*. Cambridge: Cambridge University Press.

Xia Xiangrong, 1980. *The Minning and Metallurgy History of China*. Beijing:Geological Press.

Yang Kuan, 1955. The invention of ancient Chinese iron smelting technology and development. *Journal of Literature, History & Philosophy*, 2, pp.1525-9.

Yang Kuan, 1982. *The Development of Iron and Steel Technology of Ancient China*. Shanghai: Shanghai People Press.

Ye Jun, 1975. The Study on Iron and Bronze Tools Excavated from Tonglushan Site. *Cultural Relics*, 2, pp.19-25.

Zhang Binglin, 1925. The Transition of Bronze and Iron Usage. *Huayang Guozhi*, 2, pp.1-10.

Zhao Huacheng, 1996. The discovery of iron artifacts before the mid-5th century BC and its related problems. In: School of Museology, Northwestern University eds. *The Researches of Archaeology and Cultural Relics*. Xi'an: Sanqin Press, pp.289-300.

Zhengzhou Museum, 1978. Excavations of the Iron-making site at Ancient Guxingzhen, Zhengzhou. *Cultural Relics*, 2, pp.28-43.

Authors' addresses

- Chen Jianli
  School of Archaeology and Museology, Peking University, 100871, Beijing, China
  Email: jianli_chen@pku.edu.cn
- Han Rubin
  Institute of Historical of Metallurgy and Materials, University of Science and Technology, Beijing, 100083, Beijing, China, Email: hanrubin@metall.ustb.edu.cn

Technology transfer across the Indian Ocean and South China Sea: a case study of the iron industry at Santubong, Sarawak

Brian Gilmour and Chris Doherty

Introduction

In the 1950's and 1960's the discovery of furnace remains during excavations in the coastal area of Santubong, Sarawak were interpreted as a hitherto unknown medieval iron smelting industry. Although this interpretation was later challenged a recent re-appraisal of fragmentary remains of the furnaces, backed up by some fieldwork, has shown the earlier conclusion that this was an iron smelting site (or series of sites) was substantially correct, even if relatively little was then understood about this industry.

Background

During his time as curator of the Sarawak museum in the 1950's and 1960's Tom Harrisson conducted major excavations at the coastal site of Santubong, and at neighbouring sites on the Santubong peninsula, which is dominated by the 810m high Santubong Mountain (Figs. 1 and 8). On the basis of his findings he argued that iron smelting was carried out here on a large scale, producing iron destined for export, probably to China and/or India (Harrisson and O'Connor 1969). Santubong has been identified as the centre of a region or state referred to as *Po-ni* by Chinese sources with a number of references to it over several hundred years following its first mention in the 10th Century CE. This, together with archaeological ceramic remains, suggests that this industry operated approximately from the 10th to 14th centuries, this evidence having been more recently summarised (Christie, 1988, 79).

Since Harrisson's results were published Jan Christie has argued that the reasoning behind some of his interpretations was flawed and that the evidence to hand suggested intead that this industry was more likely to have been carried out on a much smaller scale, not for export, but rather to satisfy a more limited local demand for iron (Christie, 1988). This re-interpretation was based largely on two assertions; first, that the demand for this iron was unlikely to relate to an external market because there are no mentions of the export of iron to China in the later medieval Chinese references to trade from *Po-ni*, and second, that Harrisson was mistaken in some of his archaeological interpretations that the waste material recovered by him suggested a large-scale industry.

Christie suggested that much of the material interpreted by Harrisson as iron smelting slag was more likely to have been a degraded or concreted form of lateritic iron ore mistakenly identified as slag by Harrisson, although the material itself does not seem to have been re-examined at that stage.

However a recent re-examination (by Chris Doherty) of the material – recovered during Harrisson's investigations and stored at Sarawak museum – suggests that while Harrisson may have been incorrect in some of his detailed descriptions and interpretations, his conclusion that this was a large scale iron producing centre would appear, for the most part, to have been correct.

Christie's assertion that the residual material found by Harrisson suggests the use of a primitive pit form of furnace, little more than a (bowl shaped) scoop in the ground, powered by an air blast supplied by (traditional Chinese) piston bellows via clay nozzles or tuyeres, is clearly not supported by the material surviving from Harrisson's excavations. For instance, although the tapering clay tubes found in large quantities were indeed tuyeres (as pointed out by Christie), and not crucibles (as thought by Harrisson), the material in which they were embedded represented the remains of clay walls belonging to a more sophisticated type of furnace, in this case a distinctive variety of bloomery furnace only known in southern and south-east Asia, with some of the tuyeres having been re-used in the construction of the furnace walls.

Some of the material – waste furnace debris etc – recovered by Harrisson, along with his archive, is now stored at the Sarawak Museum (of which he was Curator

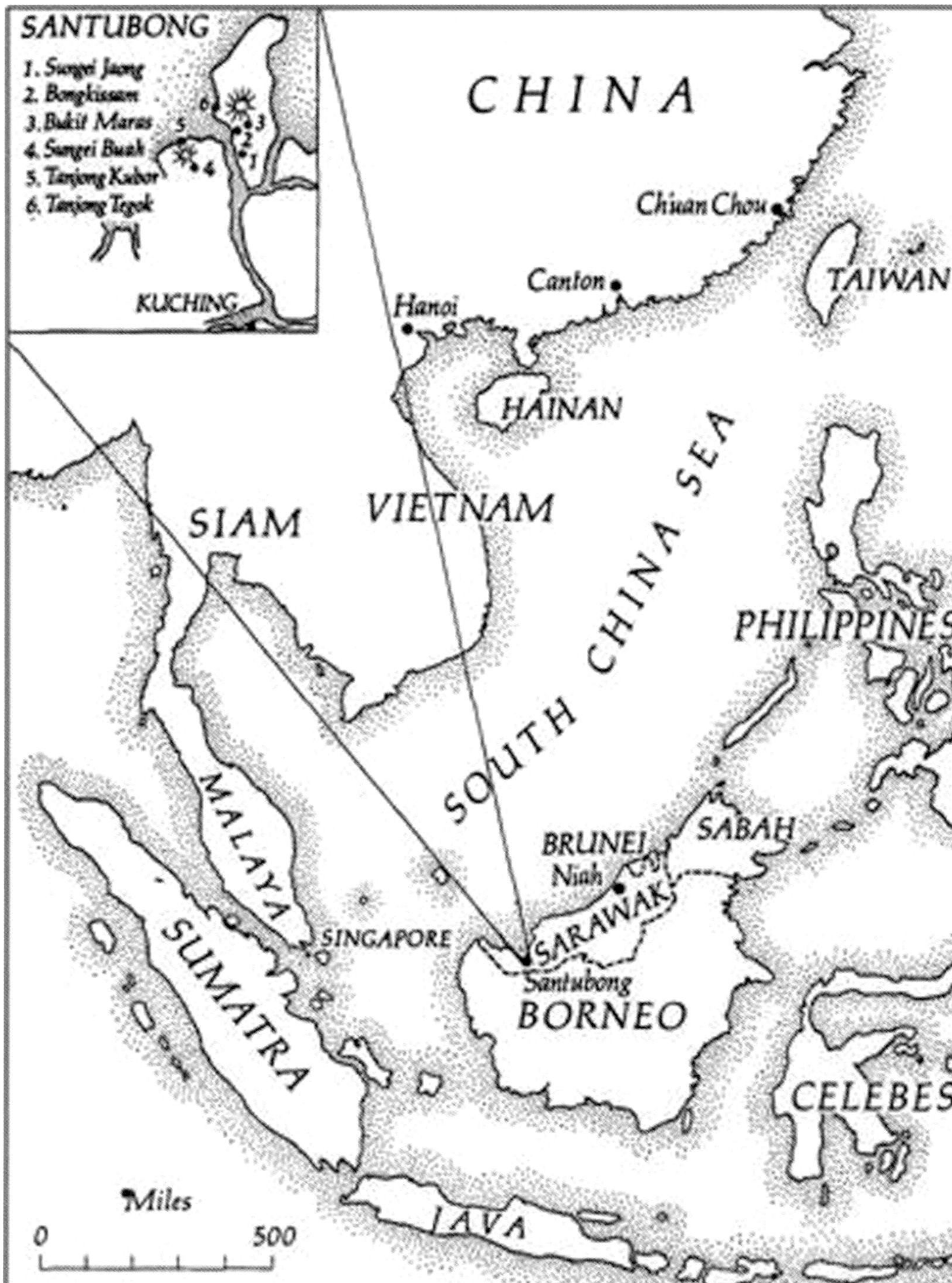

Figure 1: Location map for Santubong on the coast of western Sarawak (after Chêng 1967).

1947-1966) in nearby Kuching. The main result of the present re-examination and re-assessment of Harrisson's material has been to suggest that he may have been correct in interpreting much of it as slag, and that this may well have been present in large quantities, as he had reported. Apart from tap slag, itself not a by-product of pit furnaces, much of the debris also consisted of furnace fragments which suggest that the type of furnace represented on the site may have been a highly developed form similar to a large scale industrial type recently identified as being in use in Sri Lanka between the 7th and 11th centuries AD, but also related typologically to the *tatara* iron smelting furnace of medieval and later Japan.

We also need to re-examine the assertion that, during the 10th to 14th centuries when production at Santubong was evidently at its height, the (apparent) lack of exports of iron to China, as indicated by later medieval Chinese sources, suggests that the iron produced at Santubong was unlikely to have been produced for foreign consumption, as this seems to be based on relatively little negative evidence. There may have been no direct demand from medieval China for iron smelted at Santubong but that does not mean that there was no demand from elsewhere for this iron, or that there were no imports of iron into China from elsewhere. What we do seem to find from Chinese sources, as might perhaps be expected, is that there was some demand for more a more specialised form of iron, mainly steel. The re-appearance of just such a type of iron imported from the West (from Iran and/or India) – known as *bin* (or *pin*) – occurs in the 10th century and from then on is well known. This is described in 1368 as *bin iron* ('steely iron'):

'It is produced by the Western Barbarians. Some [types of it] have a special self-patterning, while others have a sesame-seed or snowflake patterning. When a knife or sword [made of it] is wiped clean and treated with 'gold-thread' alum [or a similar form of etchant] [the pattern] appears. Its value is greater than silver...' (Wagner 2008, 270)

In this case the steel artefacts being referred to are made of the type of ultra-high carbon crucible steel well known in medieval Iran (the probable origin of most or many of the these steel artefacts) as *fuladh* (or *fulad/pulad*) although India (and earlier Sri Lanka) were major producers either of the steel or its feed-stock (Allan and Gilmour, 2000). In India the raw steel ingots were known from the late 18th century as *wootz* (a European corruption of a south Indian, Telegu word *ukku* pronounced *wukku*, perhaps specifically for this special type of steel). Thus in any case even if iron was not being imported into China directly from Sarawak during the Middle Ages that does not necessarily mean that it was not part of a thriving export trade to other places during this time, and one with which Chinese merchants are likely to have been involved. When we start to look at what kind of well established trade network existed in this whole region in the period up to the 10th century we can perhaps see that a larger scale Santubong iron industry serving the export trade is not only perfectly logical, but also that it may have been founded as part of this trade.

Iron and steel: early production and trade in south-east Asia

We know that metals, like other traded goods, did travel widely in antiquity and in the early Middle Ages, often in the form of pigs or cakes (i.e. ingots), as well as finished objects. Seric (Chinese) iron moved westwards, while Iranian brass and steel, Indian steel (probably the crucible made steel later known in the West as *wootz*) and various other metallic products are listed as tribute in the Chinese dynastic histories. Arabic sources also record the use of *hadid al-Sini* ('Chinese iron') for mirrors. If the identification of the central Asian region of *An-hsi*, mentioned in early Chinese accounts, as Parthia is correct, then there were also seaborne links between Iran and China as early as the 3rd century CE. Once this trade route had become established, transport by ship would have been a more convenient and less costly way to carry heavy materials like the 'steely iron' mentioned in the Chinese accounts.

Under the Sasanian dynasties of the 5th and 6th centuries CE, Iran was a dominant maritime trading force at least as far east as Sri Lanka. The presence of a Nestorian Christian community at Qal'ah (the Syriac QLH) would also suggest the presence of Sasanian merchants somewhere on the Malay peninsula by the early 7th century, in a place which became an important trading centre in the medieval Islamic period.

It is perhaps significant that Qal'ah is mentioned by the Iraqi polymath Ya'cub al-Kindi, writing in the first half of the 9th century, as a producer of an 'ancient' (meaning high-quality) form of *fuladh*, a term he reserves for crucible steel. Of Kindi's three categories of 'ancient' (*'atiq*) or finest quality swords the best are said to be the Yemeni variety (Hoyland and Gilmour, 2006, 19 and 53), which we can probably assume means swords forged in the Yemen using Yemeni crucible steel. The origin of the second category of fine quality swords, the *Qala'i* is more difficult to place although it has been suggested that the name Qal'ah (Kalah) is a corruption of Kalang, an ancient port near Kuala Lumpur on the Malay Peninsula (Fatimi, 1964, 211-215). Also the presence of a Nestorian Christian community here would suggest the presence of Sasanian merchants in the early 7th century in a place which became an important trading centre on the long distance eastern trading route to China in the Islamic period (Whitehouse and Williams. 1973, 48).

Furthermore under the entry 'Qal'a' Yaqut, *Mu'jam al-Buldan* quotes from the account of a journey to China by Mis'ar ibn Muhalhil (*ca* CE 940):

> 'I returned from China to Kalah, which is the first country of India from the direction of China (in this case the Malay peninsula) and where the boats stop and go no further. In it is a great stronghold in which is the mine of Qala'i tin, found only in this stronghold, and in which are struck the Qala'i swords, which are the ancient Indian ones' (Hoyland and Gilmour 2006, 19 note 8).

Thus at present there seems to be no better candidate than Kalah/Kalang for the origin of the high quality *Qala'i* steel. This may also have been the origin, at least in part, of the steel imported into China during this period.

Sources such as Ya'qūb ibn Isḥāq al-Kindī's early 9th century sword treatise also mention the export of lesser quality *fuladh* from Sri Lanka to the sword-making centres at Khurasan and Fars in Iran, Mansura in southern Pakistan and the Yemen; Kindi also reports that *fuladh* was made there for local sword production. Evidence from archaeological investigations at Samanalawewa, unknown until recently, would suggest that Sri Lanka was a major centre for specialised, non-crucible iron and steel production between the 7th and 11th centuries CE (Juleff. 1998, p.150). The metal was smelted in wide-mouthed, wind-driven furnaces which relied on the dry monsoon season winds of that part of south-western Sri Lanka. Ancestral remains of the industry have been found, and its origins can be traced back (from archaeology and field survey) to as early as the 4th century BCE (Juleff, 1998, p.153).

The Sri Lankan industry appears to have reached its peak in the 9th century and to have ceased operating sometime during the 11th century, most likely a victim of the disruption caused by Chola invasions from southern India, which eventually brought about the end of the Sri Lankan Dry Zone civilisation under which this particular large-scale development of iron production had prospered (Juleff, 1998, p.36). Kindi's description suggests that crucible steel (probably the specialised steel of this region which became known much

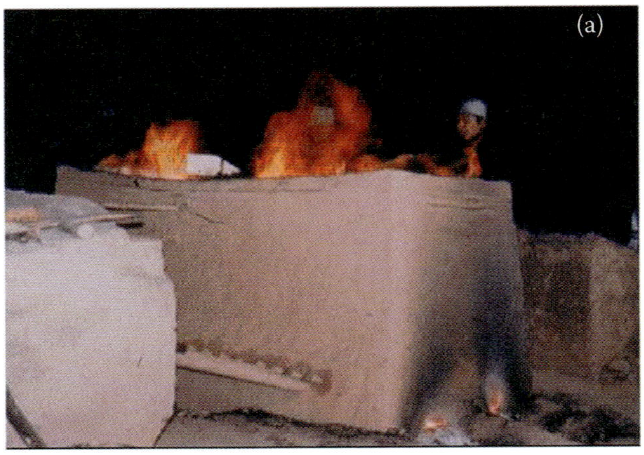

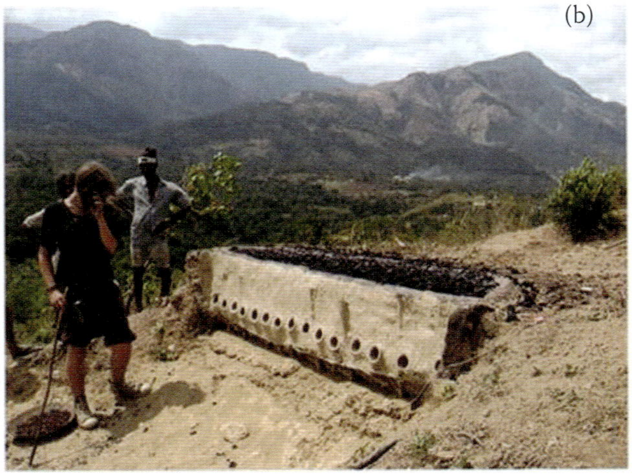

Figure 2: Showing (a) the still surviving 3.5m long Japanese *tatara* furnace being fired in January 1996, and (b) the more recently reconstructed medieval Sri Lankan, Samanalawewa furnace before being fired.

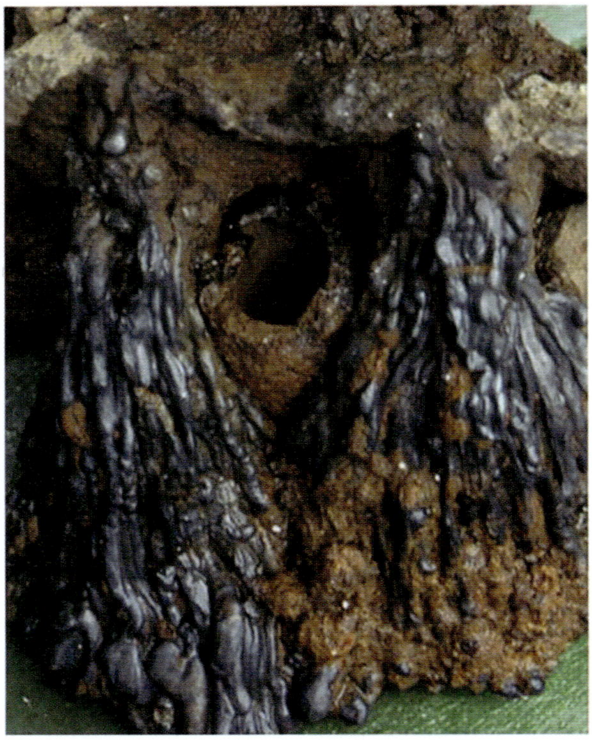

Figure 3: Fragment of the 'front' wall of one of a later (?17th century) furnace from Kuching, showing the solidified outflow of tap-slag from the subsequently blocked tuyeres, and note also the re-used vertically placed tuyere used to reinforce the furnace wall and around which the tap-slag has flowed down hill.

later to European observers as ***wootz***) had been produced in Sri Lanka as a parallel industry. Kindi makes it clear that this secondary crucible steel industry and the export of its products to Iran, probably Sind (modern south-east Pakistan) and the Mesopotamian region, was of long standing. Kindi's third category of fine quality swords, the Indian (***al-hind***) is more difficult to pin down to any particular location and may refer to both to the swords and (crucible) steel from a variety of sources. It has been argued, however, that Hindi in early Islamic times only signified the coasts and islands of India, as far east as Malaysia, rather than the mainland (Fatimi, 1963, p.122-24).

Interpretation and context of the Santubong iron industry

The furnace fragments from Santubong bear a remarkable similarity to the fragments of the distinctive wide-mouthed type found recently at Samanalawewa in Sri-Lanka, even down to the way the tuyeres from dismantled furnace walls have been re-used – placed at right angles (either vertically or horizontally) – in the construction of the new furnace wall. This at least suggests the presence at Santubong of an iron production similar to the major industry of Sri-Lanka, and possibly carried out on a similar scale. The presence of large quantities of Sung Dynasty (10th-14th Century) Chinese ceramics (as outlined by Christie) suggests that the port of Santubong, and possibly the iron industry as well, was most active in this period, at the same time and later than the Sri Lankan industry of Samanalawewa. The Sarawak/Borneo coastal region is likely to have formed part of the same extensive maritime trade network linking China, Malaysia and other parts of the 'East Indies', Sri-Lanka, India, Pakistan, Iran, Iraq and elsewhere in the Middle-East. One obvious possible eventual export destination is the Qala'i/Kalah/Kalang (crucible steel) sword making centre mentioned by Mis'ar ibn Muhalhil in the 10th century.

It may well be no coincidence that the traditional Japanese bath-shaped ***tatara*** form of iron smelting furnace (which appears to have increased steadily in size from its later medieval size, approximately 2m long, similar to the length of the Sri Lankan furnaces, to its present length of 3m+) is essentially like two Sri Lankan furnaces placed back-to back. This is explained by the fact that the Sri Lankan furnaces are wind powered from one side only, whereas the ***tatara*** furnaces are bellows powered from both sides (Fig. 2a). In each case the long side or sides of the furnace was supplied with a row of tuyeres placed fairly low down in the furnace wall. In the Sri-Lankan furnaces this part of the furnace wall, through which the blast was directed, and hence was the part most vulnerable to heat damage, was reinforced with re-used tuyeres placed both horizontally and vertically, but at right angles to their original position in the clay wall (Figs. 2b and 6). This arrangement and re-use of tuyeres is similar to that found on the Santubong fragments recovered by Harrisson (Fig. 3). Although it is impossible from the fragments alone to work out the exact form of the Santubong furnaces, none of which have yet been located, the dumps of furnace waste and slag at least indicate they were situated on the much gentler, lowermost slopes of Santubong Mountain, unlike the case in Sri Lanka at Samanalawewa.

At Samanalawewa the furnaces were wind powered, and situated on the steeper south-westerly facing hill-slopes, taking advantage of the very dry south-westerly monsoon winds that blow steadily for three months (June to August) each year in this part of southern Sri Lanka. However the air supply arrangement for the Santubong furnaces was clearly different and must have been bellows driven given that the Santubong furnaces were built on the much lower, southerly slopes of Santubong Mountain, in the lee of the Dry Monsoon winds which in any case are significantly less dry and steady than those which prevail in Sri Lanka at this time of year.

Although no actual in situ furnace remains have yet been identified, the form of the waste furnace fragments that have been recovered from Santubong suggest a very similar furnace design and operation to those excavated in Sri Lanka at Samanalawewa, even down to the way old clay tuyeres were re-used in the clay-built furnace front wall. These remains suggest a series of smelting sites scattered across the sheltered lower south-easterly slopes of Santubong Mountain. The Santubong fragments would also appear to be best interpreted or reconstructed as belonging to rectangular furnaces with a line of tuyeres near the base of at least one of the longer, roughly flat clay walls, similar in overall design to both the medieval Sri Lankan furnaces and traditional Japanese *tatara* furnace that produced a very inhomogeneous mixture of low carbon iron and medium to ultra-high carbon steel depending on the position within the furnace. A similar arrangement with the vertical placement of re-used tuyeres has also been found in fragments of furnace walls from post-medieval iron smelting furnaces from nearby Kuching (Fig. 3).

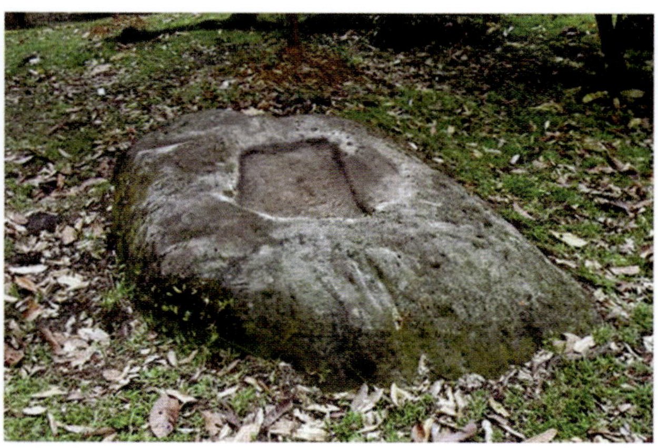

Figure 4: Boulder with a rectangular (approximately 30cm wide) trough cut into the top, the base and parts of the surrounds of which are stained a dark reddish brown, possibly from burning associated with the iron smelting here.

Unfortunately the hill-slopes at Santubong seem to be subject to continual erosion which may mean that in-situ furnace remains are hard to find, although a number of large boulders lying proud of the ground surface have been found with roughly rectangular, flat-bottomed trough-like features cut into them (Fig. 4). On their own these appear to be undatable and not necessarily connected with iron-working but if they do relate to the iron smelting carried out here they may have been working platforms perhaps for both ore preparation and bloom consolidation, although further work would be needed to establish this. Large quantities of tap slag, furnace wall debris, hammer-stones, later medieval Chinese ceramics and the like were reported to have been found both during Harrisson's investigations in the 1950's and 60's, and more recently during the investigation of a much more limited study area (approximately 200m east-west by 100m north-south) in 2006 (Buckley et al., 2007, Figs. 2 and 3).

Both the medieval Sri-Lankan iron smelting and the still surviving *tatara* iron smelting processes are highly developed forms of bloomery smelting. In the case of the long, narrow, wind-powered Samanalawewa furnaces, these were set into a south-west facing hillside to take advantage the dry monsoonal winds of that area, so that only one long clay wall was needed for the front. By contrast all four walls of the bellows-powered *tatara* furnaces were built of clay. Both the Sri-Lankan and *tatara* smelting were carried out in an annual cycle with the actual smelting only being carried out for part of the year, the rest of the time being taken up with preparation work; ore and furnace preparation, charcoal making, furnace building and so on.

Although none have yet been found it seems likely from the waste dumps that the Santubong furnaces were sited either on shallow terraces of flattish ground not far from the

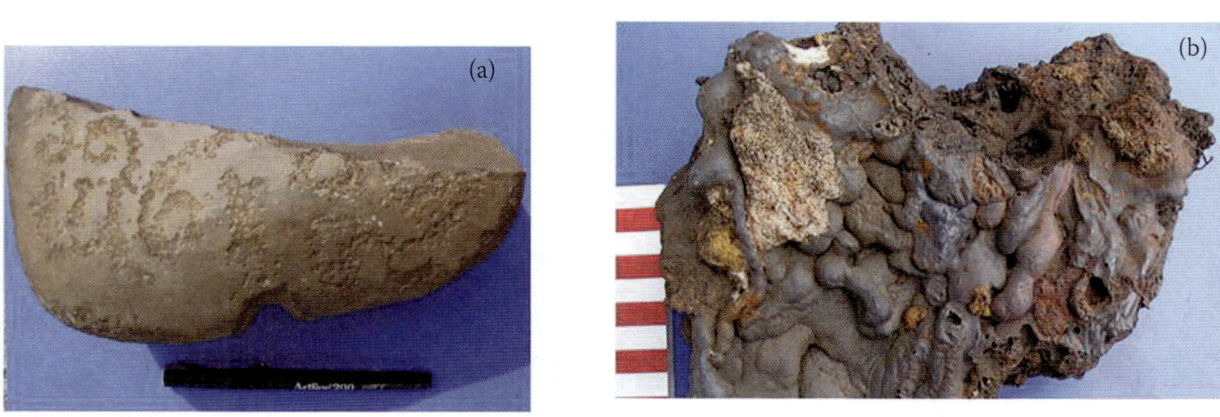

Figure 5: Typical smelting related debris found in the Santubong study area: (a) one of many fragments of decorated hammer stone (waisted for hafting) likely to be for ore preparation and, (b) one lump of a large quantity of tap slag found with structural furnace debris found in 2006 in a small study area on the lower south-west slopes of Santubong mountain (see Buckley et al., 2006).

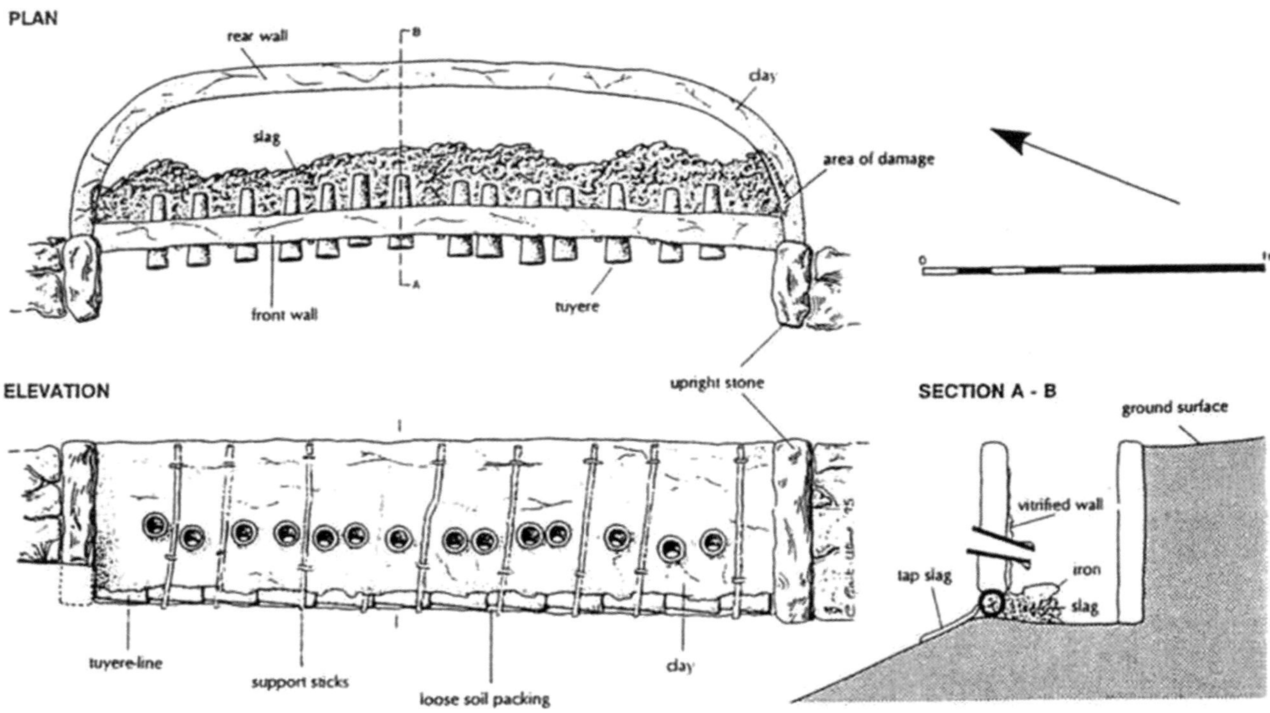

Figure 6: Reconstruction drawing of one of the medieval Sri Lankan wind driven iron smelting furnaces from Samanalawewa, set into the hill-slope here. (drawing courtesy Gill Juleff).

quayside, unlike the Samanalawewa furnaces which were built by cutting a shallow step shaped hole into the hill-slope then building the walls of the furnace so that the wide mouth faced outwards towards the prevailing wind. Thus, although the design of the Santubong furnaces were very similar in terms of wall construction to those from Sri-Lanka, rather than being wind-powered they appear to have been bellows-operated, presumably in much the same way as the surviving (and still working) NBTHK[1] *tatara* smelting furnace, (in Shimane Prefecture, western Honshu, south-west Japan; Kapp et al., 1987) itself built on a terrace on a hillside, but in this case placed at right angles so that one end of the furnace projects in the direction of the hill slope. This allows it to be built like a double-thickness Sri-Lankan furnace with a line of tuyeres built into the each of the long walls, near the base. In this case the air is fed into the tuyeres by means of bellows.

The furnace remains from Santubong suggest a highly developed form of medieval iron smelting similar to that found recently in Sri-Lanka (Fig. 6), but perhaps closer to the *tatara* process. The technological approach was certainly related to both the Sri-Lankan and Japanese industries, the knowledge of which travelled across the South China Sea as part of the working of a very widely dispersed, but highly developed trade network. From what direction this technology reached Santubong is not yet clear. The Santubong industry is perhaps best seen as a regional Malay variant of what developed, possibly from a Sri-Lankan progenitor, as a distinctive south-east Asian type of long rectangular bellows-driven iron smelting furnace. The best parallel and closest adaptation of this industry both geographically and in terms of technology would appear to be the bellows-driven rectangular furnaces still operating in 19th century Cambodia (Fig 7).

The large quantities of Chinese ceramics and correspondingly large volume of slag and furnace remains found at Santubong as well as the siting of this iron smelting indicate that this was likely to be an outward looking or export industry during the 10th to 14th century, much as Harrisson envisaged. Santubong Mountain is joined to the mainland by an isthmus on it south-eastern side (Fig. 1). The inlet on the south side of the isthmus forms a good sheltered anchorage area, one of the few along this coast, the mountain itself forming a very prominent navigational point for this part of the coast facing north into the South China Sea. This and the close proximity of both ore and fuel may have dictated why the iron smelting discovered by Harrisson was carried out here, on either side of the inlet, in the lea of the mountain in what is otherwise a damp, humid, insect-infested area.

A big change appears to have happened sometime about the 14th century with the possible end of the iron export trade indicated by the virtual disappearance of the Chinese export ceramics here. This seems to have coincided with the transfer of iron smelting further inland, in the vicinity of Kuching or beyond and may indicate a move into a smaller scale, more locally based iron industry from then on.

As Christie forcefully pointed out, Harrisson was clearly wrong in his suggestion that these were furnaces employed in crucible steel (*wootz*) making, but there is really no reason why the directly smelted or bloomery iron/steel from these furnaces should not have been exported. This iron could, in

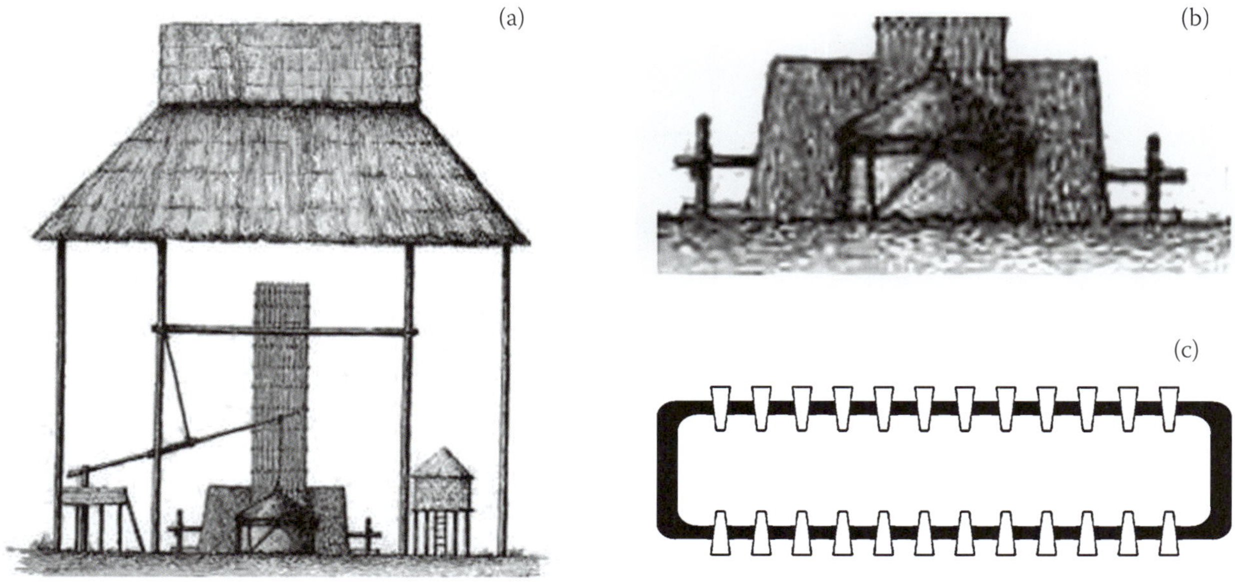

Figure 7: (a) 19th century Cambodian form of rectangular bellows driven furnace (after Maura, 1883) with (b) a slightly enlarged view of the furnace, and (c) a simplified ground plan of this type of furnace, possibly similar to the form we might expect for the medieval Santubong iron smelting furnaces.

part, have formed the feedstock for a secondary crucible steel industry either nearby or much further afield – further along one of the trade routes. Although Harrisson does not seem to have realised that the site of Santubong is situated on the edge of an iron mountain, a huge natural outcrop of lateritic iron ore, this fact would probably have soon been picked up by one of the traders plying the sea routes in this region, especially as this is such an obvious landmark on the coast of Sarawak (Fig. 8), and one with good anchorage. With abundant local supplies of wood for charcoal, as well as enriched deposits of lateritic iron (Fig. 9), there seems no reason why the iron smelting industry might not have operated on a fairly large scale during the 10th to 14th century period as suggested by the Chinese pottery found here.

Although Christie was right to point out various detailed points on which Harrisson was clearly wrong (for instance he mistook tuyeres, or air delivery nozzles, for crucibles), it seems highly likely that his conclusion, that this was a comparatively large and sophisticated iron industry, aimed at the export trade, was correct, even if he was right for the wrong reasons. There is no particular reason why the destination of this trade would have to be China, as Christie's argument would appear to suggest. It could be, for example, related to the elusive steel-making site at Qala/Kalah, famed for its quality in Kindi's day and apparently still going in the 10th Century CE, or to other production sites, perhaps (as argued by Fatimi) loosely referred to by Arab traders/travellers as Hindi or Indian, in much the same way as the East Indies were much later. One might also suggest that other markets further west along this trade route might have existed and prospered, especially once the Sri-Lankan industry was badly disrupted in the 11th century.

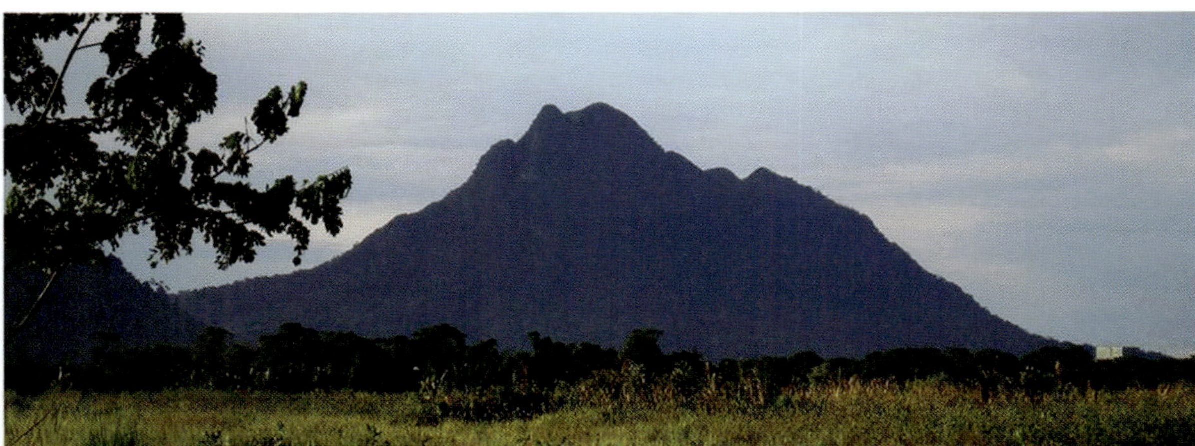

Figure 8: Santubong Mountain (810m), as seen here viewed from the south-east would have presented just as dramatic and convenient a landmark when approached across the South China Sea.

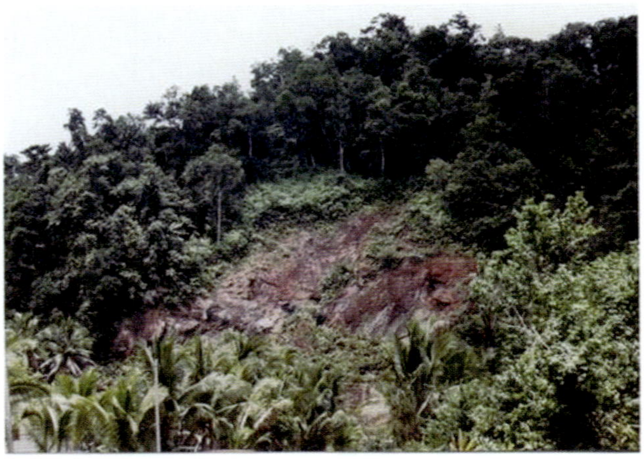

Figure 9: Iron-rich nature of the underlying ground is evident here on the lower slopes of Santubong Mountain where erosion has revealed the lateritic subsoil.

Priorities for future work on the iron industry at Santubong

Santubong could well turn out to be a pivotal site in understanding the spread and development of iron-working in the region of the South China Sea and much further afield, both in terms of how the industry was organised and operated, and also how it fitted in with the regional or international medieval trade network. Arabic and Chinese written sources may well provide more information and certainly need more careful consideration. Most of all the surviving remains of the Santubong iron industry found by Harrisson needs further thorough re-examination, much more (geophysical and land) survey work, backed up by excavation work badly needs to be carried out before this industry can be properly identified, studied and understood.

We need to try and find out what iron was being made, on what scale, and how it was used as well as why and when this took place, and how the industry was organised, all of which should be possible. In short we still need to work out the main parameters and framework for this industry. There is every reason to suppose that the results may be just as spectacular and significant as those from the Samanalawewa iron making industry in Sri-Lanka, and possibly more so given the coastal siting with good anchorage, close proximity to massive quantities of high grade iron ore and a plentiful nearby source of wood for charcoal. In short the site at Santubong has enormous potential for the rediscovery of an important early iron industry, as Harrisson seems to have sensed without really being able to work out why.

It is hoped that the work carried out so far might be seen as the first stage of a Santubong Iron Project which should undertake a complete reassessment of the all the existing material – excavated, written and other evidence – that might relate to iron smelting here. This will look at how and why the industry developed, how it functioned and fitted into the local and wider regional iron industry, economy and trade patterns. Further fieldwork and scientific analysis should lead to a better understanding of the development and nature of Santubong iron and its place in the wider iron industry of the region.

Note

1. The Nihon Bijutsu Token Hozon Kyokai or Society for the Preservation of the Japanese Art Sword.

References

Allan, J. and Gilmour, B. 2000: *Persian Steel: The Tanavoli Collection.* (Oxford University Press).

Allen, S.J. 1998: 'History, archaeology, and the question of foreign control in Early-Historic period Peninsula, Malaysia', *International Journal of Historical Archaeology*, 2, 261-289.

Buckley, R., Gnanaratnam, A., Doherty, C., Beavitt, P., and Beavitt, W. 2006: Archaeological excavations at Sungai, Santubong, Kuching, Sarawak (University of Leicester Archaeological services: unpublished report).

Chêng, T. 1968: *Archaeology in Sarawak.* (Cambridge: Heffers).

Christie, J. 1988: 'On Po-Ni: the Santubong sites of Sarawak', *The Sarawak Museum Journal*, 77-87.

Fatimi, S. 1963: 'Two letters from the Maharaja to the Khalifa', *Islamic Studies*, 2, 121-140.

Fatimi, S. 1964: 'Malaysian weapons in Arabic literature', *Islamic Studies*, 3, 199-233.

Harrisson, T. and O'Connor, S. 1969: *Excavations of the pre-historic iron industry in West Borneo.* (New York: Cornell University).

Hoyland, R. and Gilmour, B. 2006: *Medieval Islamic swords and swordmaking: Kindi's treatise "On swords and their kinds"* (edition, translation and commentary). (Oxford: Gibb Memorial Trust).

Juleff, G. 2009: 'Technology and Evolution: a root and branch view of Asian iron from protohistoric Sri Lanka to Japanese steel', *World Archaeology*, vol. 41, no. 4, 557-577.

Juleff, G. 1998: *Early iron and steel in Sri Lanka: a study of the Samanalawewa area.* (PHVZ:Mainz).

Kapp, L., Kapp, H. and Yoshihara, Y. 1987: *The craft of the Japanese sword.* (Kodansha: Tokyo, New York and London).

Moura, J. 1883: *Le Royaume du Cambodge.* (Paris: Ernest Leroux).

Wagner, D. 2008: *Ferrous Metallurgy.* Volume 5, Part II, of Joseph Needham's *Science and Civilisation in China.* (Cambridge University Press).

Whitehouse, D. and Williams, A. 1973: 'Sasanian maritime trade', *Iran*, 11, 29-49.

Author's address

- RLAHA, University of Oxford, UK.

Iron and steel making in Ancient Korea inferred from the microstructure of iron artefacts

Jang-Sik Park

ABSTRACT: Iron objects from the Korean Three Kingdoms period (ca. 300-668 AD) have been examined for the purpose of understanding iron and steel making in ancient Korea. Varying significant microstructures that were identified within particular objects in this study include: 1) a core of white cast iron surrounded by a thin steel layer of pearlite; 2) a high carbon steel interior with its carbon level gradually decreasing to that of ferrite near the surface; 3) spheroidised ultra-high carbon steel used in thin plate type objects requiring no special hardening; 4) a surface layer of white cast iron enclosing the core where the microstructure continuously varies from ultra-high carbon steel to ferrite as a function of the distance from the surface; 5) thin layers of ultra-high carbon steel embedded in the background of white cast iron; and 6) objects of almost pure iron covered with a thin surface layer of pearlite. These microstructures will be discussed in detail, with the purpose of inferring the associated engineering processes.

Keywords: Ancient Korea; iron and steel making; microstructure; iron artefacts.

Introduction

Ancient Korea is famous for manufacturing iron and steel products, as evident in the account quoted by Barnes (1999) from the Chinese history book *Weizhi*, and also from the variety and abundance of iron objects excavated. It was written in *Weizhi* that the iron articles produced near the south central coastal region were exported to the Chinese commandries occupying the northern peninsula at the time and to a few other Korean polities, as well as to Japan.

It seems that iron technology appeared in Korea during the early 3rd century BC under Chinese influence (Kim, 1992), and was adapted into a unique technological approach based on the smelting of cast iron and subsequent treatments carried out on cast iron to produce iron and steel products. The appearance of iron was a critical moment in Korean history, when many indigenous polities rose to prominence in the territories covering the Korean peninsula and southern Manchuria. They were predecessors of the early states that, with the spread of iron use, would eventually develop into Koguryo in the north and Paekche, Silla and Kaya in the south, as shown in Fig. 1 (Eckert, et al., 1990). This era is currently referred to as the Korean Three Kingdoms Period (ca. 300-668 AD), denoting the first three states that survived longer than Kaya before they were all merged into Silla, establishing in 668 AD the Unified Silla. The reference in *Weizhi* mentioned above was made concerning the Kaya iron industry, indicating its crucial role in the production and supply of iron. Throughout the entire period of their existence, the four kingdoms were in constant rivalry, and the iron industry must have been of extreme importance in their very survival. It is from this period that archaeologists have excavated the most abundant material evidence concerning ancient iron production in Korea.

Numerous archaeological excavations conducted in Korea to date have revealed that cast iron was smelted from the early period of iron use, and, within the archaeological record, iron and steel objects far exceed cast iron objects in both frequency and abundance (Kim, et al., 1994; Choi, et al., 2001). Nevertheless, little is known of the processes employed in iron and steel making. In principle, iron and steel can be produced indirectly from treating cast iron in a variety of decarburisation processes, or directly from a bloomery process, followed by the subsequent steel making by carburisation (Rostoker and Bronson, 1990; Tylecote, 1992; Wagner, 1996). It is speculated that bloom iron was smelted in ancient Korea along with cast iron (Yun, 1984). No convincing evidence, however, has been proposed either to support or refute the practice of bloomery smelting. It is necessary, therefore, to understand how the raw material for iron and steel products was produced before the iron technology established in pre-industrial Korea can be properly characterised. In his studies (Choi, et al., 2001; Park, 2003, 2005; Park and Lee, 2003; Park and Jung, 2004; Park and Mark, 2005) on iron objects from the Korean Three Kingdoms period, the author found important microstructures reflecting the processes applied for the control of the carbon content within the objects produced. This article will review these microstructures to outline steel making technologies practiced in Korea.

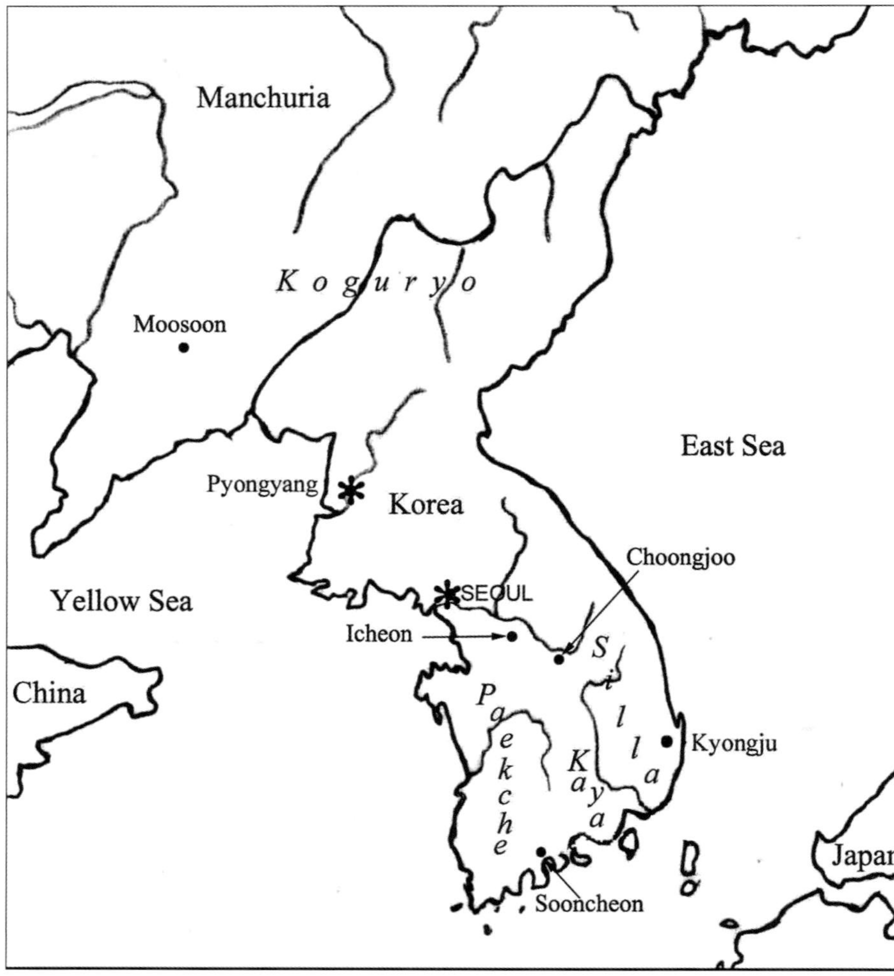

Figure 1: Political map of the Korean peninsula in the Korean Three Kingdoms Period, when the peninsula was occupied by the four rival kingdoms of Koguryo, Paekche, Silla, and Kaya. The cities indicate the sites from which the objects presented in this paper were excavated. Seoul and Pyongyang are the present-day capitals of South and North Korea.

Microstructure examination

One or more small samples were taken from each iron object for metallographic examination. The specimens were mounted and polished following standard metallographic procedures and then etched with a solution of 2 volume percent nitric acid in methanol. They were examined using an optical microscope and a scanning electron microscope (SEM). Chemical analyses using an energy dispersive spectrometer (EDS) attached to the SEM detected no other elements than iron (Fe) and carbon (C), and therefore EDS results are not presented in this paper. The carbon content in this article is based on weight fraction, and the micrographs presented are all obtained with the optical microscope, unless specified otherwise.

Figs. 2a and 2b show two arrowheads recovered from a former Koguryo military fortress (dated between 300 and 600 AD) located at Moosoon in Manchuria, as shown in Fig. 1 (Choi, et al., 2001). The arrows in the figures indicate the locations from which specimens were taken for microstructure examination. Figure 2c, a micrograph showing the structure of the object in Fig. 2a, shows white cast iron and pearlite structures in the white and dark regions, respectively. The dark layer, corresponding to the exterior of the object, has a little proeutectoid ferrite present only in the area near the surface, which indicates the increase in carbon content towards the interior. The internal region consists primarily of white cast iron of near eutectic composition, 4.3% carbon. The surface steel layer, which must have resulted from the decarburisation of white cast iron, indicates that the object was cast and then treated at elevated temperatures. Fig. 2d, showing the microstructure of the arrowhead in Fig. 2b, also contains regions of varying contrast. The dark area near the bottom centre, which corresponds to the core of the object, consists of pearlite and is surrounded by a mixture of ferrite and pearlite where the fraction of ferrite increases towards the surface. Examination of the structure in Fig. 2d at higher magnification revealed two additional aspects; the cementite phase was spheroidised and the ferrite phase was completely annealed. The structural distribution is similar in Figs. 2c and 2d, in the sense that both objects resulted from decarburisation occurring at the surface.

Fig. 3a is an iron armour scale from a private collection (Park and Mark, 2005). The object allegedly came from a former Koguryo site in Manchuria, and is almost identical in shape to other Koguryo iron scales recovered from formal excavations (Jeng and Sun, 1993). Fig. 3b is a farming hoe excavated from a military fortress (ca. 5th-6th century AD), produced in the Paekche kingdom, at Icheon – see Fig. 1 (Son, et al., 1999). These two objects are similar in two respects; they contain much more carbon than is generally required for their functional purposes and can readily be fabricated from thin iron plates. Fig. 3c shows the microstructure distribution over the entire thickness of the specimen from the area marked in the scale. The structure consists mostly of pearlite with some carbide precipitated along the grain boundaries, which is observed in steels whose carbon content is above eutectoid, 0.77%. Microstructures with somewhat lower carbon content, however, are found near the top and bottom edges. Fig. 3d, an SEM micrograph magnifying the structure of the scale, shows carbide particles or ribbons precipitated in the ferrite background. The microstructure of the hoe is shown in Figs. 3e and 3f to consist of bright cementite particles in a dark matrix of pearlite. The low magnification micrograph in Fig. 3e shows that the structure is uniform over the entire thickness. These two objects provide examples of ultra-high carbon steel being used in a situation where lower carbon steel could have been employed.

Figs. 4a and 4b present two iron objects from the former Paekche and Koguryo military fortresses (Park, 2005), located at Sooncheon and Seoul, respectively, as shown in Fig. 1. Their

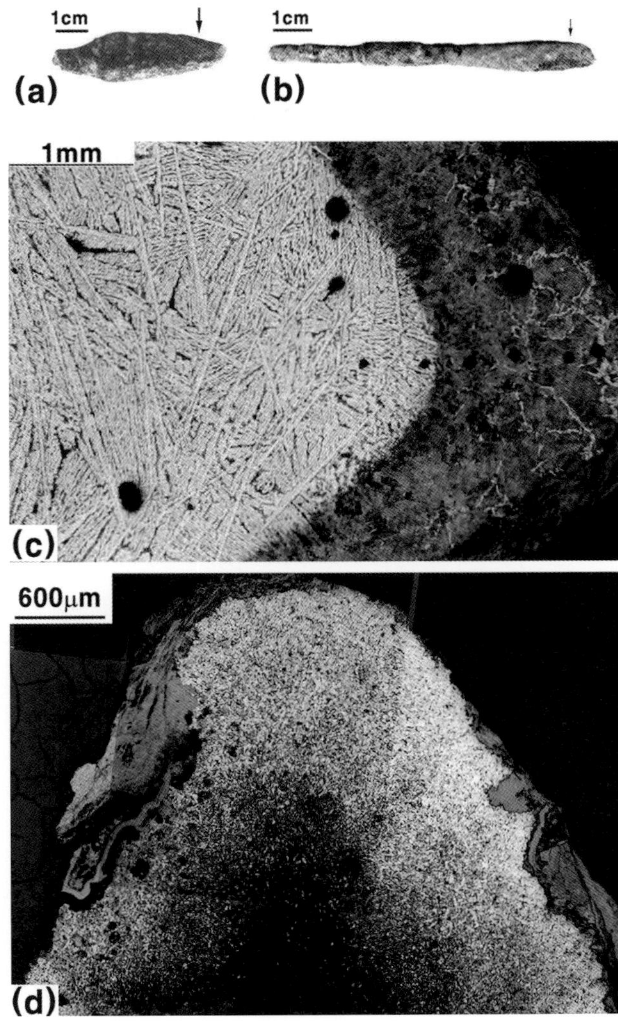

Figure 2: Arrowheads from a former Koguryo military fortress at Moosoon, shown in Figure 1, and their microstructures: (a), (b) general appearances; (c), (d) optical micrographs showing the structures of (a) and (b), respectively. The dark areas in both figures correspond to pearlite, while the white areas in (c) and (d) are filled with white cast iron eutectic and ferrite, respectively.

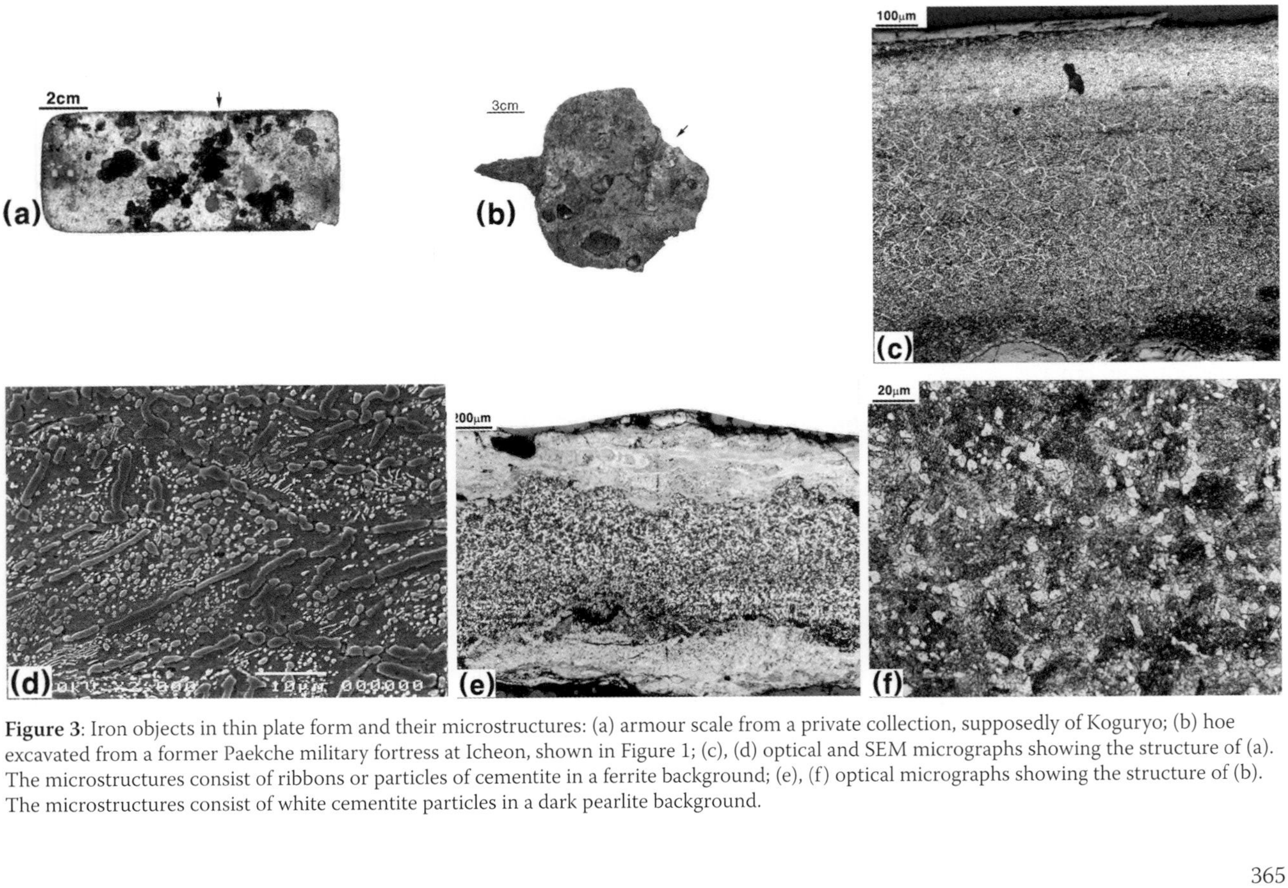

Figure 3: Iron objects in thin plate form and their microstructures: (a) armour scale from a private collection, supposedly of Koguryo; (b) hoe excavated from a former Paekche military fortress at Icheon, shown in Figure 1; (c), (d) optical and SEM micrographs showing the structure of (a). The microstructures consist of ribbons or particles of cementite in a ferrite background; (e), (f) optical micrographs showing the structure of (b). The microstructures consist of white cementite particles in a dark pearlite background.

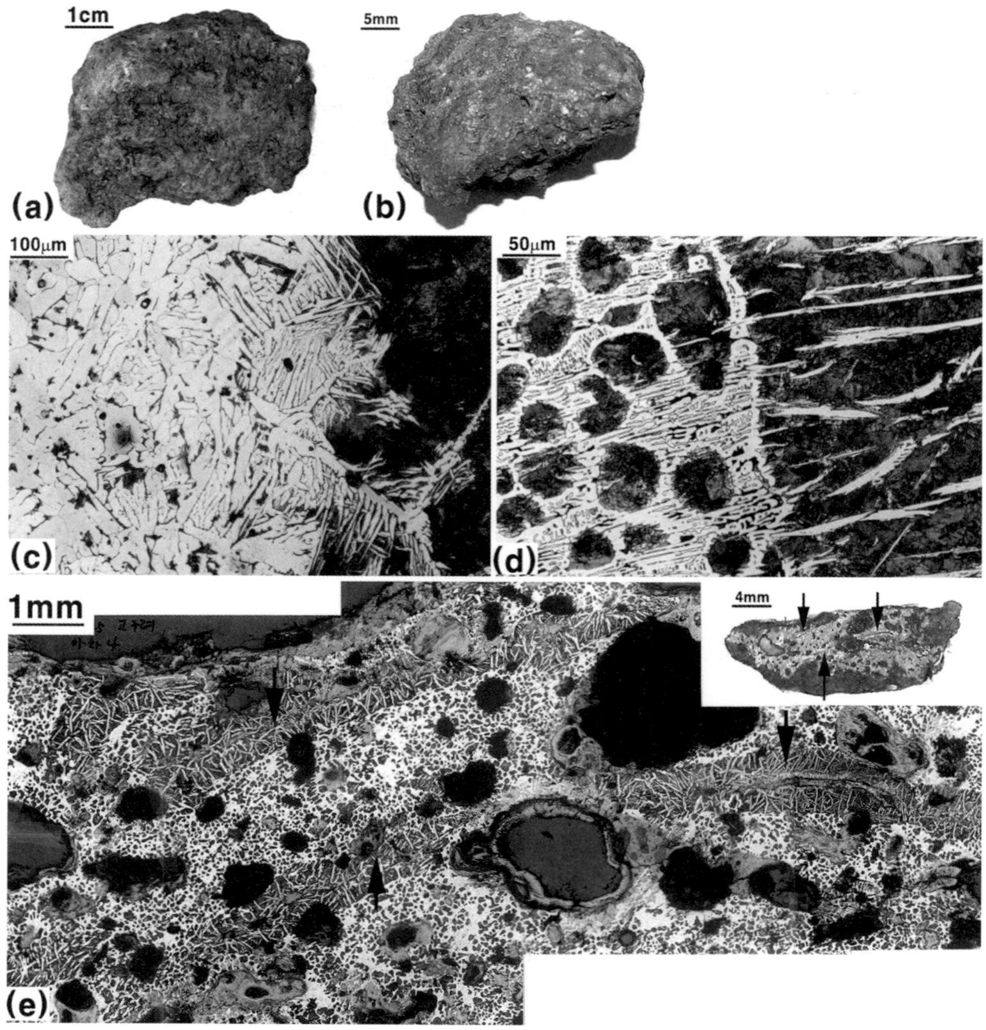

Figure 4: Anonymous iron objects and their microstructures: (a) from a former Paekche military fortress at Soonchеon, shown in Figure 1; (b) from a former Koguryo military fortress in Seoul, shown in Figure 1; (c), (d) optical micrographs showing the structure of (a). The dark areas in both figures are filled with pearlite while the white areas in (c) and (d) represent ferrite and cementite, respectively; (e) optical micrograph showing the structure of (b), which consists of thin plates of ultrahigh carbon steel (marked by the arrows) embedded in the matrix of hypoeutectic white cast iron. The dark and grey circular areas correspond to empty spaces, and the inset is a photograph showing the entire cross-section of the specimen from (b) where the arrows indicate the same areas as those in the micrograph.

appearances, suggesting no specific purposes, indicate that they were not finished products but were either discarded or reserved for later use. The Paekche object in Fig. 4a was a low carbon iron mass enclosed in a surface layer of high carbon steel or white cast iron. Fig. 4c shows the interface region between the low carbon interior of ferrite and the outer region of pearlite, and Fig. 4d shows the outermost region of white cast iron in contact with the ultrahigh carbon steel next to it. This peculiar distribution of microstructure can only be attained when a mass of low carbon iron is soaked in molten cast iron. The microstructure of the Koguryo object in Fig. 4b is shown in Fig. 4e, which covers part of the entire cross-section presented in the inset. Apart from the dark or grey holes corresponding to empty spaces or corrosion products, the structure consists of layers of ultra-high carbon steels embedded in the background of white cast iron. The arrows in the inset locate three steel layers that are clearly observable in Fig. 4e, indicated by the arrows. Structures observed at the boundary regions between steel and cast iron are qualitatively the same as those shown in Fig. 4d. This unique distribution of microstructure can only be achieved when pieces of iron or steel plates are heated in molten cast iron and then cooled at a certain rate.

Fig. 5a is an iron object in plate form (Jung, et al., 2006), and Fig. 5b an iron axe. They were excavated from two different Silla sites at Kyongju, shown in Fig. 1. The object in Fig. 5a represents a group of iron artefacts generally in the form of narrow plates, but with substantial variation in shape and size, most of which have a fan-shaped blade either on one end or on both. Based solely on their peculiar shape suggesting no specific purposes in service, they are considered semi-finished products that might have been traded as currencies or used as materials for making various iron objects with minimal treatments for shaping and hardening. Plate objects of this type were excavated in large quantities, frequently in bundles (Kim, et al., 1994), particularly from burial sites of the Silla and Kaya kingdoms. Fig. 5c, showing the microstructure of the plate object at the left end, consists of a bright interior

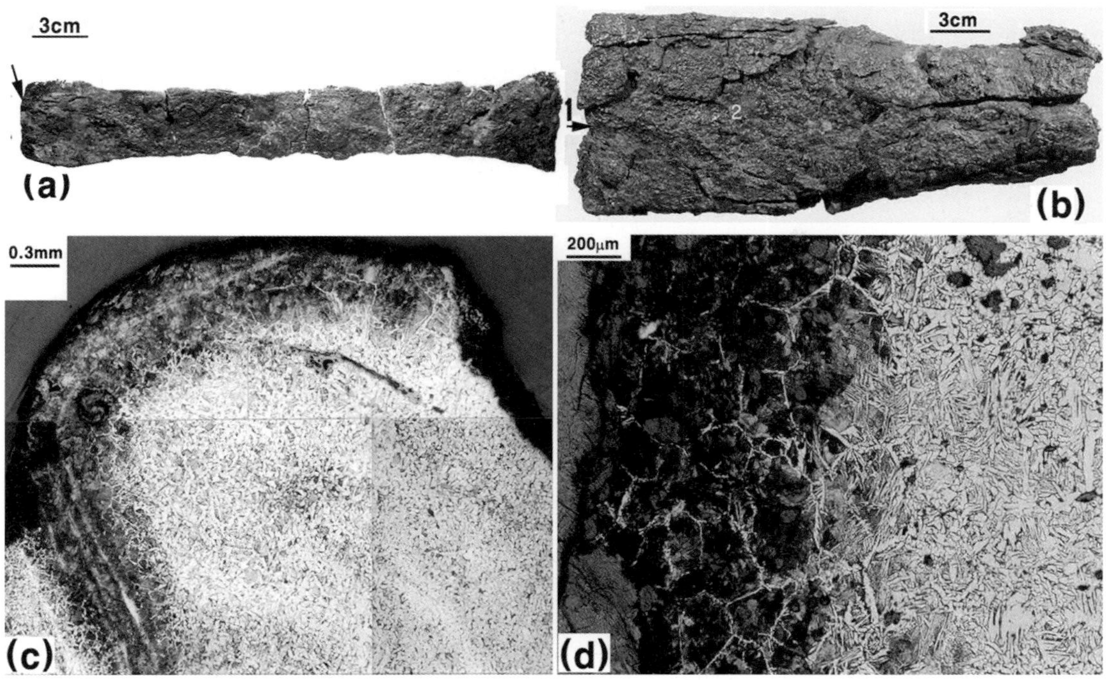

Figure 5: Iron objects from former Silla sites at Kyongju shown in Figure 1 and their microstructures: (a) anonymous object in thin plate form; (b) axe; (c), (d) optical micrographs showing the structure of (a) and (b), respectively, at the left edge marked by the arrows. The dark areas in both figures represent pearlite while the white areas correspond to ferrite.

of ferrite surrounded by a dark layer of pearlite. This unique distribution indicates that it was forged out of almost pure iron and then was subjected to a carburisation treatment in solid state. Fig. 5d, showing the microstructure of the axe at the blade marked by arrow 1 in Fig. 5b, shows a dark pearlite region in contact with the white ferrite area to the right. The pearlite region corresponds to the blade edge, while the ferrite area is facing towards the interior. The specimen taken from the body of the axe labelled 2 in Fig. 5b was found to consist mostly of ferrite. Similar to the object in Fig. 5a, this axe must also have been first forged to shape using almost pure iron and then carburised at the blade. If a plate object such as the one shown in Fig. 5a were to be used in making this axe, this could have been done with the minimal labour necessary for shaping, without the need for time-consuming carburisation.

Figs. 6a-6e are optical micrographs taken from some of the specimens examined in as-polished conditions, without etching. These micrographs aim to provide a better view of the distribution of non-metallic inclusions that are not clearly visible in the etched micrographs. Fig. 6a is a low magnified picture where the area shown in Fig. 5d is located in the middle of the left hand side. Non-metallic inclusions of varying shape and size are located in the dark areas that are randomly distributed. Fig. 6b, magnifying the area marked by the arrow, shows that they consist of grey particles of iron oxide, most probably wüstite, in a dark glassy matrix. Fig. 6c, covering roughly the same area displayed in Fig. 3c, also shows the presence of inclusions, which are well aligned and concentrated only in some horizontal planes. The inclusions in this specimen were found to consist of a single glassy phase, as illustrated in Fig. 6d, which enlarges the area marked by the arrow. One of the inclusions in Fig. 6d at the arrow is filled with pieces of the broken glassy phase. Non-metallic inclusions are almost completely absent in Fig. 6e, which covers approximately the same area shown in Fig. 3e.

Discussion

The microstructures described above illustrate a variety of processes employed in ancient Korea for the control of carbon, i.e., evidence for steel making in the manufacture of iron objects. The arrowhead in Fig. 2a presents direct evidence of white cast iron turned into steel by decarburisation at the surface. This type of structure can be obtained when the desired object is first cast into a mould, and then treated over prolonged periods of time at elevated temperatures for decarburisation. The steel layer may have been intended to overcome the problem of brittle fracture arising when white cast iron is used in the production of objects like arrowheads. The relatively thin layer indicates that the thermal treatment was stopped prematurely, either on purpose or by accident. The unique structure distribution in the second arrowhead also resulted from a prolonged thermal treatment for decarburisation, not from mixing by mechanical working. Both the spheroidised cementite and the annealed ferrite identified in its microstructure provide additional evidence for prolonged heating. Both steel and cast iron could produce a similar structure in a decarburisation treatment. In making arrowheads, however, steel can be used with no further treatment for the control of its carbon level. As such, the second arrowhead, like the first, is likely to have initially been cast

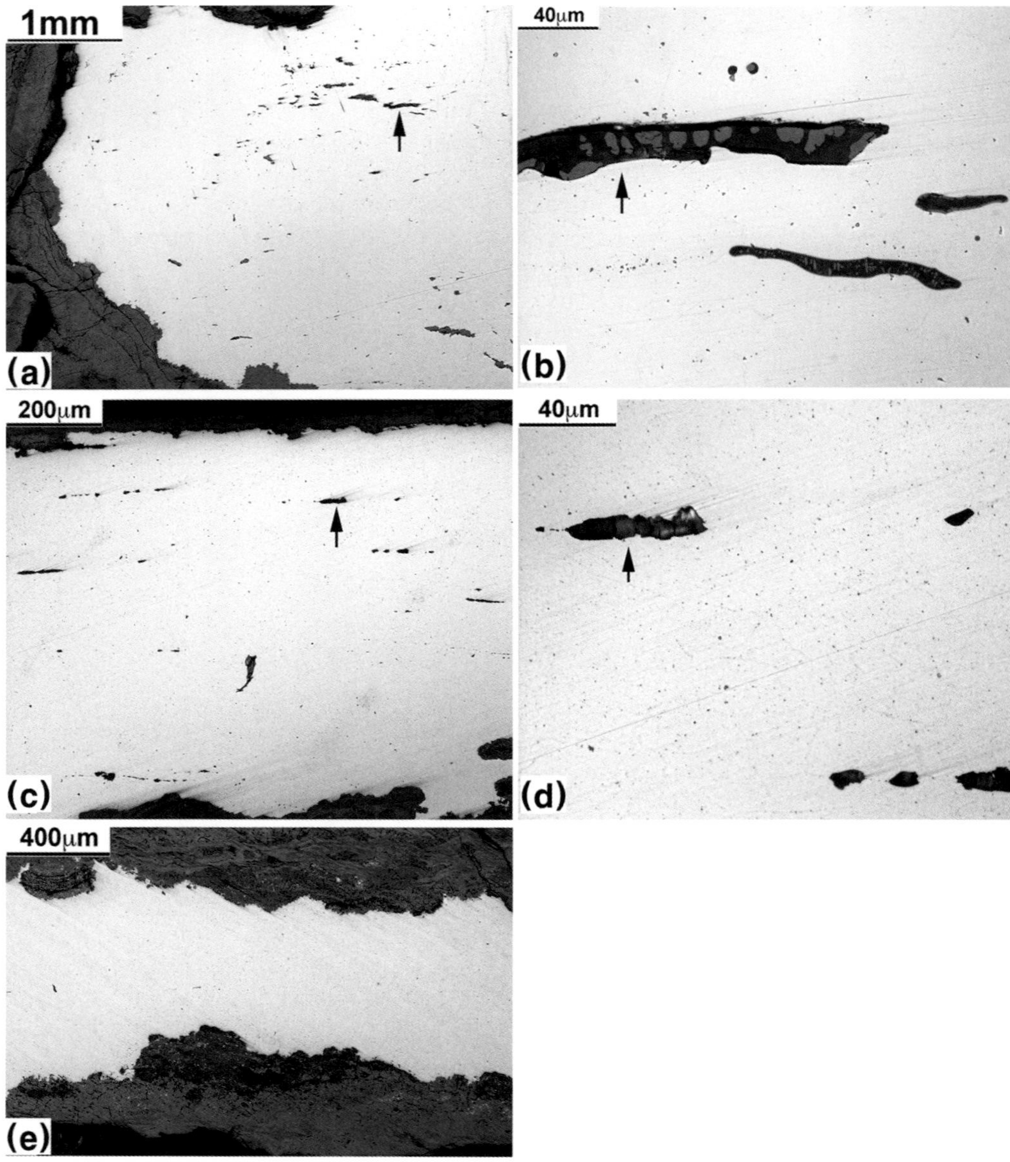

Figure 6: Optical micrographs showing the polished surface of some of the specimens examined without etching: (a) micrograph where the area shown in Figure 5d is located in the middle of the left hand side; (b) micrograph magnifying the area marked by the arrow in (a); (c) micrograph covering the same area of Figure 3c; (d) micrograph enlarging the area marked by the arrow in (c); (e) micrograph covering the same area shown in Figure 3e. The dark areas in the form of ribbons or particles in all the micrographs represent non-metallic inclusions.

from cast iron and then treated for decarburisation. If this was the case, the second arrowhead must have been given a much longer treatment at high temperatures. These arrowheads are indicative of steel making by direct decarburisation of white cast iron with no mechanical working. This technique can be used in cases where objects are first cast to shape and then decarburised, and may be employed as a method of mass production.

The use of high carbon steel in objects such as those in Figs. 3a and 3b is not necessary unless there is a specific reason for it. The spheroidised carbide particles found in both objects indicate that the high carbon content was not intended for strengthening, since they increase ductility at the sacrifice of hardness. If hardness was important, it could have readily been attained with a much lower carbon content combined with the proper thermal treatment. It is technically difficult and inefficient to produce such high carbon content through carburisation, as illustrated by referring to the objects in Figs. 5a and 5b. This high carbon content may simply arise when white cast iron is decarburised. It is unlikely that these objects were cast in pre-formed moulds and then decarburised (like the arrowheads), decarburisation being unrealistically slow to this end. Studies by the author (Park, 2004, 2008), however, showed that if forging is applied in steel making from white cast iron, the rate of decarburisation increases substantially. Steel objects produced by such thermo-mechanical treatments would naturally take the form of a thin plate. Objects in thin plate form would then be shaped from the decarburised plates with a minimal effort. In addition, forging accelerates the spheroidisation of cementite and allows the plates to be readily shaped even if their carbon content is extremely high. The objects in Figs. 3a and 3b may then be indicative of steel making from white cast iron with the aid of forging, an aspect that distinguishes the process from the one used to make the arrowheads.

The indistinct objects illustrated in Figs. 4a and 4b contain white cast iron in their microstructure. Figs. 4d and 4e both portray a process involving liquid cast iron in contact with ultra-high carbon steel remaining in solid state. Unlike those identified or alluded to in the previous objects, steel making here employs both cast iron and low carbon iron as starting materials, which are then treated at temperatures above the eutectic temperature of the Fe-C system, but below the liquidus of steel. This process is reminiscent of a steelmaking technology described in ancient Chinese texts quoted by Needham (1980) and also by Wagner (2008, pp.255-267). In this unique steelmaking method, molten cast iron is in contact with pieces of low carbon iron so that carbon atoms can diffuse between them, and thereby both are made into steel at the same time. Needham named this process cofusion, and suggested that it appeared in China no later than the 5th century AD, while direct decarburisation of white cast iron had been used as early as the 2nd century BC. The objects in Figs. 4a and 4b both represent steel making involving cast iron and low carbon iron, the former maintaining a liquid state during the process. The difference is that the first object apparently came from a low carbon iron mass treated with molten cast iron only at the surface, while the other resulted from liquid cast iron penetrating the gaps between strips of iron. The latter configuration should be much more effective at averaging carbon content.

The objects in Figs. 5a and 5b must have been made following an identical process. Low carbon iron was first forged to shape and was then given a carburisation treatment for hardening. Similar objects can also be made by welding the high carbon edge into the low carbon body. In this case, it would be technically more efficient for the welding operation to be carried out during the fabrication of final products than on intermediaries. These two objects are then representative of steel making by carburisation in solid state, which can best be adopted where low carbon iron is readily available, such as in ancient Korea. It is of significance that this specific technology has been verified far more frequently than all the others combined (Park and Lee, 2003; Park and Jung, 2004; Jung, et al., 2006). This is in agreement with archaeological evidence showing that the use of low carbon iron far exceeded that of cast iron in Korea (Kim, et al., 1994). No scientific evidence, however, has been reported of large-scale iron making from cast iron. Instead, at a recent excavation of a medieval iron production site at Choongjoo (shown in Fig. 1), more than a thousand tons of slag were discovered. This slag was characteristic of bloomery smelting, being mainly fayalitic with some wüstite, and no evidence for smelting of cast iron was identified. In addition, Park and Rehren (2011) reported evidence of large-scale bloomery smelting practiced at an iron-producing site (dated between the 2nd and 3rd century AD) in Korea. Due to Chinese influence, it is generally accepted (and with no serious effort to probe other possibilities) that the Korean Iron Age began around 300 BC and was based on a technological framework similar to the Chinese one (Yun, 1984; Taylor, 1989; Kim, 1992). Cast iron apparently played an important role in ancient Korea, but probably not as much as it did in China. Further research is required for the better understanding of historical iron metallurgy in Korea.

It is significant to note that non-metallic inclusions are nearly absent in Fig. 6e, which was taken from one of the two specimens indicative of steelmaking from white cast iron. In the other specimen, however, some inclusions are visible (see Fig. 6c), but they are distinguished from those present in the specimen indicative of steelmaking by carburisation (see Fig. 6a). In view of the growing body of evidence for bloomery smelting being widely practiced in ancient Korea, particularly in the south (see Park and Rehren, 2011), low carbon iron may have been produced directly from the bloomery process as well as indirectly from treating cast iron. Then the difference in the density and distribution of inclusions may be associated with the different raw materials used in steelmaking, i.e., cast iron versus bloomery iron. Inclusions, however, cannot be employed as a diagnostic difference between the raw materials, since their density and distribution depend inevitably on the degree of thermo-mechanical treatments applied and the skill of the ironworkers.

Summary

This study has presented material evidence from the Korean Three Kingdoms period showing a variety of engineering processes employed for steel making. They include steel making from white cast iron with and without mechanical working, steel making involving both cast iron and low carbon iron, and steel making by carburisation of low carbon iron. The processes involving cast iron either in solid or liquid state are in agreement with the descriptions in ancient Chinese texts (Needham, 1980; Wagner, 2008), and are inclusive of almost every known technology practiced in East Asia, where a unique technological tradition based on cast iron was established from as early as the 4th century BC (Wagner, 1996). It is important, however, to note that the iron technology in ancient Korea was based on the production of low carbon iron and the subsequent carbon control by carburisation, as attested by the frequency and abundance of artefacts excavated. It has yet to be determined whether low carbon iron was produced directly from bloomery smelting or indirectly from treating cast iron, or from some unknown processes.

Acknowledgements

The author acknowledges the Seoul National University Museum, the Kyongju National Research Institute of Cultural Properties, the Research Institute for Buried Cultural Properties at Dankook University, and the Sooncheon National University Museum for offering the artefacts for examination. Detailed comments from an anonymous reviewer helped to improve the manuscript considerably. The time and effort the editor spent on improving the English expression is gratefully acknowledged. This work was supported by the Korea National Research Foundation (NRF-2011-0029808).

References

Barnes, G.L., 1999. *The Rise of Civilization in East Asia: The Archaeology of China, Korea and Japan*. London: Thames and Hudson.

Choi, J.T., Jang, E.J. and Park, J.S., 2001. *Iron technologies of the Three Kingdoms Period in Korea*. Seoul: Seoul National University Museum, Research Series Vol. 10. (In Korean)

Eckert, C.J., Lee, K.B., Lew Y.I., Robinson, M. and Wagner, E.W., 1990. *Korea, Old and New: A History*. Seoul: Ilchokak Publishers for Korea Institute at Harvard University.

Jeng, T.H. and Sun, R.J., 1993. *Monographs on Koguryo*. Yanji: Yanbian University Press. (In Chinese)

Jung, Y.D., Jung, W.S., Choi, K.J. and Park, J.S., 2006. Iron Technology as Observed in Forged Iron Artifacts from the Silla Tomb Complex at Wolsanni Kyongju. *Journal of the Korean Institute of Metals and Materials*, 44, pp.359-66. (In Korean, with English title and abstract)

Kim, J.G., Kim, D.H., Choi, B.H., Park, Y.B., Lee, J.S., Yoon K.I., Cho Y.J. and Shin, C.S., 1994. *The Great HwangNam Tomb: Excavation Report on the South Tomb (Text and Photos)*. Kyongju: Kyongju National Research Institute of Cultural Properties. (In Korean)

Kim, W.Y., 1992. *Introduction to Korean Archaeology*. Seoul: Ilchi Publishing Co. (In Korean)

Needham, J., 1980. The Evolution of Iron and Steel Technology in East Asia and Southeast Asia. In: T.A. Wertime and J.D. Muhly eds. *The Coming of the Age of Iron*. New Haven and London: Yale University Press, pp.507-41.

Park, J.S., 2003. Traditional Korean sword making as estimated from the metallurgical microstructures of a privately owned sword. *Bulletin of the Metals Museum*, 36, pp.85-98.

Park, J.S., 2004. Traditional Japanese sword making from a Tatara ingot as estimated from microstructural examination. *Journal of the Iron and Steel Institute of Japan (ISIJ)-International*, 44, pp.1040-8.

Park, J.S., 2005. Steel making in the ancient Korean Kingdom of Koguryo. *Materials Characterization*, 54, pp.409-16.

Park, J.S., 2008. The key role of forging in ancient steelmaking from cast iron. *Materials Characterization*, 59, pp.647-50.

Park, J.S. and Lee, S.J., 2003. Estimation of the Iron Technology Applied in Making the Iron Sword Excavated from Gyejoksanseong Fortress. *The Paekche Yonku*, 37, pp.81-101. (In Korean, with English title and abstract)

Park, J.S. and Jung, Y.D., 2004. Iron Technology of Silla as Observed in the Microstructure of a Sword and a Spear Excavated from the Hwangnam Great Tomb. *Journal of the Korean Institute of Metals and Materials*, 42, pp.307-13. (In Korean, with English title and abstract)

Park, J.S. and Mark, E.H., 2005. The use of white cast iron in ancient Korea. *Institute for Archaeo-Metallurgical Studies*, 25, pp.9-13.

Park, J.S. and Rehren, Th., 2011. Large-scale 2nd to 3rd century AD bloomery iron smelting in Korea. *Journal of Archaeological Science*, 38, pp.1180-90.

Rostoker, W. and Bronson, B., 1990. *Pre-Industrial Iron – Its Technology and Ethnology*. Archeomaterials Monograph 1. Philadelphia, PA: Archeomaterials.

Son, B.K., Park, K.S., Kim, B.H., Hwang, J.W. and Cheong, S.K., 1999. *The 1st Excavation Report on the Ichon Seolbong Fortress*. Seoul: Research Institute for Buried Cultural Properties at Dankook University. (In Korean)

Taylor, S., 1989. The Introduction and Development of Iron Production in Korea: A Survey. *World Archaeology*, 20, pp.422-33.

Tylecote, R.F., 1992. *A History of Metallurgy*. London: The Institute of Materials.

Wagner, D.B., 1996. *Iron and Steel in Ancient China*. Leiden: E.J. Brill.

Wagner, D.B., 2008. *Science and civilisation in China / Joseph Needham. Volume 5, Chemistry and chemical technology. Part 11: Ferrous metallurgy*. Cambridge: Cambridge University Press.

Yun, T.S., 1984. *Metallurgical Studies of Early Iron Age Artifacts of Korea*. Seoul: Korea University Press. (In Korean)

Author's address

- Department of Metallurgical Engineering, Hongik University, Jochiwon Sejong, 339-701, Korea.
  e-mail) jskpark@hongik.ac.kr

Protohistoric iron weapons and tools from a burial site in West Central Thailand

Anna Bennett

ABSTRACT: The present study focuses on the iron objects from the 53 graves excavated during the 1981-1982 and 1984-1985 seasons at Ban Don Ta Phet, a protohistoric cemetery in West Central Thailand. In the course of these excavations, over 1,000 iron objects were recorded, including 350 iron tools and weapons, which were in a very good state of preservation and readily identifiable. X-radiography, metallographic observations, analyses of grain size, and hardness determinations have been employed to document fabrication techniques.

Keywords: Metallography; Protohistoric Southeast Asia; Ban Don Ta Phet.

Introduction

In Southeast Asia, the first appearance of iron in the archaeological record occurred in the second half of the first millennium BC. In Thailand, its initial appearance seems to have occurred in the west central area closer to the borders of Myanmar and subsequently in the north-eastern part of the country (Higham and Higham, 2009, Table 2), at about the same time as in southern Vietnam (Higham, 2002, p.168). Once iron appeared, it seems to have quickly replaced bronze for the manufacture of tools and weapons; bronze being retained for ornamental objects and utensils. This rapid replacement occurred at about the same time as gold, semiprecious stone polishing, and glass working first appeared; suggesting that the techniques of iron working might quite plausibly have been introduced to Southeast Asia via the Indian or Chinese merchants who were seeking raw materials, such as the plentiful supply of gold ores (Bennett, 2010). Chinese texts mention the conquest of the Upper Thai-Malay Peninsula by the Funan Kingdom, in the Mekong Delta, southern Vietnam; and these early rulers are also reported to have dispatched more than 25 missions to China between 226 and 649 AD, bearing precious gifts. The earliest written evidence of maritime trade between India and Southeast Asia comes from an ancient Chinese text known as "The History of the Former Han Dynasty 206 BC-25 AD", which was written in 32-92 AD and which contains a passage describing Han envoys sailing from the modern province of Guangdong to India with the aim of buying luxury goods such as pearls and opaque glass. Another passage in the same text describes the Han court's failed attempt to pacify the south-western barbarians (the "Xinanyi") in order to open the overland trade route leading to Daxia (Bactria), while still another account written before 91 BC clearly indicates that the Sichan – Yunnan – Burma – India overland trade route was already in use in the 2nd century BC (Laichen, 1997).

Iron from Ban Don Ta Phet

The number of iron artefacts recovered at Ban Don Ta Phet (BDTP), an early 4th or early 3rd century BC – depending on the interpretation of the calibration[1] – cemetery located in west central Thailand[2], far exceeds those found at any other protohistoric or even early historic site in Thailand. In the 1981-1982 and 1984-1985 seasons of excavations, the total number of iron objects identified was in excess of a thousand, including 310 tools and weapons which were in a very good state of preservation and readily identifiable (Table 1). The material is important because it dates to a period of profound social, political, and economic change throughout Southeast Asia, China, and India. Trade and cultural exchange between mainland Southeast Asia and India during this period is clearly

Table 1: Table showing the types of iron artefacts recovered during the 1980-81 and the 1984-85 excavations (Glover, I. C. 1990a).

Category	1980 - 81	1984 - 85	Total
Fishhook	2	3	5
Harpoon	8	6	14
Hollow-base arrowhead	4	5	9
Nails (minimum numbers)	3	1	4
Socketed arrowhead	2	4	6
Shaft-hole axe	2	2	4
Billhook	13	1 6	29
Socketed tool	36	52	88
Socketed spearhead	2	5	7
Tanged	4	19	23
Tanged arrowhead	2	7	9
Tanged blade	18	15	33
Tanged rod	4	5	9
Tanged spearhead	13	7	20
Ferule	2	0	2
Fragments (minimun)	27	21	48
Total	**142**	**168**	**310**

demonstrated by the occurrence of certain pieces of jewellery and ornaments made from semi-precious stones and probably by some of the bronze bowls excavated at BDTP (Glover et al., 1984; Glover, 1989; Bennett and Glover, 1992; Ray, 1994; Glover, 1996; Bellina and Glover, 2004; Bellina and Silapanth, 2006; Glover and Bennett, 2012). The cemetery appears to correspond to a period of transition between the traditional inhumation burials found throughout the Southeast Asian Bronze Age and the later Indianised secondary burial practices, which had been fully adopted with the complete Indianisation of Southeast Asia by the early centuries of the current era[3]. The burial deposits at the site were laid out so as to simulate primary inhumation and the skeletons were in varying stages of completeness when reburied. Skull fragments were often deposited under incomplete bronze bowls which had been purposely damaged, iron spearheads were sometimes ritually 'killed', and beads which had been deposited loose showed no evidence of having been strung as necklaces or sewn onto clothing (Glover, 1990 p.154).

The distribution of the iron artefacts among the graves suggests that at this time iron was closely linked to status. Don Ta Phet is typical of other Iron Age sites in Thailand, in that there is a range of mortuary differentiation where the dead may be ranked according to the variety and complexity of the grave goods, suggesting stratification in power relationships. Statistical analysis of the variety and complexity of the grave goods (Woods, 2001) indicates that there were very few rich burials, a large number of medium wealth burials, and a few very poor burials. There was an overall similarity throughout the graves in terms of the distribution of the bronze vessels, stone and glass beads, earthenware vessels and iron tools and weapons, but only the wealthier graves contained socketed iron spearheads. The iron agricultural tools, such as the socketed and tanged digging sticks and the billhooks, were closely associated with the distribution of spindle whorls and even though spindle whorls occurred in 31 out of 53 burials from the 1980/81 season and are one of the most common types of objects found at the site, no burial contained spindle whorls and socketed spearheads (Woods, 2001)[4]. Woods (2001) has suggested that the graves may have been organised by family links or other forms of group/clan allegiance and it is noteworthy that in the three rows located in the northern area of the site, each row contains only a single burial with a bent socketed spearhead – the first row consists of burials 265, 261 and 260, where 260 contains the spearhead; the second row consists of burials 225, 403, 263, 255 and 259, where 263 contains the spearhead; and the third row consists of burials 222, 463 and 220, where 463 contains the spearhead. The southernmost burials of each row contain spindle whorls.

Some of the iron objects were bent in half or 'killed' before burial. Blades and spearheads are the two major artefact types that were mutilated, indicating that this was a ritual largely reserved for the more prestigious objects rather than the more utilitarian ones. The objects which were bent were 2 out of 23 arrowheads, 15 out of 41 blades, 1 out of 92 digging sticks, 1 out of 8 rods, and 13 out of 23 spearheads. The ritual killing of the spearheads which involved folding the blades back on themselves prior to placing them in the graves raises questions as to whether the spearheads were used prior to the mutilation or were manufactured purely for ritual purposes and were never intended for use, as practiced in the European Viking period. The mutilation may also have been undertaken as a way of preventing looting by grave robbers or may have been associated with religious beliefs.

Prior to the present examination, only one metallographic examination of an iron object from Thailand appears to have been published (Wheeler and Maddin, 1976, p.46), and prior to the 1980s, Southeast Asian iron was still sometimes being referred to as cast; an idea undoubtedly derived from the Sino-centric hypothesis that has long regarded China as the innovating centre largely responsible for stimulating changes in the prehistoric way of life of Southeast Asian populations and developments in metal technology. Current research – both archaeological and technological – points to an equally strong, if not stronger link to the Indian sub-continent at this time, with iron working coming into Myanmar, Thailand, and probably southern Vietnam from India; where the iron technology from eastern Turkey and northern Syria first appeared in the northwest of the country around 1200 BC before spreading southwards throughout the subcontinent between 1000-800 BC (Possehl and Gullapalli, 1999, pp.153-154). More recently, Tripathi (2006, 2007, present volume) and Singh and Tewari (present volume) have asserted that iron from India can now be dated to the earlier part of the second millennium BC, suggesting that India "was an independent centre for the development of the working of iron about 1800 BC, if not earlier".

The present study focuses on the 310 iron tools and weapons that were excavated from the 53 graves during the 1981-1982 and 1984-1985 excavation seasons. Although the objects, which included both tools and weapons, for agricultural and hunting purposes, were standardised to a large degree, they exhibited a wide range of shapes and sizes. The agricultural tools included both socketed and tanged digging sticks, flat chisels, and billhooks; while the hunting tools included fish hooks, knives, harpoons, arrowheads, and tanged and socketed spearheads. Billhooks, blades, and digging sticks were the most commonly represented tools (Fig. 1). The larger socketed and tanged digging sticks clearly demonstrate a wide variation, which does not appear to be wholly due to the adaptations of the tools for different functions. The non-standardisation of these tools may suggest that the objects were forged in a large workshop, with many individuals undertaking the smithing, or alternatively that they were forged in a variety of smaller independent workshops. The very small socketed digging sticks were probably not intended for agricultural use, but may have rather been used for carpentry.

Due to long term preservation problems of this highly vulnerable material, the iron objects from Don Ta Phet were carefully documented using photography, illustration, and X-radiography. It was acknowledged that resources allowing for long-term preservation were not available, and permission was granted by the Thai Fine Arts Department to undertake a sampling programme. Thirty objects were chosen for study either because they were highly representative of a category of object or because they were highly unusual. Up to three

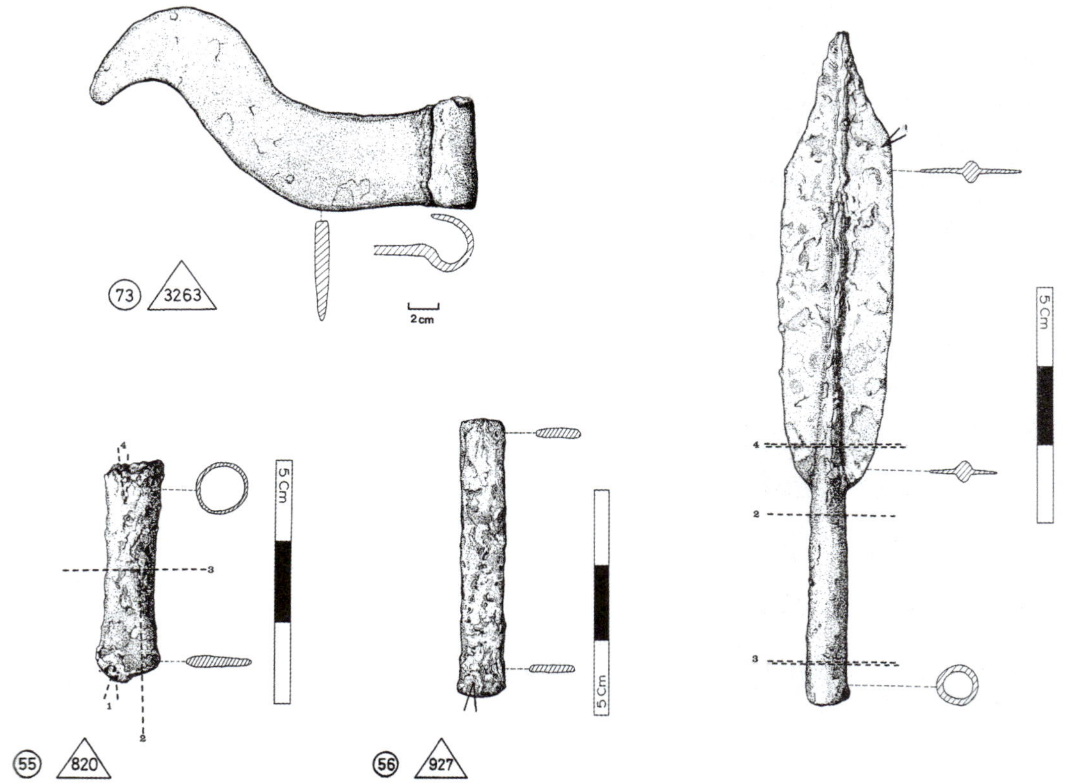

Figure 1: Drawing of a billhook, a socketed digging stick, a tanged digging stick and a socketed spearhead from BDTP. The areas sampled for metallography are indicated.

or four metallographic samples were removed from a single object[5].

The central aim of the study was to document the technology used to produce the iron objects. Due to the considerable number of objects available, it was also possible to investigate the degree to which the manufacture was standardised and consistent within each category of object. Information relating to whether objects which appear visually similar were indeed manufactured in a consistent manner, and, conversely, whether objects which are visually different, such as the socketed and tanged digging sticks, were manufactured in different ways, would help our understanding of the specialisation of ironsmiths at this time. An ability to identify differences in the level of skill used in the manufacture of the more sophisticated and rarer weapons, such as the socketed spears as compared to the simpler and more common digging sticks, was also a consideration informing the choice of the objects to be sampled. Furthermore, the location of the samples themselves was determined with specific attention, so as to be able to document any evidence of heat treatments and/or hardening of the working edges.

Metallographic examinations also afforded the possibility to investigate whether the objects showed evidence of having been used over prolonged periods prior to burial; a type of information which, due to the corrosion of iron objects, is often precluded from a surface examination.

Regarding the ritual mutilation of some of the objects, a relevant question was whether this mutilation was undertaken in the cold state, which might suggest that there was no access to a forge, or in the hot state, which might suggest ready access to and familiarity with smithing (Fig. 2).

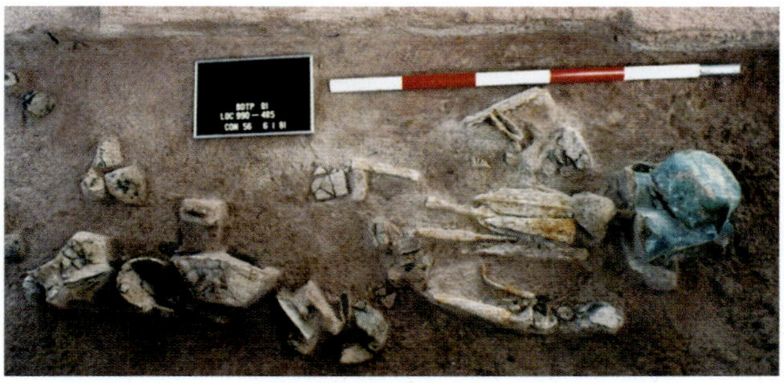

Figure 2: Photograph of re-interred primary inhumation 56 (6181) from BDTP showing a socketed spearhead which has been ritually killed by folding the blade back on itself.

Still further questions focused on determining if there was any evidence as to whether the smelting and the smithing were both undertaken at a central location, with the finished objects being traded to the surrounding areas, or whether the iron had been imported as easily transportable billets which were then divided into smaller sections to be locally forged into the desired tools and weapons.

In the present study, permission had been granted to allow generous sampling of a considerable number of the iron objects, but this is rarely possible. An important part of the present study was therefore to determine the optimum number and the position of samples for metallographic studies.

Socketed digging stick/axe, object no. 55 (820) weight: 170g; length: 13cm

X-ray examination indicated that the digging sticks were made from one piece of metal. A drawing of one of the digging sticks, indicates the position of the samples (Fig. 1). Sample 1, removed from the working edge, was mounted to expose the longitudinal direction. The main portion of the section has a uniform structure with a 0.3% carbon content, consisting of small α grains surrounded by pearlite. Traces of austenite grain boundaries and Widmänstatten plates are indicative of rapid cooling from a temperature above the upper critical point and of reheating and maintaining the iron at a low temperature, which has caused degeneration of the pearlite to spherical carbides. There is a line of oxide running the length of the section, and about one third of the thickness from the surface and the grain size on either side is quite different, suggesting that it is a weld rather than a fracture penetrated by corrosion. Beyond the weld, the carbon content gradually decreases towards the external surface and the grain size increases to ASTM 8.

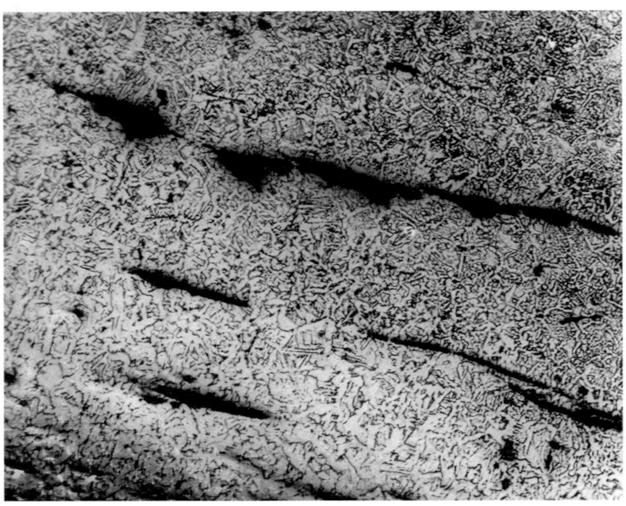

Figure 3: Photomicrograph of the sample removed from digging sticks 55 (820) showing the slag stringers in the decarburiazed surface which were more deformed. Magnification x 35.

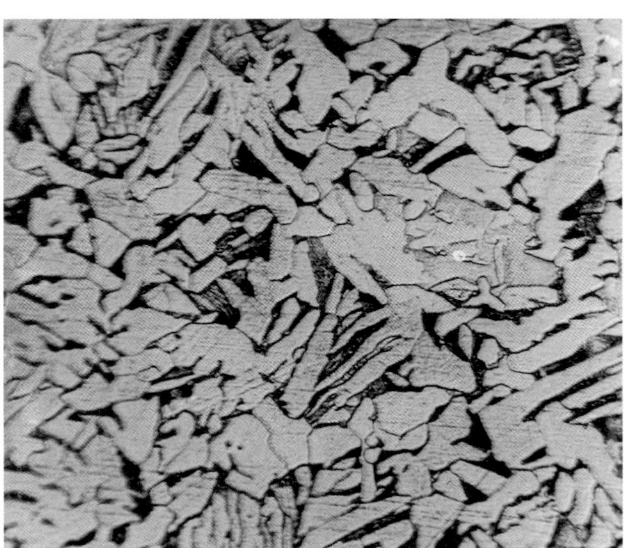

Figure 4: Photomicrograph of the sample removed from digging stick 55 (820) showing lamellar pearlite and irregular α grains indicative of air-cooling. Magnification x 225.

The presence of this weld suggests that the digging stick may have been formed by rolling over the iron billet in the form of a forge welded tubular section and then collapsing the working end, thus leaving a tapered tubular socket. However, sample 2, a longitudinal section along the centre of the object extending 4.5cm from the tip (Fig. 3) and sample 3, a transverse section through the object (Fig. 1), revealed that the socket had been made by lap welding the flattened end of a piece of iron. The apparent weld in sample 1 is due to a surplus of metal which has been folded back over one corner of the tool's working edge during forging. Sample 4, taken from the socket, has a structure similar to that observed in sample 3 and a hardness of HV/121-134 (based on 5 readings).

As with the other digging sticks that were sampled for metallurgical examination, forging appears to have been undertaken largely on one surface, which has become decarburised in the oxidising conditions of the forge. Working appears to have been carried out at a temperature above 850°C and the tools were fairly rapidly air cooled from this temperature (Fig. 4). The structure at the working edges of the tools suggests that the rate of cooling was accelerated by the reduced thickness of the metal in these areas. During the final stages of manufacture, the working ends were hammered and reheated to below 723°C. The reheating time was sufficiently short so that the pearlite in the worked areas has spheroidised, while that in the unworked areas remained unaltered. The slag inclusions in the decarburised areas are notably more deformed and are elongated along the length of the objects, clearly indicating the direction of forging. The hardness values are consistent with iron of this composition and do not suggest any hardening treatments.

Flat chisel, object no. 56 (927) weight: 175g; length: 18.5cm

A sample of 1cm in length was removed from the working edge of the chisel as shown in Fig. 1, and was mounted to expose its longitudinal direction. The regularly aligned slag stringers running the length of the sample suggest that the original iron billet was made of piled strips. The overall metal structure has a carbon content of 0.15% and an ASTM grain size of 9-10, which increases to ASTM 1 at the tip of the chisel and along one surface. These very large α grains, which are accompanied by surface decarburisation, occurred during forging and are the result of abnormal grain growth, which is typical of working at low temperatures (Fig. 5). As with the other types of objects from BDTP, the smith appears to have paid no regard to the amount of decarburisation which occurred during the forging, either at the surface of the object or at the working edge. The hardness values (based on averages of 5 readings), which were HV/146 in the main portion of the sample and HV/137 at the decarburised edge, are higher than expected for this carbon content and may reflect a degree of work-hardening, possibly during use of the tool.

Socketed sickle / Billhook, *object no. 46 (547) weight: 169g; length: 15cm*

X-ray examination indicated that the billhooks had been forged from one piece of metal, with a greater thickness occurring near the shaft socket. The objects were sampled in three areas; sample 1 was taken as a cross-section of the blade through the area of the cutting edge and was mounted to expose the lateral direction. The working edge is composed of acicular ferrite with a high carbon content in the form of spheroidised carbides (Fig. 6), a structure indicative of rapid cooling from a high temperature, followed by reheating to and maintaining at a temperature below 723°C. A small section of the working edge and the sample removed from the blunt end of the tool are composed of large ragged ferrite grains (ASTM 4). The hardness of the cutting edge (HV/167) is twice that at the blunt end (HV/86). Sample 2 is a cross-section through the area of greater metal thickness near the socket, which shows that the shaft has been forged by lapping over a tongue of metal and forge welding it onto the blade, although the welding in some of the billhooks is not complete. The main portion of sample 2 consists of equiaxed α grains (ASTM 4-5) surrounded by lamellar pearlite which show the beginnings of spheroidisation. The carbon content in these areas is 0.2%, while the external surfaces and those on either side of the weld have been completely decarburised and the α grains are significantly larger (ASTM 1). Sample 3 was removed from the socket and the hardness values based on an average of three readings are HV/116 in the high carbon areas, HV/88 in the decarburised outer edges, and HV/81 in the decarburised edges nearest the weld.

From the metallography it would appear that a right-handed smith, forging a billet of low carbon steel, shaped the cutting edge on the side facing him when holding the billhook in his left hand. The tool was then worked from the other face, holding the worked end by its thicker edge while lapping over a tongue of metal and forge-welding it on to the blade to form the shaft socket. All four original external surfaces became decarburised during the forging. The absence of the decarburised zone along the cutting edge of the blade is probably due to the tool having been repeatedly sharpened during use. The hardness of the working edge ranged from HV/87 at the blunt end of the blade to HV/161 at the cutting edge and the metallographic structures suggest that the tool was quickly air cooled, with the blade cooling more rapidly than the thicker socketed end. The cutting edge was then reheated to a temperature below 723°C and this may be associated with the sharpening of the working edge during use. The other billhooks examined, objects nos. 73 (3263) and 43 (350), had similar structures, although because of corrosion it was not possible to determine whether the blades had been used and/or sharpened.

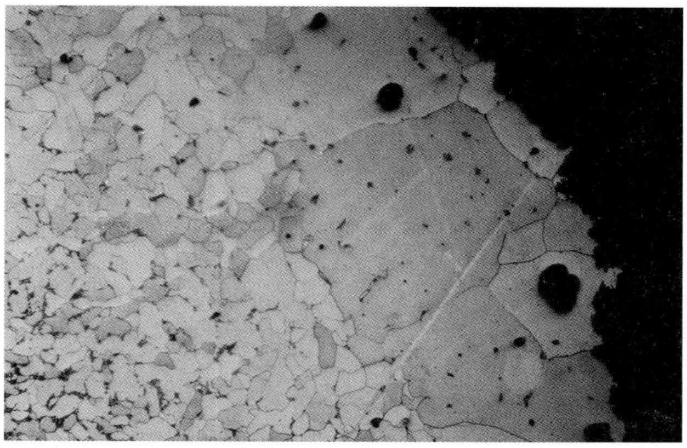

Figure 5: Photomicrograph of the sample removed from flat chisel 56 (927) showing very large alpha grains accompanied by surface decarburization. Magnification x 80.

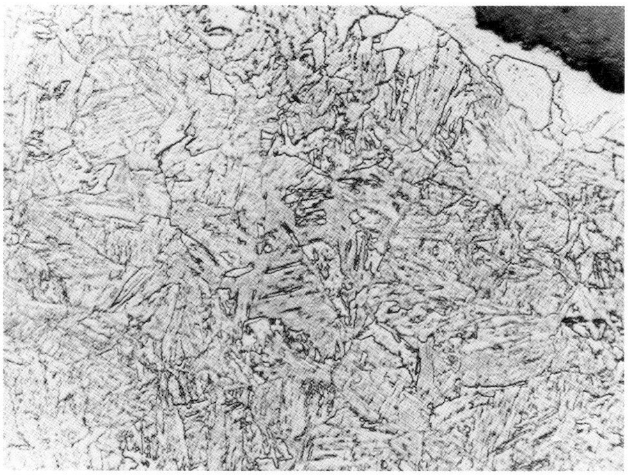

Figure 6: Photomicrograph of the sample removed from the working edge of billhook 46 (547) showing acicular ferrite with a high carbon content in the form of spheroidised carbides. Magnification x 100.

Socketed arrowhead, *object no. 44 (442) weight: 20g; length: 9.5cm*

The arrowhead, object no. 44 (442) (weight: 20g; length: 9.5cm) has a flat blade shaped like an inverted V, with a rear fin on each side. The socket is very shallow and this would have allowed the wooden shaft to fall away when the arrow hit its target. The shape of the arrow suggests that it was used for hunting.

X-ray examination indicated that the metal was almost entirely corroded. The entire object was mounted in resin in order to locate any small patches of uncorroded metal and, although no metal remains at the junction of the socket and the point, the compact magnetite corrosion has preserved the detail of the structure, confirming that the arrowhead was forged from one piece of metal. The areas of uncorroded metal are composed of equiaxed ferrite grains, which are smaller in the socket area (ASTM 7) than in the blade area. After the point and fins had been shaped, the remaining metal was folded around a former and collapsed at one end to leave a tapered tubular socket.

Harpoon/Fish hook, *object no. 56 (924) weight: 43g; length: 16cm*

X-ray examination indicated that the harpoon, had been forged from one piece of metal. Sample 1, a cross-section across the shaft, showed that the harpoon was forged from a sheet of low carbon steel, which was folded to increase its thickness, and then shaped into a cylindrical bar, tapering at both ends. The barbs were made by cutting into the metal bar and lifting pieces of metal to form the barbs. The metallographic sample is composed of alternate layers containing 0.5% and 0.05% carbon respectively, arranged in a V shape, with slag inclusions running in the same direction. In the higher carbon bands the structure varied from small equiaxed α grains (ASTM 10) to more acicular grains, indicative of air-cooling towards the surface. A small amount of surface decarburisation can be noted with the pearlite in this area being partially spheroidised; this in turn suggests that a limited degree of forging, possibly associated with the making of the barbs, was carried out at a temperature below 723°C. The hardness measurements ranged from HV/95 in the low carbon areas to HV/161 in the higher carbon areas.

***Socketed spearhead**, object no. 56 (919), weight: 650g, length: 42.5cm*

X-ray examination indicated that the spearhead (Fig. 1) was made of one piece of metal with the pronounced midrib running into a shallow socket, with no obvious weld. The midrib is very straight, even and steep, and would have been difficult to forge without the use of anvil swage blocks.[6] The spearhead might have been made by rolling over the billet to form a forge welded tubular section, which was then completely collapsed at one end but retained its tubular form in the socket portion. Alternatively, the billet might have been flattened out to form the blade at one end, while the other was folded to form the socket. In both cases the very regular shape of the socket and the tight lap weld suggest that a former may have been used.

Sample 1, removed from the blade edge, and sample 2, removed from the solid portion of the socketed area, were both composed of equiaxed α grains and a small amount of pearlite with some decarburisation at the extreme outer edges where the grains were ragged and considerably larger. Examination of samples 3 and 4, cross-sections through the entire socketed area and through the blade area respectively, show that the spearhead had been forged from a billet of piled carburised wrought iron (Fig. 7). The piled structure consists of layers with a 0.3% carbon content in a matrix of α grains alternating with layers of coarser ferrite and elongated slag filaments which have become contorted during the forging. The layers were of similar composition to those in the sample cut from the socket, although the carbon content was slightly higher and rose to 0.6% in some areas. The piled structure continued to the edges of the blade and this resulted in a central layer with a 0.6% carbon content being sandwiched between lower carbon areas (Fig. 8). Partial spheroidisation of the pearlite in the blade area and the presence of Neumann bands along the slip planes of the grains at the side of the midrib (Fig. 9) suggest that some hammering had occurred at a temperature below 500°C during the final stages of manufacture. At this temperature the α grains failed to recrystallise and the pearlite in the worked areas spheroidised, while remaining unaltered in the outer areas. This structure is typical of material which has been air cooled from a temperature above 850°C, with the blade cooling more rapidly than the thicker midrib and socketed areas. The carbon present in the blade (Fig. 8) was undoubtedly derived from the original billet and the carbon content was probably produced by accident rather than by design.

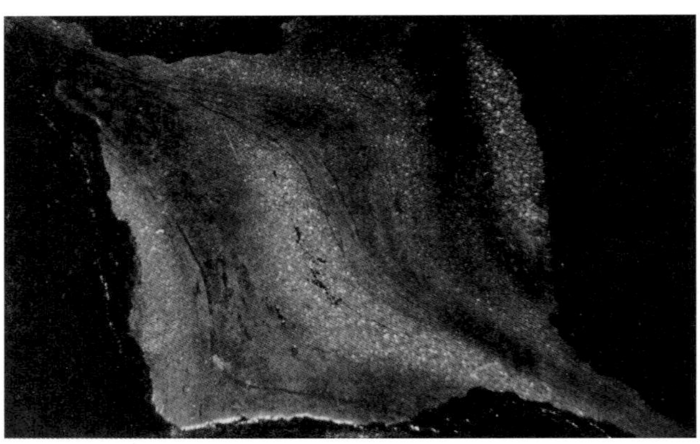

Figure 7: Photograph showing the cross section through the blade of the socketed spearhead 56 (916) showing that the spearhead had been forged from a billet of piled carburized wrought iron. The elongated slag filaments have become contorted during to forging.

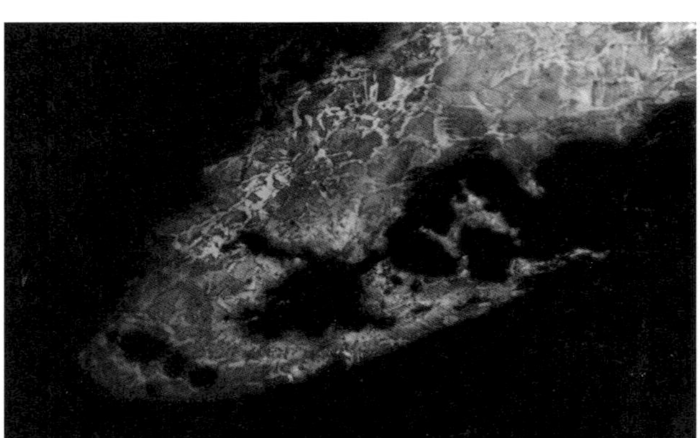

Figure 8: Photomicrograph of the sample removed from the blade of the socketed spearhead 56 (916) showing that the piled structure continued to the edges of the blade and this resulted in a central layer with a 0.6% carbon content being sandwiched between lower carbon areas. Magnification x 90.

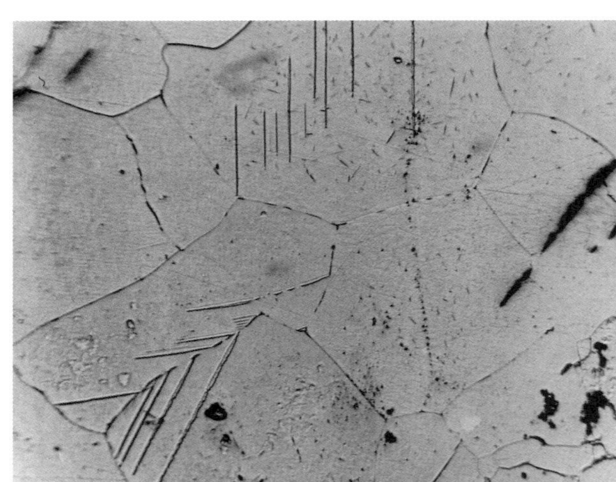

Figure 9: Photomicrograph of the sample removed from the blade of the socketed spearhead 56 (916) showing that partial spheroidisation of the pearlite and the presence of Neuman bands along the slip planes of the grains at the side of the midrib. Magnification x 180.

A 'killed' tanged spearhead*, object no 56 (923) weight: 30g; length: 10cm*

The object was in a fragmentary condition, although it was evident that the blade had been bent back in a ritual 'killing'. Samples were removed from various fragments and mounted to expose the lateral direction. Sample 1, removed from the tang, contains numerous slag inclusions and is composed of small equiaxed α grains (ASTM 9-10) with a small amount of lamellar pearlite. The hardness values, based on an average of four readings, are HV/154 in the lower carbon areas and HV/167 in those with a slightly higher carbon content. Sample 2, which was removed from an unbent area of the blade, also contains numerous slag inclusions. The central area of the sample is composed of small α grains (ASTM 9-10) and has a 0.2-0.3% carbon content. Towards the surface the grains are larger, ragged, and permeated by degenerated pearlite (Fig. 10). The outer edges are all decarburised, although one surface is more depleted in carbon than the other. The hardness values, which range from HV/180 in the core to HV/106 at the decarburised edge, suggest that the blade was not hardened, although some low temperature forging was carried out which would have removed any evidence of hardening. Sample 3, from the bent area of the blade, has a similar structure and hardness to sample 2, indicating that there is no evidence of any cold hammering and that some forging was associated with the 'killing' of the blade.

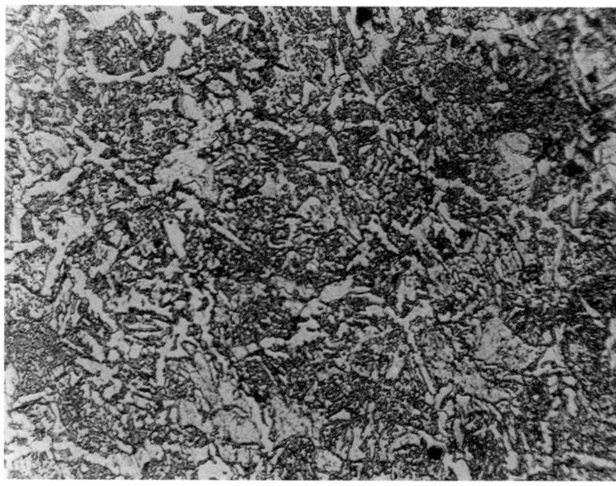

Figure 10: Photomicrograph of the sample removed from 'killed' tanged spearhead 56 (923) showing degenerated pearlite. Magnification x 190.

Summary and conclusions

What seems clear from the present study is that there was no attempt to carburise the working edges of the tools and weapons; rather a certain amount of decarburisation during forging was accepted. Furthermore, it is clear that during use, the soft wrought iron present at the working edge of the tools rapidly diminished, exposing the harder, more carbon-rich material.

The method of forging appears to be consistent both within each category of object and throughout the range of different implements. The more sophisticated objects, such as the socketed spearheads, appear to have been manufactured with a greater degree of skill than the simpler and more common digging sticks, but the technology employed remains the same. The tools and weapons examined were all forged from good quality bloomery iron billets which were composed of a piled structure consisting of layers of carbon free ferrite and layers containing up to 0.7% carbon, composed of ferrite surrounded by pearlite (Figs. 3 and 7). The carbon present in the finished objects was undoubtedly derived from the carbon used as a reducing agent during the smelting operation and there was no suggestion of any attempt to carburise the working edges of the tools or weapons. The regularly aligned slag stringers running the length of the objects indicate the direction of working (Fig. 3). In some of the more complicated shapes, such as the spearheads, the orientation of the layers and the elongated slag filaments have become contorted during forging (Fig. 7).

During the hot working some decarburisation occurred at the surface of the objects, due to the oxidising conditions in the forge. The location and depth of the decarburised surfaces could sometimes be used to indicate the manner in which the objects were held in the forge. Hammering was most frequently undertaken from one side and this caused the hammered surface to become decarburised (Fig. 5). Where both surfaces had become decarburised, one surface was invariably more depleted in carbon than the other. The extreme outer edges of these decarburised surfaces were sometimes composed of distorted, ragged ferrite grains with spheroidised pearlite at the grain boundaries. This structure is typical of material which has been forged in a falling temperature and left to cool very slowly from about 450°C – possibly in hot ashes or at the edge of the smith's fire. In other samples, the austenite grain boundaries and Widmanstätten plates indicate that the iron objects were forged at a temperature above 850°C and that the objects were subsequently rapidly air-cooled from this temperature. The working edges of the tools and weapons are invariably thinner than the main body of the objects and these would have cooled more rapidly. During the final stages of manufacture these edges were reheated for a short period and worked at a temperature below 723°C. The reheating time appears to have been sufficiently short to cause the pearlite in the worked area to have spheroidised, while leaving that in the unworked areas intact.

The smiths appear to have paid no attention to the extent of the decarburisation due to oxidation, which occurred at the surface of the objects during forging. Most of the working edges sampled indicate that the tools were either intentionally sharpened so as to completely remove the decarburised soft wrought iron or that the objects were simply used over sufficient time to allow its complete removal from the working edge. The fact that the decarburised surface could be removed from the working edges to expose the harder, more carbon-rich material, allowed the smith to ignore the decarburisation which occurred during forging.

Although it has not been possible to confirm that there was no intentional work hardening of the working edges, the hardness values of HV/165 – HV/191 in the main portion of the sections and HV/102 – HV/125 in the lower carbon areas did not suggest that any significant work hardening had been undertaken. However, a certain amount of cold working and a corresponding increase in the hardness of the metal has occurred in some instances, where hardness values of HV/140 in the decarburised areas are higher than those in the body of the sample. This increase in hardness may reflect a degree of work- hardening, possibly during use of the tool. The metal samples removed from the billhooks indicate that the smiths managed to increase the hardness of the working edges and completely remove the decarburised zones along the cutting edges of the blades. In billhook no. 46 (547), the hardness of the cutting edge was increased from HV/86 at the blunt end to HV/167, and in object no. 73 (3263) it was increased to HV/186.

In view of the large amount of iron recovered at Don Ta Phet, it is tempting to envisage a local smelting and smithing centre. However, the site has not produced any evidence of associated settlements or of any manufacturing sites. The iron minerals which are found abundantly in the surrounding area in the form of lateritic stones have an extremely low iron content and there has been considerable debate as to whether such an ore resource could have proved sufficient for successful smelting (Pryce, et al., 2006; Pryce and Natapintu, 2008). Significant iron ore deposits are available in the Wong Prachan Valley in the central plains of Thailand (Bennett, 1989; Pigott, et al., 1997), and Myanmar has extensive resources of both the iron oxides haematite and magnetite and the hydroxides goethite and limonite; however, to date, no Iron Age smelting sites have been identified in Thailand or indeed anywhere within mainland Southeast Asia.

The metallographic structures of the samples removed from the bent blades of the ritually 'killed' spearheads indicate that the bending back of the blades was undertaken in the hot state. This suggests that the people undertaking the mutilation had a knowledge of and access to some forging facilities. However, this does not necessarily imply that any forging took place at Don Ta Phet, since the cemetery is a secondary burial site and the objects may have been ritually 'killed' before being brought to the area for deposition. Given its geographical location, close to the Myanmar border and on the route to the Three Pagodas Pass, it is possible that the iron objects were imported along this route. Socketed iron spearheads which are very similar in appearance to those excavated at BDTP have been found in the Samon Valley in Myanmar (Htin, 2007). The characteristic iron billhooks (Fig. 1), which were present in 23 out of the 53 excavated graves, have been found during survey work further west in Ratchburi Province, and still other examples have been found by local tin miners, towards the Myanmar border (Bennett and Glover, 1992; Glover and Bennett, 2012). There are further examples in museum collections in the Kanchanaburi and Ratchaburi provinces (Glover, 1990 pp.160-165) and one in the collection of the National Museum of Chumphon, which comes from the contemporary early river port site of Khao Sam Kaeo in eastern peninsular Thailand; a site which has provided clear evidence of iron smithing activity in the protohistoric period, although there is still no evidence of any smelting activity (Pryce, et al., 2008). Ethnographic parallels show that in many societies, smelting and smithing were not carried out in the same production centre, but rather that the customer supplied the raw material and the smith contributed his skill as payment in return for part of the raw material as payment (Marschall, 1968, p.150). The raw iron billets may therefore have been imported and the objects forged locally according to particular requirements. Alternatively, the iron objects themselves may have been imported along with some of the more sophisticated and rarer artefacts, such as the high tin bronze bowls (Bennett and Glover, 1992; Glover and Bennett, 2012).

Notes

1. BM-2106R (context 67) 2257 ± 230 BP
   OxA-2102 (context 60) 2348 ± 70 BP
   OxA-2013 (context 78) 2390 ± 80 BP
   OxA-2014 (context 59) 2276 ± 70 BP
2. The site is located 8 km north of Phanom Tuan and 1.5km west of the road between Kanchanaburi and U-Thong and was discovered accidentally in 1975 during the building of a local school. The site has been the subject of several large excavation campaigns, the first in the mid 1970s (Chin You-di, 1976; Glover, ND; Glover, et al., ND), two in the 1980s undertaken by a British Thai team (Glover, 1990), and subsequently in 2000 by the Thai Fine Arts Department.
3. With the exception of the Hill Peoples.
4. Glover (1990) originally published one of the graves as containing a spindle whorl and a socketed spear, but context 220 was subsequently separated into two burials – 220 and 463 (Woods, 2001).
5. The samples were embedded in polyester mounting resin, cured with a peroxide catalyst. They were then polished on wet silicon carbide papers of successively finer grit, from 180 to 600. Final polishing was carried out on lap wheels, using suspensions of 7μm and 2μm alumina in distilled water or 6μm and 1μm diamond pastes.
   The samples were examined microscopically to observe any slag inclusions or welds. They were then etched with 2% Nital (2% concentrated nitric acid in industrial methylated spirits) for about 10 seconds, to reveal the metallographic structure. It was found that some surface flow of metal occurred during polishing and this had a tendency to distort the metallographic structures. This could be reduced by using only light pressure, but the polishing time was thereby increased, with a danger of surface pitting. Satisfactory results were obtained by repolishing on the final lap wheel after etching, thus removing the distorted surface layer. The samples were then reetched.
   The finished samples were examined at magnifications ranging from 10x to 1200x using an Olympus microscope and a Vickers microscope. Photomicrographs were produced of areas selected to show variations in carbon content and evidence of heat treatments. An estimation of the carbon content was made on the basis of the amount of lamellar pearlite present. Grain size was determined by comparing photomicrographs at 100x with ASTM standards (The American Society for Testing and Materials).
   Hardness determinations were made by the Vickers method. This measured resistance to the penetration of a diamond pyramid applied under a 200 g load for 15 seconds. The size of the indentation was inversely proportional to the hardness. Hardness, a property which has no simple theoretical foundation, is a more or less arbitrary number, which can be used for the purposes of comparison and provides a useful indication of mechanical properties. The number of hardness measurements carried out on each sample was dependent on observed variations in structure.
6. Similar to an anvil, but with an impression of the desired shape for the smith to work his hot metal into.

References

Bellina, B. and Glover, I.C., 2004. The archaeology of early contact with India and the Mediterranean world, from the fourth century BC to the fourth century AD. In: I.C. Glover and P. Bellwood eds. *Southeast Asia: From prehistory to history*. London: Routledge Curzon, pp.68-89.

Bellina, B. and Silapanth, P., 2006. Khao Sam Kaeo and the Upper Thai Peninsula: Understanding the mechanism of early trans-Asiatic trade and cultural exchange. In: E.A. Bacus, I.C. Glover and V.C. Pigott eds. *Uncovering Southeast Asia's Past – Selected Papers from the Tenth Biennial Conference of the European Association of Southeast Asian Archaeologists*. Singapore: National University Press, pp.379-92.

Bennett, A. (2010). Gold in early Southeast Asia. *Proceedings of workshop AURUM, the authentication and analysis of goldwork*. ArcheoSciences, 33, pp.99-107.

Bennett, A. and Glover, I.C., 1992. Decorated high tin bronze bowls from Thailand's Prehistory. *British Archaeological Reports*, 561, pp.109-19.

Chin You-di, 1976. *Ban Don Ta Phet: preliminary excavation report, 1975-76*. Bangkok: National Museum.

Glover, I.C., Alvey, B., Pisit Charoenwongsa and Narawat M. Kamnounket, 1984. The cemetery at Ban Don Ta Phet, Thailand: preliminary results from the 1980-81 excavation season. In *South Asian Archaeology 1981*, B. Allchin (ed.) Cambridge, University Press, pp.319-30.

Glover, I.C., 1989. *Early trade between India and Southeast Asia: a link in the development of a world trading system*. Centre for Southeast Asian Studies Occasional Paper 16, Hull: University of Hull.

Glover, I.C., 1990. Ban Don Ta Phet: the 1984-85 excavation. In: I.C. Glover and E.A. Glover eds. *South Asian Archaeology, 1986*. Oxford: British Archaeological Reports 561, pp.139-84.

Glover, I.C., 1996. The Southern Silk Road: Archaeological evidence for early trade between India and Southeast Asia. In: Amara Srisuchat ed. *Ancient Trades and Cultural Contacts in Southeast Asia*. Bangkok: Office of the National Culture Commission, pp.57-95.

Glover, I.C. and Bennett, A., 2012. The high tin bronzes of Thailand. In: J.G. Douglas, P. Jett and J. Winter eds. *Scientific research in Ancient Metallurgy: Proceedings of the Fifth Forbes Symposium at the Freer Gallery of Art*. London: Archetype Publications, pp.101-14.

Higham, C.F.W., 2002. *Early cultures of mainland Southeast Asia*. Bangkok: River Books.

Higham, C.F.W. and Higham, T., 2009. A new chronological framework for prehistoric Southeast Asia, based on a Bayesian model from Ban Non Wat. *Antiquity*, 83, pp.125-41.

Htin, Y., 2007. *Iron Age burials in the Samon Valley, Upper Burma*. Chiang Mai: Siam Ratana Ltd.

Laichen, Sun, 1997. Chinese historical sources on Burma: A bibliography of primary and secondary sources. *Journal of Burma Studies*, 2, pp.1-116.

Marschall, W., 1968. Metallurgie und frühe Besiedlungsgeschichte Indonesiens. *Ethnologica*, 4, pp.29-263.

Pigott, V.C., Weiss, A. and Natapintu, S., 1997. Archaeology of copper production: Excavations in the Khao Wong Prachan Valley, Central Thailand. In: R. Ciarla and F. Rispoli eds. *South-east Asian Archaeology 1992. Proceedings of the Fourth International Conference of the European Association of Southeast Asian Archaeologists. Rome, 28th September – 4th October 1992*. Rome: Istituto Italiano per l'Africa e l'Oriente, pp.119-57.

Possehl, G. and Gullapalli, P., 1999. The Early Iron Age in South Asia. In: V.C. Pigott ed. *The archaeometallurgy of the Asian old world*. Philadelphia: University Museum Monograph 89 and Symposium Series Vol. VII, The University of Pennsylvania, pp.153-75.

Pryce, T.O. and Natapintu, S., 2008. The plausibility of smelting iron from laterite: Technical notes from Ban Khao Din Tai archaeological site and Ban Prasat geological deposit, Thailand. Asian Perspectives 48(2): 249-264.

Pryce, T.O., Bellina-Pryce, B. and Bennett, A., 2006. The development of metal technologies in the Upper Thai-Malay Peninsula: Initial interpretation of the archaeometallurgical evidence from Khao Sam Kaeo. *Bulletin de l'Ecole Française d'Extrême-Orient*, 93, pp.295-315.

Ray, H.P., 1994. *The winds of change – Buddhism and the maritime links of early South Asia*. Delhi: Oxford University Press.

Tripathi, V.B., 2006. Iron technology, its impact and survival in the middle Ganga Plain. *Pragdhara*, 16, pp.109-23.

Tripathi, V.B., 2007. Towards the Wootz: iron and steel technology in India. *Indian Journal of the History of Science*, 42, pp.403-25.

Wheeler, T.S. and Maddin, R., 1976. The techniques of the early Thai metalsmith. *Expedition*, 18, pp.38-47.

Woods, M., 2001. A spatial and statistical analysis of Ban Don Ta Phet. MA Dissertation, School of Oriental and African Studies, University of London.

Author's address

- Conservation and Technical Services - Brussels
  atnbennett@analyzeart.com

Scientific approaches to technology

Geochemical survey and metalworking: a case study from Exmoor, southwest Britain

Chris Carey and Gillian Juleff

ABSTRACT: A series of geochemical surveys were undertaken on Exmoor as part of a multifaceted investigation into the rich ironworking history of this landscape. Exmoor was created by semi-metamorphosed sedimentary lithologies, which contain a range of iron ore sources. The Romano-British iron smelting site at Sherracombe Ford was one of several production complexes to exploit these ores and excavation demonstrated evidence of iron smelting and primary iron smithing. Site investigation strategies utilised a range of techniques, including geochemical survey, gradiometer survey, and excavation. This paper describes one geochemical survey applied at an intra-site level to investigate a large compacted smithing floor and a multiple-use furnace location. The discussion is focused on how geochemical surveys can be conducted on sites with archaeometallurgical remains and how such data can be analysed and interpreted. Significantly, geochemical and gradiometer surveys are compared and the datasets integrated, the resulting plot indicating a significant potential for the application of these technologies on ironworking sites.

Keywords: Geochemical survey, gradiometry, Exmoor, ironworking.

The archaeological background

The geological evolution of the southwest peninsula of Britain has created several areas of upland topography rich in metalliferous resources, such as the upland granites of Dartmoor and Bodmin Moor, and the semi-metamorphosed lithologies of Exmoor. This upland topography has in part defined a distinctive regional archaeological record that is often closely associated with the exploitation of these rich ore sources, such as speculation of Bronze Age exploitation in Cornwall (Penhallurick, 1997), potential evidence of Bronze Age exploitation on Exmoor (Juleff and Bray, 2007), the famous excavation of an Iron Age smelting site at Kestor, Dartmoor (Fox, 1954) and more recent research on Romano-British exploitation of iron ores on Exmoor (Riley and Wilson-North, 2001).

Within the southwest, archaeological evidence for the transition from a prehistoric to a typical Romano-British cultural landscape, with the exception of the city of Exeter on the Exe estuary, has proved difficult to identify. The counties of Devon and Cornwall contain numerous small enclosures that can date from the Bronze Age (Holsworthy, Devon; Green, forthcoming) through the Iron Age (Rudge, Devon; Todd, 1998) into the Romano-British period (Timberscombe; Wilson-North, forthcoming) and the post-Roman period (Dunkeswell, Devon; Silvester, 1980), and are interpreted as indicative of a continuity in settlement and landscape use through millennia. The presence of a more 'Romanised' landscape is however seemingly absent, with only a few Romano-British houses, roads or settlements identified to date (Griffith and Quinnell, 1999).

The upland region of Exmoor on the northern seaboard of the peninsula, now designated a National Park, contains significant numbers of these prehistoric enclosures and also has rich metalliferous resources dominated by iron ores but also including copper and silver/lead in localised areas. Across the moors are numerous archaeological sites presenting evidence of iron smelting, with excavation demonstrating that such sites often contain well preserved and in situ slag heaps, such as Sherracombe Ford, Brayford, and Clatworthy; and also a possible prehistoric/Romano-British mine at Roman Lode (Fig. 1). While the slag heaps at Sherracombe Ford (NGR: 272000, 136650) had been known and recorded as probably medieval on the basis of their size, the full scale of the site was not recognised until 1996, when the Exmoor National Park Authority (ENPA) carried out a qualitative assessment survey of early iron working evidence in its area of responsibility. Two charcoal samples taken from exposed sections of the slag heaps at that time returned radiocarbon dates that indicated for the first time that smelting at the site took place far earlier, in the Late Iron Age and Romano-British period.

Against this background, the Exmoor Iron project, funded by English Heritage and operated through ENPA, the National Trust and Exeter University, undertook two seasons of excavations at the site (2002 and 2003), concentrating primarily on one of the two major slag heaps with its associated working platform (platform A2: Fig 2). The objectives were to investigate the creation of the working platform and its use, and to examine the nature of the technological processes in operation at the site. The strategies adopted included open-area excavation to examine spatial and functional relationships, vertical sections traversing the platform and

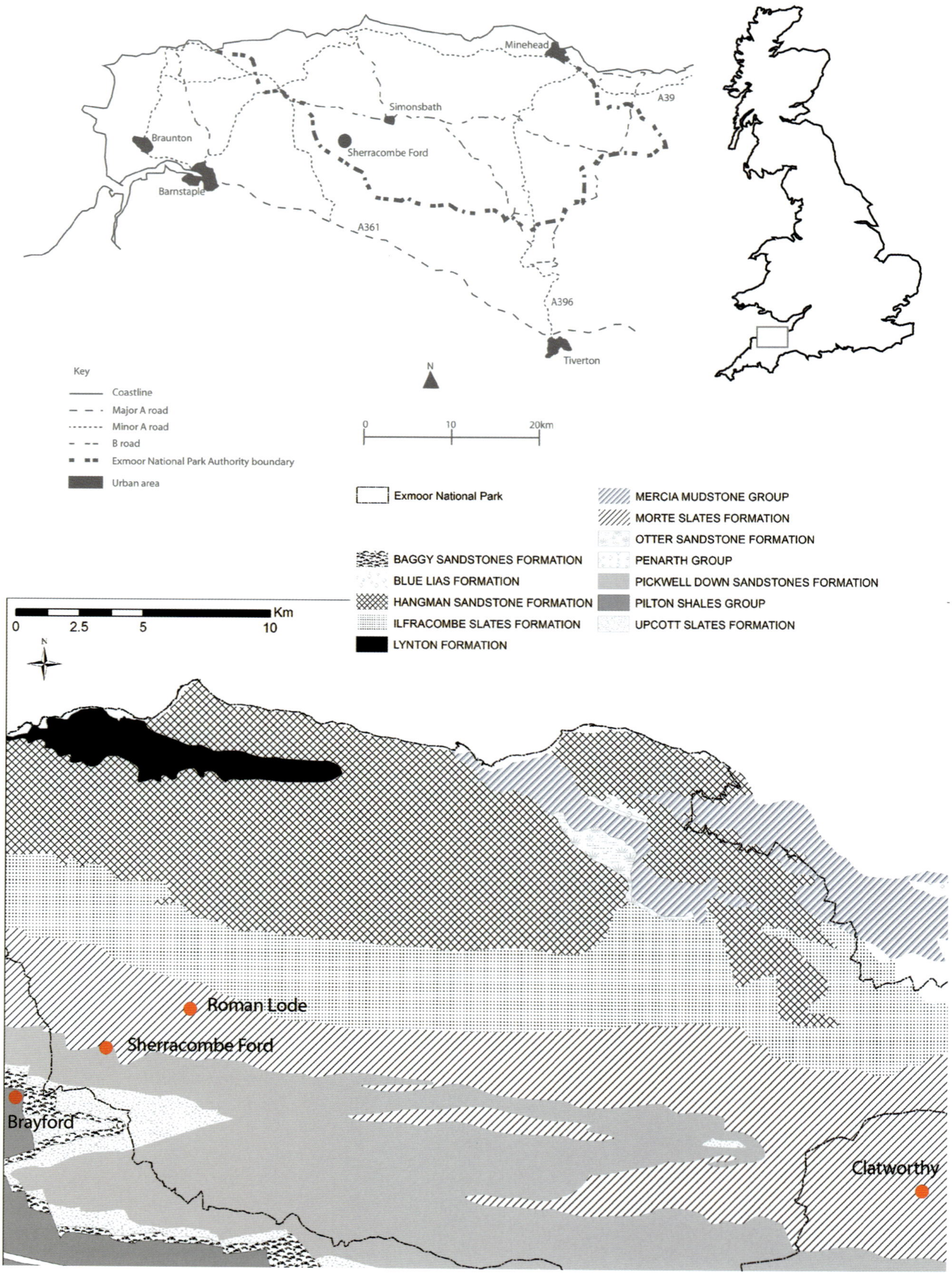

Figure 1: The location of the study area and Sherracombe Ford, relative to other significant Romano-British ironworking sites on Exmoor and the solid geology.

the slag heap to examine stratigraphy and site chronology, and quantitative sampling to characterise technologies and processes. Gradiometry survey was used as an exploratory technique and both gradiometry and geochemical surveys were used in the early stages of site investigation. This paper describes the evolution of the geochemical survey techniques applied at the site and describes the value of geochemical survey when used in combination with more conventional archaeological methods of gradiometer survey and excavation.

Introduction: geochemical survey

Geochemical survey has yet to develop as a standard archaeological technique. It has been used in an *ad-hoc* manner at a variety of different levels, ranging from the landscape (Aston, et al., 1998) to the site-specific (Linderholm and Lundberg, 1994). At one end of the spectrum it has been used in a univariate form to measure phosphate concentration (Proudfoot, 1976), whilst at the other end of the spectrum it has been applied as a multivariate analytical tool (Entwhistle, et al., 2000). Ambitious claims have been made about what geochemical survey can achieve, from site detection on a landscape scale (Jackson, 2001) through to analysis of intra-site activity (Cook, et al., 2003). However, there has been little consensus amongst practitioners regarding methodological approaches and achievable objectives, which has produced little critical development in application.

The basic premise of archaeological geochemical survey is that the activities of human beings can load geochemical deposition into their local, regional and global environments (Heron, 2001). Through detailed analysis of cations within the soil samples, it is possible to detect geochemical deposition resulting from human activity as part of the archaeological record. Traditionally this has been achieved through collecting soil samples and subsequent laboratory analysis. Recent advances, such as PXRF (portable X-ray fluorescence) allows the taking of measurements of metallic elements within the field (e.g. Craig, et al., 2007), offering considerable savings in both time and money, making the process more analogous to conventional methods of archaeological geoprospection.

Archaeological geochemical researchers have yet to fully and systemically investigate the relationship between geochemical deposition and human activities, and mechanisms of deposition and retention. However, Middleton and Price (1995) successfully demonstrated different activity zones within houses based on geochemical deposition, whilst Cook et al. (2003) defined Romano-British metalworking within buildings, both examples being local level spatial surveys. However, it is worth noting that archaeological geochemical deposition can occur on both regional and global scales, as shown by Hong et al. (1994, 1996) analysing Romano-British lead (Pb) and copper (Cu) pollution in Greenland ice cores transported through air borne pollution, and by Thorndycraft et al. (1999) analysing regional tin (Sn) pollution in alluvial sediments within valley floors.

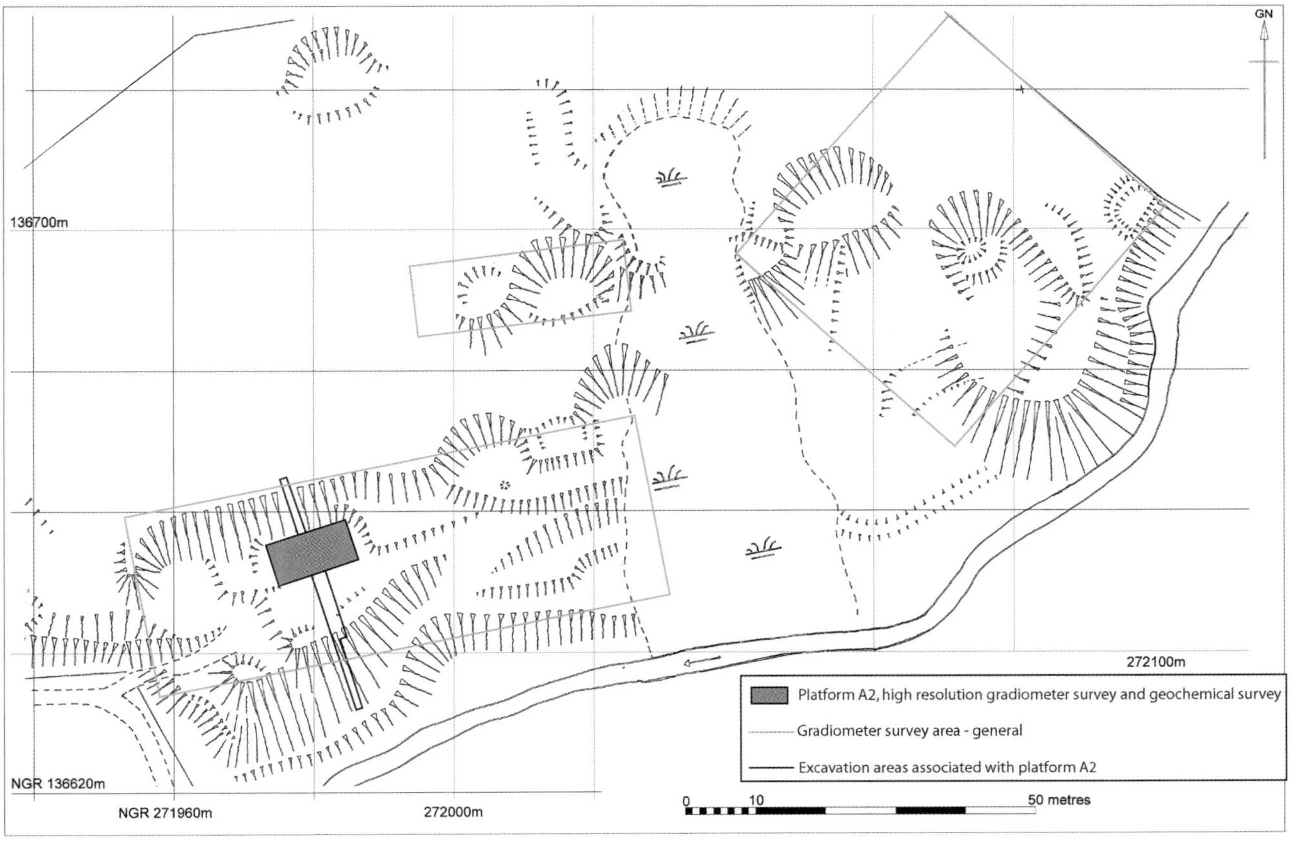

Figure 2: Earthwork plan of the Sherracombe Ford Romano-British ironworking complex, with key aspects of site investigation defined (drawn by Rob Wilson-North).

In this study, geochemical survey has been applied at an intra-site level and is used to investigate areas/features within a site. Sampling intervals are small (analogous to geophysical survey intervals), with the aim of detecting large geochemical signatures caused by metalworking. The small sample interval creates a more representative sample population. For example, a 1m sample interval allows one sample to represent $1m^2$ of archaeology. A 10m sample interval (as routinely employed on landscape scale survey, e.g. Jackson, 2001) is used to represent an area of $100m^2$. Such large sample intervals are clearly unsuitable for intra-site analysis and in general raise issues about the representative nature of a geochemical survey plot. In using geochemical survey to detect past metallurgical activity, the following assumptions are made:

1. All human activities deposit chemical elements and compounds into their immediate environment.
2. Much of this geochemical deposition is ephemeral and undetectable to the archaeological geochemist.
3. Metalworking will deposit chemical elements into the ground in the vicinity of the activity.
4. Some of this deposition from metalworking will be persistent within the soil profile.
5. This persistent deposition can be recognised against the background variance within a survey area/activity area.

The deposition of the metallic elements from metalworking is greatest in the immediate vicinity of the location where the metallurgical activity took place and can be deposited in the form of macro-artefacts (e.g. slag) and also micro-artefacts (e.g. microscopic particulates and gaseous deposition). Moving away from the activity focus, the level of geochemical enhancement decreases. In addition, some features found in the archaeological record are impermeable to geochemical deposition, e.g. a stone wall. Therefore, as well as identifying activity areas, geochemical survey has the potential to identify associated infrastructure, such as walls, through abrupt changes in geochemical concentrations.

This geochemical survey utilised a multivariable approach in both data capture and analysis. From individual soil samples, multiple elements were measured and intervariable correlation via Principal Components Analysis (PCA) was used to treat the data and identify correlations which were used to infer metalworking areas. Comparison of the original variables to the PCA allowed the identification of important elements, specifically defining impurities associated with the ore sources exploited.

Materials and methods

Field method: Gradiometer survey

The gradiometer survey carried out across platform A2 used a Geoscan FM36 gradiometer, using a 0.5m sample interval, walking 0.5m traverses. Following initial interpretation, several important geophysical anomalies thought to represent in situ remains from iron smelting related activities were further refined using a 0.25m sampling interval on 0.25m traverses and a 0.1m sampling interval on 0.1m traverses. All data was processed in Geoplot and exported into ArcGIS (v. 8.3) via Surfer (v. 8.0). All gradiometer surveys were undertaken by Substrata (Ross Dean).

Field method: Excavation and geochemical survey

The excavation strategies adopted were guided by the results of the gradiometer surveys. A large open-area trench of 12x6m positioned across the platform was the focus of excavation. Geochemical survey was undertaken during the excavation at a sample depth of c. 50cm within the excavation trench. Initial excavation exposed a blanket of colluvial 'hillwash' covering the archaeological remains. This was excavated to reveal the upper archaeological contexts and it was at this depth the geochemical samples were taken, c. 0.5m BGL (Below Ground Level). The excavation trench was cleaned by trowel, to give a coherent surface for geochemical sampling. Samples were taken using a trowel cleaned in distilled water and sample sizes were c. 10g. When collecting soil samples using a trowel, the sample will generally be context specific, i.e. the soil comes from a shallow scrape at one defined depth, which in most cases will lie within a single context. It is important to note that the extensive compacted smithing floor and furnace location, two of the central features of the excavations on the platform, were not fully exposed at the level at which the geochemical sampling took place.

Laboratory method: Geochemical survey

All samples were freeze dried after collection. After drying, each sample was homogenised by sieving through a 2mm stainless steel sieve. The element analyses required an acid digestion of the sediment sample, allowing some matrix decomposition and cation exchange. All samples were subjected to a pseudo-total acid digestion, a nitric acid (HNO_3) and hydrochloric acid (HCl) procedure which displaces the loosely bound cations within the samples but does not break down the silicate structure of the sample. This extraction procedure is less influenced by the underlying geology and soil structure, as the silicate lattice of the soil structure is not destroyed (Van Loon, 1985).

Atomic absorption spectrophotometry (AAS) was used to measure element concentrations via a Unicam 939 atomic absorption spectrometer with the Department of Geography sediment laboratories at the University of Exeter. AAS provides a relatively quick and simple method of univariable analysis. The detection limits are good (between 1 and 100ppm, depending on the element measured) and the analytical precision is high. Soil organic content was measured in each sample using Loss on Ignition at 450°C.

Data analysis and display

Principal components analysis (PCA) was used to analyse and transform the multi-element geochemical datasets, due to its ability to factorise the data matrix into (soil) sample related and (elemental) variable specific information, thereby visualising relationships between soil samples and elements through factor scores and loading plots. PCA takes a group of variables measured over a sample or population of observations, and looks at the interrelationships between them (Johnston, 1978), attempting to summarise data from a large number of variables in a smaller dimensional space, allowing data display in graphical or numerical forms. The PCA will produce a new principal component that is positioned through correlation of the original variables. Where metalworking is present and a survey is conducted on a close interval basis, it can be expected that a strong level of inter-variable correlation will exist. Therefore, when metalworking is present, a principal component can define the geochemical signature caused by deposition of metallic residues. By examining the factor scores it is possible to identify which of the samples had the most significant influence on the positioning of the component. Such samples will have high geochemical loadings and a high degree of inter-variable correlation, and are the sediments likely to have been contaminated by metalworking residues.

A GIS system was used to visualise and examine the geochemical data (MAPINFO, v. 6.5). Within a GIS environment, interpolation functions are generally used to produce a surface model from a dataset of *z* values. By using an interpolation function it is possible to create a continuous sub-surface from the original geochemical sample points that represents the concentrations of an element across a survey area or transformed data values, such as factor scores from a PCA. An Inverse Distance Weighted (IDW) function was used, which calculates the value of grid cells over the mapping area, with each data point in the calculation weighted by its distance from the centre of the cell. Further data modelling was undertaken within ArcGIS (v. 8.3) to allow synthesis of the geochemical data with excavation and gradiometer data sets, and data modelling and visualisation in a pseudo 3D environment.

Results

The excavation of platform A2 revealed a range of well-preserved and significant ironworking features, including a large furnace setting and smithing floor. The furnace setting survived to a height of 1.5m and preserved evidence of at least 10 episodes of reconstruction and use, set within a stone-revetted hollow and connected by means of a heavily trampled ramp to an extensive smithing floor (Fig. 3). The smithing

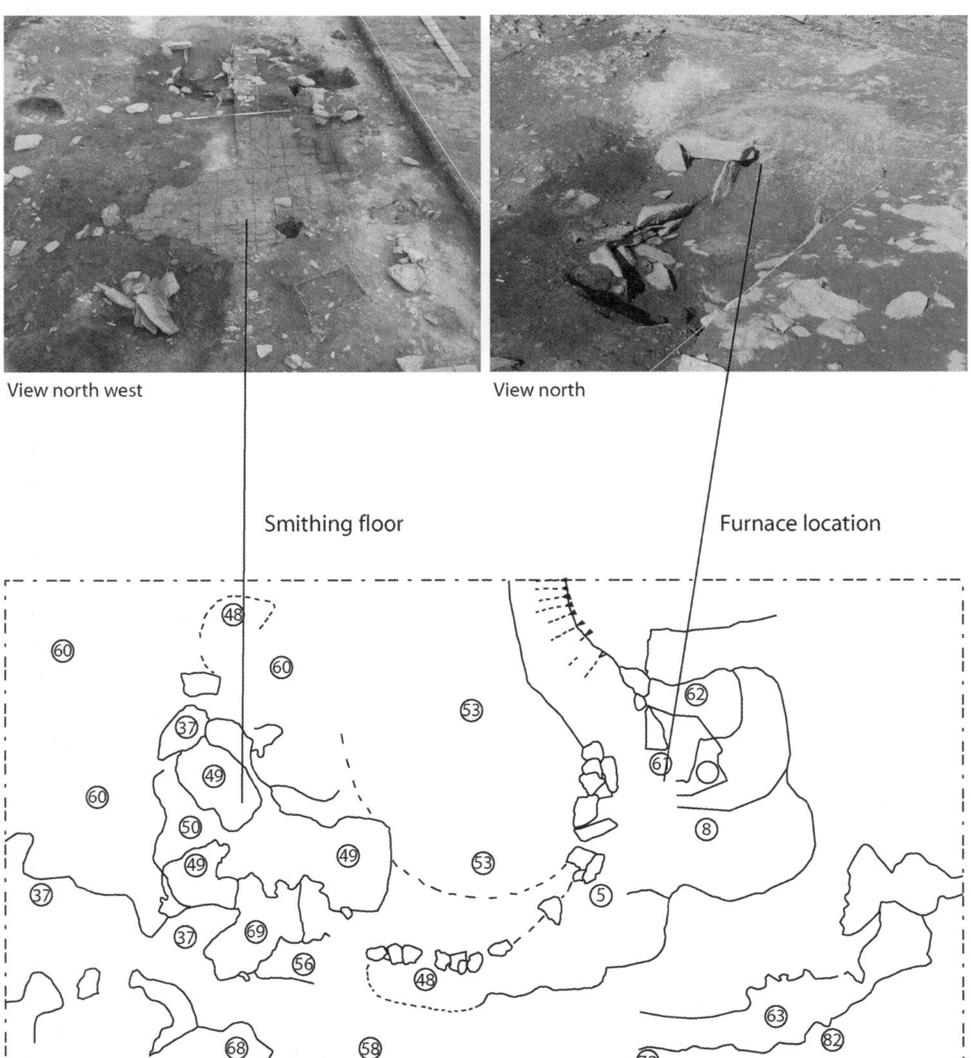

Figure 3: A schematic drawing of some of the excavation results on platform A2, with the furnace location and the smithing floor highlighted.

floor is exceptional, reaching in places 0.4m in depth of solid compacted residues formed during the primary smithing of hot iron blooms extracted from the adjoining furnace. On the platform, the remains of several other furnaces that had been truncated by successive re-cutting of the platform to reconfigure the working area were found.

The unwanted slag and waste materials from the processes operating on the platform were dumped downslope on the accreting slag heap. Excavation through the heap revealed the cyclic nature of deposition. The contexts which formed the slagheap could be categorised into three generic types:

1. Re-deposited 'natural' soil, often stony with little other material.
2. Technological debris layers, containing abundant slag, furnace lining, and charcoal. Many of these contexts were very loosely packed with large cavities but little infilling soil matrix.
3. Mixed layers with varying quantities of the 1 and 2 type of layers in them. The technological debris in these layers is generally smaller in size and quantity than in 2.

Returning to the geoprospection surveys over the platform, gradiometry identified the smithing floor in the 0.5m interval survey as anomalies GA2 and GA3 (Fig. 4). However, the exact shape of the smithing floor was not defined, and the pre-excavation interpretation of these two anomalies was as possible furnaces. It is interesting that the smithing floor produced two apparently distinct magnetic dipoles. These have subsequently been interpreted as two areas of heavily concreted smithing slag within the floor. The furnace location was clearly identifiable as the large dipole GA1.

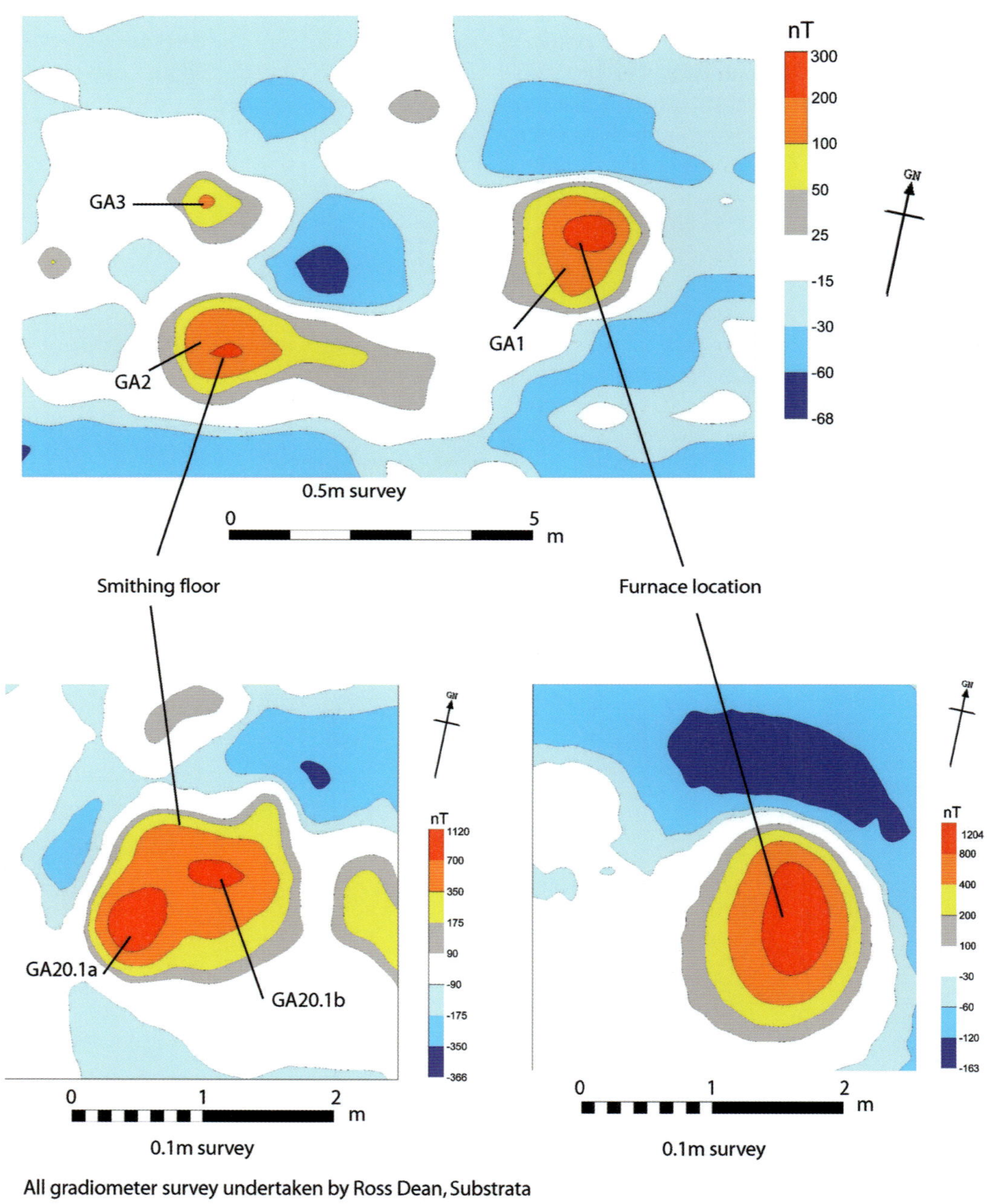

Figure 4: The gradiometer survey results from the platform A2 survey.

Geochemical survey on platform A2 produced a PCI that accounted for 35% of the original variance in the dataset and a PCII that accounted for a further 19%. The location of PCI was heavily influenced by the elements Fe, Mn and Cu, with only Cr and Mg displaying a negative association with PCI, and Zn and K both having a weak positive association. The plot produced from the PCI factor scores clearly identifies the smithing floor (excavation contexts 49, 50, 69, 56 and 37; Fig. 5). These contexts all contain hammerscale, charcoal, and semi-molten slag that had been hammered out off the raw iron as it was forged and then trampled underfoot by the working blacksmiths to form a thick concreted surface. The area of higher PCI factor scores also extends over context 60, which had distinctive heavy black staining but did not contain large amounts of the compacted smithing slag.

The second major geochemical anomaly defines the location of the furnace setting. The location represents the boundary between two different sediment matrices and is defined by a clear geochemical boundary. The sediment units producing this geochemical change can be divided into two groups. The first is composed of contexts 8, 5 and 61, and represents the furnace location and the soil matrix that has infilled the furnace location. The second is the soil profile behind the furnace that represents the packing and the filling at the back of the furnace location. The furnace location is not defined through high PCI factor scores but through a geochemical difference between two areas, both of which have a lower geochemical loading than the smithing floor.

The geochemical definition of the smithing floor provides an interesting comparison with the 0.5 m gradiometer survey, redrawn as a simple polyline file (Fig. 5). The smithing floor gradiometer dipole (GA2) did not display a classic dipole form, and the magnetic response effectively identified an area of high Fe concentrations related to heating, which could be features such as hammerscale, slag deposition, or a furnace. The form of the archaeological feature is not apparent. In contrast, the geochemical image gives almost the exact definition of the smithing floor. However, the furnace setting is defined exceptionally well by the gradiometer dipole GA1 and correlates well with the geochemical change.

The examination of the individual elements adds further information. The Fe concentrations (Fig. 6) produce a good definition of the contexts that comprise the smithing floor (49, 50, 37 and 69) and also the furnace setting (8, 61 and 62). The Fe concentrations are still elevated in context 60 surrounding the smithing floor, although not as high as on the smithing floor. The Cu concentrations also produce a good definition of the smithing floor (contexts 49, 50, 37 and 69) and the furnace setting (Fig. 6), also with weaker concentrations over context 60. The Cu complexes and compounds in the smithing floor would have originated from the original ore material exploited for smelting. Further elements followed this same depositional pattern, including Mn (Fig. 7), which also displayed a high positive association with PCI.

Of further interest are the interrelationships between different metal elements and the magnetic responses recorded by the gradiometer survey (Fig. 7). Although already viewed as a simple line drawing, the interrelationships of these modelled surfaces can be viewed within a quasi 3D environment. Visually, the communality between data groups is obvious,

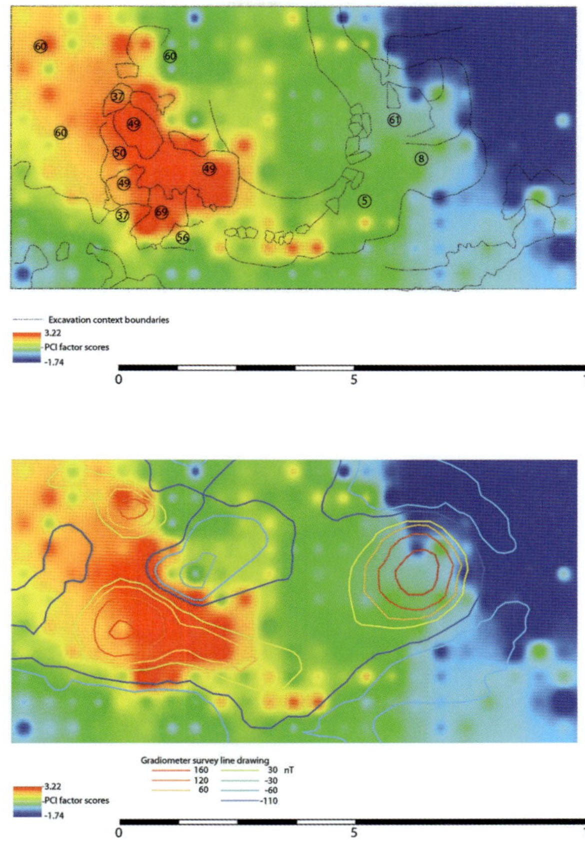

Figure 5: The geochemical survey results from the PCI factor scores and their relationship to the gradiometer survey results.

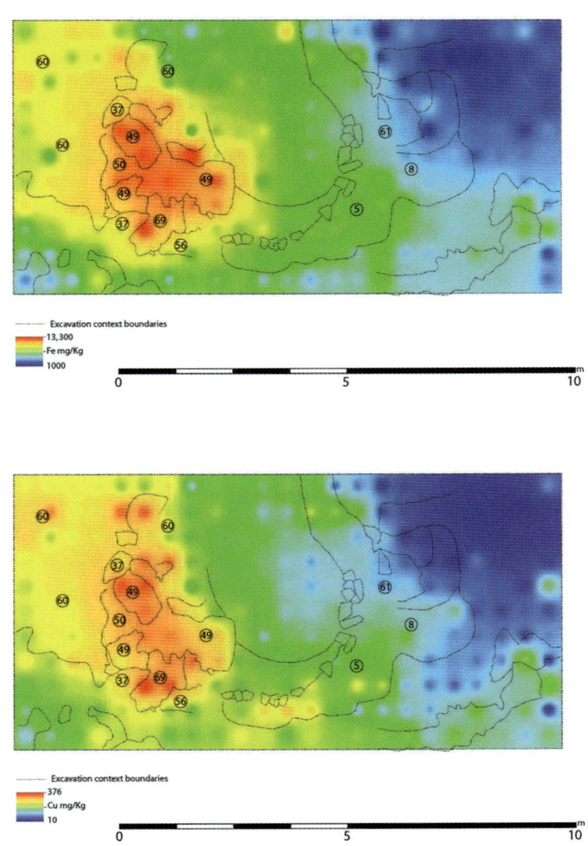

Figure 6: The geochemical plots of the Fe concentrations (upper image) and Cu concentrations (lower image).

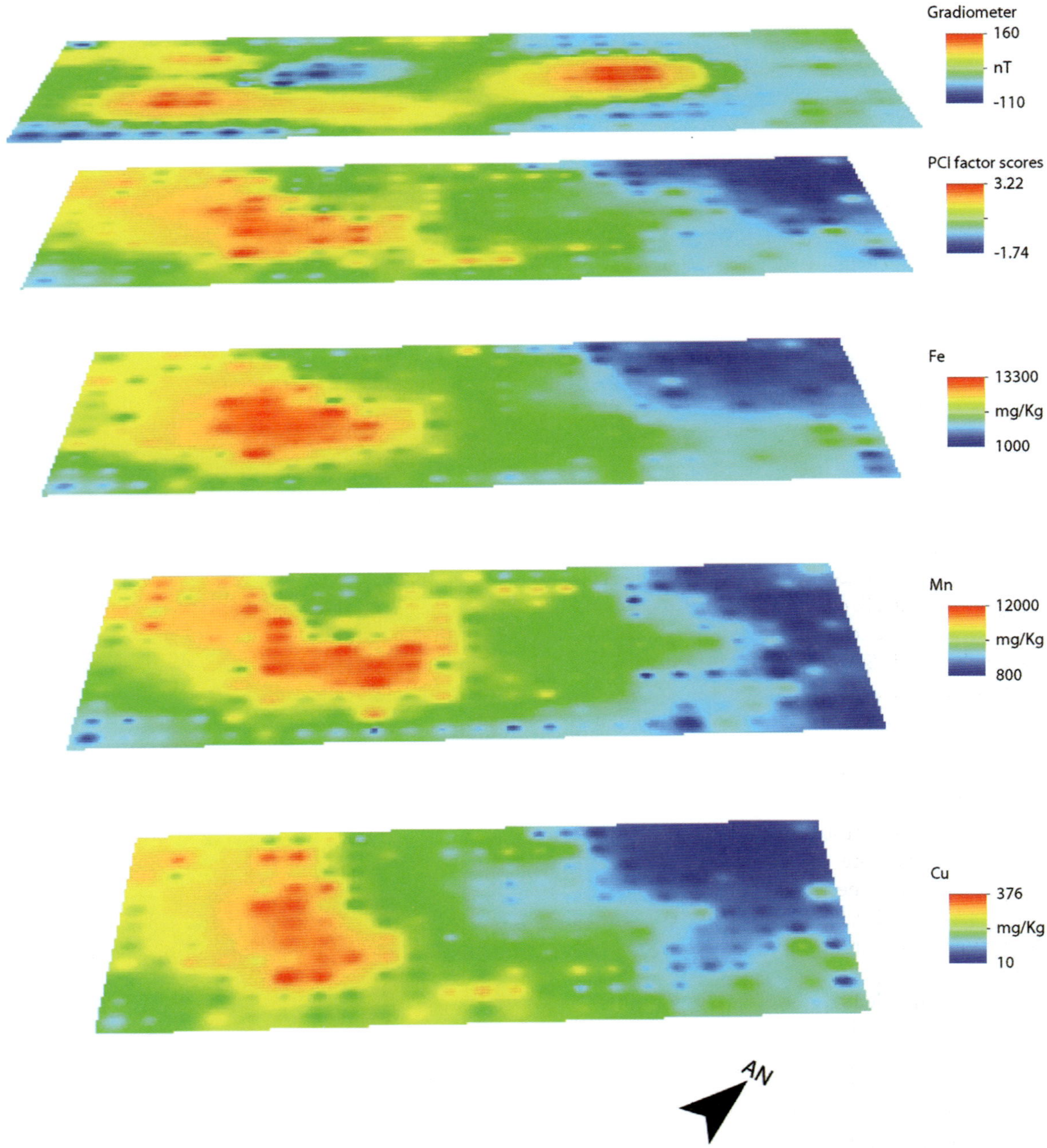

Figure 7: The visual inter-relationship of the gradiometer results, PCI factor scores and Fe, Cu and Mn concentrations.

with the smithing floor a dominant feature in all of the plots, and with the furnace location most precisely defined by the gradiometer survey (nT) and the manganese concentrations.

In particular, examination of the Fe concentrations and the gradiometer nT scores demonstrates a close visual correlation, with the nT responses effectively representing a combination of the Fe concentration and its state of oxidation or reduction. The relationship of the Fe concentrations to the magnetic nT responses were subjected to a further PCA analysis using raw (unstandardised) data. The resulting PCI accounted for 61.4% of the original variance in the dataset, only removing c. 38% of uncorrelated variance. The resultant plot of the contribution of the factor scores succinctly defines both the smithing floor and the furnace setting, demonstrating a high level of correlation with the excavated remains (Fig. 8).

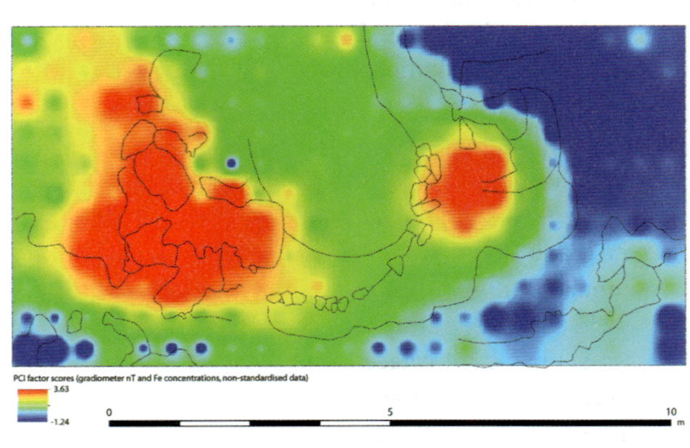

Figure 8: A data plot showing the fusion of the Fe concentrations and the gradiometer nT responses, through a PCA analysis of these two variables and plotting of the PCI factor scores.

Discussion

The results from this study demonstrate the considerable potential of using close interval, intra-site, high-resolution geochemical survey to investigate metalworking remains in combination with gradiometer survey. The PCI factor scores defined a series of anomalies that correlated very closely with the excavated archaeometallurgical features. The smithing floor was a physically large feature within the survey area which had a strong geochemical signature. These two factors, the area it covered relative to the overall survey size and the level of its geochemical enhancement, made this feature clearly identifiable in the geochemical results. Undoubtedly, the high level of correlation between Fe, Mn and Cu, and the subsequent positioning of the PCI, is primarily defined by samples from the smithing floor. The furnace setting was identified as a 'negative' geochemical feature based on the PCI factor scores, with the position of PCI derived from the metallic residues of the smithing floor. This 'negative' feature was the junction between two types of sediment matrices.

The presence of high concentrations of Cu and Mn, strongly correlating with the high Fe concentrations, suggests that both these elements were impurities within the ores being smelted at Sherracombe Ford. Small amounts of chalcopyrite are commonly found within the iron ore deposits in the Devonian rocks on southern Exmoor, and Mn is frequently noted as 'wads' within the ore bodies (Edwards, 2000). Therefore, a mode of deposition is suggested where Cu was deposited onto the smithing floor by the removal of furnace slag trapped within the iron blooms through primary smithing. Certain elements, such as Mn, are valuable additions to the iron smelting process because they act as flux, replacing iron in the slag (Tylecote, 1986, p.127). It is possible that Mn bearing ores were deliberately sought at Sherracombe Ford for smelting, potentially delivering higher iron bloom yields. Alternatively, Mn could have been added separately to the furnace as a flux, although this is considered less likely.

The geochemical and gradiometer surveys produced a high level of correlation. The furnace setting stands out as an example of the combination of these two techniques. The metallic component of the furnace setting had been removed. The location is identifiable as a weak geochemical signature caused by the change in soil structure. The preservation of the furnace setting had left intact the clay surround that had witnessed multiple firings. This produced a strong dipole identified by the gradiometer survey. Likewise, the smithing floor was identifiable as a series of high nT responses, whilst the geochemical survey provided the plan of the structure, identified through strong geochemical enhancement.

The integration and synthesis of the iron concentrations with the gradiometer data undoubtedly refines the definition of anomalies on the platform, which can be reconciled with archaeological features. The combination of the two datasets defines the furnace setting and the smithing floor with a high level of accuracy. In fact, when compared to the excavation schematic, the combined geoprospection data is moving to a level of context mapping, within a pre-excavation environment. Of course, with multi-context environments in a 3D world which have vertically accreted overtime, the mapping of contexts will be limited to the depth to which the geoprospection data is gathered. Nonetheless, this survey highlights the potential for geochemical survey to be combined with gradiometer data to map contexts on a pre-excavation basis and also to aid in the interpretation of contexts in a post excavation phase.

Conclusion

The investigation of the ironworking remains at Sherracombe Ford is significant for two reasons. Firstly, new methods of site investigation were explored, specifically, the integration of geochemical data with gradiometer data and their comparison with excavation results. The methodological development of compressing multi-element data using PCA and expressing this as spatial data plots has made the technique analogous to more conventional methods of geophysical survey. This provides a more accessible tool for archaeologists to use and understand.

By virtue of the fact that this geochemical survey methodology has been applied on an archaeometallurgical site, there are high geochemical loadings in the immediate site environment caused by the metalworking processes. This demonstrates geochemical survey to be a suitable tool to investigate such residues contained within archaeological contexts. This survey represents the first step in using a robust geochemical survey methodology to investigate metalworking sites. There is a potential to identify features that are ambiguous upon excavation by virtue of their geochemical signature, e.g. small Bronze Age furnaces; to identify remains and features on a pre-excavation or in lieu of excavation basis, or to help inform excavation, as in this case, and also to provide an environmental perspective on the archaeology of pollution, e.g. lead residues.

From an archaeological viewpoint, the investigations at Sherracombe Ford have revealed extensive evidence for Romano-British iron smelting in the uplands of Dumnonia, an area of scant Romano-British archaeological remains. The discovery of these sites has challenged conventional models of landscape evolution in the southwest of Britain and modified the archaeological perspective of Romano-British control and settlement.

Acknowledgements

All the excavation on platform A2 was undertaken by the Exmoor Iron project, under the supervision of Dr. Gill Juleff. All the gradiometer surveys at the site were undertaken by Substrata (independent geophysical consultancy) and were commissioned by the Exmoor Iron project. The earthwork survey was drawn by Rob Wilson-North (ENPA). Thanks go to English Heritage for funding this research.

References

Aston, M., Martin, M.H. and Jackson, A.W., 1998. The potential for heavy metal soil analysis on low status archaeological sites at Shapwick, Somerset. *Antiquity*, 72, pp.838-47.

Cook, S.R., Fulford, M. and Pearson, C., 2003. Soil geochemical analyses as an indication of metalworking at the excavation of a house in the Roman city of Silchester (UK). *Journal de Physique IV*, 107, pp.319-21.

Craig, N., Speakman, R.J., Popelka-Filcoff, R.S., Glascock, M.D., Robertson, J.D., Shackley, M.S and Aldenderfer, M.S., 2007. Comparison of XRF and PXRF for analysis of archaeological obsidian from southern Perú. *Journal of Archaeological Science*, 34, pp.2012-24.

Edwards, R.A., 2000. *Exmoor Geology*. Tiverton: Exmoor Books.

Entwhistle, J.A., Abrahams, P.W. and Dodgshon, R.A., 2000. The Geoarchaeological Significance and Spatial Variability of a Range of Physical and Chemical Soil Properties from a Former Habitation Site, Isle of Skye. *Journal of Archaeological Science*, 27, pp.287-303.

Fox, A., 1954. Celtic fields and farms on Dartmoor, in the light of recent excavations at Kestor. *Proceedings of the Prehistoric Society*, 20, pp.87-102.

Green, T., forthcoming. Excavations at the hillslope enclosure at Holworthy, Parracombe, North Devon. *Proceedings of the Devon Archaeological Society*.

Griffith, F. and Quinnell, H., 1999. Settlement, c. 2,500 BC to c. AD 600. In: R. Kain and W. Ravenhill eds. *Historical atlas of the southwest*. Exeter: University of Exeter Press, pp.62-8.

Heron, C., 2001. Geochemical prospecting. In: D.R. Brothwell and A.M. Pollard eds. *Handbook of archaeological sciences*. Chichester: John Wiley, pp.565-73.

Hong, S., Candelone, J.P., Patterson, C.C. and Boutron, C.F., 1994. Greenland ice evidence of hemispheric lead pollution two millennia ago by Greek and Roman civilisations. *Science*, 265, pp.1841-3.

Hong, S., Candelone, J.P., Patterson, C.C. and Boutron, C.F., 1996. History of ancient copper smelting pollution during the Roman and medieval times recorded in Greenland ice. *Science*, 272, pp.246-9.

Jackson, A.W., 2001. *Analysis of soil chemical residues and other soil associated with past human activity*. PhD Thesis, Department of Archaeology and Anthropology, University of Bristol.

Johnston, R.J., 1978. *Multivariate statistical analysis in geography*. London: Longman.

Juleff, G. and Bray, L., 2007. Minerals, metals, colours and landscape: Exmoor's Roman Lode in the Early Bronze Age. *Cambridge Archaeological Journal*, 17, pp.285-96.

Linderholm, J. and Lundberg, E., 1994. Chemical Characterisation of Various Archaeological Soil Samples Using Main and Trace Elements Determined by Inductively Coupled Plasma Atomic Emission Spectrometry. *Journal of Archaeological Science*, 21, pp.303-14.

Middleton, W.D. and Price, T.D., 1996. Identification of Activity Areas by Multi-Element Characterisation of Sediments from Modern and Archaeological House Floors Using inductively Coupled Plasma-Atomic-Emission Spectroscopy. *Journal of Archaeological Science*, 23, pp.673-87.

Penhallurick, R., 1997. The evidence for prehistoric mining in Cornwall. In: C.J. Salter ed. *Prehistoric Extractive Metallurgy in Cornwall*. Truro: Cornwall County Council.

Proudfoot, B., 1976. The analysis and interpretation of soil phosphorus in archaeological contexts. In: D.A. Davidson and M.L. Shackley eds. *Geoarchaeology: earth science and the past*. London: Duckworth, pp. 93-113.

Riley, H. and Wilson-North, R., 2001. *The Field Archaeology Of Exmoor*. Swindon: English Heritage.

Silvester, R.J., 1980. An excavation at Dunkeswell. *Proceedings of the Devon Archaeological Society*, 38, pp.53-66.

Thorndycraft, V.R., Pirrie, D. and Brown, A.G., 1999. Tracing the record of early alluvial tin mining on Dartmoor, UK. In: A.M. Pollard ed. *Geoarchaeology: exploration, environments, resources*. Geological Society Special Publications 165. London: Geological Society, pp.91-102.

Todd, M., 1998. A Hillslope Enclosure at Rudge, Morchard Bishop. *Proceedings of the Devon Archaeological Society*, 56, pp.133-52.

Tylecote, R.F., 1986. *The Prehistory of Metallurgy in the British Isles*. London: Institute of Metals.

Wilson-North, R., forthcoming. Excavation at Timberscombe hillslope enclosure, West Somerset. Internal Exmoor National Park report.

Van Loon, J.C., 1985. *Selected Methods of Trace Metal Analysis: Biological and Environmental Samples*. Chichester: John Wiley.

Authors' addresses

- Chris Carey
  Research Fellow: University of Birmingham, School of Archaeology and Antiquities and director Carey Consulting. Flat 4, Higher Threave, Westhill, Braunton, Devon. EX33 1AR.
  E-mail: C.carey@bham.ac.uk or chriscareyconsulting@gmail.com (corresponding author)
- Gillian Juleff
  Senior Lecturer, Department of Archaeology, Laver Building, North Park rd, University of Exeter. EX4 4QE.
  g.juleff@ex.ac.uk

Three ores, three irons and three knives

Quanyu Wang and Peter Crew

ABSTRACT: Experimental iron working provides the opportunity for a systematic study of the bloomery process from ore to finished product. In this project samples taken from three refined iron bars and knives, made by the experimental smelting of a bog ore and two different Northants ores, were studied to provide comparative data on the metal types, the slag inclusions and the perceived and actual qualities of the different irons. The microstructure and composition of the metals and slag inclusions were examined using metallographic microscopy and scanning electron microscopy equipped with energy dispersive X-ray spectrometry (SEM-EDX). The microhardness of the samples was also measured.

The metals have different compositions and microstructures, reflecting both the ores used and the way they were smelted and refined. All of the metals studied are heterogeneous, with variable carbon and phosphorus contents and with different types of slag inclusions. Arsenic-rich layers were found on the surface of all the bars but not in the weld lines of the knives. The phosphorus contents do not reflect the composition of the ores used. There are significant differences between what was revealed by metallography compared to the initial assessment of the quality of the knives and to their effectiveness in use.

Keywords: Bloomery iron, bar iron, knife, phosphorus, experimental smelting, slag inclusions, metallography.

Introduction

The archaeometallurgist, looking at samples under the microscope, is a long way from the heat of the furnace and the noise of the anvil. Experimental iron working can go some way towards bridging this gap by providing samples of iron smelted and refined under known conditions, allowing a more secure interpretation of the metallurgical structures and slag inclusions and thus providing a corpus of information to assist in the interpretation of archaeological samples.

The quality of bloomery iron and steel depends largely on the skill of the smelter in producing a good bloom, which can be refined to usable stock iron, and then on the skill of the blacksmith in forging useful objects. But there are many gaps in our knowledge, as experimental iron smelters, working on an occasional basis, cannot begin to emulate the skills of ancient iron workers, who practiced their craft on a regular basis. Even experienced modern blacksmiths can find it difficult to work with bloomery iron and even more difficult to assess its properties. There is still a lot to learn.

The genesis of this project was a request from a student at Leicester University, who was carrying out a comparative study of cut marks on the bone from the hillfort at Danebury, Hampshire, with those made by flint flakes and iron knives (Knight 2003). Three knives, each smelted from different ores to produce blades with ostensibly different phosphorus and carbon contents, were made for the experimental butchering of a pig, to provide comparative data. Samples of these three iron bars were used for another project at the British Museum to examine the corrosion characteristics of different metal types (Wang and Crew 2007). These iron bars and blades were examined in further detail for the project presented in this paper, the aim being to compare the metallographic results with the assessment of the properties and qualities of the three different iron blades.

The experiments

The experiments were carried out over several years at Plas Tan y Bwlch, the Snowdonia National Park Study Centre. Experiment XP27 (1989) was one of a long series of experiments smelting bog ores in a programme of research linked to excavations of the prehistoric sites at Bryn y Castell and Crawcwellt and, more recently, of the medieval site at Llwyn Du (Crew and Charlton 2007). XP57 (1993) and XP65 (1995) were part of a programme of smelting other types of ore from the main early iron producing regions of Britain, including Northants, the Forest of Dean, Cumbria and East Yorkshire. The iron from these two experiments was chosen for this project as it was made from different ores from the Jurassic Ridge series, which may have been one of the sources of the Danebury iron (Crew 1995). Analyses of the ores used are presented in Table 1 and the furnace type, operating conditions, products and yields for XP27 have been described in detail (Crew 1991).

The experiments were carried out in similar furnaces, made of the same clay and using similar beech charcoal. Thus any variation in the resulting iron would be due primarily to the different ores used, combined with differences in the operating conditions of the smelts and the degree of refining required. The technique for refining blooms requires considerable adjustment depending on the nature of the bloom, the amount

Table 1. Analyses of the ores used for XP27, River Perry bog ore and XP65, Harlaxton ore, by Vincent Serneels (XRF); XP57, Newton West Pit ore by Ian Devine (AAS).

Wt.%	*SiO_2*	*TiO_2*	*Al_2O_3*	*Fe_2O_3*	*MnO*	*MgO*	*CaO*	*K_2O*	*P_2O_5*	*Total*
XP27	7.87	0.11	2.16	80.19	2.85	0.12	2.79	0.21	2.93	99.2
XP65	20.71	0.19	3.87	72.92	0.32	0.33	0.5	0.68	0.82	100.3
XP57	13.9	0.31	5.2	74.4	0.18	1.08	3.63	0.15	1.39	100.2

ppm	*Zr*	*Sr*	*Rb*	*Ba*	*V*	*As*	*Zn*	*Cu*	*Ni*	*Cr*	*Co*
XP27	31	35	12	995	19	389	75	12	25	28	31
XP65	74	153	21	210	254	227	341		124	143	105
XP57						160	380	10	130		60

of slag contained, the fluidity of the slag and the composition of the metal. Because of our lack of experience in forging bar iron, the final smithing has mostly been carried out by a series of professional blacksmiths, each of whom had different backgrounds and skills and with differing reactions to the problems of forging bloomery iron.

XP27 was the first occasion on which a whole bloom was fully refined to bar iron (Crew 1991). Some 8kg of bog ore was smelted at an air rate of 230 litres/minute and at a 1:1 ore to charcoal ratio. This resulted in a bloom of 1700g, a 40% yield from the iron in the ore, which was refined to a square-sectioned billet of 690g (15% yield). The final product was a reproduction currency bar of 516g (12% yield) made by Heinz Denig, using a power hammer. The quality of the bar seemed to be quite reasonable, though a fine sand flux was used to help close up the fine cracks which developed as the bar was reduced in section. The bar has been studied in some detail and proved to be a composite material with a steel core (0.2 to 0.7% C and with up to 0.4% P), zones of phosphoric iron (up to 0.9%) and ferritic iron, with surface enrichment of arsenic up to 5% (Crew and Salter 1993). The smelting slags have also been analysed in some detail (Serneels and Crew 1997; Crew 2000).

XP65 was carried out in a new furnace of slightly larger diameter. 8kg of Harlaxton ore was smelted with 200 litres air/minute, again at a 1:1 ratio. The bloom of 1450g (37% yield) was not well consolidated, probably because only a short post-heat was used. The initial refining was done without difficulty to provide a good billet of 660g, an almost identical yield to that from XP27. The final refining was done by Richard Powell to produce a currency bar of only 156g (4% yield). Because of the smith's lack of experience with bloomery iron, the final bar forging took a long time with far too many heats and with considerable losses to scale. All the forging was carried out with a low fire and at relatively low temperatures, except for one full forge weld, and no flux was used at any stage. Although the final currency bar was small, the iron was very clean and had no cracks.

XP57 was one of a series of experiments smelting larger quantities of a self-fluxing Northants ore with slag tapping and hot bloom removal, with a variety of charcoal to ore ratios. In this case a low 1:2 ratio was used to smelt 16kg of ore at a blowing rate of 300 litres/minute. There were five small runs of tap slag (1.5kg in total). The result was a large bloom of 3275g (39% yield), which seemed to be of good quality, though it still contained a lot of slag. The refining was carried out by Hector Cole, with some difficulty, both in removing the slag and closing up persistent cracks, and much of the work was carried out at welding heat. Three long sessions were required for the early refining, each producing a smithing hearth cake, though the square-sectioned bar of 930g was a reasonable yield of 11%, due to the smith's skill. The intention was to forge this to a large currency bar, but as the iron was reduced in size, transverse cracks began to appear, especially towards the end of a heat when the iron was cooling. Attempts were made to improve the iron by carrying out a series of full forge welds. This was partially successful, but the bar finally fell apart into five pieces. At the time, this was thought to be a result of high melting point slag inclusions which had not been removed during the refining. One of these pieces was then successfully welded into a piece of ferritic medieval iron and drawn down to a 350g composite bar, without any signs of cracking. No flux was used or needed.

The knife smithing

These bars were subsequently used in 1999 to make a series of knives, two from each bar. This work was carried out by Hector Cole, in his own forge, using charcoal fuel. At the same time part of a phosphorus-rich (≈ 1%) currency bar, from the 9th century AD hoard found in Krakow, Poland, was refined by forge welding and used to make another knife (Nosek 1991). Most of the knives took 11-12 heats and a total of 15-20 minutes to make. The times taken are a reflection of the forgeability of each iron, though the number of heats taken and the losses of 26 to 28% were remarkably similar. The longer time (35 minutes) and greater number of heats (17) for the second knife from XP27 were because the first tang broke off, possibly due to phosphorus in the grain boundaries, and a new tang had to be forged. The XP27 metal type was known from previous study and in comparison the XP57 iron was thought to contain less carbon and phosphorus, but more than in the XP65 metal. Despite its high phosphorus content and the supposed difficulties of forging this type of iron (Vega et al., 2003a, 2003b), the easiest bar to work was that from Krakow, which gave a knife with a good edge.

The whole of the knife blades were then ground to remove the surface oxide layer, mainly to reveal any faults. Those from XP65 were clean and those from XP27 had a few fine longitudinal cracks caused by slag inclusions. The XP57 knives had transverse cracks in the core, visible only on the backs, but these had mostly been contained by the ferritic casing and there were no cracks in the cutting edge. All the knives were then sharpened by hand using whetstones (from Bryn y Castell) which took considerably longer than the forging. Finally, one example of each type was mounted in an antler handle, for use in the butchery experiments.

Assessments of the quality of the knives were made at the stages of manufacture and use, as following:

- XP27 was difficult to forge and had a good edge;
- XP65 was soft and very easy to forge, but had a poor edge;
- XP57 was easy to forge with care and had an excellent edge.

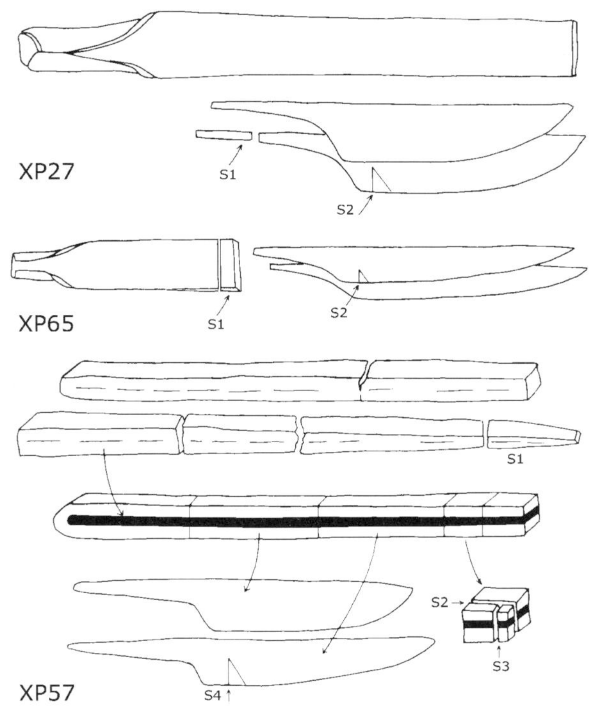

Figure 1: Drawing of the bars and knives, showing the locations of the samples.

Sampling

The samples from XP27 studied here include a section from the broken tang of the knife (S1) and a section from the knife blade (S2). The samples from XP65 were taken from end of the currency bar remaining after the knife forging (S1) and a section from one knife blade (S2). Several samples from XP57 were taken to study the iron at different stages of forging, including S1 from the end of the cracked bar, two sections from the sandwich bar, made by welding part of the broken bar into a piece of medieval ferritic iron, S2 from the long axis and S3 from the traverse section; and finally a sample S4 was taken from one of the knives (Fig. 1).

Analytical techniques

The metallographic samples were taken using an Isomet diamond saw. The sections were mounted in epoxy resin, ground and then polished using diamond paste to a finish of 1μm. After examination of the corrosion layers, the polished sections were etched using a 2% solution of HNO_3 in alcohol (nital) to reveal the metallographic structure of the metals.

A Zeiss AXIOVERT 100A microscope was used for the metallographic study of the metals. A Hitachi S-3700N Variable Pressure SEM with an Oxford INCA Energy system was used for elemental analysis and maps. The analyses were run at an accelerating voltage of 20 kV at low vacuum of 50Pa and working distance of 10mm. The detection limits for each element are variable but are typically 0.1-0.3%. The relative precision (reproducibility) is about 2% for the major elements, 10% for concentrations in the range 5-20% and it deteriorates as the detection limits are approached. Microhardness tests were carried out using a Vickers M55 Diamond Pyramid Microhardness tester on the polished surface of the metal.

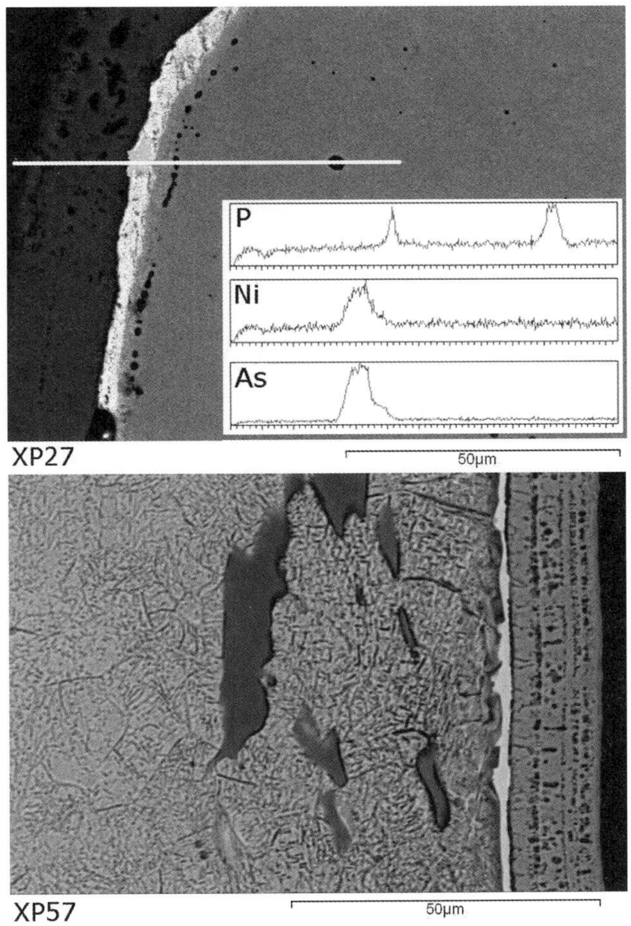

Figure 2: Arsenic-rich layers: Top: on the edge of the XP27 bar iron (S1), with linescans for P, Ni and As. *Bottom*: on the edge of the XP57 bar iron (S1).

Analytical results and discussion

All the samples from the bar irons have a thin arsenic (As)-rich layer on the surface underneath the smithing oxide layer (Fig. 2). This arsenic-rich layer can be porous and nickel (Ni) is also present in the layer. The contents of both arsenic and nickel vary between these samples with XP65 having the highest

Table 2: Compositions of slag inclusions analysed by SEM/EDX

XP27	*inclusion type*	SiO_2	TiO_2	Al_2O_3	FeO	MnO	MgO	CaO	K2O	SO_3	P_2O_5	As_2O_3	BaO
S1 knife tang	glassy slag	42.9	0.6	9.2	17.6	11.6	1.5	10.7	1.8	0.4	3.2	0.1	0.3
S1 knife tang	crystalline slag	31.2	0.4	7.0	34.3	8.3	0.8	8.0	1.6	0.6	7.2	nd	0.3
S1 knife tang	crystalline slag ?	37.0	0.5	8.1	25.9	10.0	1.2	9.4	1.7	0.5	5.2	0.1	0.3

XP65	*inclusion type*	SiO_2	TiO_2	Al_2O_3	FeO	MnO	MgO	CaO	K2O	SO_3	P_2O_5	As_2O_3	BaO
S1 currency bar	matrix of fayalitic inclusion	37.1	0.6	14.5	30.2	0.3	0.3	7.5	1.9	0.2	6.7	0.2	0.1
S1 currency bar	fayalite crystals	30.8	nd	0.4	62.8	0.6	2.3	0.5	nd	0.1	1.7	0.1	0.1
S1 currency bar	glassy slag	27.0	0.2	7.3	53.6	0.3	0.7	2.8	1.0	0.2	6.1	0.4	nd

XP57	*inclusion type*	SiO_2	TiO_2	Al_2O_3	FeO	MnO	MgO	CaO	K2O	SO_3	P_2O_5	As_2O_3	BaO
S1 cracked bar	glassy slag	38.5	0.8	15.8	6.4	0.4	2.8	27.9	2.1	0.4	4.4	0.1	nd
S2 welded bar	glassy slag	36.2	0.7	14.2	4.6	0.3	3.0	30.9	1.7	0.4	7.4	0.2	0.1
S3 welded bar	glassy slag	43.1	0.9	18.7	4.0	0.4	3.0	26.8	1.9	0.3	0.7	nd	0.1
S3 welded bar	matrix of wustitic inclusion	33.3	0.2	6.4	49.6	0.3	2.7	4.0	2.5	0.3	0.3	nd	0.2
S3 welded bar	dark phase of two phase slag	22.8	nd	2.7	69.9	0.2	0.9	0.6	0.6	nd	2.0	nd	nd
S4 knife blade	glassy slag	42.1	0.9	18.4	3.5	0.4	3.3	27.2	2.1	0.3	1.5	0.2	0.1

XP57	*P and Ca rich inclusions*	SiO_2	TiO_2	Al_2O_3	FeO	MnO	MgO	CaO	K2O	SO_3	P_2O_5	As_2O_3	BaO
S1 cracked bar	A1, A8, A9 dark globular phase	0.1	nd	nd	5.9	nd	0.8	45.9	0.5	nd	47.0	nd	nd
S1 cracked bar	A10 dark matrix	10.6	nd	1.8	43.2	0.1	0.9	12.9	0.4	1.8	28.0	0.1	0.1
S2 welded bar	A11, A12 dark globular phase	27.5	0.7	8.9	7.1	0.2	1.9	37.4	2.0	0.6	13.4	0.2	0.1
S2 welded bar	A13, A14 light matrix	29.7	nd	24.7	1.9	nd	3.2	38.6	0.4	0.1	1.0	0.3	0.1
S4 knife blade	A15, A16 light globular phase	1.4	nd	0.5	8.7	nd	0.8	42.3	0.4	0.2	45.3	0.1	0.2

(11.0% As and 8.1% Ni), XP27 having the lowest amount (4.5% As with trace of Ni), and XP57 being in between (5.8% As and 1.4% Ni). These levels are a considerable enhancement of the low values detected in the ores. The accumulation of arsenic at the surface of iron or steel undergoing oxidation is well known, with the layer being thicker when oxidation took place in the austentic range (Tylecote and Thomsen 1973).

The arsenic-rich layer varies in each of the metal bars, which is 6, 3 and 2μm for XP27, XP57 and XP65, respectively. The thickness of the arsenic-rich zone partly depends on the state of forging, as this enriched zone can break away, resulting in very high values of arsenic, nickel and sometimes cobalt in the hammer-scale (Serneels and Crew 1997). Neither the XP57 nor XP65 knives had arsenic-rich layers in the weld lines.

Metallographic studies revealed that all of these samples have heterogeneous carbon and phosphorus contents. Slag inclusions of different types were found to be present, including glassy, crystalline as well as calcium and phosphorus rich inclusions. Slag analysis was carried out using SEM-EDX and the results are shown in Table 2. The details of the analytical results for each sample are reported and discussed below.

XP27

The XP27 bar iron has a ferritic-pearlitic structure with the carbon content varying from almost nil to about 0.4%. Phosphorus was detected in some areas up to 0.2%. Another sample from the original bar of XP27 has already been studied (Crew and Salter 1993) which revealed variable carbon contents from almost nil to 0.7% as well as variable phosphorus contents in different areas of the sample, as mentioned above.

The XP27 knife is similar in microstructure from the cutting edge to the middle of the blade, though it has a very uneven distribution of carbon from one surface to the other. Fig. 3 shows that one surface has a ferritic structure, with a small amount of pearlite, containing about 0.1% carbon and the other surface has a structure close to the eutectoid composition (c. 0.6%). The differences in carbon content were reflected in the microhardness tests, which showed values varying from 221Hv in the low-carbon region to 369Hv in the high carbon region.

The phosphorus contents seem to be higher at the cutting edge than in the middle of the blade. Analyses by SEM-EDX showed an average of 0.4% phosphorus (with the maximum being 0.6%) at the cutting edge and 0.2% phosphorus (with the maximum being 0.5%) in the middle of the blade. The difference in carbon and phosphorus contents between the bar sample studied here, the sample studied by Salter, and the knife sample shows the heterogeneous nature of the metal.

Both glassy slag and slag with wüstite crystals are present in XP27; the glassy slag inclusions were found to be in the regions of carbon-rich metal and the slag with wüstite crystals was mainly present in ferritic metal, as described by Paynter (2006). Manganese is present at relatively high levels within the glassy slag, which is typical for iron smelted from a bog ore. There seems to be a crystalline-like structure in some of the glassy slag inclusions, which showed some difference in composition (Table 2).

XP65

The XP65 bar iron was found to be a phosphoric iron with a low carbon content (Fig. 4a). SEM-EDX analysis showed that it contains phosphorus in the range of 0.2-0.7%. The slag inclusions are very elongated, and were found to be three types, including glassy slag, slag with wüstite crystals and slag with fayalite crystals. The fayalite slag inclusions are much larger than the wüstite slag inclusions. A crack on one side of the bar (80μm) has a wüstite-type of slag, which is probably a result of the forge-weld being incomplete and the slag not being fully removed.

The XP65 knife was found to have a similar structure to that of the bar iron, with an uneven distribution of carbon and phosphorus (Fig. 4b). The grain size varies greatly with the largest ones being at the cutting edge (Fig. 5a). Analyses by SEM-EDX showed an average of 0.4% phosphorus (with the maximum being 0.8%) at the cutting edge and 0.5% phosphorus (with the maximum being 0.9%) in the middle of the blade.

There is a weld line, revealed by a high concentration of pearlite, running from the middle of the blade to the cutting edge, though it is not visible towards the cutting edge. An SEM-EDX linescan across the weld line (Fig. 5b) shows a sharp carbon peak and a more diffuse nickel peak. It is of some interest that arsenic was not present in the weld line. The linescans also indicate the presence of carbide on some grain boundaries and phosphorus enrichment in slag inclusions.

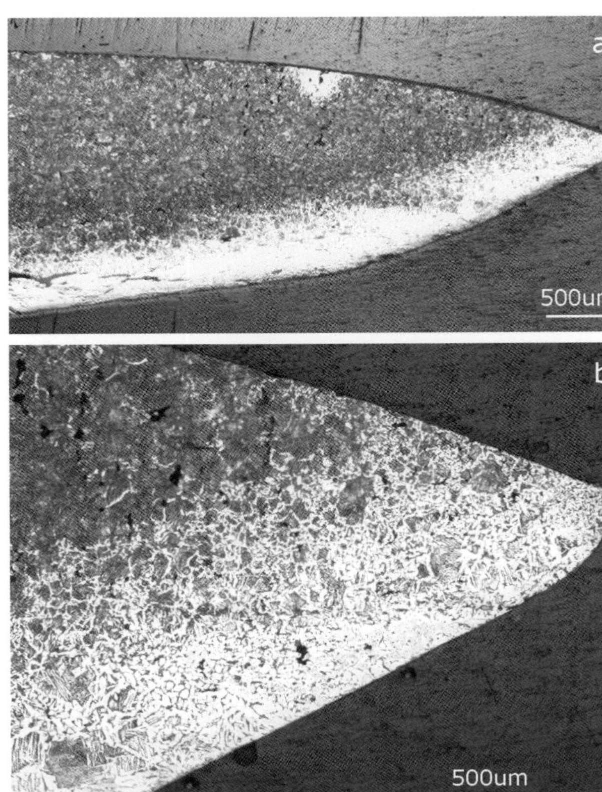

Figure 3: Photomicrographs of the XP27 knife section (S2), showing an uneven distribution of carbon.

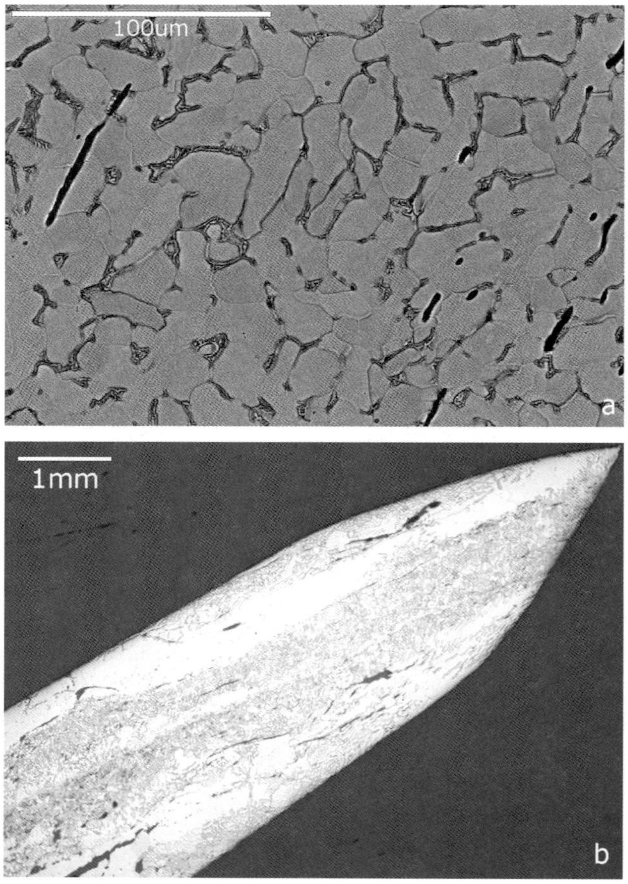

Figure 4: *a:* Backscattered electron image of the XP65 bar iron (S1), showing a phosphoric iron with a low carbon content. *b:* Photomicrograph of the XP65 knife section (S2), showing elongated slag inclusions and an uneven distribution of carbon.

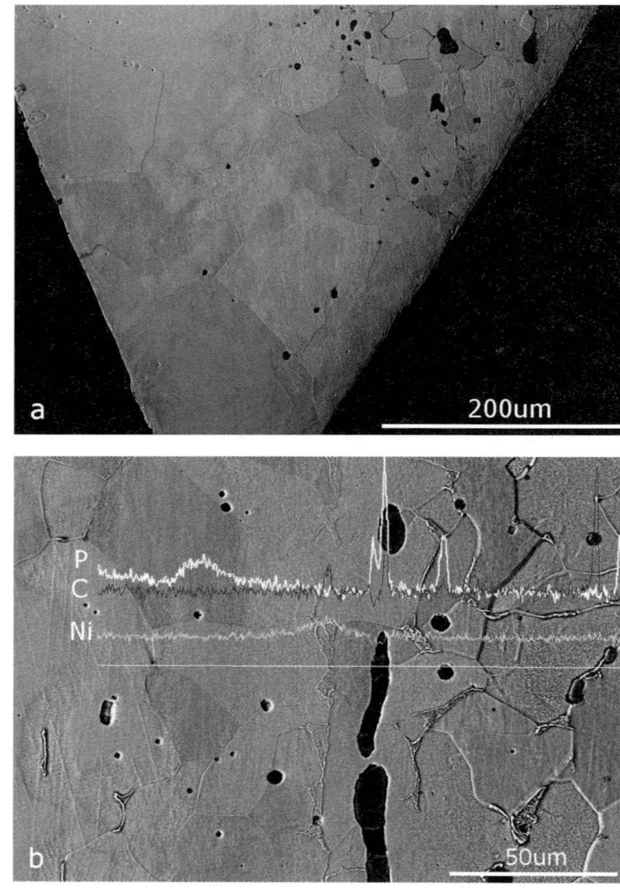

Figure 5: *a:* Backscattered electron image of the knife edge of XP65 (S2), showing large grains. *b:* Backscattered electron image of the weld line in the XP65 knife (S2), with EDX linescans for phosphorus, carbon and nickel.

XP57

Sample S1, from the end of the cracked bar, showed a ferritic structure (Fig. 6a) which was probably due to decarburisation during the extensive forging, with very elongated slag inclusions. Sample S4, taken from the middle of the knife made from the sandwich bar, showed that the carbon content decreases from the centre (about 0.2%) to the cutting edge (nearly pure ferrite), which again suggests decarburisation due to forging.

The boundaries between the medieval iron and the XP57 iron are shown clearly as a band at low magnification after etching (Fig. 6b). The welding was done so well that only a different phase or inclusions from the metal was observed on the boundaries at higher magnifications, which was proved by SEM-EDX to be pearlite. The SEM-EDX linescan revealed that phosphorus is high in some slag inclusions.

The slag inclusions in XP57 are very variable. There seem to be four main types of inclusion in S1, including wüstite slag, fayalite slag, glassy slag and slag containing mainly phosphorus and calcium (Fig. 7). The phosphorus and calcium rich slag inclusions are dense in some areas and they vary in composition, e.g. the P_2O_5 content in the inclusion in S1 is higher than that in S2 (Table 2). It seems that slag inclusions with wüstite crystals decrease in quantity and size with extent of forging, as the proportion of wüstite gets lower

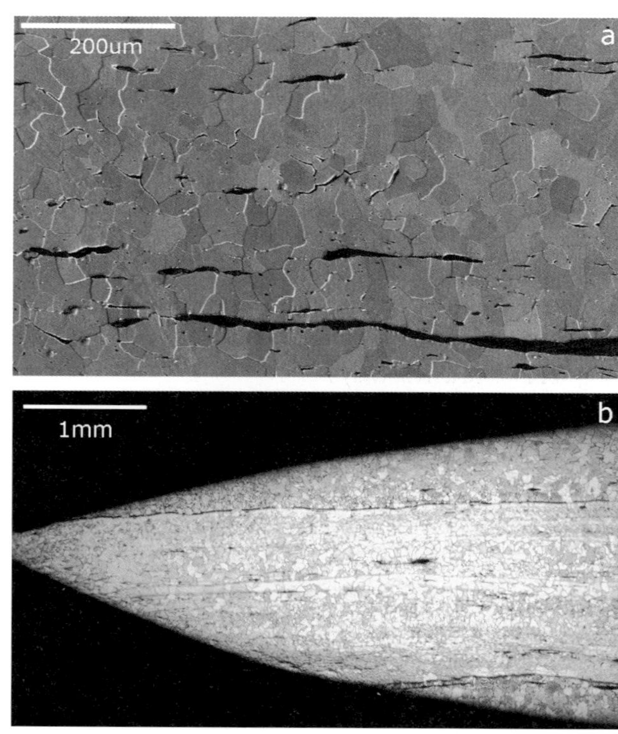

Figure 6: *a:* Backscattered electron image of the XP57 bar iron (S1), showing a ferritic structure with elongated slag inclusions. *b:* Photomicrograph of the XP57 knife edge (S4) showing the boundaries of the medieval and experimental iron.

in the sample from the sandwiched bar S3 and they almost disappear in the S4 sample from the final knife. The slag inclusions containing mainly phosphorus and calcium were found in all of the samples taken from the different stages of the experiment XP57. In sample S4 they were concentrated near the boundaries of the XP iron and the medieval iron (Fig. 8).

It has been suggested that CaO in slag can derive from added flux or lining from the smith's hearth (Piaskowski 1989) but it is now recognised that most CaO enrichment is due to the influence of the fuel ash (Crew 2000, 2007). In the case of XP57 no fluxes were used, though the ore has relatively high CaO. The composition of the phosphorus and calcium rich slag inclusions is close to $Ca_5(PO_4)_3(OH)$ but it is unlikely to be apatite, which has not been found by μXRD in slag inclusions. It is possible they may be $Ca_3(PO_4)_2$ though this is normally found only in iron from the indirect process (Dillmann and Fluzin 1997; Dillman *pers. comm.*). These slag inclusions are unlikely to be the direct cause of the cracking but might have provided the initial nucleating site of the crack if present at a high density (Salter *pers. comm.).* The presence of these unusual high Ca/P inclusions is difficult to explain, but it seems possible that they are due to P migrating from the metal as a result of the localised de-phosphorising influence of the slag inclusions (Vega et. al., 2003a). This may have been due to the high basicity of these inclusions (%CaO + %K2O / %SiO2) combined with the prolonged forging at high temperature. This type of inclusion is not yet understood and needs to be studied further.

An important question is the cause of the cracking of the XP57 bar. As all the slag inclusions seem to have been well deformed during forging, they are probably not a cause. In the central zone of Fig. 6a there are features around the grain boundaries which might be micro-cracks. One possible reason for these could be a result of the forging of the phosphoric metal at too high a temperature (Vega et al., 2003a, 2003b). With a phosphorus content of between 0.2 and 0.6% the iron must be forged at less than 900°C to avoid cracking caused by the $\acute{\alpha} + \gamma$ structure, because of the differences in the mechanical properties of the austenite and ferrite. Even with a phosphorus content as low as 0.1% the cooling iron must pass through a narrow $\acute{\alpha} + \gamma$ domain which may be sufficient to induce crack formation. This agrees with the observations made at the time of forging, that the cracks appeared as the iron was cooling and being hammered at relatively low temperatures.

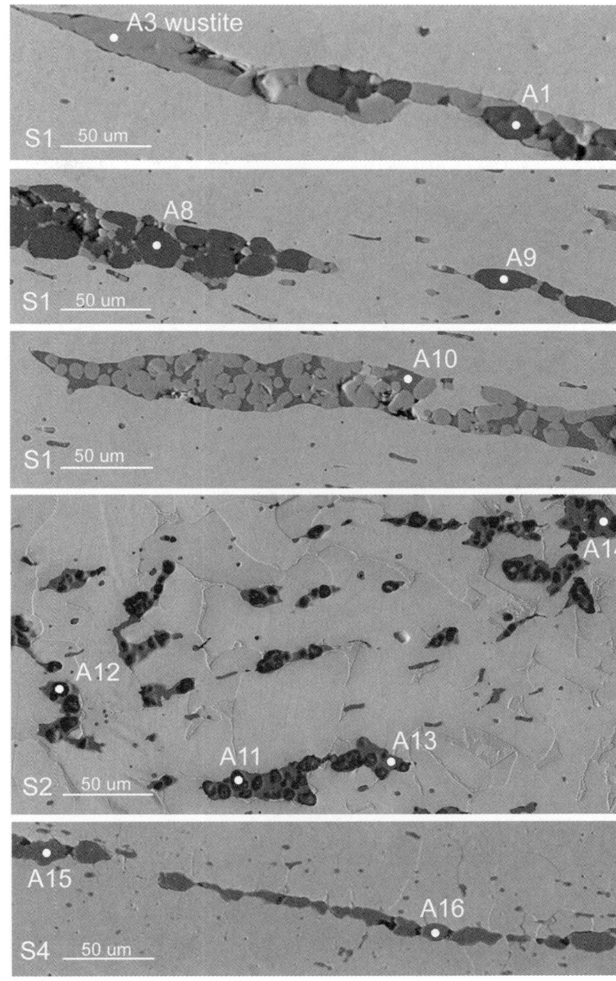

Figure 7: Composite backscattered electron images of samples from XP57, showing the locations of spot analyses of the slag inclusions.

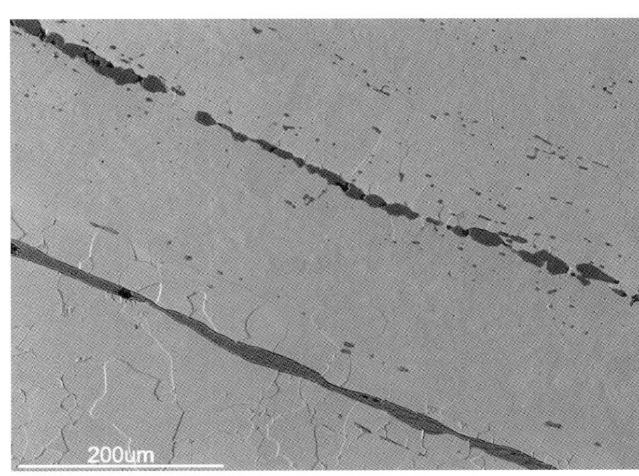

Figure 8: Backscattered electron image of the XP57 knife (S4), showing slag inclusions concentrated near the boundary of the experimental and the medieval iron.

Comparison of the three metals

As described above the metals from these samples all have heterogeneous microstructures, which are reflected in the variation of carbon and phosphorus contents (Table 3). It is not possible to give an accurate composition for the metals as carbon cannot be quantified by SEM-EDX, but these data give an indication of the composition and confirm the heterogeneity of the metals.

The phosphorus content in the metals does not reflect that given by the analysis of the ores (Table 1). Hedges and Salter (1979, 174) report that a low phosphorus content in the metal was found to be associated with a low phosphorus content in the slag as well as in the slag inclusions in the metal. The partitioning of phosphorus between the metal and slag seems to be very sensitive to the blowing rate used in smelting (Crew and Salter 1993, Table 1) which might explain the discrepancies in these experiments. However, the metal from XP65 appears to have the highest phosphorus

Table 3: Estimated carbon content based on the microstructure, phosphorous content by SEM-EDX analyses (data normalised, carbon not included and other trace elements not listed here; the balance is Fe), and microhardness of the metals

Sample	Estimated carbon content (wt%)	Phosphorus content (wt%)	Microhardness (Hv)
XP27 S1 knife tang	< 0.4	0.1	161 ± 29
XP27 S2 knife blade centre	< 0.8	0.2	320 ± 69
XP27 S2 knife blade cutting edge	< 0.8	0.4	245 ± 12
XP65 S1 currency bar	< 0.2	0.5	265 ± 14
XP65 S2 knife blade centre	< 0.2	0.5	215 ± 57
XP65 S2 knife blade cutting edge	< 0.2	0.4	208 ± 38
XP57 S1 cracked bar	almost nil	< 0.1	202 ± 14
XP57 S4 knife blade core (sandwiched in between medieval iron)	< 0.2	N/A	162 ± 3

content, despite being smelted from ore with the lowest P_2O_5 and ostensibly at the lowest blowing rate. A detailed examination of the smelting slags from these experiments would be necessary to clarify this problem. The XP57 metal has phosphorus only in some areas around the detection limit (0.1%), from an ore with a moderate P_2O_5 and a higher blowing rate, but this may be partially a result of the high CaO in the ore and slag inclusions, with the consequent de-phosphorisation of the metal, resulting in the inclusions with unusually high phosphorus and calcium.

The relatively high manganese content in the slag inclusions in XP27 reflects the nature of the ore. Silicon was detected in small concentrations in the metal for all the samples, which may reflect localised hotter and more reducing conditions in the blowing hole zone.

The microhardness tests (Table 3) generally reflect the nature of the analysed samples. The XP65 bar metal has a higher value than that of XP27 and XP57, because of its higher phosphorus content. XP57 S1 has a higher value than the other three samples from this experiment, which might be due to the cold hammering, which also perhaps contributed to the breaking of the bar. The variation of the hardness between the different samples of the XP27 metal is a reflection of the variation in the carbon content, with the knife being harder than the bar metal. The XP65 and XP57 knives, however, have a lower hardness than their bar metals.

However, the quality of the cutting edges of the knives, revealed by the butchery experiments, stands in marked contrast to the recorded microhardness values. Both the XP27 and XP57 knives were very efficient at cutting the flesh and bone of the pig, whereas the XP65 knife would not even cut the skin. There are several possible explanations for this, the most obvious being that the single samples analysed do not accurately reflect the overall quality of the knives. There is considerable variation in the reported hardness levels of phosphoric iron, which is partly due to the carbon content (Salter and Ehrenreich 1984, Fig. 10.10), though for a given carbon/phosphorus content there can be a very wide variation in the hardness. The large grain size and low carbon content at the cutting edge of the XP65 knife may account for its poor cutting quality. In contrast, the XP57 knife was effective in use, despite its similar low carbon content and very low phosphorus. This might suggest that carbon and phosphorus contents alone are not sufficient to characterise the quality of bloomery irons and it is probable that other trace elements, especially arsenic, may have had a more significant influence than is generally recognised (Tylecote and Thomsen 1973; Piaskowski 1984). A detailed study of this question is long overdue.

Conclusions

The metals are very heterogeneous with different microstructures and compositions between samples from different stages of smithing, in different areas of the same sample and with even greater differences between different areas of the same knife. The variation in carbon content is partly due to decarburisation increasing with forging. An arsenic-rich layer was found on the surface of all bar irons but not in the weld lines of the final knives.

Although only a limited number of the slag inclusions were analysed, it is clear that they are very variable, particularly in XP57, and more comprehensive analysis would be required to fully understand them. In particular, further study is needed on the calcium and phosphorus rich slag inclusions. The cracking of the XP57 bar, and the lack of cracking in the XP65 bar with a higher phosphorus content, is perhaps a confirmation of the need to forge different phosphoric irons at different temperatures.

Perhaps the most interesting question, in the context of this study, is the difference between the initially perceived types and properties of the iron, the metal types revealed by examination of the samples from the stages of forging and the differences in quality found in practical use. This is probably primarily due to the inherent variability of each metal type which cannot be fully characterised from limited sampling.

It is clear from this study that the examination of large, and preferably multiple, samples would be necessary to obtain an adequate impression of the range of variability in a metal, with clear implications for the study of archaeological objects and assessment of their quality. The quality of bloomery iron depends very much on the smelting and refining techniques used, which may not always be recognisable from the structure and composition of the final metal.

Acknowledgements

Thanks are due to all who helped with the experiments, especially to Hector Cole for forging the knives. Thanks also to Janet Lang, Chris Salter and Philippe Dillmann for their perceptive comments on the problems raised by this material.

References

Crew, P., 1991, 'The experimental production of prehistoric bar iron', *Historical Metallurgy* 25, 21-36.

Crew, P., 1995, 'Aspects of the iron supply', in Cunliffe, B., *Danebury, an iron age hillfort in Hampshire. Volume 6: A hillfort community in perspective*. Council for British Archaeology, Research Report 102, 276-284.

Crew, P., 2000, 'The influence of clay and charcoal ash on bloomery slags', in Tizzoni, C.C. and Tizzoni, M. (eds.) *Il Ferre nelle Alpi*, Bienno, 38-48.

Crew, P., 2007, 'CaO enhancement in slags: the influence of fluxing or of fuel ash?', in Crew, P. and Crew, S. (eds) Abstracts of *Early Ironworking in Europe II: archaeology, technology and experiment*. Plas Tan y Bwlch, Maentwrog, 50-52.

Crew, P. and Charlton, M., 2007, 'The anatomy of a furnace...and some of its ramifications', in La Niece. S., Hook, D. and Craddock, P. (eds) *Metals and Mines*, Archetype Publications, London, 219-225.

Crew, P. and Salter, C.J., 1993, 'Currency Bars with Welded Tips', in: Espelund, A., (ed.), *Bloomery Ironmaking during 2000 years, Volume III*. University of Trondheim, 11-30.

Dillmann, P. and Fluzin, P., 1997, 'Experimental archaeology: the contribution of micro-analysis of inclusions to the determination of the iron-working process', in Crew, P. and Crew, S., (eds.) Abstracts of *Early Ironworking in Europe: archaeology and experiment*. Plas Tan y Bwlch, Occasional Paper No. 3, 88-91.

Hedges, R.E.M. and Salter, C.J., 1979, 'Source determination of iron currency bars through analysis of the slag inclusions', *Archaeometry* 21, 161-175.

Knight, S., 2003. Butchery and intra-site spatial analysis of animal bone: a case study from Danebury Hillfort, Hampshire, England. PhD thesis, School of Archaeology and Ancient History, University of Leicester.

Nosek, E., 1991, 'Forging of high phosphorus iron', in Bielenin, K (ed.), *From Bloom to Knife*; *Materialy Archeologiczne* 26, 67-70.

Paynter, S., 2006, 'Regional variation in bloomery smelting slag of the Iron Age and Romano-British periods', *Archaeometry* 48, 271-292.

Piaskowski. J., 1984, 'Das Vorkommen von Arsen in antiken und frümittelalterlichen Gegenständen aus Renneisen', *Zeitschrift fur Archäeologie* 18, 213-226.

Piaskowski, J., 1989, 'Phosphorus in iron ore and slag, and in bloomery iron', *Archaeomaterials* 3, 47-59.

Salter, C. J. and Ehrenreich, R., 1984. 'Iron Age metallurgy in central southern Britain', in Cunliffe, B. and Miles, D., (eds), *Aspects of the Iron Age in Central Southern Britain*, Oxford University Committee for Archaeology, Monograph 2, 146-61.

Serneels, V. and Crew, P., 1997, 'Ore-slag relationships from experimentally smelted bog-iron ore', in Crew, P. and Crew, S., (eds.), *Abstracts of Early Ironworking in Europe: archaeology and experiment*. Plas Tan y Bwlch, Occasional Paper No. 3, 78-82.

Tylecote, R. F. and Thomsen, R., 1973, 'The segregation and surface enrichment of arsenic and phosphorus in early iron artefacts', *Archaeometry* 15, 193-198.

Vega, E., Dillmann, P., Fluzin, P., 2003a, 'Contribution à l'étude de fers phosphoriques en sidérurgie ancienne', *La Revue d'Archéométrie* 26, 197-208.

Vega, E., Dillmann, P., L'Heritier, M., Fluzin, P., Crew, P. and Benoit, P., 2003b, '*Forging of phosphoric iron. An analytical and experimental approach*', *Archaeometallurgy in Europe*, AIM Milan, Italy, 337-346.

Wang, Q. and Crew, P. (2007), Experimental investigation of the effects of relative humidity on corrosion of bloomery iron. Poster presented at The ICOM-CC Metal WG METAL'07 conference, 17th -21st September, 2007, Amsterdam.

Authors' addresses

- Quanyu Wang
  Corresponding author
  Department of Conservation and Scientific Research, The British Museum, Great Russell Street, London WC1B 3DG, UK.
  Email: qwang@thebritishmuseum.ac.uk
- Peter Crew
  Pen Cefn, Penrhyndeudraeth, Gwynedd LL48 6DR, UK.
  Email: crew.peter.susan@gmail.com

Iron and processes in Scandinavian blacksmithing workshops from the Iron Age to the 14th century

Arne Jouttijärvi

ABSTRACT: The present paper is based on the study of material from several iron smithing workshops in Scandinavia, dated to a period from the Iron Age to the 14th century AD. It has been found that blacksmiths used a number of key processes, including primary smithing (cleaning the bloom from most of the remaining smelting slag and producing bars or billets), welding pieces of iron and/or steel together, and final shaping of the tools or weapons. These processes can be identified by the chemical analysis of hammerscale and slag-spheres, and this methodology is explored within this paper. The workshops excavated as part of this study can be divided into three categories:

1. Simple workshops, probably farm-smithies, producing simple tools from available pieces of iron which show no sign of cleaning and welding.
2. Regular smithies, probably often village smithies, cleaning locally produced iron and making simple tools, with or without very limited use of welding.
3. More specialised workshops, producing advanced tools or weapons from bar iron and steel, employing the welding process to a larger degree.

Within northern Europe, slag inclusions (SI) can be distinguished based on differences in the local geology, making it possible to estimate the possible origin (within a broad area) of a piece of iron. The results of source determination of numerous Danish iron objects are briefly outlined below.

Keywords: Iron, hammerscale, smithies, processes, slag inclusions, source determination.

Hammerscale and slag spheres as indicators of smithing techniques

The present paper is based on the study of material from several smithing workshops in Scandinavia, dated to a period from the Iron Age to the 14th century AD, such as the knife illustrated in Fig. 1.

The work of the blacksmith consists of three main processes: cleaning the bloom iron, welding together pieces of iron and/or steel, and forming the object. These processes all result in the formation of hammerscale, and the cleaning and welding processes will also produce slag spheres. Although superficially similar, the scales and spheres will have different physical and chemical characteristics dependant on the process of origin. It is therefore possible, on the basis of chemical analysis, to establish which type of processes took place within a given workshop from the archaeometallurgical record.

Slag inclusions identified within a laminated knife can be used to explain some of the differences that can result from different processing of the material. The majority of the inclusions within such a knife will consist of slag produced during the smelting of the iron, and will therefore resemble bloomery slag, being fayalitic with a relatively large content of impurities from clay minerals, lime, and other gangue materials.

The slag inclusions are isolated by the surrounding iron, and the composition of these inclusions remains basically unchanged by the processes carried out on the iron (although some change can be induced by the presence of carbon in the iron, which will change to some extent the content of iron oxide in the slag).

A simplified way of looking at the slags is to consider only the content of iron oxide (FeO) and silicon oxide (SiO_2), as shown in Fig. 2. The remains of smelting slag are represented by circles.

Some iron objects are of course made from more than one piece of metal, and in these another type of slag inclusion will be present, because the welding areas will contain slags composed of iron oxide and the remains of the flux used in the welding process. From the investigations carried out to date, it is known that in Denmark the only type of flux used from the Iron Age (300 BC) until the medieval period was pure silica (either as pure silica sand or in the form of burned flint). Combined with the pure iron oxide formed on the surface of the metal, this process formed a fayalitic slag of greater purity than the smelting slag. In Fig. 3 the welding slags are represented by diamonds. The types of slag found

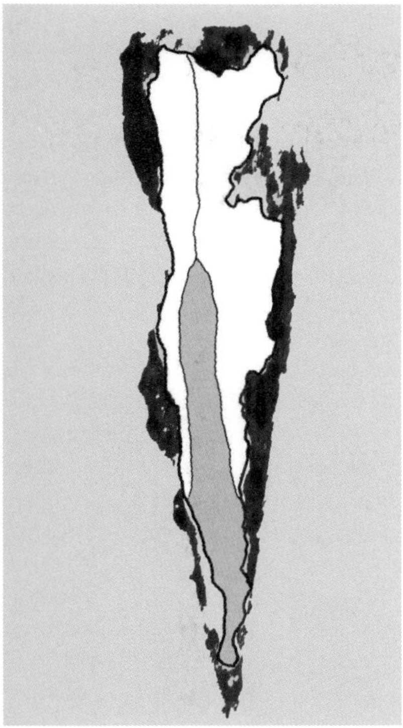

Figure 1: Cross-section of a Viking Age knife from Funen; white = iron, grey = steel, black = corrosion.

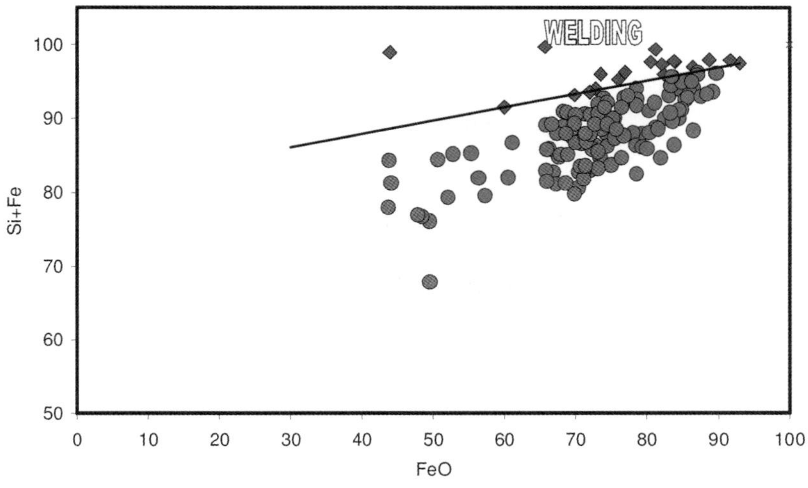

Figure 2: Iron oxide and combined iron and silicon oxide in slag inclusions in iron objects.

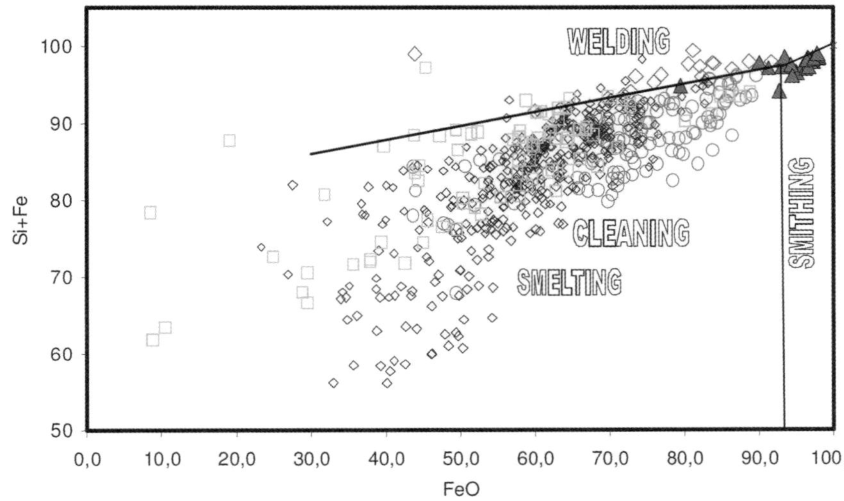

Figure 3: Compositions of hammerscales, slag spheres and slags indicating different processes.

inside the iron object are also represented in the residue left in the blacksmith's workshop. During the refining of bloom iron, flakes and droplets of what is essentially smelting slag are deposited around the anvil and larger amounts of slag are collected at the bottom of the hearth. These slag scales and spheres are of a composition similar to the smelting slags found within the iron, although they may have taken up oxides from the fuel ash and the molten clay of the forge. This is especially true for the slag left within the hearth.

Most of the slag formed during the welding operation will also form scales and spheres, and these will be chemically very similar to the welding slags found within the iron. When analysing hammerscales, one will find that, when seen in cross section, slag scales from both cleaning and welding show a structure characteristic of material that has solidified from a once molten slag. It will however be clear that a third type of scale exists, that which was never molten but was precipitated in a reaction between the surface of the solid iron and the oxygen in the atmosphere. These are the actual hammerscales, formed exclusively by the action of heating iron in the fire, and typically thinner and with a more even surface than the slag scales, from which they also differ by having a metallic bluish-black colour, as opposed to the dull grey-black of the slag. These hammerscales will consist almost entirely of iron oxide and are to be found in the far right side of Fig. 3, marked as triangles. According to this distinction, three regions can be established within the diagram, defining the residue of the three main processes used within a smithy.

The formation of hammerscale from smithing and welding without the use of flux has been described in detail, and experiments have shown that a small number of spheres can also be formed by excessive heating in the smithing process (Dungworth and Wilkes, 2007).

The examination of the residue from excavated smithies reveals that they may be classified into different types on the basis of the processes used. One type showed only slight signs of bloom-cleaning and welding. This type of workshop probably represented farm-smithies, where simple tools and fittings were made from available pieces of iron. The example shown in Fig. 4 is from an Iron Age site in southern Norway, called Rødbøl. The remains of ironworking here only consisted of a layer coloured black by charcoal and containing varying amounts of hammerscale and only very few slag spheres. More than 60% of the hammerscales had their origin in simple smithing.

The layer was systematically sampled and the physical distribution of hammerscale, pieces of vitrified clay, and larger pieces of charcoal revealed that three or four separate small workshops consisting only of a ground-level hearth and an anvil had been in operation at this site. There was no sign of any building or shed having been erected over the places of work, nor was there any sign of a local iron production.

Another type of smithy was encountered in the remains of a 13th century building at Klosterbakken, in the southern part of Jutland. Here, no traces of the floor were left, and the analyses were based on soil samples from the remnants of post holes.

As seen in Fig. 5, the majority of scales (approximately 65%) were slag scales originating from the cleaning of blooms, and the remainder primarily consisted of hammerscales from smithing.

This workshop seems to have been somewhat more advanced than the one previously mentioned, employing both the cleaning of blooms and the smithing of finished objects, although not using welding to any significant extent. The blacksmith working at Klosterbakken probably produced relatively simple tools and fittings for the farmers in the nearby villages, acquiring his iron from a local production based on bog iron ore.

As soil samples from all post-holes were analysed, it could be demonstrated that the workshop included two forges, one used exclusively for the refining of bloom iron and the other for the actual smithing.

A third type of workshop seemingly produced more advanced tools or weapons requiring the welding of pieces of iron or steel. The example shown in Fig. 6 is one of two smithies at Guldager, near Esbjerg, in the western part of Jutland. Over 60% of the hammerscale found in this smithy originated from welding.

Apparently the raw material for these smithies was bar iron, as there is very little evidence of the cleaning of blooms.

As it has been briefly mentioned above, the physical distribution of different types of smithing waste can be of great

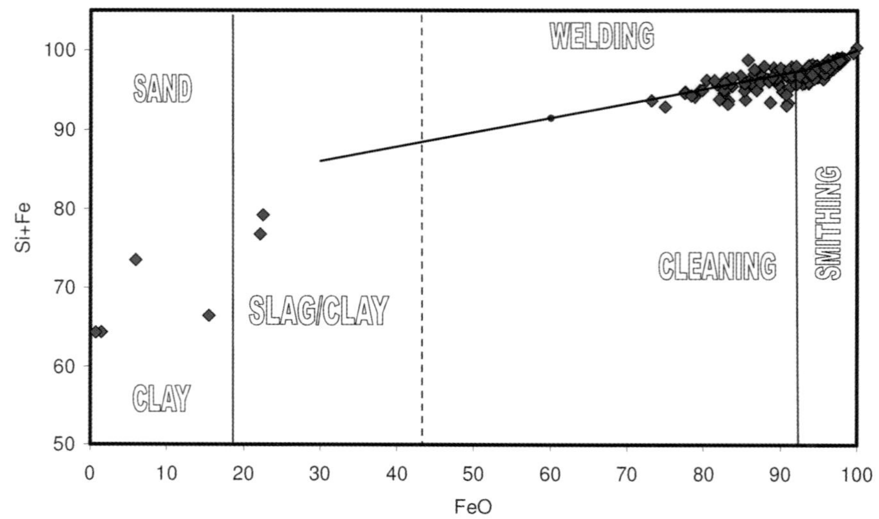

Figure 4: Hammerscale from smithy at Rødbøl, Southern Norway, indicating that only simple smithing took place at the site.

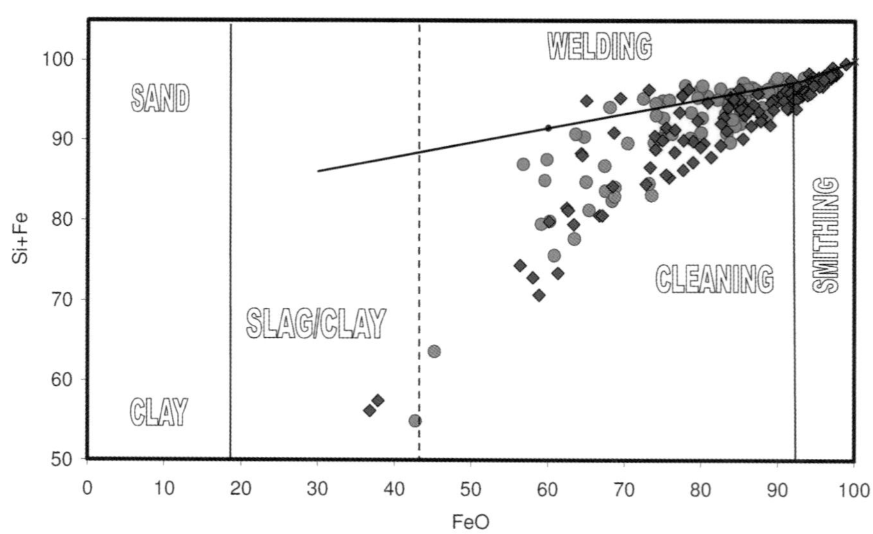

Figure 5: Hammerscale and slag spheres from a smithy at Klosterbakken, Southern Jutland, indicating a village smithy refining locally produced iron and producing simple tools and fittings.

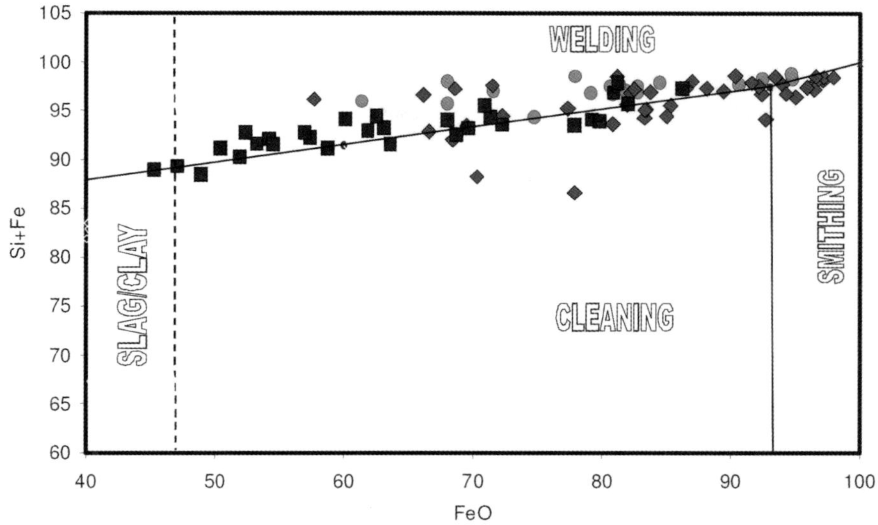

Figure 6: Hammerscale, slag spheres and slag from a smithy at Guldager, Western Jutland, indicating an advanced workshop producing complicated tools or weapons, requiring welding of iron and steel bars.

importance in the process of interpreting the archaeological remains of a workshop. Even when the archaeological remains are sparse, it is usually possible to identify the location of a forge and anvil, and to get an indication as to the position of the central working area and rubbish heaps, based on the patterning and concentrations of hammerscale and associated waste. When floor layers are present, systematic sampling can yield much more detailed information on the structure and organisation of the work in the smithy. It might even be possible to trace the movements of the artisans, based on the dragging of hammerscale and other material from the surroundings of the anvil to other parts of the workshop.

Provenancing using slag inclusion analysis

As mentioned earlier, slags produced during the cleaning of iron blooms are essentially smelting slag, and will therefore resemble the slags from smelting sites, although they will show a reaction with ashes from the charcoal. Slag inclusions found within iron artefacts will resemble bloomery slags to a greater extent, as they have been shielded from reaction with the surrounding environment by the iron metal.

Within Scandinavia and the northern part of continental Europe, a number of slag types can be distinguished based on differences in the local geology. It is therefore possible to estimate the possible origin of a piece of iron or the slag cleaned in a workshop on the basis of chemical analyses of slag inclusions or slags from primary smithing (Ilkjær, et al., 1994; Jouttijärvi, 1994, 1995; Buchwald and Wivel, 1998, Høst-Madsen and Buchwald 1999). Previous studies have primarily dealt with more local issues and have not attempted a broader geographical overview (see Hedges and Salter, 1979; Dillmann and L'Heritier, 2007; Blakelock, et al., 2009). On the basis of the main elements, the origin probably cannot be traced to a single production site, but to broader areas.

The main problem when assessing slag analysis is that slags from the same smelting site or slag inclusions within a single piece of iron often show a pronounced scatter. To a large degree, this will however be due to variations in the iron oxide content arising due to varying conditions within the smelting furnace. This dispersion can be diminished by ignoring the iron oxide and instead only looking at the composition of the gangue material. When dealing with bog iron ore this seems a sensible approach, because the gangue material is indicative of the local geology, whereas the content of iron (and manganese) oxide is a result of the reduction and oxidation processes, which vary on a much smaller scale than the overall geology. It could therefore be argued that even the manganese oxide should perhaps be left out.

Recalculating the analyses without the iron oxide tends to result in the separation of otherwise overlapping compositions of slags from different geographical areas, as exemplified in Fig. 7, showing the phosphorous and calcium oxide contents of slag inclusions in the three pieces of iron and steel used in the manufacture of the knife shown in Fig. 1.

As the bloomery slag and hence the slag inclusions are not solely composed of gangue material from the iron ore but also contains a certain amount of oxides derived from the melting of furnace clay and from the ashes of the charcoal, comparing the composition of the slag inclusions directly with the iron ores will be of limited use. It seems much more productive to compare them with well documented bloomery slag, representing both the actual types of ore chosen by the prehistoric smelters and the influence of the clay and fuel. Over the last 20 years, a few thousand analyses of bloomery slag from northern Europe have been collected from unpublished and published sources (Oelsen and Schürmann, 1954; Bachmann, 1982; Pleiner and Princ, 1984; Pleiner, 1987; Martens, 1988; Hjärthner-Holdar, 1993; Serneels, 1993, Yalcin and Hauptmann, 1995; Kronz, 1997; Joosten, et al., 1998; Spazier, 2000; Heimann, et al., 2001; Ritzdorf, 2003; Bauermeister, 2005; Buchwald, 2005), but hopefully many more will be added in the future. The following is meant as a brief outline of some of the differences encountered in the analysis of the data, and is not supposed to be an adequate description of a method.

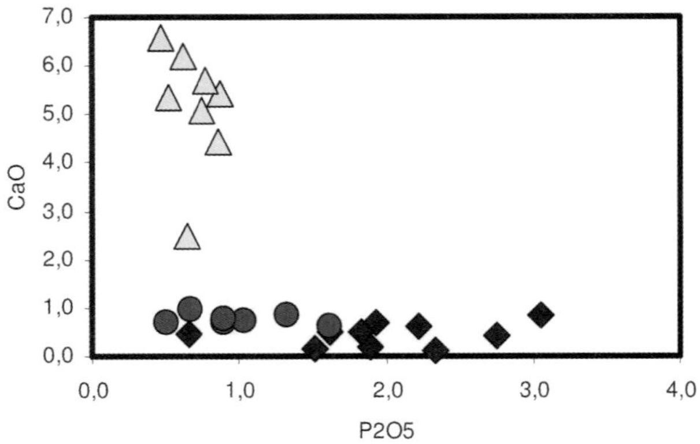

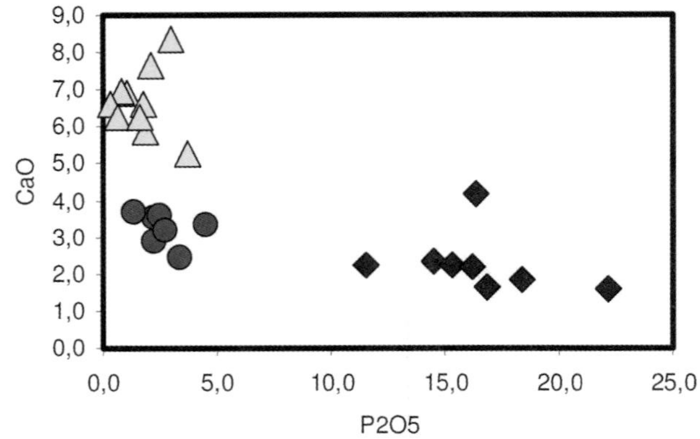

Figure 7: Uncorrected (left) and corrected (right) analyses of slag inclusions from the three pieces of iron and steel used in the knife shown in Figure 1.

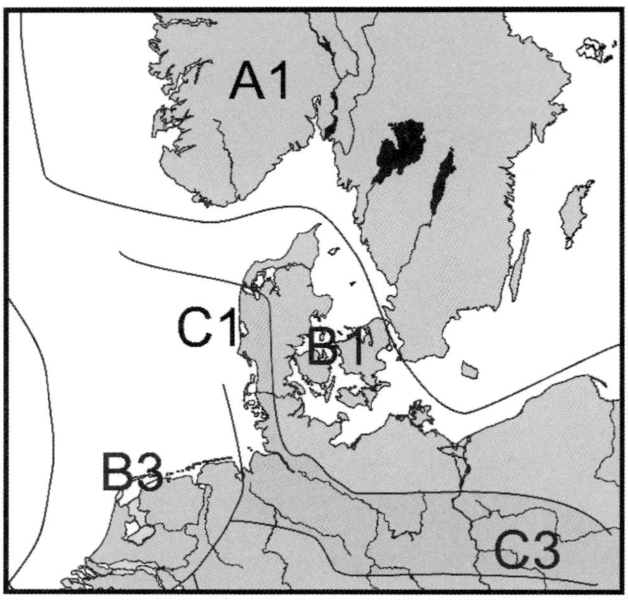

Figure 8: Tentative geographical areas to which the origin of a piece of iron might be assigned on the basis of slag inclusion analysis.

The differences seen in the bloomery slags are primarily due to large scale differences in geology between a number of geographical areas.

The North European Plain is characterised by moraine and underlying limestone, generally providing the ores, and hence the slag, with a high content of calcium, magnesium and phosphorous oxides.

This area can be further divided by the main stationary line of the last ice age. South and west of this line, the moraine is older and depleted of calcium and magnesium, as can be seen from the diagram in Fig. 9. The labels refer to the geographical areas shown in Fig. 8.

The bedrock area of Norway and Sweden (A1) overlaps both the B and C types of areas, but can be separated from the B-type areas by looking instead at the aluminium and magnesium oxides (Fig. 10). Considering these oxides also makes it easy to distinguish between B1 (eastern Denmark and the area bordering the southern coast of the Baltic Sea) and B2 (northern France).

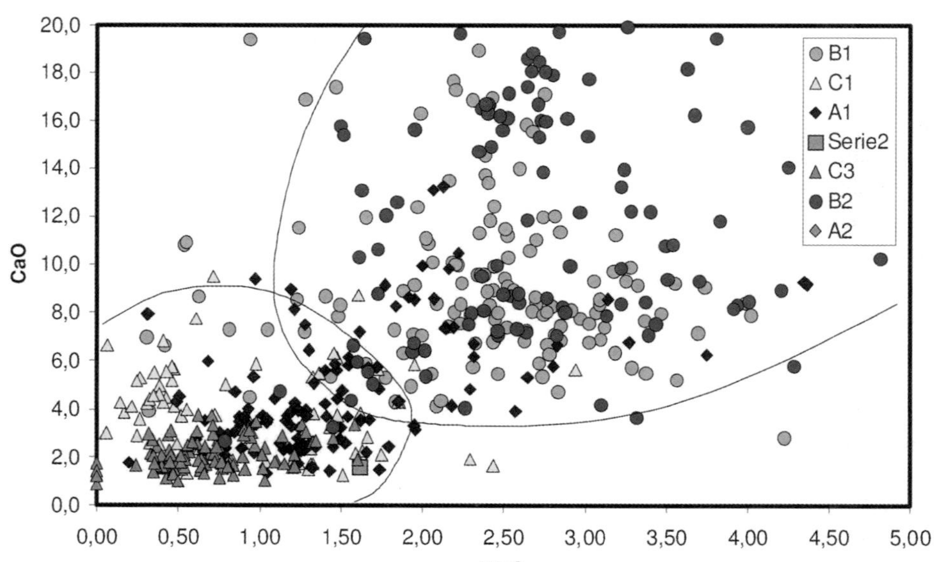

Figure 9: MgO and CaO contents of smelting slags from different geographical areas. The labels refer to the areas shown in Figure 8.

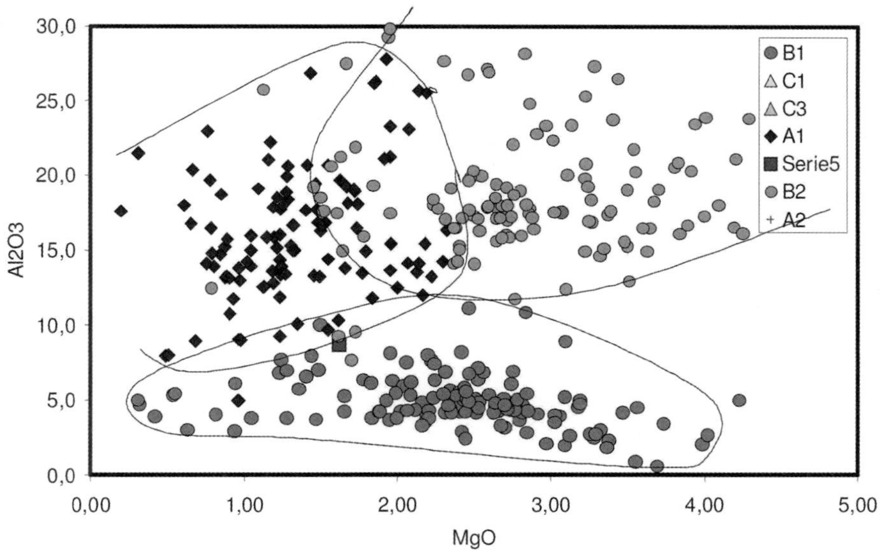

Figure 10: MgO and Al_2O_3 contents of smelting slags from different geographical areas. The labels refer to the areas shown in Figure 8.

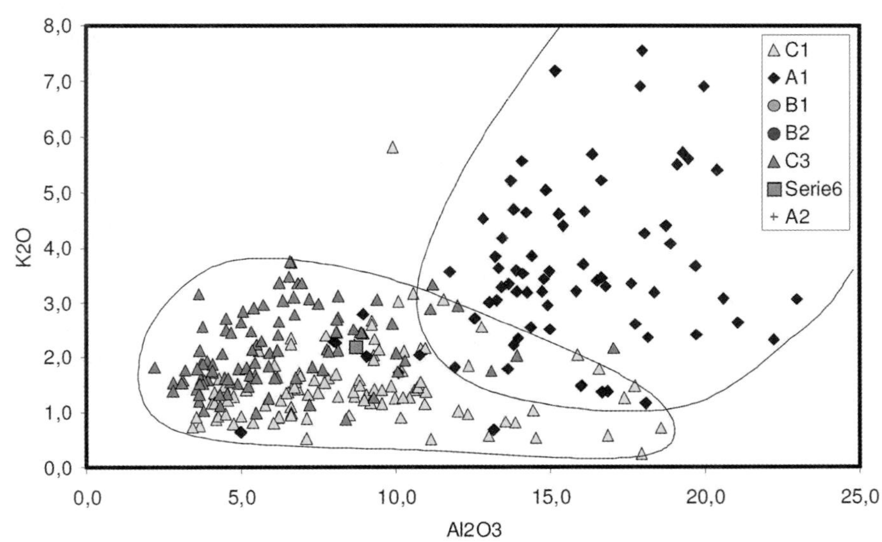

Figure 11: Al_2O_3 and K_2O contents of smelting slags from different geographical areas. The labels refer to the areas shown in Figure 8.

The A-type area can also be distinguished from the C-type parts of the North European Plain, this time according to the contents of aluminium and potassium oxides (Fig. 11).

The C1 (Western Jutland) and C2 (Northern Germany) areas do not form two clearly defined groups but rather indicate a continuous increase in potassium and calcium oxide content from north to south.

Slag inclusion analysis has also been employed as a routine method when examining iron objects, providing an indication of the origin of the iron. Using data from between 300 to 500 objects (mainly iron knives), a rough overview of the production and distribution of iron in Denmark during a period from the pre-Roman Iron Age to the 15th century AD is provided below.

The bloomery process was introduced in Denmark in the pre-Roman Iron Age, and the earliest signs of iron smelting are dated to around 300 BC. Slag inclusion analyses on items from this period indicate that almost 70% of the iron used came from local productions, and that the remaining part was almost exclusively imported from the continent. It seems that iron was primarily imported in the form of prestigious items, or imported iron was used in the form of occasional pieces of scrap. The iron smelting sites from this period consist of a few rather small furnaces found in association with settlements. It can be presumed that the demand for iron was limited at this time, and a village was able to be more or less self-sufficient in this respect.

In the Roman Iron Age, the record of iron finds is dominated by iron produced in the western part of Jutland, a finding in complete agreement with the fact that archaeological iron smelting sites have only been found in this area. It seems that the iron production was concentrated in large smelting sites, often consisting of thousands of furnaces. Currently analyses indicate that this iron probably reached no further east than Funen. Local productions may have existed in Zeeland, although archaeological evidence to support this hypothesis has yet to be found.

After around 700 AD, the industrial scale smelting sites in Jutland disappear, and there is only very limited archaeological evidence of any iron smelting taking place before the 11th century. Analyses of Viking Age items, notably knives, do however indicate that a limited production still took place. Apparently the production was carried out once again on a very small and local scale, only producing iron to meet the needs of the local community. Around 70% of the iron consumed did have its origin outside what is now Denmark, and it seems to have arrived both as prestigious items, such as swords, and in the form of slightly worked blooms, known as 'clots' in later written sources. Another type of imported metal was steel, used to produce cutting edges and primarily acquired from Central Europe or Norway/Sweden.

From written sources, it is known that iron production continued in the central part of Jutland until around 1600 AD, when the latest registrations of taxes paid in 'clots' are encountered. Only a few analyses of iron implements from this period have been undertaken, but future analyses may shed some light on the transition from a society at least partly able to fulfil its own need for iron to a total dependence on imported metal produced in the blast furnaces of Germany, Sweden, and Norway.

Conclusion

Through macroscopic observation and analyses of the chemical composition of hammerscale, slag spheres, and other forms of smithing debris, detailed information on the processes employed in prehistoric smithies can be obtained. Systematic sampling and analysis of the physical distribution of debris might yield information about the arrangement of forges, anvils and other functions in the workshops, not obtainable by purely archaeological investigation.

Large scale variations in the chemical composition of bloomery slags caused by differences in geology can be used for assessing the geographical origin of iron artefacts. Provenancing is however limited to larger geographical areas, but, if supplemented by trace element or isotope analyses, more detailed information might be acquired.

References

Bachmann, H.G., 1982. *The identification of slags from archaeological sites.* London: Institute of Archaeology Special Publication No. 6.

Bauermeister, J., 2005. *Archäometallurgische Untersuchung von Eisenschlacken aus Holtland / Kreis Leer.* Göttingen: Georg-August-Universität Geowissenschaftliches Zentrum.

Blakelock, E., Martinón-Torres, M., Veldhuijzen, H.A. and Young, T., 2009. Slag inclusions in iron objects and the quest for provenance: an experiment and a case study. *Journal of Archaeological Science,* 36, pp.1745-57.

Buchwald, V.F., 2005. *Iron and steel in ancient times.* Copenhagen: Det Kongelige Danske Videnskabernes Selskab.

Buchwald, V.F. and Wivel, H., 1998. Slag analysis as a method for the characterization and provenancing of ancient iron objects. *Materials Characterization,* 40, pp.73-96.

Dillmann, P. and L'Heritier, M., 2007. Slag inclusion analyses for studying ferrous alloys employed in French medieval buildings: supply of materials and diffusion of smelting processes. *Journal of Archaeological Science,* 34, pp.1810-23.

Dungworth, D. and Wilkes, R., 2007. *An investigation of hammerscale.* English Heritage Research Department Report 26. Portsmouth: English Heritage.

Hedges, R.E.M. and Salter, C.J., 1979. Source determination of iron currency bars through analysis of the slag inclusions. *Archaeometry,* 21, pp.161-75.

Heimann, R.B., Kreher, U., Spazier, I. and Wetzel, G., 2001. Mineralogical and chemical investigations of bloomery slags from prehistoric (8th century BC to 4th century AD) iron production sites in Upper and Lower Lusatia, Germany. *Archaeometry,* 43, pp.227-52.

Hjärthner-Holdar, E., 1993. *Järnets och järnmetallurgins introduktion i Sverige* (The introduction of iron and iron metallurgy to Sweden). Uppsala: Societas Archaeologica Upsaliensis.

Høst-Madsen, L. and Buchwald, V.F., 1999. The characterization and provenancing of ore, slag and iron from the iron age settlements at Snorup. *Historical Metallurgy,* 33, pp.57-67.

Ilkjær, J., Jouttijärvi, A. and Andresen, J., 1994. *Illerup Ådal, Proveniensbestemmelse af jern fra Illerup Ådal - et Pilotprojekt.* Jysk Arkæologisk Selskab. Aarhus

Joosten, I., Jansen, J.B.H. and Kars, H., 1998. Geochemistry and the past: estimation of the output of a Germanic iron production site in the Netherlands. *Journal of Geochemical Exploration,* 62, pp.129-37.

Jouttijärvi, A., 1994. Om muligheden for proveniensbestemmelse af jern. *Aarbøger,* 1992, pp.183-91.

Jouttijärvi, A., 1995. Kan man proveniensbestemme jern?, *Medeltida danskt järn ; Forskning i Halmstad 1.* ed Sven-Olof Olsson, Högskolan i Halmstad, pp. 93-107.

Kronz, A., 1997. *Phasenbeziehungen und Kristallisationsmechanismen in fayalitischen Schmelzsystemen – Untersuschungen an Eisen- und Buntmetallschlacken.* PhD Thesis, University of Mainz.

Martens, I., 1988. *Jernvinna på Møsstrond i Telemark, en studie i teknikk, bosetning og økonomi.* Oslo: Universitetets oldsaksamling.

Oelsen, W. and Schürmann, E., 1954. Untersuchungsergebnisse alter Rennfeuerschlacken. *Archiv für das Eisenhüttenwesen,* 25, pp.507-14.

Pleiner, R., 1987. Eine eingetiefte Eisenverhüttungswerkstatt der älteren Römischen Kaiserzeit in Lovosice, Nordböhmen. *Archeologické rozhledy,* 39, pp.75-116.

Pleiner, R. and Princ, M., 1984. Die latĚnezeitliche Eisenverhüttung und die Untersuchung einer Rennschmelze in Msec. *Památky Archeologické,* 75, pp.133-80.

Ritzdorf, H., 2003. *Die römerzeitliche Eisenmetallurgie in der Eifel.* PhD Thesis, Philosophischen Fakultät der Rheinischen Friedrich Wilhelms Universität zu Bonn.

Serneels, V., 1993. *Archéométrie des scories de fer: recherches sur la sidérurgie ancienne en Suisse occidentale.* Lausanne: Cahiers d'archéologie romande 61.

Spazier, I., 2000. Das germanische Eisenverhüttungszentrum Wolkenberg in der Niederlausitz/Südbrandenburg mit über 1000 Rennöfen. *Arbeits- und Forschungsberichte zur sächsischen Bodendenkmalpflege,* 42, pp.317-31.

Yalcin, U. and Hauptmann, A., 1995. Archäomettalurgie des Eisens auf der Schwabischen Alb. In: M. Böhm, A. Hauptmann and M. Kempa, *Beiträge zur Eisenverhüttung auf der Schwäbischen Alb.* Forschungen und Berichte zur Vor- und Frühgeschichte in Baden-Württemberg, Band 55. Stuttgart: Kommissionsverlag Theiss, pp.269-309.

Author's address

- Heimdal-archaeometry, Skovledet 30, DK-2830 Virum, Denmark heimdal@archaeometry.dk

Iron? Which iron? Methodologies for metallographic and slag inclusion studies applied to ferrous reinforcements from Auxerre Cathedral

Maxime L'Héritier, Philippe Dillmann, Sylvain Aumard, and Philippe Fluzin

ABSTRACT: Comparing data between studies of ancient ferrous artefacts is difficult due to the heterogeneous nature of the iron (e.g. carbon and phosphorus content, slag inclusions, welding lines, etc). The first aim of this paper is therefore to present a complete methodology designed to produce comparable results for the analysis of ferrous artefacts. It includes a methodology based on slag inclusion analyses used to distinguish ironmaking processes and to determine if artefacts have the same origin (i.e. reduction system). In addition, this paper provides an example of the use of this methodology using iron armatures from Auxerre Cathedral. The research has found that the ferrous artefacts described in this paper represent the earliest iron objects produced from finery ever found on a French monument, and that iron was being produced for use in construction using the finery process as early as the late 14th century. Historical studies support this conclusion, with blast furnaces and the finery process known to have been used, for example in Burgundy, at this time.

Keywords: Ferrous reinforcement, slag inclusion, metallography, methodology.

Introduction

In France, over the last decade, interdisciplinary studies have shown that large amounts of iron had been used for the construction of many medieval buildings. Among them are the Palais des Papes in Avignon (Dillmann, et al., 2003, 2004; Bernardi and Dillmann, 2005), the churches of Rouen (L'Héritier, et al., 2005, 2010; L'Héritier, 2009) and Troyes (L'Héritier, 2009; L'Héritier, et al., 2010), the cathedrals of Amiens and Beauvais (Taupin, 1996; Dillmann, 2009; Lefebvre, 2009), and the dungeon of Vincennes castle (Clément-Charpentier, 2001; Chapelot, 2003). These studies have led to a complete reconsideration of common assumptions about the use of ferrous materials in late medieval architecture.

Yet, there is much more to learn from these armatures, as they provide a great deal of well dated iron artefacts for the late Middle Ages. During this precise period in Europe, unprecedented technical changes occurred regarding iron production: blast furnaces and fineries blossomed and progressively replaced traditional bloomery furnaces. In the northern region of the French kingdom, it is widely accepted that both processes coexisted between the end of the 14th and the 16th century (Benoit, 1994).

Restoration works in the nave of Auxerre Cathedral presented archaeologists and archaeometrists with a unique opportunity to access ancient material and therefore provide new insights into medieval construction processes (Aumard, 2009). In this context, an initial set of iron armatures were sampled and analysed by the authors. As no records are available for this period, such studies offer a unique opportunity to better understand the place of iron in the 14th century construction campaigns. Parallel to the study of iron armatures in French monuments, our laboratory has recently developed several methods for distinguishing ironmaking processes based on the analysis of slag inclusions embedded in the metallic matrix (Dillmann and L'Héritier, 2007). As the earliest references to the introduction of blast furnaces and fineries in France are linked to two regions surrounding Burgundy (Champagne to the north and Nivernais to the south) and the city of Auxerre (Belhoste and Léon, 1996; Rouillard, 2003), the above mentioned method based on slag inclusion (SI) analysis will prove very efficient here.

This paper presents a complete methodology designed to produce comparable results for the analysis of ferrous artefacts: first with regards to the metallic structure, and second concerning SI analysis, including a method that allows distinguishing ironmaking processes and underlining common or different origins. Following this methodological overview, the results of recent metallographic analyses of 24 iron samples taken during the restoration of the nave of Auxerre Cathedral are provided, to demonstrate the applicability of this methodology.

Methodological considerations

Most ancient ferrous artefacts are very heterogeneous and, when performing a metallographic analysis, providing a complete and accurate description always proves difficult. Moreover, comparing results between different studies is never easy, as no clear standardisation for gathering and presenting data is available so far, each study using its own parameters and criteria. Yet, every archaeometallurgist always bears in mind the same simple questions of what, why, how and where. What kind of ferrous alloy am I dealing with? Was it intended to be this way? How was it produced and where does it come from?

Through these questions arise three essential notions: quality, production, and provenance, with quality being the most accessible, as well as the hardest factor to define. Indeed, many parameters can change the quality of the metal and therefore its mechanical behaviour, such as carbon and phosphorus content, the presence of slag inclusions, forging treatments, and so on. Hence, metallographic descriptions have to be as objective and precise as possible. However, while accumulating numbers does not define quality, the best definitions are often provided through a comparison with other artefacts or objects. The need for some kind of standardisation is therefore essential to counter problems related to comparing data when the original analytical criteria can vary significantly. The questions of production and provenance are considerably different as they rely on purely analytical results and are therefore not linked to subjectivity: conclusions in such cases rather depend on the accuracy of the method. The following methodology first deals with an objective definition of quality prior to describing the analytic means used for the determination of ironmaking processes and provenance.

Metal structure

No detailed protocol for the analysis of ferrous artefacts will be presented here, as it can be found elsewhere (Fluzin, 1983; Fluzin, et al., 2000, 2004). Because ancient ferrous artefacts are usually not homogeneous, where possible a longitudinal section should be sampled through the entire artefact in order to evaluate heterogeneities. Otherwise, representative sampling should be performed at different locations on the artefact. Then, the analysis consists in classical metallographic preparation with adapted etching such as Nital and Oberhoffer's reagent to reveal the metallographic structure. Observations always have to consider different scales: macroscopic and microscopic for the average and local compositions. As mentioned above, carbon, phosphorus, slag inclusions, and all forging treatments used by the smith have an influence on the quality, i.e. mechanical behaviour of the final object. Each of these parameters has to be precisely described.

Unlike in contemporary steels, carbon very scarcely shows a homogeneous distribution in ancient ferrous artefacts. In the bloom, the metal is usually more or less carburised, depending on local thermodynamic conditions within the furnace. Even very carburised materials have a heterogeneous carbon content. This is why assessing the carbon content of an object correctly is problematic. Here, we follow the methodology proposed by Pagès et al. (2011), which is based on two parameters: the proportion of carburised zones (> 0.2 wt%) and their average carbon content. Practically, the surface of each zone is estimated with a 0.2 wt% interval by optical microscopy on metallographic cross-sections after Nital etching, and a weighted carbon content is then defined as follows:

$\%C = (n_1\% \times 0.02 + n_2\% \times 0.1 + n_3\% \times 0.3 + n_4\% \times 0.5 + n_5\% \times 0.7 + \text{etc.})/100.$

where n_1 is the proportion of the ferritic surface, n_2 is the proportion of the surface with a C content between 0.02 wt% and 0.2 wt%, n_3 between 0.2 wt% and 0.4 wt%, etc. According to these two parameters, artefacts can then be classified into different categories: ferritic iron, heterogeneous metal (mix of iron and steel), composite steel, and steel (Table 1).

Phosphorus also has to be taken into account. Several studies clearly demonstrate that a relatively low P content in a ferrous matrix (more than 0.1 wt%) can drastically change its mechanical behaviour (Stewart, et al., 2001) and therefore its commercial or technical value. Moreover, it seems that forging phosphoric iron required specific technical gestures (Vega, et al., 2002, 2003). Thus, this material has to be identified in order to be considered separately. Specific ghosting or 'ghost structures' appear after Nital etching on polished surfaces for P contents between 0.1 and 0.6 wt%. They can be more easily detected after etching with Oberhoffer's or Klemm's etchants (Piccardo, et al., 2004). A precise evaluation of the P content however requires Wavelength Dispersive Spectrometry (WDS) coupled to a microprobe analysis[1], which is not likely to be performed routinely. Proposing a graphical chart for C content such as that put forth by Pagès et al. (2008; 2011) seems therefore pointless. Iron will be considered as not phosphoric when no ghosting is visible (noted '-'). When ghost structures cover up to 10% of the surface, it is 'slightly phosphoric' (noted '-/+'), up to 50% 'fairly phosphoric' ('+'), and above 50% 'phosphoric' ('+++').

Both phosphorus and carbon harden iron. For a complete description, micro-hardness analyses should also be performed in each area of interest (ferritic, phosphoric, carburised, etc.).

Prior to the introduction of the Bessemer and Martin processes in the 19th century, which permitted high-scale production of molten steel by decarburisation of pig iron, more or less slag inclusions remained embedded in by- or final ferrous products, depending on the refining stage, the initial amount of slag in the bloom, and also on the final shape of the artefact. These SI are a very useful means of distinguishing smelting processes and provenances. Also, regarding mechanical behaviour, they can drastically modify the quality of the metal. Moreover, it is completely impossible to estimate *ab initio* the mechanical behaviour of a given artefact if it contains numerous SI, as was demonstrated by performing uniaxial tensile tests on samples taken from bars

Table 1: Classification of ferrous materials based on their carbon content (Pagès, et al, 2008)

Type	% of the surface > 0.2 wt% C	Weighted carbon content
Ferritic	0 – 20	< 0.3
Heterogeneous	20 – 40	< 0.6
Composite steel	40 – 70	< 0.8
Steel	70 – 100	> 0.8

reinforcing the Vincennes castle dungeon in France, dating from the 14th century (L'Héritier, et al., 2005; L'Héritier, 2007; Dillmann, 2009). Evaluating the surface proportion of SI and porosities is therefore crucial for estimating the quality of by-products. In the present study, these characteristics were measured using optical microscopy. First, a black-and-white micrograph of the entire unetched sample has to be taken with good resolution (1 pixel = 2-5μm). Then, after contrast enhancement, the proportion of SI and porosities can be easily obtained using the particle counting module of any image processing software (ImagePro was used in this study). Regarding the SI proportion, the absolute accuracy of the calculation can be estimated thanks to the perimeter value (in pixels), and is usually comprised between 0.5 and 1%, mostly depending on the average size and number of SI. It should be noted that SI and porosities cannot be automatically differentiated with this method. However, for iron armatures, as well as for most final iron products, the latter are always very minor as compared to SI. Finally, the image processing software also provides various measurements, such as SI density (number of SI per mm^2) and average SI size. The combination of these three features gives a precise idea about the repartition and size of SI in the studied sample.

Lastly, microstructures related to forging treatments such as carburising, quenching, tempering, and welding have to be detected (for a good description of the identification of welding and cementation see Piaskowski, 1984). Various operations should also be distinguished, for example simple welding by compacting a porous or heterogeneous material; welding after folding and hammering; welding of different metal pieces. The most significant parameter to consider is the textural and structural homogeneity.

Slag inclusion analyses

During the reduction of iron ore, depending on thermodynamic conditions, compounds are likely to have three different behaviours in the ore/furnace lining/charcoal ashes/flux system: (i) elements that are not reduced and remain in the slag under oxidised form, (ii) elements that are mainly reduced and can be retrieved in the metal, and (iii) elements that are more or less reduced and pass into both the metal and the slag.

Concerning the bloomery process, all elements that are mainly concentrated in the slag will end up in the SI of the metallic artefacts, preserving their respective ratios. On the contrary, during the indirect process, all these elements will be eliminated during smelting in the blast furnace. Only elements reduced in the cast iron and oxidised during fining can reflect the contribution of the ore in the SI.

Regarding major elements, the Non Reduced Compounds (NRC) ending up in the slag during the bloomery process are the following: MgO, Al_2O_3, SiO_2, K_2O, and CaO. Analysis of a significant number of SI per artefact (at least 40) and filtering of the raw results is required to identify SI originating from flux during forging, but also local enrichment effects mainly due to small SI. The principle is to express an average weighted content corrected by the surface of each analysed inclusion (the detailed acquisition procedure is presented in Dillmann and L'Héritier, 2007). A careful metallographic examination of every zone in the artefact is necessary in order to locate possible different compositions (carbon, phosphorus, etc.) and welding lines. The data treatment consists of calculating 'weighted contents' by taking into account the ratio between the analysed inclusion area and the total analysed surface:

$$\%E^* = \sum_{i=1}^{n} (\%E_i \times \frac{S_i}{S_T})$$

where:

$\%E^*$: weighted content of the considered element or oxide

$\%E_i$: mass content of the element or the oxide in the i slag inclusion (SI)

S_i: Surface of the SI i where the analysis is performed

S_T: Total surface of the analysed SI

n: Total number of inclusions

This weighted content is indicated in the following with an asterisk (*) to distinguish it from the 'normal' contents (e.g. $\%Al_2O_3^*$).

More recently, a procedure for analysing trace elements in slag inclusions and extracting provenance information was proposed by Desaulty et al. (2008, 2009) and Leroy et al. (2011, 2012).

The distinction between the bloomery and indirect process by SI analysis is based on the fact that elements do not have the same behaviour in both processes. Moreover, in the blast furnace, elements such as Al, K or Mg passing mainly in the slag during smelting are eliminated from the operating chain. They are therefore strongly depleted in SI formed during fining mainly by oxidation of elements present in the cast iron. This was verified for both major and trace elements. On the contrary, other elements such as phosphorus, which could be concentrated in cast iron during ore reduction and then mainly re-oxidised during fining, will present particularly high contents in the SI. It has to be noted that pollution from charcoal and lining can occur during the processes. For that reason, there is an overlapping zone in the diagram proposed to distinguish the processes, in which bloomery and finery cannot be discriminated. Finally, some elements (e.g. Ca, Sr, Ba, Rb, etc.) entering the slag by pollution from the hearth lining, specific flux or charcoal during fining cannot be used to identify this distinction. Only major elements can be used for the calculations proposed by Dillmann and L'Héritier (2007) to distinguish the two processes.

As mentioned above, some elements, including most major elements which completely pass in the slag, such as aluminium, silicon, potassium, and calcium, are likely to be brought into the system by external pollution during smelting. Therefore, they cannot be used to determine the chemical signature of the initial ore. Nevertheless, these NRC ratios can be used to define a so-called 'workshop signature', meaning that, if the craftsman always uses the same ore, lining and charcoal, the NRC ratio of major elements is not likely to change. This was verified by experimental smelting using different types of ores (Dillmann, et al., 2005; Dillmann and L'Héritier, 2007). Thus, if in a corpus of artefacts originating from a homogeneous archaeological or historical context all the NRC ratios are the same, it is reasonable to assume that they could have the same origin. Evidently, major element analysis cannot provide further information about this origin. If the NRC ratio is not the same, two explanations can be proposed: (i) the artefacts have different provenances and (ii) they have the same provenance but a change of parameters (for example charcoal, ore, lining, etc.) occurred between the smelts.

Other elements, although they pass both into the slag and the metal depending on the thermodynamic conditions, are likely to provide indications about the initial ore. For average weighted contents in SI with over 4% MnO, the use of a high manganese ore can be considered. This was verified on archaeological artefacts coming without any doubt from high Mn ores (Dillmann, et al., 2006; Leroy, et al., 2007). Similarly, if the P_2O_5 average weighted content in SI exceeds 5%, then it doubtlessly comes mostly from the ore and not from the charcoal ashes. Thus, these two compounds can be used to distinguish artefacts with SI presenting sharply different average contents. If not, artefact provenance cannot be differentiated by these elements.

The metallic reinforcements of the Auxerre Cathedral

Context and samples

A total of 24 iron armatures could be sampled among all the ferrous reinforcements discovered in Auxerre Cathedral during the restoration works. All come from the 14th century construction campaigns in the eastern parts of the nave. Nevertheless, they belong to two different groups.

The first 11 samples are pins used in a chapel's window tracery. This window was first built in the northern aisle of the nave in the early 14th century. When the chapels were added 60 years later, the master-builders decided to dismantle the former traceries and reuse them in the newly set windows. The stones were sawed and reattached with new pins. In both construction phases, the masons indifferently used lead or mortar to seal the pins. Yet, the distinction between the two phases is very easy to make due to the tool marks printed in the lead or mortar: typical stonemason's hammer marks for the first phase and very distinct saw marks for the second one.

The second group includes 10 pins and 3 cramps coming from the pinnacles, abutments and flying buttresses of the southern part of the nave, built mainly in the late 14th century. However two cramps from the flying buttress cornering the transept are slightly earlier (ca. 1340-1350; see Table 2).

Pins from the tracery of the fourth northern chapel

Observations performed on these eleven pins show relatively low carburised structures (Fig. 1; Table 4). Most specimens plot in the ferritic or heterogeneous zones (Fig. 2). Only one (AUX 70) is inside the 'composite steel' area and can be considered as fairly carburised.

At least three of the eleven samples (AUX 67a, AUX 73a, AUX 75a) contain locally significant quantities of phosphorus revealed by the presence of ghost structures. Yet, only AUX 73a can be considered as phosphoric iron, as ghosting largely exceeds 10% of its total surface. Welding lines were observed on two cross-sections (AUX 73a and AUX 75a). All samples

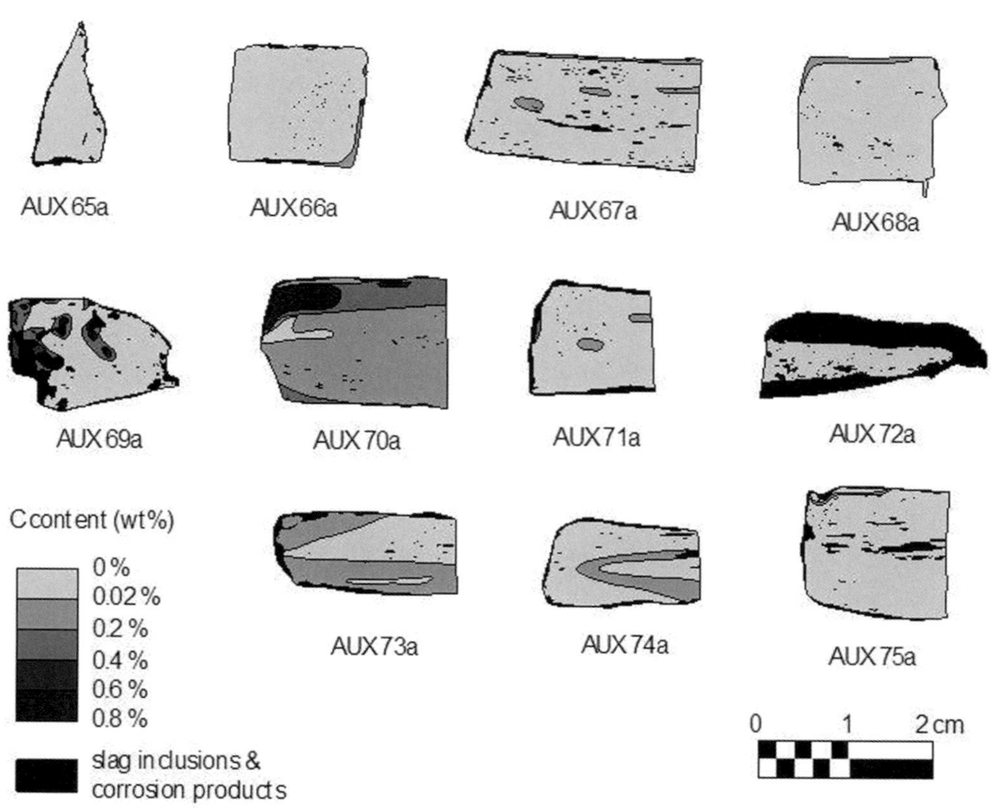

Figure 1: Metallographic observations on the cross-sections of the pins from the tracery of the fourth northern chapel.

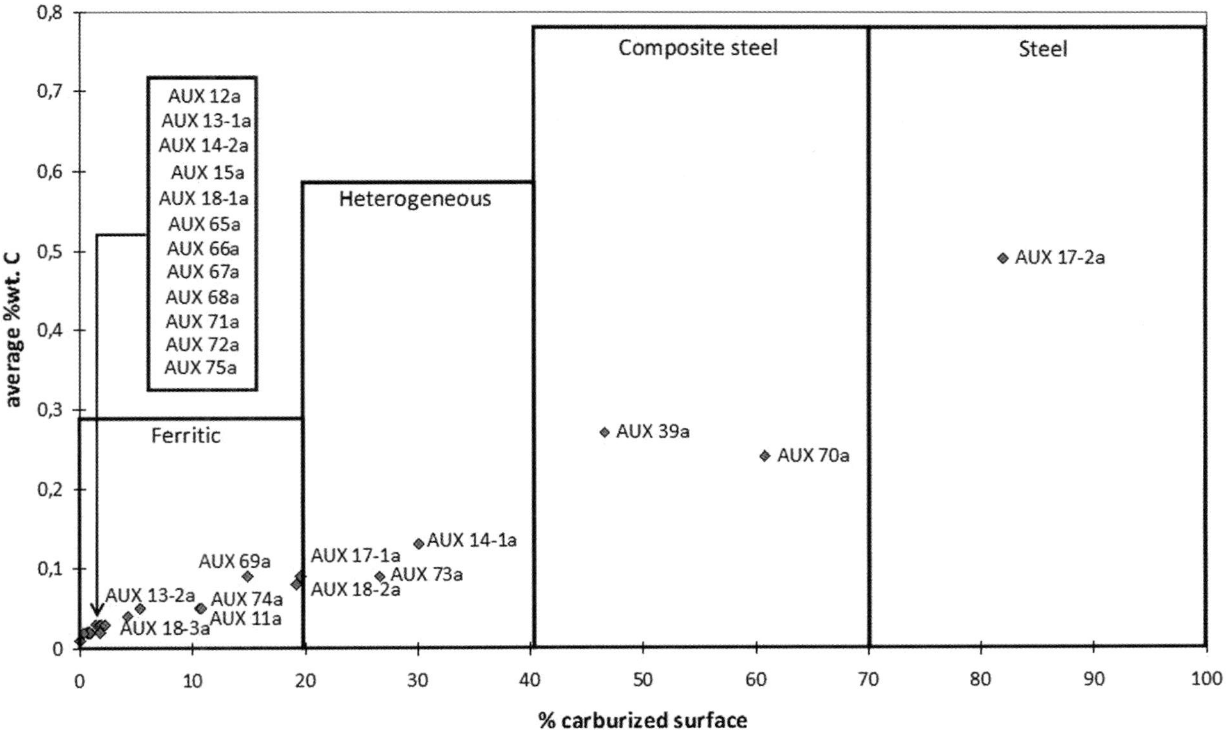

Figure 2: Carbon content of the different samples according to Pagès et al (2008; 2011).

Table 2: List of samples from Auxerre Cathedral

Number	Type	Location	Dating
AUX 11	Pin	Flying buttress abutment, 5th-6th bay south, southern pinnacle	ca. 1370-1380
AUX 12	Pin	Flying buttress mullion, 3rd-4th bay south	ca. 1380-1390
AUX 13-1	Cramp	Flying buttress, 5th-6th bay south	ca. 1380-1390
AUX 13-2	Pin	Flying buttress banister, 5-6th bay south	ca. 1380-1390
AUX 14-1	Pin	Flying buttress mullion, 6th bay south (transept)	ca. 1340-1350
AUX 14-2	Pin	Flying buttress mullion, 6th bay south (transept)	ca. 1340-1350
AUX 15	Pin	Flying buttress abutment, 5th-6th bay south, southern pinnacle	ca. 1370-1380
AUX 17-1	Cramp	Abutment top florets, 3rd-4th bay south (repair?)	ca. 1380-1390
AUX 17-2	Cramp	Abutment top florets, 3rd-4th bay south (repair?)	ca. 1380-1390
AUX 18-1	Pin	Flying buttress mullion, 3rd-4th bay south	ca. 1380-1390
AUX 18-2	Pin	Flying buttress mullion, 3rd-4th bay south	ca. 1380-1390
AUX 18-3	Pin	Flying buttress mullion, 3rd-4th bay south	ca. 1380-1390
AUX 39	Pin	Flying buttress mullion	ca.1380-1390
AUX 65	Pin	4th northern chapel's tracery: joint 9 / 15	ca.1310-1320
AUX 66	Pin	4th northern chapel's tracery: joint 1 / 3	ca.1310-1320
AUX 67	Pin	4th northern chapel's tracery: joint 14/15	ca.1310-1320
AUX 68	Pin	4th northern chapel's tracery: joint 18/19	ca. 1310-1320
AUX 69	Pin	4th northern chapel's tracery: joint 6 / 19	ca. 1310-1320
AUX 70	Pin	4th northern chapel's tracery: joint ?	ca. 1310-1320
AUX 71	Pin	4th northern chapel's tracery: joint 11 / 7	ca. 1310-1320
AUX 72	Pin	4th northern chapel's tracery: joint 3 / 5	ca.1370-1380
AUX 73	Pin	4th northern chapel's tracery: joint 3 / 1	ca. 1370-1380
AUX 74	Pin	4th northern chapel's tracery: joint 7 / 3	ca. 1310-1320
AUX 75	Pin	4th northern chapel's tracery: joint 17 / 18	ca. 1310-1320

have variable quantities of non-metallic slag inclusions and porosities (Fig. 3). Some are fairly clean, with only 2 to 4% of their surface filled with SI. Others are much dirtier: SI and porosities can reach up to 6-7% and even 12% for AUX 72a (Fig. 4). Yet, even in these dirtiest samples, SI usually remain quite small (< 1000μm^2 on average). Only AUX 75a has much bigger SI (8000μm^2 on average).

NRC ratios in SI were quite homogeneous for 8 samples: AUX 66 to AUX 72 (Fig. 5) and AUX 75a. This homogeneity indicates that these artefacts are not made of recycled metal but come from a single, unique piece of metal. Therefore, the welding lines detected on AUX 75a are only the consequence of folding during forging. On the contrary, different NRC ratios are visible in SI of samples AUX 73a (Fig. 5) and AUX 74a. Moreover, these ratios differ on each side of the welding line observed on the cross-sections. Thus, these samples are clearly made from different metallic pieces forged together in a recycling process.

In the following, SI average compositions are used to try to specify provenance groups for the pins from the northern chapel's tracery. It must be noted that no sample has a significant P_2O_5 or MnO average weighted content indicating a specific origin linked to the use of respectively phosphorous or manganiferous ores (Table 3).

Yet, NRC ratios (Fig. 6) reveal at least two different groups for the first phase pins (joints with stonemason's hammer marks): the first one with pins AUX 65, 67 and 70, the second one with AUX 66, 68, 69 and 75. AUX 71 has a slightly higher $\%K_2O^*/\%CaO^*$ ratio than the second group samples: it could belong to a third group. These observations suggest that the pins used for the first construction phase of the nave windows come from at least two (probably three) different reduction systems and therefore do not come from the same workshop. It is not possible to provide further interpretations concerning these different origins because they could be associated with a number of possibilities. For example, accounting books of the

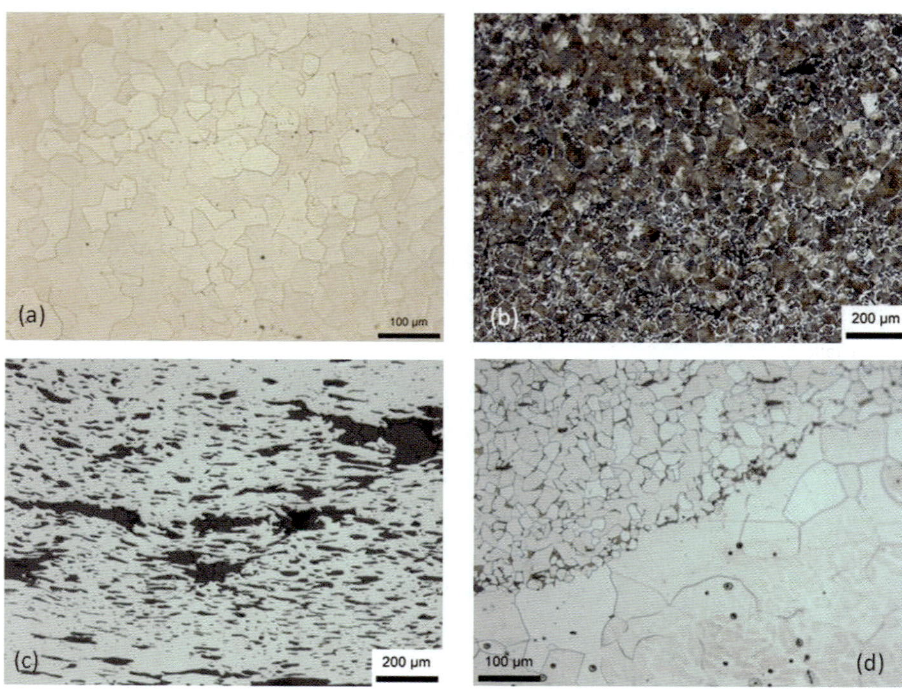

Figure 3: Optical microphotograph on samples from the fourth chapel of the nave (group 1): (a) Ferritic zone, section AUX 65a, Nital etching; (b) Carburised zone with 0.75 wt% C, sample AUX 70a, Nital etching; (c) Slag inclusion in the metallic matrix, sample AUX 67; (d) Welding line with ghosting, sample AUX 73a, Nital etching.

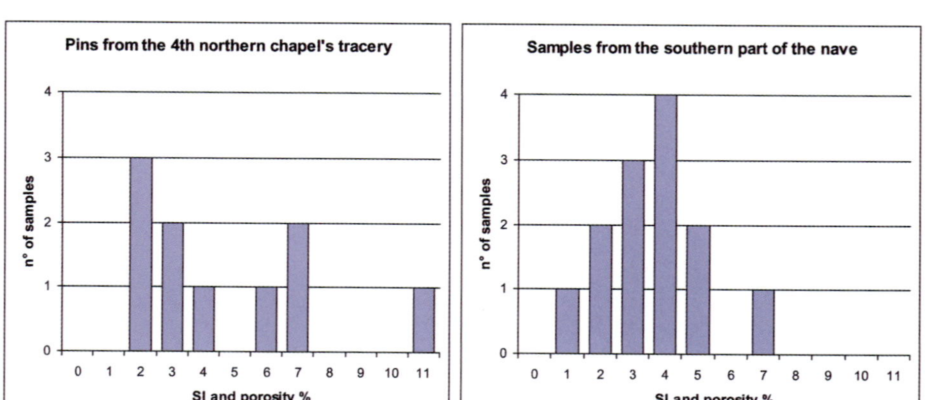

Figure 4: Repartition of SI and porosity percentages in samples coming from the tracery of the fourth northern chapel.

Table 3: Average weighted composition of slag inclusions for all samples from Auxerre Cathedral

	Na_2O*	MgO*	Al_2O_3*	SiO_2*	P_2O_5*	SO_3*	ClO*	K_2O*	CaO*	TiO_2*	V_2O_5*	Cr_2O_5*	MnO*	FeO*
AUX 11a	< 0.5	< 0.5	8.2	32	0.9	< 0.5	< 0.5	1.6	2.7	< 0.5	< 0.5	< 0.5	< 0.5	53
AUX 12a	< 0.5	< 0.5	0.9	16	10	0.8	< 0.5	< 0.5	1.7	< 0.5	< 0.5	< 0.5	< 0.5	69
AUX 13-1a	< 0.5	< 0.5	8.8	33	1.1	< 0.5	< 0.5	1.3	1.5	< 0.5	< 0.5	< 0.5	0.7	53
AUX 13-2a	< 0.5	< 0.5	10	34	1.2	< 0.5	< 0.5	1.3	1.7	< 0.5	< 0.5	< 0.5	3.2	48
AUX 14-1a	< 0.5	0.70	14	43	0.9	< 0.5	< 0.5	1.4	2.2	< 0.5	< 0.5	< 0.5	< 0.5	37
AUX 14-2a	0.5	< 0.5	9.7	29	1.2	1	< 0.5	1.7	2.3	< 0.5	< 0.5	< 0.5	< 0.5	54
AUX 15a	< 0.5	< 0.5	8.3	33	1.6	< 0.5	< 0.5	2	3.5	< 0.5	< 0.5	< 0.5	< 0.5	50
AUX 17-1a	< 0.5	< 0.5	5.5	29	3.8	1.6	< 0.5	1.6	2.9	< 0.5	< 0.5	< 0.5	0.6	55
AUX 17-2a	< 0.5	0.92	13	51	0.8	< 0.5	< 0.5	4	6.8	< 0.5	< 0.5	< 0.5	< 0.5	24
AUX 18-1a	< 0.5	< 0.5	1.2	18	12	1.2	< 0.5	< 0.5	2	< 0.5	< 0.5	< 0.5	< 0.5	64
AUX 18-2a	< 0.5	< 0.5	1.3	24	7.8	0.8	< 0.5	0.9	1.9	< 0.5	< 0.5	< 0.5	0.7	61
AUX 18-3a	< 0.5	< 0.5	8.4	33	0.9	< 0.5	< 0.5	1.5	1.9	< 0.5	< 0.5	< 0.5	< 0.5	53
AUX 39a	< 0.5	< 0.5	12	41	0.6	< 0.5	< 0.5	1.6	2.6	< 0.5	< 0.5	< 0.5	< 0.5	41
AUX 65a	< 0.5	1.0	5.6	34	4.1	0.8	< 0.5	2.7	6.7	< 0.5	< 0.5	< 0.5	0.9	45
AUX 66a	< 0.5	< 0.5	5.6	24	1.8	0.6	< 0.5	1.5	2.3	< 0.5	< 0.5	< 0.5	< 0.5	64
AUX 67a	< 0.5	1.2	7.3	43	2	< 0.5	< 0.5	3.2	6.8	< 0.5	< 0.5	< 0.5	1.1	34
AUX 68a	< 0.5	0.7	6.4	29	1.8	< 0.5	< 0.5	2.2	3.1	0.92	< 0.5	< 0.5	< 0.5	55
AUX 69a	< 0.5	0.5	6.3	31	4	0.5	< 0.5	2.5	3.9	< 0.5	< 0.5	< 0.5	1.0	50
AUX 70a	< 0.5	1.1	10	52	1.2	0.5	< 0.5	2.7	5.8	< 0.5	< 0.5	< 0.5	2.5	24
AUX 71a	< 0.5	1.1	8.7	42	0.8	< 0.5	< 0.5	4.2	4.7	< 0.5	< 0.5	< 0.5	1.5	35
AUX 72a	< 0.5	< 0.5	7.1	29	1.1	< 0.5	0.6	1.4	1.8	< 0.5	< 0.5	< 0.5	2.3	56
AUX 73a	< 0.5	1.1	7.5	26	8.1	1	< 0.5	2.4	9.5	< 0.5	< 0.5	< 0.5	< 0.5	42
AUX 74a	< 0.5	< 0.5	6.7	31	3.2	0.7	< 0.5	1.1	2.2	< 0.5	< 0.5	< 0.5	0.5	55
AUX 75a	< 0.5	0.7	6.4	29	3.5	0.6	< 0.5	2.1	3	< 0.5	< 0.5	< 0.5	1.0	54

Table 4: Summary of the metallographic observations for all samples from Auxerre Cathedral

Sample	Av. wt% C	% C zones	Type of alloy	Ghost structures	% SI and porosities	SI density (nb SI / mm^2)	Av. SI size (μm^2)	Welding lines	Homog. NRC	Reduction process
AUX 11a	0.05	10.6%	Ferritic	-	2.5%	3000	10	-	YES ?	Bloomery
AUX 12a	0.02	1.0%	Ferritic	-	5.5%	1000	50	-	YES	Indirect
AUX 13-1a	0.02	0.6%	Ferritic	-	4%	6000	5	-	YES ?	Bloomery
AUX 13-2a	0.05	5.3%	Ferritic	-/+	4%	2000	20	-	YES	Bloomery
AUX 14-1a	0.13	30.1%	Heterogeneous	-	2%	500	30	-	YES ?	Bloomery
AUX 14-2a	0.03	1.4%	Ferritic	-	5.5%	6000	10	-	NO	Bloomery
AUX 15a	0.02	0.7%	Ferritic	-	4.5%	10000	5	-	YES	Bloomery
AUX 17-1a	0.08	19.2%	Ferritic	-	4%	1000	40	1	NO	Bloomery ?
AUX 17-2a	0.49	82.0%	Steel	-	7%	2000	40	-	NO	Bloomery
AUX 18-1a	0.02	0.8%	Ferritic	-/+	4.5%	500	70	1	YES	Indirect
AUX 18-2a	0.09	19.6%	Ferritic	+	2.5%	1000	20	1	NO ?	Indirect ?
AUX 18-3a	0.04	4.2%	Ferritic	-/+	3%	1000	30	-	NO	Bloomery
AUX 39a	0.27	46.6%	Composite steel	-	4%	500	50	-	YES	Bloomery
AUX 65a	< 0.02	0.0%	Ferritic	-	2%	500	30	-	YES	Bloomery
AUX 66a	0.03	1.7%	Ferritic	-	2%	500	30	-	YES	Bloomery ?
AUX 67a	0.03	1.9%	Ferritic	-/+	7%	1000	60	-	YES	Bloomery
AUX 68a	0.02	1.8%	Ferritic	-	2%	500	30	-	YES	Bloomery
AUX 69a	0.09	14.9%	Ferritic	-	-			-	YES	Bloomery
AUX 70a	0.24	60.8%	Composite steel	-	3%	1000	30	-	YES	Bloomery
AUX 71a	0.03	2.2%	Ferritic	-	7.5%	500	110	-	YES	Bloomery
AUX 72a	< 0.02	0.0%	Ferritic	-	12%	1000	120	-	YES	Bloomery
AUX 73a	0.09	26.6%	Heterogeneous	+	3.5%	500	60	2	NO	Bloomery
AUX 74a	0.05	10.7%	Ferritic	-	4%	1000	40	1	NO	Bloomery
AUX 75a	0.02	0.3%	Ferritic	-/+	6%	8000	10	1	YES ?	Bloomery

Palais des Papes in Avignon reveal that a "Maître des ferrures" (iron master) used to buy ferrous armatures from different locations for the palace's yard (Bernardi and Dillmann 2005).

Fig. 7 shows the abacus allowing distinguishing ironmaking processes in which all 11 pins are plotted. They all are in the bloomery process area, which is not surprising according to their dating.

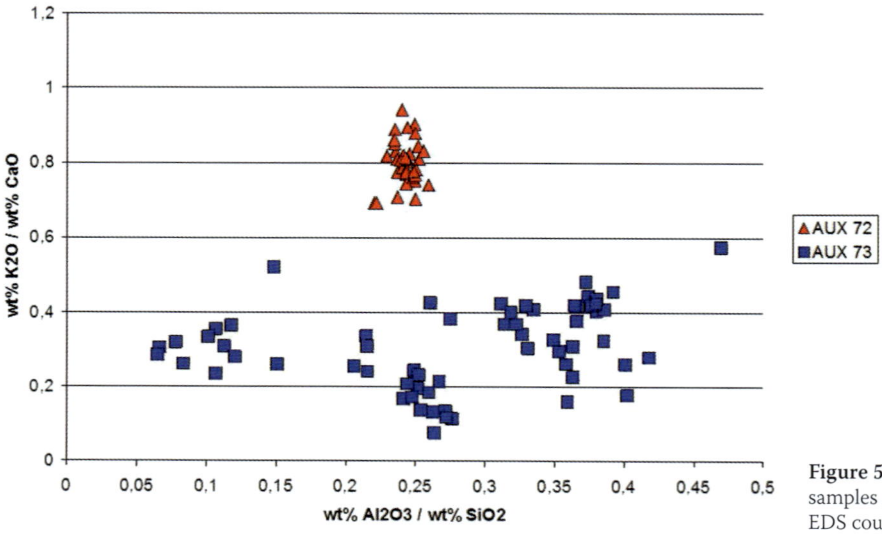

Figure 5: NRC ratios for slag inclusions of samples AUX 72 and AUX 73 (1 pt = 1 inclusion), EDS coupled to SEM.

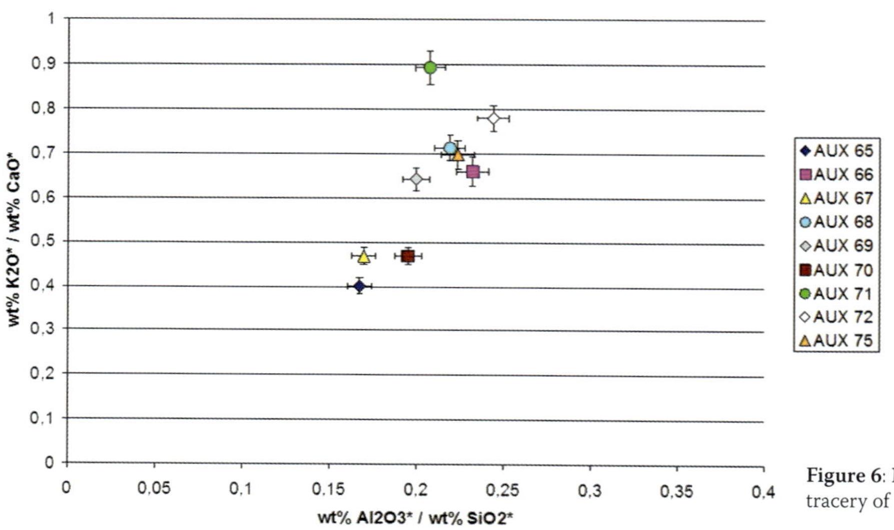

Figure 6: NRC ratios for the samples from the tracery of the fourth northern chapel.

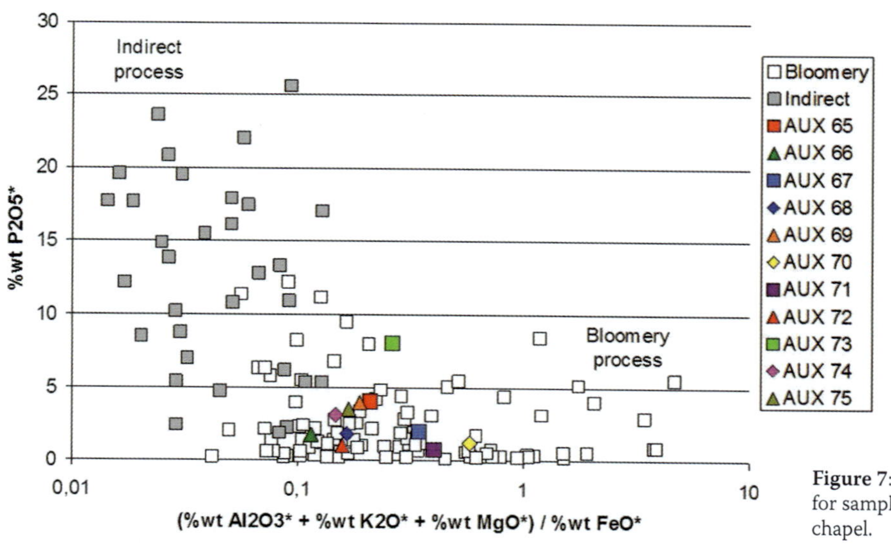

Figure 7: Distinction of the ironmaking processes for samples from the tracery of the fourth northern chapel.

Pins and cramps from the southern part of the nave

Observations performed on the cross-sections of this second set of samples (Figs. 2 and 8; Table 4) reveal that most sections are not very carburised. Ten samples plot in the 'ferritic' area as defined by Pagès et al (2008) (Fig. 2). Very local cementation features probably due to contact with charcoal in the smithy are sometimes visible. On the contrary, following this classification, AUX 17-2a can be considered as steel. This section shows a quenched structure which is probably not voluntary. Ghosting is locally visible on samples AUX 13-2a, AUX 18-1a, 18-2a, and 18-3a. Only AUX 18-2a can however be considered as fairly phosphoric (Fig. 9). On this sample, phosphorous ferrite and pearlite zones have almost the same micro-hardness, respectively 175 and 180 $Hv_{25gf/30s}$, whereas ferrite without phosphorus reaches only 125 Hv. On AUX 39a, steel zones go up to 220 Hv. This illustrates the similar hardening effect of both carbon and phosphorus on iron. Typical welding lines can be observed on samples AUX 17-1a, AUX 18-1a and 18-2a. SI and porosities range from 2 to 7% of the analysed surfaces, depending on the samples. Their repartition shows almost a Gaussian plot around 4% (Fig. 4). The dirtiest samples are equally characterised by a low density (< 10 SI/mm^2) of very big SI (> 6000µm^2) or by a high density (> 40 SI/mm^2) of small SI (< 2000µm^2). However, these features do not seem to be linked either with the type of object, or with their dimensions.

In seven samples, NRC ratios are homogeneous for all SI, indicating that the artefacts were obtained from a single piece of metal, without recycling. On the contrary, for the six other specimens, different groups of NRC ratios are evident. When within a sample these groups can be linked to areas separated by welding lines, then the artefacts were certainly forged from different pieces of recycled metal. This is particularly the case for AUX 18-2, whose inclusions also show a difference of MnO* and P2O5* contents at each side of the welding lines.

The high MnO* contents in some samples (AUX 13-2 for example) and the variation of NRC ratios prove the use of iron from different origins and reduction systems.

Considering the ironmaking processes, the SI average weighted compositions reveal that most of the artefacts come from the bloomery process (Fig. 10). Nevertheless, two of them (AUX 12 and AUX 18-1) clearly plot in the indirect process area of the abacus. This latter observation is crucial and sheds new light on the appearance of the indirect process in French Burgundy.

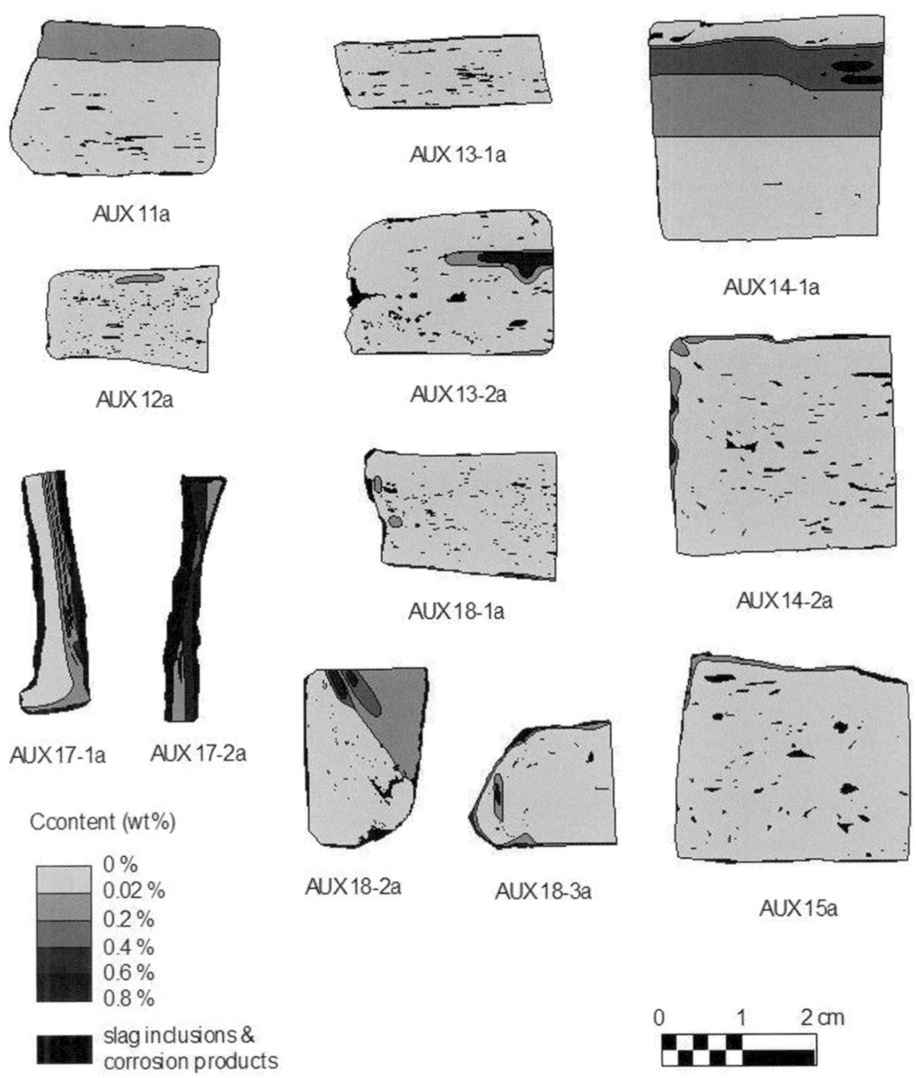

Figure 8: Metallographic observations on the cross-sections of the second set of samples.

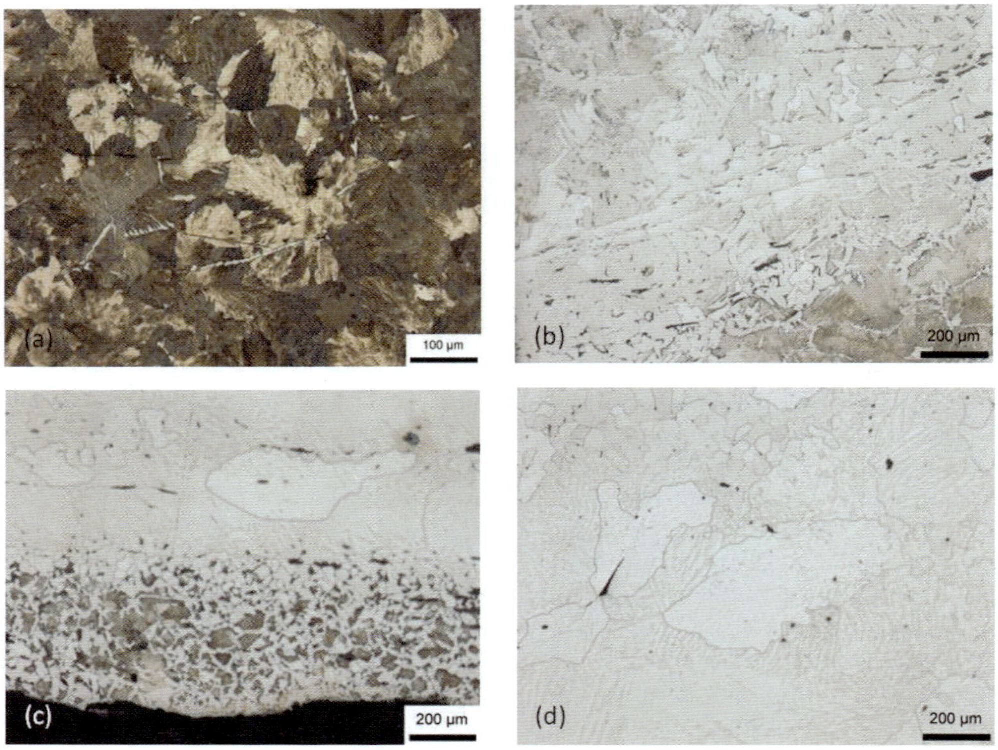

Figure 9: Microphotograph from samples from the southern part of the nave: (a) Quenching structure (martensite), sample AUX 17-2a, Nital etching; (b) Welding line, sample AUX 18-2a, Nital etching; (c) Cementation, sample AUX 18-1a, Nital etching; (d) Phosphorous ferrite showing ghosting, sample AUX 18-2a, Nital etching.

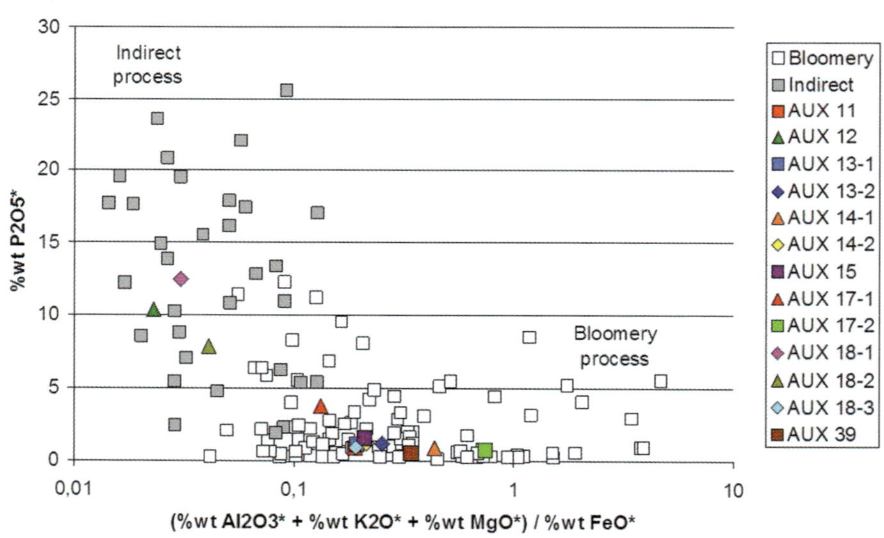

Figure 10: Distinction of the ironmaking processes for samples coming from the southern part of the nave.

Iron reinforcements employed in Auxerre Cathedral exclusively come from the bloomery process until the middle of the 14th century. During the late 14th century building campaigns, several artefacts seem to have been made with iron coming from pig iron refining. The reliable dating of these samples is provided by their location in the building studied according to archaeological and architectural considerations (Aumard, 2009). Among these observations, it was noted that the lead joints embedding indirect process artefacts are exactly the same as for artefacts made of bloomery iron, therefore proving that they are contemporary. In addition, pin AUX 18-3a (produced by the bloomery process) is located in the mullions of the same flying buttress as AUX 18-1a (produced by the indirect process). These artefacts made by the blast furnace and finery process are, to our knowledge, the oldest ever found on a monument of the French Kingdom. These results provide archaeological evidence for the rare testimonies about the blossoming of blast furnaces and fineries in Nivernais and south Champagne, two regions surrounding the town of Auxerre, during this period (Belhoste and Léon, 1996;

Rouillard, 2003). For the south of Champagne, J. Rouillard discovered in historical written sources the existence of a blast furnace in the Othe forest at Valcon sur l'Ancre (Aube), linked to a finery located in Estissac, on the border of the river Vanne from 1377 (Rouillard, 2003). Although bloomery iron is the main ferrous material used at that time in the building yard of Auxerre Cathedral, the discovery of a few artefacts made of finery iron is a very important result, because it brings material evidence for the diffusion of this process in the east of the French Kingdom in the late 14th century.

Lastly, regarding quality, the iron composition is clearly not adjusted to function, which means that there seems to be no alloy optimisation by the master mason. The variability of iron types (ferritic, steel, phosphoric, etc.) recorded in identical artefacts such as pins used in the same tracery can neither be linked to provenance nor to iron making workshops, which could be differentiated using SI analyses. This random quality rather seems to illustrate the variability of composition of the iron bloom. It can therefore be compared with most results obtained so far on medieval building yards (L'Héritier, et al., 2005, 2010; Dillmann, 2009), with the sole exception of the Palais des Papes in Avignon (Dillmann, et al., 2004).

Conclusion

This paper aimed to present an integrated methodology for studying ferrous artefacts and producing comparable results between different studies (carbon content, phosphorus content, slag inclusion cleanliness, etc.). It also detailed a methodology based on slag inclusion analyses to determine on the one hand if artefacts have the same origin (in terms of reduction system), and, on the other hand, the ironmaking process. Additionally, this methodological approach was applied to the 14th century ferrous reinforcements from Auxerre Cathedral. Using this methodology, the oldest ferrous artefacts made by the indirect process in this region (and in France) were discovered, and the first archaeometric evidence was produced to support recent historical studies which suggest the blossoming of blast furnaces and fineries in the Auxerre area during the late 14th century.

Note

1. For higher contents, Energy Dispersive Spectrometry (EDS) coupled to a Scanning Electron Microscope (SEM) analysis on the metallic matrix can be performed.

References

Aumard, S., 2009. L'emploi du plomb dans la cathédrale Saint-Etienne d'Auxerre. In A. Timbert ed. *L'homme et la matière. L'emploi du fer et du plomb dans l'architecture gothique.* Paris: Picard, pp.159-72.

Belhoste, J.-F. and Léon, P., 1996. Naissance d'une sidérurgie moderne aux confins du Berry (fin du XIVe - XVe siècles). *Mélanges Jean-Yves Ribault, Cahiers d'Archéologie et d'Histoire du Berry, Hors-série*, pp.45-52.

Benoit, P., 1994. *Les mutations techniques et scientifiques de la fin du Moyen Age et de la Renaissance.* PhD Thesis, University of Lille.

Bernardi, P. and Dillmann, P., 2005. Stone Skeleton or Iron Skeleton: The Provision and Use of Metal in the Construction of the Papal Palace at Avignon in the Fourteenth Century. In R.O. Bork ed. *De Re Metallica. The Uses of Metal in the Middle Ages.* Aldershot: Ashgate, pp.297-316.

Chapelot, J., 2003. Le donjon et son enceinte. Une résidence royale. *Les Dossiers d'Archéologie - Vincennes*, 289, pp.60-73.

Clément-Charpentier, S., 2001. Le rôle des éléments métalliques dans la conception du donjon de Vincennes. In O. Chapelot ed. *Du projet au chantier. Maîtres d'ouvrages et maîtres d'oeuvre aux XIVe-XVIe siècles.* Paris: Ecoles des Hautes Etudes en Sciences Sociales, pp.405-31.

Desaulty, A.M., Mariet, C., Dillmann, P., Joron, J.L. and Fluzin, P., 2008. A Provenance Study of Iron Archaeological Artefacts by Inductively Coupled Plasma-Mass Spectrometry Multi-elemental Analysis. *Spectrochimica Acta Part B – Atomic Spectroscopy*, 63, pp.1253-62.

Desaulty, A.M., Dillmann, P., L'Héritier, M., Mariet, C., Gratuze, B., Joron, J.L. and Fluzin, P., 2009. Does it Come from the Pays de Bray? Examination of an Origin Hypothesis for the Ferrous Reinforcements Used in French Medieval Churches Using Major and Trace Element Analyses. *Journal of Archaeological Science*, 36, pp.2445-62.

Dillmann, P., 2009. De Soissons à Beauvais: le fer des cathédrales de Picardie, une approche archéométrique. In A. Timbert ed. *L'Homme et la Matière: l'emploi du plomb et du fer dans l'architecture gothique.* Paris: Picard, pp.93-111.

Dillmann, P. and L'Héritier, M., 2007. Slag Inclusion Analyses for Studying Ferrous Alloys Employed in French Medieval Buildings: Supply of Materials and Diffusion of Smelting Processes. *Journal of Archaeological Science*, 34, pp.1810-23.

Dillmann, P., Bernardi, P. and Fluzin, P., 2003. Use of Iron for the Building of Medieval Monuments. The Palais des Papes in Avignon and Other French Buildings. In Associazione Italiana di Metallurgia ed. *Archaeometallurgy in Europe.* Milan: AIM, pp.199-208.

Dillmann, P., Bernardi, P. and Fluzin, P., 2004. Iron in Medieval Monuments. Metallographic Analysis of Irons Coming from the Palais des Papes in Avignon. *La Revue d'Archéométrie*, 27, pp.183-92.

Dillmann, P., Aranda, R., L'Héritier, M. and Fluzin, P., 2005. Forging of Iron-Steel Artifacts: Archaeology, Experiments and Archaeometry. In *1st International Conference of Paleosiderurgy and Industrial Heritage Recovery. Iron, History and Heritage.* San Sebastian: INASMET, pp.311-24.

Dillmann, P., Tereygeol, F. and Verna, C., 2006. Premières analyses métallographiques des produits sidérurgiques trouvés sur le site médiéval de Castel-Minier (Aulus-les-Bains, 09). *La Revue d'Archéométrie*, 29, pp.7-14.

Fluzin, P., 1983. Notions élémentaires de sidérurgie. In N. Echard ed. *Métallurgies africaines. Nouvelles contributions.* Paris: Musée de l'Homme, pp.13-44.

Fluzin, P., Ploquin, A. and Serneels, V., 2000. Archéométrie des déchets de production sidérurgique. Moyens et méthodes d'identification des différents éléments de la chaîne opératoire directe. *Gallia*, 57, pp.1-158.

Fluzin, P., Ploquin, A. and Dabosi, F., 2004. Approches métallurgiques et archéométriques. In M. Mangin ed. *Le Fer.* Paris: Errance, pp.113-74.

L'Héritier, M., 2007. Le fer dans l'architecture, méthodologie d'étude, exemples de Troyes et de Rouen. *Monumental*, semestriel 2, pp.98-102.

L'Héritier, M., 2009. L'utilisation du fer dans l'architecture gothique. Approche méthodologique à travers les villes de Rouen et Troyes.

In A. Timbert ed. *L'homme et la matière. L'emploi du fer et du plomb dans l'architecture gothique*. Paris: Picard, pp.61-73.

L'Héritier, M., Juhin, A., Dillmann, P., Aranda, R. and Benoit, P., 2005. Utilisation des alliages ferreux dans la construction monumentale du Moyen Age. État des lieux de l'avancée des études métallographiques et archéométriques. *La Revue d'Archéométrie*, 29, pp.117-32.

L'Héritier, M., Dillmann, P. and Benoit, P., 2010. Iron in the Building of Gothic Churches: Its Role, Origins and Production Using Evidence from Rouen and Troyes. *Historical Metallurgy*, 44, pp.21-35.

Lefebvre, E., 2009. Les tirants de fer de la cathédrale Notre-Dame d'Amiens. In A. Timbert ed. *L'emploi du fer et du plomb dans l'architecture gothique*. Paris: Picard, pp.141-7.

Leroy, S., Desaulty, A.-M., Dillmann, P., Tereygeol, F., Verna, C. and Bertrand, L., 2007. First Examinations of Ore, Slag and Iron Artefacts from the Iron-making Site of Castel-Minier (Aulus-les-Bains, 09) (France). In Associazione Italiana di Metallurgia ed. *Archaeometallurgy 2007: 2nd international conference, 17-21 June 2007, Aquileia, Italy: proceedings*. Milan: AIM, Electronic Publication.

Leroy, S., Simon, R., Bertrand, L., Foy, E., Williams, A. and Dillmann, P. 2011. First examination of slag inclusions in medieval armours by confocal SR-μ-XRF and LA-ICP-MS. Journal of Analytical Atomic Spectrometry, 26, (5), pp.1078-87.

Leroy, S., Cohen, S.X., Verna, C., Gratuze, B., Téreygeol, F., Fluzin, P., Bertrand, L. and Dillmann P. 2012. The medieval iron market in Ariège (France). Multidisciplinary analytical approach and multivariate analyses. Journal of Archaeological Science, 39, pp.1080-93Pagès, G., Long, L., Fluzin, P. and Dillmann, P., 2008. Réseaux de production et standards de commercialisation du fer antique en Méditerranée : les demi-produits des épaves romaines des Saintes-Maries-de-la-Mer (Bouches-du-Rhône). *Revue archéologique de Narbonnaise*, 41, pp.261-83.

Pagès, G., Dillmann, P., Fluzin, P. and Long, L., 2011. A Study of the Roman Iron Bars of Saintes-Maries-de-la-Mer (Bouches-du-Rhône, France). A Proposal for a Comprehensive Metallographic Approach, *Journal of Archaeological Science*, 38, pp.1234-52.

Piaskowski, J., 1984. Proposals for a Standardization of the Criteria for Determining Technological Processes in Early Iron and Steel Metallurgy. In B.G. Scott ed. *The Crafts of the Blacksmith*. Belfast: Union Internationale des Sciences Prehistoriques et Protohistoriques, Comité pour la Sidérurgie Ancienne in conjunction with the Ulster Museum, pp.169-78.

Piccardo, P., Ienco, M.G., Balasubraman, R. and Dillmann, P., 2004. Detecting Non-uniform Phosphorus Distribution in Ancient Indian Iron by Colour Metallography. *Current Science*, 87, pp.650-3.

Rouillard, J., 2003. *L'homme et la rivière: histoire du bassin de la Vienne au Moyen Âge (XIIe-XVIe siècles)*. PhD Thesis, Paris I Sorbonne, Paris; 1199p.

Stewart, J.W., Charles, J.A. and Wallach, E.R., 2001. Iron-Phosphorus-Carbon System Part I – Mechanical Properties of Low Carbon Iron-Phosphorus Alloys. *Materials science and technology*, 16**,** 275-282.

Taupin, J.-L., 1996. Le fer dans les cathédrales. *Monumental*, 13, pp.18-27.

Vega, E., Dillmann, P. and Fluzin, P., 2002. Contribution à l'étude du fer phosphoreux en sidérurgie ancienne. *La revue d'Archéométrie*, 26, pp.197-208.

Vega, E., Dillmann, P., L'Héritier, M., Fluzin, P., Crew, P. and Benoit, P., 2003. Forging of Phosphoric Iron. An Analytical and Experimental Approach. In Associazione Italiana di Metallurgia ed. *Archaeometallurgy in Europe*. Milan: AIM, pp.337-46.

Authors' addresses

- Maxime L'Héritier
  Université Paris 8, 2 rue de la Liberté, 93526 Saint-Denis cedex
- Philippe Dillmann
  LAPA UMR5060 CNRS and UMR3299 CNRS, CEA Saclay, 91191 Gif sur Yvette cedex, France
- Philippe Fluzin
  LMC UMR5060 CNRS, UTBM, 91191 Belfort cedex, France
- Sylvain Aumard
  Centre d'Etudes Médiévales d'Auxerre, 3, Place du Coche d'Eau, 89000 AUXERRE, France

Measuring variation in iron smelting slags: an empirical evaluation of group-identification procedures

Michael F. Charlton, Peter Crew, Thilo Rehren, and Stephen J. Shennan

ABSTRACT: The chemistry of ironworking slag exhibits high levels of spatiotemporally distributed variation because of differences in available resources, technological traditions, and post-depositional processes. Archaeometallurgists have devised numerous classificatory and grouping methods to break up this variation into meaningful units for investigating technological and archaeological questions. As the discipline matures, it is appropriate to ask whether or not the unit construction techniques in use are adequate for their respective tasks. This paper reviews the common practices for creating groups and classes from slag chemistry, paying special attention to their application in different research contexts. Methods are compared and evaluated against experimental ironworking slag and published data from sites in Europe. Results indicate that most of the units used to measure variation in slag chemistry perform reasonably well. However, multivariate analytical strategies generally make the most efficient use of the available data relative to the time spent analysing it.

Keywords: Slag chemistry, systematics, oxide ratios, multivariate statistics.

Introduction

Slag is unquestionably one of the most important artefacts encountered by researchers of extinct iron production technologies. Slag provides a fossilised thermochemical history of operations within a single iron production cycle and varies according to differences in furnace design, furnace operation, resource usage, and production stages. Since this same information is often missing or incomplete in the metal itself, the analysis of slag morphology, microstructure, and chemistry are essential for the effective modelling of extinct ironmaking systems (see Bachmann, 1982). Slag inclusion chemistry, in particular, is becoming increasingly recognised for its potential to source objects made of bloomery iron to the region or site where the metal was extracted (e.g., Buchwald and Wivel, 1998; Coustures, et al., 2003, 2006; Dillmann and L'Héritier, 2007; Blakelock, et al., 2009; Desaulty, et al., 2009; Leroy, et al., 2011).

Most problems in the archaeology of iron production require that the variation in slag characteristics be measured in some meaningful way. The necessary starting point for this procedure is the construction of nominal scale classifications and/or groups of slag based on morphological, microstructural, or chemical attributes. The identification and description of chemical clusters is typically based on ratio scale measures of oxide and trace element concentrations, while morphological and microstructural groups are more often generated by nominal scale measures of formal attributes. Ordinal and ratio scale measures of slag morphology and microstructure are also possible but rarely used (e.g., Gordon, 1997). Once such units are constructed, be they classes or empirical groupings, other measures of variation may be employed to quantify the attributes of their members (by calculating standard deviations, coefficients of variation, etc.) or compare groups (by conducting *t*-tests, analysis of variance, linear discriminant analysis, etc.). The particular attributes used to define or identify classes or groups may be chosen for their perceived relevance to theoretical or descriptive purposes.

The distinction between classes and groups is subtle, yet significant. Following definitions given by Dunnell (1971) and O'Brien and Lyman (2000), a class is an ideational unit with an explicitly defined set of attributes and attribute states that are necessary and sufficient for empirical specimens to be considered representative of that class. Because classes are not empirical units, they have distributed representation across space and time. This property makes them well-suited to measuring nominal scale variation, and applicable to a wide range of theoretically derived models. Groups, in contrast, are empirical collections of things that display similarity as a consequence of class membership or some measurement, such as an oxide weight percent or oxide ratio. Their definitions are fluid, determined solely by the presence of an empirical sample and changing with the addition and subtraction of new specimens. Groups have specific locations in space and time and are unreliable measures of nominal scale variation. This is not to say that non-classificatory techniques designed to identify groups within empirical samples are not without value. They have an important role to play in recognising patterned differences between empirical samples, such as iron production systems from different times and places. They describe, but cannot explain the existence of the observed patterns.

This paper reviews the methods used by archaeometallurgists to describe and group bloomery iron production slag with respect to their investigative goals. Those methods intended to have analytical significance are evaluated across multiple data sets in order to assess their ability to discern meaningful variation in slag. While not discounting the importance of morphology and microstructure, this paper focuses strictly on groups and classes constructed from ratio scale chemical attributes.

The chemical systematics of bloomery slag

Archaeometallurgists study iron production slag for two main reasons: 1) to investigate and explain engineering processes; and 2) to investigate and explain technological variation and diversity. Both lines of inquiry require systematics, or procedures for classifying and grouping observed phenomena. Assuming that iron production slag can be safely distinguished from slag derived from other processes (see Bachmann, 1982; Tylecote, 1987), it must be categorised according to its particular technological origins (i.e., bloomery furnace, blast furnace, puddling hearth, finery, smithy, or subcategories thereof). The chemical systematics for identifying such groups is not well developed, and grouping generally relies on the weight of interferences drawn from morphological, microstructural and contextual analysis (but see Gordon, 1997). The chemical systematics for slag derived from bloomery systems, while more mature, still lacks robust definitions.

When confronted with an assemblage of bloomery iron production residues obtained from a single context, the starting assumption must be that it represents a relatively synchronous deposit of material stemming from a solitary ironmaking system. The validity of this assumption ultimately rests on associated chronological indicators and stratigraphic coherency. Variation within the assemblage is caused by: 1) biases imposed on slag derived from different stages of production; 2) the heterogeneous character of slag derived from single production cycles; 3) more or less random differences between production cycles; and 4) the use of multiple production recipes during assemblage accumulation. Slag from different stages of production (smelting and smithing at its most general level) is often distinguished on the basis of morphological criteria alone (Serneels and Perret, 2003). For cases where smithing and smelting slag morphologies grade into one another, the two may be discriminated chemically. Charlton (2007, pp.127-132) provides a model for discriminating smelting from smithing slag based on the latter's chemical tendency to be enriched in fuel ash and ceramic based oxides, but further elaboration is beyond the scope of this paper. It is hereafter assumed that specimens of smelting and smithing slag have been successfully allocated into distinct groups.

All smelting slag from a bloomery system is derived from the chemistry of its parent materials and will follow the general formula:

$$\mathbf{C}_{slag} = \mathbf{C}\ [(\mathbf{O}-\mathbf{R}_{FeO}) + \mathbf{A} + \mathbf{L} + \mathbf{F} - \mathbf{V}] \qquad \text{Formula 1}$$

where **C** is oxide composition; and the weight percent contributions to composition by ore, reduced oxides, fuel ash, furnace lining, flux, and volatile elements are expressed as **O**, **R**, **A**, **L**, **F**, and **V**, respectively. The bulk chemistry of slag derived from a single unchanging recipe is expected to swarm around a mean. Variation around the mean represents sampling errors across heterogeneous specimens and random differences between smelts. The bulk chemistry of slag from different recipes should cluster such that the variation between recipes is greater than the variation within them.

The presence of multiple compositional groups in a slag sample derived from multiple phases, sites or regions may indicate the use of different mineral resources and fuels (Buchwald and Wivel, 1998), furnace design and operating parameters (Humphris, et al., 2009), and change over time (Charlton, 2007, 2009; Charlton, et al., 2010). Interpretations for the groups will vary from case to case as a consequence of their multivariate structure and, most importantly, the method of analysis. As Bishop and Neff (1989) note, techniques of statistical data analysis are both structure-revealing and structuring. The interpretative value of an analysis depends heavily on well-reasoned assumptions about the underlying structure of the data and an informed choice of technique for revealing it. Several analytical techniques have been used to identify groups within samples of bloomery smelting slag, including ternary diagrams, bivariate plots and histograms of oxides and oxide ratios, and multivariate statistics. These techniques are described and compared below.

Ternary diagrams

Ternary phase diagrams are a favoured tool for archaeometallurgists investigating the thermochemical behaviour of slag as well as for identifying the presence of multiple slag groups. Ternary diagrams are triangle-shaped graphs which display the compositions of three variables. Each side of the equilateral triangle represents the composition of one variable from 0% at one corner to 100% at the other. Any point within the graph must be defined by compositions of the three variables which have been normalised and summed to 100%. A ternary phase diagram also includes information about the relationships between chemistry, temperature, and solidification of distinct compounds (see Bachmann, 1982, pp.10-13).

One of the earliest attempts to investigate variation in slag chemistry using ternary diagrams is Morton and Wingrove's (1972) study of medieval bloomery slag, which sought to define chemical constrains imposed by ore selection and examined evidence of technological change. Using the SiO_2-anorthite ($CaAl_2Si_2O_8$)-FeO phase diagram, they showed that slag derived from rich ore tended to plot in the fayalite portion of the graph, while that from lean ores concentrated in the hercynite region. The authors also hypothesised that the shift in medieval slag towards the SiO_2 side of the graph was the result of better iron recovery from the ore.

The choice of the SiO_2-anorthite-FeO ternary was based on the authors' contention that the abundance of CaO and Al_2O_3 in slag chemistry would give rise to an anorthite phase. Its inclusion as an end member to the ternary would therefore be more accurate for determining slag physical properties and representing slag in chemical space. Fells (1983) found that the SiO_2-anorthite-FeO ternary was a valid means of representing slag chemistry, but not effective for predicting physical properties or mineralogy. Indeed, as a means of demonstrating diversity in slag samples (for either technological or resource differences), it often performs no better than the more standard FeO-SiO_2-Al_2O_3 ternary. Fig. 1 shows the same separation between the slag groups graphed in Morton and Wingrove's (1969, 1972) SiO_2-anorthite-FeO ternary graphs using the FeO-SiO_2-Al_2O_3 ternary system. Serneel's (1993) data for slag from western Switzerland and Charlton's (2007) data for slag from northwest Wales also show clear differences in groups belonging to different techno-cultures and regions when projected into FeO-SiO_2-Al_2O_3 ternary diagrams.

Plots of oxides and oxide ratios

Exploratory searches for compositional patterns are also sought by comparing oxides and oxide ratios that are related to ore composition. Fells (1983) reasoned that Al_2O_3, MnO, MgO, TiO_2, and V_2O_5 were the most useful oxides for inferring an ore source based on the analysis of slag chemistry, and systematically graphed the oxides against each other in a series of bivariate scatter plots. Her results demonstrated that the oxide pairs were capable of differentiating groups of slag, but that no single pairing was ideal. Paynter (2006) conducted a similar analysis using P_2O_5, MnO, FeO, Al_2O_3, K_2O, TiO_2, MgO, and CaO, and arrived at the same conclusion. Rather than plotting the oxide values of individual slag specimens, however, Paynter graphed only site averages.

Oxide ratios are used primarily by researchers interested in sourcing metals to the site of their primary production using slag inclusions, but are occasionally used to describe production slag as well (e.g., Allen, 1988). Certain oxide ratios are expected to remain constant from ore to slag to slag inclusions in metal, even though actual oxide concentrations will change as iron is extracted and volatile elements are burned away. Error in this prediction is the result of variable geology and unaccounted-for inputs from additive fluxes, furnace walls, and fuel ash. Buchwald (2005) suggests that the ratio of SiO_2 to Al_2O_3 (what he terms an F-value) often proves the most useful ratio for sourcing slag relative to ore, though he also finds value in others, including SiO_2/FeO, Al_2O_3/CaO, and K_2O/MgO. An experimental investigation by Blakelock et al. (2009) challenges the importance of SiO_2/Al_2O_3, since different ores were shown to have similar ratios. Paynter (2006) used a graph of CaO/MgO vs. Al_2O_3/TiO_2 to compare smelting slag to slag inclusion chemistry data published by Hedges and Salter (1979). While sometimes using bivariate scatter plots, Buchwald (2005) more often employs histograms of oxide ratio frequencies to describe slag from particular regions.

Plots of oxides and oxide ratios, regardless of the preferred method, all have their utility. Examples are shown in Figs. 2-4, comparing their performance for Serneels' (1993) slag sample from western Switzerland and Charlton's (2007) slag sample from northwest Wales. Discrimination between slag from different regions and sites is evident in each technique. Graphs of oxide ratios provide the most information and are generally more effective at identifying groups. The reason for the good performance of both techniques, although not explicitly stated by the authors, is rooted in formula 1.

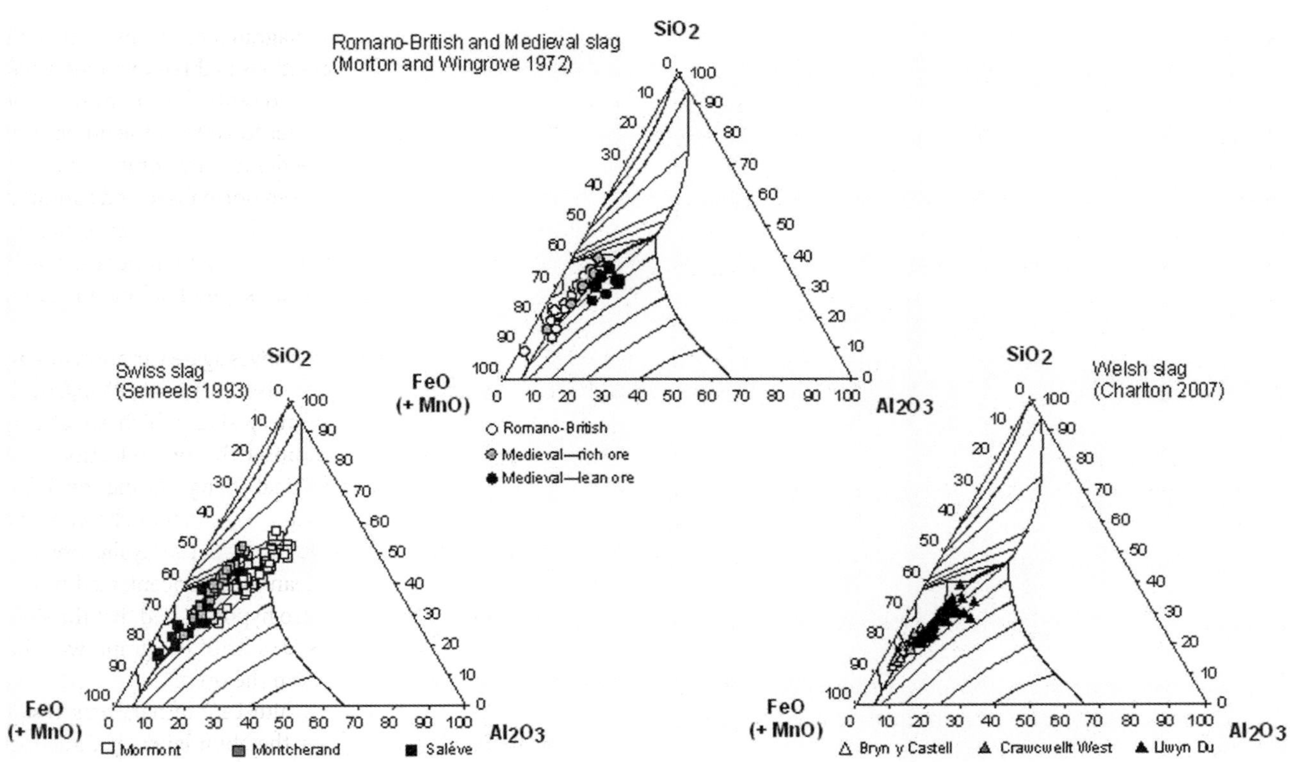

Figure 1: Employment of the FeO-SiO_2-Al_2O_3 ternary phase diagram to measure variation in slag groups from different regions.

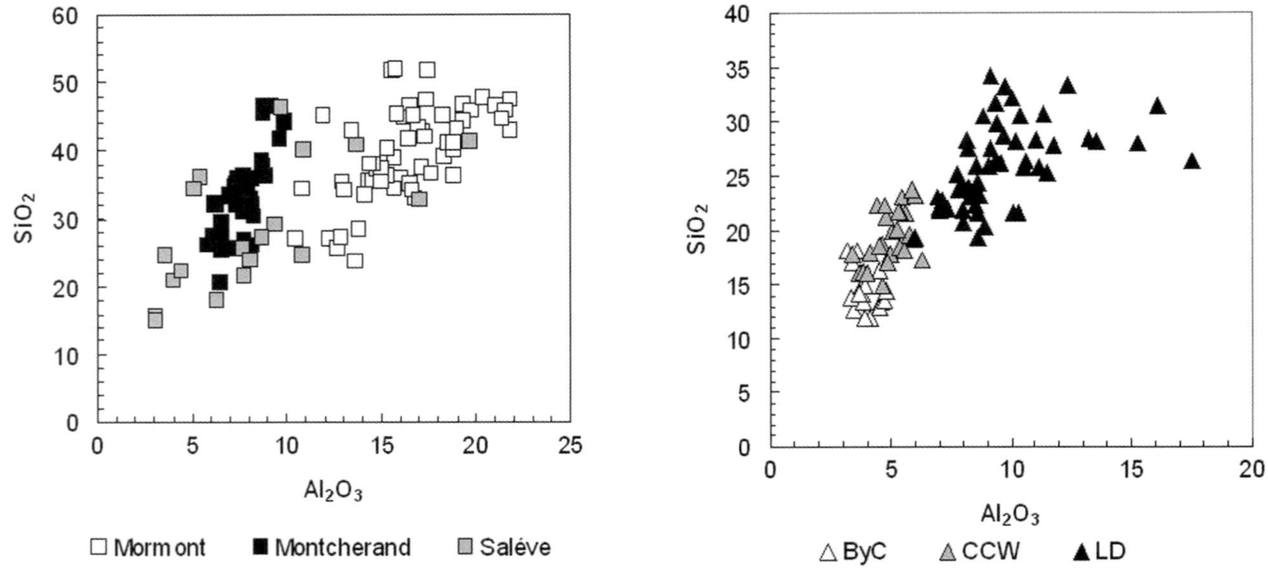

Figure 2: Scatter plot comparing Al_2O_3 vs. SiO_2 wt% measures for regional and site specific slag groups from Switzerland and Wales, respectively.

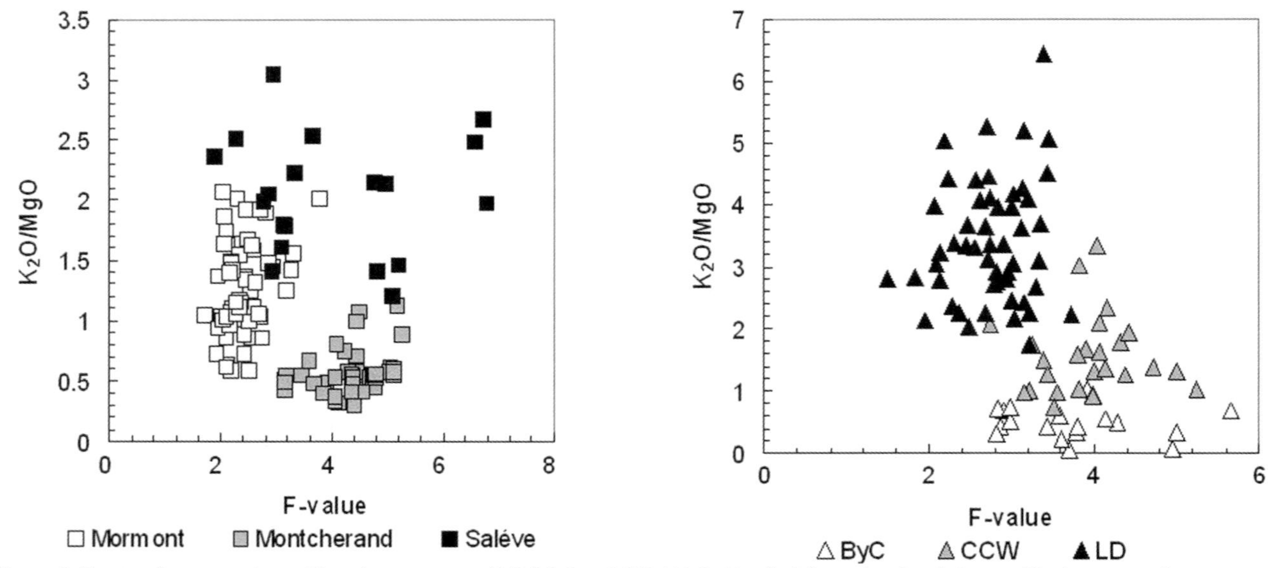

Figure 3: Scatter plot comparing oxide ratio measures – K_2O/MgO and SiO_2/Al_2O_3 (F-value) for regional and site specific slag groups from Switzerland and Wales, respectively.

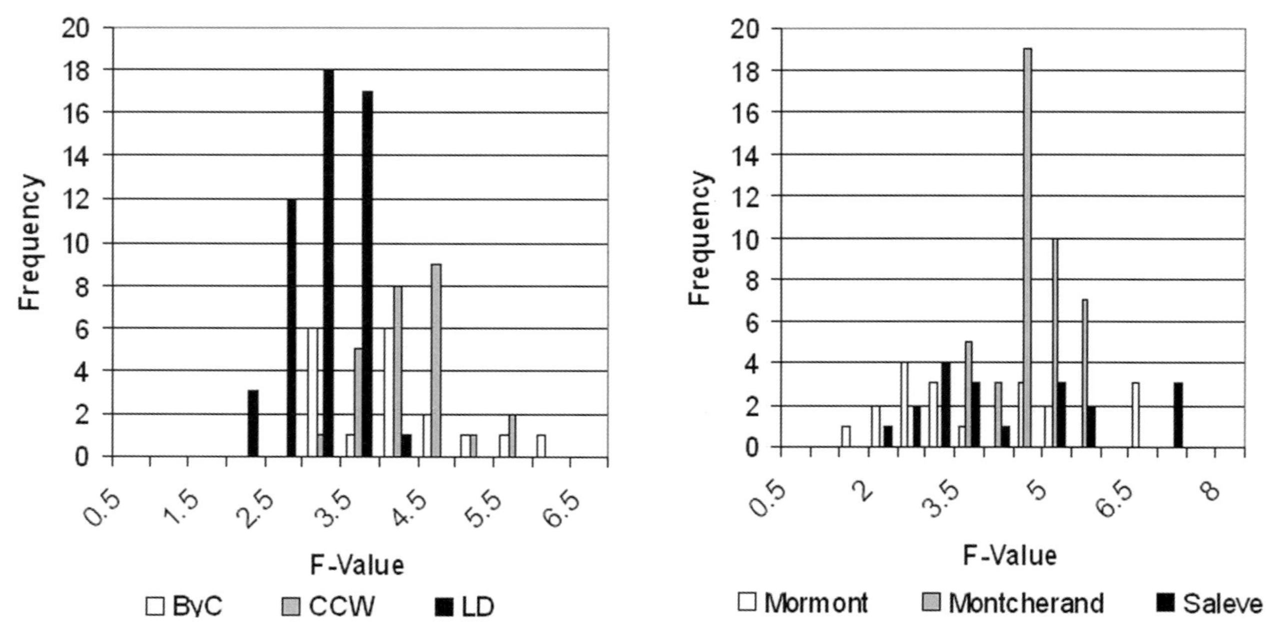

Figure 4: Histogram comparing SiO_2/Al_2O_3 (F-value) measures for regional and site specific slag groups from Switzerland and Wales, respectively.

Table 1: Parent material contribution to slag chemistry.

	MgO	Al_2O_3	SiO_2	P_2O_5	SO_3	K_2O	CaO	TiO_2	MnO	FeO	SrO	BaO
ore	●	●	●	●			●	●	●	●	●	●
clay		●	●			●		●				●
fuel ash	●			●	●	●	●				●	

The elements chosen by the authors are usually specified as being related to differences in ore chemistry. However, the contribution of all parent materials will also impact the final chemistry of the slag, sometimes significantly (Crew, 2000). Any systematic difference in the relationship between variables will be detected in plots of oxides and oxide ratios. The major weaknesses of these approaches lie in their general inability to sum up all the data in one or a few graphs, and a failure to define differences between resource variability and technological variability in smelting slag. A partial explanation of these weaknesses can be found in Table 1, which shows the primary oxides imparted to slag chemistry by ore, clay, and fuel ash, most of which occur in more than one parent material.

Principal Component Analysis – a model-based approach

Multivariate statistical techniques, such as Principal Component Analysis (PCA), offer one path to overcoming the weakness inherent in the examination of oxides and oxide ratios for determining group structure. PCA is a data reduction technique that compresses most of the significant variation found in a large multivariate data matrix (often containing nine or more variables) into a small number of new uncorrelated variables (usually between one and three), called principal components. This way, a slag chemistry measured across twelve oxides can be plotted holistically in one or two graphs (as opposed to sixty-six using bivariate oxide plots), with no or minimal loss of information. Data for individual specimens are projected onto the new principal component (PC) axes and can be related directly to the original variables, called loadings, which are represented as vectors. The structure of the loadings relative to the specimen scores can also provide important information. Loading vectors that are approximately 0° relative to one another are positively correlated (r = 1), while those 90° relative to one another are uncorrelated (r = 0), and those that are 180° relative to one another are negatively correlated (r = -1). Uncorrelated oxides and oxide ratios plotted against each other will generate group structures similar to that of the PC axes. Additional techniques such as cluster analysis or kernel density estimation may also be employed to help define data structure.

Smelting slag created by different recipes will tend to form clusters in multivariate space, whose relative positions reflect compositional differences in ore, clay, and fuel ash chemistry, as well as variation in furnace design and operation. Variation in the relative proportions of these parent materials in the slag also influences the cluster's position. Using the general model of parent material contributions to slag chemistry in Table 1, Charlton (2007) constructed a contingency table of all positive oxide correlations. Inferences about the causes of any group distinguished according to these correlations were modelled for each cell (Table 2). Strong positive correlations between K_2O and MgO, for example, lead to the inference of a group formed through higher relative contributions from fuel ash. Similarly, a strong positive correlation between SiO_2 and Al_2O_3 would suggest that groups containing larger amounts of these oxides were formed from higher relative contributions of technical ceramic from the furnace walls. Finally, there are also cases where the inference may be less certain. Strong positive correlations between P_2O_5 and CaO can suggest groups either with higher ore or fuel ash contributions, depending on the specific nature of the parent materials.

The archaeological relevance of these correlations varies from case to case. In one instance, slag groups defined by higher fuel ash contributions may reflect differences in fuel to ore ratios in the recipe. In another instance, it might suggest the use of a charcoal with higher ash content. Slag groups marked by higher concentrations of clay-related oxides that are also negatively correlated with FeO and P_2O_5 could indicate variation in the richness of utilised ores, variation in the refractory behaviour of the furnace walls/linings, or differences in redox conditions of the furnace. Deciding between these or other cases of multiple competing hypotheses requires independent lines of evidence, such as charcoal species identification and ore characterisation.

In terms of systematics, model-based PCA is a method that rests somewhere in between simple pattern recognition techniques and true classification. Like descriptive grouping exercises, results depend strictly on the specimens used in the analysis. However, model-based PCA does more than reduce data; it also provides a meaningful coordinate system which specifies the necessary and sufficient conditions for group membership relative to other groups as a consequence of oxide correlations. This coordinate system can therefore be broken up into a series of units which act as locally defined classes.

Data from three iron smelting experiments provide an ideal means of determining the efficacy of the model provided in Table 2. XP93, XP94, and XP95 were conducted with similar furnace configurations, operating parameters, clays, and fuels. The most significant source of variation was the ore (see Crew and Charlton, 2007). XP93 used Beckermet hematite – a very hard ore with an unusually high barium sulphate ($BaSO_4$) content (here represented as BaO). XP94 used a rich goethite ore from the Forest of Dean, and XP95 used a crushed bog ore. XP94 also incorporated a modified air supply system with larger diameter blowhole and deeper nozzle penetration.

A PCA of the experimental slag chemistries resulted in the formation of three distinct groups. XP93 is grouped by a high

Table 2: Contingency table for modelling resource-delimited constraints defined by positive oxide correlations.
O = Ore; FA = Fuel Ash; FL = Furnace Lining.

	MgO	Al_2O_3	SiO_2	P_2O_5	SO_3	K_2O	CaO	TiO_2	MnO	FeO	SrO	BaO
MgO		O	O	O-FA	FA	FA	FA-O	O	O	O	O-FA	O
Al_2O_3	O		FL	O		FL	O	O-FL	O	O	O	O-FL
SiO_2	O	FL		O		FL	O	FL	O	O	O	O-FL
P_2O_5	O-FA	O	O		FA	FA	O-FA	O	O	O	O-FA	O
SO_3	FA			FA		FA	FA				FA	
K_2O	FA	FL	FL	FA	FA		FA	FL		FL	FA	FL
CaO	FA-O	O	O	O-FA	FA	FA		O	O	O	FA-O	O
TiO_2	O	O-FL	FL	O		FL	O		O	O	O	O-FL
MnO	O	O	O	O			O	O		O	O	O
FeO	O	O	O	O		FL	O	O	O		O	O-FL
SrO	FA-O	O	O	O-FA	FA	FA	FA-O	O	O	O		O
BaO	O	O-FL	O-FL	O		FL	O	O-FL	O	O	O	

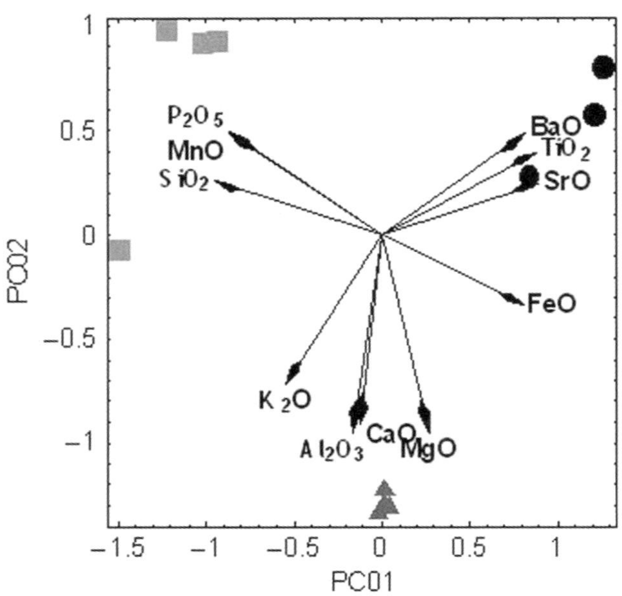

Figure 5: Graph of XP93, XP94, and XP95 slag chemistry in PC space. ● = XP93; ▲ = XP94; and ■ = XP95.

positive correlation of TiO_2, SrO, and BaO; XP94 by MgO, CaO, and Al_2O_3; and XP95 by P_2O_5 and MnO. These relationships are visible in a graph of the slag scores on the first two PC axes (Fig. 5). According to the model, the best hypothesis for XP93 is that the slag chemistry has been constrained by the ore. The same is true for XP95, where an ore constraint is the only available hypothesis. The situation with XP94 is more ambiguous. The strength of the relationship between Al_2O_3 and CaO renders an ore-based constraint the most likely hypothesis, but the correlations between MgO, K_2O, and CaO suggest fuel ash as another, perhaps complementary, possibility. Indeed, the distinct air supply system used in XP94 resulted in a significantly faster fuel burn – 24kg/hr versus 13kg/hr in XP93 (Crew and Charlton, 2007). This change in burn rate may have increased the fuel ash component of the slag.

In order to compare the performance of the model-based PCA with other techniques, a PCA was applied to Serneels' (1993) analyses of ironmaking slag and associated resources in western Switzerland. Analyses were carried out separately on slag and ore chemistries from Saléve, Mormont, and Montcherand (Fig. 6). While the groups are not as distinct as

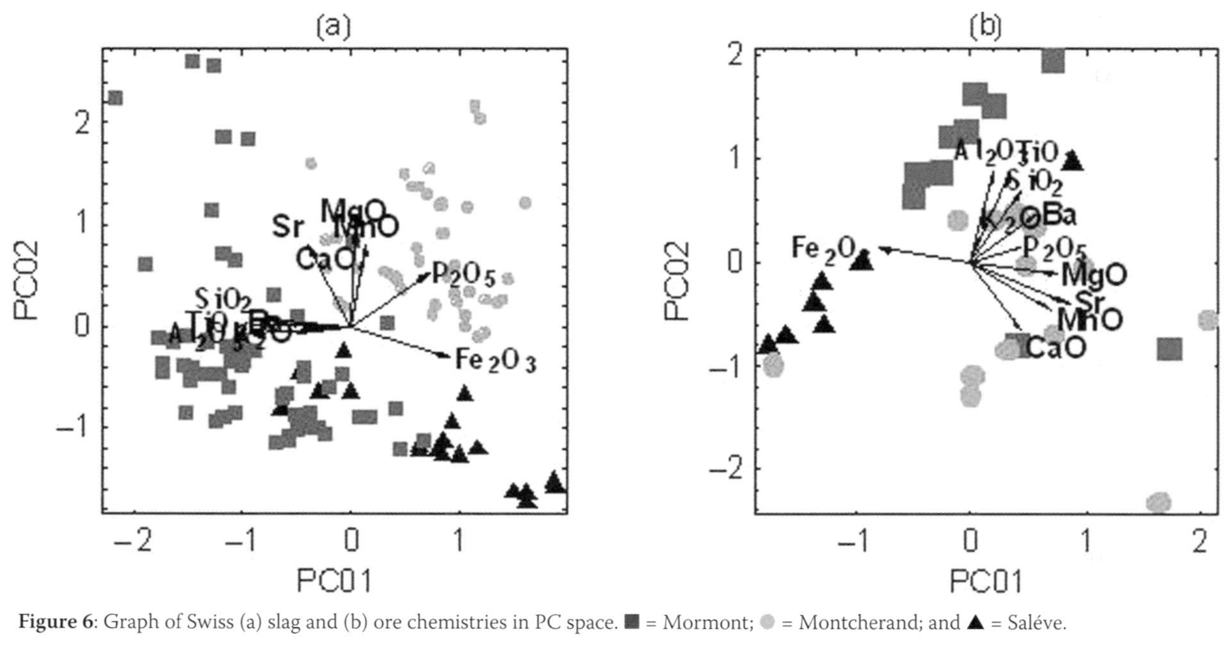

Figure 6: Graph of Swiss (a) slag and (b) ore chemistries in PC space. ■ = Mormont; ● = Montcherand; and ▲ = Saléve.

in the graph of oxide ratios, it is clear that the slags do show strong regional clustering. Saléve slags are noted for higher concentrations of Fe_2O_3 relative to Al_2O_3, TiO_2, and SiO_2. These slags overlap significantly – albeit in clustered fashion – with those from Mormont. Mormont slags display the opposite distributional pattern, being higher in Al_2O_3, TiO_2, and SiO_2 relative to Fe_2O_3. While the generalised model predicts that the major difference between the two regions is variation in ore and clay sources, a technological hypothesis can also be posited that ironmakers at Saléve produced less metallic iron per unit of ore than those in Mormont. This hypothesis is based on the observation that Saléve samples contain higher levels of Fe_2O_3 relative to typical gangue or clay oxides. Slags from Montcherand are more clearly discriminated by a combination of ore and fuel ash related oxides, suggesting that both parent materials may be constraining the chemical composition of the slag. The PCA of regional ore compositions (Fig. 6b) revealed that they are structured much the same way as the regional slags. This represents strong supporting evidence for the set of hypotheses stating that variation in ore source provides the primary explanation for interregional variability in slag chemistry in western Switzerland. The observation that Saléve and Mormont slags show internally clustered distributions suggests that other factors may also be affecting group structure.

The orientation of the loading vectors in a PCA can also indicate what oxide or oxide ratios might be worth examining for specific problems. Uncorrelated oxides and oxide ratios plotted against each other will generate group structures similar to that of the PC axes. Thus, from observing the loading arrangement on Fig. 6a, it is possible to create a new graph of Al_2O_3 vs. MgO which should reveal similar or perhaps better separation between groups (Fig. 7).

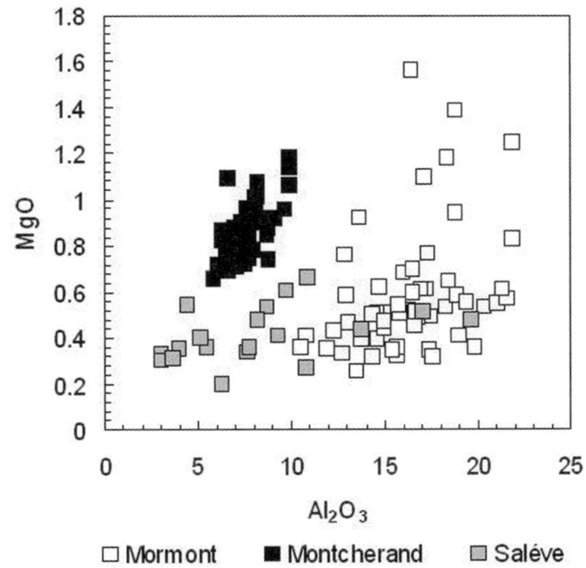

Figure 7: Graph of Al_2O_3 vs. MgO showing clear discrimination between the regional groups of Swiss slag, as suggested by the loading arrangements in Figure 6a.

Linear Discriminant Analysis

All of the methods discussed so far are referred to as unsupervised pattern recognition techniques, meaning that they are intended to identify unknown structures in the data. Supervised methods, thus called because they assume that the underlying structure is known, assign specimens to groups based on previously established patterns (see Baxter, 2006 for more details). One of the most common supervised pattern recognition techniques, Linear Discriminant Analysis (LDA), constructs a series of linear functions for a set of variables such that the distance is maximised between a set of *a priori* defined groups. The groups are defined by empirical samples of specimens called training sets, with known locations in space or some other relevant dimension. The resulting functions can then be applied to new datasets of unknown origin to determine their most likely group affiliation. A similar procedure is also possible with PCA under certain restricted conditions, though discrimination between groups is much poorer.

LDA has previously been used to group slag, usually in order to relate it to particular species of ore. Hedges and Salter (1979), in particular, used LDA to group the chemistry of slag inclusions from currency bars. That they were successful is not surprising, given that LDA requires that samples be weighted according to *a priori* group allocations. A more correct use of LDA would have required the analysis of known samples of ore, or perhaps smelting slag from different regions to use as training sets. The functions produced by this analysis could have then been applied to the unknown slag inclusions to generate probabilities of ore or smelting slag group identity. Such a strategy was used by Fells (1983). In this case, Fells used ten oxide ratios determined for three samples of ore to define her groups, then applied them to a series of excavated slag and ore. While Fells' analysis yielded good results, her work has, to the authors' knowledge, never been followed up. Stéphanie Leroy (2010), however, has developed an experiment-based approach to LDA to trace the origins of slag inclusions to particular ore bodies. Her results, though still tentative, provide a potentially powerful framework for iron object provenance studies. The application of LDA to problems surrounding iron object provenance holds great promise, but its limited development makes it impossible to directly compare its performance with other methods.

Identifying multiple smelting recipes in a single context

A final problem for identifying chemical groups of bloomery slag concerns the coexistence of multiple smelting recipes in the same context. Unlike the exploration of interregional or inter-site variation, groups from a single context cannot typically be defined by patterns of oxide correlations or *a priori*. Rather, some means of revealing the chemical structure of slag from individual contexts must be implemented. PCA provides a useful starting point, although its sole function now becomes the construction of a reduced dimensional coordinate system with which to describe the slag. If there are multiple groups present, PCA may expose visibly identifiable clusters within the data. More often, however, the results are

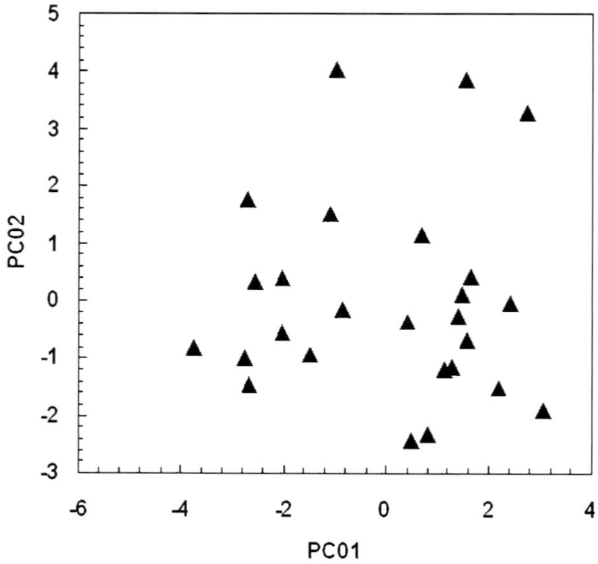

Figure 8: PCA of slag chemistry from the final production phase at Llwyn Du.

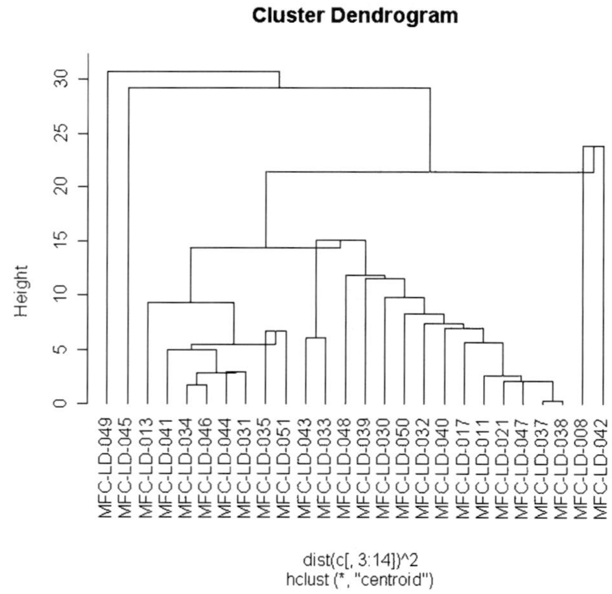

Figure 9: Dendrogram arising from a hierarchical cluster analysis of PC scores calculated from Llwyn Du slag chemistry.

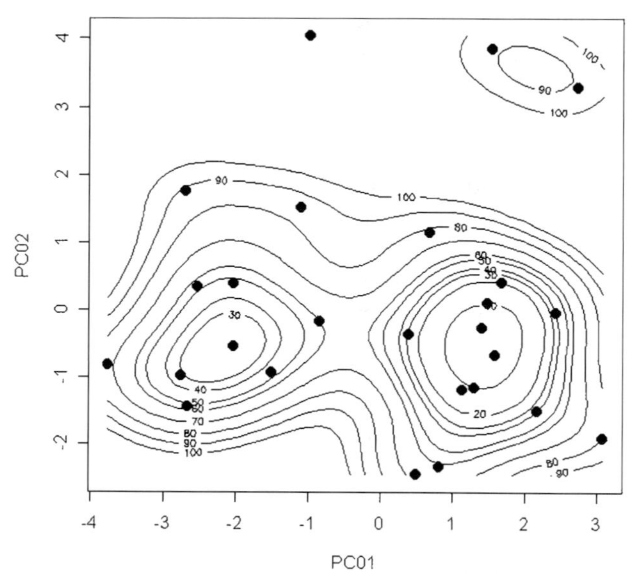

Figure 10: Bivariate kernel density estimation of the data structure in the Llwyn Du PCA. Two primary groups can be identified along with two small outlying groups.

ambiguous and open to observer bias and interpretation. Hierarchical Cluster Analysis and Kernel Density Estimation are pattern recognition techniques that provide objective and repeatable methods of identifying clustered data (Charlton, 2007). Hierarchical clustering methods begin with a distance or similarity matrix for all included specimens, based on all included variables. Variables should typically be standardised (i.e., transformed so that they exhibit unit variance and a mean of zero) prior to analysis in order to prevent those with larger variance from biasing the analysis. The most similar specimens are grouped together and the matrix recalculated successively until all individuals are joined in a single group. The result is generally represented as a dendrogram. A number of different clustering algorithms are available, using different similarity measures or criteria for grouping. Each algorithm is associated with advantages and disadvantages that must be weighed in accordance with the problem being investigated. The centroid method is used in the example provided below, and is capable of reliably detecting clusters in data that are assumed to be structured in distinct multivariate bunches of approximately the same size and shape (see Shennan, 1997; Krzanowski, 2000 for more details). The main advantages in using cluster analysis over visually defined groups are that the methods can repeatedly identify structure across large numbers of variables and avoid the problem of observer bias.

Kernel Density Estimation (KDE) is a technique for estimating a continuous frequency distribution from a sample of cases. Rather than placing cases into a series of adjacent categories like a histogram, a KDE is created by placing a 'bump' at each point and then summing the resulting heights across the axis (Baxter, et al., 1997; Baxter, 2003). The bumps are actually probability distribution functions with user-defined bandwidths. The result is an estimation of the population density suggested by the sample. Univariate KDEs are a replacement option for smoothed histograms. Their advantages compared to histograms include a body of theory for selecting appropriate bandwidths according to sample size and the lack of a defined origin. Bivariate KDEs operate in exactly the same way and are useful for highlighting point concentrations in scatter plots, revealing the structure of large data sets, and identifying groups (Baxter, et al., 1997). Because bivariate KDEs are actually a three dimensional surface, it is possible to display them as a series of contours defined in terms of the percentage of the total point density.

A PCA of smelting slag chemistry from the late medieval Welsh bloomery at Llwyn Du suggested the presence of two groups (Fig. 8). However, such a determination was visually ambiguous. A hierarchical cluster analysis using the centroid method was run across all PC axes. The resulting dendrogram suggested the presence of two closely spaced groups and four outliers (Fig. 9). A KDE was also constructed for the first two PC axes, also showing the presence of two density clusters in the data (Fig. 10). Results arising from both cluster analysis and KDE lend confidence that two groups are present, and can be used to assign specific specimens to each.

Discussion and conclusions

Systematics defines the rules by which units are constructed within a discipline, providing it with an observational language (Dunnell, 1995). Unit construction is the most fundamental task undertaken by archaeologists and archaeometallurgists alike. It places phenomena into a system of meaning that proves useful for some analytical purpose or otherwise promotes communication.

Archaeometallurgists have invested significant effort into constructing units to describe and, more rarely, classify slag in terms of its morphology, microstructure, and chemistry. Classes, because they are explicitly defined units, can be generated for the purpose of building and testing hypotheses. The ability to construct classes is important if questions related to variation in iron production are to be addressed scientifically through the analysis of slag chemistry. Morton and Wingrove (1972) classified smelting slag according to a specimen's location on the SiO_2-anorthite-FeO phase diagram. Their hypothesis was that slags plotting in the hercynite region were derived from lean ores and slags plotting in the fayalite region were derived from rich ores. While the generality of their classification scheme is open to debate, Morton and Wingrove's use of it allowed them to derive hypotheses from a model relating ore chemistry to slag chemistry. A major hurdle in using ore chemistry to classify slag has been the interference introduced by differential patterns of resource use and furnace operation. Model-based multivariate analyses provide a way forward, by generating locally defined classes based on patterns of oxide correlations in utilised resources (Charlton, 2009; Charlton, et al., 2010). Models could be applied to the use of oxide and oxide ratio plots, though none have been explicitly formulated to date.

Most other attempts to develop units based on slag chemistry, including those by Fells (1983) and Paynter (2006), resulted in the construction of groups and not clearly defined classes. Grouping methods, as mentioned above, are superb pattern recognition devices that are essential for problems relating to provenance. Importantly however, all grouping techniques, without regard to simplicity or complexity and including true classification, impose structure on data which may or may not be relevant to a particular problem. A comparison of the techniques used to identify slag compositional groups from different sites and regions shows that all perform reasonably well and may be employed for specific purposes. Ternary plots, while valuable in engineering analysis, seem to reveal the least amount of grouped structure. Bivariate scatter plots of oxides and oxide ratios can reveal significant data structure, although searches through all possible comparisons take significant time. PCA is the most effective method for revealing the structure of slag compositional groups, because it reduces analytical search time by identifying the most significant sources of variability in a data matrix with relatively few plots. PCA results can help guide decisions for additional analyses of oxides and oxide ratios when deemed necessary. LDA may prove important for sourcing slag inclusions in iron objects to particular geological regions or smelting sites, and deserves renewed consideration.

This review makes it clear that archaeometallurgists working with iron production slag have, largely through trial and error, developed a functioning set of procedures for constructing units that do the work they are intended to do – usually to recognise differences in iron production systems. Still, there is much work to be done. Though active participants in the broader aims of archaeometry, it is apparent that most archaeometallurgists have not widely explored multivariate procedures that are now common practice in the study of archaeological ceramics, glass, and lithics. The reasons for this are unclear. Such methods, as demonstrated here, can be of great value to problems related to sourcing and identifying technological diversity. More importantly, archaeometallurgy is increasingly called on to make socially relevant theoretical contributions to archaeology and ethnohistory. Questions relating to cultural transmission, innovation, and socioeconomic biases on production can only be answered scientifically through careful attention to systematics. More focus must therefore be given to constructing classes capable of addressing such problems, with slag as well as other materials.

Acknowledgements

The authors wish to thank the organisers and sponsors of the 'World of Iron' conference for their generous support. We also thank the many conference participants for their comments and informal discussions. Much of the research presented in this paper was conducted with the help and patience of the staff at the Wolfson Archaeological Science Laboratories housed at the Institute of Archaeology, University College London. Two anonymous reviewers also made insightful comments that led to the clarification of several points.

References

Allen, J.R.L., 1988. Chemical Compositional Patterns in Romano-British Bloomery Slags from the Wetlands of the Severn Estuary. *Historical Metallurgy*, 22, pp.81-6.

Bachmann, H.-G., 1982. *The Identification of Slags from Archaeological Sites.* Occasional Publication 6. London: Institute of Archaeology.

Baxter, M.J., 2003. *Statistics in Archaeology*. London: Hodder Arnold.

Baxter, M.J., 2006. A review of supervised and unsupervised pattern recognition in archaeometry. *Archaeometry*, 48, pp.671-94.

Baxter, M.J., Beardah, C.C. and Wright, R.V.S., 1997. Some Archaeological Applications of Kernel Density Estimates. *Journal of Archaeological Science*, 24, pp.347-54.

Bishop, R.L. and Neff, H., 1989. Compositional Data Analysis in Archaeology. In: R.O. Allen ed. *Archaeological Chemistry IV.* Advances in Chemistry Series 220. Washington DC: American Chemical Society, pp.57-86.

Blakelock, E., Martinón-Torres, M., Veldhuijzen, H.A. and Young, T., 2009. Slag inclusions in iron objects and the quest for provenance: an experiment and a case study. *Journal of Archaeological Science*, 36, pp.1745-57.

Buchwald, V.F., 2005. *Iron and Steel in Ancient Times.* Historisk-filosofiske Skrifter 29. Copenhagen: The Royal Danish Academy of Sciences and Letters.

Buchwald, V.F. and Wivel, H., 1998. Slag analysis as a method for the characterization and provenancing of ancient iron objects. *Materials Characterization*, 40, pp.73-96.

Charlton, M.F., 2007. *Ironworking in Northwest Wales: An Evolutionary Analysis*. PhD Thesis, University College London.

Charlton, M.F., 2009. Identifying iron production lineages: a case study in northwest Wales. In: S.J. Shennan ed. *Pattern and Process in Cultural Evolution*. Berkeley: University of California Press, pp.133-44.

Charlton, M.F., Crew, P., Rehren, Th. and Shennan, S.J., 2010. Explaining the evolution of ironmaking recipes – An example from northwest Wales. *Journal of Anthropological Archaeology*, 29, pp.352-67.

Coustures, M.P., Béziat, D., Tollon, F., Domergue, C., Long, L. and Rebiscoul, A., 2003. The use of trace element analysis of entrapped slag inclusions to establish ore – bar iron links: examples from two Gallo-Roman iron-making sites in France (Les Martys, Montagne Noire, and Les Ferrys Loiret). *Archaeometry*, 45, pp.599-613.

Coustures, M.P., Rico, C., Béziat, D., Djaoui, D., Long, L., Domergue, C. and Tollon, F., 2006. La provenance des barres de fer Romaines des Saintes-Maries-De-La-Mer (Bouches-Du-Rhône). *Gallia*, 63, pp.243-61.

Crew, P., 2000. The influence of Clay and Charcoal Ash on Bloomery Slags. In: C.C. Tizzoni and M. Tizzoni eds. *Iron in the Alps: Deposits, Mines and Metallurgy from Antiquity to the XVI Century. Proceedings of the International Conference (October 2-4, 1998, Bienno, Italy)*. Breno: Comune di Bienno, pp.38-48.

Crew, P. and Charlton, M., 2007. The anatomy of a furnace...and some of its ramifications. In: S. La Niece, D. Hook and P. Craddock eds. *Metals and Mines: Studies in Archaeometallurgy*..London: Archetype, pp.219-25.

Desaulty, A.M., Dillmann, P., L'Héritier, M., Mariet, C., Gratuze, B., Joron, J.L. and Fluzin, P., 2009. Does it come from Pays de Bray? Examination of an origin hypothesis for the ferrous reinforcements used in French medieval churches using major and trace element analyses. *Journal of Archaeological Science*, 36, pp.2445-62.

Dillmann, P. and L'Héritier, M., 2007. Slag Inclusion Analyses for Studying Ferrous Alloys Employed in French Medieval Buildings: Supply of Materials and Diffusion of Smelting Processes. *Journal of Archaeological Science*, 34, pp.1810-23.

Dunnell, R.C., 1971. *Systematics in Prehistory*. New York: Free Press.

Dunnell, R.C., 1995. What Is It That Actually Evolves? In P.A. Teltser ed. *Evolutionary Archaeology: Methodological Issues*. Tucson: University of Arizona Press, pp.33-50.

Fells, S., 1983. *The Structure and Constitution of Ferrous Process Slags*. PhD thesis, University of Aston in Birmingham.

Gordon, R., 1997. Process Deduced From Ironmaking Wastes and Artefacts. *Journal of Archaeological Science*, 24, pp.9-18.

Hedges, R.E.M. and Salter, C.J., 1979. Source Determination of Iron Currency Bars through Analysis of the Slag Inclusions. *Archaeometry*, 22, pp.161-75.

Humphris, J., Martinón-Torres, M., Rehren, Th. and Reid, A., 2009. Variability in single smelting episodes – a pilot study using iron slag from Uganda. *Journal of Archaeological Science*, 36, pp.359-69.

Krzanowski, W.J., 2000. *Principles of Multivariate Analysis: A User's Perspective*. Oxford: Oxford University Press.

Leroy, S., 2010. *Circulation au Moyen Age des Materiaux Ferreux Issus des Pyrenees Ariegeoises et de la Lombardie. Apport du Couplage des Analyses en Elements Traces et Multivariees*. PhD Thesis, Université de Technologie de Belfort-Montbéliard.

Leroy, S., Simon, R., Bertrand, L., Williams, A., Foy, E. and Dillmann, P., 2011. First examination of slag inclusions in medieval armours by confocal SR-μ-XRF and LA-ICP-MS. *Journal of Analytical Atomic Spectrometry*, 26, pp.1078-87.

Morton, G.R. and Wingrove, J., 1969. Constitution of Bloomery Slags: Part I: Roman. *Journal of the Iron and Steel Institute*, 207, pp.1556-64.

Morton, G.R. and Wingrove, J., 1972. Constitution of Bloomery Slags: Part II: Medieval. *Journal of the Iron and Steel Institute*, 210, pp.478-88.

O'Brien, M.J. and Lyman, R.L., 2000. *Applying Evolutionary Archaeology: A Systematic Approach*. New York: Kluwer Academic/Plenum Publishers.

Paynter, S., 2006. Regional Variations in Bloomery Smelting Slag of the Iron Age and Romano-British Periods. *Archaeometry*, 48, 271-92.

Serneels, V., 1993. *Archéométrie des Scories de Fer. Recherches sur la Sidérurgie Ancienne en Suisse Occidentale*. Cahiers d'Archéologie Romande 61. Lausanne: Cahiers d'Archéologie Romande.

Serneels, V. and Perret, S., 2003. Quantification of Smithing Activities Based on the Investigation of Slag and Other Material Remains. In *Archaeometallurgy in Europe, Volume 1, Procedings of the International Conference Milano, Italy, September 24-26, 2003*. Milano: Associazione Italiana di Metallurgia, pp.469-78.

Shennan, S.J., 1997. *Quantifying Archaeology*. Iowa City: University of Iowa Press.

Tylecote, R.F., 1987. *The Early History of Metallurgy in Europe*. New York: Longman.

Authors' addresses

- Michael F. Charlton*, Thilo Rehren, Stephen J. Shennan
  Institute of Archaeology, University College London, 31-34 Gordon Square, London, WC1H OPY, UK
  **Corresponding author: Permanent Address: 5430 127th PL SE, Snohomish, WA 98296, USA*
  E-mail: mike.charlton@ucl.ac.uk
- Thilo Rehren
  UCL Qatar, Hamad bin Khalifa University, Georgetown Building Doha, Qatar
  Email: director.qatar@ucl.ac.uk
- Peter Crew
  Pen Cefn, Penrhyndeudraeth, Gwynedd, LL48 6DR, UK
  Email: crew.peter.susan@gmail.com

Environmental considerations

Environment

Donald B. Wagner

'Deforestation caused by iron production' seems to be a *topos* of archaeologists and historians, invoked without much thought in many contexts. Three articles in this section take the question seriously and bring methodological rigour to a determination of the environmental impact of iron production in three places, one in Britain, two in Africa. Besides giving valuable insights into the local history of iron production here, they provide methodological considerations which will be important for the history and archaeology of iron production anywhere in the world.

Peter Crew and Tim Mighall have investigated the environmental impact of a Medieval ironworks in Wales over the ca. 65 years of its operation. The site is remarkably well preserved, and has been the subject of a very thorough excavation over a period of several years. A floating dendrochronology, taken together with historical sources, shows that the works operated in the years 1355-65, 1376-92, and 1407-19. The gaps of 10 and 15 years suggest the time needed for forest re-growth, and there is also a document of 1391-2 to be explained: it states that the forest here 'has been laid waste'.

Pollen analysis, however, indicates only minor variations in forest growth in this period, so that some forest management must have been practised to provide the 175-200 tons of wood per year which calculations suggest that the ironworks consumed. Peter Crew's painstaking study of thousands of pieces of charcoal indicates that two species were almost exclusively used, oak and birch. The oak used appears to have been entirely from large branches rather that stemwood, and this suggests forest management by lopping or pollarding rather than coppicing. The birch may have been coppiced, or it may not have been rigorously managed.

The explanation of the statement that the forest had been 'laid waste' could then be that lopping or pollarding created an *appearance* of forest destruction. Other explanations, taking in wider economic considerations, seem possible. For example, as the authors note, the ironworks may not have been the only users of forest resources here, and the particular forest management practices employed may have made the trees less useful as timber – a variety of environmental impact which has not yet been much discussed.

The surprisingly detailed results which come out of this study are made possible by a well-developed methodology which is clearly discussed in the article. Crew and Mighall's work will no doubt become a model for similar studies in other parts of the world.

The paper by Fyfe et al. presents a related study of a site in Devon, south-west England. Iron was produced here for about 150 years in the 2nd-3rd centuries AD. Production was under 1 ton per year, and may have consumed 200-300 tons of wood per year. There was definite selectivity in the species used for iron smelting, largely oak and hazel. There is evidence of coppicing here, and forest exploitation seems to have been sustainable. Again, a particularly thorough and well-developed methodology combines excavation data with geochemistry and pollen analysis to develop a chronology for the iron production at the site and to provide quantifications for both the resources needed and the iron produced. In light of this data, evidence for woodland management is assessed.

A study with a similar aim, but in an African context, is that of Barbara Eichhorn and her collaborators. Eight iron-producing sites of the Fiko tradition in Mali, covering an area of about 450 square km, were active for a period of well-over 800 years (until the 20th century AD), and are estimated to have consumed a total of 4.5-6 million tons of wood. This seems to have put wood consumption very near the estimated production capacity of the surrounding woodlands. Determination of the species of charcoal from the sites indicates that there was some preference for hardwoods, but that the choice of species largely reflects their frequency in the surrounding landscape. Thus the 'principle of least effort' more-or-less holds: woody species (both trees and shrubs) were consumed in direct proportion to their availability.

The assumption that the iron producers here used whatever species were nearest at hand (though with a bias toward hardwoods) makes it possible to estimate the environmental impact of the iron industry in its period of activity. It appears that there was a distinct vegetation change, with an increase of hardwood species that can re-sprout and a decrease in others. This is perhaps a result to be expected in any case, but we may wonder whether some form of forest management – coppicing, pollarding, or lopping – may have been practised, for it would have been noticed early on that some ways of harvesting wood gave faster re-growth than others. In a forest being exploited near its production capacity this would be an extremely important insight.

Ethnographic investigations reported by Edwinus Lyaya concerning the tree species selected for iron production in Tanzania sheds important light on what the ironworkers thought about their activities. Informants were asked what species were preferred for each of the three phases of iron production: smelting, refining, and smithing; and they were asked for the reasons for these preferences. Of the many species available to the ironworkers, very few were used. Considering the socio-cultural functions of these species in the wider society suggests interesting correlations, for example with male sexual powers, fertility, and maternity.

The reasons for selection of specific species are generally given, however, in terms which suggest concrete technical matters, and these need further investigation. When an informant states, for example, that two species used in smelting have the 'ability to clean and not dirty the iron,' he undoubtedly is referring to some quite specific and concrete phenomenon, but at the moment it is not really clear what this is. (Whether the informant's technical explanation is *correct* from an engineering standpoint is of course an entirely separate and less interesting question.)

We should also like to know more of the physical and biological properties of the tree species selected. Questions suggested by the archaeological studies mentioned above are: Is there a preference for hardwoods? and: Is there a preference for species which re-sprout easily?

As Lyaya points out, the selection for exploitation of only a handful of species out of thousands would seem to prevent large-scale deforestation. We may wonder whether this is not one latent function of species selection: If an iron industry practises neither forest management nor species selection, it will in time put itself out of business. Those that follow some such practice, for whatever reason, are more likely to survive and prosper in the long term.

As Danielle Arribet-Deroin notes, 'How the iron industry affected the environment during the European Middle Ages cannot be answered without quantifying the iron output.' Her paper gives a useful critical review of the range of methodologies which may help in answering this question. Especially interesting is the use of African analogies, where we have much more detailed ethnographic knowledge of the activities and organization of the iron producers than can ever be obtained for Medieval Europe.

Author's address

- Jernbanegade 9B
  DK-3600 Frederikssund, Denmark
  Email: dwag@alum.mit.edu

Fuel for iron – wood exploitation for metallurgy on the Dogon Plateau, Mali

Barbara Eichhorn, Caroline Robion-Brunner, Vincent Serneels and Sébastien Perret

ABSTRACT: African iron metallurgy is often associated with severe environmental degradation, due to the supposed overuse of wood-based resources as an energy supply. Such assumptions are rarely based on on-site archaeobotanical data, but are instead either inferred from wood consumption estimates or simply represent an uncritically imposed link between present degradation phenomena and past metallurgical activities. This paper discusses charcoal analytical results from slag heaps at sites belonging to the Fiko Tradition in the Dogon Country (Mali). These sites are characterised by their long-term use and intensive iron production. Anthracology points to the predominant use of dense woods with a high burning value. However, despite this preference, typical Sahelo-Sudanian woody vegetation is reflected in the charcoal assemblages. Significantly, variations within the charcoal sequence indicate distinct vegetation changes over time, and, furthermore, mass balance ratios support a relationship between these vegetation changes and metallurgical activities. This holds true for the mass production sites but probably not for other metallurgical traditions in the area with lower production levels.

Keywords: Wood, charcoal, iron metallurgy, Mali, West Africa, vegetation history, environment, degradation.

Introduction

In the archaeological literature, many severe ecological consequences of West African metallurgy have often been presumed (e.g. Haaland, 1980, 1985; Goucher, 1981, 1986; de Barros, 1986; Herbert and Goucher, 1987; Hahn, 1997; Thompson and Young, 1999; Pelzer, et al., 2004). It was mainly the exploitation of wood-based resources as fuel for the reduction process and the assumed related deforestation that were emphasised as the fundamental reasons, while the roles of air and soil pollution were seldom taken into consideration (Goucher, 1986). In addition to environmental degradation, economic decline and even the demise of political structures have in the past been associated with an ecological collapse partly caused by iron industries.

Goucher (1981) considered wood over-exploitation and the resulting degradation of forest areas, together with desiccation shifts and the disruption of traditional land use patterns due to European influence, to be among the primary reasons for the decline of West African iron-smelting. According to Goucher, this complex combination of factors and the ability of an increasingly industrialised Europe to compete with local iron production led to progressively more significant reliance on imported supplies. Haaland (1980, 1985) related the overuse of natural resources to internal political weakening and the collapse of the Ghana Empire in Mali.

The assertion that metallurgy was a significant cause of environmental degradation is mainly based on estimates of wood consumption. These estimates are sometimes inferred from the measured slag volume and calculations of charcoal consumption based on conversion ratios. Even if the magnitude of these approximations is correct, this approach is not sufficient to accurately determine vegetation change or landscape degradation. First of all, wood is a regenerative resource and its use may be sustainable as long it does not exceed the production rate (Von Maltitz and Scholes, 1995). West African savannahs are reproductive systems and even fields and fallows in traditional agro-forestry systems reproduce a considerable amount of wood (Clément, 1982; Ohler, 1985; Breman and Kessler, 1995; Nouvellet, et al., 2003). Secondly, in order to detect vegetation changes or degradation, archaeobotanical (e.g. charcoal analysis) or palaeobotanical (e.g. palynology) methods are indispensable, as they are the only reliable means available to reveal such alterations in the past. For several decades, charcoal analysis has been widely recognised as a useful tool for reconstructing past vegetation and for providing an idea of prehistoric wood use (Smart and Hofmann, 1988; Vernet, 1997; Neumann, 1999; Figueiral and Mosbrugger, 2000; Thiébault, 2002; Asouti and Austin, 2005; Dufraisse, 2006). Palaeoecological reconstructions based on charcoal analysis have so far not been applied to their fullest extent to West African metallurgy sites. Mainly charcoal originating from single furnace structures has been analysed – not allowing for establishing sequences – and the interpretation has been hampered by bad preservation or insignificant numbers of identifiable fragments (Rolando, 1992; Neumann and Vogelsang, 1996; Dupuy and Rolando, 1999; Kiethéga, 2009, p.269 ff.).

Fiko Tradition sites on the Western margin of the Dogon Plateau in Mali are characterised by surplus iron production and involvement in large-scale long-distance trade, as well as occupation through several centuries (Robion-Brunner, 2008;

Huysecom, et al., 2009; Robion-Brunner, 2010). Therefore, it is possible to construct a long-term perspective by combining analyses of charcoal samples retrieved from slag heaps with a series of radiocarbon dates. The most recent phases are supplemented by oral traditions about site history. Thus, anthracological sequences can be established (Eichhorn, 2012, Eichhorn and Neumann, in press; Eichhorn, et al., forthcoming) which help us to reconstruct the research area's vegetation history during the past millennium and draw conclusions regarding wood use for this technology. The results of these on-site analyses complement the off-site vegetation history data obtained at the nearby Ounjougou site complex, where the Holocene long-term archaeobotanical sequence is incomplete for the past 1500 years (Huysecom, et al., 2004; Eichhorn and Neumann, in press; Ozainne, et al., 2009). It is important to bear in mind that fluctuations in the charcoal record are not necessarily caused by metallurgical activities alone. Woody vegetation may have also been altered by intensified agriculture and a desiccation trend during the period in question (Mayor, et al., 2005; Höhn, 2007; Kahlheber and Neumann, 2007). Therefore, in order to confirm or disprove a correlation of vegetation changes (indicated by charcoal analysis) with the metallurgical activities of the Fiko Tradition, an interdisciplinary methodology has been adopted. Slag volume and mass were measured, and slag-charcoal-wood conversion rates established in order to estimate the magnitude of wood consumption for the whole area of the Fiko Tradition and throughout the entire chronology of occupation (Eichhorn, et al., forthcoming). The results were related to approximations of woody biomass reproduction in order to assess whether wood consumption was at the ecological threshold.

Site characteristics

Fiko Tradition sites are distributed over an area of approximately 450km^2 at the western margin of the Dogon Plateau. This metallurgical tradition consists of eight iron

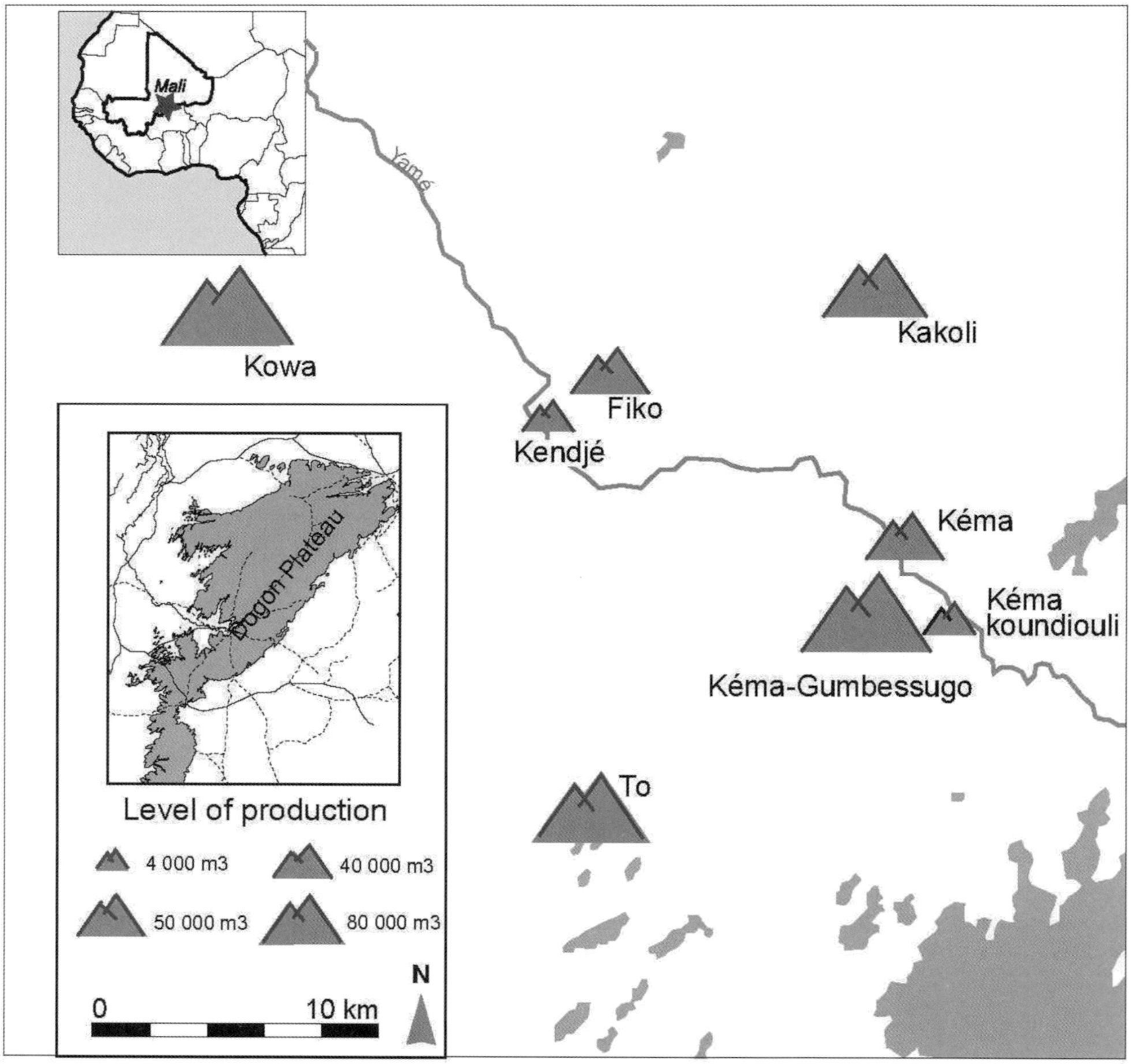

Figure 1: Map of the Fiko Tradition area, indicating the location of sites and the amount of metallurgical waste.

smelting site complexes on both sides of the Yamé River. The archaeological sites are characterised by large dumps of metallurgical waste in the form of crater-shaped mounds generally several metres high (Robion-Brunner, et al., this volume). Across this area of the plateau, the total amount of metallurgical waste is estimated at between 225,000m^3 and 300,000m^3 (Robion Brunner, 2010).

Any general considerations regarding wood consumption and sustainability outlined below refer to the whole Fiko Tradition area, while charcoal analytical results concern the sites of Fiko and Kéma-Gumbessugo (part of the Kema site complex). Both show a complex succession of slag heaps with different ages.

According to oral tradition records, the smelting activity at the Fiko site ceased during the last century, but the site had been operating on a large scale prior to that. The upper levels of the oldest slag heap are dated back to the period between the 10th and the 15th century AD. A lower layer provides an older date, from the 6th century AD; however, this date remains isolated. According to oral traditions, the end of the activity at Kéma-Gumbessugo is linked to warfare. The slag heaps near the abandoned village on the hill and those on the plain nearby are dated back to a period between the 10th and the 14th century AD. Iron production continued after displacement to Kéma-Koundiouli, which was later followed by a move to the present village of Kéma (Fig. 1; Huysecom, et al., 2009).

Environmental settings

The research area is currently located in the north of the Sudanian phytogeographic zone, at the transition to the Sahel (Fig. 2). The climate is seasonal, with a rainy season

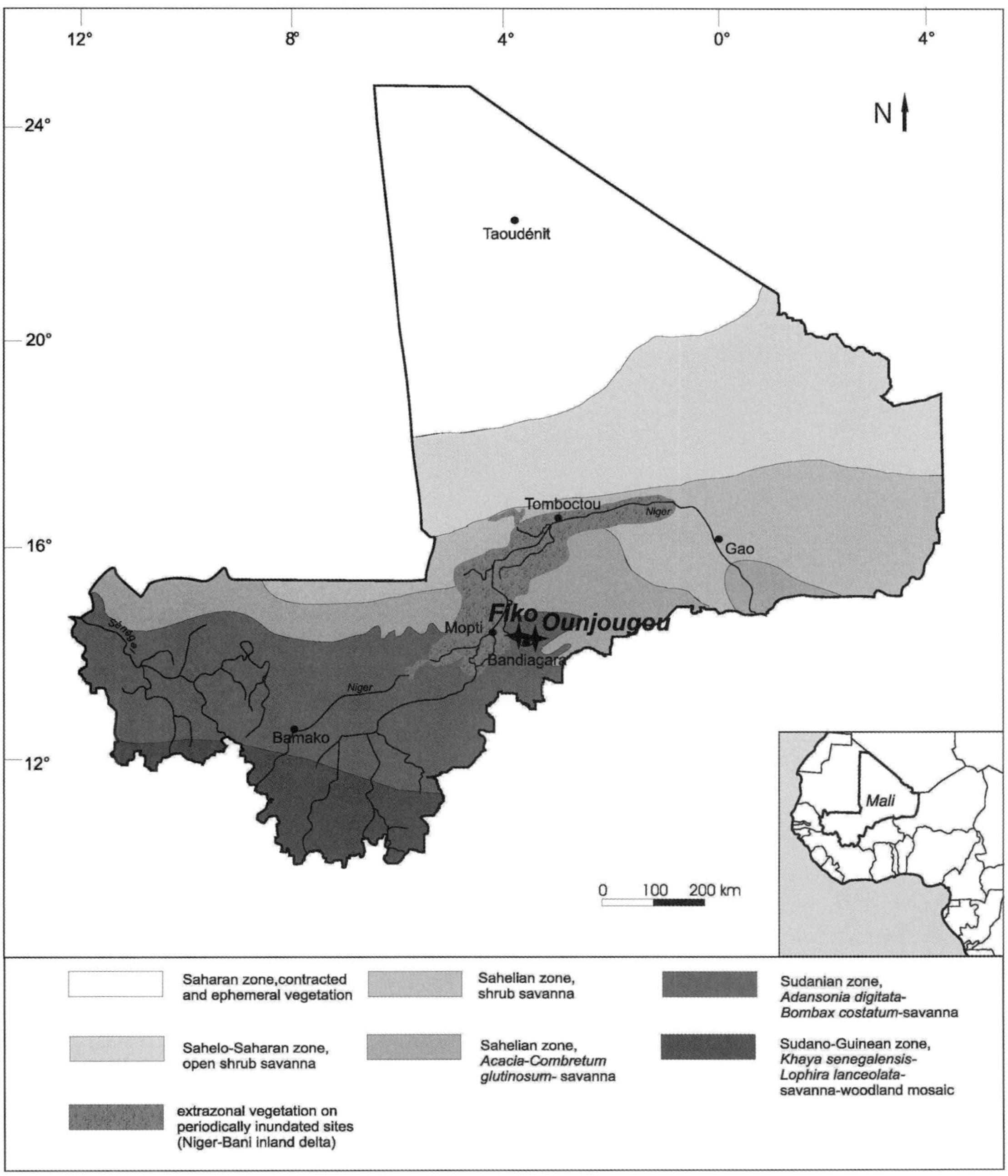

Figure 2: Major vegetation units of Mali (after Granier, 2001, modified) and the location of the Fiko and Ounjougou sites.

of three and a half months and an average annual rainfall of 500-600 mm (Coutard, 1999; Nouaceur, 2001). The Fiko Tradition sites are situated near the limit of the Sudanian *Adansonia digitata – Bombax costatum* savannahs to the south and the Sahelian *Combretum glutinosum* savannahs to the north (Fig. 2; Granier, 2001). Jaeger and Winkoun (1962) describe the natural vegetation of "favourable sites" on the Dogon Plateau as typical Sudanian savannah characterised by *Daniellia oliveri* in association with *Vitellaria parodoxa, Detarium, Parkia biglobosa, Terminalia macroptera, Khaya senegalensis, Vitex doniana, Prosopis africana, Entada africana,* and shrub species such as *Combretum micranthum* and *Guiera senegalensis.* Well-developed savannahs often occur in close proximity to semi-desert plant communities on almost barren sandstone surfaces.

At present, a cultural landscape covers the plateau and the adjacent plains, characterised by extensively cultivated park savannahs with few selected useful tree species. These agroforestry parks are locally dominated by *Faidherbia albida, Sclerocarya birrea, Lannea microcarpa, Adansonia digitata* or *Vitellaria paradoxa.* Floristically rich ravine forests with species otherwise distributed further to the south persist at the escarpment of the Dogon Plateau (Jaeger and Winkoun, 1962; Huysecom, et al., 2007).

The sites of Fiko and Kéma-Gumbessugo lie in close proximity to the Yamé River, a Niger tributary. Throughout almost the entire Holocene period, the diversity of the Yamé's gallery forest was exemplified by the extrazonal occurrence of Sudano-Guinean species such as *Syzygium guineense, Uapaca togoensis, Alchornea cordifolia* and *Lophira lanceolata,* as attested by archaeobotanical investigations at the nearby Ounjougou site complex (Ozainne, et al., 2009; Eichhorn and Neumann, Fig. 2, in press). Due to climatic deterioration and human impact, the gallery forest is today impoverished and fragmented, and characteristic species such as *Khaya senegalensis* or *Pterocarpus erinaceus* are still present but never common. Near Kéma-Gumbessugo, the palm *Borassus aethiopum* dominates the gallery forest, probably due to the fact that it is preferred by local populations due to its multiple uses.

Methodology

The applied methods have already been described in detail elsewhere (Eichhorn, et al., forthcoming), and are only briefly summarised here. Charcoal was retrieved during excavations of the Fiko and Kéma-Gumbessugo slag heaps. Visible charcoal fragments were taken directly out of the sediment which was later dry-sieved. Any fragments larger than 0.25cm were collected from the sieves. In the laboratory, charcoal taxa were identified using a number of African wood reference collections: the Frankfurt archaeobotanical laboratory, the Frankfurt DELTA/Intkey anatomical database (Neumann, et al., 2001), anatomical wood atlases and anthracological catalogues (Neumann, et al., 2001; Eichhorn, 2002; Höhn, 2005).

The burning value of a wood is directly correlated to its density or specific gravity (National Academy of Sciences, 1980, p.24; Raven, et al., 2000, p.711). Therefore, density was used as a proxy to determine whether the fuel choice was arbitrary or selective. Wood density values for each identified taxon were derived from the literature (Nygard and Elfving, 2000; Armstrong, et al., 2004; Picard, et al., 2006). Subsequently, we developed wood density classes (in 50kg x m^{-3} steps) in which all taxa were grouped. Each counted charcoal fragment was assigned to one of the classes and their respective percentages calculated (Fig. 3). Conversion rates for slag-charcoal-fresh wood were also compiled from the literature (e.g. Doat and Petroff, 1975; Haaland, 1985; Dörfler, 1995; cf. Eichhorn, et al., forthcoming). In order to determine the magnitude of annual wood reproduction for the Fiko Tradition area, we relied on data obtained from the ecological literature. Clément (1982) related the productivity of West African steppes, savannahs and woodlands to annual precipitation and site characteristics such as soil quality and human impact, and developed three productivity curves: one for sites protected from fire and other severe human impact; one for unprotected sites and medium soil quality; and one for degraded sites on poor soils. From these productivity curves, the productivity for the entire Fiko Tradition region was predicted (Table 1), assuming 500mm and 600mm of

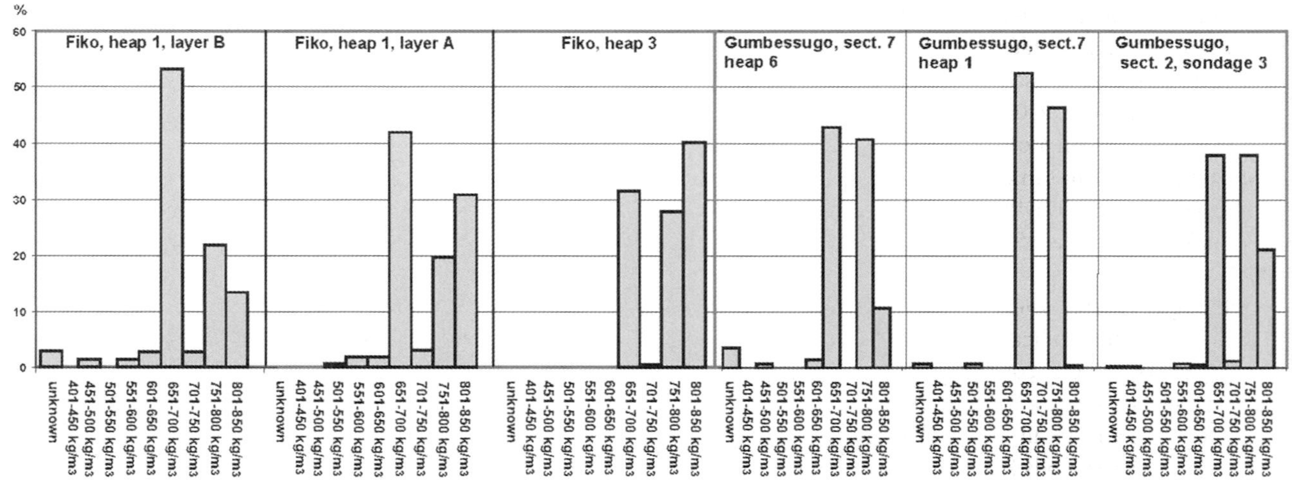

Figure 3: Percentage of charcoal fragments per wood density class at the Fiko and Kéma-Gumbessugo sites.

annual rainfall respectively. A mean value between low and medium productivity was calculated. This approach was chosen because significant human impact on the landscape was assumed for the period under consideration, due to the vast amount of land being cultivated. Shallow soils on sandstone surfaces covering a significant part of the research area were classified as minimally productive, whereas the gallery forest of the Yamé was regarded as highly productive, due to its high water balance.

A series of radiocarbon dates for Fiko and Kéma-Gumbessugo have established site chronologies, and a more or less continuous production process over a period of at least 800 years became evident (Eichhorn, 2012, Eichhorn, et al., forthcoming). Taking into consideration a single, much older radiocarbon date from Fiko, and the fact that fluctuations in iron production cannot be ruled out, estimates of different chronology scenarios are presented below (Table 1).

Table 1: Estimates of wood reproduction for the Fiko Tradition area and estimated wood consumption for the iron smelting of the Fiko Tradition (both reduction and bloom smithing in the area would approximately require the double amount of wood-based fuel).

Chronology scenario	**Annual rainfall**
number of years	500mm/ 600mm
1	5400t/7650 t
500	2,700,000t/3,825,000t
800	4,320,000t/6,120,000t
1400	7,560,000 t/10,710,000t
Wood needed for smelting	4,500,000t min
(whole sequence of metallurgy)	6,000,000t max

Fuel choice

Within the Fiko Tradition, there is no evidence of any particular preference for one or a few tree species, unlike for other iron smelting sites (Neumann and Vogelsang, 1996; Dupuy and Rolando, 1999; Neumann, 1999). On the contrary, our anthracological results show that many species were used for iron metallurgy with the dominant ones yielding high quality fuel woods. They mainly represent ubiquitous savannah and fallow trees of the region (Fig. 4 and Fig. 5). In general, the taxa *Prosopis africana*, *Terminalia* sp. and some other members of the families Combretaceae and Fabaceae are a constant presence in the charcoal samples and are represented in high proportions. Kiethéga (2009, p.249ff.) also lists a large number of species used for iron production in Burkina Faso (according to oral traditions). The same holds true for Mafa iron smelting in North Cameroon (David, et al., 1989) and ethnohistorical data confirms this aspect for north-eastern Nigeria (Sassoon, 1964).

Although distinct selectivity for single species is not evident, our results indicate that the final wood choice for the production of charcoal in the bloomery furnace was not completely arbitrary. The charcoal assemblages of both Fiko and Kema are in all samples dominated by trees and shrubs with dense woods (density >650kg/m^3, Fig. 3), which guarantee a high burning value. Light-wooded species (e.g. from the families Anacardiaceae, Burseraceae, Moraceae, Verbenaceae, Sterculiaceae and Bombacaceae) are hardly represented, although quite common in the savannahs and agroforestry parks of the area.

Vegetation change

Although some light-wooded tree and shrub species are underrepresented, the charcoal assemblages reflect the vegetation of typical northern Sudanian savannahs, woodlands and parklands quite well. This is in agreement with the 'principle of least effort' (Shackleton and Prins, 1992; Marston, 2009), which implies that people use neighbouring wood resources in a proportion corresponding to the frequency in the vegetation. The Fiko sequence (Fig. 4) shows distinct changes in the abundance of the dominant taxa in the course of the occupation. *Terminalia* sp. and *Prosopis africana* decrease, whereas *Combretum glutinosum* and *Pterocarpus* cf. *lucens* increase significantly. Among the accompanying taxa, some other changes are obvious. The shea butter tree (*Vitellaria paradoxa*), a typical resident of Sudanian agroforestry parklands, is increasingly used, and some Sudanian species are gradually replaced by Sahelian trees and shrubs, among them *Balanites aegyptiaca*. *Parinari* cf. *curatellifolia*, a species that today is only present further to the south, in areas with higher precipitation, disappears from the assemblage. These fluctuations indicate distinct vegetation changes (e.g. an increase of species that are able to re-sprout from suckers, alongside a complete retreat of other species from the area). As stated above, these changes may not necessarily be directly related to the ecological consequences of metallurgy, but might be due to other human impacts – such as intensified agriculture – or climatic oscillations.

The time span reflected in the Kéma-Gumbessugo sequence (Fig. 5) is shorter than the one corresponding to the Fiko sequence because the site was abandoned and metallurgy displaced to Kéma-Koundiouli and, later, to Kéma. Nevertheless, similar trends can be observed concerning the dominant taxa: *Terminalia* sp. and *Prosopis africana* diminish, while *Combretum glutinosum* (Fig. 6) augments. The exploitation of the gallery forest is indicated by the presence of *Khaya senegalensis*.

Estimates of wood consumption and reproduction

We can only provide rough estimates of the amount of wood that was required for the metallurgy of the Fiko Tradition (Eichhorn, 2012, Eichhorn, et al., forthcoming). The evaluated conversion rate of slag mass to fresh wood is approximately 1:20 (Table 2). For an estimated total amount of metallurgical waste of 225,000 to 300,000 tonnes, the mass of fresh wood needed would total 4,500,000 to 6,000,000 tonnes (Table 1). These numbers may vary to a certain extent as waste density (mass per volume) varies from one slag heap to another. The total amount of reproduced wood depends on the chronology (number of years) and may have varied in the past with oscillations in precipitation levels (Table 1).

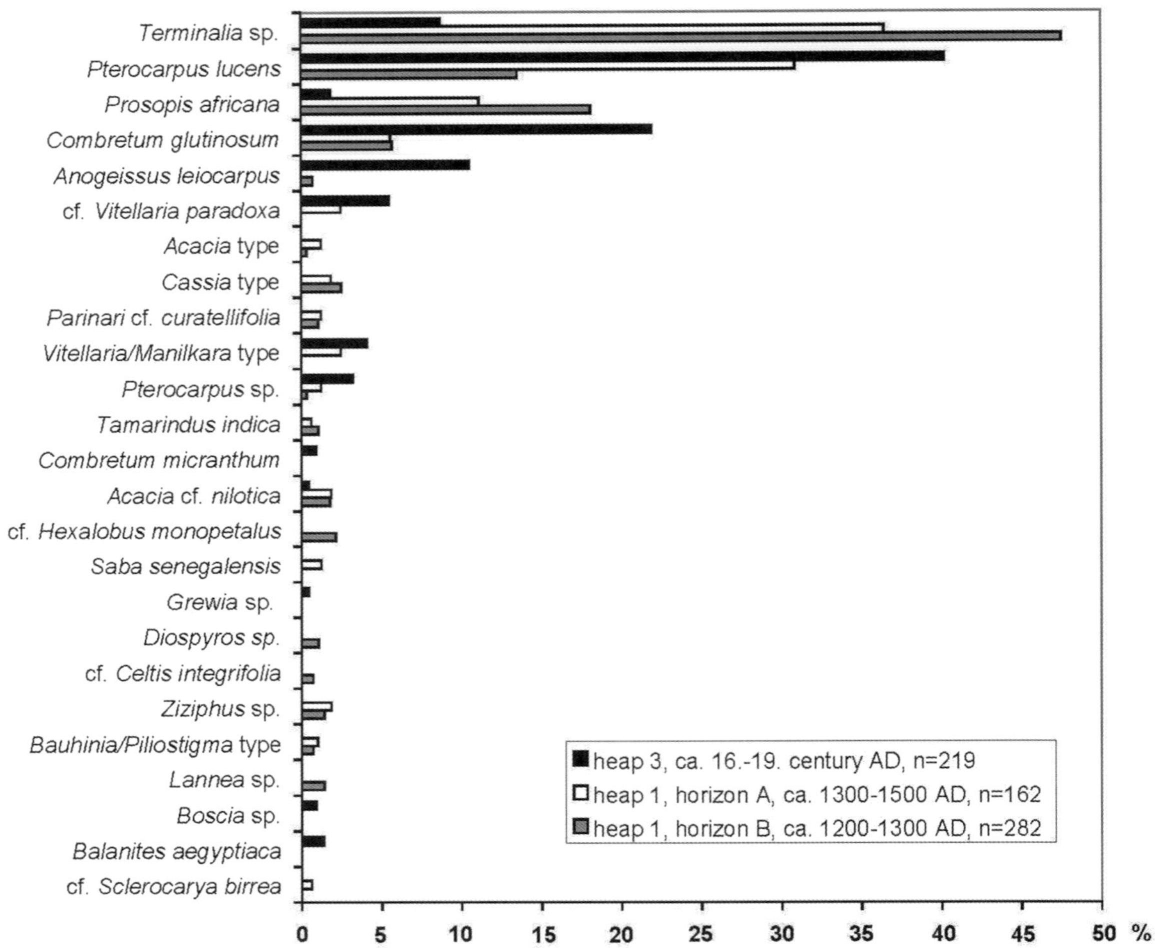

Figure 4: Charcoal analytical diagram for the Fiko site.

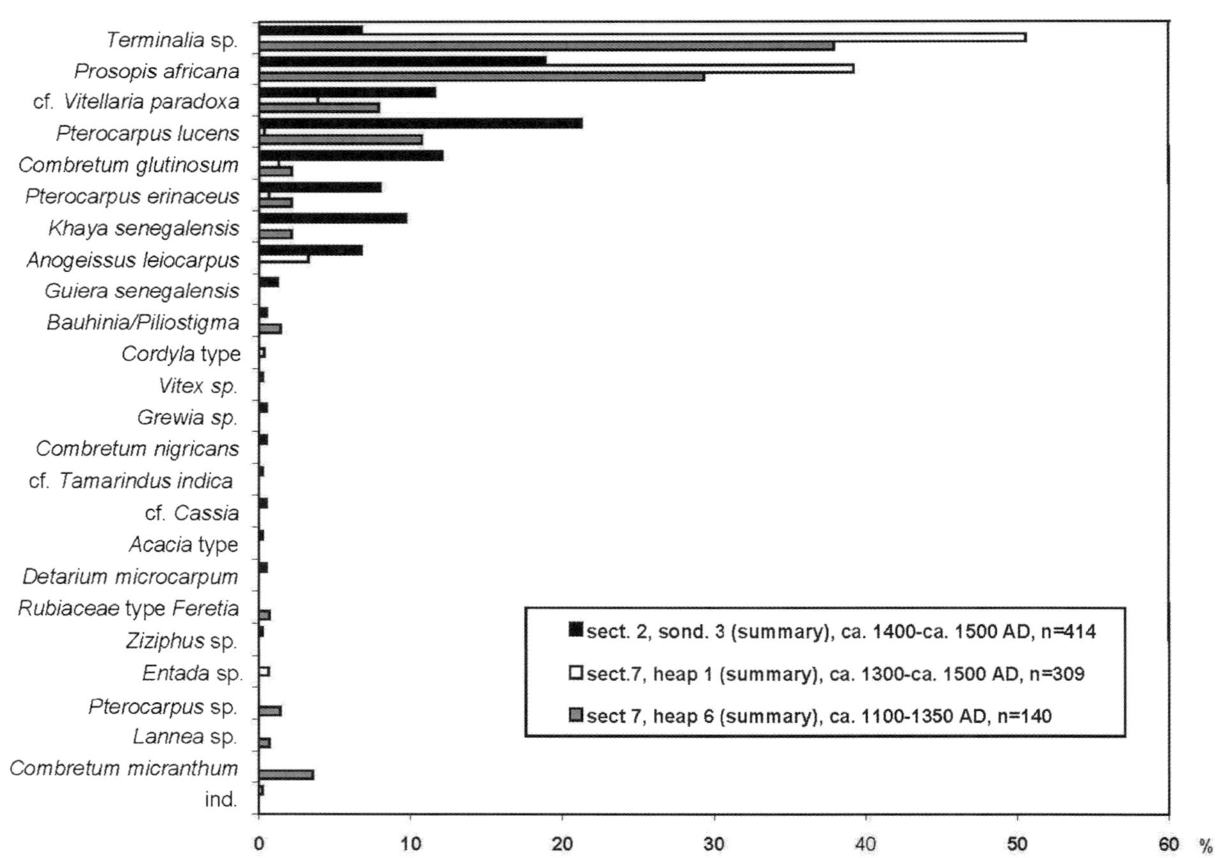

Figure 5: Charcoal analytical diagram for Kéma-Gumbessugo.

Table 2: Estimated conversion rates.

Mass balance estimates	
dry wood/fresh wood	1:1,5
charcoal/dry wood	1 : 4-5
ore/slag	1,5 : 1
ore/charcoal	1 : 2
slag/charcoal	1 : 3
slag/fresh wood	1 : ca. 20

Although these calculations can only indicate the rough magnitude of wood consumption and reproduction, they point to exploitation near the ecological threshold, especially because metallurgy was only one of the wood-consuming technologies in the area (an ecological threshold is the point where small changes in a driver, such as pollution or, in this case, wood exploitation may produce large responses in the ecosystem; Groffman, et al., 2006). Fuel used for ceramic firing and cooking, as well as construction and furniture timber have to be added to the wood amount used for iron reduction and bloom-smithing. It thus seems likely that total consumption exceeded reproduction and affected the standing stock, resulting in vegetation degradation. Even if this was not the case, the preferred use of hardwoods would have affected the composition of the vegetation. On the one hand, species with low wood density were left standing and gained relative importance in the vegetation. On the other hand, those hard-wooded species which were able to withstand cutting pressure by re-sprouting from suckers were less affected by wood exploitation than species lacking the ability to regenerate.

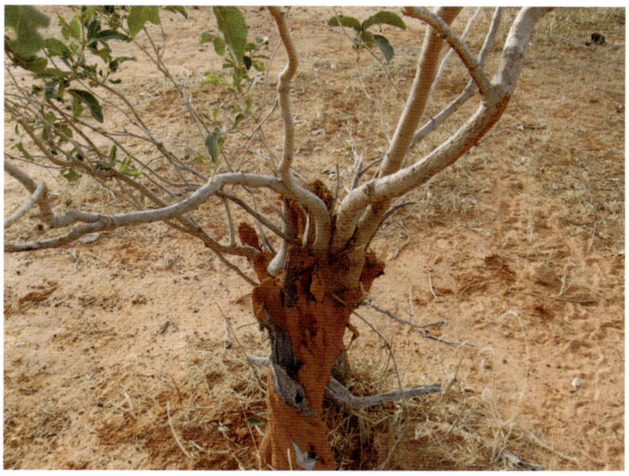

Figure 6: Cut and re-sprouted individual of *Combretum glutinosum* at Kéma-Gumbessugo.

Conclusion and outlook

An interdisciplinary archaeobotanical, metallurgical and archaeological approach has proved useful in testing the environmental impact of large-scale metallurgy on the woody vegetation of a clearly defined area with a chronology based on a series of radiocarbon dates. In contrast to earlier approaches (Haaland, 1980, 1985; Goucher, 1981, 1986; Herbert and Goucher, 1987; Hahn, 1997) which had mainly concentrated on the enormous amount of wood exploited for iron metallurgy, in this study we have also tried to integrate data concerning the reproductive capacity of woody vegetation.

Anthracological analysis carried out on charcoal samples taken from Fiko Tradition slag in the lower Yamé River Valley has revealed distinct changes in the composition of the local woody vegetation during the past millennium. Qualitative changes comprise the retreat of species currently present in the Sudano-Guinean vegetation further to the south, where higher precipitation levels prevail, and the appearance of taxa with Sahelian affinities. These changes can most likely be explained by aridification. More conspicuous are quantitative changes affecting some hardwood taxa preferred for use in the iron bloomery process. The species most prominent in the beginning gradually decline, while trees and shrubs with a high regenerative capacity such as *Combretum glutinosum* (Fig. 6) augment in the course of the occupation of the sites. Species typical of agroforestry parklands (Boffa, 1999; Kahlheber, 1999; Yaméogo, et al., 2005) also increase. Due to the high number of variables, an attempt to calculate the net wood consumption could only reveal the order of magnitude, which, however, points to exploitation near the ecological limit. It is thus most likely that at least part of the indicated vegetation changes can be directly correlated with metallurgical activities. In contrast, it is unlikely that other, much smaller-scale Dogon Country-based metallurgical traditions severely affected the surrounding woody vegetation. Further archaeobotanical investigations at other Fiko Tradition sites are currently being carried out in order to refine the picture already obtained for the lower Yamé River Valley.

Acknowledgements

We are very much indebted to the German Research Foundation (DFG), the Agence Nationale de Recherche (ANR), and the Swiss National Science Foundation (FNS) for research funding. Special thanks are due to Katharina Neumann for her support, her patience and discussions with us. We thank all the members of the project "Peuplement humain et paléoenvironnement en Afrique de l'Ouest", coordinated by Eric Huysecom, for their friendly cooperation. We are indebted to Mark Jones for language editing. Thanks are also due to the government of Mali for granting the necessary visas and permits and the "Mission Culturelle de Bandiagara" for the friendly support given to our work.

References

Armstrong, M., Lelievre, T., Reilly, D., and Robertson, B., 2004. Evaluation of the wood quality and utilisation of plantation grown *Khaya senegalensis* (African mahagony). In: D.I. Bevege et al. eds. *Prospects for high-value hardwood timber plantations in the dry tropics of northern Australia: proceedings of a workshop presented by Private Forestry North Queensland at the Centre for Tropical Agriculture, Mareeba, North Queensland, 19-21st October 2004.* Kairi: Private Forestry North Queensland, pp.1-16. (www.plantations2020.com.au/reports/pfng/acrobat2-3_armstrong_et_al.pdf).

Asouti, E., and Austin, P., 2005. Reconstructing woodland vegetation and its relation to human societies, based on the analysis and interpretation of archaeological wood charcoal macro-remains. *Environmental Archaeology*, 10, pp.1-18.

Boffa, J.-M., 1999. *Agroforestry parklands in sub-Saharan Africa*. Rome: Food and Agriculture Organization of the United Nations.

Breman, H., and Kessler, J.-J., 1995. *Woody plants in agro-ecosystems of semi-arid regions with an emphasis on the Sahelian countries*. Berlin: Springer.

Clément, J., 1982. Estimation des volumes et de la productivité des formations mixtes forestières et graminéennes tropicales. Données concernant les pays de l'Afrique francophone au nord de l'équateur et récommendations pour la conduite de nouvelles études. *Bois et Forêts des Tropiques*, 198, pp.35-58.

Coutard, S., 1999. *Étude de l'environnement géologique des sites archéologiques holocènes d'Ounjougou (Plateau dogon, Mali)*. Unpublished DEA Thesis, Laboratoire de Sciences du Sol de l'INA, Grignon.

David, N., Heimann, R., Killick, D., and Wayman, M., 1989. Between bloomery and blast furnace: Mafa iron-smelting technology in North Cameroon. *The African Archaeological Review*, 7, pp.183-208.

de Barros, P., 1986. Bassar: a quantified, chronologically controlled, regional approach to a traditional iron production center in West Africa. *Africa*, 56, pp.148-74.

Doat, J., and Petroff, G., 1975. La carbonisation des bois tropicaux. Essais de laboratoire et perspectives industrielles. *Bois et Forêts des Tropiques*, 159, pp.55-72.

Dörfler, W., 1995. Versuch einer Modellierung des Energieflusses und des Rohstoffverbrauchs während der römisch-kaiserzeitlichen Eisenverhüttung in Joldelund, Ldkr. Nordfriesland. *Probleme der Küstenforschung im südlichen Nordseegebiet*, 23, pp.175-85.

Dufraisse, A. ed., 2006. *Charcoal analysis: new analytical tools and methods for archaeology*, British Archaeological Reports International Series 1483. Oxford: Archaeopress.

Dupuy, C., and Rolando, C., 1999. Un four de métallurgie en stratigraphie à Koussané (Mali): Fouille, anthracologie et datations sur charbons. *Mémoires de la Société Préhistorique Française*, 26, pp.381-5.

Eichhorn, B., 2002. *Anthrakologische Untersuchungen zur Vegetationsgeschichte des Kaokolandes, Nordwest-Namibia*. PhD Thesis, University of Cologne, Cologne. http://kups.ub.uni-koeln.de/volltexte/2004/1178/

Eichhorn, B., 2012. Woody resource exploitation for iron metallurgy of the Fiko tradition: Implications for the environmental history of the Dogon Country, Mali. In: C. Robion-Brunner and B. Martinelli eds. *Métallurgie du Fer et Sociétés Africaines. Bilans et nouveaux paradigmes dans la recherche anthropologique et archéologique*. Cambridge Monographs in African Archaeology 81 (BAR S 2395). Oxford: Archaeopress, Hadrian Books, pp.139-148.

Eichhorn, B., and Neumann, K., in press. Holocene vegetation change and land use at Ounjougou (Mali). In: D.Q. Fuller and M.A. Murray eds. *Flora, Past Cultures and Archaeobotany in Africa*. Walnut Creek: Leftcoast Press.

Eichhorn, B., Robion-Brunner, C., Perret, S., and Serneels, V., forthcoming. *Iron metallurgy in the Dogon country (Mali): "deforestation" or sustainable use?* British Archaeological Reports International Series.

Figueiral, I., and Mosbrugger, V., 2000. A review of charcoal analysis as a tool for assessing Quaternary and Tertiary environments: achievements and limits. *Palaeogeography, Palaeoecology, Palaeoclimatology*, 164, pp.397-407.

Goucher, C., 1981. Iron is iron till it is rust: trade and ecology in the decline of West African iron-smelting. *Journal of African History*, 22, pp.179-89.

Goucher, C., 1986. *The iron industry of Bassar, Togo: An interdisciplinary investigation of African technological history*. PhD Thesis, University of Michigan, Ann Arbor.

Granier, P., 2001. Biogéographie. In: J.C. Arnaud ed. *Atlas du Mali*. Paris: Jeune Afrique, pp.22-4.

Groffman, P.M., Baron, J.S., Blett, T., Gold, A.J., Goodman, I., Gunderson, L.H., Levinson, B.M., Palmer, M.A., Paerl, H.W., Peterson, G.D., Poff, L., Rejeski, D.W., Reynolds, J.F., Turner, M.G., Weathers, K.C., and Wiens, J., 2006. Ecological thresholds: The key to successful environmental management or an important concept with no practical application? *Ecosystems*, 9, pp.1-13.

Haaland, R., 1980. Man's role in the changing habitat of Mema during the old kingdom of Ghana. *Norwegian Archaeological Review*, 13, pp.31-46.

Haaland, R., 1985. Iron production, its socio-cultural context and ecological implications. In: R. Haaland and P. Shinnie eds. *African iron working – ancient and traditional*. Oslo: Norwegian University Press, pp.50-72.

Hahn, H.-P., 1997. Eisentechniken in Nord-Togo: Kultur- und technikgeschichtliche Interpretationen. In: R. Klein-Arendt ed. *Traditionelles Eisenhandwerk in Afrika*, Colloquium Africanum 3. Köln: Heinrich-Barth-Institut, pp.129-45.

Herbert, E.W., and Goucher, C., 1987. *Resource guide for the blooms of Bandjeli : Technology and gender in West African ironmaking*. Waterton: Documentary Educational Resources.

Höhn, A., 2005. *Zur eisenzeitlichen Entwicklung der Kulturlandschaft im Sahel von Burkina Faso. Untersuchungen von archäologischen Holzkohlen*. PhD Thesis, Goethe University, Frankfurt.
http://publikationen.ub.uni-frankfurt.de/volltexte/2005/2253/

Höhn, A., 2007. Where did all the trees go? In: R. Cappers ed. *Fields of change. Progress in African Archaeobotany*. Eelde: Barkhuis Publishing, pp.35-42.

Huysecom, E., Ozainne, S., Raeli, F., Ballouche, A., Rasse, M., and Stokes, S., 2004. Ounjougou (Mali): A history of Holocene settlement at the southern edge of the Sahara. *Antiquity*, 78, pp.579-93.

Huysecom, E., Ozainne, S., Robion-Brunner, C., Mayor, A., Ballouche, A., Coulibaly, N., Eichhorn, B., Guindo, N., Keïta, D., Le Drezen, Y., Lespez, L., Neumann, K., Eichhorn, B., Rasse, M., Schaer, K., Selleger, C., Serneels, V., Soriano, S., Terrier, A., Traoré, B.D., and Tribolo, C., 2007. Peuplement humain et paléoenvironnement en Afrique de l'Ouest: résultats de la neuvième année de recherches. In: *Jahresbericht 2006*. Zürich and Vaduz: Fondation Suisse-Liechtenstein pour les recherches archéologogiques à l'étranger, pp.41-122.

Huysecom, E., Ozainne, S., Robion-Brunner, C., Mayor, A., Ballouche, A., Chaix, L., Eichhorn, B., Guindo, N., Le Drezen, Y., Lespez, L., Mezger, H., Neumann. K., Perret, S., Poudjougo, M., Rasse, M., Sanogo, K., Schneider, K., Serneels, V., Soriano, S., Soulignac, R., Traoré, B.D., and Tribolo, C., 2009. Nouvelles données sur le peuplement du Pays dogon: la onzième année de recherche du programme "Peuplement humain et évolution paléoclimatique en Afrique de l'Ouest". In: *Jahresbericht 2008*. Zürich and Vaduz: Fondation Suisse-Liechtenstein pour les recherches archéologiques à l'étranger, pp.71-183.

Jaeger, P., and Winkoun, D., 1962. Premier contact avec la flore et la végétation du plateau de Bandiagara. *Bulletin de l'Institut Fondamental d'Afrique Noire*, 24, pp.68-111.

Kahlheber, K., 1999. Indications for agroforestry. Archaeobotanical remains of crops and woody plants from medieval Saouga, Burkina Faso. In: M. Van der Veen ed. *The exploitation of plant resources in ancient Africa*. New York: Kluwer Academic/Plenum, pp.89-100.

Kahlheber, S., and Neumann, K., 2007. The development of plant cultivation in semi-arid West Africa. In: T. Denham, J. Iriarte, and L. Vrydaghs eds. *Rethinking Agriculture. Archaeological and Ethnoarchaeological Perspectives*. Walnut Creek: Left Coast Press, pp.320-46.

Kiethéga, J.-B., 2009. *La métallurgie lourde du fer au Burkina Faso. Une technologie à l'époque précoloniale*. Paris: Éditions Karthala.

Marston, J.M., 2009. Modeling wood acquisition strategies from archaeological charcoal remains. *Journal of Archaeological Science*, 36, pp.2192-200.

Mayor, A., Huysecom, E., Gallay, A., Rasse, M., and Ballouche, A., 2005. Population dynamics and paleoclimate over the past 3000 years in the Dogon country, Mali. *Journal of Anthropological Archaeology*, 24, pp.25-61.

National Academy of Sciences, 1980. *Firewood crops. Shrub and tree species for energy production*. Washington DC: US National Research Council.

Neumann, K., 1999. Charcoal from West African savanna sites: Questions of identification and interpretation. In: M. Van der Veen ed. *The exploitation of plant resources in ancient Africa*. New York: Kluwer Academic/Plenum, pp.205-19.

Neumann, K., and Vogelsang, R., 1996. Paléoenvironnement et préhistoire au Sahel du Burkina Faso. *Berichte des Sonderforschungsbereichs 268*, 7, pp.177-86.

Neumann, K., Schoch, W., Détienne, P., and Schweingruber, F.H., 2001. *Hölzer der Sahara und des Sahel. Woods of the Sahara and the Sahel. Bois du Sahara et du Sahel*. Bern: Paul Haupt.

Nouaceur, Z., 2001. Climat. In: J.C. Arnaud ed. *Atlas du Mali*. Paris: Jeune Afrique, pp. 16-9.

Nouvellet, Y., Sylla, M.L., and Kassambara, A., 2003. La production de bois d'énergie dans les jachères au Mali. *Bois et Forêts des Tropiques*, 276, pp.1-15.

Nygard, R., and Elfving, B., 2000. Stem basic density and bark proportion of 45 woody species in young savanna coppice forests in Burkina Faso. *Annals of Forest Science*, 57, pp.143-53.

Ohler, F.M.J., 1985. The fuelwood production of wooded savanna fallows in the Sudan zone of Mali. *Agroforestry Systems*, 3, pp.15-23.

Ozainne, S., Lespez, L., Le Drezen, Y., Eichhorn, B., Neumann, K., and Huysecom, E., 2009. Developing a chronology integrating archaeological and environmental data from different contexts. *Radiocarbon*, 51, pp.457-70.

Pelzer, C., Müller, J., and Albert, K.-D., 2004. Die Nomadisierung des Sahel – Siedlungsgeschichte, Klima und Vegetation in der Sahelzone von Burkina Faso in historischer Zeit. In: K.-D. Albert, D. Löhr and K. Neumann eds. *Mensch und Natur in West-Afrika*. Weinheim: Wiley VCH, pp.256-88.

Picard, N., Ouattara, S., Diarisso, D., Ballo, M., and Gautier, D., 2006. Defining units for savanna management in Sudano-Sahelian areas. *Forest Ecology and Management*, 236, pp.403-11.

Raven, P.H., Evert, R.F., and Eichhorn, S.E., 2000. *Biologie der Pflanzen*. Berlin and New York: De Gruyter.

Robion-Brunner, C., 2010. *Forgerons et sidérurgie en pays dogon. Vers une histoire de la production du fer sur le plateau de Bandiagara (Mali) durant les empires précoloniaux*. Journal of African Archaeology Monograph Series 3. Peuplement humain et paléoenvironnement en Afrique de l'Ouest, série monographique 1. Frankfurt: Africa Magna Verlag.

Robion-Brunner, C., Serneels, V., and Perret, S., this volume. Variability in iron smelting practices: confrontation of technical, cultural and economic criteria to explain the metallurgical diversity in the Dogon area (Mali).

Rolando, C., 1992. *Contribution de l'analyse anthracologique à l'étude des paléoenvironnements sahéliens*. PhD Thesis, University of Provence-Côte d'Azur, Marseille.

Sassoon, H., 1964. Iron-smelting in the hill village of Sukur, north-eastern Nigeria. *Man*, 64, pp.174-8.

Shackleton, C.M., and Prins, F., 1992. Charcoal analysis and the "Principle of Least Effort" – A conceptual model. *Journal of Archaeological Science*, 19, pp.631-7.

Smart, T.L., and Hoffman, T.S., 1988, Environmental interpretation of archaeological charcoal. In: C.A. Hastorf and V.S. Popper eds. *Current Palaeoethnobotany*. Chicago and London: University of Chicago Press, pp.165-205.

Thiébault, S. ed., 2002. *Charcoal analysis: methodological approaches, palaeoecological results and wood uses: proceedings of the Second International Meeting of Anthracology, Paris, September 2000*, British Archaeological Reports International Series 1063. Oxford: Archaeopress.

Thompson, G., and Young, R., 1999. Fuels for the furnace. Recent and prehistoric ironworking in Uganda and beyond. In: M. Van der Veen ed. *The exploitation of plant resources in ancient Africa*. New York: Kluwer Academic/Plenum, pp.221-39.

Vernet, J.-L., 1997. *L'homme et la forêt méditerranéenne de la préhistoire à nos jours*. Paris: Errance.

Von Maltitz, G.P., and Scholes, R.J., 1995. The burning of fuelwood in South Africa: When is it sustainable? *Environmental Monitoring and Assessment*, 38, pp.243-51.

Yaméogo, G., Yélémou, B., and Traoré, D., 2005. Pratique et perception paysannes dans la création de parc agroforestier dans le terroir de Vipalogo (Burkina Faso). *Biotechnology, Agronomy, Society and Environment*, 9, pp.241-8.

Authors' addresses

- Barbara Eichhorn
  Institute of Archaeological Sciences
  Department of Pre- and Protohistory
  Archaeology and Archaeobotany of Africa
  Goethe University
  Grüneburgplatz 1
  D-60323 Frankfurt
  Germany
  b.eichhorn@em.uni-frankfurt.de
- Caroline Robion-Brunner
  Laboratoire TRACES/UMR 5608
  Université de Toulouse 2 le Mirail
  Maison de la Recherche - Bâtiment 26
  5 allée Antonio Machado
  F- 31058 Toulouse cedex 9
  France
- Vincent Serneels and Sébastien Perret
  Earth Sciences – Archaeometry, Department of Geosciences,
  University of Fribourg
  Chemin du Musée 6
  1700 Fribourg
  Switzerland

Use of charcoal species for ironworking in Tanzania

Edwinus Chrisantus Lyaya

ABSTRACT: The study of the chemistry and physics of African metal production has received more attention than the biology of African metal production (tree, plant and animal species that were part and parcel of ironworking processes), because archaeometallurgists rarely attempt to explain the properties of such species; for example, accounting for their selection for metal production and processing. One possible reason for such omissions is the rarity or lack of evidence as compared to slags, furnace walls and tuyères, which are common and durable materials on ironworking sites. In order to redress this discrepancy, this paper presents an investigation of ethnographic tree species, alongside the criteria used to select such species for charcoal production, as well as their socio-cultural uses. It becomes apparent that in this part of the continent ironworkers did not practice omnivorous consumption of wood for ironworking, but instead they were species selective. Selection criteria for preferred species were dependent on both technical aspects and socio-culturally relevant parameters. It is argued here that due to species selection, it is improbable that ironworking technology led directly to total deforestation.

Keywords: Ethnographic species, traditional ironworking, bio-archaeometallurgy, omnivorous, species-selective, charcoal, deforestation, Tanzania, ironworking processes

Introduction

This paper presents an ethnographic and botanical study of tree species that were used to produce charcoal for ironworking processes in some ironworking societies in Tanzania, which include the Pare (Mwanga in Kilimanjaro), Luguru (Morogoro), Rangi (Kondoa in Dodoma), Nyamwezi (Tabora), Rongo (Geita in Mwanza), Haya (Kagera), Fipa (Sumbawanga in Rukwa), Nyakyusa (Chunya in Mbeya), and Ngoni (Ruvuma) regions (Fig. 1). The need to conduct this study is motivated by two main reasons. First, while the physical and chemical operating parameters of an iron smelting furnace are well known (Brelsford, 1949; Childs and Schmidt, 1985; Childs, 1996; Schmidt, 1997a; Buchwald and Wivel, 1998; Mucke, 2003), the properties and role of certain organic materials selected for ironworking (Todd and Charles, 1978; van der Merwe and Avery, 1987; Rehder, 1999) have received little attention. Second, the living memories concerning charcoal species are quickly disappearing from societies that were involved in ironworking in the past (e.g., Barndon, 2004), and so it is imperative to prioritise collecting such ethnographic information. Consequently, this investigation was formulated with the purpose of enriching our understanding of African bio-archaeometallurgy.

This paper has the following objectives: (1) identifying the tree species selected to make the charcoal used for ironworking; (2) an examination of the criteria for the selection of such tree species; and (3) an exploration of the socio-cultural uses of such species in particular societies.

Background information

Few published works on the biological component of African ironworking are available. The nature of the organic remains associated with ironworking may have contributed to this problem because they are rarely found at archaeometallurgical sites. However, there have been sporadic reports of the use of organic materials, such as charcoal, herbs and plants (Greig, 1937; van Noten and Raymaekers, 1988; Barndon, 1996; Childs, 1996; Schmidt, 1997a; Mapunda, 2010), as well as considerations about historic environmental conditions (Muntz, 1960; Haaland, 1985; Schmidt, 1997b; Stromquist, et al., 1999; Mapunda, 2003; Lyaya, 2008b). The following subsections discuss the ways in which organic materials have been used in metallurgy: as source of energy, for techno-medicinal purposes, and their association with environmental issues. The discussion eventually establishes significant gaps in current understandings of this bio-archaeometallurgical factor.

Tree species as source of energy

Iron smelters in Tanzania, as elsewhere in the world, used certain trees and plants for fuel purposes during iron and steel production. Charcoal was employed for the following technological functions: (1) providing energy in the form of high temperatures (van der Merwe and Avery, 1987; David et al., 1989; Rehder, 1999; Mapunda, 2002); (2) as a reducing agent (Cottrell, 1975; Avery, et al., 1988; Schmidt, 1997a; Rehder, 1999; Henderson, 2000), and (3) as a means to facilitate the

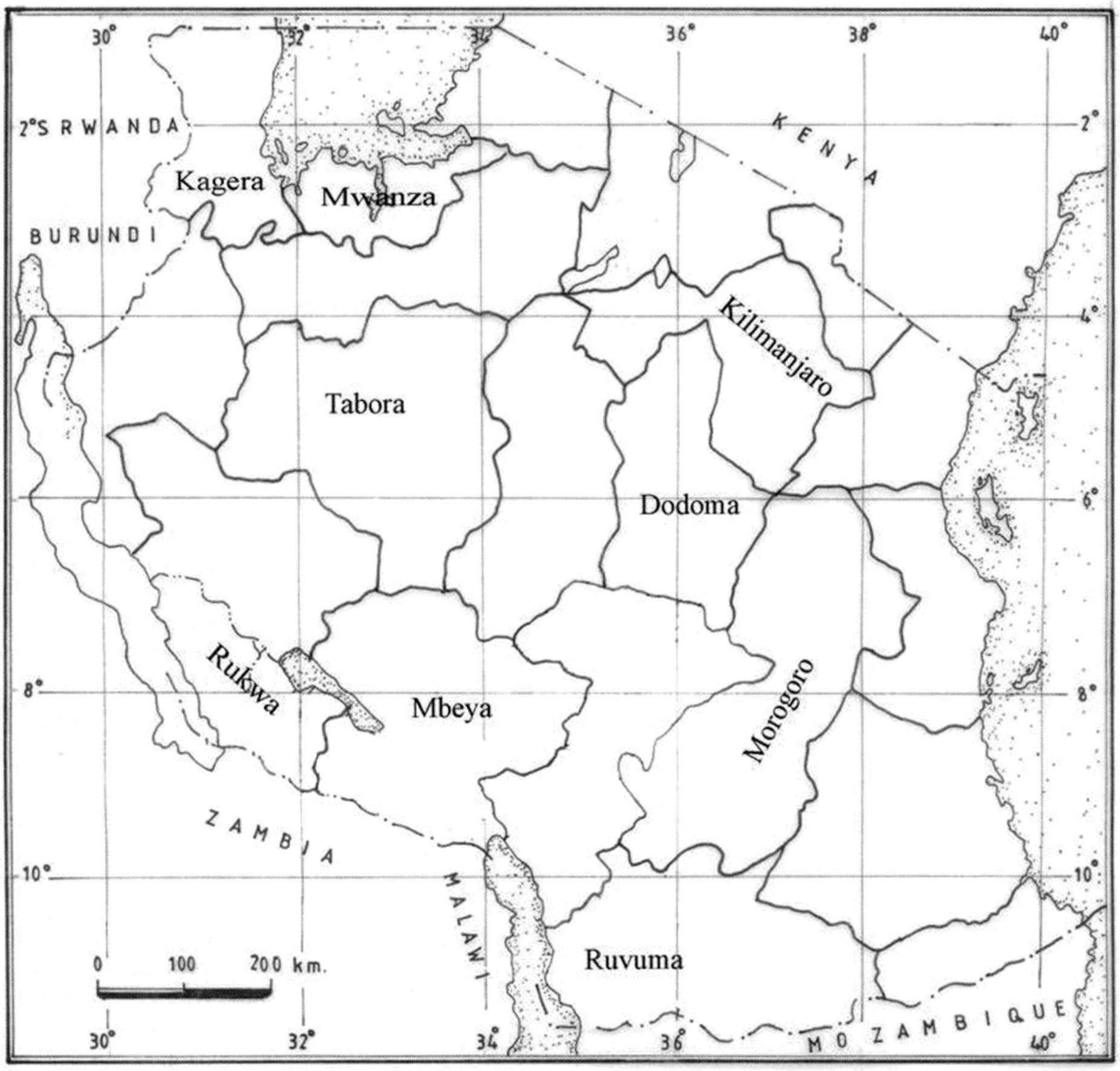

Figure 1: Map of Tanzania showing the studied ironworking regions.

formation of slag (Wertime, 1980; Schmidt, 1997a; Henderson, 2000). Other suggestions for the use of charcoal include its value as a strong, slow-burning platform within which smelting can occur, as well as ritual and ecological factors (David et al., 1989; Schmidt, 1997a, 2006; Mapunda, 2003, 2010). In addition to the employment of charcoal for iron smelting, researchers have identified species that were preferred for iron refining and forging processes respectively. As such, the tree and plant species selected for smelting charcoal were not necessarily selected for the forging process (Barndon, 2004). For example, while in Upangwa in Tanzania local tree species such as *mahato*, *mtelea*, and *mdolbole* were used to produce charcoal for iron smelting, the *mfukwa* species was specifically used for the forging process (Barndon, 2004).

Plant species for techno-magical functions

Archaeologists and archaeometallurgists suggest that the use of medicines – most of which were herbal or faunal – in ironworking activities served to ensure high quality products in many African ironworking societies (e.g., Phillipson, 1968; Wembah-Rashid, 1969, p.66; Killick, 1987; Mapunda, 1995a, p.46, 2010; Schmidt, 1997a, 2006; Barndon, 2004). For example, Barndon found that the use of *umboocha* medicines by Pangwa iron smelters of Rudewa Tanzania were intended for guarding the smelting furnace against evil intents (Barndon, 2004). It was also found that both the Ndamba of Ifakara and Bena of Njombe (both areas in Tanzania) respectively used the *Luwopola* and *Mkufwa* species as medicine to protect against mishaps at the ironworking sites (Lyaya, 2008a, 2008b). Mapunda (1995b, 2010) identified elaborate ritual practices among the Fipa of south-western Tanzania, who buried herbal materials (techno-magical) at the base of almost every furnace excavated. These materials were covered by a clay pot to protect them from burning to ash during the smelt and so losing their otherwise interminable power. Ethnographic sources claim that these materials were used for the protection of the multiple smelts going on in the furnace against evil spirits and witchcraft.

Charcoal and environmental studies

There are a growing number of studies that consider the environmental impact of iron smelting, examining charcoal consumption as a direct cause of total deforestation (e.g., Greig, 1937; Schmidt and Avery, 1983; Schmidt and Childs, 1985; Haaland, 1985; Schmidt, 1997a, 1997b; Stromquist et al., 1999). These scholars claim that iron smelters practiced omnivorous wood consumption where no selection of wood for charcoal was observed, hence upholding the belief that charcoal for iron smelting eventually led to total deforestation (e.g. Haaland, 1985; Schmidt, 1997b; Stromquist et al., 1999; Msemwa, 2001, p.49). Schmidt (1997b, p.407) writes that "the presence of charcoal from light and soft woods, as well as from wetlands trees, supports this picture of 'omnivorous' use of wood, especially surplus wood from other activities". However, it has been difficult to substantiate this claim because there are examples of ironworking societies who were selecting specific tree species for charcoal production; for example, the Rangi and Bena in Tanzania (Mapunda, 2003; Lyaya, 2008b), the Yoruba in Nigeria (Ige and Rehren, 2003, p.16), and the Lungu and Kaonde iron smelters in Zambia (Chaplin, 1961, p.53). In support of the species selection view, Mapunda (2003) argues that upon exhaustion of the preferred species, smelters may have relocated to a new area where the favoured species were available or just changed the source of the preferred charcoal species, depending on how far away the potential new source was located. For the purposes of this paper, it suffices to propose that there are two viewpoints on the relationship between ironworking technology and the environment, and both views are evaluated in the discussion section.

As mentioned above, previous researchers have acknowledged the use of flora (and fauna) in metallurgy, whether for fuel or for ritual purposes, and some have mentioned the species used (e.g., Mapunda, 1995b, 2002, 2003; Schmidt, 1997b; Ige and Rehren, 2003; Barndon, 2004). However, few authors have attempted an explanation of the properties of the species used, and from this gone on to explain their selection for use in metallurgical processes. The data below illustrates the potential of such a high-resolution study.

Methods

Following interviews with sons and daughters, grandsons and granddaughters of the people who used to be involved in iron and steel production, a survey of the tree species used for charcoal was conducted in collaboration with village governments across ironworking societies in Tanzania (see Tables 1, 2, and 3). During this survey, the majority of the species mentioned during the interviews were identified and photographed. Samples were collected for botanical classification at the Botany Department of the University of Dar es Salaam, with the assistance of Khasimu Imamu. It is acknowledged that some species may well have been forgotten during the intervening years, or simply may no longer be found.

Results

For the sake of clarity, the major findings of this study are divided into three categories: the selection of trees for the ironworking processes; the selection criteria for those species; and other socio-cultural uses of the tree species used in ironworking societies.

Selection of tree species for ironworking in Tanzania

This study has revealed that ironworking societies in Tanzania selected certain tree species for charcoal production during various ironworking processes. As outlined below, certain species were selected for iron smelting (Table 1), refining (Table 2), and smithing (Table 3).

Table 1: Charcoal species for the smelting process.

No.	Local Name	Scientific Name	Family	Zone
1	Muwanga	*Cassia siamea*	Caesalbinaceae	Morogoro
2	Msisimizi	*?*	?	Morogoro
3	Mpingo, Gembe	*Dalbergia melanoxylon*	Papilionaceae	Morogoro, Tabora, Kondoa
4	Mlama	*Bridelia cathartica*	Euphorbiaceae	Morogoro
5	Mbaazi	*Cajanus cajan*	Papilionaceae	Morogoro
6	Mlungulungu	*Cassia abbreriata*	Casalpinaceae	Morogoro
7	Mkambala	*Albizia anthelmintica*	Mimosaceae	Morogoro
8	Mkwaju	*Tamarindua indica*	Casalpinaceae	Morogoro
9	Mkalati, Ikaranyika	*Agauria salicifolia*	Ericaceae	Geita, Mwanga
10	Mgando	*Burkea africana*	Caesalpinaceae	Geita, Tabora
11	Mbanga	*Pericopsis angolensis*		Geita, Tabora
12	Kamwamba	*Monotis africanus*	Dipterocarpuceae	Rukwa
13	Umula	*Parinari curatellifolia*	Planchonex Benth	Rukwa
14	Mtomola, Mkame	*Combretum molle*	Combretaceae	Rukwa, Mwanga
15	Nansaku	*Albizia sp.*	Mimosaceae	Rukwa
16	Unzombo/Umputu	*Brachystergia manga*	Cabalpinaceae	Rukwa
17	Yunga	*Maytenus senegalensis*	Celastraceae	Rukwa
18	Nachipa	*Faurea saligna*	Proteaceae	Rukwa

cont'd...

Table 1: *Cont'd...*

No.	Local Name	Scientific Name	Family	Zone
19	Namtiti	?	?	Rukwa
20	Chimba	?	?	Rukwa
21	Chinga	?	?	Rukwa
22	Mzima	*Terminalia sericea*	?	Tabora
23	Kasanda	*Swartizia madagascansis*	?	Tabora
24	Idubilo	*Acacia albida*	?	Tabora
25	Muhanga	*Burkea africana*	Caesalpinaceae	Chunya
26	Myombo Dume	*Brachystegia boehmii*	Caesalpinaceae	Chunya
27	Myombo Jike	*Julbernarchia globiflora*	Caesaepinaceae	Chunya, Kondoa
28	Mwembe Pori	*Ozoroa insignis*	?	Chunya
29	Mpururu	?	?	Chunya
30	Mjululuzi	?	?	Kagera
31	Mshakwanyonyi	?	?	Kagera
32	Mshasha	*Sapium ellipticum*	?	Kagera
33	Mziru	*Turraea robusta*	Meliaceae	Kagera
34	Mshamako	?	?	Kagera
35	Mtoma	*Ficus sar*	Moraceae	Kagera
36	Mgege	?	?	Kagera
37	Mkaraito	*Eucalyptus sp.*	Myrtaceae	Kagera
38	Mnazi	?	?	Kagera
39	Mmo	?	?	Kagera
40	Muembe	*Mangifera indica*	Anacartiaceae	Kagera
41	Nkindo	?	?	Kagera
42	Mlulu	*Albizia gummifera*	Mimosaceae	Mwanga
43	Ng'andi	?	?	Mwanga
44	Mgunga Maji	?	?	Mwanga
45	Mgunga Chuma	?	?	Mwanga
46	Kikwata	?	?	Mwanga
47	Migunga	*Acacia tortitilis*	Mimosaceae	Kondoa
48	Michakai	*Dombeya shupangae*	Sterculiaceae	Kondoa
49	Dabili/Mitunduru	?	?	Kondoa
50	Kijame	*Acacia nilotica*	Mimosaceae	Kondoa
51	Mwanya	?	?	Kondoa
52	Milola	*Ficus sycomorus*	Moraceae	Ruvuma
53	Mikuyu	*Ficus sp.*	Moraceae	Ruvuma
54	Mikagati	*Monotes africanus*	Dipterocarpaceae	Ruvuma
55	Miyombo	*Brachystegia longifolia*	Caesalpinaceae	Ruvuma
56	Muwanga	*Pterocarpus angoleysis*	Papilionoceae	Ruvuma
57	Mivenge	*Syzygium guineense*	Myrtaceae	Ruvuma
58	Mtombati	*Dalbergia nitudila*	papilionaceae	Ruvuma
59	Misuku	*Uapaca kirkiana*	Euphorbiaceae	Ruvuma
60	Mihekela	*Terminalia mollis*	Caumbrutaceae	Ruvuma
61	Migheleghele	*Brachystegia spiciformis*	Caesalpinaceae	Ruvuma

Table 2: Charcoal species for the refining process.

No.	Local Name	Scientific Name	Family	Zones
1	Kamwamba	*Monotis africanus*	Dipterocarpuceae	Sumbawanga
2	Nansaku	*Albizia sp.*	Mimosaceae	Sumbawanga
3	Umputu	*Brachystergia manga*	Cabalpinaceae	Sumbawanga
4	Yunga	*Maytenus senegalensis*	Celastraceae	Sumbawanga

Table 3: Charcoal species for the smithing process.

No.	Local Name	*Scientific Name*	Family	Zone
1	Muwanga	*Cassia siamea*	Caesalbinaceae	Morogoro
2	Msisimizi	?	?	Morogoro
3	Mpingo	*Dalbergia melanoxylon*	Papilionaceae	Morogoro, Kondoa
4	Mfulu	?	?	Morogoro
5	Mlama	*Bridelia cathartica*	Euphorbiaceae	Morogoro
6	Mkwaju	*Tamarindua indica*	Casalpinaceae	Morogoro
7	Mkalati, Mkaranyika	*Agauria salicifolia*	Ericaceae	Geita, Mwanga, Ruvuma
8	Mgando	*Burkea africana*	Caesalpinaceae	Geita, Tabora
9	Mbanga	*Pericopsis angolensis*		Geita, Tabora
10	Kamwamba	*Monotis africanus*	Dipterocarpuceae	Rukwa
11	Umula	*Parinari curatellifolia*	Planchonex Benth	Rukwa
12	Mtomola, Mkame	*Combretum molle*	Combretaceae	Rukwa, Mwanga
13	Nansaku	*Albizia sp.*	Mimosaceae	Rukwa
14	Unzombo/Umputu	*Brachystergia manga*	Cabalpinaceae	Rukwa
15	Yunga	*Maytenus senegalensis*	Celastraceae	Rukwa
16	Namtiti	?	?	Rukwa
17	Chimba	?	?	Rukwa
18	Chinga	?	?	Rukwa
19	Mzima	*Terminalia sericea*	?	Tabora
20	Kasanda	*Swartizia madagascansis*	?	Tabora
21	Idubilo	*Acacia albida*	?	Tabora
22	Muhanga	*Burkea africana*	Caesalpinaceae	Chunya
23	Myombo Dume	*Brachystegia boehmii*	Caesalpinaceae	Chunya
24	Myombo Jike	*Julbernarchia globiflora*	Caesaepinaceae	Chunya, Kondoa
25	Mwembe Pori	*Ozoroa insignis*	?	Chunya
26	Mpururu	?	?	Chunya
27	Mlingoti	?	?	Chunya
28	Mshasha	*Sapium ellipticum*	Pax	Kagera
29	Mziru	*Turraea robusta*	Meliaceae	Kagera
30	Mkaraito	*Eucalyptus sp.*	Myrtaceae	Kagera
31	Muembe	*Mangifera indica*	Anacartiaceae	Kagera
32	Mlulu	*Albizia gummifera*	Mimosaceae	Mwanga
33	Mkithingo	?	?	Mwanga
34	Mfilifili	*Croton macrostachyus*	Euphorbiaceae	Mwanga
35	Mgunga	*Acacia tortitilis*	Mimosaceae	Mwanga, Kondoa
36	Michakai	*Dombeya shupangae*	Sterculiaceae	Kondoa
37	Dabili/Mitunduru	?	?	Kondoa
38	Kijame	*Acacia nilotica*	Mimosaceae	Kondoa
39	Mwanya	?	?	Kondoa
40	Milola	*Ficus sycomorus*	Moraceae	Ruvuma
41	Mikuyu	*Ficus sp.*	Moraceae	Ruvuma
42	Mikagati	*Monotes africanus*	Dipterocarpaceae	Ruvuma
43	Miyombo	*Brachystegia longifolia*	Caesalpinaceae	Ruvuma
44	Miwanga	*Pterocarpus angoleysis*	Papilionoceae	Ruvuma
45	Mivenge	*Syzygium guineense*	Myrtaceae	Ruvuma
46	Misuku	*Uapaca kirkiana*	Euphorbiaceae	Ruvuma
47	Mihekela	*Terminalia mollis*	Caumbrutaceae	Ruvuma
48	Migheleghele	*Brachystegia spiciformis*	Caesalpinaceae	Ruvuma
49	Mwai	*Trichilie emetica*	?	Lindi
50	Mtachi	*Cassia abbreviata*	?	Lindi
51	Mkangazi	?	?	Lindi
52	Mkunguwi	?	?	Lindi
53	Mpimbi	?	?	Lindi

The tables above illustrate that iron production cannot be associated with deforestation in the traditional ironworking regions of Tanzania. It is known, for example, that there are more than 51 tree and shrub species in Ufipa of Tanzania (Mbuya et al., 1994, pp.13-14), but only ten tree species were selected for charcoal production. Furthermore, out of 73 species documented in Uluguru, in Morogoro Tanzania (Mbuya et al., 1994, pp.22-23), ironworkers selected only eight tree species for charcoal production. It can therefore be argued that unless numerous other tree species were also being felled for different uses, this specific selection would not have caused total deforestation. Therefore, the association between indigenous iron smelting and deforestation is a simplistic one (Lyaya, 2008b).

The results also highlight that iron producers became accustomed to selecting the best species for specific processes. For example, the use of *Burkea africana* in Tabora, *Cassia abbreriata* and *Albizia anthelmintica* in Morogoro, *Dalbergia nitudila* in Ruvuma, and *Kikwata* in Mwanga was restricted to the smelting process and not employed in the smithing stage. In addition, among Fipa ironworkers, the use of the *Albizia sp.*, *Brachystargia manga* and *Monolis sp.* species was restricted to the refining process. Finally, *Agauria salicifolia* was used for the smithing process in many ironworking societies in Tanzania, such as the Rongo of Kahama (De Rosemond, 1943) and the Ndamba of Ifakara. It appears that this species is economical in terms of the rate of charcoal consumption, because it does not burn continuously, i.e. it stops or slows down the burning when one stops bellowing and vice versa. This quality was essential for the smiths, especially during lifting of the red-hot metal out of the hearth and beating it on the anvil with hammers to shape it, in these moments ironworkers saved charcoal as the fire slowed down. This information can be used to suggest that ironworking processes (smelting, refining, and smithing) were technologically, functionally (Lyaya, 2007, 2009), and bio-archaeometallurgically different. As such, there is a need for archaeometallurgists to devise a method for identifying and differentiating between these processes both in the field and in the laboratory.

It should be noted that ethnographic sources suggest that iron workers mixed compatible charcoal species to perfect performance, but that not all species could be combined successfully. For example, during interviews the author was told of the failed smelt carried out by Randi Barndon among the Fipa in the early 1990s (e.g., Barndon, 1992, 2004). It was suggested that the initial failure of the smelt was due to incorrect combination of charcoal, because when the charcoal mix was corrected, the smelt succeeded.

Criteria for the selection of charcoal species for ironworking

Regarding the selection of species, the present research has found that the selection of charcoal species for ironworking was based on several criteria. Tables 4, 5, and 6 present the list of the selection criteria.

Table 4: Criteria for the selection of charcoal species for the smelting process.

S/No.	Criteria for selection	Example tree(s)
1	Had very strong fire	Albizia gummifera
2	Had quite fire with less or no sparks	Mjululuzi
3	Had heavy charcoal	*Monotis africanus, Brachystegia boehmii*
4	Had less ash	*Burkea africana, Swartizia madagascansis*
5	Availability in nearby environment	*Cassia siamea*
6	Needed constant bellowing	*Agauria salicifolia*
7	Quality to lit and burn quickly	Mfulwe, *Cajanus cajan*
8	Ability to soften iron by something like oil?	*Tamarindua indica, Albizia sp., Dalbergia melanoxylon*
9	Ability to clean and not to dirt iron?	*Agauria salicifolia, Combretum molle*
10	Provided very long lasting fire	*Cassia siamea*
11	Able to collect all iron together?	*Albizia sp.*
12	Provided very hard charcoal	*Acacia tortitilis*
13	Ability to separate iron from ores?	*Eucalyptus sp.*, Nkindo

Table 5: Criteria for the selection of charcoal species for the refining process.

S/No.	Criterion for selection	Example Tree(s)
1	Able to clean bloom to purity	*Monotis africanus*
2	Gave heavy and strong fire	*Albizia sp., Monotis africanus*
3	Quality to collect all iron together	*Albizia sp.*

Table 6: Criteria for the selection of charcoal species for the smithing process.

S/No.	Reason for its selection	Example Tree(s)
1	Gave very strong fire	*Dalbergia melanoxylon,*
2	Gave quite fire with less or no sparks	Agauria salicifolia
3	Gave heavy charcoal	*Pericopsis angolensis, Brachystegia boehmii*
4	Had less ash	*Burkea africana, Swartizia madagascansis*
5	Needed constant bellowing	*Agauria salicifolia*
6	Quality to lit and burn quickly	Mfulwe, *Cajanus cajan*
7	Ability to soften iron by something like oil	*Albizia sp., Dalbergia melanoxylon*
8	Did not dirt iron	*Agauria salicifolia, Combretum molle*
9	Provided very long lasting fire	*Ozoroa insignis, Sapium ellipticum*
	Acted as substitutes to missing species	*Croton macrostachyus* and *Agauria salicifolia*
10	Gave very normal fire for smithing	*Parinari curatellifolia*

The tables above demonstrate that the selection of tree species for ironworking processes was based on specific qualities such as hardness or high calorific values necessary to raise temperatures for iron smelting. Apparently, ironworkers knew of the qualities of specific tree species through learnt experience passed down through generations (e.g., Schmidt, 1997a; Mapunda, 2002). As a result of their experience they knew and controlled the physical aspects of iron production, and ensured the successful manufacture of high quality iron objects. Through their understanding of the trees (and of all the other facets and materials involved in iron production), they were able to make choices that guaranteed the best outcomes. The following ethnographic examples could be considered. Fipa oral traditions maintain that the *Dalbergia melanoxylon* species were selected because they 'softened' iron, *Agauria salicifolia* species because they 'cleaned' iron, *Nansaku* species because they 'collected' iron together, and *Nkindo* species because they 'separated' iron from ores. I became a hesitant believer, because it was difficult to fully comprehend what the informants exactly meant concerning the abilities of these species. However, we can hypothesise that these tree species provided strong and perfect fire for ironworking, so that whenever they were used they led to success in the smelt, as well as resulting in good products with less or no impurities (clean iron); or it could be that these tree species naturally contained fluxing oxides (e.g., David et al., 1989; Schmidt, 1997a). It should not be surprising, therefore, that in order to take advantage of the qualities of the trees, which were possibly understood after repeated experiences, ironworkers carefully devised a working or technical belief for them and posterity. However, these oral testimonies and the validity of such properties have yet to be confirmed from a western 'scientific' perspective, and remain a focus for research.

It is noteworthy that the tendency of writers to concentrate on the technical reasons for the selection of tree species for ironworking (e.g., Cottrell, 1975; Avery et al., 1988; David et al., 1989; Rehder, 1999) is problematic, because it gives no weight to cultural beliefs in the selection of trees as presented above. There are trees with good qualities which were left out due to beliefs or for cultural reasons. For example, the author was told by informants from Upare that among the Shana and Finanga clans of the Gweno tribe in the Mwanga zone, trees which produced lots of sparks were left out regardless of their high calorific values. These tribes in particular believe that gods, the insurers of good products, are always unhappy with sparking fires. Instead, they require charcoal species that produce fires with no sparks. Another aspect supporting this argument is that when taking into consideration the importance of beliefs, the size of trees did not matter among the Fipa of Tanzania (e.g., Lyaya, 2009), for example. Because of strong belief systems, Fipa ironworkers sometimes selected small trees and left out big ones, and vice versa (e.g., Mapunda, 1995b). Based on these ethnographic examples, it seems appropriate to suggest that cultural as well as technical reasons for the selection of charcoal species are important and require due weight for the holistic understanding of this area of ironworking technology.

Socio-cultural uses of selected species

Ethnographic sources (e.g., Schmidt, 1997a, 2006; Mapunda, 2003, 2010) claim that there is a relationship between the use of charcoal tree species and society at large, because some of the species used for ironworking in Tanzania also had other socio-cultural uses (for similar observations in Malawi, see also van der Merwe and Avery, 1987).

Table 7: Socio-cultural functions of charcoal species.

No.	Species Name	Socio-cultural function
1	*Cassia siamea*	Cure for women frequent vomiting
2	*Swartizia madagascansis*	Stomach ache cure
3	*Acacia albida*	Medicines for abdominal pains
4	*Pericopsis angolensis*	Ritual medicine
5	*Mlingoti*	Cure of coughing, fever, and influenza
6	*Sapium ellipticum*	malaria and dizzying medicine, food
7	*Albizia gummifera*	Tooth brush and dental medicine
8	*Cassia abbreriata*	Cure of sexually transmitted diseases
9	*Agauria salicifolia*	Spices for traditional milk
10	*Ozoroa insignis*	Cure of diarrhoea
11	*Swartizia madagascansis*	Protection against witchcraft, insecticides in farms
12	*Dalbergia melanoxylon*	Increase of male sexual power
13	*Maytenus senegalensis*	Snake medicine
14	*Albizia sp.*	Cure for diseases such as nausea, hernia, sharp pain, bilharzias, convulsion, nose breeding and fresh wounds

Based on the list presented above (Table 7), it is clear that the majority of tree species selected for ironworking in Tanzania were also used to cure problems and diseases related to pregnancy. This information is unsurprising because African iron production processes have been associated with conception, pregnancy, and childbirth across the continent (Sutton, 1985; Nyagava, 1988; Collett, 1993; Herbert, 1993; Mapunda, 1995b, 2010; Schmidt, 1996, 1997a, 2006; Barndon, 2004). For example, *Rhynchosia resinosa* has been associated with fertility magic among the Fipa of Tanzania for a long time (Mapunda, 1995b, 2010, pp.158-160, 2011, p.172). Among the Luguru of Morogoro and Nyakikyusa of Mbeya, the *Cassia siamea* species is used to cure maternity-related problems at large. Likewise in Kagera, dizziness and nausea, as well as frequent fever and malaria problems, which are often associated with pregnant women, are essentially cured with the *Sapium ellipticum* species. Furthermore, the *Dalbergia melanoxylon* species was used to increase male sexual prowess (interview with Ernest Albert Mwanisawa, 2 September 2011). Based on such ethnographic sources, we can argue that ironworkers held the belief that because such trees cured fertility, maternity and child delivery-related problems, they could also ensure the successful delivery of an iron bloom (e.g., van der Merwe and Avery, 1987; Schmidt, 1997a; Mapunda, 2010). This illustrates the significant integration of socio-cultural beliefs with ironworking and enriches our understanding of ancient ironworking technology (e.g., van der Merwe and Avery, 1987; Collett, 1993; Herbert, 1993; Barndon, 1996, 2004; Schmidt, 1996, 1997a, 2006; Mapunda, 2003, 2010).

Discussion and conclusion

In the introduction section, we noted that this paper aimed at (1) identifying the tree species selected to make charcoal for ironworking, (2) examining the criteria for the selection of such species, and (3) exploring the socio-cultural uses of such species in respective societies. I would like to return to these objectives based on the data presented above. First and foremost, Tables 1, 2, and 3 presented the tree species preferred by ironworkers for charcoal production in some regions of Tanzania. It appears that the ironworkers did not harvest each and every tree species for charcoal, because out of the known species in their regions (see Mbuya, et al., 1994) only a few were selected.

With regard to the second objective, it appears that African ironworkers had a sound understanding of the various properties of specific charcoal wood types and mixtures of these, from both a technological and ritual perspective. Their selection criteria for charcoal for the specific stages of iron production were therefore based on both physical (such as burning time) and socio-cultural factors (such as use in fertility magic). Without denying the technical function of the species, it is noteworthy that the socio-cultural factors were critical to the smelters, as to them they represented a reality and never a myth (van der Merwe and Avery, 1987; Mapunda, 2011, p.171). To strengthen and elucidate this point, van der Merwe and Avery (1987, p.144) correctly observed that, "[iron] smelting could not be conducted without [considering socio-cultural aspects], because [these were] very important for the smelters. [In fact], it is a technology in its own right". Therefore, if we are to fully comprehend the African metal production technology, we eventually need to critically consider both technical and social aspects of the ironworking process.

Ironworkers were selective with charcoal species, therefore the association of indigenous ironworking with deforestation appears doubtful. This is because they did not practice the so-called 'omnivorous' consumption of wood (for an opposite view, see Schmidt, 1997b). Ironically, the European colonial administrators associated ironworking alone with total deforestation, in an attempt to excuse and justify the importation of metalware to Africa, which eventually, as expected, led to the demise of the indigenous African metal production technology (for details, see Kapinga, 1990; Mapunda, 2002). Although we know that iron smelting consumed large quantities of charcoal (e.g., Killick, 1991, p.64), it is difficult to believe that the cutting down of eight trees, for example, in Morogoro for ironworking charcoal would directly cause total deforestation if the other hundreds of tree species were not felled (e.g., Mbuya et al., 1994, p.22). However, there are two points to consider: (1) That the selection and removal of ironworking charcoal species could have altered the ecological system; yet, there is no environmental evidence whatsoever to prove the extinction of the species used for ironworking charcoal. For example, the bio-archaeometallurgical samples used in this study from southern Ufipa were collected from within the smelting sites' premises; this attests that the charcoal species did not undergo extinction and more likely ironworking did not utterly damage the ecological system, as some scholars would tend to believe. (2) If ironworkers are to be held responsible for deforestation, then their contribution was an indirect one, through the supply of farm implements which could have been used to clear forests much faster than stone, wooden, or bone implements (e.g. Kimambo, 1969; Mapunda, 2003) for agricultural lands (e.g., slash-and-burn type), settlements, and domestic fuel energy.

With respect to the third and final objective, we have noted that most of the selected charcoal species had additional socio-cultural uses (Table 7), and, more strikingly, that most of the species were used to heal reproduction-related problems. The symbolic association of ironworking with aspects of reproduction is a common phenomenon the world over, and Tanzania is no exception. However, what needs critical evaluation is the question regarding the originality of this association of technology and society. The question to ask is that of the two concepts, which one was the primary or original source of the association? There is no right or wrong answer in attempting to address this question, but it is critical to examine and evaluate the answers we suggest. There is an understanding that ironworkers, especially master smelters, automatically became healers of people. This is because both smelting assistants and non-smelting people trusted the techno-medicines of the smelters simply because the techno-medicines transformed stones into iron (hoes, knives, etc.). For such clients and strong believers in techno-medicines, it was not the reduction process that led to the production of metals, but the magic and strong techno-medicines of the chief smelters (Mapunda, 2010, 2011). These people strongly believed that because the techno-medicines worked for metals (turning stones into metal), they would have 'certainly' worked effectively to heal reproduction-related problems, as well as other issues. Thus, the original source of the association appears to be from ironworking to society at large. However, a second interpretation contrary to this is that society was the original source. It is difficult and impractical to believe that before ironworking there was no witchcraft, no diseases such as those related to fertility, and thus that there were no traditional healers! It could be suggested that non-ironworking healers existed and people knew and used their 'socio-medicines' to cure reproduction-related diseases and others. There is also a possibility that, by virtue of inheritance, some of these early healers could have become chief smelters, and, having turned to ironworking, they possibly re-used their 'socio-medicines' as techno-medicines, this time for a different function: assisting in the production of metal instead of a child, or protecting ironworking sites instead of households or cattle yards against mishaps and witchcraft ailments. Furthermore, those who were not lucky in turning to ironworking must have been consulted by ironworkers (who did not know the medicines before) for medicinal assistance, a transmission procedure that was always performed privately and verbally. To elucidate this point, among the Fipa of southwest Tanzania, the medicines for treating infertility and providing 'advanced bodily immunity' (*zindiko* in Swahili) were the very same techno-medicines used for ironworking, to protect against malevolent intentions (for similar observation, see Mapunda, 2010, p.160). As noted above, the problems of infertility and witchcraft existed long before ironworking began, and it is reasonable to assume that societies must have looked into such problems. One solution was to make use of trustworthy traditional healers to heal the diseases and to protect the society. Thus, the original source of the association between charcoal species and aspects related to fertility should be considered from this direction: it was knowledge originating in society at large that was transmitted to the ironworking technology. We need to learn the socio-cultural aspects associated with the ironworkers in order to fully comprehend some of the bio-archaeometallurgical and technological aspects of ironworking.

However, there is a need in the near future to test the applicability of these findings in a wider area, and across a greater time span, perhaps using multidisciplinary approaches that would include ecological and environmental concerns.

Acknowledgements

I am grateful to Prof. Bertram Mapunda of the College of Arts and Social Sciences, University of Dar es Salaam, Tanzania, who was a senior co-director of this project. Special thanks should go to the UDSM research and publication directorate for permission and funding of the field project. I would also like to thank Prof. Thilo Rehren at the UCL Institute of Archaeology for travel support, the organisers of the World of Iron 2009 conference, and the anonymous reviewers of this paper for their valuable comments.

References

Avery, D.H., van der Merwe, N.J. and Saitowitz, S., 1988. The metallurgy of the iron bloomery in Africa. In: R. Maddin ed. *The Beginning of the Use of Metals and Alloys*. Cambridge, Mass.: M.I.T. Press, pp.245-60.

Barndon, R., 1992. *Traditional Iron Working among the Fipa, An Ethnoarchaeological Study from Southwestern Tanzania*. MA Dissertation, University of Bergen.

Barndon, R., 1996. Fipa ironworking and its technological style. In: P.R. Schmidt ed. *The Culture and Technology of African Iron Production*. Florida: University Press of Florida, pp.58-73.

Barndon, R., 2004. *An Ethnoarchaeological Study of Iron Smelting Practices among the Pangwa and Fipa in Tanzania*. Oxford: Archaeopress.

Brelsford, W.V., 1949. Rituals and medicines of Chishinga ironworkers. *Man*, 49, pp.27-9.

Buchwald, V.F. and Wivel, H., 1998. Slag analysis as a method for the characterization and provenancing of ancient iron objects. *Material Characterization*, 40, pp.73-96.

Chaplin, J.H., 1961. Notes on traditional smelting in northern Rhodesia. *South African Archaeological Bulletin*, 16, pp.53-60.

Childs, S.T., 1996. Technological history and culture in western Tanzania. In: P.R. Schmidt ed. *The Culture and Technology of African Iron Production*. Florida: University Press of Florida, pp.277-320.

Childs, S.T. and Schmidt, P.R., 1985. Experimental iron smelting: the genesis of a hypothesis with implications for African prehistory and history. In: R. Haaland and P. Shinnie eds. *African Iron Working: Ancient and Traditional*. New York: Norwegian University Press, pp.121-41.

Collett, D.P., 1993. Metaphors and representations associated with precolonial iron-smelting in eastern and southern Africa. In: T. Shaw, P. Sinclair, B. Andah and A. Okpoko eds. *The Archaeology of Africa: Food, Metals, and Town*. London: Routledge, pp.499-511.

Cottrell, A., 1975. *An Introduction to Metallurgy*, 2nd edition. London: Edward Ashold Publishers Ltd.

David, N., Heimann, R., Killick, D.J. and Wayman, M., 1989. Between bloomery and blast furnace: Mafa iron-smelting technology in North Cameroon. *The African Archaeological Review*, 7, pp.183-208.

De Rosemond, C.C., 1943. Iron smelting in the Kahama district. *Tanganyika Notes and Records*, 16, pp.79-84.

Greig, R.C.H., 1937. Iron smelting in Fipa. *Tanganyika Note and Records*, 4, pp.77-80.

Haaland, R., 1985. Iron production, its socio-cultural context and ecological implications. In: R. Haaland and P. Shinnie eds. *African Iron Working: Ancient and Traditional*. Oslo: Norwegian University Press, pp.50-72.

Henderson, J., 2000. *The Science and Archaeology of Materials: An Investigation of Inorganic Materials*. London: Routledge.

Herbert, E.W., 1993. *Iron, Gender and Power: Rituals of Transformation in African Societies*. Bloomington: Indiana University Press.

Ige, A. and Rehren, Th., 2003. Black sand and iron stone: iron smelting in Modakeke, Ife, south Nigeria. *Institute of Archaeometallurgical Studies*, 23, pp.15-20.

Kapinga, V., 1990. *Kuathiriwa kwa Maendeleo ya Mwafrika: Mfano Halisi Tanzania* [How an African Development was Affected: Real Example from Tanzania]. Peramiho: Peramiho Printing Press.

Killick, D.J., 1987. Recent iron-smelting in central Malawi. *Nyame Akuma*, 28, pp.27-9.

Killick, D.J., 1991. A little-known extractive process: iron smelting in natural-draft furnaces. *Journal of the Minerals, Metals and Materials Society*, 43, pp.62-4.

Kimambo, I.N., 1969. The interior before 1800. In: I.N. Kimambo and A.J. Temu eds. *A History of Tanzania*. Nairobi: East African Publishing House, pp.14-33.

Lyaya, E.C., 2007. *The Physical Properties of Metalliferous Slag: The Case of Smelting and Smithing Processes*. MA Dissertation, University of Dar es Salaam.

Lyaya, E.C., 2008a. Archaeological field research at Ifakara, Tanzania. *Nyame Akuma*, 70, pp.11-20.

Lyaya, E.C., 2008b. Archaeological field research in Njombe, Tanzania. *Nyame Akuma*, 70, pp.21-9.

Lyaya, E.C., 2009. Revisiting African bloom refining technology in Tanzania. *Nyame Akuma*, 72, pp.30-40.

Mapunda, B.B., 1995a. Iron Age archaeology in the south-eastern Lake Tanganyika region, southwestern Tanzania. *Nyame Akuma*, 43, pp.46-57.

Mapunda, B.B., 1995b. *An Archaeological View of the History and Variation of Ironworking in Southwestern Tanzania*. PhD Thesis, University of Florida.

Mapunda, B.B., 2002. *Ufundichuma Asilia Africa Mashariki: Chimbuko Kukua na Kukoma Kwake* [Iron Technology in East Africa: Origins, Development, and Its Fall]. Peramiho: Peramiho Printing Press.

Mapunda, B.B., 2003. *Iron Technology and Deforestation: Myths and Realities*. Sussex: AFRAS, University of Sussex.

Mapunda, B.B., 2010. *Contemplating the Fipa Ironworking*. Kampala: Fontain Publishers.

Mapunda, B.B., 2011. The jack of two trades, master of both: smelting and healing in Ufipa, southern Tanzania. *African Archaeological Review*, 28, pp.161-75.

Mbuya, L.P., Msanga, H.P., Ruffo, C.K., Birnie, A. and Tengnas, B., 1994. *Useful Trees and Shrubs for Tanzania: Identification, Propagation, and Management for Agricultural and Pastoral Communities*. Nairobi: Regional Soil Conservation Unit, SIDA.

Msemwa, P., 2001. Archaeology of Upper Rufiji catchment. In: F. Chami, G. Pwiti and C. Radimilahy eds. *People, Contacts, and the Environment in the African Past*. Studies in the African Past 1. Dar es Salaam: DUP(1996) Ltd., pp.40-52.

Mucke, A. 2003. Magnetite, ilmenite and ulvite in rocks and ore deposits: petrography, microprobe analyses and genetic implications. *Mineralogy and Petrology*, 77, pp.215-34.

Muntz, A.P., 1960. Forests and iron: the charcoal iron industry of the New Jersey highlands. *Geografiska Annaler*, 42, pp.315-23.

Nyagava, S.I., 1988. *A History of the Bena to 1980*. PhD Thesis, University of Dar es Salaam.

Phillipson, D.W., 1968. Cewa, Leya, and Lala iron smelting furnaces. *South African Archaeological Bulletin*, 23, pp.102-13.

Rehder, J.E., 1999. High temperature technology in antiquity: a sourcebook on the design and operation of ancient furnaces. In: A. Hauptmann, E. Pernicka, Th. Rehren and Ü. Yalçin eds. *The Beginnings of Metallurgy*. Der Anschnitt, Beiheft 8. Bochum: Deutsches Bergbau-Museum.

Schmidt, P. R., 1996. Reconfiguring the Barongo: reproductive symbolism and reproduction among a work association of iron smelters. *In:* Schmidt, P. R. (ed.) *The Culture and Technology of African Iron Production*. Florida: University Press Florida

Schmidt, P.R., 1997a. *Iron Technology in East Africa: Symbolism, Science, and Archaeology*. Oxford: James Curry.

Schmidt, P.R., 1997b. Archaeological views of a history of landscape change in East Africa. *Journal of African History*, 38, pp.393-421.

Schmidt, P.R., 2006. *Historical Archaeology in Africa: Representations, Social Memory, and Oral Traditions*. Lanham: Altamira Press.

Schmidt, P.R. and Avery, D.H., 1983. More evidence for an advanced prehistoric iron technology in Africa. *Journal of Field Archaeology*, 18, pp.421-34.

Schmidt, P.R. and Childs, S.T., 1985. Innovation and industry during the Early Iron Age in East Africa: The KM2 and KM3 sites of Northwestern Tanzania. *The African Archaeological Review*, 3, pp.53-94.

Stromquist, L., Yanda, P., Msemwa, P., Lindberg, C. and Simonson-Forsberg, L., 1999. Utilizing landscape information to analyse and predict environmental change: the extended baseline perspective. *Ambio*, 28, pp.436-43.

Sutton, J.E.G., 1985. Temporal and spatial variability in African iron furnaces. In: R. Haaland and P. Shinnie eds. *African Iron Working: Ancient and Traditional*. Oslo: Norwegian University Press, pp.164-96.

Todd, J.A. and Charles, J.A., 1978. Ethiopian bloomery iron and the significance of inclusions analysis in iron studies. *Journal of the Historical Metallurgy Society*, 12, pp.63-87.

van der Merwe, N.J. and Avery, D.H., 1987. Science and magic in African technology: traditional iron smelting in Malawi. *Africa*, 57, pp.143-72.

van Noten, F. and Raymaekers, J., 1988. Iron smelting in central Africa. *Scientific American*, 258, pp.104-11.

Wembah-Rashid, J.A.R., 1969. Iron workers of Ufipa. *Bulletin of the International Community of Urgent Anthropological Research*, 11, pp.65-72.

Wertime, T.A., 1980. The Pyrotechnologic Background. In: T.A. Wertime and J.D. Muhly eds. *The Coming of the Age of Iron*. New Haven: Yale University Press, pp.1-24.

Author's address

- UDSM, Archaeology Unit, P.O Box 35050, Dar es Salaam, Tanzania.
  UCL Institute of Archaeology, 31-34 Gordon Square, London WC1H 0PY
  Email: edwinusl@yahoo.com

Quantifying iron production in medieval Europe: methodology and comparison with African metallurgy

Danielle Arribet-Deroin

ABSTRACT: This paper examines the output of iron and the fuel consumption from European bloomeries during the Middle Ages. The contributions of different types of evidence, both documentary and archaeological, are discussed and the rate and intensity of the production of iron at each ironworks are considered. African studies on iron metallurgy, whose methods and questions are very similar to European studies, show varied economic and social organisations. They can provide comparisons interpreting archaeological remains and formulating hypotheses, for example concerning the possibility of sporadic or permanent activity, and whether or not production lay in the hands of farmers or specialised groups.

Keywords: Quantification, bloomery, African metallurgy, medieval Europe, iron output, charcoal consumption, production rate.

Introduction

The question of how the iron industry affected the environment during the Middle Ages cannot be answered without quantifying the iron output. The relationship between bloomeries and wood resources during this period is often interpreted as being constantly under increased pressure, due to technological innovations. However, we do not yet have any reliable synthesis or global estimates for iron production during this period, such as those we have for the Romano-British one (Cleere, 1984). The question of iron output is complex, and rendered more so by the varied sources of knowledge.

In European studies, quantification often relies on the very common, but not fully discussed hypothesis that each furnace was used at an intensive rate, for example for 150 days per year (Dubois, 2000). In contrast, African researchers know that an important industrial plant could be used only two or three times a year, even in the 20th century, as illustrated in the Bwamu and Yatenga regions, both in Burkina Faso and Mali (Martinelli, 2004; Coulibaly, 2006). In both Europe and Africa, the direct process was used, and iron production is known from archaeological data, as well as from written sources in Europe and oral traditions and ethnographic records in Africa.

This paper is based on the hypothesis that studies on African metallurgy can provide some evidence, not only on the technology (Killick, 1995) and the symbolic aspects (Haaland, 2004) of medieval European iron working, but also on the rate and organisation of production. Many studies have calculated the yields in African regions in relation to the varied types of economic and social organisation, so that they could offer some clues towards a better understanding of the medieval European realities.

We shall first investigate by which methods the amount of iron produced and charcoal consumed in the medieval period can be known. Next we shall examine how the varied types of economic and social organisation of the African iron industry are interpreted and how they might lead to new interpretations of medieval European iron production.

Iron and charcoal amounts according to historical and archaeological evidence

The aim of this paper is to understand the output of each iron metallurgical district of western Europe in the Middle Ages, and gain a key insight into the use of charcoal and therefore into the environmental impact of this production. The number of production units, i.e. smelting furnaces, and the average output of each one will be evaluated as will their relation to charcoal consumption.

Written sources

Rare contracts and accounts provide some information about annual production yields at various European bloomeries dating to the end of the medieval period. The well known British accounts of the bloomeries of Tudeley, Kent, UK (Giuseppi, 1913) and Byrkeknott, Durham, UK (Lapsley, 1899), in the 14th and 15th centuries respectively (Arribet-Deroin, 2010), are very informative (Fig. 1, Table 1). These bloomeries produced an average of 5 or 5.5 blooms per week, i.e. one per day. The Byrkeknott bloomery operated almost continuously for an entire year, producing a total of 278 blooms (it should

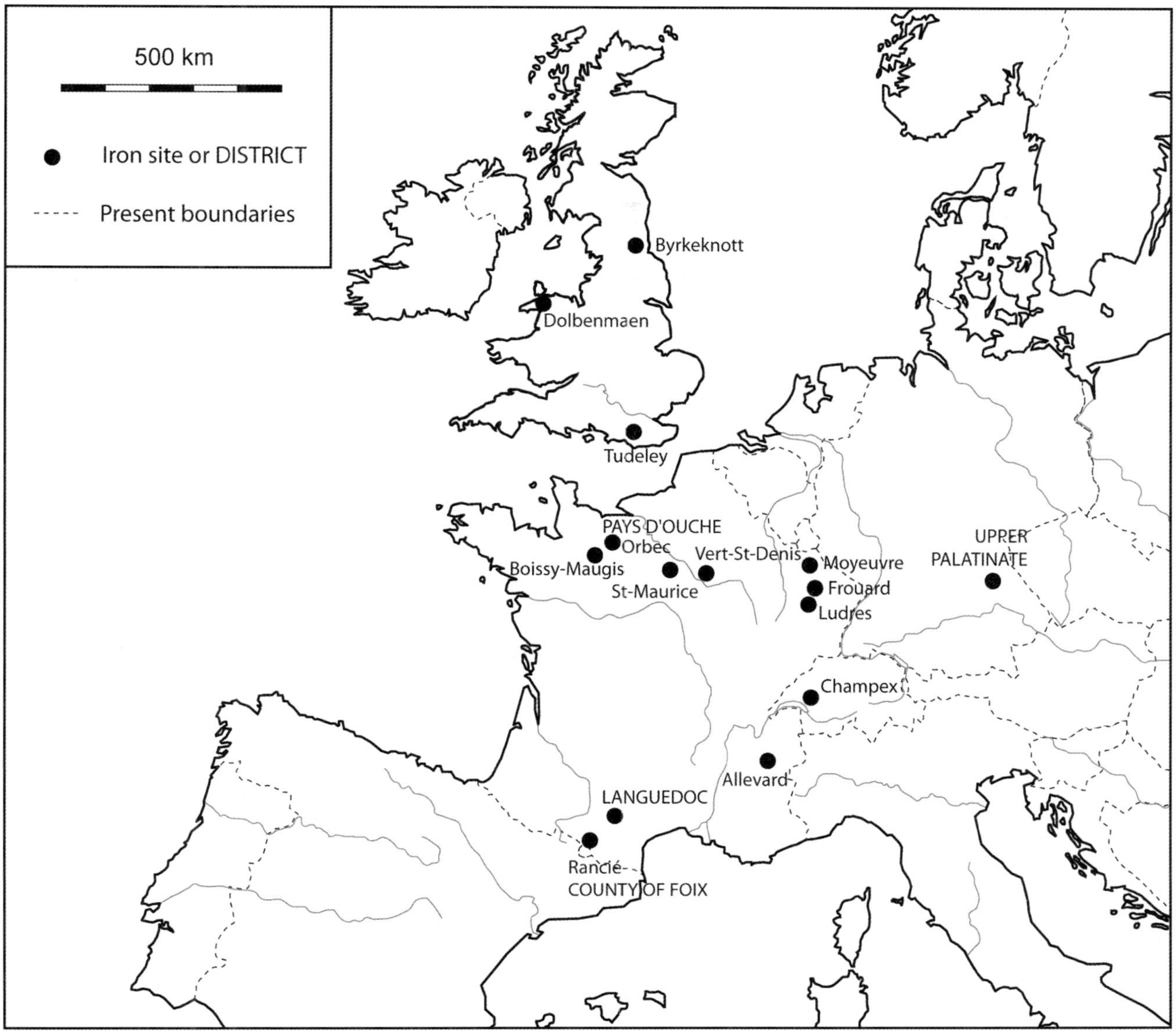

Figure 1: Studied ironworks in medieval western Europe.

Table 1: Outputs of the bloomeries of Tudeley (1329-1333, 1350-1354) and Byrkeknott (1408-1409).

	Number of weeks per year	Number of blooms per year	Number of blooms per week	Weight of each bloom in kg	Weight of iron per year in tons
Tudeley	25 to 40*	112 to 252	5.5	14-15?	1.6-3.8?
Bykeknott	49	278	5	88	28

* full years and only in the 1350s (figure unknown before)

be noted however that only 1.5 years of accounts are preserved for this site). At Tudeley, the number of weeks of labour is quite different from one year to the next, with 112 to 252 blooms being made in each full year. The bloom weight is unknown, but the hypothesis of a 30 pound bloom (Schubert, 1957, p.140) leads to an average of 1.6 to 3.8 tonnes of iron per year. At Byrkeknott, where the bloomery was water-powered, one bloom weighed 195 pounds, leading to the hypothesis that up to approximately 28 tonnes of bloom could have been produced at this site over a period of one year.

From the available documents, it is very difficult to assess charcoal usage at the bloomeries, because measurements related to charcoal were expressed in ancient units. According to new calculations (Peter Crew, personal communication), the annual consumption could have reached 14 to 28 tonnes at Tudeley and almost 130 tonnes at Byrkeknott, with a lower charcoal to iron ratio for bloomeries using water power (Table 2).

Table 2: Charcoal consumption at Tudeley (1329-1333, 1350-1354) and Byrkeknott (1408-1409).

	Weight of charcoal per bloom in kg	Ratio charcoal:iron by weight	Weight of charcoal per year in tons
Tudeley	110	7.8:1	14-28
Bykeknott	475	5.34:1	130

Some accounts only record the taxes owed by the bloomeries to the political authorities, and so these provide less information about the iron production itself. In the county of Bar (Lorraine, France) for example, a tax was paid for each working day by the ironworks of Moyeuvre (Horikoschi, 2008, p.87). The bloomeries were all located in the forest, except for a new, water-powered bloomery. This last one, built in 1323, was expected to produce twice the output of an ordinary ironworks and to consume more charcoal (Horikoschi, 2008, p.86). Although the quantity of iron was not indicated, it is possible to estimate the rate of production (Table 3) as follows: 4 to 5 days per week in 11 to 30 weeks per year in the 1320s, and a more intermittent activity in the 1340s, with 2 to 3.3 days per week. It must be noted that the number of working days and weeks for the bloomery that was water-powered and for those that were not was very similar.

The output of a bloomery is often assumed to be proportional to the taxes or payments made, but this data needs to be used cautiously, as the examples from the 14th century provided below illustrate, consumption and output can vary significantly. An average annual iron yield of 1226 kg was calculated for Dolbenmaen (north-west Wales, UK (Smith, 1995)), and one of 20 tonnes for Champex (Jura, Switzerland), produced by a water-powered bloomery (Pelet, 1988). At this last site, 7.5ha of wood are assumed to have been consumed each year, and at Orbec (Normandy, France), an area of 5ha was needed by each unpowered bloomery (Arnoux, 1993, p.313).

Documents show an intensive and high rate of production. While bloomeries were supposed to run more or less permanently, there were nevertheless great irregularities in the number of weeks of production, so that the annual output was quite different from one year to the next. According to the examples mentioned above, an unpowered bloomery could produce a few tonnes of iron and consume 15-30 tonnes of charcoal (Giuseppi, 1913; Smith, 1995), while a water-powered bloomery could produce a few tens of tonnes of iron, and consume 100 tonnes of charcoal or more (Lapsley, 1899; Pelet, 1988).

Archaeological sources

Archaeology is the main source of knowledge for European medieval ironworks, but the estimates of annual outputs remain problematic. In each case the size of the slag heap must be related to the duration of the smelting activity, which is always hypothetical and imprecise. From archaeological experiments the daily output of iron and the consumption of charcoal can be estimated. For the last 50 years, experimental archaeology has been used to estimate the daily output of iron and the consumption of charcoal for various furnace types (Leroy, et al., 2000; Crew and Crew, 2002; Crew and Charlton, 2007). The results vary (Table 4) due to a lack of standardisation within these experiments, for example the ores and furnace types used. Generally speaking, each reduction produces a bloom weighing a few kilograms (usually less than 7kg), and a greater force of air produces a larger bloom in less time (Crew and Charlton, 2007), especially with water-powered bellows (Crew and Crew, 2002).

The question of the length of each smelt and the number of smelting operations per week is significant information that remains elusive to archaeologists, not least because the bloomeries across Europe varied in their operating parameters. A further aspect of the *chaînes opératoires* of iron production not yet mentioned, but that has a significant impact on both manpower and charcoal consumption, is the forging stage, during which up to 50% of the bloom can be lost (Leroy, et al., 2000; Crew and Charlton, 2007). For example, it has been suggested that to produce a prehistoric currency bar, a charcoal to iron ratio of up to 100:1 was needed (Crew, 1991).

Archaeological surveys and excavations have provided useful information about the number and significance of smelting sites, using criteria such as the size of slag heaps and the chemical composition of slag and ore. When such data is available, an estimate of the iron yield at each site can be made (Serneels, 1993). Thus, depending on whether the amount of produced iron is large or small, each site or group of sites can be considered of more or less significance. For example, a general scheme of the different types of production sites and

Table 3: Rate of production of the bloomeries of Moyeuvre (Horikoshi 2008).

Water-powered bloomery	Unpowered bloomeries	Year	(av) Number of weeks per year	(av) Number of days per year	Number of days per week
1		1324	21	93	4.4
	8	1324*	11	51	4.6
	9	1325	24.6	102.6	4.2
	7	1326	29.4	128.4	4.4
	6	1327	27	102	3.8
1		1345-1346	44	100	2.3
	12	1345-1346	37	88	2.1

* only the half of the year (from July to December)

Table 4: Weights of iron and charcoal in some experiments

	Charcoal in kg	Iron (bloom) in kg	Ratio charcoal:iron by weight
Leroy 2000	25 to 100	1.2 to 5.5	18:1 to 27:1
	44 to 60	4 to 7	8:1 to 12:1
Crew & Charlton 2007	50	8.5	6:1
Crew & Crew 2002	140 to 290	15 to 30	8.9:1 to 12.2:1

ironworks during the early Middle Ages in Northern France (NF) has already been proposed (Cabboi, et al., 2007; Leroy, 2008). Various types of sites can be defined according to such a scheme, from less important sites to more important ones:

NF1 – ironworks in the context of rural settlements, with smithing but also smelting activities. The slag quantity is often very small, typically only a few kilograms. Such sites are relatively numerous; an example of this type of site is Saint-Maurice-Montcouronne (Ile-de-France) (Goustard, et al., 1998).

NF2 – ironworks in small districts, specialised in smelting, sometimes with forging activities. Frouard in Lorraine in the 5th and 6th centuries AD (Leroy, 1997) is an example, where a three tonne slag heap was estimated to be indicative of 1.5 tonnes of iron produced during a few tens of smelting operations.

NF3 – numerous ironworks grouped in large districts, specialised in large-scale smelting activities. In the case of Ludres (Lorraine), in the 8th-10th centuries AD (Leroy, 1997), several tonnes of slag have been found. This type of site is rare during the Middle Ages.

One of the main difficulties resides in understanding the length of the smelting activity, which is always approximate, because dating is based on ^{14}C dates. For example, at Vert-Saint-Denis near Paris, there are more than 2500 ore pits and the quantity of iron produced may have reached 150 tonnes. However, as the activity seems to have been continuous for three centuries, this would be indicative of an average output of only 500kg per year (Daveau and Goustard, 2000).

In spite of the heterogeneous sources, we can conclude that the pattern of intensive use of the furnaces by specialised manpower, which is attested at Tudeley and Byrkeknott, cannot be applied systematically to the entire medieval period. African data could help resolve this issue.

The lessons from African metallurgy

African realities are complex and varied. Some models of production, linked with labour organisations, have been proposed (Fig. 2).

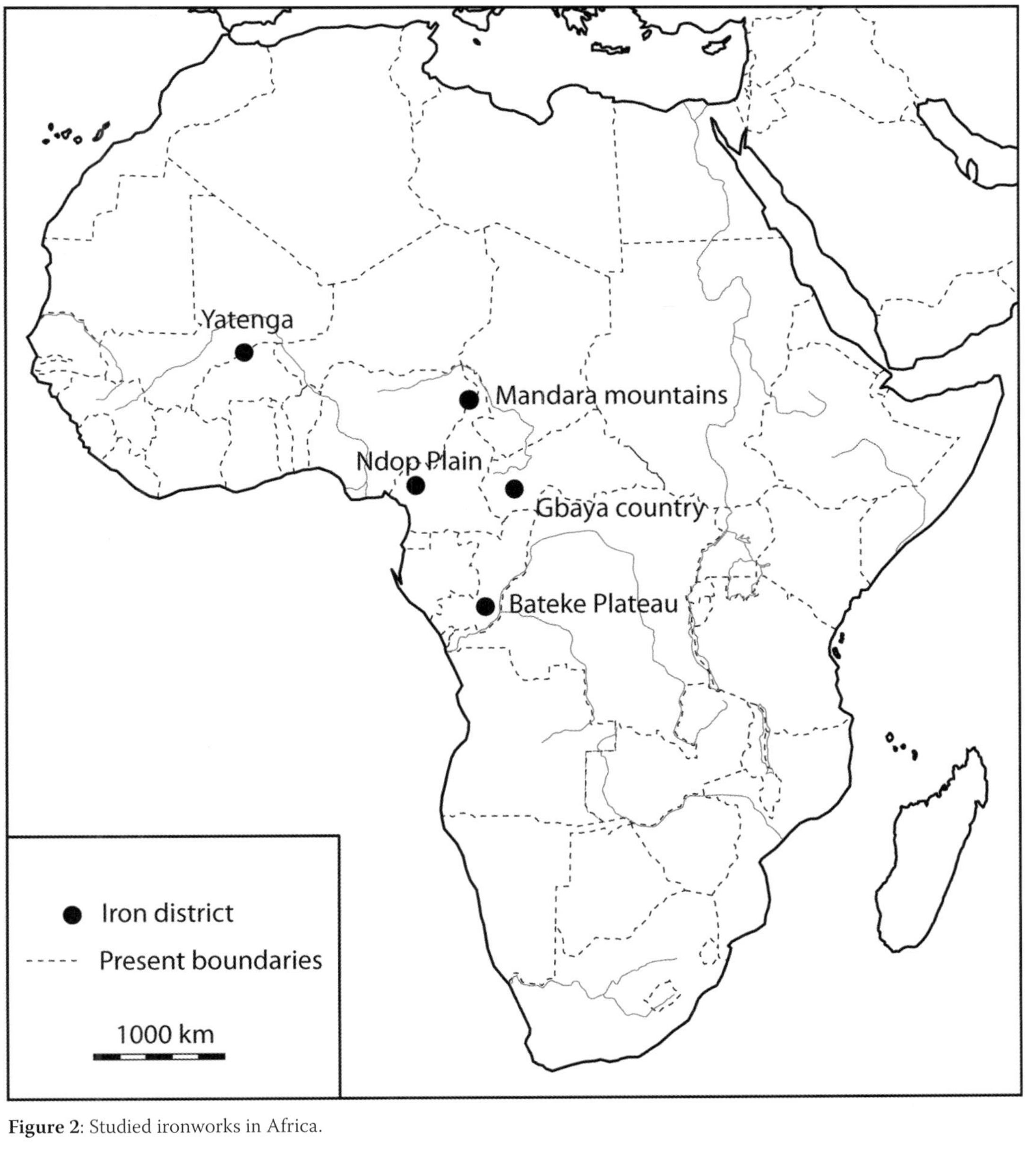

Figure 2: Studied ironworks in Africa.

Analyses of the economic organisation of the African iron production

From a study on the Bateke Plateau (spanning Gabon, Congo, and the Democratic Republic of Congo), a typology of iron production and producers has been proposed, and subsequently applied to other parts of Africa (Dupré and Pinçon, 1997, pp.146-151). The types of sites, ordered from less to more important productions, are:

AF1 – blacksmiths-smelters with a small staff, producing 50kg of new iron per year. The sites have less than $1m^3$ of slag.

AF2 – domestic industry with temporary activity according to household needs, involving co-operation of the family group, with a maximum of one reduction per month (or 12 per year); less than $1m^3$ of slag.

AF3 – occasional activities of a group involving the co-operation of several families, with a maximum of four reductions per year but certainly in several furnaces at the same time; less than $1m^3$ of slag.

AF4 – permanent domestic industry which is the main activity of a definite group, with about 50 reductions per year and a production ranging between 250 and 500kg of iron. Today these sites have between 1 to $3m^3$ of slag. The Gbaya metallurgy (Central African Republic) could have belonged to this type: smelting was the main activity of a group of smelters, but they were also hunters and farmers. They made iron to meet demand, so they worked in October before the great hunting expeditions and in March, at the beginning of agricultural work (Monino, 1983, p.295).

AF5 – intensive industry, which employed a large part of the manpower of a regional area, with 100 to 150 days of production per year and between 3 and $24m^3$ of slag visible today. The main example of this type of production is found in the Ndop Plain (Cameroon Grassfields) in the 19th century (Warnier and Fowler, 1979). In the Babungo village, an assessment has been made of an average of c. 100 tonnes of iron being produced per year, in a production involving 60 furnaces and an average of about 133 smelting operations per year in each furnace. Two thirds of the three hundred men of the village could have smelted iron, not only the owners of the ironworks, but also intermittent ironworkers.

AF6 – itinerant smelters: the amounts of iron and slag cannot be estimated in this case

Some general remarks can be made from this typology. In many cases iron working required sporadic co-operation of the smelter family or a larger group. Smelting was often a seasonal activity. The scale of the iron production can be related to the scale of trade and regional specialisation. It depended on the ore wealth, but also on the scale of the demand for iron from the society in question. In the Bateke Plateau, only 100-150g of new iron per year were required per person (Dupré and Pinçon, 1997, p.166). By contrast, in the Ndop Plain, the large outputs were related to the numerous iron objects owned by each person (Warnier and Fowler, 1979, p.348).

A social analysis of the African iron production

A link between the type of iron production and the status of blacksmiths in society has been made in the Niger Bend in West Africa (Martinelli, 2004). In a study of the Yatenga (Burkina Faso and Mali) ironworks, it was shown that iron working was the monopoly of a specialised and endogamic group of blacksmiths. These blacksmiths controlled the whole of the iron production process, from the extraction of the ore to the production of the iron objects. This social organisation was the result of an evolution related to political and technical changes: the emergence of the Yatenga Kingdom (16-18th centuries) and a new type of furnace, the '*bônga*' furnace.

More generally, two technological and social systems are described in the Niger Bend:

1) In a unitary system like Yatenga's, the entire process chain was handled by blacksmiths who had a monopoly on iron production. The society was based on closed groups, in the context of pre-colonial states.
2) In a dualist system, there were two categories of metalworkers: smelters who carried out the extraction and reduction, and blacksmiths. Smelters were also farmers and not regarded as craftsmen. This model was characteristic of the lineage societies, where political power was not centralised before the 20th century.

This scheme cannot be extended to other parts of Africa, where its pertinence might be questioned. In most areas of the Mandara mountains (northern Cameroon and eastern Nigeria) (Langlois, 2005-2006, p.193), the system could be considered as mixed, because the smelting operations were carried out by either farmers or blacksmiths, without any exclusivity. Contrary to the Yatenga case, the intensification of the production enforced the dualist trend and led to specialised areas instead of defined social groups. It can be remarked that both systems could allow large-scale iron production. In the Yatenga unitary system, each of the 1,000 to 1,500 furnaces produced 300 to 600kg of iron per year with two or three reductions, as compared to 1.5 to 2 tonnes with 100-150 reductions per year in each furnace in the Ndop Plain, a dualist system. In both cases the intensification of the production was caused by political events and accompanied by technological changes.

In spite of great differences between social organisations, studies of African iron metallurgy show the necessity of questioning the status of European ironworkers.

Hypotheses on the organisation of ironworks in medieval Europe

There are questions to be asked about the organisation of ironworks in medieval Europe: did miners, smelters and blacksmiths belong to different, privileged, and/or closed groups? Was iron production their only or main occupation and an intensive one? These issues will be examined starting from the more recent and best known period and proceeding to the previous ones.

Ironworks belonging to or encouraged by kings, lords or urban elite

During the second part of the Middle Ages, some bloomeries, such as Tudeley, belonged to lords, and some belonged to the church, such as Byrkeknott. Since the 12th century, some monastic communities (especially the Cistercians) were very active in iron production (Verna, 1995): they produced iron first for their own needs, then for trade. Since the 14th century, the great interest of the royal or seigniorial authorities is shown by the numerous contracts they concluded with entrepreneurs. These entrepreneurs could, for a payment, make use of wood and ore, and sometimes of a water course. Ironworks were generally new and water-powered, as at Moyeuvre in the county of Bar (Horikoshi, 2008), the 'moulines' in the French crown lands in Languedoc (Cabié, 1903), or in the county of Foix (Verna, 2001). There is no doubt that great output and profit were expected from such activities. Sometimes investments were made by members of the urban elite, as in the Upper Palatinate, Germany (Hirschmann *et al.*, 1987). Whether this kind of production was similar to some African type is doubtful. Such bloomeries had a permanent workforce, and the high intensity of production may be related to an increasing demand for iron, for example for use in gothic architecture (Timbert, 2009), and for producing new weapons and agricultural tools.

Iron production of rural communities

In various areas with rich ore deposits, often located in forests or mountains, the extraction and smelting of the ore were carried out by rural, community-based manpower, these groups often being given certain social privileges. In some cases, specialised smelters – known as 'férons' or 'ferrons' in French – belonged to a professional organisation called 'métiers de férons', which had a monopoly on the raw iron production in definite areas. They had their own court for professional claims and the craft was inherited. Such privileges were granted by political authorities and might have constituted a response to the concurrence of seigniorial bloomeries.

These rural producers are known in several regions of western Europe, but the best known example is the Norman one in the Pays d'Ouche (Normandy, France), because of the strength and the power of its professional court (Arnoux, 1993). There were about 200 bloomeries in this area by the 14th century. They were topographically located in areas called 'fabricas', close to the ore and situated in the forest or in villages. Each smelter had his own production unit on a piece of land, but shared collective rules and rights with the others. There is evidence of intensive iron production by groups of 'férons' during the medieval period and, as we have seen, the number of working days at Moyeuvre and the amount of wood used at Orbec were significant. Such intensity could be compared to that of the ironworks in the Ndop Plain (AF5).

In other cases, the rights related to extraction or smelting were not reserved to a specific group but belonged to an entire rural community. Two French examples are famous: Rancié in the Pyrenees (Verna, 2001) and Allevard in the Alps (Sclafert, 1926). In these valleys all the inhabitants had the right to extract ore and produce iron (or steel), but without an absolute monopoly. In both cases this was the occupation of only a small part of the population, some of whom were also farmers, while others produced iron as a seasonal occupation. The mining activity was regulated by particular institutions. Ore was smelted on the spot, or was sold to bloomeries in low valleys. This type of activity could be compared to the AF3 and AF4 types, and to productions in some areas of the Mandara mountains (Langlois, 2005-2006). In all cases there was a distinction between smelters, grouped in ore-rich areas, and blacksmiths, present in all the villages.

Which models can be used for the early medieval period?

For the early Middle Ages, the comparison with African data appears more pertinent. During this period the smelting activity often differs from the intensive and large scale production well attested to in the following centuries, but is potentially close to African models. Several types of production could be proposed (Table 5).

The NF3 type, at sites such as Ludres (Lorraine), can be closely compared to the production in the Cameroon Ndop Plain (AF5): an intensive and large-scale production, at least for purposes of regional trade. It might have been the permanent or main occupation of smelters.

More often, the sites are less significant (NF2). It could be assumed that smelting was related either to intermittent and familial activity (AF2 and AF3) or to the permanent activity of a larger group (AF4). The level of the demand and trade might be defined to allow a better comparison. The iron needs were probably high in the early Middle Ages (Leroy, 2008, p.186). Two examples show how interesting and at the same time how difficult the comparison can be.

At Vert-Saint-Denis, the assumed yield of 500kg per year produced in a few furnaces of small size, in the context of a rural settlement (Daveau and Goustard, 2000), is close to the

Table 5: Proposal of comparison between African and European ironworks.

African ironworks			European medieval ironworks	
	AF1	blacksmiths-smelters	NF1	Saint-Maurice-Montcouronne
Mandara mountains	AF2	occasional activity of a family		
	AF3	occasional activity of a group	NF2	Allevard, Rancié, Boissy-Maugis
Gbaya country	AF4	permanent industry of a group		Frouard, Vert-Saint-Denis
Ndop Plain	AF5	intensive rural industry	NF3	Ludres, Normandy 'férons'
?		intensive ironworks with investments		Tudeley, Byrkeknott
	AF6	itinerant smelters		?

African permanent domestic activity (AF4). Thus, according to African criteria, its production was a rather intensive one, with the aim to distribute beyond the village. It could have been similar to the cases of the Allevard and Rancié valleys: a production in the vicinity of an ore deposit (a modest one in this case) and a rural community able to exploit this resource, specialising in extracting and smelting.

It is known from written documents, the Boissy-Maugis 'villa' (Normandy) sent raw iron to the Saint-Germain-des-Prés abbey near Paris, to which it belonged, at the beginning of the 9th century (Hägermann, 1993, pp.117-118). About 30 people had to give approximately 33kg of iron ore per year. The proportion of the total iron output to which that figure corresponds is unknown, but it reveals a production which can be assumed to be larger than local demand. The region is known in more recent periods for its large-scale iron production (Arnoux, 1993).

When the metallurgical sites show a very small scale production (NF1), several scenarios can be proposed: this could have been the occasional activity of certain families in the rural community, like AF2, or, in contrast, the permanent activity of a specialised blacksmith, like AF1.

Conclusion

In the second part of the Middle Ages in western Europe, a model of intensive iron production seems to have been the rule: one or several tonnes of iron per year were produced by each production unit, utilising specialised manpower. Mining, charcoal production and blacksmithing were external works, and iron was produced to be sold in markets. The categories of smelters included specialised craftspeople and involved an intensive or full-time activity. Such rural communities of smelters were certainly present in the early Middle Ages.

There are a range of cases from the early Middle Ages, inspired by African metallurgy, which provide historians and archaeologists with a new point of view. It is unlikely, however, that the Yatenga's unitary system can be transposed to Europe. Generally, smelting was in the hands of farmers and the intensification of the production was linked to a specialisation of the men.

In early medieval Europe the extremely varied sizes of the sites show many possibilities, from sporadic extracting and smelting of a smith for local needs to the permanent activity of groups for large-scale trade. The separation of the tasks must also be considered, from ore extracting to blacksmithing, throughout the year. Smelting could have been seasonal work that had to be compatible with agricultural work and take into account the availability of a seasonal labour force. This is seen in some African cases (Magnuson, 1984; Nørbach, 2003) and, as such, may inspire European medievalists in their interpretations.

It appears necessary to re-examine the resource requirements and the needs of the past societies. Wood and ore availability are factors which have an impact on the potential intensity of iron production. The demand of rural and urban communities for iron, as well as the demand of the elite market, is also an important influence on the increase in output. With more reflections and archaeological studies, we may finally be able to extrapolate the amount of global iron production and its impact on the environment.

Acknowledgement

Thanks to Peter Crew for allowing me to use his re-calculations of the Tudeley and Byrkeknott data in advance of his own publication (Table 2).

References

Arnoux, M., 1993. *Mineurs, férons et maîtres de forges. Études sur la production de fer dans la Normandie du Moyen Âge, XIe-XVe siècles.* Paris: Editions du CTHS.

Arribet-Deroin, D., 2010. Pratiques et rythmes de la métallurgie du fer aux XIVe et XVe siècles: les « *bloomeries* » de Tudeley (Kent) et de Byrkeknott (comté de Durham). *Archéologie médiévale*, 40, pp.147-68.

Cabboi, S., Dunikowski, C., Leroy, M. and Merluzzo, P., 2007. Réflexions sur les formes d'organisation du travail du fer dans le nord de la France au Haut Moyen Age (l'apport des découvertes archéologiques récentes). *Medieval Europe, Paris (4e Congrès International d'Archéologie Médiévale et Moderne).* <http://medieval-europe-paris-2007.univ-paris1.fr/S.%20Cabboi%20et%20al..pdf>.

Cabié, E., 1903. Forges et moulines à fer de la Montagne Noire. *Revue du Tarn*, 20, pp.237-48.

Cleere, H., 1984. Ironmaking in the Economy of the Ancient World: The Potential of Archaeometallurgy. In: B.G. Scott and H. Cleere eds. *The Crafts of the Blacksmith.* Belfast: UISPP Comité pour la Sidérurgie Ancienne and the Ulster Museum, pp.1-6.

Coulibaly, E., 2006. *Savoirs et savoir-faire des anciens métallurgistes d'Afrique occidentale. Procédés et techniques de la sidérurgie directe dans le Bwamu (Burkina Faso et Mali).* Paris: Karthala.

Crew, P., 1991. The experimental production of prehistoric bar iron. *Historical Metallurgy*, 25, pp.21-36.

Crew, P. and Crew, S., 2002. Experimental iron working at Agorregi, 1994-1999. In: M.M. Urteaga ed. *Agorregiko burdinola eta errotak. La ferrería y los molinos de Agorregi (Aia, Gipuzkoa) II. Arkeologia esperimentala, Arquelogía experimental.* Donostia-San Sebastián: Diputación Foral de Gipuzcoa, pp.43-91.

Crew, P. and Charlton, M., 2007. The anatomy of a furnace ... and some of its ramifications. In: S. La Niece, D. Hook and P. Craddock eds. *Metals and Mines.* London: Archetype Publications in association with the British Museum, pp.219-25.

Daveau, I. and Goustard, V., 2000. Un complexe métallurgique et minier du haut Moyen Age: le site des Fourneaux à Vert-Saint-Denis (Seine-et-Marne). In: C. Domergue and M. Leroy eds. *Mines et métallurgies en Gaule. Recherches récentes.* Gallia 57. Paris: CNRS Editions, pp.77-99.

Dubois, C., 2000. Lercoul (Pyrénées ariégeoises). Un site sidérurgique du IIIe siècle de notre ère. In: C. Domergue and M. Leroy eds. *Mines et métallurgies en Gaule. Recherches récentes.* Gallia 57. Paris: CNRS Editions, pp.53-75.

Dupré, M.-C. and Pinçon, B., 1997. *Métallurgie et politique en Afrique centrale. Deux mille ans de vestiges sur les plateaux batéké Gabon, Congo, Zaïre.* Paris: Karthala.

Giuseppi, M.S., 1913. Some Fourteenth-Century Accounts of Ironworks at Tudeley, Kent. *Archaeologia*, 64, pp.145-64.

Goustard, V., Leroy, M., Serneels, V. and Daveau, I., 1998. La production sidérurgique en contexte d'habitat au VIIIe-XIIe siècles: l'apport de fouilles récentes en France et en Suisse. In P. Beck ed. *L'évolution technique au Moyen Âge, Actes du VIe congrès international d'archéologie médiévale tenu à Dijon en octobre 1996*. Paris: Editions Errance, pp.139-44.

Haaland, R., 2004. Technology, Transformation and Symbolism: Ethnographic Perspectives on European Iron Working. *Norwegian Archaeological Review*, 37, pp.1-19.

Hägermann, D. ed., 1993. *Das polyptychon von Saint-Germain-des-Prés. Studienausgabe*. Köln-Weimar-Wien: Böhlau Verlag.

Hirschmann, N., Fleisser, H. and Mahler, F., 1987. *Die Oberpfalz – Eine Europaïsches Eisenzentrum. 600 Jahre Grosse Hammereinung*. Schriftenreihe des Bergbau- und Industriemuseums Ostbayern Band 12/1 Aufsatzband. Amberg: Theuern.

Horikoshi, K., 2008. *L'industrie du fer en Lorraine XIIe-XVIIe siècles*. Langres: Editions Dominique Guéniot.

Killick, D., 1995. Variation in African Iron-smelting Practice: Implications for the Study of Prehistoric Iron Technology in Europe. In: P. Benoit and P. Fluzin eds. *Paléométallurgie du fer et cultures actes du symposium international du Comité pour la sidérurgie ancienne de l'Union internationale des sciences préhistoriques et protohistoriques, Belfort-Sévenans, Institut polytechnique de Sévenans, 1-2-3 novembre 1990*. Belfort: Vulcain, pp.59-69.

Langlois, O., 2005-2006. De l'organisation bipartite du travail du fer dans les monts Mandara septentrionaux. *Techniques et cultures*, 46-47, pp.175-209.

Lapsley, G.T., 1899. The Account Roll of a Fifteenth-Century Iron Master. *English Historical Review*, 14, pp.509-29.

Leroy, M., 1997. *La sidérurgie en Lorraine avant le haut fourneau. L'utilisation du minerai de fer oolithique en réduction directe*. Monographie du CRA 18. Paris: CNRS Editions.

Leroy, M., 2008. Les modes de production du fer au haut Moyen Age. L'exemple des ateliers sidérurgiques de Lorraine centrale. In: J. Guillaume and E. Peytremann eds. *L'Austrasie. Sociétés, économies, territoires, christianisation. Actes des XXVIe Journées internationales d'Archéologie mérovingienne (Nancy 2005)*. Nancy: Presses Universitaire de Nancy, pp.177-88.

Leroy, M., Merluzzo, P., Fluzin, P., Leclère, D., Aubert, M. and Ploquin, A., 2000. La restitution des savoir-faire pour comprendre un procédé technique: l'apport de l'expérimentation en archéologie du fer. In: P. Pétrequin, P. Fluzin, J. Thiriot and P. Benoit eds. *Arts du feu et production artisanales, XXe Rencontres internationales d'Archéologie et d'Histoire d'Antibes*. Antibes: Editions APDCA, pp.37-51.

Magnuson, G., 1984. Who were the Blacksmiths in Jämtland and Härjedalen. In: B.G. Scott and H. Cleere eds. *The Crafts of the Blacksmith*. Belfast: UISPP Comité pour la Sidérurgie Ancienne and the Ulster Museum, pp.121-30.

Martinelli, B., 2004. On the Threshold of Intensive Metallurgy: The Choice of Slow Combustion in the Niger River Bend (Burkina Faso and Mali). In H. Bocoum ed. *The Origins of Iron Metallurgy in Africa. New Light on its Antiquity: West and Central Africa*. Paris: UNESCO Publishing, pp.165-88.

Monino, Y., 1983. Accoucher du fer. La métallurgie gbaya (Centrafrique). In: N. Échard ed. *Métallurgies africaines. Nouvelles contributions*. Mémoires de la société des africanistes 9. Paris: Société des Africanistes, pp.281-309.

Nørbach, L.C. ed., 2003. *Prehistoric and Medieval Direct Iron Smelting in Scandinavia and Europe. Aspects of Technology and Science. Proceedings of the Sandbjerg Conference, 16th-20th Sept. 1999*. Acta Jutlandica 76:2, Humanities Series 75. Aarhus: Aarhus University Press.

Pelet, P.-L., 1988. Ruiner la végétation ou sauvegarder la nature, la ferrière de Champex au XIVe siècle. *Revue suisse d'histoire*, 38, pp.30-44.

Schubert, H.R., 1957. *History of the British iron and steel industry, from c.450 BC to AD 1775*. London: Routledge and Kegan Paul.

Sclafert, T., 1926. *L'industrie du fer dans la région d'Allevard au Moyen Âge*. Grenoble: Allier père et fils.

Serneels, V., 1993. *Archéométrie des scories de fer: recherches sur la sidérurgie ancienne en Suisse occidentale*. Lausanne: Cahiers d'archéologie romande 61.

Smith, K.E.S., 1995. Iron-working in North-West Wales in the Late Fourteenth Century. *Archaeological Journal*, 152, pp.246-90.

Timbert, A. ed., 2009. *L'homme et la matière. L'emploi du plomb et du fer dans l'architecture gothique. Actes du colloque de Noyon 16-17 Novembre 2006*. Paris: Picard.

Verna, C., 1995. *Les mines et les forges des Cisterciens en Champagne méridionale et en Bourgogne du Nord XIIe-XVe siècle*. Paris: AEDEH/Vulcain.

Verna, C., 2001. *Le temps des moulines. Fer, technique et société dans les Pyrénées centrales (XIIIe-XVIe siècles)*. Paris: Publications de la Sorbonne.

Warnier, J.-P. and Fowler, I., 1979. A nineteenth-century Ruhr in central Africa. *Africa (Journal of the International African Institute)*, 49, pp.329-51.

Author's address

- LAMOP UMR 8589 Université Paris 1 Panthéon-Sorbonne – CNRS
  Université Paris1 Panthéon-Sorbonne, 3 rue Michelet, 75006 PARIS FRANCE
  arribet@univ-paris1.fr

The environmental impact of Romano-British ironworking on Exmoor

Ralph Fyfe, Lee Bray, Gillian Juleff, Jessie Woodbridge and Peter Marshall

ABSTRACT: Archaeological surveys during the 1990s and early 2000s identified extensive evidence for iron working within Exmoor National Park, an upland area in southwest Britain. The archaeological evidence from these surveys demonstrates that iron extraction began in the late first millennium BC, and extended through the first half of the first millennium AD (the Romano-British period). Whilst the archaeological evidence signposts exploitation of the upland some 2000 years ago, the environmental consequences of metallurgical extraction are unclear. This paper focuses on the environmental impact of metallurgy by focussing on the largest, and best understood, iron working site on Exmoor (Sherracombe Ford). The paper draws on evidence from the excavation (cultural material and radiocarbon dates), and from off-site sequences (geochemistry and pollen analysis), to develop a chronology for iron working at Sherracombe Ford, to quantify the volumes of iron produced and the resources that would have been required (in particular the area of woodland for charcoal production), and to assess the evidence for woodland management. The chronologies suggest that Sherracombe Ford was in use for probably 225 years, based on the on-site radiocarbon evidence, and possibly as long as 300 years according to the off-site geochemical record. Estimates of total iron production are between 70-175 tonnes. The scale of woodland required for production of this volume of iron across a 300 year period is small, and could easily have been met through management of the local woodland. Pollen and charcoal evidence supports the interpretation of sustained management of woodland in the local area. Exploitation of iron in the first millennium AD in this region thus had minimal impact on the wider environment, which contradicts evidence from continental Europe of major modification of vegetation to support metallurgic activities.

Keywords: Archaeometallurgy, geochemistry, pollen, Exmoor, Romano-British, sustainability.

Introduction

Archaeological surveys and excavations in Exmoor during the late 1990s and early 2000s have, for the first time, identified extensive evidence for iron working in the region starting in the late first millennium cal BC, and extending into the Romano-British period (broadly the first half of the first millennium AD). Exmoor is an upland region in southwest Britain, and is today a National Park characterised by moorland, woodland, valleys, and farmland. Research has focused on field evidence for all stages of primary iron production, from ore extraction to smelting and initial metal refining (Fig. 1; Juleff and Bray, 2007; Brown, et al., 2009). The surveys have demonstrated a near-continuous exploitation of the upland for metallurgic products over at least the last 2000 years. Until now, explanations of landscape change and economic development of the upland have revolved around its use as an agricultural resource. In light of the recent archaeometallurgic findings, however, recognition of the role of this and other uplands as centres of extractive metal industries demands a rethink of a range of issues. These include reviewing existing ideas regarding the social and economic functioning of Exmoor in the wider region, and, from an environmental perspective, the relative importance of woodland management and clearance in relation to agrarian and industrial purposes. This paper focuses on the environmental impact of metallurgy on Exmoor by examining the likely effect of activities at Sherracombe Ford, the most extensively excavated and understood iron working site in the region, through the integration of on-site archaeological and environmental data and off-site palaeoecology and geochemistry.

This paper attempts to quantify the level of iron production at Sherracombe Ford, and characterise the scale of fuel requirements for the site. Consideration of the duration of activities is made through a combination of artefactual evidence, on-site radiocarbon dating and off-site geochemical analysis from a nearby peat system. Finally, the paper asks whether there is evidence for sustained management of the woodland resource around Sherracombe, or whether activities significantly altered the structure and nature of the surrounding landscape through unsustainable woodland clearance. There is currently no clear consensus on the impact that iron working has on the local environment, with some arguing for local, limited impacts (e.g., Mighall and Chambers, 1997) and others recognising major changes in landscape organisation and vegetation structure (e.g., Jouffroy-Bapicot, et al., 2007).

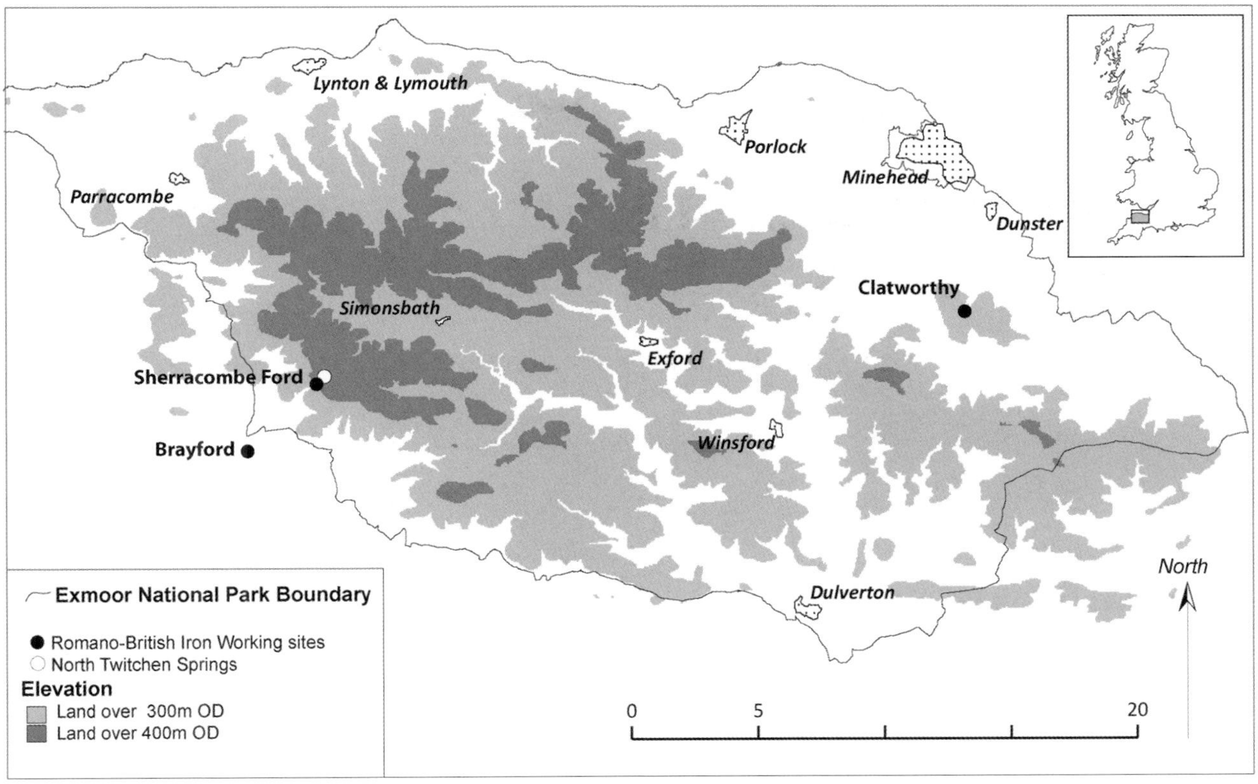

Figure 1: General location map indicating known iron working sites in Exmoor and sites mentioned in the text.

Iron working at Sherracombe Ford, Exmoor

Sherracombe Ford: nature of activities

The site of Sherracombe Ford lies on the southern fringe of Exmoor, in the parish of High Bray (British National Grid reference SS720336). This area sees the transition from high moorland to more sheltered wooded valleys and enclosed farmland. The most obvious features on the site itself are two substantial heaps of smelting waste, visible now as grassy mounds, lying close to the stream at the bottom of the valley of Sherracombe (Fig. 2). The heaps form level platforms, which were also cut back into the hillslope. While the slag heaps had been known and recorded in the local HER (Historic Environment Record) as probably medieval on the basis of their size, the full scale of the site was not recognised until 1996, when the Exmoor National Park Authority (ENPA) carried out a qualitative assessment survey of early iron working evidence in its area of responsibility. Two charcoal samples taken from exposed sections of the slag heaps at that time returned radiocarbon dates that indicated for the first time that smelting at the site took place far earlier, in the Late Iron Age – Roman period (Riley and Wilson-North, 2001). Between 2001 and 2003, the Exmoor Iron project, funded by English Heritage and operated through ENPA, the National Trust and Exeter University, undertook two seasons of excavation at the site, concentrating primarily on one of the two major slag heaps with its associated working platform. The objectives were to investigate the creation of the working platform and its use, and to examine the nature

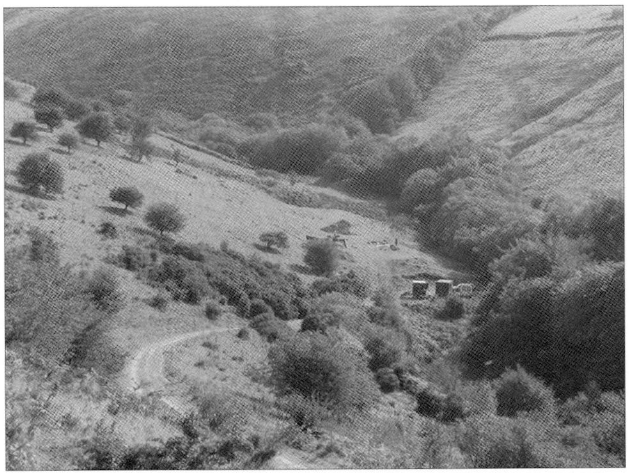

Figure 2: General overview of Sherracombe Ford during the 2001 excavation season.

of the technological processes that were in operation at the site. The strategies adopted included open-area excavation to examine spatial and functional relationships, vertical sections traversing the platform and the slag heap to examine stratigraphy and site chronology, and quantitative sampling to characterise technologies and processes.

Although the results of the fieldwork have yet to be published in full, the excavations revealed four major phases of activity. While the earliest excavated evidence for smelting occurs in Phase 1 and the remains of a furnace survived from Phase 2, the bulk of the evidence derives from Phase 3 and indicates a major intensification of activity. On the working platform, a coherent unit comprising a well-preserved, large furnace setting, preserving evidence of at least 10 episodes of reconstruction and use was set within a stone-revetted hollow and connected by means of a heavily trampled ramp to an extensive smithing floor. The latter represents an exceptional feature which in places reaches 0.4m in depth of solid compacted residues formed during the primary smithing of hot iron blooms newly extracted from the adjoining furnace. The unwanted slag and waste materials from the processes operating on the platform were dumped downslope on the growing slag heap. Excavation through these deposits revealed the cyclic nature of deposition and, through the application of a newly devised smelting waste classification scheme (Bray, 2007), this material could be sub-divided into 3 episodes composed of 11 phases, each characterised by a particular pattern of activity and distinguishable by the re-organisation of the working area, identifiable by cuts in the sequence or specific types of deposits.

Scale of iron production and fuel requirements

The scale of iron production and consequent impact on the local environment at Sherracombe Ford are based on estimates derived from a modified version of the methodology of Cleere (1976). The process is a 10-stage linear calculation (Fig. 3; Bray, 2007) that results in estimates of the volumes of fuel and woodland required for total on-site activity (stage 6); the iron ore requirements (stage 7); smithing yield (stage 8); and time and personnel requirements for production (stages 9 and 10). The focus of this paper is the impact that the metallurgical activities are likely to have had on the surrounding environment, and as such discussion will revolve around stage 6 of the process (estimates of fuel and woodland requirements). Development of the estimates relies on a range of production parameters that have been derived from excavation evidence, experimental research, and previous work (Table 1). The variables that influence

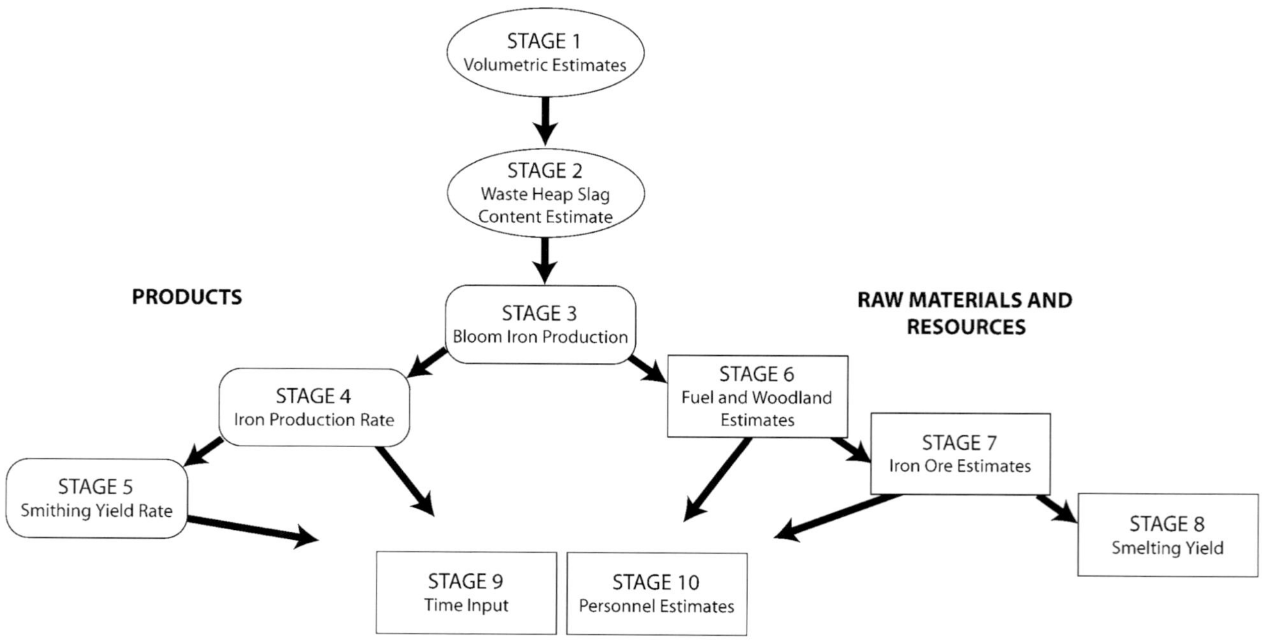

Figure 3: Ten stage linear process for calculating iron production and resource requirements (Bray, 2007).

Table 1: Parameters used in iron production estimates. Stages refer to the processes shown in Figure 3.

Stage	Description	Values	Source
2	Average waste heap density	1382 kg m^{-3}	Sherracombe Ford excavation data
2	Estimated average slag content	35.2% by weight	Sherracombe Ford excavation data
3	Bloom iron production; slag:bloom ratio	2.5 – 4:1 3:1 average	Cleere, 1971; Joosten et al., 1998; Thomas and Young, 1999
5	Iron lost during smithing	45-65% of original bloom 55% median value	Crew, 1991; Juleff, 1998; Sim, 1998; Saunder and Williams, 2002.
6	Smelting fuel charcoal:bloom ratio	9 – 16:1	Cleere, 1971; Crew, 1991; Juleff, 1998, Rehder, 2000
6	Smithing fuel charcoal:iron ratio	4 – 6:1	Crew, 1991; Sim, 1998; Saunders and Williams, 2002
6	Wood requirements wood:charcoal ratio	7 – 10:1	Cleere, 1976; Rehder, 2000
6	Woodland requirements: clear-felling	250 t ha^{-1}	Cleere, 1976; Rehder, 2000
6	Woodland requirements: new growth per year	1.6 kg m^{-2}	Whittaker and Woodwell, 1971
7	Iron ore requirements: charcoal:ore ratio	1 – 3:1	Cleere, 1971; Crew, 1991; Joosten, et al., 1997; Juleff, 1998; Saunder and Williams, 2002

the iron production process are diverse and often difficult to quantify. This makes attempts to calculate them prone to a degree of uncertainty. In addition, modern experiments that reduce some of this ambiguity also introduce another layer of uncertainty in the form of additional sources of variability, such as the relative skill levels of practitioners and differing methodologies and research agendas. Thus, it is unwise, in most cases, to assign single, definite values to any of the parameters involved in the iron production process. Instead, it is more advisable to obtain a range within which those parameters fall. The result is a calculation of iron production that is accurate perhaps to within an order of magnitude, but should only be taken to represent a minimum value, as smelting waste may have been removed from the site at any stage during the past.

Estimates of the volume of the waste heaps at Sherracombe Ford are based on extrapolation of excavation data and detailed surface survey of the site, and full details of these calculations can be found in Bray (2007); calculations based on these estimates follow the parameters given in Table 1. Waste heap A has been estimated to have a volume of 1032m^3, and waste heap B a volume of 601m^3, giving a total smelting waste of 1633m^3 for the entire site. The average waste heap density and slag content is derived from excavation data, giving a total smelting waste of 2257t, of which 794t is slag (the remainder consists of a mixture of furnace wall debris, charcoal, stone, and iron ore). The values of bloom iron production at the site are between 199 and 318t; 265t if based on the 'average' rates. If the published values for the proportion of iron lost during experimental smithing are applied to these figures, then the total iron production at Sherracombe Ford will be in the range of 70-175t iron, with the 'average' value at 119t. This is a moderate level of production when compared with other Roman production sites outside the region. At Beauport Park in the Weald (the most important iron producing area in Britain during this period), for example, Cleere (1976) estimates that a total of c. 60,000 tonnes of smelting waste were generated, of which perhaps 33,000 tonnes was slag (Bray, 2007, Table 5.23). Within the Southwest, though, production at Sherracombe Ford is large. It is one of only three sites of this magnitude currently known, with the other two also located in the Exmoor region (at Brayford and Clatworthy Reservoir; Riley and Wilson-North, 2001).

Estimation of the amount of fuel required to sustain this weight of production is based on experimental work. The weight of charcoal required for smelting falls within a range of 1790-5090t, with an average figure of 3310t based on the 'average' bloom production. Smithing activities will have required between 800 and 1810t (with an average of 1330t), meaning that the total charcoal requirements for the full life span of the site fall between 2590 and 7000t (with an average of 4640t). The ratio of wood to charcoal for charcoal production derived from experimental work give a requirement of 18,130-70,000t (average 41,190t). If this timber supply came from clear-felling, then, based on published data, between 75 and 285 hectares (average values suggest 165 hectares) of woodland were required to operate the site. Analysis of charcoal from the excavations strongly suggests woodland management in the production of the charcoal that was used at Sherracombe Ford (Gale, 2005). Whittaker and Woodwell (1971) suggest that new growth of timber in deciduous woodland is around 1.6kg m^{-2} (16t ha^{-1}), giving a total new growth area of 1130-4380ha over the life span of the site. The spatial extent of this activity, and thus its wider environmental impact, is strongly dependent on the duration of use of the site, which is discussed below. A shorter life span would have required a greater area of woodland for either clear-felling or sustained management.

The chronology of Sherracombe Ford

There are two main lines of evidence for the duration of activities at Sherracombe Ford. On-site artefactual evidence and a radiocarbon dating programme, based on material from the furnaces, features associated with the smithing floor, and charcoal from within the slag heaps interpreted as fuel from the furnaces, provide a secure understanding of when activity was taking place, although there are methodological limitations to identifying early activity (see discussion below). Off-site geochemical archives provide a second, independent check on the duration of activities at the site by preserving a stratigraphic record of atmospheric pollutants associated with smelting activities. These two data sets should be considered complementary, and are dealt with in turn here.

On-site artefactual evidence comes from the recovery of pottery within the waste heaps at the sites. Four main fabrics have been recovered from Sherracombe Ford: Black Burnished Ware (southeast Dorset type), Greywares that are probably locally produced, South Devon ware, Samian sherds, and miscellaneous coursewares. The date range of this material falls broadly into the late 2nd and the 3rd centuries AD, although it is possible that they all appeared in the early 3rd century, as certain diagnostic fabrics are missing. Radiocarbon dating of short-lived charcoal from the two slag heaps, furnaces and features associated with the smithing floor offer a second opportunity to refine the chronological framework of the metallurgic activities at Sherracombe Ford (Table 2; Fig. 4). The calibrations of the results, relating the radiocarbon measurements directly to calendar dates, are given in Table 2 and Fig. 4. All have been calculated using the calibration curve of Reimer et al. (2009) and the computer program OxCal (v4.1) (Bronk Ramsey, 1995, 1998, 2001, 2009). The calibrated date ranges cited in the table are those for 95% confidence. They are quoted in the form recommended by Mook (1986), with the end points rounded outwards to 10 years. The ranges in Table 2 have been calculated according to the maximum intercept method (Stuiver and Reimer, 1986) and those in Fig. 4 are derived from the probability method (Stuiver and Reimer, 1993). A Bayesian approach has been adopted for the interpretation of the chronology from the on-site archaeological evidence (Buck, et al., 1996). The dates of this activity can be estimated not only using the scientific dating information from the radiocarbon dates, but also by using the stratigraphic relationships between samples.

The technique used is a form of Markov Chain Monte Carlo sampling, and has been applied using the program OxCal v4.1 (http://c14.arch.ox.ac.uk/). Details of the algorithms employed by this program are available from the on-line manual or in the publications of Bronk Ramsey (1995, 1998, 2001, 2009).

Table 2: Radiocarbon samples, material and ages from Sherracombe Ford used within the chronological model (Fig. 4).

Laboratory code	Sample reference	Material and context	$\delta^{13}C$ (‰)	Radiocarbon age BP	Calibrated date (95% confidence)
OxA-12602	SF02-99	Charcoal, *Corylus* sp. roundwood from [99], the middle of the slagheap excavated in 2002	-25.6	1862 ± 30	70-240 cal AD
GU-5942	SF02-99-13	Charcoal, *Quercus* sp. roundwood from [99], the middle of the slagheap excavated in 2002	-23.0	2000 ± 50	170 cal BC – 130 cal AD
OxA-12603	SF02-63	Charcoal, *Quercus* sp. roundwood from [63], near the bottom of the slag heap excavated in 2002	-24.3	1949 ± 31	40 cal BC – 130 cal AD
GU-5941	SF02-63	Charcoal, *Quercus* sp roundwood from [63], near the bottom of the slag heap excavated in 2002	-24.2	1880 ± 70	40 cal BC – 330 cal AD
OxA-12604	SF02-24-26A	Charcoal, *Corylus* from [24], within the truncated base of a furnace	-24.8	1845 ± 31	70-250 cal AD
OxA-12605	SF02-24-26B	Charcoal, *Corylus* from [24], within the truncated base of a furnace	-24.6	1818 ± 31	90-320 cal AD
OxA-12606	SF02-84A	Charcoal, *Fraxinus* from [84], near the top of the slag heap excavated in 2002	-23.3	1883 ± 31	50-230 cal AD
OxA-12607	SF02-84B	Charcoal, *Fraxinus* from [84], near the top of the slag heap excavated in 2002	-24.3	1859 ± 30	70-240 cal AD
Beta-98972		Charcoal, unidentified from the slag heap excavated in 2002		2000 ± 50	170 cal BC – 130 cal AD
OxA-13773	505A	Charcoal, *Betula* sp. from a discrete dump of waste material within shallow pit dug into re-deposited natural shillet that makes up platform	-24.7	1876 ± 29	60-230 cal AD
OxA-13774	505B	Charcoal, *Corylus* sp. from a discrete dump of waste material within shallow pit dug into re-deposited natural shillet that makes up platform	-24.4	1916 ± 28	20-140 cal AD
OxA-13775	248A	Charcoal, *Corylus* sp. embedded in the final phase of furnace setting [327]	-27.0	1698 ± 29	250-420 cal AD
OxA-13776	248B	Charcoal, *Alnus* sp. embedded in the final phase of furnace setting [327]	-27.9	1731 ± 27	230-400 cal AD
OxA-13777	238A	Charcoal, *Betula* sp. roundwood (diam. 25 mm) from [238], the fill of circular feature [239] within a smithing floor	-26.5	1785 ± 30	130-340 cal AD
OxA-13891	238B	Charcoal, Pomoideae sp. roudnwood (diam. 25 mm) from [238], the fill of circular feature [239] within a smithing floor	-25.3	1810 ± 26	70-240 cal AD
OxA-13781	222A	Charcoal, *Alnus* sp. from [222], the fill of post-hole [223]. [223] is thought to be for a post upon which an anvil was set	-25.3	1818 ± 28	120-320 cal AD
OxA-13885	222B	Charcoal, *Corylus* sp. from [222], the fill of post-hole [223]. [223] is thought to be for a post upon which an anvil was set	-26.6	1863 ± 28	70-240 cal AD
OxA-13884	343A	Charcoal, *Quercus* sp roundwood from [343], a charcoal rich deposit sealed in a well-stratified sequence below the smithing floor	-25.8	1851 ± 26	80-240 cal AD
OxA-13778	706	Charcoal, *Quercus* sp. sapwood from [706], part of slag heap B, stratigraphically below (710)	-23.8	1825 ± 30	80-320 cal AD
OxA-13779	710	Charcoal, *Quercus* sp. sapwood from [710], part of slag heap B, stratigraphically above (706)	-22.8	1825 ± 28	90-260 cal AD
OxA-13780	701	Charcoal, *Corylus* sp. sapwood from [701], the basal context of slag heap B	-24.1	1893 ± 30	50-220 cal AD
Beta-132448		Charcoal, unidentified from slag heap B		2030 ± 50	180 cal BC – 80 cal AD

The algorithm used in the model can be derived from the structure shown in Fig. 4.

The chronological (Fig. 4) model shows good agreement between the radiocarbon dates and stratigraphy (Amodel = 90%) and provides estimates for the start of ironworking activity of 20 cal BC – 105 cal AD (95% probability; start_sherracombe; Fig. 4) or 20-75 cal AD (68% probability), and its end of 245-360 cal AD (95% probability; end_sherracombe; Fig. 4) or 260-350 cal AD (68% probability). Furthermore, the iron working activity is estimated to have lasted for 150-400 years (95% probability; Fig. 4a), and probably for 180-250 years (68% probability).

The radiocarbon evidence is not in agreement with the dating of the site suggested by the ceramics, revealing a longer duration of activities than is represented within the material culture. It is not clear to what extent this disagreement is caused by issues with both the excavation of a complex iron working site, and the taphonomic processes that create the

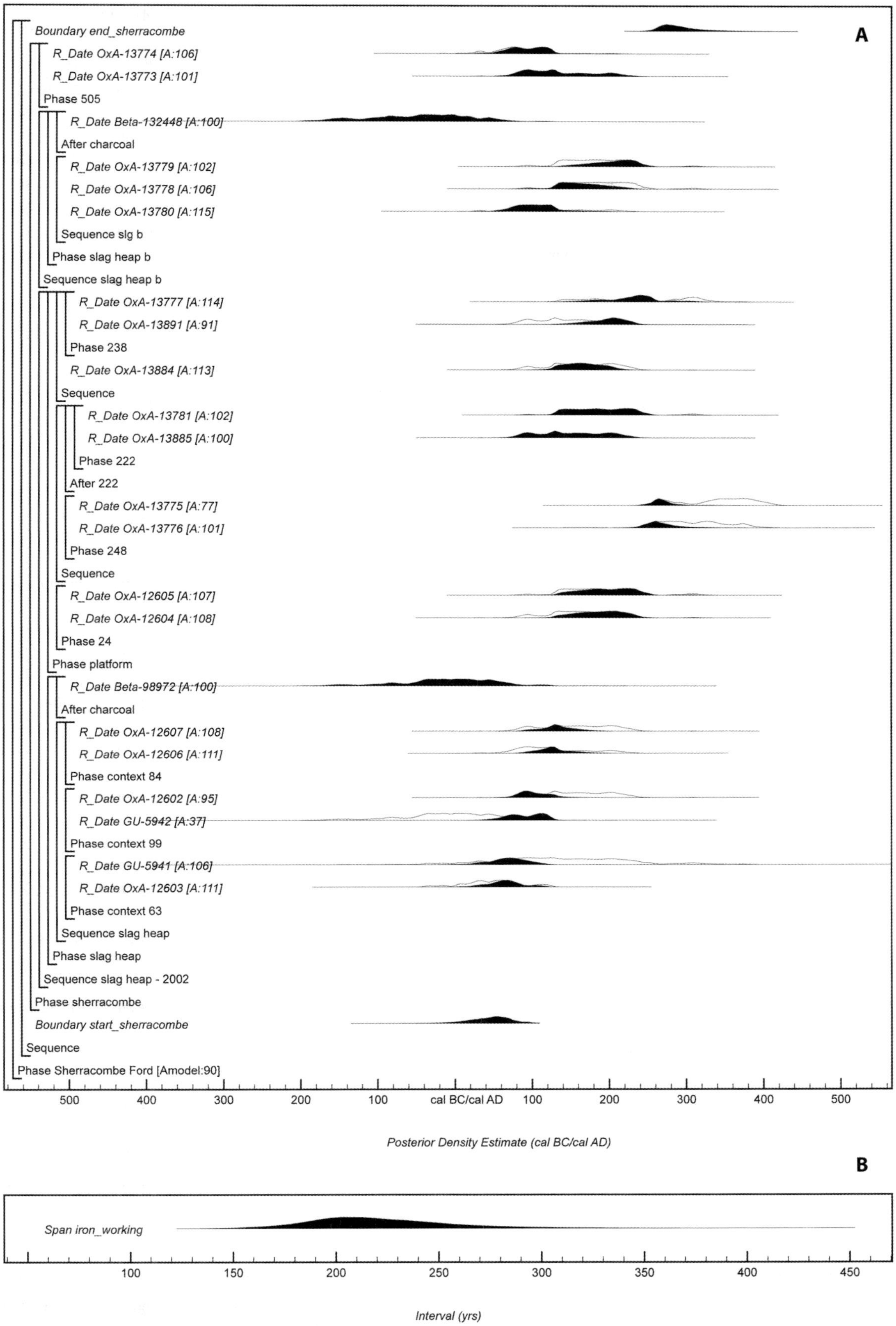

Figure 4: (a) Probability distributions of dates from Sherracombe Ford. For each of the radiocarbon dates, two distributions are plotted. The outline is the result of simple calibration; the solid based on the chronological model used. Distributions other than those relating to particular samples correspond to aspects of the model (e.g., *start_Sherracombe_Ford* is the estimated date for the start of iron working activity at the site). Figures in brackets after the laboratory numbers are the individual indices of agreement which provide an indication of the consistency of the radiocarbon dates with the prior information included in the model (Bronk Ramsey, 1995). The large square brackets down the left hand side along with the OxCal keywords define the model exactly. (b) Probability distribution for the length of time iron working activity took place at Sherracombe Ford, derived from the model in (a).

archaeological evidence itself. In the course of the working life of a site such as Sherracombe Ford, operational activities such as clearing and levelling of the working area are destructive to earlier remains. The locations of earlier furnaces are destroyed as the working area is cleaned and re-configured for new smelting and smithing installations. From a methodological perspective, it is often difficult to locate the earliest deposited material within a substantial heap of accumulated smelting debris at a production site. Individual smelting campaigns produce relatively small deposits of debris. These are discarded sequentially, both laterally and vertically, and without near-total excavation there is no guarantee that the earliest material can be identified. However, the excavations in 2002 and 2003 demonstrated that the slag heaps had coherent stratigraphy, and material for radiocarbon dating was recovered from the basal deposits. Thus, the radiocarbon dates provide the best indication of the period during which smelting activities were certainly taking place, rather than the ceramic chronologies.

The problem of destruction of the earlier record and partial recovery of evidence is not an issue in the construction of continuous geochemical archives from off-site peat sequences. Off-site geochemical analysis from sedimentary archives offers independent evidence for the timing and intensity of production based on the assumption that atmospheric pollutants associated with smelting are deposited as part of the archive (Görres and Frenzel, 1997). This approach has been employed successfully for examining the chronology and impact of early mining from a range of locations within Europe (Kempter and Frenzel, 2000; Mighall, et al., 2002; Jouffroy-Bapicot, et al., 2007). Enhanced deposition of heavy metal, and in particular lead (Pb), are thought to be reliable indicators of industrial activity within the catchment of a sequence. However, lead can be enhanced both through atmospheric deposition of pollutants, and also through the in-wash of mineral material from terrestrial sources. Concentrations of lithogenic elements such as titanium (Ti) or zirconium (Zr), whose sources would only be from terrestrial pathways, can be used to effectively remove the signal of terrestrial sources by developing ratios between Pb and the lithogenic element. Terrestrial enhancement of Pb and the lithogenic element should take place at comparable levels, and thus deviations in the ratio indicate enhanced atmospheric deposition (Shotyk, 2002). One of the additional advantages of developing independent sedimentary archives of pollutants associated with smelting is that the impact of that activity on the surrounding environment, and in particular woodland, can be inferred from the same sedimentary sequences, removing any difficulties with correlation or chronological matching.

The closest sedimentary sequence to Sherracombe Ford is a small mire towards the head of the valley within which the iron working site is located, at a distance of 600m from the smelting activities (Fig. 1). The mire, known as North Twitchen Springs, preserves just over 1m depth of peat, which has been demonstrated through radiocarbon dating to cover the last four millennia, with accumulation starting in the Early Bronze Age and continuing to recent times (Fyfe, 2012). A geochemical record has been developed for the last 2500 years from the upper part of the sequence, using contiguous 1cm samples

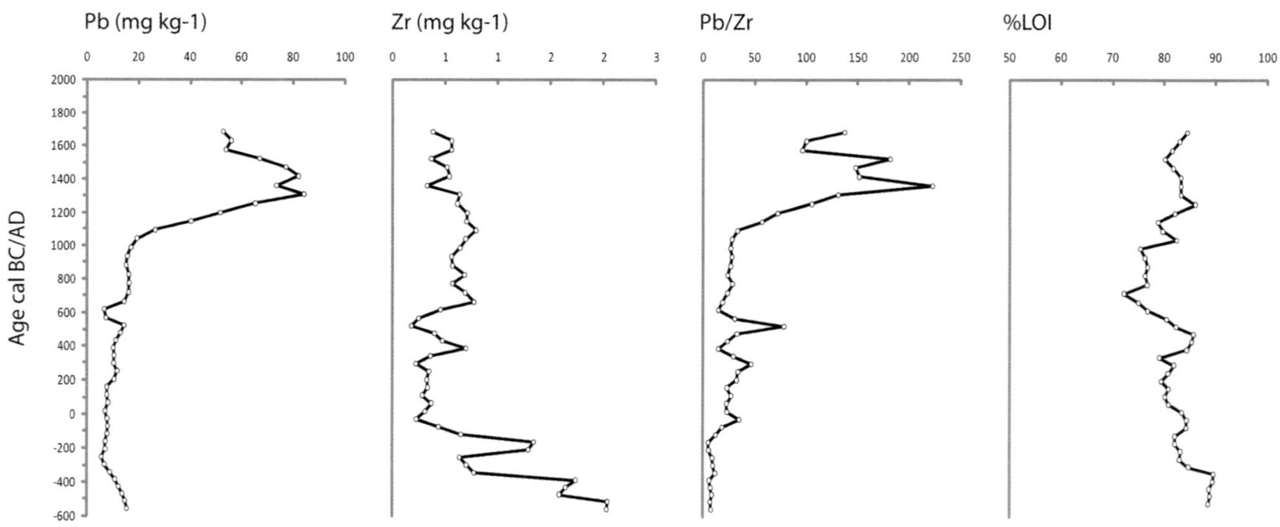

Figure 5: Geochemical data from North Twitchen Springs shown on a calibrated age time scale.

Table 3: Radiocarbon samples and ages from North Twitchen Springs.

Laboratory Code	Depth (cm)	$\delta^{13}C$ (‰)	Radiocarbon age BP	Calibrated date (95% confidence)
Beta-202087	24-26	-27.7	1430 ± 40	560-670 cal AD
Beta -202086	44-46	-27.6	2260 ± 40	400-200 cal BC
Beta -202085	64-66	-27.7	2950 ± 40	1290-1020 cal BC
Beta -202084	84-86	-26.6	3380 ± 40	1750-1540 cal BC
Beta -202083	104-106	-27.9	3620 ± 40	2120-1890 cal BC

which were air dried, ground to powder, and subjected to a nitric and hydrochloric acid digestion (Hassan, et al., 2007). The Pb and Zr concentrations were measured using ICP-MS. The accuracy of the analysis was validated through the use of certified reference material (LGC6137). In addition, loss-on-ignition values have been measured using standard methods, which provide a proxy for inwash of soil at the sampling site. These data are shown in Fig. 5 on a chronological time scale, derived from the radiocarbon dates at the site (Table 3).

The measured quantities of pollutants from off-site sequences depend on three key factors: the strength of the original pollutant source, the distance between the source of pollutants and the depositional site, and the accumulation rate at the site. Examination of the Pb data alone reveals no significant enhancement of Pb at North Twitchen Springs through the Romano-British period at all; however, the ratio of Pb/Zr indicates a clear rise in atmospheric Pb deposition starting in the 1st century BC, reaching stable elevated values in the 1st century AD and declining at the end of the 3rd century AD. A second isolated peak occurs at the start of the 6th century AD, before values fall, rising again in the later medieval period. This suggests that industrial activities were at their peak in the local area for around 300 years, between the 1st and the 3rd centuries AD. There does not appear to be any significant dip in the deposition of atmospheric pollutants through this period, suggesting continuous activity at the temporal scale of the geochemical record (samples represent c. 50 years). The local source of these pollutants is most likely to be Sherracombe Ford, owing to the proximity of the site. The other known major smelting site, at Brayford, is located 4.5km to the southwest, and atmospheric pollutants from the site are likely to have been greatly diluted at that distance. The off-site geochemistry supports the indications from the pottery chronologies that smelting was taking place at Sherracombe during the 2nd and 3rd centuries AD; however, it also allows the start of activity to be pushed back to the 1st century AD. This is also in agreement with the estimated date for the start of ironworking activity (start_Sherracombe_Ford; Fig. 4).

The environmental impact of iron working at Sherracombe Ford

The temporal duration of metallurgical activities at Sherracombe Ford is central to the consideration of the impact of iron production on the local environment, in particular on woodland resources. According to the pottery chronologies, the activity lasted around 150 years; the geochemistry and radiocarbon dates suggest it may have been as long as 300 years (although the median value from the radiocarbon dates suggests 215 years). These durations can be used to estimate the rate of metal production, charcoal requirements, and the area of woodland required per annum (Table 1). Using the figures derived above, a short (150 years) period of activity would result in between 0.46 and 1.17t of iron produced each year, requiring 14-41t of charcoal, equivalent to 98-410t of wood. This could be achieved by either clear-felling between 0.4 and 1.6ha of woodland each year, or sustained management of between 6 and 26ha of woodland. A longer period of activity (300 years) would result in between 0.23 and 0.58t of iron produced each year, requiring 7-20t of charcoal, equivalent to 48-204t of wood, from either 0.2-0.8ha of cleared woodland or 3-13ha managed woodland.

These theoretical values for woodland requirements, either through clear-felling or sustained management, can be compared with empirical data on woodland cover and management from on-site charcoal data and off-site pollen analysis from North Twitchen Springs. On-site charcoal evidence shows that a range of species were used to generate charcoal at the site, but *Quercus* (oak) and *Corylus* (hazel) were used consistently for smelting, with oak being dominant (Gale, 2005). Evidence from the industrial areas demonstrated that charcoal was prepared from various categories of wood, including narrow coppice stems (< 25mm in diameter) and wider roundwood and largewood (from wide roundwood, cordwood, or trunkwood). Some of the oak preserved clear banding in the growth rings, which may reflect local management regimes, in particular coppicing. Oak trees within hazel coppice areas accelerate their growth as soon as the hazel is cut, but slow in response to increasing competition from the regenerating stools (Gale, 2005). Thus the oak banding may reflect a short hazel coppice cycle. The species used in the production of charcoal remained consistent throughout the stratigraphy of the waste heaps, suggesting continuity in fuel procurement.

Pollen analysis from the North Twitchen Springs, combined with an understanding of the general vegetation history of Exmoor from existing palaeoecological studies (e.g., Merryfield and Moore, 1974; Francis and Slater, 1990, 1992; Fyfe et al., 2003; Fyfe 2012), allows for a wider consideration of the impact of charcoal production within a broad environmental context. The existing understanding of the vegetation history of Exmoor reveals that the upland moorland area was cleared of all woodland by the late Iron Age (Francis and Slater, 1992), and probably significantly earlier (Merryfield and Moore, 1974; Francis and Slater, 1990), resulting in an open grass and heather moorland character that may be similar to the British uplands that can be seen today. Local differences in pollen assemblages led Fyfe et al. (2003) to argue that woodland was likely to have persisted in specific topographic contexts in parts of the Exmoor landscape, namely the steep-sided combes (valleys), and "would have been suitable for management of woodland as a resource, either for timber, domestic use or charcoal production for industrial uses" (Fyfe, et al., 2003, p.231). Further empirical modelling of vegetation patterns in Exmoor using recent advances in the understanding of pollen-vegetation relationships led Fyfe (2006) to confirm this view of the distribution of Exmoor's woodland in the historic period. It seems likely, then, that woodland resources were locally available, and probably played a key role in determining the location of the iron working activities at Sherracombe Ford.

The pollen analysis from North Twitchen Springs enables a woodland history to be established within the catchment of the site, including the lower valley and Sherracombe Ford. Samples from 2cm depth intervals were prepared using standard methods and identifications were made using keys in Moore et al. (1991) and a modern reference collection. For the purposes of discussion, only selected pollen taxa are presented here (Fig. 6). Full publication of the results can be found in

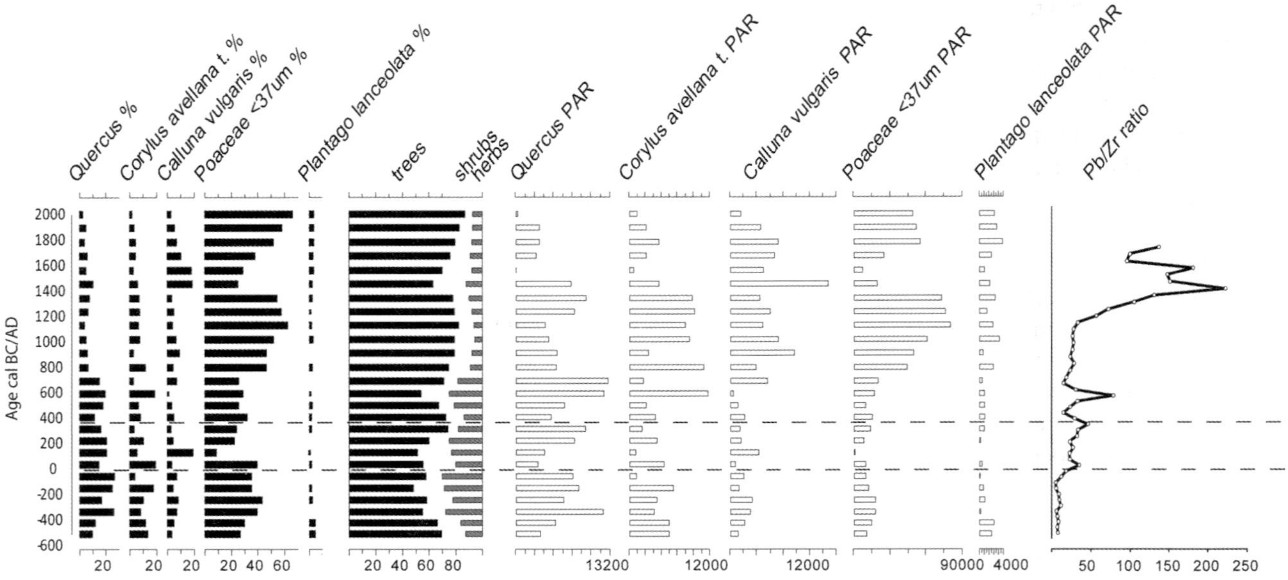

Figure 6: Simplified percentage and pollen accumulation rate diagram, North Twitchen Springs, with period of enhanced anthropogenic lead marked between dashed lines.

Fyfe (2012). The results are shown as both percentages of the total land pollen identified and pollen accumulation rates (PARs). PARs (the number of pollen grains per year) offer an absolute approach to representation of past vegetation; variable production of pollen between taxa means that the percentages of individual taxa may change in spite of no real change in the cover of the taxon in the past. This is of particular concern for woodland representation on upland margins, where changes from upland grass-heath to heather-heath may significantly alter woodland representation (Fyfe, 2006).

The summary pollen diagram (Fig. 6) indicates that there is no change in the overall representation of woodland at the start of the iron working phase indicated by the geochemistry (marked by dashed lines on Fig. 6), and that woodland representation actually increases through to the 4th century AD. Examination of the percentage records for oak and hazel indicate that oak broadly maintains its values, whilst hazel values are far more variable. This variability in hazel values might be indicative of management, if the pollen proportions represent phases of clearance and regeneration of coppice. The high values of grasses (Poaceae < 37 microns) before and after the iron working phase confirm the general pattern of a largely open upland above both Sherracombe Ford and North Twitchen Springs. The PAR data for oak is more revealing and demonstrates that the absolute influx of pollen in the 1st and 2nd centuries AD is roughly halved, which may indicate a significant clearance (or pollarding) of oak within the catchment of the site. The hazel PAR values show distinct fluctuations through the period of enhanced atmospheric Pb deposition, again possibly indicating coppice management. PARs for oak return to higher levels by the 3rd/4th century AD, suggesting recovery of oak woodland communities; the hazel values continue to fluctuate. Crucially, values for both oak and hazel are significantly higher than they are for the most recent level of the diagram, indicating that woodland was certainly more abundant in the valleys during the Romano-British period.

Interpretation of the pollen record from North Twitchen Springs requires some caution. The connection between iron working and reductions in oak and variable hazel indicative of coppicing is circumstantial, even in the light of strong support from the on-site charcoal evidence. There are a range of other possible factors that may have led to the reduction of woodland, including the use of wood for structural works or as domestic fuel. Agricultural intensification may account for woodland loss, although the very low representation of classic pastoral indicators such as ribwort plantain (*Plantago lanceolata*) probably rules this possibility out. On balance, though, it seems most probable that the patterns are a result of metal working.

Sustainable management of woodland for iron working

In summary, the evidence from off-site geochemistry seems to suggest that the iron working site at Sherracombe Ford may have operated over a period of up to 300 years from the early 1st century AD, although pottery from the extensive excavation of the site suggests a much shorter duration of activities, between the 2nd and 3rd centuries AD. Radiocarbon evidence supports the longer span of activities, but one which may be shorter than 300 years. The palynological evidence, combined with the on-site charcoal record, provides a persuasive argument for the initial clearance of some of the oak woodland and the coppicing of hazel for the production of charcoal for use within the smelting and smithing activities recorded at Sherracombe Ford. Returning to the estimations of the area required to produce sufficient charcoal through sustained management of the woodland resource discussed above, between 3 and 13ha of managed woodland would have sufficed to maintain production at Sherracombe Ford indefinitely based on the range of iron production levels derived from excavation and experimental work. This area of woodland could have easily been met within the local valley at Sherracombe. This implies a minimal overall impact on the broad environment for this scale of activities. This view echoes that of Mighall and Chamber (1997), who argued that the scale of impact around the metal working sites that they examined was localised, with little alteration of the composition of the local woodland. Other authors have taken opposite views, arguing for major modifications in plant cover during prehistoric and more recent phases of metallurgy with selective deforestation (Jouffrey-Bapicot, et al., 2007), or widespread clear-felling in advance of the establishment of coppicing cycles (Potts, 1986). In Exmoor, it is clear that Romano-British iron working took place in a landscape already substantially cleared of woodland and managed from later prehistory, according to evidence from both the palaeoecological and archaeological record (Riley and Wilson-North, 2001), meaning that the intensive exploitation of iron was not likely to have had the greatest impact on the local environment.

References

Bray, L.S., 2007. *The archaeology of iron production: Romano-British evidence from the Exmoor region*. PhD Thesis, University of Exeter.

Bronk Ramsey, C., 1995. Radiocarbon calibration and analysis of stratigraphy: the OxCal program. *Radiocarbon*, 37, pp.425-30.

Bronk Ramsey, C., 1998. Probability and dating. *Radiocarbon*, 40, pp.461-74.

Bronk Ramsey, C., 2001. Development of the radiocarbon calibration program OxCal. *Radiocarbon*, 43, pp.355-63.

Bronk Ramsey, C., 2009. Bayesian analysis of radiocarbon dates. *Radiocarbon*, 51, pp.337-60.

Brown, A.G., Bennett, J. and Rhodes, E., 2009. Roman mining on Exmoor: a geomorphological approach at Anstey's Combe, Dulverton. *Environmental Archaeology*, 14, pp.50-61.

Buck, C.E., Cavanagh, W.G. and Litton, C.D., 1996. *Bayesian Approach to Interpreting Archaeological Data*. Chichester: Wiley.

Cleere, H., 1971. Iron making in a Roman furnace. *Britannia*, 2, pp.203-17.

Cleere, H., 1976. Some operating parameters for Roman ironworks. *Bulletin of the Institute of Archaeology*, 13, pp.233-46.

Crew, P., 1991. The experimental production of bar iron. *Journal of the Historical Metallurgy Society*, 25, pp.21-36.

Francis, P.D. and Slater, D.S., 1990. A record of vegetation and land use change from upland peat deposits on Exmoor. Part 2: Hoar Moor. *Somerset Archaeology and Natural History Society Proceedings*, 134, pp.1-25.

Francis, P.D. and Slater, D.S., 1992. A record of vegetational and land use change from upland pear deposits on Exmoor. Part 3: Codsend Moors. *Somerset Archaeology and Natural History Society Proceedings*, 136, pp.9-28.

Fyfe, R.M., 2006. GIS and the application of a model of pollen deposition and dispersal: a new approach to testing landscape hypotheses using the POLLANDCAL models. *Journal of Archaeological Science*, 33, pp.483-93.

Fyfe, R.M., 2012. Bronze Age landscape dynamics: spatially detailed pollen analysis from a ceremonial complex. *Journal of Archaeological Science*, 39, pp.2764-2773.

Fyfe, R.M., Brown, A.G. and Rippon, S.J., 2003. Mid- to late-Holocene vegetation history of Greater Exmoor, UK: estimating the spatial extent of human-induced vegetation change. *Vegetation History and Archaeobotany*, 12, pp.215-32.

Gale, R., 2005. *Sherracombe Ford, Exmoor, 2002 and 2003: charcoal analysis*. Unpublished report for Exmoor Iron Project.

Görres, M. and Frenzel, B., 1997. Ash and metal concentrations in peat bogs as indicators of anthropogenic activity. *Water, Air, and Soil Pollution*, 100, pp.355-65.

Hassan, N.M., Rasmussen, P.E., Dabek-Zlotorzynska, E., Celo, V. and Chen, H., 2007. Analysis of environmental samples using microwave-assisted acid digestion and inductivity coupled plasma mass spectrometry: maximising total element recoveries. *Water, Air, and Soil Pollution*, 178, pp.232-334.

Joosten, I., Elburg, M.A., Jansen, B. and Kars, H., 1997. Calculation of the output of an early historical iron production site in the Veluwe Area, the Netherlands. In: G. de Boe and F. Verhaeghe eds. *Material Culture in Medieval Europe*. IAP Rapporten 7. Zellik: Instituut voor het Archeologisch Patrimonium, pp.43-51.

Joosten, I., Jansen, J.B.H. and Kars, H., 1998. Geochemistry and the past: estimation of the output of a Germanic iron production site in the Netherlands. *Journal of Geochemical Exploration*, 62, pp.129-37.

Jouffroy-Bapicot, I., Pulido, M., Baron, D., Galop, D., Monna, F., Lavoie, M., Ploquin, A., Petit, C., de Beaulieu, J.-L. and Richard, H., 2007. Environmental impact of early palaeometallurgy: pollen and geochemical analysis. *Vegetation History and Archaeobotany*, 16, pp.251-8.

Juleff, G., 1998. *Early Iron and Steel in Sri Lanka: A Study of the Samanalawewa Area*. Mainz: Verlag Philipp von Zabern.

Juleff, G. and Bray, L.S., 2007. Minerals, metal, colours and landscape: Exmoor's Roman Lode in the early Bronze Age. *Cambridge Archaeological Journal*, 17, pp.285-96.

Kemper, H. and Frenzel, B., 2000. The impact of early mining and smelting on the local tropospheric aerosol detected in ombrotrophic peat bogs in the Harz, Germany. *Water, Air, and Soil Pollution*, 121, pp.93-108.

Merryfield, D.L. and Moore, P.D., 1974. Prehistoric human activity and blanket peat initiation on Exmoor. *Nature*, 250, pp.439-41.

Mighall, T.M. and Chambers, F.M., 1997. Early ironworking and its impact on the environment: palaeoecological evidence from Bryn y Castell hillfort, Snowdonia, North Wales. *Proceedings of the Prehistoric Society*, 63, pp.199-220.

Mighall, T.M., Abrahams, P.W., Grattan, J.P., Hayes, D., Timberlake, S. and Forsyth, S., 2002. Geochemical evidence for atmospheric pollution derived from prehistoric copper mining at Copa Hill, Cwmystwyth, mid-Wales, UK. *The Science of the Total Environment*, 292, pp.69-80.

Mook, W.G., 1986. Business Meeting: recommendations/resolutions adopted by the twelfth international radiocarbon conference. *Radiocarbon*, 28, p.799.

Moore, P.D., Webb, J.A. and Collinson, M.E., 1991. *Pollen analysis*. Oxford: Blackwell.

Potts, R., 1986. Der pollenanalytische Nachweis extensiver Waldbewirtschaftungen in den Haubergen des Siegerlande. In: K.-E. Behre ed. *Anthropogenic indicators in pollen diagrams*. Rotterdam: AA Balkema, pp.124-34.

Rehder, J.E., 2000. *The mastery and use of fire in antiquity*. Quebec: McGill-Queen's University Press.

Reimer, P.J., Baillie, M.G.L., Bard, E., Bayliss, A., Beck, J.W., Blackwell, P.G., Bronk Ramsey, C., Buck, C.E., Burr, G., Edwards, R.L., Friedrich, M., Grootes, P.M., Guilderson, T.P., Hajdas, I., Heaton, T.J., Hogg, A.G., Hughen, K.A., Kaiser, K.F., Kromer, B., McCormac, F.G., Manning, S.W., Reimer, R.W., Richards, D.A., Southon, J.R., Talamo, S., Turney, C.S.M., van der Plicht, J. and Weyhenmeyer, C.E., 2009. IntCal09 and Marine09 radiocarbon age calibration curves, 0–50,000 years cal BP. *Radiocarbon*, 51, pp.1111-50.

Riley, H. and Wilson-North, R., 2001. *The field archaeology of Exmoor*. Swindon: English Heritage.

Saunder, L. and Williams, S., 2002. A practical treatise on the smelting and smithing of bloomery iron. *Journal of the Historical Metallurgy Society*, 36, pp.122-31.

Shotyk, W., 2002. The chronology of anthropogenic, atmospheric Pb deposition recorded by peat cores in three minerogenic peat deposits from Switzerland. *The Science of the Total Environment*, 292, pp.19-31.

Sim, D., 1998. *Beyond the bloom: bloom refining and iron artefact production in the Roman world*. British Archaeological Reports International Series 725. Oxford: Archaeopress.

Stuiver, M. and Reimer, P.J., 1986. A computer program for radiocarbon age calculation. *Radiocarbon*, 28, pp.1022-30.

Stuiver, M. and Reimer, P.J., 1993. Extended ^{14}C data base and revised CALIB 3.0 ^{14}C age calibration program. *Radiocarbon*, 35, pp.215-30.

Thomas, G.R. and Young, T.P., 1999. The determination of bloomery furnace mass balance and efficiency. In: A.M. Pollard ed. *Geoarchaeology: exploration, environments, resources*. Geological Society of London special publication 165. London: Geological Society, pp.155-64.

Whittaker, R.H. and Woodwell, G.M., 1971. Measurement of net primary productivity in forests. In: P. Duvigneaud ed. *Productivity of forest ecosystems*. Paris: UNESCO, pp.159-75.

Authors' addresses

- Ralph Fyfe* and Jessie Woodbridge
  School of Geography, Earth and Environmental Sciences, Plymouth University, Plymouth, PL4 8AA, UK
  *corresponding author: ralph.fyfe@plymouth.ac.uk, tel: 01752 585929
- Lee Bray and Gillian Juleff
  Department of Archaeology, University of Exeter, Department of Archaeology, Laver Building, North Park Road, Exeter, EX4 4QE, UK
- Peter Marshall
  English Heritage, 1 Waterhouse Square, 138-142 Holborn, London, EC1 2TN

The fuel supply and woodland management at a 14th century bloomery in Snowdonia: a multi-disciplinary approach

Peter Crew and Tim Mighall

ABSTRACT: The large amount of charcoal from the medieval bloomery at Llwyn Du has been studied in detail, revealing subtle changes in the proportion of the species used, primarily oak and birch, over some 70 years. Dendrological analysis of the oak charcoal shows that mainly branch wood was used, probably from lopping or pollarding. High resolution sampling of a peat monolith shows short-term fluctuations of the tree pollen and the microscopic charcoal. These changes can be correlated with the three phases of archaeological activity by a combination of ^{14}C and ^{210}Pb dating, together with a detailed study of the magnetic pollution preserved in the peat bog. The interpretation of the data suggests a careful use and management of the woodland, ensuring an adequate fuel supply for the bloomery. There was little overall impact on the woodland cover.

Keywords: Medieval, bloomery, iron, chronology, charcoal, pollen, woodland management.

Introduction

This paper examines the fuel supply and woodland management at a medieval bloomery in Snowdonia, north-west Wales. Llwyn Du is one of a cluster of 10 bloomeries in the Coed y Brenin area, on what was then Crown land, and there are two sites further north, at Dolbenmaen and Penmachno (Fig. 1). Crown rentals show that these bloomeries operated in the later 14th and the early 15th century (Smith 1995). Excavations carried out at Llwyn Du between 2001 and 2006 have provided the opportunity to reconstruct in some detail the production of iron over a 70 year period (Crew and Crew 2001; Crew and Charlton 2007). As part of this multi-disciplinary investigation, a palaeoenvironmental study of a monolith from a peat bog, located 50 metres from the site, has been undertaken in order to place the ironworking into its environmental context (Mighall et al. 2009). This paper presents an overview of the various strands of evidence from Llwyn Du, in order to identify the species, size and quantity of the wood and charcoal used at the bloomery, and to consider how the woodland was used and managed.

Llwyn Du is a well preserved site, with three slag dumps. Postholes suggest two phases of a large building with opposing entrances, which had a furnace at the south end and a refining hearth, with the location for a wooden anvil, in the north end (Fig. 1). The furnace had a clay and stone-cased structure, some 2 by 1.50m overall and 400mm internal diameter, with a long slag-tapping channel running into a pit. There was evidence for the position of the bellows, as a wide splay on the west side, and part of the blowing hole survived in situ (Crew and Charlton 2007).

The stratigraphy of the slag dumps and the structural evidence indicate three phases of activity. The slag dumps contain a large amount of charcoal, in the form of slag-free lenses. This indicates cyclic activity, but its periodicity cannot yet be defined. In Phase 3 the charcoal was processed and dumped in an area to the north-east, sealing a remarkable collection of medieval wood, mainly oak logs, but also including a cruck and a complete yoke (Crew and Crew 2001). The slag dumps have been carefully surveyed, giving a total volume of some 250 cubic metres. Measurement of several large samples gave a bulk density of about 1 ton per cubic metre, suggesting a total slag weight of 250 tons (Crew and Charlton 2007). This estimate is probably too high, due to the large proportion of charcoal in the slag dumps, and this question will be discussed later.

The Crown rentals indicate a theoretical level of activity at each bloomery, with iron being smelted 6 days a week for 48 weeks each year. The rent was paid in *duodena*, or six pieces of iron of 2lb each per week. However, irregular payments of the rent, with arrears being carried forward to the next annual audit, suggest that the activity was rather less intense. If the production to rent ratio was 6 times, this implies that about 50kg of raw bloom were made each week. Analysis of the ores and slags, combined with experimental work, provides a production model of 6 smelts per week, each processing some 36kg of ore in a 6 hour smelt, producing a bloom of 8.5kg, refined bar iron of 5.6kg and a total of 18kg of slag. Thus, the estimated 250 tons of slag would have resulted from some 14,000 smelts, equivalent to 280 smelts per year for 50 years (Crew and Charlton 2007). This would have required a considerable amount of charcoal, perhaps as much as 1,400 tons, or about 10,000 tons of wood, equivalent to 200 tons of wood per year.

One of the rentals, for 1391-1392, is for a bloomery on the land of the Abbey of Cymer, a local Cistercian foundation. This states that "the Queen's wood has been laid waste" and only a

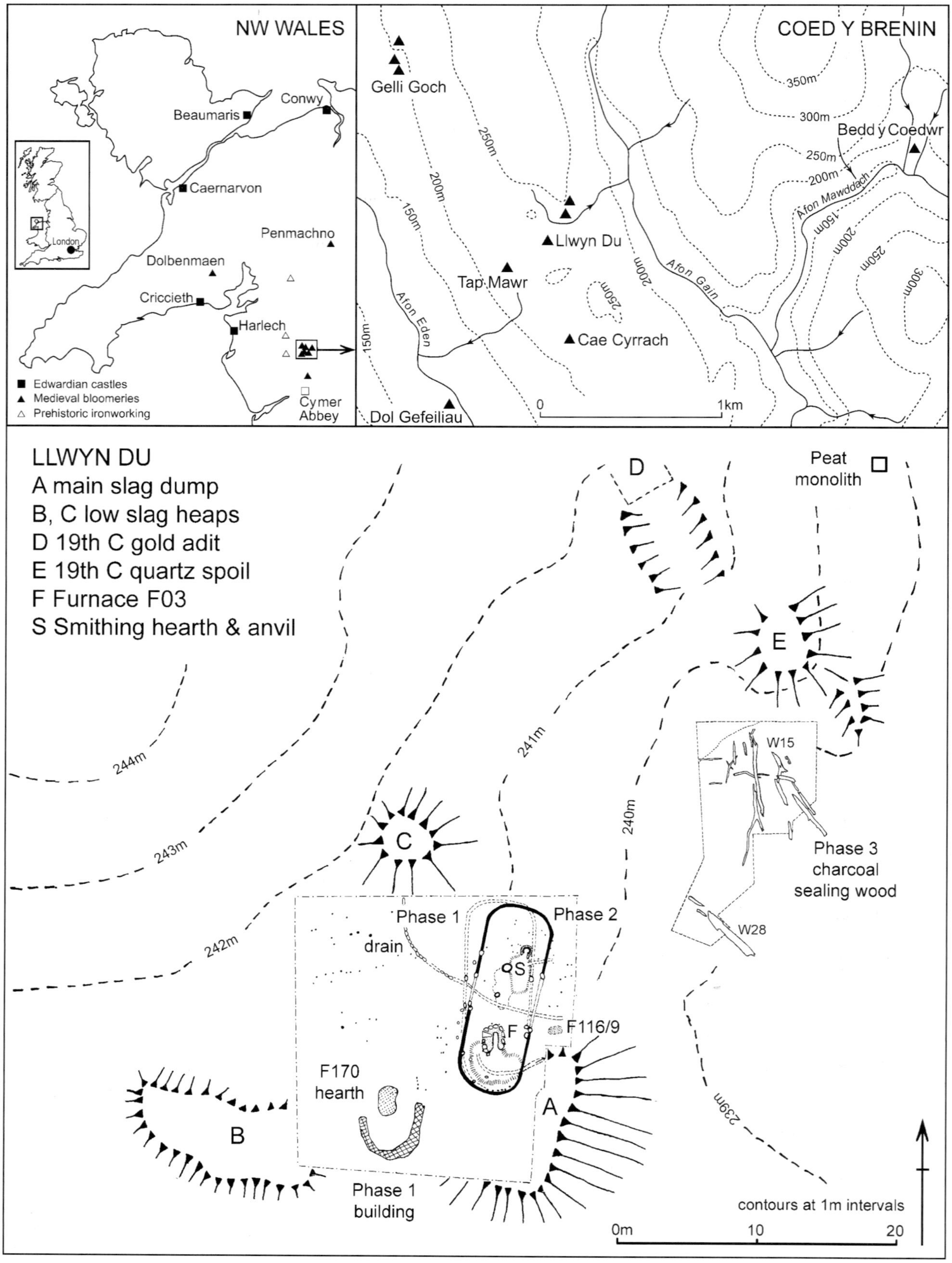

Figure 1: Map of north-west Wales, with medieval castles, medieval bloomeries, and prehistoric ironworking sites. Coed y Brenin area, with the bloomeries, contours at 50m intervals. The area west of Afon Gain was Crown land; between Afon Gain and Afon Mawddach was the grange of Cymer Abbey. Llwyn Du site plan, with excavation areas showing the Phase 1 and Phase 2 buildings, with postholes and stakeholes, the Phase 3 charcoal area sealing wood, and the peat monolith location, contours at 1m intervals.

half rent was due to the Crown, presumably for the ore (Smith 1995: 286-7). This suggests a move away from the Crown land to the bloomery at Bedd y Coedwr, which is located on the grange of Cymer Abbey (Fig. 1), to take advantage of a new woodland supply, with the other half rent presumably being paid to Cymer. The implication of this record is that there had been a significant woodland clearance, and this provided the original stimulus to examine the peat monolith, to see if this could be identified in the pollen record.

Chronology

Six radiocarbon dates were obtained from the last 10 rings of roundwood charcoal, from key stratigraphic contexts. Because of the kink in the calibration curve, all these dates give similar 2σ spans; however, as they are from a known sequence of contexts, a variable sequence model, using the constraints of the earliest and latest records, indicates that the three phases of activity occurred in the mid and later 14th century and the early 15th century.

This model can be refined using estimates for the last firing of the Phase 1 hearth (F170) and the Phase 2 furnace (F03), combining conventional archaeomagnetic dating with models of high resolution gradiometer surveys (Crew 2002). Dendrochronology can also be used. The wood gives a 166 ring site master, which does not date independently, but the charcoal deposits have samples with enough rings and bark to construct a relative sequence for the site. This can then be fitted to the chronology constructed from the radiometric dates, providing a consistent model for the site (Fig. 2).

This indicates that Phase 1 lasted from 1355 to 1365, Phase 2 from 1376 to 1392, and Phase 3 from 1407 to about 1419. The important feature of this model is that it gives estimates for the gaps between the phases of ten and fifteen years, respectively.

Charcoal species identification

Several tons of charcoal were excavated from a wide range of contexts. There were three main types of deposits: deep layers on the floor of the building, which had accumulated around the furnace during Phase 1; slag-free lenses within the slag dumps from all three phases; and the deep deposit to the north-east of the main site, interpreted as the Phase 3 charcoal processing and dumping area. Sieve analysis of several bulk samples showed that 80% of the charcoal (by weight) is less than 10mm and 98% less than 25mm. This indicates that the charcoal is dumped waste material and that the fraction which was larger than 25mm was preferred for smelting. In addition, there was a discrete dump (F116/9) in the entrance area of the building, which is interpreted as the last load of unused smelting charcoal from the very end of Phase 2. This weighed 4kg and had nearly 900 fragments of large charcoal, mostly larger than 25mm.

Over 5000 fragments of a size of 10mm or greater have been identified. The minimum reliable sample size was shown to be 100 pieces. Smaller samples were less representative, but even those with as few as 30 pieces gave a good indication of the proportions. This, however, applies only to an assemblage with two major taxa and, if there were more, the sample size would need to be larger (Davasse 1992).

Another consideration was the bias which might result from differential fragmentation. Several contexts had sufficiently large samples of the >25mm and 10-25mm size ranges for them to be analysed separately, and the results showed no significant differences. Bulk samples of small charcoal, less than 10mm in size, were also identified for the major taxa and this confirmed the proportions obtained from the larger fractions. The F116/9 deposit from the end of Phase 2 gave similar proportions to the other Phase 2 deposits, confirming the broadly representative nature of the samples from the processing waste.

The results show that 97% of all the charcoal was oak (*Quercus*) and birch (*Betula*), with only very small amounts of ash (*Fraxinus*), hazel (*Corylus*), alder (*Alnus*) and holly (*Ilex*), and very rare examples of sweet chestnut (*Castanea*) and *Pomoidiae* (Fig. 3). There are subtle variations in the proportions of oak and birch both during and between each phase. There is notably higher birch in the earliest deposits of Phase 1, but these are the least reliable data, with relatively small sample sizes. This contrasts with the Phase 1 slag lenses, which have a higher proportion of oak. In the Phase 2 slag

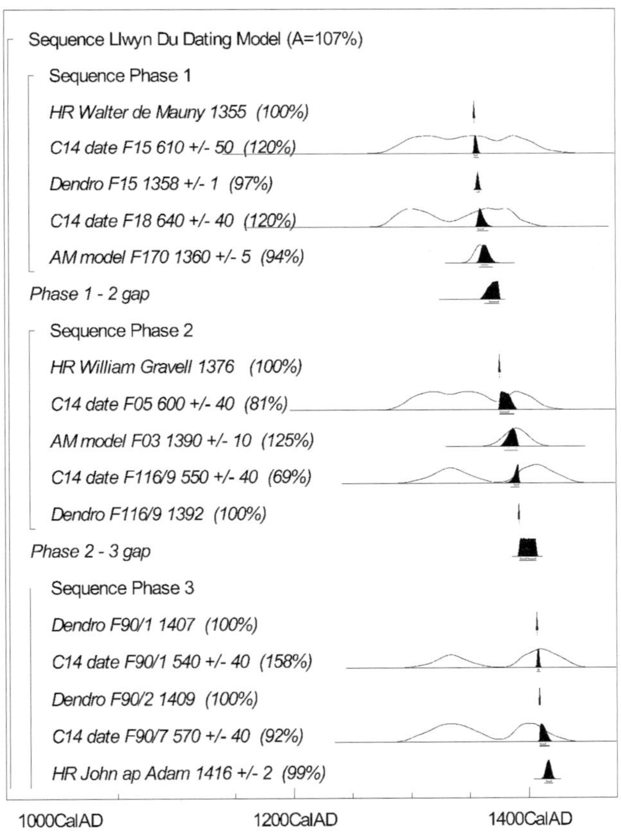

Figure 2: Llwyn Du chronology, calculated by a Bayesian model OXCAL v. 3.10 (Bronk Ramsay, 2001). 14C dates in outline are the calibrations of the individual radiocarbon results (Stuiver and Reimer, 1993). Solid distributions are probability density estimates of the calendar ages. HR dates are from the historical records, with the names of the lessees. Dendro dates are from the last 10 rings of roundwood charcoal samples with bark. AM models are estimates of the last dates of firing of the F170 hearth and the F03 furnace. The gaps between the phases are probability density estimates. Percentages in brackets are the agreement indices.

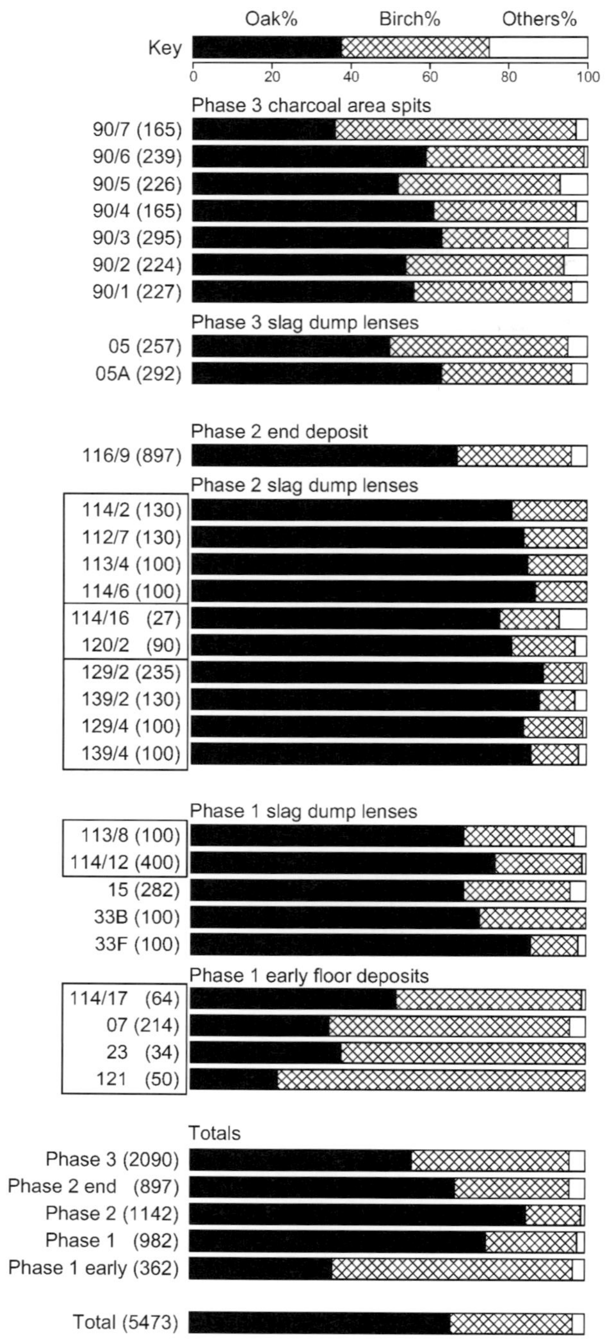

Figure 3: Llwyn Du charcoal identifications by phase and context. The contexts are approximately in stratigraphic order. Boxed context numbers are contemporary. Numbers in brackets are the sample sizes.

dump lenses, oak is dominant but birch increases slightly towards the end of the phase, especially in the final F116/9 deposit. In Phase 3, there is a higher proportion of birch throughout, with the final spit having more birch than oak.

The changing proportions of the oak and birch charcoal probably reflect the varying availability of the species in different patches of woodland. The Phase 2 deposits might indicate that oak was preferred if it was more readily available. The F116/9 deposit clearly shows that the oak and birch charcoal were mixed and used together for smelting. The limited use of other species is probably a reflection of their availability, although they may have been used for other purposes. Few samples, all of small size, could be associated unequivocally with smithing deposits, and these did not indicate a preference for any species.

The increase in the proportion of birch at the end of each phase suggests that the supply of usable oak was in decline, and the gaps in the site activity might have been to allow woodland recovery whilst iron production possibly continued elsewhere. The first abandonment of about 10 years may have been due to a decline of the woodland resource, but it could also be due to the lack of a lessee willing to take on the ironworks. The gap towards the end of the 14th century seems more likely to be due to a declining oak supply, and its timing coincides with the 1391-1392 record mentioned above. The lack of ironworking in the early 15th century is undoubtedly due to the Glyndwr rebellion, which had a major impact on this area from 1399 to 1415 (Smith and Smith 2001: 86-96). Iron production seems to have resumed at Llwyn Du by 1407, lasting until about 1419, but the documentary evidence is incomplete.

The palaeoenvironmental evidence

A peat monolith 1m deep was taken from a bog located some 50 metres to the north-east of the bloomery. Radiocarbon dating and ^{210}Pb analysis placed the iron production between 13 and 9cm, and, to correlate the specific core depth with the ironworking, mineral magnetic measurements were carried out (Mighall, et al., 2009). Fig. 4 shows gradually rising profiles for $IRM_{(0.88T)}$ and HIRM between 20.5 and 7.5cm. Three peaks in $IRM_{(0.88T)}$ and HIRM occur from 12.5 to 12cm, from 11 to 10.5cm, and from 10 to 9.5cm, although the second peak in the $IRM_{(0.88T)}$ profile is suppressed. The sampling resolution for the magnetic measurements is too coarse and the chronology too imprecise for absolute confidence, but these peaks could be the result of iron production, correlating with the three phases of activity. The gradual increase in values in the uppermost part of the monolith in $IRM_{(0.88T)}$ most likely indicates increasing levels of historical industrial pollution (Oldfield, et al., 1978, 1981).

Pollen and microscopic charcoal analysis was undertaken on the entire peat monolith using standard procedures (Barber, 1976). Samples of 5mm thickness at intervals of 1-4cm were analysed down the whole monolith to provide a long temporal record of vegetation change. Fine resolution sampling was used for the time when the iron bloomery was operational, so that each sample represented as small a period of time as possible, approximately 10 years or less. To achieve this, contiguous 2mm thick samples were taken between 13 and 9cm to reconstruct detailed vegetation changes.

Fig. 4 shows selected pollen and microscopic charcoal data from the uppermost 20cm. A gradual but staggered decrease in total arboreal pollen occurs from 18cm to the surface. This suggests that the woodland began to decline before the bloomeries became operational (18-13cm). Birch is the most affected species, but other trees, including oak, actually increase in representation. In contrast, the total arboreal pollen remained fairly stable and fluctuated around 60-70% during the period when iron production occurred (between 13 and 9cm). This suggests that the ironworkers maintained the local woodland to ensure a

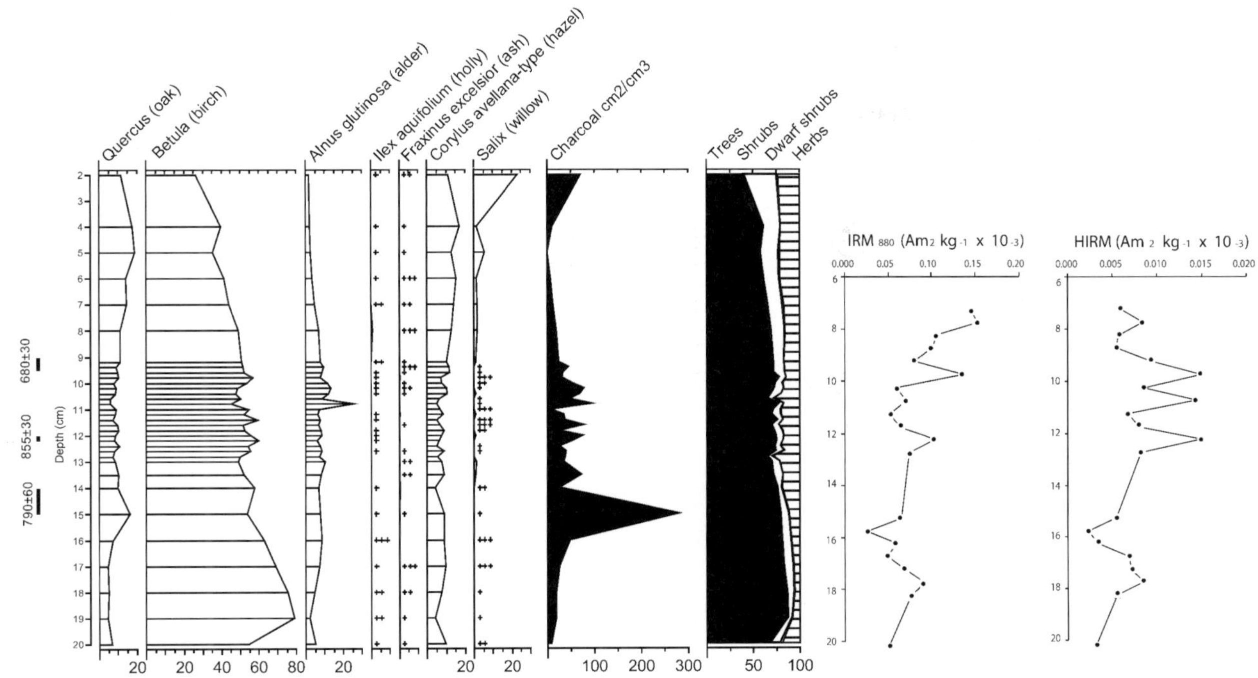

Figure 4: Selected percentages of pollen, microscopic charcoal, mineral magnetic data, and radiocarbon dates from the Llwyn Du monolith. At least 500 land pollen grains were counted for each sub-sample. Pollen was identified with the aid of keys (Fægri, et al., 1989; Moore, et al., 1991). Microscopic charcoal was counted using the method devised by Clark (1982). Mineral magnetic measurements after Mighall et al. (2009).

supply of charcoal. Birch and oak were the most abundant trees surrounding the site.

Pollen from all the major tree taxa show small-scale fluctuations between 13 and 9cm. The small decreases in oak and birch possibly represent phases of exploitation and the increases regeneration, which is consistent with woodland use and phased iron production. For example, oak decreased six times between 13 and 9.2cm. Birch has a clear decline between 12.2 and 11.8cm, and a relatively sizeable but staggered fall from 11.4 to 10.8cm. Four peaks in microscopic charcoal occurred which could have been generated by charcoal production, smelting and smithing (Fig. 4). There is some synchronicity between the declines of tree pollen, the presence of charcoal, and the peaks in mineral magnetic measurements. The first peak in HIRM and $IRM_{(0.88T)}$, between 12.5 and 12cm, coincided with a decrease in oak, alder and hazel, but birch pollen percentages increased. Birch only fell, and microscopic charcoal increased, between 12.2 and 12cm, and these patterns continued after the peaks in HIRM and $IRM_{(0.88T)}$ faded. The second peak in HIRM and, to a lesser extent, $IRM_{(0.88T)}$, from 11 to 10.5cm was coincident with lower percentages of oak, birch and hazel and a peak in microscopic charcoal. Although the tree taxa began to recover, they remained relatively low until the final peak in HIRM and $IRM_{(0.88T)}$, from 10 to 9.5cm, when microscopic charcoal also peaked again (Fig. 4). Following the abandonment of the bloomery, a slow decline in the overall tree pollen suggests that woodland clearance continued.

The evidence for woodland use and management

Recent work has shown that the simple measurement of the maximum diameter of the charcoal fragments can provide useful information to characterise a charcoal assemblage (Nelle 2002; Ludemann 2006). The diameter of each piece is estimated from the ring curvature and the ray angles. The data are presented as histograms in five diameter classes, giving a size distribution of the charcoal pieces. In addition, a nominal mean diameter (mD) can be calculated by the formula:

$$mD=(n_1 + n_2*2.5 + n_3*4 + n_4*7.5 + n_5*15)/N$$

where n_1 to n_5 are the number of fragments in each size class and N is the total number of fragments. Control data from charcoal burns with wood of known sizes shows, for example, that coppice wood gives a characteristic size distribution and mD values of 7 to 9cm (Nelle, 2002). Experimental work has demonstrated the sensitivity of the technique (Ludemann 2008), and Ludemann and Nelle (2002) have shown its value in reconstructing woodland cover and management over a large area of the Black Forest.

For Llwyn Du, over 1200 pieces of charcoal have been measured, most of which have outer rings or some sapwood, thus avoiding the difficulties which can arise with very fragmented samples (Dufraisse 2006). The size distribution of the oak charcoal and its high mD value show unequivocally that this does *not* derive from coppiced wood. The birch charcoal has a size distribution and mD value which would be more typical of coppiced wood (Fig. 5, upper two graphs). However, considering the nature of birch growth, with a dominant main stem and little usable branch wood, it is more likely that this size distribution is a result of using mainly main stem wood of relatively small diameter. This would be typical of relatively young birch trees from regeneration.

These data can be broken down by phase. The oak size distributions and mD values remain very similar through all three phases. The birch data vary more, with a higher proportion of large diameter charcoal in Phase 1 (mD 8.2) and smaller amounts in Phases 2 and 3 (mD 6.8 to 7.3).

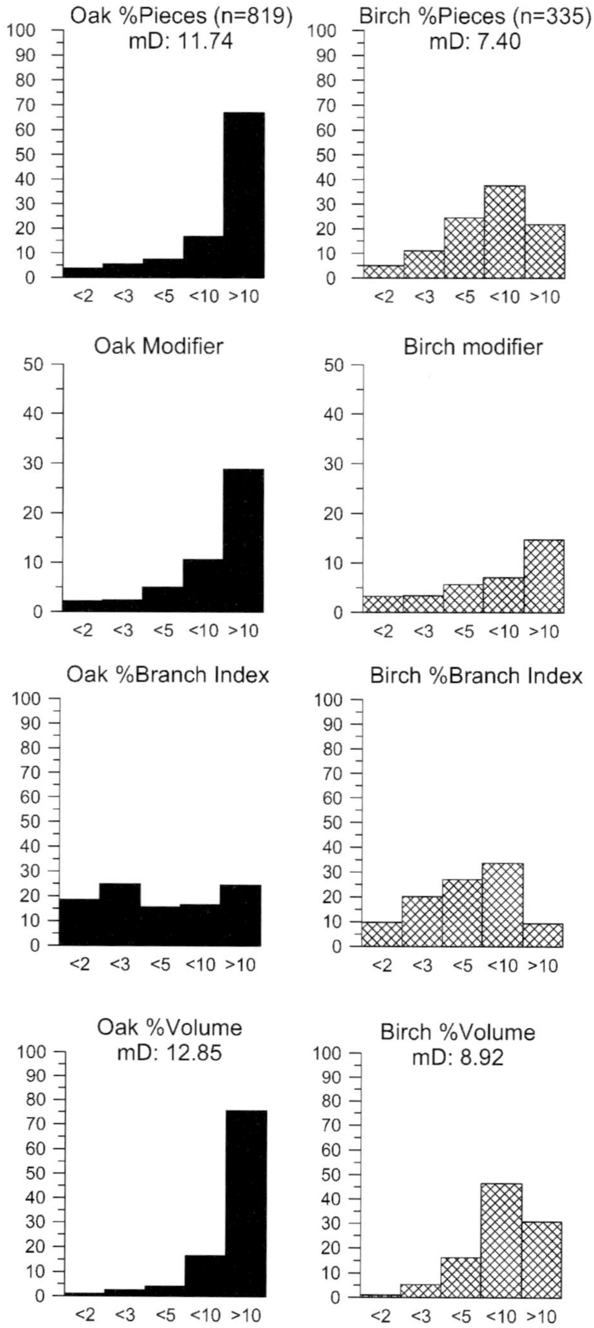

Figure 5: Llwyn Du charcoal, all phases, in diameter size classes (cm). Number of pieces, modified to branch index and to volume.

The size distribution by number of pieces has two limitations. Firstly, it assumes that each piece is of the same size and, secondly, it does not tell us anything about the sizes of the wood used, only of the charcoal fragments. The smaller diameter pieces tend to be complete, or nearly so, whereas the larger pieces are fragmented, sometimes into quite small pieces, resulting in an over-representation of the larger diameter material.

A technique has been developed to overcome this difficulty, at least in part. By measuring the maximum radius and tangent of each piece, simple trigonometry can be used to calculate a modifier using the formula:

$$m = 360/2\sin^{-1}(t/2r).$$

Thus, half and quarter sections give modifiers of 2 and 4 and, as the fragment size decreases, the modifier increases, reflecting the size of the pieces.

The number of pieces can then be divided by the total modifier for each size class, giving a new set of size distributions, which we have called the Branch Index. This is a potentially useful concept, indicating the proportion of the wood in each size class which had been carbonised (Fig. 5). The branch indices seem to indicate too much small diameter material, but, if a cross-section of the wood assemblage is drawn to scale, this shows that there is hardly enough small wood to construct a tightly packed clamp. This might suggest that the charcoal was made in pits, rather than above-ground clamps, which would require less covering material.

Finally, volume distributions per unit length can be calculated using the formula:

$$V = \Sigma[\pi(r_{max}^2 - r_{min}^2)/m].$$

This gives a size distribution of the number of pieces, weighted by the piece size (Fig. 5). The volume distributions are slightly different from, and the mD values are rather higher than, the piece distributions, but the overall pattern and relationships are similar, showing that the original piece distribution is a good approximation.

The rings have been counted and measured for about 800 pieces (Fig. 6). About 20% were complete branch wood, but only a small proportion of it is young wood with less than 20 rings, and a significant proportion of the oak has 30 to 60 rings. Incomplete pieces only give a minimum ring count, but this suggests that the oak is generally older wood, with some samples having over 100 rings, and that the birch is younger, the majority with less than 40 rings. Both the oak and birch have distributions with a positive skew. The complete pieces have a mean of 28.1 ± 11.5 for the oak and 24.4 ± 9.0 for the birch. The partial pieces have a mean of 36.9 ± 19.2 for the oak and 21.2 ± 10.4 for the birch, though these are necessarily minimum data. A tentative calculation of the number of missing early rings can be made by using a pith-distance estimate for each piece, divided by its mean ring width. This can give only an approximate estimate, as the early missing rings would certainly have been rather wider than the late rings. For the F116/9 oak, which is likely to give the most useful data, the calculated mean age is 87 ± 43 years, with nearly 40% of the pieces having more than 100 rings. Despite the inherent uncertainty of this calculation, it suggests that there was a significant proportion of mature oak available, even at the very end of Phase 2.

Fig. 6 also shows that most of the oak has a mean ring width of less than 1mm, suggesting slow growth in a closed canopy. Only a few pieces have a mean ring width over 1.6mm, but these are from samples with only a few rings. Surprisingly, the birch also has low mean ring widths, almost all being less than 1.6mm. This suggests that the birch is also from trees growing slowly in a closed canopy, rather than from regenerated trees, which would be expected to have a higher mean ring width.

It is self-evident that the smaller diameter charcoal is from branch wood, but the source of the larger diameter material needs to be established. Amongst the oak wood is a cruck (W15) which can be used to compare the growth of trunk and branch wood. The mean ring width of the trunk, with

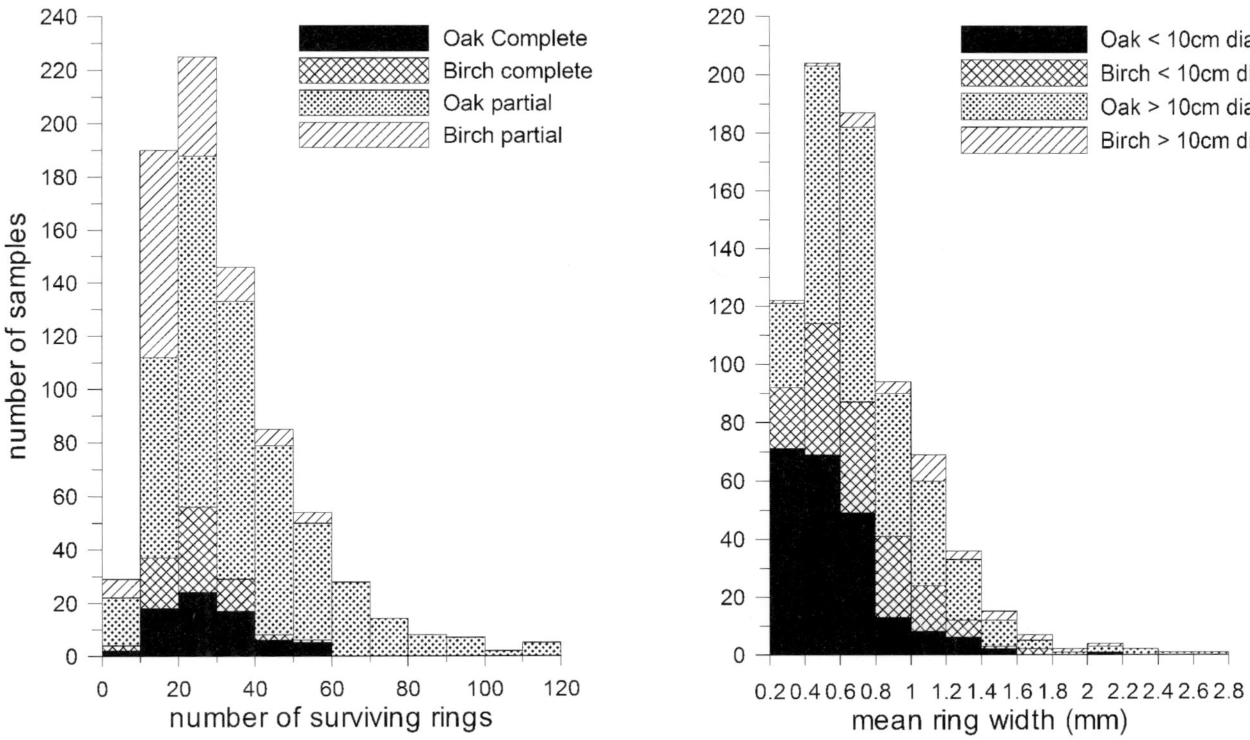

Figure 6: Llwyn Du charcoal: numbers of surviving rings (n = 837) and mean ring widths (n = 792).

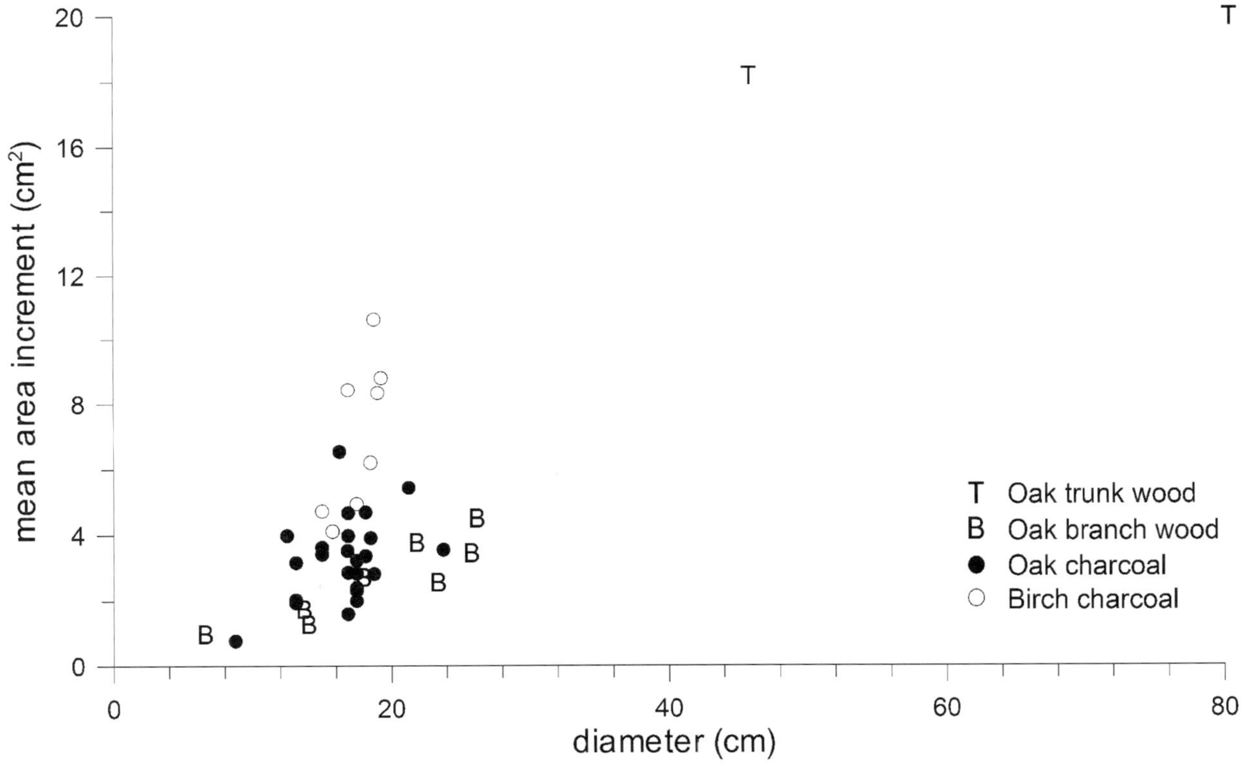

Figure 7: Llwyn Du wood and selected large diameter charcoal. Mean area increment (cm^2) plotted by radius (cm). Charcoal data adjusted to allow for shrinkage of about 20%.

80 rings, is 2.1mm, and that of the branch, with 91 rings, 0.8mm. It is more instructive to use the mean area increment (MAI), which compensates for the narrowing of the rings as the tree gets older. The W15 trunk has an MAI of 18.2cm^2 as compared to the branch MAI of 2.7cm^2, giving a much sharper discrimination. The W28 trunk, with 218 rings, has an MAI of over 20cm^2. Measurement of modern trees confirms this general pattern of trunk and branch growth.

Fig. 7 shows the MAI for all the medieval wood from Llwyn Du and for a selection of the large diameter oak and birch charcoal (adjusted to allow for shrinkage of about 20%). This indicates that the large diameter oak charcoal is primarily from branch wood, whereas the larger diameter birch is probably mostly from younger stem wood.

Discussion

The size analysis of the macroscopic charcoal shows that the oak was not being managed by coppicing and that primarily branch wood was being used. This implies that the trees were being lopped or possibly pollarded. However, the branch wood used was both relatively old and mostly of larger diameter, remaining so through all three phases. It is clear, therefore, that the oak used in the later phases is not simply regenerated timber, as the abandonment phases are both quite short. It appears that there was a careful selection of larger branches, rather than a wholesale removal of all the branches. This would give wood of an optimum size, both in terms of yield and suitability for coaling with a minimum of reduction, whilst perhaps retaining a supply of smaller branches to grow to a larger size in a relatively short time span.

In contrast, the birch used has a smaller mD value, which is probably a result of using whole younger trees, perhaps 20 to 40 years old, but again giving an optimum size range for coaling. The birch would have regenerated and grown more quickly than the oak. However, the mean ring widths of the birch suggest that most of it would have been obtained from trees growing slowly under an oak canopy, rather than from regenerated trees.

Thus, the overall picture we have is of a mature mixed oak and birch woodland being carefully used and managed, providing a near continuous supply of wood for making charcoal. It is also clear that the variations in the proportions of oak and birch are mainly a reflection of the availability of the species in the different areas of woodland being used at any one time, rather than being due to any deliberate selection (Mighall, et al., 2010).

Most of the wood used at Llwyn Du was relatively old and slowly grown material. The period of time during which the bloomery operated was too short for much of any regenerated timber to have grown to a usable size, and so the management of the woodland, whilst maintaining the tree cover, would have been primarily of benefit to future generations.

The pollen evidence is consistent with the proposed use-management regime. The lack of a significant decline argues against any major clearance of woodland, whilst the magnitude and frequency of the dips in oak pollen are consistent with lopping or pollarding. If oak was coppiced, one would expect larger and longer suppressions of the pollen, as coppiced oak can take 20 years or more before it begins to re-fruit and pollinate. Experimental work on locally pollarded oaks suggests that re-fruiting and pollen production can occur in 6 to 10 years. Pollen production would have recommenced well before the wood had re-grown to a suitable size for making charcoal. Thus, the short-term fluctuations in the oak pollen fits the hypothesis of lopped or pollarded trees, with the younger branches being left to grow and mature relatively quickly, resulting in only short gaps when individual trees were not fruiting. The dominance of birch in the pollen record suggests that there was always sufficient birch available without the need for any management. The behaviour of birch is quite different from that of oak, fruiting more rapidly and more copiously than oak, so that even the removal of relatively mature trees would result in only short term drops in the pollen production. This could also account for the asynchronicity between the fluctuations of oak and birch pollen.

The phasing of the pollen fluctuations, both between species and in relation to the site chronology, cannot be expected to reflect the woodland use directly, as there are other factors which could influence the pollen rain. These include variations in peat accumulation rate and the changing source of the tree pollen, which would have been influenced by prevailing wind, topography, distance to the peat monolith, and the density of the tree canopy. It is not possible to quantify such effects without sampling at multiple locations. Thus, although the pollen fluctuations cannot be directly related to the woodland usage, the consistently high level of the total arboreal pollen reflects the longer term maintenance of the woodland cover. The effect of lopped or pollarded trees might have created a landscape that could appear to have been cleared of useable woodland, hence leading to the 1391-1392 comment that the woodland "had been laid waste". The implication of the phrase would seem to be, therefore, not so much that the woodland had been destroyed, but that there was no longer a significant stock of wood of sufficient size for making charcoal, thus perhaps explaining the temporary move to Bedd y Coedwr and the increased use of birch in Phase 3.

There would have been a high degree of self-interest on the part of the iron workers in maintaining the woodland supply, but it is probable that there would also have been some control by the Crown, through the officer of the woodward. Direct evidence for this is rare, but a series of records survive for 1325-1326 for the Small Hundred Court for Ardudwy, held at Harlech (Lewis, 1928). These include numerous woodland offences *placite viride*, for taking wood without licence from the woodward, including the removal of cart loads of underwood, horse loads of rods, oak, hazel, ash and apple trees and even single branches of an oak tree. This demonstrates the close degree of control exercised by the woodward, no doubt in part as a way of optimising his revenue from the office. The medieval Welsh woodland economy has been discussed in detail by Linnard (1982), with examples of management practice (Fig. 8) and a remarkable record from 1351 of a coppice experiment at Hopedale, Flintshire, where part of the wood was "cut down to the ground and a part not

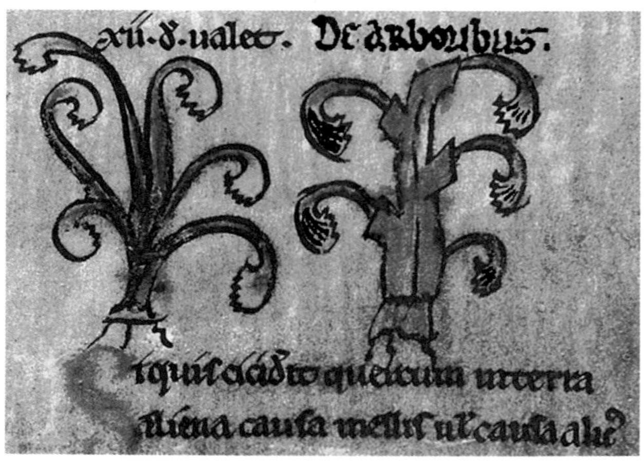

Figure 8: Coppicing and lopping in 13th century Wales (NLW MSS Peniarth 28, f22r). By permission of Llyfrgell Genedlaethol Cymru / The National Library of Wales.

so far, so that it may be possible to see which part grows best" (Linnard 1982: 38).

The 1391-1392 record shows that the Crown officials were aware of the state of the local woodland and that they were probably influential in deciding when areas should be left to recover. However, the evidence from Phase 3, with the use of a high proportion of birch, might suggest that this control had declined at the end of the 14th century, no doubt as one of the results of the Glyndwr rebellion. Perhaps the oak had been over-exploited for purposes other than ironworking.

The dating model for Llwyn Du, with two gaps of 10 and 15 years respectively, suggests that there were only some 40 years of iron working activity, and so the production model of 14,000 smelts, equivalent to 280 smelts per year for 50 years, requires revision. This model is based on the slag weight estimate of 250 tons, which is probably too high, as it does not allow for the unknown but considerable proportion of waste charcoal in the slag dumps, which would reduce the mean bulk density. If we assume that about 20% of volume of the dumps is charcoal, then the total slag weight could be 200 tons, which could be produced by 10,000 smelts, or 250 smelts per year for 40 years. This is equivalent to an annual wood requirement of about 175 tons. Unfortunately, it is impossible to estimate the area of woodland which would yield this amount of selected oak branch wood and birch, and the suggested figure of 40 tons per hectare provided by Smith (1995), based on the use of whole trees, is probably too low. A logical woodland resource area for Llwyn Du, based on local topography and excluding boggy and rocky areas, would be about 80 hectares, which was clearly sufficient to provide a carefully used woodland resource throughout the life of the bloomery.

Conclusion

The results of this study have demonstrated that the woodland at Llwyn Du was used carefully and was adequate to maintain a supply of charcoal for a bloomery working at this relatively large scale. The evidence for lopping or pollarding of the oak shows that coppicing was not the only management technique used in medieval times and that coppicing should not be invoked too readily without good evidence. However, this is the result of one detailed and fully integrated study of a site, within a well-defined archaeological framework, made possible by the exceptional quality of the evidence both from the macroscopic charcoal and the palaeoenvironmental data. There is no reason why these techniques could not be applied successfully to other metalworking sites and charcoal kilns, as in Germany, providing us with a more sophisticated picture of woodland use and management over a longer time period. It will undoubtedly show, as here, that ironworkers maintained rather than destroyed their key economic resource.

Acknowledgements

Thanks are due to Pat Denne and Bryan Dickenson for their initial work on the charcoal; to Oliver Nelle for his comments and references; and to Alison Sandison for preparing Figures 1 and 4. This project would not have been possible without the work of the excavation team, who recovered, washed and processed the charcoal with great diligence, especially Averil Downs, Kathy Laws, Brenda Telford, Sheila Rawson, Avis Reynolds, Ian Devine, Bill Jones, Bill Watkins and Terry Williams. We should like to acknowledge particularly the support of Susan Crew at all stages of the project. For their technical and scientific contributions thanks to Bruce Bevan, Ian Brooks, Mike Charlton, Ian Foster, Don Tarling, Ulf Winkler, Mick Worthington and Terry Williams. Finally, thanks to Pat Denne, Daryl Garton and Frances Lynch for their perceptive comments on earlier drafts of the paper. Funding for the project and for the radio-carbon dates was provided by the Snowdonia National Park Authority.

References

Barber, K.E., 1976. History of vegetation. In: S.B. Chapman ed. *Methods in Plant Ecology.* Oxford: Blackwell Scientific: pp.5-83.

Bronk Ramsay, C., 2001. Development of the Radiocarbon Program OXCAL. *Radiocarbon*, 43: pp.355-63.

Clark, R.L., 1982. Point count estimation of charcoal in pollen preparations and thin sections of sediments. *Pollen et Spores*, 24: pp.523-35.

Crew, P., 2002. Magnetic mapping and dating of prehistoric and medieval iron-working sites in northwest Wales. *Archaeological Prospection*, 9: pp.163-82.

Crew, P., and Charlton, M., 2007. The anatomy of a furnace...and some of its ramifications. In: S. La Niece, D. Hook and P. Craddock eds. *Metals and Mines: Studies in Archaeometallurgy.* London: Archetype: pp.219-25.

Crew, P., and Crew, S., 2001. Excavations at Llwyn Du, Merioneth, 2001. Woodland management and charcoal processing at a late 14th century ironworks. *Archaeology in Wales*, 41: pp.87-91.

Davasse, B., 1992. Aspects méthodologiques de l'anthracoanalyse des charbonnières. Histoire des forêts de la valée de Soulcem (Pyrénées ariègoises, France). In: J-P. Metaille ed. *Protoindustries et Histoires des Forêts.* Toulouse: Les Cahiers d'ISARD 3: pp.63-71.

Dufraisse, A., 2006. Charcoal anatomy potential, wood diameter and radial growth. In: A. Dufraisse ed. *Charcoal Analysis: New Analytical Tools and Methods for Archaeology*, British Archaeological Reports International Series 1483. Oxford: Archaeopress: pp.47-59.

Fægri, K., Kaland, P.E., and Krzywinski, K., 1989. *Textbook of pollen analysis.* Chichester: John Wiley and Sons.

Lewis, E.A., 1928. The proceedings of the Small Hundred Court of the commote of Ardudwy. *Bulletin of the Board of Celtic Studies*, 4: pp.153-66.

Linnard, W., 1982. *Welsh Woods and Forests: History and Utilisation.* Cardiff: National Museum of Wales.

Ludemann, T. 2006. Anthracological analysis of recent charcoal burning in the Black Forest, SW Germany. In: A. Dufraisse ed. *Charcoal Analysis: New Analytical Tools and Methods for Archaeology*, British Archaeological Reports International Series 1483. Oxford: Archaeopress: pp.61-70.

Ludemann, T., 2008. Experimental charcoal burning with special regard to anthrocological wood diameter analysis. In: G. Fiorentino and D. Magri eds. *Charcoals from the Past: Cultural and Palaeoenvironmental Implications*, British Archaeological Reports International Series 1807. Oxford: Archaeopress: pp.147-57.

Ludemann, T., and Nelle, O., 2002. *Die Wälder am Schaunisland und ihre Nutzung durch Bergbau und Köhlerei*. Baden-Württemberg: Freiburger Forstliche Forschung 15.

Mighall, T.M., Foster, I.D.L., Crew, P., Chapman, A.S., and Finn, A., 2009. Using mineral magnetism to characterise ironworking and to detect its evidence in peat bogs. *Journal of Archaeological Science*, 36: pp.130-9.

Mighall, T.M., Timberlake, S. and Crew, P., 2010. Vegetation changes in former mining and metalworking areas of Wales and Ireland during prehistoric and medieval times. In: P. Belford, M. Palmer and R. White eds *Footprints of Industry: Papers from the 300th anniversary conference at Coalbrookdale, 3-7 June 2009*. British Archaeological Reports British Series 523. Oxford: Archaeopress: pp.19-26.

Moore, P.D., Webb, J.A., and Collinson, M.E., 1991. *Pollen Analysis*. London: Blackwell Scientific Publications.

Nelle, O., 2002. Charcoal burning remains and forest stand structure – examples from the Black Forest (south-west Germany) and the Bavarian Forest (south-east Germany). In: S. Thiébault ed. *Charcoal Analysis: methodological approaches, palaeo-ecological results and wood uses*, British Archaeological Reports International Series 1063. Oxford: Archaeopress: pp.201-8.

Oldfield, F., Thompson, R., and Barber, K.E., 1978. Changing atmospheric fallout of magnetic particulates recorded in recent ombrotrophic peat sections. *Science*, 199: pp.679-80.

Oldfield, F., Tolonen, K., and Thompson, R., 1981. History of particulate atmospheric pollution from magnetic measurements in dated Finnish profiles. *Ambio*, 10: pp.185-8.

Smith, K.E.S., 1995. Ironworking in north-west Wales in the late fourteenth century. *Archaeological Journal*, 152: pp.246-90.

Smith, J.B., and Smith, Ll.B. eds., 2001. *History of Merioneth. Volume II. The Middle Ages*. Cardiff: University of Wales Press.

Stuiver, M., and Reimer, P.J., 1993. Extended ^{14}C database and revised CALIB 3.0 ^{14}C age calibration program. *Radiocarbon*, 35: pp.215-30.

Authors' addresses

- Peter Crew
  Pen Cefn, Penrhyndeudraeth, Gwynedd, LL48 6DR
  crew.peter.susan@gmail.com
- Tim Mighall
  Department of Geography and Environment, University of Aberdeen, AB24 3UF
  t.mighall@abdn.ac.uk